C0-ALP-398

Molecular Plant Breeding

In Memoriam

Norman Ernest Borlaug
(25 March 1914–12 September 2009)

Norman Borlaug was one of the greatest men of our times – a steadfast champion and spokesman against hunger and poverty. He dedicated his 95 richly lived years to filling the bellies of others, and is credited by the United Nations' World Food Program with saving more lives than any other man in history.

An American plant pathologist who spent most of his years in Mexico, it was Dr Borlaug's high-yielding dwarf wheat varieties that prevented wide-spread famine in South Asia, specifically India and Pakistan, and also in Turkey. Known as the 'Green Revolution', this feat earned him the Nobel Peace Prize in 1970. He was instrumental in establishing the International Maize and Wheat Improvement Center, known by its Spanish acronym CIMMYT, and later the Consultative Group of International Agricultural Research (CGIAR), a network of 15 agricultural research centres.

Dr Borlaug spent time as a microbiologist with DuPont before moving to Mexico in 1944 as a geneticist and plant pathologist to develop stem rust resistant wheat cultivars. In 1966 he became the director of CIMMYT's Wheat Program, seconded from the Rockefeller Foundation. His full-time employment with the Center ended in 1979, although he remained a part-time consultant until his death. In 1984 he began a new career as a university professor and went on to establish the World Food Prize, which honours the achievements of individuals who have advanced human development by improving the quality, quantity or availability of food in the world. In 1986, he joined forces with former US President Jimmy Carter and the Nippon Foundation of Japan, under the chairmanship of Ryoichi Sasakawa, to establish Sasakawa Africa Association (SAA) to address Africa's food problems. Since then, more than 1 million small-scale African farmers in 15 countries have been trained by SAA in improved farming techniques.

Dr Borlaug influenced the thinking of thousands of agricultural scientists. He was a path-breaking wheat breeder and, equally important, his stature enabled him to influence politicians and leaders around the world. His legacy and his work ethic – to get things done and not mind getting your hands dirty – influenced us all and remain CIMMYT guiding principles.

We will honor Dr Borlaug's memory by carrying forward his mission and spirit of innovation: applying agricultural science to help smallholder farmers produce more and better-quality food using fewer resources. At stake is no less than the future of humanity, for, as Borlaug said: 'The destiny of world civilization depends upon providing a decent standard of living for all'. His presence will never really leave CIMMYT; it is embedded in our soul.

Thomas A. Lumpkin
Director General, CIMMYT

Marianne Bänziger
Deputy Director General for Research and Partnerships, CIMMYT

Hans-Joachim Braun
Director for Global Wheat Program, CIMMYT

Molecular Plant Breeding

Yunbi Xu

International Maize and Wheat Improvement Center (CIMMYT)
Apdo Postal 6-641
06600 Mexico, DF
Mexico

CABI is a trading name of CAB International

CABI Head Office
Nosworthy Way
Wallingford
Oxfordshire OX10 8DE
UK

Tel: +44 (0)1491 832111
Fax: +44 (0)1491 833508
E-mail: cabi@cabi.org
Web site: www.cabi.org

CABI North American Office
875 Massachusetts Avenue
7th Floor
Cambridge, MA 02139
USA

Tel: +1 617 395 4056
Fax: +1 617 354 6875
E-mail: cabi-nao@cabi.org

A catalogue record for this book is available from the British Library, London, UK.

Library of Congress Cataloging-in-Publication Data

Xu, Yunbi.
Molecular plant breeding / Yunbi Xu.
p. cm.
ISBN 978-1-84593-392-0 (alk. paper)
1. Crop improvement. 2. Plant breeding. 3. Crops--Molecular genetics. 4. Crops--Genetics. I. Title.

SB106.147X8 2010
631.5'233--dc22

20009033246

ISBN: 978 1 84593 392 0

Typeset by SPi, Pondicherry, India.
Printed and bound in the UK by MPG Books Group.

Contents

The colour plate section can be found following p. 270

Preface

The genomics revolution of the past decade has greatly enhanced our understanding of the genetic composition of living organisms including many plant species of economic importance. Complete genomic sequences of *Arabidopsis* and several major crops, together with high-throughput technologies for analyses of transcripts, proteins and mutants, provide the basis for understanding the relationship between genes, proteins and phenotypes. Sequences and genes have been used to develop functional and biallelic markers, such as single nucleotide polymorphism (SNP), that are powerful tools for genetic mapping, germplasm evaluation and marker-assisted selection.

The road from basic genomics research to impacts on routine breeding programmes has been long, windy and bumpy, not to mention scattered with wrong turns and unexpected blockades. As a result, genomics can be applied to plant breeding only when an integrated package becomes available that combines multiple components such as high-throughput techniques, cost-effective protocols, global integration of genetic and environmental factors and precise knowledge of quantitative trait inheritance. More recently, the end of the tunnel has come in sight, and the multinational corporations have ramped up their investments in and expectations from these technologies. The challenge now is to translate and integrate the new knowledge from genomics and molecular biology into appropriate tools and methodologies for public-sector plant breeding programmes, particularly those in low-income countries. It is expected that harnessing the outputs of genomics research will be an important component in successfully addressing the challenge of doubling world food production by 2050.

What does *Molecular Plant Breeding* include?

The term 'molecular plant breeding' has been much used and abused in the literature, and thus loved or maligned in equal measure by the readership. In the context of this book, the term is used to provide a simple umbrella for the multidisciplinary field of modern plant breeding that combines molecular tools and methodologies with conventional approaches for improvement of crop plants. This book is intended to provide comprehensive coverage of the components that should be integrated within plant breeding programmes to develop crop products in a more efficient and targeted way.

The first chapter introduces some basic concepts that are required for understanding fundamentally important issues described in subsequent chapters. The concepts include crop domestication, critical events in the history of plant breeding, basics of quantitative genetics (variance, heritability and selection index), plant breeding objectives and molecular breeding goals. Chapters 2 and 3 introduce the key genomics tools that are used in molecular breeding programmes, including molecular markers, maps, 'omics' technologies and arrays. Different types of molecular markers are compared and construction of molecular maps is discussed. Chapter 4 describes common types of populations that have been used in genetics and plant breeding, with a focus on recombinant inbred lines, doubled haploids and near-isogenic lines. Chapter 5 provides an overview of marker-assisted germplasm evaluation, management and enhancement. Chapters 6 and 7 discuss the theory and practice, respectively, of using molecular markers to dissect complex traits and locate quantitative trait loci (QTL). Chapters 8 and 9 cover the theory and practice, respectively, of marker-assisted selection. Genotype-by-environment interaction (GEI) is discussed in Chapter 10, including multi-environment trials, stability of genotype performance, molecular dissection of GEI and breeding for optimum GEI. Chapter 11 provides a summary of gene isolation and functional analysis approaches, including *in silico* prediction of genes, comparative approaches for gene isolation, gene cloning based on cDNA sequencing, positional cloning and identification of genes by mutagenesis. Chapter 12 describes the use of isolated and characterized genes for gene transfer and the generation of genetically modified plants, focusing on the vital elements of expression vectors, selectable marker genes, transgene integration, expression and localization, transgene stacking and transgenic crop commercialization. Chapter 13 is devoted to intellectual property rights and plant variety protection, including plant breeders' rights, international agreements affecting plant breeding, plant variety protection strategies, intellectual property rights affecting molecular breeding and the use of molecular techniques in plant variety protection. The last two chapters (14 and 15) discuss supporting tools that are required in molecular breeding for information management and decision making, including data collection, integration, retrieval and mining and information management systems. Decision support tools are described for germplasm and breeding population management and evaluation, genetic mapping and marker-trait association analysis, marker-assisted selection, simulation and modelling, and breeding by design.

Intended audience and guidance for reading and using this book

This book is intended to provide a handbook for biologists, geneticists and breeders, as well as a textbook for final year undergraduates and graduate students specializing in agronomy, genetics, genomics and plant breeding. Although the book has attempted to cover all relevant areas of molecular breeding in plants, many examples have been drawn from the genomics research and molecular breeding of major cereal crops. It is hoped that the book can also serve as a resource for training courses as described below. As each chapter covers a complete story on a special topic, readers can choose to read chapters in any order.

Advanced Course on Quantitative Genetics: Chapters 1, 2, 4, 6, 7, 10 and 14, which cover all molecular marker-based QTL mapping, including markers, maps, populations, statistics and genotype-by-environment interaction.

Comprehensive Course on Marker-assisted Plant Breeding: Chapters 1, 2, 3, 4, 5, 8, 9, 10, 13, 14 and 15, which cover basic theories, tools, methodologies about markers, maps, omics, arrays, informatics and support tools for marker-assisted selection.

Short Course on Genetic Transformation: Chapters 1, 11, 12 and 13, which provide a brief introduction to gene isolation, transformation techniques, genetic-transformation-related intellectual property and genetically modified organism (GMO) issues.

Introductory Course on Breeding Informatics: Chapters 1, 2, 3, 4, 5, 10, 14 and 15, which cover bioinformatics, focusing on plant breeding-related applications, including basic concepts in plant breeding, markers, maps, omics, arrays, population and germplasm management, environment and geographic information system (GIS) information, data collection, integration and mining, and bioinformatics tools required to support molecular breeding. Additional introductory information can be found in other chapters.

History of writing this book

This book has been almost a decade in preparation. In fact, the initial idea for the book was stimulated by the impact from my previous book *Molecular Quantitative Genetics* published by China Agriculture Press (Xu and Zhu, 1994), which was well received by colleagues and students in China and used as a textbook in many universities. Preliminary ideas related to the book were developed in a review article on QTL separation, pyramiding and cloning in *Plant Breeding Reviews* (Xu, 1997). Much of the hopeful thinking described in this paper has fortunately come true during the following 10 years, and the manipulation of QTL has been revolutionized and become mainstream. As complete sequences for several plant genomes have become available and with more anticipated, as shown by numerous genes and QTL that have been separated and cloned individually, some of them have been pyramided for plant breeding through genetic transformation or marker-assisted selection.

I started making tangible progress on this book while working as a molecular breeder for hybrid rice at RiceTec, Inc., Texas (1998–2003). This experience shaped my thinking about how an applied breeding programme could be integrated with molecular approaches. With numerous QTL accumulating for a model crop, taking all the QTL into consideration becomes necessary. Initial thoughts on this were described in 'Global view of QTL...', published in the proceedings on quantitative genetics and plant breeding, which considered various genetic background effects and genotype-by-environment interaction (Xu, Y., 2002). Hybrid rice breeding, which involves a three-line system, requires a large number of testcrosses in order to identify traits that perform well in seed and grain production. My experience in development of marker-assisted selection strategies for breeding hybrid rice was then summarized in a review article in *Plant Breeding Reviews* (Xu, Y., 2003), which also covered general strategies for other crops using hybrids.

Moving on to research at Cornell University with Dr Susan McCouch helped me to better understand how molecular techniques could facilitate breeding of complex traits such as water-use efficiency, which is a difficult trait to measure and requires strong collaboration among researchers across many disciplines. In addition, this experience with rice as a model crop raised the issue of how we can use rice as a reference genome for improvement of other crops, which was discussed in an article published in a special rice issue of *Plant Molecular Biology* (Xu *et al.*, 2005).

With over 20 years' experience in rice, I decided to shift to another major crop by working for the International Maize and Wheat Improvement Center (CIMMYT) as the principle maize molecular breeder. CIMMYT has given me exposure to an interface connecting basic research with applied breeding for developing countries and the resource-poor. Comparing public- and private-sector breeding programmes has given me an intense understanding of the importance of making the type of breeding systems that have been working well for the private sector a practical reality for the public sector, particularly in developing countries. This has been addressed in a recent review paper published in *Crop Science* (Xu and Crouch, 2008), which discussed the critical issues for achieving this translation. My most recent research has focused on the development of various molecular breeding platforms that can be used to facilitate breeding procedures through seed DNA-based genotyping, selective and pooled DNA analysis, and chip-based large-scale germplasm evaluation, marker–trait association and marker-assisted selection (see Xu *et al.*, 2009b for further details). Thus, my career has evolved alongside the transition from molecular biology research to routine molecular plant breeding applications and I strongly believe that now is the right time for a mainstream publication providing comprehensive coverage of all fields relevant for a new generation of molecular breeders.

Acknowledgements

Assistance and professional support

The dream of writing this book could not have become reality without the wonderful support of Dr Susan McCouch at Cornell University and Dr Jinhua Xiao, now at Monsanto, who have both fully supported my proposal since 2002. Their support and consistent encouragement has greatly motivated me throughout the process. While working with Susan, she allowed me so much flexibility in my research projects and working hours so that I could continue to make progress on the writing of this book. At the same time the Cornell libraries were an indispensible source of the major references cited throughout the book. Susan's encouragement provided the impetus to keep working on the book through a very difficult time in my life. I also extend my appreciation to Dr Jonathan Crouch, the Director of the Germplasm Resources Program at CIMMYT, where I received his full understanding and support so that I could complete the second half of the book. Jonathan's guidance and contribution to my research projects and publications while at CIMMYT has significantly impacted the preparation of the book.

I would also like to thank the chief editors of the three journals for which I have served on the editorial boards during the preparation of this book: Dr Paul Christou for *Molecular Breeding*, Dr Albrecht Melchinger for *Theoretical and Applied Genetics*, and Dr Hongbin Zhang for *International Journal of Plant Genomics*. I thank them for their patience, support and flexibility with my editorial responsibilities during the preparation of the book. In addition, Drs Christou and Melchinger also reviewed several chapters in their respective fields.

My appreciation also goes to Yanli Lu (a graduate student from Sichuan Agricultural University of China) and Dr Zhuanfang Hao (a visiting scientist from the Chinese Academy of Agricultural Sciences) who helped prepare some figures and tables during their work in my lab at CIMMYT, Mexico. I would like to give special thanks to Dr Rodomiro Ortiz at CIMMYT for his consistent information sharing and stimulating discussions during our years together at CIMMYT. Finally, I would like to thank my colleagues at CIMMYT, particularly Drs Kevin Pixley, Manilal William, Jose Crossa and Guy Davenport, who provided useful discussions on various molecular breeding-related issues.

Forewords

I am greatly indebted to Dr Norman E. Borlaug, visioned plant breeder and Nobel laureate for his role in the Green Revolution, and Dr Ronald L. Phillips, Regents Professor and McKnight Presidential Chair in Genomics, University of Minnesota, who each contributed a foreword for the book. Their contributions emphasized the importance of molecular breeding in crop improvement and the role that this book will play in molecular breeding education and practice.

Reviewers

Each chapter of the book has undergone comprehensive peer review and revision before finalization. The constructive comments and critical advice of these reviewers have greatly improved this book. The reviewers were selected for their active expertise in the field of the respective chapter. Reviewers come from almost all continents and work in various fields including plant breeding, quantitative genetics, genetic transformation, intellectual property protection, bioinformatics and molecular biology, many of whom are CIMMYT scientists and managers. Considering that each chapter is relatively large in content, reviewers had to contribute a lot of time and effort to complete their reviews. Although these inputs were indispensible, any remaining errors remain my sole responsibility. The names and affiliations of the reviewers (alphabetically) are:

Raman Babu (Chapters 7 and 9), CIMMYT, Mexico
Paul Christou (Chapter 12), Lleida, Spain
Jose Crossa (Chapter 10), CIMMYT, Mexico
Jonathan H. Crouch (Chapters 13 and 15), CIMMYT, Mexico
Jedidah Danson (Chapters 7 and 9), African Center for Crop Improvement, South Africa
Guy Davenport (Chapter 14), CIMMYT, Mexico
Yuqing He (Chapter 8), Huazhong Agricultural University, Wuhan, China
Gurdev S. Khush (Chapter 1), IRRI, Philippines
Alan F. Krivanek (Chapter 4), Monsanto, Illinois, USA
Huihui Li (Chapter 6), Chinese Academy of Agricultural Sciences, China
George H. Liang (Chapter 12), San Diego, California, USA
Christopher Graham McLaren (Chapter 14), GCP/CIMMYT, Mexico
Kenneth L. McNally (Chapter 5), IRRI, Philippines
Albrecht E. Melchinger (Chapter 8), University of Hohenheim, Germany
Rodomiro Ortiz (Chapters 12, 13 and 15), CIMMYT, Mexico
Edie Paul (Chapter 14), GeneFlow, Inc., Virginia, USA
Kevin V. Pixley (Chapters 1, 4 and 5), CIMMYT, Mexico
Trushar Shah (Chapter 14), CIMMYT, Mexico
Daniel Z. Skinner (Chapter 12), Washington State University, USA
Debra Skinner (Chapter 11), University of Illinois, USA
Michael J. Thomson (Chapters 2 and 3), IRRI, Philippines
Bruce Walsh (Chapters 1, 6 and 8), University of Arizona, USA
Marilyn L. Warburton (Chapter 5), USDA/Mississippi State University, USA
Huixia Wu (Chapter 12), CIMMYT, Mexico
Rongling Wu (Chapter 1), University of Florida, Gainesville, USA
Weikai Yan (Chapter 10), Agriculture and Agri-Food Canada, Ottawa, Canada
Qifa Zhang (Chapters 8 and 12), Huazhong Agricultural University, Wuhan, China
Wanggen Zhang (Chapter 12), Syngenta, Beijing, China
Yuhua Zhang (Chapter 12), Rothamsted Research, UK

Publishers and development editors

Several editors at CABI have been working with me over the years: Tim Hardwick (2002–2006), Sarah Hulbert (2006–2007), Stefanie Gehrig (2007–2008), Claire Parfitt (2008–2009), Meredith Caroll (2009) and Tracy Head (2009). These editors and their associates have done a superb job of converting a series of manuscripts into a useable and coherent book. I thank them for their effort, consideration and cooperation.

Research grants

During the preparation of the book, my research on genomic analysis of plant water-use efficiency at Cornell University was supported by the National Science Foundation (Plant Genome Research Project Grant DBI-0110069). My molecular breeding research at CIMMYT has been supported by the Rockefeller Foundation, the Generation Challenge Programme (GCP), Bill and Melinda Gates Foundation and the European Community, and through other attributed or unrestricted funds provided by the members of the Consultative Group on International Agricultural Research (CGIAR) and national governments of the USA, Japan and the UK.

Family

It is difficult to imagine writing a book without the full support and understanding of one's family. My greatest thanks go to my wife, Yu Wang, who has given me her wholehearted and unwavering support, and to my sons, Sheng, Benjamin and Lawrence, who have retained great patience during this long adventure. And finally to my parents, for their love, encouragement and vision that unveiled in me from my earliest years the desire to thrive on the challenge of always striving to reach the highest mountain in everything I do.

Foreword

Dr Norman E. Borlaug

The past 50 years have been the most productive period in world agricultural history. Innovations in agricultural science and technology enabled the 'Green Revolution', which is reputed to have spared one billion people the pains of hunger and even starvation. Although we have seen the greatest reductions in hunger in history, it has not been enough. There are still one billion people who suffer chronic hunger, with more than half being small-scale farmers who cultivate environmentally sensitive marginal lands in developing countries.

Within the next 50 years, the world population is likely to increase by 60–80%, requiring global food production to nearly double. We will have to achieve this feat on a shrinking agricultural land base, and most of the increased production must occur in those countries that will consume it. Unless global grain supplies are expanded at an accelerated rate, food prices will remain high, or be driven up even further.

Spectacular economic growth in many newly industrializing developing countries, especially in Asia, has spurred rapid growth in global cereal demand, as more people eat better, especially through more protein-heavy diets. More recently, the subsidized conversion of grains into biofuels in the USA and Europe has accelerated demand even faster. On the supply side, a slowing in research investment in the developing world and more frequent climatic shocks (droughts, floods) have led to greater volatility in production.

Higher food prices affect everyone, but especially the poor, who spend most of their disposable income on food. Increasing supply, primarily through the generation and diffusion of productivity-enhancing new technologies, is the best way to bring food prices down and secure minimum nutritional standards for the poor.

Today's agricultural development challenges are centred on marginal lands and in regions that have been bypassed during the Green Revolution, such as Africa and resource-poor parts of Asia, and are experiencing the ripple effects of food insecurity through hunger, malnutrition and poverty.

Despite these serious and daunting challenges, there is cause for hope. New science and technology – including biotechnology – have the potential to help the world's poor and food insecure. Biotechnologies have developed invaluable new scientific methodologies and products for more productive agriculture and added-value food. This journey deeper into the genome to the molecular level is the consequence of our progressive understanding of the workings of nature. Genomics-based methods have enabled breeders greater precision in selecting and transferring genes, which has not only reduced the time needed to eliminate undesirable genes, but has also allowed breeders to access useful genes from distant species.

Bringing the power of science and technology to bear on the challenges of these riskier environments is one of the great challenges of the 21st century. With the new tools of biotechnology, we are poised for another explosion in agricultural innovation. New science has the power to increase yields, address agroclimatic extremes and mitigate a range of environmental and biological challenges.

Molecular Plant Breeding, authored by my CIMMYT colleague Yunbi Xu, is an outstanding review and synthesis of the theory and practice of genetics and genomics that can drive progress in modern plant breeding. Dr Xu has done a masterful job in integrating information about traditional and molecular plant breeding approaches. This encyclopedic handbook is poised to become a standard reference for experienced breeders and students alike. I commend him for this prodigious new contribution to the body of scientific literature.

Foreword

Dr Ronald L. Phillips

The New Plant Breeding Roadmap

The road is long from basic research findings to final destinations reflecting important applications – but it is a road that can ultimately save time and money. There may be obstacles along the way that delay building that road but they are generally overcome by careful thought and timely considerations. A new road may involve the former road but with some widening and the filling in of certain potholes. We seldom look back and think that the improvements were not useful.

The road to improved varieties by traditional plant breeding has and continues to serve society well. That approach has been based on careful observation, evaluation of multiple genotypes (parents and progenies), selection at various generational levels, extensive testing and the sophisticated utilization of statistical analyses and quantitative genetics. About 50% of the increased productivity of new varieties is generally attributed to genetic improvements, with the remaining 50% due to many other factors such as time of planting, irrigation, fertilizer, pesticide applications and planting densities.

The statistical genetics associated with traditional plant breeding can now be supplemented by extensive genomic information, gene sequences, regulatory factors and linked genetic markers. We can now draw on a broader genetic base, the identification of major loci controlling various traits and expression analyses across the entire genome under various biotic and abiotic conditions. One can anticipate a future when the networking of genes, genotype-by-environment (G × E) interactions, and even hybrid vigour will be better understood and lead to new breeding approaches. The importance of *de novo* variation may modify much of our current interpretation of breeding behaviour; *de novo* variation such as mutation, intragenic recombination, methylation, transposable elements, unequal crossing over, generation of genomic changes due to recombination among dispersed repeated elements, gene amplification and other mechanisms will need to be incorporated into plant breeding theory.

This book calls for an integration of approaches – traditional and molecular – and represents a theoretical/practical handbook reflecting modern plant breeding at its finest. I believe the reader will be surprised to find that that this single-authored book is so full of information that is useful in plant genetics and plant breeding. Students as well as established researchers wanting to learn more about molecular plant breeding will be well-served by reading this book. The information is up-to-date with many current references. Even many of the tables are packed with information and references. A good representation of international and domestic breeding is reflected through many examples. The importance of G × E interactions is clearly demonstrated. Various statistical models are provided as appropriate. The importance of defining mega-environments for varietal development is made clear. The role of core germplasm collections, appropriate population sizes, major databases and data management issues are all integrated with various plant breeding approaches. Marker-assisted selection receives considerable attention, including its requirements and advantages, along with the multitude of quantitative trait locus (QTL) analysis methods. Transformation technologies leading to the extensive use of transgenic crops are reviewed along with the increased use of trait stacking. The procurement of intellectual property that, in part, is driving the application of molecular genetics in plant breeding provides the reader with an understanding of why private industry is now more involved and why some common crops represent new business opportunities.

Molecular Plant Breeding is not like other plant breeding books. The interconnecting road that it depicts is one where you can look at the beautiful new scenery and appreciate the current view, yet see the horizon down the road.

1

Introduction

Several definitions of plant breeding have been put forward, such as 'the art and science of improving the heredity of plants for the benefit of humankind' (J.M. Poehlman), or 'evolution directed by the will of man' (N.I. Vavilov). Bernardo (2002), however, offers the most universal description: 'Plant breeding is the science, art, and business of improving plants for human benefit.'

Plants are employed in the manufacture of a multitude of products for domestic (cosmetics, medicines and clothing), industrial (manufacture of rubber, cork and engine fuel) and recreational uses (paper, art supplies, sports equipment and musical instruments) and plant breeders have therefore been driven by the challenges of meeting the ever increasing demands of the manufacturers of these products. Lewington has described the diverse uses of plants in his book *Plants for People* (2003).

Plant breeding began by the domestication of crop plants and has become ever more sophisticated. New developments in molecular biology have now led to an increasing number of methods which can be used to enhance breeding effectiveness and efficiency. This chapter includes a brief history of plant breeding together with breeding objectives and some background information relevant to the theories and technologies discussed in the following chapters of this book.

1.1 Domestication of Crop Plants

The earliest records indicate that agriculture developed some 11,000 years ago in the so-called Fertile Crescent, a hilly region in south-western Asia. Agriculture developed later in other regions. Archaeologists suggest that plant domestication began because of the increasing size of populations and changes in the exploitation of local resources (see http://www.ngdc.noaa.gov/paleo/ctl/10k.html for further details). Domestication is a selection process carried out by man to adapt plants and animals to their own needs, whether as farmers or consumers. Successive selection of desirable plants changed the genetic composition of early crops. Primitive farmers, knowing little or nothing about genetics or plant breeding, accomplished much in a short time. They did so by unconsciously altering the natural process of evolution. Indeed, domestication is nothing more than directed evolution; as a result, the process of evolution is accelerated. The key to domestication is the selective advantage of rare mutant alleles, which are desirable for successful cultivation, but unnecessary for survival in the wild. The process of selection continues until the desired mutant phenotype dominates the population. There are three important steps in the domestication process. Man not only planted seeds, but also: (i) moved seeds from their native habitat and planted them in areas to which they were perhaps not as well adapted; (ii) removed certain natural selection pressures by growing the plants in a cultivated field; and (iii) applied artificial selection pressures by choosing characteristics that would not necessarily have been beneficial for the plants under natural conditions. Cultivation also creates selection pressure, resulting in changes in allele frequency, gradations within and between species, fixation of major genes, and improvement of quantitative traits. By the end of the 18th century, the informal processes of selection practised by farmers everywhere led to the worldwide creation of thousands of different cultivars or landraces for each major crop species.

More than 1000 species of plants have been domesticated at one time or another, of which about 100–200 are now major components of the human diet. The 15 most important examples can be divided into the following four groups:

1. Cereals: rice, wheat, maize, sorghum, barley.
2. Roots and stems: sugarbeet, sugarcane, potato, yam, cassava.
3. Legumes: bean, soybean, groundnut.
4. Fruits: coconut, banana.

Certain characteristics may have been selected deliberately or unwittingly. When farmers set aside a portion of their harvest for planting in the next season, they were selecting seeds with specific characteristics. This selection has resulted in profound differences between crop plants and their progenitors. For example, many wild plants have a seed dispersal mechanism that ensures that seeds will be separated from the plants and distributed over as large an area as possible, while modern crops have been modified by selection against seed dispersal. The absence of seed dormancy mechanisms in some domesticated plants is another example. For further information see http://oregonstate.edu/instruct/css/330/index.htm and Swaminathan (2006).

It is generally believed that domestication of crop plants was undertaken in several regions of the world independently. The Russian geneticist and plant geographer N.I. Vavilov, collected plants from all over the world and identified regions where crop species and their wild relatives showed great genetic diversity. In 1926 he published 'Studies on the origin of cultivated plants' in which he described his theories regarding the origins of crops. Vavilov concluded that each crop had a characteristic primary centre of diversity which was also its centre of origin. He identified eight areas and hypothesized that these were the centres from which all our modern major crops originated. Later, he modified his theory to include 'secondary centres of diversity' for some crops. These 'centres of origin' included China, India, Central Asia, the Near East, the Mediterranean, East Africa, Mesoamerica, and South America. From these foci, agriculture was progressively disseminated to other regions such as Europe and North America. Subsequently, others including the American geographer Jack Harlan, challenged Vavilov's hypothesis because many cultivated plants did not fit Vavilov's pattern, and appeared to have been domesticated over a broad geographical area for a long period of time.

In recent years, variation in DNA fractions and other approaches have been used to study the diversity of crop species. In general, these studies have not confirmed Vavilov's theory that the centres of origin are the areas of greatest diversity, because while centres of diversity have been identified, these are often not the centres of origin. For some crops there is little connection between the source of their wild ancestors, areas of domestication, and the areas of evolutionary diversification. Species may have originated in one geographic area, but domesticated in a different region and some crops do not appear to have centres of diversity, thus a continuum of evolutionary activity is perceived rather than discrete centres.

In 1971, Jack Harlan described his own views on the origins of agriculture. He proposed three independent systems, each with a centre and a 'concentre' (larger, diffuse areas where domestication is thought to have occurred): Near East + Africa, China + South-east Asia, and Mesoamerica + South America.

Evidence gathered since that time suggests that these centres are also more diffuse than he had envisioned. After the initial phases of evolution, species spread out over large, ill-defined areas. This is probably due to the dispersal and evolution of crops associated with iterant populations. Regional and/or multiple areas of origin may prove to be more accurate than the hypothesis of a unique, localized origin for many crops. However, the probable geographic origin of many crops is listed in Table 1.1.

1.2 Early Efforts at Plant Breeding

For thousands of years selective breeding has been employed to re-engineer plants to produce traits or qualities that were considered to be desirable to consumers. Selective breeding began with the early farmers, ranchers and vintners who selected the best plants to provide seed for their next crop. When they found particular plants that fared well even in bad weather, were especially prolific, or resisted disease that had destroyed neighbouring crops, they naturally tried to capture these desirable traits by crossbreeding them into other plants. In this way, they selected and bred plants to improve their crop for commercial purposes. Although unbeknown to them, farmers have been utilizing genetics for centuries to modify the food we eat by selecting and growing seeds which produce a healthier crop that has a better flavour, richer colour and stronger resistance to certain plant diseases.

Modern plant breeding started with sedentary agriculture and the domestication of the first agricultural plants, cereals. This led to the rapid elimination of undesirable characters such as seed-shattering and dormancy and we can only speculate on how much foresight or what kind of planning based on experience was used by the first selectors of non-shattering wheat and rice, compact-headed sorghum, or soft-shelled gourds. For 10,000 years man has consciously been moulding the phenotype (and so the genotype) of hundreds of plant species as one of the many routine activities in the normal course of making a living (Harlan, 1992). Over long periods of time there was a transition from the collection of

Table 1.1. Probable geographic origins for crops.

Region	Crops
Near East (Fertile Crescent)	Wheat and barley, flax, lentils, chickpea, figs, dates, grapes, olives, lettuce, onions, cabbage, carrots, cucumbers, melons; fruits and nuts
Africa	Pearl millet, Guinea millet, African rice, sorghum, cowpea, groundnut, yam, oil palm, watermelon, okra
China	Japanese millet, rice, buckwheat, soybean
South-east Asia	Wet- and dryland rice, pigeon pea, mung bean, citrus fruits, coconut, taro, yams, banana, breadfruit, coconut, sugarcane
Mesoamerica and North America	Maize, squash, common bean, lima bean, peppers, amaranth, sweet potato, sunflower
South America	Lowlands: cassava; Mid-altitudes and uplands (Peru): potato, groundnut, cotton, maize

See http://agronomy.ucdavis.edu/gepts/pb143/lec10/pb143l10.htm for a thorough presentation on the geographic origins of crops.

wild plants for food to the selection of those to be cultivated which began to guide the evolutionary process. Now plant breeders accelerate the evolution of major crop species through skilful manipulation of breeding procedures. High-input agriculture emerged as a result of voyages of discovery and modern science.

Many traits important to early agriculturists were heritable and, therefore, could be reliably selected. However, this phase of breeding was empirical and generally not considered scientific in the modern sense because changes in these plant and animal populations were not analysed in an attempt to explain biological phenomena. At this stage of agriculture, the focus was on the practical goal of producing food rather than finding rational explanations for nature (Harlan, 1992). Ideas about heredity during the period when many early crops were domesticated ranged from mythological interpretations to near-scientific notions of trait transmission. In his Presidential Address to the American Society for Horticulture Science in 1987, Janick (1988) stated:

> The origin of new information in horticulture derives from two traditions: empirical and experimental. The roots of empiricism stem from efforts of prehistoric farmers, Hellenic root diggers, medieval peasants, and gardeners everywhere to obtain practical solutions to problems of plant growing. The accumulated successes and improvements passed orally from parent to child, from artisan to apprentice, have become embodied in human consciousness via legend, craft secrets, and folk wisdom. This information is now stored in tales, almanacs, herbals, and histories and has become part of our common culture. More than practices and skills were involved as improved germplasm was selected and preserved via seed and graft from harvest to harvest and generation to generation. The sum total of these technologies makes up the traditional lore of horticulture. It represents a monumental achievement of our forbears – unknown and unsung.

Large-scale breeding activities began very early in Europe, often under the auspices of commercial seed production enterprises. Besides selecting plants with useful characteristics, breeders also arrange 'marriages' between plants with different traits in the hope of producing fertile offspring carrying both traits. The use of artificial crosses in pre-Mendelian breeding is exemplified by the case of *Fragaria* × *ananassa* developed in the botanical garden of Paris by Duchesne, in the 17th century by crossing *Fragaria chiloense* with *Fragaria virginiana*. In England, at about the same time new cultivars of fruits, wheat and peas were being obtained by artificial hybridization (Sánchez-Monge, 1993).

Hybridization combined with selection was adopted by Patrik Sheireff in 1819 in wheat and rice where the new selections were grown along with cultivars for comparative purposes. He speculated that introduction and hybridization to be the important sources of new cultivars and stressed crossing of carefully selected parents to meet the aims of new cultivars. Although the essential elements of plant breeding were known by this time, there was still a lack of knowledge regarding the scientific basis of variation among plants. For example, the first generation of crossed materials were mistakenly expected to inevitably produce new cultivars but instead took several generations to stabilize. Many historical examples of successful plant breeding can be found in the literature, although there were still many important discoveries to be made before it could be called a technology (Chahal and Gosal, 2002).

1.3 Major Developments in the History of Plant Breeding

Plant breeders of today use various methods to accelerate the evolutionary process in order to increase the usefulness of plants by exploiting genetic differences within a species. This has been made possible by the determination of the genetic basis for developing crop breeding procedures and this in turn has a long history.

1.3.1 Breeding and hybridization

The role of reproduction in plants was first reported in 1694 by Camerarius who noticed the difference between male and female reproductive organs in maize and produced the first artificial hybrid plant. He established that seed could not be produced without the participation of pollen produced in male reproductive organs of plants. The first hybridization experiment was carried out on wheat by Fairchild in 1719 and the current technique of hybridization is largely based on the work of Kölreuter (1733–1806), a French researcher who carried out his experiments in the 1760s. Hybridization freed the breeder from the severe constraints of working within a limited population, enabled him to bring together useful traits from two or more sources, and allowed specific genes to be introduced.

By understanding the reproductive capacities of plants, plant breeders can manipulate these crosses to produce fertile offspring which carry traits from both parents. Crossing has been very valuable to plant breeders, because it allows some measure of control over the phenotype of a plant. Nearly all modern plant breeding involves some use of hybridization.

1.3.2 Mendelian genetics

It was Gregor Johann Mendel, a Moldavian monk, who in 1865 discovered the basic rules that govern heredity as a result of a series of experiments in which he crossed two cultivars of pea plants. By studying the inheritance of all-or-none variation in peas, Mendel discovered that inherited traits are determined by units of material that are transferred from one generation to another. Mendel was probably ahead of his time as other biologists of that era took 35 years to appreciate his work and plant breeding remained deprived of the deliberate application of the law of genetics until 1900 when Hugo de Vries, Carl Correns and Erich von Tschermak-Seysenegg rediscovered Mendel's work.

1.3.3 Selection

In 1859 Darwin proposed in *The Origin of Species* that natural selection is the mechanism of evolution. Darwin's thesis was that the adaptation of populations to their environments resulted from natural selection and that if this process continued for long enough, it would ultimately lead to the origin of new species. Darwin's 'Theory of Evolution through Natural Selection' hypothesized that plants change gradually by natural selection operating on variable populations and was the outstanding discovery of the 19th century with direct relevance to plant breeding.

1.3.4 Breeding types and polyploidy

Other historical developments in plant breeding include, pedigree breeding, backcross breeding (Harlan and Pope, 1922) and mutation breeding (Stadler, 1928). Natural and artificial polyploids also offered new possibilities for plant breeding. Blakeslee and Avery (1937) demonstrated the usefulness of colchicine in the induction of chromosome doubling and polyploidy, enabling plant breeders to combine entire chromosome sets of two or more species to evolve new crop plants.

1.3.5 Genetic diversity and germplasm conservation

The importance of genetic diversity in plant breeding was recognized by the 1960s and Sir Otto Frankel coined the term 'genetic resources' in 1967 to highlight the relevance and need to consider germplasm as a natural resource for the long-term improvement of crop plants. The potentially harmful effects of genetic uniformity became apparent with the epidemic of southern corn leaf blight in the USA in 1970 which destroyed about 15% of US maize in just 1 year. The National Academy of Sciences, USA, released the results of its study *Genetic Vulnerability in Major Crops* that brought into focus the causes and levels of genetic uniformity and its consequences. It was a turning point in the history of germplasm resources and the International Board for Plant Genetic Resources (IBPGR) was established in 1974, and was later renamed the International Plant Genetic Resources Institute (IPGRI) and now Biodiversity International, to collect, evaluate and conserve plant germplasm for future use.

1.3.6 Quantitative genetics and genotype-by-environment interaction

Quantitative genetics is the study of the genetic control of those traits which show continuous variation. It is concerned with the level of inheritance of these differences between individuals rather than the type of differences, that is quantitative rather than qualitative (Falconer, 1989). Several important books have been published which document the major developments in quantitative genetics and these include *Animal Breeding Plans* (Lush, 1937), *Population Genetics and Animal Improvement* (Lerner, 1950), *Biometrical Genetics* (Mather, 1949), *Population Genetics* (Li, 1955), *An Introduction to Genetics Statistics* (Kempthorne, 1957) and *Introduction to Quantitative Genetics* (Falconer, 1960).

Many of the misconceptions regarding the inheritance of quantitative traits, which include most of the economically important characters, were corrected by the classical work of Fisher (1918) who successfully applied Mendelian principles to explain the genetic control of continuous variation. He divided the phenotypic variance observed into three variance components: additive, dominance and epistatic effects. This approach has been substantially refined and applied to the improvement of the efficiency of plant breeding. Fisher also laid the foundations for scientific crop experimentation by developing the theory of experimental designs that is an essential part of any plant breeding programme. Quantitative genetics has however evolved considerably in the past two decades because of the development of plant genomics, particularly molecular markers, and other molecular tools that can be used to dissect complex traits into single Mendelian factors (Xu and Zhu, 1994; Buckler *et al.*, 2009; Chapters 6 and 7).

Genotype-by-environment interaction (GEI) and its importance to plant breeding were first recognized by Mooers (1921) and Yates and Cochran (1938). Since then, various statistical methods have been developed for the evaluation of GEI using joint linear regression, heterogeneity of variance and lack of correlation, ordination, clustering, and pattern analysis. As an important field in quantitative genetics, GEI has been receiving more attention in recent years and is covered in Chapter 10 along with molecular methods for GEI analysis.

1.3.7 Heterosis and hybrid breeding

Although early botanists had observed increased growth when unrelated plants of the same species were crossed, it was Charles Darwin who carried out the first seminal experiments. In 1877, he showed that crosses of related strains did not exhibit the vigour of hybrids. He observed heterosis, i.e. the tendency of cross-bred individuals to show qualities superior to those of both parents, in crops like maize and concluded that cross-fertilization was generally beneficial and self-fertilization injurious. In 1879, William Beal demonstrated hybrid vigour in maize by using two unrelated cultivars. The best combinations yielded 50% more than the mean of the parents. Reports by Sanborn in 1890 and McClure in 1892 confirmed Beal's earlier reports and extended the generality of the superiority of hybrids over the average of the parental forms.

1.3.8 Refinement of populations

Several different 'population breeding methods' can be used: (i) bulk; (ii) mass selection; and (iii) recurrent selection. One of the methods used for managing large populations of segregates was the 'bulk method' proposed by Harlan *et al.* (1940) for multi-parent crosses. This concept changed the breeding methodologies for self-pollinated species. Mass selection is a system of breeding in which seeds from individuals selected on the basis of phenotype are bulked and used to grow the next generation. Mass selection is the oldest breeding method for plant improvement and was employed by early farmers for the development of cultivated species from their ancestral forms.

The enhancement of open-pollinated populations of crops such as rye, maize and sugarbeet, herbage grasses, legumes, and tropical trees such as cacao, coconut, oil palm, and some rubber, depends essentially on changing the gene frequencies so that the favourable alleles are fixed, while maintaining a high (but far from maximal) degree of heterozygosity. Recurrent selection is a method of plant breeding associated with quantitatively inherited traits by which the frequencies of favourable genes are increased in populations of plants. The methodology is cyclical with each cycle encompassing two phases: (i) selection of genotypes that possess the favourable or required genes; and (ii) crossing among the selected genotypes. This leads to a gradual increase in the frequencies of the desired alleles. While recurrent selection is often successful it also has potential limitations in closed populations and this has led to numerous modifications and alternative schemes (see Hallauer and Miranda, 1988). Recurrent selection breeding methods have been applied to a wide range of plant species, including self-pollinated crops.

1.3.9 Cell totipotency, tissue culture and somaclonal variation

The discovery of auxins, by Went and colleagues, and cytokinins, by Skoog and colleagues, preceded the first success of *in vitro* culture of plant tissues (White, 1934; Nobécourt, 1939).

All the genes necessary to make an entire organism can be induced to function in the correct sequence from a living cell isolated from a mature tissue (called totipotency). Regeneration of whole plants from single cells is an important new source of genetic variability for refining the properties of plants because when somatic embryos derived from single cells are grown into plants, the plants' characteristics vary somewhat. Larkin and Scowcroft (1981) coined the term 'somaclonal variation' to describe this observed phenotypic variation among plants derived from micro-propagation experiments. When it was recognized as a genuine phenomenon, somaclonal variation was considered to be a potential tool for the introduction of new variants of perennial crops that can be asexually propagated (e.g. banana). Somaclonal variation has also been exploited by plant breeders as a new source of genetic variation for annual crops.

1.3.10 Genetic engineering and gene transfer

The discovery of the structure of DNA by Watson and Crick has enhanced traditional breeding techniques by allowing breeders to pinpoint the particular gene responsible for a particular trait and to follow its transmission to subsequent generations. Enzymes that cut and rejoin DNA molecules allow scientists to manipulate genes in the laboratory. In 1973 Stanley Cohen and Herbert Boyer spliced the gene from one organism into the DNA of another to produce recombinant DNA which was then expressed normally and this formed the basis of genetic engineering. The goal of plant genetic engineers is to isolate one or more specific genes and introduce these into plants. Improvement in a crop plant can often be achieved by introducing a single gene, and genes can now be transferred to plants using the natural gene transfer system of a promiscuous pathogenic soil bacterium, *Agrobacterium tumefaciens*. DNA can also be introduced into cells by bombardment with DNA-coated particles or by electroporation. Transgenic breeding has the potential to decrease or increase the environmental impact of agricultural practices.

The initial successes in plant genetic engineering marked a significant turning point in crop research. In the 1990s in particular, there was an upsurge of private sector investment in agricultural biotechnology. Some of the first products were plant strains capable of synthesizing an insecticidal protein encoded by a gene isolated from the bacterium *Bacillus thuringiensis* (*Bt*). *Bt* cotton, maize, and other crops are now grown commercially. There are also crop cultivars which are tolerant to or capable of degrading herbicides. Proponents stress the value of these crops in conserving tillage soil, reducing the use of harmful chemicals and reducing the labour and costs involved in crop production.

1.3.11 DNA markers and genomics

During the 1980s and 1990s, various types of molecular markers such as restriction fragment length polymorphism (RFLP) (Botstein *et al.*, 1980), randomly amplified polymorphic DNA (RAPD) (Williams *et al.*, 1990; Welsh and McClelland, 1990), microsatellites and single nucleotide polymorphism (SNP) were developed. Because of their abundance and importance in the plant genome, molecular markers have been widely used in the fields of germplasm evaluation, genetic mapping, map-based gene discovery and marker-assisted plant breeding. Molecular marker technology has become a powerful tool in the genetic manipulation of agronomic traits.

Initiated by the complete sequencing of the Arabidopsis genome in 2000 (The Arabidopsis Genome Initiative, 2000) and the rice genome in 2002 (Goff *et al.*, 2002; Yu *et al.*, 2002), the genomes of an increasing number of plants have been or are being sequenced. Technological developments in bioinformatics, genomics and various omics fields are creating substantial data on which future revolutions in plant breeding can be based.

1.3.12 Breeding efforts in the public and private sectors

Agricultural research has mainly been the responsibility of a national and/or state government department. To accelerate progress in food production especially in developing countries, international agricultural research centres were established with major emphasis on the development of high yielding cultivars. Two centres, International Rice Research Institute (IRRI), Philippines, and Centro Internacional de Mejoramiento de Maiz y Trigo (CIMMYT), Mexico, established in the 1960s, made phenomenal contributions to food production by developing shorter and higher-yielding rice, wheat and maize cultivars. Encouraged by the astonishing success of these centres and two others which were established later, the Consultative Group on International Agricultural Research (CGIAR) was established in 1971. The CGIAR now has 15 international agricultural research centres, of which eight concentrate on specific crop plants and one on genetic resources with a mission to contribute towards sustainable agriculture for food security especially in developing countries. The breeding materials developed at these centres are distributed to public and private sector research programmes for utilization in the development of locally adapted cultivars. Through National Agricultural Research Systems (NARS), these centres work in close coordination with public and private breeding programmes in each country and share their breeding technologies and stocks of germplasm.

In the USA, crop breeding, with the exception of cotton, began largely as a tax-supported endeavour with breeding programmes taking place in most State Agricultural Experimental Stations and in the United States Department of Agriculture (USDA). This pattern changed with the advent of hybrid maize when inbred lines were initially developed by public institutions and utilized to produce hybrids by private companies. With the implementation of a Plant Variety Protection Act in the USA in 1974, private breeding was expanded to include forages, cereals, soybean, and other crops. The activities of private companies contributed to the total crop breeding effort and offered a large number of cultivar options for farmers and consumers. In the USA and other industrialized countries today, the new life-science companies notably the big multinationals such as Dow, DuPont and Monsanto, dominate the application of biotechnology to agriculture, and have developed many proprietary products.

1.4 Genetic Variation

The creation of new alleles and the mixing of alleles through recombination give rise to genetic variation which is one of the forces behind evolution. Natural selection favours one phenotype over another and these phenotypes are conditioned by one or more alleles. Genetic variation is fundamental for selection, by which progress in plant breeding can be made. There are various sources of genetic variation and those described in this section are largely based on the information provided at the following web sites: http://www.ndsu.nodak.edu/instruct/mcclean/plsc431/mutation and http://evolution.berkeley.edu/evosite/evo101/IIICGeneticvariation.shtml.

1.4.1 Crossover, genetic drift and gene flow

Chromosomal crossover takes place during meiosis and results in a chromosome with a completely different chemical composition from the two parent chromosomes. During the process, two chromosomes intertwine and exchange one end of the chromosome with the other. The mechanism of crossing over is the cytogenetic base for recombination.

Gene flow refers to the passage of traits or genes between populations to prevent the occurrence of large numbers of mutations and genetic drift. In genetic drift, random variation occurs in small populations leading to the proliferation of specific traits within that population. The degree of gene flow varies widely and is dependent on the type of organism and population structure. For example genes in a mobile population are likely to be more widely distributed than those in a sedentary population, resulting in high and low rates of gene flow, respectively.

1.4.2 Mutation

A mutation is any change in the sequence of the DNA encoding a gene which leads to a change in the hereditary material when an organism undergoes DNA replication. During the process of replication, the nucleotides of a chromosome are altered, so rather than creating an identical copy of DNA strands, there are chemical variations in the replicated strands. The alteration on the chemical composition of DNA triggers a chain reaction in the genetic information of an individual. The effect of a mutation depends on its size, location (intron or exon etc.), and the type of cell in which the mutation occurs. Large changes involve the loss, addition, duplication or rearrangement of whole chromosomes or chromosome segments. Most DNA polymerases have the ability to proofread their work to ensure that the unaltered genetic material is transferred to the next generation. There are many types of mutation and the most common are listed below.

1. Point mutations represent the smallest changes where only a single base is altered. For example, a single nucleotide change may result in the change of an amino acid (aa) codon into a stop codon and thus produce a change in the phenotype. Point mutations do not usually benefit the organism as most occur in recessive genes and are not usually expressed unless two mutations occur at the same locus.

2. In synonymous or silent substitutions the aa sequence of the protein is not changed because several codons can code for the same aa, and in non-synonymous substitutions changes in the aa sequence may not affect the function of the protein. However, there have been many cases where a change in a single nucleotide can create serious problem, e.g. in sickle cell anaemia.

3. Wild-type alleles typically encode a product necessary for a specific biological function and if a mutation occurs in that allele, the function for which it encodes is also lost. The general term for these mutations is loss-of-function mutations and they are typically recessive. The degree to which the function is lost can vary. If the function is entirely lost, the mutation is called a null mutation. It is also possible that some function may remain, but not at the level of the wild-type allele, these are known as leaky mutations.

4. A small number of mutations are actually beneficial to an organism providing new or improved gene activity. In these cases, the mutation creates a new allele that is associated with a new function. Any heterozygote containing the new allele along with the original wild-type allele will express the new allele. Genetically this will define the mutation as a dominant. This class of mutation is known as gain-of-function mutations.

5. A substitution is a mutation in which one base is exchanged for another. Such a substitution could change: (i) a codon to one that encodes a different aa thus causing a small change in the protein produced; (ii) a codon to one that encodes the same aa resulting in no change in the protein produced; or (iii) an aa-coding codon to a single 'stop' codon resulting in an incomplete protein (this can have serious effects since the incomplete protein will probably not be functional).

6. Insertions/deletions (indels) produce changes by deleting or inserting sections of DNA into the 'parental' DNA sequence. Because it is usually impossible to say whether a sequence has been deleted from one plant or inserted into another, these differences are called indels. Obviously the deletion of part of a gene can seriously affect the phenotype of organisms. Insertions can be disruptive if they insert themselves into the middle of genes or regulatory regions.

7. A mutation in which one nucleotide is changed causing all the codons to its right to be altered is known as frame-shift mutation. Since protein-coding DNA is divided into codons of three bases long, insertions and deletions of a single base can alter a gene so that its message is no longer correctly parsed. As a result, a single base change can have a dramatic effect on a polypeptide sequence.

8. Mutations which occur in germ line cells including both the gametes and the cells from which they are formed are known as germinal mutations. A single germ line mutation can have a range of effects: (i) no phenotypic change; mutations in junk DNA are passed on to the offspring but have no obvious effect on the phenotype; (ii) small (or quantitative) phenotypic changes; and (iii) significant phenotypic change.

9. Mutations in somatic cells which give rise to all non-germ line tissues, only affect the original individual and cannot be passed on to the progeny. To maintain this somatic mutation, the individual containing the mutation must be cloned.

In general, the appearance of a new mutation is a rare event. Most mutations that were originally studied occurred spontaneously. Such spontaneous mutations represent only a small number of all possible mutations. To genetically dissect a biological system further, induced mutations can be created by treating an organism with a mutagenic agent.

1.5 Quantitative Traits: Variance, Heritability and Selection Index

Recent advances in high-throughput technologies for the quantification of biological molecules have shifted the focus in quantitative genetics from single traits to comprehensive large-scale analyses. So-called omic technologies have now enabled geneticists to determine how genetic information is translated into biological function (Keurentjes *et al.*, 2008; Mackay *et al.*, 2009). The ultimate goal of quantitative genetics in the era of omics is to link genetic variation to phenotypic variation and to identify the molecular pathway from gene to function. The recent progress made in humans by combining linkage disequilibrium mapping (Chapter 6) and transcriptomics (Chapter 3) holds great promises for high-resolution association mapping and identification of regulatory genetic factors (Dixon *et al.*, 2007). Information from omics research will be integrated with our current knowledge at the phenotypic level to increase the effectiveness and efficiency of plant breeding.

1.5.1 Qualitative and quantitative traits

In general, qualitative traits are genetically controlled by one or a few major genes, each of which has a relatively large effect on the phenotype but is relatively insensitive to environmental influences. Trait distribution in a typical segregating population such as an F_2 shows multi-peak distribution, although individuals within a category show continuous variation. Each individual in the population can be classified unambiguously into distinct categories that correspond to different genotypes so that they can be studied using Mendelian methods.

Quantitative traits are genetically controlled by many genes, each of which has a relatively small effect on the phenotype, but is largely influenced by environmental factors (Buckler *et al.*, 2009). Trait distribution in an F_2 population usually shows a normal or bell-shape distribution and as a result, individuals cannot be classified into phenotypic categories that correspond to different genotypes thus making the effects of individual genes indistinguishable. Quantitative genetics is traditionally described as the study of all these genes as a whole and the total variation observed in a population results from the combined effects of genetic (polygenes as a whole) and environmental factors. However, quantitative variation is not due solely to minor allelic variation in structural genes as regulatory genes no doubt also contribute to this variation. We expected polygenes to show all the typical properties of chromosomal genes both in terms of action and in transmission through meiosis.

1.5.2 The concept of allelic and genotypic frequencies

A biological population is defined genetically as a group of individuals that exist together in time and space and that can mate or be crossed to each other to produce fertile progeny. Statistically, this group is called a 'population'. Breeding populations are created by breeders to serve as a source of cultivars that meet specific breeding objectives.

At the population level, genetics can be characterized by allelic and genotypic frequencies. The allele frequencies refer to the proportion of each allele in the population, while the genotypic frequencies refer to the proportion of individuals (plants) in the population that have a particular genotype. A gene may have many allelic states. Some of the alleles of a given gene may have such marked effects as to be clearly recognized as a classical major mutant. Other alleles, though potentially separable at the DNA level, may well cause only minor differences at the level of the external phenotype. For example, one allele at a locus involved with growth hormone production could be inactive and result in a dwarf plant, while others may simply reduce or increase height by a few percent.

Allele and genotypic frequencies can be calculated by simple counting in the population. For a gene with n alleles, there are $n(n + 1)/2$ possible genotypes. The relationship between allele frequency and genotypic frequency for a single gene at the population level can be used to infer the genetic status of the gene in that population, relative to the expected equilibrium under some assumed mating system. Allele frequencies are generally not an issue in breeding populations created from non-inbred parents or from three or more inbred parents. But breeding populations in both self-pollinated and cross-pollinated crops are often created by crossing two inbred individuals.

1.5.3 Hardy–Weinberg equilibrium (HWE)

A population is in equilibrium if the allele and genotypic frequencies are constant from generation to generation. A collection of pure selfers is also at equilibrium if all are completely selfed, with $P_{A_1A_1} = p$ and $P_{A_2A_2} = q$. This implies that the allele frequency and genotypic frequency share a simple relationship:

$$P_{A_1A_1} = p^2$$
$$P_{A_1A_2} = 2pq$$
$$P_{A_2A_2} = q^2$$

or

$$(p + q)^2 = p^2 + 2pq + q^2$$

With one generation of random mating, i.e. an individual in the population that is equally likely to mate with any other individual, the above simple relationship will be obeyed. However, HWE represents idealized populations and breeders routinely use procedures that cause deviations from HWE. These procedures include the lack of random mating, the use of small population sizes, assortative mating, selection, and inbreeding during the development of progenies. Some of these procedures, such as inbreeding and the use of small population sizes, affect all loci in the population while others affect only certain loci. Suppose that two traits are controlled by different sets of loci, and a change in one trait does not affect the other. If selection occurs only for the first trait, the loci affecting that trait may deviate from HWE, but the loci for the other trait will remain in equilibrium. In large natural populations, migration, mutation, and selection are the forces that can change allelic frequencies from generation to generation.

1.5.4 Population means and variances

Theoretically, a population can be described by its parameters such as the mean and variance which depend on the probability distribution of the population. The arithmetic mean, μ, also known as the first moment about the origin, is a parameter used to measure the central location of a frequency distribution. The population variance, σ^2, also known as the second moment about the mean, provides measures of the dispersion of the distribution. If the yield trait for a cultivar that is genetically homogenous is taken as an example, the genetic effect for this cultivar population is a constant. The yield for all individuals should also be a constant provided that environmental factors do not affect the yield which is equal to the population mean. However, the yield for each individual is affected not only by its genotype but also by environmental factors such as temperature, sunlight, water, and various nutrients. As a result, individuals may have different phenotypic values, in this case yield, resulting in continuous variation among individuals. Therefore, the individual yield measures vary either positively or negatively around the population mean so that they are either higher or lower than the population mean by a certain number which is determined by its variance.

1.5.5 Heritability

The response of traits to selection depends on the relative importance of the genetic and non-genetic factors which contribute to phenotypic variation among genotypes in a population, a concept referred to as heritability. The heritability of a trait has a major impact on the methods chosen for population improvement, inbreeding, and selection. Selection for single plants is more efficient when the heritability is high. The extent to which replicated testing is required for selection depends on the heritability of the trait.

The question of whether a trait variation is a result of genetic or environmental variation is meaningless in practice. Genes cannot cause a trait to develop unless the organism is growing in an appropriate environment, and, conversely, no amount of manipulation will cause a phenotype to develop unless the necessary gene or genes are present. Nevertheless, the variability observed in some traits might result primarily from difference in the numbers and the magnitude of the effect of different genes, but that variability in other traits might stem primarily from the differences in the environments to which various individuals have been exposed. It is therefore essential to identify reliable measures to determine the relative importance of not only the numbers and magnitude of the effects of the genes involved, but also of the effects of different environments on the expression of phenotypic traits (Allard, 1999).

Heritablity is defined as the ratio of genetic variance to phenotypic variance:

$$h^2 = \frac{\sigma_G^2}{\sigma_P^2} = \frac{\sigma_G^2}{\sigma_G^2 + \sigma_E^2}$$

where σ_P^2 is phenotypic variance, which has two components, genetic variance σ_G^2 and environment variance σ_E^2. σ_E^2 can be estimated by the phenotypic variances of non-segregating populations such as inbred lines and F_1s because individuals in such a population have the same genotypes and thus, phenotypic variation in these populations can be attributed to environmental factors. σ_G^2 can be estimated using segregating populations such as F_2 and backcrosses where variance components can be obtained theoretically.

1.5.6 Response to selection

Genetic variation forms the basis for selection in plant breeding. Selection results in the differential reproduction of genotypes in a population so that gene frequencies change and, with them, genotypic and phenotypic values (mean and variance) of the trait being selected. Response to selection, or advance in one generation of selection, is measured by the difference between the selected population and their offspring population, which is denoted as *R*. Response to selection has been referred to by several different names, including genetic progress, genetic advance, genetic gain, and predicted progress or gain, and has been denoted as *R*, *GS*, *G* and ΔG.

Starting with a parental population of mean, μ, a subset of individuals is selected. The selected individuals have a mean $\bar{x}$, while the offspring of the selected population has a mean $\bar{y}$. The difference between the selected population and the original population is defined as the selection differential, and denoted by *S*, i.e.

$$S = \bar{x} - \mu$$

The response to selection, *R*, can be written as

$$R = \bar{y} - \mu$$

The relationship between *S* and R is determined by heritability,

$$R = h^2 S$$

How much of the selection differential is realized in the offspring population depends on the heritability of the trait. The heritability, h^2, in the formula can be either h_N^2 or h_B^2 (depending on whether the offspring are produced by sexual or asexual reproduction, respectively). From the above formula, $\bar{y} = \mu + h^2 S$.

The population mean of the offspring derived from the selected individuals is equal to the parental population mean plus the response to selection (Fig. 1.1). When $h^2 = 1$, the selection differential will be fully realized in the offspring population so that its mean will deviate from the parental population by *S*. When $h^2 = 0$, the selection differential cannot be realized so the offspring population mean will regress to its parental population. When $0 < h^2 < 1$, the selection differential is partially realized so that the mean of the offspring population will deviate from the parental population by $h^2 S$. It is very useful to predict the response before selection is undertaken and details of the mathematical derivation of these predictions together with the various complications encountered can be found in Empig *et al.* (1972), Hallauer and Miranda (1988) and Nyquist (1991).

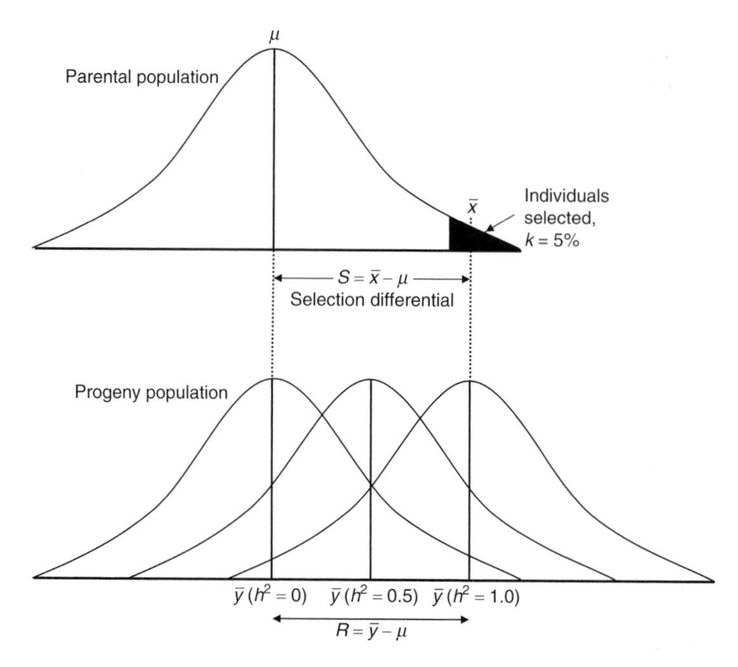

Fig. 1.1. Distribution of parental and progeny populations with a selection intensity of 5%. Because the phenotypic values of the selected plants include both a genetic and an environmental component, the progeny means depend on the heritability of the trait selected.

1.5.7 Selection index and selection for multiple traits

In most plant breeding programmes, there is a need to improve more than one trait at a time. For example, a high-yielding cultivar susceptible to a prevalent disease would be of little use to a grower. Recognition that improvement of one trait may cause improvement or deterioration in associated traits serves to emphasize the need for the simultaneous consideration of all traits which are important in a crop species. Three selection methods, which are recognized as appropriate for the simultaneous improvement of two or more traits in a breeding programme, are index selection, independent culling, and tandem selection. Independent culling requires the establishment of minimum levels of merit for each trait. An individual with a phenotype value below the critical culling level for any trait will be removed from the population. That is, only individuals meeting requirements for all traits will be selected. With tandem selection, one trait is selected until it is improved to a satisfactory level or a critical phenotypic value. Then, in the next generation or programme, selection for a second trait is carried out within the population selected for the first trait, and so on for the third and subsequent traits. A selection index is a single score which reflects the merits and demerits of all target traits. Selection among individuals is based on the relative values of the index scores. Selection indices provide one method for improving multiple traits in a breeding programme. The use of a selection index in plant breeding was originally proposed by Smith (1936) who acknowledged critical input from Fisher (1936). Subsequently, methods of developing selection indices were modified, subjected to critical evaluation, and compared to other methods of multiple trait selection.

It is generally recognized that a selection index is a linear function of observable phenotypic values of different traits. There are a number of forms of the equations available from index selection for multiple traits in grain. To construct a selection index, the observed value of each trait is weighted by an index coefficient,

$$I = b_1x_1 + b_2x_2 + \ldots + b_nx_n$$

where I is an index of merit of an individual, x_i represents the observed phenotypic value of the ith trait, and $b_1 \ldots b_n$ are weights assigned to phenotypic trait measurements represented as $x_1 \ldots x_n$. The ***b*** values are the products of the inverse of the phenotypic variance–covariance matrix, genotypic variance–covariance matrix, and a vector of economic weights. A number of variations of this index, most changing the manner of computing the ***b*** values, have been developed. These include the base index of Williams (1962), the desired gain index of Pesek and Baker (1969), and retrospective indexes proposed by Johnson *et al.* (1988) and Bernardo (1991). The emphasis in the retrospective index developments is on quantifying the knowledge experienced breeders have obtained. Baker (1986) summarized all select indexes in plant breeding developed before that time.

1.5.8 Combining ability

Combining ability is a very important concept in plant breeding and it can be used to compare and investigate how two inbred lines can be combined together to produce a productive hybrid or to breed new inbred lines. Selection and development of parental lines or inbreds with strong combining ability is one of the most important breeding objectives, no matter whether the goal is to create a hybrid with strong vigour or develop a pure-line cultivar with improved characteristics compared to their parental lines. In maize breeding, Sprague and Tatum (1942) partitioned the genetic variability among crosses into effects due to primarily either additive or non-additive effects, which correspond to two categories of combining ability, general combining ability (GCA) and special combining ability (SCA). The relative importance of GCA and SCA depends on the extent of previous testing of the parents included in the crosses. Although these concepts were developed for breeding maize, an open-pollinated crop, they are generally applicable to self-pollinated crops.

The GCA for an inbred line or a cultivar can be evaluated by the average performance of yield or other economic traits in a set of hybrid combinations. The SCA for a cross combination can be evaluated by the deviation in its performance from the value expected from the GCA of its two parental lines. If the crosses among a set of inbred lines are made in such a way that each line is crossed with several other lines in a systematic manner, the total variation among crosses can be partitioned into two components ascribable to GCA and SCA. The mean performance of a cross ($\bar{x}_{AB}$) between two inbred lines A and B can be represented as

$$\bar{x}_{AB} = GCA_A + GCA_B + SCA_{AB}$$

The GCA_A and GCA_B are the GCA of the parents A and B, respectively, and the cross of A × B is expected to have a performance equal to the sum ($GCA_A + GCA_B$) of the GCA of their parents. The actual performance of the cross, however, may be different from the expectation by an amount equivalent to the SCA. Sprague and Tatum (1942) interpreted these combining abilities in terms of type of gene action. The differences due to GCA of lines are the results of additive genetic variance and additive by additive interaction whereas SCA is a reflection of non-additive genetic variances.

1.5.9 Recurrent selection

Recurrent selection can be broadly defined as the systematic selection of desirable individuals from a population followed by recombination of the selected individuals to form a new population. The basic feature of recurrent selection methods is that they are procedures conducted in a repetitive manner, or recycling, including development of a base population with which to begin selection, evaluation of individuals from the population, and selection of superior individuals as parents that can be crossed to produce a new population for the next cycle of selection, as shown below:

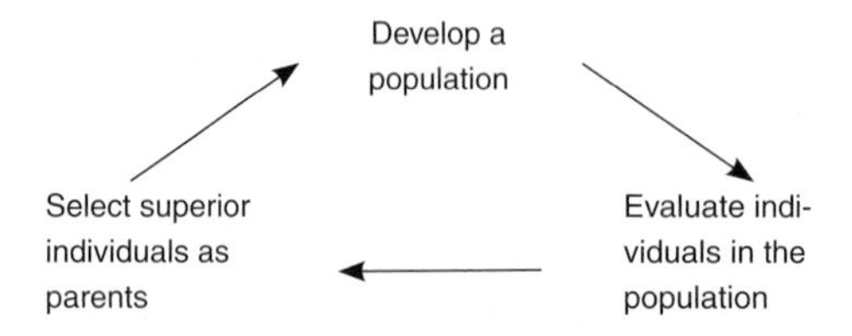

A cycle of selection is completed each time a new population is formed. The initial population that is developed for a recurrent selection programme is referred to as the base, or cycle 0, population. The population formed after one cycle of selection is called the cycle 1 population; the cycle 2 population is developed from the second cycle of selection, and so on.

Recurrent selection procedures are conducted for primarily quantitatively inherited traits. The objective of recurrent selection is to improve the mean performance of a population of plants by increasing the frequency of favourable alleles in a consistent manner in order to enhance the value of the population and to maintain the genetic variability present in the population as effectively as possible. In addition, separation of the genetic and environmental effects is an important facet of effective recurrent selection methods. The improved populations can be used as a cultivar per se, as parents of a cultivar-cross hybrid and as a source of superior individuals that can be used as inbred lines, pure-line cultivars, clonal cultivars, or parents of a synthetic line. Successful recurrent selection results in an improved population that is superior to the original population in mean performance and in the performance of the best individuals within it. Ideally, the population will be improved without its genetic variability being significantly reduced so that additional selection and improvement can occur in the future. Recurrent selection is complementary to inbred development procedures; in fact the concept of recurrent selection was developed, particularly for outcrossing crops, to rectify limitations in inbred development by continuous selfing that rapidly leads to inbreeding and allele fixation and thus inadequate opportunity for selection. There are two ways by which recurrent selection address this limitation in inbred development (Bernardo, 2002). First, recurrent selection increases the frequency of favourable alleles in the population by repeated cycles of selection. Secondly, recurrent selection maintains the degree of genetic variation in the population to allow sustained progress from subsequent cycles of selection. Genetic variation is maintained by recombining a sufficiently large number of individuals to reduce random fluctuations in allele frequencies, i.e. genetic drift.

Since the late 1950s, extensive research has been conducted to determine the relative importance of different genetic effects on the inheritance of quantitative traits for most cultivated plant species. As indicated by Hallauer (2007), quantitative genetic research has provided extensive information to assist plant breeders in developing breeding and selection strategies. Directly and/or indirectly, the principles for the inheritance of quantitative traits are pervasive in developing superior cultivars to meet the worldwide food, feed, fuel and fibre demands. The principles of quantitative genetics will have continued importance in the future.

1.6 The Green Revolution and the Challenges Ahead

The application of science and technology to crop production in the second half of the 20th century resulted in significant yield improvements for rice, wheat and maize in the developed countries, and the final result of these efforts was the Green Revolution which led to a new type of agriculture – high-input or chemical-genetic agriculture – which replaced the more traditional system. Countries involved in the 'Green Revolution', a term coined by Borlaug (1972), included Japan, Mexico, India and China among others.

By production and acreage yardsticks, agriculture has been very successful. The application of scientific knowledge to agriculture has resulted in greatly increased yields per unit land area for many of our important crops as exemplified by the 92% increase in cereal production in the developing world between 1961 and 1990. The sharp increase in human populations has been paralleled by the increase in food supply. However, yield growth rates are stagnating in some areas and, in a few cases, falling. A slowdown in the rate of yield increase of major cereals raises concern because increased yields are expected to be the source of increased food production in the future (Reeves *et al.*, 1999). On the other hand, increased national wealth resulting from economic development is not necessarily correlated with a decrease in the rate of population growth. Widespread hunger persists in a world that produces enough food.

There are many reasons for being concerned about meeting future food demands (Khush, 1999; Swaminathan, 2007). Expansion of the planet's population creates an increased demand for food and income. Other issues such as the cost of food, which may represent 60% of income in the developing world, the 800 million people who are food insecure, the 200 million children who are malnourished, and the continuing decline of available land for farming and water to irrigate crops, all indicate the need to use all the technologies available to increase productivity, assuming they can be employed in harmony with the environment. Plant breeding has generally accounted for one-half of the increases in productivity of the major crops and the future will continue to depend on advances in plant breeding. The increase in productivity has meant that large areas of land can be saved as wildlife habitats or used for purposes other than agriculture. As the availability of land and water is decreasing and populations are increasing in size, the 50% increase in food production predicted to be required over the next 25 years, poses an obvious challenge.

The danger of population growth overtaking food supplies was predicted by Malthus in 1817. The dire predictions of Malthus were forestalled, at least temporarily, by the extensive cultivation of new land and by the development of a modern agricultural science which enabled food crops to be produced at far higher yields than Malthus could have ever anticipated. However, the production of food has still not been optimized in all areas of the world.

Weather and climate profoundly affect crop production and natural events can disrupt normal climate cycles and affect agriculture. In addition, human-induced climatic change is set to accelerate during this century and this will also impact on crop production. Much of the arable land has been used for industrial purposes and land-use patterns indicate an increase in intensive farming which, however, must be sustainable.

Agricultural products are affected by abiotic and biotic stresses and one of the major challenges to the future of plant breeding is the development of cultivars and hybrids with multiple resistances or tolerances to these stresses.

The security of the food supply for an increasing world population largely depends on the availability of water for agriculture. Increasing the efficiency of water use for our major crop species is an important target in agricultural research, particularly in light of the increasing competition for limited supplies of fresh water in many parts of the world.

There are four prerequisites for greater productivity (Poehlman and Quick, 1983): (i) an improved farming system; (ii) instruction of farmers; (iii) optimization of the supply of water and fertilizers; and (iv) availability of markets. To increase crop productivity planting high-yielding cultivars must be combined with improved practices of irrigation, fertilization and pest control. Maximum crop yield will only be achieved if the improved crop cultivar receives and responds to the optimum combination of water, fertilizer and cultural practices.

1.7 Objectives of Plant Breeding

The aim of plant breeders is to reassemble desirable inherited traits to produce crops with improved characteristics. Thus far, plant breeders have mainly been concerned with bringing about a continuous improvement in the productivity of that part of the plant which is of economic importance, the stability of production through in-built resistance to pests and diseases and nutritive and organoleptic or other desired quality characters.

Many parameters and selection criteria should be included as breeding objectives. According to Sinha and Swaminathan (1984) and other sources, the major objectives of plant breeders can be summarized by the following list:

1. High primary productivity and efficient final production for each unit of cultivation and solar energy invested: to ensure that all the light that falls on a field is intercepted by leaves and that photosynthesis itself is as efficient as possible. Greater efficiency in photosynthesis could perhaps be achieved by reducing photorespiration.
2. High crop yield: plants must be selected which invest a large proportion of their total primary productivity into those areas which are commercially desirable, e.g. seeds, roots, leaves or stems.
3. Desirable nutritional value, organoleptic properties and processing qualities: the proportion of essential amino acids and the total protein in cereal grains, for example, should be increased to improve their nutritional quality.
4. Biofortifying crops with essential mineral elements that are frequently lacking in the human diet such as Fe and Zn, vitamins and amino acids (Welch and Graham, 2004; White and Broadley, 2005; Bekaert *et al.*, 2008; Mayer *et al.*, 2008; Ufaz and Galili, 2008; Naqvi *et al.*, 2009; Xu *et al.*, 2009a).
5. Modifying crop plants to generate plant-derived pharmaceuticals to supply low-cost drugs and vaccines to the developing world (Ma *et al.*, 2005).
6. Adaptation to cropping systems: including breeding for contrasting cropping, intercropping, and sustainable cropping systems (Brummer, 2006).
7. More extensive and efficient nitrogen fixation: breeding cereals that encourage the growth of increased numbers of nitrogen-fixing microorganisms around their roots to reduce the need for nitrogen fertilizer.
8. More efficient use of water whether there is a plentiful supply or dearth of water.
9. Stability of crop production by resilience to weather fluctuations, resistance to the multiple alliance of weeds, pests and pathogens, and tolerance to various abiotic stresses such as heat, cold, drought, wind, and soil salinity, acidity or aluminium toxicity.
10. Insensitivity to photoperiod and temperature: selection of crop cultivars that are insensitive to photoperiod or temperature and characterized by a high per-day biomass production would allow the development of contingency cropping patterns to suit different weather probabilities.
11. Plant architecture and adaptability to mechanized farming: the number and positioning of the leaves, branching pattern of the stem, the height of the plant, and the positioning of the organs to be harvested are all important to crop production and often determine how well plants can be harvested mechanically.
12. Elimination of toxic compounds.
13. Identification and improvement of hardy plants suitable for sources of biomass and renewable energy.
14. Multiple uses of a single crop.
15. Environmentally-friendly and stable across environments.

In conclusion, plant breeding has many breeding objectives and each of the objectives can be addressed in a specific breeding programme. A successful breeding programme consists of a series of activities as Burton (1981) summarized in six words: variate, isolate, evaluate, intermate, multiply and disseminate.

1.8 Molecular Breeding

By 2025, the global population will exceed seven billion. In the interim per-capita availability of arable land and irrigation water will decrease from year to year as biotic and abiotic stresses increase. Food security, best defined as economic, physical and social access to a balanced diet and safe dinking water will be threatened, with a holistic approach to nutritional and non-nutritional factors needed to achieve success in the eradication of hunger. Science and technology can play a very important role in stimulating and sustaining an Evergreen Revolution leading to long-term increases in productivity without any associated ecological harm (Borlaug, 2001; Swaminathan, 2007). The objectives of the plant breeder can be realized through conventional breeding integrated with various biotechnology developments (e.g. Damude and Kinney, 2008; Xu *et al.*, 2009c).

Plant breeding can be defined as an evolving science and technology (Fig. 1.2). It has gradually been evolving from art to science over the last 10,000 years, starting as an ancient art to the present molecular design-based science. With the development of molecular tools which will be discussed further in Chapters 2 and 3, plant breeding is becoming quicker, easier, more effective and more efficient (Phillips, 2006). Plant breeders will be well equipped with innovative approaches to identify and/or create genetic variation, to define the genetic feature of the genes related to the variation (position, function and relationship with other genes and environments), to understand the structure of breeding populations, to recombine novel alleles or allele combinations into specific cultivars or hybrids, and to select the best individuals with desirable genetic features which enable them to adapt to a wide range of environments.

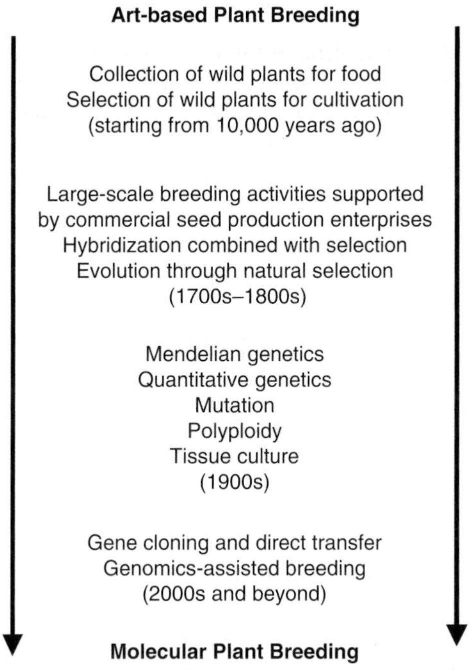

Art-based Plant Breeding

Collection of wild plants for food
Selection of wild plants for cultivation
(starting from 10,000 years ago)

Large-scale breeding activities supported by commercial seed production enterprises
Hybridization combined with selection
Evolution through natural selection
(1700s–1800s)

Mendelian genetics
Quantitative genetics
Mutation
Polyploidy
Tissue culture
(1900s)

Gene cloning and direct transfer
Genomics-assisted breeding
(2000s and beyond)

Molecular Plant Breeding

Fig. 1.2. The steps of evolution of 'plant breeding'. With the availability of more sophisticated tools, the art of plant breeding became science-based technology, molecular plant breeding.

Sequencing data for many plants is now readily available and the GenBank database is doubling every 15 months. Over 20 plant species including many important crops are in the process of being sequenced (Phillips, 2008). The next challenge is to determine the function of every gene and eventually how genes interact to form the basis of complex traits. Fortunately, DNA chips and other technologies are being developed to study the expression of multiple or even all genes simultaneously. High throughput robotics and bioinformatics tools will play an essential role in this endeavour.

New information about our crop species is expanding our capabilities to use molecular genetics. For example, we did not previously realize how similar broadly related species are in terms of their gene content and gene order. Since these species cannot usually be crossed, there was no means of assessing their relatedness. With the advent of DNA-based molecular markers, the extensive genetic mapping of chromosomes became readily possible for a variety of species. We learned that the genomes were highly similar and that this similarity allowed the prediction of gene locations among species. For example, rice has become the model or reference species for the cereals as many of the gene sequences on the rice chromosomes are shared with other cereals such as maize, sorghum, sugarcane, millet, oats, wheat and barley (Xu *et al.*, 2005). Knowing the complete DNA sequence of a model or reference genome allows genes/traits from this model to be tracked to other genomes. We have come to realize that the differences between species of plants are not due to novel genes, but to novel allelic specifications and interactions.

Since many fundamental aspects of current plant breeding procedures are not well understood, further data relating to the genetics of crop species may help to shed light on the genetic gains obtained from plant breeding. For example, in successful plant breeding programmes, the genetic base often becomes narrower rather than broader. 'Elite by elite' crosses may be the rule in these programmes. Molecular genetic markers have been widely employed to identify cryptic and novel genetic variation among cultivars and related species and used to increase the efficiency of selection for agronomic traits and the pyramid of genes from different genetic backgrounds.

Long-term selection programmes would be expected to lead to genetic fixation, however this has not been found to be the case so far and variation is still observed. Several mechanisms for *de novo* variation have been described, including intragenic recombination, unequal crossing over among repeated elements, transposon activity, DNA methylation, and paramutation. Another important feature in plant breeding whose molecular basis is not understood is heterosis although it is used as the basis for many seed-producing industries. Genomics and particularly transcriptomics are now being used to identify the heterotic genes responsible for increasing crop yields. Comprehensive quantitative trait locus-based phenotyping (phenomics) combined with genome-wide expression analysis, should help to identify the loci controlling heterotic phenotypes and thus improve the understanding of the role of heterosis in evolution and the domestication of crop plants (Lippman and Zamir, 2007), and finally to make it possible to predict hybrid performance.

Messenger RNA transcript profiling is an obvious candidate for functional genomic application to plant breeding. Although direct selection at the gene transcript level using microarray or real-time PCR may be a long-term goal, other genomic tools can be used to achieve shorter term goals with more practical applications (Crosbie *et al.*, 2006). Genetic modification of crops today involves the interfacing of molecular biology, cell and tissue culture, and genetics/breeding. The transfer of genes by cellular and molecular means will increase the available gene pool and lead to second generation biotechnology plant products such as those with a modified oil, protein, vitamin, or micronutrient content or those that have been engineered to produce compounds that can be used as vaccines or anticarcinogens.

While all these new innovations have been useful, practical plant breeding continues to be based on hybridization and selection with little change in the basic procedures. A more complete understanding of the mechanisms by which genetic and environmental variation modify yield and composition is needed so that specific quantitative and qualitative targets can be identified. To achieve this aim, the expertise of plant genomics (including various omics), physiology and agronomy, as well as plant modelling techniques must be combined (Wollenweber *et al.*, 2005) and many logistic and genetic constraints also need to be resolved (Xu and Crouch, 2008).

2

Molecular Breeding Tools: Markers and Maps

2.1 Genetic Markers

In conventional plant breeding, genetic variation is usually identified by visual selection. However, with the development of molecular biology, it can now be identified at the molecular level based on changes in the DNA and their effect on the phenotype. Molecular changes can be identified by the many techniques that have been used to label and amplify DNA and to highlight the DNA variation among individuals. Once the DNA has been extracted from plants or their seeds, variation in samples can be identified using a polymerase chain reaction (PCR) and/or hybridization process followed by polyacrylamide gel electrophoresis (PAGE) or capillary electrophoresis (CE) to identify distinct molecules based on their sizes, chemical compositions and charges. Genetic markers are used to tag and track genetic variation in DNA samples.

Genetic markers are biological features that are determined by allelic forms and can be used as experimental probes or tags to keep track of an individual, a tissue, cell, nucleus, chromosome or gene. In classical genetics, genetic polymorphism represents allelic variation. In modern genetics, genetic polymorphism is the relative difference at any genetic locus across a genome. Genetic markers can be used to facilitate studies of inheritance and variation.

Desirable genetic markers should meet the following criteria: (i) high level of genetic polymorphism; (ii) co-dominance (so that heterozygotes can be distinguished from homozygotes); (iii) clear distinct allele features (so that different alleles can be identified easily); (iv) even distribution on the entire genome; (v) neutral selection (without pleiotropic effect); (vi) easy detection (so that the whole process can be automated); (vii) low cost of marker development and genotyping; and (viii) high duplicability (so that the data can be accumulated and shared between laboratories).

Most molecular markers belong to the so-called anonymous DNA marker type and generally measure apparently neutral DNA variation. Suitable DNA markers should represent genetic polymorphism at the DNA level and should be expressed consistently across tissues, organs, developmental stages and environments; their number should be almost unlimited; there should be a high level of natural polymorphism; and they should be neutral with no effect on the expression of the target trait. Finally, most DNA markers are co-dominant or can be converted into co-dominant markers.

Table 2.1 lists the major molecular marker technologies that are currently available. Only a selection of widely-used representative types of markers will be discussed in this section. Figure 2.1 shows the molecular mechanism of several major DNA markers and the genetic polymorphisms that can be generated by restriction site or PCR priming site mutation, insertion, deletion or by changing the number of repeat units between two restriction or PCR priming sites and nucleotide mutation resulting in a single nucleotide polymorphism (SNP). There are several comprehensive reviews that cover all the important DNA markers, e.g. Reiter (2001), Avise (2004), Mohler and Schwarz (2005) and Falque and Santoni (2007). Further information regarding the application of DNA markers in genetics and breeding can be found in Lörz and Wenzel (2005). After a brief review of the classical markers, DNA markers will be discussed in more detail in this section.

2.1.1 Classical markers

Morphological markers

In the late 1800s, following his studies on the garden pea (*Pisum sativum*), G.J. Mendel proposed two basic rules of genetics,

Table 2.1. DNA markers and related major molecular techniques.

Southern blot-based markers
Restriction fragment length polymorphism (RFLP)
Single strand conformation polymorphic RFLP (SSCP-RFLP)
Denaturing gradient gel electrophoresis RFLP (DGGE-RFLP)
PCR-based markers
Randomly amplified polymorphic DNA (RAPD)
Sequence tagged site (STS)
Sequence characterized amplified region (SCAR)
Random primer-PCR (RP-PCR)
Arbitrary primer-PCR (AP-PCR)
Oligo primer-PCR (OP-PCR)
Single strand conformation polymorphism-PCR (SSCP-PCR)
Small oligo DNA analysis (SODA)
DNA amplification fingerprinting (DAF)
Amplified fragment length polymorphism (AFLP)
Sequence-related amplified polymorphism (SRAP)
Target region amplified polymorphism (TRAP)
Insertion/deletion polymorphism (Indel)
Repeat sequence-based markers
Satellite DNA (repeat unit containing several hundred to thousand base pairs (bp))
Microsatellite DNA (repeat unit containing 2–5 bp)
Minisatellite DNA (repeat unit containing more than 5 bp)
Simple sequence repeat (SSR) or simple sequence length polymorphism (SSLP)
Short repeat sequence (SRS)
Tandem repeat sequence (TRS)
mRNA-based markers
Differential display (DD)
Reverse transcription PCR (RT-PCR)
Differential display reverse transcription PCR (DDRT-PCR)
Representational difference analysis (RDA)
Expression sequence tags (EST)
Sequence target sites (STS)
Serial analysis of gene expression (SAGE)
Single nucleotide polymorphism-based markers
Single nucleotide polymorphism (SNP)

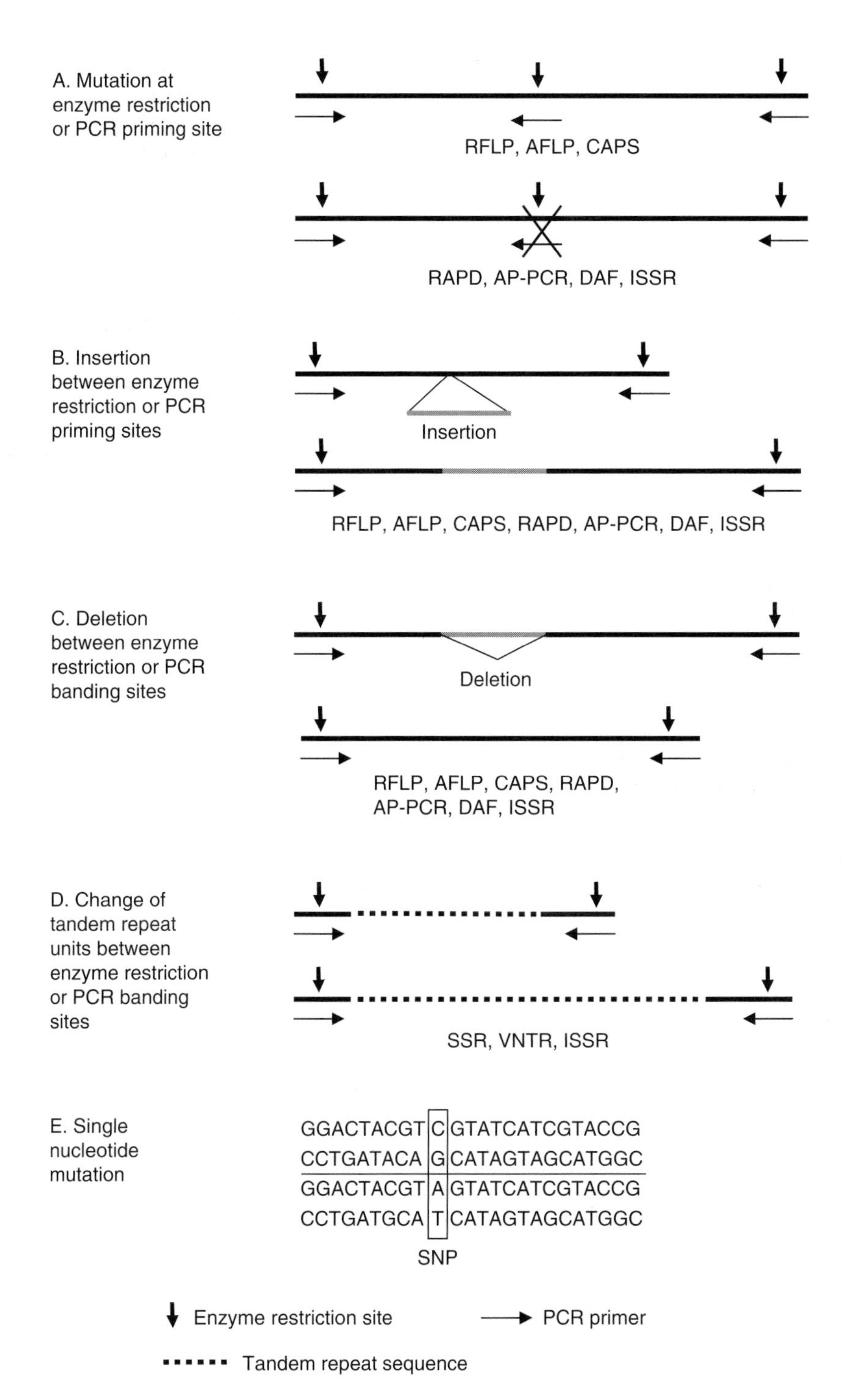

Fig. 2.1. Molecular basis of major DNA markers. Parts A–E show different ways in which DNA markers (listed below each diagram) can be generated. The cross in part A indicates that mutation has eliminated the priming site. Abbreviations: as defined in Table 2.1; VNTR, variable number of tandem repeat; CAPS, a DNA marker generated by specific primer PCR combined with RFLP; ISSR, inter simple sequence repeat.

which were later known as the Mendelian laws of equal segregation and independent assortment. Mendel selected individuals which differed in a particular trait and used them as the parental lines in a cross breeding experiment to determine the phenotype of the offspring with regard to the selected trait. The term phenotype (derived from Greek) literally means 'the form that is shown' and is used by both geneticists and breeders. The seven pairs of contrasting phenotypes studied by Mendel included round versus wrinkled seeds, yellow versus green seeds, purple versus white petals, inflated versus pinched pods, green versus yellow pods, axial versus terminal flowers and long versus short stems. The plants in the segregated populations of the pea, such as F_2 and backcross, were classified into two distinct groups depending on their phenotypes. These contrasting morphological phenotypes are the starting point for any genetic analysis and can be mapped to particular chromosomes using the Mendelian laws of inheritance and can thus be used as morphological markers of the genome and the particular trait.

Morphological markers therefore generally represent genetic polymorphisms which are visible as differences in appearance, such as the relative difference in plant height and colour, distinct differences in response to abiotic and biotic stresses, and the presence/absence of other specific morphological characteristics. A large number of variants showing particular morphological or physiological phenotypes have been generated by tissue culture and mutation breeding. Using selection techniques these variants can be genetically stabilized and then used as morphological markers.

Some genetic stocks contain more than one morphological marker, for example there are a total of over 300 morphological markers available for genetic studies in rice (Khush, 1987) and more are being created for functional genomics. Many morphological marker stocks are also available for tomato (http://www.plantpath.wisc.edu/GeminivirusResistantTomatoes/MERC/Tomato/Tomato.html), maize (Neuffer *et al.*, 1997) and soybean (Palmer and Shoemaker, 1998). Many of these markers have been linked with other agronomic traits.

Morphological markers are usually mapped by classical two- or three-point linkage tests. The linkage groups are established and the order of the markers and the relative distance between any two are determined by their recombinant frequencies. Relatively complete linkage maps have been constructed in many crop species using morphological markers and these maps provide the fundamental information for the genetic mapping of many physiological and biochemical traits.

However, it is difficult to construct a relatively saturated genetic map because of the limitation in the number of morphological markers with distinguishable polymorphisms. In addition, many morphological markers have deleterious effects on phenotypes and some are significantly affected by other factors such as environments or maturity which results in potential problems when these markers are used for genetics and plant breeding.

Cytological markers

By studying the morphology, number and structure of chromosomes from different species, particular cytogenetic features can be found, such as various types of aneuploidy, variants of chromosome structure and abnormal chromosomes. These can be used as genetic markers to locate other genes on to chromosomes and determine their relative positions, or used for genetic mapping via chromosome manipulations such as chromosome substitution.

The structural features of chromosomes can be shown by chromosome karyotype and bands. The banding patterns are indicated by colour, width, order and position, revealing the difference in distributions of euchromatin and heterochromatin. There are Q bands (produced by quinacrine hydrochloride), G bands (produced by Giemsa stain) and R bands (reversed Giemsa). These chromosome landmarks are not only useful for characterizing normal chromosomes but also for detecting chromosome mutation.

Cytological markers have been widely used to identify linkage groups within specific chromosomes and have been widely applied in physical mapping. However, because of the limited number and resolution, they have limited applications in genetic diversity analysis, genetic mapping and marker-assisted selection (MAS).

Protein markers

Isozymes are structural variants of an enzyme and while they differ from the original enzyme in molecular weight and mobility in an electric field, they have the same catalytic activity. The difference in enzyme mobility is caused by point mutations resulting from amino acid substitution such that isozymes reflect the products of different alleles rather than different genes. Therefore, isozymes can be genetically mapped on to chromosomes and then used as genetic markers for mapping other genes. Isozyme markers are based on their biochemistry and thus are also known as biochemical or protein markers.

However, their use as markers is limited. For example a total of 57 isozymes representing about 100 loci have been identified in plants (Vallegos and Chase, 1991) but for specific species only 10–20 isozymes are available so that they cannot be used to construct a complete genetic map. Each isozyme can only be identified with a specific stain which also limits their use in practice.

2.1.2 DNA markers

RFLP

Botstein *et al.* (1980) first used DNA restriction fragment length polymorphism (RFLP) in human linkage mapping and this pioneered the utilization of DNA polymorphisms as genetic markers. It is known that the genomes of all organisms show many sites of neutral variation at the DNA level. These neutral variant sites do not have any effect on the phenotype. In some cases a neutral site is nothing more than a single nucleotide difference within a gene or between genes; and in others it represents the site of a variable number of tandem repeats of 'junk DNA' present between genes. The development of RFLP markers has accelerated the construction of molecular linkage maps for many organisms, improved the accuracy of gene location, and reduced the time required to establish a complete linkage map.

The digestion of purified DNA using restriction enzymes which cut the DNA strand wherever there is a recognition site sequence (usually four to eight base pairs), leads to the formation of RFLPs which yield a molecular fingerprint that may be unique to a particular individual. If the bases are positioned at random in the genome, an enzyme having a recognition site with six bases will cleave the DNA at every 4096 bases on average (4^6). A genome of 10^9 bases could thus produce around 250,000 restriction fragments of variable length. Gel electrophoresis on such a large number of genomic DNA digestion products produces a continuous smear image. Particular fragments that are homologous between several individuals, and possibly allelic, can be separated only by means of molecular probes using the Southern technique (Southern, 1975). RFLP analysis includes the following steps (Fig. 2.2):

1. DNA isolation: a significant amount of DNA must be isolated from multiple individuals from target genotypes (parents and segregating populations, germplasm survey, garden blot, etc.) and purified to a fairly stringent degree as contaminants can often interfere with the restriction enzyme and inhibit its ability to digest the DNA.
2. Restriction digestion: restriction enzyme is added to purified genomic DNA under buffered conditions. The enzyme cuts at recognition sites throughout the genome and leaves behind hundreds of thousands of fragments.
3. Gel electrophoresis: digested products (restriction fragments) are electrophoresed on agarose gel and when visualized appear as smears because of the large number of fragments.

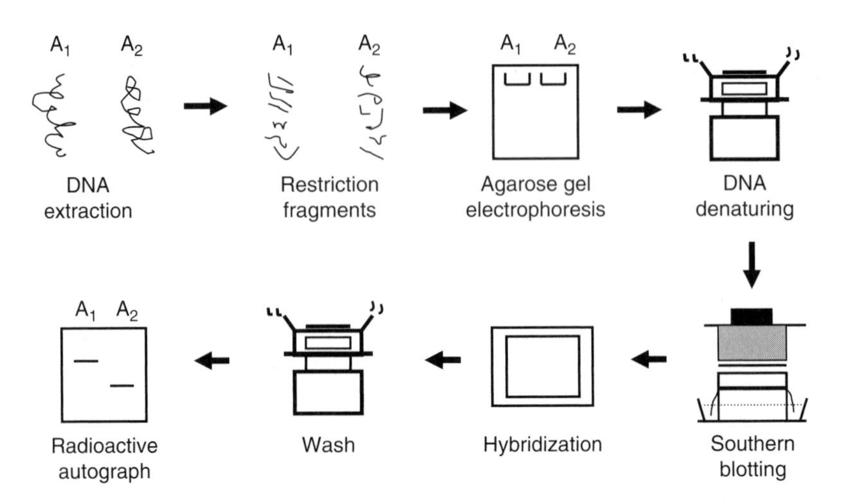

Fig. 2.2. RFLP workflow from DNA extraction to radio-autograph. Modified from Xu and Zhu (1994).

4. The agarose gel is denatured using NaOH solution and then neutralized.
5. The DNA fragments are transferred to a nitrocellulose membrane using Southern blotting.
6. Probe visualization: the membrane-bound genomic DNA is probed by hybridization using a cloned fragment of the genome of interest or a genome from a relatively close species as the probe.
7. The membrane is washed to remove non-specifically hybridized DNA.
8. In most cases the sizes of the fragments are determined by radioactive methods. The probe-restriction enzyme combinations may identify two or more differently sized fragments. Polymorphism is revealed whenever the recognized fragments are of non-identical lengths.

Differences in size of restriction fragments are due to: (i) base pair changes that result in gain and loss of restriction sites; and (ii) insertions/deletions at the restriction sites within the restriction fragments on which the probe sequence is located.

Molecular probes are DNA fragments isolated and individualized by cloning or PCR amplification. They may originate from fragmented total genomic DNA and thus contain coding or non-coding sequences, unique or repeated, of nuclear or cytoplasmic origin. They may also be complementary DNA (cDNA). The standard procedure for developing genomic DNA probes is to digest total DNA with a methylation-sensitive enzyme (e.g. *Pst*I), thereby enriching the library for single-copy sequences (Burr *et al.*, 1988). Typically, the digested DNA is size fractionated on a preparative agarose gel. DNA fragments ranging from 500 to 2000 bp are excised and eluted for cloning into a plasmid vector (e.g. pUC18). Digests of the plasmids are screened for inserts and their lengths can be estimated. Southern blots of the inserts can be probed with total sheared genomic DNA to select clones that hybridize to single- and low-copy sequences and to eliminate clones that hybridize to medium- and high-copy repeated sequences. Single- and low-copy probes are screened for RFLPs among a sample of genotypes using genomic DNAs digested with restriction endonucleases (one per assay). Typically, in species with moderate to high polymorphism rates, two to four restriction endonucleases with hexanucleotide recognition sites are tested. *Eco*RI, *Eco*RV and *Hin*dIII are widely used. In species with low polymorphism rates, additional restriction endonucleases can be tested to increase the chance of finding a polymorphism. Both the theory and the techniques for RFLP analysis in plant genome mapping have been intensively reviewed (Botstein *et al.*, 1980; Tanksley *et al.*, 1988).

Most RFLP markers are co-dominant and locus specific. RFLP genotyping is highly reproducible and the methodology is simple and requires no special instrumentation. High-throughput markers (e.g. cleaved amplified polymorphic sequence (CAPS) or insertion/deletion (indel) markers) can be developed from RFLP probe sequences. The CAPS technique, also known as PCR-RFLP, consists of digesting a PCR-amplified fragment with one or several restriction enzymes, and detecting the polymorphism by the presence/absence restriction sites (Konieczny and Ausubel, 1993).

RFLP markers are powerful tools for comparative and synteny mapping. However, RFLP analysis requires large amounts of high quality DNA and has low genotyping throughput and is very difficult to automate. Most genotyping involves radioactive methods so its use is limited to specific laboratories. RFLP probes must be physically maintained and it is therefore difficult to share them between laboratories. In addition, the level of RFLP is relatively low and selection for polymorphic parental lines is a limiting step in the development of a complete RFLP map.

RAPD

Williams *et al.* (1990) and Welsh and McClelland (1990) independently described the utilization of a single, random-sequence oligonucleotide primer in a low stringency PCR (35–45°C) for the simultaneous amplification of several discrete DNA fragments referred to as random amplified polymorphic DNA (RAPD) and arbitrary primed PCR (AP-PCR), respectively. Another related technique is DNA amplification fingerprinting (DAF) (Caetano-Anollés *et al.*, 1991). These methods differ from one another in primer length, the stringency of the conditions and the method of separation and detection of the fragments. They all can be used to identify RAPD.

The principle of RAPD consists of a PCR on the DNA of the individual under study using a short primer, usually ten nucleotides, of arbitrary sequence. The primer which binds to many different loci is used to amplify random sequences from a complex DNA template that is complementary to it (or includes a limited number of mismatches). This means that the amplified fragments generated by PCR depend on the length and size of both the primer and the target genome. Ten-base oligomers of varying GC content (ranging from 40 to 100%) are usually used. If two hybridization sites are similar to one another (at least 3000 bp) and in opposite directions, that is, in a configuration that will allow the PCR, amplification will take place. The amplified products (of up to 3.0 kb) are usually separated on agarose gels and visualized using ethidium bromide staining. The use of a single 10-mer oligonucleotide promotes the generation of several discrete DNA products and these are considered to originate from different genetic loci. Polymorphisms result from mutations or rearrangements either at or between the primer binding sites and are visible in conventional agarose gel electrophoresis as the presence or absence of a particular RAPD band. RAPDs predominantly provide dominant markers but homologous allele combinations can sometimes be identified with the help of detailed pedigree information.

RAPDs have several advantages and for this reason they are widely used (Karp and Edwards, 1997). (i) Neither DNA probes nor sequence information is required for the design of specific primers. (ii) The procedure does not involve blotting or hybridization steps thus making the technique quick, simple and efficient. (iii) RAPDs require relatively small amounts of DNA (about 10 ng per reaction) and the procedure can be automated; they are also capable of detecting higher levels of polymorphism than RFLPs. (iv) Development of markers is not required and the technology can be applied to virtually any organism with minimal initial development. (v) The primers can be universal and one set of primers can be used for any species. In addition, RAPD products of interest can be cloned, sequenced and converted into other types of PCR-based markers such as sequence tagged sites (STS), sequenced characterized amplified regions (SCAR), etc.

Reproducibility affects the way in which RAPD bands can be standardized for comparison across laboratories, samples and trials and whether RAPD marker information can be accumulated or shared. Due to frequently observed problems with reproducibility of overall RAPD profiles and specific bands, this marker class is often treated with reserve. In replication studies by Pérez *et al.* (1998), mispriming error amounted to 60%. Several factors have been shown to affect the number, size and intensity of bands. These include PCR buffers, deoxynucleotide triphosphates (dNTPs), Mg^{2+} concentration, cycling parameters, source of Taq polymerase, condition and concentration of template DNA and primer concentration. Results obtained by RAPDs are highly prone to user error and bands obtained can vary considerably between different runs of the same sample. To correct the problems that may be encountered when carrying out RAPD-PCR, it is important to bear in mind the following: (i) the concentration of DNA can alter the number of bands; (ii) RAPD profiles vary depending on the Mg^{2+} concentration and the PCR buffer provided by Taq polymerase suppliers may or may not contain Mg^{2+} ions; (iii) there are different sources of Taq polymerase and there is great variation between profiles produced using Taq polymerase obtained from different companies; (iv) there are a large number of alternative cycling times and temperatures which are equally important and depend on the type of machine used and even the wall thickness of the PCR tubes.

Generally if a PCR does not work there is likely to be something wrong with the template DNA, primers, Taq polymerase or choice of conditions. Initially it is important to try and repeat the PCR under the same conditions to ensure that there was not a simple error that resulted in the failure. In addition it is recommended that both positive and negative controls are included. A positive control with a template known to amplify well will ensure that all reagents have been added and that they are all functioning. A negative control without template DNA will reveal any contamination. In most cases if the PCR does not work and it is not clear what might be causing the problem, it is worth starting from the beginning by disposing of all the reagents used and preparing fresh ones. A careful experiment revealed that reproducibility could be improved and Taberner *et al.* (1997) reported that 3396 out of 3422 bands (99.2%) were reproducible.

On the other hand, low reproducibility is a major limitation of RAPD markers, particularly in ongoing genetic and plant breeding programmes in which the accumulated information and markers and marker data are shared between laboratories and experiments. RAPD markers may still find their applications in independent genetic diversity and phylogenetic studies that do not depend on data sharing or accumulation. As RAPD markers can be converted into other types of markers, they have a unique role in the development of target markers for crop species that have limited molecular markers available to cover the whole genome.

To overcome the problem associated RAPD analysis, Paran and Michelmore (1993) converted RAPD fragments into simple and robust PCR markers known as SCARs. This procedure increases the reproducibility of RAPD markers and also avoids the occurrence of non-homologous markers of equal molecular weight. These specific markers are obtained by introducing RAPD bands (polymorphic) into single markers which are then sequenced and specific primers are designed usually by expanding the original decamer primer sequence with 10–15 bases so that only the band of interest is amplified. In general, DNA can be isolated from agarose gels, cloned and sequenced to produce the starting DNA template for the development of a variety of PCR-based markers. The cloned and sequenced DNA fragments can then be used for the development of CAPS, single strand conformation polymorphism (SSCP) or SNP markers.

AFLP

Amplified fragment length polymorphism (AFLP; Zabeau and Voss, 1993; Vos *et al.*, 1995) is based on the selective PCR amplification of restriction fragments from a total double-digest of genomic DNA under high stringency conditions, that is, the combination of polymorphism at restriction sites and hybridization of arbitrary primers, and because of this AFLP is also called selective restriction fragment amplification (SRFA). It was perfected by the company Keygene in the Netherlands for initial use in plant improvement and has been patented. The AFLP technique combines the power of RFLP with the flexibility of PCR-based markers and provides a universal, multi-locus marker technique that can be applied to complex genomes from any source. The method is based on the identification of AFLP using selective PCR amplification of digested/ligated genomic or cDNA templates separated on a polyacrylamide gel, including restriction–ligation, pre-amplification and selective amplification (Fig. 2.3). The purified genomic DNA is first cleaved with one or more restriction endonucleases, i.e. a 6-cutter (*Eco*RI, *Pst*I and *Hind*III) and a 4-cutter (*Mse*I, *Taq*I). Adaptors of 18–20 bp and of known sequence, adapted at the sticky ends of the restriction sites, are then added to the ends of DNA fragments by a ligation reaction using T4 DNA ligase. DNA amplification is carried out using primers with the sequence specificity of the adaptor to generate a subset of fragments of different sizes (~up to 1 kb). The primer(s) also contains one or more bases at their 3' ends that provide amplification selectivity by limiting the number of perfect sequence matches between the primer and the pool of available adaptor/DNA templates. The resulting amplification products (50–400 bp size range) are typically observed by radio-labelling one of the primers followed by fragment separation on acrylamide gels to identify polymorphisms (changes in restriction sizes).

An AFLP primer is composed of a synthetic adaptor sequence, the restriction endonuclease recognition sequence and an arbitrary, non-degenerate 'selective' sequence (typically one, two or three nucleotides). In the first step, 500 ng of genomic DNA will be completely digested with two restriction enzymes, one frequent cutter (4-bp recognition site) and one rare cutter (6-bp recognition site). Oligonucleotide adaptors are ligated to the end of each restriction DNA which serve together with restriction site sequences as target sites for primer annealing, one end

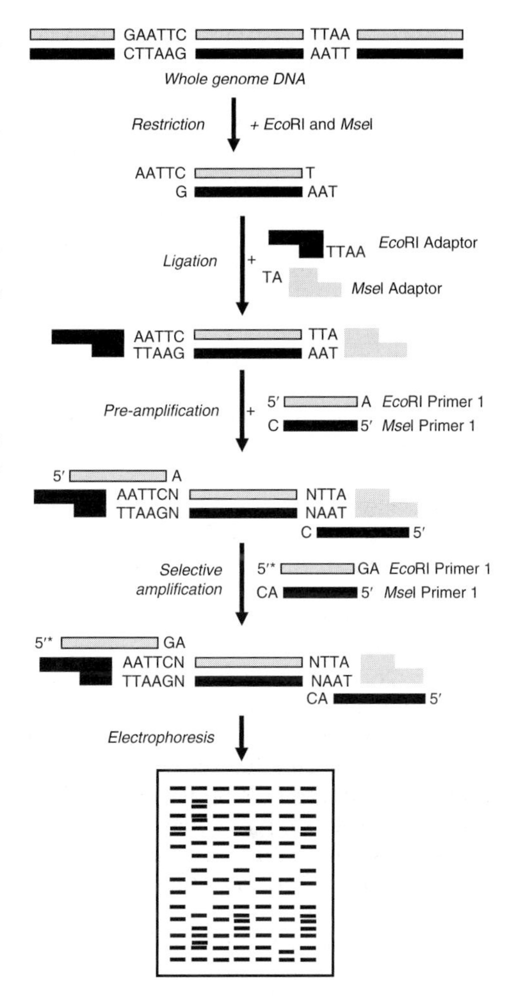

Fig. 2.3. AFLP flowchart. Adaptor DNA = short double strand DNA molecules, 18–20 bp in length, representing a mixture of two types of molecules. Each type is comparable with one restriction enzyme generated DNA end. Pre-amplifications uses selective primers, which contain an adaptor DNA sequence plus one or two random bases at the 3' end for reading into the genomic fragments. Primers for re-amplification have the pre-amplification primer sequence plus one or two additional bases at the 3' end. A tag (*) is attached at the 5' end of one of the re-amplification primers for detecting amplified molecules.

with a complementary sequence for the rare cutter and the other with the complementary sequence for the frequent cutter. In this way only fragments which have been cut by the frequent cutter and rare cutter will be amplified. Primers are designed from the known sequence of the adaptor, plus one to three selective nucleotides which extend into the fragment sequence. Sequences not matching these selective nucleotides in the primer will not be amplified so that the specific amplification of only those fragments matching the primers is achieved. The option to permutate the order of the selective bases and to recombine the primers with each other will theoretically lead to the gradual collection of all restriction fragments from a particular enzyme combination that is of a suitable size for DNA fragment analysis from a genotype. The multiplex ratio of an AFLP assay is a function of the number selective nucleotides in the AFLP primer combination, the selective nucleotide motif, GC content and physical genome size and complexity. Typically, two selective nucleotides are used for species with small genomes (1×10^8–5×10^8 bp), e.g. *Arabidopsis thaliana* L. (1×10^8 bp) and rice (*Oryza sativa* L.) (4×10^8 bp), and three selective nucleotides are used for species with large genomes (5×10^8–6×10^9 bp), e.g. maize, soybean, sunflower and many others. It is theoretically possible to use several tens of combinations of restriction enzymes at sites of four to six bases and a large number of combinations of selective bases on the amplification primers. Thus, as indicated by Falque and Santoni (2007), the restriction–amplification combinations are nearly infinite.

AFLP products can be separated in high-resolution electrophoresis systems. The number of bands produced can be manipulated by the number of selective nucleotides and the nucleotide motifs used. A well-balanced number of amplified restriction fragments ranges from 50 to150 bp. A major improvement has been made by switching from radioactive to fluorescent dye-labelled primers for the detection of fragments in gel-based or capillary DNA sequencers in which fluorescently labelled fragments pass the detector near the bottom of the gel/end of the capillary, resulting in a linear spacing of DNA fragments and therefore increasing the resolution over the whole size range (Schwarz *et al.*, 2000).

In general, AFLP assays can be carried out using relatively small DNA samples (typically 1–100 ng per individual). AFLP has a very high multiplex ratio and genotyping throughput and is relatively reproducible across laboratories. Simple off-the-shelf technology can be applied to virtually any organism with no formal marker development required and in addition, a set of primers can be used for different species. However, there are limitations to the AFLP assay. (i) The maximum polymorphic information content for any bi-allelic marker is 0.5. (ii) High quality DNA is needed to ensure complete restriction enzyme digestion. Rapid methods for isolating DNA may not produce sufficiently clean template DNA for AFLP analysis. (iii) Proprietary technology is needed to score heterozygotes and ++ homozygotes, otherwise AFLPs must be dominantly scored. (iv) AFLP markers often cluster densely in centromeric regions in species with large genomes, e.g. barley (Qi *et al.*, 1998) and sunflower (Gedil *et al.*, 2001). (v) Developing locus-specific markers from individual fragments can be difficult. (vi) AFLP primer screening is often necessary to identify optimal primer specificities and combinations otherwise the assays can be carried out using off-the-shelf technology. (vii) There are relatively high technical demands in AFLP analysis including radio-labelling and skilled manpower. (viii) Marker development is complicated and not cost-effective. (ix) Reproducibility is relatively low compared to RFLP and simple sequence repeat (SSR) markers but better than RAPD marker as AFLP reveals large numbers of bands and not all the bands will be comparable across laboratories or trials due to potential false positive, false negative and complicated gel backgrounds.

The AFLP technique can be modified so that one primer is obtained from a known multi-copy sequence to detect sequence-specific amplification polymorphisms. This approach was used successfully to generate genome-wide *Bare-1* retrotransposon-like markers in barley (Waugh *et al.*, 1997) and diploid *Avena* (Yu and Wise, 2000) as well as in lucerne by making use of consensus sequences from long terminal repeats (LTRs) of *Tms1* retrotransposon (Porceddu *et al.*, 2002). The cDNA-AFLP technique (Bachem *et al.*, 1996) which applies the standard AFLP protocol to a cDNA template, was used to display transcripts whose expression was rapidly altered during race-specific resistance reactions, for the isolation of differentially expressed genes from a specific chromosome region using aneuploids and for the construction of genome-wide transcription maps (as reviewed by Mohler and Schwarz, 2005). In addition, there are several modified AFLP techniques based on the use of endonucleases such as single endonuclease (*Msp*I) AFLP (Boumedine and Rodolakis, 1998), three endonuclease-AFLP (van der Wurff *et al.*, 2000), and second digestion AFLP (Knox and Ellis, 2001). Developments in the detection of AFLP include the replacement of radio-active detection with silver staining, fluorescent AFLP or agrarose gels for single endonuclease AFLP. Recent studies have addressed specific areas of the AFLP technique including comparison with other genotyping methods, assessment of errors, homoplasy, phylogenetic signal and appropriate analysis techniques. The study by Meudt and Clarke (2007) provides a synthesis of these areas and explores new directions for the AFLP technique in the genomic era.

SSR

Microsatellites, also known as SSRs, short tandem repeats (STRs) or sequence-tagged microsatellite sites (STMS), are tandemly repeated units of short nucleotide motifs that are 1–6 bp long. Di-, tri- and tetranucleotide repeats such as $(CA)_n$, $(AAT)_n$ and $(GATA)_n$ are widely distributed throughout the genomes of plants and animals (Tautz and Renz, 1984). One of the most important attributes of microsatellite loci is their high level of allelic variation, making them valuable as genetic markers. The unique sequences bordering the SSR motifs provide templates for specific primers to amplify the SSR alleles via PCR. Referred to as simple sequence length polymorphisms (SSLPs), they pertain to the number of repeat units that constitute the microsatellite sequence. The rates of mutation of SSR are about 4×10^4–5×10^6 per allele and per generation (Primmer *et al.*, 1996). The predominant mutation mechanism in microsatellite tracts is 'slipped-strand mispairing' (Levinson and Gutman, 1987). When slipped-strand mispairing occurs within a microsatellite array during DNA synthesis, it can result in the gain or loss of one or more repeat units depending on whether the newly synthesized DNA chain or the template chain loops out. The relative propensity for either chain to loop out seems to depend in part on the sequences making up the array and in part on whether the event occurs on the leading (continuous DNA synthesis) or lagging (discontinuous DNA synthesis) strand (Freudenreich *et al.*, 1997). SSR loci are individually amplified by PCR using pairs of oligonucleotide primers specific to unique DNA sequences flanking the SSR sequence.

Microsatellites may be obtained by screening sequences in databases or by screening libraries of clones. If no sequence is available, microsatellite markers can be developed in the following steps: construct enriched or unenriched small-insert clone library; screen it by hybridizing labelled oligo (with SSR motif of interest); sequence positive clones; design primers in single copy regions flanking SSR repeats such that the amplified fragments will be > 50 bp and < 350 bp; and identify size polymorphism on PAGE gels. For multiplexing, design primers with similar melting temperature (T_m) and a range of expected amplicon sizes to have non-overlapping groups of markers on a gel.

In rice, both an enzyme-digested (Chen, X. *et al.*, 1997) and a physically-sheared library (Panaud *et al.*, 1996) were constructed from cultivar IR36 based on size-selected DNA in the 300–800-bp range. These libraries were screened for the presence of $(GA)_n$ microsatellites by plaque and colony hybridization. A pre-sequencing screening step was used to eliminate clones where the microsatellite repeat was too near one of the cloning sites to permit accurate design of primers and to determine which end should be sequenced with priority. The basic steps include:

- PCR amplification of clone inserts and determination of their lengths before sequencing. Short and long insert clones are usually discarded.
- Selected clones are sequenced and searched for SSRs.
- Sequences within motif classes are grouped and aligned using sequence alignment software to identify redundant sequences.
- Oligonucleotide primers are designed for unique DNA sequences flanking non-redundant SSRs.
- Primers are tested and genotypes are screened for SSR length polymorphisms.

An alternative source of SSRs is to utilize expressed sequence tag (EST) and other sequence databases (e.g. Kantety *et al.*, 2002). SSRs can be identified computationally, using a BLAST query (see Simple Sequence Repeat Identification Tool available at www.gramene.org) and available genomic or EST sequences. Using this method, a total of 2414 new di-, tri- and tetra-nucleotide non-redundant SSR primer pairs, representing 2240 unique marker loci, were developed and experimentally validated in rice (McCouch *et al.*, 2002). SSR-containing sequences that consisted of perfect repeat motifs (> 24 bp in length) flanked by 100 bp of unique sequence on either side of the SSR were chosen from GenBank. Primer pairs containing 18–24 nucleotides devoid of secondary structure or consecutive tracts of a single nucleotide, with a GC content of around 50% (T_m approximately 60°C) and preferably G- or C-rich at the 3' end were automatically designed. Using electronic PCR (e-PCR) to align these designed primer pairs against 3284 publicly sequenced rice BAC and PAC clones (representing about 83% of the total rice genome), 65% of the SSR markers hit a BAC or PAC clone containing at least one genetically mapped marker and could be mapped by proxy. Additional information based on genetic mapping and 'nearest marker' information provided the basis for locating a total of 1825 designed markers along rice chromosomes.

Compared with library-derived SSRs, EST-derived SSRs are expected to display slightly fewer polymorphisms as there is pressure for sequence conservation in the coding regions (Scott, 2001). However, the availability of SSR markers from the expressed portion of the genome might facilitate their transferability across genera compared to the low efficiency of SSR markers that have been retrieved from gene-poor areas (Peakall *et al.*, 1998). This approach could be used in plant species with minimal resources and research expenditure.

Once a plant species has been completely sequenced, the entire set of available SSRs in the genome can be easily accessed through online databases. For example, the International Rice Genome Sequencing Project identified 18,828 di, tri and tetra-nucleotide SSRs that were over 20 bp in length and developed flanking primers for use as SSR markers (IRGSP, 2005). The locations of these SSRs on the physical map of rice in relation to other genetic markers can be found using the online Gramene Genome Browser (http://www.gramene.org/Oryza_sativa_japonica/index.html).

The usual method of SSR genotyping is to separate radio-labelled or silver-stained PCR products by denaturing or non-denaturing PAGE using ethidium bromide or SYBR staining although distinguishing SSRs on agarose gels is sometimes possible (Fig. 2.4). These assays can usually distinguish alleles which differ by 2–4 bp or more.

Semi-automated SSR genotyping can be carried out by assaying fluorescently labelled PCR products for length variants on an automated DNA sequencer (e.g. Applied Biosystems and Li-Cor) (Fig. 2.4). One drawback of fluorescent SSR genotyping is the cost of end-labelling primers with the necessary fluorophores, e.g. 6-carboxy-fluorescine (FAM), hexachloro-6-carboxy-flurescine (HEX) or tetrachloro-6-carboxy-fluorescine (TET). SSR length

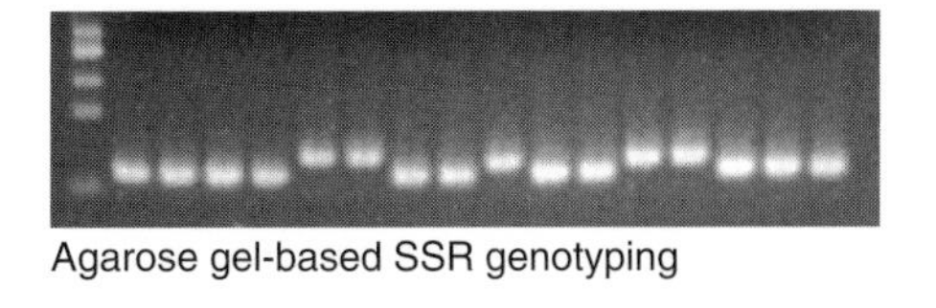

Agarose gel-based SSR genotyping

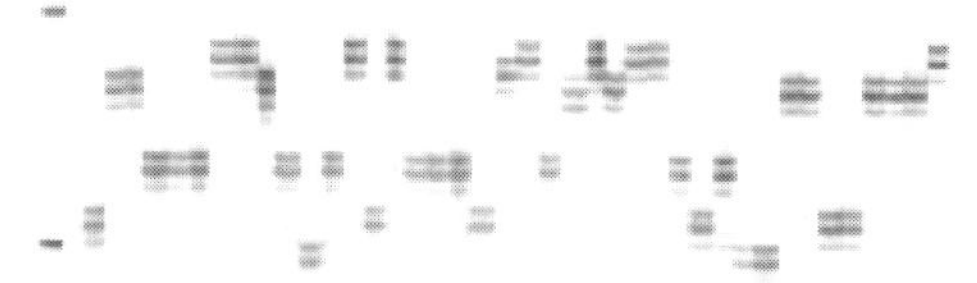

PAGE gel-based SSR genotyping

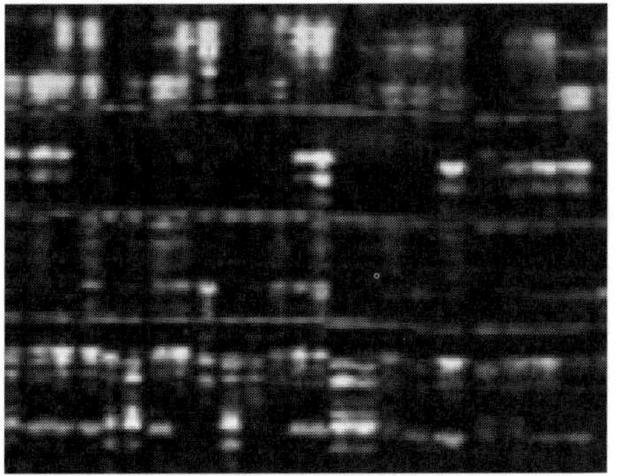

Semi-automated SSR genotyping

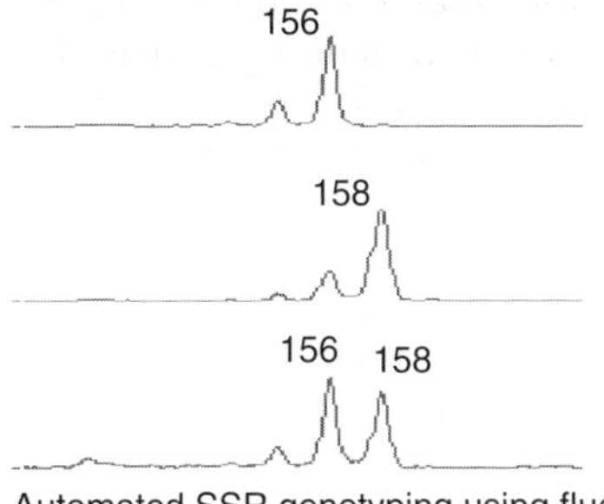

Automated SSR genotyping using fluorescent labelling

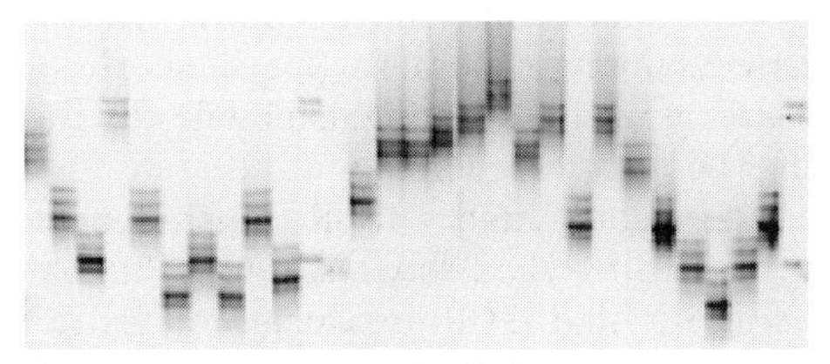

Stutter bands and multiple alleles

Fig. 2.4. Examples of genotyping systems used for SSR analysis.

polymorphisms can be also assayed using non-denaturing high pressure liquid chromatography (HPLC). SSR alleles differing by several repeat units can often be distinguished on agarose gels (Fig. 2.4).

SSRs assayed on polyacrylamide gels typically show characteristic 'stuttering'. Stutter bands are artefacts produced by DNA polymerase slippage. Typically, the most prominent stutter bands are +1 and −1 repeats (e.g. + or − 2 bp for a di-nucleotide repeat), and, if visible, the next most prominent stutter bands are +2 and −2 repeats. Stuttering reduces the resolution between alleles such that 2- or possibly 4-bp differences between alleles cannot be sharply or unequivocally distinguished on polyacrylamide gels. Figure 2.4 shows examples of different genotyping systems used for SSR analysis including multiplexing and stutter bands.

Another source of noise is the incomplete addition of non-templated adenine to PCR products thereby producing adenylated (+A) and non-adenylated (−A) DNA fragments (Magnuson *et al.*, 1996). Adding a 'pigtail' sequence (e.g. GTCTCTT) to the 5' end of the reverse primer promotes the adenylation of the 3' end of the forward strand (Brownstein *et al.*, 1996), thereby virtually eliminating the −A products and producing a more homogenous set of fragments.

SSR markers are characterized by their hypervariability, reproducibility, co-dominant nature, locus specificity and random dispersion throughout most genomes. In addition, SSRs are reported to be more variable than RFLPs or RAPDs. The advantages of SSRs are that they can be readily analysed by PCR and are easily detected on polyacrylamide gels. SSLPs with large size differences can be also detected on agarose gels. SSR markers can be multiplexed, either functionally by pooling independent PCR products or by true multiplex-PCR. Their genotyping throughput is high and can be automated. In addition, start-up costs are low for manual assay methods (once the markers have been developed) and SSR assays require only very small DNA samples (~100 ng per individual).

The disadvantages of SSRs are the labour-intensive development process particularly when this involves screening genomic DNA libraries enriched for one or more repeat motifs (although SSR-enriched libraries can be commercially purchased) and the high start-up costs for automated methods.

SNP

A single nucleotide polymorphism or SNP (pronounced *snip*) is an individual nucleotide base difference between two DNA sequences. SNPs can be categorized according to nucleotide substitution as either transitions (C/T or G/A) or transversions (C/G, A/T, C/A or T/G). For example, sequenced DNA fragments from two different individuals, AAGC**C**TA to AAGC**T**TA, contain a single nucleotide difference. In this case there are two *alleles*: C and T. C/T transitions constitute 67% of the SNPs observed in humans, and about the same rate was also found in plants (Edwards *et al.*, 2007a). In practice, single base variants in cDNA (mRNA) are considered to be SNPs as are single base insertions and deletions (indels) in the genome. As a nucleotide base is the smallest unit of inheritance, SNPs provide the ultimate form of molecular marker.

For a variation to be considered a SNP, it must occur in at least 1% of the population. SNPs make up about 90% of all human genetic variation and occur every 100–300 bases. Two of every three SNPs involve the replacement of cytosine (C) with thymine (T). This is supported by a genome-wide analysis in rice. A polymorphism database constructed to define polymorphisms between cultivars Nipponbare (from subspecies *japonica*) and 93-11 (from subspecies *indica*) contains 1,703,176 SNPs and 479,406 indels (Shen *et al.*, 2004), which equates to approximately 1 SNP/268 bp in the rice genome. Using alignments of the improved whole-genome shotgun sequences for *japonica* and *indica* rice, SNP frequencies varied from 3 SNPs/kb in coding sequences to 27.6 SNPs/kb in the transposable elements with a genome-wide measure of 15 SNPs/kb or 1 SNP/66 bp (Yu *et al.*, 2005). Based on partial genomic sequence information, SNP frequencies have been revealed in many crops, including barley, soybean, sugarbeet, maize, cassava and potato; typical SNP frequencies are also in the range of one SNP every 100–300 bp in plants (see Edwards *et al.*, 2007a for a review).

SNPs may fall within coding sequences of genes, non-coding regions of genes or in the intergenic regions between genes at different frequencies in different chromosome regions. In *Arabidopsis* the distribution of SNPs was found to be even across the five chromosomes with the exception of centromeric regions which contain few transcribed genes (Schmid *et al.*, 2003). SNPs within a coding sequence will not necessarily change the amino acid sequence of the protein that is produced due to redundancy in the genetic code. A SNP in which both forms lead to the same polypeptide sequence is termed synonymous, while if a different polypeptide sequence is produced they are non-synonymous. SNPs that are not in protein coding regions may still have consequences for gene splicing, transcription factor binding or the sequence of non-coding RNA. Of the 3–17 million SNPs found in the human genome, 5% are expected to occur within genes. Therefore, each gene may be expected to contain ~6 SNPs.

A variety of approaches have been adopted for discovery of novel SNPs in a wide range of organisms including plants. These fall into three general categories (Edwards *et al.*, 2007b): (i) *in vitro* discovery, where new sequence data is generated; (ii) *in silico* methods that rely on the analysis of available sequence data; and (iii) indirect discovery, where the base sequence of the polymorphism remains unknown. On the other hand, a large number of different SNP genotyping methods and chemistries have been developed based on various methods of allelic discrimination and detection platforms. A convenient method for detecting SNPs is RFLP (SNP-RFLP) or by using the CAPS marker technique. If one allele contains a recognition site for a restriction enzyme while the other does not, digestion of the two alleles will give rise to fragments of different length. A simple procedure is to analyse the sequence data stored in the major databases and identify SNPs. Four alleles can be identified when the complete base sequence of a segment of DNA is considered and these are represented by A, T, G and C at each SNP locus in that segment.

Sobrino *et al.* (2005) assigned the majority of SNP genotyping assays to one of four groups based on the molecular mechanisms: allele-specific hybridization, primer extension, oligonucleotide ligation and invasive cleavage. These four are described below. Chagné *et al.* (2007) added three methods to this list, sequencing, allele-specific PCR amplification, DNA conformation methods and also generalized the enzymatic cleavage method to include the invader assay and also dCAPS and targeting induced local lesions in genomes (TILLING).

1. Allele-specific hybridization (ASH), also known as allelic-specific oligonucleotide hybridization, is based on distinguishing by hybridization between two DNA targets differing at one nucleotide position (Wallace *et al.*, 1979). Allelic discrimination can be achieved using two allele-specific probes labelled with a probe-specific fluorescent dye and a generic quencher that reduces fluorescence in the intact probe. During amplification of the sequence surrounding the SNP, probes complementary to the DNA target are cleaved by the 5' exonuclease activity of Taq polymerase. Spatial separation of the dye and quencher results in an increase in probe-specific fluorescence which can be detected with a plate reader.

Under optimized assay conditions, the SNP can be detected by the difference in T_m of the two probe–template hybrids as only the perfectly matched probe–target hybrids are stable and those with one-base mismatch are unstable. To increase the reliability of SNP genotyping the probes should be as short as possible. Originally, ASH used the dot blot format in which probes are hybridized to membrane-bound genomic DNA or PCR fragments. However, the more advanced PCR-based dynamic allele-specific hybridization (DASH) method uses a microtitre plate format (Howell *et al.*, 1999). Since one of the PCR primers is biotinylated at the 5' end, the PCR products can be bound to streptavidin-coated wells and denatured under alkaline conditions. An oligonucleotide probe complementary to one allele is added to the single-strand target DNA molecules. The differences in melting curves are measured by slowly heating and observing the changes in fluorescence of a double-strand-specific, intercalating dye. The 5' nuclease or TaqMan assay, molecular beacon and the scorpion assays are all examples of ASH SNP genotyping technologies. Large-scale scanning of SNPs in a vast number of loci using allele-specific hybridization can be carried out on high-density oligonucleotide chips.

2. The Invader assay, also known as flap endonuclease discrimination, is based on the specificity of recognition and cleavage by a three-dimensional flap endonuclease which is formed when two overlapping oligonucleotides hybridize perfectly to a target DNA (Lyamichev *et al.*, 1999). The cleaved fragment may be labelled with a probe-specific fluorescent dye which fluoresces following probe cleavage due to spatial separation from the quencher. Alternatively, the flap may act as the invader probe in a secondary reaction to amplify the fluorescent signal (Invader squared) (Hall *et al.*, 2000). Third Wave Technologies Inc. (http://www.twt.com) has manufactured an Invader assay for flap endonuclease discrimination which can be carried out in solid phase using oligonucleotide-bound streptavidin-coated particles (Wilkins-Stevens *et al.*, 2001).

3. Primer extension is a term used to describe mini-sequencing, single-base extension or the GOOD assay (Sauer *et al.*, 2002). A popular method which was designed specifically for genotyping SNPs is the mini-sequencing technique (Syvänen, 1999; Syvänen *et al.*, 1990). The method forms the basis of a number of methods for allelic discrimination. The robust detection of known mutations employs oligonucleotides which anneal immediately upstream of the query SNP and are then extended by a single dideoxynucleotide triphosphate (ddNTP) in cycle sequencing reactions. The fidelity of thermostable proof-reading DNA polymerases guarantees that only the complementary ddNTP is incorporated. Several detection methods have been described for the discrimination of primer extension (PEX) products. Most popular is the use of ddNTP terminators that are labelled with different fluorescent dyes. The differentially dye-labelled PEX products can readily be detected on charge coupled device camera-based DNA sequencing instruments.

In the case of a single base extension (SBE), a primer is annealed adjacent to a SNP and extended to incorporate a ddNTP at the polymorphic site. SNaPshot (Applied Biosystems) uses differential fluorescent labelling of the four ddNTPs in a SBE reaction allowing fluorescent detection of the incorporated nucleotide. SNP-IT (Orchid Biosciences) is also based on fluorescent SBE and uses solid phase capture and detection of extension products. The GOOD assay involves extension of a primer modified near the 3' end with a charged tag to increase sensitivity to mass spectrometry detection.

Alternatives to SBE include pyrosequencing, allele specific primer extension and the amplification refractory mutation system. Real-time monitoring of PEX relies on the bioluminometric detection of inorganic pyrophosphate released upon incorporation of dNTP (Ahmadian *et al.*, 2000).

4. The oligonucleotide ligation assay (OLA) for SNP typing is based on the ability of ligase to covalently join two oligonucleotides when they hybridize next to one another on a DNA template (Landegren *et al.*, 1988). Both primers must have perfect base pair complementarity at the ligation site which makes it possible to discriminate two alleles at a SNP site. The OLA has been modified to exploit a thermostable DNA ligase, interrogate PCR templates and utilize a dual-colour detection system. OLA also gave rise to another technique, Padlock probes (Nilsson *et al.*, 1994), which uses oligonucleotide probes that ligate into circles upon target recognition and isothermal rolling-circle amplification. As reviewed by Chagné *et al.* (2007), there are several applications which have been developed to detect SNP variation using OLA, including colorimetric assays in ELISA plates, separation of the ligated oligonucleotides that have been labelled with a fluorescent dye on an automated sequencer and rolling-circle amplification with one of the ligation probes bound to a microarray surface.

DETECTION SYSTEMS. There are several detection methods for analysing the products of each type of allelic discrimination reaction: gel electrophoresis, fluorescence resonance energy transfer (FRET), fluorescence polarization, arrays or chips, luminescence, mass spectrophotometry, chromatography, etc. Fig. 2.5 summarizes the enzyme chemistry, demultiplexing and detection options in SNP genotyping.

Fluorescence is the most widely applied detection method currently employed for high-throughput genotyping in general. The use of fluorescence has been teamed with a number of different detection systems including plate readers, capillary electrophoresis and DNA arrays. In addition to fluorescence detection, mass spectrometry and light detection represent novel applications of established technology for high-throughput genotyping of SNPs.

PLATE READERS. There are many fluorescent plate readers capable of detecting fluorescence in a 96- or 384-well format (Jenkins and Gibson, 2002). Most models use a light source and narrow band-pass filters to select the excitation and emission wavelengths and enable semi-quantitative steady state fluorescence intensity readings to be made. This technology has been applied to genotyping with TaqMan, Invader and rolling-circle amplification. Fluorescence plate readers are also available which allow measurement of additional fluorescence parameters including polarization, lifetime and time-resolved fluorescence and FRET.

DNA ARRAY. Oligonucleotide arrays bound to a solid support have been proposed as the future detection platform for high-throughput genotyping. Two distinct approaches have been adopted involving ASH whereby the oligonucleotide directly probes the target and tag arrays that capture solution phase reaction products via hybridization to their anti-tag sequences.

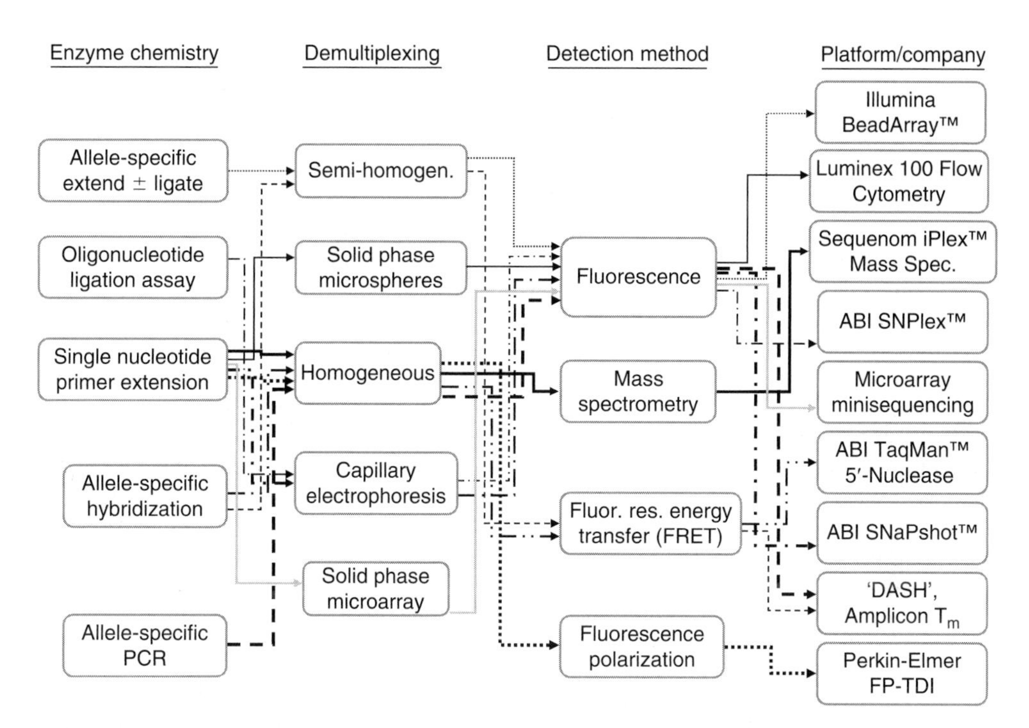

Fig. 2.5. Chemistry, demultiplexing, detection options in SNP genotyping. From Syvänen (2001) reprinted by permission from Macmillan Publishers Ltd.

The Affymetrix® Genome-Wide Human SNP Array 6.0 features more than 1.8 million markers for genetic variation, including more than 906,600 SNPs and more than 946,000 probes for the detection of copy number variation. The SNP Array 6.0 enables high-performance, high-powered and low-cost genotyping (http://www.affymetrix.com). Luminex has developed a panel of 100 bead sets with unique fluorescent labels, identifiable by flow analyser. The bead sets can be derivatized with allele specific oligonucleotides to create a bead-based array for multiplex genotyping by ASH.

Tag arrays are generic assemblies of oligonucleotides that are used to sort or deconvolute mixtures of oligos by hybridization to the anti-tag sequences. The current Affymetrix GeneChip® Universal Tag Arrays are available in 3, 5, 10 or 25 K configurations and contain novel, bioinformatically designed tag sequences that result in minimal potential for cross-hybridization.

MASS SPECTROMETRY. Many genotyping techniques involve the allele-specific incorporation of two alternative nucleotides into an oligonucleotide probe. Due to the inherent molecular weight difference of DNA bases, mass spectrometry can be used to determine which variant nucleotide has been incorporated by measuring the mass of the extended primers and this approach has been applied primarily to genotyping by primer extension using the MALDI-TOF (matrix assisted laser desorption/ionization-time of flight) mass spectrometry approach. The MALDI-TOF method is particularly advantageous for detection of PEX products in multiplex.

The polyanionic nature of oligonucleotides results in low signal to noise ratios, particularly for longer (> 40 mer) fragments. This has been addressed by specifically cleaving long probes by acidolysis of P3'-N5' phosphoramidate bonds and by a combined approach whereby the probe is digested to a very short fragment which has been derivatized to lower its charge to a single positive or negative charge.

LIGHT DETECTION. Pyrosequencing involves hybridization of a sequencing primer to a single stranded template and sequential addition of individual dNTPs. Incorporation of a dNTP into a primer releases pyrophosphate which triggers a luciferase-catalysed reaction. The genotype of a SNP is determined by the sequential addition (and degradation) of nucleotides. The light produced is detected by a charge coupled device camera and each light signal is proportional to the number of nucleotides incorporated (http://www.pyrosequencing.com), for which reason pyrosequencing is suitable for the quantitative estimation of allele frequencies in pooled DNA samples. Furthermore, pyrosequencing proved to be an appropriate method for genotyping SNPs in polyploidy plant genomes such as potato because all possible allelic states of binary SNP could be accurately distinguished (Rickert *et al.*, 2002).

There are various SNP detection systems which differ in their chemistry, detection platform, multiplex level and application; some of these will be discussed below. The reader is also referred to Bagge and Lübberstedt (2008) for further information.

The TaqMan® SNP Genotyping Assay (Applied Biosystems, Foster City, USA) is a single-tube PCR assay that exploits the 5' exonuclease activity of AmpliTaq Gold® DNA. The assay kit includes two locus-specific PCR primers that flank the SNP of interest and two allele-specific oligonucleotide TaqMan probes. These probes have a fluorescent reporter dye at the 5' end and a non-fluorescent quencher with a minor groove binder at the 3' end. Upon cleavage by the 5' exonuclease activity of Taq polymerase during PCR, the reporter dye will fluoresce as it is no longer quenched and the intensity of the emitted light can be measured. Modified probes such as locked nucleic acids, a modified nucleic acid analogue, showed better hybridization properties than standard TaqMan probes (Kennedy *et al.*, 2006). TaqMan is a simple assay, since all the reagents are added to the microtitre well at the same time in a 96- or 384-well format. Although the assay can be carried out at the monoplex or duplex level, the multiple steps can be assembled and automated so that one laboratory technician can produce 10,000 data points per day. The TaqMan platform is highly suitable for genetically modified organism tests and MAS using a few markers for a large number of samples.

The SNaPshot® Multiplex Assay (Applied Biosystems, Foster City, USA) is based on mini-sequencing, i.e. a single-base extension using fluorescent labelled ddNTPs. The system's multiplex ready reaction mix enables robust multiplex SNP interrogation of PCR-generated templates. Multiplexing can be accomplished by representing multiple SNP products spatially. This is achieved by tailing the 5' end of the unlabelled SNaPshot primers with different lengths of non-complementary oligonucleotide sequences that serve as mobility modifiers. The reactions may be carried out in 5- to 10-plex using capillary electrophoresis for data detection in a 96-well format so that one individual can generate over 10,000 data points per day. SNaPshot is suitable for MAS of several traits simultaneously and if multiple sets of 10-plex are combined, it can be used for rough mapping and marker-assisted backcrossing with several hundreds of samples and markers involved.

The SNPlex™ Genotyping System (Applied Biosystems, Foster City, USA) uses OLA/PCR technology for allelic discrimination and ligation product amplification. Genotype information is then encoded into a universal set of dye-labelled, mobility modified fragments known as Zipchute™ Mobility Modifiers, for rapid detection by capillary electrophoresis. The same set of Zipchute™ Mobility Modifiers can be used for every SNPlex™ pool regardless of which SNPs are chosen. The SNPlex™ System allows for multiplexed genotyping of up to 48 SNPs simultaneously against a single sample with the ability to detect up to 4500 SNPs in parallel in 15 min. This integrated system delivers cost-efficient, medium- to high-throughput genotyping and is suitable for various genetic and breeding applications including fingerprinting, gene mapping and MAS for both foreground and background. Both SNaPshot and SNPlex can be used with capillary electrophoresis systems as the genotyping platform which can be also used for SSR genotyping.

MassARRAY® iPLEX Gold (SEQUENOM, San Diego, USA) combines the benefits of the simple and robust single-base primer extension biochemistry with the sensitivity and accuracy of MALDI-TOF mass spectometry (see Chapter 3) detection. It uses a single termination mix and universal reaction conditions for all SNPs. The primer is extended, dependent upon the template sequence, resulting in an allele-specific difference in mass between extension products. The assays can be multiplexed up to 40 SNPs in a 384-well format allowing for throughput levels of up to 150,000 genotypes per instrument per day. MassARRAY is flexible and suitable for generating both small and large marker numbers for each sample so that it can be used for a variety of genetic and breeding purposes.

There are two major chip-based high-throughput genotyping systems, DNA microarrays developed by Affymetrix (Santa Clara, USA) and a high-density biochip assay by Illumina Inc. (San Diego, USA), both of which offer different levels of multiplexes up to several thousands or more plexes (Yan *et al.*, 2009). As an increasing number of sets of these chips become available, outsourced genotyping through companies or service centres becomes one of the options for genotyping large numbers of samples using the same set of markers (e.g. fingerprinting) to achieve high efficiency and low cost per data point.

THE FUTURE OF SNP TECHNOLOGY. A key technical obstacle in the development of microarray-based methods for genome-wide SNP genotyping is the PCR amplification step which is required to reduce the complexity and improve the sensitivity of genotyping SNPs in large, diploid genomes. The level of complexity that can be achieved in PCR does not match that of current microarray-based methods thus making PCR the limiting step in these assays (Syvänen, 2005). Highly multiplexed microarray systems have recently been developed by combining well-known reaction principles for DNA amplification and SNP genotyping.

Identification of a specific single-base change among up to billions of bases that constitute a plant species is a challenging task. PCR offers a means of reducing the complexity of a genome and increasing the copy number of the DNA templates to the levels required for the specific and sensitive detection of single-base changes. However, the design of robust PCR assays with multiplexing levels exceeding 10–20 amplicons has proven to be more difficult than initially anticipated because in multiplex PCR the number of undesired interactions between the PCR primers increases exponentially as the number of primers included in the reaction mixture increases. This interaction usually results in preferential amplification of unwanted 'primer–dimer' artefacts instead of the intended DNA templates (amplicons). Another problem in multiplex PCR is the sequence-dependent differences in PCR efficiency between the amplicons. The problems of multiplexing can be reduced to some extent by using PCR primers that are as similar to one another as possible. The multiplexing level that can be readily achieved in standard PCRs is less than that offered by current technology for producing high-density DNA microarrays. Simultaneous analysis of a reasonable amount of genomic DNA with the current detection sensitivity of microarray scanners requires an amplification step. The PCR step complicates the molecular reactions underlying the assays and introduces multiple laboratory steps into the procedures and is therefore the chief obstacle to highly multiplexed SNP genotyping.

Diversity array technology

Diversity array technology (DArT) is a novel type of DNA marker which employs a microarray hybridization-based technique developed by CAMBIA (http://www.diversityarrays.com) that enables the simultaneous genotyping of several hundred polymorphic loci spread over the genome (Jaccoud *et al.*, 2001; Wenzel *et al.*, 2004). DArT can be used to construct medium-density genetic linkage maps in species of various genome sizes. Two steps are involved: generating the array and genotyping the sample. For each sample, representative genomic DNA is prepared by restriction enzyme digestion followed by ligation of the restriction fragments to adaptors. The genome complexity is then reduced by PCR primers with complementary sequences to the adaptor and selective overhangs. Restriction generated fragments representing the diversity of the gene pool are cloned. The outcome is known as a 'representation' (typically 0.1–10% of the genome). Polymorphic clones in the library are identified by array inserts from a random set of clones. Cloned inserts are amplified using vector-specific primers, purified and arrayed on to a solid support (microarray) (Fig. 2.6A). To genotype a sample, the representation (DNA) of the sample is fluorescently labelled and hybridized against the discovery array. The array is then scanned and the hybridization signal is measured for each array spot. By using multiple labels, a representation from one sample is contrasted with that from another or with a control probe (Jaccoud *et al.*, 2001; http://www.cambia.org; http://www.diversityarrays.com). Polymorphic clones (DArT markers) show variable hybridization signal intensities for different individuals. These clones are subsequently assembled into a genotyping array for routine genotyping (Fig. 2.6B).

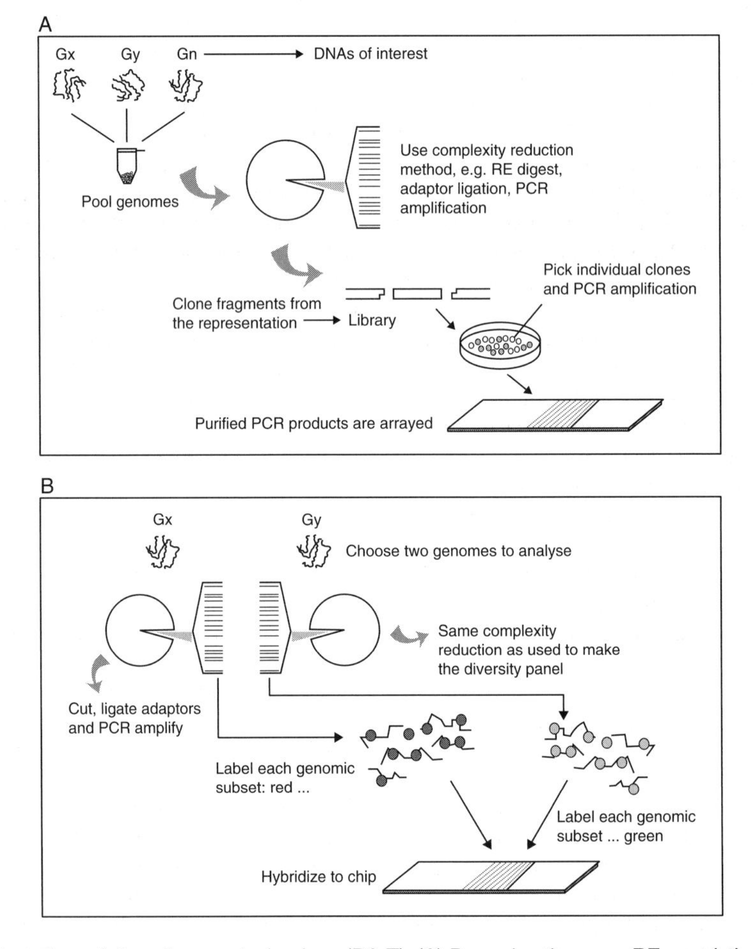

Fig. 2.6. Procedure of diversity array technology (DArT). (A) Preparing the array. RE, restriction enzyme. (B) Genotyping a sample.

DArT markers are biallelic and behave in a dominant (present versus absent) or co-dominant (two doses versus one dose versus absent) manner. DArT detects single-base changes as well as indels. It is a good alternative to currently used techniques including RFLP, AFLP, SSR and SNP in terms of cost and speed of marker discovery and analysis for whole-genome fingerprinting. It is cost-effective, sequence-independent, non-gel based technology that is amenable to high-throughput automation and the discovery of hundreds of high quality markers in a single assay. An open source software package, DArTsoft, is available for automatic data extraction and analysis. The weaknesses of this technology include marker dominance and its technically demanding nature. Also there is some concern as to whether DArT markers are randomly distributed across the whole genome, as DArT markers in barley appear to have a moderate tendency to be located in hypomethylated, gene-rich regions in distal chromosome areas (Wenzl *et al.*, 2006).

DArT technology has been successfully developed for *Arabidopsis*, cassava, barley, rice, wheat, sorghum, ryegrass, tomato and pigeon pea, while work is in progress to establish DArT in chickpea, sugarcane, lupins, quinoa, banana and coconut (http://www.diversityarrays.com). For example, a genetic map with 385 unique DArT markers spanning the 1137 cM barley genome (Wenzl *et al.*, 2004) was constructed, DArT markers along with AFLP and SSR markers were mapped on the wheat genome (Semagn *et al.*, 2006), and a cassava DArT genotyping array containing approximately 1000 polymorphic clones (Xia, L. *et al.*, 2005) is now available.

Genic and functional markers

DNA markers can be classified into random markers (RMs) (also known as anonymous or neutral markers), gene targeted markers (GTMs) (also known as candidate gene marker) and functional markers (FMs) (Anderson and Lübberstedt, 2003). RMs are derived at random from polymorphic sites across the genome whereas GTMs are derived from polymorphisms within genes. FMs are derived from polymorphic sites within genes that are causally associated with phenotypic trait variation and are superior to RMs as a result of their complete linkage with trait locus alleles and functional motifs (Anderson and Lübberstedt, 2003). The major drawback of the RMs is that their predictive value depends on the known linkage phase between marker and target locus alleles (Lübberstedt *et al.*, 1998b).

Genetic diversity at or below the species level has mostly been characterized by molecular markers that more or less randomly sampled genetic variation in the genome. RM is a very effective tool among others for the establishment of a breeding system, the study of gene flow among natural populations, and the determination of the genetic structure of GeneBank collections (Chapter 5; Xu *et al.*, 2005). RM systems are still the systems of choice for marker-assisted breeding (Xu, Y., 2003). However, 'users' of biodiversity are often not interested in random variation but rather in variation that might affect the evolutionary potential of a species or the performance of an individual genotype. Such 'functional' variation can be tagged with neutral molecular markers using quantitative trait loci (QTL) and linkage disequilibrium mapping approaches. Alternatively, DNA-profiling techniques may be used that specifically target genetic variation in functional parts of the genome.

GENIC MARKERS. A wealth of DNA sequence information from many fully characterized genes and full-length cDNA clones has been generated and deposited in online databases for an increasing number of plant species and the sequence data for ESTs, genes and cDNA clones can be downloaded from GenBank and scanned for identification of SSRs. Subsequently, locus-specific primers flanking EST- or genic SSRs can be designed to amplify the microsatellite loci present in the genes. In maize for example, gene-derived SSR markers that have been developed from genes and their primer sequences are available at www.maizeGDB.org. Genic SSRs have some intrinsic advantages over genomic SSRs because they can be obtained quickly by electronic sorting, are present in expressed regions of the genome and expected to be transferable across species (when the primers are designed from more conserved coding regions; Varshney *et al.*, 2005a). The potential use of EST-SSRs developed for barley and wheat has been demonstrated for comparative mapping in wheat, rye and rice (Yu *et al.*, 2004; Varshney *et al.*, 2005a). These studies suggested that EST-SSR markers could be used in related species for which little information is available on SSRs or ESTs. In addition, the genic SSRs are good candidates for the development of conserved orthologous markers for the genetics and breeding of different species. For example, a set of 12 barley EST-SSRs was identified that showed significant homology with the ESTs of four monocotyledonous species (wheat, maize, sorghum and rice) and two dicotyledonous species (*Arabidopsis* and *Medicago*) which could potentially be used across these species (Varshney *et al.*, 2005a).

Kumpatla and Mukopadhyay (2005) examined the abundance of SSR in more than 1.54 million ESTs belonging to 55 dicotyledonous species. They found that the frequency of ESTs containing SSR among species ranged from 2.65 to 16.82%, with dinucleotide repeats being most abundant followed by tri- or mononucleotide repeats, thus demonstrating the potential of *in silico* mining of ESTs for the rapid development of SSR markers for genetic analysis and application to dicotyledonous crops. However, EST-SSRs produce high quality markers but these are often less polymorphic than genomic SSRs (Cho *et al.*, 2000; Eujayl *et al.*, 2002; Thiel *et al.*, 2003). EST resources are also being used to mine SNPs (Picoult-Newberg *et al.*, 1999; Kota *et al.*, 2003). ESTs provide a quantitative method of measuring specific transcripts within a cDNA library and represent a powerful tool for gene discovery, gene expression, gene mapping and the generation of gene profiles. The National Center for Biotechnology Information (NCBI) database, dbEST 0900409 (http://www.ncbi.nlm.nih.gov/dbEST_summary.html) contains the largest collection of ESTs in rice, wheat, barley, maize, soybean, sorghum and potato.

Novel markers can be developed from the transcriptome and specific genes. As summarized by Gupta and Rustgi (2004), these include EST polymorphisms (developed using EST databases); conserved orthologue set markers (developed by comparing the sequences of target genomes with sequences of the closely related species); amplified consensus genetic markers (based on the known genes from model species); gene-specific tags (with primers designed using gene sequences); resistance gene analogues (with primers designed to identify consensus domains conferring resistance); exon–retrotransposon amplification polymorphism (with primers designed to combine with a long terminal repeat retrotransposon-specific primer or a randomly selected microsatellite-containing oligonucleotide); and PCR-based markers targeting exons, introns and promoter regions of known genes with high specificity.

Target region amplification polymorphism (TRAP) markers are derived from a rapid and efficient PCR-based technique which uses bioinformatics tools and EST database information to generate polymorphic markers around targeted candidate gene sequences (Hu and Vick, 2003). This TRAP technique uses two primers of 18 nucleotides to generate markers. TRAP markers are amplified by one fixed primer designed from a target EST sequence in the database and a second primer of arbitrary sequence except for AT- or GC-rich cores that anneal with introns and exons, respectively. The TRAP technique should be useful in genotyping germplasm collections and in tagging genes with beneficial traits in crop plants.

FUNCTIONAL MARKERS. Functional markers (FMs) are derived from polymorphic sites within genes causally affecting phenotypic variation. The development of FMs requires allele-specific sequences of functionally characterized genes from which polymorphic, functional motifs affecting plant phenotype can be identified. Some theoretical and application issues relevant to functional markers in wheat have been addressed (Bagge *et al.*, 2007; Bagge and Lübberstedt, 2008).

FM development requires allele sequences of functionally characterized genes from which polymorphic, functional motifs affecting plant phenotype can be identified. In contrast to RMs, FMs can be used as markers in populations without prior mapping, in mapped populations without risk of information loss owing to recombination and to better represent the genetic variation in natural or breeding populations. Once genetic effects have been assigned to functional sequence motifs, FMs derived from such motifs can be used to fix gene alleles (defined by one or several FM alleles) in several genetic backgrounds without additional calibration. This would be a major advance in the application of markers, particularly in plant breeding, for the selection of parental materials to produce segregating populations for example, as well as the subsequent selection of inbred lines (Andersen and Lübberstedt, 2003). Depending on the mode of FM characterization, they can also be used for the combination of target alleles in hybrid and synthetic breeding and cultivar testing based on the presence or absence of specific alleles at morphological trait loci. In population breeding and recurrent selection programmes, FMs can be used to avoid genetic drift at characterized loci.

A typical example is *Dwarf8* in maize which encodes a gibberellin response modulator from which FMs can be developed for plant height and flowering time. For example, nine sequence motifs in the *Dwarf8* gene of maize were shown to be associated with variation in flowering time and one particular 6-bp deletion accounted for 7–11 days difference in flowering time between inbreds (Thornsberry *et al.*, 2001). Since *Dwarf8* is a pleotropic gene (also affecting plant height) the FM from 'additional flowering time genes' should also be identified in addition to using the *Dwarf8*-derived FM. Orthologues to *Dwarf8* have been identified in wheat (*Rht1*) (Peng *et al.*, 1999), rice (*SLR1*) (Ikeda *et al.*, 2001), and barley (*sln1*) (Chandler *et al.*, 2002), and such genes have been bred into the high-yielding wheat and rice cultivars of the Green Revolution (Hedden, 2003). Altered function of alleles in these orthologous genes can reduce the response to gibberellin and consequently lead to decreased plant height. Thus, biallelic (gibberellin sensitive and insensitive) FMs can be derived for targeted and rapid cultivar breeding aimed at increasing lodging tolerance.

In this section, several widely used DNA markers have been discussed along with an overview of classical genetic markers. DNA markers have gained wide acceptance because of their genome-wide coverage and increasingly simple and easy genotyping. It can be expected that SNP markers, as the ultimate form of genetic polymorphism, will largely replace other types of markers when whole DNA sequences become available for an increasing number of plant species (e.g. Lu *et al.*, 2009; Xu *et al.*, 2009b). However, the choice of DNA markers in genetics and breeding is still highly dependent on the accessibility of geneticists and breeders to various genetic resources including the availability of DNA markers and the time and cost involved. Table 2.2 compares the five most widely used DNA markers.

2.2 Molecular Maps

The order and relative distance of genetic features that are associated with genetic variation or polymorphisms can be determined by genetic mapping. Genetic maps constructed using molecular markers can also be used to locate major genes which can then also be used as genetic markers.

2.2.1 Chromosome theory and linkage

During meiosis the parental diploid ($2n$) cell divides to produce four haploid (n) gametes. During the first meiotic division, the homologous chromosomes align and stick together in a process called synapsis. This allows spindle fibres to attach to the synapsed homologues (tetrads) and to move them as a group to the equator of the cell. As anaphase begins, the homologues can then be oriented such that they are pulled apart to opposite poles of the cell. Following

Table 2.2. Comparison of the five widely used DNA markers in plants.

	RFLP	RAPD	AFLP	SSR	SNP
Genomic coverage	Low copy coding region	Whole genome	Whole genome	Whole genome	Whole genome
Amount of DNA required	50–10 μg	1–100 ng	1–100 ng	50–120 ng	≥ 50 ng
Quality of DNA required	High	Low	High	Medium high	High
Type of polymorphism	Single base changes, indels	Single base changes, indels	Single base changes, indels	Changes in length of repeats	Single base changes, indels
Level of polymorphism	Medium	High	High	High	High
Effective multiplex ratio	Low	Medium	High	High	Medium to high
Inheritance	Co-dominant	Dominant	Dominant/ co-dominant	Co-dominant	Co-dominant
Type of probes/primers	Low copy DNA or cDNA clones	Usually 10 bp random nucleotides	Specific sequence	Specific sequence	Allele-specific PCR primers
Technically demanding	High	Low	Medium	Low	High
Radioactive detection	Usually yes	No	Usually yes	Usually no	No
Reproducibility	High	Low to medium	High	High	High
Time demanding	High	Low	Medium	Low	Low
Automation	Low	Medium	High	High	High
Development/start-up cost	High	Low	Medium	High	High
Proprietary rights required	No	Yes and licensed	Yes and licensed	Yes and some licensed	Yes and some licensed
Suitable utility in diversity, genetics and breeding	Genetics	Diversity	Diversity and genetics	All purposes	All purposes

telophase and cytokinesis, two new daughter cells are formed. Each of these daughter cells has half the chromosomes (n) of the parental cell ($2n$). The second meiotic division closely resembles mitosis with each of the nuclei generated during the first meiotic division splitting to form two more nuclei. Thus, four haploid gametes are produced.

Crossing over is the process by which homologous chromosomes exchange portions of their chromatids during meiosis, resulting in new combinations of genetic information and thus affecting inheritance and increasing genetic diversity. Genes that are present together on the same chromosome tend to be inherited together and are referred to as linked. Genes that are normally linked may be inherited independently during crossing over.

The proportion of recombinant gametes depends on the rate of crossover during meiosis and is known as the recombination frequency (r). The maximum proportion of recombinant gametes is 50% and in this case crossover between two genetic loci has occurred in all the cells. This is equivalent to the case of non-linked genes, i.e. the two loci are inherited independently.

The recombination frequency depends on the rate of crossovers which in turn depends on the linear distance between two genetic loci. Recombination frequencies range from 0 (complete linkage) to 0.5 (complete independent inheritance).

2.2.2 Genetic linkage mapping

In order to utilize the genetic information provided by molecular markers more efficiently, it is important to know the locations and relative positions of molecular markers on chromosomes. The construction of genetic linkage maps using molecular markers is based on the same principles as those used in the preparation of classical genetic maps: selection of molecular markers and genotyping system; selection of parental lines from the germplasm collection that are highly polymorphic at marker loci; development of a population or its derived lines with an increased number of molecular markers in the segregated population; genotyping each individual/line using molecular markers; and constructing linkage maps from the marker data.

The recombination frequency between two linked genetic markers can be defined in units of genetic distance known as centiMorgans (cM) or map units. If two markers are found to be separated in one of 100 progeny, those two markers are 1 cM apart. However, 1 cM does not always correspond to the same length of physical distance or the same amount of DNA. The amount of DNA per cM is referred to as the physical to genetic distance. Areas in the genome where recombination is frequent are known as recombination hot spots; there is relatively little DNA per cM in these hot spots and it can be as low as 200 kb/cM. In other areas recombination may be suppressed and 1 cM will represent more DNA and in some regions the physical to genetic distance can be up to 1500 kb/cM.

Developing mapping populations

In population development, several factors should be taken into consideration including the selection of parental lines and population types and the determination of population size.

CHOICE OF PARENTAL LINES. Four factors should be considered in selecting appropriate parental lines (Xu and Zhu, 1994):

1. DNA polymorphism: genetic polymorphism between parental lines usually depends on how closely related they are, which can be determined by criteria such as geographical distribution and morphological and isozyme polymorphisms. In general, DNA polymorphism is greater in open-pollinated species than in self-pollinated species. For example, RFLP polymorphism is very high among maize lines so that a population derived from any two inbred lines would be desirable for RFLP mapping. Genetic polymorphism is very low in tomato so that only interspecific populations are sufficiently polymorphic to allow for RFLP mapping. The level of polymorphism in rice is intermediate. In plant breeding, many novel traits have been transferred from wild species to cultivated species and such wild-cultivated crosses usually show high levels of DNA polymorphism. Several mapping populations may be needed because genetic polymorphisms that cannot be found in one population may be identified in another.

2. Purity: in the case of self-pollinated plants, the parental lines to be used for development of mapping populations should be breeding-true, i.e. homozygous at almost all of the genetic loci. Purification through further inbreeding may be needed before hybridization is carried out. Breeding-true inbred lines can be used as parents in cross-pollinated plants. For plants for which true-breeding is not possible, genetic mapping can be based on the populations derived from two heterogeneous parental lines.

3. Fertility: hybrid fertility determines whether a large segregating population can be obtained. Distant crosses are usually accompanied by abnormal chromosome pairing and recombination, segregation distortion and reduced recombination frequencies. Some distant hybrids may be partially or completely sterile so that it becomes difficult to obtain a segregating population. In this case, backcrossing populations can be used for mapping as partially sterile hybrids can be rescued by backcrossing to one of the parents.

4. Cytological features: cytological examination may be necessary in order to exclude individuals containing translocations and polyploid species containing monosomes or partial chromosomes from being used as mapping parents.

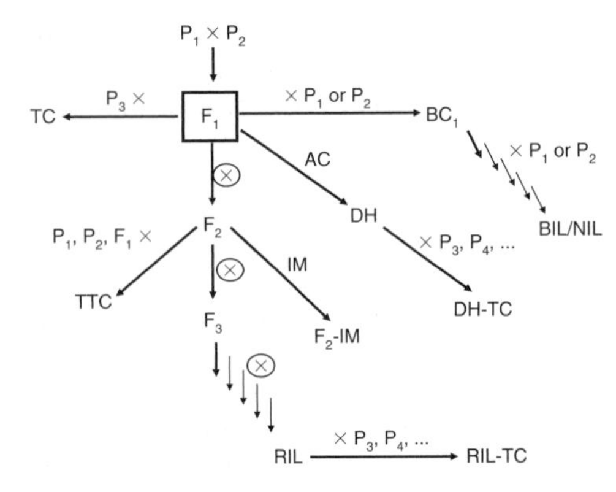

Fig. 2.7. Examples of mapping populations and their relationship. Modified from Xu and Zhu (1994). AC, anther culture; BC, backcross population; BIL, backcross inbred line; DH, double haploid; IM, intermating; NIL, near-isogenic line; RIL, recombinant inbred line; TC, testcross; TTC, triple testcross.

CHOICE OF POPULATION TYPES. There are many types of populations that can be used for genetic mapping. Figure 2.7 shows the relationship between populations derived from two or multiple parental lines. Some of these populations are discussed in detail in Chapter 4. Their use in genetic mapping is discussed below.

F_2 populations are used most frequently in linkage mapping because they are easy to develop. At each marker locus, however, 50% of the individuals in an F_2 population are heterozygous. For dominant markers, dominant homozygotes cannot be distinguished from the heterozygotes and the accuracy of mapping is therefore reduced. In order to improve the accuracy, more F_2 individuals will be needed unless co-dominant markers can be used. Another disadvantage of F_2 populations is that their genetic constitution will change during sexual reproduction so that their genetic structure is difficult to maintain. Vegetative reproduction is one method of prolonging the life of a population as exemplified by ratooning in some grass species. Tissue culture (see Chapters 4 and 12) is another method that can be used to regenerate a population without changing its constitution. Using bulked DNA from F_3 families, which are derived from F_2 individuals, is an alternative approach to prolonging the life of a population because in some crops such as rice and maize, one F_2 plant produces a large number of seeds which are sufficient for multiple plantings. By random mating within each F_3 family, an F_3 population can also be maintained.

Backcross populations (e.g. BC_1) are also frequently used in genetic mapping. BC_1 populations have only two genotypes at each marker locus which represent the corresponding gametes produced in the F_1 hybrid, an advantage over F_2 populations.

If reciprocal BC_1 populations, A × (A × B) and (A × B) × A are obtained from a cross by using the F_1 hybrid as male and female parents, respectively, the difference in recombination frequencies between male and female gametes can be compared and the former indicates the recombination frequency of male gametes while the latter indicates that for female gametes. Like F_2 populations, the genetic constitution of BC populations will change with selfing and they need to be conserved in the same way as F_2 populations. For many crop species, false hybrids may pose a problem which contributes to the inaccuracy of genetic mapping. When distant crosses are used however, backcross populations are the only populations that can be developed because of high sterility among the F_1 hybrids.

Permanent populations such as doubled haploid (DHs), recombinant inbred lines (RILs) and backcross inbred lines (BILs), which are fully discussed in Xu and Zhu (1994) and Chapter 4, provide a continuous supply of genetic material leading to the accumulation of genetic information produced in different laboratories and experiments. For major crops, there are many permanent populations available that are shared internationally with the continuous accumulation of genetic marker and phenotypic data. During population development careful attention should be paid to selection factors that could affect the segregation patterns (Xu *et al.*, 1997; Chapter 4). In some cases, distorted segregation could become very severe if selection pressure is high.

POPULATION SIZE. Achieving the maximum resolution and accuracy from genetic maps largely depends on the size of the mapping population: the larger the mapping population, the greater the accuracy of the genetic map. The research objectives dictate the size of the population. For example, the construction of marker maps requires a much smaller population than the fine mapping of QTL (Chapters 6 and 7). Construction of a high density marker map can be achieved with as few as 200 plants but fine mapping a population in order to clone a gene usually requires over 1000 plants. One alternative is to produce a relatively large population containing about 500 or more individuals from which a subset ($n \approx 150$) can then be used for the construction of a framework map as the initial step in genetic mapping. When fine mapping of a specific chromosome region is required, all the individuals in the population can be used.

With regard to the mapping power, the population size required depends on the maximum map distance that can be distinguished from random assortment and the minimum map distance at which recombination can be detected between two genetic markers (Fig. 2.8). Using a large mapping population, it is possible to map very small genetic distances between markers and also to identify weak genetic linkages. For example, one recombinant represents a 1% recombinant frequency (≈ 1 cM) for a population of 100 individuals, a 2% recombinant frequency (≈ 2 cM) for a population of 50 individuals but only a 0.1% recombination frequency (≈ 0.1 cM) for a population of 1000 individuals.

The maximum map distance that can be distinguished, *max*, can be determined as follows

$$max = r + t_{0.01,n-2}\, SE_r < 0.50 \text{ cM}$$

where n is the population size, t is Student's t parameter for a significant probability of 0.01, $n - 2$ is the degrees of freedom, SE_r is the standard error of r and r is a point estimate of recombination frequency.

The population size required also depends on the type of population. For example, more individuals from F_2 populations are required compared with BC or DH populations, because the F_2 population contains more marker genotypes and to guarantee detection of each genotype, a greater number of individuals is required. In general, F_2 population size should be doubled compared to BC in order to obtain the same mapping accuracy. Therefore, BC or DH populations are more suitable than F_2 for genetic marker mapping. The mapping power of RIL populations is in between that of F_2 and BC (DH) populations. Maximum detectable map distance and minimum resolvable map distance for F_2 and BC populations are shown in Fig. 2.9.

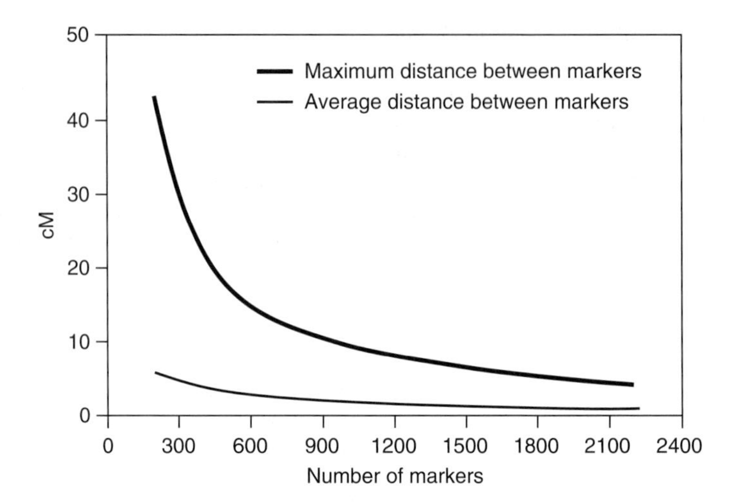

Fig. 2.8. Average and maximum distance expected between markers on a linkage map depending on number of random markers mapped for a genome with 1200 cM, e.g. 12 chromosomes of 100 cM each. The maximum distance curve is for 95% confidence level. From Tanksley *et al.* (1988) with kind permission of Springer Science and Business Media.

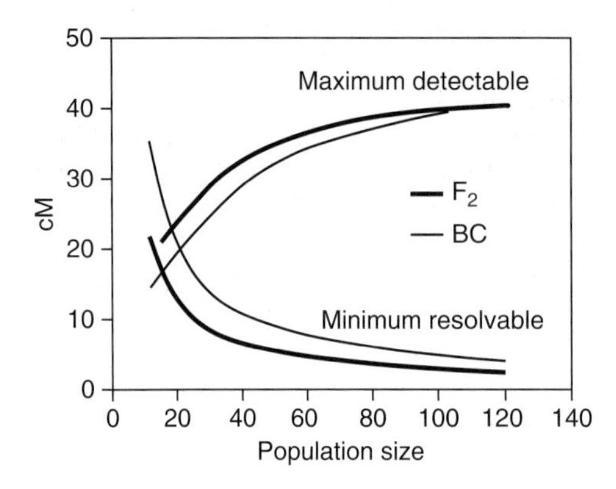

Fig. 2.9. Maximum detectable and minimum resolvable map distances between markers utilizing backcross (BC) and F_2 populations. Curves are for 99% confidence level. From Tanksley *et al.* (1988) with kind permission of Springer Science and Business Media.

Interference and mapping functions

As the genetic distance between two markers increases, the chance of double crossing over within a marker interval increases. For three linked genes, *A*, *B* and *C* with r_1 and r_2 as single crossover frequencies between *A-B* and *B-C*, the double crossover frequency between A and C can be estimated as $r_1 \times r_2$ if the two single crossovers occur independently. However, the observed frequency of double crossovers is usually lower than that expected by calculating $r_1 \times r_2$ which means that a single crossover occurring in a particular chromosome region will reduce the probability of a second single crossover occurring in its flanking regions. This phenomenon is called crossover interference.

The degree of interference can be measured by the coefficient of coincidence (*C*),

$$C = \frac{\text{Observed double crossover}}{\text{Expected double crossover}} = \frac{\text{Observed double crossover}}{r_1 \times r_2 \times n}$$

where n is the total number of individuals observed (including both recombinants and non-recombinants). When C = 0, there is complete interference and no double crossovers, this usually means that the involved chromosome region is very short. When C = 1, there is no interference, indicating that the involved chromosome region is long so that the single crossovers can occur independently.

The genetic distance estimated from the recombinant frequency will be smaller than the real distance by $2C\ r_1 r_2$ if the double crossover is not taken into account. When the genetic distance between two markers is relatively large, the adverse effect of double or multiple crossovers on the estimation of recombination frequency should be corrected. The correction can help establish a reliable function between genetic distance and recombinant frequency and this correction function is known as a mapping function.

The number of (odd) crossovers (k) in an interval defined by two genetic markers has a Poisson distribution with mean θ.

$$\Pr(\text{recombination}) = \sum_k \frac{\theta^k e^{-\theta}}{k!}$$
$$= e^{-\theta}\left(\frac{\theta}{1!} + \frac{\theta^3}{3!} + \ldots + \frac{\theta^k}{k!}\right)$$
$$= \frac{e^{-\theta}(e^{\theta} - e^{-\theta})}{2}$$
$$= \frac{1}{2}(1 - e^{-2\theta})$$

This probability is represented by r and has the following limits $0 \leq r \leq 1/2$. θ is the number of map units (M) between two markers. Assuming that $C = 1$, Haldane (1919) derived the relationship between the map distance (cM) and recombinant frequency r by solving the equation for θ:

$$\theta = -\frac{1}{2}\ln(1 - 2r)$$

which is known as Haldane's map function. When $r = 0$, $\theta = 0$ (complete linkage). When $r = 1/2$, $\theta = \infty$ (markers are unlinked), suggesting that the markers are either on the same chromosome but distant from one another or are located on different chromosomes. When $r = 22\%$, $\theta = 29$ cM.

Kosambi (1944) derived a mapping function that takes the crossover interference into account:

$$\theta = \frac{1}{4}\ln\left(\frac{1 + 2r}{1 - 2r}\right)$$

When $r = 0.22$, $\theta = 23.6$ cM. As two loci become further apart, the amount of interference allowed by the Kosambi map function decreases. For very small values of recombination (r), both Haldane and Kosambi map functions give $\theta \approx r$.

Segregation and linkage tests

With co-dominance and complete dominance models, populations F_2, BC and DH (RIL) have the segregation ratios at locus M with two alleles M_1 and M_2 shown at the bottom of the page.

Assuming two genetic loci M and N each with two alleles, M_1, M_2 and N_1, N_2 and a recombinant frequency r, genotypes and frequencies in an F_2 population derived from two parental lines, P_1 ($M_1M_1N_1N_1$) and P_2 ($M_2M_2N_2N_2$) will be as shown in Fig. 2.10.

There are three types of locus combinations between two loci M and N, depending on the dominance: (1:2:1)-(1:2:1), (3:1)-(1:2:1) and (3:1)-(3:1).

By combining the genotypes listed in Fig. 2.10, nine genotypes and their frequencies can be obtained. Similarly, we can obtain genotypes and their frequencies for (3:1)-(1:2:1) and (3:1)-(3:1) linkage combinations (Fig. 2.11).

Linkage can be determined by comparing the observed frequency for each genotype with the theoretical frequency expected from Mendelian ratios. If there are n individuals, the genotypes/phenotypes listed in Fig. 2.11 can then be identified from top to bottom as n_1 to n_9 for (1:2:1)-(1:2:1), n_1 to n_6 for (3:1)-(1:2:1) and n_1 to n_4 for (3:1)-(3:1); linkage can be determined from these observations.

Linkage detection depends on the normal segregation of the genetic loci involved, thus each locus should be tested to ensure

Population	F_2	BC	DH (RIL)
Co-dominance	**1** M_1M_1:**2** M_1M_2:**1** M_2M_2	**1** M_1M_2:**1** M_2M_2	**1** M_1M_1:**1** M_2M_2
M_1 is dominant	**3** $M_1\_$:**1** M_2M_2	**1** M_1M_2:**1** M_2M_2	**1** M_1M_1:**1** M_2M_2

F_2 gamete frequency	M_1N_1 $(1 - r)/2$	M_1N_2 $r/2$	M_2N_1 $r/2$	M_2N_2 $(1 - r)/2$
M_1N_1 $(1 - r)/2$	$M_1M_1N_1N_1$ $(1 - r)^2/4$	$M_1M_1N_1N_2$ $r(1 - r)/4$	$M_1M_2N_1N_1$ $r(1 - r)/4$	$M_1M_2N_1N_2$ $(1 - r)^2/4$
M_1N_2 $r/2$	$M_1M_1N_1N_2$ $r(1 - r)/4$	$M_1M_1N_2N_2$ $r^2/4$	$M_1M_2N_1N_2$ $r^2/4$	$M_1M_2N_2N_2$ $r(1 - r)/4$
M_2N_1 $r/2$	$M_1M_2N_1N_1$ $r(1 - r)/4$	$M_1M_2N_1N_2$ $r^2/4$	$M_2M_2N_1N_1$ $r^2/4$	$M_2M_2N_1N_2$ $r(1 - r)/4$
M_2N_2 $(1 - r)/2$	$M_1M_2N_1N_2$ $(1 - r)^2/4$	$M_1M_2N_2N_2$ $r(1 - r)/4$	$M_2M_2N_1N_2$ $r(1 - r)/4$	$M_2M_2N_2N_2$ $(1 - r)^2/4$

Fig. 2.10. Theoretical ratios in an F_2 population derived from two parents $M_1M_1N_1N_1$ and $M_2M_2N_2N_2$ with recombinant frequency r.

(1:2:1)-(1:2:1)

Genotype	Frequency
$M_1M_1N_1N_1$	$(1 - r)^2$
$M_1M_1N_1N_2$	$2r(1 - r)$
$M_1M_1N_2N_2$	r^2
$M_1M_2N_1N_1$	$2r(1 - r)$
$M_1M_2N_1N_2$	$2(1 - 2r + 2r^2)$
$M_1M_2N_2N_2$	$2r(1 - r)$
$M_2M_2N_1N_1$	r^2
$M_2M_2N_1N_2$	$2r(1 - r)$
$M_2M_2N_2N_2$	$(1 - r)^2$

(1:2:1)-(3:1)

Genotype	Frequency
$M_1M_1N_1\_$	$1 - r^2$
$M_1M1N_2N_2$	r^2
$M_1M_2N_1\_$	$2(1 - r + r^2)$
$M_1M_2N_2N_2$	$2r(1 - r)$
$M_2M_2N_1\_$	$2r - r^2$
$M_2M_2N_2N_2$	$(1 - r)^2$

(3:1)-(3:1)

Genotype	Frequency
$M_1\_N_1\_$	$3 - 2r + r^2$
$M_1\_N_2N_2$	$2r - r^2$
$M_2M_2N_1\_$	$2r - r^2$
$M_2M_2N_2N_2$	$1 - 2r + r^2$

Fig. 2.11. Genotypes and their frequencies for three linkage combinations at two loci in F_2 populations (each frequency divided by 4).

that it fits Mendelian segregation. For each of the three linkage combinations listed above, four χ^2 tests can be constructed:

- χ^2_T: general test
- χ^2_M: test to determine whether the segregation of M_1 and M_2 fits the Mendelian ratio
- χ^2_N: test to determine whether the segregation of N_1 and N_2 fits the Mendelian ratio
- χ^2_L: test to determine whether M and N loci are linked

Therefore

$$\chi^2_T = \chi^2_M + \chi^2_N + \chi^2_L$$

For linkage combination (1:2:1)-(1:2:1)

$$\chi^2_T = \frac{4}{n}\{n_5^2 + 2(n_2^2 + n_4^2 + n_6^2 + n_8^2) + 4(n_1^2 + n_3^2 + n_7^2 + n_9^2)\} - n \qquad df_T = 8$$

$$\chi^2_M = \frac{2}{n}\{2(n_1 + n_2 + n_3)^2 + (n_4 + n_5 + n_6)^2 + 2(n_7 + n_8 + n_9)^2\} - n \qquad df_M = 2$$

$$\chi^2_N = \frac{2}{n}\{2(n_1 + n_4 + n_7)^2 + (n_2 + n_5 + n_8)^2 + 2(n_3 + n_6 + n_9)^2\} - n \qquad df_N = 2$$

$$\chi^2_L = \chi^2_T - \chi^2_M - \chi^2_N \qquad df_L = 4$$

For linkage combination (3:1)-(1:2:1)

$$\chi^2_T = \frac{8}{3n}(2n_1^2 + n_3^2 + 2n_5^2 + 6n_2^2 + 3n_4^2 + 2n_6^2) - n \qquad df_T = 5$$

$$\chi^2_M = \frac{4}{3n}((n_1 + n_3 + n_5)^2 + 3(n_2 + n_4 + n_6)^2) - n \qquad df_M = 1$$

$$\chi_N^2 = \frac{2}{n}(2(n_1+n_2)^2+(n_3+n_5)^2 + 2(n_4+n_6)^2) - n \qquad df_N = 2$$

$$\chi_L^2 = \chi_T^2 - \chi_A^2 - \chi_B^2 \qquad df_L = 2$$

For linkage combination (3:1)-(3:1)

$$\chi_M^2 = \frac{1}{3n}(n_1^2 + n_2^2 - 3n_3^2 - 3n_4^2) \qquad df_M = 1$$

$$\chi_N^2 = \frac{1}{3n}(n_1^2 - 3n_2^2 + n_3^2 - 3n_4^2) \qquad df_N = 1$$

$$\chi_L^2 = \frac{1}{9n}(n_1^2 - 3n_2^2 - 3n_3^2 + 9n_4^2) \qquad df_L = 1$$

$$\chi_T^2 = \chi_A^2 + \chi_B^2 + \chi_L^2 \qquad df_T = 3$$

Similarly, three linkage combinations for BC or DH (RIL) populations can be constructed.

For example, linkage for (1:2:1)-(1:2:1) in an F_2 population as shown in Fig. 2.12 can be tested as follows

$$\chi_T^2 = \frac{4}{132}\{(56^2 + 2(6^2+5^2+4^2+3^2) + 4(27^2+1^2+0^2+30^2)\} - 132 = 165.818$$

$$\chi_M^2 = \frac{2}{132}\{2(27+6+1)^2 + (5+56+4)^2 + 2(0+3+30)^2\} - 132 = 0.045$$

$$\chi_N^2 = \frac{2}{132}\{2(27+5+0)^2 + (6+56+3)^2 + 2(1+4+30)^2\} - 132 = 0.167$$

$$\chi_L^2 = 165.818 - 0.045 - 0.167 = 165.606$$

	M_1M_1	M_1M_2	M_2M_2	Subtotal
N_1N_1	27	5	0	32
N_1N_2	6	56	3	65
N_2N_2	1	4	30	35
Subtotal	34	65	33	132 = n

Fig. 2.12. Data example used for test of linkage for (1:2:1)-(1:2:1) in an F_2 population.

We have

$$\chi_T^2 \geq \chi_{0.05(8)}^2 = 15.5$$
$$\chi_M^2 \leq \chi_{0.05(2)}^2 = 5.99$$
$$\chi_N^2 \leq \chi_{0.05(2)}^2 = 5.99$$
$$\chi_L^2 \geq \chi_{0.05(4)}^2 = 9.49$$

which indicates that both loci M and N show normal Mendelian segregation and are linked.

Maximum likelihood estimation (MLE) of recombinant frequency

To simplify, we take the linkage combination (3:1)-(3:1) (one of the alleles at each locus shows complete dominance) as an example to show how to obtain the MLE for recombination frequency. From Fig. 2.11, there are four types of phenotypes, $M_1\_N_1\_$, $M_1\_N_2N_2$, $M_2M_2N_1\_$ and $M_2M_2N_2N_2$ with theoretical frequencies p_i (i = 1, 2, 3, 4). p_i is a function of r, a parameter to be estimated, and f is a function of frequency:

$$p_i = f(r)$$

We have p_1 ($M_1\_N_1\_$) = $(3 - 2r + r^2)/4$, p_2 ($M_1\_N_2N_2$) = p_3 ($M_2M_2N_1\_$) = $(2r - r^2)/4$, p_4 ($M_2M_2N_2N_2$) = $(1 - 2r + r^2)/4$, and $\Sigma p_i = 1$.

Considering the number of individuals observed for each category, n_1, n_2, n_3 and n_4, and $\Sigma n_i = n$, they have a probability distribution of $(p_1+p_2+p_3+p_4)^n$. For a specific set of observations (n_1, n_2, n_3 and n_4), the likelihood function is:

$$L(r) = \frac{n!}{n_1!n_2!n_3!n_4!}(p_1)^{n_1}(p_2)^{n_2}(p_3)^{n_3}(p_4)^{n_4}$$
$$= \frac{n!}{n_1!n_2!n_3!n_4!}(1/4)^n(3-2r-r^2)^{n_1}(2r-r^2)^{n_2+n_3}(1-2r+r^2)^{n_4}$$

The MLE of r is $L(r)$ which can be obtained by solving the equation and setting the derivative zero

$$\frac{dL(r)}{dr} = 0$$

The natural logarithm of $L(r)$ is called *support* or *log-likelihood*. Here we have

$$\ln L(r) = C + n_1 \ln(3 - 2r + r^2) + (n_2 + n_3) \ln(2r - r^2) + n_4 \ln(1 - 2r + r^2)$$

where

$$C = \ln \frac{n!}{n_1!\, n_2!\, n_3!\, n_4!} n \ln(1/4)$$

is a constant.

The first partial derivative is the slope of a function. The slope will be zero at the maximum (global/local and/or minimum). The partial derivative is set with respect to r

$$d \ln L(r)/dr = 0$$

The partial derivative of $\ln L(r)$ is usually denoted as *score* or S

$$S = \frac{-n_1 \times 2\,(1-r)}{3-2r+r^2} + (n_2+n_3)\frac{2(1-r)}{2r-r^2} - n_4 \frac{2(1-r)}{1-2r+r^2} = 0$$

That is

$$\frac{n_1}{3-2r+r^2} - \frac{n_2+n_3}{2r-r^2} + \frac{n_4}{1-2r+r^2} = 0$$

$$\frac{n_1}{2+(1-r)^2} - \frac{n_2+n_3}{1-(1-r)^2} + \frac{n_4}{(1-r)^2} = 0$$

If $(1 - r)^2 = k$, then

$$\frac{n_1}{2+k} - \frac{n_2+n_3}{1-k} + \frac{n_4}{k} = 0$$

therefore (see equation at bottom of page) and the MLE is

$$\hat{r} = 1 - \sqrt{k}$$

According to the Rao-Cramer Unequation, the sampling variance of $\hat{r}$ is

$$\frac{1}{V_{\hat{r}}} = -E\left(\frac{d^2[\ln L(r)]}{dr^2}\right) = I$$

where $\frac{d^2[\ln L(r)]}{dr^2}$ is the secondary derivative of $\ln L(r)$ with respect to r, E is expectation, and

$$\frac{d^2[\ln L(r)]}{dr^2} = \sum_i^k \frac{n_i}{p_i^2}\left(\frac{dp_i}{dr}\right)^2 + \sum_i^k \frac{n_i}{p_i}\left(\frac{d^2 p_i}{dr^2}\right)$$

$$E\left(\frac{d^2[\ln L(r)]}{dr^2}\right) = -n\sum_i^k \frac{1}{p_i}\left(\frac{dp_i}{dr}\right)^2 + n\sum_i^k \frac{n_i}{p_i}\left(\frac{d^2 p_i}{dr^2}\right) = n\sum_i^k \frac{1}{p_i}\left(\frac{dp_i}{dr}\right)^2$$

Because $\sum_i^k \frac{d^2 p_i}{dr^2} = \frac{d}{dr}\sum_i^k p_i = 0$,

$$\frac{1}{V_{\hat{r}}} = n\sum_i^k \left[\frac{1}{p_i}\left(\frac{dp_i}{dr}\right)^2\right] = n\sum_i^k i_i = I$$

where I is the total information content and $\sum i_i = I/n$ is the information derived from a single observation.

From the above formula, the variance of $\hat{r}$ can be calculated using the information provided in Table 2.3.

To estimate k, the values of n_i listed in the table are used in the formula:

$$k = \frac{1927 \pm \sqrt{1927^2 + 8 \times 6952 \times 1338}}{2 \times 6952} = 0.7743$$

$$\hat{r} = 1 - \sqrt{0.7743} = 0.1201$$

$$V_{\hat{r}} = 1.76702 \times 10^{-5}$$

Thus,

$$\hat{r} = 0.1201 \pm \sqrt{1.76702 \times 10^{-5}} = 12.01\% \pm 0.42\%$$

$$nk^2 + (2n - 3n_1 - n_4)k - 2n_4 = 0 \quad (n = n_1 + n_2 + n_3 + n_4)$$

$$k = \frac{-(2n-3n_1-n_4) + \sqrt{(2n-3n_1-n_4)^2 + 8nn_4}}{2n}$$

Table 2.3. Calculation of the variance of recombinant frequency for two linked loci each with complete dominance.

Group	n_i	p_i	$\frac{dp_i}{dr}$	$i_i = \frac{1}{p_i}\left(\frac{dp_i}{dr}\right)^2$
$M_1\_N_1\_$	4831	$(3 - 2r + r^2)/4$	$-2(1 - r)/4$	$i_1 = \frac{(1-r)^2}{3 - 2r + r^2}$
$M_1\_N_2N_2$	390	$(2r - r^2)/4$	$2(1 - r)/4$	$i_2 = \frac{(1-r)^2}{2r - r^2}$
$M_2M_2N_1\_$	393	$(2r - r^2)/4$	$2(1 - r)/4$	$i_2 = \frac{(1-r)^2}{2r - r^2}$
$M_2M_2N_2N_2$	1338	$(1 - r^2)/4$	$-2(1 - r)/4$	$i_4 = \frac{4(1-r)^2}{4(1-r)^2} = 1$
Total	$6952 = n$	1	0	$\sum i_i = \frac{(1-r)^2}{3-2r+r^2} + \frac{2(1-r)^2}{2r-r^2} + 1$

This is an example of (3:1)-(3:1) linkage combination. Allard (1956) derived formulas for $r^\wedge$ and $V_{\hat{r}}$ for almost all possible linkage combinations and for different populations.

Likelihood ratio and linkage test

In human genetics the linkage phase (repulsion or coupling) is usually unknown thus making it impossible to calculate recombinant frequency based on the observable recombinants. As a result, likelihood ratios or odds ratios (Fisher, 1935; Haldane and Smith, 1947; Morton, 1955) have been used for linkage testing. The method is based on the comparison of the probability that observed data follow an hypothesis, for example two linked loci and the alternative hypothesis, two independent loci. The ratio of the two probabilities $L(r)/L(1/2)$ is tested as follows: $r = 1/2$ is entered into the likelihood function (see equation (a) at bottom of page).

To simplify the calculation, the log base 10 of the ratio $L(r)/L(1/2)$ known as *LOD*, is used

$$LOD = \log_{10} \frac{L(r)}{L(1/2)}$$

With $n = 6952$, $n_1 = 4831$, $n_2 = 390$, $n_3 = 393$, and $n_4 = 1338$, likelihood of odds (LOD) scores can be calculated for different r values as shown below (see (b) at bottom of page).

The result indicates that LOD scores vary with r and reach the maximum when $r = 0.12$.

If M and N are linked, $L(r)/L(1/2) > 1$, and thus *LOD* is positive. When $L(r)/L(1/2) < 1$, *LOD* is negative.

In human genetics the likelihood ratio should be greater than 1000:1, i.e. $LOD > 3$ in order to establish linkage unequivocally. The concept of the likelihood ratio is now widely used in genetic mapping of other organisms including plant species to judge the reliability of linkage estimation and to verify its existence.

$$L(1/2) = \frac{n!}{n_1!n_2!n_3!n_4!}(1/4)^n(2.25)^{n_1}(0.75)^{n_2+n_3}(0.25)^{n_4} \quad \text{(a)}$$

r	0.05	0.10	0.12	0.15	0.20	0.25	0.30
LOD	586.42	682.51	688.04	678.52	632.01	560.79	472.54

(b)

Multi-point analysis and ordering a set of markers

The methods discussed above are all based on two-point analysis using two markers at a time. However, when more than two markers from one chromosome are considered, they can theoretically be arranged in many different orders but only one particular order will match the genetic order on the chromosome and this particular order can be determined by multi-point analysis.

Consider $M_1, M_2,\ldots, M_m$ genetic markers, ordered by their real locations on a chromosome for m genetic markers, there are a total of $m!/2$ possible orders. Assume the recombinant frequency between two flanking markers, M_i and M_{i+1} is r_i. The objective is to find $r_1, r_2,\ldots, r_{m-1}$ to maximize the likelihood $L(r)$,

$$L(r) \propto p_1(r_1,r_2,\ldots,r_{m-1})^{n_1} \times p_2(r_1,r_2,\ldots,r_{m-1})^{n_2} \times \ldots \times p_m(r_1,r_2,\ldots,r_{m-1})^{n_m}$$

Using the natural logarithm, the partial derivative is then set with respect to $r_1, r_2,\ldots, r_{m-1}$. EM algorithm (Dempster *et al.*, 1977) can be used to obtain the MLE for $r_1, r_2,\ldots, r_{m-1}$, which involves multiple iteration steps of Expectation (E) and Maximization (M). The multiple steps include: (i) providing an initial set of estimates, $r^{old} = (r_1, r_2,\ldots, r_{m-1})$; (ii) using the intial estimates as the estimates of recombinant frequencies to obtain the E, i.e. the expected numbers of recombinants and non-recombinants in each marker interval; (iii) using these expected values as true values to obtain the MLE for $r^{new} = (r_1, r_2,\ldots, r_{m-1})$; (iv) repeating steps (ii) and (iii) until the MLE has converged to its maximum.

Lander and Green (1987) provided an example of the EM method for multi-point linkage analysis. Using 15 marker intervals on human chromosome 7 determined by 16 markers and initial recombinant frequencies of $r_i = 0.05$, the log-likelihood was found to be –351.45. To reduce the difference of log-likelihoods between two consective iterations to less than a given critical value (tolerance value, T = 0.01), 12 iterations were needed which resulted in convergence at log-likelihood –303.28. The probability of the observed data at the converged iteration is $10^{-303.28-(-351.45)} = 10^{48}$ times higher than that for the initial $r_i = 0.05$.

Linkage mapping in the presence of genotyping errors

As generating marker data is time consuming and expensive, maximum use should be made of the information generated. Without accounting for genotyping errors, each error in a non-terminal marker causes two apparent recombinations in the dataset. Thus every 1% error rate in a marker adds ~2 cM of inflated distance to the map. If there is an average of one marker every 2 cM, then an average of a 1% error rate will double the size of the map. There will be large distances between adjacent markers with very high error rates. These cases can be detected, either manually or automatically, and the markers removed. Such genotyping errors can be identified by simply sorting the marker data by a given linkage order to determine whether there are a large number of crossovers involved.

For the markers with low error levels that cannot be detected easily, the best strategy is to integrate error detection with map-building procedure. Cartwright *et al.* (2007) extended the traditional likelihood model used for genetic mapping to include the possibility of genotyping errors. Each individual marker is assigned an error rate which is inferred from the data as are the genetic distances. A software package, TMAP, was developed to use this model to identify maximum-likelihood maps for phase-known pedigrees. The methods were tested using a data set in *Vitis* and a simulated data set, which confirmed that the method dramatically reduced the inflationary effect caused by increasing the number of markers and resulted in more accurate orders.

Molecular maps in plants

Table 2.4 lists some representative molecular maps that have been developed for major crop plants including legumes, cereals and clonal crops, which vary in marker density,

Table 2.4. Representative genetic maps in plants.

Crop	Marker and mapping population	Map information	Reference
Azuki bean	SSR, RFLP, AFLP; 187 BC_1F_1 (JP81481 × *Vigna nepalensis*)	486 markers mapped into 11 linkage groups spanning 832.1 cM with an average marker distance of 1.85 cM, 95% genome coverage	Han *et al.* (2005)
Barley	AFLP, SSR, STS, and *vrs1*); 95 RILs (Russia 6 × H.E.S. 4)	1172 markers with a total distance of 1595.7 cM, and average marker density of 1.4 cM per locus	Hori *et al.* (2003)
	SNP, SSR, RFLP, AFLP; three DH populations	1237 markers, based on three mapping populations consisted of 1237 loci, with a total map length of 1211 cM and an average marker density of 1 locus per cM	Rostoks *et al.* (2005)
Lettuce	AFLP, RFLP, SSR, RAPD; seven inter- and intraspecific populations	2744 markers assigned to nine linkage groups that spanned a total of 1505 cM. The mean interval between markers is 0.7 cM	Truco *et al.* (2007)
Maize	SSR markers; one intermated RIL (IBM) and two immortalized F_2s	The IBM map: 748 SSR and 184 RFLP markers with a total map length of 4906 cM; two immortalized F_2 maps: 457 and 288 SSR markers with total map length of 1830 and 1716, respectively	Sharopova *et al.* (2002)
	cDNA probes; two RIL populations: IBM (B37 × Mo17) and LHRF (F2 × F252)	Framework maps: 237 and 271 loci in IBM and LHRF populations, that both maps contain 1454 loci (1056 on IBM_Gnp2004 and 398 on LHRF-Gnp2004) corresponding to 954 cDNA probes	Falque *et al.* (2005)
Oat	RFLP, AFLP, RAPD, STS, SSR, isozyme, morphological; 136 $F_{6:7}$ RIL (Ogle × TAM O-301)	426 loci (with 2–43 loci each) spanning 2049 cM of the oat genome	Portyanko *et al.* (2001)
Pearl millet	RFLP and SSR; four populations	A consensus genetic map: 353 RFLP and 65 SSR markers, marker density in four maps ranged from 1.49 cM to 5.8 cM	Qi *et al.* (2004)
Potato	AFLP markers; heterozygous diploid potato	> 10,000 AFLP loci, with marker density proportional to physical distance and independent of recombination frequency	van Os *et al.* (2006)
Rice	726 markers; 113 BC_1 (BS125 × WL02) BS125	726 markers with a total distance of 1491 cM and average marker density of 4.0 cM on the framework map, and 2.0 cM overall	Causse *et al.* (1994)
	2275 markers; 186 (Nipponbare × Kasalath) F_2	2275 markers with a total distance of 1521.6 cM, and average marker density of 0. 67 cM per locus	Harushima *et al.* (1998)
Sorghum	2590 PCR-based markers and 137 RIL (BTx623 × IS3620C)	The 1713 cM map encompassed 2926 loci	Menz *et al.* (2002)
	RFLP probes; 65 F_2 (*Sorghum bicolor* × *Sorghum propinquum*)	The *S. bicolor* × *S. propinquum* map is composed of 2512 loci, spanning 1059.2 cM, a marker per 0.4 cM	Bowers *et al.* (2003a)
Sweet potato	AFLP; (Tanzania × Bikilamaliya) F_2 population	632 (Tanzania) and 435 (Bikilamaliya) AFLP markers, with a total of 3655.6 cM and 3011.5 cM, and a marker per 5.8 cM and 6.9 cM, respectively	Kriegner *et al.* (2003)
Wheat	SSR and DArT markers; 152 RILs from a cross between durum wheat and wild emmer wheat	14 linkage groups, 690 loci (197 SSR and 493 DArT markers), spanning 2317 cM, a marker per 7.5 cM	Peleg *et al.* (2008)

and genomic coverage. For example, crops such as barley, maize, potato, rice, sorghum and wheat have high-density genetic maps while cassava, *Musa*, oat, pearl millet, sweet potato and yam have less saturated maps. The large variation in map length results from differences in the number of chromosomes and total size of the genomes as well as from the use of different numbers of markers (increasing the number of markers will generally give a larger total map length up to a certain threshold), the inclusion of skewed markers (that tend to exaggerate map distances) and the use of different mapping software (which vary in estimates of genetic distances). In addition, many published maps report more linkage groups than the basic chromosome number of that species. This is frequently the result of insufficient marker density as most saturated maps can be directly aligned with the basic chromosome complement (Tekeoglu *et al.*, 2002).

The sophistication of molecular map construction has developed from the RFLP maps of the 1980s to PCR-based markers of the 1990s to more integrated maps, as a result of the use of different types of molecular markers including genic markers, over the past decade. Linkage maps have been used in gene mapping for major genes and QTL (Chapters 6 and 7), MAS (Chapters 8 and 9) and map-based gene cloning (Chapter 11).

2.2.3 Integration of genetic maps

Integration of conventional and molecular maps

During the period 1980–1990 molecular maps were developed for many plant species. The first generation of molecular maps have been integrated with conventional genetic maps constructed using morphological and isozyme markers through cytological markers and markers shared by different maps. The 12 molecular linkage groups in rice (McCouch *et al.*, 1988) were assigned to classical linkage groups using trisomics for each of the 12 rice chromosomes. Shared markers and those which segregate in the population can be integrated with the molecular linkage map by using the same population for both conventional and molecular markers. As only very few morphological markers can segregate simultaneously in one population, integration of many of these markers requires multiple populations each with an available preliminary molecular map. If a complete linkage map for morpholgical markers is available, the positions of these markers relative to molecular markers can be inferred from the linkage relationship revealed by both morphological and molecular markers. In addition, morphological markers, including some traits of agronomic importance, can be mapped much more precisely if they are integrated with a dense molecular map and this has now become an integral step in trait and gene mapping. Integration of conventional and molecular maps has been very successful for crop plants for which relatively complete genetic linkage maps are available as a result of the use of morphological markers.

Some representative examples of such maps include rice, maize, tomato and soybean. In rice, 39 morphological markers and 82 RFLP markers were mapped together based on the segregation analysis of 19 F_2 populations derived from the crosses between *indica* cultivar IR24 and *japonica* lines with different morphological markers (Ideta *et al.*, 1996). In tomato, a number of morphological and isozyme markers were mapped with respect to RFLP markers by orienting the molecular linkage map to both morphological and cytological maps. An integrated high-density RFLP-AFLP map of tomato based on two independent *Lycopersicon esculentum* × *Lycopersicon pennellii* F_2 populations was constructed (Haanstra *et al.*, 1999), which spanned 1482 cM and contained 67 RFLP and 1175 AFLP markers. Integrated maps were also developed for maize (Neuffer *et al.*, 1997; Lee *et al.*, 2002) and soybean (Cregan *et al.*, 1999).

Integration of multiple molecular maps

For many crop plants, several molecular maps have been constructed using different populations. These populations are of variable size and structure and maps have been created using different numbers and types of markers. To build an integrated reference or consensus map, the order and genetic distance between specific markers is compared across populations and maps. Stam (1993) developed a computer program, JoinMap, for the construction of genetic linkage maps for several types of mapping populations: BC_1, F_2, RILs, DHs and outbreeder full-sib family. JoinMap can be used to combine ('join') data derived from several sources into an integrated map.

For each crop all the molecular maps developed from different populations will finally be integrated into a consensus map. This process has been very successful for several major crops and it can be expected that it will be extended to all crops when sufficient maps become available. In wheat, an SSR consensus map was constructed by fusing several genetic maps to maximize the integration of genetic mapping information from different sources (Somers *et al.*, 2004). In cotton, chromosome identities were assigned to 15 linkage groups in the RFLP joinmap developed from four intraspecific cotton (*Gossypium hirsutum* L.) populations with different genetic backgrounds (Ulloa *et al.*, 2005). In maize, two populations of intermated RILs (IRILs) were used to build a consensus map, the first panel (IBM) was derived from B73 × Mo17 and the second panel (LHRF) from F2 × F252. Framework maps of 237 loci were built from the IBM panel and 271 loci from the LHRF panel. Both maps were used to locate 1454 loci (1056 on map IBM_Gnp2004 and 398 on map LHRF_Gnp2004) that corresponded to 954 previously unmapped cDNA probes (Falque *et al.*, 2005). In barley, Wenzl *et al.* (2006) built a high-density consensus linkage map from the combined data sets of ten populations, most of which were simultaneously typed with DArT and SSR, RFLP and/or STS markers. The map comprised 2935 loci (2085 DArT, 850 other loci), spanned 1161 cM and contained a total of 1629 'bins' (unique loci). The arrangement of loci was very similar to, and almost as optimal as, the arrangement of loci in component maps created for individual populations.

Integration of genetic and physical maps

Integrated genetic and physical genome maps are extremely valuable for map-based gene isolation, comparative genome analysis and as sources of sequence-ready clones for genome sequencing projects. A well-defined correlation between the physical and genetic maps will greatly facilitate molecular breeding efforts through associating candidate genes with important biological or agronomic traits, positional cloning and comparative analysis across populations and species, and whole genome sequences, which will in turn facilitate the development of various molecular breeding tools.

Various methods have been developed for assembling physical maps of complex genomes and integrating them with genetic maps. To create an integrated genetic and physical map resource for maize, a comprehensive approach was used that included three core components (Cone *et al.*, 2002). The first was a high-resolution genetic map that provided essential genetic anchor points for ordering the physical map and for utilizing comparative information from other smaller genome plants. The physical map component consisted of contigs (sets of overlapping fingerprint clones) assembled from clones from three deep-coverage genomic libraries. The third core component was a set of informatics tools designed to analyse, search and display the mapping data. In rice, most of the genome (90.6%) was anchored genetically by overgo hybridization, DNA gel blot hybridization and *in silico* anchoring (Chen *et al.*, 2002). In wheat, the genetic–physical map relationship of microsatellite markers was established using the deletion bin system (Sourdille *et al.*, 2004). In sorghum, Klein *et al.* (2000) developed a high-throughput PCR-based method for building bacterial artificial chromosome (BAC) contigs and locating BAC clones on the genetic map in order to construct an integrated genetic and physical map. It was found that 30% of the overlapping BACs aligned by AFLP analysis provided information for merging contigs and singletons that could not be joined using fingerprint data alone. In the grasses *Lolium perenne* and *Festuca pratensis*, the physical map was integrated with a genetic map using genomic *in situ* hybridization, which was composed of 104 *F. pratensis*-specific AFLPs. The integrated map demonstrated the large-scale analysis of the physical distribution of AFLPs and variation in the relationship between genetic and physical distance from one part of the *F. pratensis* chromosome to another (King *et al.*, 2002).

An integrated genetic and physical mapping tool has been developed by the Maize Mapping Project, Columbia, Missouri, USA (http://www.maizemap.org/iMapDB/iMap.html). Contigs that were assembled by fingerprinting and the automated matching of BACs were then anchored on to IBM2 and IBM2 neighbour maps. In the Gramene database, a web-based tool, CMAP, was developed to allow users to view comparisons of genetic and physical maps (Ware *et al.*, 2002). In addition, an integrated bioinformatic tool, the Comparative Map and Trait Viewer (CMTV), was developed to construct consensus maps and compare QTL and functional genomics data across genomes and experiments (Sawkins *et al.*, 2004). All these tools can be used to build integrated maps based on shared markers and a reference map to initiate the process. The integration of genetic, cytological and physical maps is illustrated in the example shown in Fig. 3.6.

3

Molecular Breeding Tools: Omics and Arrays

The success of molecular breeding depends upon the various tools that can be used for the efficient manipulation of genetic variation. All kinds of 'omics', arrays and high-throughput technologies make it possible to carry out more large-scale genetic analyses and breeding experiments than ever before. These technologies have been incorporated into many novel genetic and breeding processes, some of which were described in Chapter 2. In this chapter, microarrays, high-throughput technologies and several aspects of genomics will be briefly discussed to provide some of the fundamental knowledge required for molecular breeding.

3.1 Molecular Techniques in Omics

Developments in molecular techniques have contributed to the various fields of 'omics', which include genomics, transcriptomics, proteomics, metabalomics and phenomics. These underlying developments include advanced gel, hybridization and expression systems, cell imaging by light and electron microscopy, high density microarrays and array experiments, and genetic readout experiments.

Using proteomics as an example, classical techniques used in proteomics involve the use of two-dimensional gel electrophoresis (2DE). The proteins can be identified by excising the spot from the gel, digesting the polypeptide into smaller peptide fragments with specific proteases, and sequencing the peptides directly or analysing them by mass spectrometry (MS). Although this method is still useful and widely used, it is limited in sensitivity, resolution, and the range of abundance of the different proteins in the sample (Zhu *et al.*, 2003; Baginsky and Gruissem, 2004). For example, abundant proteins in the sample dominate the gel whereas less abundant proteins might not be visible. New approaches involve both improved separation methods and advanced detection equipment, and several other new technologies are available for use in proteomic research (Kersten *et al.*, 2002; Zhu *et al.*, 2003; De Hoog and Mann, 2004). New detection methods and proteomic technologies are also being developed in an array format, which is increasingly being focused on protein–protein interactions, post-transcriptional modification, and elucidation of three-dimensional protein structure.

3.1.1 2-Dimensional gel electrophoresis

2DE is a form of gel electrophoresis commonly used to analyse proteins. Mixtures of proteins are separated by two properties in two dimensions in 2DE. During the early years of proteomics and until recently, profiling of protein expression relied primarily on the use of two-dimensional polyacrylmide gel electrophoresis (2D PAGE), which was later combined with MS. The basic procedure is to solubilize the protein contents of an entire cell population, tissue or biological fluid, followed by separation of the protein components in the lysate using 2DE and visualization of the separated proteins with silver staining. This approach allows only a limited display of the total protein content and can identify only the relatively abundant proteins.

2DE begins with one-dimensional electrophoresis and then separates the molecules by a second property in a direction at 90° to the first. In this technique proteins are separated in one dimension by isoelectric point and in the second dimension by mass. In one-dimensional electrophoresis, proteins (or other molecules) are separated in one dimension, so that all the proteins/molecules in one lane will be separated from one another according to the differences in a particular property (e.g. isoelectric point) between each component. The result is a gel with proteins separated out on its surface (Fig. 3.1a). The proteins can then be visualized by a variety of staining methods, the most commonly used stains are silver nitrate and Coomassie blue. By combining electrophoresis with MS, individual proteins can be profiled (Fig. 3.1b, c) and theoretical and acquired MS profiles can be matched by a database search.

An important development in 2D PAGE is the use of immobilized pH gradients

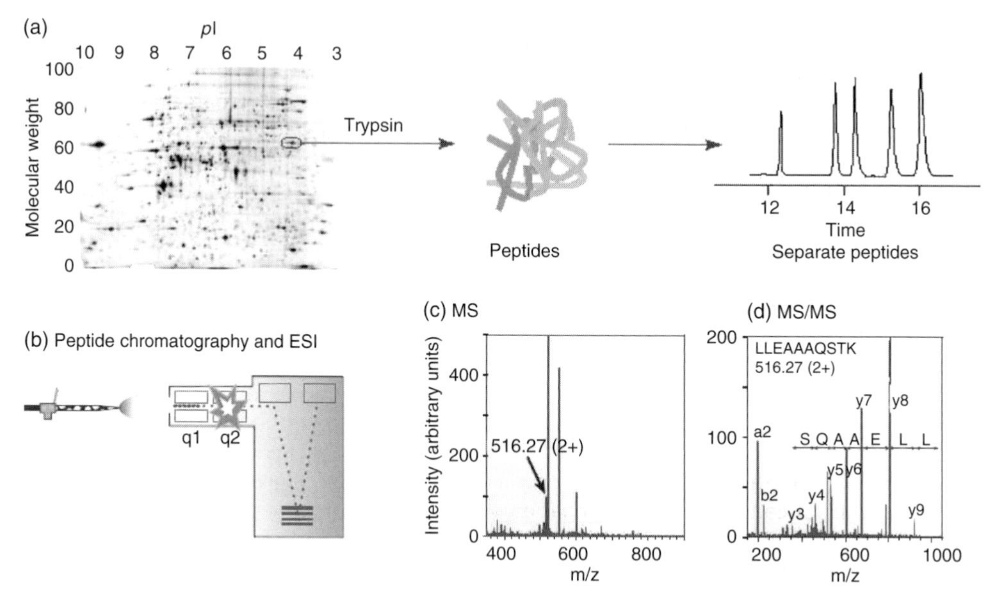

Fig. 3.1. Standard protein analysis by two-dimensional electrophoresis followed by mass spectrometry proteomics. (a) Protein is separated by two-dimensional electrophoresis: in one dimension by isoelectronic point (*pI*) and in the second dimension by mass (molecular weight). Individual peptides are obtained using trypsin to cleave peptide chains. (b) Peptides are separated by chromatography and then peptides are ionized using electospray ionization (ESI): they pass through the first quadrupole (q1) and collision chamber (q2). (c) Individual ions are separated based on their mass-to-charge (m/z) by a mass analyser. (d) From the MS spectrum, an individual peptide ion (516.27 (2+)) is selected for MS/MS analysis to produce peptide ion fragmentation patterns. Letters S, Q, A, A, E, L and L represent amino acids in the selected peptide and 'a2', b2', 'y3', etc. represent different ions.

(IPGs) in which a pH gradient is fixed within the acrylamide matrix (Gorg *et al.*, 1999). Because a wide or narrow pH range can be fixed within the gel, IPGs can be used to detect thousands of spots on a single gel with high reproducibility. A variation on this theme is the use of so-called 'zoom gels' in which the protein content of an individual sample is first fractionated into narrow pH ranges under low resolution and then each fraction is subjected to high-resolution separation by 2D PAGE. Another innovation in 2DE is differential in-gel electrophoresis (DIGE; Ünlü *et al.*, 1997) in which two pools of proteins are labelled with different fluorescent dyes. The labelled proteins are mixed and separated in the same 2DE.

Some of the main challenges facing expression proteomics, be it using 2D PAGE or any other approach, include the great dynamic range of protein abundance and a wide range of protein properties including mass, isoelectric point, extent of hyrophobicity and post-translational modifications (Hanash, 2003). Reducing sample complexity prior to analysis, for example by analysing protein subsets and subcellular organelles separately, improves the reach of 2DE or other separation techniques for the quantitative analysis of low-abundance proteins. The isolation of sub-proteome components may be combined with protein tagging to further enhance sensitivity. For example, protein tagging technologies have been implemented for the comprehensive analysis of the cell-surface proteome (Shin *et al.*, 2003).

Even with all the improvements that could be introduced, 2DE will probably remain a rather low-throughput approach that requires a relatively large amount of sample for analysis. The latter is particularly problematic when the samples to be analysed are of limited availability (Hanash, 2003). In particular, the use of laser-capture microdissection, which allows defined cell types to be isolated from tissues, yields a very small amount of protein that is difficult to reconcile with the large amounts needed for 2DE.

3.1.2 Mass spectrometry

MS is an analytical technique used to determine the composition of a physical sample by measuring the mass-to-charge ratio of the ions. It has become the method of choice for analysis of complex protein samples (Han *et al.*, 2008). MS-based proteomics has established itself as an indispensable technology for interpreting the information encoded in genomes; this has been made possible by technical and conceptual advances in many areas, most notably the discovery and development of protein ionization methods as recognized by the award of the Nobel prize for chemistry to John B. Fenn and Koichi Tanaka in 2002. Mass spectrometry instrumentation has made strides in recent years in terms of dynamic range and sensitivity (Blow, 2008).

Mass spectrometric measurements are carried out in the gas phase on ionized analytes. Mass spectrometers consist of three essential parts; the first, an ionization source, converts molecules into gas-phase ions. Once ions are created, individual ions are separated based on their mass-to-charge ratio (m/z) by a second device, a mass analyser, and transferred by magnetic or electric fields to the third, an ion detector (Fig. 3.1b, c and d). The mass analyser is central to the technology. It uses a physical property to separate ions of a particular m/z value that then strike the ion detector. The magnitude of the current that is produced at the detector as a function of time (i.e. the physical field in the mass analyser is changed as a function of time) is used to determine the m/z value of the ion. In the context of proteomics, its key parameters are sensitivity, resolution, mass accuracy and the ability to generate information-rich ion mass spectra from peptide fragments. The technique has several applications, including identifying unknown compounds by the mass of the compound molecules or their fragments, determining the isotopic composition of an element and its structure by observing the fragmentation, quantifying the amount of a compound in a sample using carefully designed methods and studying the fundamentals of gas phase ion chemistry.

There are many types of mass analysers which use static or dynamic fields and magnetic or electric fields. Each analyser type has its strengths and weaknesses. Four basic types of mass analyser used in proteomic research are: ion trap, time-of-flight (TOF), quadrupole and Fourier transform mass spectrometry (FT-MS) analyser. In ion-trap analysers, the ions are first captured or trapped for a certain time interval and are then subjected to MS or tandem MS (MS/MS) analysis. Ion traps are robust, sensitive and relatively inexpensive. A disadvantage is their relatively low mass accuracy, due in part to the limited number of ions that can be accumulated at their point-like centre before space-charging distorts their distribution and thus the accuracy of the mass measurement. The linear or two-dimensional ion trap is a recent development where ions are stored in a cylindrical volume that is considerably larger than that of the traditional, three-dimensional ion traps, allowing increased sensitivity, resolution and mass accuracy. The FT-MS instrument is also a trapping mass spectrometer, although it captures the ions under high vacuum in a high magnetic field. It measures mass by detecting the image current produced by ions cyclotroning in the presence of a magnetic field. Its strengths are high sensitivity, mass accuracy, resolution and dynamic range. In spite of the enormous potential, the expense, operational complexity and low-peptide-fragmentation efficiency of FT-MS instruments has limited their routine use in proteomic research (Aebersold and Mann, 2003). The TOF analyser uses an electric field to accelerate the ions through the same potential and then measures the time they take to reach the detector.

Techniques for ionization have been key to determining what types of samples can be analysed by MS. Electrospray ionization (ESI; Fenn *et al.*, 1989) and matrix-assisted laser desorption/ionization (MALDI; Karas and Hillenkamp, 1988) are two techniques most commonly used to volatize and ionize proteins or peptides for MS analysis while inductively coupled plasma sources are used primarily for metal analysis on a wide array of sample types. MALDI is usually coupled to TOF analysers that measure the mass of intact peptides, whereas ESI has mostly been coupled to ion traps and triple quatrupole instruments and used to generate fragment ion spectra (collision-induced spectra) of selected precursor ions (Aebersold and Goodlett, 2001). ESI creates ions by application of a potential to a flowing liquid causing the liquid to charge and subsequently spray. The electrospray creates very small droplets of solvent-containing analyte. Solvent is removed by heat or some other form of energy (e.g. energetic collisions with a gas) as the droplets enter the mass spectrometer and multiply-charged ions are formed in the process. ESI ionizes the analytes out of a solution and is therefore readily coupled to liquid-based (for example, chromatographic and electrophoretic) separation tools (Fig. 3.1). MALDI creates ions by excitation of molecules that are isolated from the energy of the laser by an energy-absorbing matrix. The laser energy strikes the crystalline matrix to cause rapid excitation of the matrix and subsequent ejection of matrix and analyte ions into the gas phase. MALDI-MS is normally used to analyse relatively simple peptide mixtures in cases where integrated liquid-chromatography ESI-MS systems (LC-MS) are preferred for the analysis of complex samples.

Key developments leading to improved detection of proteins include TOF MS and relatively non-destructive methods for converting proteins into volatile ions (Zhu *et al.*, 2003). MALDI and ESI have made it possible to analyse large molecules such as peptides and proteins. Although MALDI-TOF MS is a relative high-throughput method compared with ESI, the latter is more easily coupled with separation techniques such as LC or high pressure LC (HPLC) (Zhu *et al.*, 2003). This has provided an attractive alternative to 2DE, because even low-abundance proteins and insoluble transmembrane proteins can be detected (Ferro *et al.*, 2002; Koller *et al.*, 2002). Other MS techniques include gas chromatography–mass spectrometry (GC-MS), and ion mobility spectrometry/mass spectrometry (IMS/MS). All MS-based techniques require a substantial and searchable database of predicted proteins, ideally representing the entire genome. Protein identification is possible by comparing the deduced masses of the resolved peptide fragments with the theoretical masses of predicted peptides in the database.

Mass spectrometers are restricted in the number of ions that can be detected at any point in time. Pre-fractionation of proteins on the basis of isolation of specific cell types or subcellular organelles is often necessary to reduce the complexity (Lonosky *et al.*, 2004). Another method of fractionating a complex sample is to introduce a chromatographic technique before MS analysis. This method, referred to as multidimensional protein identification technology (MudPIT) (Whitelegge, 2002) has been used to conduct a shotgun survey of metabolic pathways in the leaves, roots and developing seeds of rice (Koller *et al.*, 2002). Compared with 2DE-MS, each method identifies unique proteins, supporting the complementary nature of the different proteomic technologies.

3.1.3 Yeast two-hybrid system

The yeast two-hybrid assay (Fields and Song, 1989) provides a genetic approach to the identification and analysis of protein–protein interactions. Yeast two-hybrid (Y2H) systems detect not only members of known complexes but also weak or transient interactions (Jansen *et al.*, 2005). The Y2H assay makes use of the molecular organization found in many transcription factors that have a DNA-binding domain and activation domains that can function independently, but when these domains are fused to two proteins that interact, the ability of the domains to control transcriptional activity is reconstituted. In this assay hybrid proteins are generated that fuse a protein X to the DNA-binding domain and protein Y to the activation domain of a transcription factor (Fig. 3.2a). Interaction between X and Y reconstitutes the activity of the transcription factor and leads to expression of a reporter gene with a recognition site for the DNA-binding domain. In the typical practice of this method, a protein of interest fused to the DNA-binding domain (the so-called bait) is screened against a library of activation-domain hybrids (prey) to select interaction partners (Phizicky *et al.*, 2003).

The key advantages of the Y2H assay are its sensitivity and flexibility (Phizicky *et al.*, 2003). The sensitivity derives in part from overproduction of protein *in vivo*, their designed direction to the nuclear compartment where interactions are monitored, the large number of variable inserts of the interacting proteins that can be examined at once, and the potency of the genetic selections. This sensitivity leads to the detection of interactions with dissociation constants around 10^{-7} M which is in the range of most weak protein interactions found in the cell and is more sensitive than co-purification. It also allows detection of certain transient interactions that might affect only a subpopulation of the hybrid proteins. Flexibility of the assay is provided by calibration to detect interactions of varying affinity by altering the expression levels of the hybrid proteins, the number and nature of the DNA-binding sites and the composition of the selection media.

Some disadvantages of the Y2H assay include the unavoidable occurrence of false negatives and false positives (Phizicky *et al.*, 2003). False negatives include proteins such as membrane proteins and secretory proteins that are not usually amenable to nuclear-based detection systems, proteins that failed to fold correctly and interactions dependent on domains occluded in the fusions or on post-translational modifications. False positives include colonies not resulting from a *bona fide* protein interaction, as well as colonies resulting from a protein interaction not indicative of an association that occurs *in vivo*.

There are several variations of the Y2H system. In the reverse Y2H system, induced *URA3* expression leads to 5-FOA being converted into the toxic substance 5-fluorouracil by Ura3p, leading to growth prohibition. Mutated or fragmented genes are created and then subjected to analysis and only loss-of-interaction mutants are able to grow in the presence of 5-FOA. In the one-hybrid system, the bait is a target DNA fragment fused to a reporter gene. Preys that are able to bind to the DNA fragment–reporter fusion will

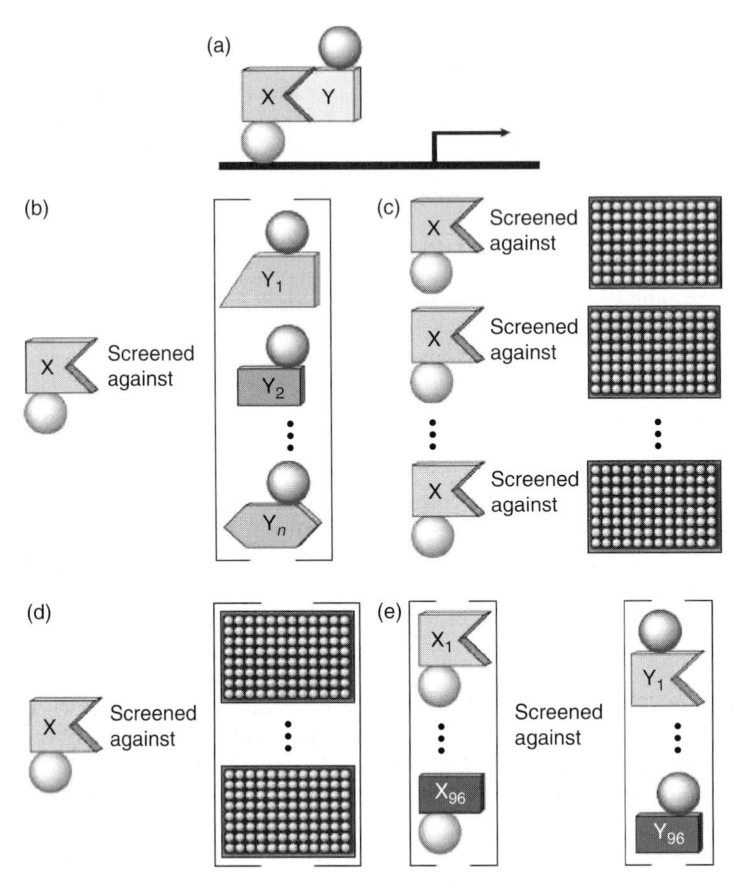

Fig. 3.2. Yeast two-hybrid approaches. (a) The yeast two-hybrid system. DNA binding and activation domains (circles) are fused to two proteins X and Y, the interaction of X and Y leads to reporter gene expression (arrow). (b) A standard two-hybrid search. Protein X, present as a DNA binding domain hybrid, is screened against a complex library of random inserts in the activation domain vector (shown in square brackets). (c) A two-hybrid array approach. Protein X is screened against a complete set of full length open reading frames (ORFs) present as activation domain hybrids (shown as yeast transformant spotted on to microtitre plates). (d) A two-hybrid search using a library of full length ORFs. The set of ORFs as activation-domain hybrids (microtitre plates in square brackets) is combined to form a low-complexity library. (e) A two-hybrid pooling strategy. Pools of ORFs as both DNA-binding domain and activation domain hybrids (in square brackets) are screened against each other. From Phizicky *et al.* (2003) reprinted by permission from Macmillan Publishers Ltd.

lead to activation of the reporter genes (*lacZ*, *HIS3* and *URA3*). In the repressed transactivator system, the interaction of bait–DNA binding domain fusion proteins and the prey–repressor domain fusion proteins can be detected by repression of the reporter *URA3*. The interaction of bait and prey enables cells to grow in the presence of 5-FOA, whereas non-interactors are sensitive to 5-FOA as a result of Ura3p production. In the three-hybrid system, the interaction of bait and prey proteins requires the presence of a third interacting molecule to form a complex. The third interacting molecule can be a protein used with a nuclear localization acting as a bridge between bait and prey to cause transcriptional activation.

Different genome-wide two-hybrid strategies have been used to analyse protein interactions in *Saccharomyces cerevisiae*. One approach involved screening a large number of individual proteins against a comprehensive library of randomly generated fragments (Fig. 3.2b). A second approach used systematic one-by-one testing of every possible protein combination using a mating assay with a comprehensive array of strains (Fig.3.2c). A third approach used a one-by-many matings strategy in which each member of a nearly complete set of strains expressing yeast open reading frames (ORFs) as DNA-binding domain hybrids was mated to a library of strains containing activation-domain fusions of full-length yeast ORFs (Fig. 3.2d). A fourth variation involved mating of defined pools of strain arrays (Fig.3.2e). Suter *et al.* (2008) reviewed the current applications of Y2H and variant technologies in yeast and mammalian systems. Y2H methods will continue to play a dominant role in the assessment of protein interactomes.

3.1.4 Serial analysis of gene expression

Serial analysis of gene expression (SAGE) is a method for the comprehensive analysis of gene expression patterns. SAGE is used to produce a snapshot of the mRNA population in a sample of interest (Velculescu *et al.*, 1995). Several variants have since been developed, most notably a more robust version, LongSAGE (Saha *et al.*, 2002) and the most recent SuperSAGE (Matsumura *et al.*, 2005) that enables very precise annotation of existing genes and discovery of new genes within genomes because of an increased tag-length of 25–27 bp. Three principles underlie the SAGE methodology: (i) a short sequence tag (originally 10–14 bp) that contains sufficient information to uniquely identify a transcript provided that the tag is obtained from a unique position within each transcript; (ii) sequence tags can then be linked together to form long serial molecules that can be cloned and sequenced; and (iii) quantitation of the number of times a particular tag is observed which provides the expression level of the corresponding transcript.

The principle of the technique is shown in Fig. 3.3: mRNAs are isolated from a tissue, and double-stranded cDNAs are synthesized from biotinylated oligo-dT primers. The DNA is cut with a frequent-cutting restriction enzyme (*Nla*III), and the 3' extremities of the double-stranded DNAs are isolated using streptavidin (which binds biotin). The double-stranded DNA is divided into two groups, the 5' extremities of which are ligated to primers A or B. These primers contain a restriction site recognized by the enzyme *Bsm*FI which cuts 20 nucleotides away from its recognition site. The two populations are then combined, ligated, amplified and sequenced. The four-nucleotide sequence CAGT (recognized by *Nla*III) allows the identification of each amplified region. The sequences obtained allow their unique identification for each gene, although the size of the sequence is very short (of the order of a dozen nucleotides), it is sufficiently adequate to identify the specific gene from which it derives by comparison with sequence databases.

SAGE can be used to identify the collection of genes transcribed in a given tissue or developmental stage. It also provides an estimate of the frequency of transcription of each identified gene because it is proportional to the frequency of the sequence in the total collection of sequences obtained. The study by Velculescu *et al.* (1995) indicated: (i) that just nine base pairs of DNA sequence are sufficient to distinguish 262,144 genes if the sequence is from a defined position in the gene; (ii) if the 9-bp sequences are placed end-to-end (concatenated) and separated by 'punctuation' then they can be sequenced 'serially' (analogous to the mechanism by which a computer transmits data); and (iii) a single sequencing reaction can yield information on 10–50 genes.

3.1.5 Quantitative real-time PCR

Real-time reverse-transcriptase PCR (RT-PCR), also known as quantitative real-time-PCR (QRT-PCR), measures PCR amplification in real time (via fluorescence) during amplification. It enables both detection and quantification (as absolute number of copies or relative amount when normalized to DNA input or additional normalizing genes) of a

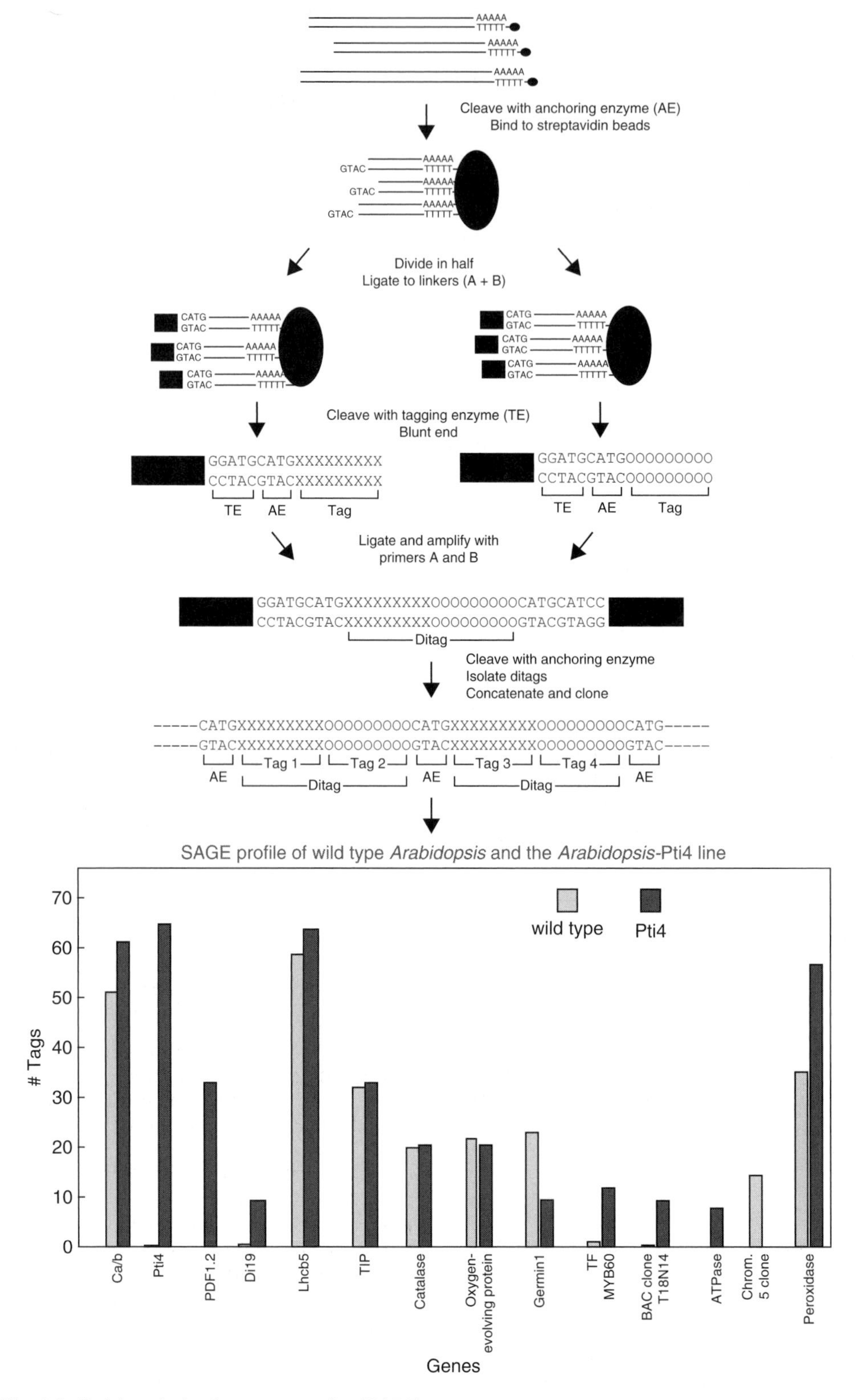

Fig. 3.3. Serial analysis of gene expression (SAGE).

specific sequence in a DNA sample. The procedure follows the general principle of PCR; its key feature is that the amplified DNA is quantified as it accumulates in the reaction in real-time after each amplification cycle. Two common methods of quantification are the use of fluorescent dyes that intercalate with double-strand DNA, and modified DNA oligonucleotide probes that fluoresce when hybridized with a complementary DNA (cDNA).

Real-time RT-PCR uses fluorophores in order to detect levels of gene expression. As mRNA becomes translated at the ribosome to produce functional proteins, mRNA levels tend to roughly correlate with protein expression. In order to adapt PCR to the measurement of RNA, the RNA sample first needs to be reverse transcribed to cDNA via an enzyme known as a reverse transcriptase. The original RT-PCR technique required extensive optimization of the number of PCR cycles, so as to obtain results during logarithmic DNA amplification, before it starts to plateau. Development of PCR technology that uses fluorophores to measure DNA amplification in real-time allows researchers to bypass the extensive optimization associated with normal RT-PCR.

In real-time RT-PCR, the amplified product is measured at the end of each cycle. This data can be analysed by computer software to calculate relative gene expression between several samples or mRNA copy number based on a standard curve (Fig. 3.4). By comparing cycles of linear amplification among target cDNAs/genes, the relative fold difference in expression can be measured as $2^{\text{cycles x} - \text{cycles y}}$. For example, comparing sample *x* linear at 37 and sample *y* linear at 27, we have 2^{37-27}, which means a 1024-fold *x* mRNA accumulation of *x* versus *y*, assuming that the sequences amplify with equal efficiency (Fig. 3.4).

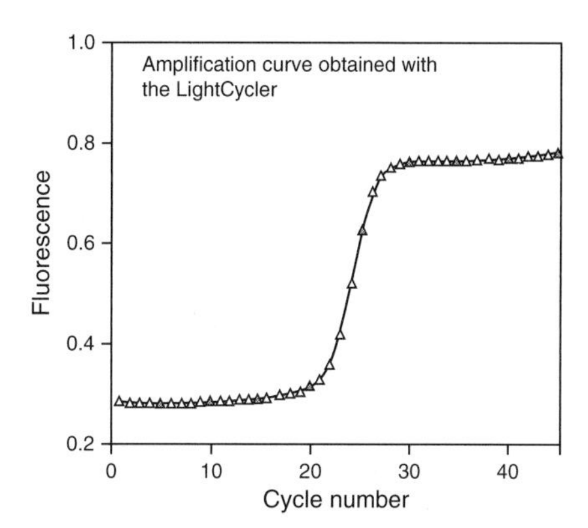

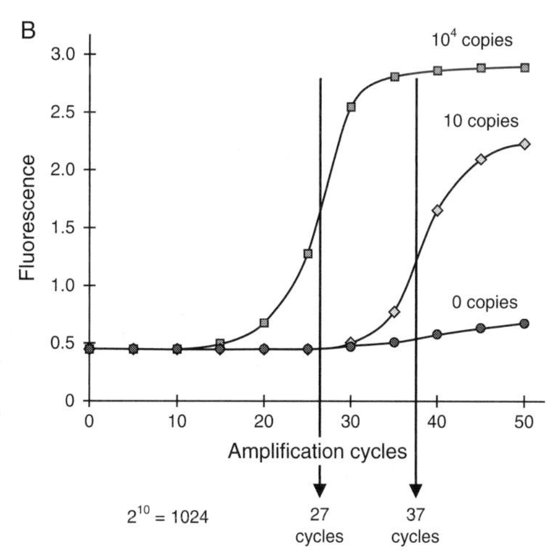

Fig. 3.4. Quantitative real-time PCR. (A) Agarose gel to show the amplification results in conventional PCR of different cycles (above) and amplification curve obtained with the LightCycler to show relative gene expression (below). (B) Relative gene expression for different mRNA copies, by which a relative fold difference in gene expression can be measured.

3.1.6 Subtraction suppressive hybridization

Suppression subtractive hybridization (SSH) (Diatchenko *et al.*, 1996) is a technique that uses PCR to quickly compare the expression of mRNA from different samples and show the relative difference in the concentration of these molecules. It can be used to enrich for differentially expressed genes. Subtracted cDNA libraries are hybridization and PCR based and result in normalization of the sample. They can be combined with full length cDNA libraries.

SSH includes the following procedures: (i) prepare cDNAs from two stages/conditions; (ii) separately digest tester (from the same source as sample to be tested) and driver cDNA (from a normal sample) to obtain shorter fragments; (iii) divide tester cDNA into two portions and ligate each to a different adaptor, while driver cDNA has no adaptors; (iv) hybridization kinetics lead to equalization and enrichment of differentially expressed sequences among single strand tester molecules; and (v) ultimately generate templates for PCR amplification from differentially expressed sequences. As a result, only differentially expressed sequences are amplified exponentially.

3.1.7 *In situ* hybridization

In situ hybridization (ISH) is a type of hybridization that uses a labelled cDNA or RNA strand (i.e. probe) to localize a specific DNA or RNA sequence in a portion or section of tissue (*in situ*) or in the entire tissue. DNA ISH can be used to determine the structure of chromosomes. Fluorescent DNA ISH (FISH) can be used to assess chromosomal integrity. RNA ISH (hybridization histochemistry) is used to measure and localize mRNAs and other transcripts within tissue sections or whole mounts.

For hybridization histochemistry, sample cells and tissues are usually treated to fix the target transcripts in place and to increase access of the probe. The probe is either a labelled cDNA or more commonly, a cRNA (riboprobe). The probe hybridizes to the target sequence at elevated temperature and the excess probe is then washed away (after prior hydrolysis using RNase in the case of unhybridized, excess RNA probe). Solution parameters such as temperature, salt and/or detergent concentration can be manipulated to remove any non-identical interactions (i.e. only exact sequence matches will remain bound). Then, the probe that was labelled with either radio-, fluorescent- or antigen-labelled bases (e.g. digoxigenin) is localized and quantitated in the tissue using autoradiography, fluorescence microscopy or immunohistochemistry. ISH can also use two or more probes labelled with radioactivity or the other non-radioactive labels to simultaneously detect two or more transcripts.

Transcriptional analysis may also be carried out by inserting a reporter gene such as *lacZ* or GFP (green fluorescent protein) downstream from the promoter under study. *lacZ* encodes β-galactosidase and its expression is detected by the blue colour obtained in the presence of X-Gal. GFP is a protein containing a chromophore which fluoresces under blue light (395 nm). These reporters are used to evaluate the expression levels and identify the tissues in which the normal gene is expressed under the chosen promoter.

3.2 Structural Genomics

Genomics is a term coined by Thomas Roderick in 1986 and refers to a new scientific discipline of mapping, sequencing and analysing genomes. Genomics is now however, undergoing a transition or expansion from the mapping and sequencing of genomes to an emphasis on genome function. To reflect this shift, genome analysis may now be divided into ‘structural genomics’ and ‘functional genomics.’ Structural genomics represents an initial phase of genome analysis and has a clear end point: the construction of high-resolution genetic, physical and transcript maps of an organism. The ultimate physical map of an organism is its complete DNA sequence.

There are an increasing number of terms ending up with -omes and -omics. Some examples include cytomics, epigenomics, genomics, immunomics, interactome, metabolomics, ORFeome, phenomics, proteomics, secretome, transcriptomics, transgenomics, etc. Genome organization, physical mapping and sequencing will be discussed in this section. For further details, readers are referred to Primrose (1995), Borevitz and Ecker (2004), Choisne *et al.* (2007) and Lewin (2007).

3.2.1 Genome organization

Major differences among various genomes

Eukaryotes have large genomes, linear chromosomes with centromeres and telomeres, low gene density disrupted by introns and highly repetitive sequences, while prokaryotes have small genomes, single and circular chromosomes (few linear) with no centromere or telomere, high gene density without introns and very few or no repetitive sequences. The genome size refers to the haploid genome since different cells within a single organism can be of different ploidy. Germ cells are usually haploid and somatic cells diploid. The size of the genome is known as the C-value and is measured by re-association kinetics. After denaturation, the rate of re-association is dependent on genome size. The larger the genome, the more repeated DNA sequences and the longer time to re-anneal, the higher the C-value. $C_0 t_{1/2}$ is the product of the DNA concentration and time required to proceed to the half way of re-association. It is directly related to the amount of DNA in the genome.

The DNA content of haploid genomes ranges from 5×10^3 for viruses to 10^{11} bp for flowering plants. Within mammals, there is only a two fold difference between the largest and smallest C-value. However, there is up to a 100-fold variation in size within flowering plants. The minimum genome size found in each phylum increases from prokaryotes to mammals (Fig. 3.5).

Among the most important food crops, rice has the smallest genome (389 Mb) (IRGSP, 2005) and wheat the largest (15,966 Mb). According to Arumuganathan

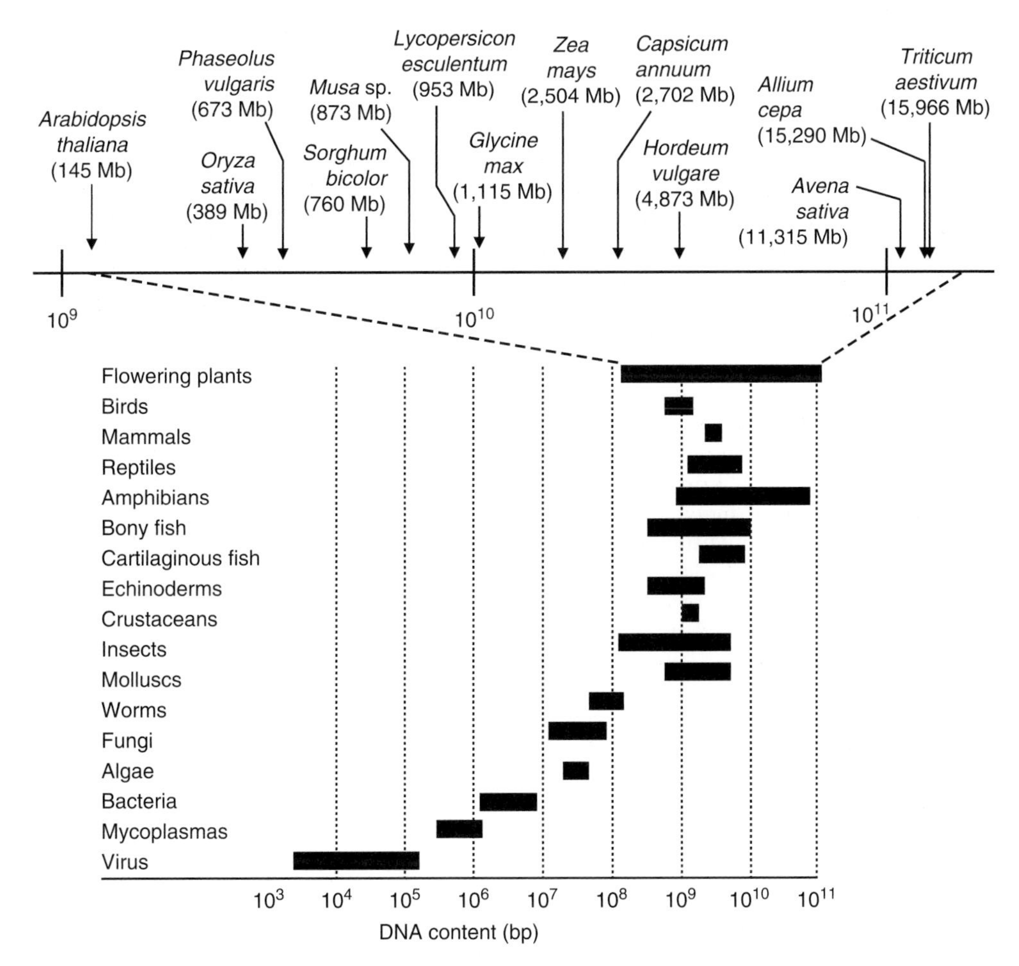

Fig. 3.5. DNA contents of organisms. Modified from Primrose (1995) and Arumuganathan and Earle (1991).

and Earle (1991), other crops can be grouped into seven classes: *Musa*, cowpea and yam (873 Mb); sorghum, bean, chickpea and pigeonpea (673–818 Mb); soybean (1115 Mb); potato and sweet potato (1597–1862 Mb); maize, pearl millet and groundnut (2352–2813 Mb); pea and barley (4397–5361 Mb); and oat (11,315 Mb).

Genome size is often correlated with plant growth and ecology and extremely large genomes may be limited both ecologically and evolutionarily. The manifold cellular and physiological effects of large genomes may be a function of selection of the major components that contribute to genome size such as transposable elements and gene duplication (Gaut and Ross-Ibarra, 2008).

Sequence complexity

Within a phylum, the number of genes in each organism is quite similar although the genome size has a 100-fold difference. It is estimated that the number of genes in flowering plants is 30,000–50,000 but the genome size variation is about 100 times (*Arabidopsis* versus wheat). This is because some large genomes contain a high percentage of repetitive DNA.

The proportions of different sequence components in representative eukaryotic genomes differ greatly. For example, the *Escherichia coli* genome consists of 100% non-repetitive sequences while tobacco contains 65% moderately repetitive and 7% highly repetitive sequences. Repetitive DNA is of two types: tandem repeats (those that are found adjacent to one another) and dispersed repeats (those that recur in unlinked genomic locations). For example, two classes of dispersed and highly repetitive DNA include SINES (short interspersed elements), i.e. shorter than 500 bases and present in 10^{5-6} copies, and LINES (long interspersed elements), i.e. longer than 5 kb and present in at least 10^4 copies per genome.

Repeated sequence families can sometimes function as regulators of gene expression. On the other hand, they can be non-functional identities (such as the so-called 'selfish DNA'). Some of the sequences are found to cause insertional or deletion mutations such as *Alu*.

3.2.2 Physical mapping

Physical mapping entails constructing a physical map which consists of continuous overlapping fragments of cloned DNA that has the same linear order as found in the chromosomes from which they are derived. A series of overlapping clones or sequences that collectively span a particular chromosomal region and form a contiguous segment is called a contig. Recommended references for physical mapping include Zhang and Wing (1997), Brown (2002), Meyers *et al.* (2004) and Lolle *et al.* (2005).

DNA libraries

Large-insert DNA libraries are one of the key components in genome research. They are especially useful for genome studies in large and complex genomes. These libraries can be used in a variety of research projects such as physical mapping of chromosomes, map-based cloning of important genes, genome organization and evolution, comparative genomics and molecular breeding programmes.

A gene or DNA library is a collection of all the genes for an organism so that there is a high probability of finding any particular segment of the source DNA in the collection. To contain a colony of bacteria for every gene, a library will consist of tens of thousands of colonies or clones. The collection is represented in the form of recombinants between DNA fragments from the organism and the vector. The library has to be *ordered* so that each clone has been placed in a precise physical location relative to others (such as in wells of microtitre plates).

Various highly efficient cloning *vectors* have been used to construct DNA libraries. Most frequently used vectors are λ phages, cosmids, P1 phages and artificial chromosomes. There are various types of artificial chromosomes including yeast artificial chromosome (YAC), bacterial artificial chromosome (BAC), binary BAC (BIBAC), P1-derived artificial chromosome (PAC), transformation-competent artificial chromosome (TAC), mammalian artificial chromosome (MAC), human artificial chromosome (HAC) and plant artificial chromosome. When the DNA is simply ligated to the vector and packaged in the phage particles, the library is said to be unamplified. In an amplified *library*, the original DNA has been subsequently increased by replication in bacteria.

Which DNA is cloned in libraries depends on the purpose of the research. *Genomic libraries* are constructed from the total nuclear DNA of an organism. In making these libraries, the DNA must be cut into clonable-size pieces as randomly as possible. Shearing or partial digestion with a frequently cutting restriction endonuclease is often used. Chromosome-specific libraries are made from the DNA of purified isolated chromosomes. A cDNA library contains a collection of cDNA clones transcribed from mRNAs collected from a specific tissue or organ at a specific growth or developmental stage under a specific environment. Therefore, a cDNA library only contains the genes that are expressed in the specific conditions. Furthermore, cDNAs do not contain introns or promoters.

Functionally, gene libraries can be classified into cloning and expression libraries. Cloning libraries are constructed by cloning vectors which contain replicons, multiple cloning sites and selection markers. Clones can be multiplied by bacterial culture. Expression libraries are constructed by expression vectors which contain specific sequences that control gene expression such as promoters, Shine-Dalgarno sequences, ATG and stop codons, etc. in addition to those contained in cloning vectors. The coding products of clones can be expressed in host cells.

cDNA libraries are often expression libraries in which clone construction is such that part or all of the encoded protein is expressed in bacteria harbouring the cloned DNA. Such expression is needed in screening libraries using antibodies or enzyme activities.

In order to be confident that virtually all regions of the genome are represented at least once in a library, considerable redundancy of cloned DNA must be included in the library. The number of DNA clones (n) needed for a certain probability (P) of finding a target clone, is calculated by the formula:

$$n = \frac{\ln(1-P)}{\ln\left(1-\frac{k}{m}\right)}$$

where k is the DNA insert size in kb and m is the haploid genome size in kb. As a rule of thumb, a library containing DNA inserts which collectively add up to three times the amount of DNA in a single gamete of the organism, will provide about 95% confidence that any DNA element in the genome is represented at least once in the library. A library that has 'five genome-equivalent' coverage (rather than three), will provide about 99% confidence of including the target element. For example, the number of BACs of an average size of 150 kb required for $5 \times$ coverage of *Arabidopsis* (m = 125,000 kb) is 3835. When DNA fragments are randomly distributed the probability of obtaining any DNA sequence from this library is no lower than 0.99.

Construction of large insert genomic libraries

Construction of large insert genomic libraries includes three steps: (i) development of the cloning vector; (ii) isolation of high molecular weight DNA; and (iii) preparation of insert DNA.

DEVELOPMENT OF LARGE-INSERT CLONING VECTORS. Developing a vector which can accommodate a large DNA fragment has been a difficult task. Ten kb is the maximum insert size of most plasmid vectors. As the insert size increases, the ligation and transformation efficiency decreases significantly.

The first such vector was the bacteriophage λ vector in which the size of the largest DNA insert is about 25 kb. This is because the fixed capacity of the phage head prevents genomes that are too long being packaged into progeny particles. Cosmids are one type of hybrid vector that replicate like a plasmid but can be packaged *in vitro* into λ phage coats. The vector can accommodate DNA inserts as large as 45 kb.

The YAC vector was developed in which an insert up to 1000 kb can be maintained. The YAC cloning system includes Tel – yeast telemeres, ARS1 – autonomously replicating sequence, CEN4 – centromere from yeast chr.4, *URA3* (Uracil) and *TRP1* (tryptophan) – yeast selection marker genes, *Amp* – ampicillin-resistance gene and Ori – origin of replication of pBR322. Although the YAC clones have played a major role in several genome projects and map-based cloning of many genes in the early 1990s, the following four problems have prevented their further use in genome studies: (i) high percentage of chimaeric clones; (ii) difficulty in DNA preparation and storage; (iii) low transformation efficiency; and (iv) instability of some inserts in yeast. In the rice cultivar Nipponbare for example, 40% of the clones in the YAC library alone were chimaeric thus limiting its use for genome sequencing or map-based cloning.

The BAC cloning system is based on the *E. coli* single copy F factor (Shizuya *et al.*, 1992). It is easy to manipulate, screen and maintain the cloned DNA. It is non-chimaeric, and has high transformation efficiency.

To facilitate gene identification in plant species, second-generation BAC vectors such as BIBAC were constructed (Hamilton *et al.*, 1996). A 150-kb human DNA fragment in the BIBAC vector was transferred into the tobacco genome by *Agrobacterium*-mediated transformation. A similar vector called TAC, was developed and used to complement a mutant phenotype in *Arabidopsis* (Liu *et al.*, 1999). Table 3.1 provides characteristics of several artificial chromosome vectors.

ISOLATION OF HIGH MOLECULAR WEIGHT DNA. Preparing quality high molecular weight (HMW) DNA (most of the DNA > 1 Mb) suitable for large insert library construction can be one of the most difficult steps in constructing a large-insert plant genomic library. There are four predominant problems involved in isolating plant nuclear DNA: (i) plant cell walls must be physically broken or enzymatically digested without damaging nuclei; (ii) chloroplasts must be separated from nuclei and/or preferentially destroyed, an important process since copies of the chloroplast genome may comprise the majority of the DNA within a plant cell; (iii) volatile secondary compounds such as polyphenols must be prevented from interacting with the nuclear DNA; and (iv) carbohydrate matrices that often form after tissue homogenization must be prevented from trapping nuclei.

Several different isolation methods have been developed. The first method was to isolate the protoplast from leaf tissue and then embed the protoplast in low-melting point agarose in the forms of a plug or bead. This method is expensive and time consuming. In addition, chloroplast DNA is not separated. The development of methods to isolate nuclei from leaf tissue has dramatically improved the procedure and quality of the HMW DNA for library construction.

Table 3.1. Characteristics of artificial chromosome vectors.

Vector	Host	Maximum size (kb)	Stability	Chimerism	DNA preparation	Plant transformation
YAC	Yeast	~1000	–	+	Difficult	No
P1	*E. coli*	~100	+	–	Easy	No
BAC	*E. coli*	~300	+	–	Easy	No
PAC	*E. coli*	~300	+	–	Easy	No
BIBAC/TAC	*E. coli* and *A. tumefaciens*	~300	+	–	Easy	Yes

PREPARATION OF INSERT DNA FOR LIGATION. The average size of DNA fragments produced by complete digestion with restriction enzymes with four- or six-base recognition sequences is too small for large insert library construction. To obtain relatively HMW restriction fragments (100–300 kb), the popular method is to partially digest the target DNA with a four-base-cut enzyme. Partial DNA digestion not only yields fragments of the desired size but also fragments the genome randomly without exclusion of any sequence.

To determine the conditions that yield a maximum percentage of fragments between 100 and 300 kb, a series of partial digestions are carried out by using different amounts of restriction enzyme for a specific digestion period. Once the optimal conditions for producing fragments between 100 and 300 kb are determined, a mass digestion using several plugs is carried out to obtain sufficient DNA for size selection. Partially digested HMW DNA is then subjected to pulsed field gel analysis.

If there is no size selection of partially digested DNA, a random library will have a preponderance of small inserts since small fragments ligate more efficiently and clones with small inserts transform with higher efficiency. Contour-clamped homogeneous electrical field (CHEF) is the most common method for separating large DNA molecules. It uses a hexagonal array of fixed electrodes and a homogeneous electrical field is generated for enhancing DNA resolution. After two-size selection using CHEF Mapper, the HMW restriction fragments must be removed from surrounding agarose before they can be used in ligation reactions. After developing the high insert library, a number of random clones can be selected to confirm the successful cloning of the inserts and the average insert size. The average insert size will then determine how many clones are needed to achieve the desired amount of genome coverage.

Physical mapping

There are five physical mapping methods: optical mapping; restriction fragment fingerprinting; chromosome walking; sequence tagged site (STS) mapping; and fluorescent *in situ* hybridization (FISH). In restriction fragment fingerprinting, individual clones are first digested with different restriction enzymes. The digested DNA is then labelled with radioactive or fluorescent dye and run on a sequence gel. The fingerprint data is collected and analysed for contig assembly. During the procedure, markers with known map position are used as probes to screen the large insert library. Clones hybridized with the same single copy marker are considered to be overlapping. PCR amplification of DNA pools using primers derived from DNA markers with known position can also be used for physical map construction. The disadvantages of this method are that it is labour intensive and filling the gaps is difficult.

STS mapping uses a sequenced tagged site (STS) which is a short region of DNA about 200–300 bases long whose exact sequence is found nowhere else in the genome. Two or more clones containing the same STS must overlap and the overlap must include the STS. There are two disadvantages to this method: it is still very labour intensive and the primer synthesis is expensive.

FISH uses synthetic polynucleotide strands that bear sequences known to be complementary to specific target sequences at specific chromosomal locations. The polynucleotides are bound via a series of linked molecules to a fluorescent dye that can be detected with a fluorescence microscope.

In addition, physical mapping can be achieved by a combination of fingerprinting, molecular linkage mapping, STS mapping, end sequencing and FISH mapping. A by-product of physical mapping is the integration of genetic, physical and sequence maps as shown in Fig. 3.6.

3.2.3 Genome sequencing

The sequencing of DNA in laboratories first began in 1978. The first genome of a multicellular eukaryote, *Caenorhabditis elegans*, was published in 1998. The rationale behind genome sequencing includes

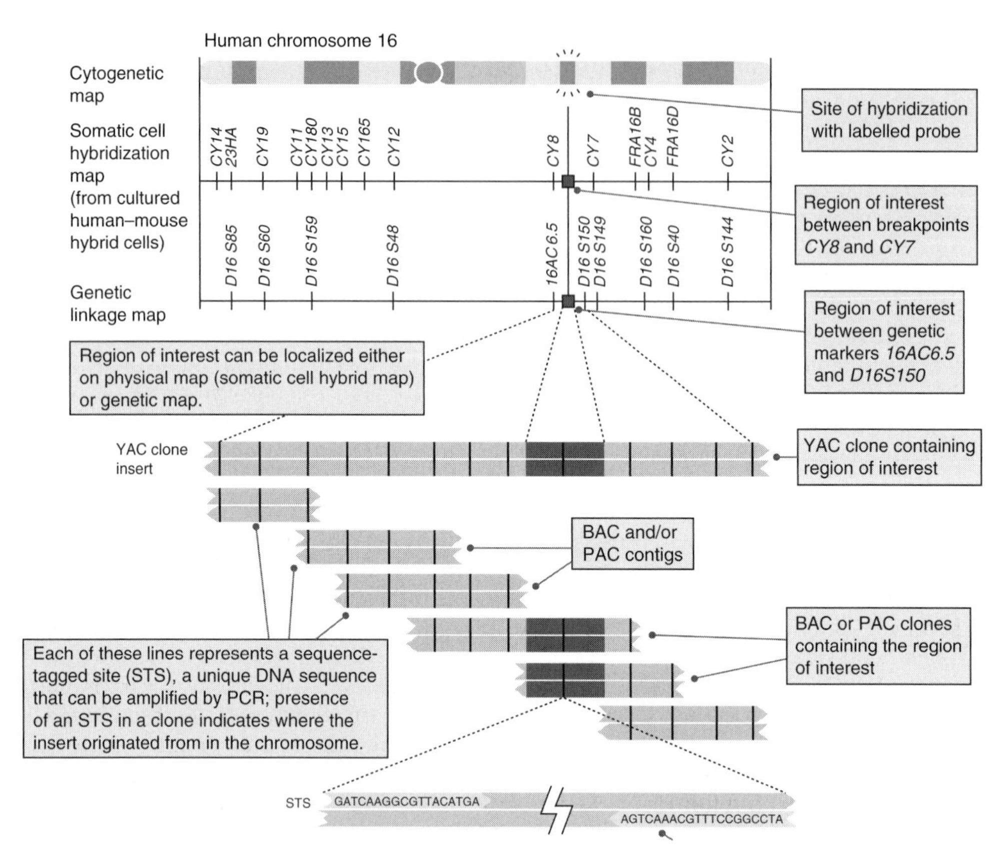

Fig. 3.6. Example of physical mapping and integration of genetic, cytological and physical maps.

identification of all the genes in the sequenced genome, elucidation of the functions and the interactions of genes in the genome, functional analysis of orthologues in related complex genomes, evolutionary analysis of genes or genomes and product development and commercial application. As the next-generation sequencing technologies continued to facilitate genome sequencing, new applications and new assay concepts (e.g. Huang *et al.*, 2009) have emerged that are vastly increasing our ability to understand genome function, including sequence census methods for functional genomics (Wold and Myers, 2008; Varshney *et al.*, 2009).

Technical developments in DNA sequencing

There are three major milestones in DNA sequencing: (i) the invention of sequencing reactions; (ii) automated fluorescent DNA sequencers; and (iii) PCR. Until the late 1970s, obtaining the DNA sequences of even five to ten nucleotides was difficult and very laborious. The development of two new methods in 1977, that of Maxam and Gilbert (chemical sequencing method) and the other by Sanger and Coulson (enzymatic sequencing), made it possible to sequence large DNA molecules. Later refinements of Sanger's chain termination method made it the preferred procedure since it has proven to be technically simpler.

The modified Sanger sequencing method or chain terminator procedure capitalizes on two properties of DNA polymerases: (i) their ability to synthesize faithfully a complementary copy of a single-stranded DNA template; and (ii) their ability to use 3'-dideoxynucleotides as substrates. Once the analogue is incorporated at the growing point of the DNA chain, the 3' end lacks a hydroxyl group and is no longer a substrate for chain elongation. Thus, the dideoxynucleotides act as chain terminators.

The development of labelling and detection techniques have contributed to an acceleration of sequencing procedures, which include ^{33}P labelled primer (1970s); ^{33}P or ^{35}S labelled primer with sharper image and lower radiation (early 1980s); and fluorescently labelled primers and dyes in four different reactions (1986). DNA sequencing became automated in the late 1980s when the primer used for each reaction was labelled with a differently coloured fluorescent tag. This technology allowed thousands of nucleotides to be sequenced in a few hours and the sequencing of large genomes then became a reality. With ABI PRISM® technology, up to four different dyes can be used to label DNA each of which can be differentiated when run together in the same lane of a gel or injected into a capillary. For DNA sequencing, this means that the four different dyes representing each of the DNA bases (A, C, G and T) can be electrophoresed together.

The improvement of polyacrylamide gel electrophoresis (in the late 1980s and early 1990s) led to high resolution, thinner gels and a sharper image. Capillary electrophoresis (CE) (1998) offers a number of performance advantages such as faster runs, small sample volumes and the ability to eliminate manual gel pouring and sample loading tasks. Walk-away automation reduces instrument-associated labour time by more than 80% over slab-gel systems. The introduction of CE resulted in the availability of automated electrophoresis instruments with much lower cost per sample (Amersham's MegaBACE and Applied Biosystems ABI3700, 3730, etc.). High-throughput sequencing can also incorporate full automation in colony picking, 96-well plasmid isolation and purification, PCR reactions, sample loading and sequence data analysis.

The new generation of high-throughput sequencing technologies promises to transform the scientific enterprise, potentially supplanting array-based technologies and opening up many new possibilities (Kahvejian *et al.*, 2008; Shendure and Ji, 2008). There are three commercial next-generation DNA sequencing systems available (Schuster, 2008) which promise vastly more sequencing capability (> 1 Gb of sequence per run) than standard capillary-based technology can produce. A high-throughput DNA sequencing technique using a novel massively parallel sequencing-by-synthesis approach called pyrosequencing was developed more recently by 454 Life Sciences (Margulies *et al.*, 2005; www.454.com). 454 Sequencing employs clonal DNA fragment amplification on beads in droplets of an aqueous–oil emulsion, followed by loading the beads into nanoscale (~ 44 μm) wells of a PicoTiterPlate which is a fibre optic chip. In each reaction cycle, one of the four deoxynucleotide triphosphates (dNTPs) is delivered to the reactor along with DNA polymerase, ATP sulfurylase and luciferase. Incorporation, which is accompanied by a chemoluminscent signal, is detected by a high-resolution charge-coupled device (CCD) sensor. 454 Sequencing is capable of sequencing roughly 100 Mb of raw DNA sequence per 7-h run with their 2007 sequencing machine, the GS FLX Genome Analyzer.

454 Sequencing allows large amounts of DNA to be sequenced at low cost compared to the Sanger chain-termination methods; G-C rich content is not as much of a problem, and the lack of reliance on cloning means that unclonable segments are not skipped; it is also capable of detecting mutations in an amplicon pool at a low sensitivity level. However, each read of the 2005 sequencing machine GS20 is only 100 bp long, resulting in some problems when dealing with highly repetitive genomes, as repetitive regions of over 100 bp cannot be 'bridged' and thus must be left as separate contigs. Also, the nature of the technology lends itself to problems with long homopolymer runs. As one of the projects using 454 sequencing, Project 'Jim' determined the first sequence of an individual, the complete genome sequence of James Dewey Watson, in May 2007.

The second high-throughput sequencing technique is Solexa™ (Illumina, Inc.; http://www.illumina.com) which depends on sequencing by synthesis. Diluted DNA templates are attached to a solid planar surface and then amplified clonally. Sequencing is performed by delivering a mixture of four differentially labelled reversible chain terminators along with DNA polymerase. The resulting signal is detected at each cycle and a new cycle can be initiated after terminator removal (Bennet *et al.*, 2005). Current average read lengths are about 30–40 bases with 1 Gb per run.

The third high-throughput sequencing technique is SOLiD™ System which enables massively parallel sequencing of clonally-amplified DNA fragments linked to beads. The SOLiD™ sequencing methodology is based on sequential ligation with dye-labelled oligonucleotides. The SOLiD™ technology provides unmatched accuracy, ultra-high throughput and application flexibility. It delivers advancements in throughput approaching 20 Gb per run. The flexibility of two independent flow cells, each capable of running 1, 4 or 8 samples, allows multiple experiments to be conducted in a single run. With unparalleled throughput and greater than 99.9% overall accuracy, the SOLiD™ System enables large-scale sequencing and tag-based experiments to be completed more cost effectively than previously possible.

There are several emerging sequencing methods: sequencing by hybridization; mass spectrophotometric techniques; direct visualization of single DNA molecules by atomic force microscopy; single-molecule sequencing strategies. The intense drive towards developing technology that can sequence a complete human genome for under US$1000 will ensure that the speed and cost of sequencing will continue to improve rapidly (Schuster, 2008). For example, a nanopore-based device provides single-molecule detection and analytical capabilities that are achieved by electrophoretically driving molecules in solution through a nano-scale pore. Further research and development to overcome current challenges to nanopore identification of each successive nucleotide in a DNA strand offers the prospect of 'third generation' instruments that will sequence a diploid mammalian genome for ~US$1000 in ~ 24 h (Branton *et al.*, 2008).

Sequencing strategies

There are two general genome sequencing strategies: (i) clone-by-clone or hierarchical sequencing (International Human Genome Sequencing Consortium, 2001); and (ii) whole shotgun sequencing (Venter *et al.*, 2001). After constructing the complete physical map, clone-by-clone sequencing can be started in any specific region. Clone-by-clone or hierarchical sequencing strategy has the following advantages: (i) the ability to fill gaps and re-sequence the uncertain regions; (ii) the ability to distribute the clones to other laboratories; and (iii) the ability to check the produced sequence by restriction enzymes. The main disadvantages are that it is expensive and time consuming for the construction of a physical map and experienced personnel are required.

The shotgun sequencing strategy consists of making small insert libraries (1–10 kb) from the genomic DNA of an organism, sequencing a large number of clones (six to eight times redundancy) and assembling contigs using bioinformatics software. It has no physical map construction and less risk of recombinant clones. It is cost effective and fast and ideal for small genome sequencing. However, it is difficult to fill gaps and re-track all the sequenced plasmids and the resulting data is less useful for positional cloning. Figure 3.7 compares the two sequencing methods.

COMBINING CLONE-BY-CLONE AND SHOTGUN SEQUENCING STRATEGIES. In 1997 The Institute of Genome Research (TIGR) launched the initiative of a whole-genome shotgun approach for the human genome. But BACs, BAC end sequences and STS markers were used extensively in assembling the sequencing data from shotgun clones. The first draft of the human genome was completed within 3 years compared with the 12 years taken by the Human Genome Project which is funded by government agencies.

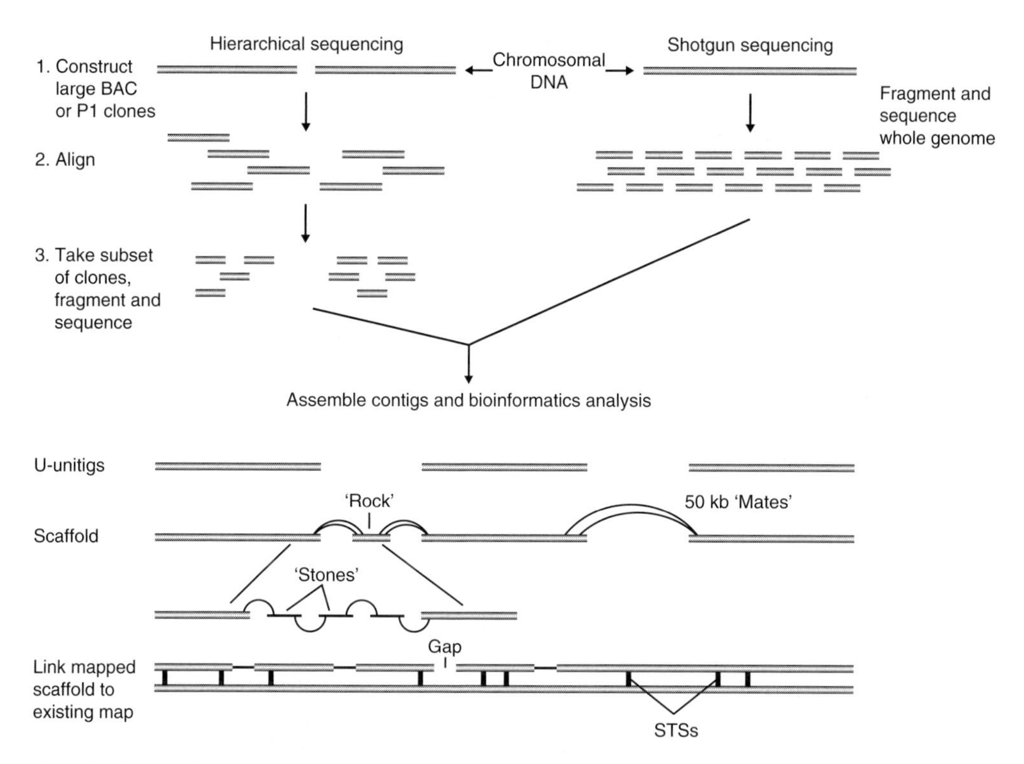

Fig. 3.7. Comparison of two sequencing strategies: assembly of a mapped scaffold. U-unitigs are assembled into scaffolds using mate-pair information to bridge gaps between two U-unitigs, and by linking unitigs to 'rock', which are less-well supported unitigs that nevertheless fit in place according to at least two independent large insert mate pairs. 'Stones' are single short contigs whose position is supported by only a single read. Gaps are filled in the finishing stage by further site-directed sequencing. Scaffolds are placed against existing genetic and physical maps by sequence tagged site (STS) matches and against the cytological map by fluorescent *in situ* hybridization (FISH).

Genome filtering strategies

The extremely large size of many crop genomes makes it difficult to decode them using the standard methods of genome sequencing such as clone-by-clone and whole-genome shotgun. Determining their complete sequences is daunting and costly. In recent years two genome filtration strategies, methylation filtration (MF) (Rabinowicz *et al.*, 1999) and C_0t-based cloning and sequencing (CBCS; Peterson *et al.*, 2002) or high C_0t (HC; Yuan *et al.*, 2003) have been suggested for selectively sequencing the gene space of large genomes. MF is based on the characteristics of plant genomes in which genes are largely hypomethylated but repeated sequences are highly methylated. Methylated DNA is cleaved when transferred into a *Mcr* + *E. coli* strain and only hypomethylated DNA is recovered. CBCS/HC separates single- and low-copy sequences including most genes from the repeated sequences on the basis of their differential renaturation characteristics. Using the MF strategy, Bedell *et al.* (2005) sequenced 96% of the genes in sorghum with an average coverage of 65% across their length. This strategy filtered out repetitive elements during the sequencing of the genome of sorghum which reduced the amount of sorghum DNA to be sequenced by two-thirds, from 735 Mb to approximately 250 Mb. Both MF and HC have been used for efficient characterization of maize gene space (Palmer *et al.*, 2003; Whitelaw *et al.*, 2003). Using high C_0t and MF, Martienssen *et al.* (2004) generated up to twofold coverage of the gene space with less than one million sequencing reads and simulations using sequenced BAC clones predicted that 5 × coverage of gene-rich regions, accompanied by less than 1 × coverage of subclones from BAC contigs, will generate a high quality mapped sequence that meets the needs of geneticists while accommodating unusually high levels of structural polymorphism. Haberer *et al.* (2005) selected 100 random regions averaging 144 kb in size, representing about 0.6% of the genome, to define their content of genes and repeats for characterizing the structure and architecture of the maize genome. Combining CBCS with genome filtration can greatly reduce the cost while retaining the high coverage of genic regions. An alternative approach is the identification of gene-rich regions on a detailed physical map and sequencing large-insert clones from these regions.

Plant genomic sequences

The first complete plant genome to be sequenced was that of *Arabidopsis*. The sequenced regions cover 115.4 Mb of the 125-Mb genome and extend into centromeric regions. The evolution of *Arabidopsis* involved a whole genome duplication followed by subsequent gene loss and extensive local gene duplications. The genome contains 25,498 genes encoding proteins from 11,000 families (The Arabidopsis Genome Initiative, 2000). *Arabidopsis* contains many families of new proteins but also lacks several common protein families. The proportion of predicted *Arabidopsis* genes in different functional categories is provided in Fig. 3.8. The complete genome sequence provides the foundation for more comprehensive comparison of conserved processes in all eukaryotes, identifying a wide range of plant-specific gene functions and establishing rapid systematic methods of identifying genes for crop improvement (Varshney *et al.*, 2009).

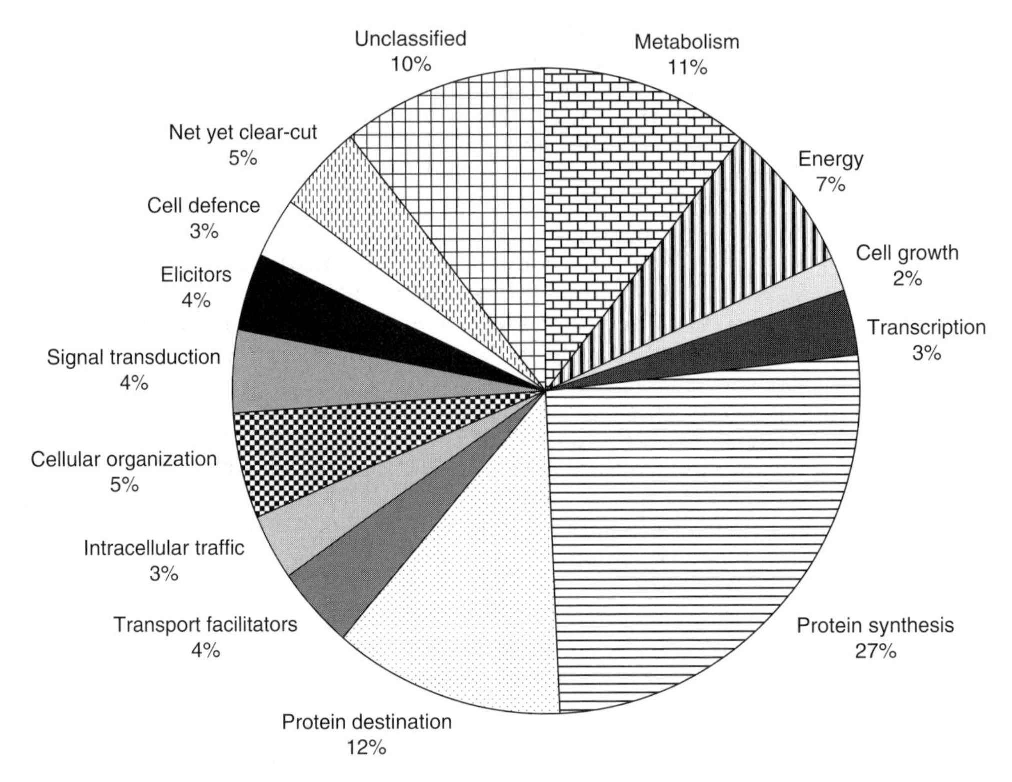

Fig. 3.8. Proportion of predicted *Arabidopsis* genes in different functional categories.

Rice was the first crop to be fully sequenced because of its importance as one of the major cereals and also because of its small genome size, small number of chromosomes (n = 12), well characterized genetic and genomic resources and availability of a large number of DNA markers and a high density genetic linkage map. Two draft sequences were completed in 2002 (Goff *et al.*, 2002; Yu *et al.*, 2002) and a complete sequence was published in 2005 (IRGSP, 2005) which is available in the National Center for Biotechnology Information (NCBI) database.

Many sequencing projects for important crop species are currently ongoing. The US Department of Energy's Joint Genome Institute (JGI) is providing funding and technical assistance to decode the genomes of several major plants, including cassava (*Manihot esculenta*), cotton (*Gossypium*), foxtail millet (*Setaria italica*), sorghum, soybean and sweet orange (*Citrus sinensis* L.) (http://www.jgi.doe.gov/sequencing/). Other plants for which there are ongoing genome sequencing projects include *Medicago truncatula* (http:///www.medicago.org/genome), *Lotus japonicum* (http://www.kazusa.or.jp), poplar, tomato (http://www.sgn.cornell.edu) and grapevine.

The International Wheat Genome Sequencing Consortium (IWGSC) has been formed to advance agricultural research for wheat production and utilization by developing DNA-based tools and resources that result from the complete sequencing of the expressed genome of common (hexaploid) bread wheat and to ensure that these tools and the sequences are available for all to use without restriction and without cost (Gill *et al.*, 2004; http://www.wheatgenome.org/). A Global *Musa* Genomics Consortium (GMGC) is decoding the *Musa* genome (http://www.newscientist.com/article.ns?id-dn1037). A Global Cassava Partnership, an alliance of the world's leading cassava researchers and developers, has proposed that sequencing the cassava genome should be a priority (Fauquet and Tohme, 2004).

To sequence the maize genome, two consortia in the USA began a pilot study: one with Jo Messing (Rutgers University), Rod Wing (Arizona University), Ed Coe (University of Missouri), Mark Vaudin (Monsanto) and Steve Rousley (Cereon); the other included Jeff Bennetzen (Purdue University), Karel Schubert and Roger Beachy (Danforth Center), Cathy Whitelaw and John Quackenbush (TIGR) and Nathan Lakey (Orion). These two pioneer programmes have been extended by a massive US programme from the National Science Foundation (NSF), USDA and the Department of Energy (DOE) led by Rick Wilson (Washington University). The sequencing strategy is a hybrid between a BAC-by-BAC approach and a whole-genome shotgun.

3.2.4 cDNA sequencing

Why cDNA sequencing

Large-scale DNA sequencing can be carried out on genomic DNA or cDNAs. There are four advantages to performing cDNA sequencing. First is the cost of sequencing a whole genome. Although DNA sequencing costs have fallen more than 50-fold over the past decade, it still costs around US$10 million to sequence three billion base pairs. It will take years to realize the goal to lower the cost of sequencing a mammalian-sized genome to US$100,000 and ultimately to cut the cost of whole-genome sequencing to US$1000 or less.

Secondly, the interpretation of the genomic sequence of eukaryotes is not straightforward in contrast to prokaryotes: coding regions are separated by non-coding regions; introns and alternative splicing occurs; one gene can lead to multiple mRNAs and gene products; a significant fraction of genomic DNA does not code for proteins (non-coding sequences).

Thirdly, cDNA sequencing helps in annotation and identification of exons and introns. Estimates of the number of human genes vary from 30,000 to 80,000. The accuracy of the *Arabidopsis* genome annotation varied from 50 to 70% in the first draft. Many *Arabidopsis* genes are still not accurately annotated.

Fourthly, sequencing cDNAs helps gain information about the transcriptome.

mRNA populations are variable among cells. The transcriptome is dynamic and constantly changing. Cells adapt to environmental, developmental and other signals by modulating their transcriptome. mRNA populations form an important level of regulation between signal perception and response. Genetically identical cells can exhibit distinct phenotypes. cDNA sequencing allows direct insight into mRNA populations and allows the dissection of the transcriptome which genomic sequencing alone does not provide. Sequencing of random cDNA clones prepared from different tissues also allows analyses of mRNA abundance.

cDNA libraries

When constructing a representative cDNA library, the source of the mRNA for the cDNA library is critical and will vary depending on the goal of the study. To estimate the diversity of mRNAs expressed in a given plant, the mRNA should represent most plant tissues and organs. On the other hand, to define the diversity of mRNAs represented in a specific tissue, organ or developmental stage, the library should be prepared from the most highly defined source feasible. As indicated by Nunberg *et al.* (1996), it is better to invest the time to harvest sufficient quantities of scarce tissue for a library rather than using materials which will contain a significant proportion of extraneous messages.

If large quantities of RNA are available, it is possible to create a plasmid library directly. This is particularly feasible since electroporation transformation efficiencies are so high. Plasmid libraries may or may not be directional and are easily arranged in an ordered array. Constructing plasmid libraries directly avoids any sequence bias, including internal deletion and *trans* recombination that may occur during the excision process.

The frequency of full-length cDNAs depends on the length of transcript (the longer the transcript the lower the frequency of obtaining full-length cDNAs). Carninci and Hayashizaki (1999) discussed the high-efficiency of full-length cDNA cloning using a cap trapper method (biotinylated cap) and thermoactivation of reverse transcriptase (cDNA synthesis at 60°C: RNA secondary structures are melted). Some normalization and subtraction methods also allow enrichment of full-length cDNAs.

For a given mRNA, multiple expressed sequence tags (ESTs) can be obtained. Depending on the extent of sampling, ESTs may or may not overlap. EST processing is needed to remove vector sequences, linker sequences, check the quality using a sequence quality filter, clean up the contaminants and chimaeric sequences and store in databases. To construct EST contigs, there are two commonly used programs: Phrap/consed and TIGR assembler. These programs generate a unigene set (contigs or Tentative Consensus): a consensus sequence for all overlapping ESTs that (supposedly) correspond to a single mRNA.

Several factors affect the quality of EST contigs: contaminating sequences, bad quality sequences, non-overlapping ESTs from the same mRNA, alternative splicing resulting in one gene with multiple mRNAs and closely related genes (chimaeric contigs). EST annotation can be carried out using similarity searches against Genbank and other databases, e.g. protein motif databases, to assign a putative function or identify functional categories. This process can be automated or manual (usually a combination of the two).

Non-random (normalized or subtracted) cDNA libraries are needed in order to overcome some of the problems with redundant ESTs in order to saturate EST databases when budget is limited or when there is a specific interest in a particular stage. Hybridization-based methods are most commonly used to decrease redundancy (reduce representation of abundant cDNAs and increase rare cDNAs). Normalized cDNA libraries are used when gene discovery is the main objective of the EST project.

cDNA sequencing

Strategies for cDNA sequencing include single-pass cDNA sequencing (ESTs), normalized cDNA libraries, subtracted cDNAs and high-throughput full-length cDNA sequencing. Single-pass cDNA can be achieved using the following steps: (i) construct cDNA libraries; (ii) randomly pick clones for sequencing (from the 5' or 3' end using vector primers); (iii) process sequences (vector/linker removal, quality control, contaminants, empty, chimaeric); (iv) construct contigs (sequences from same transcript); (v) create a unigene set; and (vi) annotate sequences. The objective of high-throughput cDNA sequencing is to obtain the full finished sequence of as many cDNAs as possible. This is necessary for complex eukaryotic genomes (human, mouse, plants). Full-length cDNA sequencing is discussed in Chapter 11 along with its use in gene cloning.

Major limitations of the cDNA sequencing approach include: (i) high redundancy of some genes in cDNA libraries; (ii) difficulty in isolating rare transcripts or developmentally-regulated genes; and (iii) the fact that some genes are not stable in *E. coli.*

3.3 Functional Genomics

The use of whole genome information and high-throughout tools has opened up a new field of research called functional genomics. Among its subdisciplines, transcriptomics (the complete set of transcripts produced in a cell) (Zimmerli and Somerville, 2005), proteomics (the complete set of proteins produced in a cell) (Roberts, 2002) and metabolomics (the complete set of metabolites expressed in a cell) (Stitt and Fernie, 2003) have been used by the plant science community. Functional genomics refers to the development and application of global (genome-wide or system-wide) experimental approaches to assess gene function by making use of the information and reagents provided by structural genomics. It is characterized by high-throughput or large-scale experimental methodologies combined with statistical and computational analysis (bioinformatics) of the results. The new information provided by all the omics disciplines will lead the plant science community to *in silico* simulations of plant growth, development and response to environmental change.

3.3.1 Transcriptomics

The transcriptome is the set of all the mRNA molecules or 'transcripts', produced in one cell or a population of cells. The term can be applied to the total set of transcripts in a given organism or to the specific subset of transcripts present in a particular cell type. Unlike the genome, which is roughly fixed for a given cell line (excluding mutations), the transcriptome can vary with external environmental conditions. Because it includes all mRNA transcripts in the cell, the transcriptome reflects the genes that are being actively expressed at any given time with the exception of mRNA degradation phenomena such as transcriptional attenuation. Transcriptomics is based on the idea that a catalogue of all the transcripts associated with a specific treatment or developmental stage provides a reasonable overview of the underlying biological processes at work. As we moved from northern blots to tiling arrays, we have advanced from a gene-by-gene world to a full genome universe. The study of transcriptomics often uses high-throughput techniques based on DNA microarray or chip technology. Suggested references for this section include Bernot (2004), Bourgault *et al.* (2005) and Busch and Lohmann (2007).

Gene expression profiling technologies provide a tool for analysing 'global' gene expression by viewing activity of all or (more typically) a substantial part of the genome at a specific time of interest. There are open and closed architecture systems for gene expression profiling. In the open architecture, all genes expressed in a tissue have the possibility of being detected (e.g. cDNA-AFLP, differential display (dd) PCR, SAGE, cDNA substraction). Advantages include the potential discovery of previously unknown genes, comprehensive coverage and the low requirements by way of equipment. Disadvantages include retrieving only a small part of the gene (since it can be laborious to clone full-length cDNA) and simple gene identification that is limited by sequences that are already in a database (otherwise the corresponding gene must be cloned).

Several alternative technologies have emerged for measuring transcript abundance in a parallel fashion. Essentially, these methods can be divided into three categories according to their underlying principle, namely PCR-, sequencing- or hybridization-based technologies. Therefore, strategies that are currently available for analysis of transcriptomes include RT-PCR (qualitative and quantitative), hybridization methods (northern blots, macroarrays, DNA microarrays, oligonucleotide microarrays), cDNA fingerprinting (differential display, cDNA-AFLP), cDNA sequencing (full-length cDNAs, subtracted cDNAs, normalized cDNA libraries, SAGE, massive parallel signature sequencing – MPSS) and combinations of the above techniques.

The most straightforward and unbiased method of analysing an RNA population is the sequencing of cDNA libraries and quantitative analysis of the resulting ESTs. Traditionally, ESTs with read-lengths of about 200–900 nucleotides have been produced by Sanger-sequencing but the associated costs have severely limited the resolution of this approach (Busch and Lohmann, 2007). Deep sequencing has become a viable alternative for unbiased large-scale expression profiling because of the development of new protocols and entirely new sequencing techniques. Non-gel-based sequencing techniques promise to deliver greatly increased throughput and a considerable cost reduction. MPSS combines *in vitro* cloning of millions of template tags on separate microbeads with ligation-mediated sequence detection. In each reaction cycle, a four-base overhang is produced on every tag to which a fluorescently labelled adaptor of defined sequence is ligated. The position and fluorescence of every microbead is monitored by a high resolution camera in each of the reaction cycles, allowing the sequences of the 17-nucleotide tags to be reconstructed (Brenner *et al.*, 2000). As indicated by Busch and Lohmann (2007), the limited length of the sequenced tags precludes the use of MPSS for *de novo* sequencing but makes it a very powerful tool for expression profiling of organisms with pre-existing sequence information. By contrast, two other high-throughput sequencing techniques as described previously, 454 and Solexa™, are ideally suited for expression-profiling purposes. Short tags are sufficient to identify a transcript unambiguously and therefore problems arising from assembling short tags into larger contigs can be ignored.

PCR product-based arrays were heavily used in the early days of global transcriptome analysis. However, the low level of standardization among laboratories, high levels of noise and experimental variation and cross-hybridization between homologous transcripts have eroded the attractiveness of these arrays. Oligonucleotide-based microarrays are now becoming the most popular technology for large-scale expression profiling because they allow the simultaneous detection of tens of thousands of transcripts at a reasonable cost. The expression level of any gene represented on the array can be deduced from the fluorescence intensity of the corresponding probe. However, microarrays only offer linear expression measurements over a range of three orders of magnitude compared to quantitative RT-PCR which has a dynamic range of five orders of magnitudes. Microarrays perform with less precision and sensitivity than other techniques when used for measuring low abundance transcripts in particular and this is manifested in their greater inter-assay variability (Busch and Lohmann, 2007). Another major limitation of microarrays designed for expression analysis is that they rely on current genome annotations, which precludes the identification of novel or very small transcription units.

Microarrays and quantitative RT-PCR have dominated expression profiling to date but deep sequencing and whole-genome tiling arrays will become increasingly important because these techniques are not limited to the detection of known transcripts. Tiling arrays, on which the entire genome is represented by evenly spaced probes, provide a novel means of transcript identification. In *Arabidopsis*, tiling arrays have been used to map transcriptionally active regions by profiling four different tissues (Yamada *et al.*, 2003).

The interaction transcriptome is the sum of all microbe and host transcripts that are produced during the interaction. The challenges in studying interaction transcriptomes include how to discriminate pathogen from host ESTs, similarity searches to genome/cDNA sequences, GC analyses and determination of hexamer frequency (windows of 6 bp). Systems genomics/transcriptomics can be used to analyse complex transcriptomes, for example the mixtures of mRNAs from different species (e.g. infected tissue, environmental samples such as soil or seawater, etc.). One challenge is to identify the species of origin in the mixtures.

3.3.2 Proteomics

Proteomics is the study of the identification, function and regulation of complete sets of proteins in a tissue, cell or subcellular compartment. Such information is crucial to understanding how complex biological processes occur at a molecular level and how they differ in various cell types, stages of development or environmental conditions (Bourgualt *et al.*, 2005). Proteomics is important as proteins are active agents in cells and they execute the biological functions encoded by genes. Sequences of genes (or genomes) and transcriptome analyses are not sufficient to elucidate biological functions. Proteomics complements transcriptomics by providing information about the time and place of protein synthesis and accumulation, as well as identifying those proteins and their post-translational modifications. Gene expression does not necessarily indicate whether a protein is synthesized, how fast it is turned over or which possible protein isoforms are synthesized (Mathesius *et al.*, 2003). In some cases, the correlation between gene expression and protein presence is as low as 0.4. First, the level of transcription of a gene gives only a rough estimate of its level of expression into a protein. An mRNA produced in abundance may be degraded rapidly or translated inefficiently, resulting in a small amount of protein. Secondly, many proteins experience post-translational modifications that profoundly affect their activities; for example some proteins are not active until they become phosphorylated. Methods such as phosphoproteomics and glycoproteomics are used to study post-translational modifications. Thirdly, many transcripts give rise to more than one protein through alternative splicing or post-translational modifications. It is generally supposed that if genomes contain tens of thousands of gene sequences, the proteome comprises several hundred thousand proteins as a result of alternative slicing and post-translational modifications. Finally, many proteins form complexes with other proteins or RNA molecules and only function in the presence of these molecules.

Proteomics has become an important approach for investigating cellular processes and network functions. Significant improvements have been made in technologies for high-throughput proteomics, both at the level of data analysis software and mass spectrometry (MS) hardware (Baginsky and Gruissem, 2006). In this section, proteomics will be briefly discussed. For further details, readers are referred to the following review articles: van Wijk (2001), Molloy and Witzmann (2002), de Hoog and Mann (2004), Saravanan *et al.* (2004), Baginsky and Gruissem (2006), Cravatt *et al.* (2007) and Zivy *et al.* (2007).

Protein extraction

Obtaining high quality protein is the first step in proteomic research. Extracting protein from plant tissue requires tissue disruption by grinding and sonication, separation of proteins from unwanted cell materials (cell wall, water, salt, phenolics, nucleic acids) by centrifugation after precipitation of proteins with acetone–trichloroacetic acid, resolubilizing protein in a solution that dissolves the maximum number of different proteins and inactivation of protease by acetone–trichloroacetic acid treatment or specific protease inhibitors. Pre-fractionation of tissue is optional for the analysis of proteins from different organelles or microsomal fractions. Solubilization requires urea or, for more hydrophobic proteins, thiourea, as a chaotrope which solubilizes, denatures and unfolds most proteins. Non-ionic zwitter detergents, e.g. 3-[3-cholamidopropyl-dimethyl-ammonio]-1-propane sulfonate (CHAPS), Triton®-X, or amidosulfobetaines are used to solubilize and separate proteins in a mixture. Sodium dodecyl sulphate (SDS) is also a strong detergent and used to solubilize membrane proteins. However, it renders a negative charge to proteins and, therefore, interferes with isoelectric focusing (Mathesius *et al.*, 2003). Reducing agents (usually dithiothreitil [DDT], 2-mercapto-ethanol or tributyl phosphine) are needed to disrupt disulfide bonds.

Protein identification and quantification

N- or C-terminal sequencing has made protein identification possible on a small scale although with limitations. Improvements in MS have made it possible to identify proteins faster, on a larger scale, using smaller amounts of protein. In addition, post-translational modifications can be determined by MS/MS analysis and proteins can be identified even when bound to other proteins in complexes. A standard technique for protein identification with MALDI-TOF MS is peptide mass fingerprinting. Protein spots in a gel can be visualized using a variety of chemical stains or fluorescent markers. Proteins can often be quantified by the intensity with which they stain. Once proteins have been separated and quantified, they can be identified. Individual spots are cut out of the gel and cleaved into peptides with proteolytic enzymes. These peptides can then be identified by MS, specifically MALDI-TOF MS. The MALDI-TOF analysis will measure very precisely (< 0.1 Da) the mass of peptides formed by this digestion. Since the cleavage sites are known, the digestion can be simulated by informatics, that is, the masses of all the peptides produced by this digestion can be calculated for all the known sequence proteins of a given organism (Zivy *et al.*, 2007). These masses will depend on the length of peptides and their composition since most amino acids have different masses. Thus, masses predicted from sequences stored in databases can simply be compared with masses effectively measured by the MALDI-TOF equipment. The greater the number of positive mass matches the more likely it is that the peptides originate from the same protein thus facilitating the rapid identification of proteins.

Protein profiling

Protein mixtures of considerable complexity can now be routinely characterized in some detail. One measure of technical progress is the number of proteins identified in each study. Such numbers can now reach the thousands for suitably complex samples. Large-scale proteomic studies are needed to solve three types of biological problem (Aebersold and Mann, 2003): (i) the generation of protein–protein linkage maps; (ii) the use of protein identification technology to annotate and, if necessary, correct genomic DNA sequences; and (iii) the use of quantitative methods to analyse protein expression profiles as a function of the cellular state as an aid to inferring cellular function.

The sequences of many mature proteins in higher eukaryotes after processing and splicing are often not directly apparent from their cognate DNA sequences. Peptide sequence data of sufficient quality provides unambiguous evidence of translation of a particular gene and can in principle, differentiate between alternatively spliced or translated forms of a protein (Aebersold and Mann, 2003). Thus, it might be tempting to systematically analyse the proteins expressed by a cell or tissue, that is, to generate comprehensive proteome maps.

The more common and versatile use of large-scale MS-based proteomics has been to document the expression of proteins as a function of cell or tissue state. Aebersold and Mann (2003) argued that to be meaningful, such data must be at least semi-quantitative and that a simple list of proteins detected in the different states is insufficient. This is because analyses of complex mixtures are often not comprehensive and therefore the non-appearance of a particular sequence in the list of identified peptides does not indicate that the peptide or protein was not originally present in the sample. Additionally, it is often impossible to prepare a certain cell type, cell fraction or tissue in completely pure form without trace contamination from other fractions. And because the ion current of a peptide is dependent on a multitude of variables that are difficult to control, this measure is not a good indicator of peptide abundance. If stable-isotope dilution has not been used, a rough relative estimate of the quantity of a protein can be obtained by integrating the ion current of its peptide-mass peaks over their elution time and comparing these 'extracted ion currents' between states, provided that highly accurate and reproducible methods are used. Increasingly, stable-isotope dilution and LC-MS/MS are used to accurately detect changes in quantitative protein profiles and to infer biological function from the observed patterns (Aebersold and Mann, 2003).

Protein–protein interactions

Protein–protein interactions occur among most proteins and there are six types of interfaces found in protein–protein interactions: domain–domain, intra-domain, hetero-oligomer, hetero-complex, homo-oligomer, and homo-complex. The analysis of protein–protein interactions can be either qualitative or quantitative. Traditional biochemical methods such as co-purification and co-immunoprecipitation have been used to identify the members of protein complexes. Proteomics-based strategies have been used to determine the composition of complexes and to establish interaction networks. The systematic, large-scale, high-throughput approaches now being taken to build maps of the interactions between proteins predicted by genome sequence information have become known as interactomics (Causier *et al.*, 2005).

There are many important characteristics of a protein–protein interaction. Obviously, it is important to know which proteins are interacting. In many experiments and computational studies, the focus is on interactions between two different proteins. However, one protein can interact with other copies of itself (oligomerization) or with three or more different proteins. The stoichiometry of the interaction is also important, that is, how many of each protein involved are present in a given reaction. Some protein interactions are stronger than others because they bind together more tightly. The strength of binding is known as affinity. Proteins will only bind to each other spontaneously if it is energetically favourable. Energy changes during binding are another important aspect of protein interactions. Many of the computational tools that predict interactions are based on the energy of interactions.

Protein interaction maps represent essential components of the post-genomic tool kits needed for understanding biological processes at a systems level. Over the past decade, a wide variety of methods have been developed to detect, analyse and quantify protein interactions, including surface plasmon resonance spectroscopy, nuclear magnetic resonance (NMR), Y2H screens, peptide tagging combined with MS and fluorescence-based technologies. Lalonde *et al.* (2008) and Miernyk and Thelen (2008) reviewed the latest techniques and current limitations of biochemical, molecular and cellular approaches for the detection of protein–protein interactions. *In vitro* biochemical strategies for identifying and characterizing interacting proteins include co-immunoprecipitation, blue native gel electrophoresis, *in vitro* binding assays, protein cross-linking and rate-zonal centrifugation. Fluorescence techniques range from co-localization to tags which may be limited by the optical resolution of the microscope, to fluorescence resonance energy transfer (FRET)-based methods that have molecular resolution and can also report on the dynamics and localization of the interactions within a cell. Proteins interact via highly evolved complementary surfaces with affinities that can vary over many orders of magnitude. Some of the techniques such as surface plasmon resonance provide detailed information regarding the physical properties of these interactions. To analyse protein complexes systematically at a sub- or full-genome level, several methods have been adapted for high-throughput screens using robotics: (i) Y2H systems; (ii) the mating-based split-ubiquitin system (mbSUS); and (iii) affinity purification of protein complexes followed by identification of proteins by MS (AP-MS).

One of the first questions usually asked about a new protein, apart from where it is expressed, is to what proteins does it bind? To study this question by MS, the protein itself is used as an affinity reagent to isolate its binding partners. Compared with two-hybrid and array-based approaches, this strategy has the advantages that the fully processed and modified protein can serve as the bait, that the interactions take place in the native environment and cellular location and that multi-component complexes can be isolated and analysed in a single operation (Ashman *et al.*, 2001). However, because many biologically relevant interactions are of low affinity, transient and generally dependent on the specific cellular environment in which they occur, MS-based methods in a straightforward affinity experiment will detect only a subset of the protein interactions that actually occur (Aebersold and Mann, 2003). Bioinformatics methods, correlation of MS data with those obtained by other methods or iterative MS measurements possibly in conjunction with chemical crosslinking (Rappsilber *et al.*, 2000) can often help to further elucidate direct interactions and overall topology of multi-protein complexes.

The ability of quantitative MS to detect specific complex components within a background of non-specifically associated proteins increases the tolerance for high background and allows for fewer purification steps and less stringent washing conditions, thus increasing the chance of finding transient and weak interactions. The same methods can be used to study the interaction of proteins with nucleic acids, small molecules and in fact with any other substrate. For example, drugs can be used as affinity baits in the same way as proteins to define their cellular targets and small molecules such as cofactors can be used to isolate interesting 'sub-proteomes' (MacDonald *et al.*, 2002).

The Y2H system has become one of the standard laboratory techniques for the detection and characterization of protein–protein interactions. It can be used to map individual amino acid residues involved in a specific protein–protein interaction. It can also be used to identify novel interactions from complex libraries of expressed proteins. The Y2H system has been widely used for determination of protein interaction networks within different organisms. In plants, the Y2H system has been successfully applied to detect interactions with phytochromes, cryptochomes, transcription factors, proteins involved in self-incompatibility mechanisms, the circadian clock and plant disease resistance (Causier *et al.*, 2005). Taken together with the recent progress made in the development of large-scale Y2H screening procedures, the time is now ripe for large-scale Y2H screens to be applied to organisms such as *Arabidopsis* and rice.

Another potential method to detect protein–protein interactions involves the use of FRET between fluorescent tags on interacting proteins. FRET is a non-radioactive process whereby energy from an excited donor fluorophore is transferred to an acceptor fluorophore that is within ~60 Å of the excited fluorophore (Wouters *et al.*, 2001). After excitation of the first fluorophore, FRET is detected either by emission from the second fluorophore using appropriate filters or by alternation of the fluorescence lifetime of the donor. Two fluorophores that are commonly used are variants of GFP: cyan fluorescent protein (CFP) and yellow fluorescent protein (YFP) (Tsien, 1998). The potential of FRET is considerable, for two reasons (Phizicky *et al.*, 2003). First, it can be used to make measurements in living cells, which allows the detection of protein interactions at the location in the cell where they normally occur in the presence of the normal cellular environment. Secondly, transient interactions can be followed with high temporal resolution in single cells. Protein interaction within the proteome might be mapped by performing FRET screens on cell arrays that are co-transferred with cDNAs bearing CFP and YFP fusion proteins.

In recent years there has been a strong focus on predicting protein interactions computationally. Predicting the interactions can help scientists predict pathways in the cell, potential drugs and antibiotics and protein functions. Proteins are large molecules and binding between them often involves many atoms and a variety of interaction types including hydrogen bonds, hydrophobic interactions, salt bridges and more. Proteins are also dynamic, with many of their bonds able to stretch and rotate. Therefore, predicting protein–protein interactions requires a good knowledge of the chemistry and physics involved in the interactions.

The principle of using hybrid proteins to analyse interactions has been extended to examine DNA–protein interactions, RNA–protein interaction, small molecule–protein interactions and interactions dependent on bridging proteins or post-translational modifications. In addition, the reconstitution of proteins other than transcription factors such as ubiquitin, has been used to establish reporter systems to detect interactions (Fashena *et al.*, 2000) and these may enable the analysis of proteins not generally suitable for the traditional two-hybrid arrays such as membrane proteins.

In the future, quantitative methods based on stable-isotope labelling are likely to revolutionize the study of stable or transient interactions and interactions dependent on post-translational modifications. In such experiments, accurate quantification by means of stable-isotope labelling is not used for protein quantification per se; instead the stable-isotope ratios distinguish between the protein composition of two or more protein complexes (Aebersold and Mann, 2003). In the case of a sample containing a complex and a control sample containing only contaminating proteins (for example, immunoprecipitation with an irrelevant antibody or isolate from a cell devoid of affinity-tagged protein), the method can distinguish between true complex components and non-specifically associated proteins.

Post-translational modifications

Proteins are converted to their mature form via a complicated sequence of post-translational protein processing and 'decoration' events. Detection of post-translational modifications is necessary, especially for phosphorylation or ubiquitinylation because they affect protein function. Phosphorylation can be detected by the use of antiphosphotyrosine antibodies on blots of 2DE or by radiolabelling proteins and detecting the labelled proteins. Glycosylation of proteins can easily be detected on gels using the periodic acid Schiff reaction. In addition, specific enzymes can be used for selective cleavage of several common post-translational modifications (Mathesius *et al.*, 2003).

Many of the post-translational modifications are regulatory and reversible which impacts biological function through a multitude of mechanisms. MS methods to determine the type and site of such modifications on single, purified proteins have been undergoing refinements since the late 1980s. In this case, peptide mapping with different enzymes is usually used to 'cover' as much of the protein sequence as possible. Protein modifications are then determined by examining the measured mass and fragmentation spectra via manual or computer-assisted interpretation. For the analysis of some types of PTMs, specific MS techniques have been developed that scan the peptides derived from a protein for the presence of a particular modification. The analysis of regulatory modifications, in particular protein phosphorylation, is complicated by the frequently low stoichiometry, the size and ionizability of peptides bearing the modifications and their fragmentation behaviour in the mass spectrometer (Aebersold and Mann, 2003). Given the difficulties of identifying all modifications even in a single protein, it is clear that at present, scanning for proteome-wide modifications is not comprehensive. One of the strategies used is essentially an extension of the approach used to analyse protein mixtures. Instead of searching a database only for non-modified peptides, the database search algorithm is instructed to also match potentially modified peptides. To avoid a 'combinatorial explosion' resulting from the need to consider all possible modifications for all peptides in the database, the experiment is usually divided into identification of a set of proteins on the basis of non-modified peptides followed by searching only these proteins for modified peptides (MacCoss *et al.*, 2002). A more functionally oriented strategy focuses on the search for one type of modification on all the proteins present in a sample. Such techniques are usually based on some form of affinity selection that is specific for the modification of interest and which is used to purify the 'sub-proteome' bearing this modification.

Many challenges remain in the large-scale mapping of post-translational modifications but it is clear that MS-based proteomics can make a unique contribution in this area. For example, systematic quantitative measurements of post-translational modifications by stable-isotope labelling would be of considerable biological interest. One of the future challenges in proteomics is to increase sensitivity to visualize low abundance proteins (e.g. regulatory proteins) as only 10% of proteins can be now visualized by 2DE. It needs a high quality database for matching sequence to MS data (or the use of MS/MS). Technical developments are needed for understanding post-translational modifications, protein complexes, protein localization and the interface with transcriptomics and metabolomics.

3.3.3 Metabolomics

Plants contain a wide diversity of low-molecular-weight chemical constituents. More than 100,000 secondary metabolites have been identified in plants and this probably represents less than 10% of nature's total (Wink, 1988). Estimates of how many metabolites occur in an individual species vary within the 5–25,000 range (Trethewey, 2005). Metabolites are the products of interrelated biochemical pathways and changes in metabolic profiles can be regarded as the ultimate response of biological systems to genetic or environmental changes (Fiehn, 2002). Plant metabolism research has experienced a second golden age resulting from synergies between genome-enabled technologies and classical biochemistry. The rapid rate at which genomics data are being accumulated creates an increased need for robust metabolomic technologies and rapid and accurate methods for identifying the activities of enzymes (DellaPenna and Last, 2008).

The metabolome refers to the complete set of small-molecule metabolites (such as metabolic intermediates, hormones and other signalling molecules and secondary metabolites) that can be found within a biological sample such as a single organism. Metabolomics is defined as the systematic survey of all the metabolites present in a plant tissue, cell and cellular compartment under defined conditions (Bourgault *et al.*, 2005). The name metabolomics was coined in the 1990s (Oliver *et al.*, 1998). The foundations of metabolomics lie in the description of biological pathways and current metabolomic databases, such as KEGG, are frequently based on well-characterized biochemical pathways. Metabolomics might be considered to be the key to integrated systems biology because it is frequently a direct gauge of a desired phenotype (Fiehn, 2002), measuring quantitative and qualitative traits such as starches in cereal grains or oils in oilseeds. Moreover, metabolomics can be correlated with genetics through genomes, transcriptomes and proteomes and therefore bypass the more traditional quantitative trait loci (QTL) approach applied to molecular plant breeding. Major recommended references for this section include Fiehn (2002), Sumner *et al.* (2003), Weckwerth (2003), Bourgault *et al.* (2005), Breitling *et al.* (2006), Schauer and Fernie (2006) and Krapp *et al.* (2007).

Targeted metabolomics involves examination of the effects of a genetic alteration or change in environmental conditions on particular metabolites (Verdonk *et al.*, 2003). Sample preparation is focused on isolating and concentrating the compound of interest to minimize detection interference from other components in the original extract. Metabolite profiling refers to a qualitative and quantitative evaluation of metabolite collections, for example those found in a particular pathway, tissue or cellular compartment (Burns *et al.*, 2003). Finally, metabolic fingerprinting focuses on collecting and analysing data from crude extracts to classify whole samples rather than separating individual metabolites (Johnson *et al.*, 2003; Weckwerth, 2003).

In stark contrast to transcriptomics and proteomics, metabolomics is mainly species-independent, which means that it can be applied to widely diverse species with relatively little time required for re-optimizing protocols for a new species. Metabolite profiling can monitor variation in the accumulation of metabolites in plant cells in culture which are ectopically expressing transcription factors, as a hypothesis-generating tool to establish the possible pathways regulated by particular regulatory proteins. The first step consists of generating a transgenic cell line expressing the regulator from a constitutive or inducible promoter. The second step is to subject extracts from transformed and control cells to various metabolic profiling approaches to determine the qualitative and quantitative differences in metabolite accumulation. A more practical approach to monitoring and purifying individual metabolites is to profile hundreds or thousands of small molecules biochemically and to screen for changes in the relative levels of these compounds. By comparing two conditions, a profile of the differences can be obtained that is then used as a blueprint to identify the individual compounds affected (Dias *et al.*, 2003). The immense chemical diversity of small biomolecules makes comprehensive metabolome screens difficult. The lack of unifying principles such as genetic codes that would assist molecule identification, comparison and causal connection is another important challenge (Breitling *et al.*, 2006).

The global study of the structure and dynamics of metabolic networks has been hindered by a lack of techniques that identify metabolites and their biochemical relationship in complex mixtures. Recent advances in ultra-high mass accuracy MS provide two advantages that can enable *ab initio* determination of metabolic networks: (i) the ability to identify molecular formulae based on exact masses; and (ii) the inference of biosynthetic relationships between masses directly from the mass spectrum. Mass spectrometers with the necessary performance parameters (mass accuracy around 1 ppm and resolution above 100,000 m/Δm) are now within the reach of many researchers and will change the way we think about metabolomics (Breitling *et al.*, 2006). The recent application of Fourier transform ion cyclotron resonance MS (FTICR-MS) to metabolomic analysis suggests a way to tackle the problem. A lower-cost alternative to high-field FTICR-MS, the Orbitrap mass analyser, promises accelerated activity in this area. These two analysers are able to achieve high resolution and mass accuracy in the 1-ppm range for biomolecular samples. In both instruments, the ionized metabolite mixture is trapped in an orbital trajectory. The frequency of their orbit depends on the mass-over-charge ratio of the ions and can be measured precisely, which is the basis of the exceptional accuracy. In FTICR-MS, trapping is achieved in a strong magnetic field which exerts a force on the charged particles that is perpendicular to their direction of motion and thus confines them to a circular path. The Orbitrap traps ions without a magnetic field and ions are trapped in a radial electrical field between a central and an outer cylindrical electrode. Theoretically, the resolving power of the FTICR-MS and Orbitrap is sufficiently high to resolve even the most complex metabolite mixtures using direct infusion.

Gas chromatography (GC)-MS or LC-MS is the tool of choice for generating high-throughput data for identification and quantification of small-molecular-weight metabolites (Weckwerth, 2003). Capillary electrophoresis (CE) is an alternative method which separates particular types of compound more efficiently and can be coupled with MS or other types of detectors. NMR, infrared (IR), ultraviolet (UV) and fluorescence spectroscopy can be used as alternative means of detection, often in parallel with MS (Weckwerth, 2003). TOF MS technology has also been used in metabolite analysis and provides a means of high sample-throughput. In the end, a combination of methods enables analysis of a broad range of metabolites.

NMR is a spectroscopic technique that exploits the magnetic properties of the atomic nucleus (Macomber, 1998). In NMR, the sample is immersed in a strong external magnetic field and transitions between the nuclear magnetic energy levels are induced by a suitably oriented radiofrequency field. In theory, any molecule containing one atom with a non-zero nuclear spin (I) is potentially visible by NMR. Considering the isotopes with a non-zero nuclear spin such as ^{1}H, ^{13}C, ^{14}N, ^{15}N and ^{31}P, all biological molecules have at least one NMR signal. There is wide variation in the sensitivity of the experiment for different nuclei, hence ^{1}H NMR remains the best choice for metabolite profiling by NMR mainly due to its natural abundance (99.8%) and sensitivity (Moing *et al.*, 2007). The NMR spectrum generally consists of a series of discrete lines (resonances) which are characterized not only by the familiar spectroscopic quantities of frequency (chemical shift), intensity and line shape, but also by relaxation times. Although less sensitive than GC or LC-MS, proton NMR spectroscopy is a powerful complementary technique for the identification and quantitative analysis of plant metabolites either *in vivo* or in tissue extracts (Krishnan *et al.*, 2005). Typically, 20–40 metabolites have been identified in metabolite profiling of plant extracts and the number of metabolites quantified can be increased with higher field strength (increasing spectral resolution) and by using microprobes for small quantity samples together with cryogenic probe heads (increasing sensitivity). One of the main advantages of ^{1}H-NMR is that structural and quantitative information can be obtained on numerous chemical species with a wide range of concentration in a single NMR experiment with excellent reproducibility.

HPLC and GC are the most widely used analytical techniques for the separation of small metabolites. GC is used to separate compounds on the basis of their relative vapour pressure and affinities for the stationary phase in the chromatographic column. It offers very high chromatographic resolution but requires chemical derivatization for many biomolecules: only volatile chemicals can be analysed without derivatization. Some large and polar metabolites cannot be analysed by GC. GC tends to give much greater chromatographic resolution than HPLC but has the disadvantages of being limited to compounds that are volatile and heat stable. A big advantage of GC is that it can be easily combined with MS, which greatly increase its utility for multicomponent profiling because of its inherent high specificity, high sensitivity and positive peak confirmation (Dias *et al.*, 2003). HPLC is a form of column chromatography used frequently in biochemistry and analytical chemistry. It is used to separate components in a mixture by using a variety of chemical interactions between the substance being analysed (analyte) and the chromatography column. Compared to GC, HPLC has lower chromatographic resolution but it does have the advantage that a much wider range of analytes can potentially be measured.

The generation of reproducible and meaningful metabolomic data requires great care in the acquisition, storage, extraction and preparation of samples (Fiehn, 2002). The true metabolic state of samples must be maintained and additional metabolic activity or chemical modification after collection must be prevented. Depending on the type of sample and the analysis performed, this can be achieved in various ways. The most common strategies are freezing in liquid nitrogen, freeze-drying, and heat denaturation to halt enzymatic activity (Fiehn, 2002). Metabolomic experiments are typically conducted by comparing experimental plants possessing an expected metabolic modification (i.e. because of the introduction of a transgene or exposure to a particular treatment) to control plants. Statistically significant changes in metabolite levels attributable to perturbations affecting the experimental plants are identified. Natural variability in metabolite levels occurs as part of normal homeostasis in plants; thus a high number of replicates is typically necessary to establish a statistically significant difference between experimental and control plants, especially if the differences between metabolite levels are subtle (Johnson *et al.*, 2003). In order to validate metabolomic studies and to facilitate data exchange, the Metabolomics Standards Initiative (MSI) has released documents describing minimum parameters for reporting metabolomic experiments. The reporting parameters encompassed by MSI include the biological study design, sample preparation, data acquisition, data processing, data analysis and interpretation relative to the biological hypotheses being evaluated. Fiehn *et al.* (2008) exemplified how such metadata could be reported by using a small case study: the metabolite profiling by GC-TOF mass spectrometry of *Arabidopsis thaliana* leaves from a knock-out allele of the gene At1g08510 in the Wassilewskija ecotype.

The large data sets and multitude of metabolites require computer-based applications to analyse complex metabolomic experiments. Ideally, such systems compile and compare data from a variety of separation and detection systems (Sumner *et al.*, 2003). Ultimately, gene functions can be predicted or global metabolic profiles associated with particular biological responses can be defined. Multivariate data analysis techniques that reduce the complexity of data sets and enable more simplified visualization of metabolomic results are currently available. These include principle-component analysis (PCA), hierarchical clustering analysis (HCA), K-means clustering and self-organizing maps (Sumner *et al.*, 2003).

Considering the natural variability in transcript, protein and metabolite levels in plants of the same genotype, correlations within complex fluctuating biochemical networks can be revealed using PCA and HCA (Weckwerth, 2003; Weckwerth *et al.*, 2004). Metabolic networks were integrated with gene expression and protein levels using a novel extraction method whereby RNA, proteins and metabolites were all extracted from a single sample (Weckwerth *et al.*, 2004).

Parallel to the development of the technologies of metabolite profiling, there has been a bewildering proliferation in the nomenclature associated with this field. At the root of the problem is that some groups have chosen to use the term metabolomics while others have opted for metabonomics. Metabolomics will be used in this book as it is derived from metabolic profiling or fingerprinting and should be a parallel terminology to transcriptomics and proteomics (Trethewey, 2005). The Human Metabolome Project led by Dr David Wishart of the University of Alberta, Canada, completed the first draft of the human metabolome consisting of 2500 metabolites, 1200 drugs and 3500 food components (Wishart *et al.*, 2007).

Schauer and Fernie (2006) assessed the contribution of metabolite profiling to several fields of plant metabolomics. As a fast growing technology, metabolite profiling is useful for phenotyping and diagnostic analyses of plants. It is also rapidly becoming a key tool in functional annotation of genes and in the comprehensive understanding of the cellular response to biological conditions such as various stresses of biotic and abiotic origin. Metabolomics approaches have recently been used to assess the natural variation in metabolite content between individual plants, an approach with great potential for the improvement of the compositional quality of crops.

3.4 Phenomics

Phenomics is a field of study concerned with the characterization of phenotypes, which are characteristics of organisms that arise via the interaction of the genome with the environment. Genomics has spawned a plethora of related omics terms that frequently relate to established fields of research. Of these terms, phenomics, the high-throughput analysis of phenotypes, has the greatest application in plant breeding.

3.4.1 Importance of phenotypes in genomics

For all sequenced organisms from the most thoroughly studied and simple bacterial cells to humans, only about two-thirds of all genes have an assigned biochemical function and only a fraction of those are associated with a phenotype. Even when phenotypes are assigned, they might represent only a partial understanding of the role of the gene. The function of a gene cannot be fully understood until it is possible to predict, describe and explain all the phenotypes that result from the wild-type and mutant forms of that gene (Bochner, 2003).

Phenotypes often cannot be predicted on the basis of the biochemical function of a gene alone because it is not clear how catalytic or regulatory activity will affect the biology of the cell or the whole organism. However, if a gene has a biological function then, for every identified gene it should be possible to define at least one phenotype. A second layer of genomic annotation could then follow in which every gene is described biologically by the phenotypes that it produces. The first step is to construct a so-called 'phenomic map' and in diploid and higher plants this will be complicated by the fact that several genes can affect gene expression and the resulting phenotypes of each other, leading to epistasis, complex traits and multifactorial stress responses (Bochner, 2003).

Advances in genetic and genomic analysis are being hindered by the slow pace at which biological (that is, phenotypic) information is being obtained, which is not keeping pace with genomic information. Bochner (1989) predicted that global phenotypic analysis would soon be needed to complement the massive amounts of genetic data being obtained and Brown and Peters (1996) called attention to 'the phenotype gap' in mouse research. The Nobel laureate Sydney Brenner, in a keynote address (at a joint Cold Spring Harbor Laboratory/Wellcome Trust Genome Informatics Conference held at Hinxton in the UK on 9 September 2002) emphasized that approaches that relied heavily on genome sequences and bioinformatics extrapolation were associated with too much noise and were becoming nonproductive. Instead, he called for a renewed focus on cellular studies and the creation of function-based cell maps in a variety of cell types by the year 2020.

However, generating phenotypic maps will not be easy. Scientists generally test and measure phenotypes one at a time, which is too slow. Almost every model system in which the genome has been sequenced has used functional genomics projects to associate the genome with the biology and this typically includes some efforts that involve phenomics. Many large-scale projects are being carried out by generally using and adapting diverse existing phenotypic technologies that range from animal autopsies to mass spectrometer analysis of cellular metabolites. A phenotype microarray technology was devised that had several attributes (Bochner, 2003): (i) it could assay about 2000 distinct culture traits; (ii) it could be used with a wide range of microbial species and cell types; (iii) it would be amenable to high-throughput studies and automation; (iv) it would allow phenotypes to be recorded quantitatively to facilitate comparisons over time; (v) it would give a comprehensive scan of the physiology of the cell; and (vi) by providing global cellular analysis, it would provide a complement to genomic and proteomic studies.

3.4.2 Phenomics in plants

The great plasticity of plant genomes in producing various phenotypes from a small amount of genetic variation has provided both challenges and opportunities for crop improvement. Detailed and systematic analysis of phenotype requires both a data repository and a means of structure interrogation. The field of phenomics developed from the phenotypic characterization of mutant plants, the descriptions of which have been published in volumes that frequently use structured ontological terms. The storage of these data in searchable databases together with the application of phenomics to high-throughput analysis, plant development and natural variation, creates the final link in the chain from the genetics of crop development to crop production (Edwards and Batley, 2004).

There is an additional need to make phenotypic data from different organisms simultaneously searchable, visible and most importantly, comparable (Lussier and Li, 2004). As an example of attempts in this field, PhenomicDB has been created as a multi-species genotype/phenotype database by merging public genotype/phenotype data from a wide range of model organisms and *Homo sapiens* (Kahraman *et al.*, 2005). To provide systematic descriptions of phenotypic characteristics of gene deletion mutants on a genome-wide scale, a public resource for mining, filtering and visualizing phenotypic data – the PROPHECY database – was established. PROPHECY is designed to allow easy and flexible access to physiologically relevant quantitative data for the growth behaviour of mutant strains in the yeast deletion collection during conditions of environmental challenges.

In plant biology, comparison of data collected by laboratories in which plants are grown under slightly different conditions can be problematic. This is especially true if the data are collected solely with reference to chronological age. Kjemtrup *et al.* (2003) described the development of a plant phenotyping platform based on a growth stage scale that will aid in the generation of coherent data. While their emphasis is on *Arabidopsis*, the principles they describe can also be applied to other plant systems. They adapted a modified version of the BBCH scale which is named after the consortium of agricultural companies that developed it (BASF, Bayer, Ciba-Geigy and Hoechst), for high-throughput phenotyping of *Arabidopsis* to collect data for both quantitative and qualitative traits spread over the developmental timeline of the plant. In the first phase of the method, data are collected enabling a series of landmark growth stages to be defined. The second phase involves the collection of detailed data for additional traits that are of particular interest at any one of these given stages. The growth stages are described as germination and sprouting, leaf development (main shoot), formation of side shoot to tillering, stem elongation or rosette growth (main shoot, shoot development), development of harvestable vegetative plant parts, inflorescence emergence (main shoot) and ear or panicle emergence, flowering on main shoot, development of fruit, ripening or maturity of fruit and seed and senescence – the beginning of dormancy.

Mutant analysis provides an alternative and typically more reliable means to assign gene function. However, this 'phenotype-centric' process, classically known as forward genetics, typically is not suitable for systematic genome-wide gene analysis, primarily due to the enormous effort required to identify each gene responsible for a particular phenotype. In spite of improvements in the cloning of genes on the basis of phenotype (such as availability of whole-genome sequences, large numbers of mapped polymorphisms and faster and cheaper genotyping technologies), it can often take over a year for a skilled scientist to move from a mutant to the affected gene. As indicated by Alonso and Ecker (2006), the combination of classical forward genetics with recently developed genome-wide, gene-indexed mutant collections is beginning to revolutionize the way in which gene functions are studied in plants. High-throughput screens using these mutant populations should provide a means to analyse plant gene functions – the phenome – on a genomic scale.

3.5 Comparative Genomics

Comparative genomics has been used to address four major research areas (Schranz *et al.*, 2007). First, all comparative analyses are based on phylogenetic hypotheses. In turn, genomics data can be used to construct more robust phylogenies. Secondly, comparative genome sequencing has been crucial in identifying changes in genome structure that are due to rearrangements, segmental duplications and polyploidy.

The alignment of multiple genomes can also be used to reconstruct an ancestral genome. Thirdly, comparative genomics data have been used to annotate homologous genes and subsequently to identify conserved *cis*-regulatory motifs. Having multiple genomes of varying phylogenetic depths has proven very useful for detecting conserved non-coding sequences. Fourthly, comparative genomics is used to understand the evolution of novel traits.

Comparative genomics provides the potential for trait extrapolation from a species where the genetic control is well understood and for which there are molecular markers to a species about which there is a limited amount of information. For example, rice is regarded as a model for cereal genomics because of its small genome. The similarity of cereal genomes in general means that the genetic and physical maps of rice can be used as reference points for exploration of the much larger and more difficult genomes of the other major and minor cereal crops (Wilson *et al.*, 1999). Conversely, decades of breeding work and molecular analysis of maize, wheat and barley can now find direct application in the improvement of rice. Comparative genomics can also be used to locate desirable alleles in gene pools close to the target crop so that transfer can achieved by conventional methods (Kresovich *et al.*, 2002).

Across plant species, genome size does not correlate with number of genes or biological complex. Physical size of genomes across plant species varies greatly, while genetic size of genomes is roughly equivalent. Large genomes usually have large physical:genetic distance ratios. Also the relationship between genes and the number of gene families is not clear. In this section, comparative maps and collinearity among related species and their implications will be discussed. Key references recommended for an overview of comparative genomics include Shimamoto and Kyozuka (2002), Ware and Stein (2003), Miller *et al.* (2004), Caicedo and Purugganan (2005), Filipski and Kumar (2005), Koonin (2005), Xu *et al.* (2005), Schranz *et al.* (2007) and Tang *et al.* (2008).

3.5.1 Comparative maps

A comparative map aligns two or more species-specific maps using common sets of markers or sequences. It requires identification of regions of sequence similarity in the genomes of different species or genera (i.e. typically, genes). Sequence similarity can be identified due to common evolutionary origins. Gene repertoire and gene order may be found conserved over larger chromosomal segments between closely related species. The long-term goals of comparative genomics are to establish relationships between map, sequence and functional genomic information across all plant species and to facilitate taxonomic and phylogenetic studies in higher plants.

Importance of comparative maps

The objective of the development of a comparative map is to identify subsets of genes that have remained relatively stable in both sequence and copy number since the radiation of flowering plants from their last common ancestor. Why are comparative maps so important? First, eukaryotic genomes are organized into chromosomes and maps summarize genetic information using chromosomes as the organizational principle. Secondly, conservation of gene identity and gene order along the chromosomes determines potential for sexual reproduction; disruption leads to speciation and major evolutionary change. Thirdly, species maps provide the context for the study of inheritance and chart the history of genetic change. Fourthly, comparative maps are the major tools for ferrying genetic information back and forth across species and genera in a systematic fashion.

Once chromosomal duplications are identified in a genome and the timing of a duplication/polyploidization event has been determined relative to angiosperm divergence nodes, ancestral gene order within the duplicated segments can be inferred. Map comparisons across divergent genera show greater conservation of ancestral gene order and gene repertoire once genome-wide duplication/gene loss within each genome is accounted for. Map comparisons between closely related species are largely unaffected because most duplications pre-date them. Comparative maps lay the groundwork for asking questions about whether specific 'linkage blocks' or gene arrangements are statistically associated with increased fitness or have a relationship between polyploidy and plant adaptation. For example, comparative linkage mapping and chromosome painting in the close relatives of *Arabidopsis* have inferred an ancestral karyotype of these species. In addition, comparative mapping to *Brassica* has identified genomic blocks that have been maintained since the divergence of the *Arabidopsis* and *Brassica* lineages (Schranz *et al.*, 2007).

An example: Arabidopsis*–tomato comparative map*

DEVELOPMENT OF *ARABIDOPSIS*–TOMATO COMPARATIVE MAP TO DETECT MACROSYNTENY. Fulton *et al.* (2002) identified over 1000 conserved orthologous sequences (COS) between tomato and *Arabidopsis* by comparison of *Arabidopsis* genomic sequence with 130,000 tomato ESTs (representing 27,000 unigenes or approximately 50% of the tomato gene content). For 1025 COS markers developed, 927 were screened against tomato DNA using Southern analysis to classify them as single, low or multiple copy, among which 85% were considered to be single or low copy (> 95% hybridization signal assigned to three or fewer restriction fragments) and 50% matched a gene of unknown function (Gene Ontology classification). A total of 550 COS markers was mapped on to the tomato genome. The size of conserved segments was generally smaller than 10 cM. Results indicated that multiple polyploidization events punctuate the evolution of *Arabidopsis* and tomato. Distinguishing orthologues from paralogues is difficult due to reciprocal loss of genes and chromosome segments following polyploidization events.

PHYLOGENETIC ANALYSIS OF CHROMOSOMAL DUPLICATION EVENTS TO DETECT MICROSYNTENY. The *Arabidopsis* genome sequence was used to analyse internal duplication events based on inferred protein matches between 26,028 genes. A total of 34 non-overlapping chromosomal segment pairs were identified consisting of 23,177 (89%) *Arabidopsis* genes (Bowers *et al.*, 2003b). To relate this 'alpha' duplication to the angiosperm family tree, all duplicated syntenic *Arabidopsis* gene pairs were compared to individual genes from pine, rice, tomato, *Medicago*, cotton and *Brassica*. It was determined whether inferred protein sequences were from duplicated syntenic gene pairs. *Arabidopsis* genes were more similar to one another than to the heterologous protein in another species.

RELATIVE AGE OF CHROMOSOMAL DUPLICATION EVENTS. It was concluded that the 'alpha' duplication event pre-dated divergence from *Brassica* about 14.5–20.4 million years ago but post-dated divergence from cotton about 83–86 million years ago.

About 50% (49–64%) of *Brassica* sequences were more similar to one duplicated *Arabidopsis* sequence than was the other *Arabidopsis* sequence to its paralogue. Only 6–19% of cotton, rice, pine, etc. sequences clustered internally to the *Arabidopsis* syntenic duplicates (Bowers *et al.*, 2003b).

POLYPLOID ANCESTRY OF MOST PLANT SPECIES. As more data accumulates, the history of angiosperms emerges as a history of genome-wide duplication followed by massive gene loss (and return to diploidy). Only 30% of *Arabidopsis* genes have retained syntenic copies in less than 86 million years since the 'alpha' duplication. In contrast, mammals appear to harbour fewer polyploidization events and less cycling of duplicated genes; 70% of human and mouse proteins show conserved synteny after 100 million years of evolution.

3.5.2 Collinearity

Orthology and paralogy

Figure 3.9 shows the concepts of orthology and paralogy. Orthologues and paralogues are two types of homologous sequence.

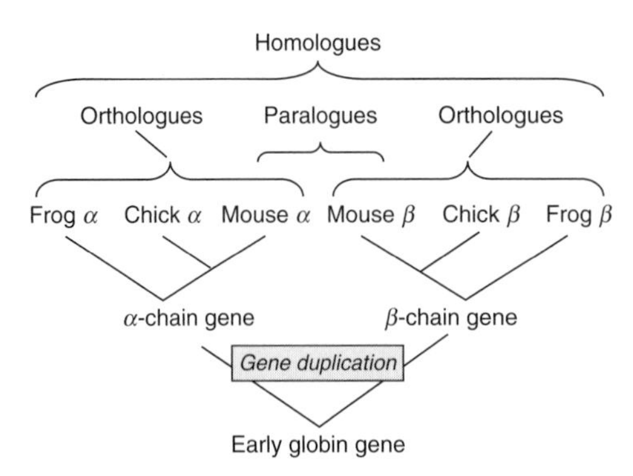

Fig. 3.9. The concepts of orthology and paralogy (from http://www.ncbi.nlm.nih.gov/Education/BLASTinfo/Orthology.html).

Orthology describes genes in different species that derive from a common ancestor. Orthologous genes may or may not have the same function. Paralogy describes genes that have duplicated (tandemly or moved to a new location) within a genome since they descended from a common ancestral gene. The word 'synteny' (from the Greek *syn*, together, and *taenie*, ribbon) refers to linkage of genes along a chromosome; currently used to indicate conservation of gene order across species. From this definition, macrosynteny means conservation of gene order across species detected at low resolution (i.e. genetic maps) while microsynteny means conservation of gene order across species analysed by high resolution (i.e. physical or sequence-based maps).

Macrocollinearity

Significant genomic collinearity in plants has been shown by comparative genetic mapping and genome sequencing, although plant genomes vary greatly in genome size and chromosome number and morphology. Comparative mapping of cereal genomes using low copy number, cross-hybridizing genetic markers has provided compelling evidence for a high level of conservation of gene order across regions spanning many megabases (i.e. macrocollinearity). Initial studies of the organization of grass genomes indicated that individual rice chromosomes were highly collinear with those of several other grass species and extensive work has shown a remarkable conservation of large segments of linkage groups within rice, maize, sorghum, barley, wheat, rye, sugarcane and other agriculturally important grasses (e.g. Ahn and Tanksley, 1993; Kurata *et al.*, 1994; van Deynze *et al.*, 1995a; Wilson *et al.*, 1999). These studies led to the prediction that grasses could be studied as a single syntenic genome. The macrocollinearity was summarized by Gale and Devos (1998) for rice and seven other cereals using what is now known as the 'circle diagram' (Plate 1). Further studies identified QTL controlling important agronomic traits which showed similarities in locations for the same or similar traits (as reviewed by Xu, 1997). Shattering and plant height are examples that were also mapped to collinear regions among grass genomes (Paterson *et al.*, 1995; Peng *et al.*, 1999). More recently, Chen *et al.* (2003) identified four QTL for quantitative resistance to rice blast that showed corresponding map positions between rice and barley, two of which had completely conserved isolate specificity and the other two had partial conserved isolate specificity. Such corresponding locations and conserved specificity suggested a common origin and conserved functionality of the genes underlying the QTL for quantitative resistance, which may be used to discover genes, understand the function of the genomes and identify the evolutionary forces that structured the organization of the grass genomes. Such findings reinforce the notion of collinearity among the cereal genomes.

This unified grass genome model has had a substantial impact upon plant biology but has not yet lived up to its potential. There are some difficulties in evaluating synteny between genomes at the macro-level (Xu *et al.*, 2005). First, the genomic marker data are very incomplete and genomic sequence data are largely lacking for many grass species. Secondly, the data are sometimes biased because the homologous DNA probes used in comparative mapping are selected for simple cross-hybridization patterns. Thirdly, many genes are members of gene families and, accordingly, it is often difficult to determine if a gene mapped in the second species is orthologous or paralogous to that in the first species. Fourthly, the collinearity of gene order and content observed at the recombinational map level is often not observed at the level of local genome structure (Bennetzen and Ramakrishna, 2002). Finally, in most early studies, no statistical analysis was used to evaluate whether the presence of a few markers in the same order on two chromosomal segments in two species occurs by chance or is truly significant.

The genome collinearity of several *Cammelineae* and *Brassicaceae* species have been recently compared to that of *A. thaliana* by comparative genetic linkage mapping and comparative chromosome painting (Schranz *et al.*, 2007). A comprehensive study identified 21 syntenic blocks that are shared by *Brassica napus* and *A. thaliana* genomes, corresponding to 90% of the *B. napus* genome (Parkin *et al.*, 2005).

Microcollinearity

Using the rice genome sequence as the reference to compare with molecular marker information of other cereals gave a result which indicated many more rearrangements than had been expected from Gale and Devos's (1998) concentric circles model. One such comparison involved more than 2600 mapped sequenced markers in maize among which only 656 putative orthologous genes could be identified (Salse *et al.*, 2004). The comparison of the wheat genetic map with the rice sequence also suggests numerous rearrangements between the two genomes with a high frequency of breakdowns in collinearity (Sorrells *et al.*, 2003). Extensive comparisons have also been made between sorghum and rice (Klein *et al.*, 2003; The Rice Chromosome 10 Sequencing Consortium, 2003). To align the sorghum physical map with the rice map, sorghum BAC clones were selected from the minimum tiling path of chromosome 3. Unique partial sequences were obtained from each BAC clone and could be directly compared with the rice sequence. This approach revealed excellent conservation between the overall structure and gene order of sorghum chromosome 3 and rice chromosome 1 but also indicated several rearrangements. Together, these studies indicate a general conservation of large syntenic blocks within cereals but with many more rearrangements and synteny breakdowns than originally anticipated.

This trend is even more obvious when synteny is analysed at the sequence level. Rearrangements may occur that involve regions smaller than a few centimorgans and would be missed by most recombinational mapping studies. Comparative sequence analysis involving large genomic segments can detect these rearrangements. Such analyses reveal the composition, organization and functional components of genomes and provide insight into regional differences in composition between related species. Recently, the sequencing of genomic segments in the cereals has enabled microcollinearity across genes or gene clusters to be investigated. Sequencing of the domestication locus *Q* in *Triticum monococcum* revealed excellent collinearity with the bread wheat genetic map (Faris *et al.*, 2003). Following the sequencing of the leaf-rust-resistance locus *Rph7* from barley, it was observed that this locus is flanked by two *HGA* genes. The orthologous locus in rice chromosome 1 consists of five *HGA* genes. In barley, only four of the five *HGA* genes are present, one is duplicated as a pseudogene and six additional genes have been inserted in between the *HGA* genes. These six genes have homologues on eight different rice chromosomes (Brunner *et al.*, 2003). The most striking rearrangement was revealed by the comparison of 100 kb around the *Bronze* locus of two maize lines. Not only does the retrotransposon distribution differ between the two lines but the genes themselves could also be different (Fu and Dooner, 2002). Comparison of the low molecular weight glutenin locus between *T. monococcum* and *Triticum durum* also revealed dramatic rearrangements: more than 90% of the sequence diverged because of retro-element insertions and because different genes are present at this locus (Wicker *et al*., 2003). Therefore collinearity can be lost very rapidly within two genomes from the same species.

With the sequencing of long regions, several studies in cereals have demonstrated incomplete microcollinearity at the sequence level. Song *et al*. (2002) identified orthologous regions from maize, sorghum and two subspecies of rice. It was found that gross macrocollinearity is maintained but microcollinearity is incomplete among these cereals. Deviations from gene collinearity are attributable to micro-rearrangement or small-scale genomic changes such as gene insertions, deletions, duplications or inversions. In the region under study, the orthologous region was found to contain six genes in rice, 15 in sorghum and 13 in maize. In maize and sorghum, gene amplification caused a local expansion of conserved genes but did not disrupt their order or orientation. As indicated by Bennetzen and Ma (2003), numerous local rearrangements differentiate the structures of different cereal genomes. On average, any comparison of a ten-gene segment between rice and a distant grass relative such as barley, maize, sorghum or wheat shows one or two rearrangements that involve genes. A simple extrapolation to the rice genome of about 40,000 genes (Goff *et al*., 2002) suggests that about 6000 genic rearrangements occurred which differentiate rice from any of the other cereals. Most of these rearrangements appear to be tiny and thus would not interfere with the macrocollinearity observed by recombinational mapping. There are exceptions however, which include chromosomal arm translocations and movements of single genes to different chromosomes (Bennetzen and Ma, 2003).

As expected, there is a high degree of gene conservation between the two shotgun-sequenced subspecies of rice, *japonica* and *indica*, which diverged more than 1 million years ago. On careful inspection, however, narrow regions of divergence can be found in these genomes (Song *et al*., 2002). These regions correspond to areas of increased divergence among rice, sorghum and maize, suggesting that the alignment of the two rice subspecies might be useful for identifying regions of cereal genomes that are prone to rapid evolution. Similar comparative analyses of *Arabidopsis* accessions have shown that both the relocation of genes and the sequence polymorphisms between accessions (in both coding and non-coding regions) are common in the *Arabidopsis* genome (The Arabidopsis Genome Initiative, 2000). Intraspecific violation of collinearity has also been identified in maize (Fu and Dooner, 2002). Han and Xue (2003) also discovered significant numbers of rearrangements and polymorphisms when comparing *indica* and *japonica* genomes in rice. The deviations from collinearity are frequently due to insertions or deletions. Intraspecific sequence polymorphisms commonly occur in both coding and non-coding regions. These variations often affect gene structures and may contribute to intraspecific phenotypic adaptations.

Implications of genome collinearity

Genomics would be much simpler if the order of genes were common (syntenic) across the major groups of plants. The usefulness of the collinearity between the genomes of model plants and important crops can be assessed by the number of failures or successes in its exploitation. For example, the analysis of the *Arabidopsis* sequence provides information that will facilitate the annotation of the rice sequence and likewise sequencing *Medicago* provides a resource for research on important crop legumes. Furthermore, the effort put into sequencing and annotating the rice genome has also been rewarded, as this annotation will be transferred to related sequences and used repeatedly in the future. The synteny between the monocots will help decipher the structure and function of the more complex genomes. A fully assembled rice sequence allows more accurate assessment of the macro- and microsynteny of rice with other cereals (Xu *et al*., 2005).

The advent of technologies for mapping genomes directly at the DNA level has made comparative genetic mapping among sexually incompatible species possible. Extensive comparative maps for marker genes have been constructed for a number of plant taxa, including species in the *Poaceae* (rice, maize, sorghum, barley and wheat), *Solanaceae* (tomato, potato and pepper) and *Brassicaceae* (*Arabidopsis*, cabbages, mustard, turnip and rape). As a result, the concept of a single genetic or ancestral map for all grasses, with species-specific modifications, is emerging (Moore *et al.*, 1995). The extensive collinearity of wheat, rye, barley, rice and maize suggests that it may be possible to reconstruct a map of the ancestral cereal genome. These conserved gene orders and the possibility of sharing DNA probes and PCR primers across species will greatly extend the power of mapping analysis by facilitating the molecular analysis of the corresponding chromosomal regions in different species and allowing information, and perhaps DNA sequences and genes, to be transferred quickly and efficiently between different species.

The challenge of finding which map, sequence and eventually functional genomic information from one species can be accessed, compared and exploited across all plant species will require the identification of a subset of plant genes that have remained relatively stable in both sequence and copy number since the radiation of flowering plants from their last common ancestor. Identification of such a set of genes would also facilitate taxonomic and phylogenic studies in higher plants that are presently based on a very small set of highly conserved sequences, such as those of chloroplast and mitochondrial genes. The conserved orthologue set of markers, identified computationally and experimentally, may further studies on comparative genomes and phylogenetics and elucidate the nature of genes conserved throughout plant evolution.

Completed genome sequences provide templates for the design of genome analysis tools in orphan species lacking sequence information. For example, Feltus *et al.* (2006) designed 384 PCR primers to conserve exonic regions flanking introns using sorghum and millet EST alignments to the rice genome. These conserved-intron scanning primers (CISP) amplified single-copy loci with 37–80% success rates; i.e. sampling most of the approximately 50 million years of divergence among grass species. When evaluating 124 CISPs across rice, sorghum, millet, Bermuda grass, teff, maize, wheat and barley, about 18.5% of them seemed to be subject to rigid intron size constraints that were independent of per-nucleotide DNA sequence variation. Likewise, about 487 conserved non-coding sequence motifs were identified in 129 CISP loci. As pointed out by Feltus *et al.* (2006), CISP provides the means to effectively explore poorly characterized genomes for both polymorphism and non-coding sequence conservation on a genome-wide or candidate gene basis and also to anchor points for comparative genomics across a diverse range of species. After the whole genomes of the major food crops have been sequenced, plant breeders will be able to access new gene tools that will facilitate the selection of outstanding individuals characterized by resistance to biotic and abiotic stresses and good seed quality, thus enabling breeders to produce new cultivars in addition to those currently available.

As a fundamental tool in biology, comparative analysis has been extended from being focused on a specific field to biology as a whole. With the growing availability of phenotypic and functional genomic data, comparative paradigms are now also being extended to the study of other functional attributes, most notably gene expression. Microarray techniques present an alternative method of studying differences between closely related genomes. Advances in microarray-based approaches (see Section 3.6) have enabled the main forms of genomic variation (amplifications, deletions, insertions, rearrangements and base-pair changes) to be detected using techniques that can easily be undertaken in individual laboratories using simple experimental approaches (Cresham *et al.*, 2008).

Tirosh *et al.* (2007) reviewed recent studies in which comparative analysis was applied to large-scale gene expression databases and discussed the central principles and challenges of such approaches. As different functional properties often co-evolve and complement one another, their combined analysis reveals additional insights. Unlike sequence-based genetic map information however, most functional properties are condition-dependent, a property that needs to be accounted for during interspecies comparisons. Furthermore, functional properties often reflect the integrated function of multiple genes, calling for novel methods that allow network-centred rather than gene-centred comparisons. Finally, one of the main challenges in comparative analysis is the integration of different data types which is becoming particularly important as additional data types are being accumulated. The lack of appropriate descriptors and metrics that succinctly represent the new information originating from genomic data is one of the roadblocks on this path. Galperin and Koller (2006) outlined recent trends in comparative genomic analysis and discussed some new metrics that have been used. This issue is related to the ontology concept and is discussed in detail in Chapter 14.

3.6 Array Technologies in Omics

It is widely believed that thousands of genes and their products (i.e. RNA and proteins) in any given living organism function in a complicated and orchestrated manner. However, traditional methods in molecular biology generally work on a 'one gene in one experiment' basis which means that the throughput is very limited and the 'whole picture' of gene function is difficult to obtain. In the late 1990s, a new technology known as a biochip or DNA microarray, attracted great interest among biologists. This technology promised to monitor the whole genome on a single array so that researchers would have a better picture of the interactions among thousands of genes at the same time.

Various terminologies have been used in the literature to describe this technology; for DNA microarrays these include, but are not limited to, biochip, DNA chip, DNA microarray and gene array. Affymetrix, Inc. owns a registered trademark, GeneChip®, which refers to its high density, oligonucleotide-based DNA arrays. However, in some articles appearing in professional journals, popular magazines and on the Internet, the term 'gene chip(s)' has been used as a general terminology that refers to DNA microarray technology. Depending on the type of molecules that are arrayed, microarrays can also be based on proteins, tissues or carbohydrates.

An array is an orderly arrangement of samples. It provides a medium for matching known and unknown molecular samples based on base-pairing (i.e. A-T and G-C for DNA; A-U and G-C for RNA) or hybridization and automating the process of identifying the unknowns. From its origin as a new technique for large-scale DNA mapping and sequencing and initial success as a tool for transcript-level analyses, microarray technology has spread into many areas by adapting the basic concept and combining it with other techniques. Microarray-based processes, either mature or under development, include transcriptional profiling, genotyping, splice-variant analysis, identification of unknown exons, DNA structure analysis, chromatin immunoprecipitation (ChIP)-on-chip, protein binding, protein–RNA interaction, chip-based comparative genomic hybridization, epigenetic studies, DNA mapping, re-sequencing, large-scale sequencing, gene/genome synthesis, RNA/RNAi synthesis, protein–DNA interaction, on-chip translation and universal microarrays (Hoheisel, 2006).

In this section, the basic procedures of arraying will be introduced and several major microarray technologies and platforms will be briefly described. The two volumes of *DNA Microarrays* (Kimmel and Oliver, 2006a, b) provide a comprehensive coverage of all the related fields from technologies and platforms to data analysis. The reader is also referred to Zhao and Bruce (2003), Amratunga and Cabrera (2004), Mockler and Ecker (2004), Subramanian *et al.* (2005), Allison *et al.* (2006), Hoheisel (2006) and Doumas *et al.* (2007).

3.6.1 Production of arrays

Complementary strands of DNA and nucleic acids in general can pair in a duplex via non-covalent binding. This fundamental characteristic is used in all DNA array techniques. Amaratunga and Cabrera (2004), Arcellana-Panilio (2005) and Doumas *et al.* (2007) describe the principles of DNA miroarray technology and how they are prepared and used. First, two terms related to microarrays, probe and target, should be introduced. The gene-specific DNA spotted on to the array is referred to as the probe and the sample to be tested that will hybridize with the probe is referred to as the target. The same probe spotted on to the array can be repeatedly hybridized with many different targets (samples). An experiment using a single DNA chip can provide researchers with information on thousands of genes simultaneously, resulting in a dramatic increase in throughput. In GeneChips (http://www.affymetrix.com/) the probe array was designed using an optimal set of oligonucleotides selected using computer algorithms and manufactured using Affymetrix light-directed chemical synthesis. Fluorescent labels were used for hybridization and detection and the Affymetrix software suite was used for data analysis and database management. Figure 3.10 illustrates a flowchart showing

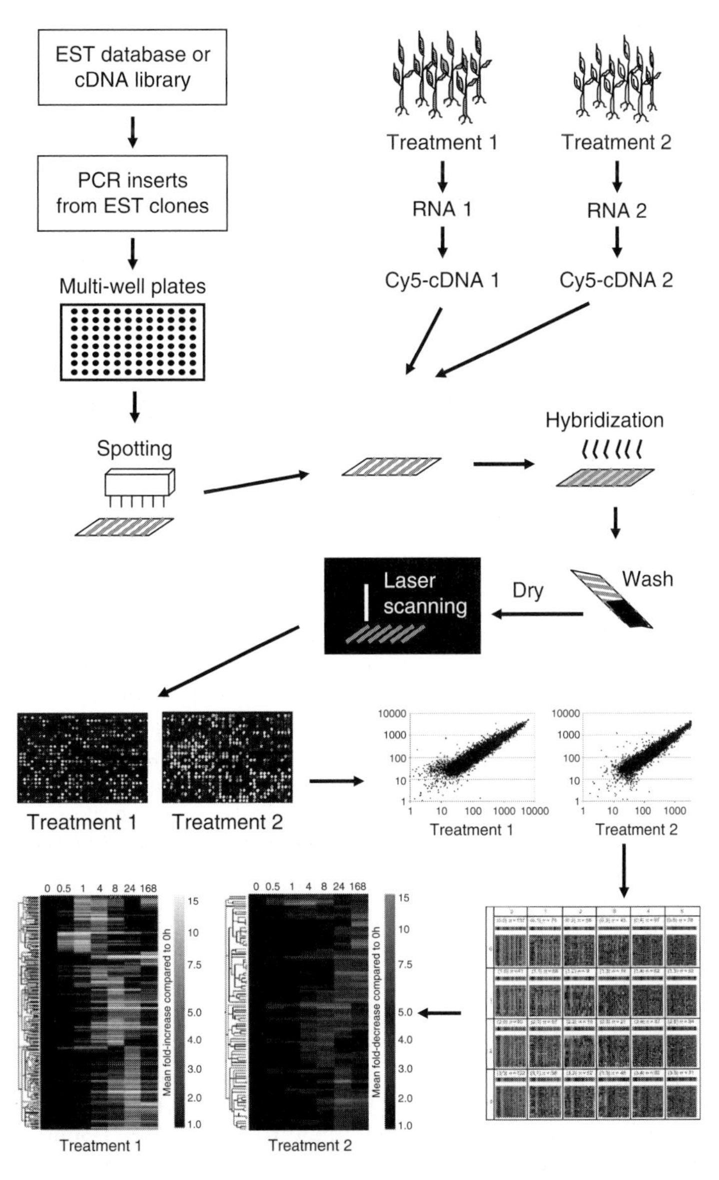

Fig. 3.10. A flowchart for a general microarray process.

a general microarray process. As DNA microarrays for whole genome expression profiling are the most mature and widely used technology, they will be used in this chapter as an example to describe the basic procedures of microarrays.

Types of arrays

An array experiment can be carried out using common assay systems such as microplates or standard blotting membranes; the arrays can be created by hand or robotics used to deposit the sample. In general, arrays are described as macroarrays or microarrays, the difference being the size of the sample spots. Macroarrays contain sample spot sizes of about 300 μm or larger and can be easily imaged with existing gel and blot scanners. The sample spot sizes in microarrays are typically less than 200 μm in diameter and these arrays usually contain thousands of spots. Microarrays also require specialized robotics and imaging equipment.

There are two main types of arrays, nylon and glass. Nylon arrays can contain up to about 1000 probes per filter. The target can be labelled using radioactive chemicals and detection of hybridization can be achieved using a phosphorimager or X-ray film. Glass arrays can hold up to about 40,000 spots per slide or 10,000 per 2 cm^2 area (limited by the capabilities of the arrayer). The target sample is labelled with fluorescent dyes and detection of hybridization requires specialized scanners.

There are two variants of the DNA microarray technology in terms of the properties of the arrayed DNA sequence of known identity. Format I: the probe cDNA (500–5000 bases long) is immobilized on a solid surface such as glass using robot spotting and exposed to a set of targets either separately or in a mixture. The development of this method, known 'traditionally' as a DNA microarray, is widely attributed to Stanford University.

Format II: an array of oligonucleotide (20–80-mer oligos) or peptide nucleic acid (PNA) probes is synthesized either *in situ* (on-chip) or by conventional synthesis followed by on-chip immobilization. The array is exposed to the labelled DNA sample and hybridized, the identity/abundance of the complementary sequences are then determined. This technology, 'historically' known as DNA chips, was developed at Affymetrix, Inc. which sells its photolithographically fabricated products under the GeneChip® trademark. Many companies are now manufacturing oligonucleotide-based microarrays using alternative *in-situ* synthesis or depositioning technologies.

Source of arrays

A collection of purified single-stranded DNA is the initial requirement. A drop of each type of DNA in solution is placed on to a specially prepared glass microscope slide by a robotic machine known as an arrayer. This process is called arraying or spotting and consists of binding a library of synthetic DNA on to a minimum surface area in a dense and homogeneous fashion. The major difference between various types of DNA arrays lies in the density of the bound probes and the manner in which these probes have been synthesized. The arrayer can quickly produce a regular grid of thousands of spots in an area the size of a dime (~ 1 cm^2), small enough to fit under the coverslip of a standard slide. The DNA in the spot is bound to the glass to prevent it from being dislodged during the hybridization reaction and subsequent wash.

The DNA spotted on to the microarray may be either cDNA (in which case the microarray is called a cDNA microarray), oligonucleotides (in which case it is called an oligonucleotide array), subgenomic regions of specific chromosomes or even the entire set of genes. The DNA spotted on to cDNA microarrays are cloned copies of cDNA that have been amplified by PCR and which correspond to the whole or part of a fully sequenced gene or putative ORF; ESTs are commonly arrayed. The selection of DNA probes to be spotted on to the microarray depends on which genes are to be studied. For plants whose genomes have been completely sequenced, it is possible to array genomic DNA from every known gene or putative ORF. To obtain sufficient DNA for arraying, each gene or putative ORF from the total genomic DNA can be amplified by PCR or each cDNA can be cloned and large numbers of identical DNA copies can be generated by growing them in bacteria.

The DNA spots on a microarray are produced either by synthesis *in situ* or by deposition of the pre-synthesized product. DNA synthesis *in situ* methods have largely been within the purview of commercial companies. In this method, 20–25-bp long gene-specific oligonucleotides are generated *in situ* on a silicon surface by combining a standard DNA synthesis protocol with phosphoramidite reagents modified with photolabile 5'-protecting groups (Doumas *et al.*, 2007). The activation for oligonucleotide elongation is achieved using a mask (Affymetrix; http://www.affymetrix.com) or maskless (NimbleGen; www.nimblegen.com) method. Alternatively, the reagents can be delivered to each spot using ink-jet technology (Agilent; http://www.agilent.com). Ongoing research and development efforts ensure the optimum design of the DNA content and continued technological advancements enable the production of increasingly higher-density arrays.

Array content

The choice of DNA type to print is fundamental. The sequence of the cDNA could be several hundred to a few thousand base pairs long. The DNA spotted on oligonucleotide arrays consist of synthesized chains of oligonucleotides corresponding to part of a known gene or putative ORF; each oligonucleotide is usually about 25–70bp long. In an oligonucleotide array, a gene is generally represented by several different oligonucleotides and they are carefully chosen for maximal specificity. Longer stretches of DNA such as those obtained from PCR of cDNA clones produce robust hybridization signals but less specificity. Short oligonucleotides (24–30nt) have greater discrimination and are also suitable for assessing single-nucleotide changes. Long oligonucleotides (50–70nt) afford an excellent compromise between signal strength and specificity and their use has increased among academic core facilities (Arcellana-Panilio, 2005). Choosing oligos corresponding to the 3' untranslated regions (3'UTR) increases the likelihood of their being specific and designing oligos close to the 3' end might also boost signal intensity.

Slide substrates

Glass microscope slides are the solid support of choice and they should be coated with a substrate that favours binding of the DNA. Development of substrates on atomically flat slide surfaces and minimum background for higher signal-to-noise ratios has contributed to the improvement of data quality (Arcellana-Panilio, 2005). Different versions of silane, amine, epoxy and aldehyde substrates which attach DNA by either ionic interaction or covalent bond formation are commercially available.

Arrays and spotting pins

The physical process of delivering the DNA to pre-determined coordinates on the array, involves spotting pens or pins carried on a print head that is controlled in three dimensions by gantry robots with sub-micron precision. A total of 30,000 features of ~ 90-μm diameter can easily be spotted on to a 25 ×75-mm slide with a maximum spotting density of over 100,000 features per slide. There are several DNA arraying technologies, including high speed robotic printing of DNA fragments on glass (usually PCR amplified cDNAs), high speed robotic printing of long oligonucleotides (70-mers; Agilent technology and many academic facilities), synthesis of oligonucleotides (25-mers) on micro-chips using photolithographic masks (Affymetrix GeneChips) and synthesis of oligonucleotides (25–70-mers) on microchips using maskless aluminium mirrors (NimbleGen GeneChips). Improvements in arraying systems have included shorter printing times and longer periods of walk-away operation. Arrayers are invariably installed within controlled-humidity cabinets to maintain an optimum environment for printing.

3.6.2 Experimental design

Careful experimental design is required to determine the type of array to run; how many replicates to use; and which samples will be hybridized to obtain meaningful data amenable to statistical analysis, upon which sound conclusions can be drawn. A biological question must first be framed and a microarray platform then chosen, followed by a decision on biological and technical replicates and the design of a series of hybridizations.

Microarray experimental design is usually governed by the aim of the experiment. An important aspect of experimental design is deciding how to minimize variation which can be thought of as occurring in three layers: biological variation, technical variation and measurement error. Replication is the easy answer to dealing with variation. To make the best use of available resources, it is important to know what to replicate and how many replicates to apply. Hybridization of two samples to the same slide is made possible by labelling each sample with chemically distinct fluorescent tags. This also provides the opportunity to make direct comparisons between samples of primary interest (Arcellana-Panilio, 2005). Using a common reference becomes more efficient when a large number of samples need to be compared. When an experiment is testing the effect(s) of multiple factors, a well-thought-out design is extremely critical so that resources are not wasted on eventually useless comparisons.

3.6.3 Sample preparation

Preparation of DNA samples for hybridization can follow general DNA extraction protocols. So here we will focus on RNA sample preparation as described by Arcellana-Panilio (2005). The sources of RNA for the samples that will be hybridized to a microarray may be obtained from different types of cells or tissues. Obtaining pure, intact RNA, free from DNA or protein contamination, is important, while the homogeneity of the RNA source itself as defined by the biological question being asked must be considered. The amount of RNA required for hybridization ranges from as little as 2–5 μg total RNA for short oligonucleotide arrays to 10–25 μg total RNA for cDNA spots and long oligonucleotide arrays. In some circumstances it becomes necessary to amplify the RNA in the sample to obtain adequate amounts for labelling and hybridization to an array.

To prepare the labelled sample, the first step is to purify mRNA from total cellular contents. There are several challenges involved: (i) mRNA accounts for only a small fraction (less than 3% of all RNA in a cell) so isolating mRNA in sufficient quantity for an experiment (1–2 μg) can be a challenge. Common mRNA isolation methods take advantage of the fact that most mRNAs have a poly-adenine, poly(A), tail. These poly(A) mRNAs can be purified by capturing them using complementary oligodeoxythymidine (oligo(dT)) molecules bound to a solid support such as a chromatographic column or a collection of magnetic beads. (ii) The more heterogeneous the cells, the more difficult it is to isolate mRNA specific to the study. (iii) Captured mRNA degrades very quickly and the mRNA has to be immediately reverse-transcribed into more stable cDNA (for cDNA microarrays). The reverse transcription reaction usually starts from the poly(A) tail of the mRNA and moves toward its head; such a reaction is described as *oligo(dT)-primed*.

3.6.4 Labelling

Before hybridization to DNA arrays or chips, the target (sample) has to be labelled to allow its subsequent detection. There are several methods that have been developed historically to detect or identify hybrid DNA molecules including the use of hydroxylapatite, radioactive labelling, enzyme-linked detection and fluorescent labelling depending on the nature of the chip, whether it is glass or nylon. In order to be able to detect which cDNAs are bound to the microarray, the sample is labelled with a reporter molecule that flags their presence. The reporters currently used in microarray experiments are fluorescent dyes known as fluors or fluorophores, chemicals that fluoresce when exposed to a specific wavelength of light. A differently coloured fluor is used for each sample so that the two samples can be differentiated on the array.

The cDNA or mRNA can be labelled either directly or indirectly. In the direct labelling procedure, fluorescently labelled nucleotide is incorporated into the cDNA products as it is being synthesized. With this method, a difference in the steric hindrance conferred by different label moieties causes some labelled nucleotides to be more efficiently used than others, producing a dye bias in which one sample is labelled at a higher level overall than the other. Cyanine 3 (Cy3) and cyanine 5 (Cy5) are large molecules that reduce reverse transcriptase efficiency of long transcripts and certain sequences. Cy3-nucleotide tends to be incorporated at a higher frequency than Cy5 although this does not necessarily translate into a better labelled target. To prevent the dye bias, the indirect labelling approach was developed where RNA is reverse transcribed in the presence of an amino allyl-modified nucleotide that enables the chemical coupling of fluorescent labels after the cDNA is synthesized. If the coupling reaction goes to completion, the frequency of labelling becomes independent of the fluorophore (Arcellana-Panilio, 2005).

The labelled sample is the target for the experiment. The number of fluor molecules that label each cDNA depends on its length and also possibly its sequence composition. For an RNA sample, either total RNA or mRNA is typically isolated and labelled using a first strand cDNA synthesis step either by direct incorporation of a fluorescent dye or by coupling the dyes to a modified nucleotide. For non-expression-based experiments, DNA rather than RNA can be labelled and hybridized to the array.

3.6.5 Hybridization and post-hybridization washes

The array holds hundreds or thousands of spots, each of which contains a different DNA sequence. If a sample contains a cDNA whose sequence is complementary to the DNA on a given spot, that cDNA will hybridize to the spot where it will be detectable by its fluorescence. In this way, every spot on an array is an independent assay for the presence of a different cDNA. Hybridization is achieved by pouring the labelled sample on to the array and allowing it to diffuse uniformly. It is then sealed in a hybridization chamber and incubated at a specific temperature for a period of time sufficient to allow hybridization reactions to complete. The experimental conditions should ensure that all areas of the array are exposed to a uniform amount of labelled sample throughout the hybridization.

Hybridizations are processed directly on the slides after target synthesis. The hybridization step is literally where everything comes together, i.e. the labelled molecules find their complementary sequences on the array and form double stranded hybrids which are strong enough to withstand stringent washes. As in the hybridization of classical Southern and northern blots, the objective is to favour the formation of hybrids and the retention of those which are specific. Hybridization conditions depend on the length of probes arrayed on the slide and need to be extensively tested before analysis. As an example, probe melting temperatures range from 42 to 70°C depending on the nature of the buffer: the presence of formamide exerts a positive effect on buffer stringency in Denhardt-type buffers which are used at 42°C, whereas Sarkosyl-based buffers are commonly used around 70°C. Exogenous DNA (e.g. salmon sperm and Cot-1 DNA) reduces background by blocking areas of the slide with a general affinity for nucleic acid or by titrating out labelled sequences that are non-specific. Denhardt's reagent (containing equal parts of Ficoll, polyvinylpyrrolidone and bovine secum albumin) is also used as a blocking agent. Detergents such as SDS reduce surface tension and improve mixing while helping to lower background at the same time. Temperature is an important factor that can be manipulated during the hybridization and post-hybridization washes of microarrays and here much can be learned from what has already been established for Northern or Southern blots. For microarrays to be useful as a means of quantifying expression the target has to be present in limiting concentrations and the probe must be present in sufficient excess so as to remain virtually unchanged even after hybridization (Arcellana-Panilio, 2005). One important feature of fluorescence detection is that it allows the simultaneous hybridization of two to several targets that have been differently labelled.

The quality of the hybridization can be assessed by spotting the sample with a set of hybridization control genes, spiking the labelled sample with a known amount of these controls prior to exposure to the array and verifying that these control genes are indeed showing up as having been hybridized.

3.6.6 Data acquisition and quantification

Once the wet phase (e.g. slide hybridization and washing off any excess labelled sample) is completed, signal detection of each of the hybridization targets can be captured, that is, the array must be scanned to determine how much of each labelled sample is bound to each spot. The signal is acquired using array scanners, either a charge-coupled device (CCD) or a confocal microscope, typically equipped with lasers to excite the fluorophores at a specific wavelength and photo-multiplier tubes to detect the emitted light. Spots with more bound sample will have more reporters and will therefore fluoresce more intensely. Whatever the scanner resolution, the microarray spot diameter needs to be five to ten times larger than the scanner resolution which can be as little as 5 μm for the most recent models. The end-product of a microarray experiment is a scanned grey scale image whose intensity measurements range from 0 to 2^{16}. The image is usually stored in a 16-bit tagged image file format (tiff, for short). The most basic scanner models offer excitation and detection of the two most commonly used fluorophores (Cy3 and Cy5) whereas higher-end models enable excitation at several wavelengths and offer dynamic focus, linear dynamic range over several orders of magnitude and options for high-throughput scanning. The objective of the scanning procedure is to obtain the best image, where the best is not necessarily the brightest (to avoid over-saturation beyond the signal range) but is the most faithful representation of the data on the slide.

Although it is only supposed to pick up the light emitted by the target cDNAs bound to their complementary spots, the scanner will inevitably pick up light from various other sources, including the labelled sample hybridizing non-specifically to the glass slide, residual (unwanted) labelled probe adhering to the slide, various chemicals used in processing the slide and even the slide itself. This extra light creates background signals. Once signal and background values are clearly defined, which is specific to each experiment, data can be extracted from the image by counting the pixels with each probe and background area and recording this in a computer readable format.

Data extraction from the image involves several steps (Arcellana-Panilio, 2005): (i) gridding or locating the spots on the array; (ii) segmentation or assignment of pixels either to foreground (true signal) or background; and (iii) intensity extraction to obtain a new value for foreground and background associated with each spot. Subtracting the background intensity from the foreground yields the true spot intensity which can be used as an approximation of relative gene expression.

3.6.7 Statistical analysis and data mining

Huge data sets are generated by microarray experiments. For example, 20 hybridization experiments with the *Arabidopsis* GeneChip generates a set of 2,624,000 data points (8200 genes × 16 oligonucleotides × 20 hybridizations). Such a massive amount of data prohibits any manual treatment. Also experimental variability is generally significant and has to be managed in order to exploit the data properly. Allison *et al.* (2006) examined five key components of microarray analysis: (i) design (the development of an experimental plan to maximize the quality and quantity of information obtained); (ii) pre-processing (processing of the microarray image and normalization of the data to remove systematic variation. Other potential pre-processing steps include transformation of data, data filtering and in the case of two-colour arrays, background subtraction); (iii) inference (testing statistical hypotheses, e.g. which genes are differentially expressed); (iv) classification (analytical approaches that attempt to divide data into classes with no prior information or into predefined classes); and (v) validation (the process of confirming the veracity of the inferences and conclusions drawn in the study).

Reproducible and reliable microarray results can be only achieved through quality control starting with data generation. Good laboratory proficiency and appropriate data analysis practices are essential (Shi *et al.*, 2008). Numerous software packages, both free and commercial, are available for quantifying microarray data. Typically, the interpreted array data will highlight a relatively small number of spots that deserve further investigation. Alternatively, the overall pattern of profiling can be used as a 'fingerprint' to characterize specific phenotypes.

The quantified data from the images are obtained in typical form of tab-delimited text files. First, dust artefacts, comet tails and other spot anomalies should be identified and flagged so that they will not enter the analysis. Pre-processing the quantified data before formal analysis includes the flagging of ambiguous spots with intensities lower than a threshold defined by the mean intensity plus two standard deviations of supposedly negative spots (no DNA, buffer and/or non-homologous DNA controls).

Interpreting the data from a microarray experiment can be challenging. Quantification of the intensities of each spot is subject to noise from irregular spots, dust on the slide and non-specific hybridization. Deciding the intensity threshold between spots and background can be difficult especially when the spots fade gradually around their edges. Detection efficiency might not be uniform across the slide, leading to excessive red intensity on one side of the array and excessive green on the other.

Data normalization addresses systematic errors that can skew the search for biological effects. One of the most common sources of systematic error is the dye bias introduced by the use of different fluorophores to label the target. Print-tip differences can also lead to sub-grid biases within the same array while scanner anomalies can cause one side of an array to seem brighter than the other. Normalization across multiple slides to remove bias can be accomplished by scaling the within-slide normalized data. In practice, examining the box plots of the normalized data of individual arrays for consistency of width can usually indicate whether normalization across arrays is required.

Spatial plots can locate background problems and extreme values. The shape and spread of scatter plots and the height and width of box plots give an overall view of data quality that can give clues about the effects of filtering and different normalization strategies. Gene expression profiling will be taken as an example for the rest of this section. Clustering algorithms are means of organizing microarray data according to similarities in expression patterns. In this case, co-expressed genes must be co-regulated, and a logical follow-up to this analysis is the search for regulatory motifs and the common upstream or downstream factors that may tie these co-expressed genes together. Treatments can be clustered based on similarity in gene expression profiles. Genes can be clustered based on similarity in expression patterns across profiles. Two mathematical approaches are often used, hierarchical or *k*-means clustering (Stanford) and self organizing maps (SOMs) (Whitehead Institute).

A strategy for identifying differentially expressed genes is to compute the *t*-statistic and correct for multiple testing using adjusted *P*-values. The B-statistic, derived using an empirical Bayes approach, has been shown in simulations to be far superior to either log ratios or the *t*-statistic for ranking differentially regulated genes (Lonnstedt and Speed, 2002). The twofold change continues to be a benchmark for researchers perusing lists of microarray data in order to validate the data by PCR, which can provide independent confirmation of the expression patterns of specific genes. However, fold change has become more of a secondary criterion for the selection of candidates for follow-up from a list of genes ranked according to more reliable measures of differential expression (Arcellana-Panilio, 2005). After preliminary data mining and statistical analysis, validation and follow-up experiments can be designed.

There are many examples of the array technologies described in this section. In yeast, 260,000 oligonucleotides corresponding to all the genes in yeast have been synthesized on to a 1.28 cm^2 chip. These chips have allowed the identification of genes expressed in various mutants under different culture conditions or at different stages of growth. Numerous genes of unknown function have thus been recognized, regulated in a manner similar to or opposite to that of genes of known function; transcription of the genome is thus incorporated into a vast combinatorial network. In plants, Affymetrix has commercialized microchips to evaluate the expression of *Arabidopsis* genes, allowing the identification of genes that are active during pathogen infection or during treatment with herbicides, fungicides or insecticides. This also facilitates the determination of which genes are transcribed in which tissues under which conditions or during which stages of development. Commercial microarrays are also available from Affymetrix for several other crop plants such as maize and tomato.

3.6.8 Protein microarrays and others

A protein chip or microarray is a piece of glass on which different molecules of protein have been affixed at separate locations in an ordered manner to form a microscopic array. Compared with DNA microarrays, the development of protein-based approaches poses technical problems for several reasons (Bernot, 2004): (i) proteins consist of 20 distinct amino acids while there are only four bases in DNA; (ii) depending on their amino-acid composition, proteins may be hydrophilic, hydrophobic, acidic or basic (while DNA is always hydrophilic and negatively charged); and (iii) proteins are often post-translationally modified (by glycosylation, phosphorylation, etc.).

Although detection of protein microarrays can be carried out using general detection methods as described above, the problem is that protein concentrations in a biological sample may be many orders of magnitude different from that of mRNAs. Therefore, protein array detection methods must have a much larger range of detection. The preferred method of detection is currently fluorescence detection. Fluorescent detection is safe, sensitive and can produce high resolution. The fluorescent detection method is compatible with standard microarray scanners but some minor alterations to software may need to be made.

Protein microarrays have been made in the following manner (Macbeath and Schreiber, 2000; Bernot, 2004). Proteins are deposited on to a support and subsequently fixed to it. Thus 1600 distinct proteins may be arranged per cm^2. These arrays are ordered so that it is known which protein is represented by any given spot. The microarrays are then incubated with other ligands (fluorescently labelled) and the result of the hybridization is analysed by confocal microscopy (it is also possible to employ radioactively labelled ligands). The protein recognized may be identified using the signal localization data obtained. The intensity of the signal obtained is proportional to the level of ligand–protein interaction.

Except for the most frequently used DNA and protein microarrays discussed above, other microarrays include those built using tissues (cells) and carbohydrates. Similar to other microarrays, a tissue chip or microarray is a piece of glass on which different tissues have been affixed, while sugar or carbohydrate microarrays include oligosaccharides, polysaccharides/glycans and glycoconjugates fixed on an array. Carbohydrates are very different from proteins in the following aspects: (i) carbohydrates are highly heterogeneous as they have a large number of different molecules determined by over 500,000 different oligosaccharides units; (ii) their synthesis is very complicated and involves a larger number of enzymes; and (iii) biological information that is stored in the various types of carbohydrates is less well understood. For these reasons, carbohydrate microarrays will be a useful tool for glycomics.

A new technology that is related to microarrays and should be mentioned is microfluidics. Microfluidics is the science and technology of systems that process or manipulate small (10^{-9}–10^{-18} l) amounts of fluids using channels with dimensions of tens to hundreds of micrometres (Whitesides, 2006). The first applications have a number of useful characteristics: (i) the ability to use very small quantities of samples and reagents and to carry out separations and detections with high resolution and sensitivity; (ii) low cost; (iii) short analysis times; and (iv) small footprints for the analytical devices. Microfluidics offers fundamentally new capabilities in the control of concentrations of molecules in space and time. In the areas of microanalysis, microfluidics offers approaches for biological analyses that require much greater throughput and higher sensitivity and resolution than were previously required. It has great potential to improve the analytical processes involved in proteomics, DNA isolation, PCR and DNA sequencing.

3.6.9 Universal chip or microarray

Most microarray platforms are designed to address a specific set of questions in a specific organism. This means that a specific microarray platform needs to be established and produced for each application. Moreover, many assays that are carried out on microarrays would work even better in a homogenous solution rather than on a solid support (Hoheisel, 2006). The establishment of 'zip-code' arrays can address these problems by separating the actual assay from the microarray hybridization (Gerry *et al.*, 1999). Such microarrays contain a set of unique and distinct oligonucleotides that are immobilized at known locations. Because they should not be complementary to any sequence in any organism and are made solely to identify the 'address' of a particular location on the microarray, they are called zip-code sequences (Fig. 3.11). The oligonucleotides are designed to have similar thermodynamic properties and thus hybridization can be carried out at one temperature and under defined stringency conditions. Instead of having to produce many different microarrays, a single design can be used for various assays.

For example, Hoheisel (2006) described a universal microarray option that involves using the L-DNA enantiomer, the mirror image form of normal D-form DNA, for the zip-code oligomer (Fig. 3.11). Because L-DNA forms a left-helical duplex, there is no cross-hybridization between L-DNA and D-DNA. However, chimaeric molecules that consist of L-form and D-form stretches can be produced by standard chemistry. Therefore, D-DNA primers are produced with an L-DNA zip-code tag that binds to the L-DNA complementary oligonucleotide on the microarray. L-DNA microarrays are stable because L-DNA is resistant to nuclease activities. Simultaneously, only the zip-code part of the molecules that is used in homogenous solution is able to hybridize to the array. Neither the D-formed primer portion nor the analyte (for example, genomic DNA or RNA preparations) will cross-hybridize with the array.

3.6.10 Whole-genome analysis using tiling microarrays

The recent explosion in available genome sequence data has made it realistic to undertake microarray analysis at the whole-genome level. Interestingly, these sequence

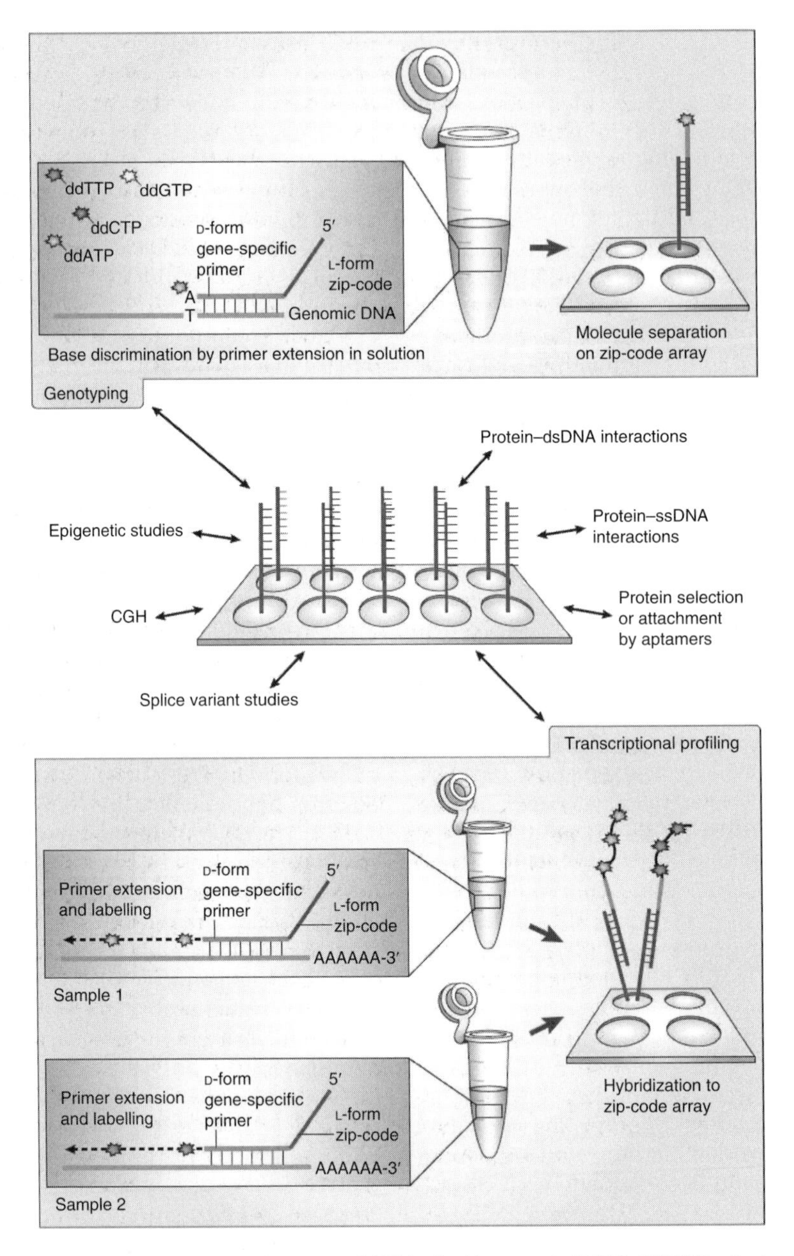

Fig. 3.11. The concept of universal microarray. dsDNA, double-stranded DNA; SSDNA, single-stranded DNA; CGH, comparative genomic hybridization.

data have led to the advent of high-density DNA oligonucleotide-based whole-genome tiling microarrays (WGAs) which can be employed to interrogate a full genome's worth of sequence data in a single experiment. This technology allows a more complete understanding of an organism's genomic content and should provide a dramatic improvement in the understanding of numerous biological processes. WGAs comprise relatively short (< 100-mer) oligonucleotide features. Furthermore, they can be created with > 6,000,000 discrete features, each comprising millions of copies of a distinct DNA sequence. For instance, the Affymetrix® GeneChip® *Arabidopsis* tiling 1.0R array (http://www.affymetrix.com) is a single array comprising over 3.2 million perfect match and mismatch probe pairs (approximately 6.4 million probes in total) tiled with 35-bp spacing throughout the complete non-repetitive *A. thaliana* genome (Zhang, X. *et al.*, 2006).

WGAs can be employed for a myriad of purposes in plants including empirical annotation of the transcriptome characterization, mapping of regulatory DNA motifs using ChIP-on-chip, novel gene discovery, analysis of alternative RNA splicing, characterization of the methylation state of cytosine bases throughout a genome (methylome) and the identification of sequence polymorphism (Gregory *et al.*, 2008). Overall, implementing standardized protocols for RNA labelling, hybridization, microarray processing, data acquisition and data normalization within the plant community will minimize sources of error and data variability between laboratories and across microarray platforms. In this way WGA analysis among a diverse set of groups will results in high-quality, easily reproducible data that will aid the research of the entire plant community.

3.6.11 Array-based genotyping

Array techniques have become increasingly popular as a tool for genome-wide genotyping since they offer an assay that is highly multiplexed at a low cost per data point. One of the earliest reports of microarray-based genotyping employed high-density WGAs produced by photolithographic synthesis (Affymetrix) for the simultaneous discovery and array of DNA polymorphisms in yeast. In genotyping assays based on microarrays, allelic variation is detected as the differential hybridization of labelled genomic DNA to individual probes or sets of probes covering identifiable genomic locations. The polymorphism of the two sequences, originating from two different cultivars or genotypes, results in differential hybridization intensity and this property associated with sequence characteristics functions is a molecular marker popularly known as single feature polymorphism (SFP). Using this approach a large number of SFPs were identified between two laboratory strains of yeast (Winzeler *et al.*, 1998).

For the larger and more complex *Arabidopsis* genome, tiling arrays were not available and hence the first experiments involved hybridization of labelled genomic DNA using Affymetrix AtGenome1 GeneChips based on available, expression-based annotation for ORFs. Despite this ORF-based focus, nearly 4000 SFPs were identified between the Columbia and Landsberg *erecta* accessions (Borevitz *et al.*, 2003). In order to determine genome-wide patterns of SFP, hybridization to the ATH1 gene expression array was used to interrogate genomic DNA diversity in 23 wild strains (accessions), in comparison with the reference strain Columbia. At < 1% false discovery rate, 77,420 SFPs with distinct patterns of variation were detected across the genome. Total and pair-wise diversity was higher near the centromeres and the heterochromatic knob region (Borevitz *et al.*, 2007). By high-density array re-sequencing of 20 diverse strains (accessions), more than 1 million non-redundant SNPs were identified (Clark *et al.*, 2007). Salathia *et al.* (2007) developed a microarray-based method that assesses 240 unique indel markers in a single hybridization experiment at a cost of less than US$50 in materials per line. The genotyping array was built with 70-mer oligonucleotide elements representing indel polymorphisms between Columbia and *Landsberg erecta*. Multi-well chips allow groups of 16 lines to be genotyped in a single experiment.

Microarray-based genotyping has recently been further developed in several crop plants. Using a high-density microarray technology pioneered at Perlegen Sciences (http://www.perlegen.com), the International Rice Functional Genomics Consortium initiated a project to identify a large fraction of the SNPs presented in cultivated rice through whole-genome comparisons of 21 rice genomes, including cultivars, germplasm lines and landraces (McNally *et al.*,

2006). Perlegen designed SNP-discovery arrays to include all possible SNP variations with multiple levels of redundancy.

Edwards *et al.* (2008) developed a microarray platform for rapid and cost-effective genetic mapping using rice as a model. In contrast to methods employing genome tiling microarrays for genotyping, the method is based on low-cost spotted microarray production, focusing only on known polymorphic features. A genotyping microarray was produced comprising 880 SFP elements derived from indels identified by aligning genomic sequences of the *japonica* cultivar Nipponbare and the *indica* cultivar 93-11. The SFPs were experimentally verified by hybridization with labelled genomic DNA prepared from the two cultivars. Using the genotyping microarrays, high levels of polymorphism were found across diverse rice accessions.

In soybean, the GoldenGate assay, which is capable of multiplexing from 96 to 1536 SNPs in a single reaction, has been tested to determine the success rate of converting verified SNPs into working arrays (Hyten *et al.*, 2008). Allelic data were successfully generated for 89% of the 384 SNP loci when it was used in three recombinant inbred line (RIL) mapping populations. Using the same system, two panels of 1536 SNP markers have been developed in maize through collaboration between Cornell, CIMMYT and Illumina, one with SNPs developed from candidate genes relevant to drought tolerance and the other with SNPs randomly distributed on the maize genome (Yan *et al.*, 2009).

4

Populations in Genetics and Breeding

Many types of populations are currently being used in genetic studies and plant breeding. The properties of a population depend on how it is developed and which parents are involved. Doubled haploids (DHs), recombinant inbred lines (RILs) and near-isogenic lines (NILs) are three important types of populations that have a long history of application in plant breeding and have been widely used in genetic mapping, gene discovery and genomics-assisted breeding since the discovery of DNA-based markers. This chapter describes in general the structure, development and utilization of these important genetic populations, based on a comprehensive discussion by Xu and Zhu (1994). More details on applications of these populations will be covered in other chapters.

4.1 Properties and Classification of Populations

Populations that are currently used in genetics and plant breeding can be classified and their properties can be described based on their genetic constitution, maintenance, genetic background and origin.

4.1.1 Genetic constitution-based classification

For two alleles, A_1 and A_2, at a specific genetic locus, A, there are three different possible genotypes, A_1A_1, A_2A_2 and A_1A_2. If a population consists of individuals that have an identical genotype (no matter whether they are homozygous, A_1A_1 or A_2A_2 or heterozygous, A_1A_2, for locus A) it is said to be homogeneous. However, if a population consists of individuals that have different genotypes (for example, some with A_1A_1 or A_2A_2, and others with A_1A_2) it is said to be heterogeneous.

Based on the above definitions, there are four types of populations:

1. Homogeneous populations with homozygous individuals: such as individuals from a cultivar of a self-pollinated species or from an inbred derived from an open-pollinated species.
2. Homogeneous populations with heterozygous individuals: such as F_1 plants derived from two homogeneous and homozygous cultivars of self-pollinated species or between two inbreds derived from an open-pollinated species.
3. Heterogeneous populations with homozygous individuals: such as pure-breeding

individuals derived from continuous selfing of a hybrid of two inbred lines or cultivars, such as RILs, where each individual is homozygous, either A_1A_1 or A_2A_2 while different individuals have different genotypes.

4. Heterogeneous populations with heterozygous individuals: such as individuals in early generations such as F_2 and F_3 derived from two inbred lines or homozygous cultivars. A set of open-pollinated cultivars of an open-pollinated species is a heterogeneous population containing heterozygous individuals.

4.1.2 Genetic maintenance-based classification

Based on whether a population can maintain its genetic constitution through selfing from one generation to another, populations can be classified into two types:

1. Tentative or temporary populations: individuals in a population, such as F_2, F_3, BC_1, BC_2, etc., have different genotypes and their genetic constitution will change with recombination resulting from selfing or inbreeding. These types of population are difficult to maintain and in most cases, the same genetic constitution can be only used once.

2. Permanent or immortalized populations: this type of populations consists of a set of pure-breeding lines derived from two parents or a common set of parents. Individuals within a line have identical genotypes, while individuals from different lines have different genotypes. Each line can serve as a segregation unit from the parental population and population structure and genetic constitution can be maintained consistently generation after generation through selfing or inbreeding processes.

4.1.3 Genetic background-based classification

Populations can be classified into those within which individuals have a nearly isogenic background and those that have a heterogeneous background. A population that consists of lines with nearly identical genetic backgrounds can be derived from genetic processes such as continuous backcrossing of a hybrid to one of its parental lines so that lines only differ for a specific target trait or locus. All other types of populations, including F_2, backcross (BC), RILs and DHs have heterogeneous backgrounds, i.e. individuals within these populations have heterogeneous backgrounds and differ not only in the target traits but also in the remainder of the traits.

4.1.4 Origin-based classification

Populations can be classified into two basic categories based on the origin of the individuals they contain: populations of natural cultivars and populations formed by planned materials among selected parents or genetic mating populations.

Populations of natural cultivars

These populations consist of a group or subset of cultivars which are selected from a large number of cultivars for specific target traits or are based on specific pedigree relationships. The variation for the target trait among groups of cultivars can be investigated and the relationship between the target trait and other traits or molecular markers can be established. For example, the genetic effect of plant height can be studied by comparing tall cultivars with short ones.

Populations formed by planned matings

Mating populations are specifically designed for genetic studies and derived from a specific genetic mating design using selected genetic stocks. There are several genetic mating designs that are widely used in genetics and breeding.

DIALLEL CROSSES. A total of n cultivars or inbred lines are selected as male and female parents to produce crosses of all possible combinations. The F_1s or F_2s derived from these crosses are then genetically analysed. The mating design is as follows:

Parent	P_1	P_2	P_3	...	P_n
P_1	×	×	×	...	×
P_2	×	×	×	...	×
P_3	×	×	×	...	×
...					
P_n	×	×	×	...	×

A full diallel analysis will include all one-way hybrids and parents while a partial or incomplete diallel analysis may contain just half the diallel without reciprocals or parents. Diallel crosses are usually used to estimate general combining ability for the parents and special combining ability for specific crosses, providing information for producing hybrids.

NORTH CAROLINA DESIGNS. There are three North Carolina designs, denoted by NCI, NCII, and NCIII. These designs are most often used in cross-pollinated crops and to study broad-based populations. Their use in self-pollinated crops usually involves many inbred lines that can reasonably be considered to represent a large, reference population, e.g. late maturing soybean adapted to a geographical belt of USA. To simplify the description, however, inbreds are taken as an example.

NCI: two inbred lines are crossed to produce F_2, and then some individuals are randomly selected from the F_2 population as males to intermate with other randomly selected females. The offspring derived from this intermating will be used in genetic analysis. The design can be described as below:

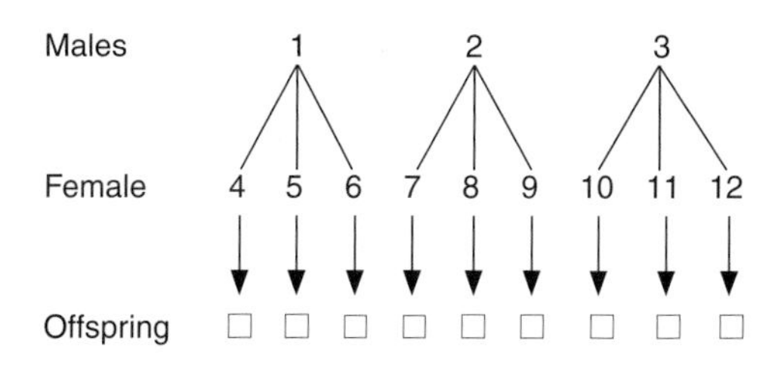

NCII: n parental lines are divided into two groups, one as males and the other as females, to produce crosses of all possible combinations.

Cultivar	1	2	3	...	n_1
n_1+1	×	×	×	...	×
n_1+2	×	×	×	...	×
n_1+3	×	×	×	...	×
...					
n_1+n_2	×	×	×	...	×

NCIII: n individuals are selected from an F_2 population to backcross with two parents, P_1 and P_2:

F_2 individual	1	2	3	...	n
P_1	×	×	×	...	×
P_2	×	×	×	...	×

TRIPLE TESTCROSS (TTC) AND SIMPLIFIED TTC (sTTC). TTC is an extension of NCIII, where n individuals ($n > 20$) are selected from an F_2 population to backcross with both parental lines, P_1 and P_2, and the F_1 ($P_1 \times P_2$):

F_2 individual	1	2	3	...	n
P_1	×	×	×	...	×
P_2	×	×	×	...	×
F_1	×	×	×	...	×

In sTTC: n cultivars or strains ($n > 20$) are selected from the germplasm pool to cross with two cultivars or strains, P_H and P_L, which show extreme phenotypes (with the highest and lowest phenotypic values), respectively.

Strain	1	2	3	...	n
P_H	×	×	×	...	×
P_L	×	×	×	...	×

The populations derived from the above genetic mating designs have been widely used in conventional quantitative genetics to study and subsequently exploit modes of gene action determining the inheritance and expression of the target traits. The reader is referred to Hallauer and Miranda (1988) and

Mather and Jinks (1982) or sections discussing quantitative genetics in plant breeding texts for details regarding the genetic information that can be derived from the study of hybrids or families formed using each of these mating designs. Some of these designs have also been used in genetic mapping of quantitative traits.

Inbreeding populations

This type of population includes segregating populations such as F_2 and F_3 populations which are derived from selfing or sibmating an F_1 hybrid, BC populations that are derived from backcrossing the F_1 to one of the parents or advanced BC populations derived by multiple backcrossings of the F_1 to one of the parents.

Populations used in genetic studies and plant breeding can be derived from any of the mating designs discussed above. For breeding purposes, the sizes of populations that will be maintained can be much smaller than those used in genetic studies because breeders only need to retain the populations with desirable traits. For genetic studies, however, geneticists need to maintain as large a population as possible and all types of segregates including those with undesirable traits.

4.2 Doubled Haploids (DHs)

Cells or plants that contain a single complete set of chromosomes are called haploid. Haploids derived from diploids are called monoploid, while haploids derived from polyploids are called poly-haploid. Diploids produced from chromosome doubling of haploids are called doubled or double haploid (DH). The DH approach has several advantages that make it useful in genetics and plant breeding. DHs can be produced via *in vivo* and *in vitro* systems. Haploid embryos are produced *in vivo* by parthenogenesis, pseudogamy or chromosome elimination after extensive crossing. The haploid embryo is rescued, cultured and chromosome-doubling produces DHs. The *in vitro* methods include gynegenesis (ovary and flower culture) and androgenesis (anther and microspore culture). Forster *et al.* (2007) reviewed various approaches for haploid production in plants. Forster and Thomas (2004) and Szarejko and Forster (2007) reviewed the use of DHs in genetic studies and plant breeding. Recent reviews on specific crop species are available for tomato (Bal and Abak, 2007) and nutraceutical species (Ferrie, 2007).

4.2.1 Haploid production

There are several approaches to haploid production. Naturally occurring haploids have been reported in a number of species including tobacco, rice and maize. In barley, the *hap* initiator gene was reported to control haploidy and spontaneous haploids were recovered at high frequency (Hagberg and Hagberg, 1980), with up to 8% haploid offspring being recovered when a cultivar that was homozygous for the *hap* allele was used as the female parent to cross with other cultivars, but none were produced from the reciprocal cross. In maize, the indeterminate gametophyte gene (*ig*) results in a monoploid embryo either from the sperm cell or the egg cell (Kermicle, 1969). Although DHs can be recovered from such spontaneous haploids, their frequencies are usually too low for genetics and breeding purposes.

With the recognition of the importance of DHs in plant breeding, extensive efforts have been made to induce haploid embryogenesis and increase the frequency at which DHs can be recovered. The benefits of DHs have already been demonstrated in many research and breeding programmes. This progress has led to DH cultivars for commercial production and DH populations for genetics and breeding studies. In barley, over 100 cultivars have been released and similar numbers of rice and rapeseed DH cultivars have been listed (Forster and Thomas, 2004). DHs have also been used successfully in recalcitrant species such as oat (Kiviharju *et al.*, 2005) and rye (Tenhola-Roininen *et al.*, 2006).

Maluszynski *et al.* (2003) edited a manual presenting a set of protocols for the production of DH in 22 major crop plant species including four tree species. The manual contains various protocols and approaches to DH production that have been successfully used for different germplasm resources in each species. The protocols describe in detail all the steps in DH production, from donor plant growth conditions, through *in vitro* procedures, media composition and preparation to regeneration of haploid plants and methods for chromosome doubling. The manual enables the researcher to choose the most suitable method for production of DH for their particular laboratory conditions and plant materials, e.g. microspore versus anther culture, wide hybridization or gynogenesis. The manual also contains information on the organization of a DH laboratory, basic DH media and associated simple cytogenetic methods for ploidy level analysis. An excellent overview of haploid induction and the application of doubled haploids is provided for *Brassicaceae*, *Poaceae* and *Solanaceae* in *Haploids in Crop Improvement II* (*Biotechnology in Agriculture and Forestry*) edited by Palmer *et al.* (2005).

There are now five methods generally applicable to the production of haploids in plants with frequencies that are useful for genetics and breeding programmes (Palmer and Keller, 2005):

- Extensive hybridization crosses followed by chromosome elimination from one parent of a cross, usually the pollination parent.
- Gynogenesis: cultured unfertilized isolated ovules and ovaries of flower buds develop embryos from cells of the embryo sac.
- Androgenesis: cultured anthers or isolated microspores undergo embryogenesis or organogensis directly or through intermediate callus.
- Parthenogenesis: development of an embryo by pseudogamy, semigamy or apogamy.
- Inducer-based approach: haploid-inducing lines are used to produce haploids.

Chromosome or genome elimination

Haploid embryos can be produced in plants after pollination by distantly related species. In most cases, normal double fertilization takes place to form a hybrid zygote and endosperm. Chromosome or genome preferential or uniparental elimination arises as a result of certain crosses; fertilization occurs but soon afterwards the genome of one parent is preferentially eliminated. Haploids can be produced by interspecific hybridization followed by chromosome elimination. In barley, this extensive hybridization method consists of crossing cultivated barley, *Hordeum vulgare* ($2n = 2x = 14$) with the wild, diploid cross-pollinated perennial *Hordeum bulbosum* ($2n$:::: $2x = 14$). Most progeny (95%) are barley haploids, while the remainder is made up by diploid hybrids. This technique, called the *bulbosum* method, has been extensively utilized for the production of haploids in barley. Haploids can also be produced in hexaploid wheat (var. Chinese Spring) by chromosome elimination following hybridization of wheat with *H. bulbosum* (both $2x$ and $4x$). A frequency of 13.7% grain set with $2x$ *bulbosum* and 43.7% grain set with $4x$ *bulbosum* were obtained (Barclay, 1975). During formation of the embryo the chromosomes of *H. bulbosum* are eliminated. The immature embryos are cultured *in vitro* and plantlets from these monoploid embryos can be induced via an efficient chromosome doubling technique to produce fertile flowers bearing homozygous hexaploid seeds.

The production of embryos as a result of wheat × maize crosses was first reported by Zenkteler and Nitzsche (1984). Laurie and Bennett (1986) cytologically examined embryos produced via this system and found maize chromosomes to be preferentially eliminated during the first three cell divisions, leaving a haploid complement of wheat chromosomes. This method was used in wheat haploid production and applied with some success in generating genetic and mapping populations (Laurie and Reymondie, 1991). Mean frequencies of fertilization, embryo formation, embryo germination and haploid regeneration of 83, 20, 45 and 8%, respectively have been reported (Chen *et al.*, 1999). Significant differences in the percentage of embryo germination and haploid regeneration were observed among crosses suggesting that the efficacy of haploid production could be improved by selection of more responsive parents. Eighty per cent of haploid plants were doubled and had a normal seed set; however, only 6% produced viable progeny. Ultimately two DH green plants per pollinated head were obtained on average. The frequency of haploid regeneration was increased from 35 to 50% in the winter 2000 study using a pre-cold treatment of embryos.

Factors that have been reported to affect the production of haploids by the chromosome elimination approach include genotype, temperature during growth (higher temperature resulting in a higher rate of elimination), genome ratio of parental lines and others. Factors affecting the efficacy of DH production in the wheat x maize system include: (i) expertise and consistency in protocol implementation; (ii) control of temperature and light regimes for optimal plant growth and reproduction in both wheat and maize; (iii) wheat F_1 genotype differences; (iv) timing of 2,4-dichlorophenoxyacetic acid (2,4-D) treatment; and (v) growth stage at which colchicine is applied.

Compared to anther culture, the wheat × maize system (sometimes called the maize pollen method) has three advantages: less genotype-dependent response, greater efficacy and less time consuming. Based on Kisana *et al.* (1993), the maize pollen method is about two to three times more efficient than anther culture. In the study by Chen *et al.* (1999), twice as many green plants were regenerated (mean = 7.54%) using the maize pollen method than anther culture. Kisana *et al.* (1993) reported that aneuploids or gross chromosomal abnormalities were not observed and confirmed that chromosome variations were not common in wheat × maize-derived plants. They also concluded that this technique could save 4–6 weeks in obtaining the same age haploid green plants.

The cross incompatibility barrier in wheat has been successfully overcome by using maize pollen. The wheat × maize technique is currently being used as an alternative to the *bulbosum* technique and anther culture for wheat haploid production. In order to use the wheat × maize system in practical breeding programmes, further enhancement of embryo formation, germination and green-plant regeneration and doubling is needed. Some green plants will die during colchicine-induced chromosome doubling and during transplantation of the colchicine-treated seedlings to the field; therefore, the final population size may be too small to represent a sufficient number of possible genotypes to make selection effective. In addition, application of 2,4-D is crucial and without it there may be no seed set or embryo formation. Of the various methods tested, the use of spikelet culture offers a practical and versatile alternative for the production of wheat polyhaploid using wheat × maize sexual crossing (Kaushik *et al.*, 2004).

SOMATIC REDUCTION AND CHROMOSOME ELIMINATION. Cases are known where either spontaneously or due to specific treatments, the chromosome number was reduced to half in the somatic tissues, a phenomenon described as somatic reduction or reductional mitosis. Early studies include that of Swaminathan and Singh (1958), who induced a haploid branch on a watermelon by irradiation of the seed used. This must have occurred by the reduction of chromosome number in the somatic tissue through an unknown mechanism (perhaps due to spindle organizer abnormalities). Similarly, in *Sorghum vulgare*, somatic tetraploid ($2n = 4x$) cells responded to colchicine treatment and gave rise to diploid cells which took over the growing point completely thus giving rise to diploid individuals. There are also a number of other chemicals such as chloramphenicol and para-fluorophenylalanine (an amino acid analogue) which have sometimes been successfully used for production of haploids in a number of materials. Elimination of parental chromosomes has also been observed in somatically-produced wide hybrids. In these cases, the elimination tends to be irregular and incomplete, leading to asymmetric hybrids or cybrids (Liu, J.H. *et al.*, 2005).

MECHANISM OF CHROMOSOME ELIMINATION. Several hypotheses have been presented to explain uniparental chromosome elimination during hybrid embryo development in plants: for example, differences in timing of essential mitotic processes due to asynchronous cell cycles or asynchrony in nucleoprotein synthesis leading to a loss of the most retarded chromosomes. Other hypotheses propose the formation of multipolar spindles, spatial separation of genomes during interphase and metaphase, parent-specific inactivation of centromeres and by analogy with the host-restriction and modification systems of bacteria, degradation of alien chromosomes by host-specific nuclease activity. Gernand *et al.* (2005) provide evidence for a novel chromosome elimination pathway in wheat × pearl millet hybrids that involves the formation of nuclear extrusions during interphase in addition to post-mitotically formed micronuclei. They found that the chromatin structure of nuclei and micronuclei was different and heterochromatinization and DNA fragmentation of micronucleated pearl millet chromatin was the final step during haploidization.

The mechanism of chromosome elimination in *Hordeum* hybrids was studied by Subrahmanyam and Kasha (1975) and Bennett *et al.* (1976) and the following conclusions were drawn: (i) normal double fertilization occurs in interspecific crosses as confirmed by cytological study; and (ii) after fertilization there is a gradual and selective elimination of *H. bulbosum* chromosomes from nuclei of endosperm as well as embryo cells so that eventually haploid embryos are produced. A sudden shortage of proteins in the developing embryo and endosperm and the better ability of *vulgare* chromosomes to form spindle attachments relative to *bulbosum* chromosomes, may be responsible for elimination of the *bulbosum* chromosomes. Other possible causes such as differences in mitotic cycle, congression during mitosis, etc. were ruled out by the authors.

It has also been demonstrated that the elimination of *bulbosum* chromosomes is under genetic control (Subrahmanyam and Kasha, 1975). The above-mentioned authors used primary trisomics and monotelotrisomics in crosses with tetraploid *H. bulbosum* and concluded that both arms of chromosome 2 and the short arm of chromosome 3 of *H. vulgare* are responsible for chromosome elimination, although their effect may be neutralized or offset if a sufficient dose of *bulbosum* chromosomes is available.

Ovary culture or gynogenesis

Ovary culture involves production of a haploid individual by culture of unfertilized ovaries to obtain haploid plants from egg cells or other haploid cells of the embryo; the process is known as gynogenesis. Under the appropriate culture conditions the unfertilized cell of the embryo sac develops into an embryo by as yet unknown mechanisms. Haploid plants generally originate from egg cells in most species (*in vitro* parthenogenesis) but in some species, e.g. rice, they arise chiefly from the synergids; in at least *Allium tuberosum* even antipodal cells produce haploid plants (*in vitro* apogamic) (Mukhambetzhanov, 1997).

Gynogenesis may occur either via embryogenesis or plantlet regeneration from callus. In rice 2-methyl-4-chlorophenoxyacetic acid (MCPA) generally leads to a small amount of protocorm-like callus formation from which shoots and roots regenerate, while picloram promotes embryo regeneration. In contrast, sugarbeet usually shows embryo development while in sunflower embryos regenerate following a callus phase. In general, regeneration from a callus phase appears, at least for the present, to be easier than direct embryogenesis.

Generally, gynogenesis has two or more stages and each stage may have distinct requirements. In rice, two stages, i.e. induction and regeneration, are recognized. During induction, ovaries are floated on a liquid medium containing low auxin levels and kept in the dark, while for regeneration they are transferred on to an agar medium containing a higher auxin concentration and incubated in the light.

Depending on the species, unfertilized ovules, ovaries or flower buds can be cultured. In some members of the *Chenopodiaceae*, *Liliaceae* and *Cucurbitaceae*, gynogenesis is the main route to DH production (Palmer and Keller, 2005). Even where anther or microscope culture is successful, gynogenetic haploids have been produced, e.g. in barley, maize, rice and wheat.

San Noeum (1976) was the first to demonstrate that gynogenesis can be induced under *in vitro* conditions. She obtained gynogenic haploids using an ovary culture of *H. vulgare*. Subsequently, success has been obtained with many species, e.g. wheat, rice, maize, tobacco, petunia, gerbera, sunflower, sugarbeets, onions, rubber, etc. About 0.2–6% of the cultured ovaries show gynogenesis and one or two plantlets, rarely up to eight, originate from each ovary.

Embryogenic frequency is low in many cases, but relatively high frequencies have been reported in some cases (Alan *et al.*, 2003; Martinez, 2003). The rate of success varies considerably with species and is markedly influenced by explant genotype so that some cultivars do not respond at all. In rice, *japonica* genotypes are far more responsive than *indica* cultivars. In most cases, the optimum stage for ovule culture is the nearly mature embryo sac, but in rice ovaries at the free nuclear embryo sac stage are the most responsive.

The culture response is still genotype dependent (Alan *et al.*, 2003; Bohanec *et al.*, 2003). Generally, for culture of whole flowers, ovary and ovules attached to placenta respond better, but in gerbera and sunflower isolated ovules give a better response. Cold pretreatment (24–48 h at 4°C in sunflower and 24 h at 7°C in rice) of the inflorescence before ovary culture enhances gynogenesis.

The composition of the culture medium and stage of embryo sac development are important considerations for successful culture (Keller and Korzun, 1996). Growth regulators are crucial in gynogenesis and at higher levels they may induce callusing of somatic tissues and even suppress gynogenesis. Growth regulator requirements seem to depend on species. For example, in sunflower growth regulator-free medium is the best and even a low level of MCPA induces somatic calli and somatic embryos. But in rice, 0.125–0.5 mg l^{-1} MCPA is optimal for gynogenesis. The sucrose level also appears to be critical; in sunflower 12% sucrose leads to gynogenic embryo production while at lower levels somatic calli and somatic embryos were also produced. Ovaries are generally cultured in the light but in some species at least, e.g. sunflower and rice, incubation in the dark favours gynogenesis and minimizes somatic callusing; in rice light may lead to the degeneration of gynogenic pro-embryos.

Ovary culture has two main limitations: (i) it is not successful in all species; and (ii) the frequency of responding ovaries and the number of plantlets per ovary is usually low. Therefore, anther culture is preferred over ovary culture; only in those cases where anther culture fails, e.g. sugarbeet and for male sterile lines, ovary culture assumes significance.

Anther culture or androgenesis

Anther culture or androgenesis is a process by which a haploid individual develops from a pollen grain. Anther culture is often the method of choice for DH production in crop plants (Sopory and Munshi, 1996). Good aseptic techniques are required but the methods are generally simple and applicable to a wide range of crops (Maluszynski *et al.*, 2003). In general, haploid plants are generated *in vitro* from the microspores contained in the anther and require chromosome doubling treatments. The number of chromosomes in haploid plants can be doubled either naturally or by colchicine treatment.

The process involved in anther culture is poorly understood. Investigations have been hampered by the presence of the sporophytic anther wall that presents direct access to the microspores contained within. This has become an important issue because although many species respond to anther culture, responsive genotypes can be a limiting factor thus making it necessary to study, understand and manipulate microspore embryogenesis in order to develop genotype-independent methods (Forster *et al.*, 2007). Many factors influence the production of anther-culture-derived plants including the physiological status of the donor plants, pre-treatment of anthers, developmental stage of the pollen, components in the medium and culture conditions such as light, temperature and humidity. The constraints associated with this approach are the selective response of genotypes to the anther culture process or medium, the high rate of albino formation and somaclonal variation. These factors have been discussed by Taji *et al.* (2002) and are summarized in the following discussion.

The genotype of the donor plant plays a significant role in determining the frequency of pollen plant production. There are genotypes extremely recalcitrant to anther culture. In rice, for example, *japonica* cultivars are much easier to culture from anthers than *indica* cultivars. Genotype-dependency is a major constraint that affects its wide application.

The culture medium plays a vital role since the requirements vary with the genotype and probably the age of the anther as well as the conditions under which donor plants are grown. The medium should contain the correct amounts and proportions of inorganic nutrients to satisfy the nutritional as well as physiological needs of the many plant cells in culture. Sucrose is considered to be the most effective carbohydrate source which cannot be substituted by other disaccharides. The concentration of sucrose also plays an important role in the induction of pollen plants. Activated charcoal is also added to the culture medium.

In addition to basal salts and vitamins, hormones in the medium are critical factors for embryo or callus formation. Cytokinins (e.g. kinetin) are necessary for induction of pollen embryos in many species of *Solanaceae*, except tobacco. Auxins, in particular 2,4-D, greatly promote the formation of pollen callus in cereals. For regeneration of plants from pollen calli, a cytokinin and lower concentration of auxin are often necessary.

Certain organic supplements added to the culture medium often enhance the growth of anther culture. Some of these include the hydrolysed products of proteins such as casein (found in milk), nucleic acids and others. Coconut milk obtained from tender coconuts is often added to tissue culture media as it contains a complex mixture of nucleic acids, sugars, growth hormones and some vitamins.

The physiological state of the parent plant plays a role in haploid production. In various plant species it has been shown that the frequency of androgenesis is higher in anthers harvested at the beginning of the flowering period and declines with plant age. This may be due to deterioration in the general condition of the plants, especially during seed set. The lower frequency of induction of haploids in anthers taken from older plants may also be associated with a decline in pollen viability. Seasonal variations, physical treatment and application of hormones and salts to the plant also alter its physiological status which is reflected in changes in the anther response to culture.

Temperature and light are two physical factors which play an important role in the culture of anthers. Higher temperatures (30°C) yield better results. Temperature shocks also enhance the induction frequency of microspore androgenesis. Frequency of haploid formation and growth of plantlets are generally better in the light. Certain physical and chemical treatments given to flower buds or anthers prior to culture can be highly conducive to the development of pollen into plants. The most significant is cold treatment.

The developmental stage of pollen greatly influences the fate of the microspore. Androgenesis occurs when a microspore or pollen is induced to shift from a gametophytic pathway to a sporophytic pathway of embryo formation. Anthers of some species (*Datura*, tobacco) give the best response if pollen is cultured at first mitosis or later stages (postmitotic), whereas in most others (barley, wheat, rice) anthers are most productive when cultured at the uninucleate microspore stage (premitotic). Anthers at a very young stage (containing microspore mother cells *m* tetrads) or a late stage (containing binucleate, starch-filled pollen) of development are generally ineffective, albeit that some exceptions are known.

Barley and rice are considered to be model cereal crops for androgenesis. The application of barley anther culture protocols to other cereals such as wheat yielded a low frequency of green plants. Although a high frequency of green plants is produced for most barley crosses, androgenesis still poses some problems that need to be addressed. There are barley genotypes which are extremely recalcitrant to microspore division and/or with a high rate of albinism. The rate of embryogenesis is still low and poorly-developed embryos are formed very frequently. New methods are needed that reduce the cost of DH production and are effective for all genotypes.

Future objectives in plant androgenesis include the development of efficient androgenesis protocols for a wide range of genotypes, a better understanding of the biological processes involved in the stress pre-treatment, the study of the influence of different micronutrients on the induction of gametic embryogenesis and possible gametophytic selection. Identification of genetic loci associated with the anther culture response process will facilitate the understanding of the mechanisms underlying androgenesis. Identification and localization of molecular markers linked to the yield of green plants per anther and the evaluation of their potential use for the prediction of the anther culture response of genotypes will also help to optimize the production of DHs.

Semigamy

Semigamy is a form of parthenogenesis and occurs when the nucleus of the egg cell and the generative nucleus of the germinated pollen grain divide independently, resulting in a haploid chimera (a plant whose tissues are of two different genotypes). Semigamy is a type of facultative apomixis in which the male sperm nucleus does not fuse with the egg nucleus after penetrating the egg in the embryo sac. Subsequent development can give rise to an embryo containing haploid chimaeral tissues of paternal and maternal origins. In cotton, the semigametic phenomenon was first reported by Turcotte and Feaster (1963), who developed the Pima line 57-4 that produced haploid seeds at a high frequency. Currently semigamy is the only feasible means for production of haploids in cotton (Zhang and Stewart, 2004).

There are many examples of DH lines developed from cultivars and intra- and interspecific hybrids between upland cotton (*Gossypium hirsutum* L.) and American Pima cotton (*Gossypium barbadense* L.) using semigamy. The semigametic trait has also been transferred into different cotton cytoplasms to facilitate rapid replacement of nuclei. Stelly *et al.* (1988) proposed a scheme called hybrid elimination and haploid production system using a cotton strain with semigamy (*Se*), lethal gene (Le_2^{dav}), virescent (v_7) and male sterility or glandless (gl_2gl_3).

Semigametic lines can produce 30–60% haploids when self-pollinated and about 0.7–1.0% androgenic haploids when used as female parents in crosses with normal non-semigametic cottons (Turcotte and Feaster, 1967). A unique feature of semigamy is that the inheritance of the gene is conveyed by both male and female gametes but expression of the trait in terms of haploid production occurs only in the female parent. As a consequence, for example, in reciprocal crosses between *SeSe* and *sese* parents, haploids will be produced only when *SeSe* or *Sese* is the female parent.

The results reported by Zhang and Stewart (2004) verified that semigamy in cotton is controlled by one gene, previously designated *Se*. The gene functions sporophytically and gametophytically resulting in an incomplete dominance mode of action. Consistent with the difference between the two parental isogenic lines, semigametic $F_{2.3}$ lines had significantly lower chlorophyll content than non-semigametic $F_{2.3}$ lines, an observation that was confirmed by a significant association between haploid production and chlorophyll content. The *Se* gene and the gene for reduced chlorophyll content could be either the same or closely linked.

Inducer-based approach

Haploid inducing lines have been used in maize to produce haploids by development of the unfertilized egg cells (Eder and

Chalyk, 2002). A haploid induction rate of up to 2.3% was detected by Coe (1959) in crosses with the inbred line Stock 6. A higher rate (about 6%) was later obtained by Sarker *et al.* (1994) and Shatskaya *et al.* (1994) in progenies of crosses between Stock 6 and Indian and Russian germplasm, respectively. Inducer lines are now available with haploid seed induction rates of 8–12% in temperate maize germplasm (Melchinger *et al.*, 2005; Röber *et al.*, 2005).

Segregation studies (Lashermes and Beckert, 1988; Deimling *et al.*, 1997) and quantitative trait loci (QTL) analysis (Röber, 1999) demonstrated that *in vivo* haploid induction in maize is a quantitative trait under the control of an unknown large number of loci. Individual QTL explained only small parts of the genetic variation.

Compared with other methods of DH production such as anther culture, the inducer-based approach is rather efficient, less dependent on the genotype and can be practised in almost every maize breeding programme without access to expensive laboratory facilities (Röber *et al.*, 2005; http://www.uni-hohenheim.de/~ipspwww/350a/linien/indexl.html).

Requirements for *in vivo* DH production in practical breeding include: (i) availability of inducer genetic stocks; (ii) high induction rate; (iii) the inducer is a good pollinator; (iv) reproducibility with reasonable seed quantities; (v) availability of a marker system that is independent of the genetic background of the female and of environmental effects and can be used for effective and unambiguous identification of haploid kernels; and (vi) availability of an artificial chromosome doubling system with high doubling rates that is safe, simple and cost-effective.

Since the late 1990s, these requirements have been partially met in maize with: (i) inducer lines (e.g. RWS and UH400 developed at the University of Hohenheim) with improved induction rates of 10% or higher; (ii) a combination of two dominant markers (anthocyanin colour of endosperm and embryo for identification of haploids and anthocyanin coloration of stalk for identification of false positives in the field); and (iii) improved chromosome doubling systems using colchicine that gave a doubling rate of greater than 10%.

A scheme to show *in vivo* haploid induction includes the following steps:

- Creating new variation by intercrossing with selected lines.
- *In-vivo* haploid induction in generation F_1.
- Chromosome doubling of haploid seedlings:
  - selection of haploid kernels;
  - germination of kernels;
  - cutting of coleoptile;
  - doubling procedure: treatment of seedlings with colchicine;
  - planting of treated seedlings in greenhouse;
  - transplanting DH plants at the three-leaf stage to the field and selfing (generation D_0); and
  - formation of testcross hybrids.
- Evaluation of testcrosses in multi-environment yield trials (two stages).

4.2.2 Diploidization of haploid plants

As described above, haploids can be produced through various approaches. Haploid plants may grow normally under *in vitro* or greenhouse conditions up to the flowering stage, but viable gametes are not formed due to the absence of one set of homologous chromosomes and consequently, there is no seed set.

The only mechanism for perpetuating the haploids is by duplicating the chromosome complement in order to obtain homozygous diploids. In pollen-derived plants duplication of chromosomes may occur spontaneously in cultures. However, the spontaneous chromosome doubling rate of haploids is usually low. In maize, for example, the rate ranges from 0 to 10% (Chase, 1969; Beckert, 1994; Deimling *et al.*, 1997; Kato, 2002). Therefore, it is necessary to diploidize the haploids by chemical means. Thus, artificial chromosome doubling (diploidization) is necessary for the efficient large-scale use of haploid plants.

Chromosome doubling is thought to occur by one or more of four mechanisms, namely endomitosis, endoreduplication, C-mitosis or nuclear fusion (Jensen, 1974; Kasha, 2005). Endomitosis is described as chromosome multiplication and separation but failure of the spindle leads to one restitution nucleus with the chromosome number doubled. It has also been called 'nuclear restitution'. Endoreduplication is duplication of the chromatids without their separation and leads to diplo-chromosomes or to polytene chromosomes if many replications occur. Endoreplication is a common feature in specialized plant cells where cells become differentiated or enlarged in cells that are very active in metabolite production. C-mitosis is a specific form of endomitosis where, under the influence of colchicine, the centromeres do not initially separate during metaphase while chromosome arms or chromatids do separate. Nuclear fusion occurs when two or more nuclei divide synchronously and develop a common spindle. Thus, two or more nuclei could result with doubled, polyploid or aneuploid chromosome numbers.

A simple procedure designed to achieve diploidization involves immersion of very young haploids in a filter-sterilized solution of colchicine (0.4%) for 2–4 days, followed by their transfer to the culture medium for further growth. In maize, the highest doubling rates are achievable by immersing 2–3-day-old seedlings in a colchicine solution as suggested by Gayen *et al.* (1994). Using an improved version of this method, Deimling *et al.* (1997) obtained doubling rates of up to 63%. The studies of Eder and Chalyk (2002) using genetically broader materials yielded an average doubling rate of 27%. Optimized methods for colchicine treatment of haploid seedlings yield average success rates of about 10% fertile diploid plants with satisfactory seed set (Mannschreck, 2004). In this procedure chromosome or gene instabilities are minimal compared to other methods of colchicine or chemical treatment.

4.2.3 Evaluation of DH lines

Randomness

Systems used to produce DH lines should not have preference to specific gametes, which means each gamete should have the same probability of developing into a haploid. Chromosome elimination using the *bulbosum* approach in barley is usually a random process and there is no significant segregation distortion associated with it. Park *et al.* (1976) and Choo *et al.* (1982) did not find any gamete preference associated with this approach by comparing the DH and single seed descent (SSD) populations. In rice, however, especially for the DH populations derived from anther culture of distant crosses, distorted segregations were found for isozymes, restriction fragment length polymorphisms (RFLPs) and morphological traits. As a result, the segregation of two types of homozygotes deviated from the 1:1 ratio for many single gene loci (Chen, Y. *et al.*, 1997).

Stability

Theoretically, DH lines have two properties: complete homogeneity and homozygosity within lines. Except for the variation that might be produced during anther culture or other generation processes, DH lines should be genetically stable and the mutation rate that can occur in DH lines should be in the same range as that of other true-breeding cultivars.

There are some reports that identified somaclonal variation associated with anther culture-derived DH lines (Chen, Y. *et al.*, 1997) and theories that account for the origin of somaclonal variation (Taji *et al.*, 2002). The variation within a DH line can be divided into two categories: (i) variation originating from the genetic heterogeneity of somatic cells of the source (haploid) plant; (ii) variation due to structural alterations of DNA and chromosomes caused by tissue culture.

Somaclonal variation as an important cause for instability of DH lines and is not restricted to, but is particularly common in, plants regenerated from callus. The variations can be genotypic or phenotypic which in the latter case can be either genetic or epigenetic in origin. Typical genetic alterations are: changes in chromosome numbers (polyploidy and aneuploidy), chromosome structure (translocations, deletions and duplications) and DNA sequence (base mutations).

GENETIC VARIATION ARISING FROM SOURCE PLANTS. The source plants used to initiate cultures are likely to be heterogeneous with respect to the state of differentiation, ploidy level and age. These explant-related factors will affect the genetic make-up of the cells produced in the culture and thus the callus arising from such a group of cells with diverse genetic make-up will inevitably lead to a mixed population of cells. Depending on the cell types from which the plants are originated, those regenerated from such a genetically mosaic callus will undoubtedly be of different genetic make-up. Taji *et al.* (2002) indicated that such genetic mosaicness seems to occur commonly in polyploid plants rather than in diploids or haploids.

GENETIC VARIATION ARISING DURING CULTURE. Although a significant degree of genetic variability can be traced to the genetically heterogeneous cell types of explant at least in polyploid species, there is substantial evidence to indicate that much of the variability observed in generated plants stems from the culture process itself. Aneuploids, polyploids or cells with structurally altered chromosomes may arise in culture. Many differentiated cells when induced to divide in culture, undergo endoduplication of chromosomes resulting in the production of tetraploid or octaploid cells with distinct phenotypes.

Various phenomena have been observed in tissue culture of various plant species which explain the production of cells with unusual ploidy levels (Bhojwani, 1990). Occurrence of multi-polar spindles due to failure of spindle formation during cell division is one of the contributing factors. Absence of spindle formation during mitosis results in the appearance of cells with doubled chromosome number while the formation of multi-polar spindles on chromosomes lagging at anaphase cause the development of cell lines with haploid, triploid or other uneven ploidy status.

Many studies have indicated that cryptic structural modification of individual chromosomes is more likely to cause somaclonal variation than modification induced by ploidy changes in many tissue-cultured plants. Chromosomal changes occurring during tissue culture include transposition of mobile genetic elements (transposons), chromosome breakage and repositioning of chromosome segments.

As summarized by Taji *et al.* (2002), several mechanisms have been proposed to explain the genetic variability that occurs in tissue culture. The most possible causes are:

1. Reduced regulatory control of mitotic events in culture: the ploidy status of plants generated from callus, cell suspension or protoplast cultures of certain species differ significantly despite the fact that the cultures originate from a highly homogenous genetic background. This indicates a lack of tight regulation of cell-cycle-related controls during cell proliferation in culture.
2. Use of growth regulators: plant growth regulators, particularly synthetic auxins such as 2,4-D, are considered to be the major cause of genetic variability in culture. For example, cytokinins at low concentrations have been shown to reduce the range of ploidy in culture while low levels of both auxins and cytokinins appear to preferentially activate the division of cytologically stable meristematic cells enabling the regeneration of genetically uniform plantlets.
3. Medium components: some of the mineral nutrients influence the establishment of genetic variability in culture. For instance, by altering the levels of phosphate and nitrogen as well as the form of nitrogen in the medium, the genetic composition (ploidy level) of the cultured cells can be controlled to a considerable extent. A marked increase in chromosome breakage has been observed in plant cell cultures grown with different levels of magnesium or manganese.

4. Culture conditions: some culture conditions, such as incubation temperatures above 35°C and long duration of culture, have been implicated in inducing genetic variability in regenerated plants.
5. Inherited genomic instability: molecular studies indicate the existence of certain regions of genome that are more susceptible to tissue-culture-induced structural alternations, although the reason for the increased susceptibility of these genomic loci known as 'hot spots' is not fully understood.

CAUSES OF EPIGENETIC VARIATION IN TISSUE CULTURE. Any culture-induced changes which are stable but not heritable have frequently been considered as epigenetic variation. However, a greater understanding of genetic and epigenetic alterations in tissue culture in the recent past has led to a clear distinction between these two types of variation. For instance, genetic mutations occur randomly and at a much lower rate than epigenetic variations. Genetic changes are usually stable and heritable. Epigenetic variation may also lead to stable traits; however, reversal can occur at high rates under non-selective conditions. Epigenetic traits are often transmitted through mitosis in a stable manner but rarely through meiosis and the level of induction of epigenetic traits is directly related to the selection pressure experienced by the cells. Epigenetic changes are generally assumed to reflect alteration in expression rather than in the information content of genes.

As Taji *et al.* (2002) summarized, the epigenetic variation observed in cultured cells or regenerated plants is mainly due to three cellular events: (i) gene amplification; (ii) DNA methylation; and (iii) increased activity of transposable elements. In plants, nearly 25% of the genome can be methylated at cytosine residues but the significance of this cytosine methylation is not apparent. It has been suggested that methylation (and demethylation) of DNA is one of the ways of controlling transcriptional activity and that this process can be affected by the tissue culture process. The non-heritable genetic variability observed in many tissue culture systems could thus be attributed to tissue-culture-induced methylation or demethylation of DNA. The activity of transposons and retrotransposons induced by tissue culture could also be responsible for some of the genetic and epigenetic variability observed in culture.

4.2.4 Quantitative genetics of DHs

DH lines that are derived randomly from an array of gametes produced by F_1 plants are very useful in quantitative genetics. Compared with diploid genetic models for populations such as F_2, F_3 or BC, there are no dominance or dominance-related epistasis effects involved in the genetic model of DH populations. As a result, additive, additive-related epsitasis and linkage effects can be investigated properly. As a permanent population, DH lines can be replicated as many times as desired across different environments, seasons and laboratories, providing endless genetic material for phenotyping and genotyping particularly for understanding the genotype-by-environment interaction. In DH populations, the additive component of genetic variance is larger than that of diploid populations such as F_2 and BC. Choo *et al.* (1985) discussed in detail the quantitative genetics associated with DH populations, including detection of epistasis, estimation of genetic variance components, linkage test, estimation of gene numbers, genetic mapping of polygenes and tests of genetic models and hypotheses. Röber *et al.* (2005) compared the expected gain from selection for DH lines and other populations and implications of epistatic effects, which is briefly described here.

Expected gain from selection

As is well known from quantitative genetics (see e.g. Falconer and Mackay (1996) and also Chapter 1), the expected gain from selection can be described by $\Delta G = i\, h_x r_G\, \sigma_y$, where i is the selection intensity, h_x the square root of the heritability of the selection criterion, r_G the genetic correlation between selection criterion and gain criterion and σ_y the standard deviation of the gain criterion. In long-term breeding programmes, the decisive gain criterion for evaluating selection progress in hybrid breeding is the general combining ability (GCA) of the improved lines. At the beginning of a breeding cycle the test units are the DH lines per se and later on in the cycle their testcrosses.

Strong selection (large *i*) leads to a small effective population size and consequently to a loss of genetic variance due to random drift. To keep this loss within certain limits, a minimum number of lines should be recombined after each breeding cycle. This number depends on the inbreeding coefficient (*F*) of the candidate lines. The number should be (2*F*) times larger for inbred lines than for non-inbred genotypes. Assuming that S_2 lines (F = 0.75) are recombined in conventional breeding, the number of DH lines (F = 1) would have to be increased 1:0.75 = 1.33-fold to preserve an equivalent level of genetic variation. This means that the selection intensity must be reduced accordingly when using DH lines.

In contrast to the selection intensity, h_x and r_G increase when using DH lines. This increase is particularly large in the first testcross stage. Neglecting epistasis, the GCA variance of inbred lines is equal to 1/2 $F\,\sigma_A^2$ (Falconer and Mackay, 1996), where σ_A^2 is the additive variance of the base population. Thus the GCA variance of DH lines is 1:0.75 = 1.33 times larger than that of S_2 lines. This leads to better differentiation among the testcrosses and consequently to higher heritability. Seitz (2005) compared three sets of S_2 and S_3 lines each with DH lines derived from the same crosses and evaluated the same testers in the same environments. On average, the estimated genetic testcross variances for grain yield (bu. acre^{-1}) amounted to 50, 94 and 124 for S_2, S_3 and DH lines, respectively.

The genetic correlation between selection and gain criterion (r_G) also increases with the degree to which the tested lines have been inbred. For example, the correlation between S_t lines and their homozygous progenies for GCA is equal to $\sqrt{F_t}$ whereas for DH lines this correlation is 1. Thus compared with S_2 lines, the correlation of DH lines is $1{:}\sqrt{0.75} = 1.15$ times stronger.

Implications of epistatic effects

Epistatic gene action may positively or negatively affect hybrid performance (Lamkey and Edwards, 1999). In most cases, epistatic effects have been reported to cause a decrease in the testcross performance of segregating generations (Lamkey *et al.*, 1995) or to penalize three-way and double crosses compared to their non-parental single crosses (Sprague *et al.*, 1962; Melchinger *et al.*, 1986). These effects are commonly referred to as 'recombinational loss' and may be explained by a disruption during meiosis of co-adapted gene arrangements assorted by prior natural and artificial selection. Marker-based analyses of QTL partially corroborate this hypothesis (Stuber, 1999). To avoid recombinational loss and still offer a chance to select for new positive interactions, a balance between recombination and fixation of gene arrangements is needed. The DH-line approach might offer the method for achieving this goal as homozygosity can be reached in one cycle of recombination when F_1 is used for DH development or in different cycles when segregating populations of different generations are used.

4.2.5 Applications of DH populations in genomics

In genetics, DHs may serve to recover recessives. Using DHs, linkage data can be obtained directly by sampling gametes as monoploids. DHs are ideal for the study of mutation frequencies and spectra. As DHs represent homozygous, immortal and true breeding lines, they can be repeatedly phenotyped and genotyped so phenotypic and genotypic information can be accumulated over years and across laboratories. In genomics, DHs are therefore ideal for studying complex traits that are quantitatively inherited which may require replicated trials over many years and locations for accurate phenotyping.

DH populations are desirable genetic materials for genetic mapping including the construction of genetic linkage maps and gene tagging using genetic markers. They can be produced relatively rapidly, requiring 1–1.5 years to become established after the initial cross and they provide an ongoing population that can be used indefinitely for mapping. QTL analysis is facilitated by using DH mapping populations and the homozygosity of DHs enables accurate phenotyping by replicate trials at multiple sites (Forster and Thomas, 2004). In addition, in DH populations, dominant markers are as efficient as co-dominant markers because linkage statistics are estimated with equal efficiency (Knapp *et al.*, 1995). DHs can be also used to increase the expression level of a transgene (Beaujean *et al.*, 1998).

Only the application of DHs for construction of genetic linkage maps will be discussed here. Assuming that the two parental lines used for production of DH populations have the genotypes P_1 (*AABB*) and P_2 (*aabb*), their F_1 will produce four types of gametes, *AB*, *Ab*, *aB* and *ab*. As a result of 'single-sex production', these gametes produce four types of haploids and by chromosome doubling will produce four types of DHs: *AABB*, *AAbb*, *aaBB* and *aabb*. When *A-a* and *B-b* are independent (not linked), the four types of DH lines are present in identical proportions: 25%. The segregation of two loci in DH population is shown in Fig. 4.1, among which *AB* and *ab* are parental gametes and *Ab* and *aB* are recombinant gametes while *AABB* and *aabb* are genotypes for parental lines and *AAbb* and *aaBB* are genotypes for recombinants. It is expected that for each molecular marker there are two parental genotypes in DH populations and in any DH line only one of the parental bands revealed by markers will show up.

A general step before map construction and gene mapping is to evaluate the DH population. In rice, 66 DH lines were derived from the F_1 between *indica* Apura and upland *japonica* Irat 177 by anther culture. Heterozygosity was found for some loci with two parental bands while non-parental alleles (or new alleles) were found for other loci. The limitations of using this DH population in genetic linkage mapping do not result from the partial heterozygosity or new alleles but from the low RFLP polymorphism identified between the parents (S.R. McCouch, Cornell University, personal communication). Only 40% of the 100 tested RFLP markers detected polymorphism. Of the markers that had been mapped on to an F_2 population, IR34583/Bulu, only 55% were polymorphic between Apura and Irat 177. However, a relatively saturated molecular map can be established if other types of molecular markers such as single sequence repeats (SSRs) or single nucleotide polymorphisms (SNPs) are used.

In barley, a DH population consisting of 113 lines was derived by anther culture from the F_1 hybrid between two spring barley cultivars, Prottor and Nudinka (Heun *et al.*,

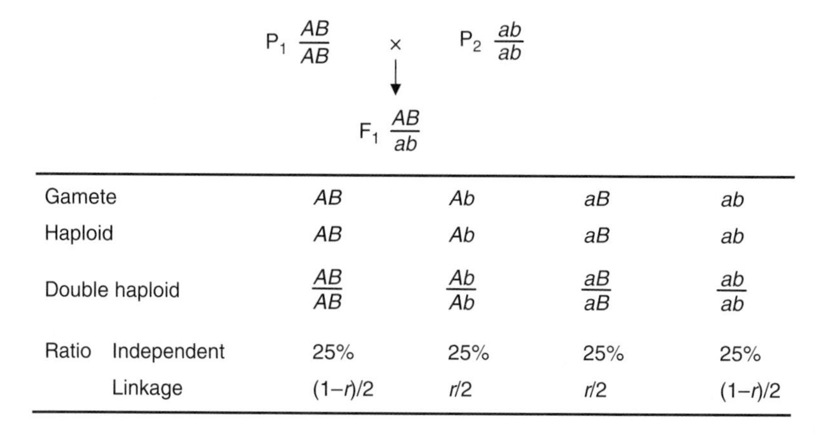

Gamete		*AB*	*Ab*	*aB*	*ab*
Haploid		*AB*	*Ab*	*aB*	*ab*
Double haploid		*AB*/*AB*	*Ab*/*Ab*	*aB*/*aB*	*ab*/*ab*
Ratio	Independent	25%	25%	25%	25%
	Linkage	$(1-r)/2$	$r/2$	$r/2$	$(1-r)/2$

Fig. 4.1. Segregation of two genetic loci in a DH population.

1991). A genetic map was constructed using 55 RFLP markers and two known genes and is the first complete molecular map to be constructed using DH populations in crops. Since then, many DH populations have been developed using the different approaches described above and have been used for map construction and genetic mapping.

4.2.6 Application of DHs in plant breeding

The benefits of DHs in plant breeding have been widely reviewed; readers should refer to Forster and Thomas (2004), Forster *et al.* (2007), and the five volumes on In Vitro *Haploid Production in Higher Plants* edited by Jain *et al.* (1996–1997).

Application of DHs in plant breeding can be described by comparison of the time required to obtain fixed inbreds relative to inbreeding, starting from a heterozygote:

Selfing of a heterozygote	Haploids of a heterozygote
Gametes: 1/2 *A* + 1/2 *a*	Gametes: 1/2 *A* + 1/2 *a*
F_2 1/4 *AA*, 1/2 *Aa*, 1/4 *aa*	chromosome doubling
F_3 1/4 *Aa*	1/2 *AA* + 1/2 *aa*
F_4 1/8 *Aa*	
F_5 1/16 *Aa*	
F_6 1/32 *Aa*	
~ 1/2 *AA* + 1/2 *aa*	

Apparently, the DH approach has a time reduction of three to four generations compared to inbreeding-based breeding. The DH approach features many logistical advantages simplifying breeding to a large extent and enabling evaluation of genetically fixed hybrid components from the very beginning of the selection process. Depending on the material, the costs and the breeding scheme adopted, the DH approach can reduce the time for development and commercialization of new inbred lines and lead to a higher expected genetic gain per unit of time.

As outlined above, DH lines extracted from a heterozygote or a segregating population represent 'immortalized', reproducible gametes that can be immediately evaluated to assess their true breeding potential for target traits. They have the following advantages and clear beneficial applications (Melchinger *et al.*, 2005; Röber *et al.*, 2005; Longin *et al.*, 2006; W. Schipprack, University of Hohenheim, personal communication):

- providing the quickest possible route to complete homozygosity;
- giving an immediate product of stable recombinants from species crosses;
- no masking effects because of the high homogeneity attained in the first generation of DH populations;
- increased performance per se due to selection pressure in the haploid phase and/or during the first generation of DHs;
- complete genetic variance accessible from the very beginning of the selection process;
- easy integration of line/hybrid development with recurrent selection;
- reduced efforts in the nursery after the first multiplication of DH lines compared to a conventional breeding nursery;
- maximum genetic variance in line per se and testcross trials;
- high reproducibility of early-selection results;
- high efficiency in stacking specific targeted genes in homozygous lines; and
- simplified logistics for seed exchange between main and off-season programmes since each line is fixed and can be represented by a single plant.

DHs have been used in plant breeding programmes to produce homozygous genotypes in a number of important species, e.g. tobacco (*Nicotiana tabacum* L.), wheat, barley, canola (*Brassica napus* L.), rice and maize (Maluszynski *et al.*, 2003), but only rarely in triticale, oat, rye and others. Research in crops such as rice, wheat and maize has shown that significant progress in haploid technology is attainable given an intensive research effort. Well-established methods in these crops have allowed major parts or whole breeding programmes to be based on DH production. Oat, triticale, wild barley, potato and cabbage are examples of crops where DH technologies are less advanced but in which hundreds of

DHs may still be obtained (Tuvesson *et al.*, 2007). In other crops, including some vegetable species and forage and turf grass species, DH methods are being developed, but applications in crop improvement are rare. The DH approach has yet to be exploited in leguminous species, predominantly due to their cultivation in developing countries and consequent paucity of research funding. Difficulties have also been posed by the small anther size and relatively low number of microspores per anther in legume crops (Croser *et al.*, 2006).

The DH technique offers an efficient tool for extracting individual gametes from heterozygous materials and transforming them into homozygous lines that can be reproduced *ad libitum* by selfing. DHs extracted from a heterogeneous population, e.g. landraces, represent 'immortalized', reproducible gametes that can be immediately evaluated to assess their true breeding potential for target traits. They can also serve as source material for breeding programmes of hybrids and synthetics. Furthermore, DH lines may be used for long-term conservation of heterogeneous germplasm resources such as landraces without the risk of genetic drift and other changes in gene frequencies, as well as for in-depth characterization of the breeding potential of each heterogeneous germplasm collection because each of the extracted DH lines can be evaluated in replicated trials in diverse environments.

With some DH methods, only a tiny fraction of the haploid seedlings will germinate and survive to the adult stage due to the uncovered genetic load and the stress in plant development exerted by colchicine treatment for chromosome doubling. Nevertheless, because the DH technique is rather simple, it is feasible to generate and identify large numbers of haploid seeds, treat them with colchicine and transplant them to the field. Hence, by starting with a sufficiently large number of haploid seeds it is possible to generate hundreds of viable DH lines with acceptable agronomic performance.

DHs are essentially important in the evaluation of diversity, because they fix rare alleles and aid the efficient selection for quantitative traits in breeding. In outcrossing species, DHs enable undesirable recessive genes to be eliminated from lines at any breeding stage (Forster and Thomas, 2004).

Development of DHs through anther culture has been very successful with many cultivars released in barley breeding worldwide and in rice breeding in China since the 1970s. The production of DHs has become the preferred tool in many advanced plant breeding institutes and commercial companies for breeding many crop species. Due to the obvious advantages of DH lines and the enhancements made in *in vivo* haploid induction in recent years, many commercial breeding companies such as Agreliant, Monsanto and Pioneer are presently adopting or are already routinely using this technology in their maize breeding programmes (Seitz, 2005). Recurrent selection for testcross performance using DHs has reduced the cycle length and improved genetic advance (Gallais and Bordes, 2007). In some companies *in vivo* haploid induction has more or less replaced conventional line development with up to 15,000 DH lines per year per breeding programme and over 100,000 DH lines per year across all programmes at costs of US$10 or less per DH line. The first maize hybrids produced using DH lines have been commercialized in the USA and Europe (W. Schipprack, University of Hohenheim, personal communication). However, the development of new, more efficient and cheaper large-scale production protocols has meant that DHs have also recently been applied in less advanced breeding programmes.

4.2.7 Limitations and future prospects

Genetics and breeding in DHs have not given the desired and expected dividends, despite the substantial investments made in haploid research since the late 1980s. Some of the widely recognized limitations of DH breeding are as follows: (i) haploids cannot be obtained in the high frequency required for selection in most important crop species; (ii) the cost–benefit ratio in DH breeding is often not favourable, thus discouraging its use despite the obvious advantages; (iii) haploids and DHs will express recessive deleterious traits and deleterious mutations may arise during the DH development process including anther culture, particularly for open-pollinated species; (iv) different ploidy levels may be available so that haploid status may need to be confirmed cytologically; alternatively, pollen culture may be necessary, which is expensive and has a relatively low success rate and is also genotype-dependent in many species; (v) doubled haploidy may also decrease genetic diversity, which is better maintained in heterozygous lines; (vi) the success of the DH method is highly genotype dependent, so is not yet suitable for all breeding programmes; (vii) some techniques, e.g. inducers in maize (especially the good ones), are proprietary and not available to all interested breeders; and (viii) health and legal concerns related to handling the chemical doubling agents.

The Third International Conference on Haploids in Higher Plants (12–15 February 2006, Vienna, Austria) highlighted the following issues that are important to future studies on DHs: (i) new methods of haploid and DH plant formation; (ii) mechanism of initiation of haploids; (iii) application of haploid cells, gametes, haploid and DH plants in fundamental and applied science; (iv) genes controlling haploid formation from female and male gametes; and (v) methods of diploidization of haploids.

4.3 Recombinant Inbred Lines (RILs)

Recombinant inbred lines or random inbred lines (RILs) are usually a part of the ultimate products of many breeding programmes and are also used as genetic materials. They can be produced by various inbreeding procedures. To help understand the whole process of development and applications of RILs, the inbreeding procedure and its effects will be discussed first.

4.3.1 Inbreeding and its genetic effects

RILs result from continuous inbreeding such as selfing or sibmating starting from an F_2 population until homozygosity is reached. There are two genetic responses to inbreeding, gene recombination and genotype homogenization. Starting from a heterozygote at a locus *A*-*a*, for example, selfing will produce three genotypes, *AA*, *Aa* and *aa*. With continuous selfing, two homozygotes, *AA* and *aa*, will not segregate, while the heterozygote *Aa* will continue to segregate producing the three genotypes. However, the proportion of heterozygotes in the population will decrease with continuous selfing and will approach zero. This process can be described as below.

Consider one locus with two alleles, *A* and *a*, underlying continuous selfing. Homozygotes will increase by 50%, while heterozygotes will decrease by 50% with each generation of selfing. At generation *t*, the proportion of heterozygotes in the population will be $(1/2)^t$, while the proportion of homozygotes will be $1 - (1/2)^t$; the homozygotes *AA* and *aa* each account for $[1 - (1/2)^t]/2 = (2^t - 1)/2^{t+1}$ (Table 4.1).

When two or more loci, for example *k* loci, are involved, successive selfing from F_1 hybrids will produce $(1/2)^{tk}$ heterozygotes and $[1 - (1/2)^t]^k = [(2^t - 1)/2^t]^k$ homozygotes at generation *t*. The more loci are involved, the longer it takes to reach homozygosity (Fig. 4.2). In the seventh generation of selfing starting from a heterozygous hybrid for example, the proportion of homozygotes will be 99% for the population with one heterozygous locus involved, 96% for the population with five heterozygous loci involved, 89% for 15 loci, 79% for 30 loci and 46% for 100 loci.

If heterozygous loci are linked, successive inbreeding can still produce a homozygous population. However, the rate of approach to homozygosity depends on the recombination frequencies between the linked loci. The lower the recombination frequency, the higher the proportion of homozygotes in the population and the more rapidly the population becomes homogenized. If the recombination frequency, *r*,

Table 4.1. Genotypes derived from a single-locus heterozygote and their frequencies in selfing generations.

Generation	Genotype			Frequency of heterozygotes	Frequency of homozygotes
	AA	*Aa*	*aa*		
0	–	1	-	1	0
1	1/4	2/4	1/4	1/2	50.0
2	3/8	2/8	3/8	1/4	75.0
3	7/16	2/16	7/16	1/8	87.5
4	15/32	2/32	15/32	1/16	93.8
5	31/64	2/64	31/64	1/32	96.9
10	1023/2048	2/2048	1023/2048	1/2048	99.9
...					
t	$(2^t - 1)/2^{t+1}$	$2/2^{t+1} = 1/2^t$	$(2^t - 1)/2^{t+1}$	$1/2^t$	$1 - 1/2^t$

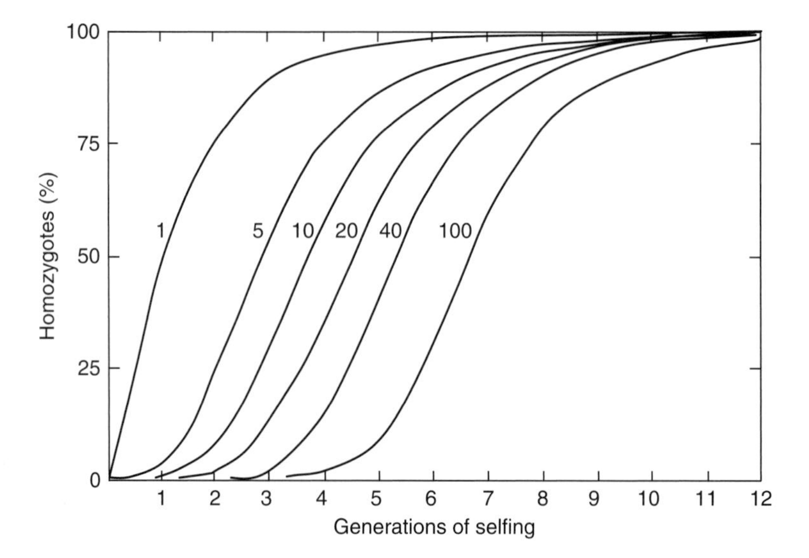

Fig. 4.2. Effects of generations and genetic loci on the proportion of homozygotes in self-pollinated populations (numbers of generations are 1, 5, 10, 20, 40, 100).

is close to zero or two loci are completed linked, the rate of homogenization will be close or equal to the rate for the population with one heterozygous locus. If r is about 50%, the rate of homogenization will be about the same as that for the population with two heterozygous loci. It can be estimated that for two linked loci and after one generation of selfing, the proportion of homozygotes will be 41% for r = 10%, 34% for r = 20%, 26% for r = 40% and 25.26% for r = 45%.

Continuous inbreeding (e.g. selfing) results in the fixation of segregation so that the genetic combinations of two parental genomes represented in individual F_2 plants are each represented by an RIL (Fig. 4.3). The genetic combinations of two parental genomes are fixed in a group of RILs.

For quantitative traits that are controlled by polygenes or multiple QTL, the mean value of the population will return to the average value of the parental lines because dominance and dominance-related epistasis will dissipate with increasing homogenization. The variance will also change with increasing homogenization but the direction of change will depend

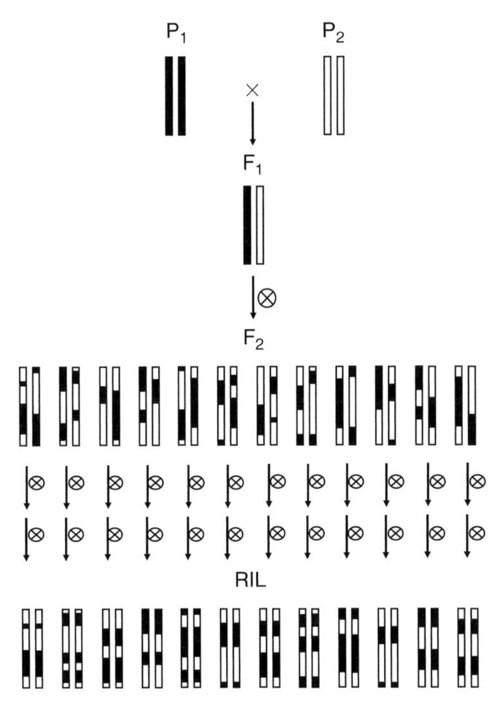

Fig 4.3. Production of RILs by successive selfings. Two parental lines, P_1 and P_2, are crossed to produce F_1. The F_1 is then selfed to produce F_2. The selfing process continues until a certain level of homozygosity is reached. The end product consists of a set of RILs, each of which is a fixed recombinant of the parental lines.

on the effect of related genes and their interactions. Figure 4.4 shows the changes in mean value and variance in RIL populations derived by SSD under different genetic models.

In animals, RILs are usually stable for up to 20 generations of sibmating. Such long-term continuous sibmating results in such a low viability that it is very hard to maintain the population. The mouse was the first animal used for genetic mapping with RILs and its RIL population is relatively small. However, the problems associated with small population size can be ameliorated by using combined information from multiple sets of RILs. In plant species by contrast, it takes about half of the time required in animals to obtain stable RILs through inbreeding. Also, maintaining RIL populations in plants can be achieved at much lower costs, making it possible to manipulate large-sized populations. For some plant species, such as tobacco and *Brassica* however, self-incompatibility prohibits the production of RILs through inbreeding.

4.3.2 Development of RILs

RILs are the products of successive inbreeding. Based on reproduction systems and the degree of inbreeding, there are several types of procedures for developing RILs.

Full-sib mating: for outbreeding organisms, the most severe inbreeding is full-sib mating, i.e. mating between the offspring of the same parents. Because outbreeding organisms are highly heterozygous, they have to be inbred for several generations to approach homozygosity. The inbreeding parents can be then used to produce progenies that will be intermated to produce the next generation of progenies. This process will continue until the progenies are highly homozygous.

Selfing: for self-pollinated plants, cultivars are genetically homozygous so they can be used to produce hybrids directly, followed by successive selfing. There are two different procedures for the management of the progenies, bulking and SSD. In the bulking method, hybrids are bulk planted and harvested until F_5 to F_8 before they are planted by families.

Single seed descent (SSD) method

The SSD method was proposed by the Canadian scientist Guolden in 1941. Starting from F_2, one or several seeds are harvested from each plant and planted to produce the next generation until F_5 to F_8. When most plants are homozygous, all the SSD seeds from each plant are harvested to produce RILs. Plant breeders use three procedures to implement the concept of SSD (Fehr, 1987).

SINGLE-SEED PROCEDURE. When the single-seed procedure is used, the size of the populations will decrease in each generation

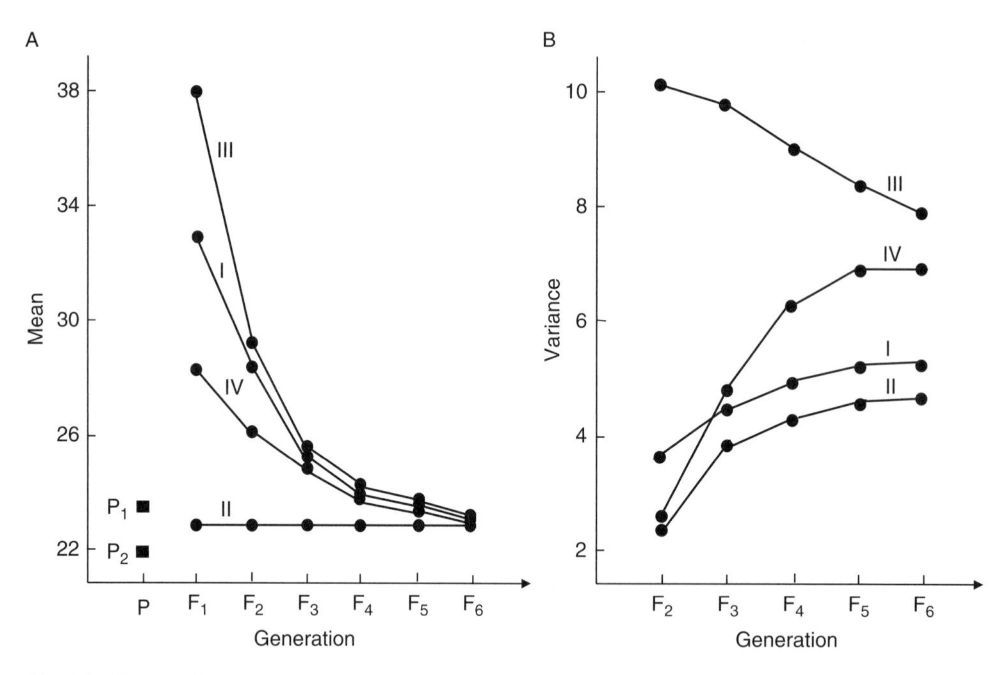

Fig. 4.4. Change of mean (A) and variance (B) in RIL populations derived by SSD. (I) Additive – increasing alleles are completely dominant. (II) Additive without dominance effect. (III) Additive – increasing alleles are completely dominant with complementary interaction. (IV) Additive – increasing alleles are completely dominant with duplicate interaction.

because of lack of seed germination or failure of plant establishment to produce seed. It is necessary to decide on the number of inbred plants that are desired in the last generation and begin with an appropriate population size in the F_2 generation. The single-seed procedure ensures that each individual in the final population traces to a different F_2 individual. However, the procedure cannot ensure that a particular F_2 will be represented in the final population because failure of any seed to germinate or generate a productive plant automatically eliminates that seed's F_2 family.

SINGLE-HILL PROCEDURE. The single-hill procedure can be used to ensure that each F_2 plant will have progeny represented in each generation of inbreeding. Progeny from individual plants are maintained as separate lines during each generation of inbreeding by planting a few seeds in a hill or row, harvesting self-pollinated seeds from the hill and planting them in another hill the following generation. An individual plant is harvested from each line when the population has reached the desired level of homozygosity.

With the single-hill procedure the identity of each F_2 plant and its progeny can be maintained during self-pollination. When the identity of an F_2 is maintained, the seed packet and hill must be properly identified with a line designation for planting and harvest.

MULTIPLE-SEED PROCEDURE. Use of the single-seed procedure requires that the size of the populations in F_2 be larger than in later generations, due to lack of seed germination or plant establishment for seed set. Usually, two samples are harvested, one for planting in the next generation and one for a reserve. Researchers sometimes bulk two or three seeds from each plant during harvest. Part of the sample is planted and the remainder is reserved. The procedure is referred to as modified SSD. The number of seeds planted and harvested each season depends on the number of lines desired from the population and the anticipated percentages of seed germination, seedling establishment and seed set.

Advantages and disadvantages of SSD procedures

Fehr (1987) summarized the merits of the SSD procedures and indicated the following advantages:

- They are an easy way to maintain populations during inbreeding.
- Natural selection does not influence the population unless genotypes differ in their ability to produce at least one viable seed each generation.
- The procedures are well suited to greenhouse and off-season nurseries where the performance of genotypes may not be representative of their performance in the area in which they are normally grown.

The disadvantages are: (i) artificial selection is based on the phenotype of individual plants, not on progeny performance, when SSD is used for cultivar development rather than genetic population development; and (ii) natural selection cannot influence the populations in a positive manner unless undesirable genotypes do not germinate or set any seed.

4.3.3 Map distance and recombinant fraction in RIL populations

Theoretically and in practice, no matter how many cycles of inbreeding are completed, some degree of heterozygosity will always exist in the RIL population. From the above discussion, we can estimate the remaining heterozygosity for each generation of inbreeding. In genetic mapping, nearly completely homozygous RILs are used. RILs have undergone several cycles of meiosis before fixation, which differs from F_2 or BC populations where only one cycle of meiosis occurs. As a result, linked genes have more opportunities to recombine in RIL populations. This property was discovered by Haldane and Waddington (1931) by studying inbreeding populations. For tightly linked loci, the number of recombinants observed in RILs is twice that observed in the populations with only one cycle of meiosis. At the beginning stage of genetic mapping, this multiple recombination in RILs makes it difficult to detect linkage. Once linkage relationships are roughly established among loci, the greater frequency of recombination makes it easy to detect non-allelism among loci. It also makes the estimation of genetic distances more accurate because the confidence interval for an estimated genetic distance is a function of recombination frequency. With the increased number of meiosis events, there are more opportunities to find recombinants between two tightly linked loci (Fig. 4.5).

In populations that have undergone only one cycle of meiosis, recombinant frequency r (%) is linearly related to map distance R (cM), as indicated by the dashed line in Fig. 4.5. In RIL populations derived from selfing, r is almost equal to $2R$ when the map distance is small, which is indicated by the solid line and formula $R = r/(2-2r)$ (Fig. 4.5). For the RIL populations derived

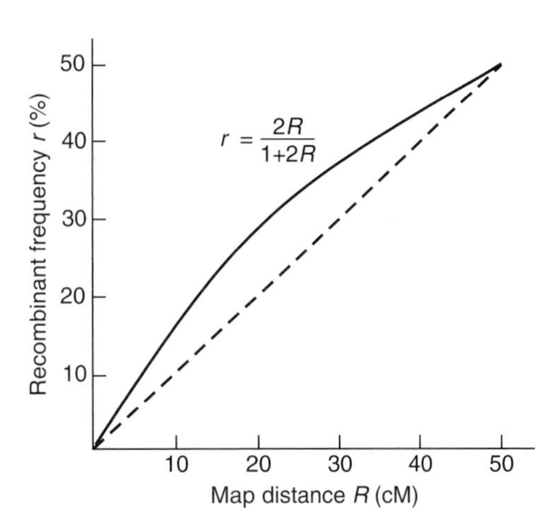

Fig 4.5. Relationship between map distance (R) and recombinant frequency (r) for RIL populations derived by continuous selfing (solid line) and for populations that have undergone only one cycle of meiosis, e.g. F_2 or BC (dashed line).

from sibmating, the skew becomes more significant with r nearly equal to $4R$ when the map distance is small.

4.3.4 Construction of genetic maps using RILs

As each RIL is inbred as a DH line and thus can be propagated indefinitely, a panel of RILs has a number of advantages for genomic studies: (i) each line needs to be genotyped only once; (ii) multiple individuals can be phenotyped from each line to reduce individual, environmental and measurement variability; (iii) multiple invasive (destructive) phenotypes can be obtained on the same set of genomes; and (iv) as recombinations are more frequent in RILs than in populations with only one meiosis, greater resolution can be achieved in genetic mapping.

In genetic mapping with RIL populations, recombinant frequency should be converted into map distance using the formula $R = r/(2-2r)$ proposed by Haldane and Waddington (1931). There are no mapping functions available for RIL populations to adjust for double crossover events as there are for populations with one cycle of meiosis as discussed in Chapter 2. When the map distance is within the range that allows confidence about linkage detection, recombinant frequency has a linear relationship with map distance (Fig. 4.5; Silver, 1985).

Non-linked loci may be linked simply due to chance. These false linkages can often be confirmed by whether a linkage detected with one marker is also judged to be linked by other markers in the same linkage group and whether the suspected linkage found in one population can also be detected in other RIL populations. Mouse geneticists discussed the case when a linkage could not be certain because of small population sizes, and Silver (1985) provided a table for the 95 and 99% confidence intervals for estimated map distances based on RILs derived from sibmating. At low rates of recombination, these intervals are relatively small when compared with those obtained from the binomial distribution for F_2 and BC_1 populations of comparable population size. According to Taylor (1978), RILs derived from sibmating were more powerful in the estimation of map distances than populations undergoing single meiosis when $R \leq$ 12.5cM. Based on Taylor's method, it can be inferred that RILs derived from self-pollination have greater influence on the estimation of map distance when $R \leq$ 23cM.

Because of the advantages of RILs, they have been receiving great attention in genomics research. Numerous RIL populations have been developed in plant species, especially in maize and rice. Burr *et al.* (1988) reported RFLP maps constructed using two maize RIL populations, T232 × CM37 and CO159 × Tx303. Among 334 mapped genetic loci, 220 were polymorphic in both populations. By comparing the map distances obtained from these two populations with each other and with publicly accepted map distances, they found that the differences could be twice as large in some cases. Although these differences were still within the range of confidence intervals, they might be due to the genetic difference in recombinant frequencies at specific chromosomal regions. In maize there is no significant polymorphism caused by chromosome rearrangement, except for chromosome 10. Therefore, it is not surprising that there was no significant difference in map distance between the two maize RIL populations. Table 4.2 provides some examples of RIL populations developed in maize (Burr *et al.*, 1988) and in rice (Xu, Y., 2002) that have been widely used for linkage mapping and gene tagging.

4.3.5 Intermated RILs and nested RIL populations

Intermated RILs

The production of RILs allows for the accumulation of recombination breakpoints during the inbreeding phase. However, the accumulation in RILs is limited by the fact that each generation of inbreeding makes the recombining chromosomes more similar to one another so that meiosis ceases to

Table 4.2. Some examples of RIL populations developed in maize (Burr *et al.*, 1988) and rice (Xu, 2002).

Species	Population	Population size	No. markers[a]
Maize	T232 × CM37	48	
	CO159 × Tx303	160	
	Mo17 × B73	44	
	PA326 × ND300	74	
	CK52 × A671	162	
	CG16 × A671	172	
	Ch593-9 × CH606-11	101	
	CO220 × N28	173	
Rice	9024 × LH422	194	141
	CO39 × Moroberekan	281	127
	Lemont × Teqing	315	217
	IR58821 × IR52561	166	399
	IR74 × J almagna	165	144
	Zhenshan 97 × Minghui 63	238	171
	Asominori × IR24	65	289
	Acc8558 × H359	131	225
	IR1552 × Azucena	150	207
	IR74 × FR13A	74	202
	IR20 × IR55178-3B-9-3	84	217

[a]Numbers of markers shown for the first generation of genetic maps and more markers have been added to many of these maps since then.

generate new recombinant haplotypes. As an alternative to RILs, Darvasi and Soller (1995) proposed randomly mating the F_2 progeny of a cross between inbred founders and using successive generations of random mating to promote the accumulation of recombination breakpoints in the resulting advanced intercross lines or intermated recombinant inbred lines (IRILs). IRIL design has obvious appeal in its union of the advantages of both IRILs and RILs and has been employed in the production of mapping populations in several species. As a result, IRILs have become interestingly popular for QTL mapping.

Breeding designs for IRILs have been investigated by Rockman and Kruglyak (2008). Their results indicated that the simplest design, random pair mating with each pair contributing exactly two offspring to the next generation, performed as well as the most extreme inbreeding avoidance schemes in expanding the genetic map, increasing fine-mapping resolution and controlling genetic drift. Circular mating designs offered negligible advantages for controlling drift and gave greatly reduced map expansion. Random-mating designs with variance in offspring number are also poor at increasing mapping resolution. It is suggested that the most effective designs for IRIL construction are inbreeding avoidance and random mating with equal contributions from each parent to the next generation.

Multi-way or nested RIL populations

Using two or more RIL populations in genetic mapping provides several advantages: (i) polymorphisms that are not detected in one population may be detected in another; (ii) weak linkage identified in one population can be confirmed or excluded by using other populations; (iii) multiple populations with shared genetic data can be combined and considered as a single population to provide more reliable results; and (iv) multiple populations provide a wide spectrum of target loci across the genome since for quantitative traits the related loci with genetic differences between the parents are almost always different from population to population.

The Complex Trait Consortium for mouse proposed the development of a large panel of eight-way RILs (Complex Trait Consortium, 2004). An eight-way RIL, also known as Collaborative Cross, is formed by intermating eight parental inbred strains followed by repeated sibling mating to produce a new set of inbred lines whose genome is a mosaic of the eight parental strains (Broman, 2005). Such a panel of RILs would serve as a valuable resource for mapping the loci that contribute to complex phenotypes in mouse and would support studies that incorporate multiple genetic, environmental and developmental variables into comprehensive statistical models of complex traits (Complex Trait Consortium, 2004). The genomes of eight founder strains are rapidly combined and are then inbred to produce finished RIL strains. Eight-way RIL strains achieve 99% inbreeding by generation 23. Each strain captures ~135 unique recombinant events. With genetic contributions from multiple parental strains including several wild derivatives, the eight-way RILs will capture an abundance of genetic diversity and will retain segregating polymorphisms every 100–200 bp. This level of genetic diversity will be sufficient to drive phenotypic diversity in almost any trait of interest. An estimated ~1000 strains will be required to guarantee high mapping resolution and detect extended networks of epistatic and gene–environment interactions. This estimate is based on the statistical power necessary to detect biologically relevant correlations among thousands of measured traits. A set of 1000 strains containing 135,000 recombinant events is a far more powerful and flexible research tool than a set of 100 strains containing the same total number of recombinant events. Mounting evidence suggests that gene–gene interactions (epistasis) are crucial in many complex disease aetiologies. A set of 1000 strains will readily support simultaneous mapping of many two-way and three-way epistatic interactions.

Similar resources have been developed in maize, *Arabidopsis* and *Drosophila*. To map much of the segregating variation in maize, 25 RIL mapping populations were created. Twenty-five diverse lines were selected to capture 80% of the nucleotide polymorphism in maize. In order to provide a uniform evaluation background, each line was crossed to a common parent, B73 (the standard US inbred), to form 25 RIL populations. Each of these RIL populations has at least 200 RILs, each descended from a unique F_2 plant, resulting in a total of 5000 RILs. Using SSD and low density planting, 88% success in advancing lines per generation was achieved. This has developed as an integrated mapping approach, called Nested Association Mapping (NAM), which exploits simultaneously the advantages of linkage analysis and association (or linkage disequilibrium, LD) mapping as discussed in Chapter 6. The power of NAM for genome-wide QTL mapping has been demonstrated by computer simulation with varied numbers of QTL and trait heritabilities (Yu *et al.*, 2008). With a dense coverage (2.6 cM) of common-parent-specific (CPS) markers, the genome information for 5000 RILs can be inferred based on the parental genome information. Essentially, the linkage information captured by the CPS markers and the LD information among loci residing between CPS markers was then projected to RIL based on parental information, ultimately allowing for genome-wide high-resolution mapping. The power of NAM with 5000 RILs allowed 30–79% of the simulated QTL to be precisely identified. In the ongoing genome sequencing projects, NAM would greatly facilitate complex trait dissection in many species in which a similar strategy can be readily applied.

4.4 Near-isogenic Lines (NILs)

Near-isogenic lines (NILs) derived from inbreeding are in most cases the product of successive backcrossing. The methods for obtaining NILs, including the genetic effects of backcrossing will be described first, followed by their applications in genomics and plant breeding.

4.4.1 Backcrossing and its genetic effects

Backcrossing is a hybridization method by which the hybrid is crossed back to one of its parental lines. The hybrid could be of any generation, although usually it starts with the F_1. Backcrossing has been widely used in plant breeding to improve one or a few major traits (target traits) that are agronomically or genetically important but are lacking in a current commercial or elite cultivars. In this case, the commercial cultivar is crossed with the germplasm (the donor) that provides the target trait and then a backcross programme starts by backcrossing the hybrid to the commercial cultivar. This process continues by selecting the progeny produced in the previous backcross that has the target trait to backcross again to the commercial cultivar, until the progeny very nearly resembles the commercial cultivar except for the target trait transferred from the donor.

The commercial cultivar used recurrently in this process is called the recurrent parent (RP), while the germplasm as the donor of the target trait is called the donor parent (DP). The final product of continuous backcrossing will be a backcross inbred line (BIL) with an almost identical genome to the RP except for the target trait/locus. The final BILs are produced by one or more generations of selfing following the final backcross. A set of BILs can be produced simultaneously by producing BILs for different target traits/genes/chromosome regions using the same recurrent parent. In the extreme case, BILs can be produced for any of the genetic loci distributed on the whole genome by marker-assisted selection (MAS), which will be discussed later.

With continued backcrossing, the proportion of the recurrent genome in the backcross progeny increases while that of donor genome decreases. When one locus is involved in controlling the target trait, the proportion of the allele from the non-recurrent parent or DP in the backcross progeny will be $(1/2)^t$, while the proportion of the allele from the RP will be $1 - (1/2)^t$, at the tth generation of backcrossing. The same is true for the whole genome (Fig. 4.6).

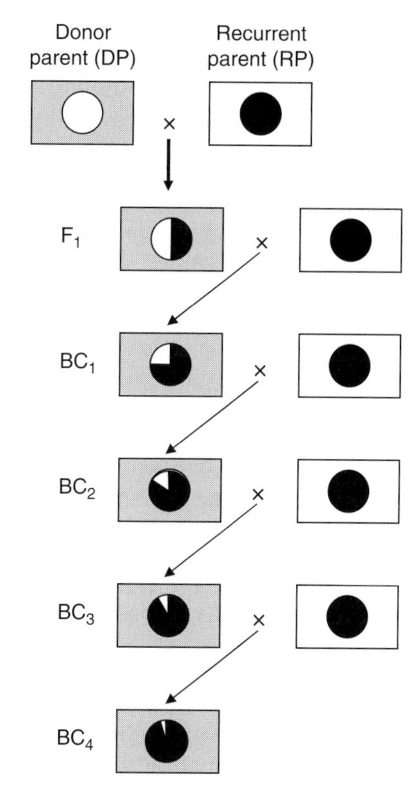

Fig. 4.6. Genetic effect of backcrossing (with perfect Mendelian segregation).

If there is no linkage among k loci that differ between the two parents, the progeny that are homozygous for the recurrent parent alleles will account for $[1 - (1/2)^t]^k$ at the tth generation of backcrossing. That is, for the fifth backcross and $k = 10$ loci, 72.8% of the backcross progeny will be homozygous and have the same genotype as the recurrent parent at these loci. Backcross-derived homogenization is different from that resulting from self-pollination where the former is homogenized towards the RP while the latter is homogenized towards the parental genotypes (both parents) and their recombinants.

Because of genetic linkage, the donor genome will be replaced by the recurrent genome at a lower rate in the genomic region around the target gene and the genes nearby than in other regions. The low rate depends on the tightness of linkage, that is, how far the target gene is from the linked genes. Under conditions of no selection, the probability of obtaining recombinants, that is, the DP alleles at non-target loci that are replaced by the RP alleles, will be $1 - (1 - r)^t$. Therefore, the more backcrosses, the greater the chance of recombination.

When the allele for the target trait is dominant, the candidate progeny are selected for further backcrossing only when they are heterozygous at the target locus and bear a high level of phenotypic resemblance to the recurrent parent. In this way, the recovery of recurrent progeny will be accelerated.

When the allele for the target trait is recessive the plants used for backcrossing can also be selfed. The same plants are selfed and backcrossed and their progeny can be planted side by side and compared to find which of the plants used for backcrossing contain the target allele based on the phenotype of the selfed progeny. If the selfed progeny is segregating for the target trait, the backcross progeny from the same plant can be used for further backcrossing. If the selfed progeny is not segregating for the target trait, the corresponding progeny from the backcross should be excluded from further backcrossing. Using linked markers to select the target trait (Chapter 8) will make it possible to continue backcrossing without the required selfing to select the plants containing recessive genes.

According to the rate of recovery of the RP genome determined by $[1 - (1/2)^t]^k$, an infinite number of backcrosses are theoretically needed to eliminate the complete genome from the DP. The currently available BILs are mostly derived from fewer than ten (usually five to six) backcrosses. As a result, specific BILs may still contain alleles from the donor at a large number of loci in addition to the target locus. This is the reason why BILs derived from repeated backcrosses are frequently called NILs.

4.4.2 Other methods for production of NILs

In addition to the repeated backcrosses, there are several other approaches to production of NILs as discussed by Xu, Y. (2002):

Selfing-derived NILs. Pairwise NILs can be developed through continuous selfing while keeping the target trait locus heterozygous. Once other genetic loci are almost all fixed, an additional generation of selfing will result in a pair of NILs that differ only at the target locus (Xu and Zhu, 1994). Selfing-derived NILs can be any combination of parental genotypes and each pair of NILs have identical genetic constitution (except for the target locus), whereas the backcross-derived NILs have the same genetic constitution as the RP.

Whole genome selection of permanent populations. With the accumulation of permanent mapping populations such as RILs and DHs described in the previous sections, it is possible to find two which are almost genotypically identical for the whole genome except for one or a few marker loci.

Mutation. Creation of a collection of single-locus mutants from a donor inbred line is a quick approach to producing a large number of NILs. For most mutants, mutation only occurs at one or few genetic loci. These mutants can be considered near-isogenic to their 'wild-type' donor and are thus known as isomutagenic lines (IMLs). IMLs have been widely used in functional genomics research for gene cloning (see Chapter 11 for details).

Chromosome substitution. Using chromosome engineering and/or MAS, whole or partial chromosome substitution lines can be created, so that each line has one chromosome or partial chromosome replaced. A set of chromosome substitution lines, also known as introgression lines, can be produced to cover the whole genome so that each represents a piece of chromosome from the donor genome.

4.4.3 Introgression line libraries

Genetic stocks of genome-wide coverage are the platform required for large-scale gene discovery (Chapter 11) and efficient marker–trait association (Chapter 6). Extensive research has been carried out worldwide to develop genome-wide genetic stocks for functional genomics research. Eshed and Zamir (1994) proposed to exploit introgression lines (ILs), also known as chromosome substitution lines (CSSLs) or IL libraries, which could be generated by repeated backcrossing and MAS with the whole genome covered by the contigs of the introgressed segments from the DP. ILs have a high percentage of the RP genome and a low percentage of the DP genome. They offer several advantages over conventional populations: (i) they provide useful stocks for highly efficient QTL or gene identification and fine mapping; (ii) they can contribute to the detection of epistatic interactions between QTL; and (iii) they can be used to map new region-specific DNA markers (Eshed and Zamir, 1995; Fridman *et al.*, 2004). Several sets of IL are now available in barley, maize, rice, soybean and wheat that contain beneficial alleles from wild relatives thus enriching the genetic diversity in the primary gene pools of these crops. These ILs when crossed with cultivars produce progenies with enhanced trait values as demonstrated by the increased yield in tomato and wheat (Gur and Zamir, 2004; Liu *et al.*, 2006). ILs are of particular value for assessing the phenotypic consequences associated with diverse donor alleles in particular genetic backgrounds and subsequently for map-based cloning of genes (Zamir, 2001). IL libraries will facilitate large-scale QTL cloning and functional genomics research into complex phenotypes by resolving four major technical difficulties (Li, Z.K. *et al.*, 2005). They allow: (i) efficient identification of QTL with a large effect on specific target phenotypes; (ii) efficient fine mapping of target QTL and determination of candidate genes underlying QTL; (iii) efficient determination and verification of functions of QTL candidate genes; and (iv) the dissection of gene networks and metabolic pathways underlying the complex phenotypes.

There are many reports of the development of ILs in various crops, but most of them involve relatively small population sizes (Dwivedi *et al.*, 2007) that are not sufficiently accurate to cover a significant part of the genetic variation. In barley, a set of 146 ILs was derived from BC_2F_6 of the cross Harrington and Caesarea (*Hordeum vulgare* ssp. *spontaneum*), covering an average 12.5% of the *H. spontaneum* genome (Matus *et al.*, 2003). In rice, there are several reports involving development of over 100 ILs. For example, 147 ILs were developed from *Oryza sativa* (Taichung 65) and *Oryza glumaepatula* reciprocal crosses containing *O. glumaepatula* or Taichung 65 cytoplasm but with entire chromosome segments from *O. glumaepatula* (Sobrizal *et al.*, 1999); 140 ILs were derived from a cross between *japonica* cv. Nipponbare, and an elite *indica* line Zhenshan 97B (Mu *et al.*, 2004); 159 ILs carrying variant introgressed segments from *Oryza rufipogon* Griff. in the background of the *indica* cultivar Guichao were developed, representing 67.5% of the *O. rufipogon* genome and a 92.4–99.9% (97.4% average) representation of the RP genome (Tian *et al.*, 2006). As additions to this list, three examples from rice and tomato will be discussed below in detail.

Rice ILs

To facilitate research into the functional genomics of complex traits in rice, Z.K. Li *et al.* (2005) developed over 20,000 ILs in three elite rice genetic backgrounds (two high-yielding *indica* cultivars, IR64 and Teqing, and a new plant type tropical *japonica*, NPT (IR68552-55-3-2)) by marker-assisted selective introgression for a wide range of complex traits, including resistance/tolerance to many biotic and abiotic stresses, morpho-agronomic traits, physiological traits, etc. A total of 195 accessions from 34 countries were used as donors in the backcrossing programme, representing different subspecies, ecotypes and gene pools. Twenty-five randomly chosen plants from each BC_1F_1 population were backcrossed with each RP to produce 25 BC_2F_1 lines. From each cross, 25 BC_2F_1 lines were planted in the following season and seed from individual plants of 25 BC_2F_1 lines from each cross were bulk-harvested to form a single bulk BC_2F_2 population. In addition, 30–50 superior high-yielding BC_2F_1 plants from each cross were further backcrossed with the RPs to produce BC_3F_1 lines and likewise BC_4F_1 lines. Similarly, BC_3F_2 and BC_4F_2 bulks were generated by bulk-harvesting seeds of all BC_3F_1 and BC_4F_1 lines from each cross. The BC bulks were then screened for their resistance or tolerance to different abiotic and biotic stresses, including drought, salinity, submergence, anaerobic germination, zinc deficiency, brown planthopper, etc. and in all cases the severity of the stresses was strong enough to kill the RPs and only the surviving BC progeny were selected. Selection for many agronomic traits, such as flowering time, plant height, plant type related traits (leaf and culm angles), grain quality parameters, yield related traits, etc., was also undertaken based on visual observation of BC bulks. Each of the selected BC progeny was tested for selected phenotypes and selfed for two or more generations to form a homozygous IL. ILs within each genetic background are phenotypically similar to their RP but each carries one or few traits introgressed from a known donor. Together, these ILs contain a significant portion of the alleles affecting the selected complex phenotypes in which allelic diversity exists in the primary gene pools of rice. A forward genetics strategy was proposed and demonstrated with examples for using these ILs for large-scale functional genomics research (Chapter 11). Complementary to the genome-wide insertional mutants, these ILs provide new methods for highly efficient gene discovery, candidate gene identification and cloning of important QTL for specific phenotypes based on the convergent evidence from QTL position, expression profiling and functional and molecular diversity analysis of candidate genes.

In another example in rice, a single-segment substitution line (SSSL) library was developed using Hua-Jing-Xian, an elite *indica* cultivar from South China as the recipient and 24 cultivars including 14 *indica* and ten *japonica* cultivars from worldwide sources as donors (Xi *et al.*, 2006). The current library consists of 1529 SSSLs with an average substitution segment length of 18.8 cM. The library covers 28705.9 cM of genome in total which is equal to 18.8 genome-equivalents. This library has been used for QTL mapping of many traits (Xi *et al.*, 2006; Liu *et al.*, 2008).

The Lycopersicon pennellii *ILs*

Using whole genome marker analysis, Eshed and Zamir (1994) developed a permanent mapping population designed for QTL analysis. This resource is composed of a tomato cultivar (*Lycopersicon esculentum* cv. M82) which includes single introgressed genomic regions from the wild green-fruited species *L. pennellii*. This congenic resource, composed of 76 ILs, provides nearly complete coverage of the wild species genome. The IL map is connected to the high-resolution F_2 map which is composed of 1500 markers (IL chromosome maps) with seed of the ILs available through the C.M. Rick Tomato Genetics Resource Center, University of California Davis.

Applications of the ILs developed in rice and tomato will be further discussed in later chapters.

4.4.4 Gene tagging strategy using NILs

Due to genetic linkage, the chromosome fragments around the target locus will be dragged into backross progeny and may be retained in the subsequent progeny. This phenomenon is called linkage drag. The basic idea behind gene tagging using NILs is to use the opportunity provided by linkage drag to identify the molecular markers located on the chromosome segments around the target locus. This can be accomplished by comparing the marker genotypes among RP, DP and NILs. When the genotype at a particular marker locus is the same for the NIL and DP but different for the RP, possible linkage between this marker and the target locus can be determined (Fig. 4.7).

Success in gene tagging using NILs relies on the assumption that there is genetic difference between DP and RP in the chromosome region flanking the target locus. Apparently, the likelihood of detecting this difference depends on the length of the chromosome regions in the DP which have been retained during the backcross; this parameter decreases with the increasing number of backcrosses. Detecting these differences also depends on the molecular polymorphism in this region in DP and RP. When DP and RP are from different species such as cultivated and wild species, there is a high probability of finding polymorphisms between them. Conversely, the probability will be low if DP and RP are genetically closely related to each other. The likelihood of detecting molecular polymorphisms between DP and RP at the target region can be increased by using large numbers of markers and/or different types of markers.

With the increasing availability of complete sets of BILs or NILs that cover the whole genome, NIL mapping strategies provide a convenient approach to tagging and isolating numerous genes. In the following section, the discussion will be focused on major gene-related issues while QTL mapping using NILs is described in Chapter 7.

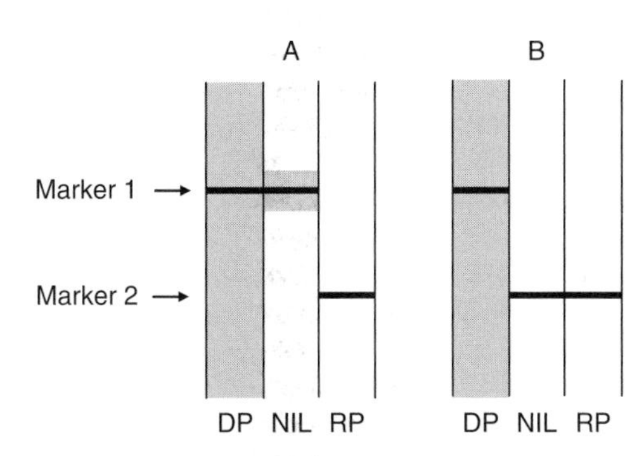

Fig. 4.7. Gene tagging using NILs. (A) Possible linkage, NIL has the same allele as DP at Marker 1 locus. (B) No linkage, NIL has different alleles from DP at both marker loci (1 and 2). NIL, near-isogenic line; DP, donor parent; R, recurrent parent.

4.4.5 Theoretical considerations in genetic mapping using NILs

One source of errors in NIL-based mapping is the possibility that the RP not only differs from the DP in the regions around the target locus but also differs at other loci distributed all over the genome. This is because for a limited number of backcrosses, t, the DP genome retained in the backcross progeny is $1/2^t$, which creates the possibility that the polymorphic markers in the retained regions could be falsely identified as being located in the target region. These false positive markers are not actually linked with the target region. As calculated theoretically by Muehlbauer *et al.* (1988), for a genome containing 20 chromosomes each of 50 cM, progeny derived from five backcrosses will retain DP alleles at four out of 100 randomly selected marker loci. Among these four retained DP loci, it is estimated that only one or two will not be linked to the target gene. This estimation is based on the assumption that there is no selection for the RP phenotype, that is, individuals that are heterozygous at the target marker loci are selected randomly to be backcrossed with RP.

Backcross introgression without selection for the RP phenotype

Assume that a plant species has n chromosomes, each of L M and the objective is to transfer the target gene (classic marker) located in the middle of one of the chromosomes from DP to RP by t backcrosses ($b = t + 1$). Suppose there are 100 polymorphic markers between DP and RP and these markers are randomly distributed over the genome. In the final backcross progeny, the proportion of the marked chromosome (M, where the target gene is located) from the DP will be

$$U_{Mb} = [2\,(1-e^{-bL/2})/b]/L$$

with variance

$$V_{Mb} = (2/L^2)\{[2 - (bL + 2)e^{-bL/2}]/b^2 - [(1/b)(1 - e^{-bL/2})]^2\}$$

With 20 chromosomes (n = 20) each 50 cM long (L = 0.5) and five backcrosses (t = 5, and $b = t + 1 = 6$), the proportion of the DP genome containing the target chromosome is

$$U_{Mb} = 0.5179 \pm 0.2498 = 51.79\% \pm 24.98\%$$

It is expected that 51.79% of the target chromosome will come from DP. As the chromosome is 50 cM in length, this proportion can be converted into genetic map distance by the following calculation

$$0.5179 \times 50\,cM = 25.90 \pm 12.49\,cM$$

In the backcross progeny, the proportion of the DP genome in the non-target chromosomes (N) is

$$U_{Nb} = (1/2)^b = 0.015625 = 1.56\%$$

with variance

$$V_{Nb} = (1/2)^b(2/L)[(1/b)(1 - e^{-bL/2})] - [(1/2)^b]^2$$

This can be converted into genetic map distance as follows:

$$0.015625 \times 50\,cM = 0.78 \pm 4.43\,cM.$$

In the backcross progeny therefore, the proportion of the DP genome (target and non-target chromosomes) in the total genome (T) will be

$$U_{Tb} = [L \times U_{Mb} + (n - 1)L \times U_{Nb}]/(nL)$$

with variance

$$V_{Tb} = [L \times V_{Mb} + (n - 1)(L \times V_{Nb})]/(nL)^2$$

Therefore

$$U_{Tb} = [(50 \times 0.5179) + (20 - 1) \times 50 \times 0.015625]/(20 \times 50) = 0.04074 \pm 0.1028$$

If NILs are used, interaction between the target QTL and other major genes/QTL can be eliminated and only epistasis between multiple target QTL needs to be considered. With removal of noise from heterogeneous backgrounds, the proportion of variance explained by the target QTL will increase and minor QTL can be identified. Without disturbance from the background effect, multiple QTL can be easily separated. Since all genotypic variation comes from the target loci, environmental effects can be estimated. In QTL cloning, NILs have been used to map the target QTL precisely by using all of these advantages.

Backcross introgression with selection for RP phenotype

The above discussion applies to backcrossing without selection for RP phenotype. In reality, however, the individuals most like the RP are always selected for backcrossing in order to obtain the introgressed lines as rapidly as possible. The effectiveness of phenotype-selected backcrossing depends on how the following requirements are met: (i) allelic differences exist not only at the target locus but also at many other loci across the genome between RP and DP; (ii) these differences control the phenotype in backcross progeny, i.e. are highly inheritable; (iii) each backcross produces enough progeny to allow individuals of different phenotypes to be identified. If these requirements are met or at least partly met, selection for the RP phenotype will help to improve the rate at which backcrossing progeny recover the RP alleles not only at the target loci but also around the target region.

Selection for the RP phenotype accelerates the replacement of the DP genome and at completion of backcrossing the average proportion of the DP genome retained in the backcross progeny will be reduced. Because the replacement of the DP genome is proportional to the number of backcrosses, the effect of selection for the RP phenotype on the replacement and retention of the DP genome will decrease as backcrossing continues.

To derive the same formulas for backcross introgression with selection for RP phenotype as those discussed for the case without RP phenotype selection, the three factors (requirements) affecting phenotypic selection mentioned above need to be determined. However, these factors vary depending on specific DP × RP crosses so that it is impossible to develop general formulas that are applicable to different crosses. Instead, Muehlbauer *et al.* (1988) provided two examples to explain the effect of selection for the RP phenotype on the retention of the DP genome for marked and unmarked chromosomes. In each case, selection for the RP phenotype significantly reduced the DP genome retained in backcross progeny.

4.4.6 Application of NILs in gene tagging

NILs have been successfully used in gene tagging for almost all crop species with available molecular marker systems and NILs. Some pioneering examples in this field include identifying molecular markers for the *Tm-2a* gene controlling resistance to tobacco mosaic virus in tomato (Young *et al.*, 1988), the *Dm*1, *Dm*3 and *Dm*11 genes for downy mildew resistance in lettuce (Paran *et al.*, 1991) and the *Pto* gene for stem rot (*Pseudomonas*) resistance in tomato (Martin *et al.*, 1991). Since then, numerous studies using NILs for gene tagging and map-based gene cloning have been reported. From these studies, some general conclusions can be drawn:

- Backcrossing significantly reduces the linkage drag around the target region.
- The more backcrosses that are carried out, the smaller the linkage drag fragment in backcross progeny and the less frequently will false positives be found between the NILs.
- Molecular markers can be used to improve the efficiency of backcrossing by significantly reducing the linkage drag and increasing the RP genome ratio.
- Linkage drag has significant influence on the recovery of the RP genome around the target region, indicating its important impact on backcross breeding programmes.
- Using multiple NILs significantly reduces the probability of reporting false positives for linked loci.
- Selection for the RP phenotype reduces the DP genome ratio during the backcross process compared to cases where the RP phenotype has not been selected.

4.5 Cross-population Comparison: Recombination Frequency and Selection

4.5.1 Recombination frequencies across populations

The recombinant frequencies in DH and RIL populations have been compared in maize (Murigneux *et al.*, 1993), wheat (Henry *et al.*, 1988) and rice (Courtois, 1993; Antonio *et al.*, 1996) using populations derived from different crosses.

DH lines are used extensively in barley breeding programmes to reduce the time required to obtain pure lines and to increase breeding efficiency. The availability of high-density linkage maps of barley makes it possible to compare the recombination frequencies in *H. bulbosum* (Hb) and anther culture-derived DH lines (see Section 4.2.1) across most of the barley genome. These methods differ in three major aspects: (i) the Hb and anther culture-derived DH lines arise from female and male recombinant products, respectively; (ii) the optimal donor plant growth conditions differ (Pickering and Devaux, 1992); and (iii) the *in vitro* culture phases are distinct: microspores evolve into embryoids that give rise to plantlets, while in the Hb method, plantlets develop from zygotic embryos. Recombinant frequency is likely to be affected by the first two features. Devaux *et al.* (1995) reported the results of an experiment comparing map distances observed in Hb and anther culture-derived DH lines obtained from an F_1 (Steptoe × Morex) hybrid. Male (anther culture-derived) and female (Hb-derived), DH populations were used to map the barley genome and thus determine the different recombination rates occurring during meiosis in the F_1 hybrid donor plants. The anther culture-derived (male recombination) population showed an 18% greater recombination rate than the Hb-derived population. This increased recombination rate was observed for every chromosome and most of the chromosome arms. Examination of linkage distances between individual markers revealed eight segments with significantly higher rate of recombination in the AC-derived population and one in the Hb-derived population. Although three out of eight segments that appeared significantly longer in the AC population are non-telomeric, the most significant increases were noted for the telomeres of the long arms of chromosomes 2 and 5.

Significant excess male recombination was also found at the most distal regions on chromosome 9 in tomato (de Vicente and Tanksley, 1991), proterminal regions of several linkage groups in *Brassica nigra* (Lagercrantz and Lydiate, 1995) and for several terminal intervals in *Pennisetum glaucum* (Busso *et al.*, 1995). In barley, most of the chromosome termini have been tagged with telomeric markers that were mapped in the Steptoe × Morex HB population (Kilian *et al.*, 1999). These telomeric markers in the AC population will need to be mapped in order to test whether an increased frequency of recombination at the telomeric regions is a general phenomenon in barley or is limited to particular chromosomal arms.

With more complete genetic maps now available for between-sex comparison both in plants and animals the tendency towards increased recombination frequency at the telomeres in male meiosis seems to be emerging as a general phenomenon. Since the differences in chiasma distribution show a similar trend, increased recombination frequency at the telomeres may be a biologically significant phenomenon. There is some evidence that sub-telomeric regions of human and cereal genomes are much more gene-rich than proximal regions. If this proves to be true for other species, it could be speculated that increased male recombination at the telomeric regions is selectively advantageous and therefore evolutionarily conserved since it results in an increased number of gene assortments in the more abundant male gametes. Male gametes are not only more abundant, but also energetically less costly and subject to selection at the pollination and/or fertilization stages.

4.5.2 Unintentional selection during the process of population development

The process by which genetic and plant breeding populations are developed may include various unintended selection pressures which result in the deviation of genotypic and allelic frequencies from Mendelian expectation. Segregation distortion has been documented in a wide range of organisms including plants. Distorted segregation can be detected with almost any type of genetic marker, including morphological mutants, isozymes and DNA markers (Table 3 in Xu *et al.*, 1997). The same allele at a given locus can be distorted in either of two directions, with an allele frequency higher or lower than the expected. For example, waxy and non-waxy rice kernels on a segregating plant could show ratios of equal to or larger or smaller than the expected ratio (3:1), with the proportion of waxy rice kernels ranging from 8.9 to 95.6% depending on the crosses (Xu and Shen, 1992d).

As reviewed by Xu *et al.* (1997), aberrant segregation ratios in plants may arise from a variety of physiological or genetic causes and may be manifested as differential transmission in either the male or female germ line or may result from post-zygotic selection prior to genotypic evaluation. Most commonly, however, skewed segregation appears to arise from male gametophytic selection, through the selective influences of the gynoecium, including genetic incompatibility, environmental effects and the differential competitive ability of genetically-variable pollen.

Selection pressure associated with DH development

The representativeness of DH lines can be severely affected by the process involved in DH development. For the DH populations derived from anther culture (male gametophytes), the observed distortion in segregation can be attributed to differential viability or lethality of pollen or to selective regeneration in *in vitro* culture and clearly not to the selective influences of the gynoecium or the differential competitive ability of pollen. The distortion in three chromosomal regions (two on chromosome 2 and one on chromosome 10) detected in the DH populations by Xu *et al.* (1997) indicated an overrepresentation of alleles from the *japonica* parent that has been proven to be easily regenerated by anther culture. The other parent is *indica* which belongs to a subspecies that is more recalcitrant to anther culture (Shen *et al.*, 1982; Yang *et al.*, 1983). It has been suggested that these regions may be associated with the preferential regeneration of *japonica* genotypes during anther culture. Yamagishi *et al.* (1996) also identified markers in several chromosomal regions that showed aberrant segregation ratios favouring *japonica* alleles in a DH population, although these markers segregated normally in the corresponding F_2 population. They concluded that these regions contained genetic factors which conferred a selective advantage on the *japonica* genotypes during anther culture.

Selective regeneration of genotypes has also been reported in other plants. Very strong distortions of single locus segregations were observed in an anther culture-derived barley population (Devaux *et al.*, 1995). Devaux and Zivy (1994) demonstrated that some markers showing distorted segregation are linked to genes involved in the anther culture response. In another barley DH population, a significant proportion (44%) of the mapped markers showed distorted segregation which was caused mainly by the prevalence of alleles from the parent that responded better to *in vitro* culture (Graner *et al.*, 1991). Although segregation distortion may arise from genetic, physiological and/or environmental causes and the relative contribution of each of these factors may differ in specific populations, much of the reported segregation distortion in anther culture-derived populations is likely to be the result of using parental genotypes that differ in their response to anther culture.

Selection pressure involved in RIL development

Deviation from randomness due to selection pressure in the production of RILs is a potential problem that needs more attention. In contrast to the populations derived from a one-step homogenization, RILs are produced by many generations of inbreeding during which plants are subjected to selection pressures generated by various environmental disturbances and competition among plants that may well occur for many years and seasons and in many locations. The distortions resulting from selection pressures involved in RIL development can be understood by comparison of multiple populations of different genetic structures derived from the same crosses and by comparison of populations produced by different approaches.

He *et al.* (2001) compared molecular marker segregations between DH and RIL populations derived by anther culture and SSD respectively from the same rice cross, ZYQ8 (*indica*) × JX17 (*japonica*). In the RIL population, 27.3% of the markers showed distorted segregation at the $P < 0.01$ level, of which 90% of the markers favoured *indica* alleles while in the DH population, 18.2% of the markers were skewed almost equally towards *indica* and *japonica* alleles. This might reflect the different types of selection pressures to which the DH and RIL populations were subjected. Eight commonly distorted regions on chromosomes 1, 3, 4, 7, 8, 10, 11 and 12 were detected in both RIL and DH populations of which seven skewed towards *indica* alleles and one towards a *japonica* allele. Five of them were located near gametophytic gene loci (*ga*) and/or sterility gene loci (*S*).

To compare the frequency and location of loci showing distorted allele frequencies between different population types (F_2, DHs and RILs), information from 53 populations with a known number of distorted markers was summarized and analysed (Xu *et al.*, 1997). In summary, RIL populations had significantly higher frequencies of distorted markers (39.4 ± 2.5%) than other population structures (DH: 29.4 ± 3.5%; BC: 28.6 ± 2.8%;

F_2: 19.3 ± 11.2%), which may indicate the cumulative effects of selection pressures during the process of RIL development. Distorted segregation in RIL populations derived via SSD represents the cumulative effect of both genetic (G) and environmental (E) factors on multiple generations and the G × E interaction becomes more pronounced with the progress of selfing. Thus, it is difficult to distinguish genetic from environmental causes of distortion in RIL populations. However, an over-representation of *indica* alleles in two chromosomal regions on chromosomes 3 and 6 was specific to one RIL population. These chromosomal regions may be associated with a selective advantage in the *indica* growth environment in which the RIL population was developed.

In contrast to DH and RIL populations where genotypic frequencies are a perfect reflection of the allele frequencies due to lack of heterozygotes, F_2 populations offer the potential to detect an advantage or disadvantage associated with the heterozygote class at specific loci, even when the parental allele frequency is normal.

Expression of distorters with low heritability will be influenced by the environment and therefore these will be detected only in experiments carried out under well-controlled conditions. Because the segregation distortion occurs either during, or just before or after meiosis, the experimental environment must be controlled during the reproductive phase of the parental lines, although the effect will only be detected in the offspring.

Genetics of selection associated segregation distortion

The genetic control of distorted segregation has been studied in rice (as summarized by Xu *et al.*, 1997) and barley (Konishi *et al.*, 1990, 1992) using morphological and isozyme markers. The genetic basis of segregation distortion may be the abortion of male or female gametes or the selective fertilization of particular gametic genotypes. Distortion at a marker locus in rice may be caused by linkage between the marker and the gene conferring lower pollinating ability, the gametophyte gene (*ga*) (Nakagahra, 1972) also referred to as a gamete eliminator or pollen killer, causing abortion of gametes (Sano, 1990). A large number of *ga* loci and sterility gene loci (*S*) have been identified using morphological markers.

If segregation distorters have high heritability, they will be detected in almost any population if the parents differ at the genetic locus in question and in almost any environment in which the population is grown. For a specific chromosomal region, the probability of a distortion locus being falsely assigned will decrease with the number of populations sharing the same distortion and with the number of markers in a cluster of distorted markers. Use of multiple populations developed in multiple environments would facilitate the detection of highly heritable genetic segregation distortion factors. Chromosomal regions associated with marker segregation distortion in rice were compared using six molecular linkage maps (Xu *et al.*, 1997). Mapping populations were derived from one interspecfic backcross and five inter-subspecfic (*indica/japonica*) crosses including two F_2 populations, two DH populations and one RIL population. Marker loci associated with skewed allele frequencies were distributed on all 12 chromosomes. Distortion in eight chromosomal regions showed the grouping of previously identified gametophyte (*ga*) or sterility genes (*S*). Three additional clusters of skewed markers were observed in more than one population in regions where no gametophytic or sterility loci had been reported previously. A total of 17 segregation distortion loci were postulated and their locations in the rice genome were estimated. Using a single F_2 cross, Harushima *et al.* (1996) identified 11 major segregation distortions at ten positions on chromosomes 1, 3, 6, 8, 9 and 10 and at least two of these segregation distortion regions (on chromosomes 1 and 3) were also detected by Xu *et al.* (1997).

A similar comparison was undertaken among four maize mapping populations using 1820 co-dominant markers (Lu *et al.*, 2002). On a given chromosome nearly all of the markers showing segregation distortion favoured the allele from the same parent.

A total of 18 chromosomal regions on the ten maize chromosomes were associated with segregation distortion. The consistent location of these chromosomal regions in four populations suggested the presence of segregation distortion regions. Three known gametophytic factors are possible genetic causes for the presence of these regions.

In *Populus* most markers exhibiting segregation distortion generally occurred in large contiguous blocks on two linkage groups and it has been hypothesized that divergent selection had occurred on the chromosomal scale among the parental species (Yin, T.M. *et al.*, 2004).

Segregation distortion loci were mapped to chromosomal regions including three regions on chromosome 5D in *Aegilops tauschii* using 194 molecular markers for an F_2 population (Faris *et al.*, 1998). Two sets of reciprocal BC populations were used to further analyse the effect of sex and cytoplasm on segregation distortion. Extreme distortion of marker segregation ratios in the chromosome 5D regions was observed in populations in which the F_1 was used as the male parent and ratios were skewed in favour of one parent. There was some evidence of differential transmission caused by nucleo-cytoplasmic interactions. This result, along with other studies, indicated that loci affecting gametophyte competition in male gametes are located on 5DL.

To map segregation-distorting loci using molecular markers, both a maximum likelihood (ML) method and a Bayesian method were developed (Vogl and Xu, 2000). ML mapping was implemented by use of an expectation-maximization algorithm and the Bayesian method was developed using the Markov chain Monte Carlo (MCMC) approach. Bayesian mapping is computationally more intensive than ML mapping but can handle more complicated models such as multiple segregation-distorting loci.

Implications for genetics and plant breeding

The phenomenon of segregation distortion is intimately linked to the probability of producing specific recombinants of interest in genetics and breeding populations. In genetics, construction of genetic maps and identification of linkage among markers and between markers and genes depends on the segregation patterns of all the markers and genes involved. In breeding, the success of obtaining specific genes, genotypes and gene combinations depends on the probability of the target genes and gene combinations occurring at a ratio expected by Mendelian segregation. To broaden the genetic base of cultivated species, breeders often undertake wide hybridization but they frequently fail to recover recombinants of interest, in part as a result of non-random survival or generation of offspring. On the other hand however, phenotypic selection during inbreeding including the backcrossing process, which breeders of course utilize, can significantly improve the probability of recovering the desired alleles.

Identification of genetic factors associated with segregation distortion will contribute to our understanding of where these genetic factors are located and how they might be managed in a breeding programme. If a target locus is known to be linked to a segregation distortion locus and is underrepresented in a desired population, the frequency of the favourable allele can be increased by using molecular markers to select for recombinants in the region of interest. To reduce the negative influence of segregation distortion in plant breeding, it is reasonable to decrease the number of generations required for stabilizing breeding lines. The production of DH populations from F_1 hybrids minimizes the number of generations required to reach homozygosity and therefore maximizes the chance of retaining desirable alleles in a population unless they are linked to segregation distortion factors that affect DH production. In wide crosses where wild alleles tend to be disproportionately lost, the frequency of rare alleles can be enhanced by adjusting the type of selection and population structure used in accordance with genetic information relating to segregation distortion, thus providing further opportunities for favourable recombination in later generations (Xu *et al.*, 1997). To understand the underlying mechanism(s) that is responsible for segregation distortion, it would be useful to develop NILs containing individual segregation distortion loci so that the effect of these factors could be evaluated systematically in different genetic backgrounds and environments. NILs would also provide material for cloning these genetic factors to permit a more in-depth characterization of their molecular structure and function.

5

Plant Genetic Resources: Management, Evaluation and Enhancement

Plant genetic resources are one of the most important tools in agricultural research and are used for the improvement of productivity and sustainability of production systems, both in the developed and in the developing world. The beginnings of the contemporary international system for germplasm conservation can be traced back to the brilliant and pioneering work of the Russian botanist, Nikolai Vavilov. He was the first, in the 1920s, to realize the importance and potential benefits to be derived from gathering plant genetic resources from around the world and organizing them into a collection. He noted that the task was not only to gather plants for the immediate breeding needs of Soviet agriculture, but also to save seeds from extinction. He also recognized that modern cultivars were replacing the local landraces and that plant research was destroying the very foundation of its own existence, thereby threatening global food security. Since the late 1950s, there has been an increasing awareness and documentation of the benefits of biodiversity and the risks associated with genetic erosion. Different methods have been developed to conserve plant genetic resources and make them available to breeders. Analytical methods including molecular markers and geographic information systems (GIS) have been developed to characterize genetic diversity and make its management, evaluation and enhancement more efficient. In this chapter, most fields related to plant genetic resources will be covered, including germplasm collection, maintenance, evaluation, enhancement, utilization and documentation. As an introduction, biodiversity and genetic diversity will be discussed first.

Biodiversity is of ecological, economic and cultural importance. Diversity within an ecosystem allows it to survive and be productive while providing an enormous range of products and services for exploitation by man. Agrobiodiversity, as a component of the total biodiversity, is important to agriculture. It helps ensure sustainability, stability and productivity (Hawtin, 1998). As recognized by the Convention on Biological Diversity, there are three interdependent levels of biodiversity: ecosystem level, species level and genetic level, each of which is influenced by, and influences the other.

As described by Hawtin (1998), ecosystem diversity can be defined as the variability between interdependent communities of species and the physical environment in which they live. Diverse agroecosystems can lead to a wide variety of enterprises on a national, regional or community scale which in turn contributes to maximizing food security, helps to increase employment opportunities and increase local or national self-reliance. Species diversity relates to the cultivars of species within an area. A diversity of crop and animal species on the community, farm or field levels can help add stability by reducing reliance on a single enterprise. Such diversity can also lead to a more efficient use of resources, for example by providing increased opportunities for nutrient recycling. Species diversity at the field level, e.g. by planting crop mixtures, can help to provide a buffer against adverse conditions, pests and diseases. In addition, diversity in agricultural enterprises allows for a more efficient use of inputs such as labour. Genetic diversity refers to the variation within a species which represents the total genetic information contained in individual plants of a species, each consisting of a unique assembly of genes constituting its evolutionary heritage. This diversity begins at the molecular level, is carried as sequences of instructions on chromosomes and provides the foundation for environmental adaptation and ultimately for the evolution of species. Genetic diversity enables species to adapt to new ecosystems and environments or changes in the current environment, by natural and/or human selection. Diversity within a crop species helps diminish the risk of losses through diseases or pests and provides opportunities for the exploitation of different features of the microenvironment by, for example the presence of diverse growth habits and rooting patterns. Such factors can contribute both to greater stability and in many circumstances, greater productivity. At the same time, genetic diversity within crops helps to provide a reservoir of genes for future crop improvement by farmers and professional plant breeders. Figure 5.1 provides an example of genetic diversity in maize kernel phenotypes which exists in the maize germplasm, although only a few of these phenotypes now exist in cultivated maize.

Of the three levels which comprise global biological diversity, genetic diversity has received the greatest attention within the agricultural community. As the raw material of future elite cultivars and an indicator of sustainability of agricultural production, the status of genetic diversity is of utmost concern for agricultural production and particularly for societies in general. This chapter will focus on genetic diversity within crops, i.e. the genetic resources that lie at the heart of sustainable agricultural development and provide the basis for the continued evolution and adaptation of crops.

5.1 Genetic Erosion and Potential Vulnerability

5.1.1 Genetic erosion

For centuries human intervention has altered the dynamic relationships among the various ecosystems used for food, feed, fuel, fibre, shelter and medicines. However, one of the most profound and irreplaceable changes that humans have wrought is the acceleration of the rates of extinction of species caused by human colonization, extension and intensification of agriculture and industrialization. For example, deforestation of tropical rainforests at the current rate, may result in the elimination of somewhere between 5 and 15% of the world's species between 1990 and 2020. Given the current estimate of about 10 million species in the world, these rates would translate into a loss of 15,000–50,000 species year^{-1} or 50–150 species day^{-1}. Thomas *et al.* (2004) estimated the total extinction of plant species in Amazonia following the maximum expected climate change leading to habitat destruction or climatic unsuitability to be 69% for species with seed dispersal and 87% for those without seed dispersal. The clearing of forests and the spread of urban areas is also resulting in the disappearance of the wild relatives of crop plants. On the other hand, as indicated by Brown and Brubaker (2002) several hundred crop and wild plant species previously used by humans, are now classified as underutilized or neglected.

The erosion in biodiversity is caused by a multiplicity of factors, including loss, fragmentation and degradation of habitats; introduction of alien species into ecological niches; overgrazing; excessive harvesting beyond the levels of natural regeneration

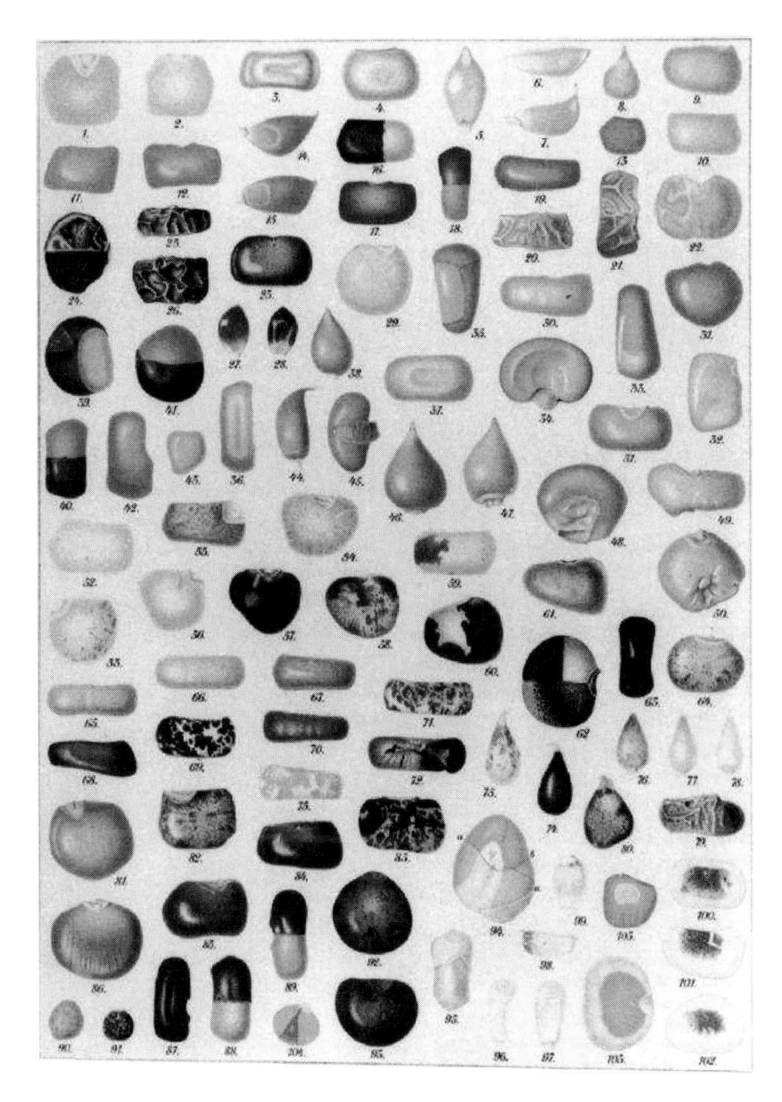

Fig. 5.1. Maize kernel phenotype. After Neuffer *et al.* (1997; the original plate is from Correns, 1901).

as in the case of trees harvested for timber; pollution of various media that sustain the biological nutrient cycles in ecosystems; deforestation and land clearance which is cited as being the most frequent cause of genetic erosion in Africa; adverse environmental conditions such as drought and flooding; introduction of new pests and diseases; population pressure and urbanization; war and civil strife; technological advances in agriculture, particularly the green revolution which resulted in the abandonment of traditional crops in favour of new ones, the settlement of new lands, changes in cultivation methods and changes in agricultural systems. Apart from the narrowing of genetic diversity by extensive mono-cropping, the Green Revolution has contributed indirectly to the loss of biodiversity by soil mining. As a result of the heavy use of fertilizers, chemical inputs and irrigation, agricultural crops which rely on Green Revolution technology have rendered the soil sub-fertile and in some cases inhospitable to other species. In addition, the phenomena of genetic drift and selection pressure produce cumulative genetic erosion that may sometimes exceed the genetic erosion actually taking place in the field (Esquinas-Alcázar, 1993). Concerns about the genetic erosion of plant genetic resources were first articulated by scientists in the mid-20th century and have since become an important part of national policies and international treaties.

Genetic erosion, or the reduction in genetic diversity in crop plants, takes on various shapes depending on one's viewpoint, including the reduction in the number of different crop species being grown and the decrease in genetic diversity (number of unrelated cultivars being grown with crop species). In addition, other organisms, both within and across agroecosystems, are increasingly taken into account when assessing biodiversity as it relates to agriculture (Collins and Qualset, 1999; Hillel and Rosenzweig, 2005). As one of the major factors for genetic erosion, the transition from primitive to 'advanced' cultivars as a result of plant breeding, is worthy of a further discussion. This has occurred by two distinct pathways: (i) selection for relative uniformity, resulting in 'pure' lines, multilines, single or double hybrids, etc.; and (ii) selection for closely defined objectives. Both processes have resulted in a marked reduction in genetic variation. At the same time there has been a tendency to restrict the gene pool from which parental material has been drawn. This is a result of the high level of productivity achieved when breeding within a restricted but well-adapted gene pool and breeding methods that have made it possible to introduce specifically desired improvements such as disease resistance and quality characteristics, into breeding stocks with a minimum of disturbance to the genotypic structure by backcrossing or transgenic approaches.

In the process of modern plant improvement, the traditional cultivars (landraces) of farmers have been replaced by modern cultivars. In the 1990s, only about 15% of the global area devoted to rice and 10% of the developing world's wheat area were planted to landraces (Day Rubenstein *et al.*, 2005). Other examples include the following: (i) of the nearly 8000 cultivars of apple that grew in the USA at the turn of the 20th century, more than 95% no longer exist; (ii) only 20% of the maize types recorded in 1930 in Mexico can now be found; and (iii) only 10% of the 10,000 wheat cultivars grown in China in 1949 remain in use (Day Rubenstein *et al.*, 2005; Gepts, 2006). The process is under way in all countries, both developed and developing and unfortunately includes some of the richest primary and secondary gene centres of several important food crops (Dodds, 1991). As the demand for uniform performance and grain quality has increased, new cultivars including hybrids are increasingly derived from adapted, genetically related and elite modern cultivars. The more genetically variable but less productive primitive ancestors have been almost excluded from most breeding programmes. In a study of pedigree relationships among 140 US rice accessions, Dilday (1990) concluded that all parental germplasm in public cultivars used in the southern USA could be traced back to 22 plant introductions in the early 1900s and those used in California could be traced back to 23 introductions. The same situation is true for soybean and wheat. Virtually all modern US soybean cultivars can be traced back to a dozen strains from a small area in north-eastern China and the majority of hard red winter wheat cultivars in the USA originated from just two lines imported from Poland and Russia (Duvick, 1977; Harlan, 1987).

5.1.2 Genetic vulnerability

Genetic vulnerability is the potentially dangerous condition which results from a narrow genetic base. One of the most tragic cases on record caused by genetic vulnerability is the Irish potato famine of the 1840s in which more than one million Irish starved to death as a consequence of a massive attack of late blight (*Phytophthora infestans*) that destroyed the Irish potato crop. The potato had been the main staple of the Irish diet for the preceding centuries. The underlying cause of the catastrophe was the narrow genetic base of the potato plants in that country; all had originated from a small quantity of uniform materials brought from Latin America in the 16th century. Other famous examples include the coffee rust epidemic in Ceylon (1868) and Southern corn leaf blight epidemic in the USA (1970).

Genetic vulnerability stems from genetic uniformity, examples of which are homozygosity (often recessive) as a result of clonal reproduction and the formation of F_1 hybrids from inbred parents (e.g. hybrid maize). The types of uniformity desired in a crop are: (i) rapid and uniform germination of seeds; (ii) nearly simultaneous flowering and maturation; (iii) stature that promotes mechanical harvest; (iv) product uniformity for taste, flavour and chemical composition; and (v) year-to-year stability of yield (Wilkes, 1993). With the substitution and consequent loss of a primitive cultivar, the genetic diversity contained in it is eliminated. To prevent such losses, samples of the replaced landraces should be adequately conserved for possible future use. The tendency to eliminate the genetic diversity contained in primitive landraces of plants jeopardizes the possible development of future cultivars adapted to tomorrow's unforeseeable needs.

As a few elite cultivars have come to dominate the major crops worldwide, genetic vulnerability has increased. As Wilkes (1993) indicated, we are now promoting a carpet of closely related dwarf-stature cultivars across the grain belts of the world. The magnitude of this potential is made clear by the fact that most of the hybrid rice planted in Asia now shares the same maternal cytoplasm and most of the high-yielding bread wheat cultivars are presently based on only three types of cytoplasm. There are many more such examples in other important crops. The burden of genetic vulnerability has been placed primarily on the shoulders of plant breeders because elements in the technology of plant breeding can be designed to minimize its impact, for example by developing synthetic or composite cultivars and multi-lines. Biotechnology, including genomics, promises the potential to both enhance and further endanger diversity. As a double-edged sword, biotechnology can enhance or jeopardize the greater utilization of genetic resources.

5.2 The Concept of Germplasm

5.2.1 A generalized concept of germplasm

Germplasm can be defined as the genetic materials that represent an organism. The expression of plant genetic resources usually refers to the sum total of genes, gene combinations or genotypes embodied as cultivars that are available for the genetic improvement of crop plants. Following the proposal of Harlan and de Wet (1971), plant genetic resources were classified into three gene pools that reflected the increasing difficulties in carrying out sexual crosses and obtaining viable and fertile progenies. Gene Pool I includes the crop species itself and its wild progenitor. Crosses within Gene Pool I can generally be made easily and the resulting progeny is viable and fertile. This gene pool corresponds closely to the biological species concept. Gene Pools II and III include other species that are less related to the crop species of interest. Crosses between Gene Pools II and III are possible but are usually more difficult to achieve. The progeny shows reduced viability and fertility. Finally, crosses between Gene Pools I and III are the most difficult. Special techniques such as tissue culture and embryo rescue must be used to obtain a progeny from these crosses. The progeny often show a severe reduction in viability and fertility. The operational definition of Harland and de Wet (1971) has been very useful because it reflects the realities of the breeding process, particularly the introduction of new genetic diversity into the populations of a breeding programme by sexual hybridization (Gepts, 2006). However, it could be argued that this definition may need to be expanded to include a Gene Pool IV based on the advances in scientific technology and increased awareness of the benefits of biodiversity in general. Availability of plant transformation techniques (as discussed in Chapter 12) has extended the reach of plant breeding beyond the limitations imposed by sexual cross-compatibility and as a result, the Gene Pool IV should include all organisms as a potential source of genetic diversity.

Classical germplasm can be defined according to reproduction systems to include seeds from sexual plants and all types of tissue such as roots, stems and other organs that can be used for reproduction in asexual plants. Therefore, germplasm is traditionally defined as a morphologically distinct biological object. Different plant species or cultivars from the same species can be distinguished from each other morphologically based on size, colour and shape. In the case of sexual plants, seeds are the major carrier of germplasm and for most plant species germplasm can be maintained and reproduced by the collection and regeneration of seeds. Seeds are of major importance in the process of germplasm management and are collected, maintained and reproduced. Evaluation and utilization of germplasm is dependent on the seeds that can be used to generate plants and on other useful organs such as root, leaves, stems or even seeds themselves.

With the development of molecular biology, the concept of germplasm has been generalized and broadened. As a carrier of genetic material, germplasm can be anything that carries genetic information required for controlling and rebuilding an organism which includes genes and their clones, chromosome segments and even pieces of functional DNA sequences. The generalization of the concept of germplasm depends on two major developments: cell totipotency (the potential for regenerating a whole plant from a single cell) and the development of the gene concept (genetic material can be traceable to a small piece of DNA that controls a biological trait and codes for a specific protein) (Xu and Luo, 2002; Fig. 5.2).

DNA as a type of genetic resource is rapidly increasing in importance. DNA from the nucleus, mitochondrion and chloroplasts are now routinely extracted and immobilized on to nitrocellulose sheets where the DNA can be probed with numerous cloned genes. With the development of PCR, specific fragments or entire genes from a mixture of genomic DNA can now be routinely amplified (Engelmann and Engels, 2002). Genetic information can be synthesized and living variants can be created or rebuilt by using DNA sequence information. These advances have led to the formation of an international network of DNA repositories for the storage of genomic DNA (Adams, 1997). The advantage of this technique is that it is efficient and simple and overcomes physical limitations or constraints. The disadvantage lies in problems with subsequent

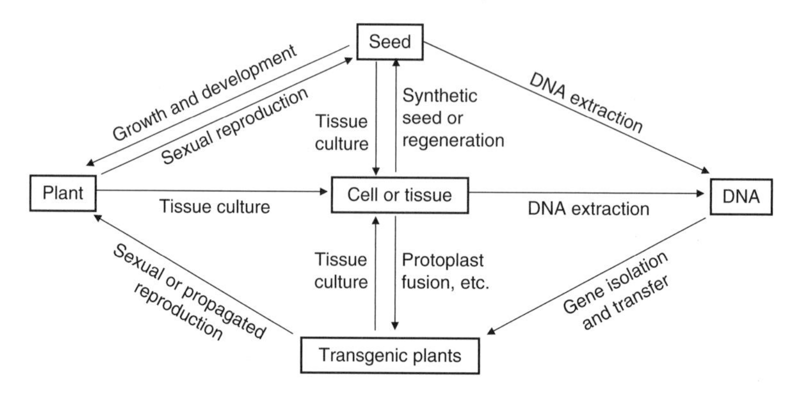

Fig. 5.2. Germplasm carriers and their conversion through biological, molecular and biotechnological approaches. Modified from Xu and Luo (2002).

gene isolation, cloning and transfer (Maxted *et al.*, 1997). A tool of potential importance is the development of collections of DNA samples *in glacie* from wild species.

Genome sequencing and the understanding of the function of all plant genes will have significant impacts on the conservation of plant genetic resources. The role that the genetic resources community might undertake in respect to conserving molecular genetic products has yet to be defined. As such, genebanks are facing new demands from the user community. There are an increasing number of resources being generated by the molecular community that are relevant to conservation work. Some genebanks might wish to store primers, probes and DNA libraries to facilitate their work, in addition to populations generated for gene isolation and plant improvement. In the future, users may want to receive functional DNA sequences, genes, clones or markers instead of seeds or other traditional means of transporting DNA. They may want a series of specific alleles rather than accessions where alleles are segregating. As Kresovich *et al.* (2002) indicated a future-oriented analysis of these possible trends and their implications is very important in order to predict and thus prepare for the changing role of genebanks and curators.

On the other hand, the concept of germplasm is no longer defined for each species or crop plant and its relatives. With the development of technologies for gene cloning and transfer as discussed in Chapters 11 and 12, the genetic barriers that used to exist among different species or genera no longer exist and genes can be exchanged freely between different families and genera or even between plants, animals and bacteria. Useful genes identified in one species can be used to modify another species. Taxa that are evolutionarily related (e.g. grasses, legumes and members of the *Solanaceae*) have strikingly similar genome organizations as discussed in Chapter 2. At the fundamental level of the gene, many sequences are highly conserved across families. Therefore, users of generic resources will acquire useful genes from repositories independent of their source (Kresovich *et al.*, 2002). As a carrier for genetic materials, therefore, germplasm is no longer limited to a specific species. Germplasm can be managed based on the classification of genes or properties across species rather than the terminology and classification of the plant. With the new technologies that have been developed in molecular biology and genomics, the gene pool for any given species has expanded well beyond the tertiary gene pool and can be taken to include any gene from any source, perhaps in the future, even to new synthetic or shuffled sequences.

As germplasm collection turns increasingly to tissues, cells and DNA, methods for collecting and preserving these materials may need to be modified or revolutionized by using diversified techniques for preservation and reproduction. For example, preservation of tissues and cells can be achieved by subculture or regeneration instead of the storage of seeds. Using cloning and transformation, DNA can be transferred into other plants to obtain transgenic plants (Fig. 5.2). Germplasm management in the future will become an integrated science that is closely linked with biotechnology including tissue culture, gene cloning and transformation, molecular marker technology and synthetic seed technology. As summarized by Taji *et al.* (2002), there are five main areas of biotechnology that can directly assist plant conservation programmes: (i) molecular biology, particularly molecular markers: assisting in germplasm collection, aiding genebank design and accession structure and assessing genome stability, genetic diversity, population structure and distribution patterns; (ii) molecular diagnostics: assessing phytosanitary status; (iii) *in vitro* culture: micropropagation, slow growth and embryo rescue; (iv) cryopreservation: long-term conservation of seed-recalcitrant species, vegetatively propagated species and biotechnological products; and (v) information technology: documentation, training, transfer technology, germplasm exchange, DNA databases, genome maps, genebank inventories and international networking. These areas will be discussed in various sections of this chapter.

5.2.2 Classical germplasm

Classical germplasm can be identified by their position in an agricultural ecosystem:

1. Commercial cultivars (cultivars in current use): these are the standardized and commercialized cultivars that have in general been created by professional plant breeders. Many of them are characterized by high productivity when subjected to intensive cultivation systems requiring heavy investment (fertilizers, irrigation, pesticides, etc.) and most by uniformity which may lead to a high degree of genetic vulnerability.
2. Advanced breeding lines: these are the materials obtained by plant breeders as intermediate products. These lines usually have a narrow genetic base because in general they have originated from a small number of cultivars or populations.
3. Landraces or traditional/primitive cultivars: these are primitive cultivars that have evolved over centuries or even millennia, have been influenced decisively by migrations and have been subjected to both natural and artificial selection. There is a large diversity between and within these cultivars that are adapted to survive in often unfavourable conditions, have low but stable levels of production and are therefore characteristic of subsistence agriculture. These primitive cultivars have been frequently used as parental lines to breed new cultivars.
4. Wild and weedy species related to cultivated species: these are either the ancestors of a cultivated crop or species that are genetically close to the crop so that genes can flow between them without much difficulty or where the genetic barriers between them can be removed by certain methods such as embryo rescue of hybrids. These germplasm resources are becoming increasingly important as genetic engineering has facilitated the transfer of genes between different plant species, families and genera.
5. Special genetic stocks: this category includes other genetic combinations such as genetic, chromosomal and genomic mutants which have been naturally or artificially produced and conserved in the collections of either geneticists or plant breeders. The material can be of value in itself or used as a tool for research.
6. Co-adapted or symbiotic organisms: in which two forms of a crop, two distinct crops or a crop and its symbiont (unrelated weeds or a legume and a nodule-forming bacteria) are grown together.

5.2.3 Artificial or synthetic germplasm

Man-made germplasm is described as artificial or synthetic and can be of the following types:

1. Organisms possessing exotic genetic materials including transgenic plants and engineered plants.
2. Organisms containing genetic modifications including variants induced by physical and chemical agents, somaclonal variants derived from tissue culture and mutants occurring naturally during the process of maintenance and production of germplasm.
3. Synthetic cultivars and species where the former includes novel cultivars derived from distant crossing with relatives such as cultivars of sorghum with maize chromosomal segments or genes and the latter includes a man-made cereal, octaploid triticale, which is derived from hybridization of wheat (*Triticum aestivum*) and rye (*Secale cereale*).
4. Variants with chromosome changes in structure and number including polyploids produced by chromosome doubling, somatic allopolyploids from cell hybridization and aneuploids which contain an abnormal chromosome number due to a missing chromosome.
5. A set of individuals with a specific genetic structure as discussed in Chapter 4, including nearly isogenic lines that show isogenic differences at specific genetic loci, recombinant inbred lines that are derived from continuous inbreeding of an F_1 hybrid and doubled haploid lines derived from female or male gametes by gynogenesis or androgenesis using techniques such as anther culture and chromosome doubling.

These types of germplasm resources have become very important materials in genetics and plant breeding.

5.2.4 *In situ* and *ex situ* conservation

As discussed above, the prospects of species extinction is forever with or without human intervention. Therefore, the best preparation for future uncertainties is to conserve as many gene pools, species and ecosystems as possible, whether they have actual or potential utility to humankind. Use for the benefit of humankind is the strongest justification for the conservation of plant genetic resources. During recent decades awareness has been raised of the importance of conserving crop gene pools to ensure that the breeder has adequate raw materials. This process is dynamic since the breeder is continually seeking new alleles and allelic combinations to improve the performance of a crop species in its target environments.

Conservation of germplasm resources goes far beyond the preservation of a species. The objectives must be to conserve sufficient diversity within each species to ensure that its genetic potential will be fully available in the future. Conservation of germplasm resources has been generalized to include all activities relating to germplasm management such as collection, maintenance, rejuvenation and multiplication, evaluation, exchange and documentation. In this section the discussion will focus on the methods for germplasm conservation.

Conservation and control are two issues broadly related to biodiversity that have a bearing on the role of biotechnology. Conservation refers to the maintenance or enhancement of biodiversity – particularly the plant species – and control refers to accessing this diversity. Two basic conservation strategies, *in situ* and *ex situ* conservation, each composed of various techniques are employed to conserve genetic diversity for various research and development programmes including plant breeding. *In situ* conservation refers to the preservation of genetic resources within the evolutionary dynamic ecosystems of their original or natural environment including conservation in nature reserves and on the farm. This type of conservation control is most suited to wild related species. *Ex situ* conservation entails removing germplasm resources (seed, pollen, sperm and individual organisms) from their original habitats and preserving them in botanical gardens or gene/seed banks. Examples of different methods of *ex situ* conservation include field genebanks, seed storage, pollen storage, *in vitro* conservation and DNA storage.

There is an obvious fundamental difference between these two strategies: *ex situ* conservation involves the sampling, transfer and storage of target taxa remotely from the collection areas whereas *in situ* conservation involves the designation, management and monitoring of target taxa where they are encountered (Maxted *et al.*, 1997). Another difference lies with the more dynamic nature of *in situ* conservation as opposed to the more static nature of *ex situ* conservation.

Each technique has its own advantages and limitations. The major drawback of *ex situ* conservation is that the evolution of species would be frozen since no further adaptation to environmental or biotic stresses indigenous to their origin can take place and that the processes of selection and continuous adaptation to those local habitats are halted. Other disadvantages are that long-term integrity of the germplasm remains in question and that high rates of mutation exist among the *ex situ* stored plants. Further drawbacks are the occurrence of genetic drift (random loss of diversity due to the fact that the samples collected and multiplied are necessarily very small) and selection pressure (the materials are usually multiplied in ecogeographical areas which differ from those in the areas where they were originally collected).

In situ techniques allow the conservation of greater inter- and intraspecific genetic diversity than is possible in *ex situ* facilities. They also permit continued evolution and adaptation to take place, whether in the wild or on the farm where selection by man also plays a critical role. For some species such as many tropical trees, it is the only feasible method of conservation. The main drawback is the difficulty in characterizing, evaluating and assessing genetic resources and susceptibility to hazards such as extreme weather conditions, pests and diseases. In addition, the monetary expense may be quite high, especially where there is pressure for alternative uses of the land. The method selected for *in situ* conservation depends on the nature of the species. Traditional crop cultivars may be conserved on the farm while undomesticated relatives of food crops may require land to be set aside as reserves (Hawtin, 1998).

In situ conservation is especially appropriate for wild species and for landrace materials on the farm while *ex situ* conservation techniques are particularly appropriate for the conservation of crops and their wild relatives (Engelmann and Engels, 2002). *In situ* conservation of biodiversity enables the preservation of the knowledge of farming systems, including biological and social knowledge associated with them. *Ex situ* conservation on the other hand, divorces the biological from the social context.

In situ and *ex situ* systems for conservation of germplasm resources should be considered as complementary and not antagonistic. The current approach is to combine both methods of conservation depending on such factors as reproductive biology, nature of the storage organs and propagules and availability of human, financial and institutional resources (Bretting and Duvick, 1997). Many major food plants produce seeds that undergo maturation drying or can be dried to low-water content due to their tolerance to extensive desiccation and can therefore be stored dry at low temperatures. Seeds of this type are known as 'orthodox' (Roberts, 1973). Storage of such orthodox seeds is the most widely practised method for *ex situ* conservation of plant genetic resources and about 90% of the 6.1 million accessions stored in genebanks are maintained as orthodox seed (Engelmann and Engels, 2002). For most species, stored seed is the most genetically relevant, i.e. it is the raw material with which the breeder works and seed propagation is an integral part of the growth cycle of the crop.

There are a significant number of crops which fall outside the category of orthodox seed for several reasons. First, some species do not produce seeds at all and consequently are propagated vegetatively; these include banana and plantain (*Musa* spp.). Secondly, some species such as potato, other root and tuber crops such as yams (*Dioscorea* spp.), cassava (*Manibot esculenta*), sweet potato (*Ipomoea batatas*) and sugarcane (*Saccharum* spp.) either have some sterile genotypes and/or do not produce orthodox seed. However, if they are capable of seed production, these seeds are highly heterozygous and are therefore of limited utility for the conservation of particular genotypes. These crops are usually propagated vegetatively to maintain the genotypes as clones (Simmonds, 1982). Thirdly, a considerable number of species, predominately tropical or subtropical in origin such as coconut, cacao and many forest and fruit tree species, produce seeds which do not undergo maturation drying and are shed at relatively high moisture content. Such seeds are unable to withstand desiccation and are often sensitive to chilling. Seeds of this type are called recalcitrant and need to be kept in moist, relatively warm conditions to maintain viability (Roberts, 1973; Chin and Roberts, 1980). Even when recalcitrant seeds are stored in an optimal manner, their lifespan is limited to weeks or occasionally months.

Other conservation methods are needed for these recalcitrant species. These include conservation as living collections in field genebanks as described above for *in situ* conservation or *in vitro* conservation either as living plantlets, plant tissue on appropriate media often under conditions of slow growth or by cryopreservation at very low temperatures, generally using liquid nitrogen. For those problem species whose seeds do not survive under conventional storage conditions largely because they cannot tolerate desiccation and die when exposed to low temperatures, the field genebank is the conventional approach to their conservation. However, there are many drawbacks to this, not least being that field genebanks cannot provide secure, long-term conservation as compared to the safety and low input requirements of a seed genebank.

5.3 Collection/Acquisition

For most species the material to be collected consists of seeds, although in some cases it may be bulbs, tubers, cuttings, whole plants, pollen grains or even tissue samples for *in vitro* culture depending on the characteristics of the species and the manner in which the material is to be conserved. Much work has been carried out on the collection and acquisition of germplasm resources worldwide. The centres of the Consultative Group on International Agricultural Research (CGIAR) have the responsibility of collecting, preserving, characterizing, evaluating and documenting the genetic resources of the cultivated and wild relatives of the cereals (barley, maize, millets, oat, rice, sorghum and wheat), legumes (Bambara groundnuts, chickpea, common bean, cowpea, faba bean, grasspea, lentil, pea, groundnut, pigeonpea and soybean), roots and tubers (Andean root and tuber crops, cassava, potato, sweet potato and yam) and *Musa* (both banana and plantain). Based on the most recently available data, over 6 million accessions are stored *ex situ* throughout the world; of these, some 600,000 are maintained within the CGIAR system and the remaining 5.4 million accessions are stored in national or regional genebanks. Nearly 39% are cereals, 15% food legumes, 8% vegetables, 7% forages, 5% fruits, 2% roots and tubers and *c.*2% oil crops (Scarascia-Mugnozza and Perrino, 2002). Approximately 527,000 accessions are stored worldwide in field (*in situ*) genebanks, of which 284,000 are in Europe, 10,000 in the Near East, 84,000 in Asia and the Pacific, 16,000 in Africa and 117,000 in the Americas (FAO, 1998). There are 1500 botanical gardens (11% private) worldwide which maintain living collections of plants. About 10% of these also have seed banks and 2% *in vitro* collections. Vegetatively propagated species, forest trees, medicinal and ornamental species, and plant genetic resources for food and agriculture which are of local significance are usually well represented.

5.3.1 Several issues on germplasm collections

How representative a collection is compared to the entire species is a major concern of germplasm collections. A breeder will usually look for 'useful' agronomic characteristics (selective sampling), whereas the population geneticist may try to collect randomly (random sampling). It should be noted that the concept of 'usefulness' is relative and may vary according to the objectives and information available to the collectors. Collections can be made more representative by analysing patterns of ecogeographic differentiation to identify related species that comprise crop gene pools, ensuring that 90% of the input is not being targeted to save only 10% of the known diversity, and planning for additional exploration and collection to amplify the collections while avoiding any duplication of effort. Since genetic erosion will not wait for approval of pending international agreements or networking arrangements, plans for the collection of germplasm should take into account the numbers of samples estimated to be required by the World Resources Institute for crop gene pools, forest species, medicinal plants, ecosystem rehabilitation and traditional underexploited plants. Molecular markers such as randomly amplified polymorphic DNA (RAPD), restriction fragment length polymorphism (RFLP), simple sequence repeats (SSRs) and single nucleotide polymorphisms (SNPs) have contributed to a better understanding of the genetic structure of gene pools and, together with techniques such as GIS, offer new potential for mapping diversity which would help to establish representative germplasm collections more efficiently.

Optimal sampling methodology during the collection of field germplasm requires a clear understanding of the genetic structure of the crop species in question. Biotechnology can help to reduce the practical impediments to efficient collecting in at least two ways. First, biochemical and molecular characterization techniques can be used to provide information about the availability of genetic diversity in a given collecting area, thereby facilitating more rational and effective sampling. Molecular markers can be used to measure the degree of divergence within species, analyse inter- and intrapopulational diversity and monitor genetic erosion within genebank collections. Secondly, *in vitro* propagation methods can be modified for application in the field to provide new ways of collecting problem materials.

For clonally propagated and recalcitrant seed-producing species, the materials collected are often bulky and heavy. Furthermore, they are often soil bearing, thereby introducing a plant health hazard. Recalcitrant seeds and vegetative explants such as shoots, suckers or tubers have a limited lifespan and may be prone to decomposition through microbial attack. In some cases, suitable materials for collection may not even be available and seed may be immature or absent as a result of grazing. However, new *in vitro* collecting techniques involve the principles of *in vitro* inoculation and culture without the cumbersome and complex conditions that normally pertain to the laboratory. This was originally explored for cacao buds and the coconut embryo and was also successfully adapted for several other materials (Withers, 1993).

The observance of adequate quarantine, disease indexing and disease eradication procedures are essential for the safe movement of germplasm from its origin to genebanks and among genebanks and users. Clonally-propagated crops present particular problems in that they are commonly collected in the form of vegetative propagules that carry a relatively high risk of disease transmission. They may accumulate systemic pathogens since they lack the pathogen filter that the seed production stage can offer. The potential for eliminating pathogens via meristem-tip culture, sometimes linked to other therapeutic processes such as thermotherapy, is now an important component of the process of introducing many clonally-propagated crops into conservation collections. The introduction of the enzyme-linked immunosorbent assay (ELISA) and other methods based on nucleic acid, biochemical and molecular technology provide new methods for detecting pathogens.

Wild relatives of our present crop plants, although agronomically undesirable, may also have acquired many desirable stress-resistant characteristics as a result of their long exposure to nature's pressures. Many recent studies using wild relatives in genetic mapping have identified 'cryptic' alleles that do not exist in cultivated plants (for details see Chapter 7) which make conservation of wild species a more important component in germplasm resources than ever before. Requirements for the development of collection strategies suitable for wild relatives has been increasing and genomic tools including molecular markers can help to identify the genetic diversity and merits that exist in the wild relatives by the methods discussed in Section 5.5.

5.3.2 Core collections

As germplasm collections of major crop plants continue to grow in number and size around the world, better access to and use of the genetic resources in collections have become important issues. Potential users require either populations representative of the diversity or accessions that describe particular agronomic characters (e.g. disease resistance, drought tolerance). In either case the managers of collections may find it difficult to meet such needs. The very size and heterogeneous structure of many collections have hindered efforts to increase the use of genebank materials in plant breeding. Recognizing this, Frankel (1984) proposed that a collection could be represented by what he termed a core collection, which would 'represent with minimum repetitiveness, the genetic diversity of a crop species and its relatives'. The accessions excluded from the core collection would be retained as the reserve collection. Construction of a core collection involves selecting approximately 10% of the germplasm accessions to represent at least 70% of the genetic variation (e.g. Brown, 1989a, b) unless the entire germplasm collection is very large, in which case less than 10% would be necessary. This proposal was further developed by Frankel and

Brown (1984) and Brown (1989a), who outlined how to achieve core coverage of the collection by using information regarding the origin and characteristics of the accessions. In terms of practical use, the three major objectives of the core collection are to set up as wide a representation as possible of the genetic diversity to be able to conduct intensive studies on a reduced set of genotypes and to attempt to extrapolate the results thus obtained to facilitate research on appropriate genotypes in the base collection (Noirot *et al.*, 2003).

The core proposal was a radical departure in thinking regarding genetic resources (Frankel, 1986). Until then, the main emphasis had been on the open-ended task of collecting as many samples as possible and securing their survival in storage, irrespective of continuing cost and use. Frankel and Brown (1984) introduced the notion of adequacy of sampling of the species range. Analysis of climatic, ecological and geographical information on the species range could be used to suggest where distinctly different environments or separated localities occurred for that species. This analysis could be checked with the available collections and used to identify places or habitats where collections had been excessive and others where further collection is warranted. In this way, a complete collection can be built up, from which a core collection can be extracted.

Using all the available data, core collections are arranged to make their entries representative of genetic diversity. The basic procedure is to recognize groups of related or similar accessions within the collection and sample from each group. Presently, in the constitution of a core collection, most researchers agree on the need for stratification prior to the sampling. In other words, the organization of the variability in groups and subgroups should be taken into account. There are clear benefits to the greater use of these more precise measures of genetic variation. Equally clearly, it is costly in human and financial resources to generate these measures so they can only be employed in a limited number of collections. Therefore, the selection of which species and which samples to include is crucial. Since the aim is to obtain the maximum amount of useful information from a limited sample, the use of core collections is an obvious approach.

A general procedure for the selection of a core collection can be divided in four steps:

- Definition of the domain: the first step in creating a core collection is defining the material that should be represented, i.e. the domain of the core collection.
- Division into groups: the second step is dividing the domain into groups which should be as genetically distinct as possible.
- Allocation of entries: the size of the core collection should be determined and the choice of number of entries per group should be made.
- Choice of accessions: the last step is the choice of accessions from each group that are to be included in the core.

Several different methods have been used to construct core collections and these aim to represent most of the genetic diversity with the fewest number of accessions possible (see for example Noirot *et al.*, 2003). Many reports have been published on the formation of core subsets. Hintum (1999) described one such system, the Core Selector to generate representative selections of germplasm accessions. Upadhyaya and Ortiz (2001) developed a two-stage strategy for developing a mini-core collection, again based on selecting 10% of the accessions from the core collection representing 90% of the variability of the entire collection. In this process, a representative core collection is first developed using all the available information on geographic origin, characterization and evaluation data. In the second stage, the core collection is evaluated for various morphological, agronomic and quality traits to select a subset of 10% accessions from this core subset (or 1% of the entire collection) that captures a large proportion (i.e. more than 80% of the entire collection) of the useful variation. At both stages in selection of core and mini-core collections, standard clustering procedures are used to separate groups of similar accessions combined with various statistical tests to identify the best representatives.

Molecular markers have been used to construct core subsets which preserve as much of the diversity present in the original collection as possible (Franco *et al.*, 2005, 2006). Genetic markers on three maize data sets and 24 stratified sampling strategies were used to investigate which strategy conserved the most diversity in the core subset as compared with the original sample (Franco *et al.*, 2006). The strategies were formed by combining three factors: (i) two clustering methods (unweighted pair-group means arithmetic (UPGMA) and Ward); based on (ii) two initial genetic distance measures; and using (iii) six allocation criteria (two based on the size of the cluster and four based on maximizing distances in the core (the D method) used with four diversity indices). The success of each strategy was measured on the basis of maximizing genetic distances (Modified Roger and Cavalli-Sforza and Edwards distances) and genetic diversity indices (Shannon index, proportion of heterozygous loci and number of effective alleles) in each core. For the three data sets, the UPGMA with D allocation methods produced core subsets with significantly more diversity than the other methods and were better than the M strategy implemented in the MSTRAT algorithm for maximizing genetic distance.

Using the advanced M strategy with a heuristic search for establishing core sets, a program known as POWERCORE has been developed (Kim *et al.*, 2007). The program supports development of core-sets by reducing the redundancy of useful alleles and thus enhancing their richness. The output of the POWERCORE has been validated using some case studies and the program effectively simplifies the generation process of core-set while significantly cutting down the number of core entries, maintaining 100% of the diversity. POWERCORE is applicable to various types of genomic data including SNPs.

Based on phenotypic evaluation of economically important traits and the use of DNA markers, studies of genetic diversity aimed at developing core collections have been reported for several plant species. Crops with cores established at the early stage include lucerne, barley, chickpea, clover, lentil, medic, groundnut, bean, pea, safflower and wheat (Clark *et al.*, 1997). Mini-core collections are reported for crops such as chickpea (Upadhyaya and Ortiz, 2001), groundnut (Upadhyaya *et al.*, 2002), pigeonpea (Upadhyaya *et al.*, 2006b) and rice (1536 accessions, D.J. Mackill, International Rice Research Institute (IRRI), personal communication). Such efforts have led to the identification of diverse germplasm with beneficial traits of significant economic value being found in barley and many legume crops (Dwivedi *et al.*, 2005, 2007; Brick *et al.*, 2006). Table 5.1 provides examples for core collections that have been established with a relatively large number of germplasm accessions included. Several types of data were used for each crop, with geographic origin usually being one of the first criteria used for selection.

In rice, methods for selecting accessions to construct a core collection were investigated based on shared allele frequencies (SAFs) and the frequency of unique RFLP and SSR alleles (Xu *et al.*, 2004; Fig. 5.3). Subsets of various sizes were selected (representing 5–50% of the US and world collections) using random selection as a control. For each sample size, 200 replications were analysed using a re-sampling technique and the number of alleles in each subgroup was compared with the total number of alleles identified in the larger collection from which the subsets were sampled. A cultivar subset (13% of the entire collection) selected on the basis of both SAFs and number of unique alleles detected, represented 94.9% of the RFLP alleles but only 74.4% of the SSR alleles. It can be expected that selection criteria based on additional sources of information will further improve the value and representativeness of core collections. This resource may serve as a source of novel alleles for genetic studies and for broadening the genetic base of US rice cultivars. In addition, the following conclusions were drawn (Xu *et al.*, 2004): (i) more samples were needed to represent

Table 5.1. Description of core collections in barley, cassava, finger millet, maize, pearl millet, potato, rice, sorghum and wheat (modified from Dwivedi *et al.*, 2007).

Crop	Description[a]	Number of accessions	Reference
Barley	USDA-ARS barley core collection	2,303	Bowman *et al.* (2001)
	Core collection	670	Fu *et al.* (2005)
Cassava	Core collection	630	Chavarriaga-Aguirre *et al.* (1999)
Finger millet	Core collection	622	Upadhyaya *et al.* (2006a)
Maize	Chinese maize core collection	1,193	Li *et al.* (2004)
Pearl millet	Core collection	1,600	http://icrtest:8080/Pearlmillet/Pearlmillet/coreMillet.html
Potato	Core collection	306	Huamán *et al.* (2000)
Rice	USDA core collection	1,801	Yan *et al.* (2004)
	IRRI core collection	11,200	Mackill and McNally (2004)
Sorghum	Core collection	3,475	Rao and Rao (1995)
Wheat	Novi Sad Core collection	710	Kobiljski *et al.* (2002)
	Chinese common wheat core collection	340	Dong *et al.* (2003)

[a] Abbreviations: IRRI, International Rice Research Institute; USDA-ARS, United States Department of Agriculture-Agricultural Research Service.

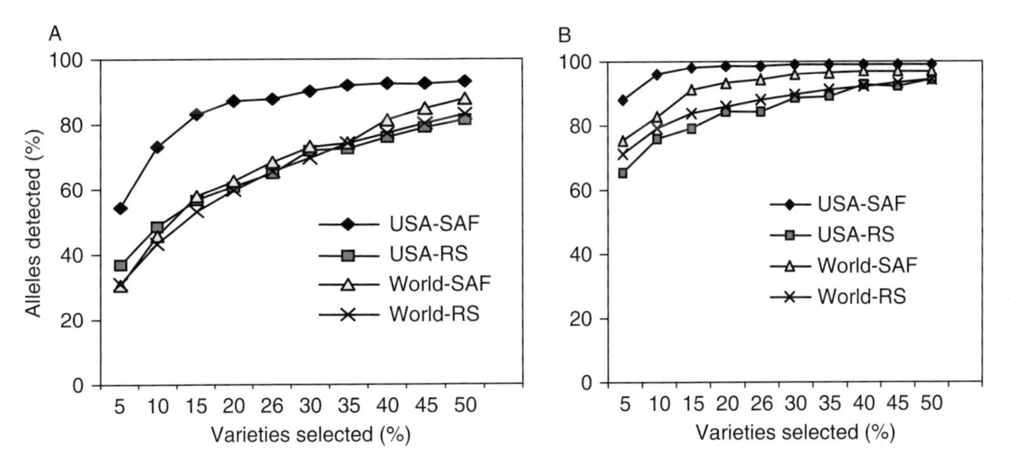

Fig. 5.3. Comparison of selection methods based on shared allele frequency (SAF) or random selection (RS) for identifying members of a core collection in rice. Proportion of RFLP (A) and SSR (B) alleles detected in US and World collections based on SAF or RS. Modified from Xu *et al.* (2004).

the world collection, which was more diverse than the US collection, which contained more pedigree-related cultivars; (ii) combining the use of SAF and unique alleles improved the representativeness of the core collection; (iii) core collections selected by SAF required fewer samples than random selection for the same level of representativeness; and (iv) more samples were needed to adequately represent genetic diversity if highly polymorphic markers were used (e.g. SSRs versus RFLPs).

The core collection concept has aroused considerable worldwide interest and debate within the plant germplasm resources community. It has been welcomed as a way of making existing collections more accessible through the development of a small group of accessions that would be the focus of evaluation and use and provide an entry point to the large collections that it aims to represent. However, a concern that still remains is that the available knowledge regarding genetic diversity in any crop is insufficient to enable a meaningful core to be developed and that the most useful characters often occur at such a low frequency that they would be omitted from any small core collection. Other concerns regarding core collections include rendering the reserve collection more vulnerable to loss, the lack of representation of rare, endemic alleles and a poor relationship with the specific needs of users (Gepts, 2006).

When molecular markers are developed from DNA sequences with unknown or no function, identical marker alleles among collections may not necessarily mean that these collections share identical functional alleles linked to the marker locus. Genetic variation for important phenotypic traits could be lost if core collections are based solely on the use of such anonymous DNA markers. As the genome sequence is deciphered and the function of many genes is determined, gene-specific markers with identified functional nucleotide polymorphisms (FNPs) will become available for many genes. Core collections of germplasm constructed using FNPs could be assembled to represent a 'core collection' of genes. As gene structure–function relationships are clarified with greater precision, it will be possible to focus attention on genetic diversity within the active sites of a structural gene or within key promoter regions. This will make it productive to screen large germplasm collections for FNPs, targeting the search for alleles that are likely to be phenotypically relevant at specific loci. From a primary collection, a user who had identified an accession or accessions of interest would move to the next level of information where clusters of germplasm known to represent a broader spectrum of diversity within a specific gene pool, or a specific trait, could be defined. The second level of investigation could be conducted using carefully designed sets of molecular markers known to target specific traits or regions of the genome. The construction of core collections using these approaches may help establish heterotic groups from which parents can be chosen to establish base populations for breeding hybrid crops.

5.4 Maintenance, Rejuvenation and Multiplication

The main task of a germplasm bank is to conserve germplasm in a state in which it can be indefinitely propagated without loss of genetic diversity or integrity. In general, the term 'base collection' is applied to collections stored under long-term conditions, whereas the term 'active collection' is used for collections stored under medium-term conditions and 'working collection' refers to breeders' collections usually stored under short-term conditions. Monitoring the health of collections, particularly of field genebanks, assessment of accession viability and rejuvenation and multiplication of collections are essential housekeeping functions. For most crops with seed as the germplasm carrier, maintenance, rejuvenation and multiplication processes have been well established. This section will focus on problem crops and the methodologies that will become increasingly important in the field.

5.4.1 *In vitro* storage techniques

In the late 1970s and early 1980s, tissue culture or *in vitro* culture techniques had begun to make an impact in plant physiological studies, vegetative propagation, disease eradication and genetic manipulation. *In vitro* storage techniques were then recognized as a way of conserving the genetic resources of problem crops and also of providing a conservation method for the emerging field of plant biotechnology. Figure 5.4 provides a flowchart for plant tissue culture from various tissues to generate plantlets. A common factor involved in plant tissue culture is the

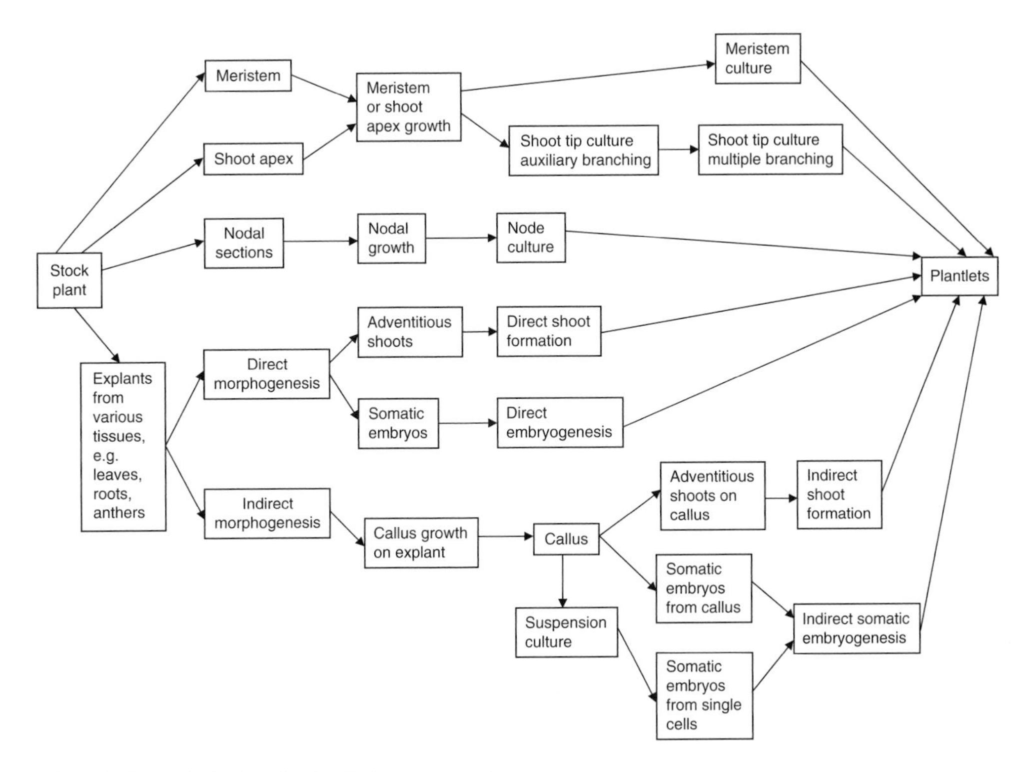

Fig. 5.4. The principal methods of micropropagation.

growth of microbe-free plant material in an aseptic environment such as on sterilized nutrient medium in a test tube.

Along with other new technologies, *in vitro* techniques are now increasing the efficiency and security of conservation for not only problem crops but many others as well. *In vitro* storage techniques have the following advantages (Dodds, 1991): (i) most *in vitro* systems posses the potential for very high clonal multiplication rates under controlled environmental conditions; (ii) by generating plants through meristem culture in combination with thermotherapy, the culture systems are aseptic, can easily be kept free from fungi, bacteria, viruses and insect parasites and are isolated from many external threats, thus ensuring the production of disease-free stocks and simplifying quarantine procedures for international exchange of germplasm; (iii) due to the miniaturization of explants they require less storage space with continuous availability and ease of shipment; (iv) in an ideal tissue-culture storage system, genetic erosion is reduced to zero; (v) by means of pollen and anther culture, haploid plants may be produced that are of use in both genetics and breeding programmes (for details see Chapter 4); (vi) they are useful in plant breeding programmes as a means of rescuing and subsequently culturing zygotic embryos from incompatible crosses that normally result in embryo abscission; and (vii) reduced expenses, both in labour and financial terms, as compared to maintaining large field collections is a further factor contributing to the use of *in vitro* collections. Other advantages of *in vitro* techniques in germplasm conservation include early maturation, e.g. in forest tree spp.; culturing and fusion of inter- or intraspecific protoplasts; transformation of nuclear and cytoplasmic structures of plant cells through insertion of foreign DNA, etc.; and production and transformation of useful natural compounds in fermenters (biosyntheses/biotransformations) (Kumar, 1993; Ashmore, 1997).

There are some examples of the application of *in vitro* storage techniques. Techniques have been developed for the collection of species that produce recalcitrant seeds and for vegetatively propagated material, which enable a collector to introduce the material *in vitro*, under aseptic conditions, directly in the field (Withers, 1995). This approach will allow germplasm collection to be made in remote areas (e.g. in the case of highly recalcitrant cacao seeds) or when the transport of the collected fruits would become prohibitively expensive (e.g. collecting coconut germplasm). Also in cases where the target species does not have seeds or other storage organs to be collected or when budwood would quickly lose viability or is contaminated, establishment of aseptic culture in the field will facilitate collection and improve its efficiency (Engelmann and Engels, 2002).

The disadvantages of *in vitro* maintenance are relatively high inputs of time and labour for culture establishment and maintenance, potential losses due to contamination or mislabelling, risk of microbial infection at each subculture, the cumulative risk of somaclonal variation with time and accidental loss through equipment failure. With certain tissue cultures, the morphogenetic potential of the cultures may decline after growth under *in vitro* conditions for an extended period of time. Prolonged maintenance of de-differentiated plant cells and tissues *in vitro* by repeated subculture is expensive, time consuming, labour intensive and often results in a reduction in morphogenic or biosynthetic capacity and changes in genetic, chromosomal or genomic composition, such as mutations, aneuploidy and polyploidy. Somaclonal variation resulting from subculture may manifest itself at the molecular, biochemical or phenotypic level. However, its extent can be limited by controlling environmental factors such as medium formulation and subculture interval, but cannot be eliminated with certainty unless metabolism is suspended. The type of explant and the preservation method can have a significant impact on survival and the extent of somaclonal variation. In general, organized cultures (meristems, shoot tips and embryos) are more stable than non-organized cultures (protoplasts, suspensions and calli). Thus, organized cultures have a better likelihood of retaining their genetic integrity during prolonged culture *in vitro*. There is a need to develop and utilize storage methods that reduce the maintenance requirements of plant cultures, while maintaining genetic, biochemical and phenotypic stability. For short- to medium-term maintenance, cell and tissue cultures may benefit from some form of growth reduction, but for long-term maintenance, growth suspension must be recommended.

In vitro culture requires strict control of environmental conditions such as medium constituents and often such conditions are not immediately applicable to a wide range of species or even to every selection within a species. Thus, culture conditions often need to be developed for each particular species, subspecies or even culture of interest. The International Board for Plant Genetic Resources (IBPGR, now known as Biodiversity International) published general recommendations for *in vitro* storage, including recommendations on the design and operation of culture facilities (IBPGR, 1986). Extensive research has been carried out to reduce growth rates by reducing the components in the culture medium and modifying the physical environment. Modification of the gaseous environment by mineral oil overlay or control of the gas balance can retard growth. However, the most practical and effective slow growth methods to date involve reducing the culture temperature and/or adding osmotic retardants to the culture medium. Cultures of many species can be maintained in this way for 6 months–2 years without the need for subculturing.

5.4.2 Cryopreservation

Efforts have been made to reduce or eliminate the risks discussed above by the development of slow growth methods for medium-term storage and cryopreservation for long-term storage (Withers, 1993; Harding, 2004) which is considered to be the most effective way of preserving *in vitro* cultures for extended periods. Cryogenic storage is achieved at below −130°C, where liquid water is absent and molecular kinetics and diffusion rates are extremely low. In practice, this refers to storage in or over liquid nitrogen (−196°C liquid phase, −150°C vapour phase). Under conditions of cryopreservation, metabolism ceases and therefore, time effectively stops. Advantages of cryopreservation include indefinite storage without subculturing, frequent viability testing or plant generation and maintenance of the biosynthetic and regeneration capacity of cultures over time. The only threats to the survival and integrity of material once conditions have been determined for safely conveying it to and from the storage temperature are accidental thawing under suboptimal conditions and free radical damage induced by background irradiation (Benson, 1990). The former risk can be minimized by appropriate equipment, backup systems and laboratory procedures. The latter can be minimized by screening and application of free radical-scavenging cryoprotectants. The success of cryopreservation depends on many factors such as the starting material and its precondition, the cryoprotectant treatment and the freezing and thawing rates. Evidence suggests that genetic stability is maintained in cryopreserved materials and any genetic damage probably occurs during the actual freezing and thawing phases, rather than during storage.

The first successes in plant cell cryopreservation were achieved in the early 1970s. Cell suspension cultures have been among the most amenable materials to this method of preservation. The procedure for these materials involves the following stages: pre-growth, cryoprotection, cooling (protective dehydration), storage, increasing the temperature, post-thaw treatment and recovery growth. Cryopreserved embryogenic suspension cultures from which plants can be regenerated are a potentially valuable tool for conservation. Because of concerns regarding genetic stability, most cryopreservation research for genetic conservation has concentrated on organized cultures such as shoot tips, shoots and zygotic embryos. Some plant materials have survived well under conventional cryopreservation while others have remained completely unresponsive. There is a large group of materials that survive to some degree at the cell and tissue level, but with some physical damage. In these cases callusing, which has the attendant risks of somaclonal variation, is employed instead. Recent efforts have increased the number of species that can be cryopreserved as shoots and have improved the quality of cryopreserved specimens. However, progress has been slow and has involved a relatively narrow range of species. Recalcitrant seeds present a dual problem in cryopreservation: they are often large and therefore prone to structural injury and they are very sensitive to dehydration. Perhaps the most promising and intriguing of the new developments in cryopreservation involves artificial seed technology. Dereuddre *et al.* (1991) pioneered a technique involving encapsulation, dehydration and cryopreservation of shoot tips or somatic embryos. These are encapsulated in an alginate gel, dehydrated in air or by incubation in a hypertonic sucrose solution and cooled rapidly. This can give much higher survival levels and result in less structural damage than conventional approaches.

5.4.3 Synthetic seeds and storage of DNA

Two other techniques may be useful for conserving genetic materials that are difficult to conserve with conventional techniques. These are the production and preservation of synthetic seeds and the preservation of nucleic acids (DNA). Storage of DNA is in principle, simple to carry out and widely applicable; however, it in no way replaces or solves the problem of the storage of germplasm since at this point whole organisms cannot be regenerated from DNA. It may be viewed instead as complementary. As techniques in biotechnological approaches to breeding progress however, it is possible to envisage a role for the discrete collection of genes governing particular traits. Furthermore, as the barriers between gene pools are reduced through biotechnology thereby facilitating the selective transfer of genes, relevant materials stored in the form of DNA may not necessarily originate from the target crop's own gene pool.

Plant preservation initiatives have by necessity, focused on the conservation of species and landraces of international agricultural importance and to a lesser extent, endangered and threatened species. Conversely, little effort has been made to systematically collect and preserve the increasing number of genotypes being developed with biotechnological applications (Owen, 1996). Since these elite cultures are being maintained by individual researchers or laboratories, they are in danger of being lost. Owen (1996) highlighted several methods for the maintenance and storage of plant germplasm, with particular emphasis on those techniques most applicable to the preservation of elite cultures used in biotechnology and the plants derived from them.

Somatic and zygotic embryos have been suggested as useful propagules for preservation. Somatic embryogenesis involves adventitious propagation and produces cultures that are convenient to handle and amenable to some technologies such as artificial seed production, which can be linked to storage by cryopreservation. Preservation of some recalcitrant species has been made possible by the observation that excised embryos behave in an orthodox manner and can be cryopreserved. Research has also been conducted to determine the feasibility of using desiccated somatic embryos or encapsulated somatic embryos. Preservation of these 'synthetic seeds' would be especially useful for the preservation of clonal lines; however, more research is needed to elucidate how to increase viability after drying and to inhibit precocious germination.

The total genetic information of a plant can be readily isolated and DNA segments can be stored in lyophilized form. Thus, it would be a useful method for the storage of genes of interest to gene transfer technologies, similar to stored DNA. Pollen storage would also be a useful adjunct for germplasm preservation of lines developed for breeding programmes; however, cytoplasmic genes may not be conserved. Pollen storage can preserve genes, but may not preserve desirable gene combinations.

5.4.4 Rejuvenation and multiplication

The loss of the germination capacity of stored seeds necessitates their periodic rejuvenation. As seed ages, before losing germination capacity, mutations increase and if rejuvenation of the material does not take place within a particular period of time, the genetic structure of the population can vary. The multiplication site should have ecological characteristics similar to those where the material was collected in order to prevent selection that can change the allelic frequencies, even eliminating those alleles most sensitive to certain soil–climate factors.

Tissue culture can be applied to mass produce carbon copies of a selected (elite) plant whose agronomic characteristics are known (Fig. 5.4). It allows propagation of plant material with high multiplication rates in an aseptic environment. Since the 1970s, *in vitro* propagation techniques, mainly based on micropropagation and somatic embryogenesis, have been extensively developed and applied to thousands of different species.

As indicated previously, *in vitro* propagation techniques can be used for rapid clonal multiplication of germplasm in the vegetative form and also for other materials such as recalcitrant seeds which are often available in relatively small numbers. Preference is generally given to propagation methods that confer the lowest risk of somaclonal variation in culture, such as shoot or meristem cultures reproduced by non-adventitious means. However, other factors must be included in the total equation; these include availability of the preferred propagation technique, ease and rate of multiplication and amenability to storage.

5.5 Evaluation

Just having thousands of germplasm accessions available is not helpful if they are not appropriately utilized. The evaluation of germplasm resources is a prerequisite for their utilization in crop improvement. Vast genetic resources are available for crop plants, but, to date, few of them have been well characterized, either phenotypically or genotypically. Of those that have been characterized, it is usually only for a small number of traits and usually only on a phenotypic level; on a genotypic level the available information on useful traits is even less abundant. However, as the need for readily available genetic and genomic information to use in targeted applications has grown so has interest in and support for molecular evaluation of genebank materials. The evaluation of a population or germplasm accession starts at the moment of collection and should never actually end. The term 'descriptor' is used increasingly often in referring to each of those characters considered important and/or useful in the description of a population or accession. Species descriptors differ according to whether they have been selected by plant breeders, botanists, geneticists or experts in other disciplines.

Genetic evaluation and utilization of seed-based germplasm has focused on the characteristics of the seed itself and on whole plants such as morphological and physiological traits, stress tolerance and quality. Most traits scored first are easy to score, often by eye. Genetic evaluation of germplasm resources has been broadened with the development of plant breeding programmes incorporating molecular biology. Morphological evaluation can be extended from several traits of economical importance to almost all traits that differ among germplasm accessions. This paradigm for germplasm characterization is based on the evaluation of phenotypic variation of entries from a genebank for a clearly defined characteristic that is recognizable in the whole plant. This approach works well when the phenotype is controlled by major genes. For traits such as yield, which is genetically controlled by many genes, it is more difficult to distinguish accessions by phenotypic evaluation alone because the same phenotype may be controlled by different genes and vice versa. As a result, exotic germplasm which is perceived to be a poor bet for the improvement of most traits based on phenotypic examination, may contain some superior genes (alleles) for the improvement of most traits, but they lie buried amid the thousands of accessions maintained in genebanks (e.g. de Vicente and Tanksley, 1993; Xiao *et al.*, 1998). Visual evaluation is not efficient enough to identify all of these features. For example, evaluation of anatomic and quality characteristics can be upgraded by using new techniques to reveal the novel differences that cannot be identified by eye.

As types of conserved germplasm extend to include tissues, cells and DNA, additional evaluation criteria and methods are needed. For example, specificity of tissues or cells can be determined by their responses to specific culture media. DNA samples can be evaluated by their physical and chemical properties such as absorbance spectra, electrophoretic separation and staining reactions. Even for seed-based germplasm, molecular biology provides many novel methods for evaluating germplasm at the cellular, chromosomal and molecular levels; for example, identification of chromosome abnormalities and differences at the DNA level. As a result, germplasm will be evaluated not only at morphological and physiological levels but also at a multidisciplinary level that includes molecular biology. Among all the available new techniques, the most feasible is molecular marker technology which is based on DNA differences and/or its integration with quantitative trait loci (QTL) analysis. As described in Chapters 6 and 7, this technique allows a more precise identification and definition of genes, alleles and the useful traits they underlie and can be used to identify and extract superior genes (alleles) from inferior germplasm, thus allowing germplasm evaluation to move from morphological and physiological levels to biochemical and molecular levels.

5.5.1 Marker-assisted germplasm evaluation

Marker-assisted germplasm evaluation (MAGE) aims to complement phenotypic evaluation by helping to define the genetic architecture of germplasm resources and by identifying and managing germplasm that contains alleles associated with traits of economic importance (Xu, Y. *et al.*, 2003). Molecular markers may allow for characterization based on genes, genotypes and genomes which provides more precise information than classical phenotypic or passport data. Molecular marker data can be used to answer questions of identity, duplication, genetic diversity, contamination and integrity of regeneration. In addition, molecular markers are extremely powerful for identifying zygosity at important loci in species which are vegetatively propagated such as potato, sugarcane, taro and sweet potato. Many features revealed by molecular markers, such as unique alleles, allele frequency and heterozygosity, mirror the genetic structure of germplasm resources at the molecular level (Lu *et al.*, 2009). On a more fundamental level, molecular marker information can lead to the identification of useful genes contained in collections and aid in the transfer of these genes into well-adapted cultivars. MAGE can play an important role in the procedures related to the acquisition/distribution, maintenance and use of germplasm (Bretting and Widrlechner, 1995; Xu, Y. *et al.*, 2003). As summarized by Xu, Y. *et al.* (2003), molecular markers can be used for: (i) differentiating cultivars and constructing heterotic groups; (ii) identifying germplasm redundancy, underrepresented alleles and genetic gaps in current collections; (iii) monitoring genetic shifts that occur during germplasm storage, regeneration, domestication and breeding; (iv) screening germplasm for novel genes or superior alleles; and (v) constructing a representative subset or core collection.

The realization of the importance of MAGE led to the formation of the Generation Challenge Programme (GCP) (http://www.generationcp.org) (Dwivedi *et al.*, 2007). The GCP aims to utilize molecular tools and comparative biology to explore and exploit the genetic diversity housed in existing germplasm collections with a particular focus on improving the drought tolerance of various cereals, legumes and clonal food crops. One of the primary goals of the GCP is the extensive genomic characterization of global crop-related genetic resources (composite collections); initially using SSR markers to determine population structure and now moving on to whole genome scans (including SNP and diversity array technology (DArT) arrays) and functional genomics analysis of subsets of germplasm (mini-composite collections). Thus, the GCP has created composite collections covering global diversity for most of the 20 CGIAR mandated crops. These consists of up to 3000 accessions or no more than 10% of the total number of available accessions for inbreeding crops and 1500 accessions for outcrossing species (where each accession must be treated as a population). It is expected that this analysis will also lead to the development of genetically broad-based mapping and breeding populations. The results from these GCP-supported projects are already starting to be made available for the benefit of the scientific community. Furthermore, the GCP is supporting a project on allele diversity at orthologous candidate (ADOC) genes that will produce and deliver a public data set of allelic diversity at orthologous candidate genes across eight important GCP crops and assess whole sequence polymorphism in a DNA bank of 300 reference accessions for each crop. This reference germplasm which has already undergone one level of genome scan, will be evaluated for traits associated with drought tolerance to test for associations between observed polymorphisms and trait variability (http://www.intl-pag.org/14/abstracts/PAG14_W264.html).

Molecular markers can be used for germplasm management in different ways. Markers with known functional alleles or associated with agronomic traits can be used to trace, select and manage these alleles or traits. Genetic markers that reveal multiple bands or represent multiple loci such as RAPD or amplified fragment length polymorphism (AFLP), are usually difficult to trace back to specific alleles/loci, so they need to be converted into markers that are locus-specific such as sequence tagged sites (STSs), SSRs or SNPs. Neutral markers or markers in unknown chromosomal regions can be used for fingerprinting and background examination. In this case any type of marker that detects a high rate of polymorphism is useful as long as it is able to reveal genome-wide polymorphism. Spooner *et al.* (2005) provided some examples of the use of molecular markers in genebank management, including assessing the level of redundancy within and between collections and the genetic integrity of accessions during the course of genebank operations such as regeneration as well as the presence and magnitude of gene flow. As discussed in Chapter 9 and by Xu, Y. (2003), an efficient MAGE system consists of several key components.

MAGE largely depends on multivariate analysis of DNA genotypes. There are several questions that need to be answered for each experiment: Which entities should be sampled? What is the nature of the genetic material to be sampled? How should heterogeneous, segregating populations be sampled? What types of variables should be measured? How many variables (e.g. markers) should be measured? Should analyses be carried out on raw multivariate data or derived genetic similarities? Among these questions, the most frequent might be how many markers are enough for a genome-wide MAGE; the answer however, depends on the questions that MAGE is expected to answer, and it also depends on the type of marker being used. Smith *et al.* (1991) used 200 RFLP markers dispersed across the maize genome to fingerprint 11 inbred lines (the genetic distance matrix comprised 55 elements). They estimated distance matrices by sampling 5–200 RFLP markers in increments of five (e.g. from 5, 10, 15, up to 200). They concluded that accuracy was sufficient with 100 or more markers. Bernardo (1993) concluded that 250 or more marker loci were needed to produce precise estimates of coefficients of co-ancestry. As a rough estimation, the number of markers that are needed to detect linkage disequilibrium between any two markers in the genome can be used to judge how many markers are needed for a genome-wide MAGE, which is apparently crop-dependent and also higher than in most MAGE projects that have been reported if genome-wide germplasm is evaluated within closely related germplasm or populations at gene or sequence level. With the development of SNP markers covering whole genomes and high-throughput array-based genotyping systems (Chapter 3; Lu *et al.*, 2009; Yan *et al.*, 2009), using markers developed from all candidate genes for all available germplasm collections can be a realistic target in the near future.

5.5.2 *In vitro* evaluation

The basis for obtaining unique plants from cell culture resulted from the observation that plant cells in culture, are genetically variable. Cell culture was envisioned as a means of preferentially selecting cell lines with a mutation to cope with a specific selective agent. Plants generated from such cell lines in many cases expressed the new traits at the whole-plant level giving enhanced agronomic value for that specific trait. There was a correlation between cell-level response to a selective agent and that of the entire plant. By using a specific selective agent to preferentially screen for a specific cell line, the number of plants involved in whole-plant level screening can be reduced. Techniques useful in the characterization and evaluation of cultures for genetic stability include nuclear cytology, isozyme analysis, DNA marker-based analysis, as well as other molecular and biochemical methods. As large populations of cells can be screened *in vitro* in a small space as compared to traditional screening of large plant populations in the greenhouse or field, this could save a vast amount of time, space, labour and money and allow for screening all year round in a controlled laboratory. Also, *in vitro* screening could reduce some of the problems conventional screening may encounter due to environmental variation and poor uniformity within a field.

Plants lose many of their distinguishing phenotypic characteristics when transferred to *in vitro* culture. Therefore, accurate records and vigilance are essential to ensure that genetic integrity is preserved. The risk of somaclonal variation can be reduced by monitoring, but frequent regeneration of plants from stored cultures and monitoring them in the field is costly and inefficient. Techniques are needed that can be applied easily and economically to cultured material. Visual monitoring *in vitro* will only detect the most gross of variants, e.g. variegated leaves or extreme dwarfism. More accurate, wide-ranging and reliable monitoring may be achieved by biochemical methods and molecular marker techniques. Minute changes at the DNA level can be detected with molecular markers that provide an ideal means of determining genetic integrity. It may become possible to detect undesirable variants in an overnight procedure at the culture stage, thereby eliminating the need for establishing *in vitro* propagated plantlets in the field for monitoring.

5.5.3 Genetic diversity

New genetic technologies, especially large-scale DNA sequencing (Chapter 3), have led to the development of molecular systematics and new methods of measuring genetic similarity and divergence in plant species and populations. It is now possible to compare organisms from the genome level (using for example, fluorescent *in situ* hybridization or FISH) down to the level of single nucleotides (DNA sequencing and SNPs). Molecular markers have been used for genetic diversity studies for the following purposes: (i) examination of genotype frequencies for deviations at individual loci and characterization of molecular variation within or between populations; (ii) construction of 'phylogenetic' trees or classification of germplasm accessions based on genetic distance and determination of heterotic groups for hybrid crops; (iii) analysis of the correlation between the genetic distance and hybrid performance, heterosis and specific combining ability; and (iv) comparison of genetic diversity among different groups of maize germplasm. Taking maize as an example, some applications in these areas can be found in Melchinger (1999), Warburton *et al.* (2002), Betrán *et al.* (2003), Reif *et al.* (2004), Xia, X.C. *et al.* (2005) and Lu *et al.* (2009). Such studies have provided useful information for genebank curation, gene identification and breeding.

Understanding the range of diversity and the genetic structure of gene pools is critical for the effective management and use of germplasm resources. The first question to ask might be about the distinctiveness of the concerned entities since the issue of what level of diversity we should actually try to maintain is still under debate. Some have argued that highly unique entities should be given preference over equally rare taxa with close relatives of abundant distribution (Vane-Wright *et al.*, 1991) while others argue that evolutionary potential is highest in species-rich groups since the ability to adapt is seemingly greater (Erwin, 1991). On the other hand, the importance of species versus subspecies, hybrids and populations has generated considerable debate about the scientific legitimacy of legal conservation units (O'Brien and Mayr, 1991). Therefore, as indicated by Hahn and Grifo (1996), the first measures to be taken with molecular methods are taxon-specific markers and estimation of the degree of differentiation between units.

Diversity studies are generally undertaken using molecular markers that are assumed to be neutral, that is, not within expressed regions of the DNA. The correlation between molecular variation and quantitative variation in expressed traits has rarely been studied in detail but is an issue that must be addressed if studies in genetic diversity are to be used more effectively in biodiversity assessment and conservation (Butlin and Tregenta, 1998). Across a large genome, such as that of maize, diversity can accumulate so that 150 million sites are commonly polymorphic. A small but important proportion of these polymorphisms is responsible for the complex variation in phenotypic traits. Molecular markers have increased our understanding of the spatial and temporal patterns of genetic variation and of the evolutionary mechanisms that generate and maintain variation. However, the direct benefit of these data to either practical biodiversity conservation or germplasm collection management is equivocal (Harris, 1999).

Several past studies have highlighted the decline of genetic diversity in modern cultivars compared to landraces or wild relatives. In maize, for example, Liu *et al.* (2003) evaluated the genetic diversity among 260 diverse maize inbred lines with 94 SSR markers and found that tropical and subtropical inbreds contain a greater number of alleles and gene diversity than temperate inbreds. It was also found that maize inbreds capture less than 80% of the alleles seen in the landraces, suggesting that landraces can provide substantial additional genetic diversity for maize breeding. After analysing over 100 maize inbred lines and teosinte accessions with 462 SSRs, Vigouroux *et al.* (2005) concluded that many alleles in the progenitor species of maize (teosinte) are not present in maize. Wright *et al.* (2005) compared SNP diversity between maize and teosinte in 774 genes and concluded that maize accessions had much less genetic diversity consistent with products of artificial selection and crop improvement. These reports in maize along with genetic mapping studies involving wild relatives in other crops, support earlier conclusions that non-adapted and wild related species contain untapped sources of new alleles for future crop breeding improvement (Tanksley and McCouch, 1997).

Factors impacting genetic diversity

The extent of polymorphism differs substantially between species and sampled loci. In a comprehensive study of variation within a maize chromosome, the diversity at 21 loci varied by 16-fold (Tenaillon *et al.*, 2001). The variation between loci may partly reflect sampling effects but selection and other factors play a more important role (Table 5.2). Although many factors influence diversity, the neutral theory of evolution suggests that the level of polymorphism (θ) should be the product of the effective population size (N_e) and the mutation rate (μ) with $\theta = 4\ N_e\mu$ (Kimura, 1969). Unfortunately, there is little empirical proof of this in plants. Background selection is likely to be one of the major factors determining nucleotide diversity and it suggests that diversity should be shaped by recombination at the intragenomic scale and by the outcrossing rate at the species level. Strong selection pressure is important in decreasing the nucleotide diversity of some plant species. During the selection of advantageous phenotypes, some crops appear to have passed through bottlenecks that substantially reduced diversity

Table 5.2. Factors that impact nucleotide diversity (reprinted from Buckler and Thornsberry (2002) with permission from Elsevier).

Factor	Correlation with diversity	Scope
Mutation rate	Positive	Often whole genome
Population size	Positive	Whole genome
Outcrossing	Positive	Whole genome
Recombination	Positive	Whole genome
Positive-trait selection	Negative	Individual genes
Line selection	Positive	Whole genome
Diversifying selection	Positive	Individual genes
Balancing selection	Positive	Individual genes
Background selection	Negative	Individual genes or whole genome
Population structure	Mixed	Whole genome
Sequencing errors	Positive	Individual genes
PCR problems	Negative	Individual genes

(Doebley, 1992). Balancing selection and/or frequency-dependent selection may also play an important role in increasing diversity at specific loci within a genome. In these selection regimes, selection favours the maintenance of multiple alleles with different effects over evolutionary time.

Measurement of diversity

The estimation of genetic similarity is vital to the formulation of optimal germplasm management strategies and lies at the core of modern plant systematics and evolutionary biology. Plant systematicists and evolutionary geneticists have developed techniques for analysing genetic similarity that may be ideally suited for addressing certain germplasm management issues. Kresovich and McFerson (1992) highlighted the important role of genetic diversity assessment in plant genetic resource management. One simple estimate of genetic diversity in a given taxon, germplasm collection or geographic region is the number of taxa included in the larger unit (e.g. the number of subspecies found in a species in a given region). Yet, the number of recognized subordinate taxa may vary substantially among taxonomic treatments as may the actual level of genetic differentiation among such taxa (Bretting and Goodman, 1989). Accordingly, diversity estimates derived from genetic marker data may be more valuable than counts of taxa for most germplasm management applications, since such estimates can be more easily compared across taxa and the focus may be on conserving genes rather than taxa.

Because the genome is assayed directly, DNA-based technologies circumvent the often poor correspondence between morphological and genetic diversity in crop species. With STSs developed from expressed sequence tags (ESTs), it is even possible to use expressed genes specific to life history stages, rather than anonymous sequence differences to assay genetic differences among accessions. Because database comparisons can often identify the functional product of an EST, the genebank manager obtains not only an indicator of genetic diversity and the relationships among accessions but also an increase in the information content of the sample accession (Brown and Brubaker, 2002).

When genetic marker data can be interpreted by a locus/allele model, allelic diversity can be described by: (i) the percentage of polymorphic loci, calculated by dividing the number of polymorphic loci by the total number of loci assayed; (ii) the mean number of alleles per locus, calculated by dividing the total number of alleles detected by the number of loci assayed; (iii) total gene diversity or average expected heterozygosity (Nei, 1973; Brown and Weir, 1983), calculated by

$$H = 1 - \sum_{i} \sum_{j=1}^{m} P_{ij}^2 / m \qquad (5.1)$$

and (iv) polymorphic information content (PIC), which was described by Botstein *et al.* (1980) to refer to the relative value of each marker with respect to the amount of polymorphism exhibited and is estimated by

$$PIC_i = 1 - \sum_{j=1}^{m} P_{ij}^2 \qquad (5.2)$$

In both Eqns 5.1 and 5.2, P_{ij} is the frequency of the jth allele at the ith of m loci. The variances of all these estimates are affected by the number of loci and by sample size – the number of progeny assayed per plant, plants assayed per population or number of populations assayed per taxon (Brown and Weir, 1983; Weir, 1990). Various theoretical and empirical studies suggest that for precise estimates, the number of loci assayed may be more critical than the sample size but that the latter should be as large as practical.

In various applications of molecular marker data, a proper choice of a similarity s or dissimilarity coefficient ($d = 1 - s$) is important and depends on factors such as: (i) the properties of the marker system employed; (ii) the genealogy of the germplasm; (iii) the operational taxonomic unit (OTU) under consideration (e.g. lines, populations); (iv) the objectives of the study; and (v) the necessary preconditions for subsequent multivariate analyses.

A wide variety of pairwise genetic similarity measures is available but only a few have been widely applied. Reif *et al.* (2005) examined ten dissimilarity coefficients widely used in germplasm surveys (Table 5.3) with special focus on applications in plant breeding and seed banks, by investigating their genetic and mathematical properties, examining the consequences of these properties for different areas of application in plant breeding and seed banks and determining the relationships between these ten coefficients. A Procrustes analysis of a published data set consisting of seven International Maize and Wheat Improvement Center (CIMMYT) maize populations demonstrated close affinity between Euclidean, Rogers', modified Rogers' (Rogers, 1972; Wright, 1978) and Cavalli-Sforza and Edwards' distance on one hand, and between Nei's standard and Reynolds's dissimilarity on the other. This study also showed that the genetic and mathematical properties of dissimilarity measures are of crucial importance when choosing a genetic dissimilarity coefficient for analysing molecular data.

Table 5.3. Dissimilarity coefficients d for allelic informative marker data. p_{ij} and q_{ij} are allelic frequencies of the jth allele at the ith locus in the two operational taxonomic units consideration, n_i is the number of alleles at the ith locus, and m refers to the number of loci.

Variable	Dissimilarity coefficient		Range
d_E	$\sqrt{\sum_{i=1}^{m}\sum_{j=1}^{n_i}(p_{ij}-q_{ij})^2}$	Euclidean	$0,\sqrt{2m}$
d_R	$\frac{1}{m}\sum_{i=1}^{m}\sqrt{\frac{1}{2}\sum_{j=1}^{n_i}(p_{ij}-q_{ij})^2}$	Rogers (1972)	0,1
d_W	$\frac{1}{\sqrt{2m}}\sqrt{\sum_{i=1}^{m}\sum_{j=1}^{n_i}(p_{ij}-q_{ij})^2}$	Modified Rogers	0,1
d_{CE}	$\sqrt{\frac{1}{m}\sum_{i=1}^{m}(1-\sum_{j=1}^{n_i}\sqrt{p_{ij}q_{ij}})}$	Cavalli-Sforza and Edwards (1967)	0,1
d_{RE}	$-\ln(1-\theta)$	Reynolds *et al.* (1983)	$0,\infty$
d_{N72}	$-\ln\frac{\sum_{i=1}^{m}\sum_{j=1}^{n_i}p_{ij}q_{ij}}{\sqrt{\sum_{i=1}^{m}\sum_{j=1}^{n_i}p_{ij}^2\sum_{i=1}^{m}\sum_{j=1}^{n_i}q_{ij}^2}}$	Nei (1972)	$0,\infty$
d_{N83}	$\frac{1}{m}\sum_{i=1}^{m}(1-\sum_{j=1}^{n_i}\sqrt{p_{ij}q_{ij}})$	Nei *et al.* (1983)	0,1

$$\theta=\frac{\sum_{i=1}^{m}\left\{\frac{1}{2}\sum_{j=1}^{n_i}(p_{ij}-q_{ij})^2-\frac{1}{2(2n-1)}\left[2-\sum_{j=1}^{n_i}(p_{ij}^2+q_{ij}^2)\right]\right\}}{\sum_{i=1}^{m}(1-\sum_{j=1}^{n_j}p_{ij}q_{ij})}$$

Germplasm classification

Germplasm can be classified on the basis of morphological traits, geographic distribution, evolutionary and breeding history, pedigree and/or genotypic diversity at the molecular level. Both categorical and quantitative data have been used for phenotype-based classification. A broad-based approach to germplasm classification will contribute to our understanding of the genetic structure of subpopulations within a species, how to identify useful gene donors and the rationale for constructing heterotic groups for hybrid breeding. A classification technique may be considered optimal if it has these characteristics (Crossa and Franco, 2004): (i) produces clusters that respond to the optimization of a target function; (ii) is linked to a technique for defining the optimum number of groups, preferably in the form of a statistical hypothesis test; (iii) helps to calculate a measure of the quality of the clusters; (iv) assigns observations to the groups, based on the probability of each observation belonging to each group; (v) uses the information available in categorical variables as well as in continuous variables; and (vi) may be extended to the problem of classification when the variables are measured in different environments. The best numerical classification strategy is the one that produces the most compact and well-separated groups, that is, minimum variability within each group and maximum variability among groups. Crossa and Franco (2004) reviewed geometric classification techniques as well as statistical models based on mixed distribution models. The two-stage sequential clustering strategy, which uses all variables, continuous and categorical, tends to form more homogeneous groups of individuals than other clustering strategies. The sequential clustering strategies can be applied to three-way data comprising genotype × environment attributes. This approach groups genotypes with consistent responses for most of the continuous and categorical traits across environments.

Patterns of genetic similarity among taxa or germplasm collections can be visualized by cluster analysis and ordination. Ideally, these two multivariate techniques are deployed together because their strengths are complementary (Sneath and Sokal, 1973; Dunn and Everitt, 1982; Sokal, 1986). In cluster analysis, taxa, germplasm collections or genetic markers are arranged in a hierarchy (called a phenogram or dendrogram) by an agglomerative algorithm according to patterns occurring in a matrix of pairwise genetic similarities as described above. The hierarchies obtained from cluster analyses are highly dependent on both the similarity measure and the clustering algorithm used. The most frequently used clustering methods involve arithmetic means (either UPGMA or weighted-pair group means arithmetic (WPGMA)) (Sneath and Sokal, 1973). One of the commercial packages which implements these and other methods is NTSYS (http://www.exetersoftware.com/cat/ntsyspc/ntsyspc.html). More recently, a comprehensive set of statistical methods for genetic marker data analysis, designed especially for SSR/SNP data analysis, PowerMarker, has been widely used for cluster analysis (e.g. Lu *et al.*, 2009). PowerMarker has options for selecting different distances and clustering methods and is free for download at http://statgen.ncsu.edu/powermarker/.

With ordination, the multidimensional variability in a pairwise, intertaxa or intermarker similarity matrix can be portrayed in one or several dimensions through eigenstructure analysis. Ordination is best suited to revealing interactions and associations among taxa or germplasm accessions described by traits that vary continuously and quantitatively. Principal component, principal coordinate and linear discriminant analyses are the ordination techniques most relevant for potential germplasm management applications.

There are numerous reports on germplasm classification using molecular markers. Only two examples will be discussed here. In sorghum, 46 converted exotic lines representing all five races and nine intermediate races of sorghum were fingerprinted using AFLP and SSR markers. A total of 453 scored marker loci were used to calculate genetic similarities between the lines. The dendrogram constructed using UPGMA grouped 31 lines into three major clusters with Jaccard coefficients greater than 0.75. The remaining 15 lines were grouped into four small sub-clusters each with two lines and seven single accession nodes (Perumal *et al.*, 2007). RFLP marker-based analysis of 236 rice cultivars identified two major groups which corresponded to the two major rice types, *indica* and *japonica*. By comparison of allele frequencies between *indica* and *japonica* cultivars, several subspecies-specific alleles were identified, with one allele existing in more than 99% of *indica* cultivars and another in more than 99% of *japonica* cultivars (Xu, Y. *et al.*, 2003).

Figure 5.5 provides an example of clustering analysis using 169 SSR markers to classify 18 US rice cultivars collected or selected before 1930 (Lu *et al.*, 2005). These cultivars were classified into three groups, which corresponded to three types of cultivars with different grain sizes, i.e. short grain cultivars in the western US rice belt (California) and medium and long grain cultivars in the southern US rice belt. These three groups of cultivars (Fig. 5.5) formed the foundation of germplasm resources for breeding short-grain temperate *japonica* and medium- and long-grain tropical *japonica* cultivars, respectively, in the USA.

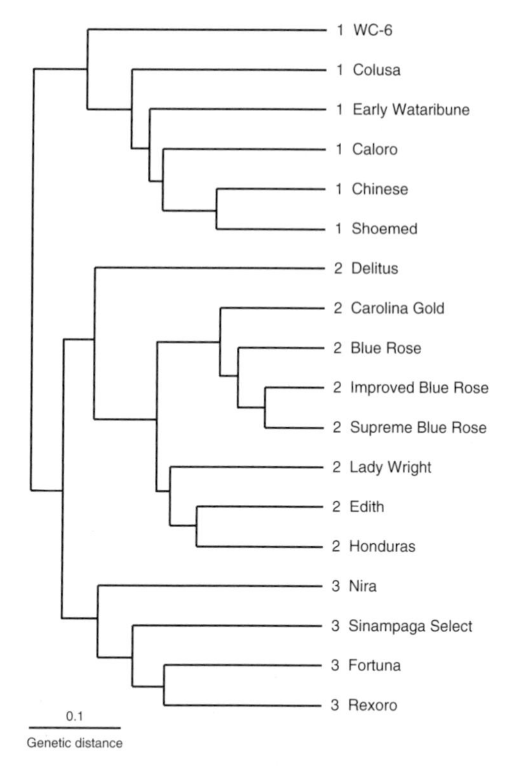

Fig. 5.5. Groups of 18 US rice cultivars collected or selected before 1930 based on 169 SSR markers using UPGMA methods and Nei's (1972) genetic distance (Lu *et al.*, 2005). Three groups (1, 2, 3) can be identified, consisting of 6, 8, and 4 cultivars, and representing short-, medium-, and long-grain US rice cultivars, respectively. From Lu *et al.* (2005) with permission.

Germplasm classification can be used to construct heterotic groups so that cultivars within each group have a high level of similarity in their genetic backgrounds. As a result, intergroup hybrids show a higher level of heterosis than within-group hybrids. Commercial maize hybrids are typically created between inbreds from opposite, complementary heterotic groups. Heterotic patterns in many crop species have been established based solely on large numbers of testcrosses and extensive breeding experience. For inbreeding species for which subspecies or subpopulation differences may be older or more pronounced than in cross-pollinating species, DNA-based markers can be used to classify germplasm accessions into different heterotic groups, each with a high level of similarity. Research results from rice, *Brassica napus*, barley and wheat indicate that DNA markers are very useful tools for the construction of heterotic groups (Xu, Y. 2003). Divergence at molecular marker loci has also been useful in assigning maize inbreds to known heterotic groups previously established in breeding programmes, and the molecular information agreed with pedigree information (Lee *et al.*, 1989; Melchinger *et al.*, 1991; Messmer *et al.*, 1993).

Two areas need further development in germplasm classification: methods of data analysis and the understanding of molecular diversity in relation to quantitative variation. Methods for the analysis of molecular data have not kept up with the sophistication of the methods of data generation (Harris, 1999). Thus, it is common to find sophisticated molecular data (e.g. AFLP) being analysed using similarity measures derived decades ago. Similarity measures and classification methods are needed specifically for handling molecular marker data from polyploid species.

Phylogenetics

One of the most important roles of genetic markers in plant germplasm management is in the elucidation of the systematic relationships within genera, tribes and families and obtaining characteristic genetic profiles of germplasm. Using the similarity measures and classification methods described above, genetic markers of all types have been instrumental in characterizing systematic and evolutionary genetic relationships and in establishing a germplasm's taxonomic identity which will probably change how the germplasm accessions are managed and utilized. As indicated by Bretting and Widrlechner (1995), clarifying evolutionary relationships among intermediate taxa may challenge the germplasm manager's judgement and acuity. Molecular taxonomy will substantially improve our knowledge of the primary, secondary and tertiary gene pools of many crops and evolutionary studies will help identify crop ancestors, past genetic bottlenecks and opportunities for introducing useful variation. It is particularly vital for germplasm management purposes to discriminate recently synthesized, naturally occurring F_1 hybrids and/or hybrid derivatives from taxonomically intermediate taxa originating from convergent-parallel evolution, clonal variation, recombinational speciation and/or the retention of intermediate ancestral traits (where the latter includes the phenomenon known as lineage sorting; Avise, 1986).

Supraspecific systematic relationships are best elucidated by phylogenetic methods. These methods can sometimes help estimate phylogenetic relationships among crops and related taxa and accordingly, may help determine whether a weedy crop relative is a crop progenitor or a feral crop derivative. As the exact systematic relationships among a crop and its relatives are better understood, germplasm conservation and utilization strategies may change tangibly. For example, since at least the mid-1980s, maize evolutionists in general have accepted the tripartite hypothesis of Mangelsdorf (originating in the 1930s and reviewed in Mangelsdorf (1974)). This hypothesis postulated that maize evolved directly from an undiscovered wild maize and that teosinte was derived from a hybrid between maize and *Tripsacum* species. During that period, substantial resources (relative to those devoted to similar programmes with teosinte) were allocated to improving maize with introgressed *Tripsacum* germplasm (Galinat, 1977). To summarize, a clear understanding of the systematic relationships among a crop and its wild relatives is vital for sound genetic resource management and for crop improvement as a whole.

Taxonomic relationships have been re-evaluated for many crop plants by using molecular markers and genomic sequences that cover some part of the genome for specific traits, attempting to replace the classical morphological survey with a point survey using data obtained from one or more marker or sequence loci. For example, studies of the genetic architecture of key yield-related components (e.g. flower and seed production, maturity and photoperiod response) will enable us to focus on areas of the genome where diversity is particularly important for this trait (Hodgkin and Ramanatha Rao, 2002). Phylogenetic studies provide a fundamental gain in genetic knowledge not only to prove that two individuals or gene copies differ but also to place them in a hierarchy of relationships based on the timing of a shared ancestor.

Phylogenetic diversity can be also estimated by whole genome analysis and genome-scale phylogenetic trees can be created. Such genome-trees can be built based on gene content, gene order, evolutionary distances between orthologues and concatenated alignments of orthologous protein sequences (see Wolf *et al.* (2003) for a review). Both the initial results and the general notion that using genome-wide information helps enhance the phylogenetic signal suggest that the future belongs to these approaches.

5.5.4 Collection redundancies and gaps

As a large number of germplasm accessions are available for each cultivated plant, many likely represent duplicate or nearly identical samples of the same cultivar while others embody those with rare alleles or highly unusual allele combinations or those where many of their genes or alleles are underrepresented in current collections. Molecular technology will help us to understand the genetic structure of existing collections and to design appropriate acquisition strategies. In particular, genetic distance can be calculated as described previously to identify particularly divergent subpopulations that might harbour valuable genetic variation complementary to that in current holdings.

Germplasm redundancy exists in many germplasm collections due to the different names given to the same cultivars or duplicate samplings of the same accessions. Duplication of germplasm among collections is substantial. Eliminating this type of duplication has often been suggested as a way of reducing the costs associated with the operation of genebanks. Lyman (1984) estimated that at least 50% of the germplasm held consists of duplicated accessions. The Food and Agriculture Organization (FAO) (1998) estimates that of the 6 million accessions stored worldwide, only between 1 and 2 million are unique. On the other hand, it has been recognized that all germplasm should be backed up in at least two different sites to avoid complete loss of the collections at any given site. Pedigree-related cultivars, sibling lines and early isogenic lines may represent another type of redundancy because they are genotypically duplicated at most genetic loci. For example, US rice cultivars M5, M301, M103, S201, Calrose, Calrose 76, CS-M3 and Calmochi-202 shared the same panel of alleles at all of the 100 RFLP loci surveyed. Each of these cultivars can be traced back to a common ancestor, Caloro. In addition, no genetic polymorphism could be detected at another 60 loci between Calrose and Calrose 76 when a more polymorphic marker type, SSR, was used (Xu, Y. *et al.*, 2004). This is probably due to the fact that they are isolines, with Calrose 76 representing a variant derived from Calrose via chemical mutagenesis. Using 15 SSR markers, Dean *et al.* (1999) assayed 19 sorghum (*Sorghum bicolor* (L.) Moench) accessions identified as 'Orange' currently maintained by the US National Plant Germplasm System (NPGS). They found that most accessions were genetically distinct, but two redundant groups were found. The variance analysis also indicated that it should be possible to reduce the number of Orange accessions held by NPGS by almost half without seriously jeopardizing the overall genetic variation contained in these holdings.

Germplasm collections can be compared for the frequencies of alleles at all genetic loci so that distinctive alleles, allele combinations and allele frequency patterns can be identified for a given population. Chromosomal regions containing loci that show the greatest changes in allele frequency between the collections can be located. The rationale for this analysis is to define genomic regions where selection gave rise to allele combinations or allele frequency patterns that distinguish a group of accessions with less diversity from those in a more diverse group. Alleles originally found in ancestral cultivars or the wild relatives may be gradually lost through domestication and breeding. Modern breeding programmes generally rely on a small number of superior accessions which results in genetic uniformity and loss of diverse alleles that could be important to future breeding programmes. Valuable lost genes or alleles can be recovered by going back to the ancestors or wild relatives of our crop species. Using 47 SSR markers, Christiansen *et al.* (2002) determined the variation of genetic diversity in 75 Nordic spring wheat cultivars bred during the 20th century. They found that some alleles were lost during the first quarter of the century whereas several new alleles were introduced in the Nordic spring wheat material during the second quarter of the century.

The allele frequencies at 100 RFLP and 60 SSR loci in rice were compared between the US and world collections and between the two major types, *indica* and *japonica*, within the world collection (Xu *et al.*,

2004). Among 34 alleles at 20 RFLP and 14 SSR loci that were found most frequently (allele frequencies are 20.4–59.5%) in the world collection, three of them were completely lost and 31 of them were underrepresented (less than 5%) in the US collection (*japonica*) while some of them were also lost or underrepresented in *indica* types. As examples, the lost alleles and the underrepresented alleles with frequencies of less than 2% are listed in Table 5.4. Selection against these alleles is clear and stems from the fact that modern US rice cultivars have been developed from a small set of germplasm introductions.

5.5.5 Genetic drifts/shifts and gene flow

To generate stocks for distribution or to maintain the seed viability, a variable accession needs to be regenerated regularly. During this process there is a risk that the genetic integrity of the accession will be compromised by genetic drift, selection or gene flow (Sackville Hamilton and Chorlton, 1997). Genetic drift is a stochastic phenomenon of fluctuations in allele frequencies in the offspring deviating from the parental population which may result in random fluctuations in allele frequencies from generation to generation or the eventual loss of alleles from the population. The change in allele frequency in one generation for diploid organisms can be quantified by $q(1 - q)/2N_e$, where q represents the frequency of allele and N_e the effective population size (Falconer, 1981). Thus, the effective population size determines the extent of genetic drift. Effective population sizes are generally smaller than actual population sizes because of unequal numbers of females and males, overlapping generations, non-random mating, differential fertility and fluctuations in population size (Falconer, 1981; Barrett and Kohn, 1991). Genetic drift may be controlled by adjusting the size of the regenerating population or developing improved regeneration methods (Engels and Visser, 2003). Genetic drift can be measured using molecular markers that are neutral and co-dominant because of the random nature of the process.

The genetic composition of populations may also change during regeneration due to selection. Selection is different from genetic drift because it does not affect all loci simultaneously and usually occurs towards particular genotypes or loci. Selection during regeneration can be inferred from strong shifts in marker allele frequencies for certain loci between parental and offspring populations (Spooner *et al.*, 2005). Maintaining genetic diversity and preventing genetic shifts are important objectives for germplasm

Table 5.4. SSR and RFLP alleles lost or underrepresented in the US collection but most frequent in the world collection (the markers designated with the 'RM' prefix are SSRs and others are RFLPs) (selected from Xu *et al.*, 2004).

			Allele frequency (%)			
Chromosome	Marker	Allele	World	USA	World *japonica*	World *indica*
1	RM259	156bp	20.4	0	0	35
1	CDO118	17kb	49.5	1.6	0	85.7
2	RM207	131bp	21.7	0	4.7	35
3	RM7	181bp	31.5	1.6	2.3	54.1
5	RM233B	138bp	41.9	1.7	2.5	74.5
7	RM11	143bp	29.6	1.6	4.7	48.4
9	RM219	216bp	21	0.9	0	35.6
9	RM257	149bp	21.5	0	0	36.5
9	RM205	123bp	46.4	1.6	4.5	77.8
9	CDO1058	4.1kb	55.9	1.6	4.5	93.8
12	RG901X	4.6kb	43.6	1.6	9.1	69.8

conservation. In open-pollinated species, deviations from random mating, primarily in the form of assortative or consanguineous matings, need to be monitored during germplasm regeneration. In maize, deviations from random mating have been widely studied, with emphasis on detailed multi-locus isozyme analyses of one or two synthetic or open-pollinated maize cultivars (Kahler *et al.*, 1984; Pollak *et al.*, 1984; Bijlsma *et al.*, 1986). In general, levels of selfing did not exceed those expected under random-mating models, but significant deviations were caused by temporal variation in the pollen pool or by gametophytic selection.

The genetic profiles of germplasm can change during the course of medium- or long-term storage. Storage effects fall into three broad categories: (i) the occurrence of mutations; (ii) the occurrence of chromosomal aberrations; and (iii) shifts in gene frequencies resulting from differential genotypic viability in heterogeneous populations. After a comprehensive review of storage effects on seeds, Roos (1988) found little evidence for heritable changes in germplasm attributable to storage-induced chromosomal aberrations and noted 'little need for concern about mutation as a significant factor in altering the composition of germplasm collections'. As indicated by Bretting and Widrlechner (1995), however, differential seed longevity can markedly reduce genetic variability over time. This is well documented by experiments involving mixtures of eight bean lines (Roos, 1984) and four seed storage protein genotypes within a cultivar of wheat (Stoyanova, 1991).

Genetic shifts can also be caused by *in vitro* culture. The genetic stability of germplasm maintained in tissue culture (*in vitro*) has historically been monitored with karyotypic markers such as chromosome number and morphology (D'Amato, 1975) because cytological variability has been considered a primary cause of somaclonal variation. Lassner and Orton (1983) reported that *in vitro* cultures of celery with identical isozyme profiles were markedly variable cytologically. More recently, *in vitro* culture has been shown to induce changes because of the mobilization of transposable elements (Jiang *et al.*, 2003; Kikuchi *et al.*, 2003; Nakazaki *et al.*, 2003). These findings should reinforce the concept that the genetic stability of *in vitro* cultures should be monitored with a battery of different genetic markers, particularly transposon-based DNA markers that collectively span the whole genome.

Germplasm accessions that are mainly self-pollinated may contain a certain level of heterogeneity providing a buffer for maintaining genetic diversity and preventing genetic shifts. Monitoring heterogeneous accessions will help develop strategies for regeneration of germplasm samples without loss of the allelic diversity provided by heterogeneity. In general, a higher level of heterozygosity was found in traditional cultivars as reported in rice by Olufowote *et al.* (1997). Genetic diversity resulting from heterozygosity or heterogeneity was also found within inbred lines from different sources in rice (Olufowote *et al.*, 1997) and maize (Gethi *et al.*, 2002). In another rice example (Xu *et al.*, 2004), a total of 120 (50.8%) of the 236 rice accessions were found to be heterozygous/heterogeneous at one or more RFLP or SSR loci and the number of heterozygous loci detected in a single rice accession ranged from 0 to 39 (25.3% of the 160 loci). These heterozygous allele patterns may indicate either seed mixtures or true heterozygosity remaining in these cultivars although all accessions had been purified before genotyping and no apparent phenotypic variation was detected.

In plant breeding, specific accessions are selected as parental lines from which to develop new cultivars based on availability of target traits or their overall performance. Some accessions have been used more frequently than others and as a result, considering cultivars that have been bred and used as a whole, genotypic selection and the change of allele frequencies at various genetic loci have occurred resulting in the loss or underrepresentation of specific genotypes, alleles and allele combinations.

Gene flow plays an important role in generating troublesome weeds that are difficult to control. Gene flow can also lead to either crop diversification (Harlan, 1965; Jarvis and Hodgkin, 1999) or genetic assimilation (Para *et al.*, 2005) or a combination of both at different times and in different locations. Molecular marker analysis can be used to monitor gene flow among cultivars developed in a long history of plant breeding (vertical flow), among pedigree-related cultivars developed in a relatively short period (horizontal flow) and among cultivated species and weeds. Tracing specific alleles or genes has become an important objective in parentage control and cultivar identification which has been used to protect the rights of breeders. Further discussion on gene flow between transgenic plants and weeds can be found in Chapter 12.

5.5.6 Unique germplasm

To broaden the genetic base of specific cultivated species, the genetic diversity within collections must be assessed in the context of the total available genetic diversity for each species. With the use of DNA profiles, the genetic uniqueness of each accession in a germplasm collection or in a population can be determined and the identity and frequency of individual alleles can be clearly described and characterized (Brown and Kresovich, 1996; Smith and Helentjaris, 1996; Lu *et al.*, 2009). A DNA bank can be developed to undertake allele mining for identifying unique germplasm containing novel alleles and allele combinations.

The sampling of exotic germplasm should emphasize the genetic composition rather than the appearance of something very different. Accessions with DNA profiles most distinct from that of modern germplasm are likely to contain the greatest number of novel alleles (different from those already present in the elite gene pool). Marker analysis could be used to identify accessions harbouring rare or novel alleles so that the functional significance of the resident genes can be determined using both traditional crossing and sequence-based genomics approaches. Considering the allele frequency profiles across all cultivars and germplasm accessions will give us some idea of which germplasm may retain or contain the rare genes/alleles and further phenotypic characterization may determine whether these alleles will be important to our future breeding programmes. The germplasm that holds unique alleles may contain unique genetic variation required for trait improvement. For example, 15 (6.4%) of the 236 rice accessions examined by Xu *et al.* (2004) contained unique alleles (those present in only one of the cultivars) for at least one RFLP locus and 81 (34.3%) rice accessions had unique SSR alleles. The germplasm accessions identified as having unique alleles also had unusual geographic origins with high genetic diversity and could have potential use in the exploitation of heterosis and novel alleles for agronomic traits.

The degree of genetic similarity between any two cultivars can be calculated as the proportion of shared alleles. The most similar accessions share alleles at almost all marker loci while the least similar accessions have few or no alleles in common. When evaluating genetic similarity, shared allele frequencies (SAFs) can be averaged over all possible pairs of cultivars in a sample. A smaller average similarity indicates a greater genetic difference with respect to the rest of the cultivars in the collection. Based on the averaged SAF, the most diverse accessions can be selected to represent cultivars that host the least-frequent alleles and are genotypically most different from other accessions. From 236 rice cultivars, Xu *et al.* (2004) selected the 16 most diverse accessions (with SAF < 50%) based on RFLP markers and 49 accessions based on SSR markers. Most of these selections, such as Caloro, Cina, Badkalamkati, DGWG and TN1, were ancestral cultivars that had been used as parents in breeding programmes more than 40 years previously; none of the selections includes lines from the US collection which has a much narrower genetic basis.

Genetic mapping studies involving interspecific crosses have identified novel/superior alleles originating from phenotypically unfavorable distant relatives that enhance the performance of modern cultivars (Xiao *et al.*, 1998; Moncada *et al.*, 2001; Brondani *et al.*, 2002; Nguyen *et al.*, 2003;

Thomson *et al.*, 2003). These novel alleles are present in germplasm collections but have not been previously identified because they are hidden in the inferior phenotype. The valuable alleles identified from *Oryza rufipogon* that increase yield in commercial cultivars have been used to improve the best hybrids that have been commercialized in China since before 1989 (Xiao *et al.*, 1998) and new hybrids containing these *O. rufipogon* introgressions demonstrate more than a 30% yield advantage over previous Chinese hybrids (Yuan, 2002). The novel genes/alleles identified from a germplasm collection can also be utilized in transformation experiments by one of the methods described in Chapter 12 if sexual transfer is impossible or too slow, especially during the phase of testing new genes and alleles in a common genetic background. Utilization of genetic resources in plant breeding is the major task of plant breeders, a topic which is discussed in detail in Chapters 7 and 9.

5.5.7 Allele mining

There are several options for identifying or capturing diversity that might not exist in the germplasm pool of existing breeding lines: allele mining, transformation, mutation breeding, use of landraces or synthetic polyploids and wide crossing (Able *et al.*, 2008). Allele mining, which is important for utilizing novel alleles hidden in genetic diversity, will be discussed here.

Molecular and functional diversity of crops genomes can be characterized by allele mining, identification of distinct 'haplotypes' for different inbred lines, single feature polymorphism (SFP) analysis, discovery of nearly identical paralogues (NIPs; Emrich *et al.*, 2007) and determination of their evolutionary implications. In general, there are two approaches that have been elaborated for allele mining: re-sequencing (e.g. Huang *et al.*, 2009) and EcoTILLING (discussed in Chapter 11) (Comai *et al.*, 2004). Whole genome genotyping using gene-based markers can be used as the foundation of the re-sequencing method. Allele mining from germplasm collections is in its infancy currently facing the fundamental challenge to establish which of the various alleles present is functionally different from the wild type and where possible to identify which new alleles beneficially influence the target trait. Methods to ascertain allele function include marker-assisted backcrossing (MABC), transformation, transient expression assays and association analysis using an independent set of germplasm for association mapping from that used to identify the original allele. As more of these studies are carried out, it is hoped that the growing database of comparisons between sequence variation and phenotype will allow bioinformaticians to identify patterns that can form the basis of future predictive methods. The current rate limiting factor for the effective use of outputs from allele mining in breeding programmes is that there is insufficient information on the relationship between SNP variation and changes in phenotypes that may be useful for breeders. However, the resources and tools necessary to perform *in silico* trait targeted selection of the outputs from allele mining are becoming available. Thus, proof-of-concept projects are now being carried out in model organisms in order to study the relationships between SNP haplotypes and changes in phenotypes. This has already led to the development of predictive tools that can identify those SNPs with a high probability of conferring deleterious phenotypes. However, the next big step in this area is the development of bioinformatics tools to compare sequence variation with protein and functional domain variation or with public databases including associated phenotype data, in order to predict which sub-selections of SNP haplotype variants have the maximum likelihood of providing beneficial phenotypic variation in the target trait. It is likely that SNPs in promoter and non-coding regions will also be important for predictive phenotype analysis.

The same methodology used in association mapping may also be used for allele mining of the diverse core subsets of germplasm being created from breeder's lines, genebank accessions and wild relatives. Once a gene of interest is positively identified (via association mapping or any other technique) and the sequence determined, the same gene can be re-sequenced (entirely or in part) in all the individuals in the subset (Huang *et al.*, 2009; Vashney *et al.*, 2009). Changes in the DNA sequence corresponding to new alleles of this locus will be identified in this manner and individuals carrying the new alleles can be evaluated for the target trait to determine the associated change in phenotype and value for subsequent use in breeding programmes. These alleles may not ever have been found via simple phenotypic screens, either because it is not possible to grow and measure every plant in a large germplasm collection under all possible environmental conditions, because its effect may be masked in an unsuitable genetic background or because its effect may be so small that it will not be found unless specifically sought in carefully controlled phenotypic screens (not generally possible on a very large scale).

5.6 Germplasm Enhancement

Although evaluation data are vital to the active use of conserved germplasm, this information does not automatically guarantee active use. Most genebank accessions are landraces that have been grown by traditional farmers for centuries. They have undergone countless cycles of selection for adaptation to biotic and abiotic stresses. Therefore, they may have specific value for modern crop improvement. Unadapted, wild germplasm is valuable and often offers great potential for improvement of important characteristics; however, breeders have usually been unwilling to access this valuable resource because of the detrimental effect of other genes carried along with the selected gene by linkage drag.

The term 'pre-breeding' is often used to designate the phase between evaluation and breeding. Many programmes that aim to facilitate the utilization of plant germplasm include the process of pre-breeding, also known as 'development breeding' or 'germplasm enhancement'. Duvick (1990) defined 'pre-breeding' as making particular genes more accessible and usable to breeders by adapting 'exotic' germplasm to a local environment without losing its essential genetic profile and/or introgressing high-value traits from exotic germplasm into adapted cultivars. Although the end-products of pre-breeding are usually deficient in certain desirable characters, they are attractive to plant breeders due to their greater potential for direct utilization in a breeding programme when compared to the original source(s). Germplasm enhancement or converting unadapted germplasm to a usable form is a key to modern crop improvement. Thus, germplasm curators should not expect large germplasm requests from breeders who are searching for enhanced or improved germplasm with specific traits rather than the raw, often unadapted germplasm that are the principle components of major genebanks. Effective alliance with germplasm enhancement specialists and experimental biologists (molecular geneticists, physiologists, biochemists, pathologists and entomologists) would enhance the use of unadapted germplasm.

Take the common bean (*Phaseolus vulgaris* L.) as an example for crop pre-breeding. Evaluation of wild common bean accessions has shown resistance to insects and diseases and higher N, Fe and Ca content in the seeds, which will ultimately contribute to improvements in nutritional quality and yield (Acosta-Gallegos *et al.*, 2007). In this situation, the pre-breeding efforts will be enhanced by: (i) information on gene pool origins, classification of syndrome traits, molecular diversity and mapping data of the wild forms; (ii) indirect screening for biotic and abiotic stresses; and (iii) marker-assisted selection.

5.6.1 Purification of germplasm collections

Off-types in germplasm collections are a potential quality problem. Traditionally, off-types are defined as individual plants that are phenotypically different from the typical plants or the plants developed by breeders. They may result from mechanical mixtures, outcrossing, mutation or residual genetic variation. From the point of view of germplasm conservation, the off-types could be mixed with the real type and when the proportion of the off-types is sufficiently high within accessions, they can dominate the collection and cannot be differentiated from the typical plants. From the point of view of germplasm utilization, the presence of off-type plants will reduce the uniformity of the crop and thus reduce its productivity and quality. Phenotypic off-types can be easily rogued if there are not too many and if they can be distinguished from the real type by phenotype. In addition to the off-types that are visible phenotypically, many off-types are genetically different from the typical plants but are difficult to distinguish visually. The presence of genotypic off-types may impose a more severe effect on germplasm and could be one of the reasons for genetic drift and cryptic loss of germplasm accessions. Both genotypic and phenotypic off-types may be exaggerated by multiplication. Molecular technology provides a powerful mechanism for distinguishing both the phenotypic and the genotypic off-types from the typical plants. Marker–trait associations and high-resolution molecular markers such as SSRs and SNPs could be used to distinguish two plants with very similar genetic backgrounds. With ten or more co-dominant molecular markers for example, breeders can identify distinct off-types from their breeding populations and hybrid seed bulks and obtain detailed genotypic information such as the source of the off-type genotypes and proportion of the off-types to typical plants. A selection and purification decision can then be made to refine the germplasm collection and breeding materials.

Heterogeneity existing in a germplasm collection reduces the potential of utilization and reduces the interest of plant breeders. The first step in genetic enhancement for this kind of germplasm is purification by selecting typical plants to obtain true-breeding genotypes. This is extremely important for wild relatives of self-pollinated crops.

5.6.2 Tissue culture and transformation in germplasm enhancement

A new level of enthusiasm and activity in plant cell culture research developed in the early 1970s with reports of plant cell lines resistant to amino acid analogues, nucleotide analogues, antibiotics and plant pathogen toxins. The real excitement of these reports from plant breeding was the potential to generate crop plant germplasm that expressed new sources of resistance to herbicides, plant pathogens and mineral and salt stresses that could not be obtained through conventional breeding methodologies. The manipulation of plant cells, tissues and organs *in vitro* is producing an increasing number of unique clones of industrial, biochemical, genotypic and agronomic importance. Examples include regenerable genotypes, transformants, haploids, polyploids, mutants, isogenic lines, somaclonal variants, somatic hybrids and secondary product-producing cultures. In addition, a wide array of industrial chemicals is derived from plants, including flavours, pigments, gums, resins, waxes, dyes, essential oils, edible oils, agrochemicals, enzymes, anaesthetics, analgesics, stimulants, sedatives, narcotics and anticancer agents. The ultimate strategy was to put the most advanced breeding germplasm into cell culture and obtain either by selection or somaclonal variation a derived line improved by the addition of one new trait.

Advances in tissue culture and molecular biology have opened new avenues for the precise transfer of novel genes into crop plants from diverse biological systems (plants, animals and microorganisms) which were previously not feasible. The development of efficient procedures for culture of somatic cells, pollen, protoplasts and for plant regeneration from a large number of plant species combined with a broad-suite of tools, including improved DNA vector systems based on Ti and Ri plasmids of *Agrobacterium*, direct DNA transfer methods, transposable elements, series of promoters, marker genes and a large number of cloned genes, have made gene transfer more precise and directed (for details see Chapter 12). As a result of these developments, transgenic plants have been produced in many plant species with foreign genes inserted for a wide range of traits. These developments have resulted in a large number of germplasm/cultivars with enhanced agronomic traits.

5.6.3 Gene introgression in germplasm enhancement

Up until now the primitive cultivars and related wild populations have been a fruitful and sometimes the sole, source of genes for pest and disease resistance, adaptations to difficult environments and other agricultural traits. The proportions of unadapted and adapted genomes and/or genotypes persisting in enhanced germplasm will differ according to the particular goals of the enhancement programme. Incorporation programmes seek to increase genetic diversity by maximizing the proportion of unadapted genome/genotypes that is retained. In contrast, when introgressing adapted germplasm with high-value traits, only the requisite high-value genes should be transferred. Finally, yield enhancement efforts identify whichever proportion of the unadapted and adapted genome/genotypes that optimizes the yield of the desired end product.

Gene introgression from wild species through molecular marker-assisted selection will be discussed in detail in Chapters 8 and 9. Only two examples that relate to germplasm enhancement will be given here. Isozyme, RFLP and morphological markers diagnostic for chromosomes of one of the wild-weedy relatives of tomato facilitated efforts to introgress wild genomic segments into elite tomato-breeding germplasm (DeVerna *et al.*, 1987, 1990). As a result, the elite germplasm received wild genomic segments that improved horticulturally valuable traits to increase the yield (Rick, 1988). Notably, as a result of this programme, modern tomato cultivars may be more genetically diverse than heirloom, vintage cultivars (Williams and St Clair, 1993).

Stuber and Sisco (1991) and collaborators introgressed into a maize inbred line adapted to Iowa, yield-enhancing genomic segments from an inbred line adapted to Texas by: (i) identifying the favourable segments through yield trials coordinated with molecular (RFLP and isozyme) marker genotyping; and (ii) transferring into the Iowa line, with the help of molecular marker genotyping, only the favourable segments from the Texas line. Although favourable segments were identified in field trials conducted in the diverse environments of North Carolina, Iowa and Illinois, both the recipient and the donor lines could be considered somewhat alien to the primary breeding site for this programme in North Carolina. Nevertheless, these two examples do represent cases in which genetic markers apparently facilitated yield enhancement successfully. With numerous genes identified from wild relatives of crop species by molecular markers, marker-assisted gene introgression will be used increasingly in germplasm enhancement.

The importance of the various base broadening procedures that are being explored at present should be emphasized. Genetic resources workers need to collaborate with plant breeders in the development of procedures that allow effective testing of new materials and their introduction into improvement programmes in a systematic manner. As indicated by Hodgkin and Ramanatha Rao (2002), both introgression and incorporation programmes will be needed.

5.7 Information Management

Information management has become increasingly important in plant breeding and germplasm conservation as a large amount of data is accumulating. Since breeding-related information is described in detail in Chapter 14 only issues related to germplasm management will be discussed in this section.

5.7.1 Information system

There are two areas in which the rapid developments of the past few years have had a major effect on plant genetic resources work: molecular genetics as discussed in Chapters 2 and 3 and information technology. Information generated throughout germplasm conservation activities must be stored in an easily accessible form. Dissatisfaction with the quantity, quality and availability of information on the accession level is the most frequent concern expressed by genebank clients (Fowler and Hodgkin, 2004). The information situation has improved in the last few years. The CGIAR System-wide Information Network for Genetic Resources (SINGER) provides access to data on the plant collections held in trust by the Future Harvest Centres while the US Genetic Resources Information Network (GRIN) DA system and the European EURISCO system provide access to data on collections held in the USA and Europe, respectively. The information revolution enables us to manage and process the very large amounts of data generated in various areas and the Internet potentially provides global access to that data. Information management has always had a central place in plant genetic resources conservation. The need to identify, record and communicate information about accessions has led to the development of a substantial infrastructure with relatively highly developed database structures and information management systems. The information revolution will profoundly affect our understanding of the organisms we conserve. All scientists involved in germplasm collection, conservation and utilization would benefit from professionally built, deployed and maintained information resources for their favourite organisms.

Application of GIS technologies to the management of information on global plant diversity is one of the greatest achievements made in plant genetic resources conservation since the late 1950s (Kresovich *et al.*, 2002). A GIS system is a database management system that can simultaneously handle digital spatial data and associated non-spatial attribute data. Spatial or location data are acquired via geographic positioning system devices which are now quite inexpensive and have become part of the obligatory equipment for field explorations. In addition, an increasing body of geo-referenced data has become available, i.e. data associated with coordinate and altitude information. These geo-referenced data include both biological (e.g. landcover, cattle density) and non-biological (e.g. climate, topography, soil and human activity) data. The non-spatial attributes are any biological, including genetic, data associated with the individual accessions collected. Thus, GIS is a tool designed to visualize and analyse spatial patterns in genetic data in relation to ecological data; it is also a hypothesis-generating tool for investigating processes that shape genomes. With a free mapping program, DIVA-GIS, we can create grid maps of the distribution of biological diversity to identify 'hotspots' and areas that have complementary levels of diversity (Hijmans *et al.*, 2001; http://www.diva-gis.org/). Furthermore, information generated by GIS analysis can help in conserving and using genetic diversity as effectively and efficiently as possible (Greene and Guarino, 1999; Jarvis *et al.*, 2005).

Examples of GIS applications, as summarized by Gepts (2006), include: (i) study of isolation by distance and its effect on the genetic structure of gene pools by comparing genetic and geographic distances; (ii) linking diversity and environmentally heterogeneity; (iii) determination of species distribution and areas of greatest diversity; (iv) identification of germplasm with specific adaptation; (v) predicting the distribution of species of interest and identifying new areas for germplasm exploration; (vi) planning germplasm exploration trips by identifying highly diverse areas, ecologically dissimilar areas, under-conserved areas and areas containing threatened species, timing of the exploration and additions to passport data; (vii) designing zoning plans for *in situ* conservation integrated with socio-economic and indigenous knowledge data; and (viii) establishment of core collections (e.g. those based in part, on environmental variables such as length of the growing season, photoperiod, soil types and moisture regimes).

5.7.2 Standardization of data collection

Issues related to data standardization become more prominent with the accumulation of data from different collections. Genetic resources documentation systems have been developed in many different places and using very different approaches. With the increasing international collaboration among genebanks, the different approaches used by individual genebanks has made it increasingly difficult to exchange information and the need to formulate international formats for the documentation of crop germplasm became apparent (Hazekamp, 2002). To assist in this task, the International Plant Genetic Resources Institute (IPGRI) started the production of crop descriptor lists in 1979 and more than 80 descriptor lists have been produced. In 1996, IPGRI went one step further and defined a set of multi-crop passport descriptors (Hazekamp *et al.*, 1997). These descriptors aim to provide consistent coding schemes for a set of common passport descriptors for all crops and will therefore facilitate the accumulation of passport data into multi-crop information systems.

Another issue will be the need to ensure that there are common vocabularies, also called ontologies, where different databases are being linked and that there are reasonably straightforward ways of linking them (Sobral, 2002). Several genebanks mentioned data standardization as obstacles to fuller use of passport data. The lack of standardization in taxonomic treatments seriously hampers the exchange of basic biological data. Not only are genebanks reliant on a wealth of basic biological data to fulfil their collection acquisition and maintenance tasks, access of users to genebank collections is equally hampered by inconsistent taxonomies. Scientific names are notoriously inconsistent among the collection databases of different organizations (Hulden, 1997). Not only do taxonomic treatments frequently change as a result of new research, but also conflicting treatments coexist for some species. This creates major problems when trying to exchange basic biological information (Hazekamp, 2002).

The Species 2000 and Integrated Taxonomic Information System 'Catalogue of Life' (http://www.sp2000.org/) aims to create a comprehensive catalogue of all known species of organisms on Earth by 2011. Rapid progress has been made recently and the 2006 Annual Checklist contains 884,552 species, approximately half of all known organisms. This database is a valuable asset and will provide a unified structure through which a wealth of basic biological information can be linked. It is essential for genebanks to adhere to such taxonomic standards not only to facilitate data exchange between genebanks but also to ensure effective linkage with related biological disciplines. While initiatives such as the Species 2000 project will go a long way towards rationalizing the use of different taxonomic treatments, there is still a need for adequate taxonomic capacity within national programmes to apply such taxonomic standards correctly (Hazekamp, 2002).

5.7.3 Information integration and utilization

Multiple conservation approaches rely on good documentation systems integrated across programmes and based on individual commodities. For instance, samples can be sorted when it is known who stores what and redundancies can be identified. Site collection data and characterization data enable unique materials to be identified, provide knowledge of where unique landraces or primitive cultivars originated so that they can be conserved and multiplied in areas with similar environments and facilitate decisions on what should justifiably be conserved by whom. It is necessary to provide a framework for a more integrated approach to biology where information from widely different sources can be brought together to help understand crop plant performance and diversity and the forces that are responsible for the patterns we observe. To integrate information generated in different areas, it is important for all researchers to follow general rules for reporting their genotyping and phenotyping results. One direction for the use of a germplasm database is to combine information collected from global research efforts. Facilitating cross talk among currently existing genome databases specializing in sequence information or expression data with germplasm databases documenting phenotypic and genotypic variation will add value to all sources of information.

Integration of molecular information is important but it is even more important (and challenging) to integrate it with phenotypic information obtained at the level of the whole organism, the latter typically provided through germplasm repositories and mutant stock centres. This need is clear because organisms are more than the sum of their parts. In addition, it is within this organism context that data can be transformed into useful information and perhaps, knowledge. Databases from genebanks will be linked with those from botanical gardens and protected areas to support conservation planning. Sequence information from molecular databases will be compared with expressed sequence tagged sites (ESTs) obtained from diversity studies to target markers associated with potentially useful traits (Hodgkin and Ramanatha Rao, 2002). Once integration of molecular information with organism (phenotypic) information is achieved in a robust manner then it is very likely that the resulting system will modify the way we think about germplasm conservation and enhancement.

Comprehensive DNA fingerprinting of crop gene pools, including as many cultivars, hybrid parents and progenies as possible, is the first step for using molecular marker technology in germplasm enhancement. DNA fingerprinting data may be stored as alleles and as scanned images of gels and autoradiographs. These data must be integrated with both phenotypic information and passport and pedigree information. A database of DNA marker alleles for the elite gene pool of a crop provides information on specific DNA polymorphisms that is needed to design, execute and analyse genetic mapping experiments targeted at specific traits or specific crosses. The same database serves as a classification tool describing the overall levels and patterns of variation within the crop gene pool and illustrating subdivisions within a gene pool such as heterotic groups. Such information is useful in making predictions about the performance of new cultivars and hybrids or selecting parents for crosses that are likely to yield new gene combinations or afford an optimal degree of performance. Plant geneticists and breeders can use the data from a germplasm evaluation project as a guide in choosing the most efficient crosses for genetic studies and breeding. For example, a genotyping project of n accessions theoretically provides polymorphism surveys for $n \times (n - 1)/2$ possible cross combinations. With an increasing number of markers surveyed on a variety of germplasm accessions and as more data flow into the database from multiple sources, it will be increasingly possible to determine the genetic constitution and genetic relationships among a wide range of parental lines, cultivars and wild relatives. This also provides the foundation for developing hypotheses based on association genetics and haplotypic patterns to relate agronomically important phenotypes to the presence or absence of specific molecular alleles. The most ambitious project would be to genotype all available germplasm accessions using markers from all candidate genes of a genome. As a result, all germplasm-related efforts could then be based on the whole genome.

An efficient approach to the screening of germplasm involves the ability to rapidly create a nested series of core collections based on information about geographic, phenotypic and genotypic diversity stored in a database. The construction of such a system would require a large-scale effort to provide genotypic information using a standard set of markers that could serve as a reference point. As new markers and marker systems were developed, they could be overlaid on to the essential framework of diversity established previously. An increasingly powerful information system could be developed if data models were made explicit and the data structures were modular so that new types of genetic information could be readily incorporated as they became available (Chapter 14). By accumulating historical information in a systematic manner, germplasm collections would rapidly gain value because they could be screened computationally for essential molecular and phenotypic characteristics of interest.

Databases for whole genomic sequences for several important species, both dicots and monocots, are available allowing directed discovery of genes in higher plants and classification of alleles present in a wide range of breeding germplasm. As indicated by Sorrells and Wilson (1997), the identification of the genes controlling a trait and knowledge of their DNA sequence would facilitate the classification of variation in a germplasm pool based on gene fingerprinting or characterization of variation in key DNA sequences. Classification of functional sequence variants within genes such as FNP at a large number of targeted loci would substantially reduce the amount of work required to determine their relative breeding value and lead to the identification of superior alleles.

5.8 Future Prospects

Germplasm collection, management and evaluation are complex and endless endeavours. Protecting plant genetic diversity poses political, ethical and technical challenges (Esquinas-Alcázar, 2005). Germplasm curators face serious problems, including insufficient operational funds, needs for new germplasm acquisitions, genetic erosion during genebank management, lack of research opportunities and a high turnover of personnel. On the other hand, there are also several major constrains to the efficient use of conserved germplasm which include: (i) insufficient stock of viable seed for distribution; (ii) high cost of seed production and distribution; (iii) lack of relevant characterization and evaluation data; (iv) reluctance of breeders to use unadapted germplasm; (v) plant quarantine regulations; and (vi) obstacles posed by legislation on plant breeders' rights and intellectual property rights.

Developments in recent years have heightened the public nature of the debate over fair access and benefit sharing, and have strengthened the role of global institutes and instruments in protecting these public goods as well as ensuring access to them and the information associated with them. The Global Crop Diversity Trust and the Challenge Programme on Unlocking Genetic Diversity (now known as the Generation Challenge Programme) are two of these developments (Thompson *et al.*, 2004). The Global Crop Diversity Trust has been established as an international fund whose goal is to support the conservation of crop diversity over the long term. The establishment of the Trust involves a partnership between the FAO and the CGIAR. The Trust aims to match the long-term nature of conservation needs with long-term secure and sustainable funding by creating an endowment that will provide a permanent source of funding for crop diversity collections around the world. The CGIAR has taken the initiative for a large research programme that aims to use molecular tools to unlock the genetic diversity in genebank collections for transfer to breeding programmes. This Challenge Programme brings together advanced research institutes, national programmes from developing countries and many of the CGIAR institutes. CIMMYT, IRRI and IPGRI were the founders of this Challenge Programme. Apart from advancing state-of-the-art techniques for molecular characterization of germplasm, one of the main thrusts of the programme is developing molecular toolkits and information systems for crops other than the 'big five' model plant species (rice, *Arabidopsis*, *Medicago*, wheat and maize), including crops such beans, cassava, banana and millets.

Although molecular markers have been recognized as the most useful tools among molecular techniques, the high cost per data point is a bottleneck that is currently limiting the extensive use of molecular markers in germplasm management. The cost of molecular genotyping depends on marker type and its capacity in high-throughput analysis. For example, the cost of SNP analysis is now about US$0.20–0.30 per genotype, with a cost of only a few cents per genotype expected in the coming years (Jenkins and Gibson, 2002). With well-established marker systems and sequencing facilities, genotyping with SSR markers costs about US$0.30–0.80 per data point, depending on marker multiplexing and the number of markers genotyped for each sample (Xu *et al.*, 2002). There are several ways to reduce the cost of genotyping. First, increasing the throughput using automated genotyping and data-scoring systems can help increase the daily data output (Coburn *et al.*, 2002). Secondly, the optimization of marker systems, including facilities and personnel, will result in less cost per data point. Now there are powerful new techniques for screening thousands of plants for sequence variation in any particular gene which is known to be of importance in a breeding programme. Detection of SNPs can reduce a collection of many thousands of accessions to some tens of plants with sequence changes in the gene of interest. These can then be screened for their phenotypic characteristics and used where appropriate (Peacock and Chaudhury, 2002).

Genomic research has helped establish an information flow from molecular markers to genetic maps to sequences to genes and to functional alleles. Apparently, however, there is still a gap between sequence-based information targeting genes and alleles and breeding-related information targeting germplasm, pedigrees and phenotypes. Phenotypic evaluation still provides the foundation for the functional analysis of many genes even when a complete genomic sequence becomes available. Integration of breeding-related phenotypic evaluation with the high-throughput evaluation of mutants in a genomics context will hasten progress towards understanding the functions of all plant genes in the years ahead. This information will enhance the efforts to use MAGE effectively to achieve a substantial increase in the efficiency of germplasm management in the years to come.

With new genotyping methods such as the locus-specific microarray and resequencing, it is now possible to scan variation at a locus from many accessions on a single chip as described in Chapter 3. Consequently, the breadth of genetic information from thousands of DNA polymorphisms and the depth of phenotypic measurements hold promise for identifying marker–trait correlations through linkage disequilibrium-based association genetics. The current QTL cloning procedures are time consuming; for example, in species that have two growing seasons per year, it may take 5 years to produce the population needed for fine-scale mapping. With thousands of genes evaluated for QTL effects, a more efficient approach is needed to complement map-based cloning. This role may be fulfilled by the application of association tests to naturally occurring populations (Buckler and Thornsberry, 2002). This process, which can be called association mapping or linkage disequilibrium mapping within a sample of known pedigree (described in detail in Chapter 6), exploits related individuals that differ for a particular trait to establish which region of the genome is associated with the phenotype among the population members. In order to apply this method to mapping genes using a plant genetic resources collection, the following prerequisite resources will be required: (i) a dense set of molecular markers; (ii) passport and phenotypic data; (iii) information on population structure; and (iv) a sample with contrasting genotypes for the trait of interest (Kresovich *et al.*, 2002). For several reasons, there is great enthusiasm at present about the promise of linkage disequilibrium-based association studies for uncovering the genetic components of complex traits in humans: dense SNP maps across the genome, elegant high-throughput genotyping techniques, simultaneous comparison of groups of loci, statistical measures for assessing genome-wide significance and phenotypic insights as the basis for comparative genomic studies among different human groups are all available. These conditions have already been or will soon be satisfied in some plant species as well, and association studies have been reported in many plant species including maize, rice, barley and *Arabidopsis*. These studies bring together the power of genomics with the richness of crop germplasm collections and promise to provide new insights into the genetic bases of domestication and productivity of our major food crops.

In the age of genomics, new attention is being focused on the value of germplasm resources, including whole plants, seeds, plant parts, tissues and clones from distinct species and synthetic germplasm and all types of mutants. The ultimate goal of germplasm conservation is to maintain diversity of the genes and gene combinations (Xu and Luo, 2002). Information regarding germplasm resources is being increasingly extracted from studies involving the relationship between genome sequence and its biological and evolutionary significance in the context of genetic resources. This information can be translated across species in a comparative context and thus the effective management of germplasm resources today involves both the practical management of seeds, tissues, clones, cells and mutant stocks and the effective management of large reservoirs of electronic information that helps us decipher the value and meaning of the genetic information contained within each germplasm accession. An important question will be whether the increased use of molecular tools in genebanks will make genebanks into true 'banks of genes' and whether, for example, associated sequence data will be freely available.

A user who has identified an accession or accessions of interest from a primary collection would then move to the next level of information, where clusters of germplasm known to represent a broader spectrum of diversity within a specific gene pool (a subspecies or ecotype within a species) or a specific trait (resistance to diseases and pests) could be defined. The second level of investigation could be conducted using carefully designed sets of molecular markers known to target specific traits or designed to provide haplotype data for specific regions of the genome. With the availability of the genome sequences from rice and other plants, accelerated efforts are underway to determine the function of all genes in a plant. As gene structure–function relationships are clarified with greater precision, it will be possible to focus attention on genetic diversity within the active sites of a structural gene or within key promoter regions. This will make it productive to screen large germplasm collections for FNPs, targeting the search for alleles that are phenotypically relevant and have high breeding value.

Locating useful genes in collections will require an integrated approach that brings together information from molecular studies and other areas. This might include using an extensive set of molecular markers for diversity studies, analysis of the extent of linkage disequilibrium and identification of areas of the genome where important genes may occur combined with more conventional approaches using passport data, GIS and collections (Kresovich *et al.*, 2002). These techniques could together provide an optimum set of candidate accessions for phenotypic and genetic analyses. In all cases, efficient characterization methods will remain an essential component of any plant genetic resources studies.

Finally, the domestication stories of maize and tomato should provide a warning to all curators and users of genetic resources that major phenotypic differences between accessions do not always mean that there are equally extensive genetic differences. In addition, significant contributions to agronomically desirable traits may result from the regulation, both spatial and temporal, of gene expression rather than from differences in amino acid sequences or protein structures. The challenge for curators now is to interpret how this knowledge affects not only genetic resource conservation in general, but where and how to look for alleles that will be useful for genetic diversity characterization and plant breeding.

6

Molecular Dissection of Complex Traits: Theory

As discussed in Chapter 1, quantitative variation in phenotype can be explained by the combined action of many discrete genetic factors or polygenes, each having a rather small effect on the overall phenotype and being influenced by the environment. The contribution of each quantitative locus at a phenotypic level is expressed as an increase or decrease in trait value and it is not possible to distinguish the effect of various loci acting in this manner from one another based on phenotypic variation alone. Furthermore, the effect of particular environmental variables is also expressed as a quantitative increase or decrease in the final trait value. The same amount of total genetic variation can be produced by allelic variation at many loci, each having a small effect on the trait or at a few loci having a larger effect. As both genetic and environmental factors contribute in the same positive or negative manner to trait value, it is generally not possible, from the phenotypic distribution of the trait alone, to distinguish the effect of genetic factors from those of environmental factors as sources of variation in traits. Therefore, breeding for quantitative traits tends to be a less efficient and time-consuming process.

Tools for directed genetic manipulation of quantitative traits have undergone a crucial revolution since the late 1980s with the development of molecular markers (Chapter 2). As a result, interchange between molecular biology and quantitative genetics, which have developed independently for many years, has become apparent since the 1990s (Peterson 1992; Xu and Zhu, 1994). Since then, high-density molecular maps have been constructed in many crops and genome-wide mapping and marker-based manipulation of genes affecting quantitative traits has become possible. Traits which have been improved largely by conventional breeding and genetically analysed by biometrical methods in the past can be manipulated now using molecular markers. Location and effect of the genes controlling a quantitative trait can be determined by marker-based genetic analysis. A chromosomal region linked to or associated with a marker which affects a quantitative trait was defined as a quantitative trait locus (QTL) (Geldermann, 1975). A QTL that has a large effect and can explain a major part of total variation can be analysed genetically as a major gene in most cases. In this chapter, our discussion is devoted to the genes with relatively small effects.

Classical quantitative genetics is based on the statistical analysis of the mean and variance of the trait of interest (Chapter 1). It would be preferable to move away from statistical considerations of variances to directly examine individual QTL. Although oit is possible to examine candidate loci, it is more likely that we can work indirectly with QTL via linked marker loci. Early QTL studies were based on manipulations of whole chromosomes, including substitution of one chromosome from one inbred line into another. The approach was refined to apply to small segments of chromosomes, delineated first by morphological markers (Thoday, 1961).

As large numbers of molecular markers become available and thus the whole genome mapping of quantitative traits become feasible, Xu (1997) proposed the concept of separating, pyramiding and cloning QTL, describing how multiple quantitative trait loci (QTL), either clustered together or dispersed in different chromosomes, can first be separated or dissected by molecular marker-assisted QTL mapping and selection and then pyramided into one genetic background, either by marker-assisted selection (MAS) or transformation of cloned multiple QTL, to create transgressive progeny in plant breeding. At about the same time, *Molecular Dissection of Complex Traits* became an attractive title for a multi-author book (Paterson, 1998).

The basic method used in Mendelian genetics to find linkage relationship is to classify individuals based on phenotypes and then compare the proportions of these groups with the theoretical ratio expected from independent loci and estimate the recombination fraction. QTL mapping is establishing linkage between marker loci and QTL. The fundamental principle is the same, i.e. classifying individuals. Depending on what criterion is used for classification of individuals, there are two major types of QTL mapping methods: marker-based analysis and trait-based analysis.

1. Marker-based analysis: methods for locating chromosomal regions or loci affecting quantitative traits or QTL based on their linkage relationships to Mendelian marker loci was first presented by Thoday (1961) and applied in experimental and agricultural species. These studies are all based on differences generated by QTL linked to the marker locus in the mean value of a quantitative trait between marker genotypes in a segregating population such as F_2, BC and DH between two inbred lines (Chapter 4). If a marker is linked to a QTL, marker and QTL alleles co-segregate to some extent in the progeny. As a result, the frequencies of QTL genotypes will be different among marker genotypes (Fig. 6.1) and hence the distribution (e.g. the mean and variance) of the quantitative trait will vary over the marker genotypes. Marker-based analysis of linkage proceeds by testing for phenotypic differences among marker genotypes (Soller and Beckmann, 1990).

Before the discovery of molecular markers, marker-based analysis utilized the data from single markers (e.g. Sax, 1923). Here, only one or a few markers could be analysed in an experiment because the number of markers available at that time was limited. Further, most were morphological or biochemical markers making it impossible to construct a complete linkage map by using single or even multiple populations. With the development of high-density molecular maps, it became apparent that simple (one-locus) marker-based

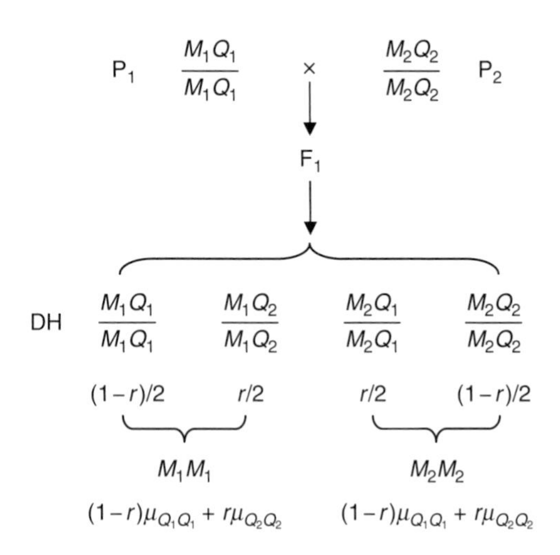

Fig. 6.1. The frequency distribution of QTL genotypes, Q_1Q_1 and Q_2Q_2, within two marker genotypes, M_1M_1 and M_2M_2, in a double haploid (DH) population. r is the recombination frequency between the marker and QTL loci. When $r = 0.5$ (there is no linkage between them), the frequencies of Q_1Q_1 and Q_2Q_2 are the same between the two marker genotypes, which means that there is no phenotypic difference between the two marker genotypes.

analysis alone could not fully utilize the genetic information harboured in complete linkage maps for QTL mapping. To fully exploit the potential of complete linkage maps to locate QTL more efficiently and accurately, many QTL mapping approaches have been developed using multiple markers simultaneously.

2. Trait-based analysis: an alternative approach to TB analysis is to examine marker allele frequencies in the lines originating from a segregating population but selected for specific phenotypes (Stuber *et al.*, 1980, 1982). In such populations, selection, especially for the phenotypic extremes, would be expected to change the allelic frequencies of segregating plus or minus alleles at QTL thus affecting the trait in question. Although variation in quantitative traits is continuous in a population, two extremes of phenotypes could be distinguishable if the intermediate phenotypes are excluded. The frequency of plus alleles increases in the high extreme and the frequency of minus alleles increases in the low extreme (ref. Fig. 7.6B). Hitchhiking effects between such QTL alleles and nearby marker alleles would be expected to generate corresponding changes in the allelic frequencies of the coupled marker alleles. Consequently, marker loci at which allele frequencies differed significantly in the high and low extremes would be considered to be in linkage with QTL having an effect on the trait under selection. In this way, the number of segregating QTL affecting the trait under selection and their general map locations could be determined. The deviation of allele frequency from the Mendelian ratio in each of the two extremes can be tested by the χ^2 statistic. This method is called trait-based analysis (Keightley and Bulfield, 1993). As less statistical but more practical issues are associated with trait-based analysis, utilizing this method is discussed in Chapter 7 along with selective genotyping and pooled DNA analysis.

This chapter discusses theoretical or statistical aspects of QTL mapping. To understand better the story behind QTL mapping, the reader should have some basic knowledge of statistics, experimental design, linear modelling and the theory of probability. Although the readers who have little background in these fields are recommended to focus on Chapter 7 for more practical concepts of molecular dissection of complex traits, they will still benefit by scanning through each section of this chapter. On the other hand, however, it is a challenge to have all the statistical issues in this field fully described in a single chapter. The following references are highly recommended for a full coverage of QTL mapping statistics: Xu and Zhu (1994), Lynch and Walsh (1997), Liu (1998), Sorensen and Gianola (2002) and Wu *et al.* (2007). Furthermore, there are several websites which provide free access to statistical genomics and QTL mapping courses (e.g. http://www.stat.wisc.edu/~yandell/statgen/course/).

6.1 Single Marker-based Approaches

Even though trait variation is often known to be genetic, the number and location of the genes controlling this variation is generally unknown. On the other hand, marker genotypes can be scored precisely. If there is an association between marker type and trait value, it is likely that a trait locus is close to that marker locus. Therefore, the simplest analyses consider each marker locus in turn.

6.1.1 Assumptions

Assume two inbred parental lines, P_1 and P_2 and their F_1, a marker locus M with two alleles, M_1 and M_2 linked with a QTL (Q) with two alleles Q_1 and Q_2 and the recombination fraction between M and Q is r with the trait values normally distributed, we have

$$P_1(Q_1Q_1M_1M_1): Y \sim N(\mu_{Q_1Q_1}, \sigma^2)$$

$$P_2(Q_2Q_2M_2M_2): Y \sim N(\mu_{Q_2Q_2}, \sigma^2)$$

$$F_1(Q_1Q_2M_1M_2): Y \sim N(\mu_{Q_1Q_2}, \sigma^2)$$

For the populations derived from this cross, there are three marker types in F_2 (M_1M_1,

M_1M_2 and M_2M_2), two marker types in the double haploid (DH) (M_1M_1 and M_2M_2) or in the backcross (BC) (M_1M_1 and M_1M_2 for BC derived from backcrossing F_1 to P_1 and M_1M_2 and M_2M_2 for BC derived from backcrossing F_1 to P_2). Similarly, QTL genotypes for each population can be obtained. Tables 6.1, 6.2 and 6.3 provide the genotypic frequencies, means and variances for each marker type and QTL genotypes in populations F_2, DH and BC.

In most cases, we assume that trait variances are homogeneous over the different QTL genotypes, so that $\sigma^2_{Q_1Q_1} = \sigma^2_{Q_1Q_2} = \sigma^2_{Q_2Q_2} = \sigma^2$ for F_2 and we also make this same assumption for other populations such as BC, DH and recombinant inbred lines (RILs).

For the convenience of discussion, frequencies for genotypes in F_2 and DH (BC) populations are expressed by a matrix:

$$p_{ij} = \begin{bmatrix} (1-r)^2 & 2r(1-r) & r^2 \\ r(1-r) & 1-2r+2r^2 & r(1-r) \\ r^2 & 2r(1-r) & (1-r)^2 \end{bmatrix}$$

$(i, j = 1, 2, 3 \text{ for } F_2)$

and

$$p_{ij} = \begin{bmatrix} 1-r & r \\ r & 1-r \end{bmatrix}$$

$(i, j = 1, 2 \text{ for DH or BC})$

Table 6.1. Genotypic frequencies, means and variances in marker and QTL genotypes in the F_2 population.

	Genotypic frequency					
Marker genotype	Q_1Q_1	Q_1Q_2	Q_2Q_2	Sample mean	Sample variance	Sample size
M_1M_1	$(1-r)^2$	$2r(1-r)$	r^2	$\mu_{M_1M_1}$	$\sigma^2_{M_1M_1}$	$n_{M_1M_1}$
M_1M_2	$r(1-r)$	$1-2r+2r^2$	$r(1-r)$	$\mu_{M_1M_2}$	$\sigma^2_{M_1M_2}$	$n_{M_1M_2}$
M_2M_2	r^2	$2r(1-r)$	$(1-r)^2$	$\mu_{M_2M_2}$	$\sigma^2_{M_2M_2}$	$n_{M_2M_2}$
Mean	$\mu_{Q_1Q_1}$	$\mu_{Q_1Q_2}$	$\mu_{Q_2Q_2}$			
Variance	$\sigma^2_{Q_1Q_1}$	$\sigma^2_{Q_1Q_2}$	$\sigma^2_{Q_2Q_2}$			

Table 6.2. Genotypic frequencies, means, and variances in marker and QTL genotypes in the DH population.

	Genotypic frequency				
Marker genotype	Q_1Q_1	Q_2Q_2	Sample mean	Sample variance	Sample size
M_1M_1	$1-r$	r	$\mu_{M_1M_1}$	$\sigma^2_{M_1M_1}$	$n_{M_1M_1}$
M_2M_2	r	$1-r$	$\mu_{M_2M_2}$	$\sigma^2_{M_2M_2}$	$n_{M_2M_2}$
Mean	$\mu_{Q_1Q_1}$	$\mu_{Q_2Q_2}$			
Variance	$\sigma^2_{Q_1Q_1}$	$\sigma^2_{Q_2Q_2}$			

Table 6.3. Genotypic frequencies, means and variances in marker and QTL genotypes in the BC population derived by backcrossing to P_1.

	Genotypic frequency				
Marker genotype	Q_1Q_1	Q_1Q_2	Sample mean	Sample variance	Sample size
M_1M_1	$1-r$	r	$\mu_{M_1M_1}$	$\sigma^2_{M_1M_1}$	$n_{M_1M_1}$
M_1M_2	r	$1-r$	$\mu_{M_1M_2}$	$\sigma^2_{M_1M_2}$	$n_{M_1M_2}$
Mean	$\mu_{Q_1Q_1}$	$\mu_{Q_1Q_2}$			
Variance	$\sigma^2_{Q_1Q_1}$	$\sigma^2_{Q_1Q_2}$			

6.1.2 Comparison of marker means

Backcross design

Now we take the BC population as an example to show how to detect marker–trait association through comparison of means for the different marker classes.

The genotypic array for the BC population derived from backcrossing F_1 to P_1 is

$$\frac{1-r}{2}Q_1Q_1M_1M_1 + \frac{r}{2}Q_1Q_2M_1M_1 + \frac{r}{2}Q_1Q_1M_1M_2 + \frac{1-r}{2}Q_1Q_2M_1M_2$$

For a BC population derived from backcrossing F_1 to P_2, the two marker genotypes are M_1M_2 and M_2M_2 and the two QTL genotypes are Q_1Q_2 and Q_2Q_2, with other items unchanged.

Only the marker genotypes can be directly observed and BC individuals have two classes of segregates: marker types M_1M_1 and M_1M_2. The trait distributions in these two classes are

$$M_1M_1\text{: } Y\sim(1-r)\,N(\mu_{Q_1Q_1},\sigma^2) + rN(\mu_{Q_1Q_2},\sigma^2)$$

$$M_1M_2\text{: } Y\sim rN(\mu_{Q_1Q_1},\sigma^2) + (1-r)N(\mu_{Q_1Q_2},\sigma^2)$$

The means and variance of the two mixture distributions are

$$\mu_{M_1M_1} = (1-r)\mu_{Q_1Q_1} + r\mu_{Q_1Q_2}$$

$$\mu_{M_1M_2} = r\mu_{Q_1Q_1} + (1-r)\mu_{Q_1Q_2}$$

$$\sigma^2_{M_1M_1} = \sigma^2_{M_1M_2} = \sigma^2 + r(1-r)(\mu_{Q_1Q_1} - \mu_{Q_1Q_2})^2$$

The expected difference in average trait values is

$$\mu_{M_1M_1} - \mu_{M_1M_2} = (1-2r)(\mu_{Q_1Q_1} - \mu_{Q_1Q_2})$$

Therefore, only when $r = 0.5$, i.e. there is no linkage between M and Q,

$$\mu_{M_1M_1} - \mu_{M_1M_2} = 0$$

If $r < 0.5$, there is linkage and $\mu_{M_1M_1} \neq \mu_{M_1M_2}$. The smaller r (and hence and the tighter the linkage), the bigger the difference between $\mu_{M_1M_1}$ and $\mu_{M_1M_2}$. The difference reaches the maximum, $\mu_{M_1M_1} - \mu_{M_1M_2} = \mu_{Q_1Q_1} - \mu_{Q_1Q_2}$, when $r = 0$, i.e. M and Q are completely linked. In this case, all differences between marker genotypes can be attributed to the effect of the putative QTL.

The difference between marker genotypes can be tested by the t-test statistic

$$t = \frac{\tilde{\mu}_{M_1M_1} - \tilde{\mu}_{M_1M_2}}{\sqrt{s^2\left(\frac{1}{n_{M_1M_1}} + \frac{1}{n_{M_1M_2}}\right)}}$$

The quantity s^2 is a pooled estimate of the variance within each marker class of BC individuals. The higher the t value, the more significant the difference and the closer the linkage is between M and Q.

The above discussion can be extended to DH populations where $\mu_{M_1M_2}$ is replaced by $\mu_{M_2M_2}$ and $\mu_{Q_1Q_2}$ by $\mu_{Q_2Q_2}$.

F_2 design

For an F_2 population, there are ten trait–marker genotypes with three distinguishable marker classes. The trait distributions are

$$M_1M_1\text{: } (1-r)^2\,N(\mu_{Q_1Q_1},\sigma^2) + 2r(1-r)\,N(\mu_{Q_1Q_2},\sigma^2) + r^2N(\mu_{Q_2Q_2},\sigma^2)$$

$$M_1M_2\text{: } r(1-r)\,N(\mu_{Q_1Q_1},\sigma^2) + [r^2+(1-r)^2]\,N(\mu_{Q_1Q_2},\sigma^2) + r(1-r)\,N(\mu_{Q_2Q_2},\sigma^2)$$

$$M_2M_2\text{: } r^2\,N(\mu_{Q_1Q_1},\sigma^2) + 2r(1-r)\,N(\mu_{Q_1Q_2},\sigma^2) + (1-r)^2\,N(\mu_{Q_2Q_2},\sigma^2)$$

The trait means of these three marker classes are

$$\mu_{M_1M_1} = (1-r)^2\mu_{Q_1Q_1} + 2r(1-r)\mu_{Q_1Q_2} + r^2\mu_{Q_2Q_2}$$

$$\mu_{M_1M_2} = r(1-r)\mu_{Q_1Q_1} + [r^2+(1-r)^2]\mu_{Q_1Q_2} + r(1-r)\mu_{Q_2Q_2}$$

$$\mu_{M_2M_2} = r^2\mu_{Q_1Q_1} + 2r(1-r)\mu_{Q_1Q_2} + (1-r)^2\mu_{Q_2Q_2}$$

The trait variances of these three classes are

$$\begin{aligned}\sigma^2_{M_1M_1} &= \sigma^2 + 2r(1-r)[(\mu_{Q_1Q_1} - \mu_{Q_1Q_2}) \\ &\quad - r(\mu_{Q_1Q_1} + \mu_{Q_2Q_2} - 2\mu_{Q_1Q_2})]^2 \\ &\quad + r^2(1-r)^2(\mu_{Q_1Q_1} + \mu_{Q_2Q_2} - 2\mu_{Q_1Q_2})^2 \\ \sigma^2_{M_1M_2} &= \sigma^2 + r(1-r)[(\mu_{Q_1Q_1} - \mu_{Q_1Q_2})^{2^2} \\ &\quad + (\mu_{Q_2Q_2} - \mu_{Q_1Q_2})^2 \\ &\quad - r^2(1-r)^2(\mu_{Q_1Q_1} + \mu_{Q_2Q_2} - 2\mu_{Q_1Q_2})] \\ \sigma^2_{M_2M_2} &= \sigma^2 + 2r(1-r)[(\mu_{Q_2Q_2} - \mu_{Q_1Q_2}) \\ &\quad - r(\mu_{Q_1Q_1} + \mu_{Q_2Q_2} - 2\mu_{Q_1Q_2})]^2 \\ &\quad + r^2(1-r)^2(\mu_{Q_1Q_1} + \mu_{Q_2Q_2} - 2\mu_{Q_1Q_2})^2\end{aligned}$$

For an additive trait, i.e. $\mu_{Q_1Q_1} + \mu_{Q_2Q_2} - 2\mu_{Q_1Q_2} = 0$, the three F_2 variances are in general equal. Then

$$\sigma^2_{M_1M_1} = \sigma^2_{M_1M_2} = \sigma^2_{M_2M_2} = \sigma^2 + 2r(1-r)\delta^2$$

where $\delta^2 = (\mu_{Q_1Q_1} - \mu_{Q_1Q_2})^2 = (\mu_{Q_2Q_2} - \mu_{Q_1Q_2})^2$.

The hypothesis of no linkage between marker and trait loci can be tested by comparing the three marker class means. Under this hypothesis, the three marker means and variances will be equal regardless of the degree of dominance.

For an F_2 population we can construct two *t*-tests to test marker additive and dominance effects separately. Let $\tilde{\mu}_{M_1M_1}$, $\tilde{\mu}_{M_1M_2}$, $\tilde{\mu}_{M_2M_2}$ be the observed trait means of the groups of individuals with marker genotypes M_1M_1, M_1M_2 and M_2M_2 for a marker in an F_2 population with corresponding samples size $n_{M_1M_1}$, $n_{M_1M_2}$ and $n_{M_2M_2}$ and variances $\sigma^2_{M_1M_1}$, $\sigma^2_{M_1M_2}$ and $\sigma^2_{M_2M_2}$. Recall the additive and dominance effects as defined in Chapter 1. To test marker additive effect, the test statistic is

$$t_1 = \frac{\tilde{\mu}_{M_1M_1} - \tilde{\mu}_{M_2M_2}}{\sqrt{S^2\left(\dfrac{1}{n_{M_1M_1}} + \dfrac{1}{n_{M_2M_2}}\right)}} \qquad (6.1)$$

with

$$S^2 = \frac{(n_{M_1M_1} - 1)\sigma^2_{M_1M_1} + (n_{M_2M_2} - 1)\sigma^2_{M_2M_2}}{n_{M_1M_1} + n_{M_2M_2} - 2}$$

To test marker dominance effect, the test statistic is

$$t_2 = \frac{\tilde{\mu}_{M_1M_2} - (\tilde{\mu}_{M_1M_1} + \tilde{\mu}_{M_2M_2})/2}{\sqrt{S^2\left(\dfrac{1}{n_{M_1M_2}} + \dfrac{1}{4n_{M_1M_1}} + \dfrac{1}{4n_{M_2M_2}}\right)}} \qquad (6.2)$$

with (see equation at bottom of page).

6.1.3 ANOVA

The test of marker–trait association can also be carried out by analysis of variance (ANOVA) to analyse between marker classes for each marker. If there is no linkage between marker locus M and QTL, no marker–trait association will be found. As a result, the means of marker genotypes are equal for F_2, i.e.

$$\hat{\mu}_{M_1M_1} = \hat{\mu}_{M_1M_2} = \hat{\mu}_{M_2M_2}$$

If the individual groups of marker genotypes are considered as independent samples, phenotypic differences among marker genotypes can be tested by one-way ANOVA with unequal sample sizes. The ANOVA model is

$$y_{ij} = \mu + \tau_i + \varepsilon_{ij}$$

where y_{ij} is the phenotypic value for the *j*th individual within the *i*th marker genotype and μ is the phenotypic mean of the mapping population. We therefore have

$$\begin{aligned}\tau_i &= \hat{\mu}_i - \mu \\ \varepsilon_{ij} &= y_{ij} - \hat{\mu}_i\end{aligned}$$

with $i = 1, \ldots, k$ (F_2: $k = 3$; BC (DH): $k = 2$) and $j = 1, \ldots, n_i$. We have an ANOVA table

$$S^2 = \frac{(n_{M_1M_1} - 1)\sigma^2_{M_1M_1} + (n_{M_1M_2} - 1)\sigma^2_{M_1M_2} + (n_{M_2M_2} - 1)\sigma^2_{M_2M_2}}{n_{M_1M_1} + n_{M_1M_2} + n_{M_2M_2} - 3}$$

as shown in Table 6.4. A significant treatment effect implies linkage to a segregating QTL.

6.1.4 Regression approach

Instead of a *t*-test or ANOVA, the trait value can be regressed on marker genotype. Using BC populations as an example, for the *j*th individual,

$$Y_j = \beta_0 + \beta_{YX} X_j + \varepsilon_j$$

where the indicator variable X_j takes the values 1 or 0 according to whether the individual has marker genotype M_1M_1 or M_1M_2. The variances and correlations among variables Y and X are:

$$\mu_X = \frac{1}{2},\ \mu_Y = \frac{1}{2}(\mu_{Q_1Q_1} + \mu_{Q_1Q_2})$$

$$\sigma_X^2 = \frac{1}{4},\ \sigma_Y^2 = \sigma^2 + \frac{1}{4}\delta^2$$

$$\sigma_{XY} = \frac{1}{4}(1-2r)\delta$$

$$\rho_{XY} = (1-2r)\sqrt{1+4\sigma^2/\delta^2}$$

So, the regression coefficient for Y on X is

$$\beta_{YX} = (1 - 2r)\delta$$

This method uses one marker at a time to test whether this marker is significantly associated with the quantitative trait under investigation, using the statistic

$$t = \frac{\hat{\beta}_{YX} - 0}{\hat{s}_{\beta_{YX}}}$$

From this statistic, we know that the environmental error will affect $\hat{s}_{\beta_{YX}}$ but not $\hat{\beta}_{YX}$ so that reducing the environmental error by controlling environmental factors will improve the QTL mapping effect. The major drawback of such linear model-based (e.g. ANOVA, regression) single marker approaches is that they do not indicate which side of the marker the QTL is located nor how far it is from the marker.

6.1.5 Likelihood approach

For a normal variable, $Y \sim N(\mu,\sigma^2)$, the likelihood for the parameters (see Chapter 2

Table 6.4. One-way ANOVA for quantitative trait values among marker genotypes. *df*, degree of freedom; *SS*, sum of squares; *MS*, mean square; *EMS*, expected mean square.

Source	*df*	*SS*	*MS*	*EMS*
Between genotype	df_g	SS_g	MS_g	$\sigma_e^2 + n_0\,\sigma_\tau^2$
Error	df_e	SS_e	MS_e	σ_e^2
Total	df_T			

$$df_T = \sum_i n_i - 1,\ df_t = k - 1,\ df_e = df_T - df_t$$

$$SS_T = \sum_i \sum_j y_{ij}^2 - (\sum_i \sum_j y_{ij})^2 / \sum_i n_i$$

$$SS_t = \sum_i \frac{(\sum_j y_{ij})^2}{n_i} - \frac{(\sum_i \sum_j y_{ij})^2}{\sum_i n_i}$$

$$SS_e = SS_T - SS_t$$

$$n_0 = \frac{(\sum_i n_i)^2 - \sum_i n_i^2}{(\sum_i n_i)(k-1)}$$

for a basic description of the likelihood method) is

$$L(\mu,\sigma^2) = \frac{e^{-(Y-\mu)^2/(2\sigma^2)}}{\sqrt{2\pi\sigma^2}}$$

If Y_{1i} and Y_{2i} are the trait values for the ith individuals in BC marker classes M_1M_1 and M_1M_2, then the likelihood from all $n_{M_1M_1}$ and $n_{M_1M_2}$ backcross individuals is shown in the equation at the bottom of the page. The hypothesis of no linkage can be tested by the likelihood ratio statistic

$$\lambda = \frac{L(\tilde{\mu}_{Q_1Q_1}, \tilde{\mu}_{Q_1Q_2}, \hat{\sigma}^2, r = 0.5)}{L(\tilde{\mu}_{Q_1Q_1}, \tilde{\mu}_{Q_1Q_2}, \hat{\sigma}^2, \hat{r})}$$

The estimates of $\mu_{Q_1Q_1}$, $\mu_{Q_1Q_2}$, σ^2 will be different for r being estimated or set to 0.5.

6.2 Interval Mapping

An indication that the recombination fraction r is less than 0.5 from single-marker analyses is confounded by the size of the effects of locus Q, since it is actually the product $(1-r)\delta$ that is being tested for departure from zero. A marker close to a QTL of small effect will give the same signal as a marker some distance from a QTL of large effect. Lander and Botstein (1989) developed a QTL mapping method known as interval mapping, by using two flanking markers. van Ooijen (1992) described this method in more detail to make it more understandable, while Xu *et al.* (1995) extended the statistics issues to DH populations.

6.2.1 Assumptions

Suppose the two parental lines used to produce a mapping population (F_2, BC or DH) have genotypes $M_1M_1Q_1Q_1N_1N_1$ and $M_2M_2Q_2Q_2N_2N_2$, respectively, at two marker loci M (with alleles M_1 and M_2) and N (with alleles N_1 and N_2) and a QTL, each with two alleles. The QTL is located between two marker loci. Its genetic distances to M and N are θ_{MQ} and θ_{QN}, respectively. The genetic distance (cM) between markers M and N is θ. θ can be converted into recombinant frequency, r, by

$$r = \frac{1}{2}\tanh(2\theta) = \frac{1}{2}\frac{e^{2\theta} - e^{-2\theta}}{e^{2\theta} + e^{-2\theta}}$$

Theoretically we have $0 < r_{MQ} < r$. When there is no crossover interference, $r = r_{MQ} + r_{QN} - 2r_{MQ}r_{QN}$.

Hypotheses include

H_0: $r_{MQ} = r_{QN} = 0.5$ QTL unlinked to markers
H_1: $\min(r_{MQ}, r_{QN}) < 0.5$ QTL linked to markers

or

H_0: $\min(r_{MQ}, r_{QN}) > r_{MN}$ QTL exterior to interval
H_1: $\min(r_{MQ}, r_{QN}) < r_{MN}$ QTL interior to interval

Under the assumption of no interference, when Q is interior to MN, the event of no recombination between M and N is equivalent to no recombination in both intervals MQ and QN or no recombination in either interval:

$$(1 - r_{MN}) = (1 - r_{MQ})(1 - r_{QN}) + r_{MQ}\, r_{QN}$$
$$r_{MN} = r_{MQ} + r_{QN} - 2r_{MQ}r_{QN}$$
$$(1-2r_{MN}) = (1 - 2r_{MQ})(1 - 2r_{QN})$$

where r_{MN} is known, there is only one independent unknown recombination fraction.

Assume the phenotypic values of n measured individuals or families/lines in the mapping population is $y = \{y_1, y_2, \ldots, y_n\}$

$$L = \prod_{i=1}^{n_{M_1M_1}} \left[\frac{1-r}{\sqrt{2\pi\sigma^2}} \exp\left(\frac{-(Y_{1i} - \mu_{Q_1Q_1})^2}{2\sigma^2} \right) + \frac{r}{\sqrt{2\pi\sigma^2}} \exp\left(\frac{-(Y_{1i} - \mu_{Q_1Q_2})^2}{2\sigma^2} \right) \right]$$
$$\times \prod_{i=1}^{n_{M_1M_2}} \left[\frac{r}{\sqrt{2\pi\sigma^2}} \exp\left(\frac{-(Y_{2i} - \mu_{Q_1Q_1})^2}{2\sigma^2} \right) + \frac{(1-r)}{\sqrt{2\pi\sigma^2}} \exp\left(\frac{-(Y_{2i} - \mu_{Q_1Q_2})^2}{2\sigma^2} \right) \right]$$

and the genetic effects of three QTL genotypes, Q_1Q_1, Q_1Q_2 and Q_2Q_2 follow normal distribution $N(\mu_{Q_1Q_1},\sigma^2)$, $N(\mu_{Q_1Q_1},\sigma^2)$ and $N(\mu_{Q_2Q_2},\sigma^2)$. Therefore, the effect of QTL on the quantitative trait can be described by the mixture of these three normal distributions, with proportions P_{m1}, P_{m2} and P_{m3}, respectively, for the marker locus M. P_{m1} is zero for the BC population derived from backcrossing F_1 to P_2, P_{m2} is zero for DH populations and P_{m3} is zero for the BC population derived from backcrossing F_1 to P_1. The probability density function for the phenotypic value of an individual or line is

$$f(y_i|m_i;r_M) = \sum_{q=1} P_{mq} f_q(y)$$

where m is the marker genotype; r_M is the recombinant frequency between the marker M and QTL and P_{Mq} is the probability of QTL genotype ($q \in \{1,2,3\}$ for an F_2 population), which depends on the marker genotype m and r_M and

$$f_q(y) = \frac{1}{\sqrt{2\pi\sigma^2}} \exp\left[-\frac{(y-\mu_q)^2}{2\sigma^2}\right]$$

which is the probability density function for a normal distribution with mean μ_q and variance σ^2, with $\mu_1 = \mu_{Q_1Q_1}$, $\mu_2 = \mu_{Q_1Q_2}$ and $\mu_3 = \mu_{Q_2Q_2}$ for an F_2 population.

6.2.2 Likelihood approach

For two specific markers M and N, the F_2 and BC (DH) populations have nine and four marker genotypes, respectively, and any individual in the population must have one of these genotypes. For the individuals or lines with a specific marker genotype, the sum of the probabilities for three QTL genotypes, Q_1Q_1, Q_1Q_2 and Q_2Q_2, is $P_{m1} + P_{m2} + P_{m3} = 1$. From the genotypic frequencies provided in Tables 6.5 and 6.6, probabilities for three QTL genotypes can be obtained. The combined probability density function or likelihood function for all individuals/lines in a population can be expressed as:

$$L = L(\mu_q, \mu_j, \sigma^2, y_1, y_2, \ldots y_n)$$

$$= \prod_{i=1}^{n} f(y_i|m_i;r_M) = \prod_{i=1}^{n}\sum_{q=1} P_{mq} f_q(y)$$

$$= \prod_{i=1}^{n}\sum_{q=1} P_{mq} \frac{1}{\sqrt{2\pi\sigma^2}} \exp\left[-\frac{(y_i-\mu_q)^2}{2\sigma^2}\right]$$

Table 6.5. The expected genotypic frequencies in the F_2 population (each frequency × 4).

Genotype	Q_1Q_1	Q_1Q_2	Q_2Q_2
$M_1M_1N_1N_1$	$r_M^2 r_N^2$	$2r_M(1-r_M)r_N(1-r_N)$	$(1-r_M)^2(1-r_N)^2$
$M_1M_1N_1N_2$	$2r_M^2 r_N(1-r_N)$	$2r_M(1-r_M)[r_N^2+(1-r_N)^2]$	$2(1-r_M)^2 r_N(1-r_N)$
$M_1M_1N_2N_2$	$r_M^2(1-r_N)^2$	$2r_M(1-r_M)r_N(1-r_N)$	$(1-r_M)^2 r_N^2$
$M_1M_2N_1N_1$	$2r_M(1-r_M)r_N^2$	$2[r_M^2+(1-r_M)^2]r_N(1-r_N)$	$2r_M(1-r_M)(1-r_N)^2$
$M_1M_2N_1N_2$	$4r_M(1-r_M)r_N(1-r_N)$	$2[r_M^2+(1-r_M)^2][r_N^2+(1-r_N)^2]$	$4r_M(1-r_M)r_N(1-r_N)$
$M_1M_2N_2N_2$	$2r_M(1-r_M)(1-r_N)^2$	$2[r_M^2+(1-r_M)^2]r_N(1-r_N)$	$2r_M(1-r_M)r_N^2$
$M_2M_2N_1N_1$	$(1-r_M)^2 r_N^2$	$2r_M(1-r_M)r_N(1-r_N)$	$r_M^2(1-r_N)^2$
$M_2M_2N_1N_2$	$2(1-r_M)^2 r_N(1-r_N)$	$2r_M(1-r_M)[r_N^2+(1-r_N)^2]$	$2r_M^2 r_N(1-r_N)$
$M_2M_2N_2N_2$	$(1-r_M)^2(1-r_N)^2$	$2r_M(1-r_M)r_N(1-r_N)$	$r_M^2 r_N^2$

Table 6.6. The expected genotypic frequencies in the DH (BC) population (each frequency × 2).

Genotype	$Q_1Q_1(Q_1Q_2)$	Q_2Q_2
$M_1M_1N_1N_1(M_1M_2N_1N_1)$	$r_M r_N$	$(1-r_M)(1-r_N)$
$M_1M_1N_2N_2(M_1M_2N_2N_2)$	$r_M(1-r_N)$	$(1-r_M)r_N$
$M_2M_2N_1N_1(M_2M_2N_1N_2)$	$(1-r_M)r_N$	$r_M(1-r_N)$
$M_2M_2N_2N_2(M_2M_2N_2N_2)$	$(1-r_M)(1-r_N)$	$r_M r_N$

where m_i is the marker genotype for the ith individual/line, with a total of n individuals/lines in the population.

Maximum likelihood estimates (MLEs) of parameters μ_j and σ^2 are those for maximizing the above likelihood function. In order to maximize the function, we take the logarithm for the likelihood,

$$\ln L = \ln\left(\prod_{i=1}^{n} f(y_i | m_i; r_M)\right)$$
$$= \ln \frac{n}{\sqrt{2\pi\sigma^2}} + \sum_{i=1}^{n} \ln \sum_{q=1} P_{mq} \exp\left[-\frac{(y_i - \mu_q)^2}{2\sigma^2}\right] \quad (6.3)$$

If we define

$$W_q(y_i | m_i; r_M) = \frac{P_{mq} f_q(y_i)}{f(y_i | m_i; r_M)}$$

thus the probability that QTL genotype is q, when an individual or line has phenotype y and marker genotype m, is determined by W_q.

Set the derivative of Eqn 6.3 as zero and solve the equation to obtain

$$\hat{\mu}_q = \frac{\sum_{i=1}^{n} [W_q(y_i | m_i; r_M) \times y_i]}{\sum_{i=1}^{n} W_q(y_i | m_i; r_M)} \quad (6.4a)$$

$$\hat{\sigma}^2 = \frac{1}{n} \sum_{i=1}^{n} \sum_{q=1} [W_q(y_i | m_i; r_M) \times (y_i - \mu_q)^2] \quad (6.4b)$$

When QTL is located between the two marker loci, Eqn 6.4 has no explicit solution. However, it can be solved using EM iteration method (Dempster *et al.*, 1977). The E (expectation) step in the EM method is to obtain the expectations for unknown missing data by using known data (y and m) and initial approximations of, for example for a F_2 population, $\lambda \in (\mu_1, \mu_2, \mu_3, \sigma^2)$(using the average phenotypes of quantitative trait, x_1, x_2 and x_3, for individual/line groups of marker genotypes and the sample variance of the population, s^2, as initial values). The M (maximization) step is to maximize the likelihood function (Eqn 6.3) to obtain a new cycle of λ values, by using the initial values of λ and the expectation obtained for missing data. E and M steps are processed alternatively by using the new λ to replace the old λ, until the likelihood function (Eqn 6.3) does not increase (the difference between the two iterations is less than a predetermined critical value).

Under the null hypothesis H_0: $\mu_i = \mu_L$ ($i \neq L$) (there is no linked QTL), the likelihood function becomes

$$L_0 = L(\mu_P, \sigma_P^2, y_1, y_2, \ldots y_n) = \prod_{i=1}^{n} f(y_i) \quad (6.5)$$

where

- $\hat{\mu}_P = \frac{1}{n} \sum_{i=1}^{n} y_i$ is the average mean of the mapping population;
- $\hat{\sigma}_P^2 = \frac{1}{n} \sum_{i=1}^{n} (y_i - \mu_P)^2$ is the variance of the mapping population; and
- $f(y_i) = \frac{1}{\sqrt{2\pi\sigma^2}} \exp\left[-\frac{(y_i - \mu_P)^2}{2\sigma^2}\right]$ is the normal density function with mean μ_P and variance σ_P^2.

The statistic for the likelihood ratio test of the alternative hypothesis (at least one QTL exists at this location) can be converted into a likelihood of odds (LOD) score,

$$LOD = \log_{10}\left(\frac{L(\mu_j, \sigma^2, y_1, y_2, \ldots y_n)}{L_0(\mu_P, \sigma_P^2, y_1, y_2, \ldots y_n)}\right)$$

For an interval bracketed by the two markers M and N, a LOD score is calculated for every scanning position. LOD scores obtained for all marker intervals located on the same chromosome form a likelihood profile to show possible position(s) of QTL associated with the quantitative trait. This method uses two flanking markers at a time to test whether there is any QTL locating at the interval bracketed by the two markers. For a specific interval, the test is carried out at any point by moving a step from one marker to the other. After completion of the test for the interval, the test moves to the next two flanking markers. The LOD score does not provide a test for the presence of a QTL between the two markers and so is not a formal test of a QTL within the interval. Instead the LOD compares the likelihood of the QTL being at the position characterized by recombination fractions r_{MQ}, r_{QN} against the likelihood that it is at some position unlinked to the interval.

The amount of support for a QTL at a particular map position is often displayed graphically through use of likelihood (or profile) maps (Fig. 6.2), which plot the likelihood-ratio statistic (or a closely related quantity) as a function of the map position of the putative QTL. Lander and Botstein (1989) plotted the LOD scores defined by Morton (1955).

Empirically, a QTL is claimed when the LOD is larger than a critical value predetermined (for example, 2 or 3) or generated by permutation. The location of the QTL should be the chromosome region that corresponds to the highest likelihood map if LOD scores in several flanking marker intervals are larger than the critical values. The two-LOD support interval determined by the range of the highest LOD minus two LOD provides an empirical confidence interval for the range of QTL location (Fig. 6.2). Simulation studies show that a two-LOD interval is close to the 95% interval.

6.3 Composite Interval Mapping

Most of the single-QTL methods can be extended to multiple QTL by conditioning additional marker loci and using conditional probabilities for multi-locus genotypes. This approach has been used to develop explicit models for two or three linked QTL (e.g. Knapp, 1991; Haley and Knott, 1992; Martinez and Curnow, 1992; Jansen, 1996; Satagopan *et al.*, 1996). Kearsey and Hyne (1994), Hyne and Kearsey (1995) and Wu and Li (1994, 1996) also proposed a very simple

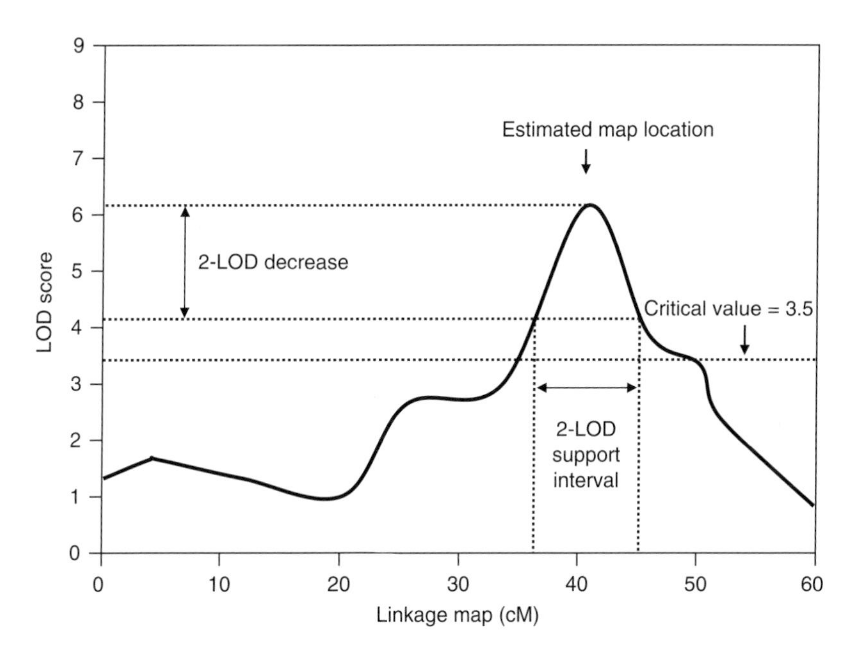

Fig. 6.2. Hypothetical likelihood map for the marker–QTL association on a linkage map in internal marker analysis. A QTL is indicated if any part of the likelihood map exceeds a critical value. In such cases, the estimated QTL location is the value of centimorgans giving the highest likelihood. Approximate confidence intervals for QTL position (two-LOD support intervals) are often constructed by including the set of all centimorgan values giving likelihoods within two-LOD scores of the maximum value.

regression based method that simultaneously considers all the markers on a single chromosome for locating multiple linked QTL. Wright and Mowers (1994) and Whittaker *et al.* (1997) showed how positional information for linked QTL can be extracted from the regression coefficients of a standard multiple regression incorporating several linked markers (see Lynch and Walsh (1998) for a detailed discussion of these topics). Here a composite interval mapping (CIM) developed by Zeng (1993, 1994) and Jansen and Stam (1994) is described following Zeng (1998).

6.3.1 Basis

The interval mapping methods described above and other methods for single QTL are not suitable when there are multiple QTL closely linked on one chromosome. When two QTL are closely linked in coupling phase, i.e. two increasing (or decreasing) alleles are linked together, the marker between the two QTL will show the highest *t*- or *F*-value for single marker analysis because of the additivity of two linked QTL. As a result, a false QTL will be declared to be located between the two real QTL, which is called a 'ghost QTL'. The same is true for interval mapping. Interval mapping gives results that can be confounded by the presence of additional QTL outside the interval being considered.

Ideally, when we test an interval for a QTL, we would like to have our test statistic independent of the effects of possible QTL at other regions of the chromosome to avoid 'ghost' QTL. If such a test can be formulated, we can simplify mapping for multiple QTL from a multiple dimensional search problem to a one-dimensional search problem, as the test for each interval is independent and for each marker interval we can effectively consider the possibility of the presence of only a single QTL. This test can be constructed by using a combination of interval mapping with multiple regression. Largely because of the linear structures of the locations of genes on chromosomes, multiple regression analysis has a very important property in that the partial regression coefficient of a trait on a marker is expected to depend only on those QTL which are located on the interval bracketed by the two neighbouring markers and to be independent of any other QTL if there is no crossing-over interference and no epistasis (Stam, 1991; Zeng, 1993).

6.3.2 Model

CIM is an extension of interval mapping with some selected markers also fitted in the model as cofactors to control the genetic variation of other possibly linked or unlinked QTL. Using the appropriate unlinked markers can partly account for the segregation variance generated by unlinked QTL, while the effects of linked QTL can be reduced by including markers linked to the interval of interest. To test for a QTL on an interval between adjacent marker M_i and M_{i+1} in particular, we extend the model

$$y_j = \mu + b^* x_j^* + e_j$$

to

$$y_j = \mu + b^* x_j^* + \sum_k b_k x_{jk} + e_j$$

where y_j is the trait value of the *j*th individual in a population, b^* is the effect of the putative QTL, x_j^* refers to the putative QTL and x_{jk} refers to those markers selected for genetic background control.

6.3.3 Likelihood analysis

The likelihood function is specified as

$$L(b^*, \mathbf{B}, \sigma^2) = \prod_{j=1}^{n} p_{1j}\phi\left(\frac{y_j - \mathbf{X}_j\mathbf{B} - b^*}{\sigma}\right) + p_{0j}\phi\left(\frac{y_j - \mathbf{X}_j\mathbf{B}}{\sigma}\right)$$

where $\mathbf{X}_j\mathbf{B} = \mu + \sum_k b_k x_{jk}$. The MLEs of the various parameters can be found in a similar manner as for interval mapping. For b^* see Eqn a at the bottom of the page. Setting this derivative to zero gives

$$\sum_{j=1}^{n} P_j(y_j - \mathbf{X}_j\mathbf{B} - b^*) = 0$$

where P_j is calculated by Eqn b (see bottom of page). This leads to the solution given by Zeng (1994) as

$$\hat{b}^* = \sum_{j=1}^{n}(y_j - \mathbf{X}_j\hat{\mathbf{B}})P_j \Big/ \sum_{j=1}^{n} P_j$$
$$= (\mathbf{Y} - \mathbf{X}\mathbf{B})'\mathbf{P}/c$$

where $c = \sum_{j=1}^{n} p_j$, $\mathbf{Y} = \{y_j\}_{n\times 1}$, $\mathbf{P} = \{P_j\}_{n\times 1}$, and a prime denotes transposition.

Differentiating the log-likelihood with respect to $\mathbf{B}$

$$\frac{\partial \ln L}{\partial \mathbf{B}} = \sum_{j=1}^{n}[P_j\mathbf{X}_j'(y_j - \mathbf{X}_j\mathbf{B} - b^*) + (1 - P_j)\mathbf{X}_j'(y_j - \mathbf{X}_j\mathbf{B})]/\sigma^2$$

Expressed in matrix notation, the equation $\partial \ln L/\partial \mathbf{B} = 0$ becomes

$$\mathbf{X}'(\mathbf{Y} - \mathbf{X}\mathbf{B}) = \mathbf{X}'\mathbf{P}b^*$$
$$\hat{\mathbf{B}} = (\mathbf{X}'\mathbf{X})^{-1}\mathbf{X}'(\mathbf{Y} - \mathbf{P}\hat{b}^*)$$

Differentiating the log-likelihood with respect to σ^2:

$$\frac{\partial \ln L}{\partial \sigma^2} = \sum_{j=1}^{n}[P_j(y_j - \mathbf{X}_j\mathbf{B} - b^*)^2 + (1 - P_j)(y_j - \mathbf{X}_j\mathbf{B})^2]/(2\sigma^4) - n/(2\sigma^2)$$

Setting this derivative to zero leads to the solution

$$n\hat{\sigma}^2 = (\mathbf{Y} - \mathbf{X}\hat{\mathbf{B}})'(\mathbf{Y} - \mathbf{X}\hat{\mathbf{B}}) - \hat{b}^{*2}c$$

6.3.4 Hypothesis test

The hypotheses to be tested are $H_0{:}b^* = 0$ and $H_1{:}b^* \neq 0$. The likelihood function under the null hypothesis is

$$L(b^* = 0, \mathbf{B}, \sigma^2) = \prod_{j=1}^{n}\phi\left(\frac{y_j - \mathbf{X}_j\mathbf{B}}{\sigma}\right)$$

with the MLEs

$$\hat{\hat{\mathbf{B}}} = (\mathbf{X}'\mathbf{X})^{-1}\mathbf{X}'\mathbf{Y}$$
$$\hat{\hat{\sigma}}^2 = (\mathbf{Y} - \mathbf{X}\hat{\hat{\mathbf{B}}})'(\mathbf{Y} - \mathbf{X}\hat{\hat{\mathbf{B}}})/n$$

The likelihood ratio (LR) test statistic is

$$LR = -2\ln\frac{L(b^* = 0, \hat{\hat{\mathbf{B}}}, \hat{\hat{\sigma}}^2)}{L(\hat{b}^*, \hat{\mathbf{B}}, \hat{\sigma}^2)} \quad \text{or}$$

$$LOD = -\log_{10}\frac{L(b^* = 0, \hat{\hat{\mathbf{B}}}, \hat{\hat{\sigma}}^2)}{L(\hat{b}^*, \hat{\mathbf{B}}, \hat{\sigma}^2)}$$

Like Lander and Botstein's interval mapping, this test can be performed at any position in a genome. Thus it gives a systematic strategy to search for QTL in a genome. As the test statistic is almost independent for each interval, a test on each interval is more likely to test for a single QTL only.

6.3.5 Selection of markers as cofactors

Which markers should be added as cofactors has no single solution as the question

$$\frac{\partial \ln L}{\partial b^*} = \sum_{j=1}^{n}\frac{p_{1j}\phi([y_j - \mathbf{X}_j\mathbf{B} - b^*]/\sigma)}{p_{1j}\phi([y_j - \mathbf{X}_j\mathbf{B} - b^*]/\sigma) + p_{0j}\phi([y_j - \mathbf{X}_j\mathbf{B}]/\sigma)}\frac{y_j - \mathbf{X}_j\mathbf{B} - b^*}{\sigma^2} \qquad \text{(a)}$$

$$P_j = \frac{p_{1j}\phi([y_j - \mathbf{X}_j\mathbf{B} - b^*]/\sigma)}{p_{1j}\phi([y_j - \mathbf{X}_j\mathbf{B} - b^*]/\sigma) + p_{0j}\phi([y_j - \mathbf{X}_j\mathbf{B}]/\sigma)} \qquad \text{(b)}$$

depends on the number and positions of underlying QTL, the information that is not available a priori. Suppose the interval of interest is delimited by markers i and $i + 1$. Additional markers $i - 1$ and $i + 2$ as cofactors account for all linked QTL to the left of marker $i - 1$ and to the right of marker $i + 2$. Thus, while these cofactors do not account for the effect of linked QTL in the intervals immediately adjacent to the one of interest, they do account for all the linked QTL.

The number of cofactors should not exceed $2\sqrt{n}$ where n is the number of individuals in the analysis (Jansen and Stam, 1994) or alternatively it can be determined automatically by F-to-enter or F-to-drop criterion in the forward or backward stepwise regression analysis. A first approach would be to include all unlinked markers showing significantly marker–trait association (detected, for example, by stand single-marker regression). If several linked markers from a single chromosome all show significant effects, one might just use the marker having the largest effect. A related strategy is to first perform a multiple regression using all markers unlinked to the region of interest and then eliminate those that are not significant.

In the computer program designed for CIM, QTL CARTOGRAPHER, a two-step procedure for practical data analysis was implemented. In the first step, n_p markers that are significantly associated with the trait are selected by (forward or backward) stepwise regression. In the second step (mapping step), for each testing interval, except of the markers for the putative QTL, two markers that are at least W_s cM away from the test interval (one for each direction) are first picked up to fit in the model to define a testing window for blocking other possible linked QTL effects on the test. Then, those selected n_p markers that are outside of the testing window are also fitted into the model to reduce the residual variance.

The accuracy of locating QTL provided by CIM is at the cost of reduced statistical power because markers selected as cofactors around the test interval will pick up some effect of the QTL that is located in the test interval. Therefore, markers that are close to the interval under test are not suitable as cofactors. To solve this problem, only markers that are some distance away from the test interval can be selected. Because the size of this test window depends on test intervals, different sizes should be tested to find a suitable window size for each test interval.

6.3.6 Inclusive composite interval mapping

In Zeng's (1993, 1994) algorithm, the QTL effect at the current testing position and regression coefficients of the marker variables used to control genetic background are estimated simultaneously in an expectation and conditional maximization (ECM) algorithm. Thus, the same marker variable may have different coefficient estimates as the testing position changes along the chromosomes. The algorithm used in CIM cannot completely ensure that the effect of QTL at the current testing interval is not absorbed by the background marker variables, which may result in biased estimation of the QTL effect.

A modified algorithm called inclusive composite interval mapping (ICIM) was proposed by Li *et al.* (2007). In ICIM, marker selection is conducted only once through stepwise regression by considering all marker information simultaneously and the phenotypic values are then adjusted by all markers retained in the regression equation except the two markers flanking the current mapping interval. The adjusted phenotypic values are finally used in interval mapping. The modified algorithm has a simpler form than that used in CIM but a faster convergence. ICIM retains all the advantages of CIM over interval mapping and avoids the possible increase of sampling variance and the complicated background marker selection process. Extensive simulations using two genomes and various genetic models indicated that ICIM has increased detection power, reduced false detection rate and less biased estimates of QTL effects. ICIM has been extended to map digenic interacting

QTL (Li *et al.*, 2008). Windows-supported software, ICIMAPPING, was developed for using the ICIM mapping approach and is available at http://www.isbreeding.net.

6.4 Multiple Interval Mapping

Genetic mapping approaches involving multiple QTL have been developed. In general, there are three different approaches: (i) maximum likelihood using EM include multiple interval mapping (MIM) (Kao and Zeng, 1997) and sequential testing to search model space; (ii) multiple imputation (Sen and Churchill, 2001) uses Bayesian log posterior odds and sequential testing and pairwise plots to search; and (iii) Markov chain Monte Carlo (MCMC) (Satagopan *et al.*, 1996) employs Markov chain sampling to search model space. In this section, we will focus on MIM based on Kao and Zeng (1997) and Kao *et al.* (1999).

MIM is a multiple-QTL oriented method combining QTL mapping analysis with the analysis of genetic architecture of quantitative traits through a search algorithm to search for number, positions, effects and interaction of significant QTL. Using markers for simultaneous multiple QTL analysis was suggested first by Lander and Botstein (1989), although the idea was pursued only with a very limited scope. Bayesian statistics via MCMC for mapping QTL is also based on multiple QTL, particularly when it is combined with a reversible-jump process, which will be discussed in the Section 6.7.

MIM consists of four components:

- an evaluation procedure to analyse the likelihood of the data given a genetic model (number, position and epistasis of QTL);
- a search strategy to select the best genetic model (among those sampled) in the parameter space;
- an estimation procedure to estimate all parameters of interest in the genetic architecture of quantitative traits (number, positions, effects and epistasis of QTL; genetic variances and covariances explained by QTL effects) given the selected genetic model; and
- a prediction procedure to estimate or predict the genotypic values of individuals based on the selected genetic model and estimated genetic parameter values for MAS.

6.4.1 Multiple interval mapping model and likelihood analysis

For m putative QTL, the model of MIM is specified as

$$y_i = \mu + \sum_{r=1}^{m} \alpha_r x_{ir}^* + \sum_{r \neq s \subset (1, \ldots, m)}^{t} \beta_{rs}(x_{ir}^* x_{is}^*) + e_i \qquad (6.6)$$

where:

- y_i is the phenotype value of individual i;
- i indexes individuals of the sample ($i = 1, 2, \ldots, n$);
- μ is the mean of the model;
- α_r is the marginal effect of putative QTL r;
- x_{ir}^* is an indicator variable denoting genotype of putative QTL r (defined by 1/2 or –1/2 for the two genotypes), which is unobserved but can be inferred from marker data in sense of probability;
- β_{rs} is the epistatic effect between putative QTL r and s;
- $r \neq s \subset (1, \ldots, m)$ denotes a subset of QTL pairs that each shows a significant epistatic effect, because if all pairs of m QTL are fitted in the model, the model can be over parameterized;
- m is the number of putative QTL chosen by either their significant marginal effects or significant epistatic effects;
- t is the number of significant pairwise epistatic effects; and
- e_i is a residual effect of the model assumed to be normally distributed with mean zero and variance σ^2.

As the genotypes of an individual at many genomic locations are not observed (but marker genotypes are), the model contains missing data. So the likelihood function of the data given the model is a mixture of normal distributions

$$L(\mathbf{E}, \mu, \sigma^2) = \prod_{i=1}^{n} \left[\sum_{j=1}^{2^m} p_{ij} \phi(y_i | \mu + \mathbf{D}_{ij}\mathbf{E}, \sigma^2) \right] \quad (6.7)$$

The term in brackets is the weighted sum of a series of normal density functions, one for each of 2^m possible multiple-QTL genotypes. p_{ij} is the probability of each multi-locus genotype conditional on marker data; **E** is a vector of QTL parameters (αs and βs), $\mathbf{D}_{ij}$ is a vector of the genetic model design specifying the configuration of x^*'s association with each α and β for the jth QTL genotype (see Kao and Zeng, 1997); and $\phi(y_i | \mu, \sigma^2)$ denotes a normal density function for y with mean μ and variance σ^2.

Thus the probability density of each individual is a mixture of 2^m possible normal densities with different means $\mu + \mathbf{D}_{ij}\mathbf{E}$ and mixing proportions p_{ij} which are calculated from marker information.

The procedure to obtain MLEs using an EM algorithm has been described by Kao and Zeng (1997). In the $[t + 1]$th iteration, the E-step is

$$\pi_{ij}^{[t+1]} = \frac{p_{ij}\phi(y_i | \mu^{[t]} + D_{ij}E^{[t]}, \sigma^{2[t]})}{\sum_{j=1}^{2^m} p_{ij}\phi(y_i | \mu^{[t]} + D_{ij}E^{[t]}, \sigma^{2[t]})} \quad (6.8)$$

The M-step is shown in Eqns 6.9–6.11 (see bottom of page) where E_r is the rth element of **E** and D_{ijr} is the rth element of $\mathbf{D}_{ij}$.

These equations can be expressed in a general form in matrix notation as (Kao and Zeng, 1997)

$$\mathbf{E}^{[t+1]} = diag(\mathbf{V})^{-1}[\mathbf{D}'\mathbf{\Pi}'(\mathbf{Y} - \mu) - nondiag(\mathbf{V})\mathbf{E}^{(t)}] \quad (6.12)$$

$$\mu = \frac{1}{n}\mathbf{1}'[\mathbf{Y} - \mathbf{\Pi}\mathbf{D}\mathbf{E}]$$

$$\sigma^2 = \frac{1}{n}[(\mathbf{Y} - \mu)'(\mathbf{Y} - \mu) - 2(\mathbf{Y} - \mu)'\mathbf{\Pi}\mathbf{D}\mathbf{E} + \mathbf{E}'\mathbf{V}\mathbf{E}]$$

with

$$\mathbf{V} = \{\mathbf{1}'\mathbf{\Pi}(D_r \# D_s)\}_{r,s=1,\ldots,w} \text{ and } \mathbf{\Pi} = \{\pi_{ij}\}$$

where # denotes the Hadamard product, which is the element-by-element product of corresponding elements of two same order matrices and ' denotes transposition of a matrix or vector.

The two forms (Eqns 6.9 and 6.12) are actually somewhat different for computation. Equation 6.12 implies the update of **E** as a vector in one step and Eqn 6.9 implies the update of each element in **E** in turn (always using the most recently updated

$$E_r^{[t+1]} = \frac{\sum_i \sum_j \pi_{ij}^{[t+1]} D_{ijr}[(y_i - \mu^{[t]}) - \sum_{s=1}^{r-1} D_{ijs}E_s^{[t+1]} - \sum_{s=r+1}^{w} D_{ijs}E_s^{[t]}]}{\sum_i \sum_j \pi_{ij}^{[t+1]} D_{ijr}^2} \quad (6.9)$$

$$\mu^{[t+1]} = \frac{1}{n}\sum_i \left(y_i - \sum_j \sum_r \pi_{ij}^{[t+1]} D_{ijr} E_r^{[t+1]} \right) \quad (6.10)$$

$$\sigma^{2[t+1]} = \frac{1}{n}\left[\sum_i (y_i - \mu^{[t+1]})^2 - 2\sum_i (y_i - \mu^{[t+1]}) \sum_j \sum_r \pi_{ij}^{[t+1]} D_{ijr} E_r^{[t+1]} + \sum_r \sum_s \sum_i \sum_j \pi_{ij}^{[t+1]} D_{ijr} D_{ijs} E_r^{[t+1]} E_s^{[t+1]} \right] \quad (6.11)$$

values for other parameters). Equation 6.9 is more stable than Eqn 6.12 numerically: Eqn 6.12 can lead to divergence in certain cases and Eqn 6.9 can always lead to convergence, although at a slightly slower pace (Z.-B. Zeng, North Carolina State University, personal communication).

Note on the meaning and difference between p_{ij} and π_{ij}: p_{ij} is the probability of each multi-locus QTL genotype conditional on marker genotype and π_{ij} is the probability of each multi-locus QTL genotype conditional on marker genotype and also phenotypic value.

The test for each QTL effect, say E_r, is performed by a likelihood ratio test conditional on other selected QTL effects (see equation at bottom of page).

For given positions of m putative QTL and $m + t$ QTL effects, the likelihood analysis can proceed as outlined above. Now the task is to search and select the best genetic model (number, positions and interaction of QTL) that fits the data well.

6.4.2 Model selection

Pre-model selection

As the evaluation of the MIM model is computationally intensive, it is important to select a good pre-model for MIM analysis. The following procedure can be used. First, select a subset of significant markers. Then, use the results from marker selection to perform CIM to scan the genome for candidate positions. Finally, evaluate and test each parameter in the pre-model under MIM and drop any non-significant estimate in a stepwise manner.

Model selection using multiple interval mapping

After the first evaluation of the pre-model, perform the following stepwise selection analysis to finalize the search for a genetic model under MIM.

1. Begin with a model that contains m QTL and t epistatic effects.
2. Scan the genome to search for the best position of an $(m + 1)$th QTL and then perform a likelihood ratio test for the marginal effect of this putative QTL. If the test statistic exceeds the critical value, this effect is retained in the model.
3. Search for the $t + 1$ epistatic effect among the pairwise interaction terms not yet included in the model and perform the likelihood ratio test on the effect. If LOD exceeds the critical value, the effect is retained in the model. Repeat the process until no more significant epistatic effects are found.
4. Re-evaluate the significance of each QTL effect currently fitted in the model. If LOD for a QTL (marginal or epistatic) effect falls below the significant threshold conditional on other fitted effects, the effect is removed from the model. However, if the marginal effect of a QTL that has significant epistatic effect on other QTL falls below the threshold, this marginal effect is still retained. This process is performed in a stepwise manner until the test statistic for each effect is above the significance threshold.
5. Optimize estimates of QTL positions based on the currently selected model. Instead of performing a multi-dimensional search around the regions of current estimates of QTL positions (which is an option), estimates of QTL positions are updated in turn for each region. For the ith QTL in the model, the region between its two neighbour QTL is scanned to find the position that maximizes the likelihood (conditional on the current estimates of positions of other QTL and QTL epistasis). This refinement process is repeated sequentially for each QTL position until there is no change in the estimates of QTL positions.

$$LOD = \log_{10} \frac{L(E_1 \neq 0, \ldots, E_{m+t} \neq 0)}{L(E_{1\neq 0}, \ldots, E_{r-1} \neq 0, E_r = 0, E_{r+1} \neq 0, \ldots, E_{m+t} \neq 0)}$$

6. Return to step 2 and repeat the process until no more significant QTL effects can be added into the model and estimates of QTL positions are optimized.

Stopping rules

An important issue associated with model selection is a stopping rule for the model search algorithm or criterion for comparing different models. In regression analysis with model selection, the stopping rules are usually decided by minimizing the final prediction error (FPE) criterion or information criteria (IC).

The FPE criterion is

$$S_k = (n + k)RSS_k / (n - k)$$

where RSS_k is the residual sum of squares and k is the number of parameters fitted in the model. The IC of the general form is

$$IC = -2[\log L_k - kc(n)/2] \quad (6.13)$$

where L_k is the likelihood of data (Eqn 6.7) given a genetic model with k parameters and $c(n)$ is a weighting function of the sample size (examples given below). This is approximately equivalent to

$$IC = \log [RSS_k/n] + kc(n)/n$$

in regression analysis.

The IC criteria can be related to the F-to-enter statistic (for regression analysis) or LR-to-enter statistic (for likelihood analysis) in the stepwise selection procedure. It was shown (Miller, 1990: p. 208) that Eqn 6.12 leads to the F-to-enter statistic for regression analysis at the minimum (see Eqn 6.14 at bottom of page) provided that $c(n)/n$ is small. As $LR = n \log (SSR_k/SSR_{k+1})$ in the setting of regression analysis, Eqns 6.13 and 6.14 imply that the LR-to-enter statistic for likelihood analysis at minimum is

$$LR_k = -2\log\frac{L_k}{L_{k+1}} \le n\log(c(n)/n+1) \approx c(n)$$

The criterion is basically defined by the choice of the penalty $c(n)$. Using $c(n) = 2$ as suggested by Akaike (1969) would mean that the final threshold in LOD is 0.43. $c(n)$ can take a variety of forms, such as: $c(n) = \log(n)$, which is the classical Bayes information criterion (BIC); $c(n) = 2$, which is the Akaike information criterion (AIC) (Zou and Zeng, 2008).

In reference to QTL analysis on markers, Broman (1997) suggested using $c(n) = \delta \log n$ and recommended δ be between 2 and 3. For $n = 100$–500, the threshold in LOD would be 2–2.7 for $\delta = 2$ and 3–4 for $\delta = 3$. However, this argument is still rather arbitrary and does not relate to the genetic length of the linkage map, number of markers and linkage groups or the distribution of markers.

6.4.3 Estimating genotypic values and variance components of QTL effects

Given estimates of the QTL parameters, the genotypic values of an individual can be estimated. This estimation is complicated by the fact that QTL genotypes are not observed directly, rather only marker genotypes are observed. Thus, the estimation for an individual is the weighted mean of all possible genotypic values, weighted by the probability ($\hat{\pi}_{ij}$) of each QTL genotype conditional on both the marker and the phenotypic data. From Eqn 6.10, this estimation equation is

$$\hat{y}_i = \hat{\mu} + \sum_{j=1}^{2^m}\sum_{r=1}^{m+t} \hat{\pi}_{ij} D_{ijr} \hat{E}_r$$

$$\frac{SRR_k - SRR_{k+1}}{SRR_{k+1}/(n-k-1)} \le (n-k-1)(e^{c(n)/n}-1) \approx 2c(n)\left(1-\frac{k+1}{n}\right) \quad (6.14)$$

where the first summation is over all possible 2^m QTL genotypes and the second summation is over all effects of the model (m main effects and t epistatic effects). $\hat{\mu}$ is the MLE of μ obtained from Eqn 6.10 at the equilibrium of the final model and $\hat{E}_r$ is the MLE of QTL effect E_r obtained from Eqn 6.9. $\hat{\pi}_{ij}$ is the MLE of π obtained from Eqn 6.8.

To predict the genotypic values of quantitative traits based on marker information only, we need to use

$$\hat{y}_i = \hat{\mu} + \sum_j \sum_r p_{ij} D_{ijr} \hat{E}_r$$

as $\hat{\pi}_{ij}$ is a function of phenotype y_i which is unavailable in early selection.

The genetic variances and covariances explained by each QTL effect can be estimated directly from the likelihood analysis. Applying the EM algorithm, Eqn 6.12 leads to

$$\hat{\mathbf{E}} = \hat{\mathbf{V}}^{-1}\mathbf{D}'\hat{\mathbf{\Pi}}'(\mathbf{Y} - \hat{\mu})$$

This implies

$$\hat{\sigma}^2 = \frac{1}{n}[(\mathbf{Y} - \hat{\mu})'(\mathbf{Y} - \hat{\mu}) - \hat{\mathbf{E}}'\hat{\mathbf{V}}\hat{\mathbf{E}}]$$

or Eqn 6.15 at the bottom of the page where $\bar{y} = \Sigma_{i=1}^{n} y_i/n$ and $\bar{D}_r = \Sigma_{i=1}^{n}\Sigma_{j=1}^{2^m}\hat{\Pi}_{ij}D_{ijr}/n$.

In this form $\hat{\sigma}^2$ is expressed as a difference between the MLE of total phenotypic variance $\hat{\sigma}_p^2$ (the first part of Eqn 6.15) and that of the genetic variance $\hat{\sigma}_g^2$ (the second part of Eqn 6.15). $\hat{\sigma}_g^2$ can be further partitioned into the equation at the very bottom of the page. $\hat{\sigma}_{E_r}^2$ estimates genetic variance due to the QTL effect E_r and $\hat{\sigma}_{E_r,E_s}^2$ estimates genetic covariance between QTL effects E_r and E_s.

It is convenient and informative to combine the variance due to each QTL effect with half of the covariances between this QTL effect and other effects, and report this variance component as the variance component explained by this QTL effect

$$\hat{\sigma}_r^2 = \hat{\sigma}_{E_r}^2 + \frac{1}{2}\sum_{s \neq r} \hat{\sigma}_{E_r,E_s}$$

Whereas $\hat{\sigma}_{E_r}^2$ estimates the variance of the rth QTL effect in linkage equilibrium (in which $\sigma_{E_r,E_s} = 0$), $\hat{\sigma}_r^2$ estimates the contribution to the total variance in the current population with linkage disequilibrium. Estimates of these variances, covariances and variance components can be given as a ratio of the total phenotypic variance. Note that $\hat{\sigma}_g^2/\hat{\sigma}_p^2$ is the coefficient of determination (R^2) of the MIM model. Note also whereas $\hat{\sigma}_{E_r}^2$ is always positive, $\hat{\sigma}_r^2$ is not necessarily positive.

$$\begin{aligned}\hat{\sigma}^2 &= \frac{1}{n}\left[\sum_{i=1}^{n}(y_i - \hat{\mu})^2 - \sum_{r=1}^{m+t}\sum_{s=1}^{m+t}\sum_{i=1}^{n}\sum_{j=1}^{2^m}\hat{\pi}_{ij}D_{ijr}D_{ijs}\hat{E}_r\hat{E}_s\right] \\ &= \frac{1}{n}\left[\sum_{i=1}^{n}(y_i - \bar{y})^2 - \sum_{r=1}^{m+t}\sum_{s=1}^{m+t}\sum_{i=1}^{n}\sum_{j=1}^{2^m}\hat{\pi}_{ij}(D_{ijr} - \bar{D}_r)(D_{ijs} - \bar{D}_s)\hat{E}_r\hat{E}_s\right]\end{aligned} \quad (6.15)$$

$$\begin{aligned}\hat{\sigma}_g^2 &= \sum_{r=1}^{m+t}\left[\frac{1}{n}\sum_{i=1}^{n}\sum_{j=1}^{2^m}\hat{\pi}_{ij}(D_{ijr} - \bar{D}_r)^2\hat{E}_r^2\right] + \sum_{r=2}^{m+t}\sum_{s=1}^{r-1}\left[\frac{2}{n}\sum_{i=1}^{n}\sum_{j=1}^{2^m}\hat{\pi}_{ij}(D_{ijr} - \bar{D}_r)(D_{ijs} - \bar{D}_s)\hat{E}_r\hat{E}_s\right] \\ &= \sum_{r=1}^{m+t}\hat{\sigma}_{E_r}^2 + \sum_{r=2}^{m+t}\sum_{s=1}^{r-1}\hat{\sigma}_{E_r,E_s}\end{aligned}$$

6.5 Multiple Populations/Crosses

6.5.1 Experimental designs

There are many different types of populations available in genetics and breeding programmes (Chapter 4). However, only very few of them, as discussed previously, have been exploited for QTL mapping. These crosses/populations, derived from divergent inbred lines, populations and species, provide potential opportunities for more convenient QTL mapping and better integration of mapping with plant breeding programmes.

One of the most frequently used crosses are BCs with only two genotypes at a locus which are simple to analyse. Another common cross, the F_2, which has three genotypes at a locus, can be used to estimate both additive and dominance effects. Compared to the BC, it is more complex for data analysis particularly for multiple QTL with epistasis, while it provides more opportunities and information to examine genetic structure or architecture of QTL and has more power than the BC.

Some less frequently used crosses are: (i) F_3, F_4, etc. derived by selfing or random mating from F_2. The random mating increases recombination and expends the length of the linkage map and thus increases the mapping resolution (estimation of QTL position). (ii) Repeated BCs that are derived by continuous backcrossing to one of the parental lines and the end-product would be the near-isogenic lines (NILs). (iii) Multi-way crosses, derived by crossing a hybrid with another hybrid or an inbred/cultivar, which results in a population with three or four parental lines involved. (iv) Multiple crosses derived from a complicated mating design such as NCII and diallel design (Chapter 4).

With all available crosses and populations, QTL mapping can be based on multiple populations derived from the same inbred parents or a hybrid, or on multiple crosses derived from different parental lines. In the former case, there are two possible alleles segregating in the populations for the diploid species, while in the latter case, more than two possible alleles will be segregating, depending how many parental lines are involved. On the other hand, all crosses or populations could be derived from outbred parents, which results in an uncertain coupling/repulsion phase and is heterozygous for some or all parental lines.

For QTL mapping with crosses from segregating populations, similar model and analysis procedures can be used as inbred crosses, but with more complicated analysis. The probability of the allelic origin for each genomic point from observed markers needs to be estimated. This type of population has low power for QTL analysis because QTL alleles may not be preferentially fixed in the parental populations and it makes power calculations more difficult.

For multiple crosses developed from different heterogenous parental lines, half-sib or full-sib relationships may exist in some of the individual plants used for mapping. Half sibs can be analysed based on the segregation of one parent, which is similar to the backcrossing model and analysis. This type of population is less powerful for QTL detection because there is more uncontrollable variability in the other parents. The allelic effect differences are only analysed for one parent, not for those between widely differentiated inbred lines, populations and species. Generally the relevant heritability is low for QTL analysis. For full sibs, there are four genotypes at a locus; allelic substitution effects can be estimated for male and female parents and their interaction (dominance). Information for QTL analysis is double that for half sibs and this type of population should be more powerful.

6.5.2 QTL for multiple crosses

QTL analysis with multiple crosses can be achieved separately for each cross, which is simple but inefficient with less power. Multiple crosses derived from different parents have more power because more individuals are involved and there are thus more informative markers. These types of crosses can be used to study the effect of QTL under different genetic backgrounds such as genotype by cross interaction and epistatic interaction. For all multiple crosses, a more reasonable analysis would be the combined analysis over crosses. In this way, crosses created or evaluated at different times can be combined and multiple projects in a team or across research groups can be related and shared. Disadvantages include more complications for analysis with few software packages available and having to account for the multiple related crosses (where individuals may be correlated to each other both genotypically and phenotypically). QTL mapping approaches have been developed for four-way crosses (Xu, 1996) and crosses derived from multiple inbred lines (Liu and Zeng, 2000).

Broman *et al.* (2003a) discussed how to combine multiple crosses in QTL analysis. For crosses with founders unrelated to each other, a naïve sum of separate LODs by cross can be used assuming a different gene action in different crosses, or combined analysis can be used for independent crosses. For crosses with related founders, QTL analysis depends on genetic relationships within and between crosses. With constant genetic covariance within a cross, all individuals have the same genetic relationship and combined analysis has no effect on single cross analysis. However, genetic covariance may differ between crosses, depending on the expected number of alleles shared by identity by descent (IBD). It should be noted that covariance across multiple crosses is not constant. In these cases, combined analysis will provide results that are different from single cross analysis. The problems with multiple cross analysis can be fixed simply by the introduction of blocking factors for crosses as a random effect for genetic relationships. This addresses the constant covariance with each cross and different covariances between crosses, which provides an appropriate recombination model for crosses to relate the recombination rate to distance and common phenotype model across all crosses to allow cross by genetic effect interactions.

Ignoring polygenic effects will result in a biased additive effect estimate, detecting dominance when it does not exist and increased variance and thus bias QTL results, although location estimate is unbiased. Statistical power can be increased by combining crosses, which is important when several related crosses are created. The threshold idea for testing and loci intervals has been extended to multiple crosses (Zou *et al.*, 2001).

Jannick and Jansen (2001) developed a method to map epistatic QTL by identifying loci with strong interaction between QTL and genetic background. The approach requires large populations derived from multiple related inbred-line crosses. The method is applied to simulate DH populations derived from a diallel among three inbred parents. This approach allows detection of QTL involved not only in pairwise but also higher-order interaction and does so with one-dimensional genome searches.

The North Carolina Experimental III (design NCIII in Chapter 4), originally designed by Comstock and Robinson (1952), is the first complex design that was exploited for QTL mapping. In NCIII, the experimental units are produced from BC matings of F_2 plants to the two parental lines from which the F_2 was derived. Additive and dominance components of variance can be estimated with nearly equal precision under the assumption of diploidy, biallelic and equal gene frequencies and absence of linkage and epistasis. Cockerham and Zeng (1996) extended Comstock and Robinson's ANOVA to include linkage and epistasis for F_2 and F_3 progenies and developed orthogonal contrasts for QTL mapping using single-marker ANOVA. Melchinger *et al.* (2007) demonstrated the exceptional features of NCIII for identification of QTL contributing to heterosis. They defined a new type of heterotic gene effect, denoted as the augmented dominance effect d_i^*, which is equal to the net contribution of QTL_i to mid-parent heterosis (MPH). It comprises the dominance effect d minus half the sum of additive × dominance epistatic interactions with genetic background. The novelty of their approach is that QTL that significantly contribute to MPH are identified and both dominance and epistasis are accounted for.

An elegant experimental design that can provide a test of significance for the presence of epistasis is the triple testcross (TTC) design (Chapter 4), proposed by Kearsey and Jinks (1968), which is an extension of NCIII. In the TTC design, testcrosses are produced not only with the two parental lines but also with the F_1 derived from them. For every progeny from a segregating population, e.g. F_2 plant or RIL, three sets of data can be generated: (i) the average parental testcross performance; (ii) the difference between the parental testcross performances; and (iii) the deviation of testcross progenies with the F_1 from the mean of the parental testcrosses. Kearsey *et al.* (2003) and Frascaroli *et al.* (2007) presented experimental results from QTL analyses based on the TTC design with data from *Arabidopsis* and maize, respectively. Melchinger *et al.* (2008) gave genetic expectations of QTL effects estimated with the TTC design in the presence of epistasis. With the TTC design, dominance × additive epistatic interactions of individual QTL with the genetic background can be estimated with one-dimensional genome scans. They demonstrated that the limitation of NCIII in the analysis of heterosis to separate QTL main effects and their epistatic interactions with all other QTL can partially be overcome with the TTC design. They also presented genetic expectations of variance components for the analysis of TTC progeny tested in a split-plot design, assuming digenic epistasis and arbitrary linkage. Kusterer *et al.* (2007) used the theory to study heterosis for biomass-related traits in *Arabidopsis*.

6.5.3 Pooled analysis

Very often, more than two mapping populations are studied for the same or related traits. QTL analysis on pooled data from multiple mapping populations was suggested by Lander and Kruglyak (1995). Pooled analysis provides a means for evaluating, as a whole, evidence for the existence of a QTL from different studies and examining differences in gene effect of a QTL among different populations.

Walling *et al.* (2000) extended least square interval mapping (Haley *et al.*, 1994) to analysis of combined data from seven porcine populations, while Li, R. *et al.* (2005) extended the Bayesian QTL analysis method (Sen and Churchill, 2001) to analysis of combined data from four mouse populations. The former (Walling *et al.*, 2000) is simple and computation and general statistical software such as SAS is applicable. The latter (Li, R. *et al.*, 2005) adopted a new QTL analysis method and this requires special software. Some earlier studies (Rebai and Goffinet, 1993; Xu, 1998; Liu and Zeng, 2000) also developed QTL analysis methods for data which may be produced from several populations.

Guo, B. *et al.* (2006) provided an example of pooled analysis of data from multiple QTL mapping populations. Least square interval mapping was extended for pooled analysis by inclusion of populations and cofactor markers as indicator variables and covariate variables separately in the multiple linear models. The general linear test approach was applied to the detection of QTL. Single population-based and pooled analyses were conducted on data from two $F_{2:3}$ mapping populations, Hamilton (susceptible) × PI 90763 (resistant) and Magellan (susceptible) × PI 404198A (resistant), for resistance to cyst nematode in soybean. It was demonstrated that where a QTL was shared among populations, pooled analysis showed increased LOD values for the QTL candidate region over single population analyses. Where a QTL was not shared among populations, however, the pooled analysis showed decreased LOD values for the QTL candidate region over single population analyses. Pooled analysis on data from genetically similar populations may have a higher power of QTL detection relative to single population-based analyses. An important issue emerges from such pooled analyses: because of this dilution effect, a QTL with strong effects, but existing in only one or few populations, may become undetectable if a large number of populations are pooled.

6.6 Multiple QTL

6.6.1 Reality of multiple QTL

A multiple QTL model is designed to: (i) effectively search over the 'space' of genetic architecture for the number and positions of loci, gene action (additive, dominance, epistasis); (ii) select 'best' or 'better' model(s) including what criteria to use and where to draw the line; and (iii) estimate 'features' of model such as means, variances and covariances, confidence regions and marginal or conditional distributions (Broman *et al.*, 2003a).

The multiple QTL approach should have several advantages relative to single QTL approaches. First, statistical power and precision can be improved so that the number of QTL detected will increase and better estimates of loci (less bias, smaller intervals) will be provided. Secondly, the inference of complex genetic architecture including patterns and individual elements of epistasis can be improved; means, variances and covariances can be estimated appropriately and the relative contributions of different QTL can be assessed. Thirdly, estimates of genotypic values can be improved with less bias (more accurate) and smaller variance (more precise).

Is there any limit of estimation for QTL? As indicated by Bernardo (2001), the reasonable number of QTL for an efficient MAS is 10. A larger number such as 50 is too big. Phenotype is a better predictor than genotype when there are a large number of QTL. Increasing sample size does not give multiple QTL any advantage. Also it is hard to select many QTL simultaneously because there are 3^m possible genotypes to choose from when a trait is controlled by m QTL. Genetic linkage between QTL, i.e. multi-collinearity, will lead to correlated estimates of gene effects and the precision of each effect drops as more predictors are added. There is a need to balance bias and variance. A few QTL can dramatically reduce bias while many predictors (QTL) can increase variance. Finally, estimation of QTL parameters depends on sample size, heritability and environmental variation.

What can we do with the QTL below the limits of detection? There is a problem of selection bias: QTL of modest effect can sometimes be detected but their effects are biased upwards when detected (Beavis, 1994). To avoid sharp in/out dichotomy, caution should be taken about only examining the 'best' model and the probability that a QTL is in the model should be considered. Building m detected loci into the QTL model will directly allow uncertainty in genetic architecture and model selection over number of QTL.

6.6.2 Selecting a class of QTL models

There are many parameters to be considered when selecting a class of QTL models (Broman *et al.*, 2003a): (i) number of QTL, single QTL or multiple QTL of known or unknown number; (ii) location of QTL with known positions and widely spaced (no two QTL within a marker interval) or arbitrarily close; (iii) gene action including additive and/or dominance effects, epsitatic effects (four combinations for diploid species – *aa*, *ad*, *da*, *dd*; more combinations for species with higher levels of ploidy) and phenotypic distribution (normal, binomial, Poisson, etc.).

Consider a phenotype normally distributed with

$$\Pr(Y \mid Q,\theta) = N(G_Q,\sigma^2)$$

Typical assumptions that are required for building a model are: (i) normally-distributed environmental variation, i.e. residuals e (not Y!) give a bell-shaped histogram; (ii) genetic value G_Q is a composite of m QTL, i.e. Q = $(Q_1, Q_2, \ldots, Q_m)$; and (ii) genetic effect uncorrelated with environment. That is,

$$Y = \mu + G_Q + e,\ e \sim N(0, \sigma^2)$$
$$E(Y \mid Q,\theta) = \mu + G_Q,\ \mathrm{var}(Y \mid Q,\theta) = \sigma^2$$
$$\theta = (\mu, G_Q, \sigma^2)$$

Considering multiple QTL, the genotypic value can be partitioned (assuming no epistasis) as

$$G_Q = \theta_{Q(1)} + \theta_{Q(2)} + \ldots + \theta_{Q(m)}$$
$$\text{or } G_Q = \sum_j \theta_{Q(j)}$$

Thus genetic variance can be partitioned as

$$\text{var}(G_Q) = \sigma_G^2 = \sum_j \sigma_{G(j)}^2,$$
$$\sigma_{G(j)}^2 = \text{var}(\theta_{Q(j)})$$

with partitioned heritability h^2

$$h^2 = \frac{\sigma_G^2}{\sigma_G^2 + \sigma^2} = \sum_j \frac{\sigma_{G(j)}^2}{\sigma_G^2 + \sigma^2}$$

With many optional models for selection, alternative QTL models should be compared. The comparison can be based on the residual sum of squares (RSS), information criteria such as Bayes information criteria (BIC) and Bayes factors (Broman *et al.* 2003a).

1. Comparing models can be based on the RSS, which has a nice property in that it never increases as the model grows in size. The goal is to obtain a small RSS with the 'simplest' model.
2. Classical linear models that can be used for comparing models include mean squared error (MSE), Mallow's C_p and adjusted R^2.
3. Models can be compared based on re-sampling techniques, which include bootstrap (re-sampling with replacement from data), cross validation (repeatedly dividing data into estimation and test sets) and sequential permutation tests which are conditional on the QTL already in the model and stops when added QTL are not significant.
4. There are some information criteria for comparing models built on RSS and likelihoods, which included Akaike information criteria (AIC), Bayes/Schwartz information criteria (BIC), BIC-delta (BIC_δ) and Hannon–Quinn information criteria (HQIC).

6.6.3 Multiple QTL with epistasis

When a trait is controlled by multiple QTL, it is very possible that some epistasis may occur between loci. With two QTL involved, there are four types of epistasis, *aa*, *ad*, *da* and *dd*. With more than two loci involved, there would be higher-order epistasis.

Considering genetic models with epistasis, genotypic values can be partitioned with epistasis as

$$G_Q = \theta_{Q(1)} + \theta_{Q(2)} + \theta_{Q(1,2)}$$

This genetic variance can be partitioned accordingly,

$$\text{var}(G_Q) = \sigma_G^2 = \sigma_{G(1)}^2 + \sigma_{G(2)}^2 + \sigma_{G(1,2)}^2$$

For 2-QTL interactions

$$G_Q = \sum_j \theta_{1Qj} + \sum_j \theta_{2Qj}$$

where $\theta_{1Qj} = \theta_{Q(j1)}$, $\theta_{2Qj} = \theta_{Q(j1,\,j2)}$; $j_1, j_2 = 1, \ldots, m_j$.

With an extra subscript k to keep tracking the order of loci, the genetic variance is partitioned as

$$\sigma_G^2 = \sigma_{1G}^2 + \sigma_{2G}^2$$

$$\sigma_{kG}^2 = \sum_j \sigma_{kGj}^2,\ \sigma_{kGj}^2 = \text{var}(\theta_{kQj})$$

Considering m QTL ($m > 2$) with higher order epistasis, it sums over order k and over QTL index j,

$$G_Q = \sum_k \sum_j \theta_{kjQ}$$

$$\theta_{kjQ} = \theta_{(j1,\,j2,\,\ldots,\,jk)Q}$$

Genetic variance is partitioned as

$$\sigma_G^2 = \sum_k \sigma_{kG}^2,\ \sigma_{kG}^2 = \sum_j \sigma_{kGj}^2,$$
$$\sigma_{kGj}^2 = \text{var}(\theta_{kQj})$$

With so many parameters, a large sample size is needed for even modestly reasonable estimates.

QTL mapping incorporating multiple QTL with epistasis have received much attention with various statistical approaches developed (e.g. Doebley *et al.*, 1995; Jannink and Jansen, 2001; Boer *et al.*, 2002; Carlborg and Andersson, 2002; Yi and Xu, 2002; Yang, 2004; Baieri *et al.*, 2006; Alvarez-Castro and Carborg, 2007). Examples of QTL mapping software that can handle multiple QTL with epistatic effects are QTL CARTOGRAPHER and MULTIQTL.

6.7 Bayesian Mapping

A coherent approach to statistical modelling is provided by the Bayesian paradigm, which has been applied successfully in various contexts (Malakoff, 1999), including problems in genetics (Shoemaker *et al.*, 1999; Huelsenbeck *et al.*, 2001; Sorensen and Gianola, 2002; Xu, S., 2003). In Bayesian analysis everything is treated as an unknown variable with a prior distribution. A variable can be classified into one of two classes: observables and unobservables. The observables include data (phenotypic values, marker scores and pedigrees, etc.). The unobservables include parameters. Generally, this approach includes a careful consideration of the structure of the problem at hand, which then culminates in a model (likelihood) and in prior beliefs of unobservables expressed in a form of a probability distribution. Given the likelihood and prior beliefs, Bayesian machinery then delivers exactly the relevant information through the posterior probability distribution of the unobservables. The prior use can range from non-informative to very informative distributions and should reflect the knowledge available (e.g. derived from the earlier studies or from existing theory). The simulation (integration) method can be used to generate an approximate sample from the posterior distribution. The sampling algorithm tailored for a specific model is called the MCMC sampler.

6.7.1 Advantages of Bayesian mapping

Bayesian methodology has become popular in QTL mapping because of the availability of simulation-based MCMC algorithms. MCMC provides an approach for achieving a number of analytic goals that are otherwise difficult to achieve (Xu, S., 2002). Bayesian mapping allows the use of prior knowledge of QTL parameters and the posterior variances and credibility intervals for the estimated QTL parameters is automatically obtained. With MCMC approaches, it is possible to perform linkage analysis with any number of marker loci, multiple-trait loci and multiple genomic segments. At the same time, these approaches allow the use of pedigrees of arbitrary size and complexity. In addition to mapping the loci, Bayesian reversible-jump MCMC approaches allow one to estimate the number of loci and associated individual-locus model parameters as well as covariate effects in a joint linkage and oligogenic segregation analysis. This is particularly useful when multiple contributing loci are considered, but the number is unknown. It is advantageous to be able to estimate the number of contributing loci, rather than to fix this number a priori. The compromise made in order to achieve these goals lies in the overall approach, which is based on statistical sampling rather than exact enumeration of all possible underlying but unobserved genotypes.

6.7.2 Bayesian mapping statistics: a brief overview

With observed phenotypic trait values, markers and linkage map data (Y, X) and unknown quantitative genotype (Q), we can study the unknowns (θ, λ, Q), where λ is QTL location and θ their genetic effects.

$$Q \sim \Pr(Q \mid Y_i, X_i, \theta, \lambda)$$

Genotypes Q for every individual at m QTL can be sampled and their positions, marginal effects and epistatic effects θ can be tested.

The properties of the posterior distribution can be studied by using prior distributions that are independent between QTL and by drawing samples from posterior probabilities. The conditional posterior probability for multiple imputation or MCMC is shown in the equation at the bottom of the page.

To construct a Markov chain around posterior distribution, we need posterior probability as a stable distribution of the Markov chain. In practice, the chain tends towards stable distribution. The MCMC algorithm starts with given values of the parameters in the prior distributions and the initial values for all the unknowns generated from their prior distributions

$$(\lambda,Q,\theta,m) \sim \Pr(\lambda,Q,\theta,m \mid Y,X)$$

and m-QTL model components from full conditionals are updated with the following updating steps:

- update genetic effects θ given genotypes and traits;
- update locus λ given genotypes and marker map; and
- update genotypes Q given traits, marker map, locus and effects.

This generates the following chain of estimates:

$$(\lambda, Q, \theta, m)_1 \rightarrow (\lambda, Q, \theta, m)_2 \rightarrow \ldots \rightarrow (\lambda, Q, \theta, m)_N$$

To ensure that the chain mixes well, the initial values may have low posterior probability at the period of burn-in (initial iterations of the MCMC process that are used to locate the sampler in this part of the sample space).

After the burn-in period, realizations of (λ, Q, θ, m) are sampled from the chain and stored. Once enough realizations have been sampled, empirical posterior distributions for parameters in (λ, Q, θ, m) can be created from the posterior sample.

From full conditionals for a model with mQTL, it is hard to sample from joint posterior probability

$$\Pr(\lambda,Q,\theta \mid Y,X) = \Pr(\theta)\Pr(\lambda)\Pr(Q \mid X,\lambda)\Pr(Y \mid Q,\theta)/\text{constant}$$

But it is easy to sample parameters from full conditionals as following:

$$\Pr(\theta \mid Y,X,\lambda,Q) = \Pr(\theta \mid Y,Q) = \Pr(\theta)\Pr(Y \mid Q,\theta)/\text{constant} \text{ (for genetic effects)}$$

$$\Pr(\lambda \mid Y,X,\theta,Q) = \Pr(\lambda \mid X,Q) = \Pr(\lambda)\Pr(Q \mid ,\lambda)/\text{constant} \text{ (for QTL locus)}$$

$$\Pr(Q \mid Y,X,\lambda,\theta) = \Pr(Q \mid X,\lambda)\Pr(Y \mid Q,\theta)/\text{constant (for QTL genotypes)}$$

6.7.3 Bayesian mapping methods

When fully structuring the gene mapping problem in the Bayesian framework, the types of models considered to be suitable (e.g. no epistasis) need to be defined a priori. A prior opinion concerning the plausible values of the model dimension (the number of parameters) in addition to plausible values of the parameters themselves then need to be incorporated. This includes the prior distributions attached to the number of influential genes (QTL) and to their effects, which together reflect the prior beliefs concerning sensitivity towards small gene effects. In general, the MCMC analysis requires specific prior or proposal, distributions and involves many iterations. In order for the MCMC process to provide useful estimates, it is necessary for the sampler to move around the sample space successfully.

Bayesian mapping was initiated by Hoeschele and VanRaden (1993a,b) and subsequently developed by Satagopan

$$\Pr(\theta,\lambda,Q \mid Y,X) = \frac{\Pr(Q \mid X,\lambda)\Pr(Y \mid Q,\theta)\Pr(\lambda \mid X)\Pr(\theta)}{\Pr(Y \mid X)}$$

et al. (1996) and Sillanpää and Arjas (1998, 1999). Since then, various Bayesian mapping methods have been developed for different models and genetic systems, including the reverse-jump MCMC Bayesian method (Green, 1995; Satagopan *et al.*, 1996; Sillanpää and Arjas, 1998, 1999; Sillanpää and Corander, 2002), model selection framework (Yi, 2004; Yi *et al.*, 2005, 2007) and the shrinkage estimation (SE) method (Xu, S., 2003; Zhang and Xu, 2004; Wang, H. *et al.*, 2005). Wu and Lin (2006) concluded that the SE method allows analytical strategies for QTL mapping to expand to whole-genome mapping of epistatic QTL by use of all markers. However, the number of variables involved is so large that the computation time is too long. To solve this problem, Zhang and Xu (2005) proposed the penalized maximum likelihood (PML) method. Yi and Shriner (2008) reviewed Bayesian mapping methods and associated computer software for mapping multiple QTL in experimental crosses. They compared and contrasted the various methods to clearly describe the relationship between them.

Bayesian shrinkage estimation (BSE) method

With BSE, the number of effects that can be handled can be larger than the number of observations. The BSE method has been extended to map multiple QTL (Zhang and Xu, 2004; Wang, H. *et al.*, 2005) and epistatic QTL.

Assuming m QTL, Q_1, Q_2, ... and Q_m, the model for the quantitative trait value can be written as

$$y_i = b_0 + \sum_{j=1}^{q} x_{ij} b_j + e_i$$

where y_i is the quantitative trait value for individual i, b_0 is the mean, b_j is the main effect of Q_j, x_{ij} is coded as 1/2 or −1/2 if the genotype of Q_j is $Q_j\,Q_j$ or $Q_j\,q_j$ ($j = 1, \ldots, m$), and m is not equal to the number of markers for multiple-marker analysis but the number of marker intervals for multiple QTL analysis. The BSE method allows spurious QTL effects to be reduced towards zero, while QTL with large effects are estimated with virtually no shrinkage. To do this, each marker effect is allowed to have its own variance parameter, which in turn has its own prior distribution so that the variance can be estimated. Henceforth, prior distributions for all parameters are firstly assumed, i.e. $p(b_0) \propto 1$, $p(\sigma_e^2) \propto 1/\sigma_e^2$, $p(b_j) = N(0,\sigma_j^2)$ and $p(\sigma_j^2) \propto 1/\sigma_j^2$ ($j = 1, \ldots, q$); then conditional posterior distributions (CPD) for all parameters and hyperparameters are deduced, i.e. CPD for b_j is $N(\hat{b}_j, s_j^2)$ where

$$\bar{b}_j = \left(\sum_{i=1}^{n} x_{ij}^2 + \sigma_e^2 / \sigma_j^2 \right)^{-1} \sum_{i=1}^{n} x_{ij} \left(y_i - b_o - \sum_{k \neq j}^{q} x_{ik} b_k \right)$$

and

$$s_j^2 = \left(\sum_{i=1}^{n} x_{ij}^2 + \sigma_e^2 / \sigma_j^2 \right)^{-1} \sigma_e^2$$

The CPD for σ_j^2 is an inverted chi-square distribution; and finally, we sample observations of all parameters from the corresponding CPD. When the sampling chain converges to the stationary distribution, the sampled parameters actually follow the joint posterior distribution. When the sample of a single-parameter is considered, this univariate sample is actually the marginal posterior sample for this parameter. Therefore, the number, positions and effects of QTL can be estimated.

Provided the *j*th QTL is false (i.e. effect size is zero), the estimate of σ_j^2 will tend to zero and the mean and variance of the posterior distribution for b_j regress to zero so that the sampled observations of b_j are close to zero. Note that updating the variance σ_j^2 for the *j*th QTL is important because this either overcomes the shortcomings of the fixed ridge parameter in ridge regression or reflects the information of the data. If $b_j \to 0$ in the formula $\sigma_j^2 = b_j^2/\chi_{v=1}^2$, then $\sigma_j^2 \to 0$; however, dividing b_j^2 by a chi-square variable allows σ_j^2 a chance to recover because $\chi_{v=1}^2$ can be very small by chance.

Bayesian shrinkage analysis was used to develop a QTL model for mapping multiple QTL for dynamic traits (such as growth trajectories) under the maximum likelihood framework (Yang and Xu, 2007). The growth trajectory was fitted by Legendre polynomials. The method combines the shrinkage mapping for individual quantitative traits with the Legendre polynomial analysis for dynamic traits. The multiple-QTL model was implemented in two ways: (i) a fixed-interval approach where a QTL is placed in each marker interval; and (ii) a moving-interval approach where the position of a QTL can be searched in a range that covers many marker intervals. Simulation showed that the Bayesian shrinkage method generated much better signals for QTL than the interval mapping approach.

Model selection

A composite model space approach was proposed by Yi (2004) for mapping multiple non-epistatic QTL and extended by Yi *et al.* (2005) to epistatic QTL mapping for continuous traits. The key advantage of this approach is that it provides a convenient way to reasonably reduce the model space and to construct efficient algorithms for exploring the complicated posterior distribution. Yi *et al.* (2007) proposed a Bayesian model selection approach of genome-wide interacting QTL for ordinal traits in experimental crosses. They first developed a Bayesian ordinal probit model for multiple interacting QTL on the basis of the composite model space framework and then used this framework to develop an efficient MCMC algorithm for identifying multiple interacting QTL for ordinal traits.

Penalized maximum likelihood (PML) method

Integrating the shrinkage estimation with maximum likelihood (ML) method can decrease the running time. PML is different from the ML method because the function to be maximized is a penalized likelihood function rather than a likelihood function. Penalized likelihood is similar to the posterior distribution of the parameters, with the prior distribution of the parameters serving as the penalty. So PML method depends on the prior distribution. It estimates means and variances of prior distributions of QTL effects together with QTL effects, i.e. QTL effects can be estimated by using the equation shown at the bottom of the page. If, $\sigma_j^2 \to 0$ then $\hat{b}_j \to \mu_j$. Additionally, $\hat{\mu}_j = \hat{b}_j/(\mu + 1)$, so $\hat{b}_j \to 0$. This explains the reason why the estimate of a false-QTL effect is close to zero. Note that the PML method can select variables in the estimation of parameters, handle a model with the number of considered effects ten times larger than the sample size (Zhang and Xu, 2005; Hoti and Sillanpää, 2006) and be a refined method of mapping QTL (Yi *et al.*, 2006) because of small residual variance at the beginning of parameter estimation. However, the PML method cannot detect epistasis between nearby markers because of their multi-collinearity. For the real data analysis, two approaches are available for epistatic analysis. With the PML method along with the variable-interval approach, whole-genome mapping of epistatic QTL may be carried out by the use of all markers or the BSE method along with the variable-interval approach can be used to map epistatic QTL.

MCMC and especially the Gibbs sampler allows for the efficient exploration of very complex likelihood surfaces and calculation of Bayesian posterior distributions. For these reasons, Walsh (2001) predicted that the next 20 years will likely be marked by a strong influx of Bayesian methods replacing their likelihood counterparts. In contrast to classical methods, the Bayesian MCMC approach necessitates more human

$$\hat{b}_j = \left(\sum_{i=1}^{n} x_{ij}^2 + \sigma_e^2 / \sigma_j^2 \right)^{-1} \left[\sum_{i=1}^{n} x_{ij} \left(y_i - b_0 - \sum_{k \neq j}^{q} x_{ik} b_k \right) + \mu \sigma_e^2 / \sigma_j^2 \right]$$

effort and care to ensure that the simulation produces a representative sample from the posterior distribution. This requires careful monitoring of the convergence and the mixing properties of the MCMC sampler.

6.8 Linkage Disequilibrium Mapping

QTL mapping methods most frequently used so far are largely based on segregating populations derived from two parental lines, although some of them may be modified to use multiple populations simultaneously. Linkage disequilibrium (LD) or association mapping to be discussed in this section can be exploited to identify QTL using collections of germplasm, cultivars and all available genetic and breeding materials, by which molecular dissection of complex traits can be more closely linked up with plant breeding programmes.

LD is also known as gametic phase disequilibrium, gametic disequilibrium and allelic association. Simply stated, LD is the 'non-random association of alleles at different loci'. It is the correlation between polymorphisms (e.g. single nucleotide polymorphisms (SNPs)) that is caused by their shared history of mutation and recombination. The terms linkage and LD are often confused. Linkage refers to the correlated inheritance of loci through the physical connection on a chromosome, whereas LD refers to the correlation between alleles in a population. The confusion occurs because tight linkage may result in high levels of LD. For example, if two mutations occur within a few bases of one another, they undergo the same pressures of selection and drift through time. Because recombination between the two neighbouring bases is rare, the presence of these SNPs is highly correlated and the tight linkage will result in high LD. In contrast, SNPs on separate chromosomes experience different selection pressures and independent segregation; these SNPs thus have a much lower correlation or level of LD.

This section focuses on the basic concepts of LD mapping. Important references for this section include Jannink and Walsh (2002), Flint-Garcia *et al.* (2003), Breseghello and Sorrels (2006b), De Silva and Ball (2007), Mackay and Powell (2007), Oraguzie *et al.* (2007), Zhu *et al.* (2008), Buckler *et al.* (2009), Myles *et al.* (2009) and Yu *et al.* (2009).

6.8.1 Why linkage disequilibrium mapping?

Allele association between marker loci and association between marker alleles and phenotypes can be designated as marker–marker association and marker–trait association, respectively (Xu, Y., 2002). As discussed previously, the objective of linkage mapping is to identify simply inherited markers in close proximity to genetic factors affecting QTL. This localization relies on processes that create a statistical association between marker and QTL alleles and that selectively reduce that association as a function of the distance between the marker and QTL. Recombination in meiosis that leads to DHs, F_2 or RILs reduces the association between a given QTL and markers distant from it. Unfortunately, derivation of these populations (Chapter 4) requires relatively few meiosis, such that even markers that are far from the QTL (e.g. 10 cM) remain strongly associated with it. Such long-distance association hampers precise localization of the QTL. One approach to fine mapping is to expand the genetic map, for example, through the use of RILs and advanced intercross lines (Chapter 7).

Although designed segregating populations are easy to create, they come with a number of disadvantages (Malosetti *et al.*, 2007). First, the amount of segregating genetic variation within the population is limited, because at most two alleles per locus can segregate in a diploid species, where in the absence of allele polymorphisms between the parents no QTL can be identified. Secondly, the genetic backgrounds within which mapping studies take place are generally not representative of the backgrounds used in elite germplasm (Jannink *et al.*, 2001). In order to increase the genetic polymorphism, the parental lines are usually selected from highly diverse germplasm. Thirdly, the relatively low number of generations after maximum LD, where the maximum LD is reached in the F_1, implies a reduced number of sampled meioses within designed populations (typically a few hundred), leading to relatively long stretches of chromosome being in LD. Consequently, the characteristic size of confidence intervals for QTL locations is between 10 and 20 cM (Darvasi *et al.*, 1993). In addition, germplasm resources and breeding populations that have been accumulating in breeding programmes with available phenotypic information cannot be used so that genetic mapping and breeding are usually two separate, independent procedures.

LD mapping takes advantage of events that created association in the relatively distant past. Assuming many generations and therefore meioses have elapsed since these events, recombination will have removed association between a QTL and any markers not tightly linked to it. LD mapping thus allows for much finer mapping than standard biparental cross approaches. At a fundamental level, both LD and linkage rely on the co-inheritance of adjacent DNA variants, with linkage capitalizing on this by identifying haplotypes that are inherited intact over several generations and LD relying on the retention of adjacent DNA variants over many generations. Thus, LD studies can be regarded as very large linkage studies of unobserved, hypothetical pedigrees (Cardon and Bell, 2001). LD analysis has the potential to identify a single polymorphism within a gene that is responsible for the difference in phenotype and is perfectly suited for sampling a wide range of alleles from germplasm collections with high resolution (Flint-Garcia *et al.*, 2003). A less obvious additional attractive property is that LD mapping approaches offer possibilities for QTL identification in polyploidy crops with hard to model segregation patterns (Malosetti *et al.*, 2007).

For marker–trait association, differences in both phenotype and allele frequency can be identified in a group of cultivars that are derived from a common ancestral gene pool (Xu and Zhu, 1994). The procedure is regarded as an initial screening for identification of QTL (Bar-Hen *et al.*, 1995; Virk *et al.*, 1996). The development of saturated linkage maps and highly informative microsatellite and SNP markers in plants makes it possible to systematically survey marker–trait association on a whole-genome scale. Compared with transmission-based linkage mapping, LD mapping provides more opportunities for breeding applications since hundreds of germplasm accessions that are useful as parents in breeding are involved. An important asset of LD mapping strategies is the straightforward utilization of large amounts of historical phenotypic data that are available for mapping efforts at no or little extra costs, especially when evaluation of the trait is time and money consuming, as is the case with mean yield, adaptability and stability. As an increasing number of germplasm accessions are evaluated with molecular markers and phenotyped for agronomic traits, it is essential to consider using the LD mapping approach to map genes or at least to provide a pre-screen for linkage-based genetic mapping (Xu, Y., 2002).

6.8.2 Measurement of linkage disequilibrium

A variety of statistics have been used to measure LD. Delvin and Risch (1995) and Jorde (2000) reviewed the relative advantages and disadvantages of each statistical approach. Here, we introduce the two most common statistics for measuring LD: r^2 and D'. Consider a pair of loci with alleles A and a at locus one and B and b at locus two, with allele frequencies π_A, π_a, π_B and π_b, respectively. The resulting haplotype frequencies are π_{AB}, π_{Ab}, π_{aB} and π_{ab}. The basic component of all LD statistics is the difference between the observed and the expected haplotype frequencies,

$$D_{ab} = (\pi_{AB} - \pi_A \pi_B)$$

The distinction between these statistics lies in the scaling of this difference (Flint-Garcia *et al.*, 2003).

The first of the two measures, r^2, also described in the literature as Δ^2, is calculated as

$$r^2 = \frac{(D_{ab})^2}{\pi_A \pi_a \pi_B \pi_b}$$

It is convenient to consider r^2 as the square of the correlation coefficient between the two loci. However, unless the two loci have identical allele frequencies, a value of 1 is not possible. Statistical significance (P-value) for LD is usually calculated using either Fisher's exact test to compare sites with two alleles at each locus or multifactorial permutation analysis to compare sites with more than two alleles at either or both loci.

Alternatively, the LD statistic D' (Lewontin, 1964) is calculated as

$$|D'| = \frac{(D_{ab})^2}{\min(\pi_A \pi_b, \pi_a \pi_B)} \text{ for } D_{ab} < 0$$

$$|D'| = \frac{(D_{ab})^2}{\min(\pi_A \pi_B, \pi_a \pi_b)} \text{ for } D_{ab} > 0$$

D' is scaled based on the observed allele frequencies, so it will range between 0 and 1 even if allele frequencies differ between the loci. D' will only be less than 1 if all four possible haplotypes are observed; hence, a presumed recombination event has occurred between the two loci.

The statistics r^2 and D' reflect different aspects of LD and perform differently under various conditions. Figure 6.3 presents three scenarios of how linked polymorphisms may exhibit different levels of LD (Flint-Garcia *et al.*, 2003). Figure 6.3A shows an example of absolute LD, where the two polymorphisms are completely correlated with one another. An instance when absolute LD can develop is when two linked mutations occur at a similar point in time and no recombination has occurred between the sites. In this case, the history of mutation and recombination for the sites is the same. Both r^2 and D' have a value of 1 in this scenario. Figure 6.3B shows an example of LD when the polymorphisms are not completely correlated, but there is no evidence of recombination. One way this type of LD structure can develop is when the mutations occur on different allelic lineages. This situation can reflect the same recombinational history but different mutational histories. This is the situation in which r^2 and D' act differently, with D' still equal to 1, but where r^2 can be much smaller. Figure 6.3C shows an example of polymorphisms in linkage equilibrium. If the sites are linked, then equilibrium could be produced by a recombination event between the two sites. In this case, the recombinational history differs for the various haplotypes but the mutational history is the same. Hence, both r^2 and D' will be zero.

Although neither r^2 nor D' perform extremely well with small sample sizes and/or low allele frequencies, each has distinct advantages. Whereas r^2 summarizes both recombinational and mutational history, D' measures only recombinational history and is therefore the more accurate statistic for estimating recombination differences. However, D' is strongly affected by small sample sizes, resulting in highly erratic behaviour when comparing loci with low allele frequencies. This is due to the decreased probability of finding all four allelic combinations of low frequency polymorphisms even if the loci are unlinked. For the purpose of examining the resolution of association studies, the r^2 statistic is preferred, as it is indicative of how markers might correlate with the QTL of interest.

There are two common ways to visualize the extent of LD between pairs of loci (Flint-Garcia *et al.*, 2003). LD decay plots are used to visualize the rate at which LD declines with genetic or physical distance (Fig. 6.4). Scatter plots of r^2 values versus genetic/physical distances between all pairs of alleles within a gene, along a chromosome or across the genome are constructed. Alternatively, disequilibrium matrices are effective for visualizing the linear arrangement of LD between polymorphic sites within a gene or loci along a chromosome (Plate 2). It should be noted that LD decay is unpredictable. Both plot types highlight the

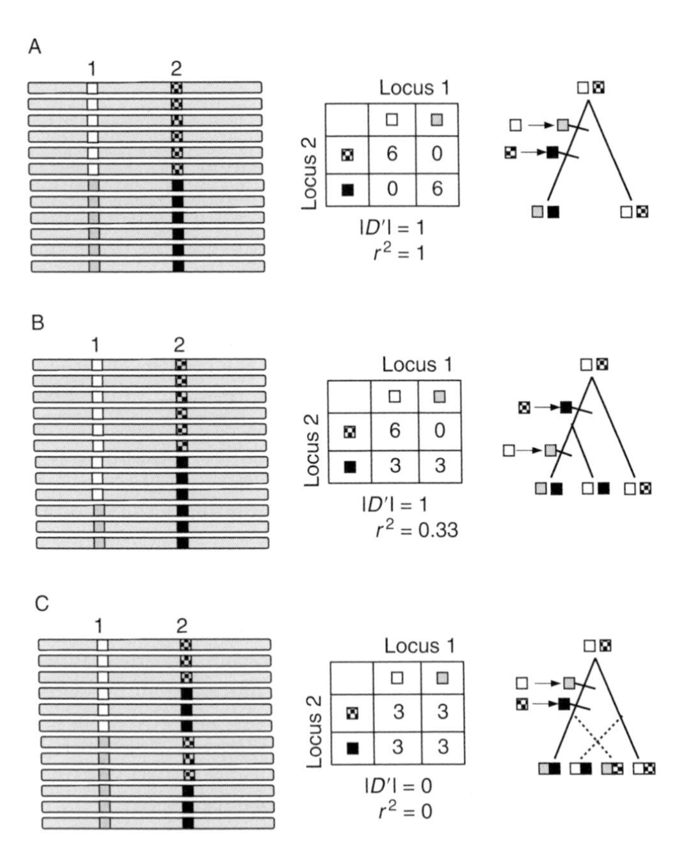

Fig. 6.3. Hypothetical scenarios of linkage disequilibrium (LD) between linked polymorphisms caused by different mutational and recombinational histories demonstrating the behaviour of the r^2 and D' statistics. Images in the left column represent the allelic states of two loci. The middle column represents the 2 × 2 contingency table of haplotypes and the resulting r^2 and D' statistics. The right column represents a possible tree responsible for the observed LD present. (A) An example of absolute LD, where the two polymorphisms are completely correlated with one another. (B) An example of LD when the polymorphisms are not completely correlated, but there is no evidence of recombination. (C) An example of when polymorphisms are in linkage equilibrium. Modified from Rafalski (2002).

random variation in LD owing to a variety of forces discussed below.

The limits of linkage analysis and LD mapping when they are used alone can be overcome by a joint mapping strategy as demonstrated by Wu and Zeng (2001) in which a random sample from a natural population and the open-pollinated progeny of the sample were analysed jointly. The joint linkage and LD mapping strategy was extended to map QTL segregating in a natural population (Wu *et al.*, 2002b). The extension allows for simultaneous estimates of a number of genetic and genomic parameters including the allele frequency of QTL, its effects, its location and its population association with a known marker locus.

6.8.3 Factors affecting linkage disequilibrium

In a large, randomly mated population with loci segregating independently, but in the absence of selection, mutation or migration, polymorphic loci will be in linkage equilibrium (Falconer and

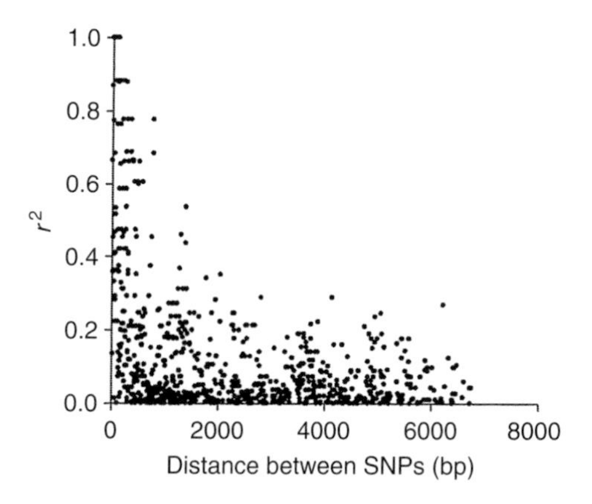

Fig. 6.4. Linkage disequilibrium (LD) decay plot of *shrunken 1* (*sh1*) in maize. LD, measured as r^2, between pairs of polymorphic sites is plotted against the distance between the sites. For this particular gene, LD decayed within 1500 bp. Data from Remington *et al.* (2001).

Mackay, 1996). Mutation provides the raw material for producing polymorphisms that will be in LD. Recombination is the primary force that eliminates both linkage and association over generations and the main phenomenon that weakens intrachromosomal LD, whereas interchromosomal LD is broken down by independent assortment. The rate for recombination to erode LD is slow between closely linked loci. For example, for loci that are 1 cM apart, more than 50% of the initial disequilibrium remains after 50 generations (Falconer and Mackay, 1996). However, LD decays with time but not in the case of LD beyond 5–10 cM except due to epistasis. A variety of mechanisms generate LD, including linkage, selection and admixture, several of which can operate simultaneously. Some common mechanisms are summarized from Jannink and Walsh (2002), Flint-Garcia *et al.* (2003) and Mackay and Powell (2007).

1. Founder effect: when populations are expanded from a small number of founders, the haplotypes present in the founders will be more frequent than expected under equilibrium. Three special cases are noteworthy. First, genetic drift affects LD by this mechanism in that a population experiencing drift derives from fewer individuals than its present size. Secondly, by considering an individual with a new mutation as a founder, we see that its descendants will predominantly receive the mutation and loci linked to it in the same phase. Linkage marker alleles will therefore be in LD with the mutant allele. Finally, an extreme case arises in the F_2 population derived from the cross of two inbred lines. Here, all individuals derived from a single F_1 founder genotype and association between loci can be predicted on the basis of their mapping distance.

2. Mutation: immediately after a mutation occurs, it is in LD with all other loci: the new mutation only occurs on a single haplotype. In successive generations, recombination causes LD to decay as new haplotypes are created, but this process takes a long time for closely linked markers. Most of the polymorphisms we observe are old: many generations are required for allele frequencies to rise to a frequency at which we detect them. Therefore, most pairs of polymorphic loci show little LD originating from mutation unless closely linked.

3. Population structure: the presence of subgroups in the sample in which individuals are more closely related to each other than the average pair of individuals taken at random in the population. Substructure is a common cause of covariance of polygenic effects because relatives tend to share marker and gene alleles genome-wide. LD arises in structured populations when allelic frequencies differ at two loci across subpopulations, irrespective of the linkage status of the loci. Admixed populations, formed by the union of previously separate populations into a single panmictic one, can be considered a case of structured population where substructuring has recently ceased. As gene flow between individuals of genetically distinct populations is followed by intermating, an admixture results in the introduction of chromosomes of different ancestry and allele frequencies.

4. Selection: this changes allele frequencies at QTL determining the selected trait, which causes LD between the selected allele at a locus and linked loci. This process, called hitchhiking, generates LD among markers around the selected locus. Moreover, selection for or against a phenotype controlled by two unlinked loci (epistasis) may result in LD despite the fact that the loci are not physically linked. Negative LD will occur between loci affecting a trait in populations under stabilizing or directional selection as a result of the Bulmer effect. Positive LD will occur between loci affecting a trait under disruptive selection. When loci interact epistatically, haplotypes carrying the allelic combination favoured by selection will also have higher frequencies than expected.

5. Mating patterns: population mating patterns can strongly influence LD. Generally, LD decays more rapidly in outcrossing species as compared to selfing species (Nordborg, 2000). This is because recombination is less effective in selfing species, where individuals are more likely to be homozygous, than in outcrossing species. LD breaks down rapidly with random mating (Pritchard and Rosenberg, 1999).

6. Genetic drift: population size plays an important role in determining the level of LD. In small populations, the effects of genetic drift result in the consistent loss of rare allelic combinations which increase LD levels. When genetic drift and recombination are in equilibrium,

$$r^2 = \frac{1}{1+4Nc}$$

where N is the effective population size and c is the recombination fraction between sites (Weir, 1996). Therefore, LD can be created in populations that have recently experienced a reduction in population size (bottleneck) with accompanying extreme genetic drift (Dunning *et al.*, 2000). During a bottleneck, only few allelic combinations are passed on to future generations. This can generate substantial LD. The activities of plant breeders themselves can result in bottlenecks – the introduction of a new disease resistance or agronomic trait might result in a period of breeding in which a small number of parental lines are used extensively, generating some degree of LD.

7. Migration: if two populations, differing in allele frequency, are brought together, LD is created. Less extreme population admixture or migration also generates LD.

6.8.4 Methods for linkage disequilibrium mapping

The transmission disequilibrium test and derivatives

The first and most robust method for distinguishing QTL–marker associations arising from LD between closely linked markers from spurious background associations is the transmission disequilibrium test (TDT) (Spielman *et al.*, 1993). Neither linkage alone nor disequilibrium alone (i.e. between unlinked markers) will generate a positive result hence the TDT is an extremely robust way of controlling for false positives.

The single progeny in each family is usually selected for an extreme phenotype. Parents and progeny are genotyped, but only parents heterozygous at the marker locus are included in the analysis. From each parent, one allele must be transmitted to the progeny and one is not. Over all families a count is made of the number of transmissions and non-transmissions. In the absence of linkage between QTL and marker, the expected ratio of transmission to non-transmission is 1:1. In the presence of linkage it is distorted to an extent that depends on the strength of LD between the marker and QTL. The distortion is tested in a chi-squared test. Power depends on the strength of LD and on the effectiveness of selection of extreme progeny in driving segregation away from expectation (Mackay and Powell, 2007).

This elegant test is extremely robust to the effects of population structure, particularly in human genetics, but is susceptible to an increase in false positive results generated by genotype error and biased allele calling (Mitchell and Chakravarti, 2003). This risk can be reduced by modelling genotype errors and missing data in the analysis or by comparing the transmission ratio for extreme phenotypes with that for control individuals or for the opposite extreme. The TDT has been extended to study haplotype transmissions, quantitative traits, the use of sib pairs rather than parents and progeny and information from extended pedigrees.

In crops, parental and progeny lines are usually separated by several generations of gametogenesis rather than by one. In this case, the TDT is still valid, but might no longer be so robust: the process of breeding might itself distort segregation patterns. A family-based association test that is applicable to plant breeding programmes has been proposed by Stich *et al.* (2006). The authors point out that for candidate gene studies, this method is more cost-effective than the alternative methods described below given that no additional control markers are required. However, some power will be lost because only progeny derived from F_1s known to have a heterozygous marker genotype are informative. Laird and Lange (2006) reviewed TDT and other family-based association tests.

Structured association

Structured association provides a sophisticated approach to detecting and controlling population structure (Pritchard *et al.*, 2000a,b; Falush *et al.*, 2003; Mackay and Powell, 2007). To deal with non-functional, spurious associations between a phenotype and an unlinked candidate gene caused by the presence of population structure and unequal distribution of alleles within sub-populations, several methods have been proposed. Pritchard *et al.* (2000b) proposed a method of testing association that depends on the inferred ancestries of individuals. Ancestries were inferred by a Bayesian method proposed by Pritchard *et al.* (2000a). Thornsberry *et al.* (2001) extended this method to deal with a quantitative trait and studied a candidate gene for the control of flowering time in maize.

The computer program STRUCTURE (http://pritch.bsd.uchicago.edu/software/structure2_1.html; Pritchard *et al.*, 2000a) uses computationally intensive methods to partition individuals into populations given molecular marker data. Many individuals or lines will not belong uniquely to one population, but will be the descendents of crosses between two or more ancestral populations. STRUCTURE also estimates the proportion of ancestry attributable to each population. Following allocation of individuals to populations, the test for association is carried out in a model fitting exercise. Here, the principle is that variation attributable to population membership is accounted for first, using estimates of population membership from STRUCTURE and then the presence of any residual association between the marker and phenotype is tested. For example, to test for association between a quantitative trait and a microsatellite, the trait is first regressed on the estimated coefficients of population membership and then on the marker – coded as a factor as if in an analysis of variance (Aranzana *et al.*, 2005). Alternatively, the groups can be integrated as an extra factor or a set of covariables in a statistical model relating phenotype to genotype (Thornsberry *et al.*, 2001; Wilson *et al.*, 2004).

As a valid alternative to the use of STRUCTURE, classical multivariate analysis methods can be used to classify genotypes. In that case a matrix of genetic/genotypic distances is calculated from molecular marker information and used as input for clustering and/or scaling techniques (Ivandic *et al.*, 2002; Kraakman *et al.*, 2004). For collections of cultivars and breeding lines, genotypic relationships as obtained from the pedigree or from similarities in neutral marker profiles (Yu *et al.*, 2006) can be translated into distances that are subsequently analysed by cluster analysis. The groups detected by such a cluster analysis can be interpreted as representing population structure and form an approximation to the original relationships between the genotypes as present before the grouping. The identified groups can be used as a kind of correction factor in association analyses.

Principal component analysis

A method termed EIGENSTRAT (Price *et al.*, 2006) is based on principal component analysis (PCA) across a large number of biallelic control markers with a genome-wide distribution. The PCA summarizes the variation observed across all markers into a smaller number of underlying component variables. These can be interpreted as relating to separate, unobserved, subpopulations from which the individuals in the dataset (or their ancestors) originated. The loadings of each individual on each principal component describe the population membership or the ancestry of each individual. However, these estimates are not ancestral proportions (values can be negative) in the same way that estimates of ancestry from STRUCTURE are. The loadings are used to adjust individual candidate marker genotypes (coded numerically) and phenotypes for their ancestry. The adjusted values are independent of estimated ancestry so a statistically significant correlation between an adjusted candidate marker and adjusted phenotype is therefore evidence of close linkage of a trait locus to the marker.

The EIGENSTRAT approach is similar to that of structured association but is less dependent on assessing the number of ancestral populations. Although each principal component is attributed to a separate population, the analysis is robust to the number included in the analysis, provided this is sufficiently large to capture all true population effects.

EIGENSTRAT was developed for analysing human datasets, which have high-density genotyping and low levels of population differentiation. Many crops have much higher levels of population differentiation than those found in human data sets and often only low densities of markers are available. EIGENSTRAT, unlike structured association, will not readily handle multi-allelic markers. However, a microsatellite with ten alleles could be coded as ten biallelic loci, all in complete LD. An analysis of human data showed that EIGENSTRAT was little affected by LD among > 3 million SNPs. The method shows great promise but additional research is required to establish its suitability for crops.

Mixed models

Parisseaux and Bernardo (2004) show how to integrate the pedigree-based relationship matrix into a QTL mapping analysis within a mixed-model framework. Yu *et al.* (2006) proposed the QK mixed-model LD mapping approach that promises to correct for LD caused by population structure and family relatedness. In this approach, both a marker-based relationship matrix (**K**) and a factor representing population structure are included in a mixed model for association analysis of a single trait in a single environment. The population structure matrix **Q** was calculated by the software STRUCTURE, which gives for each individual under consideration the probability of membership in each subpopulation.

Malosetti *et al.* (2007) proposed a LD approach based on mixed models with attention to the incorporation of the relationships between genotypes, whether induced by pedigree, population substructure or otherwise. Furthermore, they emphasized the need to pay attention to the environmental features of the data as well, i.e. adequate representation of the relations among multiple observations on the same genotypes. They illustrated their modelling approach using 25 years of Dutch national cultivar list data on late blight resistance in the genetically complex crop of potato. As markers, they used nucleotide binding-site markers, a specific type of marker that targets resistance or resistance-analogue genes. To assess the consistency of QTL identified by their mixed-model approach, a second independent data set was analysed. Two markers were identified that are potentially useful in selection for late blight resistance in potato.

Malosetti *et al.* (2007) showed how powerful and flexible a mixed-model framework can be for association mapping in plant species. They illustrated the use of mixed models by analysing two independent data sets on resistance to late blight, where the second data set served as an empirical check on the QTL identified in the first set. The approach can be implemented in any statistical package with extended facilities for mixed-model analyses.

Stich *et al.* (2008) evaluated various methods for LD mapping in the autogamous species wheat using an empirical data set, determined a marker-based kinship matrix using restriction maximum-likelihood (REML) estimate of the probability of two alleles at the same locus being identical in state but not identical by descent, and compared the results of LD approaches based on adjusted entry means (two-step approaches) with the results of approaches in which the phenotypic data analysis and the association analysis were performed in one step (one-step approaches). On the basis of the phenotypic and genotypic data of 303 soft winter wheat inbreds, their results indicated that the ANOVA approach was inappropriate for LD mapping in the germplasm set examined. Their observations suggested that the QK methods proposed by Yu *et al.* (2006) are appropriate for LD mapping not only in allogamous species such as humans and maize, but also in the autogamous species wheat. LD mapping approaches using a kinship matrix estimated by REML were more appropriate for LD mapping than the QK method proposed by Yu *et al.* (2006) with respect to: (i) the adherence to the nominal α-level; and (ii) the adjusted power for detection of QTL. They showed that the data set could be analysed by using two-step approaches of the proposed LD method without substantially increasing the empirical Type I error rate in comparison to the corresponding one-step approaches.

Quantitative inbred pedigree disequilibrium test

Stich *et al.* (2006) conducted a study to: (i) adapt the quantitative pedigree disequilibrium test to typical pedigrees of inbred lines produced in plant breeding programmes; (ii) compare the newly developed quantitative inbred pedigree disequilibrium test (QIPDT) with the commonly employed logistic regression ratio test (LRRT), with respect to the power and Type I error rate of QTL detection; and (iii) demonstrate the use of the QIPDT by applying it to flowering data of European elite maize inbreds. QIPDT and LRRT were compared based on computer simulations modelling 55 years of hybrid maize breeding in Central Europe. Furthermore, QIPDT was applied to a cross-section of 49 European elite maize inbred lines genotyped with 722 AFLP markers and phenotyped in four environments for several days to anthesis. Compared to LRRT, the power to detect QTL was higher with QIPDT when using data collected routinely in plant breeding programmes. Application of QIPDT to the 49 European maize inbreds resulted in a significant ($P < 0.05$) association located at a position for which a consensus QTL was detected in a previous study.

Bayesian methods

Bayesian methods based on the MCMC algorithm have been developed for mapping multiple QTL as discussed in Section 6.7. Those based on Bayesian variable selection (e.g. Yi *et al.*, 2003;Yi, 2004; Sillanpää and Bhattacharjee, 2005) are advantageous in that they can be implemented via a simple and easy-to-use Gibbs sampler and can be extended to the whole genome LD mapping (Kilpikari and Sillanpää, 2003).

Iwata *et al.* (2007) proposed an approach that combines a Bayesian method for mapping multiple QTL with a regression method that directly incorporates estimates of population structure and multiple QTL. The efficiency of the approach in simulated- and real-trait analyses of a rice germplasm collection was evaluated. Simulation analyses based on real marker data showed that the model could suppress both false-positive and false-negative rates and the error of estimation of genetic effects over single QTL models, indicating statistically desirable attributes over single QTL models.

6.8.5 Applications of linkage disequilibrium mapping

There are many reports on LD mapping in various plants (e.g. Thornsberry *et al.*, 2001; Kraakman *et al.*, 2004; Breseghello and Sorrels, 2006a; Auzanneau *et al.*, 2007; Crossa *et al.*, 2007; Brown *et al.*, 2008; D'hoop *et al.*, 2008; Raboin *et al.*, 2008; Weber *et al.*, 2008; Buckler *et al.*, 2009; Chan *et al.*, 2009; McMullen *et al.*, 2009; Stich *et al.*, 2009). Earlier attempts at establishing association between traits and markers across germplasm collections concerned rice, oats, maize, sea beet and barley. In rice, Virk *et al.* (1996) predicted the value for six traits using multiple linear regression. In oats, Beer *et al.* (1997) found associations between markers and 13 quantitative traits in a set of 64 landraces and cultivars. In maize, Thornsberry *et al.* (2001) found associations between *Dwarf8* polymorphisms and flowering time. In sea beet, Hansen *et al.* (2001) mapped the bolting gene, using AFLP markers in four populations. In barley, Igartua *et al.* (1999) concluded that marker–trait associations for heading date, found in mapping populations, were to some extent, maintained in 32 cultivars. Ivandic *et al.* (2003) found association between markers and the traits of water-stress tolerance (chromosome 4H) and powdery mildew resistance in 52 wild barley lines. Chromosome 4H is, according to Forster *et al.* (2000), known for many loci involving abiotic stress tolerance, including salt tolerance, water use efficiency and adaptation to drought environments.

Using 237 rice accessions collected from around the world and genotypic data for 100 restriction fragment length polymorphism (RFLP) and 60 simple sequence repeat (SSR) marker loci and phenotypic data for 12 traits, a stronger marker–marker association was found in the cultivar groups that had greater genetic variation or closer pedigree relationship (Xu, Y., 2002). Markers within linkage groups showed stronger allelic association than markers between linkage groups. The statistical associations, however, could not be interpreted solely from genetic linkage. Comparison of marker–trait association in different cultivar groups demonstrated that both phenotypic variation and pedigree relationship among rice accessions strongly influenced the association detection. A highly consistent allele–trait association was revealed among multiple alleles at a given locus. Several chromosomal regions as hot spots for marker–trait associations have been assigned to QTL clusters. Several highly consistent allele–trait associations were revealed among multiple alleles at specific loci.

The same data set was used to evaluate the potential of discriminant analysis, a multivariate statistical procedure, to detect candidate markers associated with agronomic traits (Zhang *et al.*, 2005). Model-based methods revealed population structure among the lines. Marker alleles associated with all traits were identified by discriminant analysis at high levels of correct percentage classification within sub-populations and across all lines. Associated marker alleles pointed to the same and different regions on the rice genetic map when compared to previous QTL mapping experiments. Results suggested that candidate markers associated with agronomic traits can be readily detected among inbred lines of rice using discriminant analysis combined with other methods.

Using 236 AFLP markers and 146 modern two-row spring barley cultivars, associations between markers were found for markers as far apart as 10 cM (Kraakman *et al.*, 2004). Subsequently, for the 146 cultivars the complex traits mean yield, adaptability (Finlay–Wilkinson slope) and stability (deviations from regression) were estimated from the analysis of cultivar trial data. Regression of those traits on individual marker data disclosed marker–trait associations for mean yield and yield stability. Many of the associated markers were located in regions where earlier QTL were found for yield and yield components. In tetraploid potato, LD mapping has been successfully applied to studying disease resistances for which candidate genes were defined (Gebhardt *et al.*, 2004; Simko *et al.*, 2004a,b).

Historical multi-environmental trial data provides comprehensive phenotypic data for LD mapping and modelling genotype-by-environment interaction. Crossa *et al.* (2007) reported a comprehensive study using historical wheat data. Mapped diversity array technology (DArT) markers (Chapter 2) were used to find association with resistance to stem rust, leaf rust, yellow rust and powdery mildew, plus grain yield in five historical wheat international multi-environment trials from the International Maize and Wheat Improvement Center (CIMMYT) conducted from 1970 to 2004. Two linear mixed models were used to assess marker–trait associations incorporating information on population structure and covariance between relatives. Several LD clusters bearing multiple host plant resistance genes were found. Most of the associated markers were found in genomic regions where previous reports had found genes or QTL influencing the same traits. In addition, many new chromosome regions for disease resistance and grain yield were identified. Phenotyping across up to 60 environments and years allowed modelling of genotype-by-environment interaction, thereby making possible the identification of markers contributing to both additive and additive-by-additive interaction effects.

As whole genome sequences become available for more and more plant species and thus sequence-based markers covering whole genomes, genome-wide association (GWA) studies are becoming popular in genetic studies to replace the candidate gene-based approach. This boom follows a long germination period during which the necessary concepts, resources and techniques were developed and assembled (Kruglyak, 2008). With the completion of the initial wave of GWA scans in humans, McCarthy *et al.* (2008) reviewed each major step in the implementation of a GWA scan, highlighting areas where there is an emerging consensus over the ingredients for success and those aspects for which considerable challenges remain.

In plants, it is apparent that molecular marker-based germplasm evaluation will produce a large data set that can be explored for LD studies for crops with different levels of LD (e.g. Lu *et al.*, 2009), with the development of highly informative DNA markers (e.g. SNPs) and high-throughput genotyping technology. The number of SNPs that are required for GWA obviously depends on the genomic extent of LD because genotyped SNPs must be spaced sufficiently densely to be in LD with most of the variants that are not genotyped. In sugarbeet, LD extended up to 3 cM (Kraft *et al.*, 2000), while in some *Arabidopsis* populations LD even exceeded 50 cM (Nordborg *et al.*, 2002). In contrast, in maize LD had already diminished after 2000 bp (Remington *et al.*, 2001). The marker density in many plant species will allow effective GWA.

6.9 Meta-analysis

The explosion of interest in QTL mapping has led to numerous studies in plants, each based on its own experimental population(s). Each experiment is limited in size and usually restricted to a single population or a cross, planted in a specific environment(s). Therefore, QTL effects that can be detected are also limited. One direction for QTL analysis is to combine information from several studies – for example, by meta-analysis of the results of QTL studies (Goffinet and Gerber, 2000) or joint analysis of the raw data (Haley, 1999) as discussed in Section 6.5.3.

Efforts to combine findings from separate studies have a long history. Glass (1976) proposed a method to integrate and summarize the findings from a body of research. He called the method meta-analysis. Since then, meta-analysis has become a widely accepted research tool in a variety of disciplines (Hedges and Olkin, 1985). Meta-analysis involves the application of standard statistical principles (hypothesis testing, inference) to situations where only summary information is available (e.g. published reports) and not the source unit record data. Well-conducted meta-analysis allows for a more objective appraisal of the evidence, which may lead to resolution of uncertainty and disagreement. Meta-analysis makes the literature review process more transparent, compared with traditional narrative reviews where it is often not clear how the conclusions follow from the data examined (Smith and Egger, 1998). The application of meta-analysis to QTL detection is recent (Goffinet and Gerber, 2000; Hayes and Goddard, 2001). The combining of the results across studies can provide a more precise and consensus estimate of the location of a QTL and its effect as compared with any single study. However, there are many challenges in combining the results of QTL mapping across studies, including differences in marker density, linkage map, sample size, study design, as well as statistical methods used. One aspect that might transcend the meta-analysis problem and benefit the whole field of QTL detection and location is the reliability of the principal parameters which characterize QTL: position, confidence interval, R^2 and LOD score (Hanocq *et al.*, 2007). These parameters are critical to the meta-analysis process but are often only partially reported in research papers.

6.9.1 Meta-analysis of QTL locations

We followed the method described by Goffinet and Gerber (2000). In summary, with a total of m published reports of a QTL on a particular chromosome, the statistical question is to decide on whether these reports represent a single QTL, two QTL, etc. up to m separate QTL (one for each publication).

Assessment of the number of QTL can be made on the basis of a likelihood ratio test, AIC or adjusted AIC, as in the method outlined by Goffinet and Gerber (2000). This involves selecting from among the best-fitting models with 1, 2, ..., m distinct QTL. As a result each published QTL can then be allocated to its respective consensus QTL. Note that usually, only the latest paper in a publication series on the same study population was included, to avoid duplication of the same QTL report. For a publication to be included in meta-analysis, it ideally provides the interval map (test statistic profile). As well as providing the estimate of the QTL location ($\hat{d}_i$), the interval map also enables estimation of the standard error for the QTL location, $\sigma_i = se(\hat{d}_i)$, after conversion of the test statistic to a (approximate) log-likelihood ($\ln L$) scale. It has been suggested that the standard error can be estimated from the curvature (Fisher information) of the log-likelihood profile at the estimated map position

$$\sigma_i = [-\partial^2 \ln L/\partial d^2 |_{d=\hat{d}_i}]^{-1/2}$$

In particular, the curvature is estimated by fitting a local quadratic near the maximum of $\ln L$ and determining the coefficient of the quadratic term. These standard errors are used to construct a weighted estimate of QTL location, the weights being inversely proportional to the squared standard errors ($w_i = \sigma_i^{-2}$).

For studies that did not include an interval map, average standard errors

$$\bar{\sigma} = \sqrt{(1/m)\sum_{i=1}^{m} \sigma_i^2}$$

can be computed based on the studies where interval maps were available.

6.9.2 Meta-analysis of QTL maps

Integration of genetic maps and QTL by iterative projections on a reference map is now widely used to position both markers and QTL on a single and homogeneous consensus map (e.g. Arcade *et al.*, 2004; Sawkins *et al.*, 2004). Comparison of multiple QTL mapping experiments by alignment to a common reference or consensus map offers a more complete picture of the genetic control of a trait than can be obtained in any one study. In order to study QTL congruency, Goffinet and Gerber (2000) proposed an original approach based on a meta-analysis strategy. Etzel and Guerra (2003) developed a meta-analysis based on an approach to overcome the between-study heterogeneity and to refine both QTL location and the magnitude of the genetic effects. Both the methods of Goffinet and Gerber (2000) and Etzel and Guerra (2003) are limited to a small number of underlying QTL positions (from one to four for the former and only one for the latter) which is a serious limitation for a whole genome study of QTL congruency. Even if the average number of QTL per experiment is around four in plants (Kearsey and Farquhar, 1998; Xu, Y., 2002; Chardon *et al.*, 2004), it would be expected that more than four genes can be involved in the trait variation on a single chromosome.

A meta-analysis of flowering time and related traits in maize from 22 QTL detection studies concluded that a total of 62 different QTL are likely to be involved in the variation of these traits, whereas on average four to five QTL were detected in single-population analyses (Chardon *et al.*, 2004). To remove these impediments, Veyrieras *et al.* (2007) developed a new two-stage meta-analysis procedure in order to integrate multiple independent QTL mapping experiments with the aim of creating a global framework to evaluate the homogeneity of both genetic marker and QTL mapping results from literature and public databases. First, it implements a new statistical approach to merge multiple distinct genetic maps into a single consensus map which is optimal in terms of weighted least squares and can be used to investigate recombination rate heterogeneity between studies. Secondly, assuming that QTL can be projected on the consensus map, MetaQTL, a computational and statistical package developed for the whole-genome meta-analysis of QTL mapping experiments, offers a new clustering approach based on a Gaussian mixture model to decide how many QTL underlie the distribution of the observed QTL. Contrary to existing methods, MetaQTL offers a complete statistical process to establish a consensus model for both the marker and the QTL positions on the whole genome.

6.9.3 Meta-analysis of QTL effects

After estimating the consensus QTL position using the above approach, a meta-analysis can be conducted for the effect size for each consensus QTL. Suppose that for a consensus QTL, the QTL allelic substitution effects (a) differ from sire to sire and assume that $a \sim N(0,\sigma_A^2)$. The purpose behind this meta-analysis is to estimate the variance of these effects, σ_A^2. Next assume that for each sire in the available studies, the estimate of the QTL allelic substitution effect, a_i, is $\hat{a}_i$ with corresponding standard error $\varsigma_i = se(\hat{a}_i)$ and variance ς_i^2, $i = 1, 2, ..., n$, where n is the number of sires. To model the imprecision of $\hat{a}_i$ estimating a_i, we assume that $\hat{a}_i \mid a_i \sim N(a_i,\varsigma_i^2)$ and consequently, the unconditional distribution of estimated effects will be $\hat{a}_i \sim N(0,\varsigma_i^2 + \sigma_A^2)$. As also considered by Hayes and Goddard (2001), there are two other features that need to be modelled in the meta-analysis. First, since it is to a certain extent arbitrary which sire allele is labelled as having a positive effect, we will ignore the sign and condition on $a_i > 0$ and $\hat{a}_i > 0$. Secondly, only 'significant' QTL tend to be published (resulting in potential publication bias), so we assume that $\hat{a}_i > c$ where c is the 'threshold' QTL effect that just reaches 'publication level'. With these constraints, the probability density function, $h(\bullet)$, for the observed QTL effects will be

$$h(\hat{a}_i \mid a_i > c) = n_i(\hat{a}_i)/[1 - N_i(c)],\ \hat{a}_i > c$$

where say,

$$n_i(y) = \frac{1}{\sqrt{2\pi(\varsigma_i^2 + \sigma_A^2)}} \exp\left(-\frac{y^2}{2(\varsigma_i^2 + \sigma_A^2)}\right)$$

is the normal probability density function and $N_i(y) = \int_{-\infty}^{y} n_i(t)dt$ the corresponding cumulative normal distribution function. So there are two parameters to be estimated, σ_A^2 and c and this is achieved by an ML procedure.

For those papers where ζ_I was not reported, the average value ($\bar{\varsigma}$) is computed in a similar way to that of $\bar{\sigma}$. However, because the different studies were conducted under different conditions, there was a large variation in the phenotypic standard deviation across studies for a particular trait. Consequently, both the effect estimates and their standard errors were re-scaled by dividing by their reported phenotypic standard deviations (where reported) or by appropriate consensus standard deviations used for international evaluations where this was not reported. Consequently, the consensus estimate of σ_A^2 will be the proportion of the phenotypic variance explained by the consensus QTL.

6.9.4 Examples of meta-analysis

Meta-analysis of all identified QTL promises to contribute to our understanding of fundamental questions and to expedite crop improvement. Khatkar *et al.* (2004) reviewed the results of QTL mapping in dairy cattle. Based on the information available in the public domain, they developed an online QTL map for milk production traits. To extract the most information from these published records, a meta-analysis was conducted to obtain consensus on QTL location and allelic substitution effect of these QTL. The meta-analysis indicated a number of consensus regions, the most striking being two distinct regions affecting milk yield on chromosome 6 at 49 cM and 87 cM explaining 4.2 and 3.6% of the genetic variance of milk yield, respectively. Outputs from such analyses highlight the specific areas of the genome where future resources should be directed to refine characterization of the QTL.

To identify the genome regions of bread wheat involved in the control of earliness and its three components: photoperiod sensitivity, vernalization requirement and intrinsic earliness, Hanocq *et al.* (2007) carried out a QTL meta-analysis to examine the replicability of QTL across 13 independent studies and to propose meta-QTL. QTL were projected on to the reference map using the BIOMERCATOR 2.0 software (Arcade *et al.*, 2004). To assess the reliability of this projection, five variables were calculated to assess QTL projection quality for each QTL: (i) the percentage of QTL confidence interval (CI) included in the linked region; (ii) N_m (the number of common markers characterizing a QTL CI region, i.e. within and flanking it); (iii) local map density (which is computed as the local average distance of the N_m markers on the projected map); (iv) maximum gap size or the size of the largest interval between adjacent markers when considering the N_m markers; and (v) the weighted standard deviation standardized to 100 cM (which evaluates the heterogeneity in homothetic coefficients for intervals within and flanking a QTL CI region). Chromosomes of groups 2 and 5 had greater control over the incidence of earliness as they carry the known, major genes *Ppd* and *Vrn*. The other four chromosome regions played an intermediate role in control of earliness.

In cotton, a total of 432 QTL involving cotton fibre quality, leaf morphology, flower morphology, resistance to bacteria, trichome distribution and density and other traits that were mapped in one diploid and ten tetraploid interspecific cotton populations, was aligned using a reference map which consisted of 3475 loci in total and was depicted in a CMAP resource (Rong *et al.*, 2007). Meta-analysis of polyploidy cotton QTL showed unequal contributions of sub-genomes to a complex network of genes and gene clusters implicated in lint fibre development. QTL correspondence across studies was only modest, suggesting that additional QTL for the target traits remain to be discovered. Crosses between closely-related genotypes differing by single-gene mutants yield profoundly different QTL landscapes, suggesting that fibre variation involves a complex network of interacting genes. Meta-analysis linked to synteny-based and expression-based information provides clues about specific genes and families involved in QTL networks.

Munafò and Flint (2004) described how meta-analysis works and considered whether it will solve the problem of underpowered studies or whether it is another affliction visited by statisticians on geneticists. A crucial question for any meta-analysis is the degree of heterogeneity that exists between the individual studies, which is perhaps, not surprisingly common. Ioannidis *et al.* (2001) conducted a meta-analysis of 370 studies addressing 36 genetic associations. They found that significant between-study heterogeneity is frequent and that the results of the first study often correlate only modestly with subsequent research on the same association. It has been argued that meta-analysis is analogous to averaging the characteristics of apples and orange (Hunt, 1997) and consequently, its outcome is meaningless. Another concern in meta-analysis is publication bias that can exist when non-significant findings remain unpublished, thereby artificially inflating the apparent magnitude of the effect. The concern is not new and was raised in the late 1950s in relation to psychiatric and psychological research in humans (Sterling, 1959).

As indicated by Munafò and Flint (2004), meta-analysis has been successful in revealing unexpected sources of heterogeneity, such as publication bias. If heterogeneity is adequately recognized and taken into account, meta-analysis can confirm the involvement of a genetic variant, but it is not a substitute for an adequately powered primary study.

6.10 *In Silico* Mapping

As an alternative to designed mapping experiments using an F_2 or BC mapping population, *in silico* mapping was developed to detect genes by simultaneously exploiting existing phenotypic, genotypic and pedigree data available in breeding programmes and genomic databases. Grupe *et al.* (2001) were the first to use this approach to investigate whether chromosomal regions regulating quantitative traits (QTL intervals) could be computationally predicted with the use of the mSNP database and available phenotypic information obtained from mouse inbred strains. The phenotypic and genotypic information was analysed *in silico* to identify candidate QTL intervals. Ability of the computational method to correctly predict QTL intervals was evaluated and 19 of 26 experimentally verified QTL intervals for ten phenotypic traits were correctly identified. *In silico* mapping can eliminate many months to years of laboratory work required to generate, characterize and genotype intercross progeny, reducing the time required for QTL interval identification to milliseconds when a large number of related data become available.

6.10.1 Pros and cons

As massive amounts of phenotypic data for different traits have accumulated in public and private plant breeding programmes in major crop species, *in silico* mapping in plants has become possible and attractive. Compared with designed mapping experiments, *in silico* mapping has several advantages (Grupe *et al.*, 2001; Parisseaux and Bernardo, 2004). First, *in silico* mapping exploits larger populations than designed mapping experiments. In maize, for example, thousands of experimental hybrids are evaluated each year (Smith *et al.*, 1999). In contrast, the small populations (e.g. fewer than 500 progenies) often used in designed mapping experiments lead to a low power for detecting QTL (Melchinger *et al.*, 1998), overestimation of QTL effects (Beavis, 1994) and imprecise estimates of QTL location (van Ooijen, 1992; Visscher *et al.*, 1996). Secondly, phenotypic data used for *in silico* mapping are obtained through more extensive testing under multiple, diverse environments. An experimental maize hybrid is typically evaluated in 20 environments; those that are eventually released as cultivars are evaluated in up to 1500 location–year combinations (Smith *et al.*, 1999). The use of many environments permits the sampling of a sufficient set of QTL × environment interactions. Thirdly, the hybrids and inbreds tested typically represent a wide sample of the germplasm and genetic backgrounds. In contrast, only a narrow genetic background is exploited in designed mapping experiments that use F_2 or BC populations. Fourthly, the data used for *in silico* mapping are already available without extra cost.

Offsetting these advantages are three main complications to *in silico* mapping (Parisseaux and Bernardo, 2004). First, the performance data are highly unbalanced:

the same set of hybrids or inbreds are evaluated in a different set of environments, as some hybrids or inbreds that fail to perform well are discarded and those that perform well are subjected to more testing. Secondly, the hybrids or inbreds do not comprise a single homogenous population. Any *in silico* mapping procedure would therefore have to account for pedigree relationships and differences in the genetic backgrounds among tested hybrids or inbreds. Thirdly, few crops have enough data available for *in silico* mapping.

6.10.2 Mixed-model approach

The usefulness of *in silico* mapping has been explored via a mixed-model approach in maize (*Zea mays* L.) to determine whether the procedure gave results that were repeatable across populations (Parisseaux and Bernardo, 2004). Multi-location data were obtained from the 1995–2002 hybrid testing programme of Limagrain Genetics in Europe, which included: (i) multi-location phenotypic data for 22,774 single-cross hybrids; (ii) SSR marker data at 96 loci for the 1266 parental inbreds of the single-cross hybrids; and (iii) pedigree records for the 1266 parental inbreds which were classified into nine different heterotic groups.

Using a mixed-model approach, the general combining ability effect associated with marker alleles in each heterotic pattern was estimated. The numbers of marker loci with significant effects – 37 for plant height, 24 for smut (*Ustilago maydis* (DC.) Cda.) resistance and 44 for grain moisture – were consistent with previous results from designed mapping experiments. Each trait had many loci with small effects and few loci with large effects. For smut resistance, a marker in bin 8.05 on chromosome 8 had a significant effect in seven (out of a maximum of 18) instances. For this major QTL, the maximum effect of an allele substitution ranged from 5.4% to 41.9%, with an average of 22.0%. It is concluded that *in silico* mapping via a mixed-model approach can detect associations that are repeatable across different populations.

Because of differences in the germplasm used, the numbers of QTL identified through *in silico* mapping were not directly comparable with those previously detected through designed mapping experiments. On the one hand, the wide range of germplasm sampled with *in silico* mapping enhances the detection of many QTL. On the other hand, mapping populations are often developed by crossing two parents that are widely divergent for a trait, e.g. susceptible parent and resistant parent for smut. A diverse mapping population also enhances the detection of many QTL. In the largest QTL mapping study published in maize (976 families from an F_2 population, genotyped with 172 markers and evaluated in 19 environments), Openshaw and Frascaroli (1997) detected 36 significant markers for plant height and 32 for grain moisture (data for smut resistance were absent). This result for plant height (36 QTL) was consistent with the number of significant markers detected for plant height (37) via *in silico* mapping. The number of significant markers (44) for grain moisture was larger than that detected by Openshaw and Frascaroli (1997), perhaps because of a wider range of maturities sampled in the *in silico* mapping germplasm than in the single F_2 population used by Openshaw and Frascaroli (1997). For smut resistance, Lübberstedt *et al.* (1998a) detected 19 significant markers across four populations, whereas Kerns *et al.* (1999) detected 22 significant markers in one population. These previous results were consistent with the number of significant markers (24) detected for smut resistance in *in silico* mapping.

6.10.3 Statistical power

It has been shown that the heritability and genetic architecture (e.g. number of QTL and distribution of effects) of the trait and resources available for QTL mapping (e.g. sample size and number of markers) affect the statistical power of designed QTL mapping experiments as discussed in this chapter. These genetic and non-genetic factors are also expected to affect the power of *in silico* mapping via a mixed-model approach.

The statistical power of the *in silico* mapping method was evaluated via a mixed-model approach in hybrid crops (Yu *et al.*, 2005). Simulation mimicked a two-stage breeding process in maize, with inbred development and hybrid testing. First, two opposite heterotic groups were considered, each having a total of $n_1 = n_2 = 112$ inbreds developed from different ancestral inbreds. Secondly, it was assumed that $n = 600$ or 2400 hybrids, among all potential single-cross hybrids ($112 \times 112 = 12,544$) between the two heterotic groups, had data available from multi-location performance trails. The number of inbreds in each heterotic group and the number of hybrids with available phenotypic data were chosen to agree with the empirical data of Parisseaux and Bernardo (2004).

A total of 64 simulation experiments was conducted. These 64 experiments had contrasting values of six different parameters: level of initial LD ($t = 10$ or 20 generations of random mating), significance level ($\alpha = 0.01$ or 0.0001), number of QTL ($l = 20$ or 80), heritability ($H = 0.40$ or 0.70), number of markers ($m = 200$ or 400) and sample size ($n = 600$ or 2400 hybrids). For each experiment, 50 runs were conducted with different locations of QTL and markers on the genetic map and different inbreds and hybrids.

It was found that the average power to detect QTL ranged from 0.11 to 0.59 for a significance level of $\alpha = 0.01$ and from 0.01 to 0.47 for $\alpha = 0.0001$. The false discovery rate ranged from 0.22 to 0.74 for $\alpha = 0.01$ and from 0.05 to 0.46 for $\alpha = 0.0001$. As with designed mapping experiments, a large sample size, high marker density, high heritability and small number of QTL led to the highest power for *in silico* mapping via a mixed-model approach. The power to detect QTL with large effects was greater than the power to detect QTL with small effects. It is concluded that gene discovery in hybrid crops can be initiated by *in silico* mapping. Finding an acceptable compromise, however, between the power to detect QTL and the proportion of false QTL would be necessary.

In plant breeding programmes, the phenotypic data are highly unbalanced and the inbreds and hybrids have a pedigree structure. *In silico* mapping via a mixed-model approach accommodates unbalanced data, pedigree relationships and different heterotic groups of parental inbreds by fitting relevant terms in the mixed model. Furthermore, the relative effects of the QTL are measured by the regression coefficients of the significant markers and the approximate positions of the QTL are indicated by the location of the significant markers.

As with other QTL mapping methods, the results from *in silico* mapping should be followed by fine mapping at the target regions, sequence analysis and functional tests of gene effects (Glazier *et al.*, 2002). In hybrid crops for which multiple heterotic groups exist, *in silico* mapping via a mixed-model approach can be applied to different heterotic patterns. Subsequently, the markers or the genomic regions that show a repeatable association with the trait of interest across different populations can be considered as the prime targets for further analysis (Parisseaux and Bernardo, 2004). Cross validation by conducting *in silico* mapping in multiple heterotic patterns would result in better control of the overall false discovery rate and provide increased confidence for conducting further investigation in putative QTL regions.

6.11 Sample Size, Power and Thresholds

6.11.1 Power and sample size

There are two types of errors that can be made when carrying out a statistical test. A false positive (a Type I error) occurs when the null hypothesis is rejected when in fact it is correct. We control for this by setting a low significance level α for a test (the probability of a false positive). The other source of error is a false negative (a Type II error), i.e. failing to reject the null hypothesis when in fact it is false. The power of a test is defined to be the probability that the null hypothesis is rejected when it is indeed false. Hence if β is the probability of a false negative, the power is $1 - \beta$. The discussion in this section is based on Broman *et al.* (2003a) and Zeng's presentation at the Plant and Animal Genome XI meeting, 2003 (http://statgen.ncsu.edu/zeng/QTLPower-Presentation.pdf).

First a simple case (a point for departure) is one marker and one QTL for F_2. Assume that the QTL genotypic effects for Q_1Q_1, Q_1Q_2 and Q_2Q_2 are a, d and $-a$, respectively.

The marker effects can be tested, referring to Eqn 6.1, by

$$t_1 = \frac{\mu_{M_1M_1} - \mu_{M_2M_2}}{\sqrt{\dfrac{\sigma^2}{n/4} + \dfrac{\sigma^2}{n/4}}} = \frac{(1-2r)2a}{\sqrt{8\sigma^2/n}} \quad (6.16)$$

and referring to Eqn 6.2, by

$$t_2 = \frac{\mu_{M_1M_2} - (\mu_{M_1M_1} + \mu_{M_2M_2})/2}{\sqrt{\dfrac{\sigma^2}{n/2} + \dfrac{\sigma^2}{n} + \dfrac{\sigma^2}{n}}} = \frac{(1-2r)2d}{\sqrt{4\sigma^2/n}} \quad (6.17)$$

Note that $\mu_{M_1M_2}$ does not contribute to the test in Eqn 6.16; adding $\mu_{M_1M_2}$ in Eqn 6.16 does not increase the efficiency of the test unless $|d| \geq a/2$ (but see below for the calculation of sample size required with dominance).

When n is large, the observed difference $\hat{t}$ is approximately normally distributed and the power $1 - \beta$ to detect the difference (for one-tailed test) is

$$1 - \beta = \Pr[\hat{t} > z_\alpha \textit{ with } \hat{t} \sim N(t,1)] = 1 - \Phi(z_\alpha - t)$$

where z_α is the z critical value of the test with $(1 - \alpha)$ confidence under the null hypothesis $t = 0$ and $\Phi(x)$ is the standard normal cumulative distribution function.

For given α and β, the sample size n required for the test is

$$n_1 = 8\left[\frac{z_\alpha + z_\beta}{(1-2r)2a/\sigma}\right]^2 \quad (6.18)$$

for additive effect, and

$$n_2 = 4\left[\frac{z_\alpha + z_\beta}{(1-2r)d/\sigma}\right]^2 \quad (6.19)$$

for dominance effect.

Factors determining the required sample size

1. If the test is two-tailed (the usual case), z_α should be replaced by $z_{\alpha/2}$.
2. For interval mapping the required sample size can be reduced by a factor of $(1 - r^*)$ where r^* is the recombination frequency between an interval of two marker loci. Example: if r^* is about 0.23 for a 30-cM interval, then, $(1 - 2r)^2$ in Eqns 6.18 and 6.19 can be replaced by $(1 - r^*) = 0.77$ to account for the worst case scenario where a QTL is located in the middle of an interval ($r \approx r^*/2$).
3. In the test, if many unlinked markers are used for controlling genetic background, most of genetic variance in the population can be removed from the residual variance (the idea of CIM) and σ_r^2 may be roughly approximated by the environment variance σ_e^2. The overall heritability of the trait matters enormously.
4. For a systematical search for QTL in a genome, the Type I error α for each test should be substantially lower to account for increased false positive probability in an overall search. In most cases, the use of $\alpha^* = 0.001$ (a very conservative level) for each individual test should be sufficient to ensure an overall false positive rate of less than 5%.

The relevant sample size can be calculated as

$$n_1 \cong \frac{8}{0.77}\left[\frac{z_{\alpha^*} + z_\beta}{2a/\sigma_e}\right]^2$$

for additive effect.

Now it remains to determine the likely magnitudes of $2a/\sigma_e$. Suppose that a QTL contributes to a proportion f of the genetic variance σ_g^2 in an F_2 population. Assuming that no other genes are linked to the QTL and ignoring the dominance ($d = 0$),

$$\frac{(2a)^2}{8\sigma_e^2} = f\sigma_g^2/\sigma_e^2$$

σ_g^2/σ_e^2 is an unknown quantity. For example, assuming $h_{F2}^2 = \sigma_g^2/(\sigma_g^2 + \sigma_e^2) = 0.6$ means

$$\frac{\sigma_g^2}{\sigma_e^2} = 1.5 \text{ and } \frac{(2a)^2}{\sigma_e^2} = 12f$$

Given that $\alpha^* = 0.001$ and $\beta = 0.1$ ($z_{0.001} + z_{0.1} = 3.09 + 1.28 = 4.37$), the required sample sizes for detecting leading QTL for f = 0.01, 0.02, 0.05, 0.1, 0.2, 0.3, 0.4 and 0.5 are n = 1653, 826, 330, 165, 82, 55, 41 and 33.

Effects of dominance

Depending on the degree of the dominance effect, the sample size required for detecting a dominance effect may need to be substantially increased. Dominance does not, however, affect the calculation of the power detecting QTL. For example, suppose $d = a$. In this case we may use

$$t_3 = \frac{\mu_{M_1\_} - \mu_{M_2M_2}}{\sqrt{\frac{\sigma^2}{3n/4} + \frac{\sigma^2}{n/4}}} = \frac{(1-2r)2a}{\sqrt{16\sigma^2/(3n)}}$$

But because of dominance

$$\frac{3(2a)^2}{16} = f\sigma_g^2$$

Thus as long as f, the proportion of the genetic variation attributed to the QTL, is fixed, the required sample size for the test is unchanged.

Effect of linkage: multiple linked QTL

There are two issues that need to be considered:

1. Detection of QTL on the chromosome: for two linked QTL, if the model is misidentified (two QTL analysed as one), the power to identify the 'one QTL' is based on the joint effect of QTL (a weighted sum). If the two QTL are in coupling linkage, the joint effect is aggregated and thus power is increased. If the two QTL are in repulsion linkage, the joint effect is reduced. Thus power is decreased and it can be very low. However, if the model can be identified correctly (searching for two QTL or conditional searching), the issue is about separating linked QTL and the power to identify repulsion-linked QTL is not necessarily very low.
2. Separating linked QTL (identifying both QTL): the required sample size is increased by a factor (Zeng, 1993)

$$\frac{\sigma_i^2}{\sigma_{i\cdot j}^2} = \frac{1/4}{r(1-r)}$$

where σ_i^2 is the variance of marker i and $\sigma_{i\cdot j}^2$ is the variance of marker i conditional on marker j.

The values for these factors corresponding to the recombinant frequency, r, between the two QTL are shown at the bottom of the page.

r	0.5	0.4	0.3	0.2	0.15	0.1
$\frac{1}{4r(1-r)}$	1	1.04	1.19	1.56	1.96	2.78

r	0.09	0.08	0.07	0.06	0.05	0.04	0.03	0.02	0.01
$\frac{1}{4r(1-r)}$	3.05	3.40	3.84	4.43	5.26	6.51	8.59	12.76	25.25

QTL detection and power calculation depend on QTL mapping analysis procedure: CIM is more powerful than simple interval mapping; MIM is more powerful than CIM.

The power of the test can be increased by combining information from multiple related traits, multiple crosses and multiple environments. The genetic structure becomes more complex in this case and so does the statistical analysis. But, there are definite advantages in the joint multiple trait analysis for QTL identification (Jiang and Zeng, 1995) and of course for hypothesis testing (pleiotropy) and parameter estimation.

There are many factors that determine how large a sample size is required for a specific QTL mapping experiment. The sample size depends on the heritability of the trait of interest (any knowledge or guess), how large an effect of a QTL (as a minimum) is expected to be detected (for example, detect a QTL that explains 5% variation), what is the likely complexity of the genetic architecture of QTL and how many QTL, distribution of effects, epistasis, etc.

6.11.2 Cross validation and sample size

Cross validation (CV) is a re-sampling technique that samples from a genetic cross (e.g. F_2 intercross) and divides a large-size sample into several subsamples (e.g. $k = 5$). As an example, cross validation that was used for a sample size study (Melchinger *et al.*, 2004) will be discussed in this section.

Melchinger *et al.* (2000) compiled a literature survey based on 45 published QTL studies in crops encompassing 34 complex traits. Sample sizes ranged between 60 and 380 with a median of 150. In most studies only a small number of QTL (median of six) were detected which generally explained a surprisingly large proportion (50% and more) of the genetic variance. Although these findings seem to contradict Fisher's (1918) infinitesimal model upon which quantitative genetics is based, Beavis (1998) conjectured from simulation results that different conclusions might be drawn if larger experimental populations were evaluated. Experimental QTL studies in plant species have been inadequate for drawing inferences about numbers, magnitudes and distribution of QTL for most quantitative traits. Unless large numbers of progeny are evaluated for QTL, MAS will have minimal impact on plant breeding (Gimelfarb and Lande, 1994a) and new breeding strategies based on evaluation of large numbers of progeny will be necessary to realize the potential of MAS.

To test this hypothesis, the largest QTL experiment available in plants conducted by Pioneer Hi-Bred® in maize was analysed in detail by Schön *et al.* (2004). This study consisted of testcrosses of 976 $F_{4:5}$ lines derived from the cross of two elite lines. These materials together with testcrosses of their parents were evaluated in 16 environments. The $F_{4:5}$ lines were assayed with 172 RFLP markers covering the entire genome. With the entire data set (comprising $N = 976$ genotypes and $E = 16$ environments) the number of detected QTL confirmed the infinitesimal model of quantitative genetics (e.g. 30 QTL detected with LOD ≥ 2.5 for plant height, explaining 61% of the genetic variance).

For studying the effects of sample size as well as genotypic and environmental sampling on the outcome of QTL analyses, the entire data set was partitioned into smaller data sets with $N = 488$, 244, 122 and $E = 16$, 4, 2. After randomization of genotypes and environments, the partitioning of the experimental data reference population, $P_{ED}(N, E)$, was repeated to obtain a total of 120 different small data sets for given values of N and E. Within each $P_{ED}(N, E)$, heritabilities were estimated and QTL analyses were performed for each data set with LOD = 2.50 and 3.21. Fivefold CV accounting for genotypic sampling was applied by subdividing each small data set into five genotypic samples. Four genotypic samples were used as the estimation set for localization of QTL and estimation of their effects. The fifth sample was used as a test set to obtain asymptotically unbiased estimates of the proportion of the genotypic variance explained by QTL in each test set. For each small data set five different estimation sets and corresponding test sets are possible. By randomization of the genotypes assigned to the five subsamples 120 estimation sets and test sets were generated and averaged for estimation of parameters. In the estimation set step, some (four) subsamples are used to predict QTL loci λ and effects θ. Heritability h^2 can be also predicted by

$$h^2 = \frac{\sigma_g^2}{\frac{\sigma^2}{re} + \frac{\sigma_{ge}^2}{e} + \sigma_g^2}$$

In the test set step, other (one) subsamples can be tested by using loci from estimation sets to predict effects θ and the proportion of genotypic variance explained.

Reducing N from 976 to 244 or even to 122 decreased the average number of detected QTL (n_{QTL}) by more than half irrespective of E (Fig. 6.5). By comparison, reducing E from 16 to 4 or 2 had a much smaller effect on n_{QTL}. In all instances, a large variation in n_{QTL} was observed among different data sets especially for smaller values of N and E. Although n_{QTL} decreased with smaller data sets, the estimates of genetic variance explained remained almost the same due to a tremendous increase in the bias. This illustrates that QTL effects obtained from smaller sample sizes are usually highly inflated, leading to an overly optimistic assessment of the prospects of MAS. Moreover, inferences about the genetic architecture (number of QTL and their effects) of complex traits cannot be achieved reliably with smaller sample sizes.

6.11.3 Confidence interval of QTL location

In genome scans to detect QTL or targeted scans to attempt to replicate previously reported QTL, estimation of the position of a QTL is usually achieved with low precision, even in controlled crosses from inbred lines, but the accuracy of the position is important, either for subsequent introgression or fine mapping (e.g. Lynch and Walsh, 1998). Darvasi and Soller (1997) presented empirical predictions of the confidence interval of QTL location for dense marker maps in experimental crosses. They showed from simulation results for BC and F_2 populations from inbred lines that the 95% confidence interval was a simple function of sample size and the effect of the QTL:

$$CI95_{BC} = 3000/[n(a + d)^2] \text{ for BC}$$

$$CI95_{F_2} = 1500/(na)^2 \text{ for } F_2$$

where a and d are the additive and dominance effects of the QTL in residual standard deviation units, respectively, and n the sample size. The results from Darvasi and Soller (1997) can be used in combination with power studies before an experiment to assess the expected confidence region of a QTL given its effect and the sample size of the experiment. After an experiment is conducted and a QTL is detected, using either the LOD drop-off method (Fig. 6.2; Lander and Botstein, 1989) or a bootstrap procedure (Visscher *et al.*, 1996; Talbot *et al.*, 1999) to estimate the confidence interval of a QTL.

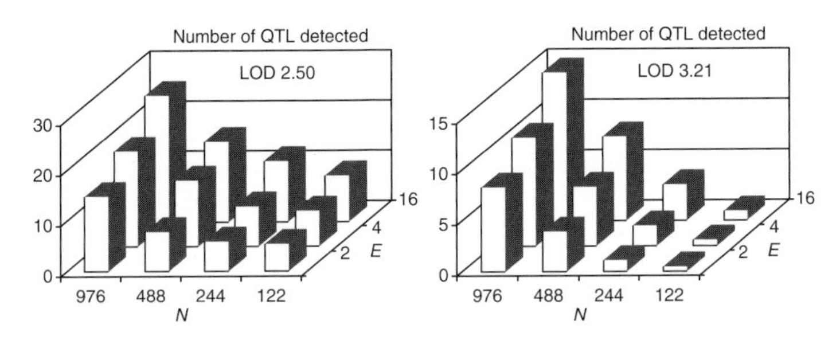

Fig. 6.5. Average number of QTL (n_{QTL}) detected in the estimation set of 120 different data set and partitionings $P_{ED}(N, E)$ using standard cross validation and LOD = 2.50 and 3.21 for plant height. From Melchinger *et al.* (2004) with kind permission of Springer Science and Business Media.

Visscher and Goddard (2004) derived by theory simple equations that can be used to predict any confidence interval and give expressions for the 95% interval. This confidence interval in centimorgans is

$$CI(1-\beta) \approx (200x)X_{1-\beta}/(nd^2)$$

with $x = 2$ for F_2 and $x = 4$ for BC and $X_{1-\beta}$ the threshold of a central chi-squared distribution with 1 degree of freedom corresponding to a cumulative density of $(1 - \beta)$. For example, for BC and F_2 populations, the 95% confidence interval are predicted as

$$CI95_{BC} \approx (200)(4)(3.84)/(nd^2) = 3073/(nd)^2$$

and

$$CI95_{F_2} \approx (200)(2)(3.84)/(nd^2) = 1537/(nd)^2$$

The prediction of the CI as a function of the proportion of the variance explained by the QTL (q) is

$$CI \approx 200X_{1-\beta}(1-q^2)/(nq^2)$$

For example, the 95% CI for a QTL that explains 35% of the variation in either a BC or F_2 population is

$$CI95 \approx (200)(3.84)(1-0.35)/(nq^2) = 499/(nq)^2$$

A general form of the prediction of the CI for dense marker maps that also applies to other population structures is

$$CI(1-\beta) \approx 200X_{1-\beta}/\lambda$$

where λ is the non-centrality parameter of a chi-squared test for the presence of a QTL at the true QTL location.

6.11.4 QTL thresholds

QTL thresholds for interval mapping

Interval mapping scans across loci at location λ in a genome to find the evidence for QTL with large LOD (λ). The relationship between LOD and LR is LOD (λ) = 0.217 × LR. Setting a genome-wide LOD threshold can protect against one or more false positives and roughly adjust the size of each individual test. LOD distribution under the null hypothesis at any particular location λ depends on design (1 degree of freedom for BC, 2 degrees of freedom for F_2) with

$$\frac{LOD(\lambda)}{0.217} = LR \sim \chi_1^2 \text{ or } \chi_2^2$$

Some point-wise P-values for different levels of LR and LOD are provided in Table 6.7.

Genome-wide threshold

Assume a dense marker map with markers everywhere. LR test statistics are correlated and correlation drops off quickly with distance but there is no correlation for unlinked markers.

Based on Lander and Bostein (1989), a genome-wide threshold value, t, can be determined using the Ornstein–Uhlenbeck process as follows

$$\Pr(\max_{\lambda \text{ in genome}} LR(\lambda) > t) = \alpha \approx (C + 2Gt)\alpha_t$$

with $Pr(\chi_1^2 > t) = \alpha_t$, where C = number of chromosomes; G = length of genome in centimorgans; t = genome-wide threshold value; and α_t = corresponding point-wise significance level.

Figure 6.6 shows LOD thresholds for different point-wise and genome-wise P-values in BC and F_2 populations (Broman *et al.*, 2003a).

Permutation and thresholds

It is possible to derive the distribution of any test statistic under an appropriate null

Table 6.7. LR, LOD and point-wise *P*-values.

		P-value	
LR	LOD	1 d.f.[a]	2 d.f.
10	1	0.0319	0.1
31.6	1.5	0.0086	0.0316
100	2	0.0024	0.0024
1,000	3	0.0002	0.001
10,000	4	< 0.0001	0.0001

[a] d.f., degree of freedom.

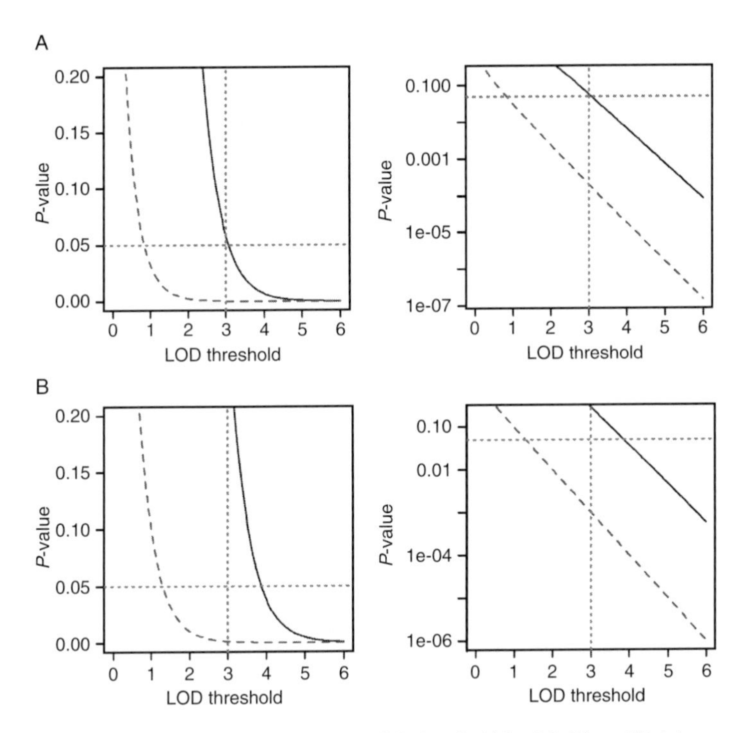

Fig. 6.6. Point-wise and genome-wide *P*-values and LOD threshold for BC (A) and F_2 intercross (B).

hypothesis by 'shuffling' the quantitative trait values among the individuals in the data set. If there is a QTL effect at specific location(s) in the genome, there will be an association between the trait values and the point of analysis on the genetic map. If there is no QTL present in the genome, or it is unlinked to the point of analysis, there is no marker–trait association (i.e. exactly the situation described under the null hypothesis). Permutation tests resolve the problem of finding a significance threshold by simulating a large number of permutations (say 1000) of the observed data set (marker observations are shuffled with respect to traits) so that distribution of the test statistic (LOD score) can be estimated under a null hypothesis as no relationship between traits and markers. This distribution determines how large the LOD values obtained from a particular data set by chance can be.

Shuffling trait values among individuals in the data set represents the situation under the null hypothesis (Churchill and Doerge, 1994), i.e. randomness. Shuffling trait values does not change the summary statistics such as the number of individuals, mean and variance of the individuals. Estimating significance threshold values by permutation includes four steps (Doerge and Churchill, 1996):

1. Hold the genetic map fixed (i.e. keep the marker information from a sampled individual intact. If the individual has m and y, the elements of the vector m should be kept together and the trait values y should be shuffled over these).
2. Shuffle the trait values.
3. Analyse the shuffled data set by applying a t-test, likelihood ratio test or calculation of LOD score.
4. Store the test statistic from each analysis point of step 3 in an analysis matrix.

Repeat steps 2–4 N times.

Threshold values can be created for comparison-wise (per marker), chromosome-wise (chromosome specific) and experiment-wise (experiment specific). In the computer software developed for the

CIM method, the permutation method is incorporated into the model so that empirical thresholds for declaring significant QTL can be calculated.

6.11.5 False discovery rate

Declaring the presence of a QTL always carries some risk that such a declaration is false. The risk can be judged by the false discovery rate (FDR), which represents the probability that a QTL is false, given that a QTL has been declared. A high FDR can result in false leads and wasted resources in characterizing and exploiting genes for quantitative traits, as well as confuse the QTL literature and databases. Knowledge of the magnitude of the FDR would be helpful for designing QTL mapping experiments and for properly interpreting their results.

Take an example as given by Bernardo (2004). Suppose that out of 64 independent markers, 60 are unlinked to QTL in a mapping population. Of these 60 markers, three are incorrectly declared to be linked to a QTL (i.e. false positives, Fig. 6.7) and 57 are correctly declared to be unlinked to QTL (i.e. true negatives, Fig. 6.7). The comparison-wise significance level or Type I error rate, denoted by α_C, is equal to (number of false positives)/(number of false positives + number of true negatives). In the example in Fig. 6.7, α_C is equal to 3/(3 + 57) = 0.05. Studies to map QTL have differed in the significance levels used. Some investigators have used stringent significance levels of $\alpha_C \cong 0.0001$, as suggested by a permutation test to control the experiment-wise error rate (Churchill and Doerge, 1994), whereas other investigators (Openshaw and Frascaroli, 1997) have used a relaxed significance level of $\alpha_C = 0.1$.

Regardless of the significance level used, a misconception is that α_C is equal to the proportion of false positives among all declared marker–QTL linkages. In other words, if 20 QTL have been declared at a significance level of $\alpha_C = 0.05$, a misconception is that only 20 × 0.05 = 1 of the 20 declared QTL is false. The false discovery rate is defined as the probability that a QTL is false, given that a QTL has been declared (Benjamini and Hochberg, 1995; Fernando *et al.*, 2004); it is equal to (number of false positives)/(number of false positives + number of true positives). In the example in Fig. 6.7, the FDR is equal to 3/(3 + 1) = 0.75 rather than 0.05. The FDR can therefore be much greater than α_C (Fernando, 2002).

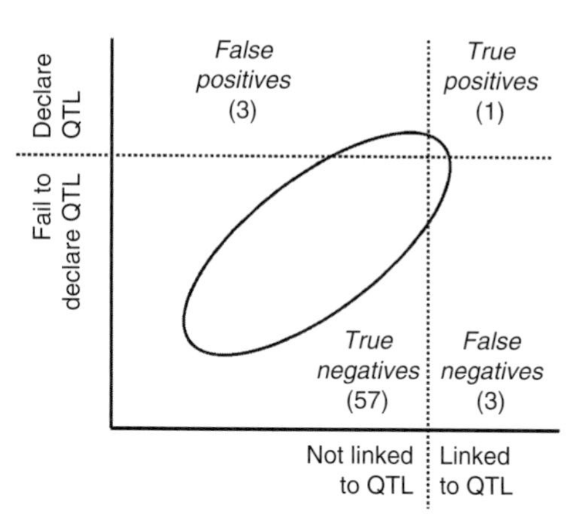

Fig. 6.7. Outcomes of a significance test for detecting QTL. From Bernardo (2004) with kind permission of Springer Science and Business Media.

The simulation study by Bernardo (2004) was undertaken to determine the FDR in an F_2 mapping population, given different numbers of QTL, population sizes and trait heritabilities. Markers linked to QTL were detected by multiple regression of phenotype on marker genotype. Phenotypic selection and marker-based recurrent selection were compared. The FDR increased as α_C increased. Notably, the FDR was often 10–30 times higher than the α_C level used. Regardless of the number of QTL, heritability or size of the genome, the FDR was ≤ 0.01 when α_C was 0.0001. The FDR increased to 0.82 when α_C was 0.05, heritability was low and only one QTL controlled the trait. An α_C of 0.05 led to a low FDR when many QTL (30 or 100) controlled the trait, but this lower FDR was accompanied by a diminished power to detect QTL. Larger mapping populations led to both a lower FDR and increased power. Relaxed significance levels of $\alpha_C = 0.1$ or

0.2 led to the largest responses to marker-based recurrent selection, despite the high FDR. To prevent false QTL from confusing the literature and databases, a detected QTL should, in general, be reported as a QTL only if it was identified at a stringent significance level, e.g. $\alpha_C \cong 0.0001$. In conclusion, the question of 'what proportion of declared QTL in plants are false?' cannot be answered definitely because QTL studies have used different significance levels, traits differ in the number of underlying QTL and experiments have used different types of mapping populations (e.g. BC instead of F_2 populations).

As a potential alternative to the FDR, Chen and Storey (2006) proposed a generalized version of genome-wide error rate (GWER). Rather than guarding against any single false positive linkage from occurring, the generalized GWER allows the researcher to guard against exceeding more than k false positive linkage, where k is chosen by the user. For example, if we set $k = 1$, then the goal for $GWER_k$ is to prevent more than one false positive linkage from occurring. The user can apply this significance criterion at the appropriate value of k or at several values of k. $GWER_k$ allows the user to provide a more liberal balance between true positives and false positives at no additional cost in computation or assumptions.

6.12 Summary and Prospects

QTL mapping has evolved from single marker-based, two flanking marker-based, to multiple marker-based approaches and finally to all-marker-based whole genome approaches. It started by using simple and well-characterized F_2 or RIL populations derived from biparental lines and now extends to any population including those derived from multiple parents or randomly selected materials. As an increasing number of populations, maps and QTL information have been accumulating worldwide, it has become increasingly important to use all available data for meta-analysis, pooled analysis and *in silico* mapping.

QTL mapping has also been moving from single trait-based analysis to integrated analysis of multiple traits or even thousands of expression 'traits' simultaneously. Methods for some specific traits including triploidy endosperms and dynamic traits across different developmental stages and methods for complicated genetic effects including epistasis and genotype-by-environment interaction (Chapter 10) have also been developed.

Statistical methods have been developed for almost all types of complicated situation that would be encountered in the genetic dissection of complex traits. However, most methods still remain at the stage of theory, published by statisticians who keep moving forward to new publication opportunities or optimizing their methodologies with endless efforts, leaving few feasible options for geneticists. It should be noted that the best method means nothing but statisticians' games unless it can be translated into user-friendly software.

Practically, there might be no general method that can meet all requirements. This is because, on the one hand, the simplest methods (such as phenotypic comparison of alternative marker genotypes) are the best for simple traits; and on the other hand, modelling or simulation, no matter how many parameters can be incorporated into the model, might be too complicated for complex traits that interact with various environments. Ultimately, dissection of complex traits will rely on continuous research effort in applied QTL mapping and integrated utilization of various information and materials that has been accumulating, including genetic and breeding materials (populations, structured materials), molecular markers (sequences and genes) and various phenotypic data collected across environments (years, seasons and locations).

Commercial breeding programmes grow thousands of progenies per year derived from multiple, related and unrelated crosses and evaluate them for many agronomically important traits in diverse environments. With the use of high-throughput facilities (DNA sequencers, DNA microarrays, protein chips, etc.), these materials can be assayed in parallel with the use of genomic tools. Thus, the current limitations of classical QTL mapping studies may soon be overcome by pedigree-based and/or haplotype-based QTL mapping approaches (Jannink *et al.*, 2001; Jansen *et al.*, 2003). The main ideas include: (i) the exploitation of pedigree and phenotypic data, routinely collected in applied plant breeding programmes, for QTL mapping; and (ii) the sampling of full QTL variation present in a wide range of germplasm, which allows the breeders to search for the best alleles ('allele mining') present in elite materials as well as in genetic resources. On the other hand, molecular dissection of complex traits will depend on the utilization of both linkage- and LD-based methods (Manenti *et al.*, 2009; Myles *et al.*, 2009).

7

Molecular Dissection of Complex Traits: Practice

In recent years, quantitative trait loci (QTL) research has been attracting many scientists resulting in numerous publications every year. From the viewpoint of practice, however, one should consider the whole picture of QTL from single QTL to multiple QTL, single traits to trait complexes, homogeneous to heterogeneous genetic backgrounds and static mapping to dynamic mapping. Advances in molecular dissection of complex traits could answer the following questions: How many genes are involved in genetic control of each quantitative trait in a segregating population? Can we separate closely linked QTL into single units? How can we compare QTL across different genetic backgrounds and developmental stages? How can we handle multiple traits and expression QTL? Issues related to these questions will be addressed in this chapter. Some theoretical considerations will be also discussed as complementary to Chapter 6. For general resources, readers may refer to Xu (1997), Liu (1998), Lynch and Walsh (1998), Paterson (1998), Flint and Mott (2001), Xu, Y. (2002), Collard *et al.* (2005), Gibson and Weir (2005) and Wu and Lin (2006).

QTL mapping practice will usually include creating, genotyping and phenotyping mapping populations, generating genetic linkage maps and establishing marker–trait association. QTL mapping requires three major data sets, which are for molecular marker, phenotype and linkage map, plus genetic mapping software that provide an appropriate mapping procedure with user-friendly output. The linkage map can be specific to the mapping population or is inferred from physical positions available for the molecular markers.

7.1 QTL Separating

Most quantitative traits can be genetically associated with molecular markers that are located in different chromosomal regions. These regions represent either separate single QTL or multiple closely linked QTL. The number and distribution of multiple QTL on chromosomes determines their manipulability in genetics and breeding. Generally, multiple QTL affecting a specific trait have four possible distributions on chromosomes (Xu, 1997; Fig. 7.1): (i) independent QTL – genes are independently distributed on each chromosome; (ii) loosely linked QTL – genes are located on the same chromosome but separated by large distances so that they recombine with high frequency and can be easily separated; (iii) clustered QTL – genes are closely linked or clustered in a specific chromosomal region so that

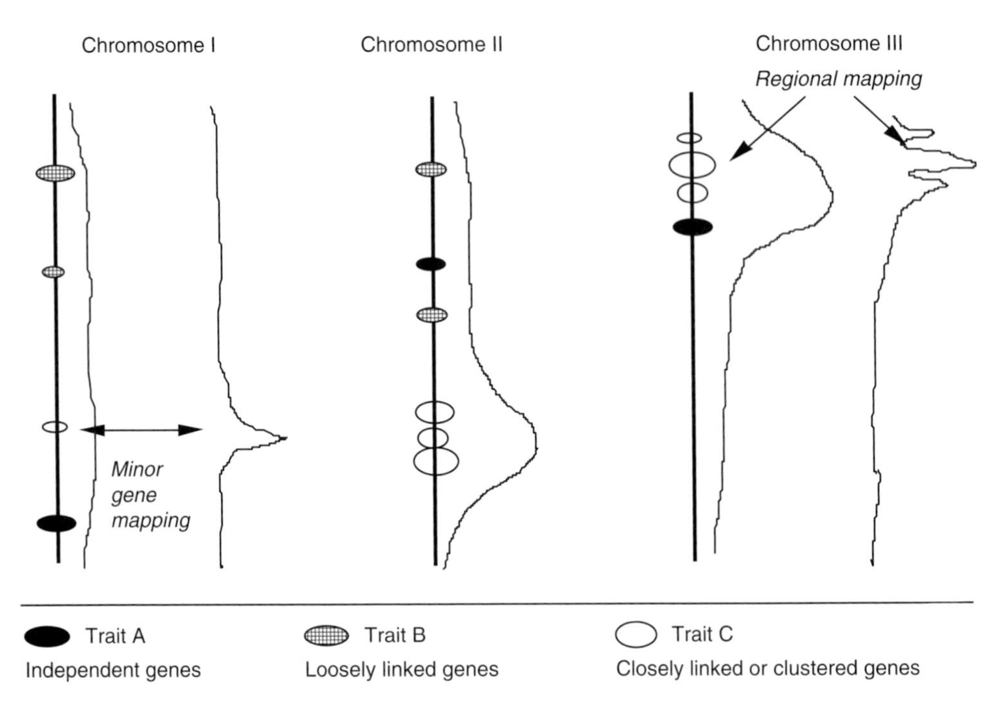

Fig. 7.1. Models for QTL distribution. Three traits (A, B and C) are used as examples for independent QTL (Trait A), loosely linked QTL (Trait B), closely linked or clustered QTL (Trait C, Chromosome II and III), and mixed model (Trait C). Detectable QTL are indicated by circles, and their effects are represented by the circle sizes. Likelihood maps for Trait C are given at the right side of each linkage map, and for Chromosomes I and III, two likelihood maps are given to show expected results from minor QTL mapping and regional mapping. From Xu (1997). This material is reproduced with permission of John Wiley & Sons, Inc.

they behave as one gene with major effect; and (iv) mixed distribution – for a specific trait, QTL have a combined distribution of the three models above.

Because of continuous variation in quantitative traits, QTL genotypes cannot be easily determined by inspecting the distribution of trait phenotype alone. This is one of the fundamental problems of quantitative genetics. Historically important genetic parameters, e.g. genetic variances and heritability (Chapter 1), summarize the effects caused by all QTL but do not provide information to distinguish the effects of individual QTL. In order to understand the genetic structure of QTL and ultimately clone them, multiple QTL affecting the same trait must be mapped on to chromosomes and their effects must be well separated. Theoretically, multiple QTL can be separated into single manipulable factors by different methods including mapping and selection approaches.

7.1.1 Mapping approaches

In theory, molecular-marker-based QTL mapping can be used to draw inferences about QTL allelic differences. Normally, however, it is difficult to determine whether the effect detected with a particular molecular marker is due to one QTL with large effect or linked QTL each with relatively small effect. For this reason, the term QTL usually describes a region of a chromosome defined by linkage to a marker gene (Tanksley, 1993). Using a mapping procedure, one can partition multiple QTL into single manipulable units and determine whether a QTL is comprised of one or more genes (Fig. 7.1). This strategy depends on both the resolution of molecular maps and the mapping power for the QTL with small effect and requires improvement of the statistical power in QTL analysis, saturation of molecular maps and optimization of population structures.

Fine mapping

It is generally acknowledged that a typical higher plant genome includes 10,000–100,000 genes, scattered through a total of 10^8–10^{10} bp of DNA. Consequently, 0.1% of the genome would include an average of 10–100 genes. Several genes lying close together, each with a small effect on a trait, could appear to be a single QTL of large effect (Michelmore and Shaw, 1988; Paterson *et al.*, 1988). Reducing the size of the regions identified as containing QTL through fine mapping has been envisioned as an initial step in identifying single QTL that ultimately could be manipulated using transformation (recombinant DNA) technology (Stuber, 1994a; Tanksley *et al.*, 1995). Current strategies for mapping QTL depend on comparing the means of recombinant and non-recombinant classes. Given the practical size of segregating populations (n = 200–300) and the marker density of molecular maps most frequently used for QTL studies, preliminary mapping resolution of QTL has been limited to approximately 10–20 cM – inadequate for distinguishing between single gene and multi-gene composition. To reveal just what lies at a 'locus', techniques with much higher resolution are necessary.

Conventional mapping populations such as backcross (BC) and F_2 have limitations in fine mapping of QTL due to the lack of sufficient recombinational events even in large populations. Therefore, an alternative approach for fine mapping is to exploit populations whose derivations are based on multiple cycles of recombination. These populations include recombinant inbred lines (RILs) (Burr and Burr, 1991) and advanced intercrossing lines (AILs) (Darvasi and Soller, 1995) or intermated recombinant inbred lines (IRILs) (Liu *et al.*, 1996). As described in Chapter 5, RILs are produced by continually selfing or sibmating the progeny of individual members of an F_2 population until virtually homozygosity is achieved. Considering the fact that more and more RILs have been accumulated in breeding programmes and such populations can be exploited for the mapping of one or more traits, the RIL approach is practical for most crops. AILs are initiated by a cross between two inbred lines and derived by sequentially and randomly intercrossing each generation until advanced generations are attained. For these two kinds of populations, many recombinational events required for fine mapping of QTL are accumulated in a single relatively small population over the course of many generations. Due to more opportunity for meiotic recombination, they have the advantage of possibly distinguishing more closely linked QTL. For example, with the same population size and QTL effect, the 95% confidence interval of a QTL map location of 20 cM in the F_2 is reduced fivefold after eight additional random mating generations (F_{10}; Darvasi and Soller, 1995). RILs have similar effects on the resolution power of QTL mapping. It is worth noting that increases in recombinational events will reduce the effect due to the QTL associated with any particular marker and thus, these populations are more suitable for fine mapping of QTL of moderate and large effects. In another approach, Paterson *et al.* (1990) suggested that recombinant individuals could be identified in primary generations and selectively multiplied in subsequent generations so that the recombinant classes occur at near equal frequency with the non-recombinant classes, increasing the power for statistical comparisons among the classes.

The benefits of using designs such as RILs other than the conventional F_2 and BC, where genotyping and phenotyping could be done on the same set of individuals, include reducing cost and environmental variance and taking advantage of the changes in population structures of other RIL populations. Kao (2006) proposed a statistical method considering the differences in population structures between different RIL populations on the basis of a multiple-QTL model to map for QTL in different designs.

The proposed method has the potential to improve the resolution of genetic architecture of quantitative traits and can serve as an effective tool to explore the QTL mapping study in the system of RIL populations. Martin and Hospital (2006) described the non-independence of multiple recombinations arising in RIL recombination data even though there may be no interference in each meiosis. They also provide formulas for interference tests, gene mapping and QTL detection in RIL populations.

A new genetic map of maize, ISU–IBM Map4, that integrates 2029 existing markers with 1329 new indel polymorphism (IDP) markers has been developed using IRILs from the intermated B73 × Mo17 (IBM) population (Fu *et al.*, 2006). The mosaic structures of the genomes of 91 IRILs, an important resource for identifying and mapping QTL and expression QTL (eQTL), were defined. When this RIL population was evaluated in four environments for resistance to southern leaf blight (SLB) disease caused by *Cochliobolus heterostrophus* race O (Balint-Kurti *et al.*, 2007), four common SLB resistance QTL were identified in all environments, two in bin 3.04 and one each in bins 1.10 and 8.02/3. A comparison was made between SLB QTL detected in two populations, independently derived from the same parental cross: the IBM advanced intercross population and a conventional RIL population. Several QTL for SLB resistance were detected in both populations, with the IBM providing between 5 and 50 times greater mapping resolution.

Population size and mating design are two important aspects to be given adequate consideration during the development of IRILs. Although random intermating of F_2 populations has been suggested for obtaining precise estimates of recombination frequencies between tightly linked loci, Frisch and Melchinger (2008) in a recent simulation study showed that sampling effects due to small population sizes in the intermating generations have abolished the advantages of random intermating that were reported in previous theoretical studies considering an infinite population size. They also propose a mating scheme for intermating with planned crosses that yields more precise estimates than those under random intermating.

Minor QTL mapping

In most QTL identification studies, rather stringent threshold probability levels have been set so that there is a low risk in making Type I errors (i.e. false positives). Thus, only those QTL with sufficiently large phenotypic effects to be detected statistically can be identified while QTL with smaller effects will fall below the threshold of detection (cf. Fig. 7.1). When multiple QTL are located in the same chromosomal region, the ones with smaller effects cannot be detected in most instances. This 'overshadow' effect of major QTL over minor QTL makes the molecular marker approach biased towards the detection of QTL of large phenotypic effects. It should be pointed out that these major QTL would be ones with high heritabilities, easily manipulated through traditional breeding practices and may already be fixed in many breeding lines. There are accumulated data from numerous QTL studies to establish definitely that QTL affecting a number of quantitative traits are distributed throughout the genome and certain chromosomal regions appear to contribute greater effects than others. More surprising has been the finding that in many instances a large proportion of quantitative variation can be explained by the segregation of a few major QTL. It is not uncommon to find individual QTL that can account for more than 20% of the phenotypic variation in a population (Table 1 of Tanksley, 1993) and values as high as 85.7% (Lin *et al.*, 1995) (a major gene with distinct bimodal distribution) have been reported for a single major QTL. It may be reasonable, therefore, to use marker technology as a means for placing greater emphasis on those QTL showing only relatively minor effects (minor QTL).

Detectability of a trait locus is severely limited by its genetic background. A straightforward background effect is 'dilution', i.e. the more QTL alleles that exist, the smaller the relative contribution of a given locus (Frankel, 1995). The smallest effects a QTL can have and still be detected by the marker method depend on a number of factors (Tanksley, 1993; Chapter 6), which include:

1. Map distance: the closer a QTL is to a marker, the smaller the QTL effect and still be detected statistically. This relationship indicates that the power of QTL mapping can be improved with the saturation of molecular maps. Now, many high-density linkage maps providing high quality reference maps for QTL mapping have been available for many crop plants.
2. Sample size: the larger the sample (population) size, the more likely the effects of smaller QTL will reach statistical significance. This relationship indicates that the detection of QTL with a relatively small effect largely depends on the size of the mapping population. Using a typical sample size ($n < 500$), two or more genes closely linked (within 20 cM) will usually be detected as a single QTL (i.e. they cannot be distinguished as separate QTL when mapped with the interval approach of two flanking markers). In maize using an F_2 population size of 1700 individuals and probability threshold of 0.05, a QTL contributing as little as 0.3% of phenotypic variance was reported (Edwards *et al.*, 1987). In experiments with smaller sample sizes and higher probability thresholds, QTL that explain less than 3% of the phenotypic variance are not normally detected. The bias towards detecting QTL with larger effects means that it is unlikely that one will ever detect, map and characterize all of the QTL affecting a character in any single segregating population.
3. Heritability: the larger the environmental effect on the character (i.e. low heritability), the less likely a QTL will be detected. Estimates of heritability can be improved by controlling environmental error. Permanent mapping populations such as RILs, double haploids (DHs) and advanced backcrossing populations or near-isogenic lines (NILs) can be used to improve the mapping power by replicate phenotyping in different environments (years, seasons or locations).
4. QTL threshold: higher probability thresholds for declaring a QTL effect significantly reduce the chances of spurious QTL being reported, but also reduce the chances of detecting QTL with smaller effects. This relationship indicates that development of QTL mapping methods which can improve the mapping power with a specific sample size would benefit the separation of minor QTL. Based on the concept of permutation tests, Churchill and Doerge (1994) described a method to determine an appropriate threshold value for declaring significant QTL effects, providing an alternative approach to the likelihood of odds (LOD) drop-off method (Lander and Botstein, 1989). The conditional empirical threshold and the residual empirical threshold yield critical values that can be used to construct tests for the presence of minor QTL effects while accounting for effects of known major QTL (Doerge and Churchill, 1996). Now the permutation method has been widely used in various QTL mapping approaches and some QTL mapping software such as QTL CARTOGRAPHER (http://statgen.ncsu.edu/qtlcart/WQTLCart.htm) provide the permutation function for the statistical methods incorporated.

In rice, a total of 15 QTL for heading date (*Hd1–Hd3, Hd3b–Hd14*) were identified in several populations derived from crosses between Nipponbare, a rice cultivar from Japan, and Kasalath, a rice cultivar from India (as reviewed by Yano *et al.*, 2001). Nine of these have been mapped as single Mendelian factors and studies have shown that *Hd1*, *Hd2*, *Hd3a*, *Hd3b*, *Hd5* and *Hd6* are involved in day-length response (reviewed by Uga *et al.*, 2007). Using an extremely late heading (202 days to heading) cultivar Nona Bokra from India and *japonica* cultivar Koshihikari (105 days) from Japan, QTL analysis identified 12 QTL on seven chromosomes. The Nona Bokra alleles of all QTL contributed to an increase in heading date. Comparison of chromosomal locations between heading date QTL detected between these two cultivars and 15 QTL identified from Nipponbare × Kasalath populations revealed that eight of the heading date QTL were nearby the *Hd1*, *Hd2*, *Hd3a*, *Hd4*, *Hd5*, *Hd6*, *Hd9* and *Hd13*. The results suggested that the strong photoperiod sensitivity in Nona Bokra was generated mainly by the accumulation of additive effects of particular alleles at previously identified QTL (Uga *et al.*, 2007). This also indicates that multiple QTL for complex traits like extremely late-heading can be dissected by QTL mapping.

Regional mapping

The most common method of placing molecular markers on a linkage map is by random cloning of genomic/cDNA sequences, by PCR-based detection of polymorphism, or by single nucleotide polymorphism (SNP) markers based on chip technology (Chapter 3), followed by linkage analysis. This whole genome mapping method is extremely useful and can be used to construct both low and high resolution maps of complex genomes. Nevertheless, this approach is limited when one is interested in targeting a particular chromosomal region. A majority of random markers will ultimately be mapped outside of a target interval and as the interval size decreases the odds of any new randomly generated marker being placed within it decreases (Tanksley *et al.*, 1995).

Two strategies have been proposed to target the chromosomal region of interest (regional mapping; Xu, 1997) and have proven effective for identifying from a large number of markers the few that reside near a targeted locus. Both involve the use of genetic stocks that are (almost) genetically identical, except in the regions flanking the targeted gene. The first strategy uses NILs, which are generated by introgression (Wehrhahn and Allard, 1965). As discussed in Chapter 4, the inbred lines differ at the targeted locus or region. If the donor parent and the recurrent parent are sufficiently divergent, it is possible to detect polymorphisms between the pairs of NILs. The marker that detects such polymorphisms will likely be linked to the target gene. As early examples, Young *et al.* (1988) used NILs and pools of restriction fragment length polymorphism (RFLP) probes to detect new markers within the *Tm-2a* region of tomato. Using a similar strategy, Martin *et al.* (1991) were able to use random PCR amplification on NILs to isolate new markers near the tomato *Pto* disease resistance locus. Now NILs have been widely used in map-based gene cloning through fine mapping.

With the accumulation of permanent mapping populations such as RILs and DHs, it is possible to select lines that are almost genotypically identical in the whole genome except for only one or a few marker loci. Combined with phenotypic similarity, this information can be used to obtain NILs for qualitative or quantitative traits (Xu, 1997).

The second strategy, referred to as DNA pooling or the bulked segregant analysis (BSA), which is discussed later in this chapter, relies on the use of segregating populations (Michelmore *et al.*, 1991; Giovannoni *et al.*, 1991), which does not require highly specialized genetic stocks. This strategy is derived from the concept of selective genotyping based on selection for contrasting phenotypes or bracket DNA markers.

Both regional mapping methods described above, when used in conjunction with high-volume DNA marker technology, permit one to screen thousands of loci and selectively identify those adjacent to the gene of interest in a specific chromosomal region and are very suitable for analysis of clustered QTL. Moreover, these regional mapping approaches can be accomplished without having a genetic map for the species. For major genes, the efficiency of the NIL and BSA approaches have been demonstrated in plants and early examples of success include Young *et al.* (1988), Michelmore *et al.* (1991), Giovannoni *et al.* (1991), Schüller *et al.* (1992), Mackill *et al.* (1993) and Pineda *et al.* (1993).

To solve the issue associated with the masking effects of major QTL and epistatic interactions of multiple QTL involved in RIL-based QTL mapping, Keurentjes *et al.* (2007a) empirically compared the QTL mapping power of a genome-wide NIL population with an already existing RIL population derived from the same parents in *Arabidopsis thaliana*. By analysing and mapping QTL affecting six developmental traits with different heritability, overall, QTL with smaller effects could be detected in the NIL population more easily than in the RIL population, although the localization resolution was lower. In general, population size is more important than the number of replicates to increase the mapping power of RILs, whereas for NILs several replicates are absolutely required.

In an effort to identify putative candidate genes underlying drought tolerance in rice, Nguyen *et al.* (2004) developed several expression sequence-based markers using BSA for saturation mapping of QTL regions. Thirteen of the markers were localized in the close vicinity of the targeted QTL regions. In rice, substitution mapping of a flowering-time QTL associated with transgressive variation has separated a previously located QTL, *dth1.1*, into at least two sub-QTL (Thomson *et al.*, 2006). The QTL *dth1.1* was associated with transgressive variation for days to heading in an advanced BC population derived from the *Oryza sativa* cultivar Jefferson and an accession of the wild rice relative *Oryza rufipogon*. A series of NILs containing different *O. rufipogon* introgressions across the target region were constructed to dissect *dth1.1* using substitution mapping. In contrast to the late-flowering *O. rufipogon* parent, *O. rufipogon* alleles in the substitution lines caused early flowering under both short and long day-lengths and provided evidence for at least two distinct sub-QTL: *dth1.1a* and *dth1.1b*. Potential candidate genes underlying these sub-QTL included genes with sequence similarity to *Arabidopsis GI, FT, SOC1* and *EMF1* and *Pharbitis nil PNZIP*. Evidence from families with non-target *O. rufipogon* introgressions in combination with *dth1.1* alleles also detected an early flowering QTL on chromosome 4 and a late-flowering QTL on chromosome 6 and provided evidence for additional sub-QTL in the *dth1.1* region.

7.1.2 Screening for allele dispersion

When multiple QTL control a trait, their alleles of positive or negative effect (increasing or decreasing trait value) tend to be dispersed among genetic stocks, with positive alleles at one or some loci but negative alleles at others. The phenomenon where QTL alleles of similar effect are dispersed among genetic stocks is referred to as allele dispersion. However, a genetic stock may contain (associate) all the alleles of similar (positive or negative) effect at the multiple QTL; this is referred to as allele association. For traits naturally selected towards intermediate phenotype, alleles of similar effect at multiple loci are more likely dispersed than associated. In natural or breeding populations with neutral selection, there are many genetic stocks that have positive alleles at some loci but negative alleles at others. The extreme phenotypes of quantitative traits come from the association of QTL alleles while the intermediate phenotype usually indicates allele dispersion. Therefore, different QTL alleles with similar effect can be identified from the existing populations. On the other hand, if one has allele-associated stocks in hand, the QTL alleles could be separated by selection for different genotypes. Allele dispersion differs from linkage equilibrium in two respects. First, allele dispersion refers to independent or linked loci controlling the same trait; while in linkage equilibrium the related genetic loci are usually supposed to be genetically linked and control different traits. Secondly, allele dispersion represents a situation in which any two non-genetically related genotypes (strains) within a species show allelic differences at the same genetic locus, while linkage equilibrium represents a situation in which a constant gene frequency has been reached in a given population derived from two related strains.

Genetic stocks with dispersed QTL alleles usually show similar phenotype, making it difficult to identify genetic differences only by phenotypic evaluation. When these stocks are used as the parents to produce segregating populations, however, a part of the progeny will have transgressive phenotypes, i.e. they are phenotypically outside of the range of the parents, because these progeny associate all alleles of similar effect with the result of recombination of different QTL alleles. Positive and negative transgressive individuals will arise from the associations of positive and negative alleles, respectively. Transgression caused by dominance and/or overdominance can also be excluded by successively selfing the transgressive individuals to determine if they maintain the same phenotype in advanced generations. If no significant epistasis can be detected by biometrical genetic analysis, the transgressive segregation in populations derived from two genetic stocks provides evidence for allele dispersion. Based on the additive-dominance model, Xu and Shen (1992a) suggested three methods to screen the separable QTL alleles by detection of allele (gene) dispersion, including: (i) testing the homogeneity between F_2 phenotypic variance and the environmental variance estimated from phenotypic variances of non-segregating populations (P_1, P_2 and F_1); (ii) testing the differences of means among F_1, F_2, $F_1 \times P_1$ and $F_1 \times P_2$; and (iii) comparing the genetic parameters such as gene effects and genetic variances estimated from the cross derived by intermating transgressive individuals of two kinds with those estimated from the cross of the original stocks.

Classical genetic analysis provides some examples for allele dispersion. The first example in plants may come from *Nicotiana rustica*. The allelic differences for final height, flowering time and related characters were largely dispersed between two cultivars (genotypes) 1 and 5 (Jinks and Perkins, 1969, 1972; Perkins and Jinks, 1973) with 127 and 103 cm of final height and 77 and 72 days of flowering time (days after sowing), respectively. Among the random sample of 82 inbred lines derived from the cross between these two cultivars, transgressive lines were found and two of them, B2 and B35, were the shortest and tallest in final height (92 and 144 cm, respectively) and the earliest and latest to flower (70 and 84 days, respectively). The simultaneous analysis of the two contrasting crosses (1 × 5 and B2 × B35) indicated the allele dispersion in the original cultivars (Jayasekara and Jinks, 1976). Another example is rice tiller angle (the angle between the main stem and its tillers). Transgressive segregation was found in the two crosses derived from four *indica* rice cultivars with similar tiller angle and the extreme strains with largest and smallest tiller angles were obtained by successively selfing the transgressive individuals (Xu and Shen, 1992b). Comparative genetic analysis of two contrasting crosses (from the original cultivars and from the corresponding extreme strains) revealed that two loci were responsible for the genetic difference of tiller angle in each pair of the original cultivars, and alleles of similar effect were dispersed in the original cultivars but associated in the extreme strains (largest tiller angle strains having all the positive alleles and smallest tiller angle strains having all the negative alleles). The second cycle of crossing between the extreme strains derived from different original crosses revealed further transgression. Biometrical genetic analysis and selection response indicate that four loci controlled the total variation of tiller angle in four original cultivars, each cultivar carrying two positive alleles at only one locus (Xu *et al.*, 1998).

Allele-dispersion can also be identified based on QTL mapping results. In QTL mapping, phenotypic difference between parents is not necessary for detection of QTL. In most cases where no parental difference is found, QTL are still detected, which could be due to the complementary patterns of positive and negative allelic effects. QTL mapping can provide information about the genetic constitution of each segregate in the mapping population so that one can infer which individual carries desirable alleles and then separate the multiple QTL by selection for individuals with different allele combinations. For example, if four QTL are inferred to control a trait, the allelic constitution can be determined for all individuals and each QTL. Therefore, one can easily screen the individuals carrying the positive allele at each of the four QTL. Because of allele dispersion, it is not likely that one QTL mapping experiment using any single population can detect all the QTL affecting a given trait. Therefore, independent experiments tend to reveal different QTL or QTL alleles. Comparison of QTL effects and mapping positions may result in separation of the multiple QTL. However, this approach largely depends on the precision of QTL mapping results available. So far, numerous QTL affecting the same traits have been identified in most crops and they are different in locations and effects. Variation among investigations and among populations can logically be expected for the following reasons (Smith and Beavis, 1996): (i) different polymorphisms in the populations studied; (ii) different number and location of polymorphic regions affecting the trait; (iii) environmental effects or genotype–environment interaction; and (iv) small sample size. With the development of highly polymorphic DNA markers such as simple sequence repeats (SSRs), the first reason will become less important. Permanent populations can be phenotyped at different seasons, years or locations, reducing the environmental effect on QTL identification. Using relatively large sample size, combined with highly polymorphic markers, permanent population and replicate phenotyping, will help to determine whether the variation of QTL mapping comes from different QTL constitutions of populations or not. It is of interest to note that 'cryptic' factors were frequently uncovered (e.g. Stuber, 1995; Ragot *et al.*, 1995), indicating the possibility of QTL dispersion. For example, genetic factors contributing to high grain yield and tall stature in maize occasionally have been associated with marker alleles from low-yielding, short-statured parental lines (Edwards *et al.*, 1992). A wild rice species with low yield potential contains genes that may significantly increase the productivity of the high-yielding cultivated rice (Xiao *et al.*, 1996a). There are numerous recent examples available to support these early reports.

As observed in rice QTL mapping, on the average, about four QTL are identified for each trait (Table 7.1; Xu, Y., 2002), the same as the average obtained for 176 trial–trait combinations as reviewed by Kearsey and Farquhar (1998). When QTL identified for the same trait are summarized over different projects/populations, this number becomes much larger. For example, rice plant height has been mapped using 13 populations, with 63 QTL reported. Some of the QTL are allelic to each other, i.e. they were mapped to the same chromosomal region or intervals of less than 15 cM. After elimination of possible allelic QTL, the total number of QTL for plant height is reduced to 29, with up to five QTL existing on a chromosome (Xu, Y., 2002). The QTL *qPH1-1*, which corresponded to a major semi-dwarf gene *sd-1* and *qPH8-1* were each detected in six populations. QTL *qPH2-2* and *qPH3-3* were each detected in five populations. Over 50 major

Table 7.1. The number of QTL identified in rice using permanent mapping populations. See Xu, Y. (2002) for all references.

Population	Population size	Number of markers	Number of traits	Number of QTL
IR64/Azucena DH	105–135	146–175	56	215
Zhaiyeqing 8/Jingxi 17 DH	132	137–243	35	115
9024/LH422 RIL	194	141	25	74
CO39/Moroberekan RIL	143–281	127	14	121
Lemont/Teqing RIL	255–315	113–217	8	46
IR58821/IR52561 RIL	166	399	5	28
IR74/Jalmagna RIL	165	144	5	18
Nipponbare/Kasalath BIL	98	245	4	19
Zhenshan 97/Minghui 63 RIL	238	171	3	6
Asominori/IR24 RIL	65	289	2	17
Acc8558/H359 RIL	131	225	1	11
IR1552/Azucena RIL	150	207	1	4
IR74/FR13A RIL	74	202	1	4
IR20/IR55178-3B-9-3 RIL	84	217	1	4
Overall	65–315	113–399	161	682

genes for dwarf and semi-dwarf mutants have been found (Kinoshita, 1995) and 14 of them were linked to molecular markers (Huang *et al.*, 1996; Kamijima *et al.*, 1996), with 13 of them (93%) co-localized with plant height QTL. More plant-height QTL will likely be co-localized with major loci, as more major loci are linked with molecular markers. These co-localizations support Robertson's (1985) hypothesis that alleles for qualitative mutants are simply 'lost-function' alleles at the same loci underlying quantitative variation. Until QTL are mapped to higher degrees of precision and/or cloned, however, it would be difficult to prove that the particular QTL actually correspond to known loci defined by macromutant alleles and which QTL are allelic to each other. The QTL allelism test and the determination of the major-gene and QTL correspondence depend on the availability of high-density molecular maps with a common set of markers shared among researchers.

With the generalization of this concept, non-allelic alleles can be searched for among an entire set of related species. The high incidence of transgressive segregation in interspecific crosses tells us that individuals that do not exhibit a particular trait often carry superior/hidden alleles that condition that trait. It is likely that non-allelic alleles usually would be present in other strains but missed due to limitations in genetic analysis. By using the entire set of the related species, it thus may be possible to identify all of the genes involved in a given trait or physiological process because the genes phenotypically hidden in one species may not be hidden in another species (Bennetzen, 1996).

7.2 QTL for Complicated Traits

7.2.1 Trait components

Many quantitative traits are a complex consisting of different or related components or subtraits. For example, there are six direct component effects upon days to flowering in legume seed crops. Four are genetic factors, which regulate three plant processes and one plant characteristic and two are environmental factors, which modulate the ongoing gene actions of one or more of the three plant processes. These six direct components are the rate of node and leaf development, change from node to flower, vernalization requirement, node number at the minimal days to flowering, impacting photoperiod and impacting temperature (Wallace, 1985). In most crops, the yield trait is comprised of several yield components and oil or protein content is related to many compounds and amino acids. In rice, low fertility controlled by polygenes can be partitioned into several components including male and female sterility, or ovary and pollen abortion so that polygenes can be divided into several single genes with different effects and thus can be handled with ease. Dissecting or partitioning a complex trait into separate components can benefit both QTL mapping and cloning.

7.2.2 Correlated traits

Trait correlation may arise from either pleiotropic effects of single genes or from tight linkage of genes affecting different traits. Correlated traits often share some QTL mapped to similar chromosomal regions. In most *Poaceae*, increased plant height often correlates with late flowering. Comparative data support the possibility that different, closely linked genes, rather than a single gene, account for correlated traits. In sorghum, two of the three QTL affecting flowering time were associated with height QTL and two major QTL were mapped within overlapping 90% confidence intervals, explaining 85.7% and 54.8% of phenotypic variation, respectively, and showing similar gene action (dominance/additive = 0.72 and 0.73) (Lin *et al.*, 1995). Many pairs of independent discrete mutations affecting height and flowering are closely linked in corresponding locations of wheat (*Ppd1* and *Rht8*) (Worland and Law, 1986; Hart *et al.*, 1993) and rice (*Se-1*/*Se-3* and *d-4*/*d-9*) (Kinoshita and Takahashi, 1991; Causse *et al.*, 1994).

Correlated traits have been analysed separately in QTL mapping without using correlation information and thus correlation between traits will affect the mapping of any single trait involved. Considering the correlation between different physiological characters, a polygenic complex controlling a physiological trait can be manipulated to a large extent by cloning only one or a few QTL from this complex. That is, all functions of a polygenic complex affecting multiple physiological characters can be started and pushed by making one of the functions into a highly efficient system. This was observed in growth-hormone-transformed mice (Palmiter *et al.*, 1983): when growth hormone was produced largely by inserting additional copies, all other components could respond appropriately to this change, although growth hormone is only one of the components affecting development. It seems that QTL controlling closely related quantitative traits could be manipulated to show linked response by only manipulating one or a few of these QTL.

Multivariate analysis of complicated traits can be used to investigate the structure of a genetic system that includes allelic variation at multiple loci, intermediate phenotypes and their relationships. Jiang and Zeng (1995) proposed a method for QTL detection based on a multivariate normal model with unconstrained covariance structure. Alternatively, dimension reduction techniques, such as principal component analysis, can be applied to a set of correlated traits. Multivariate QTL analyses can provide enhanced power and resolution in QTL mapping when traits are highly correlated and share common genetic determinants (Korol *et al.*, 2001). Mapping studies that investigate clusters of related phenotypes often reveal a network of genetic effects, in which each phenotype is influenced by multiple loci (heterogeneity) and different phenotypes share one or more loci in common (pleiotropy). The complexity of observed QTL networks will vary depending on the traits and the power of the study design. It is also likely that physiological interactions independent of genetic factors may result in correlated phenotypic response. Li *et al.* (2006) introduced a method for the analysis of multi-locus, multi-trait genetic data that provides an intuitive and precise characterization of genetic architecture and they showed it is possible to infer the magnitude and direction of causal relationships among multiple correlated phenotypes. They illustrated the technique using body composition and bone density data from mouse intercross populations. The identification of causal networks sheds light on the nature of genetic heterogeneity and pleiotropy in complex genetic systems.

Genetic correlation can be understood at gene expression levels. Coordinated regulation of gene expression levels across a series of experimental conditions provides valuable information about the function of correlated transcripts. In order to annotate gene function and identify potential members of regulatory networks, Lan *et al.* (2006) explored correlation of expression profiles across a genetic dimension, namely genotype segregating in a panel of 60 F_2 mice derived from a cross used to explore diabetes in obese mice. They first identified 6016 seed transcripts for which they observed that gene expression is linked to a particular region of the genome. Then they searched for transcripts whose expression is highly correlated with the seed transcripts and tested for enrichment of common biological functions among the lists of correlated transcripts. They found and explored the properties of 1341 sets of transcripts that share a particular 'gene ontology' term. Thirty-eight seeds in the G protein-coupled receptor protein signalling pathways were correlated with 174 transcripts, all of which are also annotated as G protein-coupled receptor protein signalling and 131 of which share a regulatory locus on chromosome 2. They noted that many of these findings would have been missed by simple eQTL analysis without the correlation step. Trait correlation combined with linkage mapping is more sensitive compared with linkage mapping alone.

7.2.3 Qualitative–quantitative traits

Many economically important quantitative characters, including plant height, pest and disease resistance and grain quality, in plant populations exhibit the combined effects of both major genes and polygenes. In other words, many traits are influenced by both qualitative and quantitative genes or by a major–minor gene system, showing bimodal distribution (Fig. 7.2; for examples see Jiang *et al.*, 1994 and Lin *et al.*, 1995). These traits can be defined as quantitative–qualitative traits (QQT) (Mo, 1993a,b) or semi-quantitative traits (Stuber, 1995). Separation of the major gene effects from the polygenic effects is important for understanding the whole genetic system of these kinds of traits and for mapping and cloning of the genes involved.

Jiang *et al.* (1994) used Elston's (1984) model of mixed major locus and polygenes with modification and extension to obtain reliable information necessary for assessment of the use of major genes in a breeding programme. As indicated by Xu (1997), QTL with larger effects will mask the QTL with smaller effects in the same or nearby locations so that the latter cannot be easily detected if they are mapped simultaneously. In order to eliminate the phenotypic effect of the 'major gene' from the 'residual' (error) term in the analysis of the significance of other QTL, Lin *et al.* (1995) tried two approaches, one by adjustment of the phenotypic value of individual plants for the major gene effect and one by use of the 'fix QTL' algorithm in MAPMAKER/QTL to fix the QTL effect with the largest LOD. As indicated by the authors, such approaches incur a risk that a fraction of the 'residual' (error) variance will be removed from the experimental model, due to chance correlation with the fixed parameters, artificially reducing the remaining 'error' term and increasing the likelihood of false positives. Doerge and Churchill (1996) used the conditional empirical threshold and the residual empirical threshold to search for multiple QTL. Once a major QTL has been detected, its

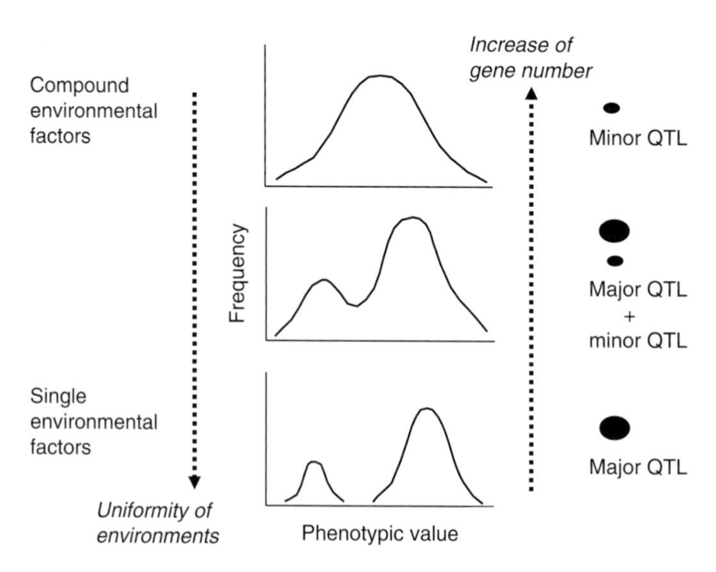

Fig 7.2. Relationship between phenotypes, genes, and environments. Discrete phenotypic distribution for qualitative traits arises from major genes, bimodal distribution for qualitative–quantitative traits from the joint effect of a major QTL (with dominant effect) and some minor QTL, and normal distribution for typical quantitative traits from many minor QTL. With partition and uniformity of environments, some continuously distributed traits can be converted into a bimodal of discretely distributed traits. From Xu (1997). This material is reproduced with permission of John Wiley & Sons, Inc.

phenotypic effects can be accounted for in the search for secondary QTL. This method is suitable for unlinked multiple QTL and/or QTL residing on different chromosomes. Mapping methods suitable for major–minor genes warrant further research.

7.2.4 Seed traits

The improvement of seed yield and quality is one of the most important objectives in cereal breeding. As a major storage organ of cereal seeds, endosperms provide humans with proteins, essential amino acids and oils. An understanding of the inheritance of endosperm traits is critical for the improvement of yield potential and seed quality. Genetic behaviour in triploid endosperms is very different from that of the maternal plants that supply the components for grain growth and development. Thus, methods suited for genetic analysis of traits in maternal plants (diploids for most cereal crops) cannot directly be used for endosperm traits (Xu, 1997). Based on triploid models, biometrical methods have been proposed for conventional genetic analysis of endosperm traits (Gale, 1975; Bogyo *et al.*, 1988; Mo, 1988; Foolad and Jones, 1992; Pooni *et al.*, 1992; Zhu and Weir, 1994). Any analytical method for endosperm traits needs to combine a QTL analytical method developed for diploid maternal plants with a triploid model proposed for conventional genetic analysis.

On the other hand, the genetic system controlling endosperm traits may be much more complicated than that which controls the traits of maternal plants. Because maternal plants provide seeds with a portion of their genetic material and almost all the nutrients required for growth and development, seed traits are genetically affected by both the seed nuclear genes and the maternal nuclear genes. In addition, cytoplasmic genes may also affect some seed traits through their indirect effects on the biosynthetic processes of chloroplasts and mitochondria. To understand endosperm traits with biological accuracy, one should take into consideration maternal genetic effects and cytoplasmic effects along with the direct genetic effects of seeds (Xu, 1997). As seeds initiate a new generation that differs from their maternal plants, some seed traits should be considered as a generation advanced over their maternal plants. Since the DNA used in most molecular analyses has been extracted from leaves or tissues of maternal plants, genetic analysis of endosperm traits should be based on the DNA extracted from both maternal plants and endosperm tissues in order to understand the relative contribution of the different genetic factors to the variation of endosperm traits.

Many years after Xu's (1997) advocacy, several articles were published detailing the unique difference associated with triploid traits and some statistical methods have been developed with consideration of the trisomic inheritance of the endosperm and the generation difference between the mapping population and the endosperm (Wu *et al.*, 2002a,c; Xu, C. *et al.*, 2003; Kao, 2004; Cui and Wu, 2005; Wang, X. *et al.*, 2007). In general, the proposed triploid-based methods use the marker information either from only the maternal plants or from both the maternal plants and their embryos for mapping endosperm traits and provide better detection power and estimation precision than diploid-based methods. The genetic models are also developed to handle epistatic effects (Cui and Wu, 2005) and to use bulked grain samples (Wang, X. *et al.*, 2007).

Zheng *et al.* (2008) conducted QTL analysis on maternal and endosperm genome for three cooking quality traits (amylose content, gel consistency and gelatinization temperature) in rice using a genetic model with endosperm and maternal effects and environmental interaction effects. The results suggested that a total of seven QTL were associated with cooking quality of rice, which were subsequently mapped to chromosomes 1, 4 and 6. Six of these QTL were also found to have environmental interaction effects.

As we discussed earlier, several studies have shown that maternal genotypic variation could greatly influence the estimation of the direct effects of QTL underlying endosperm traits. Recently, Wen and Wu (2008) proposed methods of interval mapping of endosperm QTL using seeds of F_2 or BC_1 (an equal mixture of $F_1 \times P_1$ and $F_1 \times P_2$ with F_1 as the female parent) derived from a cross between two pure lines ($P_1 \times P_2$). The most significant advantage of these experimental designs is that the maternal effects do not contribute to the genetic variation of endosperm traits and therefore the direct effects of endosperm QTL can be estimated without the influence of maternal effects. In addition, these experimental designs could greatly reduce environmental variation because a few F_1 plants grown in a small block of field will produce sufficient F_2 or BC_1 seeds for endosperm QTL analysis. More recently, He and Zhang (2008) proposed mapping endosperm trait loci (ETL) and epistatic ETL (eETL) as an efficient way to genetically improve grain quality using an alternative random hybridization design. Using a penalized maximum likelihood method, the endosperm trait means of random hybrid lines together with known marker genotype information from their corresponding parental F_2 plants were used to estimate efficiently and without bias the positions and all of the effects of eETL. This new method may enable us to map triploid eETL in the same way as diploid quantitative traits in future.

7.3 QTL Mapping across Species

For species connected by parallel genome mapping, as described in Chapter 3, it should be possible to compare the map positions of QTL for the same or similar characters. In this case, breeders might be able to predict the positions of important QTL (e.g. for growth rates in animals or yield in plants) in one species based on mapping studies from the others. Coincidence of map positions would support the hypothesis that loci underlying natural quantitative variation have been conserved during long periods of evolutionary divergence (i.e. they are orthologous genes). The collinearity of genomes among related species provides a central tool for the determination of the relative 'allelism' of genes in different species (Bennetzen, 1996).

Many QTL mapping studies have been published for species connected by comparative linkage maps, which can be used to infer some conclusions regarding the hypothesis of conserved QTL among divergent species. Perhaps the first evidence for orthologous QTL comes from comparative mapping in mung bean and cowpea (Fatokun *et al.*, 1992), where the researchers showed that the single most important QTL for determining seed weight in these two distinct species mapped to the same locus in both genomes and that the chance occurrence of such coincidental mapping is very unlikely. Lin *et al.* (1995) and Xiao *et al.* (1996c) discussed the putative orthologous QTL across grass species. Despite of different chromosome numbers and ploidy levels, homoeologous relationships among rice, maize, wheat, oats and barley chromosomes have been defined by using common anchor probes (Ahn and Tanksley, 1993; Ahn *et al.*, 1993; van Deynze *et al.*, 1995a, b). This information allows the comparison of locations of QTL affecting the same or corresponding traits in different species. For the QTL documented, some of them show similarities in locations for the same or similar traits. As an example, take flowering traits (days to heading, flowering and anthesis). The QTL close to CDO1081 on rice chromosome 3 coincides with a similar QTL on the homoeologous chromosomes 1 (Stuber *et al.*, 1992) and 9 (Koester *et al.*, 1993; Veldboom *et al.*, 1994) of maize, chromosome 4 of barley (Hayes *et al.*, 1993) and chromosome 5 of hexaploid oat (Siripoonwiwat, 1995). Across 15 maize populations studied by seven groups of researchers, 55 QTL or mutants affecting flowering time were reported. A total of 26 (47%) are clustered in five regions that span 12.1% of the maize genome. One flowering QTL reported in sorghum (Pereira *et al.*, 1994), three QTL in rice (Li *et al.*, 1995), three discrete mutants in wheat (Hart *et al.*, 1993) and one in barley (Laurie *et al.*, 1994) were included. Such a coincidence of QTL map positions among these distinct species suggests that this kind of locus can be traced back to the last common ancestor of these species. Paterson *et al.* (1995) indicated that in sorghum, rice and maize, three similar phenotypes (seed size, disarticulation of the mature inflorescence and day-neutral flowering) are largely determined by a small number of QTL that correspond closely in the three taxa, which impels the comparative mapping of complex phenotypes across large evolutionary distances.

Further studies identified QTL controlling important agronomic traits that showed similarities in locations for the same or similar traits (i.e. Fatokun *et al.*, 1992; Lin *et al.*, 1995; Xiao *et al.*, 1996c; for a review, see Xu, 1997). Shattering and plant height are examples that were also mapped to collinear regions among grass genomes (Paterson *et al.*, 1995; Peng *et al.*, 1999). Chen *et al.* (2003) identified four QTL for quantitative resistance to rice blast that showed corresponding map positions between rice and barley, two of which had completely conserved isolate specificity and the other two had partial conserved isolate specificity. Such corresponding locations and conserved specificity suggested a common origin and conserved functionality of the genes underlying the QTL for quantitative resistance.

In forest trees, a comparative genetic and QTL mapping was performed between *Quercus robur* L. and *Castanea sativa* Mill., two major forest tree species belonging to the *Fagaceae* family (Casasoli *et al.*, 2006). Oak EST-derived markers (sequence tagged sites, STSs) were used to align the 12 linkage groups of the two species. Fifty-one and 45 STSs were mapped in oak and chestnut, respectively. These STSs, added to SSR markers preciously mapped in both species, provided a total number of 55 orthologous molecular markers for comparative mapping within the *Fagaceae* family. Homologous genomic regions identified between oak and chestnut allowed comparison QTL positions for three important adaptive traits. Co-location of the QTL controlling the timing of bud burst was significant between the two species. However, conservation of QTL for height growth was not supported by statistical tests. No QTL for carbon isotope discrimination was conserved between the two species. Putative candidate genes for bud burst can be identified on the basis of co-locations between EST-derived markers and QTL.

Schaeffer *et al.* (2006) reported a strategy for consensus QTL maps that leverages the highly curated data in MaizeGDB, in particular, the numerous QTL studies and maps that are integrated with other genome data on a common coordinate system. In addition, they exploited a systematic QTL nomenclature and a hierarchical categorization of over 400 maize traits developed in the mid-1990s; the main nodes of the hierarchy are aligned with the trait ontology at Gramene, a comparative mapping database for cereals. Consensus maps are presented for one trait category, insect response (80 QTL); and two traits, grain yield (71 QTL) and kernel weight (113 QTL), representing over 20 separate QTL map sets of ten chromosomes each.

The use of anchor markers has enabled detection of possible orthologous QTL by comparing QTL across cereals or construction of phylogenetic relationships. Although it is unclear how many claimed orthologous QTL are real, detection of QTL that are common across cereals at least indicates that the same QTL could be identified from very different genetic backgrounds.

In a significant across species study, Campbell *et al.* (2007) identified a set of evolutionarily conserved and lineage-specific rice genes, which is termed conserved *Poaceae*-specific genes (CPSGs) reflecting the presence of significant sequence similarity across three separate *Poaceae* subfamilies. Using the rice genome annotation, along with genomic sequence and clustered transcript assemblies from 184 species in the plant kingdom, they have identified a set of 861 rice genes that are evolutionarily conserved among six diverse species within the *Poaceae* yet lack significant sequence similarity with plant species outside the *Poaceae*. It was interesting to note that the vast majority of rice CPSGs (86.6%) encode proteins with no putative function or functionally characterized protein domain and for the remaining CPSGs, 8.8% encode an F-box domain-containing protein and 4.5% encode a protein with a putative function. On average, the CPSGs have fewer exons, shorter total gene length and elevated GC content when compared with genes annotated as either transposable elements (TEs) or those genes having significant sequence similarity in a species outside the *Poaceae*. At the genome level, syntenic alignments between sorghum (*Sorghum bicolor*) and 103 of the 861 rice CPSGs (12.0%) could be made, demonstrating an additional level of conservation for this set of genes within the *Poaceae*.

7.4 QTL across Genetic Backgrounds

Phenotypic expression of quantitative traits is affected to a great extent by internal genetic background. That is at least in part because gene action is not always independent between QTL controlling different traits and between QTL and the corresponding major genes or other major genes. Difficulty in obtaining consistent results from different QTL mapping experiments for the same trait can be partly attributed to the action of the heterogeneous genetic background from which mapping populations are derived, besides the reasons mentioned in the Section 7.1.2.

7.4.1 Homogeneous genetic backgrounds

Populations developed for QTL analysis can be very heterogeneous in genetic backgrounds, with hundreds or thousands of genes segregating simultaneously, or very homogeneous with only a target gene segregating. Homogeneous or isogenic backgrounds such as NILs can be created through the five approaches as described in Chapter 4 and Xu, Y. (2002). Tanksley and Nelson (1996) proposed an advanced BC QTL analysis in which a hybrid F_1 is backcrossed to the recurrent parent for several times to obtain advanced generations (e.g. BC_2, BC_3). If additional BC generations proceeded with whole genome selection, this approach can be used to create QTL-NILs. The uniformity of genetic background among QTL-NILs and their donors should permit straightforward phenotypic evaluation and facilitate QTL mapping.

In maize, a set of 89 NILs was created using marker-assisted selection (MAS) (Szalma *et al.*, 2007). Nineteen genomic regions, identified by RFLP loci and chosen to represent portions of all ten maize chromosomes, were introgressed by backcrossing three generations from donor line Tx303 into the B73 genetic background. NILs were genotyped at an additional 128 SSR loci to estimate the size of introgression and the amount of background introgression. Tx303 introgressions ranged from 10 to 150 cM in size with an average of 60 cM. Across all NILs, 89% of the Tx303 genome is presented in targeted and background introgressions. A parallel experiment of testcrosses of each NIL to the unrelated inbred, Mo17, was conducted in the same environments to map QTL in NIL testcross hybrids.

In *Arabidopsis*, a genome-wide coverage NIL population was developed by introgressing genomic regions from the Cape Verde Islands (Cvi) accession into the Landsberg *erect* (Ler) genetic background (Keurantjes *et al.*, 2007b). QTL mapping power of the new population was empirically compared with an already existing RIL population derived from the same parents. For that, QTL affecting six developmental traits with different heritability were analysed and mapped. Overall, in the NIL population smaller-effect QTL than in the RIL population could be detected although the location resolution was lower. Furthermore, the effect of population size and of the number of replicates on the detection power of QTL affecting the developmental traits was estimated. In general, population size is more important than the number of replicates to increase the mapping power of RILs, whereas for NILs several replicates are absolutely required. These analyses are expected to facilitate experimental design for QTL mapping using these two common types of segregating populations.

QTL have been fine mapped by applying a mapping strategy based on analysis of large progenies derived from NILs. This approach requires the construction of highly inbred lines involving many generations prior to generating the cross needed for fine mapping. Instead of homogenizing the complete genetic background, as in the NIL approach, Peleman *et al.* (2005) have chosen to focus specifically on the loci involved in expression of the phenotype. The strategy involved simultaneous fine mapping of QTL already at the F_2 stage rather than producing inbred lines prior to fine mapping. The main principle of the approach is the selective genotyping and phenotyping of only those plants that yield information on the map position of the QTL. Such plants are selected after a first rough-scale mapping by standard methods (e.g. 200 F_2 individuals). After identification of the QTL for the trait of interest, a larger part of the population (e.g. 1000 F_2 plants) is screened with markers flanking the QTL to identify sets of QTL isogenic recombinants (QIRs). QIR plants carrying a recombination event in one QTL while they are homozygous at all other QTL are most informative. The trait complexity can thus be reduced to a monogenic trait as plants with all but one QTL having an identical homozygous genotype are selected. These QIRs are subsequently genotyped with sufficient markers at the recombinant QTL region to precisely map the recombinant event within the QTL-bearing interval. Phenotyping other QIRs becomes more reliable by reducing the trait complexity as these plants are nearly isogenic for all QTL that affect the trait. Peleman *et al.* (2005) demonstrated that for fine mapping oligogenic traits, homogenizing the background genome is not required. The method was demonstrated by fine mapping a QTL responsible for erucic acid content in rapeseed. For quantitative traits that are controlled by many QTL each with relatively small effects, progeny test and background selection with markers to cover the whole genome would be required in order to minimize the genetic background effect and to confirm the phenotype for the recombinants selected.

7.4.2 Heterogeneous genetic backgrounds

Although genetic distances and order of DNA markers are comparable among very different crosses, QTL mapping using different populations derived from the same cross has identified very different QTL. Only some QTL are common across populations of different structures, such as DHs and RILs derived from a single cross (He *et al.*, 2001) where there is an identical set of genes segregating. Heterogeneous genetic backgrounds can also come from various crosses derived from different cultivars. Genetic materials with heterogeneous genetic backgrounds can be used to estimate epistasis, detect non-allelic QTL and discover multiple alleles. QTL mapping for the same traits using different populations can be illustrated using seed dormancy in barley as an example, where QTL were compared across seven RIL populations and one DH population derived from crosses including 11 cultivated strains and one wild barley strain showing the wide range of seed dormancy levels (Hori *et al.*, 2007). Linkage maps were constructed based on EST, SSR, RFLP and morphological markers, each map consisting of 82–1114 markers (Table 7.2). Using these populations, a total of 38 QTL clustered around 11 regions were identified on the barley chromosomes except chromosome 2H among eight populations. The QTL at the centromeric region of the long arm of chromosome 5H was identified in all populations with different degrees of dormancy depth and period (Fig. 7.3).

Considering several populations derived from diverse parental materials increases the probability that a QTL will be polymorphic in at least one population. To go beyond comparison of results between populations, some authors have proposed jointly analysing the different populations. This can be done first for independent

Table 7.2. Summary of linkage map information for eight permanent mapping populations in barley (from Hori *et al.* (2007) with kind permission of Springer Science and Business Media).

Population	Number of markers						Total map length (cM)
	Morpho-logical	EST	SSR	RFLP	AFLP	Total	
Haruna Nijo × H602 DH (DHHS)	4	1055	35	16		1110	1362.7
Russia 6 × H.E.S.4 (RHI)	3	75	34	3	1134	1249	1595.7
Mokusekko 3 × Ko A (RIA)	2	102		1		105	1233.5
Harbin 2-row × Khanaqin 7 (RI1)	4	81				85	1217.0
Harbin 2-row × Turkey (RI2)	4	29	45	1	328	407	1377.1
Harbin 2-row × Turkey 45 (RI3)	2	80				82	1103.8
Harbin 2-row × Katana (RI4)	4	90				94	1208.2
Harbin 2-row × Khanaqin 1 (RI5)	2	76	32			110	1078.3

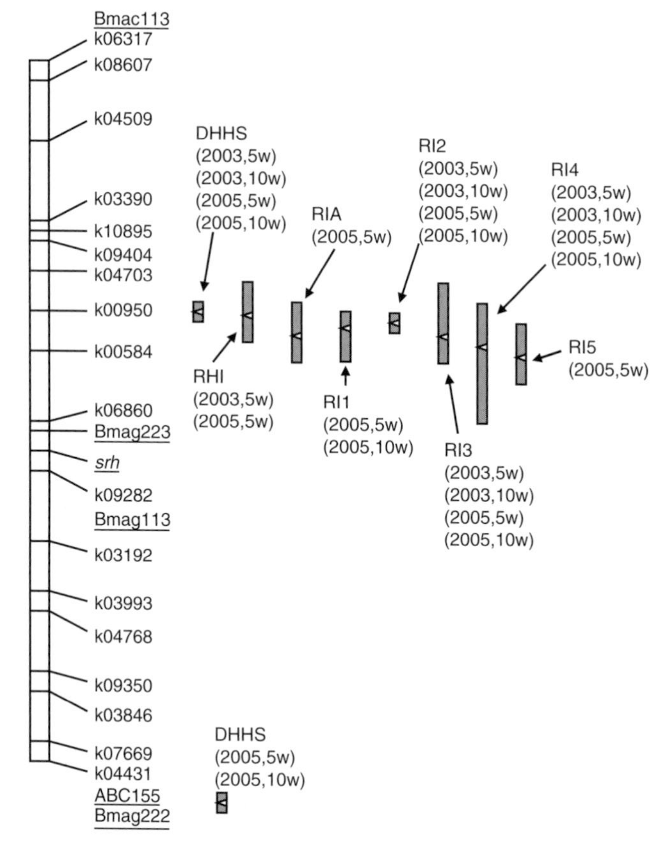

Fig. 7.3. A consensus barley linkage map based on eight mapping populations (for codes see Table 7.2) and positions of QTL for seed dormancy at 5 and 10 weeks (5w and 10w, respectively) after ripening in 2003 and 2005. Linkage groups are oriented with short arms from the top. The anchor loci including SSR, RFLP and morphological markers are indicated with under line. QTL positions are indicated by grey boxes. Peaks of the significant marker intervals as indicated by triangles in boxes. Only chromosome 5H is included, which shows large-effect QTL near the centromere on the long arm in all of the populations. From Hori *et al.* (2007) with kind permission of Springer Science and Business Media.

populations (no known pedigree relationship between the parents of the different populations) (Muranty, 1996; Xu, 1998). In this case, QTL effects are nested (in the statistical sense) within populations and the number of parameters to be estimated increases with the number of populations. Also, the lack of connections between populations does not allow global comparison of the effects of all QTL alleles segregating in the different populations. An alternative approach is therefore to develop connected populations (common parents among populations). Under the assumption of additivity, considering identical allelic effects over populations rather than nesting effects within populations reduces the total number of parameters and, consequently increases the power of QTL detection (Rebai and Goffinet, 1993; Jannink and Jansen, 2001). In such an analysis, the effects of alleles segregating are estimated simultaneously, which facilitates a global comparison. This is of particular interest to identify the parental origin(s) of favourable allele(s) at each QTL.

QTL for six quality traits in tomato (fruit weight, firmness, locule number, soluble solid content, sugar content and titratable acidity) were studied in order to investigate their individual effect and their stability over years, generations and genetic backgrounds (Chaïb *et al.*, 2006). Three sets of genotypes corresponding to three generations were compared: (i) an RIL population containing 50% of each parental genome; (ii) three BC_3S_1 populations segregating simultaneously for the five regions carrying fruit quality QTL, but almost fully homozygous for the recipient genome on the eight chromosomes carrying no QTL; and (iii) three sets of QTL-NILs (BC_3S_3 lines) which differed from the recipient line only in one of the five chromosome regions. Eight of the ten QTL detected in RILs were recovered in the QTL-NILs with the genetic background used for the initial QTL mapping experiment, with the exception of two QTL for fruit firmness. Several new QTL were detected. In the two other genetic backgrounds, the number of QTL in common with the RILs was lower, but several new QTL were also detected in advanced generations.

7.4.3 Epistasis

Importance of epistasis

The importance of epistasis to the genetic control of quantitative traits has been debated. As one of the early supports to the importance of epistasis, Eshed and Zamir (1996) reported that QTL epistasis is a significant component in determining phenotypic values using tomato NILs. For the five yield-associated traits, 20–40% of the 45 dichromosome segment combinations were epistatic, which is much higher than would be expected by chance alone. The detected epistasis was predominantly less-than-additive, i.e. the effect of double heterozygotes was smaller than the sum of the effects of the corresponding single heterozygotes. Several other studies showed that the epistatic variance can account for a large proportion of the genetic variance of quantitative traits (Carlborg *et al.*, 2005; Malmberg and Mauricio, 2005; Malmberg *et al.*, 2005). Epistatic interaction among loci could contribute substantially to the variation in complex traits (Carlborg and Haley, 2004; Marchini *et al.*, 2005).

In contrast, after a review of most studies conducted at that time Tanksley (1993) suggested that strong epistatic interactions are the exception and not the rule for naturally occurring polygenes. These conclusions are supported to some extent by the few studies in which individual QTL have been genetically identified by introgression from other QTL in NILs and have been shown to continue producing their same individual effects (De Vicente and Tanksley 1993; Eshed and Zamir 1995) and also by a recent report on negligible interaction using a barley DH population (Harrington × TR306) developed by the North American Barley Genome Mapping Project for QTL mapping, which consisted of 145 lines and 127 markers covering a total genome length of 1270 cM. These DH lines were evaluated in ~25 environments for seven quantitative traits: heading, height, kernel weight, lodging, maturity, test weight and yield. Xu and Jia (2007) applied an empirical Bayes method that simultaneously estimates 127 main effects for all markers and main-effect QTL (single marker) and the largest epistatic effect (single pair of markers) explained ~18 and 2.6% of the phenotypic variance, respectively. On average, the sum of all significant main effects and the sum of all significant epistatic effects contributed 35 and 6% of the total phenotypic variance, respectively. Epistasis seems to be negligible for all the seven traits. They also found that whether two loci interact does not depend on whether or not the loci have individual main effects. This invalidates the common practice of epistatic analysis in which epistatic effects are estimated only for pairs of loci of which both have main effects.

The contradicting reports may result from the fact that QTL mapping studies and analytical methods have not been able to detect epistasis and thus the conclusions could be biased, preferentially identifying genes that have large effects and/or act independently (Xu, 1997). This argument is supported by the results that QTL with large effects are detected in very different crosses and environments. The second reason is that ordinal QTL analysis was made with populations segregating for the whole genome simultaneously so that it may be difficult to detect an interaction in a specific combination of QTL genotypes. For example, Yano *et al.* (1997) predicted an interaction between the two largest QTL, *Hd1* and *Hd2*, for heading date. But the existence of another QTL, *Hd6* and its interaction could not be detected in their primary population (F_2), where many epistatic interactions could exist in so-called minor QTL. Successful examples for detection of epistatic interactions by using primary populations seem to be related to population sizes and structures, quantitative traits, the number of existing QTL and QTL effects. The more QTL involved, the more difficult is the detection of significant differences for individual QTL. Although using a large population size may help to detect epistatic interactions, it increases experimental errors, due to increasing difficulty in managing such a population effectively.

Statistical methods for epistatic QTL

Methods for mapping QTL with epistatic effects are still premature. Some methods utilize models including a single epistatic effect at a time (Holland, 1998; Malmberg *et al.*, 2005), while others apply a model selection strategy that searches for multiple epistatic effects (Carlborg *et al.*, 2000; Yi *et al.*, 2003, 2005; Baieri *et al.*, 2006). Xu (2007) developed an empirical Bayes method that can simultaneously estimate main effects and all individual markers and epistatic effects of all pairs of markers. Recently, epistatic QTL analysis has been extended to the genome-wide level. Such an example is that Yi *et al.* (2007) proposed a Bayesian model selection approach of genome-wide interacting QTL for ordinal traits in experimental crosses. Stich *et al.* (2007) examined a genome-wide QTL mapping strategy using genome sequence information of RILs that were generated from several crosses of parental inbreds. The SNP haplotype data of B73 and 25 diverse maize inbreds were used to simulate the production of various RIL populations. Higher power to detect three-way interactions was observed for RILs derived from optimally allocated distance-based designs than from nested designs or diallel designs. The power and proportion of false positives to detect three-way interactions using a nested design with 5000 RILs were for both the 4-QTL and 12-QTL scenario of a magnitude that seems promising for their identification. To find an optimal model for the epistatic effect, Bayesian model selection (George and McMulloch, 1993), by taking advantage of Markov chain Monte Carlo (MCMC) sampling, is a more efficient algorithm than both the exhaustive and the heuristic searches. The simulation experiments conducted by Xu (2007) showed that the MCMC-based methods performed satisfactorily when the sample size was 600. The empirical Bayes method is more robust to small sample size than the MCMC-based full Bayes methods. Considering that most

QTL studies reported so far have sample sizes less or much less than 600, a larger mapping population has to be created in order to use the MCMC-based full Bayes methods developed so far for QTL mapping where epistatic effects are involved.

Population strategies for epistatic QTL studies

QTL interaction has been analysed using different types of plant materials including a series of chromosomal substitution lines or QTL-NILs. If NILs are used, interaction between the target QTL and other major genes/QTL can be eliminated and only epistasis between multiple target QTL needs to be considered. With removal of noise from heterogeneous backgrounds, the proportion of variance explained by the target QTL will increase and minor QTL can be identified.

Epistatic interactions between QTL and the genetic background can be addressed using connected designs of multiple populations, provided the mating design contains 'loops' (in the simplest case, three populations derived from three parents A × B, B × C and A × C). In such designs, epistasis can be tested through the comparison between: (i) a 'connected additive' model where the allele effects at a QTL are assumed to be identical in the different populations; and (ii) an 'hierarchical' model where allele effects are nested within populations, which accounts for possible interactions with the genetic background. Such an analysis tests for consistency of allelic effects over populations and therefore permits evaluation of the contribution of QTL-by-genetic-background epistatic effects to variation in QTL results observed among populations, relative to that of other factors such as allelic relationships between parental inbreds and statistical noise. Tests for epistasis in connected designs following this principle have been proposed by several authors (Rebai *et al.*, 1994; Charcosset *et al.*, 1994; Jannink and Jansen, 2000, 2001). One of the advantages of these tests, when compared to testing only for digenic interactions, is to enable the detection of epistatic interactions of higher order (Charcosset *et al.*, 1994). The statistical properties of QTL-by-genetic-background interaction tests in the case of a single digenic interaction has been analysed by means of simulations (Jannink and Jansen, 2001) and the result showed that it was possible to identify the two QTL involved by using an appropriate statistical test and also proposed guidelines for the interpretation of the sign of the QTL-by-genetic-background interaction effects. For more complex situations, the results are less predictable. Several digenic epistatic interactions that involve a given QTL may add up if similar in sign, yielding a significant interaction with the genetic background whereas none of them were significant. They may also cancel out each other if opposite in signs and lead to no detectable interaction with the genetic background. It is therefore interesting to compare both types of interactions.

Blanc *et al.* (2006) presented results from six connected F_2 populations of 150 $F_{2:3}$ families each, derived from four maize inbreds and evaluated for three traits of agronomic interest using the MCQTL software (Jourjon *et al.*, 2005). This software permits the joint analysis of multiple populations using a composite interval mapping method based on a linearized regression model (Haley and Knott 1992; Charcosset *et al.* 2000). They first detected QTL in each population independently (single-population analysis), secondly on the whole design without taking into account connections (multi-population disconnected analysis), and then on the global design using connections (multi-population connected analysis). Lastly, they tested for digenic interactions and for locus-by-genetic-background interactions, estimated the contribution of epistasis to the variation of the traits studied and checked if epistatic interactions could explain discrepancies among the analyses. The joint estimation of the different parental allele effects in a connected model allowed them to identify, for each QTL, the parental inbred line(s) that carried the most interesting allele(s). Taking into account the connections between populations increased the number of QTL detected and the accuracy of QTL position estimates. Many epistatic interactions were detected, particularly for grain yield QTL (R^2 increase of 9.6%). Allelic relationships and epistasis both contribute to the lack of consistency for QTL positions observed among populations, in addition to the limited power of the tests.

Melchinger *et al.* (2008) derived quantitative genetic expectations of QTL effects obtained from one-dimensional genome scans with the triple testcross (TTC) design and pairwise interactions between marker loci using two-way analyses of variance (ANOVA) under the F_2- and the F_∞-metric model. It was demonstrated that the TTC design can partially overcome the limitations of the design III in separating QTL main effects and their epistatic interactions in the analysis of heterosis, and that dominance × additive epistatic interactions of individual QTL with the genetic background can be estimated with a one-dimensional genome scan.

7.4.4 Multiple alleles at a locus

Two-parent derived populations in diploid crops have only two alleles segregating at each locus. Identification of multiple alleles requires comparison of populations derived from different crosses. To distinguish QTL alleles identified in one cross from those in another, all mapped alleles must be accurately sized and documented.

As an example of multiple alleles at a locus, rice amylose content, mainly controlled by the *wx* gene, can be taken as an example. Wide variation in amylose content occurs and cultivars with different amylose content, varying from waxy (0–2%), very low (3–9%), low (10–19%) and intermediate (20–25%) to high (> 25%), have been selected in breeding programmes. Conventional genetic studies using cultivars with different amylose contents revealed transgressive segregation in F_2s in almost all possible parental combinations (Pooni *et al.*, 1993). A polymorphic microsatellite was identified in the *wx* gene (Bligh *et al.*, 1995) located 55 bp upstream of the putative 5'-leader intron splice site. Ayres *et al.* (1997) determined the relationship between polymorphism at that locus and variation in amylose content. Eight *wx* microsatellite alleles were identified from 92 long-, medium- and short-grain US rice cultivars, which explained 85.9% of the variation. The amplified products ranged from 103 to 127 bp in length and contained $(CT)_n$ repeats, where n ranged from 8 to 20. Average amylose content in cultivars with different alleles varied from 14.9 to 25.2%. Using more diverse rice germplasm accessions ($n = 243$), Zeng *et al.* (2000) identified 15 alleles at the *wx* locus, using microsatellite class and G-T polymorphism, resulting in a total of 16 alleles identified so far. Now the question is whether the multiple alleles identified at the waxy locus can be associated to QTL alleles and whether the case can be extended to other traits or genetic loci.

Using molecular markers with multiple alleles in QTL mapping will help identify multiple QTL alleles. QTL studies using different populations have identified some common QTL. It is necessary, however, to further clarify whether they identified common or different alleles at those QTL. Reporting the sizes of associated alleles and using allele-rich markers in QTL studies will provide information required for this clarification, with the assumption that each marker allele has a corresponding QTL allele.

7.5 QTL across Growth and Developmental Stages

Classical breeding methods rely heavily on end-point measurements of agricultural productivity, which are influenced by different parameters and consequently different genes, in different environments. If more specific measures of agricultural productivity can be identified, for example, physical or chemical properties of the plant which relate directly to productivity under a particular environmental stress, it will be much more feasible to identify the underlying genes by mapping. However,

1

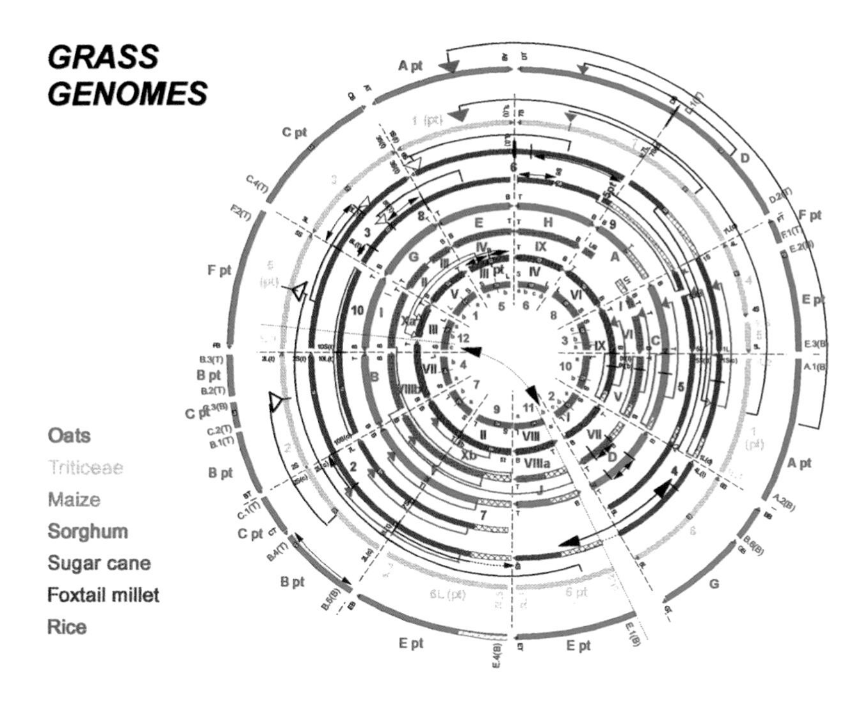

2

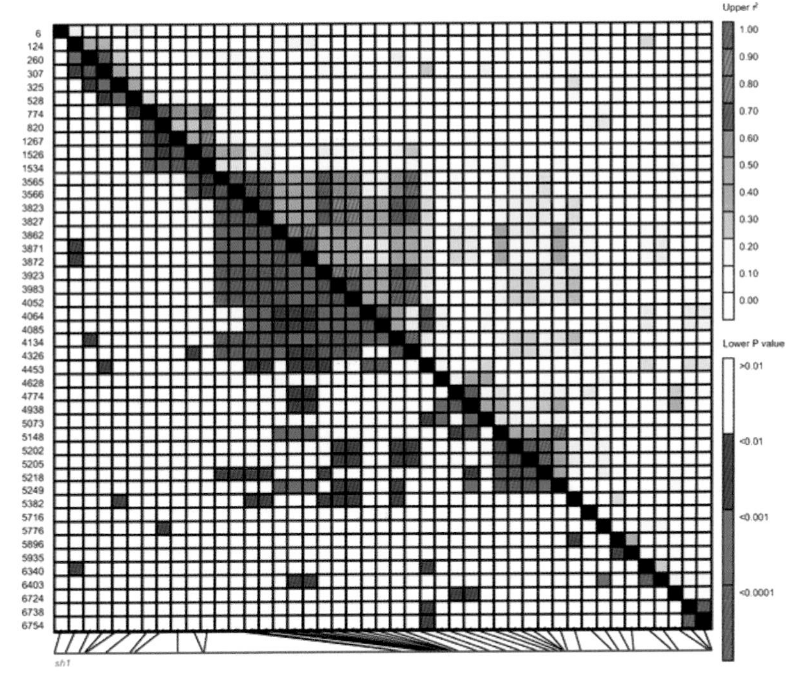

Plate 1. Circle diagram for comparative genomics in cereals. (From Gale and Devos (1998) © 1998 National Academy of Sciences, USA.)

Plate 2. Disequilibrium matrix for polymorphic sites within *sh1*. Polymorphic sites are plotted on both the *x*-axis and the *y*-axis. The pair-wise calculation of linkage disequilibrium (LD) (r^2) is displayed above the diagonal with the corresponding *P*-values for Fisher's exact test displayed below the diagonal. Coloration is indicative of the corresponding *P*-value or r^2 values from the bars on right. Notice that some blocks of LD do persist over larger distances within the gene, which do not necessarily correspond to tight linkage. (Reproduced with permission of Annual Reviews Inc., from Flint-Garcia *et al.* (2003); permission conveyed through Copyright Clearance Center, Inc.)

3

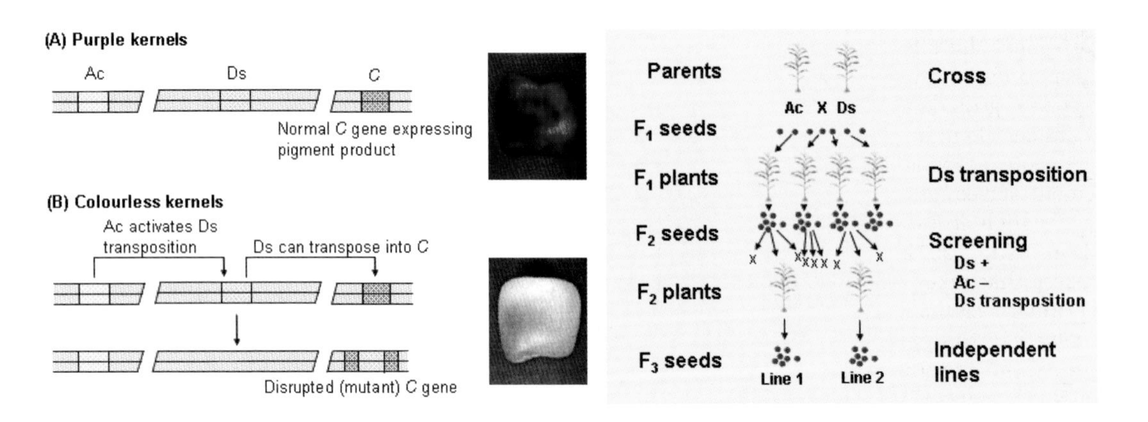

4

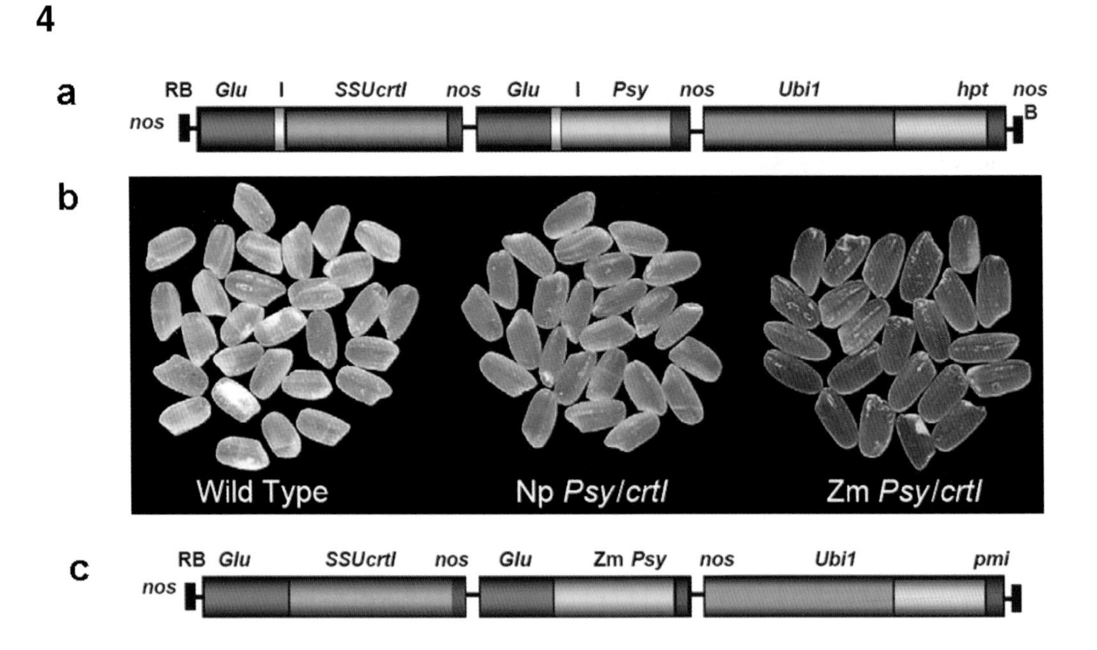

Plate 3. Ac/Ds tagging system in plants.

Plate 4. Carotenoid enhancement of the rice endosperm by transformation with *psy* orthologues and *crtI*. (a) Schematic diagram of the T-DNAs used to generate transgenic rice plants. The T-DNA comprise the rice glutelin promoter (*Glu*) and the first intron of the catalase gene from castor bean (I), *E. uredovora crtI* functionally fused to the pea RUBISCO chloroplast transit peptide (*SSUcrtI*) and a phytoene synthase from each of five plant species (*psy*), with a nos terminator, as well as a selectable marker cassette comprising the maize polyubiquitin (*Ubi1*) promoter with intron, hygromycin resistance (*hpt*) and nos terminator. (b) Photograph of polished wild-type and transgenic rice grains containing the T-DNA (as above) with the daffodil *psy* (Np) or maize *psy* (Zm) showing altered colour due to carotenoid accumulation. (c) Schematic diagram of the T-DNA in pSYN12424 used to develop 'Golden Rice 2'. *pmi* is maize phosphomannose isomerase gene. (From Paine *et al.* (2005) reprinted by permission from Macmillan Publishers Ltd.)

measures of agricultural productivity usually reflect the effects of many genes, acting at different times during the lengthy period of growth and development of the organism. Genetic expression of quantitative traits associates closely with development stages and may have one specific stage which is most suitable for identification (Xu, 1997). For any biochemical process that takes place in growth, possibly much less than 0.1% of the operations involved in the process would lead to the activity and the final output of a plant cell (Peterson, 1992). Therefore, developmental research on quantitative traits is important not only for quantitative genetic analysis by conventional methods, but also for determination of the best stage for identification by molecular approaches. In rice, for example, genetic analysis on tiller number was made at different growth stages and differential phenotypic expression was found, indicating that evaluation and selection of tiller number should be at the peak-tillering stage (Xu and Shen, 1991). At this stage, phenotypic differences in tiller number among genetic materials was maximized and suitable for distinguishing different genotypes. In general, intensive research in developmental genetics of quantitative traits is fundamental for QTL mapping and cloning but limited work has been done in most crops.

7.5.1 Dynamic traits

Quantitative traits that can be measured repeatedly during the development of life are called longitudinal traits in humans, but more often called dynamic traits in animals and plants. Some genes control the phenotypic values of the dynamic traits at fixed time points and others may alter the transitions of the phenotypes between consecutive time points. The growth pattern of a dynamic trait is called the growth trajectory (Yang and Xu, 2007). On the other hand, the normal functions of biological process are strongly correlated with the genes that control them. Several studies in animal models have identified various so-called circadian clock genes and clock-controlled transcript factors through mutants. Circadian rhythms in plants are poorly understood and they can be determined as complex dynamic traits.

Some dynamic traits do not fit a normal distribution as well as a typical quantitative trait because of their extremely non-linear nature. These traits are best characterized in terms of sudden transitions to qualitatively distinct phenotypic states as opposed to quantitative extensions of previous states, and are called time-to-event or time-to-failure traits. The most notorious of such traits is probably the death of the organism and many other traits such as flowering can be interpreted in this way. In a typical time-to-event (or time-to-failure) experiment one follows a sample over time and records the times (e.g. hours, days) at which the event occurs to a given individual. The resulting phenotypic distribution is usually right-skewed.

7.5.2 Dynamic mapping

Many agriculturally and biomedically important traits undergo predictable changes in genetic time. These changes are in part driven by the temporal regulation of the genes or QTL underlying these phenotypes. To understand genetic expression at different developmental stages, dynamic mapping has been proposed (Xu, 1994, 1997; Xu and Zhu, 1994). Xu, Y. (2002) summarized three approaches to dynamic analysis or time-related mapping using phenotypic data collected across developmental stages. One is based on analysis of trait values measured at each observation time (e.g. Bradshaw and Stettler, 1995; Plomion *et al.*, 1996; Price and Tomos, 1997; Verhaegen *et al.*, 1997), from which the accumulated effect of a QTL, from the beginning of ontogenesis to each observation time, can be estimated. This is called effect-accumulation analysis or unconditional QTL mapping (Yan *et al.*, 1998a). The second approach is to analyse trait-value increments observed at sequential time intervals (e.g. Bradshaw and Stettler, 1995; Plomion *et al.*, 1996; Verhaegen *et al.*, 1997), from which the incremental or net effect of a QTL at each time interval can be estimated. This is called effect–increment analysis or conditional QTL mapping (Yan *et al.*, 1998a). Phenotypic data collected at different growth stages or time intervals can be analysed either separately or jointly. Compared to separate analysis, joint analysis can synthesize all the information from different times or time intervals to give a comprehensive estimate of each QTL position, according to which a corresponding complete expression (or expression rate) curve of each QTL can be estimated (Wu *et al.*, 1999). In practice, both separate and joint analyses should be conducted. A third approach to looking at the QTL over time is to do a multivariate analysis based on fitting the parameters of the growth curve (animal breeders call this general approach 'random regression').

The significant advantage of dynamic mapping is that it provides a quantitative framework for testing the interplay between genetic (inter)actions and the pattern of development in a time course. Dynamic mapping constructs a setting for precisely estimating and predicting a number of fundamental events in the genetic control of development (Wu *et al.*, 2004), which include: (i) the timing of a QTL to turn on and off to affect growth in a time course; (ii) the duration of the dynamic genetic effect of a QTL; (iii) the magnitude of the genetic effect of a QTL on maximal growth rate; and (iv) the pleiotropic effect of the growth QTL on other developmental traits related to growth processes.

In general, there are four types of QTL effects in time-to-event experiments (as modified from Wu and Lin (2006) and Johannes (2007)). These include: (i) early-acting: QTL expressed at the early stage of the developmental process but not during the rest of the process; (ii) late-acting: QTL expressed only at the late stage of the developmental process; (iii) inversely acting: QTL highly expressed at the early stage but with low expression at the late stage, or vice versa; and (iv) proportionally acting: QTL either expressed with a proportionally increased or decreased rate or with a consistent rate.

As an early example in dynamic QTL mapping, Yan *et al.* (1998a,b) used rice IR64/Azucena DHs to study the developmental characteristics of QTL for tiller numbers and plant height by conditional and unconditional interval mapping, in combination with phenotyping these traits every 10 days after transplanting. They concluded that many QTL identified at the early stages were undetectable at the final stage. Conditional mapping identified more QTL than unconditional mapping. Temporal patterns of gene expression changed with developmental stages. Genes at a specific genomic region might have opposite genetic effects at various growth stages. For chromosomal regions significantly associated with plant height, conditional QTL were found only at one to several specific periods and no QTL for plant height was continually active during the entire period of growth.

7.5.3 Statistical methods for dynamic mapping

Several dynamic mapping methods developed (Ma *et al.*, 2002; Wu, W. *et al.*, 2002; Wu *et al.*, 2004; Wu and Lin, 2006) have made it possible to test interesting hypotheses about the quantitative genetic control of the rate of change in the phenotype as well as the time specificity of the genetic effects. To be informative, these latter methods require that a measured trait value can be obtained from the same individual at different time points and that the phenotype can be described as a process unfolding along a continuous trajectory.

The biological and statistical advantages of dynamic mapping result from joint modelling of the mean-covariance structures for developmental trajectories of a complex trait measured at a series of time points. While an increased number of time points can better describe the dynamic pattern of trait development, significant difficulties in performing dynamic mapping arise from prohibitive computational times required as well as from modelling the structure of a high-dimensional covariance matrix. An efficient approach for applying dynamic mapping to high-dimensional data is through dimensional reduction, i.e. the transformation that brings data from a high- to low-order dimension. Zhao *et al.* (2007) developed a statistical model for dynamic mapping of QTL that govern the developmental process of a quantitative trait on the basis of wavelet dimension reduction. By breaking an original signal down into a spectrum by taking its averages (smooth coefficients) and difference (detail coefficients), they used the discrete Haar wavelet shrinkage technique to transform an inherently high-dimensional biological problem into its tractable low-dimensional representation within the framework of dynamic mapping constructed by a Gaussian mixture model. The wavelet-based parametric dynamic mapping holds great promise as a powerful statistical tool to unravel the genetic machinery of developmental trajectories with large-scale high-dimensional data.

To be informative, methods based on the test of time specificity of the genetic effects require that a measured trait value can be obtained from the same individual at different time points and that the phenotype can be described as a process unfolding along a continuous trajectory. Johannes (2007) developed the idea of time-varying QTL effects in the context of time-to-event analysis. An extension of the Cox model (EC model) (Therneau and Grambsch, 2000) was applied to an interval-mapping framework. In its simplest form, this model assumes that the QTL effect changes at some time point t_0 and follows a linear function before and after this change point. The approximate time point at which this change occurs is estimated. Using simulated and real data, the mapping performance of the EC model was compared to the Cox proportional hazards (CPH) model, which explicitly assumes a constant effect. The results showed that the EC model detects time-dependent QTL, which the CPH model fails to detect. At the same time, the EC model recovered all of the QTL the CPH model detects. It was concluded that potentially important QTL may be missed if their time-dependent effects are not accounted for.

The most cumbersome issue in multiple QTL mapping for dynamic traits is how to determine the optimal number of QTL. To do this, variable selection via stepwise regression is commonly used in maximum-likelihood mapping. Reversely-jump Markov chain Monte Carlo (RJ-MCMC) is the corresponding variable selection procedure used in Bayesian analysis. However, RJ-MCMC is shown to be subjected to poor mixing and slow convergence to the stationary distribution. Variable selection by Bayesian shrinkage analysis and stochastic search are more efficient than RJ-MCMC (reviewed in Yang and Xu, 2007). In these methods, no variable selection is conducted in an explicit manner; rather, a treatment similar to variable selection is made implicitly by shrinking the effects of excessive QTL to zero. Yang, R.Q. *et al.* (2006) developed an interval-mapping procedure to map QTL for dynamic traits under the maximum-likelihood framework. They fitted the growth trajectory by Legendre polynomials. The method was intended to map one QTL at a time and the entire QTL analysis involved scanning the entire genome by fitting multiple single-QTL models. Yang and Xu (2007) proposed a Bayesian shrinkage analysis for estimating and mapping multiple QTL in a single model. The method is a combination between the shrinkage mapping for individual quantitative traits and the Legendre polynomial analysis, an extensively used linear growth model in animals, for dynamic traits. Simulation study showed that the method generated a much better signal for QTL than the interval-mapping approach.

Although various statistical methods have been developed to meet the requirements of dynamic mapping for different trait categories, effectiveness and efficiency of these methods needs further studies. Application of these methods to QTL mapping needs full support of user-friendly mapping software.

7.6 Multiple Traits and Gene Expression

Plant breeders manage numerous phenotypes simultaneously in order to develop a suitable breeding product for a suitable environment. Geneticists face the same challenge when handling many transcripts in genetic mapping for gene expression. In terms of the complexity and variability, both phenotypes handled by breeders and transcripts handled by geneticists belong to a same category: multiple traits. However, discussion in this section will focus on gene expression. All issues discussed in this section can be found in the corresponding discussion on multiple quantitative traits in Chapter 1.

7.6.1 Features of gene expression

An emerging approach is to ask whether the parameters of gene activity at the level of transcription regarding additivity, heritability and complexity parallel those of classical phenotypic traits (Gibson and Weir, 2005). There are six features that characterize gene expression studies. First, there is now a reasonable expectation that for any tissue from any organism sampled under a particular set of environmental conditions, 10–50% of the transcripts will be found to vary as a result of heritable differences (Stamatoyannopoulos, 2004).

Secondly, numerous examples were observed of non-additivity of transcription including over- and under-dominance (F_1 with higher or lower expression, respectively, than either parent), parent-of-origin, maternal and reciprocal F_1 effects, indicating an unexpected complexity to the mapping of genotype on to the transcriptional phenotype. Similar results have been observed in studies targeting specific candidate genes in maize (Auger *et al.*, 2005) and wheat (Sun *et al.*, 2004) and in a massively parallel signature sequencing (MPSS) analysis of hybrid oysters (Hedgecock *et al.*, 2002).

Thirdly, genetic complexity of transcript levels is reflected by the QTL number and effect size and also many forms of genetic complexity. Direct evidence for genetic complexity of transcript levels comes from detecting multiple QTL for at least some expression traits. Even the detected QTL typically explain only a minority of trait variation (Rockman and Kruglyak, 2006). In yeast, the median phenotypic effect of a detected QTL was 27% of genetic (inheritable) variance explained and only 23% of traits had a QTL that explained > 50% of the genetic variance (Brem and Kruglyak, 2005; Fig. 7.4). Whereas visible trait variation is often described by several QTL that collectively account for up to half of the genetic variance and individually rarely > 20% of it, eQTL accounting for 25–50% of transcriptional variation are prevalent (as summarized by Gibson and Weir, 2005). It is clear that major-effect QTL are more prevalent than many investigators would have expected.

Fourthly, transcriptional variation is probably highly polygenic. It is important to recognize that even in the cases where a major-effect eQTL explains half of the genetic variance for transcript abundance, the other half remains to be accounted for and in most cases will be caused by undetected loci. Because conservative thresholds of detection are required to adjust for the extraordinarily large number of comparisons involved in a genome-wide linkage scan for several thousand transcripts (the so-called 'multiple comparison problem'), most true eQTL remain undetected. Based on a yeast data set studied by Brem and Kruglyak (2005) transcription is more often likely to be highly polygenic than monogenic: only 3% of highly heritable transcripts are consistent with single locus inheritance, 18% suggest control by two loci and > 50% require at least five loci under an additive model (Fig. 7.5). They also argue that more than half of the transcripts show transgressive segregation (transcript abundance in F_2 progeny falls outside the range of both grandparents) and that > 15% are better explained by models that include epistatic interaction. Clearly, the landscape of gene expression in yeast is genetically complex and it can be expected that it will be anything but more complex in higher eukaryotes.

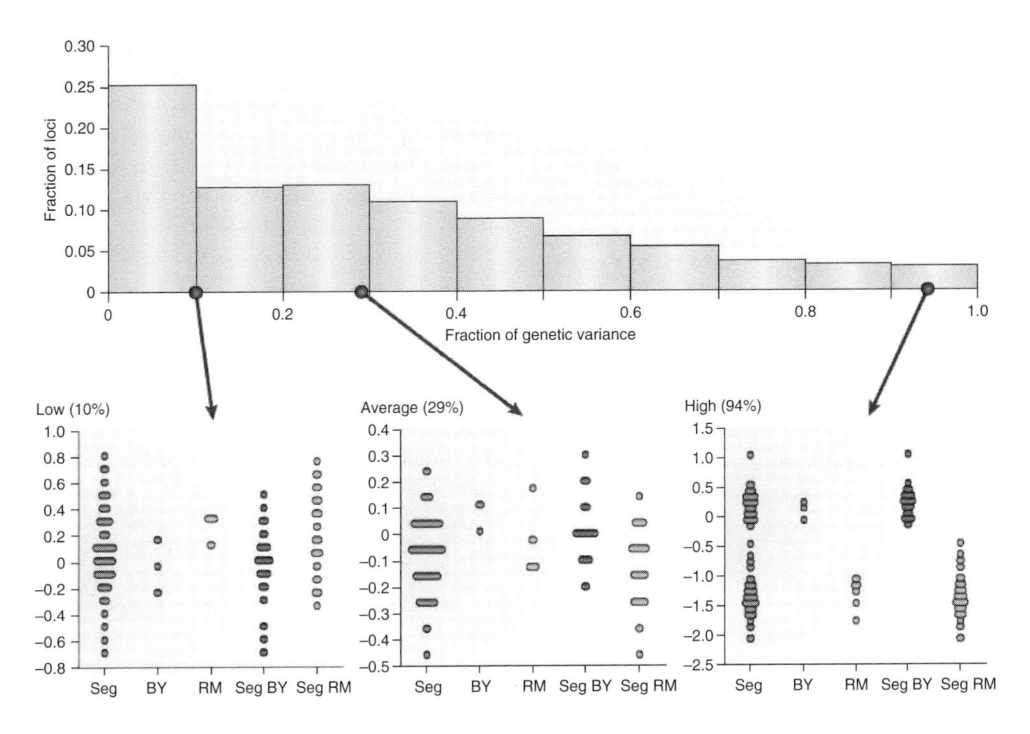

Fig. 7.4. Most gene expression traits are affected by multiple loci. Each bar represents the fraction of QTL that explain a percentage of genetic variance in the range on the *x*-axis. For each trait with significant linkage(s), only the single most significant QTL is included. Data are derived from the first table in Brem and Kruglyak (2005). The panels below the plot show examples of QTL that explain, from left to right, low (10%), average (29%) and high (94%) percentages of genetic variance. In each panel, the left-most column shows the relative expression of the corresponding genes in all 112 segregants (Seg), the next two columns show the expression in replicates of the two parent strains (BY, RM), and the last two columns show the expression in the segregants that inherit the QTL alleles for the first and second parent strains (Seg BY, Seg RM). From Rockman and Kruglyak (2006) reprinted by permission from Macmillan Publishers Ltd.

Fifthly, up to one-third of eQTL are *cis* acting. If an eQTL is mapped to a genomic region where the expression trait (eTrait) gene is located, it may suggest the *cis*-regulatory mechanism for the eQTL, i.e. certain sequence variations around the gene region of the eTrait, may directly influence the transcript abundance of the gene. In most of these cases, intuition suggests that the eQTL effect is likely to be caused by polymorphism in the regulatory region of the gene, namely sequence variation in the binding sites for transcription factors. Otherwise, mapping results may indicate *trans*-acting regulations, i.e. the variation of an eTrait is affected by sequence polymorphisms in other genes.

Sixthly, most of eQTL studies detect eQTL 'hotspots' that explain variation for multiple transcripts. The straightforward interpretation of these cases is that the eQTL identifies a regulatory gene that co-regulates as many as 25 downstream targets (but see de Koning and Haley, (2005) and Pérez-Enciso (2004) for a more critical interpretation).

7.6.2 Examples of eQTL in plants

Dissection of the genetics underlying gene expression combines large-scale microarray analyses of expression profiles and conventional QTL mapping of the same segregating population. In this analysis the expression profiling is considered a quantitative phenotype affected by multiple

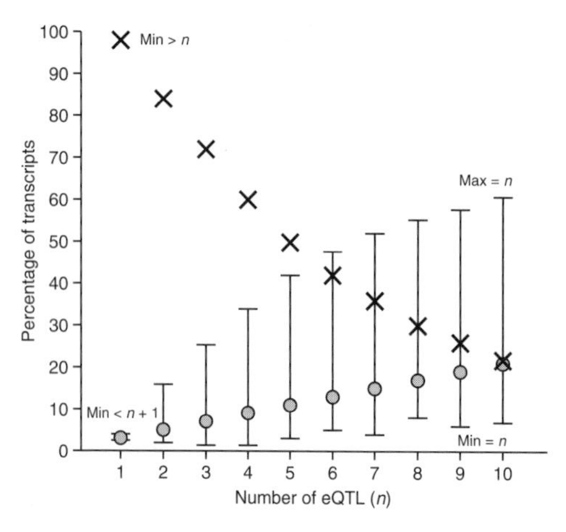

Fig. 7.5. Inference of polygenic regulation from eQTL analysis. A plot of the data from Brem and Kruglyak (2005) showing the range of complexity of transcriptional regulation in yeast inferred by their likelihood analysis. The error bars indicate the range of the percentage of differently expressed transcripts in the F_2 segregation that are predicted to be regulated by n genes indicated on the x-axis. The large Xs indicate the minimum number of transcripts regulated by more than n eQTL: for example, at least 20% of transcripts are predicated to have more than ten eQTL. The circles place a low limit on the number of transcripts regulated by up to n eQTL: for example, at least 10% of transcripts are regulated by four or fewer eQTL. Reprinted from Gibson and Weir (2005) with permission from Elsevier.

genes and environmental factors (Jansen and Nap, 2001). This approach has facilitated the identification of genomic regions or eQTL associated with transcript variation in co-regulated genes and when correlated with phenotypic data from a quantitative character, has successfully identified candidate genes by co-localizing gene eQTL and trait QTL (Brem *et al.*, 2002; Klose *et al.*, 2002; Wayne and McIntyre, 2002; Schadt *et al.*, 2003; Rockman and Kruglyak, 2006; Keurentjes *et al.*, 2007b).

Plants exhibit massive changes in gene expression during morpho-physiological and reproductive development as well when exposed to a range of biotic and abiotic stresses. These have been observed as differences in transcriptional profiles in many crops. Variation in transcript abundance is now being associated with gene expression using eQTL analysis in an increasing number of crops. For example, Kirst *et al.* (2004) dissected the genetic and metabolic network underlying variation in growth in an interspecific BC population of eucalyptus. QTL analysis of transcript levels of lignin-related genes showed that their mRNA abundance is regulated by two genetic loci, coordinating genetic control of lignin biosynthesis. These two loci co-localize with QTL for growth, suggesting that the same genomic regions are regulating growth and lignin content and composition. Using a high-density oligonucleotide array and phenotypically divergent rice accessions and their transgressive segregants, Hazen *et al.* (2005) measured the expression of approximately half of the genes in rice (~21,000) to associate changes in stress-regulated gene expression with QTL for osmotic adjustment (OA), which is a known mechanism of drought tolerance. A total of 662 transcripts were observed to be expressed differentially between the parental lines. Only 12 genes were induced in the low OA parent (CT9993) at moderate dehydration stress levels while over 200 genes were induced in the high OA parent (IR62266). Sixty-nine genes were upregulated in all high OA lines and nine of those genes were not induced in any of the low OA lines, of which four could be annotated as follows: sucrose synthase, a pore protein, a heat shock protein and a late embryogenesis abundant (LEA) protein. Previous conventional QTL mapping using the same two rice accessions showed that the parental genotypes differed for five of the OA QTL, that two of these QTL are syntenic with other cereal drought stress QTL (Zhang *et al.*, 2001) and a major OA QTL in the same genomic region on rice chromosome 7 is also reported in a different cross (Lilley *et al.*, 1996). Of the 3954-probes that correspond to this part of the chromosome, few showed a differential expression pattern between the high and low OA lines. Thus, these preliminary results demonstrate the power of integrating quantitative analysis of gene expression data with genetic map information to identify genetic and metabolic networks that would not have been identified through conventional QTL analysis.

Guo, M. *et al.* (2006) applied genome-wide transcript profiling to gain a global picture of the ways in which a large proportion of genes are expressed in the immature ear tissues of a series of 16 maize hybrids that vary in their degree of heterosis. Key observations include: (i) the proportion of allelic additively expressed genes is positively associated with hybrid yield and heterosis; (ii) the proportion of genes that exhibit a bias towards the expression level of the paternal parent is negatively correlated with hybrid yield and heterosis; and (iii) there is no correlation between the over- or under-expression of specific genes in maize hybrids with either yield or heterosis. The relationship of the expression patterns with hybrid performance is substantiated by analysis of a genetically improved modern hybrid (Pioneer® hybrid 3394) versus a less improved older hybrid (Pioneer® hybrid 3306) grown at different levels of plant density stress. The proportion of allelic additively expressed genes is positively associated with the modern high-yielding hybrid, heterosis and high-yielding environments, whereas the converse is true for the paternally biased gene expression. The dynamic changes of gene expression in hybrids responding to genotype and environment may result from differential regulation of the two parental alleles. Their findings suggested that differential allele regulation may play an important role in hybrid yield or heterosis and provide a new insight to the molecular understanding of the underlying mechanisms of heterosis.

Recently, Keurentjes *et al.* (2007b) described the results of genome-wide expression variation analysis in an RIL population of *A. thaliana* and for many genes variation in expression could be explained by eQTL. The nature and consequences of this variation are discussed based on additional genetic parameters, such as heritability and transgression and by examining the genomic position of eQTL versus gene position, polymorphism frequency and gene ontology. Besides, the authors have developed an approach for genetic regulatory network construction by combining eQTL mapping and regulatory candidate gene selection.

Most eQTL mapping studies to date have searched for eQTL by analysing gene expression traits one at a time. As thousands of expression traits are typically analysed, this can reduce power because of the need to correct for the number of hypothesis tests performed. In addition, gene expression traits exhibit a complex correlation structure, which is ignored when analysing traits individually. To address these issues, Biswas *et al.* (2008) applied two different multivariate dimension reduction techniques, the singular value decomposition (SVD) and independent component analysis (ICA) to gene expression traits derived from a cross between two strains of *Saccharomyces cerevisiae*. In total, 21 eQTL were found, of which 11 were novel and both *cis* and *trans*-linkages to the meta-traits were observed. These results demonstrated that dimension reduction methods are a useful and complementary approach for probing the genetic architecture of gene expression variation.

As we have discussed earlier, a range of biological and statistical tools enable research on natural variation to move from simple reductionistic studies focused on individual genes to integrative studies connecting molecular variation at multiple loci with physiological consequences. Hansen *et al.* (2008) provides a comprehensive review focusing on recent examples that demonstrate how expression QTL data can be used for gene discovery and to untangle complex regulatory networks. The latter is also briefly discussed in Chapter 10.

7.7 Selective Genotyping and Pooled DNA Analysis

As introduced in Chapter 6, replacing individual genotyping by selecting only the individuals from the high and low tails of the population distribution (selective genotyping) or DNA analysis in pools of the selected individuals (pooled DNA analysis) was proposed for QTL analysis and for testing of linkage between markers and a major gene. This concept is referred to as 'tail analysis' (Hillel *et al.*, 1990; Dunnington *et al.*, 1992; Plotsky *et al.*, 1993), 'bulked segregant analysis' (Giovannoni *et al.*, 1991; Michelmore *et al.*, 1991), or 'selective DNA pooling' (Darvasi and Soller, 1994) and is an effective solution to reduce costs associated with genotyping large mapping populations. As reducing the size of a QTL mapping population will decrease the detection power (Charcosset and Gallais, 1996) and also increase the QTL confidence interval, as well as the risk of detecting false QTL, selective genotyping can save much more cost than using smaller population sizes while maintaining the same mapping power as the large populations. Take a large population with 500 individuals and select 25 individuals from each tail, which means that selective genotyping will only cost 10% (= 2 × 25/500) of the total cost required for genotyping the whole population. When pooled DNA analysis is used, two tails can be genotyped as two individuals, which brings the genotyping cost down to 0.4% (= 2/500) of the total cost. Apparently, the bigger the original population size, the more saving there will be on all related costs including genotyping.

7.7.1 Major gene-controlled traits

Major gene-controlled traits can be selectively genotyped through bulked segregant analysis (Fig. 7.6A; Xu and Crouch, 2008). Briefly, individuals are selected so that two groups of individuals have contrasting phenotype, i.e. resistant versus susceptible plants, or early versus late maturity, with a randomized genetic background for other traits. Molecular markers that detect polymorphism between the two groups may be linked to the character and this possibility can then be tested in a segregating population. In order to isolate additional markers near a gene for which the genetic map location is known, individuals that are homozygous across a target interval can be selected from a segregating population based on known bracket markers. DNA from these individuals is then combined into two pools: one homozygous for one parental type and one for the other at the two marker loci. The result is that each DNA pool is homozygous at all loci within and adjacent to the target region. However, the homozygous target region differs between the two pools in parental origin, thus providing the basis for selection of polymorphic markers specific to the target region. When pooled DNA samples are subsequently utilized as templates for random primer amplification via PCR, polymorphism should result only if the primer primes within or adjacent to the target interval. This polymorphism can also be detected by probing with other molecular markers.

The genotyping in this approach becomes very simple since it relies on just two DNA pools each of plants from one or other of the phenotypic extremes (Giovannoni *et al.*, 1991; Michelmore *et al.*, 1991). The pools that have been used in plants usually consist of 10–15 individuals taken from as large a population as possible. This approach has already been successfully used in many plants (e.g. Barua *et al.*, 1993; Hormaza *et al.*, 1994; Villar *et al.*, 1996; van Treuren, 2001; Zhang *et al.*, 2002).

7.7.2 Quantitative traits

When selective genotyping is used for quantitative traits (Fig. 7.6B; Lander and Botstein, 1989; Xu and Crouch, 2008), the changes of marker allele frequencies between two tails can be used for detection marker–trait association as suggested by Stuber *et al.* (1980, 1982). Due to the hitchhiking effect, selection would change frequencies of markers that are more linked to the QTL for the selected trait (Lebowitz *et al.*, 1987). Darvasi and Soller (1992) have shown that selective genotyping led to a marked decrease in the number of individuals genotyped for a given power. This approach can be bidirectional if the two tails of the distribution are considered or unidirectional if only one tail is considered. The latter is more suitable for traits subjected to strong selection of unfavourable environments. Several applications of this method have already been reported for QTL detection or QTL validation in

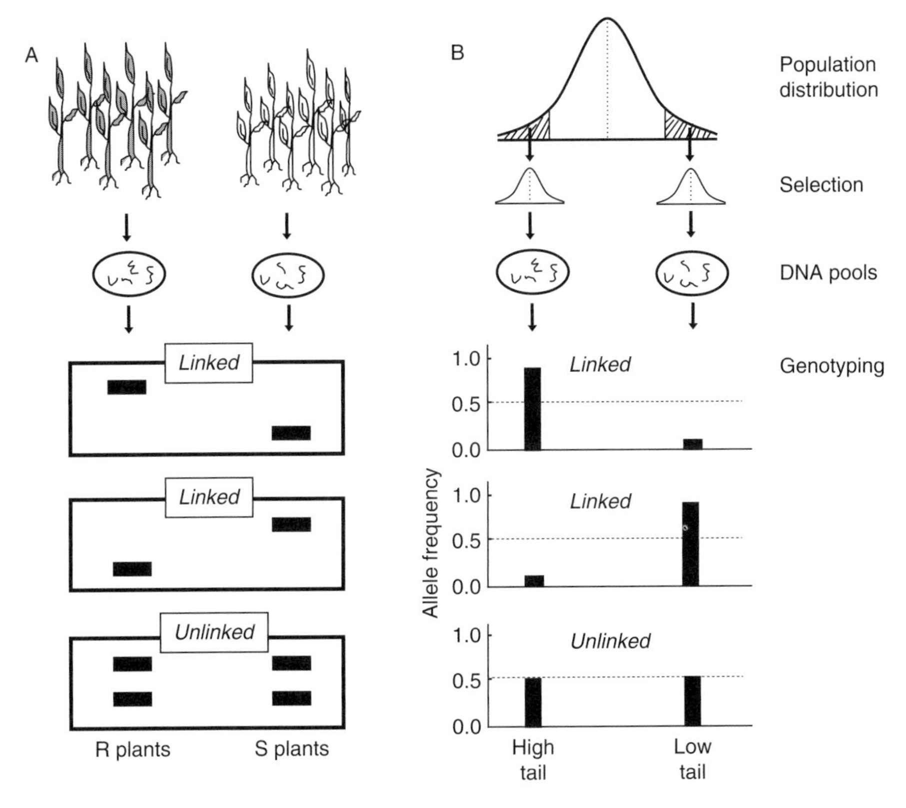

Fig. 7.6. Selective genotyping and pooled DNA analysis. (A) Pooled analysis using disease resistant (R) and susceptible (S) plants as an example. DNA pools are constructed from R and S plants selected from a mapping population and then genotyped by molecular markers. When the two DNA pools show different alleles at a specific marker locus, the marker is linked with the disease response, while when both pools show the same heterozygous genotype, the marker is unlinked with the disease response. (B) Pooled DNA analysis using extreme plants selected for a target quantitative traits from two tails of a normal distribution in the mapping population. Marker–trait linkage is revealed by allele frequency at specific marker loci. When allele frequencies are significantly different between two pools at a marker locus, the marker is linked with the target traits, while when the allele frequencies are very close to each other (each approximately to 0.5), the marker is unlinked with the target trait. In both A and B, assume that the marker is dominant and reveals polymorphisms between the parental lines that are used to derive the mapping population.

plants (Foolad and Jones, 1993; Zhang, L.P., *et al.* 2003; Wingbermuehle *et al.*, 2004; Coque and Gallais, 2006). It is especially useful for genes that have large effects on the trait of interest. It can also be used for traits controlled by a few major-effect QTL (Quarrie *et al.*, 1999). Furthermore, in maize, with two cycles of recurrent selection on phenotype from a population of F_4 independent families, Moreau *et al.* (2004) have shown that the significant changes in marker allele frequency were for a marker locus located in the vicinity of the detected QTL. An additional interest of the marker frequency approach is that it makes it possible to use DNA pooling of selected individuals to estimate the frequencies needed for the tests (Darvasi and Soller, 1994).

7.7.3 The power of selective genotyping and pooled DNA analysis

There are several problems associated with pooled or bulked DNA analysis in plants as summarized by Xu and Crouch (2008). These include: (i) a relatively small number of markers has been used to try to cover the whole genome with the assumption that the recombinant frequencies are consistent across the genome and genes of interest can be readily identified within a marker density of 15–25 cM; (ii) contrasting individuals have been selected from a relatively small population size so that the phenotypic difference between the pools may be only big enough for identification of large-effect genes/QTL; (iii) when allele signal is judged by a gel-based genotyping system, allele frequency in each pool cannot be quantified accurately and the allele signal generated by a small percentage of individuals in the pool cannot be detected and thus, the genetic difference between the pools can be only scored as presence and absence; and (iv) because of the above reasons, a relatively small number of individuals (about 15) is included in each pool to guarantee that the real associated markers will not be missed, at a cost of a high level of false positives (marker–trait is not really associated to each other but still indicated so statistically). The false positive markers have to be eliminated by a whole population validation step with all putative markers.

Simulation studies have been carried out by Xu *et al.* (2008) and Sun *et al.* (2009) using QTL IciMapping (available at http://www.isbreeding.net), an integrated computing package for common QTL mapping methods including single marker analysis, traditional interval mapping (Lander and Botstein, 1989), and inclusive composite interval mapping for additive (Li *et al.*, 2007) and interacting (Li *et al.*, 2008) QTL. Several parameters associated with selective genotyping were simulated based on the assumption that phenotypic extremes from two tails of a recombinant inbred population can be reliably selected and that they can be genotyped either individually so that the allele frequency in each tail can be inferred or genotyped using bulked DNA from each tail so that the allele frequencies can be estimated based on the relative signal strength of two DNA pools. Simulated parameters include total population size (ranged from 200 to 3000), tail population size (15–100 plants in each tail, equivalent to 13–50% of selection rate), number of QTL (1–5), marker density (1–15 cM), QTL effect (explaining 1–20% of phenotypic variation), two linked QTL and two QTL with epistatic interaction. One hundred simulation runs were carried out for each scenario from which the power of QTL detection and the mean LOD score were then calculated.

Comparative analysis of two selective genotyping strategies (Fig. 7.7) indicated that conventional selective genotyping (Fig. 7.7A, Strategy A, where relatively small total and tail population sizes were used with a low density of marker coverage), resulted in the detection of only one marker in the target region with an average LOD score of 3.94 and power of detection of 67%. In contrast, Strategy B (Fig. 7.7B), where large total and tail population sizes were used along with a high density of marker coverage, resulted in the detection of multiple markers around the target region with the highest having a LOD score of 10.37 and a power of detection of 98%.

When various QTL effects (responsible for 1–20% of the total phenotypic variation), tail sizes (15–100) and total population sizes (200, 500, 1000 and 3000) (Fig. 7.8) were used in the simulation analysis, the power of QTL detection indicated the optimum total and tail population sizes required for detection of small QTL. To identify QTL explaining 15% of the phenotypic variation with a 95% or higher power of QTL detection, will require a population size of 200 or more with a minimum tail size of 15, which matches most reported cases of successful use of bulked DNA analysis. However, to detect QTL of small effect, ranging from 3 to 10% of the phenotypic variation, 50–100 individuals needed to be selected from each tail of a population with 1000 individuals, in order to have a 95% power of QTL detection (Fig. 7.8). The simulation analysis also indicated that the power of detection would not change when multiple QTL (two to five) are involved but they are independent of each other. The simulation also indicated that selective genotyping can be also used to

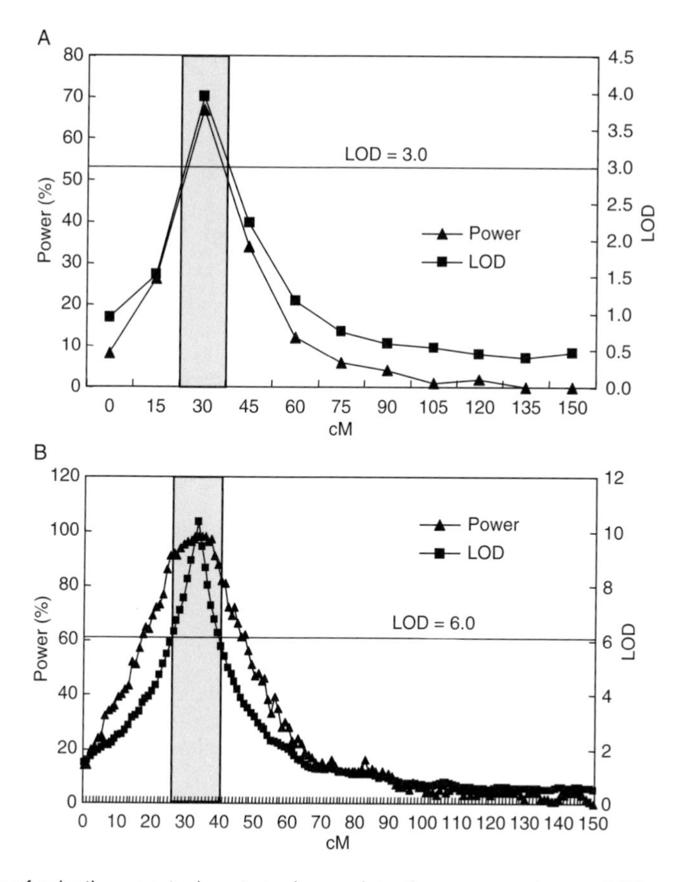

Fig. 7.7. Effects of selective genotyping strategies on detection power and mean LOD score around the target region (15 cM, grey area) assuming the QTL explain 10% of phenotypic variation. (A) Strategy A: population size = 200, tail size = 15, marker density = 15 cM, resulting in only one marker showing positive in the target region with an LOD score = 3.94 and power of detection = 67%, which has been widely used in conventional bulked DNA analysis. (B) Strategy B: population size = 500, tail size = 30, marker density = 1 cM, resulting in multiple markers showing positive in the target region with LOD = 10.37 and power of detection = 98%, which is proposed for selective genotyping-based fine mapping.

separate linked QTL at a distance of 25 cM and detect epistatic QTL.

A study of the optimization of selective genotyping without assumption on the QTL effect such as a negligible contribution r_p^2 of the QTL to the phenotypic variance, was developed, with selection either of both tails (bidirectional genotyping or BSG) or only one tail (unidirectional genotyping or USG) (Gallais *et al.*, 2007). For a given population size of phenotyped plants the optimal proportion selected for selective genotyping is around 30% for each tail. For the same investment as in ANOVA, by investing more in phenotyping than in genotyping when the cost ratio of genotyping to phenotyping is higher than 1, the optimal proportion selected appears to be between 10 and 20% for each tail. It is mainly affected by the cost ratio and decreases when the cost ratio increases. At this optimum, BSG is competitive with ANOVA, or even more powerful, when the cost ratio is higher than 1. USG can also be competitive when the cost ratio is higher than 2. Using experimental data from two populations of about 300 F_4 inbred families of maize, it was verified that BSG at the optimum gives the same results as

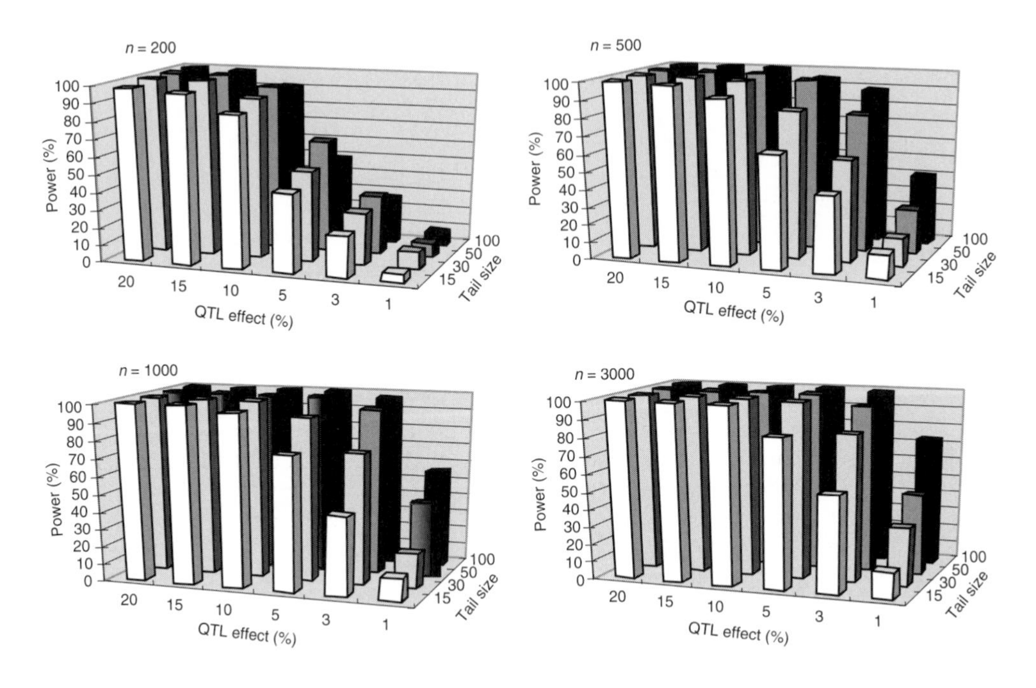

Fig. 7.8. Detection power of selective genotyping under various QTL effects (1–20%, percentage phenotypic variation explained by identified QTL), tail sizes (15–100) and total population sizes (200–3000). A total of 100 permutations were implemented for each case and each combination.

ANOVA or is better whereas USG is less powerful or equivalent.

7.7.4 Use of selective genotyping and pooled DNA analysis

Replacement of entire population genotyping

A general conclusion that can be drawn from the simulation analyses is that selective genotyping can be used to replace the entire population genotyping approach in almost all cases including QTL with relatively small effects as well as QTL with epistatic interactions or link QTL. It is recommended that for selective genotyping for QTL of different effects, the population sizes would be: 20 individuals from each tail of a population with 200 individuals for large QTL (15% or larger), 50 individuals from each tail of a population with 500–1000 individuals for medium-size QTL (3–10%) and 100 individuals from each tail of a population with 3000–5000 individuals for small QTL (0.2–3%).

'All-in-one plate' – genetic mapping of all target traits in one step

A large number of trait-specific genetic and breeding materials, with novel properties including inbreds/cultivars with extreme phenotypes, eternal/fixed segregating populations (e.g. RILs, DHs, NILs, introgression lines (ILs)), genetic stocks (e.g. single-segment substitution lines (SSSLs)) and mutant libraries, have been developed and maintained across the world for most crops. These are valuable directly for the purpose they were developed but also offer a novel resource for genetic mapping and gene discovery when used collectively. These materials have often been phenotyped in multiple environments due to their permanently fixed genetic composition. By collecting phenotypic extremes from currently available genetic and breeding materials and utilizing selective genotyping and pooled DNA analysis, it is theoretically possible that one 384-well plate could be designed to cover almost all major gene/QTL-controlled agronomic traits of importance in a crop species (Xu *et al.*, 2008; Sun *et al.*, 2009).

Genome-wide association mapping

Developments in SNP genotyping technologies and methodologies recently reported in human genomics have made it possible now to carry out genome-wide linkage-disequilibrium-based association mapping in human beings by using an integrated technology package including selective genotyping, pooled DNA analysis and microarray-based SNP genotyping with 100,000 markers (Sham *et al.*, 2002; Meaburn *et al.*, 2006; Yang, H.-C. *et al.*, 2006a). This system has the power to estimate allele frequencies and identify unique alleles from a pooled DNA sample of several hundreds of individuals. If this approach is successfully translated to plants it will resolve many of the constraints of pooled DNA analysis. The high frequency of false positive markers that would be detected when substantially fewer plants are used in each pool could be avoided if a pooled DNA can be formed using many more plants selected from a large population. However, optimizing SNP genotyping systems for pooled DNA analysis is considerably more complicated than for SSR markers and suffers a much higher level of redundancy. Where this has been achieved in human genomics, it required at least half a million SNPs as a starting point in order to identify 100,000 optimized SNPs suitable for pooled analysis. This density of SNP markers is available in rice and maize and in due course other crops when whole genome sequences are generated.

Genome-wide association mapping may provide a shortcut to discovering functional alleles and allelic variations that are associated with agronomic traits of interest. Selective genotyping and pooled DNA analysis can be extended to using inbred lines with extreme phenotypes selected from various collections of germplasm. This is in principal similar to linkage-disequilibrium-based association mapping but using selected phenotypic extremes. For association mapping of quantitative traits governed by a large number of minor genes that interact with each other and the environment, selective genotyping will face the same challenges as experienced with linkage-based QTL mapping using entire population genotyping.

Integration with selective phenotyping

The selective phenotyping method involves preferentially selecting individuals to maximize their genotypic dissimilarity. Selective phenotyping is most effective when prior knowledge of genetic architecture allows focus on specific genetic regions (Jin *et al.*, 2004; Jannink, 2005) and specific allele combinations. As genotyping becomes cheaper, it may be more efficient to first carry out low density genotyping of the whole population in order to identify the most informative subset of individuals in terms of minimum level of relatedness between individuals plus optimum subpopulation structure and allele representativeness. Then carry out precision phenotyping of this subset, particularly for the traits that are difficult or expensive to evaluate. And then finally carry out dense whole genome genotyping of the individuals from the tails of the phenotypic distribution. In this way, the total number of individuals to be phenotyped and genotyped may not change, but the power of the analysis will be dramatically increased. This approach could also be achieved for traits where phenotypic extremes can be easily identified by using a simple screening method, for example abiotic stress tolerance where a large number of plants/families can be eliminated easily under stress conditions through visual scoring. As the original population can be selected under a strong environmental stress to eliminate a large proportion of the plants, only the most stress tolerant and probably the most stress sensitive plants too, are selected for genotyping. Following selective genotyping of the individuals with extreme phenotypes, precision phenotyping of the resultant subset of individuals can be carried out using physiological component and surrogate traits. High-density planting and selection at early stages of plant development, combined with selective phenotyping

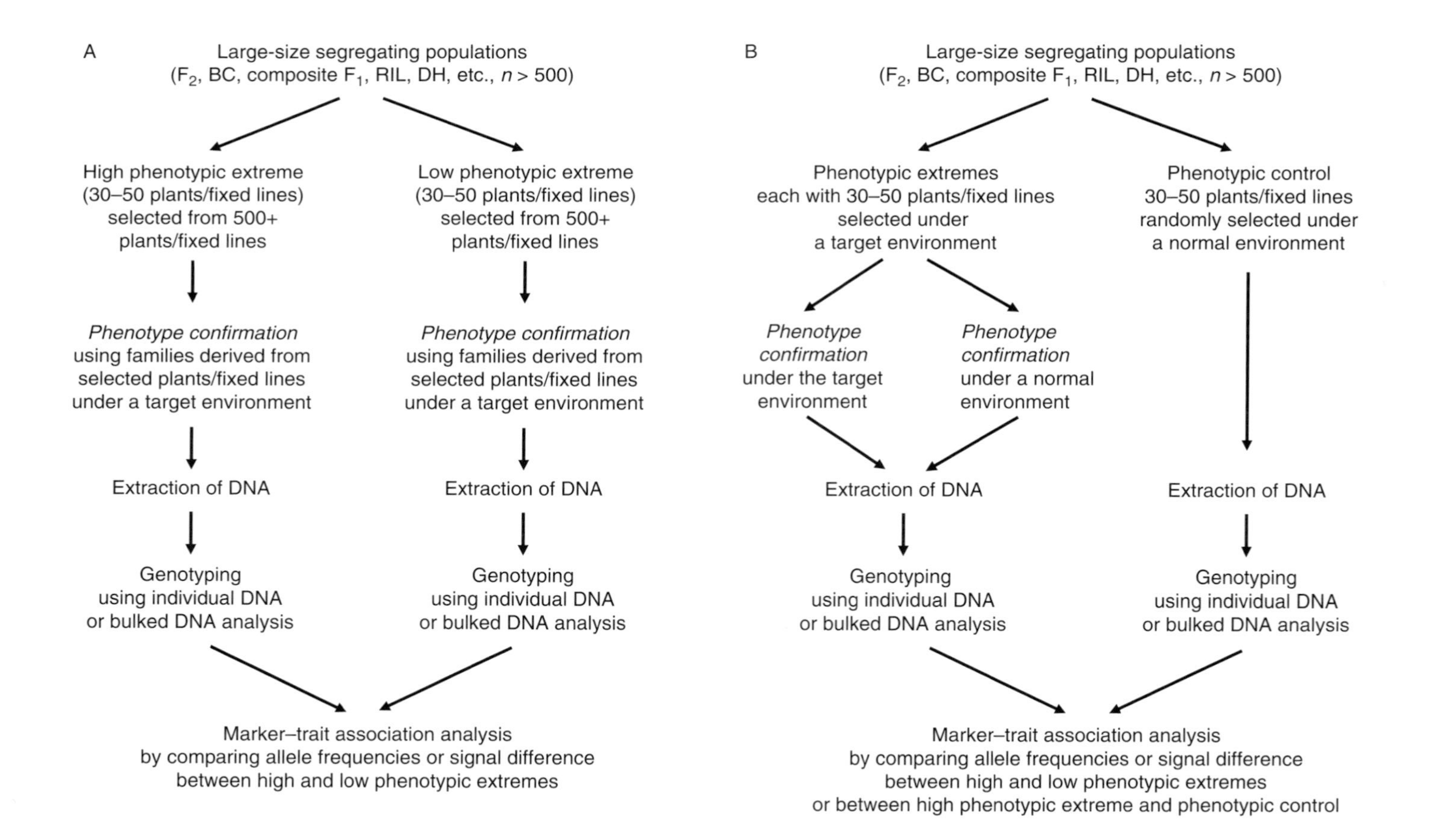

Fig. 7.9. Flowchart for large-scale selective genotyping and genetic mapping, including: selection of phenotypic extremes from large-size segregating populations, phenotype confirmation, DNA extraction, genotyping and marker–trait association analysis. (A) A procedure for most target traits which can be scored phenotypically for all individuals/fixed lines, and then high- and low-phenotypic extremes are selected for further analysis. (B) A procedure particularly suitable for abiotic and biotic stress tolerance where only the phenotypic extreme for tolerance is available under a target environment and comparison is made between the extreme and the phenotypic control that is randomly selected from the individuals/fixed lines under a normal environment. From Xu and Crouch (2008) with permission.

and genotyping should also be investigated as a potential option for some traits in order to allow one to work with more plants/families at the same cost (Xu and Crouch, 2008). Where the target trait is influenced by planting density or strong selection pressure this will clearly confound the ability to make genetic gain. However, many major-gene controlled traits can be investigated in this way without much disturbance.

Figure 7.9 shows this method for detection of marker–trait association for stress tolerance and other traits that can be selected for phenotypic extremes under a target environment. It can be inferred that phenotypic extremes or extremely stress tolerant plants are those with an accumulation of favourable alleles from multiple loci, each with small to large effects, so that genetic mapping will identify the genetic regions with relatively large accumulative effect on the target trait. In this case, lessons learnt from long-term selection of protein and oil content in maize may be highly instructive in this endeavour (Dudley and Lambert, 2004), particularly regarding the success of marker-assisted recurrent selection (MARS) to accumulate favourable alleles at numerous loci. In this approach, pyramiding of minor genes can be achieved using MARS to accumulate minor QTL where decades of breeding efforts have resulted in the fixation of all major genes.

It can be expected that selective genotyping and pooled DNA analysis, which have been widely used with mixed success in genetic mapping, will become increasingly important in genetic mapping and MAS and will gradually replace entire population genotyping in many cases. Selective genotyping will greatly facilitate and improve genetic mapping and marker-assisted breeding procedures in general. As genome-wide selective genotyping become possible, an effective information management and data analysis system will be required to make full use of the potentialities of selective genotyping in genetics, genomics and plant breeding.

8

Marker-assisted Selection: Theory

Conventional breeding procedures have been limited by the fact that the sources of genetic variation amenable to its manipulations are restricted in practice to those available within the gene pool of a single species. In addition, conventional breeding manipulations involve crossing entire genomes, relying on independent assortment and recombination to produce superior recombinants; and attempting to identify these from among many segregation products. Even for alleles with a major and clear-cut phenotypic effect this can involve many generations and the task may be almost impossible when desired and undesired alleles are closely linked. In principle, these limitations can be overcome by the genetic engineering procedures such as recombinant DNA methodologies. These procedures have enabled the cloning of genes with defined activities and their introduction in a rapid and highly specific manner across species boundaries. Thus, all kinds of life now become available as a source of useful alleles and the mixing of entire genomes is avoided (Beckmann and Soller, 1986a). The basic limitation in genetic engineering applications at present seems to be the public acceptance of genetically modified organisms (GMOs) derived from gene cloning and transformation approaches as described in Chapters 11 and 12.

In marker-assisted breeding the plant breeder takes advantage of the association between agronomic traits and allelic variants of genetic, mostly molecular, markers. The general idea behind marker-assisted breeding is as follows. Before a breeder can utilize linkage-based associations between traits and markers, the associations have to be assessed with a certain degree of accuracy and thus marker genotypes can be used as indicators or predictors of trait genotypes and phenotypes. When the alleles in question are few in number and have major effects on phenotype, such as a single gene-based disease resistance, the assessment of association is straightforward: mapping a monogenic trait goes along with the mapping of markers, as described in Chapter 2, while introduction of the desired alleles into the cultivar can be carried out readily by the classical breeding procedures of crossing, backcrossing, selfing and selection. In both cases, breeders depend on a clear relationship between genotype and phenotype to monitor the presence of the desired alleles in the populations of concern. For quantitative traits, however, a reliable assessment of trait–marker association requires large-scale field experiments as well as statistical techniques, known as quantitative trait loci (QTL) mapping as described in Chapters 6 and 7. Once marker–trait associations have reliably been assessed, the breeder is able to monitor the transmission of trait genes via closely linked markers, thus enabling 'genotype building', i.e. construction of desired genotypes by deliberate crossing and selection, using the marker genotype as a selection criterion.

The potential value of genetic markers, linkage maps and indirect selection in plant breeding has been known for over 80 years. Since the advent of DNA marker technology in the 1980s, it has dramatically enhanced the efficiency of plant breeding. In the past 20 years, a number of breeding companies have, to varying degrees, used markers to increase the effectiveness of selection in breeding and to significantly shorten the development time of cultivars (Dwivedi *et al.*, 2007). Now, advances in automated technology enable a new approach in marker-assisted breeding, called 'Breeding by Design'. The advances in applied genomics and the possibility to generate large-scale marker data sets provide us with the tools to determine the genetic basis for all traits of agronomical importance. Also, methods for assessing the allelic variation at these agronomically important loci are now available. This combined knowledge will eventually allow the breeder to combine favourable alleles at all these loci in a controlled manner, leading to superior cultivars (Peleman and van der Voort, 2003).

Changing concepts and molecular approaches provide opportunities to develop rational and refined breeding strategies. Knowledge about map position and allelic variation at agronomically important loci in concert with available, easy-to-assay molecular markers have made possible the design of superior cultivars. Compared to phenotypic assays, as summarized from Xu, Y. (2002), Peleman and van der Voort (2003), Xu, Y. (2003) and Xu and Crouch (2008), DNA markers offer great advantages to accelerate the cultivar development time as a result of the following.

1. Increased reliability: the outcome of phenotypic assays is affected, among others, by environmental factors, the heritability of the trait, the number of genes involved, the magnitude of their effects and the way these loci interact. Hence, error margins on the measurement of phenotypes tend to be significantly larger than those of genotyping scores based on DNA markers.
2. Increased efficiency: DNA markers can be scored at seedling stage or even based on seed before germination. This is especially advantageous when selecting for traits which are expressed only at later stages of development, such as traits associated with flower, fruit and seed. By selecting at the seedling stage or based on seed DNA, considerable amounts of time and space can be saved.
3. Reducing costs: there are ample traits where the determination of the phenotype costs more than the performance of genotyping using a PCR assay or hybridization. In a high-throughput setting, the material and consumable cost for a PCR assay will typically not exceed US$2. In comparison, the growth of a tomato or pepper plant to full maturity in a heated greenhouse will cost approximately US$20. Every plant that can be rejected before planting, particularly for those with the seed that is big enough for single seed-based DNA extraction, will in such settings save a considerable amount of money.

The use of DNA markers for indirect selection offers greatest benefits for quantitative traits with low heritability as these are the most difficult characters to assess in field experiments. Obviously, the development of marker-assisted assays for such traits is difficult and costly due to the extensive phenotypic assays required for such traits. However, once the knowledge exists to estimate the parameters which determine the trait of interest, a well-designed experimental set up will result in the availability of marker-assisted selection (MAS) tools, which can reduce to a major extent future application of phenotypical assays (Peleman and van der Voort, 2003). As described in previous chapters, molecular marker technology will help identify favourable alleles for agronomic traits, associate these alleles with specific molecular markers and introduce them from one genetic background to another through MAS. Theoretical considerations in MAS will be discussed in this chapter and the practical issues in MAS will be discussed in Chapter 9.

8.1 Components of Marker-assisted Selection

Key issues in successful deployment of molecular markers in MAS are as follows as summarized from Xu, Y. (2003) and Mohler and Singrün (2004):

1. Markers should co-segregate or map as close as possible to the target gene (e.g. less than 2 cM), in order to have low recombination frequency between the target gene and the marker. Accuracy of MAS will be improved if, rather than a single marker, two markers flanking the target gene are used. Ideally, gene-based markers that are developed from the sequence of the target gene, or functional markers that reveal functional differences associated with the target gene, are more preferred as segregation between the marker and the target gene will no longer exist or will be reduced to a minimum.
2. For unlimited use in MAS, markers should display polymorphism between genotypes that have and do not have the target gene.
3. Cost-effective, simple and high-throughput markers are required to ensure genotyping power needed for the rapid screening of large populations. Hybridization-based non-PCR markers that can reveal difference from DNA samples directly would be more preferred.

In addition, marker-assisted background selection depends on molecular markers that are well characterized and distributed over the whole genome. It is most desirable if gene-based markers are used for both marker-assisted foreground and background selection. In this case, a core set of markers can be established for both purposes so the same markers can be used for foreground selection in some crosses but background selection in others.

As summarized by Xu, Y. (2003), there are five key components that are required for efficient MAS, including: (i) suitable genetic markers and their characterization; (ii) high-density molecular maps; (iii) established marker–trait associations for traits of interest; (iv) high-throughput genotyping systems; and (v) functional data analysis and delivery.

8.1.1 Genetic markers and maps

Desirable DNA markers for MAS should meet the following requirements: detection of high frequency of polymorphism, co-dominance, abundance, whole genome coverage, high duplicability, suitability for high-throughput analysis and multiplexing, technical simplicity, cost effectiveness, requirement of a small amount of DNA and user-friendly (such as suitability for different genotyping systems and facilities). Among all these requirements, co-dominance is the most important for MAS in breeding hybrid crops, because the two parental inbreds and hybrid combinations can be distinguished unambiguously by co-dominant markers. Single nucleotide polymorphism (SNP) markers have great potential for super high-throughput analysis using chip technology and become available in more and more plant species, while simple sequence repeat (SSR) markers are more widely used across different crops. For all types of DNA markers available so far, SSR satisfies all the requirements. As estimated from a draft rice sequence, the density of SSRs in the genome is approximately one SSR per gene. These markers can be shared internationally through Internet-distributed primer sequences. SSR markers can be genotyped manually using agarose or polyacrylamide gels and ethidium bromide or silver staining, or in highly automated facilities. SSR markers can be multiplexed doing PCR and for multiple-sample loading on gels using fluorescent labelling. DNA extracted from a small piece of leaf or from single dry seeds will be enough to run several hundreds of markers. As more and more genes are cloned, it will be possible to develop molecular markers based on sequences within the gene of interest. Intragenic markers provide several advantages over gene-linked markers. First, there is no recombination between the marker and the gene, or intergenic recombination. Secondly, multiple alleles can be tagged and distinguished. An example is the SSR marker in rice, RM190, derived from a microsatellite sequence with a splice site in the waxy gene, which is responsible for amylose synthesis, an important grain quality trait for rice. As more and more cloned genes become available, functional markers developed from the target gene will be the best choice in MAS.

Selection of target chromosomal regions based on associated markers (foreground selection) and selection of genetic background for one of the parental genomes (background selection) may require different markers. Markers for foreground selection must be genetically mapped and associated with agronomic traits. Genetic markers revealing multiple bands or representing multiple loci are usually difficult to trace back to the specific allele/locus known to be associated with the trait, particularly when the population used for MAS is different from the population used for mapping. These types of markers include randomly amplified polymorphic DNAs (RAPDs) and amplified fragment length polymorphisms (AFLPs), which as such are not good for foreground selection. To use marker–trait associations based on these markers, it is best to use markers that are more locus-specific, such as sequence tagged site (STS) or SSR markers. For background selection, any type of markers that display a high rate of polymorphism is useful. Background selection does not require the use of mapped markers as long as they can reveal genome-wide polymorphism and the revealed differences can be traced back to their parents. As indicated previously, however, it is desirable to develop a core set of gene-based markers for both foreground and background selection so this set of markers can be used as universal markers across different populations.

The efficiency of MAS largely depends on how well markers are linked to the target trait. Construction of a high-density genetic map using high-throughput molecular markers is the first step to a large-scale MAS programme. A reference map is required for each crop or species based on a permanent segregating population that can be shared internationally allowing the placement of additional markers on the same map. This map should be constructed using markers that are friendly to users. There are two reasons why we need a high-density molecular map. First, a minimum requirement for MAS based on marker–trait association includes a three-marker system: one marker co-segregating with the trait for foreground selection and the other two flanking a target region for recombinant selection. Since the target gene can be located in any region of the genome, a dense map is required for identifying this triplet at any position in the genome. Even if a genic or functional marker is available for the target trait, the triplet is still needed for selection against the donor genome around the target when marker-assisted gene introgression is involved. Secondly, markers identified using mapping populations may not be polymorphic in breeding populations derived from other parental lines. To guarantee that the three-marker system will work for other breeding populations, many markers have to be identified around the target region. For the crop species with physical maps and whole genome sequences available, molecular markers will be available to cover the entire genome so that markers targeting specific genes and genomic regions can be selected from the marker and/or sequence databases. Using less expensive array-based genotyping systems, a core set of array-based of molecular markers can be established for all genetic mapping and genome-wide MAS for different populations.

8.1.2 Marker characterization

It is not enough to just have thousands of genetic markers in hand. To use molecular markers efficiently, they have to be characterized for many features, including: number of alleles; polymorphism information content (PIC); allelic difference (e.g. allele sizes and their range); allele feature (e.g. haplotypes) in standard or control cultivars; signal strength under specific genotyping conditions; background or noise signal; PCR or hybridization conditions; chromosome location (flanking markers and genetic distances); and information required for multiplexing.

Characterization of molecular markers helps to identify markers close to the genes of importance to breeding programmes and to evaluate germplasm and breeding materials. A core set of molecular markers should be characterized for each plant species and these markers should be evenly distributed on all chromosomes and suited for multiplexing. Many crop plants have now established core-set markers and have been used for evaluation of germplasm accessions, construction of heterotic pools and MAS (see Xu, Y. (2003) for an example in rice). Many efforts have been contributed to characterize array-based markers and optimize genotyping systems.

Allele number

The number of alleles at a marker locus is related to the genetic diversity that can be revealed by a particular marker. The more alleles at a locus, the higher the degree of diversity that can be revealed and the more efficiently closely related lines can be distinguished. SNP markers show polymorphism by two different nucleotides, usually displaying two different alleles across germplasm. Restriction fragment length polymorphism (RFLP) markers have many fewer alleles per locus compared to SSR markers. As a typical example, an average of 2.7 RFLP alleles and 11.9 SSR alleles per locus has been reported by Xu *et al.* (2004) based on a large set of germplasm accessions.

Polymorphism information content (PIC) value

The relative informativeness of each marker can be evaluated based on its PIC value, as described by Botstein *et al.* (1980), which reflects the amount of polymorphism and is a function of the number of alleles and allele frequencies at any given locus. PIC value is used to refer to the relative value of each marker with respect to the amount of polymorphism exhibited, which can be estimated by $PIC_i = 1 - \sum_{j=1}^{n} P_{ij}^2$, where P_{ij} is the frequency of the jth allele for marker i and the summation extends over n alleles (Weir, 1990; Anderson *et al.*, 1993). The calculation is based on the number of alleles detected by a marker at a given locus and the relative frequency of each allele in the tested accessions. In rice, the average PIC value was almost twice as high for SSRs (0.66) as for RFLPs (0.36) (Xu *et al.*, 2004).

Informative markers

Based on PIC values and the number of alleles detected, a set of highly informative markers can be selected such that the same amount of genotyping information can be obtained by surveying fewer molecular markers. Selected markers should be evenly distributed throughout the genome. As an example, a group of 24 RFLP/SSR markers was selected from 236 accessions × 160 markers rice data set as a set of highly informative markers for preliminary fingerprinting of rice germplasm and breeding populations (Xu *et al.*, 2004).

In addition to the requirements discussed previously for a marker system, there are some other requirements for a marker and a core set of markers. Taking SSR markers as an example, a useful marker should have many alleles per locus (> 10), high PIC value (> 0.8), suitable difference in allele sizes (4–10 bp between any two alleles), strong signal for detection, less background or noise signal and high replicability or reliability. As SNP markers have only two alleles at each locus, the informative marker sets rather than individual markers should be used and their informativeness can be judged by their haplotypes. Also allelic polymorphism should be considered for the markers developed from different regions of a gene for candidate gene-based, gene-based or functional markers. No matter which type of markers is used, a useful set of markers should provide whole genome coverage, even distribution on each chromosome and high potentiality for multiplexing or high-throughput genotyping.

8.1.3 Validation of marker–trait associations

Establishment of highly significant marker–trait associations is one of the prerequisites for MAS, which is discussed in Chapters 6 and 7. Demonstrated linkages between target traits/genes and molecular markers are traditionally based on genetic mapping experiments and it is important to confirm that these associations are consistent in mapping and breeding populations. As an example, Knoll and Ejeta (2008) validated three QTL for early-season cold tolerance in sorghum using two other populations and they found that all three associated markers were shown to retain influence in the different genetic backgrounds.

In many cases, however, genetic mapping results obtained from specific crosses cannot be used for MAS for the same traits in different crosses. There are three reasons for this phenomenon. First, quantitative traits are usually controlled by many genes. Genes are only segregating at the loci where two parents are genetically different and thus can be mapped using the population derived from these two parents. For a randomly selected mapping population, the parents will have a strong chance to share identical alleles at some of the genetic loci. There is a high probability that segregating genes in any breeding population could be different from the genes already mapped. Secondly, multiple alleles at a locus work in the same way to complicate MAS, because mapping parents could have alleles that are different from those of breeding populations. Interaction among these multiple alleles will modify marker–trait associations when different allele combinations are considered. Thirdly, genotype-by-environment interaction could make the establishment of marker–trait association depend on specific environments. Thus, QTL markers identified using a single mapping population may not be automatically used directly in unrelated populations without marker validation and/or fine mapping (Nicholas, 2006). The marker–trait association must be validated, in representative parental lines, breeding populations and phenotypic extremes before it can be used for routine MAS, particularly for QTL with relatively small effects. In a proportion of cases, markers will lose their selective power during this validation step. In these cases, there is a need to identify new markers (through fine mapping or candidate gene analysis) around the target locus in order to find marker–trait associations that are shared across different breeding populations. Shortcuts in this process may be possible through cross comparison with dense maps or by screening candidate gene markers for species where these are available. By finding several markers within a single gene, it is much more likely that the parents of any breeding population will be polymorphic for at least one of them, thus allowing breeders to track the alleles donated from each parent throughout the breeding process, speeding MAS and backcrossing in any cross. In the better studied species, up-to-date marker–trait association and associated SNP markers can be routinely accessed through gene cloning and fine mapping reports.

MAS has been successfully applied to date for monogenic and oligogenic traits controlled by major genes and for QTL of large effect influencing complex traits, particularly in the private sector (Dwivedi *et al.*, 2007). For more precise genotypic selection of complex traits such as the minor-gene controlled abiotic stress tolerance, more closely linked markers, preferably gene-based markers, or even better, functional nucleotide polymorphic markers (Rockman and Wray, 2002; Andersen and Lübberstedt, 2003), need to be developed. This should be combined with precision phenotyping in order to maximize the power of detection and minimize the chance of false negative.

There are many reasons why close marker–trait associations are required:

(i) chromosomal location associated with the trait must be reduced to a manageable piece of DNA if cloning of specific genes is necessary; (ii) to identify all the related genes for a specific trait, a high-density genetic map is required because the fewer markers are used, the smaller proportion of genetic factors contributing to that trait will be sampled; (iii) large genetic distances between markers and target traits will contribute to the rapid decrease of MAS efficiency after several successive cycles of selection; and (iv) to minimize linkage drag involved in gene introgression, closely linked markers around the target region are needed.

QTL mapping presumes accurate phenotypic scoring methods, something that can be difficult to optimize and even more difficult to keep consistent for months or years. Just a few mis-scored individuals can totally confound QTL discovery and placement (Young, 1999). This is also true for fine mapping of major genes for map-based cloning, where mis-scoring of several plants in a population with thousands of individuals will result in a large error (up to 1 cM) in estimating genetic distances. High levels of accuracy are required to dissect a chromosomal region associated with a given trait and narrow down the candidate region to a single contig or several mega bases, that is, a set of clones that can be assembled into a linear order.

8.1.4 Genotyping and high-throughput genotyping systems

To make marker-based technology practical for breeding applications, an automated genotyping system is required. As an ultimate marker type, SNPs have gained wide acceptance as genetic markers for use in linkage and association studies, especially for human genetics and many crop plants as well. High-throughput SNP genotyping has great potential for many applications, including MAS on the basis of whole genome approaches. This has led to a requirement for high-throughput SNP genotyping platforms. Development of such a platform depends on coupling reliable chemical assays with an appropriate detection system to maximize efficiency with respect to accuracy, speed and cost. With current technology platforms (e.g. Illumina), one lab can deliver throughputs in excess of one million data points per day, with an accuracy of > 99%, at a cost of US$0.06–0.10 per data point using array-based SNP genotyping system. In order to meet the demands of the coming years, however, genotyping platforms need to deliver throughputs in the order of one million genotypes per day at a cost of only a few cents per genotype (instead of per data point). In addition, DNA template requirements must be minimized such that hundreds of thousands of SNPs can be interrogated using a relatively small amount of genomic DNA. Released whole genomic sequences in model and crop plants including *Arabidopsis*, rice and maize have been used to develop gene-based SNPs for other related species.

8.1.5 Data management and delivery

To handle the daily data flow from the laboratory to the breeder and integrate information from molecular markers, genetic mapping and phenotyping, many informatics tools are needed. Decision support tools required in molecular breeding are fully discussed in Chapter 15 so only data management and delivery are briefly described here as a component of MAS.

For efficient data management and delivery, it is important for all researchers to follow general rules through all these procedures. A standard reporting system is also critical for comparative genomics, QTL allelism tests, data sharing and mining and the correspondence between major genes and QTL. As discussed by Xu, Y. (2002), a standard system for marker–trait association should include associated alleles and allele characterization such as allele sizes, gene effects, variation explained by each gene or all genes in the model, gene interaction if more than one gene is identified and genotype-by-environment interaction if more than one environment is involved. Genetic information should be shared and combined with data generated in plant breeding, for example, germplasm diversity, mapping populations, pedigrees, graphical genotypes, mutants and other genetic stocks.

With thousands or even millions of data points flowing out of a laboratory daily, timely scoring and delivery of the results to breeders are basic requirements for a high-efficiency breeding system. Well-trained assistants for genotyping and scoring, coupled with research scientists who can analyse data in meaningful ways, are the key components for a data management and delivery system. A laboratory with well-equipped facilities has to be also well equipped with qualified personnel and software required for data integration, manipulation, analysis and mining. Timely delivery of data to the breeder is also equally important, because in many cases the time window the breeder can use for selection is very limited. With the high-throughput genotyping and data management systems currently available, it takes 5–10 days to generate and analyse data including activities ranging from leaf tissue harvesting to DNA extraction, genotyping, data scoring, analysing, summarizing and reporting. The number of data points that can be generated and thus the number of plants that can be handled in a week in a lab depends on the level of high throughput and the availability of genotyping facilities.

8.2 Marker-assisted Gene Introgression

Gene introgression involves the introduction of a target gene into a productive, recipient line or cultivar. It can be used in both backcrossing and intercrossing programmes. By using DNA markers to identify recombinants, introgressed chromosome segments might be 'trimmed' to minimal size, reducing the extent to which the recurrent genotype is disrupted by undesirable alleles closely linked to the target trait. At the same time, genetic background can be selected genome-wide to minimize the donor genome content (DGC).

Historically, Tanksley and Rick (1980) and Tanksley (1983) considered the use of isozyme markers to speed the introgression of a trait controlled by a gene of major Mendelian effect from an exotic resource population to a cultivar. In this case, the general problem is to eliminate the exotic donor genome as rapidly as possible by replacing it with the recipient cultivar genome while retaining the gene of interest from the donor. This is generally accomplished by a larger number of backcross (BC) generations. Tanksley and Rick (1980) pointed out that if the chromosomes of the donor strain carry isozyme or other markers differentiating them from the recipient chromosomes, the number of BC generations required can be reduced dramatically by selecting against the donor marker alleles. Because of the general lack of isozyme or other markers differentiating cultivars from one another, Tanksley and Rick (1980) and Tanksley (1983) proposed this scheme as primarily of use for introgression from wild species to cultivar. With the availability of informative molecular markers to differentially mark the whole genome, however, this technique has been used for cultivar-to-cultivar introgression within a species.

Efficiency of MAS in gene introgression is affected by many factors including population size, genome size, marker–gene linkage intensity and number of markers. Stam (2003) raised several issues that are relevant to the design of an introgression-breeding programme:

- What amount of variation in DGC is to be expected in generations BC_1, BC_2, etc.?
- To what extent does this depend on the number of chromosomes and the genome size?
- What population size is required to guarantee, with 90% certainty, that at least one individual in a BC_1 has a DGC of less than e.g. 0.30?
- If markers are flanking the target gene, what are the optimal population sizes in successive generations to ensure

that the donor segment dragged along with the target gene is smaller than the segment bracketed by these flanking markers?
- Does it pay to increase the number of markers for background selection? If so, to what extent does this depend on population sizes used and/or genome size?
- If a certain pre-set goal, e.g. less than 0.05 DGC, is to be achieved in a given number of generations, should population size in successive generations be constant or is it better to vary population size over generations?
- If the number of generations is not a limiting factor, but the total number of plants to be genotyped is, then what is the optimal distribution of plant numbers over generations?
- Do the same guidelines for optimal transfer of a single target gene also hold for the transfer of multiple genes?

Some of these issues on gene introgression have been also addressed by a number of authors, using an analytical approach, numerical methods, computer simulations or a combination thereof (Hospital *et al.*, 1992; Hospital and Charcosset, 1997; Hospital, 2001; van Berloo *et al.*, 2001; Stam, 2003). As a special case, Frisch (2004) discussed the issues related to introgression of a recessive gene, where recurrent backcrossing without the aid of molecular markers requires progeny tests in each BC generation in order to determine whether a plant is a heterozygous carrier of the recessive gene or not.

8.2.1 Marker-assisted foreground selection

There are several approaches to using molecular markers to select an associated target gene or allele (foreground selection). Foreground selection can be used for gene introgression from one genetic background to another and pyramiding multiple genes/alleles to a genotype from multiple donors as well. For a specific target gene or allele, foreground selection can involve one to several markers. The simplest way is to use one closely linked marker (on either side of the target locus). The most complicated approach is to integrate foreground selection with background selection using multiple markers for the target locus and many others for covering the entire genome, this is referred to as 'whole genome selection' in this book and differs from genome-wide selection which will be discussed later in this chapter. The most frequently used approach is to use a triplet, marker–target–marker. Depending on how close the linked markers are to the target, the population sizes required for identification of particular genotype, the cost and efficiency related to foreground selection compared to phenotypic selection varies significantly. For example, a two-genetic locus model with one marker and one target locus involved can be simplified as selection for a single gene-based marker when the marker is developed from the target gene.

Selection using single markers

The reliability of foreground selection largely depends on the genetic distance between the markers and the target gene. If only one marker, located on one side of the target gene, is used in selection, the linkage between the marker and the gene has to be very tight in order to have relatively high selection efficiency. Suppose a marker locus (M/m) is linked with the target locus (Q/q) with recombination frequency of r and the F_1 has genotype MQ/mq, where Q is the target allele to be selected; when M is linked to Q, Q can be selected on the basis of M. The probability that the Q/Q genotype can be obtained through selection of marker genotype M/M, that is, the probability for selecting the correct individuals, is

$$P_1 = (1 - r)^2 \quad (8.1)$$

From Fig. 8.1, the probability for selecting the correct individuals decreases rapidly with the increase of recombination frequency. In order to have over 90% probability, the recombination frequency between

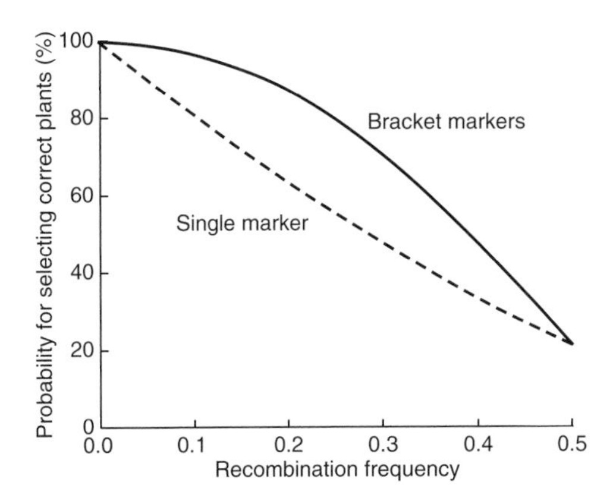

Fig. 8.1. Relationship between the recombination frequency between marker and target gene and the probability for selecting correct plants based on linked markers.

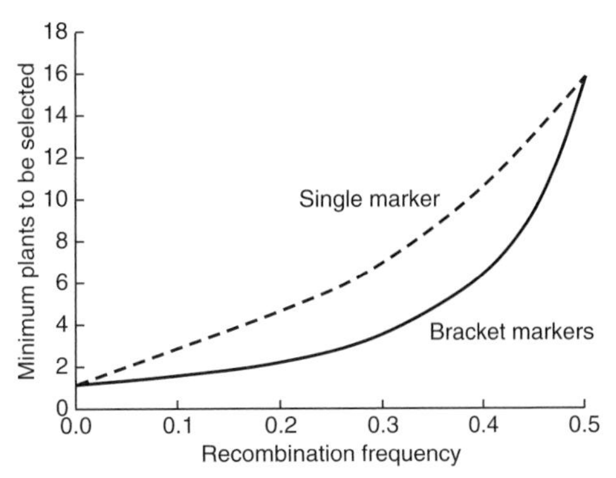

Fig. 8.2. Relationship between the recombination frequency between marker and target gene and the minimum plants that should be selected in MAS. Assume the target gene is between the middle of two markers, i.e. $r_1 = r_2$, when bracket markers are considered.

the marker and the target gene must be less than 0.05. When r is larger than 0.10, the probability reduces to the below 80%. However, if we just want to have at least one selected individual with the target genotype, MAS is still very helpful even if the linkage is very loose. If the probability for getting at least one desired individual with the target genotype is P_2, the minimum sample size required to obtain the target genotype M/M can be obtained from the probability function of the binomial distribution as

$$n = \log(1 - P_2)/\log(1 - P_1) \quad (8.2)$$

Frisch *et al.* (1999b) provided the probabilities for getting the target genotype, P_2, which are not only defined by factors associated with the target gene and its flanking markers but also by the condition that the complete chromosome region between a flanking marker and the nearest telomere consists entirely of the recurrent parent genome.

Figure 8.2 indicates the relationship between the minimum sample size required and the recombination frequency when $P_2 = 0.99$. Even if the recombination frequency is as high as 0.3, only seven plants that have M/M genotype are needed in order to have 99% probability that at least one of them has the target genotype. In selection without using molecular markers, which is equivalent to non-linkage between the marker and the target gene ($r = 0.5$), at least 16 plants are needed.

Selection using bracket markers

By monitoring markers flanking the target locus and the recipient alleles at the flanking markers, the length of intact donor chromosome segment around the target gene can be reduced efficiently (Tanksley *et al.*, 1989). This rationale can be used to determine the population size in a BC programme such that the recombinants between the target gene and the flanking markers can be found with a high probability.

To reduce false positives in MAS, flanking markers or multiple markers around the region should be used simultaneously. A three-marker system, with three markers located on a chromosome block, will be desirable in this case (Zhang and Huang, 1998). The marker in the middle, preferably intragenic or co-segregating with the gene, will be used to indicate the presence of the target gene in the selection process. The marker on each side will be used to indicate the absence of the chromosome segment from the donor parent (negative selection), that is, selection for recombination between the target gene locus and the marker loci. As more and more genes have been cloned, the marker in the middle could be developed from the cloned gene. This system will be very useful when the target gene is only available in a wild species and linkage drag is associated with the chromosome segment to be introgressed.

Suppose there are two marker loci (M_1/m_1 and M_2/m_2) which are located on each side of the target gene (Q/q) with recombination frequencies r_1 and r_2 and the F_1 has genotype M_1QM_2/m_1qm_2. The F_1 will produce two gametes with the marker genotype M_1M_2, one of which is the parental type containing the target allele (M_1QM_2) and the other is the double-crossed containing non-target allele (M_1qM_2). Because the frequency of double crossing is very low, the double-crossed gamete is very rare. As a result, the probability of making the correct selection for the target allele Q based on the presence of M_1 and M_2 is very high. Under no interference, the probability of obtaining the target genotype Q/Q based on selection of M_1M_2/M_1M_2 in the F_2 generation is

$$P_1 = (1-r_1)^2\,(1-r_2)^2/[(1-r_1)(1-r_2)+r_1r_2]^2 \quad (8.3)$$

When the target gene is located in the middle of two flanking markers, i.e. $r_1 = r_2$, the probability of making the right selection is minimized. Figures 8.1 and 8.2 show the relationship between the minimum number of plants required and r_1 (or r_2), when $P_2 = 0.99$ and $r_1 = r_2$. Selection efficiency is much higher using two flanking markers than using one marker. With interference between single crossovers (as is generally the case), the frequency of actual double crossovers is lower than the expected value (which assumes no interference). Therefore, the actual probability for making the right selection based on flanking markers should be higher than the theoretical expectation.

The population size required to generate (in a single BC generation) a high probability of obtaining at least one plant recombinant between the target gene and both flanking markers is greater than the reproductive rate for most crop species. For example, for a flanking marker distance of 5 cM on each side of the target gene, about 4000 individuals are required to find a double recombinant with a probability of 0.99 (Frisch *et al.*, 1999b). Therefore, Frisch (2004) proposed a sequential strategy to find an individual with recombination between the target gene and one flanking marker in generation BC_1 and a recombinant between the target gene and the second flanking marker in generation BC_2 (also see Fig. 8.4b for further explanation of this strategy).

Table 8.1 gives the optimum population size n_1 in generation BC_1 and corresponding expected population size $E(n_2)$ in generation BC_2 such that the expected total number of individuals $E(n) = n_1 + E(n_2)$ required to introgress one gene with a minimum number of individuals in a two-generation BC programme is minimized. The values depend on the map distances d_1 and d_2 between the target gene and two flanking markers (Frisch, 2004).

Table 8.1. The expected total number of individuals $E(n) = n_1 + E(n_2)$ required to introgress one gene with a minimum number of individuals in a two-generation BC programme (from Frisch (2004) with kind permission of Springer Science and Business Media).

	Map distance d_2 (cM)				
Map distance d_1(cM)	4	6	8 $n_1/E(n_2)$[a]	12	16
4	143/252	136/186	130/155	123/128	117/117
6		91/167	88/135	83/105	79/93
8			66/125	63/94	60/80
12				48/83	41/68
16					32/62

[a]n_1 is the optimum population size in generation BC_1 and $E(n_2)$ the corresponding expected population size in generation BC_2.

Selection using multiple markers for multiple targets

MAS provides opportunities for simultaneous selection of multiple traits/genes using multiple markers. In some cases, multiple pathogen races or insect biotypes must be used to identify plants for multiple resistances, but in practice phenotypic selection may be difficult or impossible because different genes may produce similar phenotypes that cannot be distinguished from each other. Marker–trait association can be used to simultaneously select multiple resistances from different disease races and/or insect biotypes and pyramid them into a single line through MAS.

For example, to find a restorer for cytoplasmic male sterility (CMS) in rice through testcrossing and progeny test, a candidate male plant has to be testcrossed with a CMS line to find out if it has fertility restorability based on the fertility of testcross progeny. However, sterility in testcross progeny could result from the absence of either restorability genes or wide compatibility genes or both when an intersubspecific cross is involved. MAS using multiple markers could be used to distinguish the two different types of sterility. As another example, consider phenotypic selection for multiple traits in rice, such as thermal-sensitive genic male sterility (TGMS), amylose and wide compatibility. Candidate plants must be tested in two different environments where TGMS can be identified. Each plant must be testcrossed with wide compatibility testers, following up with a progeny test in the next season. At the same time, a relatively large amount of seed must be harvested for amylose measurement. While conventional selection methods require a delay until a large number of seeds are available and a reasonable level of homozygosity is reached, in MAS only a leaf harvested at any growth stage in any segregating population is required, with the availability of associated markers for these traits.

As genetic mapping information accumulates from different mapping populations, it may be possible to establish a complete profile for all the genes associated with a specific trait or trait category. The multiple-marker approach can be used to select the best trait/gene combinations based on selection for each of the target loci whose position in the genome is known. It is possible to select the best cassette for any traits and/or trait combinations.

When single chromosomes are distinguishable, partial genome selection or whole chromosome selection are possible as an alternative to whole genome selection so that the other chromosomes remain unchanged. MAS could be focused on a chromosomal region/arm if it is separable from the rest of the genome. Genes controlling the same traits or trait category may cluster in some specific chromosomal regions, which are called gene blocks. Regional mapping strategies (Xu, 1997; Monna *et al.*, 2002), combined with a high-density genetic map, can help construct high-density regional maps that target gene blocks for separation of closely linked genes.

8.2.2 Marker-assisted background selection

In a BC programme, molecular markers can be used for indirect selection for the presence of a favourable allele (Tanksley, 1983) and for selection against the undesirable genetic background of the donor genotype (Tanksley *et al.*, 1989). Selection for the remainder of the genome excluding the target gene(s), i.e. genetic background, is called background selection (Hospital and Charcosset, 1997).

The background selection is aiming at the whole genome. In a segregating population, each chromosome represents a random combination of two parental chromosomes. So we have to know the parental combination of each chromosome in order to do whole genome selection, i.e. the entire genome has to be covered by molecular markers. For an individual plant, we can infer the parental origin for each marker allele across the whole genome when genotypes at all marker loci are known and thus we can infer the parental combination for each chromosome.

Concept of graphical genotypes

In breeding programmes, it is important to consider the complete genome of individuals, in addition to specific target genes. In sexually reproducing organisms, segregating progeny contain chromosomes that are mosaic of chromosomal pieces derived from their parents. Knowledge of the molecular marker genotype at one specific locus or haplotype across several closely linked loci yields information about the parental origin of alleles at that particular site in the genome. Knowledge of the molecular marker genotypes of many linked loci throughout the entire genome yields an estimate of the exact composition of an individual's chromosomes in terms of its parents. In other words, information about linked points in a genome permits deduction of a continuous genotype, which can be displayed graphically (Fig. 8.3). Graphical genotypes provide a clear picture of the genomic structure of each plant, which facilitates MAS, and are particularly useful for background selection. The selection is first made for foreground to retain the target gene and background selection is then made among the plants already selected for foreground.

In any meiosis, zero, one, or more crossover events may occur between a given pair of homologues. When crossover has occurred, a complete description of an individual's genome would include information on changes in allelic constitution due to recombination, as well as information on the locations where crossover events occurred (Young and Tanksley, 1989a). In transmission genetics, individuals are routinely described by their genotype at one or more genetic loci of interest. The description is generally alphabetic or numerical in nature and provides precise information on the derivation and allelic constitution at the specific loci. High-density molecular maps can be used to determine the genotype of an individual at thousands of loci and thus it is possible to deduce the most probable genetic constitution for regions of interest or the entire genome in a given individual. The graphical genotype, which portrays molecular data in a graphical form, has a number of advantages over numerical

a 1: 1122222222; 2: 22222; 3:111122; 4: 333333; 5: 22333; 6: 11111; 7: 333332; 8: 22111; 9:3333; 10: 1233322 11: 111; 12: 23333

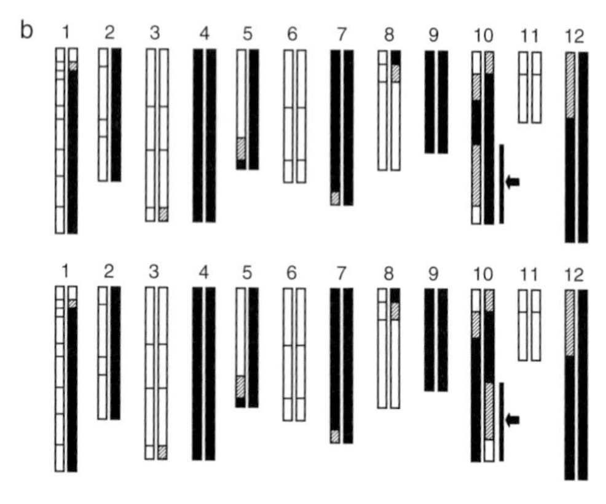

Fig. 8.3. Graphical genotype for an individual from a tomato F_2 population derived from a cross between *Solanum esculentum* and *Solanum pennellii* (also known as *Lycopersicon esculentum* and *Lycopersicon pennellii*). (a) Numerical RFLP data presented in order along the 12 chromosomes in the genome: 1, homozygous for *S. esculentum*; 2, heterozygous; 3, homozygous for *S. pennellii*. (b) Graphical genotypes derived from the numerical RFLP data shown in a. White intervals indicate segments derived from *S. esculentum*; blackened intervals from *S. pennellii*; and striped intervals indicate segments containing a crossover event. Two homologues of each chromosome pair are shown side by side. Two isomeric graphical genotypes of equal likelihood are shown that differ only in the region noted by the arrow and thick line. From Young and Tanksley (1989a) with kind permission of Springer Science and Business Media.

genotypes. It is similar to cytological karyotypes in describing an entire genome in a single graphic image, but different in that graphical genotypes would be inferred from molecular marker data and thus would show the genomic constitution and parental derivation for all points in the genome.

To develop a graphical genotype, molecular marker data, obtained in a numerical form, need to be transformed into an easily interpretable and accurate graphic image. Young and Tanksley (1989a) developed the mechanics for conveying RFLP data in the form of a graphical genotype, applied this concept to BC and F_2 populations of tomato and discussed several issues relating to the potential power and application of graphical genotypes. The term RFLP markers used in their paper can be extended to include genotypes derived from all types of molecular markers that are co-dominant and haplotypes derived from di-allelic markers such as SNPs. This concept (developed on the basis of structured populations such as BC and F_2) can be extended to all populations including natural populations that consist of germplasm accessions or cultivars.

Requirements for deducing graphical genotypes

In order to construct a graphical genotype, certain conditions must be met. First, a well populated or high density, molecular map, for the entire genome of the species must be available. This map should consist of a large number of markers that cover the entire genome with at least one marker every 10 cM or less. In addition, it is also necessary that the *cis–trans* configuration for the molecular markers be known in order to prepare a graphical genotype. In populations derived from inbred lines, such as breeding populations consisting of BC or F_2 progeny, the *cis–trans* configuration can be inferred simply by the knowledge of the breeding scheme. In more complex situations, complete molecular marker data must be obtained for three generations in order to prepare graphical genotypes for individuals in the third generation. In humans, for example, molecular marker data must be determined for grandparents and parents in order to develop graphical genotypes for the children in the pedigree. Without this knowledge of *cis–trans* configuration, molecular marker data from some regions of the genome may have more than one possible graphical genotype, all of which are equally likely to be correct.

Assumptions employed in developing graphical genotypes

The primary assumption required for the development of graphical genotypes is that the genotype of a region between two molecular markers is inferred from the genotypes of the markers that delimit the interval. When inferring the graphical genotype of an interval from the genotypes of the marker endpoints, there are often alternative configurations that will satisfy the available marker data. Young and Tanksley (1989a) used the most likely configurations to develop a graphical genotype. Thus, simple configurations requiring the fewest number of crossover events were utilized in developing a graphical genotype, while alternative configurations that require one or more multiple crossover events are not. In practice, this means that if two consecutive loci have the same genotype, the genotype of the segment between the markers is inferred to be that of two flanking markers. When two adjacent loci have different marker genotypes, it is inferred that a crossover event had taken place somewhere between the two loci.

Since the genotype of a non-recombinant interval is inferred from the genotype of its marker endpoints, double crossovers (or other even numbers of crossovers) in a given interval will falsify this inference and the likelihood of double crossovers increases by the square of the probability of a crossover between the adjacent molecular markers. Thus, for any interval, the probability that the inferred genotype will be correct is $1 - r^2$, where r is the probability of a crossover event between adjacent molecular markers. For the total genome, the probability that there are no incorrect intervals is

$$P_t = \prod_{n=1}^{\text{Total intervals}} (1 - r_n^2) \quad (8.4)$$

This equation considers only double crossovers and assumes interference between crossovers to be negligible. As an example, consider an organism with a total genome size of 1000 cM in which molecular markers are evenly spaced over the entire genome. The expected proportion of the genome which is described correctly by the graphical genotype is calculated by first determining the probability of 0, 1, 2,... intervals that are incorrectly described for a given spacing of molecular markers. These probabilities, along with the spacing size, are then used to determine the expected length of the genome correctly inferred, which is then divided by the total genome size to yield the expected proportion of the genome that is accurately portrayed by the graphical genotype. With molecular markers spaced every 10 cM, an inferred graphical genotype will have a probability of only 30% of being exactly correct for all regions (i.e. no incorrect intervals). However, this same graphical genotype will be accurate in describing the genome constitution for over 99% of the genome. Even when the spacing between molecular markers increases to 30 cM, the inferred graphical genotype will be accurate for approximately 95% of the genome. Apparently, as the number of manageable and available molecular markers becomes unlimited compared to the number when the concept was proposed, the correct probability will be improved significantly.

Cis–trans ambiguity happens in an F_2 population when heterozygous loci are separated by a stretch of one or more homologous loci. In this situation, two equally likely graphical genotypes are possible that differ in the *cis–trans* configuration of the flanking heterozygous regions (see Fig. 5 of Young and Tanksley, 1989a). Calculations based on the Poisson distribution indicated that only 6% of a genome consisting of ten chromosomes of 100 cM each will be ambiguous. The utility of graphical genotypes in F_2 populations will not generally be seriously impaired by *cis–trans* ambiguities.

Application of graphical genotypes

A graphical representation of a genotype deduced from RFLP data for a randomly selected individual from a tomato F_2 population provided by Young and Tanksley (1989a) is shown in Fig. 8.3. Note that it is not only possible to see which portions of each set of homologues are derived from each parent, but also the regions in which crossovers took place.

Using graphical genotypes, plants can be selected that not only contain the gene(s) of interest, but also have the highest probability that the rest of the genome will return to that of the recurrent parent with additional crossing. Although the concept of the graphical genotype was proposed a long time ago, it has been widely used in different fields of genomics. It has been used, as described in Chapter 4, for selection of genome-wide introgression lines as a library to cover all traits and the whole genome segment by segment. As molecular marker data increase exponentially with the availability of high-throughput genotyping systems, the concept of the graphical genotype and its derivatives have received more attention and are widely used in MAS, near-isogenic line (NIL) construction, introgression line library development and association mapping. As numerous points in the genome can be covered by markers, graphical genotypes can be simplified by displaying them using the physical positions of markers rather than the intervals determined by flanking markers.

8.2.3 Donor genome content in BC generations

DNA marker-based whole genome selection or background selection can be used to accelerate recovery of a recurrent genotype in the backcrossing process for improving parental lines. The basic principle of background selection (as opposed to 'foreground selection' on the target gene) is that in any given BC generation the actual DGC varies around the theoretical mean value.

Once QTL alleles of interest in the resource (donor) parent have been identified by linkage to resource-specific marker alleles, repeated backcrosses to the cultivar (while choosing in each cycle only those backcrossing progeny carrying the exotic QTL-linked marker alleles) will allow the effective introgression of the linked quantitative alleles from the donor into the cultivar. Depending on the number of alleles to be introgressed, it may be possible to expedite matters by actively selecting against exotic marker alleles (and hence against the associated chromosomal regions) that are not in linkage to introgressed alleles.

Table 8.2 shows the frequency of a favourable allele after one to six BC generations, with and without selection for a

Table 8.2. The frequency of a favourable allele after a given number of BC generations, with and without selection for a linked marker allele or for a pair of markers bracketing the favourable allele (marker bracket) and the proportion of recipient genome recovered with and without MAS against the remaining exotic genome. From Beckmann and Soller (1986a) by permission of Oxford University Press.

Number of BC generations	MAS for favourable alleles (frequency of favourable alleles)			MAS against remainder of exotic genome (proportion of recipient genome recovered)	
	None	Single marker[a]	Marker bracket[b]	None	Full marker coverage[c]
1	0.25	0.81	0.92	0.75	0.85
2	0.12	0.73	0.88	0.88	0.99
3	0.06	0.66	0.85	0.94	1.00
4	0.03	0.59	0.82	0.97	1.00
5	0.02	0.53	0.78	0.98	1.00
6	0.01	0.48	0.75	0.99	1.00

[a]Proportion of recombination between marker allele and linked favourable allele, 0.10.
[b]Proportion of recombination between the two markers of the bracket, 0.40.
[c]Two markers per chromosome.

linked marker allele. Also shown are results if selection is for a pair of marker alleles bracketing the QTL to be introgressed. Suppose that the proportion of recombination between marker allele and linked favourable allele is 0.10 when a single marker is used and the proportion of recombination between the two markers of the bracket is 0.40. The comparison of interest is the frequency of the introgressed alleles after three generations of marker-assisted backcrossing (MAB) (single marker, 0.66; marker bracket, 0.85) compared to the frequency of the introgressed allele after five to six unassisted BC generations (0.01). In the former case, the introgressed allele will have an immediate effect on cultivar value and can be rapidly brought to fixation by selfing or selection.

With two markers per chromosome used for MAS against the remainder of the exotic genome, the proportion of recipient (recurrent) genome recovered in BC_2 will be equal to that obtained in BC_6 without MAS (Table 8.2). This result is also given Fig. 8.4 and is well recognized by many authors (e.g. Tanksley *et al.*, 1989; Hospital *et al.*, 1992; Frisch *et al.*, 1999a, 2000). Therefore, selection based on markers that distinguish between donor and recurrent parent genome may considerably accelerate the recovery of the recurrent parent genome.

If only two background selection markers on the target chromosomes are used (assuming direct selection for the target gene), the distances d_1 and d_2 between the target gene and markers can be chosen such that the expected DGC on the target chromosome is minimized if both markers are fixed for the recipient alleles (Hospital *et al.*, 1992) by applying

$$d_1 = d_2 = \frac{1}{2}\ln(1+2\sqrt{s}) \quad (8.5)$$

where s is the proportion of selected BC_1 individuals. This approach is based on the assumption of an infinite population size and the optimum properties only hold true if two markers in the carrier chromosome of the target gene are used (Frisch, 2004).

8.2.4 Linkage drag in gene introgression

When transferring a single gene from a donor into the genetic background of a recurrent parent by repeated backcrossing, genetic linkage will cause fragments of the donor genome surrounding the target gene to be 'dragged' along, which is called 'linkage drag', a persistent problem in plant breeding for gene introgression. Small donor genome fragments, not linked to the target gene,

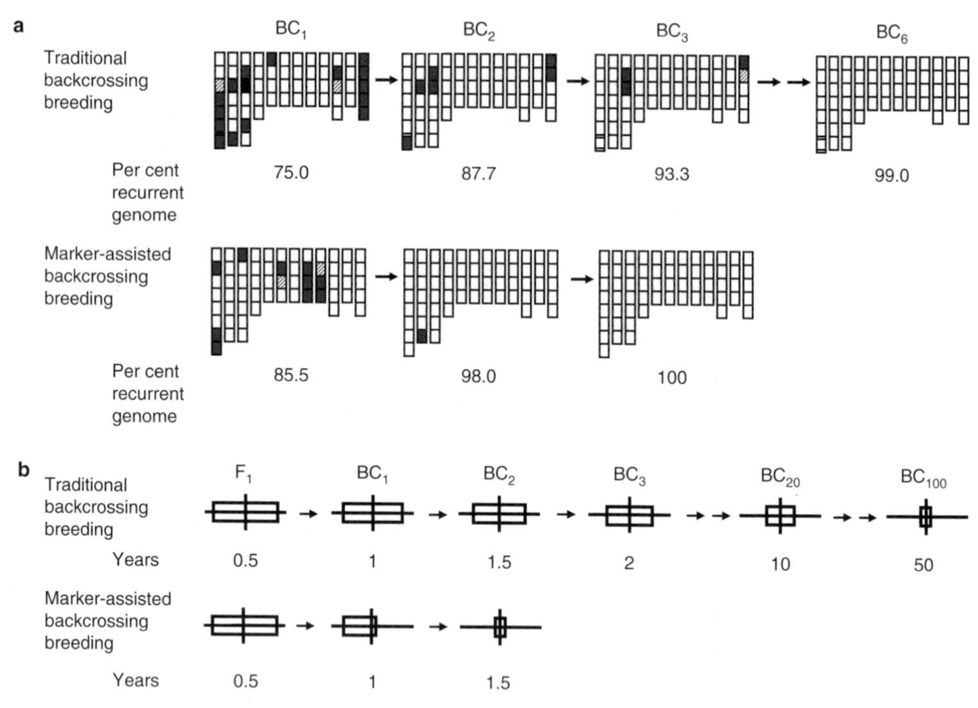

Fig. 8.4. Comparison of traditional and marker-assisted backcrossing breeding (assuming that co-dominant markers are used). (a) Rate to return to recurrent parent genotype in regions of genome unlinked to gene(s) being introduced. (Top) Traditional backcrossing breeding. Graphical genotypes were generated for randomly selected individuals from various BC generations derived from a single BC_1 individual by computer simulation. Only one homologue of each of the 12 tomato chromosomes is shown (the other homologue can be derived exclusively from the recurrent parent). Darkened regions indicate donor genome segments, striped regions indicate segments in which crossovers occurred, and white regions indicate recurrent genome segments. Each interval is 20 cM in length. The numbers beneath each graphical genotype indicate the percentage of the genome derived from the recurrent parent. The average number of generations required to return to the recurrent genome, as estimated from 20 independent simulations, was 6.5 ± 1.7 generations. (Bottom) Graphical genotypes of individuals from marker-assisted backcrossing breeding programme showing return to the recurrent parentage in only three generations. In each BC generation, 30 progeny were generated and the best (in terms of percentage recurrent parent genome) was used as the parent for the next BC generation. (b) Expected linkage drag around a selected gene held heterozygous during backcrossing. (Top) Traditional backcrossing breeding. (Bottom) MAS for plants carrying chromosomes with recombination near the selected gene. Markers tightly linked to the gene of interest are used to identify individuals with crossovers within 1 cM on one side of the selected gene in BC_1. These recombinant individuals are then backcrossed to the recurrent parent and other tightly linked markers are used to select recombinants within 1 cM on the other side of the target gene in BC_2. The expected number of years to obtain a given level of linkage drag (for a typical crop with a generation time of 0.5 years) is shown below. From Tanksley *et al.* (1989) reprinted by permission from Macmillan Publishers Ltd.

may also end up in the recipient's genetic background. The removal of linked segments occurs in a complex fashion that was described by Hanson (1959) and further elaborated by Stam and Zeven (1981). Their work showed that it takes many generations to remove the linked donor segments. For example, even after 20 BCs, one expects to find a sizable piece (10 cM) of the donor chromosome still linked to the gene being selected (Stam and Zeven, 1981), which is shown in Fig. 8.4b. In practice, this region may be larger or smaller than the expected value owing to the large variance associated with the expected value and because a breeder inevitably practices selection among the progeny. In most plant genomes, 10 cM is enough DNA to contain hundreds of genes. Therefore, backcrossing results in the transfer, not only of gene(s) of interest, but also of additional linked genes from the donor. This phenomenon can often result in a new cultivar modified for characters other than those originally targeted. Not surprisingly, many examples of 'linkage drag' are known in which undesirable traits that are closely linked to a target gene are carried along during the breeding programme, particularly when an exotic germplasm is involved.

In addition to linkage drag, unlinked DNA from the donor parent must also be removed during a BC breeding programme. In order to obtain a better idea of the relative importance of linked versus unlinked donor segments in BC breeding, a simple curve was derived from the works of Hanson (1959) and Stam and Zeven (1981) to compare the amount of foreign DNA due to these two sources as a function of the number of BC generations (Young and Tanksley, 1989b). The results of this analysis demonstrated that for a hypothetical genome of ten chromosomes of 100 cM each, the proportion of unlinked DNA derived from the donor genome is greater than that of remaining linked DNA only in the first four BC generations. After this time, the proportion of donor DNA due to linkage drag far exceed unlinked DNA by a factor of 50 and in the 20th BC generations, linked donor DNA exceeds unlinked by a factor of more than 10^5. This simple analysis clearly emphasizes the importance of linkage drag as the prominent problem in BC breeding programmes.

In a traditional BC programme, the linked segments usually remain large for many generations not because recombination had not occurred in these regions, but because there is no effective way to identify recombinant individuals. In classical breeding it is usually only by chance that such recombinants are occasionally selected which contribute to a reduction in the size of the donor segment. With high-density molecular maps it is possible to directly select individuals that have experienced recombination near the gene of interest. In approximately 150 BC plants there is a 95% chance that at least one plant will have experienced a crossover within 1 cM on one side or the other of the gene being selected. Molecular markers allow unequivocal identification of these individuals (Young and Tanksley, 1989b). With one additional BC generation of 300 plants, there would be a 95% chance of a crossover within 1 cM of the other side of the gene, generating a segment surrounding the target gene of less than 2 cM. This would have been accomplished in two generations with molecular markers, while it would have required, on average, 100 generations without molecular markers (Fig. 8.4b). It is apparent that the ability to select for desirable recombinants in a region of interest is a function of the number of markers mapped in that region, as well as the number of plants assayed. As plant molecular maps become more saturated, the efficiency of selecting recombinants will increase.

Peleman and van der Voort (2003) provided an example of linkage drag that happened in gene transgression of lettuce. In the 1990s, Keygene was involved in a marker-assisted breeding approach that led to the development of a novel lettuce cultivar resistant to the aphid *Nasonovia ribisnigri* (Jansen, 1996). This aphid is a major problem in field-grown lettuce areas in Europe and California causing reduced and abnormal growth in addition to spread of viral diseases. Resistance to this aphid could be introgressed from a wild relative of lettuce, *Lactuca virosa*, by repeated backcrossing. However, despite many rounds of backcrossing the new product was of extremely poor quality, bearing yellow leaves and a greatly reduced head. This could either have been caused by a pleiotropic effect of the resistance gene or by linkage drag, a negative trait closely linked to the positive trait of interest. Marker analysis eventually demonstrated that the reduced quality was caused by linkage drag. In this case, the linkage drag was recessive, only visible in the homozygous state, thereby seriously increasing the difficulty to select for recombinations based on the phenotype.

It was decided to use DNA markers flanking the introgression to pre-select for individuals that are recombinant in the vicinity of the gene. More than a thousand F_2 plants were screened this way, leading to the selection of some 100 individuals bearing a recombination or even double recombinations in the vicinity of the gene. Only those individuals needed to be phenotyped for both the resistance and, at the F_3 level, for the absence of the negative characteristics. This approach eventually led to the selection of an individual bearing recombination events very close to each side of the gene thereby removing the linkage drag. The results demonstrated that the (recessive) linkage drag was due to tightly linked factors on both sides of the resistance gene. As indicated by Peleman and van der Voort (2003), this result would have been very hard to obtain by classical selection methods.

8.2.5 Effect of genome size on gene introgression

The influence of genome size on the distribution of DGC in BC generations has important consequences for the attainable rate of donor genome substitution. The distribution of DGC in a BC_1 generation is shown in Fig. 8.5 for three genome sizes (haploid number of chromosomes × map length): small: 5 × 100 cM; medium: 10 × 100 cM; large: 15 × 150 cM (Stam, 2003). The important feature that can be observed is that the variance in DGC decreases as genome size (total centimorgans) increases.

From the tabulated cumulative distribution of Fig. 8.5 the probability of less than a given DGC can be read. For example, the probability that DGC is less than 0.35 equals 0.21, 0.12 and 0.06 for the small, medium and large genome, respectively. From these probabilities one can calculate the population size required to ensure that with e.g. 90% certainty at least one plant will occur with less than a given DGC. Let the threshold DGC be x and let the corresponding probability be p_x. Then from Stam (2003) the required minimum population size N satisfies

$$1 - (1 - p_x)^N > P_C \quad (8.6)$$

where P_C is the pre-set level of certainty.

For the three genome sizes, the population size required is given in Table 8.3 to find with 90% probability at least one or at

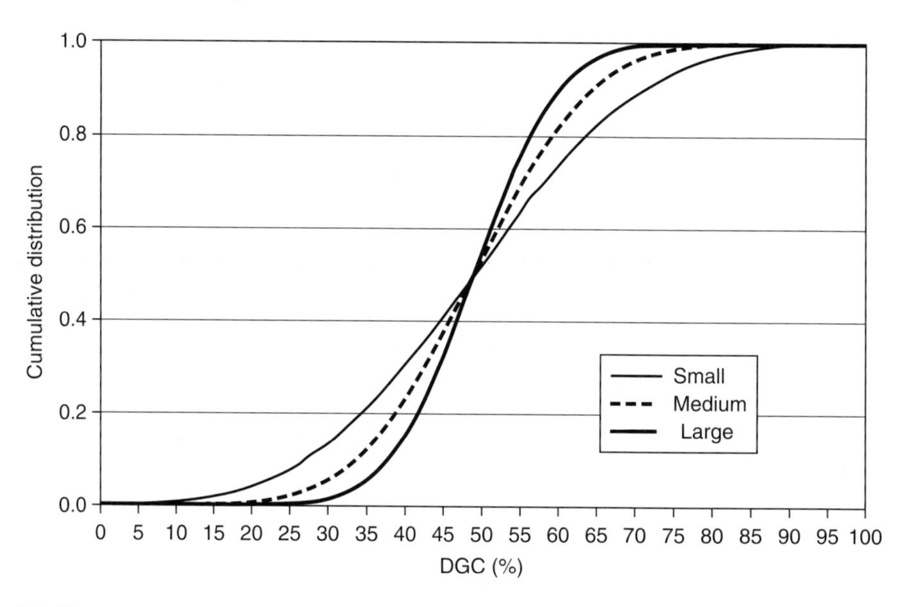

Fig. 8.5. The cumulative distribution of donor genome content (DGC) in a BC_1 generation for a small, medium or large genome. Results based on 50,000 replicate simulation runs. After Stam (2003).

Table 8.3. Population sizes (expressed as number of individuals) required in a BC_1 to obtain (probability 0.90) at least one or at least two plants with less than a certain donor genome content (DGC) with a small, medium or large genome (Stam, 2003).

	At least one			At least two		
DGC	Small	Medium	Large	Small	Medium	Large
< 0.45	5	5	6	8	9	11
< 0.40	7	9	14	14	16	25
< 0.35	10	18	40	17	31	68
< 0.30	16	41	169	28	69	285
< 0.25	28	111	822	48	187	>1000

least two plants with less than a given DGC in a BC_1 generation, which tells the importance of genome size. For example, for DGC to be less than 0.40 in at least one plant, a large genome requires approximately a twofold larger population size as compared to a small genome (14 versus 7). As DGC decreases, this tendency increases rapidly as up to tenfold for DGC less than 0.30 and thirty times for DGC less than 0.25. From these simple calculations, the price to be paid for a rapid decline of DGC in a large genome is twofold (Stam, 2003): (i) the larger the genome size, the more markers (the more marker data points per plant) required; and (ii) the larger the genome size, the larger population size required to attain a given rate of donor genome substitution.

When multiple BC generations are considered, there are selection strategies on population sizes. Employing increasing, constant, or decreasing population sizes from generations BC_1 to BC_3 in a simulation study had little effect on the recurrent parent genome values of the selected BC_3 plants (Frisch *et al.*, 1999a). For example, allocating a total of n = 300 plants such that 100 plants are generated in each of generations BC_1 to BC_3, (ratio n_1:n_2:n_3 = 1:1:1) resulted in a lower 10% percentile of the recurrent parent genome (Q10) of 97.4%, while various ratios from 3:2:1 on the one extreme to 1:3:9 on the other resulted in Q10 values of 97.3 and 97.4%, respectively. In contrast, employing a large population size in generation BC_1 multiplies the number of marker data points required for the marker-assisted BC programme. For example, only 2650 marker data points were required for n_1:n_2:n_3 = 1:3:9, while 5000 or even 7250 marker data points were required for ratios 1:1:1 and 3:2:1, respectively. However, in a multi-stage selection for a quantitative trait, large populations in early generations are advantageous because when high selection intensity is applied, a large selection gain is expected due to the large segregation variance (Frisch, 2004).

8.2.6 Background selection at carrier chromosome

Donor genome substitution is most important and at the same time most difficult, for chromosomes that carry the target gene(s). Suppose that the target gene is flanked by two markers at map distances d_1 and d_2 which can be used for background selection as described previously. Within a given number of generations the introgressed segment must be smaller than the segment covered by d_1–d_2. Then, given a pre-set probability of reaching this goal (with a 99% success rate), what are the optimal population sizes in successive BC generations?

The answer to this question has been given by Hospital and Decoux (2002) and can readily be obtained with the software package POPMIN (http://moulon.inra.fr/~fred/pro grams). Table 8.4 provides three important features based on the results from POPMIN. First, the smaller the segment (interval)

Table 8.4. Optimum population sizes (expressed as number of individuals) required in successive BC generations to achieve with 99% certainty two markers flanking the target gene becoming detached in at least one plant of the last BC generation. Σ*N*, accumulated number of plants. (Σ*N*), average accumulated number of plants; this is less than Σ*N* because with a certain probability the goal may be reached before the final generation. Figures indicated in 'configuration' column are distances in centimorgans (cM). *T*, target locus; *d*1, *d*2, flanking markers (Stam, 2003).

Configuration	Generation	BC_1	BC_2	BC_3	Σ*N*	(Σ*N*)
*d*1-10-*T*-10-*d*2	2	62	100	–	162	(137)
	3	25	36	76	137	(74)
*d*1-5-*T*-5-*d*2	2	118	200	–	318	(289)
	3	48	70	149	267	(149)

bracketed by the markers, the more plants are required because rare recombinants are less likely to occur in smaller populations. Secondly, population size should increase as generations proceed as two-sided detachment (crossover) is in most cases a two-stage process. If no detachment (crossover) occurs at any side in a given generation, more plants are required in the generation(s) thereafter. Thirdly, allowing more generations (three versus two) to achieve the goal requires fewer plants to be grown and genotyped in total, indicating a trade-off between speed and cost (total sample size) of the introgression programme.

In addition, the POPMIN software also allows the user to specify the initial genotype at both markers and the target locus. Given an initial condition of BC generation, e.g. BC_1, the user can optimize population sizes in the following BC_2, BC_3 etc. Conversely, if no single recombinant has been obtained in a given BC generation, an increase of the originally planned population sizes in generations thereafter is needed.

In terms of the relative importance of background selection for the carrier chromosomes and the remainder genome, Hospital (2002) considered background selection on carrier chromosomes to be more important than on non-carriers and thus assigned different weights to carrier and non-carrier markers. Frisch and Melchinger (2001) considered 'multi-stage' selection of markers: after selection of the target gene(s), one selects plants based on carrier markers and finally, from the obtained subset, one selects based on non-carrier markers.

8.2.7 Whole genome selection for genetic background

The question arises about the number of markers (per chromosome) that should be used for whole genome selection for genetic background and how this depends on genome and/or population sizes. Several authors (see e.g. Hospital and Charcosset, 1997; Frisch *et al.*, 1999a,b) have shown that in a moderately sized population from which the 'most promising' plant is selected for further backcrossing, an increase in the number of markers per chromosome beyond two is hardly rewarding (Table 8.5). An increase from 1 to 8 markers reduces DGC in relative sense (from 0.13 to 0.07 in BC_2), but the absolute effect is limited. However, when rapid progress requires using larger population sizes, especially in the case of a large genome (where larger population sizes are required anyway), the situation is different (Table 8.6).

Table 8.5. Average decrease of DGC in a BC programme with a medium genome size and single target gene. Each chromosome has 1, 2 or 8 markers, uniformly distributed over the chromosome. One plant out of 50 is selected in each generation for backcrossing. The selected plant satisfies the following conditions: (i) it carries the target allele; and (ii) it has the smallest number of markers of donor signature. Results based on 5000 replicate simulation runs (Stam, 2003).

Number of markers	BC_1	BC_2	BC_3
1	0.34	0.13	0.07
2	0.31	0.09	0.04
8	0.30	0.07	0.02

Table 8.6. Average DGC attained in BC_2 for small and large genome sizes with various population sizes and number of markers per chromosome (Stam, 2003).

Number of markers	Population size	Genome size	
		Small	Large
2	50	0.082	0.121
	200	0.079	0.095
	400	0.078	0.088
8	50	0.040	0.100
	200	0.021	0.067
	400	0.019	0.055

Two general conclusions can be drawn (Stam, 2003): (i) For a small genome with few markers per chromosome, increasing the population size makes little sense. When many markers are available, however, an increase of population size does reduce DGC, but hardly so beyond $N = 200$. (ii) For a large genome, increasing the population size is beneficial, irrespective of the number of markers per chromosome. Obviously, with increasing genome size more independent recombination events are required to attain a given reduction in DGC, which in turn demand larger populations for their discovery.

Whole genome selection for background will help reduce the DGC. The question about what final level of DGC is 'acceptable' cannot easily be answered in general terms. When only relying on estimated DGC, based on markers, one still runs the risk that after finalization a tiny donor fragment contains a few 'wild type' genes that confer an undesirable trait. Especially in a rapid cycling introgression programme that hardly allows phenotypic selection for general agronomic performance, undesired donor traits may unexpectedly turn up despite an expensive and theoretically powerful BC scheme (Stam, 2003). On the other hand, desirable DGC levels across different BC populations largely depend on the genetic difference between the donor and recurrent parents. In many cases, unsaturated backcrossing with selection for the target gene may be enough particularly when the donor parent is also a commercial cultivar.

8.2.8 Multiple gene introgression by repeated backcrossing

Since little additional effort is required to screen with multiple molecular markers after sampling and DNA extraction, one could consider adding many genes simultaneously to a cultivar through MAB. For example, batteries of disease resistance genes could be added in a few generations, as opposed to the many generations required with traditional breeding. The ability to rapidly adjust existing cultivars should allow breeders to more quickly respond to market demands, as well as unexpected environmental pressures, such as the appearance of new pathogens.

With marker assisted introgression, allele frequencies for the introgressed alleles are sufficiently high that two of three alleles could be readily introduced and brought to fixation in a given breeding cycle (Beckmann and Soller, 1986a). Without MAS, many BC progeny will have to be screened for the introduced trait, due to the extreme rarity of BC progeny carrying desired exotic alleles.

Several authors have considered optimization aspects of multiple gene transfer by repeated MAB (van Berloo *et al.*, 2001; Frisch and Melchinger, 2001; Hospital, 2002; Stam, 2003). Frisch (2004) discussed the introgression of two dominant genes. It is clear that, roughly speaking, the effects of population size, genome size and total number of markers on the efficiency of recurrent parent genome recovery are similar to those for single gene transfer. As an example, Table 8.7 shows the effect of population size for the introgession of three target genes in a genome of medium size using eight markers per chromosome for background selection (Stam, 2003). With multiple targets an increase of population size enhances the efficiency. However, the average DGC decrease in the BC population more apparently when the population sizes increase from small to medium such as from 50 to 100.

The number of target genes does affect the answer to the question whether a given total number of plants should be distributed

Table 8.7. Average DGC in BC_2 and BC_3 in an example of the simultaneous introgression of three target genes in a genome of medium size, using eight markers per chromosome for background selection. A single plant was selected for further backcrossing, carrying the three target alleles and having the smallest number of markers of donor signature. Averages based on 1000 replicate simulation runs (Stam, 2003).

Population size	BC_2	BC_3
50	0.18	0.09
100	0.14	0.06
200	0.11	0.04
400	0.09	0.03

over two or three generations. Comparison of average DGC attained with a total of 900 plants, distributed over two or three BC generations with medium genome size and eight selection markers per chromosome (Stam, 2003), showed that three BC generations each with 300 plants is more effective than two BC generations each with 450 plants. The average DGC for the former is 0.010 for one target gene and 0.036 for three target genes while for the latter these two numbers are 0.023 and 0.083, respectively. Again, there is a trade-off among time (how many generations involved), cost (how many data points to generate per generation) and efficiency (how soon the recurrent parent genome can be recovered).

A complication arising with multiple QTL transfer is the uncertainty about the exact location of QTL. Hospital and Charcosset (1997) investigated the optimal location of markers to be used in foreground selection. This optimization process should also consider the relative economic importance of the target traits for the multiple QTL to be introgressed.

8.3 Marker-assisted Gene Pyramiding

Agricultural productivity is the result of growing superior genotypes in an environment which allows them to express their superiority (Boyer, 1982). Increasing genetic value relies on increasing the frequency of favourable genes controlling that trait. To create a superior genotype, the breeder must assemble many genes which work well together and, for a specific trait, assemble the alleles with similar effects from different loci. This process is called pyramiding, by which different QTL alleles can be recombined and the true-breeding lines associating alleles of similar (positive or negative) effect can be selected (Xu, 1997; Fig. 8.6). Related techniques include effectively identifying the individuals with favourable allele combinations, assembling different alleles into a common genetic stock to produce new genotypes and determining the joint effects of alleles at different loci. In the words of Allard (1988): 'Emphasis was therefore shifted…to a particulate approach…determining the individual effects of single marker loci on adaptive change, then determining the joint effects of pairs of loci.'

8.3.1 Gene-pyramiding schemes

If all genes cannot be fixed in a single step of selection, it is necessary to cross again selected individuals with incomplete, but complementary, sets of homozygous loci (Xu *et al.*, 1998). However, such strategies are limited to small numbers of target loci. To accumulate more loci in a single genotype by selection on markers, Hospital *et al.* (2000) proposed a marker-based recurrent selection (MBRS) method using a QTL complementary strategy in a randomly mating population. When evaluating this method using simulations with 50 detected QTL in a population of 200, they found that the frequency of favourable alleles went up to 100% in ten generations when markers were located exactly on the QTL, but up to only 92% when marker–QTL distance was 5 cM. The reduced efficiency in the latter case comes from the probability of 'losing' the QTL during the breeding scheme because of recombination between the markers and QTL. This effect becomes more severe with increasing duration of the breeding scheme because of the accumulation of meiosis;

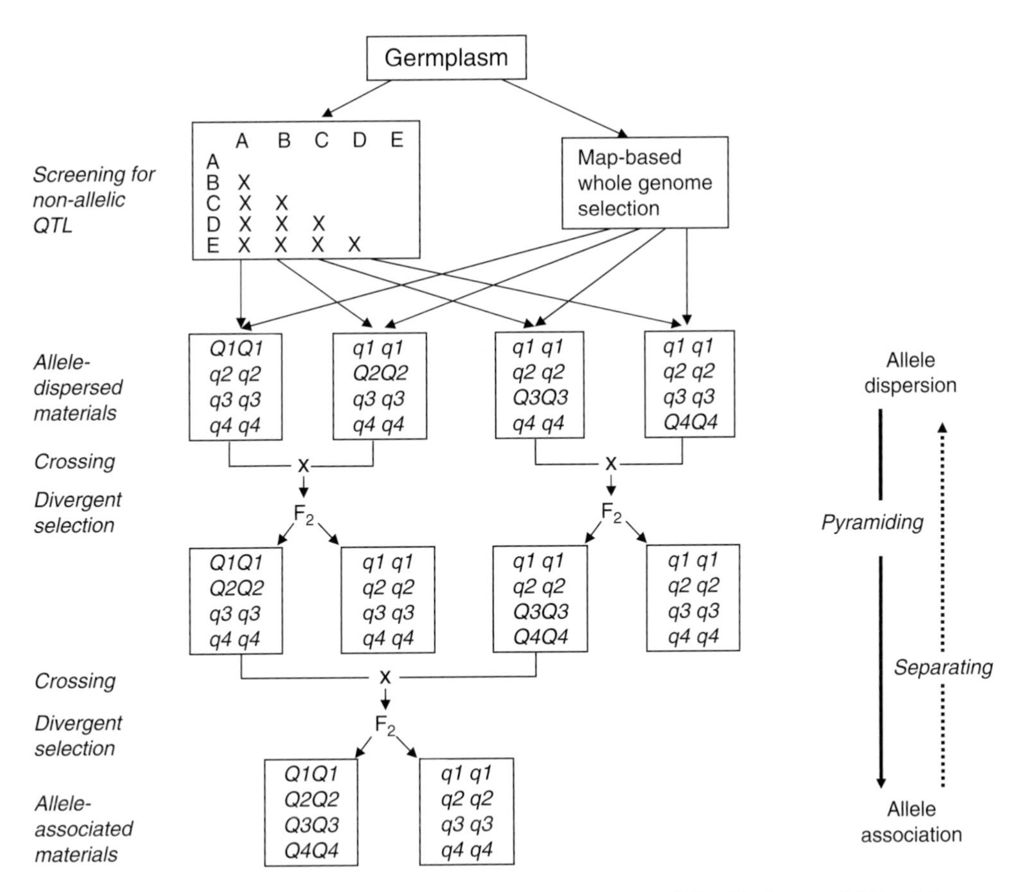

Fig. 8.6. A procedure for QTL separating and pyramiding. Non-allelic QTL with dispersed QTL alleles are identified by observation of transgressive segregation and map-based whole genome selection, and then recombinants are obtained by divergent phenotypic selection from crosses derived from non-allelic QTL materials. Two cycles of cross-selection are exemplified to pyramid non-allelic QTL at four loci (*Q1-q1*, *Q2-q2*, *Q3-q3* and *Q4-q4*). QTL separating is a reverse process of pyramiding, in which allele-associated materials are used as parents to produce a segregating population (F_2) and intermediate phenotype is selected in order to get allele-dispersion individuals. From Xu (1997). This material is reproduced with permission of John Wiley & Sons, Inc.

hence, it is important to cumulate and fix the target genes as rapidly as possible. Hospital *et al.* (2000) concluded that the optimization of pair-wise crosses between selected individuals is the most efficient way to decrease the duration of the breeding scheme under the constraint of at constant cost.

Servin *et al.* (2004) developed a general framework to optimize breeding schemes to accumulate identified genes from multiple parents into a single genotype (gene-pyramiding schemes). This section will introduce the theory developed by these authors on marker-assisted gene pyramiding.

Definition

To accumulate into a single genotype the genes that have been identified in multiple parents, assume we have n loci of interest and a set of founding parents labelled $\{P_i, i \in [1,\ldots, n]\}$ with P_i being homozygous for the favourable alleles at the ith locus and homozygous for unfavourable alleles at the remaining $n - 1$ loci. We assume that the recombinant fractions between the loci are known and we want to derive the ideal genotype (ideotype) that is homozygous for the favourable allele at all n loci.

As shown in Fig. 8.7, the gene-pyramiding scheme has two parts. The first part is called a *pedigree* and is aimed at accumulating all target genes in a single genotype (called the *root genotype*). The second part is called the *fixation steps*, which aims at fixing the target genes into a homozygous state, that is, to derive the ideotype from the root genotype. A pedigree can be represented by a binary tree with n leaves corresponding to the n founding parents and $n - 1$ nodes. Each node of the tree is called an intermediate genotype and has two parents. Each intermediate genotype, which is a particular genotype selected from among the offspring, becomes a parent in the next cross. Denote the gametes (subsets of genes) passed on from the parents to the intermediate genotype as s. Take $H_{(s_1)(s_2)}$ as an example, the intermediate genotype must produce and pass on to its offspring a gamete carrying all the favourable alleles in s_1 and s_2.

There are many possible procedures that can be used to fix the root genotype, one of which is to generate a population of doubled haploids (DHs) as described in detail in Chapter 4. Using the DH procedure, the ideotype can be developed in just one additional generation after the root genotype is obtained, plus one more generation for seed increase to produce large populations. The fixation steps using the DH procedure can be outlined as follows.

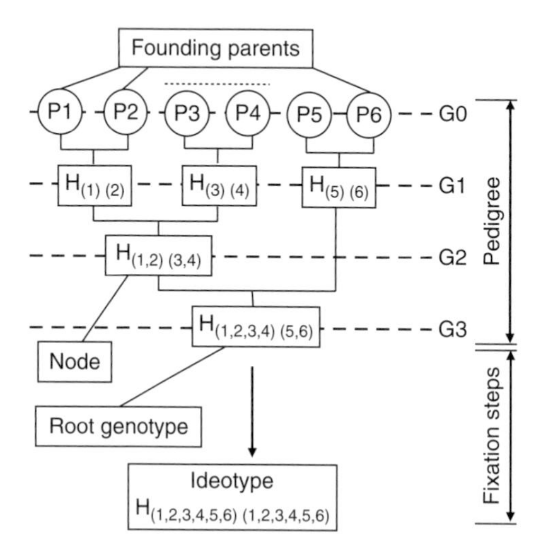

Fig. 8.7. Example of gene pyramiding scheme cumulating six target genes. The *pedigree* part is aimed at cumulating one copy of all target genes from founding parents in a single genotype (*root genotype*). The *fixation step* is to derive the *ideotype* from the *root genotype* which fixes the target genes into a homozygous state. From Servin *et al.* (2004) with permission.

First, obtain a genotype carrying all favourable alleles in coupling, namely $H_{(1,2,\ldots,n)(B)}$ by crossing the root genotype with a blank parent (denoted as $H_{(B)(B)}$) containing none of the favourable alleles. This guarantees that the linkage phase of the offspring is known and that the $H_{(1,2,\ldots,n)(B)}$ genotype can be identified without ambiguity.

Second, self $H_{(1,2,\ldots,n)(B)}$ to give the ideotype in one generation.

Pedigree height

The number of generations a pedigree spans is called the *pedigree height*, denoted h. If the fixation steps span two generations, the complete gene-pyramiding scheme spans $h + 2$ generations. A pedigree is of maximum height when just one cross is performed at each generation (involving an intermediate genotype H and a founding parent). This type of pedigree is called a cascading pedigree. Conversely, a pedigree is of minimum height when the maximum number of crosses is performed at each generation. The height n of a pedigree cumulating n genes satisfies

$$\lceil Log_2(n) \rceil \leq h \leq n - 1 \tag{8.7}$$

where $\lceil x \rceil$ denotes the smallest integer larger than or equal to x.

Number of pedigrees

The number of pedigrees accumulating n genes is the number of binary trees with n labelled leaves. The root genotype of a pedigree accumulating n target genets comes from the cross of two parents carrying, respectively, p and $n - p$ (non-overlapping) target genes, where $(1 \leq p \leq n - 1)$. Let $N(p)$ be the number of subpedigrees cumulating p specified genes. Summing up over all possible values of p, the number $N(p)$ of pedigree cumulating n genes can be computed via

$$N(n)=\frac{1}{2}\sum_{p=1}^{n-1}\binom{n}{p}N(p)N(n-p) \quad (8.8)$$

The factor ½ is there to ensure that the crossing of two given parents is counted only once. This recursion can be solved (see the Appendix in Servin *et al.*, 2004) and leads to

$$N(n)=\prod_{k=2}^{n}(2k-3)=(2n-3)(2n-5)\ldots 1 \quad (8.9)$$

for the total number of pedigrees cumulating the n genes. The total number of pedigrees increases very fast with the number of loci considered. For example, when $n = 3, 4, 5, 6, 7$, the total numbers of pedigrees are 3, 15, 105, 945 and 10,395, respectively.

Gene transmission probability through a pedigree

Given the recombination fractions between loci, we can compute the probability that an intermediate genotype $H_{(s_1)(s_2)}$ passes on to its offspring the set of genes s that is the union of s_1 and s_2. If denoted by $v(s)$ the total number of genes in the set s, we have $v(s) = v(s_1)+v(s_2)$. Let $\{a_i\}$ be the genes in set s ranked according to their position on the genetic map, so that $s = (a_1,a_2,\ldots,a_{v(s_1)+v(s_2)})$. Let $r_{x,y}$ be the recombinant fraction between x and y. The probability that a gamete generated by $H_{(s_1)(s_2)}$ contains the set s of genes is

$$P(H_{(s_1)(s_2)}\rightarrow s)=\frac{1}{2}\prod_{i=1}^{v(s)-1}\pi(i,i+1) \quad (8.10)$$

where $\pi(i, i+1) = r_{a_i,a_{i+1}}$, if genes a_i and a_{i+1} are in different subsets and $\pi(i, i+1) = (1 - r_{a_i,a_{i+1}})$, otherwise.

Note that other target genes might be on the map, located between the a_i's, but not belonging to the set s; recombinations between these genes do not matter here. As an example illustrating Eqn 8.10, consider the genotype $H_{(1,3)(2,5,6)}$. The probability that it passes the set (1, 2, 3, 5, 6) is (see Eqn 8.11 at bottom of page). Knowing these probabilities, the overall probability of obtaining the root genotype of a given pedigree is the product, over all the pedigree's nodes (other than the root node), of the probabilities calculated as in Eqn 8.10.

Minimum population sizes necessary to obtain ideotype

Lets call P_f and P_m the probabilities computed as in Eqn 8.10 that each parent of a given node passes on its particular subset of genes. From these probabilities we can compute the population size N needed to get the intermediate genotype at this node with a probability of success γ. The probability that none of the N offspring has the right genotype is $(1 - P_fP_m)^N$; identifying this with $1 - \gamma$ gives

$$N=\frac{\ln(1-\gamma)}{\ln(1-P_fP_m)} \quad (8.12)$$

where ln denotes the natural logarithm. From Eqn 8.12, the population sizes required at each node can be computed. Now the overall probability of success of the pedigree is the product of the probabilities of success at each of its nodes. Similarly, the population sizes required for the fixation steps can be computed. The nodes associated with combining two founding parents always pass on their target genes. Let p be the number of other nodes in the breeding scheme; if they all have the some probability of success γ as considered here, then the overall probability of success of the gene-pyramiding scheme is γ^p. The sum of all population sizes needed

$$P(H_{(1,3)(2,5,6)})\rightarrow(1,\ 2,\ 3,\ 5,\ 6)=\frac{1}{2}(r_{1,2})(r_{2,3})(r_{3,5})(1-r_{5,6}) \quad (8.11)$$

in the gene-pyramiding scheme (pedigree and fixation steps) is denoted by N_{tot}. The target of the population sizes to be handled at any node or step during the whole gene-pyramiding scheme is N_{max}.

A case study

Servin *et al.* (2004) developed a computer program to build all pedigrees leading to the ideotype for a given number n of genes. Given the $r_{i,j}$ values, the program determines the gene transmission probabilities and the cumulated population size N_{tot} for each pedigree followed by the fixation steps.

A case study was provided by Servin *et al.* (2004) for cumulating four genes, which might be frequently used in accumulating major-gene controlled traits such as disease resistance. The 15 possible pedigrees for accumulating four genes located on a single chromosome were generated with the assumption that the recombination fraction between adjacent loci are the same and correspond to 20 cM using Haldane's mapping function (Haldane, 1919). As the recombination fraction is the same for all pairs of adjacent loci, some gene-pyramiding schemes have the same transmission probability or population sizes, which we would note is not true for almost all practical cases.

Figure 8.8 shows the three schemes, each representing multiple gene pyramiding schemes with the same accumulated population size that necessitate the smallest N_{tot}. The population sizes were computed so that the probability of success of each scheme was 0.99. In the scheme based on a cascading pedigree (Fig. 8.8a), there are four nodes for which the probability of obtaining the intermediate genotype is not 1. The probability of success used at each of those nodes was thus $0.99^{1/4} = 0.9975$. In the two other schemes, the number of such nodes is three, so that the probability of success used at each of these nodes was $0.99^{1/3} = 0.9967$. The hybrids between founding parents are obtained with a probability of 1, so that the population required at the corresponding nodes was assumed to be one individual.

Figure 8.8a shows a gene-pyramiding scheme involving a cascading pedigree. It spans five generations ($h = n - 1 = 3$ for the pedigree height, plus two generations for the fixation steps) and requires the smallest cumulated population size ($N_{tot} = 325$) of all the schemes. The two other best schemes last four generations ($h = \mathrm{Log}_2(n) = 2$ for the pedigree height plus two generations for the fixation steps). The scheme that necessitates the next smallest N_{tot} (= 961) is the one represented in Fig. 8.8b. It cumulates loci 1 and 4 on one subpedigree and 2 and 3 on the other, before generating the $H_{(1,2,3,4)(B)}$ genotype. The population sizes needed for this gene-pyramiding scheme are large at all nodes when compared to the cascading type. The gene-pyramiding scheme represented in Fig. 8.8c necessitates an even larger N_{tot} (= 1001) because a huge population size is needed to produce the root genotype $H_{(1,2)(3,4)}$; conversely, the population size needed to produce the $H_{(1,2,3,4)(B)}$ genotype is much smaller ($N = 97$).

Xu *et al.* (1998) provided a practical example to accumulate four loci controlling tiller angle using phenotypic selection following both a scheme similar to the cascading pedigree and the other similar to the schemes described by Fig.8.8b, c.

Although the cascading pedigree-based gene-pyramiding scheme needs the smallest population size to combine all favourable alleles, it takes more generations to derive the root genotype, compared to other schemes. When genotyping cost is a more important limiting factor than the time required for delivery of breeding products, the cascading pedigree scheme should be chosen. However, when quick development of a root genotype becomes more important, particularly in the private breeding sector, a couple of generations less would make a huge difference in market-share competition.

Theoretically the method described above can be extended to the schemes involving many genes. As more genes are involved, the pedigree height (the number of generations a pedigree spans) increases and so does the cumulative population size. The difference of population sizes between difference schemes will also increase. As the number of genes increases, the population size needed in each generation would

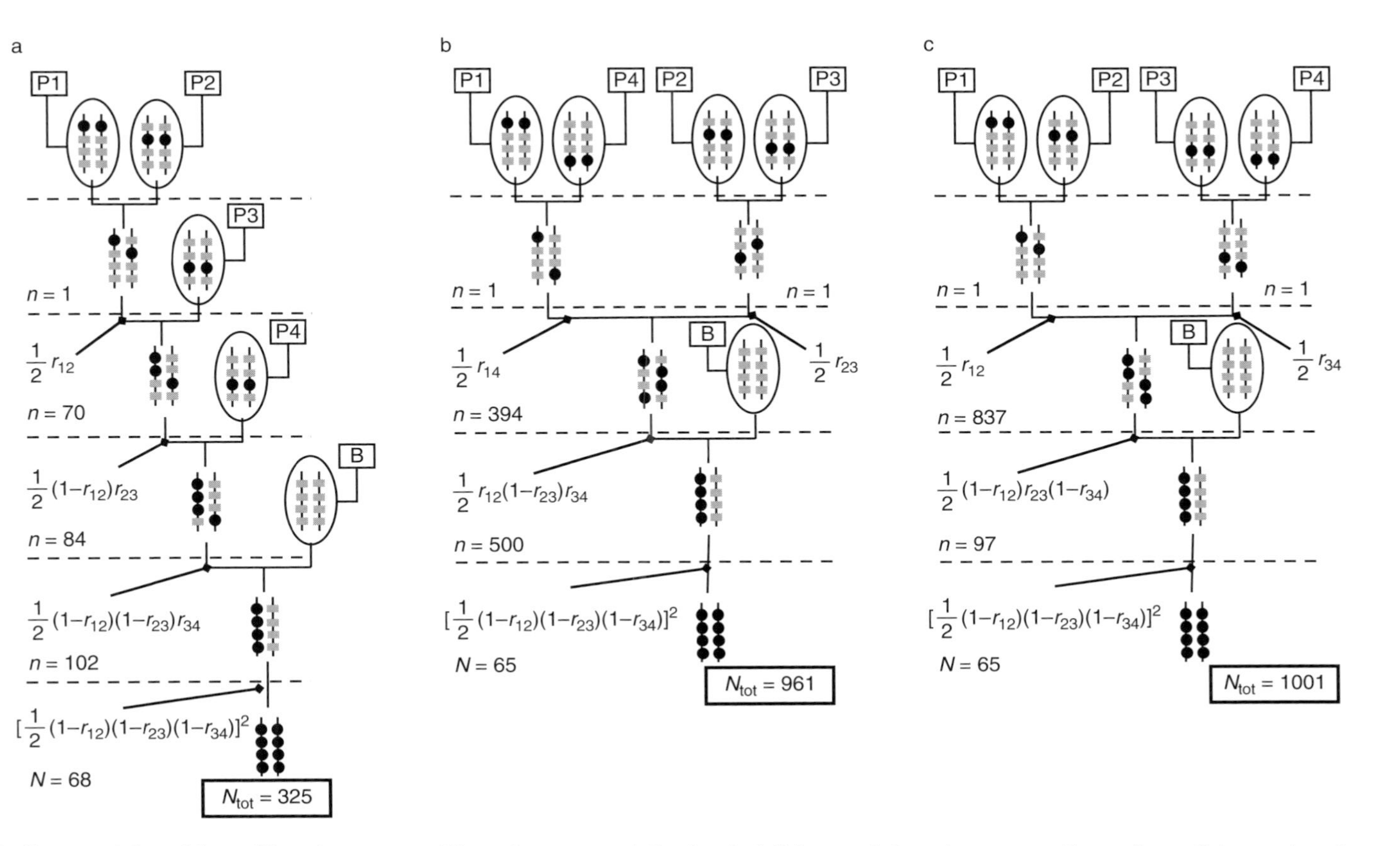

Fig. 8.8. Representation of three different gene-pyramiding schemes cumulating four loci. Scheme a is based on a cascading pedigree. Schemes b and c differ by the order of crosses of the founding parents. The target genes are represented by solid circles and other genes by shaded boxes. At each node the transmission probabilities of the targeted genes from parent to offspring are given. When the probability is equal to one, it is not indicated. The population sizes needed at each node (N) and the cumulated population sizes (N_{tot}) are provided. From Servin *et al.* (2004) with permission.

become so large for some schemes that it would be practically impossible. As a result, the cascading pedigree-based scheme would become the only choice, although it would take many more generations to derive the root genotype.

8.3.2 Crossing and selection strategies

Different crossing and selection strategies may require vastly different population sizes to recover a target genotype with the same certainty even when the same parents are used (Bonnet *et al.*, 2005). Determination of the most efficient strategy has the potential to dramatically decrease the amount of resources (plants, plots, marker assays and labour) required to combine a set of target alleles into an idea genotype (ideotype). Considerable efficiency gains can be achieved if plant breeders are able to choose the most appropriate cross (e.g. single cross, BC or top-cross) and best MAS methods.

In using markers, several scenarios are commonly faced by breeders: (i) pyramiding alleles at multiple loci including consideration of most appropriate cross type; (ii) minimizing marker screening costs by sequential culling; (iii) use of incomplete linked markers to combine target alleles; and (iv) combining alleles linked in repulsion in crosses segregating for other unlinked target alleles. Wang *et al.* (2007) used population genetic theory to establish general rules for the numbers of markers required, the best crossing strategies and the level of inbreeding to maximize the efficiency of marker implementation where there is no recombination between marker and allele of interest.

Comparing biparental, back- and top-crosses

If n loci differ between two parents with favourable alleles at n_1 loci in the first parent P_1 and n_2 in the second parent P_2, then relative proportions of the target genotype in DH or recombinant inbred line (RIL) populations derived from F_1, P_1BC_1 (backcrossed to P_1) and P_2BC_1 (backcrossed to P_2) are

$$f_{F_1} = \left(\frac{1}{2}\right)^n$$
$$f_{P_1BC_1} = \left(\frac{3}{4}\right)^{n_1}\left(\frac{1}{4}\right)^{n_2} \quad (8.13)$$
$$f_{P_2BC_1} = \left(\frac{1}{4}\right)^{n_1}\left(\frac{3}{4}\right)^{n_2}$$

The three proportions were used as a guide as to whether a BC reduced population size and to indicate which parent should be used as the recurrent parent.

If the target alleles are dispersed among three parents, i.e. P_1, P_2 and P_3, a top-cross (or three-way cross), e.g. $(P_1 \times P_2) \times P_3$ is required to combine all alleles. If each parent carries different alleles, the alleles contributed by parents P_1 and P_2 in the first cross will be present at frequencies of 0.25 following a top-cross with P_3 and the alleles contributed by P_3 will each have a frequency of 0.5. If n_1, n_2 and n_3 are the number of target favourable alleles in the three parents, respectively, under the condition of no selection, the expected proportion of individuals with the target genotype in the DH or RIL population is

$$f_{TC} = \left(\frac{1}{4}\right)^{n_1+n_2}\left(\frac{1}{2}\right)^{n_3} = 2^{n_3-2n} \quad (8.14)$$

where $n = n_1+n_2+n_3$. Equation 8.14 is used to determine the order in which to cross parents to minimize the population sizes required in a top-cross.

Minimizing the total number of marker assays with sequential culling

In a population of N individuals to be screened sequentially with markers at n independent loci and where only those with the target genotype are retained for screening with the next marker, the total number of marker assays (M) required to identify the target genotype at all loci can be calculated according to

$$M = N + Nf_1 + Nf_1f_2 + \ldots + Nf_1f_2 \ldots f_{n-1} \quad (8.15)$$

where f_1, f_2, ... f_n are the proportion of individuals retained after screening with each marker. For any set of markers, M will be minimized if the marker with the lowest retained fraction f (or highest culling rate) is used first, followed by the next lowest and so on. The total cost (C) of marker assays can be determined from Eqn 8.15 by inclusion of the cost of each assay

$$C = Nc_1 + Nf_1c_2 + Nf_1f_2c_3 + \ldots + Nf_1f_2 \ldots f_{n-1}c_n \quad (8.16)$$

where c_1, c_2, ... c_n are the cost of the marker assays. The total cost, C, is minimized when

$$\frac{c_1}{1-f_1} < \frac{c_2}{1-f_2} < \ldots < \frac{c_n}{1-f_n}.$$

It should be noted that the analytic expression for the cost of sequential culling ignores the costs of plant/line handling (tagging, leaf sampling, etc.) and DNA extraction, which are fixed with total sample size and cannot be reduced by sequential culling. If these fixed costs are major parts of the expense for genotyping, the order of markers used in the sequential culling may become less important. As high-throughput genotyping systems have been established for using all markers for all samples to make the genotyping most cost-effective overall, the order of markers used in the sequential culling may become less important.

Enrichment of favourable alleles at early generations

When many (unlinked) markers have to be selected, the frequency of a target homozygous genotype will be low and a large population size will be required. For example, in the F_2 of a biparental cross between two inbreds segregating at five unlinked loci, the frequency of the target genotype is $0.25^5 = 0.00098$ and the minimum population size (Eqn 8.2) to recover at least one target genotype is 4714 ($\alpha = 0.01$). If selection is made among homozygous lines (i.e. DH or RIL populations) from the same cross, the frequency of the target genotype is $0.5^5 = 0.03125$ with a minimum population size of only 146 ($\alpha = 0.01$), i.e. the target genotype is more readily recovered with a smaller population size if selection is delayed until greater homozygosity has been reached.

For more segregating loci, population sizes quickly increase even in DH or RIL populations. For example, in a biparental population with eight unlinked segregating loci, the frequency of the target genotype in a homozygous population is $0.5^8 = 0.0039$, the minimum population size 1777. In these instances, Bonnet *et al.* (2005) proposed a two-stage selection strategy. The first stage is 'F_2 enrichment', where F_2 individuals carrying the entire set of target alleles in either homozygous or heterozygous form are selected. F_2 enrichment takes advantage of the high expected frequency of carrier (either homozygous or heterozygous) at each locus of 0.75. The value of the technique can be seen in a population segregating at 12 loci, where the frequency of genotypes selected in an F_2 enrichment step is $0.75^{12} = 0.031676$, resulting in the minimum population size of 144 F_2 generations, compared to the frequency of $0.25^{12} = 5.960464 \times 10^{-8}$ and a population size of > 77 million to identify a single homozygous individual in the F_2.

After F_2 enrichment, the frequency of each of the 12 target alleles in the selected population is increased from 0.5 to 0.67. The second step is to generate a population of more or less homozygous lines from the selected F_2. The frequency of the target genotype in DH/RIL populations generated from the enriched F_2 will have been increased from 0.5^{12} to 0.67^{12}, resulting in a decrease in minimum population size from 18,861 to 596. Thus, with enrichment, both the F_2 and the DH/RIL populations are of a more practical size for breeding.

The allele enrichment can be done for more than one generation when multiple-generation selection is involved. Enrichment at two selection stages (e.g. in F_2 and F_3) always requires greater assay numbers than simple F_2 enrichment (Wang *et al.*, 2007). As indicated by Bonnet *et al.* (2005), F_2 enrichment increased the frequency of selected alleles, allowing large reductions in minimum population size for recovery of target genotypes (commonly around 90%) and/or selection at a greater number of loci. So the gain from another cycle of allele enrichment selection in F_3 following enrichment in F_2 is at best minor and often results in a small net increase in minimum population size.

For a top-cross of three adapted wheat lines from an existing breeding programme, simulation of changes in allele frequencies at nine target genes (seven unlinked) showed that population size was minimized with a three-stage selection strategy in the F_1 generation of the top-cross (TCF1), the F_2 generation of the top-cross (TCF2) and DHs. Enrichment of allele frequencies in TCF2 reduced the total number of lines screened from > 3500 to < 600. Eight of the genes were present at frequencies > 0.97 after selection (Wang *et al.*, 2007).

8.3.3 Gene pyramiding for different traits

The methods discussed above are for pyramiding genes affecting a specific trait. However, aggregating favourable genes from different traits in a genotype has long challenged plant breeders. The principles discussed above can be used in the same way to accumulate QTL alleles controlling different traits. A distinct difference in concept is that alleles at different trait loci to be accumulated may have different favourable directions, i.e. negative alleles are favourable for some traits but positive alleles are favourable for others. Therefore, one may need to combine the positive QTL alleles of some traits with the negative alleles of others to meet breeding objectives. Marker-assisted gene pyramiding is also important when considering multiple traits, as in phenotypic selection each of these traits has to be tested in different environments, different developmental stages or different stages of a breeding programme.

Attention should be paid to trait correlation when one practises pyramiding of alleles for different traits. Positive correlation will facilitate the pyramiding process involved in selection for alleles with the same favourable direction, but impede the process of selection for QTL alleles with different favourable directions and vice versa for negative correlation. If correlation results from pleiotropic effects of a marker gene rather than linkage, it is difficult, if not impossible, to select towards the direction opposite to the correlation.

8.3.4 Marker-assisted recurrent selection versus genome-wide selection

Recurrent selection is considered as one of the selection approaches to combine favourable alleles distributed among different sources of germplasm. There are various new versions of recurrent selection available with which molecular marker information is incorporated. The key advantages of these new versions are the availability of genetic data for all progeny at each generation of selection, the integration of genotypic and phenotypic data and the rapid cycling of generations of selection and information-directed matings at continuous nurseries.

Marker-assisted recurrent selection (MARS) was proposed in the 1990s (Edwards and Johnson, 1994; Lee, 1995 Stam, 1995) which uses markers at each generation to target all traits of importance and for which genetic information can be obtained. Genetic information is usually obtained from QTL analyses performed on experimental populations, which includes QTL locations and effects. When the QTL mapping is conducted based on a biparental population, both parents often contribute favourable alleles. As a result, the ideal genotype is a mosaic of chromosomal segments from the two parents. The goal of MARS is to obtain individuals with as many accumulated favourable alleles as possible. However, the ideal genotype, defined as the mosaic of favourable chromosomal segments from two parents, will usually never occur in any F_n population of realistic size (Stam, 1995). As discussed previously, a breeding scheme to produce or approach this ideal genotype based on individuals of the experimental population could involve several successive generations of crossing individuals (Stam, 1995; Peleman and van der Voort, 2003) and would therefore constitute what is referred to as MARS or genotype construction. This idea can be extended to situations where favourable alleles come from more than two parents. Please note that MARS can also start without any QTL information while selection can be based on significant marker–trait association established during the MARS process.

All simulation studies revealed that MARS was generally superior to phenotypic selection in accumulating favourable alleles in one individual (van Berloo and Stam, 1998, 2001; Charmet *et al.*, 1999) and MARS was between 3% and almost 20% more efficient than phenotypic selection (van Berloo and Stam, 2001). The advantage of MARS over phenotypic selection was greater when the population under selection was larger or more heterozygous including BC_1 or F_2 populations.

Through simulation, Bernardo and Yu (2007) assessed the response due to MARS compared with genome-wide selection and to determine the extent to which phenotyping can be minimized and genotyping maximized in genome-wide selection. By their definition, MARS refers to the improvement of an F_2 population by one cycle of MAS (i.e. based on phenotypic data and marker scores) followed by three cycles of marker-based selection (i.e. based on marker scores only) in an off-season nursery (Johnson 2001, 2004). The marker scores are typically determined from about 20 to 35 markers that have been identified, in a multiple regression model, as significantly associated with one or more traits of interest (Edwards and Johnson, 1994). Genome-wide selection refers to marker-based selection without significant testing and without identifying a subset of markers associated with the trait (Meuwissen *et al.*, 2001). The effects on the target trait (i.e. breeding values) of all genotyped markers distributed across the genome are fitted as random effects in a linear model. The trait values are then predicted as the sum of an individual's breeding values across all the genotyped markers and selection is subsequently based on these genome-wide predictions. In this book, the term 'genome-wide selection' will be only used for this specific situation.

Let's consider genome-wide selection and MARS as depicted in Fig. 8.9. Bernardo and Yu (2007) simulated genome-wide selection by evaluating DHs for testcross performance in Cycle 0, followed by two cycles of selection based on markers. Cycle 0 is evaluated during the regular growing season when phenotypic measurements are meaningful. Cycles 1 and 2 of genome-wide selection and MARS are conducted in an off-season nursery, where phenotypic evaluations were not meaningful but where three generations can be grown in 1 year. For Cycle 0, genome-wide selection and MARS can be considered to be either involved in the F_2 plants or the production of DHs. Response to marker-based selection is greater with DHs than with F_2 plants.

Assuming that individuals are genotyped for N_M markers and breeding values associated with each of the N_M markers were predicted and were all used in genome-wide selection, Bernardo and Yu (2007) found that across different numbers of QTL (20, 40 and 100) and levels of heritability, the response

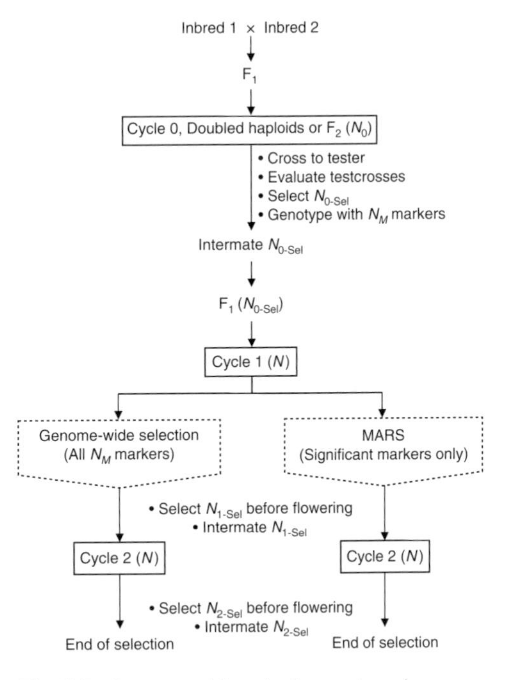

Fig. 8.9. Genome-wide selection and marker-assisted recurrent selection (MARS). From Bernardo and Yu (2007) with permission.

to genome-wide selection was 18–43% larger than the response to MARS. Regardless the heritability and the number of QTL, response to genome-wide selection were smallest when N_M = 64 markers were used. A minimum of N_M = 128–256 polymorphic markers should be used in genome-wide selection in maize and more markers should be used for complex traits that have, at the same time, a high heritability. In contrast, response to MARS were largest with N_M = 64 or 128 markers. Genome-wide selection is most useful for complex traits that are controlled by many QTL and have low heritability. Responses to selection were maintained when the number of DHs phenotyped and genotyped in Cycle 0 was reduced and the number of plants genotyped in Cycles 1 and 2 was increased. Such schemes that minimize phenotyping and maximize genotyping would be feasible only if the cost per marker data point is reduced to about US$0.02. As availability of large numbers of SNP markers in many crop plants and array-based cheap genotyping systems, genome-wide selection, as a brute-force and black-box procedure that exploits cheap and abundant molecular markers, is superior to MARS in plants. Please note that in genome-wide selection, one does not need any QTL information. Rather, one uses a general regression approach in a test set to obtain an estimate of breeding value from a very dense marker set and then selects on this marker set.

8.4 Selection for Quantitative Traits

The most significant distinction of quantitative inheritance is that there is no corresponding (simple) relationship between genotype and phenotype, although conventional plant breeding is based on selection of phenotypes. This is the major reason why the efficiency of conventional plant breeding is often low. Therefore, the main objective of MAS should be for quantitative traits according to their importance and necessity. In principle, the methodologies developed for qualitative traits in MAS are also applicable to quantitative traits. However, more factors should be taken into consideration when quantitative traits are involved. First, QTL mapping so far has provided limited results so that there is no trait for which all related QTL have been located precisely. Therefore, it is very difficult, if not impossible, to make a comprehensive selection for any specific trait. It is also a complicated issue to simultaneously select for multiple QTL. Secondly, epistasis will affect both the efficiency and the final products of MAS. Thirdly, there are certain genetic correlations among quantitative traits so MAS for one trait may also modify other correlated traits. Therefore, it is much more difficult to apply MAS to quantitative traits.

8.4.1 Selection based on phenotypic values

The theoretical basis for phenotypic selection is that phenotypic value is an approximate estimate of genotypic value and thus, selection based on phenotypic value can be considered approximately as a selection based on genotypic value. The higher the relatedness between phenotypic and genotypic values, the higher the efficiency of phenotypic selection.

It should be noted that, under random mating, only a fraction of the total genotypic value, namely the component contributed by additive effects, can be transmitted from one generation to the next and therefore, only selection for the additive component of genotypic value is effective. More precisely, the more closely the additive effect of an individual resembles its phenotypic value, the higher the efficiency of phenotypic selection. In animal breeding, the additive value of an individual is often known as its breeding value. The relatedness of phenotypic value to the additive effect depends on narrow-sense heritability ($h^2 = \sigma_A^2/\sigma_P^2$), where σ_A^2 and σ_P^2 are additive genetic variance and phenotypic variance, respectively. The higher h^2, the greater is the relatedness between phenotypic value and additive effect. When $h^2 = 1$, the phenotypic value is equal to the additive effect value. The efficiency of phenotypic selection increases as the narrow-sense heritability increases.

In conventional plant breeding, improvement of quantitative traits has been relying on direct selection. Direct selection is to select the individuals with extreme phenotype (one with either the largest or the smallest phenotypic value) in each generation so that the population mean changes towards the direction of selection. As discussed in Chapter 1, the efficiency of direct selection can be determined by response to selection (R) or genetic advance (ΔG), which is defined as the difference between the population mean of the progeny derived from the selected individuals ($\bar{y}$) and the population mean of the original or parental population (μ), i.e. $\Delta G = \bar{y} - \mu$ (Fig. 1.1). The higher the genetic advance, the higher the efficiency of selection. Apparently, genetic advance is positively proportional to heritability. Under a given heritability, genetic advance depends on selection rate (the proportion of selected individuals to the total individuals in the population). The smaller the selection rate, the larger the selection intensity (the difference of population means between the selected individuals and the original population) and the larger the genetic advance.

8.4.2 Selection based on marker scores

From discussion above, under random mating, selection should be much more efficient if it can be based on additive effects. The key issue here is how to estimate the additive effect for each plant. Theoretically, it can be estimated through marker–QTL analysis. By primary QTL analysis, however, it is difficult to detect all QTL and estimate their effects precisely and thus the estimate of additive effect is only an approximate with a potentially large estimation error. In order to obtain an accurate estimate for an individual's additive effect, it is necessary to map each QTL precisely. At present, MAS can be only processed based on approximate additive effects.

Most QTL mapping methods can be used to obtain additive effects (see Chapter 6 for details). In practice, however, a more convenient and efficient method is required. Here MAS method proposed by Lande and Thompson (1990) will be described, which is based on marker–trait regression and has been widely accepted. Under the additive-effect model, the marker–trait regression equation is

$$y = \mu_0 + \sum_{i=1}^{N} a_i x_i + \varepsilon \qquad (8.17)$$

where y is the phenotypic value of an individual, μ_0 is the model mean, a_i is the additive effect of marker i, x_i is the category variable of marker i (with values of 1, 0 and −1 for marker genotypes MM, Mm and mm), ε is random environmental error and N is the number of markers. By step-wise regression, markers with significant effects on the target trait and thus most probably linked to QTL can be selected and additive effect estimates ($\hat{a}_i$) can be used to calculate marker score for each plant:

$$m = \sum_{i=1}^{n} \hat{a}_i x_i \qquad (8.18)$$

where n is the number of markers selected. Marker score m is an approximation of the additive effect and the extent of approximation depends on the proportion of additive genetic variance explained by selected markers (σ_M^2) to the total additive genetic variance (σ_A^2), that is, $p = \sigma_M^2/\sigma_A^2$. The higher the value of p is, the better m is as a predictor of an individual's additive genetic value. Only when $p = 1$, is m equal to an individual's additive effect. Selection based on marker scores is called marker-score selection.

Both marker score and phenotypic value are the approximations of additive effect and their degrees of approximation depend on p and h^2, respectively. So selection efficiencies of these two methods depend on the relative magnitudes of p and h^2. That is, selection based on marker scores might not be more efficient than selection based on phenotype, depending on whether p is larger than h^2.

Let's now discuss the direct selection. Denote genetic advances obtained through marker-score selection and phenotypic selection as ΔG_M and ΔG_P, respectively. Under the same selection rate, the relative efficiency of these two methods is

$$RE_{MP} = \frac{\Delta G_M}{\Delta G_P} = \sqrt{\frac{p}{h^2}} \tag{8.19}$$

which indicates that the relative efficiency is determined by the relative magnitudes of p and h^2. It can be inferred that for traits with relatively low heritability the relative efficiency is high. The lower the heritability is, the higher the relative efficiency. For the traits with relatively high heritability, the efficiency of phenotypic selection will be high enough so that there is no necessity for marker-score selection. In addition, marker-score selection may be less efficient than phenotypic selection because of estimation errors of marker scores.

Although selection based on marker scores has relatively higher efficiency when heritability is low, the power for detecting QTL will be decreased and the sample error of marker scores will increase and thus the efficiency of selection based on marker scores will decrease if the heritability is too low (Moreau *et al.*, 1998). Under low heritability, therefore, it is necessary to increase population size and use low thresholds for declaring QTL in order to improve the power for QTL detection and to decrease the estimate error of marker scores (Gimelfarb and Lande, 1994a; Hospital *et al.*, 1997; Moreau *et al.*, 1998).

If the estimates of marker scores are reliable, selection based on marker scores will have significant genetic advance in early generations. However, genetic advance will often decrease with the advance of generations and disappear in three to five generations so that no further significant genetic advance is possible (Edwards and Page, 1994). There are two reasons for this phenomenon. First, genetic recombination breaks the linkage relationship between the marker and QTL. Secondly, favourable alleles with minor effects are lost during the selection while the unfavourable alleles are fixed (become homozygous) at the rate that is faster in marker-score selection than in phenotypic selection (Hospital *et al.*, 1997). The first issue can be solved by constant re-evaluation and screening for markers that have significant effects on the trait. If molecular markers are re-evaluated and selected for associations in each generation, selection efficiency will be improved significantly (Gimelfarb and Lande, 1994a). However, this approximation will increase the cost for molecular marker analysis. It should be more reasonable if this re-evaluation and selection can be performed every two to three generations (Hospital *et al.*, 1997).

8.4.3 Index selection

As discussed above, both marker score and phenotypic value are approximates of additive effect, each containing only partial information of additive effects that could complement each other. If marker score and phenotypic value can be combined, selection based on the integrated information should have higher efficiency. Therefore, Lande and Thompson (1990) proposed that a selection index should be constructed using marker score and phenotypic value:

$$I = b_z z + b_m m \tag{8.20}$$

which can be optimized by choosing the weight coefficients b_z and b_m to maximize the rate of improvement in the mean phenotype per generation.

The selection method based on the selection index is called index selection. In the above equation, z is the phenotypic value and m is the marker score. The optimal weight coefficients b_z and b_m are

$$b_z = \frac{\sigma_G^2 - \sigma_M^2}{\sigma_P^2 - \sigma_M^2} = \frac{(1-p)h^2}{1-ph^2} \tag{8.21}$$

and

$$b_m = \frac{\sigma_P^2 - \sigma_G^2}{\sigma_P^2 - \sigma_M^2} = \frac{1-h^2}{1-ph^2} \tag{8.22}$$

respectively.

The selection index also approximates to the additive effect, to which extent depends on its heritability (Knapp, 1998):

$$h_I^2 = \frac{(1-p)h^2}{1-ph^2} + \frac{p(1-h^2)}{h^2 - 2ph^2 + p} \quad (8.23)$$

From this equation, the higher the selection index heritability (h_I^2), the better predictor of additive effect the selection index becomes and the higher the selection efficiency. When $p = 0$, $h_I^2 = h^2$, i.e. index selection is equivalent to phenotypic selection. For a given h^2, h_I^2 increases as p increases and it increases dramatically when h^2 is low. h_I^2 increases rapidly when $0 < p \leq 0.5$ (Fig. 8.10). This indicates that low heritability has a strong effect on marker score and in this case MAS should have a higher impact on selection.

For directional selection, the relative efficiency of index selection and phenotypic selection can be expressed as (Lande and Thompson, 1990):

$$RE_{IP} = \frac{\Delta G_I}{\Delta G_P} = \sqrt{\frac{p}{h^2} + \frac{(1-p)^2}{1-ph^2}} \quad (8.24)$$

where ΔG_I is the genetic advance of selection index. Fig. 8.11 shows how RE_{IP} changes with p under different levels of h^2. For a given p, RE_{IP} increases with the decrease of h^2, that is, MAS is more efficient when heritability is low; while RE_{IP} increases with the increase of p but the increase rate becomes slow when h^2 is high. When h^2 reaches to an intermediate level ($h^2 = 0.5$), index selection has no apparent advantage. When $h^2 = 1$, RE_{IP} does not change with p, with a constant value of 1, indicating that in this case molecular markers do not provide any extra information so that MAS has no positive contribution at all.

The relative efficiency of index selection and marker-score selection can be expressed as

$$RE_{IM} = \frac{\Delta G_I}{\Delta G_M} = \frac{RE_{IP}}{RE_{MP}} = \sqrt{1 + \frac{h^2(1-p)^2}{p(1-ph^2)}} \quad (8.25)$$

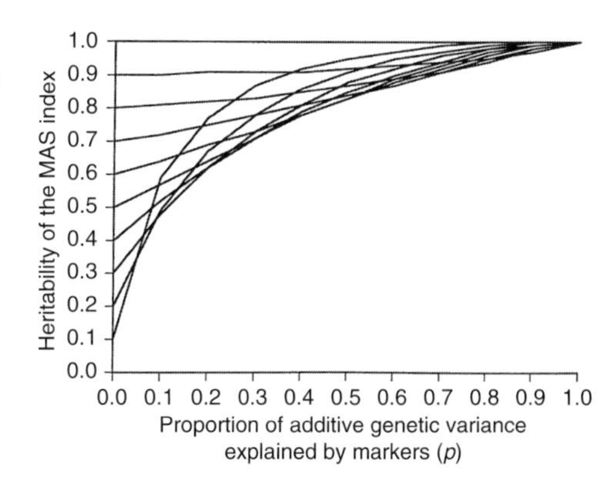

Fig. 8.10. Relationship between heritability of the MAS index (h_I^2) and the proportion of the additive genetic variance $p = \sigma_M^2/\sigma_G^2$ with p ranging from 0.0 to 1.0 and heritability (h^2) ranging from 0.1 to 1.0, where σ_M^2 is the additive genetic variance associated with markers and σ_G^2 is the additive genetic variance. From Knapp (1998) with permission.

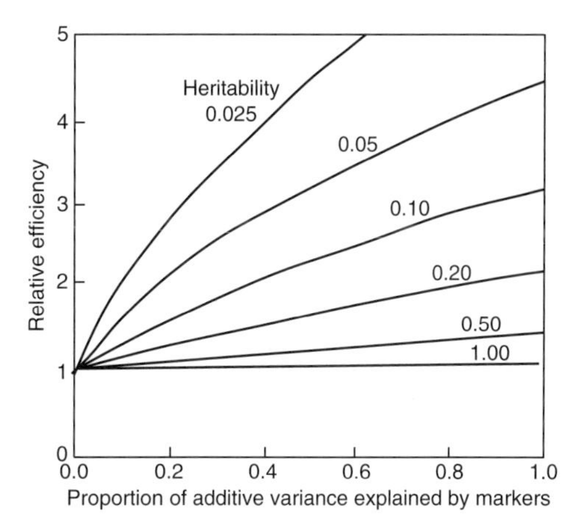

Fig. 8.11. Efficiency of MAS in the improvement of a single trait relative to traditional individual selection index with the same selection intensity, assuming very large sample sizes. Relative efficiency is plotted as a function of the proportion of the additive genetic variance in the trait significantly associated with the marker loci, for various values of the heritability of the trait. From Lande and Thompson (1990) with permission.

which indicates that no matter what values p and h^2 take, there is $RE_{IM} \geq 1$. Therefore, index selection always has higher selection efficiency than marker-score selection, which has been proven by computer simulation (Whittaker *et al.*, 1997) and is different from the situation where selection is based on marker scores.

Index selection depends on both phenotypic value and marker score. Therefore, the factors that affect the efficiency of marker-score selection will also affect the efficiency of index selection. Computer simulation (Gimelfarb and Lande, 1994a,b, 1995) indicated that index selection is more efficient than phenotypic selection at least for the first several generations, but this advantage disappears very quickly with the advance of generations. In advanced generations, index selection might be less efficient than the phenotypic selection. This could happen in advanced generations: when the degree of the additive effect explained by marker score is not as good as that by phenotypic value (i.e. $p < h^2$), while sampling errors of weight coefficients in Eqn 8.20 amplify the relative importance of marker score so that the proportion of additive genetic variance explained by the selection index is not as good as that by phenotypic value ($h_1^2 < h^2$). Therefore, both marker-score and index selection have advantage only at the early stages of selection and, in advanced generations phenotypic selection works better.

Although index selection utilizes more genetic information and thus it is more efficient than marker-score selection, it costs more and needs more work in order to gain the extra information for phenotypic value. Furthermore, measurement of phenotypic value is limited to the stage at which the trait is expressed, which cancels the advantage that MAS can be done at any stage. In addition, when the measurement of phenotype needs to be progeny tested, the cycle of index selection becomes longer so that the advantage of index selection may not offset this disadvantage. For example, hybrid maize yield has to be measured by progeny test and each cycle of index selection will take 2 years, while four cycles of marker-score selection can be done in 2 years (Edwards and Page, 1994). Although marker-score selection has lower genetic advance per cycle compared to index selection, it has higher genetic advance per unit time because more cycles can be done in a given period. Based on this result, Hospital *et al.* (1997) proposed a selection strategy with one-generation of index selection and several generations of marker-score selection alternatively. In the generation of index selection, a relatively larger population is required for re-evaluation and selection based on molecular markers, in order to maintain the reliability of marker–trait regression. Conversely, relatively smaller populations can be used in the generation of marker-score selection.

8.4.4 Genotypic selection

Both marker-score selection and index selection depend on genotypic value or more specifically the additive components of genotypic value rather than the genotype itself. Therefore, both selection methods are for selection of genotypes through genotypic values, which is indirect and they have no virtual difference from phenotypic selection. This is not exactly the concept of MAS which has been proposed and expected. Because genotypic value is the result of genotypic expression, different genotypes may have the same genotypic value, that is, a genotypic value could match up with many genotypes. There is a loss or degeneracy of genetic information from genotype to genotypic value. This information degeneracy will result in a low efficiency of selection and the loss of some favourable QTL alleles with relatively smaller effects. The more the QTL involved in selection, the higher the chance that favourable alleles will be lost. Therefore, a more efficient selection method should be that based on the genotype itself (which is called genotypic selection) as MAS for the qualitative trait. More specifically, each target QTL is selected based on its two flanking markers, a single closely linked marker, or a gene-based marker.

Currently, genotypic selection of quantitative traits is now limited by the availability of QTL that have been fine mapped. For most quantitative traits, only QTL with large effects have been mapped on a relatively rough scale, leaving a lot of minor

QTL non-detectable. To improve the efficiency and reliability of MAS, markers that are flanking the target QTL should be tightly linked. However, if the target region is too small, it might not contain the target QTL because the primary QTL mapping is not so accurate. It is important to develop flanking markers that bracket the target QTL with high confidence in order to have high selection efficiency.

As discussed in the previous section for selection of qualitative traits, it is better to use three linked markers in selection and the best positions of these markers will be determined by the confidence interval of the QTL. The middle marker should be tightly linked to, located exactly at, or identical to, the QTL, which should be bracketed by two flanking markers. The optimized window size for the target region determined by the flanking markers is positively proportional to the confidence interval of the QTL. The larger the QTL confidence interval, the bigger the target region bracketed by the flanking markers is required for guaranteeing that the QTL is located in the target region.

As indicated by Hospital and Charcosset (1997), using position-optimized markers to follow the target QTL in BC breeding, favourable alleles at four independent QTL can be transferred from the donor parent to the recurrent parent, with a population consisting of several hundreds of individuals. If there is linkage between QTL and QTL are precisely mapped or larger population sizes are used, more QTL can be transferred simultaneously.

8.4.5 Integrated marker-assisted selection

As discussed previously, marker–trait association identified in one population has to be validated before it can be used for MAS in other populations. One of the best ways to avoid the marker the validation step is to integrate genetic mapping with MAS, that is, marker–trait associations identified from a breeding population will be used for MAS of the same population. This is critical for quantitative traits that are genetically controlled by many genes and interact with environments. Advanced backcrossing QTL (AB-QTL) analysis, proposed by Tanksley and Nelson (1996) to accelerate the process of molecular breeding, is one of the approaches that can be used for this purpose. Stuber *et al.* (1999) discussed their effort to test a marker-based breeding scheme for systematically generating superior lines without any prior identification of genes in the donor sources. Identifying and mapping of genes in the donor is a bonus obtained when the derived NILs are evaluated. This method is somewhat similar to AB-QTL analysis. Other approaches include using associations identified in F_2 populations to select the subsequent self-pollinated populations.

The AB-QTL strategy postpones QTL mapping until the BC_2 or BC_3 generation. The delay of QTL analysis offers advantages for QTL characterization such that the probability is reduced for the detection of QTL displaying epistatic interactions among donor alleles due to their overall low frequency. In fact, there will be a higher probability of detecting additive QTL which still function in a near-isogenic background. During the generation of BC_2 or BC_3 populations, negative selection is being exercised to minimize the occurrence of unfavourable donor alleles. The advantage of focusing on the BC_2 or BC_3 population is that they offer sufficient statistical power for QTL identification on the one hand and on the other hand provide sufficient similarity to the recurrent parent to select for QTL-NILs in a short time span (within 1–2 years). By use of QTL-NILs, the QTL discovered can be verified and the NILs may serve directly either as improved cultivars or as a parent cultivar in case of hybrid crops (Peleman and van der Voort, 2003).

The AB-QTL approach can be exploited for pyramiding QTL alleles. Each time that AB-QTL analysis is applied, the map positions of donor QTL affecting key traits will likely be discovered so that QTL mapping information derived from AB-QTL analysis is cumulative. Based on this knowledge, as indicated by Tanksley and Nelson (1996), it would be straightforward to combine favourable donor QTL alleles detected in one experiment with non-allelic QTL affecting the same trait from other experiments in which a different donor parent was used. In this way, it should be possible to pyramid all non-allelic QTL with similar effects detected within a given species or across the related species, if they act without much influence of epistasis.

The AB-QTL approach has been successfully used to identify markers for QTL contributing to fruit size, shape, colour and firmness together with soluble solids and total yield in tomato. On this basis, QTL-marker associations were identified in one BC generation and immediately applied in the subsequent BC generation some 6 months later (Tanksley *et al.*, 1996). In rice, a series of advanced backcrossing populations have been developed through collaborations between Cornell University and breeders around the world to identify and introgress trait-enhancing alleles from wild species into high-yielding elite cultivars. The first such study employed a cross between the wild rice relative *Oryza rufipogon* and the Chinese *indica* hybrid 'V20'/'Ce64' (Xiao *et al.*, 1998). Although the *O. rufipogon* accession was phenotypically inferior for all 12 traits studied, transgressive segregation was observed for all traits and 51% of the QTL detected had beneficial alleles from *O. rifupogon*. By MAS and field selection, an excellent CMS restorer line ('Q661') carrying one of the QTL for yield components has been developed. Its hybrid, 'J23A'/'Q661', out-yielded the check hybrid by 35% in a replicated trial for the second rice crop in 2001 (Yuan, 2002). A second QTL study used an advanced BC population between the same *O. rufipogon* accession and the upland *japonica* rice cultivar 'Caiapo' and identified beneficial QTL alleles from *O. rufipogon* for 56% of the trait-enhancing QTL detected (Moncada *et al.*, 2001). A third study employed the *O. rufipogon* in a cross with the long-grain 'Jefferson', a US tropical *japonica* cultivar and the *O. rufipogon* allele was favourable for 53% of the yield and yield component QTL (Thomson *et al.*, 2003). There are several ongoing projects to introgress these favourable alleles from the *O. rufipogon* accession into cultivated rice.

8.4.6 Response to marker-assisted selection

MAS for traits controlled by major genes will receive a strong response. However, the response to selection, or genetic advance, for quantitative traits will depend on several factors: linkage between markers and genes, trait heritability, gene effects, gene interactions, population size, the number of plants selected and the breeding scheme. In classical selection theory, the expectation, genetic variance and heritability of the target trait are required, as well as the covariance between the target trait and selection criterion in the case of indirect selection. In backcrossing without selection, as described in Chapter 4, the expected donor genome proportion in generation BC_n is $1/2^{n+1}$. In backcrossing with selection for the presence of a target gene, Stam and Zeven (1981) derived the expected donor genome proportion on the carrier chromosome of the target gene. Their results were extended to a chromosome carrying the target gene and the recurrent parent alleles at two flanking markers (Hospital *et al.*, 1992) and to a chromosome carrying several target genes (Ribaut *et al.*, 2002a).

An example in Lande and Thompson (1990) demonstrated that on a single trait the potential selection efficiency by using a combination of molecular and phenotypic information, compared to standard methods of phenotypic selection, depends on the heritability of the trait, the proportion of additive genetic variance associated with marker loci and the selection scheme. As discussed previously, the relative efficiency of MAS is greatest for traits with low heritability if a large fraction of the additive genetic variance is associated with marker loci. Limitations that may affect the potential utility of MAS in applied breeding programmes include: (i) the level of linkage disequilibrium in the populations, which affects the number of marker loci needed; (ii) sample size needed to detect trait loci with low heritability; and (iii) sample errors in the estimation of relative weights in the selection indices.

Frisch and Melchinger (2005) developed a theoretical framework for MAS for the genetic background of the recurrent parent in a BC programme to predict the response to selection and give criteria for selecting the most promising BC individuals for further backcrossing or selfing. The approach dealt with selection in generation n of the BC programme, taking into account pre-selection for the presence of one or several target genes, the linkage map of the target gene(s) and markers and the marker genotype of the individuals used as non-recurrent parents for generating BC generations.

Response to selection R is defined as the difference between the expected donor genome proportion μ in the selected fraction of a BC_n population and the expected donor genome proportion μ' in the unselected BC_n population:

$$R = \mu - \mu' \qquad (8.26)$$

Prediction of the response to selection can be employed to compare alternative scenarios with respect to population size and required number of markers. This application was illustrated by the example of a BC_1 population using model genomes close to maize (ten chromosomes of length 2 M) and sugarbeet (nine chromosomes of length 1 M) with markers evenly distributed across all chromosomes, a target gene located 66 cM from the telomere on a chromosome and one individual is selected as the non-recurrent parent of generation BC_2.

The expected response to selection for maize ranged from ~5% of the donor genome (20 markers, 20 plants) to 12% (120 markers, 1000 plants) and for sugarbeet it ranged from ~7% to 15% (Fig. 8.12). To obtain a response to selection of ~10% with 60 markers, a population size of 180 is required in maize, corresponding to ~180/2 × 60 = 5400 marker data points (MDP). By comparison, in sugarbeet a population size of 60 is sufficient, resulting in only 30% of the MDP required for maize. The result indicates that the efficiency of MAB in crops with smaller genomes is much higher than that in crops with larger genomes.

Using > 80 markers in maize (corresponding to a marker density of 25 cM) or > 60 markers in sugarbeet (marker density 15 cM) resulted in only a marginal increase of the response to selection, irrespective of the population size employed (Fig. 8.12). Increasing the population size up to 100 plants resulted in substantial increase in response to selection on both crops and using even larger populations still improves the expected response to selection. Frisch and Melchinger (2005) concluded that increasing the response to selection by increasing the number of markers employed is possible only up to an upper limit that depends on the number and length of chromosomes. In contrast, increasing the response to selection by increasing the population size is possible up to population sizes that exceed the reproduction coefficient of most crop species.

An optimum criterion for the design of MAS in a BC population can be defined by the expected response to selection reached with a fixed number of MDP. For a fixed number of MDP in sugarbeet, designs with large populations and few markers always reached larger values of response to selection than designs with small populations and many markers (Fig. 8.12). For maize, the same trend was observed for 500 and 1000 MDP, while for a larger number of MDP the optimum design ranged between 40 and 50 markers. Therefore, in BC_1 populations of maize and sugarbeet and a fixed number of MDP, MAS is, within certain limits, more efficient for larger populations than for higher marker densities.

In theory, MAS is proposed to be more efficient than phenotypic selection when the heritability of a trait is low, where there is tight linkage between QTL and markers (Dudley, 1993; Knapp, 1998), with larger population sizes (Moreau *et al.*, 1998) and in earlier generations of selection before recombinational erosion of marker–trait associations (Lee, 1995). Edwards and Page (1994) proposed that the distance between markers and QTL was the factor that most

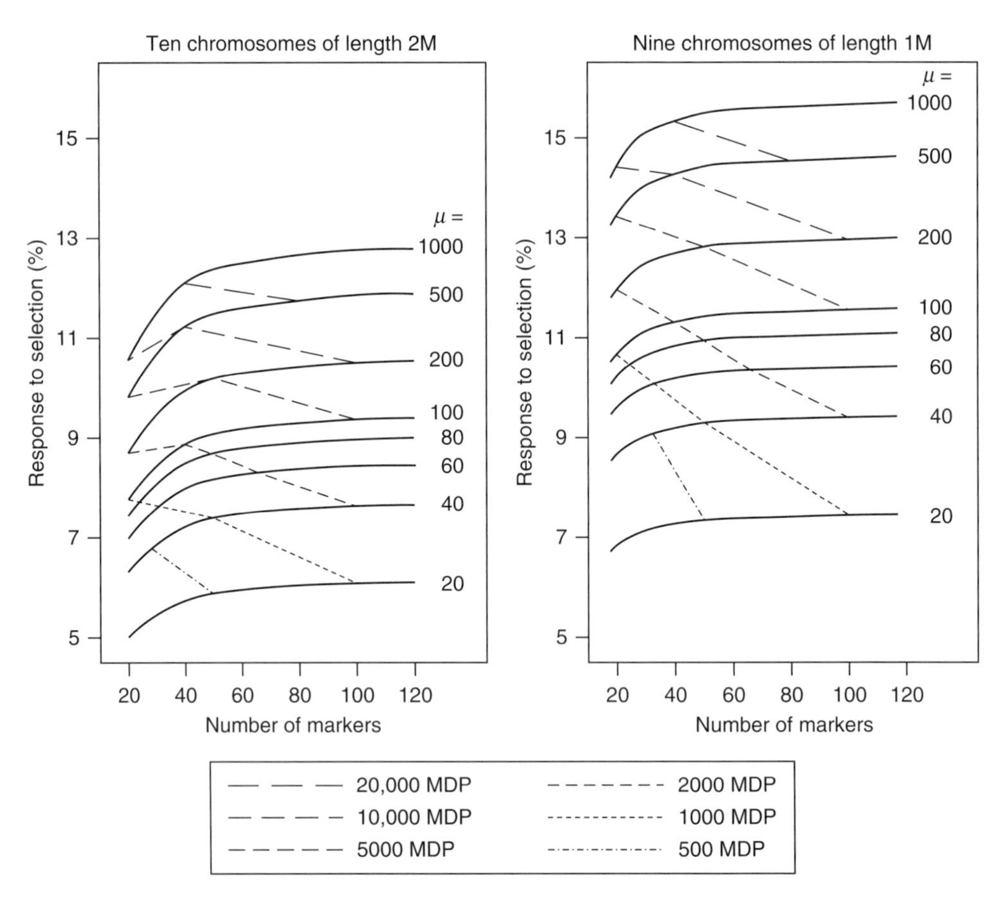

Fig. 8.12. Expected response to selection throughout the entire genome and expected number of required marker data points (MDP) when selecting the best out of μ = 20, 40, 60, 80, 100, 200, 500 and 1000 BC_1 individuals. Model of the maize genome with ten chromosomes of length 2 M (left-hand side). Model of the sugarbeet genome with nine chromosomes of length 1 M (right-hand side). From Frisch and Melchinger (2005) with permission.

limited genetic gains from MAS. Yousef and Juvik (2001a) reported an empirical experiment that provided equivocal results regarding the relative efficiency of MAS and phenotypic selection in enhancing economically important quantitative traits in sweet corn. MAS and phenotypic selection were applied to three $F_{2:3}$ base populations with either the *sugary 1* (*su1*), *sugary enhancer 1* (*se1*), or *shrunken 2* (*sh2*) endosperm mutations. One cycle of selection was applied to both single and multiple traits such as seedling emergence. Selection efficiencies were evaluated on the basis of gains over one cycle. Among 52 paired comparisons between MAS and phenotypic selection composite populations, MAS resulted in significantly higher gain than phenotypic selection for 38% of the comparisons, while phenotypic selection was significantly greater in only 4% of the cases. The average MAS and phenotypic selection gains, calculated as percent increase or decrease from the randomly selected controls, was 10.9% and 6.1%, respectively.

Recognizing that small mapping populations are not adequate for QTL mapping is the first and most important realization needed in the research community (Young, 1999). Scientists must understand that simply demonstrating that a complex trait can be dissected into QTL and mapped to approximate genomic regions using DNA markers is not enough. Projects need to utilize better scoring methods, larger population sizes, multiple replications and environments, appropriate quantitative genetic analysis, various genetic backgrounds and, whenever possible, independent verification through advanced generations or parallel populations (Melchinger *et al.*, 1998; Utz *et al.*, 2000; Schön *et al.*, 2004). Only then will sufficient experimental evidence be in place for a successful MAS programme.

'What if we knew all the genes for a quantitative trait in hybrid crops?' This was asked by Bernardo (2001), when working on the prediction of hybrid performance through computer simulation. With maize as a model species, he found through trait and gene best linear unbiased prediction (TG-BLUP) that gene information is most useful in selection when few loci (e.g. ten) control the trait. With many loci ($\geq$ 50), the least square estimates of gene effects become imprecise. Gene information consequently improves selection efficiency among hybrids by only 10% or less and actually becomes detrimental to selection, as more loci become known. Bernardo further indicated that increasing the population size and trait heritability to improve the estimates of gene effects also improves phenotypic selection, leaving little room for improvement of selection efficiency via gene information. He thought genomics is of limited value in selection for quantitative traits in hybrid crops. Epistatic interactions, which were assumed absent in his study, would make the estimation of gene effects even more difficult. It is unknown whether methods other than TG-BLUP or multiple regression would substantially enhance the usefulness of gene information in selection.

8.5 Long-term Selection

As one of the most powerful tools available to biology, selection is used in the plant and animal sciences to develop improved crop cultivars and livestock breeds. Selection is also used in laboratory species to test many of the assumptions of the underlying quantitative genetic models and to test the limits of selection itself. The power of selection is best presented by the selection responses that have been observed in two important agricultural species. US maize yield increased from a pre-1930 average of 1.6 t ha^{-1} (26.1 bushels $acre^{-1}$) to an average of 8.6 t ha^{-1} (134.7 bushes $acre^{-1}$) for the 5 year period from 1998 to 2002, a fivefold increase over 70 years (http://www.usda.gov/nass/). Of course, not all of the increase is due to selection, but studies have consistently shown that genetics can account for 50% of the increase. Milk yield in Holsteins had increased from 5870 kg in 1957 to 11,338 kg in 2001, representing a doubling in milk yields over 44 years (http://aipl.arsusda.gov/dynamic/trend/current/trndx.html). There is evidence that the genetic trend continues to increase with time in Holsteins. Molecular techniques have provided novel tools to analyse the final selection product and reveal the change of genetic structures with the progress of selection experiments.

Including long-term selection in this chapter is justified by a concept 'reversed breeding-to-genetics', which starts with a selection programme to pyramid favourable alleles from various sources of germplasm to create transgressive variation and great selection response followed by genetic analysis (usually marker-assisted evaluation) to identify the genes and alleles associated with the selection response. As it will take years to pyramid genes and alleles from multiple sources through genetic mapping and MAS, the 'reversed breeding-to-genetics' approach can be used to exploit cumulated novel alleles and genes by taking advantage of the availability of plant materials that have been accumulating in genetic and breeding programmes. Combining with the strategy of selective genotyping revised by Xu *et al.* (2008) and Sun *et al.* (2009), it could be more realistic by starting with selection to pyramid alleles followed by genetic analysis to identify the genes, compared to the 'genetics-to-breeding' approach by which genes/QTL are mapped in separate genetic analyses and then to pyramided by MAS. In this section, we will discuss the long-term selection experiments in plants and marker-assisted evaluation of the selection results, which can be considered the 'reversed breeding-to-genetics' approach.

8.5.1 Long-term selection in maize

There are several long-term selection experiments in maize (Duvick *et al.*, 2004; Hallauer *et al.*, 2004; Dudley and Lambert, 2004). The most well-known is the selection for oil and protein contents which has been running for over 100 generations. The detail about this experiment can be found in the special volume of *Plant Breeding Reviews* (Volume 24, Part 1, 2004). Only some significant procedures and results will be summarized here.

Procedure

The long-term selection experiment for oil and protein content in maize was initiated at the University of Illinois by C.G. Hopkins before the rediscovery of the Mendelian laws (Hopkins, 1899). The most recent update on this long-term selection can be found from Dudley and Lambert (2004). Although the original goal was to produce agriculturally valuable crops by increasing the oil and protein content of the kernels, the results are also quite remarkable from a theoretical viewpoint. One of the most interesting results was that the continued selection did not deplete the variability. Truly the results were not in full compliance with the simple Mendelian expectation.

In 1896, Hopkins initiated selection in the open-pollinated maize cultivar 'Burr's White' (Hopkins, 1899). He analysed 163 ears for oil and protein concentration. The 24 ears highest in protein, the 12 ears lowest in protein, the 24 ears highest in oil and the 12 ears lowest in oil were selected to initiate the Illinois High Protein (IHP), Illinois Low Protein (ILP), Illinois High Oil (IHO) and Illinois Low Oil (ILO) strains, respectively. Both forward and reverse selection has been conducted at different times in the experiment. The forward phase of the experiment was divided into four segments as follows:

- Segment 1. Generations 0–9, mass selection based on chemical composition. Numbers of ears analysed and selected varied but approximately 20% of the ears analysed were selected. Each strain was grown in a separate but isolated field.
- Segment 2. Generations 10–25. 120 ears per strain were analysed and 24 were saved. Seed from each ear was planted ear-to-row. Alternate rows were detasselled and 20 ears were analysed from each of the six highest yielding rows. Four ears were saved per row.
- Segment 3. Generations 26–52 in IHP and ILP; generations 26–58 in IHO and ILO. Twelve selected ears were arbitrarily divided into two lots (A and B) of six ears. Seed within each lot was bulked and planted in the nursery. Silks in lot A were pollinated by a bulk sample of pollen from 15–20 plants in lot B while silks in lot B were pollinated with pollen from lot A. Thirty ears from each lot were analysed and the 12 most extreme of the 60 ears analysed were saved.
- Segment 4. Generations 53–90 in IHP and ILP; 59–90 in IHO and 59–87 ILO. The selection procedure was the same as in segment 3 but 90–100 kg of N fertilizer ha^{-1} were added to the soil. Only 87 generations were completed in ILO because of difficulties with seed set and seed quality which cause a loss of some generations.

Following 48 generations of forward selection, reverse selection was initiated in each of the four strains to form four new strains: Reverse High Protein (RHP), Reverse Low Protein (RLP), Reverse High Oil (RHO) and Reverse Low Oil (RLO) (Figs 8.13 and 8.14). The objective was to determine the extent of residual variability available for selection. The selection procedure was the same as in the forward strains except that selection was for low protein in IHP, high protein in ILP, etc. Following seven generations of selection in RHO, selection was against reverse to initiate the Switchback High Oil (SHO)

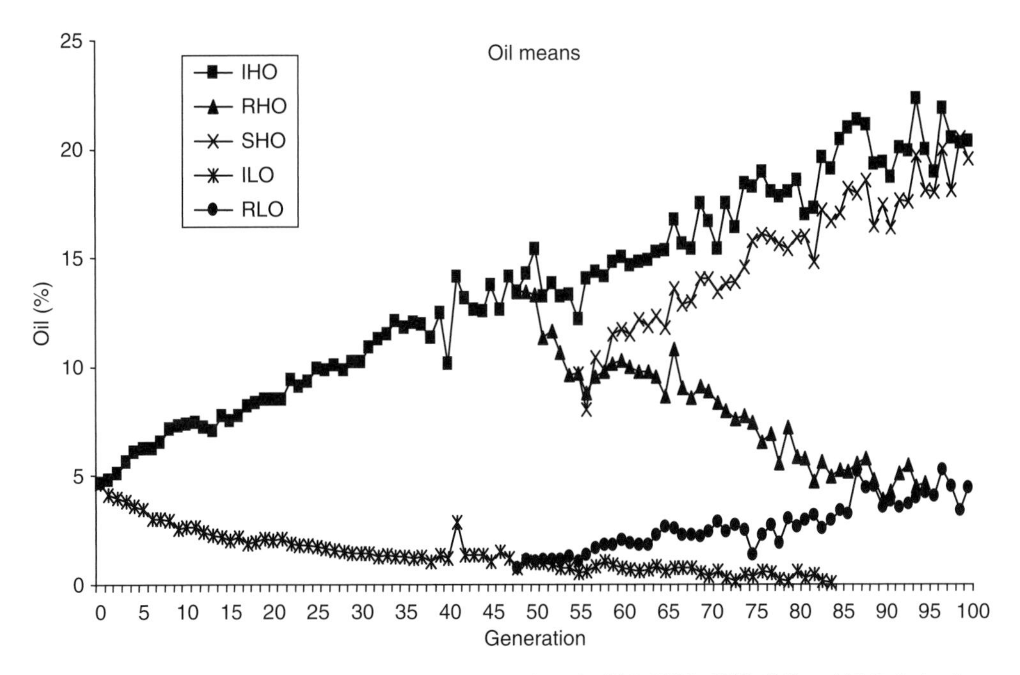

Fig. 8.13. Mean oil percentage plotted against generations for IHO, RHO, SHO, ILO and RLO derived from 100 generations of selection. From Dudley and Lambert (2004). This material is reproduced with permission of John Wiley & Sons, Inc.

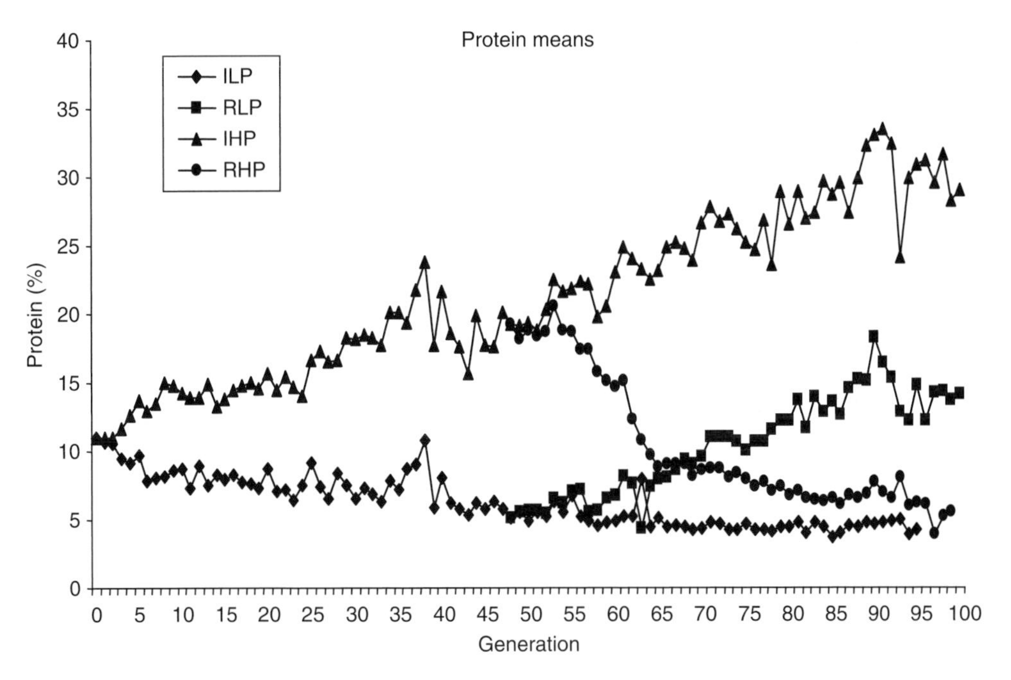

Fig. 8.14. Mean protein percentage plotted against generations for IHP, RHP, ILP and RLP derived from 100 generations of selection. From Dudley and Lambert (2004). This material is reproduced with permission of John Wiley & Sons, Inc.

strain. Beginning in generation 90 of ILP, a new strain called Reverse Low Protein 2 (RLP2) was initiated by selection for high protein in ILP. This strain was initiated to determine whether genetic variability still existed that could be exploited by selection after an apparent lack of progress in ILP for nearly 35 generations. The selection procedures were the same as in the regular and reverse selection strains, and have been described in detail (Dudley *et al.*, 1974; Dudley and Lambert, 1992).

Limits to selection

Response over all generations is presented by means for each generation plotted against generation number for all strains (Figs 8.13 and 8.14). The data upon which the figures were based is available in details in Dudley and Lambert's (2004) Appendix Tables 5.A1–5.A5.

One of the objectives of this experiment was to determine the limits to selection for oil and protein in maize. The question has been answered for low oil and low protein in that progress ceased when oil became so low it was no longer measurable with the analytical tools available. Protein apparently reached a lower limit after approximately 65 generations when no further progress was possible with the selection methods used. This low limit is likely physiological in nature.

Three types of evidence suggested an upper limit has not been reached for oil in IHO and SHO. Significant genetic variation still exists in generation 98 and has not changed since generation 65 as results from the per se evaluation trials showed significant increases in oil during the last five generations measured. Thus 100 generations of selection have not eliminated genetic variability and an upper limit has not been reached for either IHO or SHO.

For IHP, the results are not clear. Data from the evaluation trials indicated no significant increase in protein since generation 88. Genetic variance in generation 98 was not significantly different from that in generation 68. Also there is no apparent viability problem in IHP as in experiments with other species where viability problems caused progress to cease, significant genetic variability was found in strains which had plateaued. Thus, it is not clear whether an upper limit has been reached for protein in IHP.

Based on significant progress in the reverse selection strains, genetic variance had not been exhausted at generation 48. The results from RLP2 are inclusive as to whether exploitable genetic variance for high protein still existed at generation 90 in ILP. Results from the per se evaluation trials suggest that progress is being made, but the generation data do not confirm this result.

One unusual result of reverse selection occurred in RHP. All the gain from the first 48 generations of selection was dissipated by the next 15 generations of selection (Fig. 8.14). The progress per generation for these 15 generations was approximately 0.68% per generation a rate at least three times that in any other segment of any strain for protein.

Explanation of progress

For oil, total gain in IHO is approximately four times the total gain in ILO. For protein, the gain in IHP is approximately three times the gain in ILP. Gain in the high direction for both oil and protein is greater from generations 49 to 100 than from generations 0 to 48. In contrast, nearly 90% of the gain from selection in ILO and ILP came in the first 48 generations of selection. Given that the lower limit to selection for ILO and ILP is near zero and the upper limit could approach 100%, this greater gain in the high direction is not surprising. The gain between generations 48 and 100 for IHO (9.7% oil) is similar to that in RHO (9.0% oil) and the gain in IHP (12.5%) protein from generations 48 to 100 is similar to that in RHP (12.2% protein). The gain in RLO from generations 48 to 100 is nearly ten times that in ILO and the gain in RLP over the same generations is 13 times that in ILP.

These results are consistent with the gene frequency estimates assuming a model with a relatively larger number of genes affecting the traits, each with relatively similar effects and additive gene action. The frequency of favourable alleles (q) in the original population was estimated as approximately 0.2; therefore, greater gain for higher oil or protein should be possible than lower values. When reverse selection was initiated, q was estimated to be approximately 0.5 in both IHP and IHO. Thus, selection in either direction should be possible and the total possible change should be approximately the same in either direction. The switchback selection occurred at a gene frequency of 0.35, which could allow greater progress in the high direction than in the low, as was observed.

By evaluating the results from selection of 48 generations, Leng (1962) suggested four possible genetic interpretations: (i) accidental outcrossing; (ii) favouring of heterozygotes in selection; (iii) high rate of mutation of the 'chemical genes' concerned; and (iv) release of variability by some unknown means. He immediately dismissed these interpretations because: (i) pollination has been under strict control throughout the long-term study. (ii) Favouring heterozygosity cannot be ruled out; however 'The rapid response to reverse selection in all four strains, if it were attributed to residual heterozygosity alone, would have required the level of heterozygosity to have remained at nearly the same level through 48 generations of successful selection. This appears highly improbable.' (iii) Since 'all four strains are relatively uniform and show no evidence of being highly mutable... mutation is not considered a likely explanation.' (iv) A plausible mechanism is that continued recombination plays a role.

However, as indicated by Dudley (1977), it is possible to explain all the progress by segregation of a relatively large number of genes (n), each at a relative low frequency (q) in the original population. The number of additive genetic standard deviations of progress possible for a given value of n and q as calculated by Dudley (1977) based on theory derived by Robertson (1970) is shown in Table 8.8 for a sample of values of q and n. For the progress of 21 σ_A made in IHO and 18 σ_A in IHP, gene frequency needs

Table 8.8. Ultimate limits to selection, measured as number of σ_A with varying values of n (number of loci segregating) and q (frequency of favourable alleles) (from Dudley (1977) with permission).

	q				
n	0.1	0.25	0.5	0.75	0.9
10	13	8	3	3	2
50	30	17	10	6	3
100	42	24	14	8	5
200	60	35	20	12	7

to be low, approximately 0.25, and $n > 50$. Such values are consistent with estimates of q of approximately 0.2 for both oil and protein and n of 54 and 123 for oil and protein obtained, respectively. Although these results suggest all the progress would be explained by segregation of a large number of genes in the original population, mutation cannot be eliminated as a possible source of some of the variation upon which selection continues to operate.

Goodnight (2004) and Eitan and Soller (2004) suggested epistasis as an important factor to explain the negative or positive heterosis for oil and protein observed in the crosses involving the long-term selection strains, supporting the hypothesis that additive × additive espitasis was important. Further evidence comes from the Design III study of Moreno-Gonzalez *et al.* (1975) where crosses of both the F_2 and the F_6 of the cross of IHO × ILO back to the parents exhibited negative heterosis for oil. This hypothesis is further supported by the presence of significant negative heterosis for protein in the crosses of IHP × RLP and IHP × ILP (Dudley *et al.*, 1977).

Dudley *et al.* (1974) suggested part of the continued response in IHP could be due to a change in environment because the addition of N-fertilization in generation 53 increased response per generation from 1.4 to 1.6 g kg^{-1} protein per cycle. The increase of available N fertilizer presumably allowed alleles for higher protein to be expressed and selected.

Finally, Walsh (2004) argued that mutation was a necessary assumption to explain the result of the long-term selection experiment.

He indicated that gain based on mutational variance is expected to exceed that from gain based on residual segregation from the original population after about 46 generations for oil and 33 for protein. Although per-locus mutation rates are typically very small, for a wide range of traits the mutational variance introduced in each generation is on the order of 1/100th of the environmental variance. This can be quite significant after 10–20 generations. Keightley (2004) reviewed selection experiments in inbred lines and concluded that mutational variance was important in selection response. However, as indicated by Dudley (2007), neither Walsh nor Keightley considered the effects of epistatic interactionon selection response. They concluded that epistasis may be an important factor in explaining long-term response to selection, which has been supported by the results from the crosses of IHO × ILO and IHP × ILP for epistatic interactions as more epistatic interactions were significant than expected by chance and the number of markers associated only with significant epistatic effects ranged from 46.3 to 72.2% of the total number of significant markers detected (Dudley, 2008). I would rather suggest that both large numbers of loci at low gene frequency in the original population and their recombination and epistatic interaction in the long-term selection lines should have contributed to the long-term selection response.

Marker-assisted evaluation

Response to phenotypic selection can be evaluated and associated genes can be identified using molecular markers. The Illinois long-term selection experiment on maize oil and protein contents (Dudley and Lambert, 1992, 2004) and marker-assisted evaluation (Goldman *et al.*, 1993) provides such an example. The long-term divergent selection response can be attributed to the accumulative action of alleles with similar effect that had been dispersed among the individuals of the original population (Xu, 1997), while *de novo* mutations may be an alternative explanation for this divergence, as indicated by selection for bristle number in *Drosophila* (Mackay, 1995). The selection strains offer a unique opportunity to investigate the genetic basis of kernel chemical traits and have been used to produce maize populations to map the QTL responsible for the selection response (Goldman *et al.*, 1993). By using 90 genomic and cDNA clones distributed throughout the maize genome to detect RFLPs between IHP and ILP strains, 22 loci distributed on ten chromosome arms were significantly associated with protein concentration and clusters of three or more significant loci were detected on chromosome arms 3L, 5S and 7L, suggesting the presence of QTL with large effects at these locations. A multiple linear regression model consisting of six significant loci on different chromosomes explained over 64% of the total variation (Goldman *et al.*, 1993). These significant QTL associations can be used to account for the long-term selection response and the protein content difference between the IHP and ILP strains. It can be expected that the longer the selection proceeds, the bigger the difference of protein content will be in the resulting selection strains and thus the potential to detect additional QTL, as long as the populations continue to respond to selection. This expectation can be tested by QTL mapping using the crosses from the IHP and ILP strains derived from different cycles of selection.

Instead of using the extreme divergence of the parents to create mapping populations, Wassom *et al.* (2008) identified kernel QTL in a genetic background more relevant to practical breeding by using 150 BC_1-derived S_1 lines (BC_1S_1s) from IHO and recurrent parent B73. Oil, protein and starch were measured in BC_1S_1s and in Mo17-top-cross hybrids. Multiple regression models with 3–9 QTL detected for each trait by composite interval mapping explained 46.9, 45.2 and 44.3% of phenotypic variance for oil, protein and starch, respectively, in BC_1S_1s and 17.5, 22.9 and 40.1% for oil, protein and starch, respectively, in the testcross hybrids.

Laurie *et al.* (2004) used an association study to infer the genetic basis of dramatic changes that occurred in response to selection for changes in oil concentration. The study population was produced by a cross between the high- and low-selection lines at generation 70 when the oil concentrations were estimated for IHO as 16.7% and for ILO as 0.4%, followed by ten generations of random mating and the derivation of 500 lines by selfing. These lines were genotyped for 488 genetic markers and the oil concentration was evaluated in replicated field trials. As a single admixture event between IHO and ILO created linkage disequilibrium (LD) between genes with different allelic frequencies and the ten-generation random mating eliminated essentially all association between unlinked markers and most of those between loosely linked markers, the population can be used for LD mapping. Three methods of analysis were tested in simulations for ability to detect QTL. Using the most effective method-model selection in multiple regression, ~ 50 QTL were detected which accounted for ~ 50% of the genetic variance, suggesting that > 50 QTL are involved. The QTL effect estimates are small and largely additive. About 20% of the QTL have negative effects (i.e. not predicted by the parental difference), which is consistent with hitchhiking and small population size during selection. The large number of QTL detected accounts for the smooth and sustained response to selection throughout the 20th century.

Mikkilineni and Rocheford (2004) characterized RFLP variant frequencies in two cycles (65 and 91) of IHP, ILP, RHP and RLP. As revealed by RFLPs, considerable variation at the DNA level was maintained in the Illinois long-term selection protein strains even after 91 generations of selection. Only one locus was observed with a unique RFLP variant detected in just one of the four strains. Although only 35 RFLP loci were looked at, it does not appear there was much variation that might potentially be attributable to mutation. The inbreeding values calculated from the RFLP data from cycles 65/69 and 91 were lower than those calculated on the strains before molecular marker data were available. Maize undergoes inbreeding depression and thus there may have been some natural selection within the selection strains for more vigorous and more heterozygous plants. Also, the effective population size, due to the system used for bulking of pollen from multiple tassels and using the bulked pollen to fertilize many ears, may be larger than previously calculated, contributing to less inbreeding than estimated earlier (Walsh, 2004). The lower levels of inbreeding observed for the reverse strains than the forward strains suggest the change in direction of selection for protein levels may have contributed to maintenance of heterozygosity over generations in the reverse strains. There were trends in the variant frequencies in the forward and reverse strains that are consistent with response to selection.

All the RFLP loci selected to assay on the strains based on association with QTL for protein contents in IHP × ILP-derived mapping populations showed frequency trends consistent with response to selection in one or both of the reverse strains. Only one RFLP locus that showed a trend has not been identified as a QTL. The selection of probes based on previous QTL associations most likely increased the probability of identifying loci with variant frequency trends. These probes are more likely to reveal variants that respond to reverse selection, if they were not fixed by cycle 48 when the reverse selection was initiated. These loci are therefore good candidates to look for changes in variant frequencies in response to reverse selection. Eight probes (23%) showed reverse trends for the RHP strain. Twelve probes (34%) showed trends for the RLP strain. One probe (3%) showed a trend for just the RHP strain. Five probes (14%) showed trends for just the RLP strain. Seven probes (20%) showed trends in common for both the strains. All seven loci that displayed trends in both directions were associated with QTL in IHP × ILP mapping populations (Goldman *et al.*, 1993, 1994; Dijkhuizen *et al.*, 1998).

RFLP genotypic and variants frequency difference among cycle 90 of the oil strains, IHO, ILO, RHO and RLO, were also determined (Sughroue and Rockeford, 1994). A high degree of variant polymorphism was found among the four oil strains and many RFLP loci were still segregating within the oil strains after 90 generations of selection. RFLP variant trends consistent with response to directional selection were detected in comparisons among the four oil strains.

Application to plant breeding

The total gain from selection, both in absolute value and in number of additive genetic standard deviations, is well beyond what might have been expected from the distribution of oil and protein values in the original population. Likewise, they are well beyond what has been possible by selection for agronomic traits such as grain yield. To illustrate the possible increases in maize grain yield if selection for yield was as effective as for oil and protein, estimates of grain yield and σ_A from two maize synthetics, RSSSC (a stiff-stalk synthetic) and RSL (a Lancaster derivative) obtained in Illinois were used. The original means were 6.66 t ha^{-1} for RSL and 9.23 t ha^{-1} for RSSSC. Assume a gain of 24 σ_A, the approximately average of what was observed for oil and protein. The gain would be 33.28 t ha^{-1} for RSL and 27.44 t ha^{-1} for RSSSC or a yield at the limit of 39.94 t ha^{-1} for RSL and 36.68 t ha^{-1} for RSSSC. Assuming some heterosis, the ultimate yield would be around 43.96 t ha^{-1}. These values are not unreasonable when the fact that a yield of over 31.4 t ha^{-1} was reported in Iowa in 2002. As indicated by Dudley and Lambert (2004), these results suggest the existence of more genetic variability and more plasticity in the maize genome than is usually expected. They also suggest that limits to selection for yield have not been reached. To mark the importance of the long-term selection experiment for protein and oil in maize at the University of Illinois, the conference titled 'Long-term Selection: A Celebration of 100 Generations of Selection for Oil and Protein in Maize' was held on 17–19 June 2002 in Urbana, Illinois.

8.5.2 Divergent selection in rice

Xu *et al.* (1998) reported a divergent selection experiment in rice. Transgressive segregation of tiller angle was found in two rice F_2 populations, 5002 × Zhu-Fei 10 and HA79317-7 × Zhen-Nong13. By divergent selection for tiller angle in each F_2 population, two types of true-breeding extremes were obtained, one with larger tiller angle and the other with smaller tiller angle. Transgression of tiller angler was confirmed in the two extreme crosses (Xu and Shen, 1992b). For loci contributing to variation in tiller angle, the alleles of similar effect were proved to be dispersed in the original parents but associated (pyramided) in the extreme selections. By crossing two extreme strains each derived from one original cross, new transgression was found in the F_2 and then two types of extremes were obtained by the second cycle of divergent selection. By crossing the second-cycle extremes with each other and the third cycle of divergent selection for larger tiller angle, all positive alleles from the four original parents were pyramided. The transgression in each original cross can be explained by the complementary action of the genes, which had been dispersed between the original parents and complemented each other when they were pyramided in the extreme strains (Xu *et al.*, 1998). Since this transgression was observed in replicate experiments (Xu and Shen, 1992b,c), it is unlikely that the results are due to the mutation events as reported for experiments on divergent selection for bristle number in *Drosophila* (Mackay, 1995).

We would expect genetic fixation with long-term selection programmes. However, selection experiments discussed above for maize for high and low protein or oil and in *Drosophila* for bristle numbers (Yoo, 1980) show no indication of genetic fixation from long-term selection resulting in remarkable changes in phenotype. Frequent identification of large-effect QTL, as reviewed by Tanksley (1993), Kearsey and Farquhar (1998) and Xu, Y. (2002), makes steady and sustained selection response puzzling: alleles of large effects should be fixed rapidly, after which no further response would be seen. Barton and Keightley (2002) named two factors that might explain this apparent paradox. First, QTL-mapping experiments underestimate the number of QTL and overestimate their effects. Secondly, mutation generates alleles of large effect, which can be picked up quickly enough by selection to sustain a continuing selection response. Several mechanisms have been described that can create *de novo* variation, including intragenic recombination, unequal crossing over among repeated elements, transposon activity, DNA methylation and paramutation. Barton and Keightley (2002) listed several factors that make it difficult to estimate the true numbers and effects of loci influencing a quantitative trait. Hyne and Kearsey (1995) pointed out that in a typical experiment (heritability ~ 40%, ~ 300 F_2 individuals), no more than ~ 12 QTL are ever likely to be detected, which is supported by empirical data on the numbers of QTL detected in plants as reviewed by Tanksley (1993), Kearsey and Farquhar (1998) and Xu, Y. (2002). Both Beavis (1994) and Utz and Melchinger (1994) indicated that unless samples are large (> 500, for example), the effects of statistically significant QTL are substantially overestimated. As the results from RFLP-based evaluation of long-term selected maize lines, the numbers of QTL discovered in various experiments have begun to get close to the numbers of genes required to explain the long-term selection response.

Studies of this type would not be possible without the availability of the long-term selection strains. This fact points to the importance of maintaining longer-term selection programmes so that these kinds of genetic stocks are available for various types of studies. Maintenance of long-term breeding materials is becoming more challenging in an era frequently focused on short-term genomic-based experiments funded by short-term competitive grants. However, it is genetic stocks developed by public sector long-term breeding and selection programmes that frequently facilitate many of these studies at molecular level.

9

Marker-assisted Selection: Practice

Developments in genomics have provided new tools for discovering and tagging novel alleles and genes useful for improving target traits and for manipulating those genes in breeding programmes through marker-assisted selection (MAS), i.e. selection of phenotypic traits indirectly using markers that are closely linked to the traits or are developed from the gene-related sequences. Plant breeding will benefit from MAS through: (i) more effectively identifying, quantifying and characterizing genetic variation from all available germplasm resources (Tanksley *et al.*, 1989; Tanksley and McCouch, 1997; Gur and Zamir, 2004); (ii) tagging, cloning and introgressing genes and/or quantitative trait loci (QTL) useful for enhancing the target trait through genetic transformation and molecular marker technologies (Dudley, 1993; Gibson and Somerville, 1993; Paterson, 1998; Peters *et al.*, 2003; Gur and Zamir, 2004; Peña, 2004; Holland, 2004; Salvi and Tuberosa, 2005); and (iii) manipulating (differentiating, selecting, pyramiding and integrating) genetic variation in breeding populations (Stuber, 1992; Xu, 1997; Collard *et al.*, 2005; Francia *et al.*, 2005; Varshney *et al.*, 2005b; Wang, J. *et al.*, 2007). MAS can also have significant utility in plant breeding programmes through assisting PVP (plant variety protection) and DUS (distinctness, uniformity and stability testing) processes (CFIA/NFS, 2005; Heckenberger *et al.*, 2006; IBRD/World Bank, 2006). If the marker loci are sufficiently close on genetic or physical maps then reasonably good inferences may be made about the cultivar's haplotype. Such information is used to establish identity, resolve disagreements related to germplasm ownership and acquisition, enforce laws intended to encourage genetic diversity of the hybrids and avoid using inbreds that contain transgenes which may violate regulatory considerations and restrictions. These are often the very first applications of genomics in private sector breeding programmes, which is discussed in Chapter 13.

Using molecular markers in plant breeding programmes has been widely discussed (Beckmann and Soller, 1986a; Paterson *et al.*, 1991; Dudley, 1993; Stuber, 1994a; Xu and Zhu, 1994; Lee, 1995; Hospital and Charcosset, 1997; Xu, Y., 2002, 2003; Eathington *et al.*, 2007; Bernardo, 2008; Collard and Mackill, 2008; Xu and Crouch, 2008; Xu *et al.*, 2009b, d). Using rice and other cereal crops as examples, Xu, Y. (2003) provided a comprehensive review on the MAS system, germplasm evaluation, hybrid prediction and seed quality control. Much has happened in maize breeding since Stuber and Moll (1972) first reported that selection for grain yield in maize had resulted in changes in allele frequencies at several isozyme loci throughout the genome. In so doing, they essentially laid the grounds for MAS in maize. Indeed, if phenotypic selection (PS) could produce a change in marker allele frequencies, then why could deliberately altering marker allele frequencies at specific loci not produce predictable phenotypic changes for one or several traits?

The success of MAS depends on location of the markers with respect to genes of interest. Three kinds of relationships between the markers and respective genes could be distinguished. (i) The molecular marker is located within the gene of interest, which is the most favourable situation for MAS and in this case, it could be ideally referred to as gene-assisted selection. While this kind of relationship is the most preferred one, it is also difficult to find this kind of marker. (ii) The marker is in linkage disequilibrium (LD) with the gene of interest throughout the population. LD is the tendency of a certain combination of alleles to be inherited together. Population-wide LD can be found when markers and genes of interest are physically close to each other. Selection using these markers can be called LD-MAS. (iii) The marker is in linkage equilibrium with the gene of interest throughout the population, which is the most difficult and challenging situation for applying MAS.

The efficiency of MAS depends on many factors associated with how the underlying marker–trait associations (MTAs) were identified, including: the size of the mapping population, the nature of the phenotyping, the design and analysis of the experiment, the number of markers used, the distance between marker loci, the genomic region containing the desired QTL and the proportion of additive genetic variance explained by the marker, the selection method and the experimental design. The efficiency of MAS also depends on many factors associated with its application, including: the crop and breeding system, the molecular breeding process and the nature of the genotyping pipeline. For private breeding programmes, MAS has offered several attractive features, most of which are related to time and resource allocations.

As conventional breeding systems attempt to combine more and more target traits, there tends to be an overall loss of breeding gain and an increase in the duration of breeding cycles (time to generate a new product). Hence, MAS offers potential to greatly improve the overall pace, precision and impact of the breeding progress by assembling target traits in the same genotype more precisely, with less unintentional losses and in fewer selection cycles.

9.1 Selection Schemes for Marker-assisted Selection

MAS is most useful for traits where phenotypic evaluation is expensive or difficult, particularly for those polygenic traits with low heritability that are highly affected by the environment. It is also useful to break linkages between the target traits and undesirable genes in so-called marker-accelerated backcross breeding. MAS may also offer the opportunity to address goals not possible through conventional breeding, such as pyramiding different sources of disease resistance that have similar phenotypes. Indirect selection based on marker genotype rather than phenotype can be used to accelerate the speed and increase the precision of genetic progress, reduce the number of generations and when integrated into optimized molecular breeding strategies, it can also lower costs of selection. Xu, Y. (2002) discussed six situations that are most suitable for MAS. These include selection without testcrossing or a progeny test; selection independent of environments; selection without laborious fieldwork or intensive laboratory work; selection at an earlier breeding stage; selection for multiple genes and/or multiple traits; and whole genome selection.

9.1.1 Selection without testcrossing or progeny test

In plant breeding, many traits need testcrossing and a progeny test for unambiguous identification. Typical examples include male-sterility restorability, wide compatibility, heterosis and combining ability. In testcrossing, each candidate plant will be crossed to testers and then its genotype will be inferred from a progeny test in the next season. Each candidate plant must be harvested and maintained separately and only the plants with the target trait will advance to the next level. Testcrossing may continue for several generations until the selected plants reach a certain level of homozygosity. Using MAS, testcrossing and/or a progeny test can be eliminated since the target trait can be identified from the candidate plant itself, based on marker genotypes, saving laborious testcrossing and time-consuming progeny tests.

9.1.2 Selection independent of environments

Many traits must be screened in specific or controlled environments where they can be fully expressed. For example, photoperiod or temperature sensitivity can only be identified by comparison of their phenotypes in two distinct photoperiod or temperature conditions. For identification of insect/disease resistance, plants must be inoculated artificially or naturally. For abiotic resistance, such as drought, salinity and submergence tolerance and lodging resistance, selection in traditional breeding programmes can only be done when the specific stress is present. To measure responses to agrochemicals, such as herbicides and plant growth regulators, these chemicals must be applied to plants at the right stage under suitable environments. MAS has made it possible to perform indirect selection for these traits.

9.1.3 Selection without laborious field or intensive laboratory work

Many important traits are phenotypically invisible or unscorable by visual observation and must be measured in the laboratory using sophisticated equipment or facilities, or a large number of samples are required, which means that it cannot be measured until late generations when a relatively large amount of seed becomes available for each selection entry. Chemical and physical properties of grain are examples that fall under this category. Traits such as tissue cultivability need laborious laboratory work for testing each sample. Using MAS, a piece of leaf harvested at any growth stage of plants or even a piece of endosperm will be enough for accurate measurement of all the traits mentioned above, once associated markers have been identified.

9.1.4 Selection at an early breeding stage

Traits that are only measurable at or after the reproductive stage would be good candidates for MAS. For example, grain quality can only be tested using mature seeds. Yield heterosis and yield potential must be measured after harvest or/and in advanced generations. For forest and fruit trees, many traits have to wait up to several years until adult stages for phenotyping. MAS can be made at any stage and in any generation, so that breeders do not need to maintain a large number of candidate plants generation after generation (year after year). A recent summary report on wheat MAS (Kuchel *et al.*, 2008) indicated that the integration of MAS for specific target genes, particularly at the early stages of a breeding programme, is likely to substantially increase genetic improvement.

9.1.5 Selection for multiple genes and multiple traits

In some cases, multiple pathogen races or insect biotypes must be used to identify plants for multiple resistances, but, in practice, this may be difficult or impossible, because different genes may produce similar phenotypes that cannot be distinguished from each other. MTA can be used to select multiple resistances simultaneously.

Consider selection for multiple traits – for example, temperature-sensitive genic male sterility (TGMS), amylose and wide compatibility in rice. Candidate plants must be tested under two different environments where TGMS can be identified. Each plant must be testcrossed with wide-compatibility testers, following up with a progeny test in the next season. At the same time, a large amount of seed must be harvested for amylose measurement. Thus, using PS methods, we must wait until a large number of seeds are available and a reasonable level of homozygosity is reached.

9.1.6 Whole genome selection

MAS can also be practised at the whole genome level. Whole genome selection can be used to eliminate the donor genome in backcross breeding or to get rid of linkage drag when a wide cross is involved. Combined with MAS for multiple traits, whole genome selection allows the breeder to transfer multiple traits through backcrossing simultaneously.

High density molecular maps can be used to determine the genotype of an individual at many, sometimes thousands of loci and make it possible to deduce the most favourable genetic constitution for various regions throughout the entire genome in a given individual. By portraying molecular data in a graphical form, as discussed in Chapter 8, a graphical genotype can be inferred to show the genomic constitution and parental derivation for all points in the genome (Young and Tanksley, 1989a), which opens up the possibility of conveniently analysing quantitative traits in map-based whole genome selection. As an extension of this concept, the graphical genotype can be described for QTL and used to identify from mapping populations the desirable individuals with a favourable combination of different QTL alleles or with association of all alleles of similar effects across the whole genome.

9.2 Bottlenecks in Application of Marker-assisted Selection

To analyse the bottlenecks that may limit the application of MAS in plant breeding, it is necessary to have a brief overview of the current status of MAS. Several private companies have been routinely using MAS in breeding programmes, benefiting from their long-term basic research programmes and the availability of all the components of MAS. It is certainly a big investment for a breeding company/institution to start from scratch to run an efficient and fully operational MAS-based breeding programme. In contrast to conventional breeding schemes, the methods and design of infrastructure needed to support MAS have been the areas of greatest change. In order to utilize MAS, companies had to make significant investments to assemble or modify various aspects of infrastructure such as methods to detect DNA polymorphism, manage information, or analyse and track samples, software to relate genotype with phenotype and off-season or continuous nurseries (Ragot and Lee, 2007). These components had to be integrated with each other and with breeding activities, which meant that scientists needed to learn how and when MAS provided a comparative advantage over other methods when taking into account time and cost components.

MAS has been applied in the private sector for crops of great commercial interest including maize, soybean, canola, sunflower and vegetables. MAS in maize cultivar development aims at recovering an ideal genotype defined as a mosaic of favourable chromosomal segments from the parents (referred to as genotype construction). More specifically MAS in maize has been used to simultaneously select for multiple traits (selection based on marker information only) such as yield, biotic and abiotic stress resistance and quality attributes (Ragot *et al.*, 2000; Eathington, 2005; Eathington *et al.*, 2007), several of which are polygenic in nature. Although there is very limited information on successful breeding product delivery, the first commercial products of molecular breeding (rather than limited MAS) have been released from multinational breeding companies. The first molecular breeding maize hybrids developed by Monsanto entered the US commercial portfolio in the 2006 cropping season and as estimated by 2010, over 12% of the commercial crop in the USA will be derived from molecular breeding (Fraley, 2006).

MAS has also been used to some extent in the public sector for plant breeding through gene introgression and gene pyramiding, particularly for major-gene controlled disease resistance and for the crops of less interest to the private sector (for a review, see Dwivedi *et al.*, 2007). William *et al.* (2007b) reported the use of MAS in the International Maize and Wheat Improvement Center (CIMMYT) wheat breeding programmes. Large MAS programmes have been developed to help wheat breeding in Australia with MAS extensively used for at least 19 genes, or chromosome regions for cultivar development in Australian wheat breeding programmes (Eagles *et al.*, 2001). During the last few years, remarkable progress in implementation of MAS strategies for cultivar development has been achieved by the MAS Wheat Consortium in the USA and 80 MAS projects were completed and over 300 additional backcrossing programmes are attempting to incorporate 22 different disease and pest resistance genes and 21 alleles favouring bread-making and pasta quality (Dubcovsky, 2004). With all the efforts above and other MAS breeding programmes in the public sector worldwide, however, there are very few documented releases or registrations of new cultivars. Some examples available, as shown in Table 9.1, include two rice cultivars released in the USA, Cadet and Jacinto, with unique cooking and processing quality traits (http://www.ars.usda.gov/is/AR/archive/dec00/rice1200.pdf). In Indonesia, two rice cultivars, Angke and Conde, released possessing resistance to bacterial blight, produced 20% greater yield over IR64 (Bustamam *et al.*, 2002). In common bean, USPT-ANT-1 was registered as an anthracnose-resistant pinto bean germplasm line which contained

Table 9.1. Examples of released crop cultivars developed through marker-assisted selection.

Crop	Trait	Breeding product	Reference
Common bean	Resistance to anthracnose	Resistance incorporated in Pinto bean cultivar, USPT-ANT-1 containing *Co-4*^{2} gene that confers resistance to all known North American races of anthracnose in USA	Miklas *et al.* (2003)
Pearl millet	Downy mildew	The parental lines of the original hybrid (HHB 67) were improved for downy mildew resistance through MAS and conventional backcross breeding, and new hybrid HHB 67-2 with improved resistance to downy mildew released in India	Navarro *et al.* (2006)
Rice	Bacterial blight	Angke and Conde, possessing resistance to bacterial blight, produced 20% greater yield over IR64 and released in Indonesia	Bustamam *et al.* (2002)
Rice	Amylose content	Cadet and Jacinto with unique cooking and processing quality traits released in USA	http://www.ars.usda.gov/is/AR/archive/dec00/rice1200.pdf

the *Co-4*^{2} gene conferring resistance to all known North American races of anthracnose in the USA (Miklas *et al.*, 2003). In pearl millet, the parental lines of the original hybrid (HHB 67) were improved for downy mildew resistance through MAS and conventional backcross breeding and a new hybrid HHB 67-2 with improved resistance to downy mildew was released in India (Navarro *et al.*, 2006). Quality protein maize from an extra-early single cross maize hybrid, Vivek Maize Hybrid-9, which was developed through MAS for the *opaque2* gene (Babu *et al.*, 2005) has recently been released in India.

The limited success in breeding product delivery in MAS can be further illustrated by the numbers of publications that have been generated on QTL mapping and MAS since the discovery of the first generation of DNA markers (Xu and Crouch, 2008). The term 'marker-assisted-selection' first appeared over two decades ago (Beckmann and Soller, 1986b) and was initially focused on the potential uses. A decade later, the community term became increasingly interested in application of the genes tagged by molecular markers and the term appeared in over 100 journal articles in 1995 (Fig. 9.1). However, the first real article on application of MAS in plant breeding using DNA markers is probably the one published by Concibido *et al.* (1996) for soybean cyst nematode resistance. The volume of publications on the development and to a lesser extent application of markers for assisting plant breeding has increased dramatically during the last decade. As a result, the number of articles containing the term 'marker-assisted-selection' climbed to over 1000 in 2003 (Fig. 9.1). There is limited targeted public sector funding to support the large-scale validation, refinement and application of MAS in field breeding. This can be seen from the number of articles with the term 'marker-assisted-selection' (1390 in 2004) compared to the number of articles with the term 'quantitative trait locus' or 'quantitative trait loci' (1250 in 1998 and 4440 in 2005, Fig. 9.1). Most of articles on MAS result from either investments from donors with a scientific mandate or academic institutions with a specific interest in showing promising applications of MAS in plant breeding. To convert promising publications into practical application in field breeding requires breaking through many practical, logistical and genetical bottlenecks (Xu and Crouch, 2008). This includes developing simple, quick and cheap technical protocols for sampling, DNA extraction

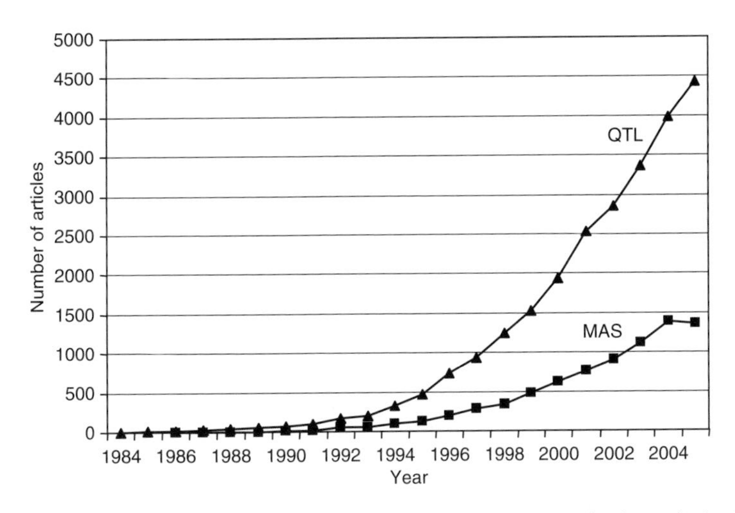

Fig. 9.1. The numbers of articles with the terms QTL ('quantitative trait locus' or 'quantitative trait loci') and MAS ('marker-assisted-selection') by years from Google Scholar, 4 August 2007). From Xu and Crouch (2008) with permission.

and genotyping that remain reliable and precise when routinely applied at high-throughput. This also includes developing tailored sample and data tracking and management systems plus powerful decision support tools to ensure effective integration of genotyping into breeding programmes. Xu and Crouch (2008) discussed the bottlenecks associated with translation of MAS from publications to practice, particularly in public sector breeding programmes. William *et al.* (2007a) provided technical, economic and policy considerations on MAS in crops based on lessons from the experience at an international agricultural research centre. In principle, effective MAS systems are the result of the following activities:

- Developing DNA extraction and tissue sampling and tracking systems appropriate for large-scale field trials.
- Establishing a platform for molecular data generation, management and analysis that meet the needs of plant breeding.
- Developing analytical methods for synthetic cultivar development, heterotic group construction and hybrid prediction using molecular marker information.
- Exploiting genetic and breeding materials including populations, hybrids, open-pollinated populations, landraces under selection and synthetic cultivars from ongoing breeding programmes.
- Validating MTAs using any population that is genotyped genome-wide for marker-assisted backcrossing (MABC) and phenotyped for target traits, leading to the update and refinement of the marker set.
- Optimizing MAS systems and refining MAS breeding programmes by improving the following procedures: high-throughput sampling, DNA extraction and genotyping, environment control and characterization, precision phenotyping, integration of diversity analysis, genetic mapping and MAS, and data generation, interpretation and delivery systems.

9.2.1 Effective marker–trait association

QTL publications have been increasing tremendously in the past two decades as shown in Fig. 9.1 and involving almost all crop plants and all types of agronomic traits (as reviewed by Dwivedi *et al.*, 2007). However, reports of QTL mapping to date have tended to be based on individual small to moderately sized mapping populations screened with a relatively small number of markers, providing relatively low resolution of MTAs (Xu, Y., 2002, 2003; Salvi and Tuberosa, 2005). Very few of the QTL reported have been utilized in plant breeding through MAS. Thus the community is investing a large amount of money and labour in generating a lot of publications with little impact on applied plant breeding. One of the approaches to effective MTA is selective genotyping and pooled DNA analysis discussed in Section 7.7.

Some inherent limitations to MAS are related to the estimates of QTL position and genetic effects and the rates of false positives and negatives. Confidence intervals for QTL are typically 10–15 cM; a genetic region that should not be a major barrier for implementing MAS although it could become a limitation to achieving genetic gain by preventing the selection of desired recombination events. The advent of association mapping and a growing pool of candidate genes should provide some resources needed to minimize problems related to the estimation of QTL position. The genetic effects of QTL are overestimated for many reasons, some of which are linked to experimental designs for phenotyping or population development while others are inherent to the process of QTL detection (Lee, 1995; Beavis, 1998; Melchinger *et al.*, 1998; Holland, 2004).

9.2.2 Cost-effective and high-throughput genotyping systems

Private corporations have established or are developing the capacity to produce hundreds of millions of data points per year in service laboratories, distinct from research units. Besides, smaller 'biotech' companies are developing technologies that could reduce the cost of each marker data point to a mere few US cents (Ragot and Lee, 2007). Without considering any other cost in MAS, however, the current cost associated with DNA extraction alone is already a big burden for many plant breeding programmes in terms of sample-based cost, especially in the early stages when few assays are required on each sample. So a great effort will be needed first to minimize the cost associated with each step of DNA extraction including sampling, labelling, reagents and plastic consumables.

PCR amplification is a necessary and also expensive step for all PCR-based markers. Multiplexing PCR primers has been an approach to significantly reduce the PCR related cost but it takes a lot of effort to optimize the protocol for suitable multiplex marker sets. Multiplexed PCR primers work well for genetic diversity analysis. When they are used for genetic mapping and MAS, however, they have to be optimized and even redesigned for each specific cross or population because there is no universal marker set that contains markers that are polymorphic across all crosses or populations.

Another significant cost related to MAS is the step of marker detection after PCR amplification, which can be significantly different from one assay type to another. When screening PCR-based markers by agarose gel electrophoresis, which is considered more suitable for MAS of single traits, gel preparation and electrophoresis and scoring time for a 50–200-sample-gel can take as long as 3–4 h. Using microtitre plates or dot blot detection of allele-specific gene-based markers offers substantially higher throughput and lower costs than gel-based assays. However, those systems are not suitable for large-scale MAS using large numbers of markers for both genetic background selection and multiple target traits. Effective and efficient marker genotyping systems for large-scale MAS depend on a high-throughput detection system that works with a large number of markers. In general, developing and optimizing such a detection system is time-consuming and also expensive.

Continual improvement in the capability of laboratories to generate molecular data has come through the development of new types of markers allowing increasing automation. However, this has tended to come with the negative consequence of an increase in the cost of equipment required to achieve high-throughput low-cost genotyping and in turn, the capacity to see molecular genotyping achieve impacts at the scale of modern plant breeding programmes. Due to the large up-front costs of assembling infrastructure and personnel for genotyping, it is unlikely that individual national marker laboratories could produce data points in a cost-efficient manner. In advanced laboratories and in animal and human research, this has led to an increased tendency towards centralization and in particular, a shift to an out-sourcing mode of operation. Therefore, the actual genotyping might be most efficiently and effectively carried out through regional hubs and/or out-sourcing services. Collard *et al.* (2008) discussed genotyping systems that might be suitable for different situations and breeding programmes including gel- and non-gel-based genotyping systems for remote breeding station laboratories and capillary- and array-based genotyping systems for regional hub laboratories.

9.2.3 Phenotyping and sample tracking

Once a high-throughput system has been established for DNA extraction, PCR amplification and marker detection, the bottleneck will be the phenotyping that is required for MTA before MAS and sample tracking that is required for a large number of plants and families during MAS. Phenotyping has been considered as critical in the era of post-genomics and is now receiving greater attention than ever. Precision and global phenotyping of a large number of plant samples is very expensive and time-consuming and is the limiting factor that affects the accuracy of genetic mapping and the power of MAS. Private corporations have realized the need for such high precision phenotyping as can be seen from their active recruiting of trait-specific phenotyping scientists often located in targeted areas where the trait of interest can be more easily measured (e.g. positions dedicated to drought tolerance and located in arid regions of the world) (Ragot and Lee, 2007). Beyond laboratories, plant handling is becoming a bottleneck to high-throughput protocols. High-throughput facilities have to be established and equipped at continuous nursery sites potentially to handle millions of plants per year.

The level of heritability of measured traits depends on whether the phenotyping can be repeated across different seasons, locations and environments. Clustering target locations into mega-environments and comparing these with selection at different locations has been used to understand how the target environments for a breeding programme differentiate the germplasm with respect to yield and other agronomic traits (e.g. Rajaram *et al.*, 1994; Lillemo *et al.*, 2004; Chapter 10). Cross-population and environment comparison of phenotyping will determine how the MTAs identified under one environment can be used for selection under another. In this case, well characterized environments and well established selection criteria are essential prerequisites for the development of a reliable precision phenotyping system. Precision and high-throughput multi-locational phenotyping, together with effective sampling and data acquisition systems being developed for many traits, provides the potential to develop a phenomics-based protocol for trait-specific breeding programmes. This will not only help understand the phenotypic profile a plant possesses but also improve the precision of genetic mapping and thus MAS for the target phenotype.

Tracking samples from the field to the harvest bags to DNA plates for DNA extractions, PCR amplification and marker detection and then tracing back to the field plants selected based on the genotyping is a time-consuming and error-prone step, which translates into a large proportion of cost as a whole for MAS. As plant breeders always work with a large number of plants and populations and some crop species cannot be as easily organized in the field as others, to facilitate the sample collecting and tracking, sampling tracking will finally determine whether MAS can be processed in a high-throughput manner and thus whether MAS is practicable on a large scale.

9.2.4 Epistasis and genotype-by-environment interaction

Genetic effects related to epistasis are either poorly estimated or ignored by breeding programmes (Holland, 2001; Crosbie *et al.*, 2006). Such assessments of genetic effects will inflate predictions of genetic gain. The relative merit of MAS will depend on the nature of predictions, actual results and costs of alternative methods.

The importance of genotype-by-environment interaction (GEI; discussed in detail in Chapter 10), as a bottleneck in marker-assisted breeding (MAB), has been recognized because it affects both the power of QTL detection and the response to MAS. To evaluate QTL by environment interaction, precision phenotyping at multiple location/environment trials is required. Selection of suitable locations for phenotyping and accurate estimation of QTL effects across environments are two factors that determine whether the QTL identified can be used for effective MAS. Also in MTA, either through linkage mapping or LD mapping, QTL-by-environment interaction effects should also be incorporated in to the statistical model for MTA.

9.3 Reducing Costs and Increasing Scale and Efficiency

Highly abundant single nucleotide polymorphism (SNP)-based genic markers provide great potential for increasing scale and efficiency and thus reducing the cost of MAS because genotyping can be automated. Developments of high-throughput genotyping platforms are largely driven by human and animal research and applications. However, there are many commonalities among MAB of livestock and human health diagnostics, which will provide many important spillovers for molecular plant breeders. The feasibility of marker-assisted approaches for plant breeding is heavily influenced by the relative cost (in time and money) compared with conventional breeding.

There are several ways to reduce the MAS cost. First, high-throughput analysis using automated genotyping and data scoring systems will help increase the daily data output. Secondly, using the same sample for selection of multiple traits will reduce the trait-based cost. Thirdly, selection at an early stage of plant development or before planting and an early stage of the breeding process will minimize the number of plants that need to be retained so that the overall breeding cost will come down. Fourthly, optimization of MAS systems, including facilities and personnel, will result in less cost per data point.

While not truly an inherent limitation of the methods involved, one unavoidable limitation of MAS is the cost of assembling and integrating the necessary infrastructure and personnel. These can be substantial and beyond the means of many programmes. For such programmes, implementation of MAS could lead to a delusional or unbalanced reallocation of resources from vital activities such as high-quality phenotypic evaluation and selection in the target environment (Ragot and Lee, 2007). Currently, only the largest maize breeding programmes in a given market or region have the scale of sales and diversity of products that can justify and support MAS and withstand some of the financial burdens of establishing and replacing components of the system (e.g. changes in the methods and platforms for detecting DNA polymorphisms).

The economic story from DNA sequencing may tell us what we can expect in terms of cost reduction in marker genotyping. Sequencing cost per finished base was US$10 in 1990, but was reduced to US$1 in 1996, US$0.10 in 2002 and US$0.01 in 2006, which is a thousand times cheaper than in 1990. The cost of genotyping using molecular markers depends on marker type and its capacity in high-throughput analysis. With well-established marker systems and sequencing facilities, genotyping with simple sequence repeat (SSR) markers costs about US$0.30–0.80 per data point, depending on marker multiplexing and the number of markers genotyped for each sample (Xu *et al.*, 2002). For example, the lowest cost of SNP analysis in maize is now about US$90 for 1536 samples.

9.3.1 Cost–benefit analysis

Cost–benefit analysis will help us understand which components in the system need to be improved and where the bottlenecks for large-scale application of MAS are, as preliminary outputs in this area have been achieved in maize (Dreher *et al.*, 2003; Morris *et al.*, 2003) and wheat (Kuchel *et al.*, 2005). This analysis needs to be constantly updated as new genotyping systems become available and new optimizations are implemented in respective genotyping labs. Since many factors that can reduce cost may influence genetic gain, it is essential that cost–benefit analysis modules be integrated into those facilitating the genetic modelling and simulation of different breeding systems (Wang *et al.*, 2003, 2004; Wang, J. *et al.*, 2007).

The economic merit of MAS could include situations in which molecular costs are more than offset by savings in phenotypic evaluation. If molecular costs are in addition to, not in place of, phenotypic costs, the economic merit of MAS will become questionable and more difficult to evaluate. In other cases, the ability to select early offsets the extra costs that are associated with MAS. Detailed cost–benefit analysis of various elements of DNA marker development and application, including the cost of the required genotyping platforms and professional expertise, needs to be assessed at the earliest possible stage. This is particularly important at this time when most public plant breeding programmes are not adequately funded or poorly equipped to reach a critical threshold of marker assay throughput.

Comparative studies exist about the benefit of MAS versus PS (van Berloo and Stam, 1999; Yousef and Juvik, 2001a). The benefit depends on the heritability of the trait and the population size. When the heritability is high, the cost involved in genotyping many plants may not outweigh the expected benefits from PS. As calculated for recombinant inbred lines (RILs), a benefit can be expected with a range of heritability of 0.1–0.3 (van Berloo and Stam, 1998). If the value is less than 0.1, it is not possible to detect the QTL with the accuracy required to rely on flanking markers for selection.

In considering the use of molecular markers to improve the mean value of a generic trait in a population formed by crossing two inbred lines, Moreau *et al.* (2000) show that MAS will be more cost-effective than PS when the ratio of phenotypic to genotypic evaluation costs exceeds a critical level. A comparison between MAS and conventional greenhouse screening of common beans for resistance to common bacterial blight showed that the cost of MAS is about one-third less than that of the greenhouse test (Yu *et al.*, 2000).

A study designed to compare the cost-effectiveness of conventional breeding and MAB in maize was carried out in Mexico at CIMMYT (Dreher *et al.*, 2003; Morris *et al.*, 2003). This study proceeded in two stages. In the first stage, costs associated with use of conventional and MAS methods for maize breeding were estimated using a spreadsheet-based budgeting approach (Dreher *et al.*, 2003). The costs of initially developing molecular markers linked to the trait of interest were not considered; the analysis assumed that suitable molecular markers were already available. Field operations and laboratory procedures required for conventional and MAS breeding projects were identified and costed out. Sensitivity analysis was then performed to determine how the costs of field operations and laboratory procedures are likely to change with improvements in research protocols and/or fluctuations in prices of key inputs. This information was used to compare the cost of using conventional screening methods and MAS to achieve a well-defined breeding objective – identification of plants carrying a mutant recessive form of the *opaque2* gene in maize that is associated with Quality Protein Maize (QPM). In addition to generating empirical cost information that will be of use to CIMMYT research managers, the first stage of the study produced four important insights, which can be applied in general to other cases of MAS. First, for any given breeding project, detailed budget analysis will be needed to determine the cost-effectiveness of MAS relative to PS methods. Secondly, direct comparisons of unit costs for phenotypic and genotypic analysis provide useful information for research managers, but in many cases, technology choice decisions are not made solely on the basis of cost. For example, time considerations will often be critical, since genotypic and phenotypic screening methods may differ in their time requirements. Even when 'real time' requirements are similar, for applications in which phenotypic screening requires samples of mature grain, genotypic screening often can be completed much earlier in the plant growth cycle. Thirdly, the choice between conventional and MAS methods may be complicated even further because the two are not always direct substitutes. Using molecular markers, breeders may be able to obtain more information about what is going on at the genotypic level such as genetic background than they can obtain using phenotypic screening methods. Fourthly, when used with empirical data from actual breeding programmes, budgeting tools are needed to improve the efficiency of existing protocols and to inform decisions about future technology choices.

In the second stage of the study (Morris *et al.*, 2003), the costs associated with the use of conventional and MAS methods at CIMMYT were compared for a particular breeding application: introgressing an elite allele at a single dominant gene into an elite maize line (line conversion). At CIMMYT, neither method shows clear superiority in terms of both cost and speed: conventional breeding schemes are less expensive, but MAS-based breeding schemes can be completed in less time. For applications involving tradeoffs between time and money, relative profitability can be evaluated using conventional investment theory. Private firms, which can raise operating capital by drawing on corporate cash reserves, floating shares in the stock market, or borrowing in commercial credit markets, have been actively implementing MAS to maximize the net benefits generated by their breeding programmes (also profits) by opting for technologies that allow them to bring new products into the market faster, even if these technologies are more costly to implement. In contrast, public plant breeding programmes, which are more likely to face capital constraints in the sense that they are usually required to operate within their budget allocation, have been much slower to implement MAS. Public breeding programmes can maximize the returns to their limited resources by sticking to lower-cost PS methods, even though this means that breeding projects will take longer to complete.

For many plant breeding projects, the relative attractiveness of PS versus MAS will not be in doubt. When switching between PS and MAS implies a trade-off between time and money, the cost-effectiveness of DNA markers depends critically on four parameters: (i) the relative cost of phenotypic versus genotypic screening; (ii) the time savings achieved using MAS; (iii) the size and temporal distribution of benefits associated with accelerated release of improved germplasm; and (iv) the availability to the breeding programme of operating capital. All four of these parameters can vary significantly between breeding projects, suggesting that detailed economic analysis may be needed to predict in advance which selection technology will be optimal for a given breeding project (Morris *et al.*, 2003).

9.3.2 Seed DNA-based genotyping and MAS system

DNA extraction currently represents the single largest cost in most MAS pipelines and often presents the rate limiting step for scale-up of the whole process. Development and optimization of a non-destructive single-seed-based DNA extraction system will play a significant role in enhancing MAS efficiency, particularly for traits expressed late in the cropping season. Compared to MAS using DNA extracted from leaves and other tissues, seed-based DNA genotyping has many advantages, including: (i) identification of desirable genotypes and discard of undesirable genotypes before planting; (ii) increasing the speed of breeding cycles by selecting genotypes during the off season; (iii) reducing the time-consuming and error-prone sample collecting step that currently involves harvesting leaf tissue from plants in the field or glasshouse which then need to be retraced when the genotyping data is released; and (iv) saving land because only selected genotypes (seeds) are planted. Although DNA extraction from single dried seed has been studied in many plant species, most reports focus on destructive protocols. To develop a comprehensive and operational system for MAS using single-seed-based and non-destructive DNA extraction, the extracted DNA must have a high quality compared to leaf-tissue DNA so as not to confound the PCR amplification and detection process. Similarly, the quantity of DNA should be large enough for whole genome genotyping and DNA extraction should be high-throughput, while sampled seeds should maintain a high level of germination.

A seed DNA-based genotyping system that is feasible for crop species with relatively large seeds has been developed in CIMMYT for molecular breeding in maize (Gao *et al.*, 2008). An optimized genotyping method using endosperm DNA sampled from single maize seeds was developed (Fig. 9.2), which can be high-throughput and is generally applicable to different types of kernels. The seed DNA-based genotyping method involved excising endosperm pieces from imbibed maize seeds, then grinding the pieces into powder in a 96-tube plate using a tissue shaker to improve efficiency. Sampled seeds were stored in two 48-well plates as a unit for facilitating trace data from desirable genotypes to corresponding candidate seeds. Using the seed DNA-based genotyping method, the DNA extraction process and following genotyping can be

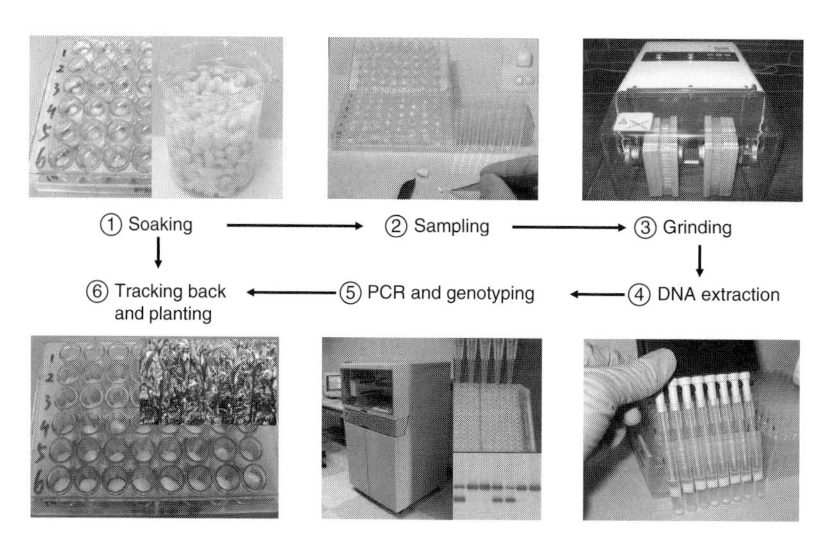

Fig. 9.2. Flowchart of large-scale seed DNA-based genotyping system. From Gao *et al.* (2008) with kind permission of Springer Science and Business Media.

done in 96-well plates using regular extraction buffers, the DNA quality is functionally comparable with that of leaf DNA and the DNA amount extracted from 30 mg of endosperm is sufficient for up to 200–400 agarose gel-based markers and several million chip-based SNP markers. By comparing endosperm and corresponding leaf DNA of an F_2 population, genotyping errors caused by pericarp contamination and hetero-fertilization averaged 3.8% and 0.6%, respectively, depending on the SSR markers used. Endosperm sampling did not affect germination rates under controlled conditions, while under field conditions the germination rate, seedling establishment and normalized different vegetative index (NDVI) were significantly lower than that of controls for some genotypes. Careful field management could compensate for these slight effects on germination and seedling establishment. Seed DNA-based genotyping lowered costs by 24.6% compared to leaf DNA-based genotyping due to reduced field plantings and labour costs.

As seed DNA-based genotyping can be processed before planting, for example selecting on F_2 seeds harvested from an F_1 plant, it is possible to select desirable genotypes before planting. This has a potentially large impact on breeding programmes, from changing population sizes and selection pressures to differences in field design and strategies in MAS. Over several breeding cycles, this is likely to result in accelerated gain and improved efficiency. Another advantage of seed DNA-based genotyping is that genotyping can continue until at least a minimum number of desirable genotypes are identified. This means that target genotypes can be identified by genotyping populations as small as possible, saving the cost of sampling all available plants in the field while avoiding the risk that no desirable genotypes can be found with available plants in the field (as there is no way to go beyond the plants that have been planted), compared to leaf DNA-based genotyping. For example, a theoretical proportion of homozygotes at n target loci in an F_2 population is $(1/4)^n$ and thus for three loci, 1/64 plants in the population will have the desirable genotypes. As seed DNA-based genotyping can stop at any stage once the suitable amount of target seeds that carry desirable genotype has been identified, the number of seeds that have to be genotyped could be much less than, or in the worst case equal to, the number of plants that have to be planted in the field. For leaf DNA-based genotyping, to ensure a 99% probability of obtaining at least one desirable genotype, a minimum number of plants that have to be planted is log(1 – 0.99)/log(1 – 1/64) = 292. As the number of target loci increases, the minimum number of plants that have to be planted in the field will go beyond the capacity of most current breeding programmes. All these factors have significant impacts on procedures, methods and strategies for MAS that have been well established for the leaf DNA-based genotyping, simplifying the process and improving breeding efficiency. An important next step is a comprehensive modelling and analysis of all aspects of this genotyping method, as have been done for leaf DNA-based MAS under the assumption that selection is made after planting since the pioneer study by Lande and Thompson (1990). This also needs to incorporate both negative factors such as hetero-fertilization, endosperm triploidy and potential pericarp contamination and positive factors such as reduced labour time and selection of desirable genotypes before planting.

There are several issues to be considered before the seed DNA-based genotyping developed in maize can be extended to other crops (Gao *et al.*, 2008). First, crops should have relatively large seeds with at least 8–10 mg of tissue that can be sampled for DNA extraction in order to meet the requirement of single-seed-based genotyping, particularly for agarose gel-based genotyping. Secondly, seed texture and tissues (endosperm from monocots and cotyledons from dicots) should be suitable for sampling, or the seed can be soaked without significantly affecting germination rate when the dry seed is too hard for excising. Thirdly, the pericarp contamination can be neglected because it is at a relatively low level or the pericarp can be removed easily during sampling. Finally, a suitable DNA extraction protocol may need to be developed for each specific crop and can be used for crop seeds with a high percentage of specific chemical compositions such as fat, protein and starch. It can be expected that this approach would, to a great extent, replace leaf DNA-based genotyping that has been used in many crops for intellectual property protection, transgene detection, genetic testing for cultivar purity and hybridity, gene mapping, genetic diversity analysis and MAS.

9.3.3 Integrated diversity analysis, genetic mapping and MAS

Genetic mapping and MAS usually involve multiple consecutive steps from development of mapping populations, genetic mapping and marker validation to MAS. New multipurpose methodologies are emerging that facilitate the integration of genetic diversity analysis, MTA analysis, MAS validation and application within a single breeding programme context. These methodologies rely on utilization of multiple approaches such as LD analysis using a set of diverse genotypes, advanced backcross QTL (AB-QTL) mapping (Tanksley and Nelson, 1996) and 'mapping-as-you-go' (MAYG) (Podlich *et al.*, 2004), so that various steps in the process can be integrated. In the MAYG approach, estimates of QTL allele effects are continually revised by remapping new elite germplasm generated over cycles of selection, thus ensuring that QTL estimates remain relevant to the current set of germplasm in the breeding programme. The integration of genetic mapping and MAS offers two major advantages: (i) ability to carry out MTA analysis using breeding populations directly rather than having to follow time-consuming development of genetic populations; and (ii) combining MTA and its validation. This saves time, both in the process itself but also in the generation of the necessary genetic materials. However, perhaps most importantly, the common use of end-user relevant genetic materials throughout the process is likely to dramatically reduce the level of redundancy that is commonly experienced when transferring outputs from genetic studies and validating them in breeding populations.

9.3.4 Developing breeding strategies for simultaneous improvement of multiple traits

Strategy development for multiple trait improvement will include understanding the correlation between different traits (including the interaction between component traits of a very complex trait such as drought tolerance); genetic dissection of the developmental correlation between multiple traits; understanding of genetic networks for correlated traits; and construction of selection indices with multiple traits. Much progress has already been made in this area which is relevant to drought tolerant crops, e.g. in maize (Edmeades *et al.*, 2000; Bänziger *et al.*, 2006) and wheat (Babar *et al.*, 2006, 2007). A MAS kit can be developed to include markers associated with a set of key major-gene controlled traits plus markers evenly covering the whole genome for marker-assisted background selection. Several thousands of well-selected SNP markers can be fitted into a single chip and they can be updated and ultimately replaced by gene-based and functional markers as more and more genes are identified and functionally characterized for traits of economic importance. Selection for multiple traits can be completed in one step as long as the population is large enough to allow desirable individuals to combine different traits. However, the number of trait loci that can be manipulated in one step is limited as the population size required to cover the recombinants increases exponentially with the increase of the number of traits/loci. To manipulate multiple genes/traits that are beyond the population sizes that are amenable, a two-stage selection strategy involved two generations proposed by Bonnet *et al.* (2005) and simulated by Wang, J. *et al.* (2007) can be employed, as discussed in Chapter 8. In this approach, individuals are selected first by all target markers for both homozygous and heterozygous forms to obtain a subset of population that contain higher frequencies of the target alleles so that a much smaller population size is required in the following generation to obtain the homozygotes at the target loci.

9.4 Traits Most Suitable for MAS

With currently available molecular markers and genotyping systems, some traits are more suitable for MAS than others. Xu, Y. (2002) evaluated various traits and listed those most suitable for MAS, which include traits requiring testcrossing or progeny testing, environment-dependent traits and seed and quality traits.

9.4.1 Traits requiring testcrossing or progeny testing

Cytoplasmic male sterility and fertility restoration

Many important crop species, including rice, sorghum and sunflower, depend on cytoplasmic male sterility (CMS) and its fertility restoration for hybrid seed production. A large amount of testcrossing and progeny testing is involved in breeding CMS lines and their restorers. Testcrossing can start as early as with the F_2 generation. F_2 plants will be selected first for other agronomic traits and selected plants are testcrossed to maintainer lines for CMS-maintaining ability or to restorer lines for restorability. The testcross progeny will be planted the following season for fertility observation. Only the plants with complete sterile testcross progeny (for CMS) or completely fertile testcross progeny (for restorability) will be moved to the next breeding procedure. MAS can be used to replace testcross and progeny testing if markers for fertility restorability are developed. Xu, Y. (2003) listed restorability genes that have been associated with molecular markers in 12 crop species including maize, rice, sorghum, wheat, barley, rye, sunflower, oilseed rape, sugarbeet and onion. Some of them have been cloned (e.g. Desloire *et al.*, 2003; Koizuka *et al.*, 2003; Komori *et al.*, 2004; Wang, Z. *et al.*, 2006) and many more cloning studies would be expected. Cloned genes provide an opportunity of developing genic or functional markers for selection of fertility restorability.

Outcrossing

Evolutionary change in plant mating systems from outcrossing (cross-pollination) to inbreeding (self-pollination) has occurred frequently throughout the history of flowering plants and has been described as the most common evolutionary trend in angiosperm reproduction (Stebbins, 1957, 1970). For example, wild rice is frequently cross-pollinated, while cultivated rice is self-pollinated. Many characters involved in mating system evolution, such as sizes of floral organs or amount of pollen produced, are quantitative in nature. Hybrid seed production depends on the improvement of outcrossing-related traits and for self-pollinated crops, it might involve a reconstruction (or recovery) of the outcrossing mating system (Xu, Y., 2003).

Various techniques to produce hybrids have been developed depending on the crop, including hand emasculation, roguing of staminate plants in dioecious lines, use of gynoecious or highly female lines, CMS and genetic male sterility, protogyny, or self-incompatibility (Janick, 1998). The rate of outcrossing is often the limiting factor determining whether a hybrid has potential for commercialization: seed cost and price are both largely dependent of how easy it is to produce high-quality hybrid seed that both seed providers and farmers accept. Maize was particularly suitable for hybrid breeding because of monoecism and the simple emasculation techniques practised in breeding that allowed for easy inbreeding and outbreeding (Simmonds, 1979). The necessity of high seeding rates in highly self-pollinated crops such as rice and wheat introduces an economic problem: seed production costs must be low enough and yield of hybrids in the farmers' fields must be high enough that farmers can profit from purchase and use of hybrid seed and companies can profit from their production and sale (Goldman, 1999).

Yield of hybrid seed is determined by many variables, both genetic and environmental. In productive, favourable environments, seed yield from seed set through cross-pollination can approach those of conventional self-pollinated cultivars in wheat (Lucken, 1986) or might be up to 80% of inbred lines in rice (Yuan and Chen, 1988; Lu *et al.*, 2001). The breeder's approach to high, stable seed production is: (i) to identify those plant and flower features that affect cross-pollination; (ii) to find variation for these traits; and (iii) to incorporate genes for favourable expression of traits into parental lines (Lucken, 1986). Considering all hybrid cereal crops with the CMS system, measurements for increased outcrossing rate will include choice of favourite climate conditions for seed production; ensuring flowering synchronization of the two parents; providing a suitable pollen source; developing male sterile lines with desirable outcrossing traits; supplementary pollination; and adjustment of flowering habit and stigma characteristics using growth regulators such as gibberellic acid (Xu, Y., 2003).

Many plants are naturally self-pollinated. Their floral structure is adapted for inbreeding. Breeding parental lines may need to completely convert the floral structure and make them suitable for outcrossing. Outcrossing in rice depends on the capacity of stigmas to receive alien pollen and the capacity of anthers to emit much pollen to pollinate other plants in the proximity (Oka, 1988). Linkage between long exerted stigma and undesirable agronomic traits in wild rice species is quite strong and needs to be broken to incorporate these traits into selected genotypes. On the other hand, using the gene *eui* (elongated upmost internode) to correct the panicle enclosure associated with CMS has been used in China for high-yielding seed production with the minimized gibberellic acid application. This gene has been cloned (Zhu *et al.*, 2006) and hopefully the gene transfer can be facilitated by MAS.

The floral structure of wheat is considered to be oriented towards cross-pollination (Wilson, 1968). However, a close examination of its floral traits clearly indicated that wheat is less suited, in its present form, to cross-pollination than crops such as maize, sorghum and rye (Wilson and Driscoll, 1983). After review of the status of hybrid wheat, Lucken and Johnson (1988) indicated the need for acquiring more knowledge about genetic variation of floral biology, including: (i) spike and flower morphology; (ii) pollen dispersal, buoyancy, durability and vigour; (iii) stigma accessibility, receptivity and durability; and (iv) development of selection screens for these traits.

Many factors affecting outcrossing provide opportunities for MAS. However, there are very few investigations on genetic mapping of traits related to outcrossing. Grandillo and Tanksley (1996) examined anther length in a backcross between *Lycopersicon esculentum* and *Lycopersicon pimpinellifolium*. They found two QTL affecting this trait, on chromosomes 2 and 7, which accounted for only 24% of the phenotypic variation. Georgiady *et al.* (2002) investigated traits that distinguish outcrossing and self-pollinating forms of currant tomato, *L. pimpinellifolium*. A total of five QTL were found involving four traits: total anther length, sterile anther length, style length and flowers per inflorescence. Each of these four traits had a QTL of major (> 25%) effect on phenotypic variance. In rice, some genetic mapping projects have been undertaken that target outcrossing. Two QTL for the rate of exserted stigma in the RILs derived from the cross between an *indica* cultivar, Peikuh, and a wild rice, W1944 (*Oryza rufipogon* Griff.; Uga *et al.*, 2003). Nine QTL for the frequency of stigma exsertion were detected in the RILs derived from the cross between a *japonica* cultivar, Asominori, and an *indica* cultivar, IR24 (Yamamoto *et al.*, 2003). A further QTL analysis was conducted using the F_2 population between a *japonica* cultivar, Koshihikari, and a breeding line showing exserted stigma selected from the backcross population between IR24 as a donor and *japonica* cultivars (Miyata *et al.*, 2007). A highly significant QTL (qES3), which had been predicted in the RILs of IR24, was confirmed at the centromeric region on chromosome 3. qES3 increases about 20% of the frequency of the exserted stigmas at the IR24 allele and explains about 32% of the total phenotypic variance. A QTL near-isogenic line (NIL) for qES3 increased the frequency of the exserted stigma by 36% compared to that of Koshihikari in a field evaluation.

It is anticipated that MAS will provide a powerful tool to help fix the outcrossing-related issues in crops that are naturally self-pollinated but have great potential in hybrid breeding. Testcrossing required traits, such as stigma longevity and receptivity and labour-intensive traits, such as pollen load, can be selected much more easily through linked markers.

Wide compatibility

Hybridization barriers exist in distant crosses of many crop species to some extent. Because the parents are not genetically compatible, hybrids derived from intersubspecific crosses such as *indica* × *japonica* in rice are partially or completely sterile with seed set less than 30%. Some intermediate cultivars have little or no hybrid barrier with either *indica* or *japonica*, which can be called wide-compatible cultivars. The 'wide compatibility' trait can be thus defined as the ability to make intersubspecific hybrids fertile.

To identify wide compatibility and transfer the related genes to other genetic backgrounds, testcrossing and progeny testing are required, as for fertility restoration. In rice, several sets of testers were carefully selected for this purpose. A lot of work is involved in testcrossing and progeny testing to find out the cultivars or plants with wide compatibility. Molecular marker-assisted identification of wide compatibility genes have been reported (Wang G.W., *et al.*, 2005, 2006; Zhao *et al.*, 2006), which will accelerate and facilitate the breeding process by eliminating or minimizing testcrossing and progeny testing. Wide compatibility has been selected in rice using associated SSR markers.

Heterosis

Exploitation of heterosis or hybrid vigour to increase crop yields started early in the 20th century with maize. From inbreeding a number of crop plants including maize, George H. Shull developed a perspective on heterosis that he outlined in a 1908 publication entitled 'The composition of a field of maize'. Hybrids provide many advantages in a crop production system. The principal benefit is increased yield. In open-pollinating species, one of the most often overlooked benefits is uniformity, an element which has allowed for the rapid expansion of production in many crop plants such as the vegetables. Additional benefits may include stress tolerance and pest resistance and other performance characteristics. Breeders of hybrid crops can react faster and with more options to meet changing markets, customer needs and production demands. Other advantages of hybrids include the ability to combine useful dominant genes available in different inbred lines, to optimize the expression of genes in the heterozygous state and to produce unique traits.

Xu, Y. (2003) discussed four features of hybrid breeding associated with hybrid prediction, including selection for hybrid performance, seed production and commercialization and grain production. Hybrid performance depends on genes and their interactions and combinations. Selection for hybrid performance in breeding programmes is based on testcrossing and progeny testing. That is, we breed hybrids through selection of parental lines with desirable agronomic traits. To associate the parental phenotype with hybrid performance, breeders have to cross their candidate breeding lines with several testers and from the hybrid progeny, to determine if the candidates contain the genes required for hybrids and whether the parental combinations produce useful hybrids. This indirect selection, based on testcrossing and progeny testing, is time-consuming and very expensive. Furthermore, the association between the parental line and hybrid from one cross cannot be used to make a prediction about other associations.

A cross of two extremely low-yielding inbreds can give a hybrid with good mid-parent or high-parent heterosis but poor performance, whereas a cross of two high-yielding inbreds might exhibit less mid-parent or high-parent heterosis but nevertheless produce a hybrid with good performance. High-yielding hybrids owe their yield not only to heterosis but also to other heritable factors that are not necessarily influenced by heterosis. For effective selection, one needs to know the relative importance of each genetic contribution – of heterosis and non-heterosis – in individual hybrids (Duvick, 1999). MAS for heterosis could become possible, as discussed later and will facilitate breeding processes through associated markers.

9.4.2 Environment-dependent traits

For some traits, phenotypic expression largely depends on specific environments. Typical examples of environment-dependent traits include photoperiod/temperature sensitivity, environment-induced genic male sterility (EGMS) and abiotic and biotic stresses. MAS is particularly useful to such traits as they can be selected under any environments through associated markers. Before MAS, however, MTA has to be established under the environments where the phenotype can be expressed, in most cases, under controlled environments. Controlled environments can be compared with each other or with natural environments. If two environments mainly differ in one macro-environmental variable, they are considered contrasting or near iso-environments (NIEs) and the standard plot-to-plot variation and other residual micro-environmental effects can be neglected (Xu, Y., 2002). If the two environments are from experiments in different years or locations, it is assumed that location and year effects do not confound the effect of the macro-environmental factor.

Some traits need to be measured under NIEs, where plants respond differently. In such cases, one environment imposes much less stress on plants than the other, for example, two environments with normal and high temperatures. The effect of the stress environment can be measured by comparing it to a much-less-stress or non-stress environment. A relative trait value is then derived from two direct trait values measured in each environment to ascertain the sensitivity of plants to the stress. If different plants have an identical phenotype under the much-less-stress environment (this is not true for a segregating population in most cases), the direct trait value in the stress environment can be used to measure sensitivity. When both environments impose little stress on plants and the plants respond differently, however, relative trait values should be used (Xu, Y., 2002).

Photoperiod/temperature sensitivity

A typical example for environment-dependent traits is photoperiod sensitivity in many plant species that can only be measured in NIEs, one with short day-length and the other with long day-length. Plants start to flower when specific photoperiod and/or temperature conditions are met. In hybrid crops, flowering synchronization of two parents is one of the factors influencing hybrid seed production and thus the economic advantage over the inbred lines/cultivars. To understand photoperiod and temperature responses, hybrids and their parents must be planted in a variety of environments or NIEs. Genetic study of these responses will finally characterize the parental photoperiod-thermo response pattern and its effect on their hybrids and thus make hybrid photoperiod-thermo response predictive.

Using a rice double haploid (DH) population between 'Zhaiyeqing 8' and 'Jingxi 17', days-to-heading and photo-thermo sensitivity were investigated in two environments (Beijing and Hangzhou, China) that differ mainly in day-length and temperature (Xu, 2002). Four chromosomal regions were significantly associated with days-to-heading in either or both locations, whereas a different locus on chromosome 7 (G397A-RM248) was significantly associated with photo-thermo sensitivity, indicating that the photo-thermo sensitivity QTL was independent of the QTL for days-to-heading. By evaluating days-to-flowering of individual 'CO39'/ 'Moroberekan' RILs under 10h and 14h day-lengths and greenhouse conditions, Maheswaran *et al.* (2000) identified 15 QTL for days-to-flowering. Only four of them were also identified as influencing response to photoperiod.

Different QTL have been identified using direct and relative trait values and in rice, days-to-heading and photoperiod are often controlled by different QTL as discussed above. On the other hand, direct and relative traits could share some QTL. That means days-to-heading and photoperiod sensitivity are genetically related to some extent because both traits are related to the basic vegetative growth that plants must achieve to flower. There are QTL mapping studies undertaken in NIEs, but QTL were mapped using trait values scored in each environment rather than using relative measures. The traits themselves were mapped rather than the relative response measured under the NIEs. In rice, numerous QTL for days-to-heading or -flowering have been mapped using molecular markers but very few of them have been tested under both long- and short-day conditions. Using an F_2 between *japonica* 'Nipponbare' and *indica* 'Kasalath', Yano *et al.* (1997) identified two major and three minor QTL for heading date. Three of them (*Hd1*, *Hd2* and *Hd3*) were identified later as photoperiod sensitivity genes by testing the QTL-NILs under different day-lengths (Lin *et al.*, 2000) and one of them (*Hd1*) was cloned (Yano *et al.*, 2000).

Environment-induced genic male sterility

Male sterility can be induced by specific environmental factors. An EGMS was first discovered in rice by Shi (1981) from 'Nongken 58', a *japonica* cultivar. The mutant 'Nongken 58S' is sterile when the days are long (> 13.5 h) but becomes fertile when days are short (< 13.5 h). Thus, fertility conversion is triggered by the length of photoperiod. EGMS has also been reported in pepper, tomato, wheat, barley, sesame, pea, rape and soybean.

The dependency of male sterility on temperature or photoperiod-temperature interaction requires two different environments in the breeding and selection process. Breeding populations have to be planted in one environment where the plants will be sterile to make sure of the presence of sterility genes and in another where the plants will be fertile to confirm the fertility conversion and produce seeds. Using associated molecular markers, confirmation of fertility conversion involving two environments can be avoided. Genetic mapping studies in rice have laid a foundation for MAS in breeding EGMS lines. To facilitate incorporation of the *tms2* gene in rice, an SSR marker, RM11, located on chromosome 7, was identified and found to be useful in identifying heterozygous fertile plants in F_2 populations and F_3–F_4 progenies for selection of progenies in advance (Lu *et al.*, 2004). Lang *et al.* (1999) reported that PCR-based markers were 85% accurate in identifying *tms3* in the juvenile stage.

Biotic and abiotic stresses

Breeding of insect and disease resistance and tolerance to abiotic stresses has become a worldwide issue. To identify insect/disease resistance, plants must be inoculated artificially or naturally, or in specific environments where the stress exists. Artificial inoculation is impractical when the insects/diseases are under quarantine control. On the other hand, evaluation of plant response to different insects/diseases or different biotypes/strains/races of the same stress agents is very difficult, if not impossible, using traditional screening methods.

In traditional breeding programmes, selection for tolerance to abiotic stress such as salinity, drought and submergence tolerance and lodging resistance can only be done in specific environments that are either present at specific locations or created at well-controlled environments. Selection for these traits is considered most difficult in breeding programmes.

For effective MAS, development of suitable selection criteria is critical for both MTA and the following MAS, particularly for abiotic stresses. Taking drought tolerance in rice as an example, current knowledge on physiology suggests that drought tolerance depends on one or more of the following components: (i) the ability of roots to exploit deep soil water to provide for evapotranspirational demand; (ii) the capacity for osmotic adjustment that allows plants to retain turgidity and protect meristems from extreme desiccation; and (iii) control over non-stomatal water loss from leaves (Nguyen *et al.*, 1997). These components are generally applicable to other cereal crops. A large number of QTL had been identified in rice for osmotic adjustment, dehydration tolerance, abscisic acid accumulation, stomatal behaviour, root penetration index, root thickness, total root number, root length, total dry root weight, deep root dry weight and root pulling force (Zhang *et al.*, 1999). In maize, grain yield under drought stress is negatively correlated with the anthesis-silking interval (ASI), the difference in days between pollen shedding and silk emergence. A short ASI means rapid silk extrusion because time to anthesis is little affected by drought.

9.4.3 Seed and quality traits

Seed traits

As a major storage organ of crop seeds, endosperms provide humans with proteins, essential amino acids and oils. An understanding of the inheritance of endosperm traits is critical for the improvement of seed quality. Genetic behaviour in triploid endosperms is very different from that of the maternal plants that supply assimilates for grain growth and development. Thus, methods suited for genetic analysis of traits in maternal plants (diploids for most cereal crops) cannot directly be used for endosperm traits (Xu, 1997). Any genetic analytical method for endosperm traits needs to combine a genetic method developed for diploid maternal plants with a triploid model proposed for conventional genetic analysis.

The genetic system controlling endosperm traits may be much more complicated than that which controls traits of the plant per se. Because the plant provides seeds with a portion of their genetic material and almost all the nutrients required for growth and development, seed traits are genetically affected by both the seed nuclear genes and the maternal nuclear genes. In addition, cytoplasmic genes may also affect some seed traits through their indirect effects on the biosynthetic processes of chloroplasts and mitochondria. To understand endosperm traits with biological accuracy, one should take into consideration maternal genetic effects and cytoplasmic effects along with the direct genetic effects of seeds. As seeds initiate a new generation that differs from their maternal plants, some seed traits should be considered as one generation advanced over their maternal plants. Genetic analysis of endosperm traits should be based on the DNA extracted from both maternal plants and endosperm tissues in order to understand the relative contribution of the different genetic factors to the variation of endosperm traits (Xu, 1997). In many cases, all endosperm traits have been treated the same as other traits of the plant, with few reports (Tan *et al.*, 1999) that considered the generation advancement issue.

Hybrid seed traits

Although F_1 plants are uniform, seeds borne on them represent the F_2 seed generation and are expected to segregate for some grain characteristics. Major determinants of grain quality, for example in cereal crops, are milling; grain size, shape and appearance; and cooking and eating characteristics. Some grain tissues are of maternal origin and some result from fertilization and union of genetically diverse gametes. For example, the lemma and palea of the rice hull are maternal tissues. Seed size and shape are determined by the shape and size of hulls and the latter is determined by the genotype of F_1 plants. As a result, all F_2 seeds borne on F_1 plants have nearly identical dimensions even though the parents could have very different seed sizes. Endosperm is triploid tissue resulting from the union of one male nucleus with two female nuclei. If the parents differ in endosperm traits, these traits among F_2 grains on F_1 plants show clear-cut segregation (Kumar and Khush, 1986; Tan *et al.*, 1999). Single seed analysis of a rice hybrid, 'Shanyou 63', indicated that the amylose content for seeds on a F_1 plant could range from 8% to 32% when two parents had 15.8% and 27.2% amylose content, respectively. A similar situation was reported for barley. If the parents differ significantly in malting quality characters, the grain produced by barley hybrids will be heterogeneous and heterozygous for characters critical to the malting process (Ramage, 1983).

Quality traits

Many quality traits, including seed traits discussed above, are genetically controlled by multi-genic loci, or by multiple alleles at a locus because of the triploidy of the endosperm. As a result, the same phenotypes may come from different genetic factors or different alleles from the same locus. PS for the same trait values may not result in the same alleles or genes fixed in parents.

On the other hand, almost all quality traits are only measurable at or after the reproductive stage. MAS will help distinguish different genetic loci that contribute to the same quality traits. Methods for non-destructive extraction of DNA from single dry seeds, as discussed in Section 9.3 and Xu *et al.* (2009d), provide an opportunity for selection of seed traits so that selection can be processed before planting. Early-stage selection also provides more opportunities for selection of traits with relatively low heritability. MAS could be used for early-stage quality tests or DNA-based quality tests, whereas such tests would be delayed in a conventional breeding programme because a relatively large amount of seeds is required.

Genetic contribution to quality comes from both parents, but one of them could be more important than another in some specific situations. Endosperm properties might be affected more by female parents due to the maternal effect, or more by male parents due to the xenia effect. The composition and development of the kernels can be changed by the nature of pollen. This was first shown by Kiesselbach (1926) as the change of a sweet corn endosperm into a starchy endosperm after pollination of a sweet corn female by a flint endosperm male. Large xenia effects were observed for sorghum malt quality in the F_1 but this was entirely lost in the F_2 generation (Wenzel and Pretorius, 2000). Curtis *et al.* (1956) observed that the germ is markedly influenced in weight, oil and protein content by both the seed parent and the pollen parent of corn, with a pronounced maternal effect.

9.5 Marker-assisted Gene Introgression

As discussed in Chapter 8, major applications of MAS in plant breeding are the transfer of novel alleles from a donor to elite germplasm and to pyramid all favourable alleles from different sources into one genetic background. In the former case, marker-assisted background selection is used to eliminate the donor genetic background, while in the latter case the background selection may not be necessary depending on whether the recipient is best commercialized or not. Although MAS has been widely used in the private sector for breeding of both major gene controlled traits and MARS for quantitative traits, it has limited application in the public sector because of the reasons described in the previous sections. Some significant applications of MAS in plant breeding have been made in several Consultative Group on International Agricultural Research (CGIAR) centres. As an example, molecular markers are being used to facilitate selection at CIMMYT for a set of traits that have low heritability but high economic value or cannot be effectively screened in Mexico on a seasonal basis. These traits include parameters associated with root health, foliar diseases and factors associated with quality (Table 9.2). All MAS is tightly coupled with the existing field breeding operations. The application of markers begins by characterizing the cross block materials. Parental material is first characterized with markers for known genes to identify the parents with favourable alleles, which are then selectively combined in crosses (William *et al.*, 2007b). There are numerous examples available in MAS including its application in gene introgression and pyramiding for both qualitative and quantitative traits. Several recent reviews provided a good coverage in general methods (Xu, 2003) and applications in several major crops (Dwivedi *et al.*, 2007). Reviewing all the details for all traits and crops is beyond the scope of this chapter. In this section, an overview is provided for marker-assisted gene introgression.

Table 9.2. List of markers along with the chromosomal location of the target genes, currently in use for MAS at CIMMYT (from William *et al.* (2007b) with kind permission of Springer Science and Business Media).

Trait	Gene	Marker type	Chromosome
Resistance to *Heterodera avenae*	*Cre1*	STS	2BL
Resistance to *H. avenae*	*Cre3*	STS	2DL
Crown rot	Qtl-2.49	SSR	1DL
Flour colour/*Pratylenchus neglectus*	*Rlnn-1*	STS	7BL
Boron tolerance	*Bo-1*	SSR	7BL
Russian wheat aphid	*Dn2*	SSR	7D
Russian wheat aphid	*Dn4*	SSR	1D
Hessian fly	*H25*	SSR	4A
Stem rust	*Sr24*	STS/SSR	3DL
Stem rust	*Sr25*	STS	7DL
Stem rust	*Sr26*	STS	6AL
Stem rust	*Sr38*	STS	2AS
Stem rust	*Sr39*	STS	2B
Durable leaf and brown rust	*Lr34*/*Yr18*	STS	7DS
Swelling volume	GBSS-null	STS	4B
Grain hardness	Hardness	STS	5ABD
Dough strength	*Glu1BX*	STS	1BS
Barley yellow dwarf virus	*BDV2*	STS	7DS
Agronomy	*Rht-B1b*	STS	4B
Agronomy	*Rht-D1b*	STS	4D
Agronomy	*Rht8*	SSR	2D
Pairing homologue	*ph1b*	STS	5B
High protein gene	*Gpc-B1*	STS	6B

9.5.1 Marker-assisted gene introgression from wild relatives

Novel alleles and genetic diversity widely exist in wild relatives of cultivated plants. Wild crop relatives are traditionally looked upon as potential sources of gene(s) for various agronomic traits including resistance to many pests and diseases that are not available in cultigens, thus making them a valuable resource for gene transfer in cultivated species (Tanksley and McCouch, 1997). Both conventional crossing and selection and molecular breeding (MAS and transgenics) have been used to transfer pest and disease resistances from wild relatives to cultivated crop species. Resistance gene(s) from wild relatives have facilitated large-scale cultivation of crops in disease or pest endemic regions of the world, i.e. bacterial blight and grassy stunt virus in rice, bacterial blight in maize and potato and nematodes in many crops. Wild relatives are usually inferior to modern cultivars with respect to yield and seed quality. However, trait-enhancing alleles have been identified and introgressed into cultivated species from wild species through marker-assisted novel allele discovery. The successful transfer of improved fruit yield and processing quality in tomato (Rick, 1974; de Vicente and Tanksley, 1993; Fulton *et al.*, 1997; Bernacchi *et al.*, 1998a,b; Fridman *et al.*, 2000; Yousef and Juvik, 2001b) led to the realization that wild relatives can contain beneficial genes (in addition to resistance to biotic stresses) associated with yield and seed quality, although these are often phenotypically masked by deleterious genes and are thus difficult to identify and transfer through PS and breeding.

Concerns about reduced genetic diversity among commercial hybrids and depletion of genetic diversity in gene pools used in breeding may be partially alleviated by successful implementations of MAS. MABC may revive interest in using essentially untapped exotic germplasm as a source of favourable alleles for improvement of elite cultivars (Ragot and Lee, 2007). Very small and targeted chromosomal segments of exotic origin can be introgressed into elite inbred lines with limited risk of carrying along undesirable characteristics. Such an approach could be beneficial in many crops although no accounts of its implementation have been reported despite the many years of reports of its successful use in tomato (Tanksley *et al.*, 1996; Bernacchi *et al.*, 1998a,b; Robert *et al.*, 2001), rice (Xiao *et al.*, 1998) and soybean (Concibido *et al.*, 2003).

Wild relatives of rice within the genus *Oryza* are not only a rich source of information on the origins of variation within the genus but also a viable source of a wide variety of agronomically important germplasm for future breeding in rice and other cereals as well. To fill the gulf between national research programmes and breeding applications in developing countries, an international programme, the Generation Challenge Programme – Cultivating Plant Diversity for the Resource Poor (www.generationcp.org), has been established to begin to characterize and utilize a wide spectrum of germplasm collections. Molecular markers have been proven particularly useful for accelerating the backcrossing of a gene or QTL from exotic cultivars or wild relatives into an elite cultivar or breeding line (Tanksley and Nelson, 1996). Favourable genes or alleles from wild species of rice have been detected after backcrossing to elite cultivars (Xiao *et al.*, 1998; Moncada *et al.*, 2001; Septiningsih *et al.*, 2003; Thomson *et al.*, 2003). Similarly, this approach can identify alleles from exotic cultivars that result in improved phenotype, even though the parent may not possess inferior phenotype for this trait (Tanksley and McCouch, 1997; Xu, 1997, 2002). McCouch *et al.* (2007) summarized results from a decade of collaborative research using advanced backcross populations derived from *O. rufipogon* L. to: (i) identify QTL-associated improved performance in cultivated rice *Oryza sativa* L.; and (ii) to clone genes underlying key QTL of interest. They demonstrated that AB-QTL analysis is capable of: (i) successfully uncovering positive alleles in wild germplasm that were not obvious based on the phenotype of the parent; (ii) offering an estimation of the breeding value of exotic germplasm;

(iii) generating NILs that can be used as the basis for gene isolation and also as parents for further crossing in a cultivar development programme; and (iv) providing gene-based markers for targeted introgression of alleles using MAS.

Development of exotic genetic libraries, (also known as CSSL, chromosome segment substitution line; IL, introgression line; or CL, contig line) is another approach to enhance utilization of wild relatives to expand crop gene pools. These genetic stocks provide a well characterized potential resource for uplifting the yield barriers through pyramiding beneficial loci and fixing of positive heterosis. For example, when tomato introgression lines carrying three-independent yield-promoting genomic regions were pyramided, the progenies produced more than 50% greater yield compared to controls (Gur and Zamir, 2004). Yoon *et al.* (2006) reported that several rice lines outperformed Hwaseongbyeo (approximately 1 t ha^{-1} increase in grain yield). Several grain characteristics – including grain weight, were improved after crossing an advanced introgression line containing *Oryza grandiglumis* segments, HG101 (very similar to Hwaseongbyeo) with Hwaseongbyeo. The above examples demonstrate that wild relatives contain desirable alleles for agronomic traits even though their effect is phenotypically not evident in wild relatives. It is important that more emphasis should be given to exploit wild relatives to identify yield enhancing alleles to further raise the yield potential of crop cultivars.

Using AB-QTL analysis, yield and grain-quality enhancing alleles from wild relatives have been successfully introgressed in rice, wheat, barley, sorghum, common bean and soybean. Dramatic yield advantages have been reported in rice, for example, through the introduction of two yield-enhancing QTL alleles (*yld1.1* and *yld2.1*) from *O. rufipogon* (AA genome) into 9311 (one of the top performing parental lines used in the production of super hybrid rice in China). This contributed in excess of 20% yield increases in rice; i.e. about 1 t ha^{-1} gain in yield in some of the newly bred cultivars, largely because of increases in panicle length, panicles per plant, grains per plant and grain weight. These improved lines with 9311-type genetic backgrounds are being used to raise the existing yield potential of super hybrid rice in China (Liang *et al.*, 2004). *O. grandiglumis* (allotetraploid, CCDD genome species) is another wild relative contributing positive alleles for increased grain yield in rice. In contrast, only 6–8% increase in grain yield was reported when positive alleles from *Hordeum spontaneum* were introgressed into barley. Wild relatives also contributed positive alleles for improved grain characteristics in rice (long, slender and translucent grains and grain weight), wheat (grain weight and hardness) and barley (grain weight, protein content and some malt quality traits). Of particular interest is a locus for grain weight, *tgw2*, which contributed positive alleles from *O. grandiglumis* that are independent from undesirable effects of height and maturity (Yoon *et al.*, 2006). In a similar study, Ishimaru (2003) identified a grain weight QTL, *tgw6*, responsible for increased yield potential without any adverse effects on plant type, or grain quality in the Nipponbare genetic background. Similarly, alleles from *Glycine soja* conveyed 8–9% increase in grain yield and improved the protein content in soybean (Concibido *et al.*, 2003).

9.5.2 Marker-assisted gene introgression from elite germplasm

Unquestionably, the most pervasive and direct use of MAS by the private sector has been with backcrossing of transgenes into elite inbred lines, the direct parents of the commercial hybrids, particularly in maize (Ragot *et al.*, 1995; Crosbie *et al.*, 2006). Currently, the most widely deployed transgenes and combinations thereof (i.e. gene stacks) are for resistance to herbicides or insects (e.g. *Ostrinia* and *Diabrotica*). As the commercial maize crop of any region, maturity zone, market or country is not yet uniform or homogeneous for any transgene, maize breeders have elected to develop near-isogenic versions (transgenic and non-transgenic) of elite inbreds and commercial hybrids in order to satisfy combinations of licensing agreements, agronomic practices, regulatory requirements, market demands and product development schemes (Ragot and Lee, 2007). This has required companies to have two parallel maize breeding programmes, transgenic and non-transgenic. In this manner, MABC of transgenes and to a lesser degree, of native genes and QTL for other traits, has expedited the development of commercial hybrids. Unless regulatory issues change dramatically, MABC will remain the preferred means of delivering transgenes to the market.

MABC clearly provides the information needed to reduce the number of generations of backcrossing, to combine (i.e. 'stack') transgenes, 'native' genes or QTL into one inbred or hybrid quickly and to maximize the recovery of the recurrent parent's genome in the backcross-derived progeny. In several private breeding programmes, MABC has enabled the number of backcrossing generations needed to recover 99% of the recurrent parent genome to be reduced from six to three, reducing the time needed to develop a converted cultivar by 1 year (Crosbie *et al.*, 2006; Ragot *et al.*, 1995). As a line derived by MABC can be made to be very similar to the original non-converted line, most of its attributes, including agronomic performance, can be assumed to be equal or similar to those of the original line.

Marker-assisted gene introgression is thought to be promising in rice because a number of rice cultivars are widely grown for their adaptation, stable performance and desirable grain quality. Chen *et al.* (2000) used such an approach to transfer the bacterial blight resistance gene *Xa21* into Minghui 63, a widely used parent for hybrid rice production in China. Ahmadi *et al.* (2001) used a similar approach to introgress two QTL controlling resistance to rice yellow mottle virus into the cultivar IR64. Such approaches, however, can only sample a small number of accessions.

Using PCR-based DNA markers for tracking the *RB* gene in potato breeding populations, several marker-positive selected lines showed resistance to late blight. *RB* has also been cloned and transformed into Katahdin, a highly susceptible potato cultivar. The Katahdin transformed plants with *RB* showed broad-spectrum resistance against a wide range of late blight isolates (Song *et al.*, 2003). Clearly, by having the full sequence of the target gene, it should be possible to develop a highly efficient low cost assay system for this trait. The best example of the use of MAS in commercial barley breeding is the barley yellow mosaic virus complex where a variety of different markers have been developed for selection of the *rym4* and *rym5* resistance genes and one, the SSR Bmac0029, is used by many European winter barley breeders (Rae *et al.*, 2007). The cloning of the *rym4/5* locus (Stein *et al.*, 2005) provides the basis of a diagnostic marker for *rym4/5*-based virus resistance.

As reviewed by Dwivedi *et al.* (2007), MAS coupled with backcross and pedigree breeding methods and field evaluation has led to reports in the literature of genetic enhancement for resistance to bacterial blight (*Xa21*), gall midge (*Gm-6t*) and brown plant hopper (*Bph1* and *Bph2*) in rice; to leaf rust (*Lr19*, *Lr51* and *Yr15*) in wheat; to yellow dwarf virus (*Yd2*), stripe rust (*Yr4*) and powdery mildew (*mlo-9*) in barley; and to downy mildew (major QTL) in pearl millet. The progenies showed the same resistance level as the donor parental lines both in greenhouse and field evaluations.

9.5.3 Marker-assisted gene introgression for drought tolerance

The International Rice Research Institute (IRRI) has several drought-tolerance breeding programmes using identified QTL and MAS. QTL affecting root parameters were identified using a rice DH population derived from the cross IR64 × Azucena. An MABC programme was started to transfer the alleles of Azucena (upland rice) at four QTL for deeper roots on chromosomes 1, 2, 7 and 9 from selected DH lines into 'IR64' (elite rice cultivar) (Shen *et al.*, 2001). The backcross progeny were selected strictly on the basis of their genotype at the marker loci in the target regions up to the BC_3F_2, from which BC_3F_3 NILs were developed and compared to 'IR64' for the target root traits. Of the three tested NILs carrying target 1 (QTL on chromosome 1), one had significantly improved root traits over IR64. Three of the seven NILs carrying target 7 (QTL on chromosome 7) alone, as well as three of the eight NILs carrying both targets 1 and 7, showed significantly improved root mass. Four of the six NILs carrying target 9 (QTL on chromosome 9) had significantly improved maximum root length.

Steele *et al.* (2006) initiated MABC to improve drought tolerance into Kalinga III, an upland *indica* cultivar. After five backcrosses and conducting over 3000 marker assays (2548 restriction fragment length polymorphism (RFLP) and 700 SSR) on 323 plants, the NILs were developed and evaluated for root traits. The target segment on chromosome 9 (RM242-RM201) significantly increased root length under both irrigated and drought stress environments. Azucena alleles at the locus RM248 (below the target root QTL on chromosome 7) delayed flowering. However, selection for the recurrent parent allele at this locus produced early-flowering NILs that are suited to upland environments in eastern India.

Anthesis-silking interval (ASI) is an important trait associated with drought tolerance in maize. Ribaut *et al.* (1996, 1997) initiated a major MAB programme to transfer five genomic regions involved in the expression of a short ASI from Ac7643 (a drought tolerant line) to CML247 (an elite tropical breeding line). Five genomic regions were transferred using flanking PCR-based markers. Seventy of the best BC_2F_3 (i.e. S_2 lines) lines were crossed with two testers, CML254 and CML274. These hybrids and the BC_2F_4 families derived from selected BC_2F_3 plants were evaluated for 3 years under drought stress conditions. The best five MABC-derived hybrids yielded, on average, at least 50% more than the control hybrids under water stress conditions (Ribaut *et al.*, 2002b; Ribaut and Ragot, 2007). However, this difference became less marked when the intensity of stress decreased: for a stress inducing less than 40% yield reduction, performance of testcross hybrids resulting from MAS was no better than the 'original' version of CML274.

A major QTL on linkage group 2 (LG2) is associated with increased grain yield and harvest index under terminal stress in pearl millet cultivar PRLT 2/89-33 (Yadav *et al.*, 2002). The performance of QTL MAS-derived top cross hybrids (TCH) was compared with that of field-based TCH. Progenies with the best overall ability to maintain under terminal stress environments were used to generate the TCH and these were compared with randomly mated TCH made from randomly selected progenies from the entire population (irrespective of performance under terminal drought stress). In both cases progenies were selected irrespective of the presence or absence of favourable alleles at the putative drought tolerance QTL and evaluated across 21 environments (non-stress, terminal stress and gradient stress). The QTL MAS-derived hybrids were significantly, but only modestly, higher yielding both in full and in partial terminal stress environments. However, this advantage under stress was at the cost of lower yield of the same hybrids under non-stressed environments. The QTL MAS-derived hybrids flowered earlier and had limited effective basal tillers, low biomass and high harvest index. All these traits are similar to that of the drought tolerant parent thus confirming the effectiveness of the putative drought tolerant QTL on LG2 (Bidinger *et al.*, 2005).

9.5.4 Marker-assisted gene introgression for quality traits

Rice

Rice amylose content, mainly controlled by the *wx* gene, is a good example of MAS. Ayres *et al.* (1997) determined the relationship between polymorphism at that locus and variation in amylose content. Eight *wx* microsatellite alleles were identified from 92 long-, medium- and short-grain US rice cultivars. When used as predictors of amylose content, these eight alleles explained an average of 85.9% of the variation. The amplified products ranged from 103bp to 127bp in length and contained $(CT)_n$ repeats, where n ranged from 8 to 20. Average amylose content in cultivars with different alleles varied from 14.9% to 25.2%. Although the mircrosatellite marker was located in the intron of the waxy gene, a complete association between marker alleles and amylose contents still depends on fully understanding other genes involved in the starch synthesis.

To improve the most widely grown hybrid rice, Zhou *et al.* (2003) successfully introduced the wx-MH fragment from the restorer line Minghui 63 into the male sterile line Zhenshan 97B, which was subsequently transferred to Zhenshan 97A, using MAS in three generations of backcrossing followed by one generation of selfing. The introduction of this fragment has greatly improved the cooking and eating quality of inbred lines and their resultant hybrids, with the agronomic performance essentially the same as the original maintainer line and resultant hybrid. Liu *et al.* (2006) used MAS to introgress the *Wx-T* allele (conferring intermediate amylose content and thus good quality) into two widely used maintainers (Longtefu and Zhenshan 97) and their relevant male-sterile lines to generate improved *indica* hybrids. The resulting maintainer lines and hybrids showed improved cooking and eating quality with no significant alterations in their agronomic traits.

Rice with low glutelin content is suitable for patients affected by diabetes and kidney failure. The *Lgc-1* locus confers low glutelin in the rice grain, located on chromosome 2 between flanking markers (Miyahara, 1999). This trait has been successfully incorporated into *japonica* rice with 93–97% selection efficiency using SSR2-004 and RM358 markers (Wang, Y.H. *et al.*, 2005). Additionally, grain quality traits such as 1000-seed weight, kernel length/breadth ratio, basmati type aroma and high amylose content have been combined with resistance to bacterial blight using MABC breeding (Ramlingam *et al.*, 2002; Joseph *et al.*, 2003).

Wheat

Sun *et al.* (2005) used a novel sequence tagged site (STS) marker for improving polyphenol oxidase (PPO) activity in bread wheat. Breeding wheat cultivars with low PPO activity is the best approach to reduce undesirable darkening of bread wheat based end-products, particularly for Asian noodles. Based on the sequences of genes conditioning PPO activity during kernel development, 28 pairs of primers were developed. One of these markers, *PPO18*, mapped to chromosome 2AL, can amplify a 685-bp and an 876-bp fragment in the cultivars with high and low PPO activity, respectively. QTL analysis indicated that the PPO gene co-segregated with the STS marker *PPO18* and is closely linked to *Xgwm312* and *Xgwm294* on chromosome 2AL, explaining 28–43% of phenotypic variance for PPO activity across three environments. A total of 233 Chinese wheat cultivars and advanced lines were used to validate the correlation between the polymorphic fragments of *PPO18* and grain PPO activity. The results showed that *PPO18* is a co-dominant, efficient and reliable molecular marker for PPO activity and can be used in wheat breeding programmes targeting noodle quality improvement.

Maize

The endosperm of the maize seed has several distinct regions that have different physical properties. The aleurone is the outer layer of the endosperm, composed of specialized cells that secrete hydrolytic enzymes during germination. Beneath the aleurone are starchy endosperm cells filled with starch and storage proteins, thus creating two distinct regions – the 'vitreous' or glassy endosperm and the 'starchy' endosperm. The vitreous endosperm transmits light, whereas the starchy endosperm does not. Typically, the endosperm is ~90% starch and 10% protein (Gibbon and Larkins, 2005). Normal maize protein is deficient in two essential amino acids (lysine and tryptophan), has a high leucine:isoleucine ratio and biological value (Babu *et al.*, 2004). A naturally occurring recessive mutant gene *opaque2*, observed first in a Peruvian maize landrace, gives a chalky appearance to the kernels and has improved protein quality due to increased levels of lysine and tryptophan in the endosperm (Mertz *et al.*, 1964). However, this trait appears to be associated with inferior agronomic traits such as brittleness and increased susceptibility to insect pests. With the discovery of 'modifier genes' that alter the soft, starchy texture of the endosperm, maize breeders developed hard endosperm *o2* mutants designated as 'Quality Protein Maize' (QPM) (Prasanna *et al.*, 2001; Nelson, 2001; Xu *et al.*, 2009d) which have the phenotypes and yield potential of normal maize but maintain the increased lysine content of *o2*. *Opaque2* is a recessive trait but due to the effect of the modifiers, QPM behaves as a quantitative trait. Using SSRs and backcross breeding, Babu *et al.* (2005) developed maize lines that had twice the amount of lysine and tryptophan as compared to local cultivars and recovered up to 95% of the recurrent parent genome in two backcross generations.

Barley

Malt is a major raw material for the production of beer. Characters that affect malting quality include malt extract content, α- and β-amylase activity, diastatic power, malt β-glucan content, malt β-glucanase activity, grain protein content, kernel plumpness and dormancy, all are quantitatively inherited and variously influenced by the environment (Zale *et al.*, 2000). There are a few barley cultivars with good malt quality that brewers are reluctant to change from due to their concerns about the resultant changes in flavour and brewing procedures. For example, the goal of the US Pacific Northwest barley breeding programme is to produce high yielding NILs that maintain traditional malting quality characteristics but transfer QTL associated with yield, via MABC, from the high yielding cv. Baronesse to the North American two-row malting barley industry standard cv. Harrington. Schmierer *et al.* (2004) targeted the Baronesse chromosome 2HL and 3HL fragments presumed to contain QTL that affect yield. Using backcross breeding and QTL/marker information, they identified a NIL (00–170) that when evaluated for yield over 22 environments and for malt quality over six environments, produced yield equal to Baronesse while maintaining a Harrington-like malt quality profile. Other studies have also reported the development of lines with improved malt quality: white aleurone colour and high α-amylase content (Ayoub *et al.*, 2003) and high in β-glucan and fine-coarse difference (Igartua *et al.*, 2000).

9.6 Marker-assisted Gene Pyramiding

Gene pyramiding is the process that brings the genes or alleles dispersed in different cultivars into a cultivar/genotype. QTL pyramiding is an important strategy for rebuilding the outputs from reductionist genomics research into whole traits of value for crop improvement. Genes can be pyramided through pedigree breeding by crosses involving multiple parental lines containing different favourable alleles or MABC to introgress those alleles into the same genetic background. One of the approaches for the pedigree method is to use NILs. Once the desirable QTL have been detected, then NILs are generated for each QTL in a common elite genetic background and the effect of each QTL individually evaluated. The selected NILs containing the most important QTL for the target trait are subjected to pairwise crosses to pyramid two or more QTL for one or more target traits. For example, in rice QTL for increased grain number (*Gn1*) and QTL for reduced plant height [*Ph1*(*sd1*)] were pyramided in the Koshihikari background producing a 23% increase in grain yield while reducing the plant height by 20% compared with Koshihikari (Ashikari *et al.*, 2005).

9.6.1 Gene pyramiding for major genes

The great opportunity offered by MAS to select superior lines based on genotype rather than phenotype becomes clearly obvious, particularly in the case of combining different simple inherited resistance genes of large effects for a given pathogen in a single genotype (gene pyramiding). Gene pyramiding is particularly important for disease resistance breeding. It is a useful approach to the durability or level of pest and disease resistances, or to increase the level of abiotic stress tolerance. Genes controlling resistance to different races or biotypes of a pest or pathogen and genes contributing to agronomic or seed quality traits can be pyramided together to maximize the benefit of MAS through simultaneous improvement of several traits in an improved genetic background. Gene pyramiding multiple genes for resistance to different diseases can offer great financial rewards through extending the lifespan of new cultivars. Such an approach has been used for the backcross transfer of QTL for downy mildew resistance in pearl millet (Witcombe and Hash, 2000).

Many reports have been available for marker-assisted gene pyramiding although few of them result in the release of commercial cultivars. Table 9.3 lists some representative examples from barley, common bean, rice, soybean and wheat, most of which are for major genes. Gene pyramiding includes combinations of genes for resistance to multiple races of the same disease, genes for resistance to different diseases and genes for disease and insect resistance. In rice, three blast resistance genes have been pyramided into one cultivar. First, three blast resistance genes (*Pi-2*, *Pi-1* and *Pi-4*) were mapped on to rice chromosomes 6, 11 and 12. Gene pyramiding started with three NILs, each carrying one of the genes. After two cycles of crossing and selection, a plant containing the three was obtained and it has been used as a source for these genes in plant breeding. An integrated breeding programme including MAS was used to improve an elite hybrid rice, 'Shanyou 63', a cross between 'Zhenshan 97' and 'Minghui 63'. Xa_{21}, a wide-spectrum bacterial blight resistance gene, was introduced into the restorer 'Minghui 63' by MAS and a *Bt* gene that is toxic to stem borer was introduced into 'Minghui 63' through transformation. An allele at the *Wx* locus from 'Minghui 63' was transferred by MAS to 'Zhenshan 97' to improve cooking and eating quality of the hybrid, resulting in a new version of 'Zhenshan 97' with medium amylose content, soft gel consistency and high gelatinization temperature. The pyramiding of *Bt*, Xa_{21} and *wx* genes created an improved 'Shanyou 63' (He *et al.*, 2002). Other successful examples in rice include improved pyramided lines and cultivars containing gene combinations for bacterial blight, blast, brown plant hopper, yellow stem borer and sheath blight. In wheat, powdery mildew (*Pm2*, *Pm4a*, *Pm6*, *Pm8* and *Pm21*) pyramided lines and those with resistance to *Fusarium* head blight (six QTL), orange blossom midge (*Sm1*) and leaf rust (*Lr21*) were bred through MAS. Resistance to barley mild mosaic virus and barley yellow mosaic virus complex and stripe rust has been separately incorporated through MAS in barley. Many of these pyramided lines showed enhanced resistance to pests and diseases, some even out-yielded the controls under high disease or pest pressure in field conditions. In legumes, resistances to rust and anthracnose (QTL) have been combined in common bean.

9.6.2 Gene pyramiding through marker-assisted recurrent selection

Marker-assisted recurrent selection (MARS) schemes and infrastructure have been developed for 'forward breeding' of native genes and QTL for relatively complex traits such as disease resistance, abiotic stress tolerance and grain yield (Ribaut and Betrán, 1999; Ragot *et al.*, 2000; Ribaut *et al.*, 2000; Eathington, 2005; Crosbie *et al.*, 2006). Eathington (2005) and Crosbie *et al.* (2006) reported that the rates of genetic gain achieved through MARS in maize were about twice those of PS in some reference populations. Marker-only recurrent selection schemes have been implemented for a variety of traits including grain yield and grain moisture (Eathington, 2005), or

Table 9.3. Examples of marker-assisted gene pyramiding for resistance to biotic stresses in crops.

Crop and target trait	Gene	Breeding scheme	Marker	MAS product	Reference
Barley yellow mosaic virus and barley mild mosaic virus	*rym4, rym5, rym9* and *rym11*	Simple and complex crosses using double haploids	RAPDs and SSRs	DHs carrying *rym4, rym9* and *rym11* and those with *rym5, rym9* and *rym11*	Werner *et al.* (2005)
Barley stripe rust	QTL (1H, 4H and 5H)	Backcross derived introgression lines	SSRs	Introgression lines carrying 1H, 4H or 5H individually or in combinations	Richardson *et al.* (2006)
Common bean rust (*Uromyces appendiculatus*) and anthracnose (*Colletotrichum lindemuthianum*)	Nine major genes each for rust and anthracnose	Three backcrosses	RAPDs	Lines combining resistance to rust and anthracnose	Faleiro *et al.* (2004)
Rice bacterial blight (BB)	*Xa4, xa5, xa13* and *Xa21*	Pedigree breeding	RFLPs	Pyramided lines showing broader spectrum of resistance to BB	Huang *et al.* (1997)
	Xa7 and *Xa21*	Pedigree breeding	6 PCR-based markers	Pyramided lines showing stronger resistance to BB than lines with single genes	Zhang, J. *et al.* (2006)
Rice bacterial blight (BB), stem borer (SB), blast, and brown planthopper (BPH)	*Xa21* and *Xa7*; *Bt* (SB); *Pi1, Pi2, Pi3*; and *Qbph1* and *Qbph2*	Pedigree breeding	AFLP 1415, STS P3, M5, 248, RM144, RM224 and *Pi2*	Improved 'Minghui 63' showing broader resistance to BB and combined resistance to BB and SB, and improved 'Zhenshan 97' showing better resistance to BPH	He, Y. *et al.* (2004)
Rice bacterial blight (BB), yellow stem borer (YSB), sheath blight (SB) (*Rhizoctonia solani*)	*Xa21, Bt* and *RC7 chitenase* (Sb)	Pedigree breeding	Pc822 (*Xa21*), *Bt* and *RC7* chitinase	Lines carrying three genes resistant to BB, YSB and SB	Datta *et al.* (2002)

(*Continued*)

Table 9.3. Continued

Crop and target trait	Gene	Breeding scheme	Marker	MAS product	Reference
Rice blast (BL) [*Magnaporthae grisea* (Herbert) Borr. (anamorphe *Pyricularia oryza* Cav.)]	*Pi1, Piz-5* and *Pita*	Pedigree breeding	RFLPs	The pyramided lines showing better resistance to blast	Hittalmani *et al.* (2000)
Rice blast (BL) and bacterial blight (BB)	*Piz-1* and *Piz-5* (BL) and *Xa21* (BB)	Pedigree breeding	RZ536 and r10 (BL) and *Xa21* (1.4 kb fragment of pC822)	The pyramids showing enhanced resistance to BL and BB	Narayanan *et al.* (2004)
Soybean corn earworm (CEW) (*Helicoverpa zea* Boddie)	QTL and Bt (*cry1Ac*)	Three backcrosses	Nine SSRs	The pyramid lines with a detrimental effect on larval weights and on defoliation by CEW	Walker *et al.* (2002)
Soybean corn earworm and soybean looper (*Pseudoplusia includens*)	*cry1Ac* and QTL (PI 229358)	Two backcrosses	Six SSRs and sequence-specific primers *cry1Ac*	Lines carrying *cry1Ac* and QTL alleles resistant to three lepidopteran pests	Walker *et al.* (2004)
Wheat Fusarium head blight (FHB) (*Fusarium graminearum*), orange blossom midge (*Sitodiplosis mosellana*) and leaf rust (*Lr21*)	Six FHB QTL, *Sm1* for midge and *Lr21* for leaf rust	Two backcrosses	gwm533, gwm493 and wmc808	Resistant progenies containing chromosome segments FHB, *Sm1* and *Lr21*	Somers *et al.* (2005)
Wheat powdery mildew (*Erysiphe graminis* DC. *F. tritici* Em. Marchal)	*Pm2, Pm4a, Pm6, Pm8* and *Pm21*	Pedigree breeding	RAPD and SCAR markers[a]	Lines with *Pm2* and *Pm4a* immune to powdery mildew	Wang *et al.* (2001)

[a]RAPD, randomly amplified polymorphic DNA; SCAR, sequence characterized amplified regions.

abiotic stress tolerance (Ragot *et al.*, 2000) and multiple traits are being targeted simultaneously. Selection indices were apparently based on ten to probably more than 50 loci, these being either QTL identified in the experimental population where MARS was being initiated, QTL identified in other populations, or genes. Marker genotypes are generated for all markers flanking QTL included in the selection indices (Ragot *et al.*, 2000). Plants are genotyped at each cycle and specific combinations of plants are selected for crossing, as proposed by van Berloo and Stam (1998). Several, probably three to four, cycles or MARS are conducted per year using continuous nurseries. Results reported in these recent communications about private MARS experiments (Ragot *et al.*, 2000; Eathington, 2005) are in sharp contrast to those in earlier publications (Openshaw and Frascaroli, 1997; Moreau *et al.*, 2004). As summarized by Ragot and Lee (2007), this selection response can be attributed to: (i) rather large sizes of the populations submitted to selection at each cycle; (ii) use of flanking versus single markers; (iii) selection before flowering; (iv) increased number of generations from one to four generations per year; and (v) lower cost of marker data points.

9.7 Marker-assisted Hybrid Prediction

Hybrid performance largely depends on general combining ability (GCA) of the parental lines and the specific combining ability (SCA) between the parents. GCA is defined as an attribute of an inbred line and is measured as the average performance of all hybrids made with that inbred line as a parent. The higher the GCA of an inbred, the higher the average performance of its hybrids. SCA is defined for specific combinations of parents and is measured by the deviation of the hybrid performance from the expected performance as estimated from the GCA of the parents. As a result, hybrid performance is determined by its parents' GCA and the cross's SCA. Hybrid performance can be measured by the heterosis, the performance of a hybrid over their parental lines.

Suppose a breeder has 100 inbreds from heterotic group 1 and 100 inbreds from heterotic group 2. There are 10,000 possible (group 1 × group 2) single crosses. For developing new hybrids, there are 495,000 possible (group 1 F_2) × (group 2 tester) combinations and 495,000 possible (group 1 tester) × (group 2 F_2) combinations, if testcrossing starts from the F_2. Due to limited resources, breeders are unable to test all combinations in all environments of interest but may test a limited set of single crosses and F_2 × tester combinations. Typically, < 1% of the maize single crosses tested by a breeder eventually become commercial hybrids (Hallauer, 1990). Therefore, predicting hybrid performance has always been a primary objective in all hybrid-breeding programmes. Methods for predicting the performance of single crosses would greatly enhance the efficiency of hybrid breeding programmes. Development of a reliable method for predicting hybrid performance and/or heterosis without generating and testing hundreds or thousands of single cross combinations has been the goal of numerous studies using marker data and combinations of marker and phenotypic data, particularly in maize and rice.

9.7.1 Genetic basis of heterosis

QTL for heterosis

Heterosis is a complex physiological phenomenon affected by many factors. Yield is the most important trait in crop-based heterosis analysis. Understanding the genetic basis of heterosis is the fundamental basis for hybrid prediction. Several different hypotheses have been proposed for the explanation of heterosis. Among these hypotheses, arguments focused on the dominance hypothesis (Davenport, 1908) and the overdominance hypothesis (East, 1908; Shull, 1908), both of which are based on describing the genetic effects of single loci. Recent studies have indicated that epistasis plays an important role in genetic control of both quantitative traits and heterosis. The dominance hypothesis proposes that heterosis results from the cancellation of effects from deleterious recessive alleles, contributed by one parent, by dominant alleles contributed by the other parent in the heterozygous F_1. This hypothesis emphasizes the contribution of the dominance to heterosis. The overdominance hypothesis assumes that a specific heterozygous combination of alleles at a single locus is superior to either of the homozygous combinations of the parental alleles at that locus. With development of molecular markers, QTL mapping in rice and maize addressed the classical models by breaking down heterosis into Mendelian factors and assessing their modes of inheritance (Stuber *et al.*, 1992; Xiao *et al.*, 1995; Yu *et al.*, 1997a; Li, Z.K. *et al.*, 2001; Luo *et al.*, 2001; Hua *et al.*, 2002, 2003; Lu, H. *et al.*, 2003). The evidence showed that both dominance and locus-specific overdominance have a role in heterosis, with some involvement of epistasis, although the relative contribution of each of these mechanisms is still unclear. Crow (1999, 2000) provided a historical review on the dominance and overdominance hypotheses. Xu, Y. (2003) and Lippman and Zamir (2007) provided a review on all possible hypotheses including epistasis.

In many investigations, genes for yield per se and genes for yield-related heterosis have been confounded with each other. Reports in the 1990s on dominance (Xiao *et al.*, 1995), overdominance (Stuber *et al.*, 1992) and epistasis (Yu *et al.*, 1997a) were based on the use of yield and yield components per se to measure hybrid performance without use of parental lines as a control to derive values for the mid-parent or better-parent heterosis. The method of measurement will identify genes for yield and yield components rather than genes for heterosis. For open-pollinated species like maize, which has severe inbreeding depression, it is very difficult (if not impossible) to do side-by-side comparisons of the F_1 hybrids with their parents. But, theoretically, this comparison is absolutely necessary if heterosis rather hybrid performance needs to be measured (Xu, Y., 2003).

Several investigations were reported later on for genetic analysis of heterosis per se in rice. Li, Z.K. *et al.* (2001) investigated the genetic basis of heterosis in rice using 254 RILs derived from a cross between 'Lemont' (*japonica*) and 'Teqing' (*indica*) and two backcross and two testcross populations derived from crosses between the RILs and their parents plus two testers (Zhong 413 and IR64). As a result, most QTL associated with decreased grain yield and biomass, or with heterosis in rice appeared to be involved in epistasis and about 90% of the QTL contributing to heterosis appeared to be overdominant. Hua *et al.* (2002, 2003) designed a mating scheme that generated a fixed or 'immortalized' F_2 population, using a population of 240 RILs derived from the Zhenshan 97 × Minghui 63 cross. In this design, crosses were made between the RILs chosen by random permutations of the 240 RILs. In each round of permutation, the 240 RILs were randomly divided into two groups and lines in the two groups were paired at random without replacement to provide parents for 120 crosses. Three rounds of such random permutations, including 360 crosses, resulted in two conclusions. First, all kinds of genetic effects, including single-locus heterotic effects caused mostly by overdominance and all three forms of digenic interactions (additive by additive, additive by dominance and dominance by dominance) appeared to play a role in the genetic basis of heterosis in the 'immortalized F_2' population. However, the QTL were not fine mapped, leaving open the possibility that, as in maize, the single-locus effects were due to pseudo-overdominance, rather than true overdominance. Secondly, single-locus heterotic effects and dominance-by-dominance interaction could, together, adequately account for the genetic basis of heterosis in the F_1 hybrid.

To assess the importance of loci with overdominant (ODO) effects in expression of heterosis, Semel *et al.* (2006) employed NILs, carrying single marker-defined chromosome segments from distantly related wild species *Solanum (Lycopersicon) pennellii*, to partition heterosis into defined genomic regions, eliminating a major part of the genome-wide epistasis. They detected 841 QTL for 35 diverse traits. NILs showing greater reproductive fitness are characterized by the prevalence of ODO QTL, which were virtually absent for the non-reproductive traits. ODO results from true ODO due to allelic interactions of a single gene or from pseudo ODO involving linked loci with dominant alleles in repulsion. In their study, although they detected dominant and recessive QTL for all phenotypic traits but ODO only for the reproductive traits indicates that pseudo ODO is unlikely to explain heterosis in NIL, thus they favour the true ODO model, a single functional Mendelian locus, involved in heterosis.

Gene expression analysis of heterosis

Using serial analysis of gene expression (SAGE), Bao *et al.* (2005) surveyed transcripomes in panicles, leaves and roots of a super-hybrid rice (*LYP9*) in comparison to its parental inbred cultivar genotypes (*93-11* and *PA64s*). They identified 595 upregulated and 25 downregulated tags in *LYP9* that were related to enhancing carbon- and nitrogen-assimilation, including photosynthesis in leaves, nitrogen uptake in roots and rapid growth in both roots and panicles. They found massive complementation at the transcript level that further suggests that the underlying mechanisms of heterosis may not be as simple as have been reported from studies of a small number of genes (Birchler *et al.*, 2003).

Yao *et al.* (2005) used an interspecific hybrid between common wheat (*Triticum aestivum* L., $2n = 42$, AABBDD) line 3338 and spelt (*Triticum spelta* L., $2n = 42$, AABBDD) line 2463, which is highly heterotic both for aerial growth and for root-related traits. In their research they included an expression assay using modified suppression subtractive hybridization (SSH) to generate four subtracted cDNA libraries between the wheat hybrid and its parental genotypes. Of the 748 non-redundant cDNAs obtained, 465 cDNAs had high sequence similarity to GenBank entries in diverse functional categories, such as metabolism, cell growth and maintenance, signal transduction, photosynthesis, response to stress, transcription regulation and others. They further confirmed the expression patterns of 68.2% SSH-derived cDNAs by reverse Northern blot, while semi-quantitative RT-PCR exhibited similar results (72.2%). This suggests that the genes differentially expressed between hybrids and their parents are involved in diverse physiological pathways, which may contribute to heterosis in wheat.

Maize inbred lines B73 and Mo17 produce a heterotic F_1 hybrid. Based on analysis with a 13,999 cDNA microarrays, Swanson-Wagner *et al.* (2006) compared global patterns of gene expression in seedlings of the hybrid (B73 × Mo17) with those of its parental genotypes. A total of 1367 expressed sequence tags (ESTs) were observed to be significantly differentially expressed, using an estimated 15% false discovery rate as cut off. All possible modes of gene action were observed, including additivity, high- and low-parent dominance, underdominance and overdominance. A total of 1062 of the 1367 ESTs (78%) exhibited expression patterns that are not statistically distinguishable from additivity while the remaining 305 ESTs exhibited non-additive gene expression. About 181 of the 305 non-additive ESTs exhibited high-parent dominance, 23 ESTs showed low-parent dominance, while 44 ESTs displayed underdominance or overdominance. These results suggest that multiple genetic mechanisms, including overdominance, contribute to heterosis. This contrasts with previous studies that reported heterosis was due to gene action of only a small set of maize genes (Song and Messing, 2003; Guo *et al.*, 2004; Auger *et al.*, 2005). Further analysis of allelic variation in gene expression in the maize hybrid and its parental lines (B73 and Mo17) identified a subset of 27 genes that are differentially expressed in parental lines. When the transcriptional contribution of each allele from the inbred line was analysed in the hybrid, the majority of the differential expression was observed to be due to *cis*-regulatory variation and not due to differences in *trans*-acting regulatory factors. This suggest a predominance of additive expression and a lack of epistatic effects, as genes subject to *cis*-regulatory variation are expected to be expressed at mid-parent, or additive, levels in the hybrids (Stuper and Springer, 2006). Using a 57,000 maize gene-specific long-oligonucleotide microarray containing about 32,000 genes to study the differential gene expression between a maize hybrid and its parental genotypes (B73 and Mo17), Scheuring *et al.* (2006) revealed that at least 800 genes were expressed at two- to tenfold higher levels in the hybrid than the parent genotypes. Using Massively Parallel Signature Sequencing (MPSS), an open-ended mRNA profiling technology, of nearly 400 allelic signature tag pairs, Yang, X. *et al.* (2006) found 60% of the genes expressed in meristems of the hybrid were significantly different in allele-specific transcript level as compared to the parental genotypes. This suggests an abundance of *cis*-regulatory polymorphisms affecting hybrid meristem gene expression. Furthermore, when comparing the expression of the same allele in the hybrid versus inbred parents, they found 50% of the genes expressed at a significantly different level. Such differences in expression are likely to be attributed to the effect of *trans*-acting factors that differ between the hybrid and inbreds. While *cis*-regulatory variation predicts additive expression, *trans*-regulation may result in non-additive expression in the hybrid. Thus, studying the effect of transcript regulation at an allele-specific level provides a different level of understanding of gene regulation than focusing on overall expression in the hybrid.

As indicated by Lippman and Zamir (2007), however, differences in methodology aside, a fundamental problem in these studies is that they cannot associate novel expression patterns in hybrids with any heterotic phenotypes. As too many loci have been revealed to differ between two parental lines, the key issue is to understand further which really matter with the expression of heterosis. This is very much like the situation where experienced breeders can tell how many traits could be different between two parental lines they are working with but they cannot tell how the differences contribute to production of a heterotic hybrid. Mapping expression QTL (eQTL) and testing whether there is an association between eQTL and phenotypic QTL should be the next logical step. However, it can be expected that numerous eQTL will be identified across different species and populations, which would repeat the history of phenotypic QTL mapping where numerous QTL have been identified but nothing can be confirmed for their heterotic effects. Further research is required to identify the QTL that genetically control heterosis and their interactions in gene networks associated heterotic effects across the whole genome.

Prospects on genetic basis of heterosis

Heterozygosity and its related gene interactions are the primary genetic basis for explanation of heterosis because the hybrid is heterozygous across all genetic loci that differ between the parents. Thus, the degree of heterosis depends on which loci are heterozygous and how within-locus alleles and inter-locus alleles interact with each other (Xu, Y., 2003). Interaction of within-locus alleles results in dominance, partial dominance, or overdominance, with a theoretical range of dominance degree from zero (no dominance) to larger than 1 (overdominance). Interaction of inter-locus alleles results in epistasis. Genetic mapping results have indicated that most QTL involved in heterosis and other quantitative traits had a dominance effect. As statistical methods that can estimate epistasis more efficiently become available, epistasis has been found more frequently and proven to be a common phenomenon in the genetic control of quantitative traits including heterosis (Xu, Y., 2003). With so many genetic loci involved, it is unlikely that there is no interaction at all between any pair of them.

It can be concluded that two different types of allele interaction, both within-locus and inter-locus, play an important role in the genetic control of heterosis. Contribution of a specific locus to heterosis could be due to any single type of these interactions. When multiple loci are involved which were not taken into account in the early 1900s, various combinations of within-locus and inter-locus interactions (especially dominance-by-dominance interaction) could contribute to the genetic control of heterosis. For a specific cross and specific trait, heterosis might be explainable by any single type of these interactions (Xu, Y., 2003). For different crosses, species, or traits, however, their heterosis has to be explained by the dominance of different degrees in combination with all possible inter-locus interactions, as indicated by Goldman (1999). A full understanding of heterosis will depend on cloning and functional analysis of all genes that are related to heterosis. This process would be very similar to that for understanding disease resistance genes that functionally appear much simpler than heterosis.

9.7.2 Heterotic groups

Heterotic groups are the backbone of successful hybrid breeding. In most cases, breeding for heterosis without knowledge of heterotic patterns has proven to be a hit-or-miss approach (Jordaan *et al.*, 1999). The concept of heterotic groups or heterotic pools was first developed in maize, based on the observation that inbreds selected out of certain populations tended to produce better performing hybrids when crossed to inbreds from other groups (Hallauer *et al.*, 1988). This recognition resulted from the systematic crossing of thousands of inbred lines from different source populations and evaluation of the hybrids (Havey, 1998). In the review of capturing heterosis in forage crop cultivar development, Brummer (1999) indicated that the key to successful semi-hybrid production is to keep heterotic groups separate, only intercrossing them for testing and release. Breeding highly heterotic hybrids largely depends on selection of desirable parents as a prerequisite for most hybrid breeding programmes and thus depends on genetic diversity in the germplasm resources available to plant breeders. Therefore, construction or development of heterotic groups has been one of the key components in hybrid breeding for many crops. Introgressing exotic germplasm is often suggested as an approach to increase genetic differences between opposing heterotic populations, thereby potentially increasing heterotic response. An understanding of heterotic relationship between populations is needed to exploit exotic germplasm intelligently. Melchinger and Gumber (1998) reviewed the development of heterotic groups in five major crops with different pollination systems: allogamous maize and rye; partially allogamous faba bean and oilseed rape; and autogamous rice.

A possible explanation for heterotic groups is that populations of divergent genetic backgrounds have unique allelic diversity that may have arisen from founder effects, genetic drift, or the accumulation of unique allelic diversity by mutation or selection. Significantly greater heterosis could result from this genetic diversity by specific interallelic interactions (overdominance), repulsion-phase linkage among loci showing dominance (pseudo-overdominance) (Havey, 1998) and/or inter-locus interaction (epistasis). Apparently, the most obvious potential heterotic groups are either geographically separated populations or separate subspecies and ecotypes. Melchinger and Gumber (1998) recommended the following criteria for the identification of heterotic groups and patterns in descending order of importance: (i) high mean performance and large genetic variance in the hybrid population to ascertain future selection response; (ii) high per se performance and good adaptation of both or at least one of the parental heterotic groups; (iii) low inbreeding depression in the source materials for the development of inbreds; and (iv) a stable CMS system without deleterious side effects, as well as effective restorers and maintainers, if hybrid breeding is based on CMS.

Construction of heterotic groups based on hybrid performance

With large numbers of inbred or open-pollinated lines or populations available, it is not feasible in most crops to make diallel crosses and produce sufficient F_1 seed for multi-environment field-testing. Therefore, Melchinger and Gumber (1998) suggested a multi-stage procedure to identify heterotic groups, which consists of the following steps: (i) grouping the germplasm based on genetic similarity; (ii) selection of representative genotypes (e.g. two or four lines or one population) from each subgroup for producing diallel crosses; (iii) evaluation of diallel crosses among the subgroups together with parents in replicated field trials; and (iv) selection of the most promising cross combinations as potential heterotic patterns using the identification criteria. If established heterotic patterns are available, using selected elite genotypes from them as testers for the production and evaluation of the germplasm to be classified is recommended. Based on the testcross performance, populations or lines having similar combining ability and heterotic response could be merged to constitute a new independent heterotic group, if they behave differently from the existing heterotic groups; however, if their behaviour is similar to an existing heterotic group, they could be merged with it to enlarge its genetic base. Heterotic patterns in many crop species have been established solely based on the large numbers of testcrosses and breeding experience, without the use of molecular markers.

Ron Parra and Hallauer (1997) reviewed heterotic patterns used in the major maize production regions of the world. Some patterns have had importance in specific production regions. Others have been exploited on several continents, for example, the heterotic patterns based on 'Reid Yellow Dent' (RYD) and 'Lancaster Sure Crop' (LSC) from the temperate USA and Tuxpeño and Estación Tulio Ospina from tropical Mexico and South America. Two heterotic groups from which inbreds commonly are selected and used to produce superior maize hybrids are Iowa Stiff Stalk Synthetic (BSSS) and derivatives of LSC (Darrah and Zuber, 1986; Gerdes and Tracy, 1993). Although both populations are primarily comprised of southern dent germplasm, LSC has more northern flint germplasm than BSSS (Smith, 1986; Gerdes and Tracy, 1993). With genetically balanced sets of crosses, inter-group hybrids out-yielded the respective intra-group hybrids by 21% in RYD × LSC crosses (Dudley *et al.*, 1991) and by 16% in flint × dent crosses (Dhillon *et al.*, 1993). In both studies, the percentage of increase in heterosis for yield of inter-group over intra-group crosses was about twice as large as for the hybrid yield itself. Most heterotic grouping reports are for maize, with only very few on other crops including summer squash (Anido *et al.*, 2004) and rapeseed (Qian *et al.*, 2007).

Rice might be the only crop where hybrids are widely grown but very few studies on heterotic groupings have been reported. Heterosis in rice has been utilized largely through CMS. Fortunately, rice breeders in China identified the restorers for CMS from geographically distant rice cultivars from South-east Asia and used them in hybrid rice breeding. This resulted in high levels of heterosis among intra-subspecies (*indica* × *indica*) hybrids. A large-scale screening of diverse CMS maintainers and restorers provided some clue as to heterotic pattern. Three ecotypes from different subspecies, *indica*, *japonica* and *javanica*, have different morphological and physiological characteristics and ecogeographical distribution and, therefore, serve as a basis for defining distinct heterotic groups (Xu, Y., 2003). As summarized by Yuan (1992), heterosis for grain yield in crosses among the three rice ecotypes has the following trend: *indica* × *japonica* > *indica* × *javanica* > *javanica* × *japonica* > *indica* × *indica* > *japonica* × *japonica*. This mirrors the current situation of heterotic pools in rice. It is well known to hybrid rice breeders that a high level of heterosis results from crosses between CMS lines bred in China and restorer lines derived from South-east Asian *indica* cultivars, which is the heterotic pattern for *indica* × *indica* hybrids.

Construction of heterotic groups using molecular marker information

Molecular markers have been playing an increasingly important role in the construction of heterotic groups since the 1990s. Most reports are focused on maize, wheat, barley and canola. Because marker-based groupings reflect the genetic differences among parental lines, they can contribute to parental improvement and to effective selection for heterotic hybrids. In general, heterotic groups constructed on the basis of marker information match up very well with pedigrees, but have the advantage that missing historical information, such as the incomplete pedigree information or ambiguous pedigree, will not affect the marker-based method.

In maize, different types of molecular markers have been successfully used to differentiate heterotic groups with results that are consistent with pedigree-based grouping (Mumm and Dudley, 1994; Liu *et al.*, 1997; Peng *et al.*, 1998; Wu *et al.*, 2000; Menkir *et al.*, 2004). Based on heterosis and combining ability analyses using cultivars from different heterotic groups, Peng *et al.* (1998) proposed seven heterotic patterns for the utilization of maize heterosis. Divergence at molecular marker loci has been useful in assigning maize inbreds to known heterotic groups previously established in breeding programmes and the molecular information agreed with pedigree information (Lee *et al.*, 1989; Melchinger *et al.*, 1991; Messmer *et al.*, 1993). Side-by-side phenotypic evaluation of a sequence of successful maize hybrids produced by Pioneer Hi-Bred International, Inc., representing each decade from the 1930s to present, provides a description of the phenotypic changes for a number of key traits that the breeders have directly or indirectly changed. Genetic fingerprints of the inbred parents of these hybrids provide a description of the genotypic changes that have occurred in association with the sustained breeding effort (Fig. 9.3; Cooper *et al.*, 2004). Important phases can be identified over this period of breeding. Initially double-cross hybrids (1920s–1960s) were developed. From the 1960s there was a relatively rapid transition to the use of single-cross hybrids, the foundation of which was the organization of the maize germplasm into heterotic groups, represented in this example by the Stiff Stalk Synthetic (SS) and the Non Stiff Stalk Synthetic (NSS) groups (Fig. 9.3).

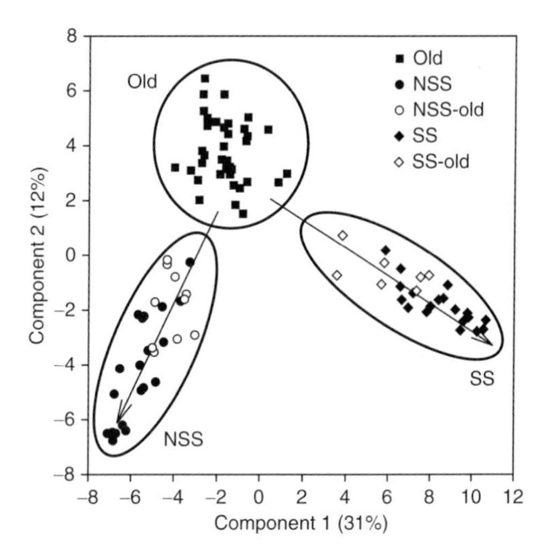

Fig. 9.3. A plot of the inbred scores on the first two principal components from analysis of SSR marker profiles of the parents of the maize hybrids (SS, Stiff Stalk Synthetic inbred line; NSS, Non Stiff Stalk Synthetic inbred line). The large boundaries distinguish three main groups of lines: Old, the old inbred lines used before the formation of the heterotic groups; the other two groups represent SS and NSS inbred lines. The arrows indicate the direction of the progression of inbred improvement in the SS and NSS heterotic groups. From Cooper *et al.* (2004) with permission.

Using 160 RFLP markers and 21 wide-compatibility cultivars and three *indica* and three *japonica* cultivars, Zheng *et al.* (1994) constructed a dendrogram tree and discussed the potential of wide compatibility in hybrid breeding using *indica* × *japonica* crosses. Based on diallel crosses among eight *indica* lines representing the parents of the best-performing commercial rice hybrids grown in China, Zhang *et al.* (1995) studied molecular divergence and hybrid performance. Their results suggest the existence of two heterotic groups within *indica*, one comprised of rice strains from southern China and the other comprised of strains from South-east Asia. Using two types of molecular markers, RFLPs and amplified fragment length polymorphisms (AFLPs), Mackill *et al.* (1996) obtained similar grouping results. Using RAPD and SSR markers, Xiao *et al.* (1996b) separated the ten parental lines into two major groups that correspond to *indica* and *japonica* subspecies. These results and the results from barley (Melchinger *et al.*, 1994) and wheat (Sun *et al.*, 1996; Ni *et al.*, 1997) also supported the conclusion that DNA markers are very useful tools for construction of heterotic groups.

Future direction

It is evident from the review of various studies that adapted populations, isolated either by time and/or space, are the most suitable candidates for promising heterotic patterns. Genetic diversity can be related to geographic origin of parental lines. The geographical variation can be related to ecological and environmental variations that, in turn, dictate survival fitness, created by spontaneous and induced genetic variation in natural and directed-selection situations. Consequently, the parental lines derived from different geographic origins are considered to have more genetic diversity than those derived from the same geographic origin. During internationalization of plant breeding efforts and massive exchange of unimproved and improved germplasm throughout the world attention needs to be paid to avoid the negative effect of using distant crosses that might mix up heterotic groups existing among cultivars of different geographic origins. For example, breeding wide-compatible inbred cultivars as a bridge for harnessing *indica*/*japonica* heterosis in rice has reduced heterosis compared to what would be expected from crosses between typical *indica* and *japonica* cultivars (Xu, Y., 2003).

Heterotic groups should not be considered as closed populations, but should be broadened continuously by introgressing unique germplasm to warrant medium- and long-term gains from selection. Heterotic groups consisting of poorly utilized and unadapted germplasm should be enhanced through joint public–private breeding ventures. Different phenotypes may or may not reflect divergent genetic backgrounds. Phenotypically different populations may possess the same genetic background and divergent phenotypes may be conditioned by allelic differences at relatively few loci (Havey, 1998). MAS can be useful in creating, maintaining and improving heterotic groups. As discussed above, marker-based grouping of germplasm and breeding populations will help establish heterotic groups that hold maximum genetic diversity between groups but minimum diversity within groups. Identification of marker alleles that are specific to each heterotic group will help keep them genotypically separated. MAS can be used to improve the existing heterotic groups through introgressing target genes from one heterotic group or outsource germplasm to another with minimum linkage drag from the donor.

9.7.3 Marker-assisted hybrid prediction

It is reasonably believed that heterosis originates, in some way, from the genetic differences or heterozygosity between the parents. Theoretically, hybrid performance is equal to the average parental performance plus heterosis. In the past several decades, hybrid prediction has been largely based on the evaluation of genetic diversity among parental lines. It has been expected that understanding the relationship between heterozygosity/parental difference and heterosis would help predict hybrids. The development of molecular marker techniques has provided new tools for hybrid prediction and DNA markers have been used extensively in investigating correlations between parental genetic distance (GD) and hybrid performance.

Genome-wide heterozygosity and hybrid prediction

The relationship between parental genetic divergence and hybrid performance was first studied in maize. Variability for molecular markers generally agreed with pedigree information and assignment (based on hybrid performance) to known heterotic groups (Smith, O.S. *et al.*, 1990; Dudley *et al.*, 1991; Melchinger *et al.*, 1991); however, variability at molecular marker loci was ineffective in predicting specific hybrid performance from crosses among maize inbreds (Lee *et al.*, 1989; Melchinger *et al.*, 1992). Some reports indicated high correlation between hybrid performance/ heterosis and parental GDs or the degree of heterozygosity (Lee *et al.*, 1989; Smith, O.S. *et al.*, 1990; Stuber *et al.*, 1992; Reif *et al.*, 2003), while others revealed very weak correlations (Godshalk *et al.*, 1990; Dudley *et al.*, 1991). Correlations between single-cross performance and molecular marker diversity for unrelated parental inbreds have been too low to be of any predictive value (Godshalk *et al.*, 1990; Melchinger *et al.*, 1990; Dudley *et al.*, 1991), which is also supported by the result from sorghum (Jordan *et al.*, 2004). Molecular-based GD estimates also failed to predict superior hybrid performance in oat (Moser and Lee, 1994), soybean (Gizlice *et al.*, 1993), chickpea (Sant *et al.*, 1999) and pepper (Geleta *et al.*, 2004). A recent large-scale experiment in maize also supported this unpredictability. Using three sets of six sister-line inbred lines, each set being highly related and derived from a common parent cross and 45 sister-line hybrids generated by a partial diallel, Lee, E.A. *et al.* (2007) re-examined the relationship between degree of relatedness, genetic effects and heterosis in maize. The three sets of sister lines ranged between 47 and 77% identical-by-descent, creating a series of lines that potentially vary in gene frequency. They reported three relevant findings regarding heterosis for grain yield: (i) substantial genome-wide heterozygosity is not a requirement for the expression of heterosis; (ii) there is not a consistent relationship between degree of relatedness and the magnitude of heterosis; and (iii) the presence of non-additive genetic effects is not a requirement for the manifestation of heterosis.

Hybrids are more predictable within than between heterotic groups

Correlations between heterozygosity/GD and hybrid performance/heterosis varied for hybrids between lines that belong to the same heterotic group (within-group hybrids). In maize, correlations of GD with F_1 performance and heterosis were significant and positive for all traits of within-group hybrids, flint × flint crosses, but not for the subset of flint × dent and dent × dent crosses (Boppenmaier *et al.*, 1993). This was supported by Benchimol *et al.* (2000) using 18 tropical maize inbred lines where correlations of parental GDs with single crosses and their heterosis for grain yield were higher for line crosses from the same heterotic groups than the crosses from different heterotic groups. In rice, Xiao *et al.* (1996b) reported that yield potential and its heterosis showed significantly positive correlations with GD for *indica* × *indica* or *japonica* × *japonica* crosses, but the correlations were not significant for *indica* × *japonica* crosses. It was confirmed by Zhao *et al.* (1999) that very little correlation was detected in intersubspecific crosses using diallel crosses derived from 11 elite rice cultivars. In other cases, however, weak or no correlation was found for within-group hybrids. Examples include weak or no significant associations of GD with F_1 performance and mid-parent heterosis in soybean (Cerna *et al.*, 1997), wheat (Martin *et al.*, 1995) and US long-grain rice cultivars (Saghai Maroof *et al.*, 1997). These results may be due to the low levels of heterosis in these cultivar groups.

Based on results from various studies in maize, Melchinger (1993) summarized the relationship between parental GD and mid-parent heterosis (MPH) in a schematic representation. For crosses among related lines, there exists a tight association between GD and MPH for yield characters because both measures are a linear function of co-ancestry, *f*, and thus decrease with increasing *f*. For intra-group crosses, the correlation *r*(GD, MPH) is generally positive, too. This can be explained by hidden relatedness between some parents considered to be unrelated based on their pedigree and the presence of the same linkage phase between QTL and marker loci in the maternal and paternal gametic arrays of intra-group hybrids, which results in a positive covariance between GD and MPH (Charcosset *et al.*, 1991). In contrast, no significant association between both measures exists for inter-group hybrids. In this case, the maternal and paternal gametic arrays may differ in the linkage phase for many QTL–marker pairs; as a consequence, positive and negative terms cancel each other in their net contribution to covariance (GD, MPH), resulting in a low or zero correlation (Charcosset and Essioux, 1994).

Heterosis-associated markers and hybrid prediction

It has been common practice in most studies to determine GD or heterozygosity estimates from a set of DNA markers chosen for good coverage of the entire genome but not for linkage to genes influencing heterosis of the target trait. Theoretical investigations (Charcosset *et al.*, 1991) and computer modelling (Bernardo, 1992) demonstrated that with intra- and inter-group crosses the correlation between GD and MPH is expected to decrease if genes influencing heterosis are not closely linked to markers used for calculation of genetic estimates and vice versa if markers employed for calculation of GDs are not linked to genes controlling the trait. Hence, increasing the marker density alone will not necessarily improve the ability to predict MPH by GD estimates; rather, markers must additionally be selected for tight linkage to genes affecting heterosis of the target trait in the germplasm under study. This is corroborated by comparison of results obtained with 209 AFLPs versus 135 RFLPs (Ajmone Marsan *et al.*, 1998) and a study by Dudley *et al.* (1991). Using these associative loci will help establish strong correlations between heterozygosity and heterosis. However, allelic differences at marker loci do not assure allelic differences at linked loci for heterosis. For a limited number of markers to be useful as predictors for hybrid performance, the effects of alleles at the loci linked to specific marker alleles must be ascertained (Stuber *et al.*, 1999).

Zhang *et al.* (1994) proposed two statistical parameters, general and specific heterozygosity, to measure genotypic heterozygosity. The former is the heterozygosity calculated from the GDs between the parents using all possible markers and the latter is that from using marker loci that are significantly associated with the traits of interest revealed by single factorial analysis of variance. The results from rice indicated that there was a weak correlation between general heterozygosity and heterosis but a significant correlation between specific heterozygosity and heterosis for yield and biomass.

Favourable allele combination and hybrid prediction

Heterogenic gene combinations may not always lead to heterosis and heterosis may ultimately depend upon the balance between favourable and unfavourable interactions of genes. It is reasonably inferred that heterosis could be caused by specific gene combinations derived from the two parents. Those genes may simultaneously produce different genetic effects in different genetic backgrounds. So, for parental improvement and hybrid prediction, investigating the specific gene combinations that contribute to heterosis should be more important than studying any single gene or QTL. Using 99 half-diallel rice hybrids derived from nine CMS lines and 11 restorer lines, Liu and Wu (1998) found that four favourable alleles and six favourable heterotic patterns on the parental lines significantly contributed to the heterosis of their hybrids for grain yield, whereas six unfavourable alleles and six unfavourable heterotic patterns significantly reduced heterosis. They suggested that optimal hybrids with superior grain yield could be developed by assembling those favourable alleles into and removing the unfavourable alleles from their parental lines.

Conclusions and prospects

There are several conclusions that can be drawn from the numerous investigations on the relationships between heterozygosity and GD with hybrid performance and heterosis. First, the higher the heterozygosity between the parents, the stronger the heterosis is. Secondly, using more markers alone will not improve the prediction.

Thirdly, prediction is possible using markers known to be associated with hybrid performance or heterosis if the association is used to predict performance of a hybrid derived from the same heterotic pattern. Fourthly, genetic variation (the presence of heterosis) is a prerequisite for prediction. Fifthly, the relationship of heterozygosity with heterosis and with hybrid performance will be different if the two involve different genes (Xu, Y., 2003). The last conclusion was supported by results of Zhu *et al.* (2001) that heterosis was highly significant but hybrid performance was not when 57 rice accessions from six ecotypes and their hybrids were genotyped by 48 SSR and 50 RFLP markers. It is anticipated that prediction could be possible if heterozygosity is derived from specific marker loci that are associated with heterosis and hybrid performance and all possible associated loci have been identified and their effects and interactions clearly defined.

Considering the fact that only heterotic crosses are of commercial importance and of interest to the breeder, the practical value of the genetic distance approach for prediction of heterosis and hybrid performance is limited (Vuylsteke *et al.*, 2000). This is true for some crop species like maize. For rice, however, the reproductive barrier between the two subspecies, *indica* and *japonica*, has enforced a limitation on the utilization of *indica/japonica* heterosis, although the use of the wide-compatibility gene(s) has had a great impact on the limitation. Hybrid breeding for *indica* rice has been based on crosses within the *indica* group. The strong relationship between the heterozygosity at marker loci and heterosis within the *indica* group as reported before (Xiao *et al.*, 1996b) indicates that GD estimates based on molecular markers could be very useful in assigning *indica* cultivars into different subgroups for hybrid *indica* rice development.

Screening for heterosis-related molecular markers as suggested by Melchinger *et al.* (1990b), using specific heterozygosity proposed by Zhang *et al.* (1994) and identifying favourable combinations of allele and heterotic patterns (Liu and Wu, 1998) are among the approaches that could be exploited further to improve the prediction of hybrid performance/heterosis using molecular markers. Understanding genetic variation among cultivars to be tested and identifying markers associated with heterosis and heterosis-related traits are two important components in hybrid prediction. We should keep in mind that marker–heterosis associations identified in one cross may not be suitable for selection in others because heterosis could be controlled by many genes and each cross has different genes and gene combinations in action.

Despite their low values, the inbred-hybrid yield correlations were positive. They indicated a tendency for high-yielding inbreds to produce high-yielding hybrids. Hybrid breeding is always accompanied by the improvement of parental lines. Modern maize inbreds, grown at today's high density, can yield nearly as much as hybrids of the 1930s (Duvick, 1984; Meghi *et al.*, 1984). Duvick (1999) has suggested that if as much effort had been put into improvement of open-pollinated varieties (OPVs) as has been devoted to hybrid improvement over the years, the gap between the best hybrids and the best OPVs might be less than what it currently is. Some authors even argue that OPVs might be superior to hybrids (Lewontin and Berlan, 1990), but their assumption is not backed up by data.

The potential application of DNA markers in hybrid breeding depends very much upon whether divergent heterotic groups have been established or not and upon crop species. If well-established heterotic groups are unavailable, marker-based GD estimates can be used to avoid producing and testing crosses between closely related lines. Furthermore, crosses with inferior MPH could be discarded prior to field-testing based on prediction. Another potential application exists. If new lines of unknown heterotic pattern or inbreds developed from crosses between parents from different heterotic groups (e.g. commercial hybrids) are to be evaluated for testcross performance, GD estimates could assist the breeder in the choice of appropriate testers for evaluating the combining ability of the lines.

9.8 Opportunities and Challenges

Plant breeding has generally accounted for one-half of the increases in productivity of the major crops and the future will continue to depend on its advances. However, the rate, scale and scope of uptake of genomics in crop breeding programmes have continually lagged behind expectations. This is little different to the adoption of quantitative genetics, mechanization and computerization during the last century. This is partly due to the long product development cycle in plant breeding and in turn the long-term nature of feedback from the market regarding the impact of any changes in the cultivar development pipeline. Opportunities and challenges we are facing in MAS will be discussed in this section.

9.8.1 Molecular tools and breeding systems

The prerequisite for increasing accessibility of MAS to breeders is developing a highly efficient breeding system, particularly for resource-limited plant breeding programmes in developing countries. Several strategies can be used to establish such a system through the use of MAS, including: (i) selection at early breeding stages to eliminate most segregants, particularly for highly inheritable traits; (ii) selection at early developmental stage using high-selection pressure and an optimized selection rate, particularly for large-size plants; (iii) one-step selection for multiple traits using high-throughout genotyping; (iv) utilization of cost-effective genotyping systems; (v) highly efficient phenotyping, sample tracking and data acquisition; (vi) development and utilization of quick fixation and stabilization approaches; and (vii) genotyping once and phenotyping multiple times. To increase accessibility of MAS to breeders, the most important thing is to build skills and capacity in developing countries and to develop decision support tools to facilitate MAB programmes. Over the next decade, MAS technologies will become cheaper and easier to apply on a large scale, MAS can be carried out for all genes related to important target traits and using information from genotyping of all germplasm in the breeding system.

9.8.2 Crop-specific issues

The bottlenecks in MAS could be specific to crops except for those discussed in the previous section. For example, a possible limitation of MAS with maize is the structure and content of various gene pools. Examples of maize gene pools would include European flint and dent germplasm, US dents and various heterotic groups within each of these and other larger pools. Surveys with DNA markers have established differences among such groups of germplasm (Smith and Smith, 1992; Niebur *et al.*, 2004). In addition, the efficacy for MAS in relatively complex populations such as synthetics and OPVs has not been investigated.

For open-pollinated crops, breeding for complex traits is limited by an additional bottleneck that there is no standardized protocol available for MAS that can be automatically applied to the various breeding systems required for development of inbred, hybrid, population and synthetic cultivars, where the material at many stages in the breeding process is highly heterogeneous and highly heterozygous. This is very different from breeding systems for inbred crops such as wheat that almost always start and end with inbred lines (Koebner and Summers, 2003) and rice that may start with inbreds and ends with inbreds or inbred-based hybrids (Xu, Y., 2003). Thus, MAS efforts in open-pollinated crops can consist of two simultaneous approaches, one using the MTAs that have been identified previously and the other based on an integrated genetic diversity analysis, MTA analysis and MAS approach to discover, validate and apply new marker associations all in the same breeding populations albeit at different generations. It is a challenge to make MAS applicable from the earliest possible stages of the breeding programme while giving the flexibility to sequentially improve the power of the MAS as data accumulates and information is integrated through subsequent breeding processes.

9.8.3 Quantitative traits

Traditionally the heritability of quantitative traits was the most common predictor of genetic gains for different plant breeding methods. DNA markers may be used today to accelerate and enhance overall breeding methods by combining DNA marker and phenotyping data in a selection index. Geneticists and plant breeders need to deal with linkage disequilibrium while using MAS in recurrent selection, especially when using polymorphic markers arising from mapping populations, which tend to be from diverse parents and thus may not be relevant for target breeding materials. The power of MAS will also continue to rely heavily on the accuracy and precision of phenotyping and the characterization and evaluation of germplasm in the field. Issues such as the error term to test for the significance of a QTL, detecting small effects with narrow genetic variance, or the number of QTL not related to genetic variance or divergence of parents are all under-researched areas that need priority attention by geneticists. Addressing these issues will allow plant breeders to define the optimum number of individuals/lines and markers to be used in their MAS programmes.

Plant breeders are ready to apply MAS for quantitative traits when the genetic gain and time or cost efficiency from doing so are clearly higher than through PS methods. Initial emphasis in this area should be on traits for which a robust cost-effective phenotyping system is not available. To quickly reach this stage requires a paradigm shift in strategy among the marker–trait identification community: from efforts to identify all QTL influencing the target trait to a focus on identification of a few QTL having the largest effect on the target trait. QTL of major effect may be easier to detect (in the right genetic material) and, be less influenced by GEIs and genetic background effects. Of great importance will be a shift away from analysis of entire genetic populations to an emphasis on selected individuals with extreme phenotypes from relevant breeding populations and genetic stocks and likely, pooled DNA analysis using the selected individuals (Xu and Crouch, 2008). Of equal importance will be a shift from linked markers to diagnostic gene-based markers, which will generally be SNP-based and thus readily scalable for high-throughput haplotyping.

9.8.4 Genetic networks

The potential for MAS to contribute to improvements in crops should increase in parallel with our understanding of the relationships among genomes, the environment and phenotypes. Candidate transgenes will be developed on a regular basis and their contributions to crop improvement will be realized in the most efficient manner with MAS. Likewise, the identification of candidate native genes and their gene products and functions and of other DNA sequences (e.g. micro-RNA (miRNA), matrix attachment and regulatory regions), will improve the power of methods such as association mapping and genome scans to assess their genotypic value in the context of defined reference populations of significance to plant breeding.

Plants exhibit massive changes in gene expression during morpho-physiological and reproductive development as well as when exposed to a range of biotic and abiotic stresses. A new field of genetics of global gene expression has emerged based on the application of traditional techniques of linkage and association analysis for the thousands of transcripts measured by microarrays. Dissecting the architecture of quantitative traits in this way connects DNA sequence variation with phenotypic variation and is improving our understanding of transcriptional regulation and regulatory variation (Rockman and Kruglyak, 2006).

9.8.5 Marker-assisted selection in developing countries

There are many additional factors that will affect the application of MAS in developing countries. Building the necessary skills among national programme staff and ensuring those programmes have possession or access to sufficient capacity is an essential prerequisite.

Several crop-specific biotechnology networks have been established in Asia, Africa and Latin America during the 1980s and 1990s. Many of these covered a wide range of activities including upstream research and capacity building. Unfortunately, in some cases major donors have pulled out from further funding of such networks. However, all these networks still present an excellent basis for the development of molecular breeding communities of practice that can be used to validate, refine and apply new technologies in national breeding programmes. Conversely in other crops, conventional breeding networks have sufficiently matured to become prime candidates for the introduction of MAS systems and other molecular breeding approaches. However, many of these breeding programmes are not receiving international development assistance or are significantly under-funded, which seriously threatens their long-term impact. Molecular breeding consortia accessing joint venture genotyping hubs or commercial service providers appear to be an increasingly realistic option where those facilities can provide the right quality, quantity and timeline of service to fit the given breeding system.

Capacity building will upgrade the skills of participating plant breeders and improve the understanding of plant breeding and associated molecular technologies among the broader community. As many molecular techniques become sufficiently routine, there will be many opportunities for scientists to profitably shift their attention to experimental design, analysis and interpretation – as opposed to their current predominant time contribution to data generation.

Policy options for research, development and diffusion of the products of MAS in developing countries depend on the development objectives and priorities of the agricultural sector, its various subsectors and cross-cutting activities dealing with science and technology. Dargie (2007) discussed the policy considerations and options for developing and implementing MAS programmes and projects for developing countries. He considered three categories of countries with different capacities of facilities and personnel. (i) Countries with high-quality personnel and facilities for phenotypic evaluation and selection and in molecular biology: through the establishment of centralized centres of excellence and sectoral/subsectoral institutions, they have the potential to develop and validate molecular markers and apply MAS routinely. (ii) Countries with reasonable capacities for phenotype evaluation and selection and some capacities to apply molecular marker methods: these countries have less comprehensive breeding programmes and therefore can cover fewer species. Using regional centres of excellence, such as Bioscience eastern and central Africa (BecA), is an option to implement MAS in their breeding programmes. (iii) Countries with limited capacities in phenotypic evaluation and selection and no capacities to apply molecular techniques: their options are to partner with institutions of the CGIAR system and other advanced institutions in developed and developing counties and import cultivars and advanced breeding lines developed by these institutions through MAS that contain the needed traits. Establishment of molecular breeding community of practice will help upgrade scientific and technical expertise in molecular biology itself and in linking molecular and phenotypic approaches through species and theme-specific networks, workshops, training courses, scientific visiting, etc., to implement MAS across these types of developing countries.

10

Genotype-by-environment Interaction

Genotype is defined as an individual's genetic make-up – the nucleotide sequence of DNA that is transmitted from parents to offspring as discussed in Chapter 2. The phenotypic expression of a genotype depends on environments that may be defined as the sum total of circumstances surrounding or affecting an organism or a group of organisms. Cultivars of a crop as genotypes, when grown under a wide range of conditions, are exposed to different soil types, fertility levels, moisture contents, temperatures, photoperiods, biotic and abiotic stresses and cultural practices. As gene expression may be modified, enhanced, silenced, or timed by the regulatory mechanisms of the cell in response to internal and external factors, the genotypes (cultivars) may specify a range of phenotypic expression that are called the norm of reaction, or plasticity, which is simply the expression of variability in the phenotype of individuals of identical genotype (Bradshaw, 1965). As a result, one cultivar may have the highest yield in some environments and a second cultivar may excel in others. Changes in the relative performance of genotypes across different environments are referred to as genotype-by-environment interaction (GEI). GEI must be explained from the environmental part, from the genotypic part and from both simultaneously. Environmental characterization using data from Geographic Information System (GIS) tells us the actual environmental factors such as maximum temperature, minimum temperature, precipitation, sun radiation, etc.

GEI can rise due to changes in the genotype, the environment, or both. It is ubiquitous, occurring for virtually every aspect of plant growth and development and touching every discipline of biological science. A large proportion of biological research in agricultural science is concerned with study of GEI. Scientists are increasingly aware that much scientific inference is conditional because of GEI. This awareness has led to greater interest and therefore advances, in our understanding of the factors influencing plant growth and development. In turn, there has been considerable improvement in the performance of many crop species (Cooper and Byth, 1996). Nevertheless, we are far from developing an adequate understanding of the factors influencing adaptation, even for our major agricultural species. Consequently, there is considerable opportunity for improvement in strategies of plant breeding.

The relative performance of genotypes across environments determines the importance of an interaction. There is no GEI when the relative performance among genotypes remains constant across environments. In Fig 10.1a, cultivar A has the same yield superiority over cultivar B across two environments (E1 and E2). No GEI is present

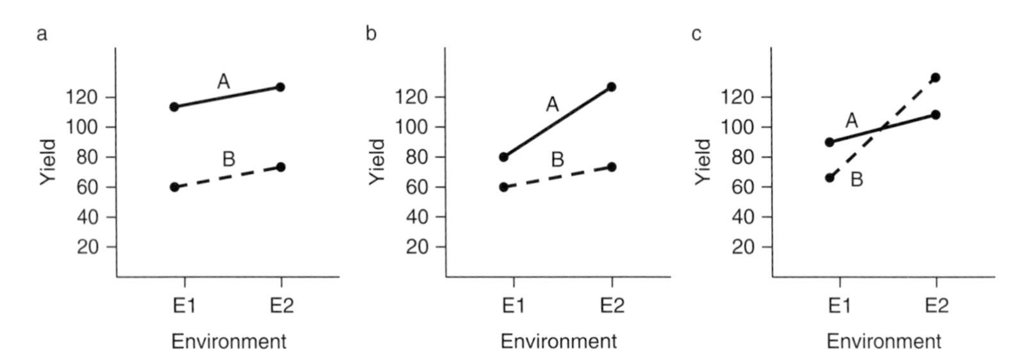

Fig. 10.1. The relative performance of two cultivars (A and B) in two environments (E1 and E2). (a) No GEI is present. (b) GEI is present but does not alter genotypic ranking. (c) GEI is present and alters genotypic ranking. Modified from Allard and Bradshaw (1964).

because the yield differential between the cultivars is 50 units in both environments – proportionality is maintained, that is, the difference between any two genotypes in any two environments is the same. GEIs can occur in two ways. (i) The difference among genotypes can vary without any alternation in their rank, which is referred to as non-crossover interaction. In Fig 10.1b, a GEI is present because cultivar A yields 20 units more than cultivar B in environment E1 but 50 units more in environment E2. (ii) The rank among cultivars change across environments, which is referred to as crossover interaction (COI). In Fig 10.1c, cultivar A is more productive in environment E1, but cultivar B is more productive in environment E2. The most important GEI for the plant breeder is the COI caused by changes in rank among genotypes.

Existence of GEI has significant influence on the efficiency of crop improvement via plant breeding, largely because they confound comparisons among genotypes with the environment of test and complicate the definition of breeding objectives. It is argued that to overcome these constraints to crop improvement we need to develop an understanding of the differences in plant adaptation associated with the differences in performance and in particular the GEI. GEI is of interest to plant breeders for several reasons (Fehr, 1987). (i) The need to develop cultivars for specific purposes is determined by an understanding of GEI. Unique cultivars may be required for different row spacings, soil types or planting dates. (ii) The potential need for unique cultivars in different geographical areas requires an understanding of GEI. The importance of this interaction can determine if division of a large geographical area into subareas is needed and justified for testing new genotypes and recommending cultivars to crop producers. (iii) Effective allocation of resources for testing genotypes across locations and years is based on the relative importance of genotype × location, genotype × year and genotype × location × year interactions. (iv) The response of genotypes to variable productivity levels among environments provides an understanding of their stability of performance. An understanding of the genotype stability across environments helps in determination of their suitability for the fluctuations in growing conditions that are likely to be encountered.

There are several key areas in GEI study: (i) methodology for effective environmental characterization and classification; (ii) strategies for partitioning GEIs into repeatable and non-repeatable components; (iii) experimental evidence to quantify the relative efficiencies of direct selection for target traits and indirect selection strategies based on crop physiological principles; (iv) integrated utilization of multi-environment trial data, pedigree information and genotypic data of cultivars; and (v) determination of genetic loci responsible for GEI and molecular dissection of GEI components. Discussion in this chapter is mainly based on several important references including Feher (1987), Romagosa and Fox (1993), Knapp (1994), Xu and Zhu (1994), Cooper and Hammer (1996), Bernardo (2002), Chahal and Gosal (2002), Kang (2002), Crossa *et al.* (2004), Cooper *et al.* (2005), van Eeuwijk *et al.* (2005) and Yan *et al.* (2007).

10.1 Multi-environment Trials

A major objective in plant breeding programmes is to assess the suitability of individual crop genotypes for agricultural purposes across a range of agro-ecological conditions. Appropriate experimental procedures are required to understand and determine the importance of GEI. For this purpose breeders conduct so-called multi-environment trials (METs). In a MET, a set of genotypes is evaluated across a number of environments that hopefully represent the target environment to select widely or specifically adapted genotypes. As an example, Table 10.1 provides a MET data set for 18 winter wheat cultivars tested at nine Ontario locations in 1993 from Yan *et al.* (2007). The performance of genotypes in METs is analysed by statistical models developed to describe and interpret genotype-by-environment data (GED). The statistical analysis should provide estimates for parameters that indicate both how well genotypes perform on average across the environmental range and how well they perform in specific environmental conditions.

10.1.1 Experimental design

An understanding of the steps involved in the design, implementation, analysis and interpretation of METs can be useful. Planning of any experiment begins with a statement of the concept or hypothesis to be evaluated, sometimes phrased in the form of a question. Is the relative performance among genotypes different with conservation tillage versus conventional tillage? Do genotypes respond differently to high versus low rates of inorganic nitrogen fertilization? The breeder may have a hypothesis about the answer to the question on the basis of practical experience. It is critical that the hypothesis should not be regarded as factual, an attitude that can bias the interpretation of the experimental results. A MET that involves multiple genotypes, years and locations is usually required. The GEI is considered to be absent if all genotypes perform similarly across all the environments, i.e. total variation is explained only by main effects of environments and genotypes.

The empirical mean response, $\bar{y}_{ij}$, of the ith genotype ($i = 1,2,\ldots,I$) in the jth environment ($j = 1,2,\ldots,J$) with r replications in each of the $I{\times}J$ cells is expressed as

$$\bar{y}_{ij} = \mu + \tau_i + \delta_j + (\tau\delta)_{ij} + \bar{\varepsilon}_{ij} \qquad (10.1)$$

where μ is the grand mean over all genotypes and environments, τ_i is the additive effect of the ith genotype, δ_j is the additive effect of the jth environment, $(\tau\delta)_{ij}$ is the non-additivity, GEI, of the ith genotype in the jth environment and $\bar{\varepsilon}_{ij}$ is the (average) error assumed normally and independently distributed, i.e. $NID(0, \sigma^2/r)$, where σ^2 is the within-environment error variance, assumed to be constant.

Except for μ, all the terms in Eqn 10.1 are usually treated as random effects. To provide a complementary framework for a genetic interpretation of the observed trait variation, we can also consider the trait phenotypic variation as the combination of a 'genetic signal' component $[\tau_i + (\tau\delta)_{ij}]$, an 'environmental context' component (δ_j) and an 'environmental noise' component ($\bar{\varepsilon}_{ij}$). For the variance-covariance (VCOV) structure of the error term, $\bar{\varepsilon}_{ij}$, various choices are possible, the simplest being that $\bar{\varepsilon}_{ij}$ is independently identically normally distributed. The terms of Eqn 10.1 can also be considered as fixed effects depending on the sampling methods used and the general purpose of the study. For example, if 'environment' refers to locations, then they may be considered a fixed effect when they are not randomly chosen from all possible sites in an area, while if the environment

Table 10.1. Mean yield (Mg ha^{-1}) of 18 winter wheat cultivars (G1–G18) tested at nine Ontario locations (E1–E9) in 1993 (from Yan *et al.* (2007) with permission).

	Test environments									
Genotypes	E1	E2	E3	E4	E5	E6	E7	E8	E9	Mean
G1	4.46	4.15	2.85	3.08	5.94	4.45	4.35	4.04	2.67	4.00
G2	4.42	4.77	2.91	3.51	5.70	5.15	4.96	4.39	2.94	4.31
G3	4.67	4.58	3.10	3.46	6.07	5.03	4.73	3.90	2.62	4.24
G4	4.73	4.75	3.38	3.90	6.22	5.34	4.23	4.89	3.45	4.54
G5	4.39	4.60	3.51	3.85	5.77	5.42	5.15	4.10	2.83	4.40
G6	5.18	4.48	2.99	3.77	6.58	5.05	3.99	4.27	2.78	4.34
G7	3.38	4.18	2.74	3.16	5.34	4.27	4.16	4.06	2.03	3.70
G8	4.85	4.66	4.43	3.95	5.54	5.83	4.17	5.06	3.57	4.67
G9	5.04	4.74	3.51	3.44	5.96	4.86	4.98	4.51	2.86	4.43
G10	5.20	4.66	3.60	3.76	5.94	5.35	3.90	4.45	3.30	4.46
G11	4.29	4.53	2.76	3.42	6.14	5.25	4.86	4.14	3.15	4.28
G12	3.15	3.04	2.39	2.35	4.23	4.26	3.38	4.07	2.10	3.22
G13	4.10	3.88	2.30	3.72	4.56	5.15	2.60	4.96	2.89	3.80
G14	3.34	3.85	2.42	2.78	4.63	5.09	3.28	3.92	2.56	3.54
G15	4.38	4.70	3.66	3.59	6.19	5.14	3.93	4.21	2.93	4.30
G16	4.94	4.70	2.95	3.90	6.06	5.33	4.30	4.30	3.03	4.39
G17	3.79	4.97	3.38	3.35	4.77	5.30	4.32	4.86	3.38	4.24
G18	4.24	4.65	3.61	3.91	6.64	4.83	5.01	4.36	3.11	4.48
Mean	4.36	4.44	3.14	3.49	5.68	5.06	4.24	4.36	2.90	4.19

refers to years then they can be considered as randomly chosen. If years and locations are typically representing a normal combination of years and locations they can be perfectly considered as random effects.

The genotypes chosen for an assessment of possible interactions are an important consideration in designing the experiment. Some analyses of GEI are not based on an experiment specifically designed for that purpose, particularly the assessment of the importance of interactions with locations and years. Instead, breeders utilize data from test genotypes including cultivars, hybrids, populations and experimental lines that have been evaluated over locations and years as a part of normal testing programmes.

It is desirable to have at least two replications in each location and year to obtain an estimate of experimental error so that it is possible to test the significance of the interactions of interest. Any additional replications will allow a more reliable estimate of the experimental error. However, sometimes resources are not available for replicating all genotypes so that only some entries are replicated. In this case, it is an augmented design; it is a perfectly legitimate design, although the precision is lower.

10.1.2 Basic data analysis and interpretation

For all MET data, basic analyses should include the calculation of mean values, determination of the statistical significance of the sources of variation and estimation of appropriate variance components. The sources of variation in an experiment are partitioned into main effects and their interactions (Table 10.2). The mean squares for the sources of variation are determined and appropriate *F*-tests are conducted to assess the probability that a source of variation is significant. Components of variance can be calculated for the main effect of the genotypes and their interactions with the locations and years. Standard errors can be computed for each variance component.

Data interpretation includes the statistical significance of various variation sources

Table 10.2. Analysis of variance for experiments in an annual crop with different numbers of locations and years (from Johnson *et al.* (1955) with permission).

Sources of variation	Degrees of freedom	Expected mean squares
One location in 1 year		
Replications	$r-1$	–
Genotypes	$g-1$	$\sigma_e^2 + r(\sigma_g^2 + \sigma_{gl}^2 + \sigma_{gy}^2 + \sigma_{gly}^2)$
Error	$(r-1)(g-1)$	σ_e^2
One location in 2 or more years		
Years	$y-1$	–
Replications in years	$y(r-1)$	–
Genotypes	$g-1$	$\sigma_e^2 + r(\sigma_{gy}^2 + \sigma_{gly}^2) + ry(\sigma_g^2 + \sigma_{gl}^2)$
Genotypes × years	$(g-1)(y-1)$	$\sigma_e^2 + r(\sigma_{gy}^2 + \sigma_{gly}^2)$
Error	$y(r-1)(g-1)$	σ_e^2
One year at two or more locations		
Locations	$l-1$	–
Replications in locations	$l(r-1)$	–
Genotypes	$g-1$	$\sigma_e^2 + r(\sigma_{gy}^2 + \sigma_{gly}^2) + rl(\sigma_g^2 + \sigma_{gl}^2)$
Genotypes × locations	$(g-1)(l-1)$	$\sigma_e^2 + r(\sigma_{gy}^2 + \sigma_{gly}^2)$
Error	$l(r-1)(g-1)$	σ_e^2
Two or more locations in 2 or more years		
Years	$y-1$	–
Locations	$l-1$	–
Replications in years and locations	$yl(r-1)$	–
Years × locations	$(y-1)(l-1)$	–
Genotypes	$g-1$	$\sigma_e^2 + r\sigma_{gly}^2 + ry\sigma_{gl}^2 + rl\sigma_{gy}^2 + ryl\sigma_g^2$
Genotypes × years	$(g-1)(y-1)$	$\sigma_e^2 + r\sigma_{gly}^2 + rl\sigma_{gy}^2$
Genotypes × locations	$(g-1)(l-1)$	$\sigma_e^2 + r\sigma_{gly}^2 + ry\sigma_{gl}^2$
Genotypes × years × locations	$(g-1)(y-1)(l-1)$	$\sigma_e^2 + r\sigma_{gly}^2$
Error	$yl(r-1)(g-1)$	σ_e^2

and their practical implications. The genotype × location interaction measures the consistency of performance among genotypes at different locations. The consistency of performance of genotypes in different years is indicated by the genotype × year interaction. The genotype × location × year interaction measures the consistency of the genotype × location interaction across years. For all of these mentioned interactions, an examination of mean values is necessary to determine if a significant interaction is due to a change in rank among genotypes or to changes in the differences among genotypes without rank change (ref. Fig. 10.1).

Genotype × location interaction

Wide fluctuations in the rank of genotypes across test locations suggest that it may be desirable to develop genotypes for different locations through independent selection and testing programmes. The cost of establishing independent programmes for different geographical areas is substantial; therefore, the decision can be difficult. Before establishing independent breeding programmes, the breeder should make a detailed examination of the environmental factors responsible for the genotype × location interaction. As suggested by Fehr (1987), if the differences among locations are due to soil type or other factors that are consistent from year to year, independent programmes may be appropriate. Temporary differences among locations associated with unusual climate conditions would not justify this.

Another consideration in determining the implications of genotype × location interaction is that fluctuations in rank may not preclude selection of superior genotypes for multiple locations. Assume that a group of genotypes are divided into three classes: good, intermediate and poor. A genotype ×

location interaction could be caused by fluctuations in rank among genotypes within the three classes, but not among classes. Such an interaction would be unlikely to justify the establishment of breeding programmes for independent locations, at least for the initial stages of testing.

Genotype × year interaction

An inconsistent ranking among genotypes grown in different years is in some regards more difficult to deal with than a genotype × location interaction. A breeder does not have the option of establishing independent breeding programmes for different years (Fehr, 1987). The primary option available is to identify genotypes that exhibit superior performance on the average across years. This involves the testing of genotypes in several years before selection of one for release as a cultivar. To reduce the length of time for genetic improvement, multiple locations in 1 year often are used as a substitute for years. The substitution is only effective when the range of climate conditions among locations in single years is comparable to that among years.

Genotype × year × location interaction

This interaction can first be used to test if the genotype × location interaction is repeatable across years and thereby mega-environments can be established. It can be used secondly when there are fluctuations in the ranking of genotypes associated with individual location–year combinations. Here the breeder must identify genotypes with superior average performance over locations and years. When METs are performed across several years, the interaction is referred to as a three-mode (three-way) data array, in which the modes are genotypes, locations and years. By extension of a two-way additive main effect and multiplicative interaction (AMMI) mode to a three-way mode, Varela *et al.* (2006) offered us a natural approach for assessing the response in locations and years or for studying the multi-attribute response of genotypes in environments. The three-way mode was applied to two data sets. Data set 1 comprised genotype (25) × location (4) × sowing time (4) interaction with eight traits measured. The structure of data set 2 is genotype (20) × irrigation regimes (4) × year (3) on grain yield. Their results showed that the three-way AMMI analysis gave sensible and useful information that have otherwise been unavailable to the breeder in relation to the differential responses of genotypes in different locations and in several years and the different relationship between locations in different years.

10.2 Environmental Characterization

The objectives of GED analysis (i.e. MET data for a single trait) should include three major aspects: (i) mega-environment analysis; (ii) test-environment evaluation; and (iii) genotype evaluation (Yan and Kang, 2003), all of which are associated with environmental characterization. Yan *et al.* (2007) use the yield data listed Table 10.1 as an example to illustrate the three aspects of bi-plot analysis. When supplemental information (e.g. data on environmental or genotypic covariates) is available, a fourth aspect, which is to understand the causes of genotype main effect (G) and GEI (Yan and Kang, 2003; Yan and Tinker, 2006) can be included as described in Section 10.3.2.

Environmental characterization involves definition of the key factors which influence both performance level and the relative performance of genotypes in an experiment, as well as assessment of the relevance of these factors to the target environments. This provides a basis for understanding the results from individual experiments and predicting their application to elsewhere. This can be extended to include the sociological factors that influence the utilization of cultivars by farmers. In the majority of METs conducted there is no clear definition of the environmental challenge. Further, in many METs there is no measurement of how well the test environments match those of the target environments. Different categories of sites may be identified to assist environmental characterization. Benchmark sites, which are intensively monitored, could be useful for gaining an understanding of the mixture of environment encountered in the production system. Cooper and Hammer (1996) listed three strategies for environmental characterization: (i) direct measurement of environmental variables during an experiment, which is possible but is both time consuming and resource intensive and, therefore, costly; (ii) quantitative analytical approaches based on statistical methods and simulation models; and (iii) utilization of reference and probe genotypes for specific environmental factors. It should be noted that statistical methods discussed in the next section can also be used for environmental characterization.

10.2.1 Classification of environments

Every factor that is a part of the environment has the potential to cause differential performance, i.e. GEI. Environmental factors can be classified as either predictable or unpredictable factors (Allard and Bradshaw, 1964). Predictable factors are those that occur in a systemic manner or are under human control, such as soil types, planting dates, plant densities, fertilizer rates, tillage practices, crop rotation patterns. Unpredictable factors are those that fluctuate and cannot be artificially controlled, including rainfall, temperature and relative humidity. From an applied perspective it may also be useful to distinguish between environmental factors that can be manipulated by the farmer and those that cannot. In METs, management factors are generally considered as part of the environment and often are not explicitly distinguished from climatic, temporal and regional factors. They are considered to be sampled in combination with the different locations and years in the METs. However, wherever possible plant breeders can and often do separate, out of the pool of things called environment, factors identified to be repeatable and important in the target environments. Predictable factors can be evaluated individually and collectively for their interaction with genotypes. Studies have been made of genotype × soil type, genotype × row spacing, genotype × planting date, genotype × plant population and genotype × fertilization interactions.

To maximize grower's yields, the growing region often has to be subdivided into relatively homogeneous mega-environments and appropriate genotypes deployed for each of these mega-environments. A mega-environment is defined as a portion (not necessarily contiguous) of a crop species' growing region with a fairly homogeneous environment that causes similar genotypes to perform best. For several reasons, identifying mega-environments has attracted much attention (Gauch and Zobel, 1997). First, interest has grown in providing adapted materials for marginal environments which are stressed in a variety of ways that typically generate large GEIs so genotypes that win in highly favourable environments may rank poorly there. Secondly, greater concern about long-term soil conservation and about lowering pesticide and fertilizer usages has stimulated a greater diversity of management practices, hence creating a variety of mega-environments within major crop regions previously managed in a much more uniform manner. Finally and more generally, most plant breeders feel that they are exploiting rather than ignoring the potential for yield increases that resides in GEIs. Genotype evaluation and test-environment evaluation become meaningful only after the mega-environment issue is addressed.

Mega-environments are broad, usually international and frequently transcontinental, which can be defined by similar biotic and abiotic stresses, cropping system requirements, consumer preferences and, for convenience, by a volume of production of the relevant crop sufficient to justify its attention. For example, 'tropical lowland, late-maturing, white dent' maize with relevant disease resistances occupies 3.8 million ha across 18 countries (Gauch and Zobel, 1997). This definition encompasses environmental, genotypic, geographical and even economic aspects of mega-environments. Subdivision of a crop's growing region into several mega-environments implies more work for plant breeders and seed producers, but it also implies higher heritabilities, faster progress for plant breeders and potentially stronger competitiveness for seed producers. On the other hand, the necessary and sufficient condition for mega-environment division is a repeatable which-won-where pattern rather than merely a repeatable environment-grouping pattern (Yan and Rajcan, 2002; Yan and Kang, 2003). Mega-environments are used to allocate resources in a breeding or research programme, to rationalize germplasm and information exchanges between breeding programmes (allowing even small programmes to progress by focusing on the most promising material), to increase heritabilities within relatively well-defined and predictable environments, to increase the efficiency of testing and breeding programmes and to target genotypes to appropriate production areas. Many other terms, however, have essentially the same meaning, such as agro-climatic or eco-geographic regions.

Appropriate mega-environment analysis should classify the target environment into one of three possible types (Table 10.3). Type 1 is the easiest target environment one can hope for, but it is usually an overoptimistic expectation. Type 2 suggests opportunities for exploiting some of the GEI. Such opportunities should not be overlooked if they exist, which is the whole point of mega-environment analysis and GEI analysis. Type 3 is the most challenging target environment and, unfortunately, also the most common one. Statistical methods for grouping environments involves classification procedure, ordination procedures (i.e. using coordinates in a graph to depict relationships among environments), or the joint use of a classification and an ordination procedure (DeLacy and Cooper, 1990).

Clustering analysis, which can be used for environment classification, usually involves the creation of hierarchical groups of environments just as the same as described for germplasm classification in Chapter 5. A given environment is more similar to an environment in the same cluster than to an environment in a different cluster, in terms of genotype rankings, rather than the physical factors of the environments per se. The clustering procedure requires some measure of dissimilarity or distance between environments. Data from METs are typically unbalanced because the genotypes and locations often vary from year to year. But the statistical distance between two environments can be determined from the performance of the subset of genotypes that are grown in both environments (DeLacy and Cooper, 1990). The distance measures for genotypes summarized by Lin *et al.* (1986) can be used for environment clustering, whereas Ouyang *et al.* (1995) measured the distance between j and j' as

Table 10.3. Three types of target environments based on mega-environment analysis (from Yan *et al.* (2007) with permission).

	With crossover GEI	No crossover GEI
Repeatable across years	Type 2: target environment consisting of multiple mega-environments. Strategy: select specifically adapted genotypes for each mega-environment. A single year multilocation trial may be sufficient.	Type 1: target environment consisting of a single, simple mega-environment. Strategy: test at a single test location in a single year suffices to select for a single best cultivar.
Not repeatable across years	Type 3: target environment consisting of a single but complex mega-environment. Strategy: select a set of cultivars for the whole region based on both mean performance and stability based on data from multiyear and multilocation tests	

$$D_{jj'} = \frac{1}{g}\sum_{i=1}^{g}\left(\frac{P_{ij}-\mu_j}{s_j} - \frac{P_{ij'}-\mu_{j'}}{s_{j'}}\right)^2$$
$$= \frac{1}{g}\sum_{i=1}^{g}(t_{ij}-t_{ij'})^2 \qquad (10.2)$$

where g is the number of genotypes grown in both j and j'; μ_j (or $\mu_{j'}$) is the mean of all the genotypes in environment j (or j'); and s_j ($s_{j'}$) is the phenotypic standard deviation among all the genotypes in environment j (or j'). When all genotypes grown in environment j are also grown in environment j', Eqn 10.2 can be rewritten as (Ouyang *et al.*, 1995)

$$D_{jj'} = 2\left(1-\frac{1}{n}\right)\left(1-r_{jj'}\right) \qquad (10.3)$$

where $r_{jj'}$ is the correlation between environment j and environment across genotypes. This implies that the distance between two environments is $D_{jj'} = 0$ if the performance of genotypes are perfectly correlated in the two environments, i.e. $r_{jj'} = 1$. In contrast, the distance approaches $D_{jj'} = 2$ if $r_{jj'} = 0$. The distance approaches a maximum of $D_{jj'} = 4$ when COIs occur and $r_{jj'}$ approaches −1.

As described in Chapter 5, several methods are available for joining individual environments and environment clusters together to form a (new) cluster on the basis of $D_{jj'}$. A common procedure is the average linkage method, also called the unweighted pair-group method using arithmetic averages (UPGMA) method, in which the distance between two clusters is equal to the average distance between an environment in the first cluster and an environment in the second cluster. The cluster diagram or dendrogram is used to graphically illustrate the groups of environments (Fig. 10.2). The cluster diagram indicates the hierarchical clustering of environments and the average distances at which they are joined. The clusters based on $D_{jj'}$ from METs are often consistent with geographical groupings (Bernardo, 2002). For example, Ouyang *et al.* (1995) partitioned the 90 counties in Iowa on the basis of the performance of seven maize hybrids grown in a total of 2006 environments. Cluster analysis partitioned the counties into a northern group and a southern group, although two south-eastern Iowa counties were clustered with the northern Iowa group (Fig. 10.2). The north–south groups are consistent with

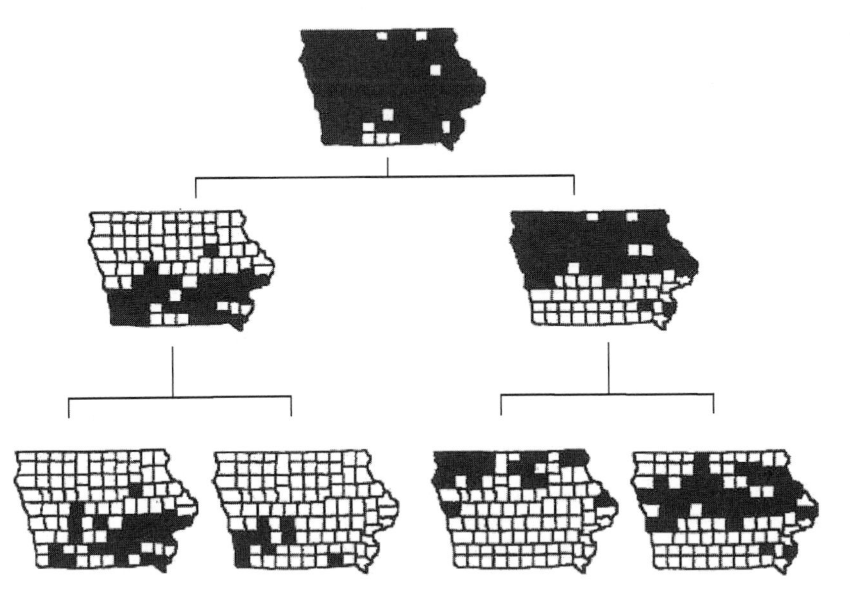

Fig. 10.2. Cluster analysis of Iowa counties. Adapted from Ouyang *et al.* (1995); original figure provided by Rex Bernardo.

differences in days to maturity between the high altitudes and the low altitudes. The Iowa counties were further subdivided into a south-eastern cluster, a south-western cluster, a northern cluster and a central cluster.

In the which-won-where view of the genotype main effect (G) plus genotype-by-environment interaction (GGE) bi-plot (Fig. 10.3) based on the data in Table 10.1, the nine environments fell into two sectors with different winning cultivars. Specifically, G18 was the highest yielding cultivar in E5 and E7 (but only slightly higher than several other cultivars with markers in close proximity to G18) and G8 was the highest yielding cultivar in the other environments. This crossover GE suggests that the target environments may be divided into different mega-environments.

The effectiveness of a cultivar evaluation system largely depends on the genetic correlation between genotype performance in METs and in the target population of environments (TPE). Plant breeders have favoured classifications based on the similarity of cultivar discrimination in trials. However, these efforts frequently fail to provide adequate assessments of the TPE, since they require long-term performance data, which are not normally collected due to high cost. To describe the TPE, Löffler *et al.* (2005) performed crop simulations for each US Corn Belt Township for the period 1952–2002, using standard Crop Environment Resource Synthesis (CERES)-Maize model inputs. To classify METs, input data were collected at or near the trial sites. Grain yield and biotic stress data for model confirmation were collected from 18 hybrids grown in replicated trials in 266 environments in 2000–2002. On the basis

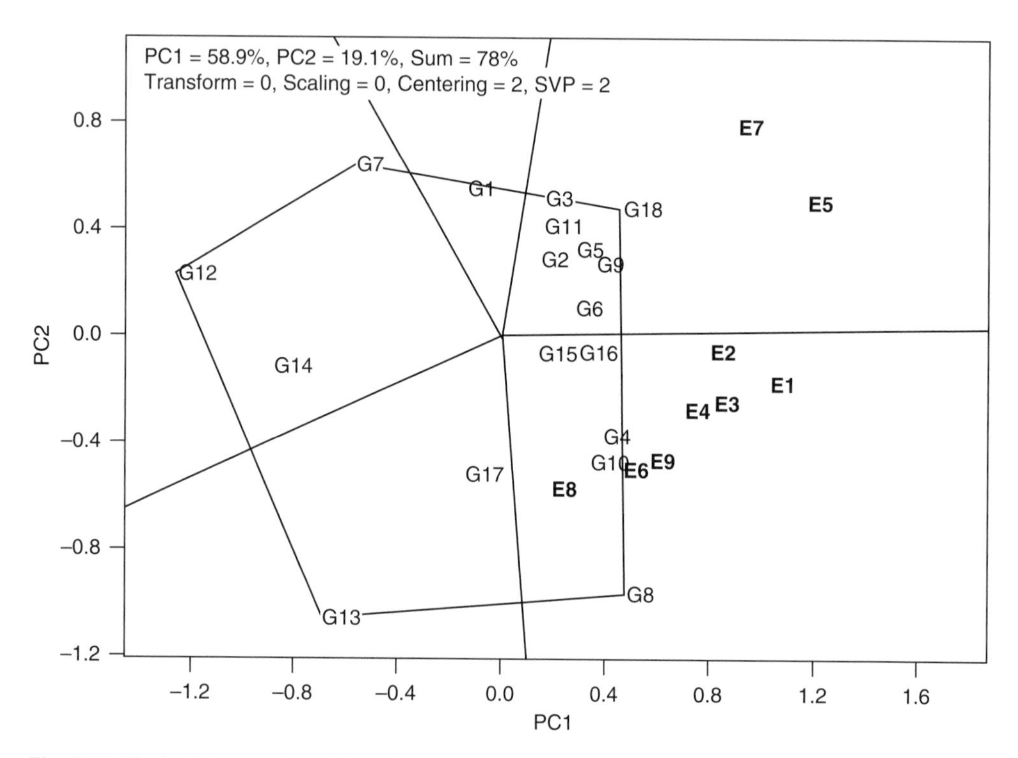

Fig. 10.3. The 'which-won-where' view of the GGE bi-plot based on the G × E data in Table 10.1. The data were not transformed ('Transform = 0'), not scaled ('Scaling = 0'), and were environment-centred ('Centering = 2'). The bi-plot was based on environment-focused singular value partitioning ('SVP = 2') and therefore is appropriate for visualizing the relationships among environments. It explained 78% of the total G + GE. The genotypes are labelled as G1–G18 and the environments are labelled as E1–E9. From Yan *et al.* (2007) with permission. PC, principal component.

of prevailing conditions during key growth stages and observed patterns of GEI, six major environment classes (EC) were identified. The relative frequency of each EC varied greatly from year to year and significant hybrid × EC interaction variance was observed. This environmental classification system provided a useful description of some of the features of both the TPE and the MET. Knowledge of the spatial (locations) and temporal (years) distributions of ECs that influence the incidence of GEI can be used to improve cultivar performance predictability in the US Corn Belt TPE.

Subdivision of a crop's growing regions into several mega-environments could be avoided if genotypes could be found with yield superiority throughout the region, that is, cultivars bred in favourable environments would also perform best in different or unfavourable environments. However, one can hardly expect a single cultivar or a hybrid to flourish the world over, under all environments and management practices. A cultivar planted outside its mega-environment frequently suffers yield reductions. Furthermore, even if the breeding goal is wide adaptation (rather than mega-environment directed breeding), it would still be the best strategy to identify several mega-environments and place a test location in each to select wide adaptation. It has been a normal practice that multinational breeding companies have their programmes established to target specific eco-geographic regions.

10.2.2 GIS and environment characterization

Modern plant breeding programmes increasingly use information from different sources, including geographic information provided by GIS (http://www.gis.com). GIS integrates hardware, software and data for capturing, managing, analysing and displaying all forms of geographically referenced information. GIS allows us to view, understand, question, interpret and visualize data in many ways that reveal relationships, patterns and trends in the form of maps, globes, reports and charts. A GIS can be viewed in three ways. *The Database View*: a GIS is a unique type of database of the world – a geographic database (geodatabase). It is an 'information system for geography'. Fundamentally, a GIS is based on a structured database that describes the world in geographic terms. *The Map View*: a GIS is a set of intelligent maps and other views that show features and feature relationships on the earth's surface. Maps of the underlying geographic information can be constructed and used as 'windows into the database' to support queries, analysis and editing of the information. *The Model View*: a GIS is a set of information transformation tools that derive new geographic data sets from existing data sets. These geo-processing functions take information from existing data sets, apply analytic functions and write results into new derived data sets. To utilize GIS data more effectively, several software packages have been developed. ESRI software offers scalable solutions for researchers at National Agricultural Research Services (NARS), universities and international research centres. From field-based products like ArcPad to the server level Spatial Database Engine (ArcSDE), data can be collected and managed. The Internet Map Server (ArcIMS) allows research sites separated by great geographical distances to be connected in real time and ArcGIS provides all of the necessary tools to analyse the spatial components of agricultural data sets.

There is a growing need to classify production environments by combining biophysical criteria with socio-economic factors. Geospatial technologies, especially GIS, are playing a role in each of these areas and spatial analysis provides unique insights. Use of GIS to characterize wheat production environments is described by Hodson and White (2007) by drawing from examples at the International Maize and Wheat Improvement Center (CIMMYT). Since the 1980s, the CIMMYT wheat programme has classified production regions into mega-environments based on climatic, edaphic and biotic constraints. Advances in spatially disaggregated data sets and GIS tools allow mega-environments to be characterized and mapped in a much more quantitative manner. The combination of improved crop distribution data and key biophysical data at high spatial resolutions also permits exploring scenarios for disease epidemics, as illustrated for the stem rust race Ug99. Availability of spatial data describing future climate conditions may provide insights into potential changes in wheat production environments in the coming decades. Increased availability of near real-time daily weather data derived from remote sensing should further improve characterization of environments, as well as permit regional-scale modelling of dynamic processes such as disease progression or crop water status. Below are some examples where plant breeding research is benefiting from implementing a spatial aspect to environment characterization.

The first example is to use GIS parameters in grouping sites to ensure that breeders choose as many variable sites as possible to represent the target region. The present mega-environments in the Southern African Development Community (SADC) countries are confounded within each country, which limits the exchange of germplasm among them. A study was undertaken to revise and group similar maize-testing sites across the SADC countries that are not confounded within each country (Setimela *et al.*, 2005). The study was based on 3 years (1999–2001) of regional maize yield trial data and GIS parameters from 94 sites. Sequential retrospective (Seqret) pattern analysis methodology was used to stratify testing sites and group them according to their similarity and dissimilarity based on mean grain yield. The methodology used historical data, taking into account imbalances of data caused by changes over locations and years, such as additions and omission of genotypes and locations. Cluster analysis grouped regional trial sites into seven mega-environments, mainly distinguished by GIS parameters related to rainfall, temperature, soil pH and soil nitrogen with an overall $R^2 = 0.70$. This analysis can reveal challenges and opportunities to develop and deploy maize germplasm in the SADC region faster and more effectively.

The second example is to use GIS parameters to determine the *Striga*-prone areas in Africa. *Striga* is an obligate parasitic weed that attacks cereal crops in sub-Saharan Africa. In western Kenya, it has been identified by farmers as their major pest problem in maize. A new technology, consisting of coating seed of imidazolinone resistant (IR) maize cultivars with the imidazolinone herbicide, imazapyr, has proven to be very effective in controlling *Striga* on farmer fields. To help extension agents and seed companies to develop appropriate strategies, the potential for this technology was analysed by combining different data sources into a GIS (De Groote *et al.*, 2008). Superimposing secondary data, field surveys, agricultural statistics and farmer surveys made it possible to clearly identify the *Striga*-prone areas in western Kenya. By extrapolation over the maize area in the zone, total potential demand for IR-maize seed is estimated at 2000–2700 t year^{-1}. Similar calculations, but based on much less precise data and expert opinion rather than farmer surveys or trials, gives an estimate of the potential demand for IR-maize seed in Africa as 153,000 t year^{-1}.

The third example is to classify maize growing environments based on drought related parameters. GEIs in southern African maize growing environments result from factors related to maximum temperature, season rainfall, season length, within-season drought, subsoil pH and socio-economic factors that result in sub-optimal input application. The difficulty of choosing appropriate selection environments has restricted breeding progress for abiotic stress tolerance in highly variable target environments. Bänziger *et al.* (2006) applied cluster analysis to the most prominent GEIs and grouped trial sites into eight mega-environments mainly distinguished by season rainfall, maximum temperature, subsoil pH and N application. GIS information available for season rainfall, maximum temperature and subsoil pH (Hodson *et al.*, 2002) was used to map maize mega-environments (Table 10.4; Fig. 10.4). Classification by maximum temperature distinguished different elevations: mega-environments A–E corresponding to the mid-altitudes; mega-environments F

Table 10.4. Characteristics of maize mega-environments in southern Africa as identified through sequential retrospective pattern analysis of multi-environment trials (reprinted from Bänziger *et al.* (2006) with permission from Elsevier).

Maize mega-environment	Maximum temperature (°C)	Season precipitation (mm)	Subsoil pH (water)	Area in southern Africa (10^3 ha)	Area in southern Africa (%)
A	24–27	> 700	< 5.7	46,282	18.2
B	24–27	> 700	> 5.7	28,826	11.4
C	24–30	< 700		48,291	19.0
D	27–30	> 700	< 5.7	17,166	6.8
E	27–30	> 700	> 5.7	49,589	19.6
F	> 30	> 700		17,146	6.8
G	> 30	< 700		38,403	15.1
H	< 24			7,897	3.1

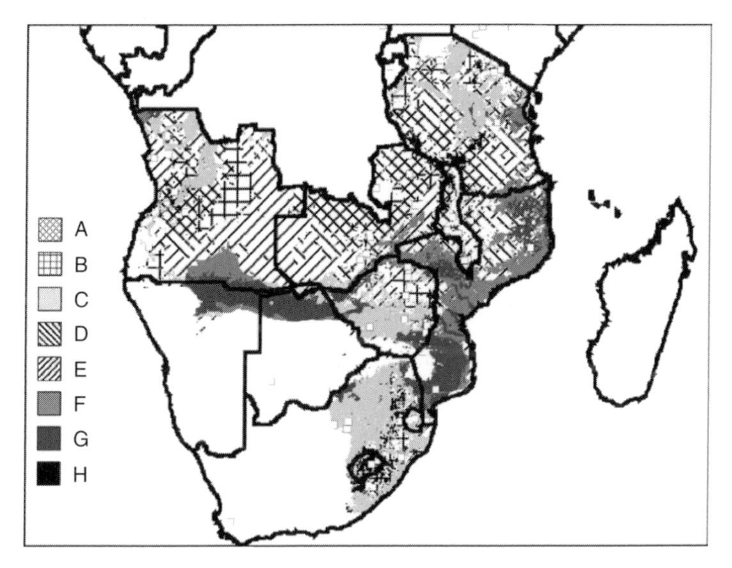

Fig. 10.4. Maize mega-environments in southern Africa delineated by combinations of maximum temperature, season precipitation and subsoil pH. Table 10.4 gives details of the eight environments A–H. White areas with rainfall < 400 mm were excluded from the analysis. Squares indicate trial sites used for defining mega-environments. Climatic and edaphic data were from Hodson *et al.* (2002). Reprinted from Bänziger *et al.* (2006) with permission from Elsevier.

and G to the lowlands; and mega-environment H to the highlands. However, they also seem to be related to disease incidence, with leaf diseases such as *Cercospora zeae-maydis*, *Puccinia sorghi* and *Exserohilum turcicum* the most prevalent in mega-environments A and B, downy mildews occurring in mega-environments F and G and *Puccinia polysora* and *Helminthosporium maydis* likely occurring mostly in mega-environment F. Most trials with suboptimal N application clustered with trials in mega-environments D and E, which may be indicative of less fertile soil types in those areas or may simply be coincidental.

10.2.3 Selection of locations for testing

The purpose of test-environment evaluation is to identify test environments that effectively identify superior genotypes for a mega-environment. An 'ideal' test environment should be both discriminating of the genotypes and representative of the mega-environment. The selection of locations for the evaluation of a quantitative character involves a number of considerations. Locations generally are chosen to represent the area where a new cultivar is to be grown commercially. The cost of transporting machinery and personnel may influence the distance of a location from the main research centre, when the testing is largely based on a mechanized system. The availability of suitable land may be a factor when the size of the test area is large. The test environments should be evaluated for being, or not being, representative of the target environment and for their power to discriminate among genotypes.

A primary consideration in site selection is the diversity of environments that can be obtained within a year. This is particularly important when widely adapted cultivars are desired. A breeder will attempt to test at locations that have environments as diverse as those that would be encountered at one location in 2 or more years (Fehr, 1987). Selection of locations for testing can be based on analysis of variance (ANOVA), correlation and cluster analyses as those used for selection and evaluation of genotypes tested in METs as will be discussed in the next section.

Developing specific cultivars for each subregion of a target region, instead of widely adapted cultivars, may exploit positive genotype-by-location (GL) interactions to increase crop yields. With reference to the Algerian durum wheat (*Triticum durum* Desf.) region, Annicchiarico *et al.* (2005) performed a study aimed at: (i) comparing AMMI versus joint regression modelling of GL effects; (ii) verifying the reliability of a GIS-based definition of two subregions that extended the site classification on the basis of GL effects as a function of long-term winter mean temperature; and (iii) comparing wide versus specific adaptation in terms of observed and predicted yield gains. Twenty-four cultivars from international centres in Europe and North Africa were evaluated across 3 years in a total of 47 environments by randomized complete block designs with four replications per trial. Results indicated that the AMMI + cluster analysis and pattern analysis classified test locations consistently and in good agreement with the GIS-based subregion definition. Under the hypothesis of six selection environments assigned to subregions in proportion to their size (three sites in each of 2 years) for late stage selection, specific adaptation provided 2–7% greater gains than wide adaptation over the region at similar costs. The advantage of specific adaptation was much larger (39% determined on the basis of observed gains) for the smaller, stressful inland subregion, where specific adaptation may also enhance food security.

Repeatable GL interaction revealed in METs can be exploited by site-specific cultivar recommendations. There is uncertainty, however, on methods for defining recommendations and extending results to non-tested locations. With reference to durum wheat in Algeria, Annicchiarico *et al.* (2006) compared methods for defining the best pair of cultivars for local recommendation based on: (i) observed data; (ii) joint regression-modelled data; (iii) AMMI-modelled data; (iv) factorial regression-modelled data; (v) AMMI modelling interfaced with a GIS; and (vi) factorial regression modelling interfaced with a GIS. The last two methods extended the recommendations to all sites in a GIS as a function of long-term climatic data. GIS-based recommendations implied a slight yield decrease relative to those based on conventional modelling. However, they allowed for about 9% higher yields than those of most-grown cultivars, while enlarging the scope for site-specific recommendations and assisting national seed production and distribution systems.

10.3 Stability of Genotype Performance

In general, there are two concepts of stability for genotype performance, static and dynamic. Static stability, also referred to as the biological concept of stability, implies that a genotype has a stable performance across environments with no among-environment variance, i.e. a genotype is non-responsive to increased levels of inputs. Dynamic stability implies that a genotype's performance is stable, but for each environment, its performance corresponds to the estimated or predicted level, which is also referred to as the agronomic concept of stability. Lin *et al.* (1986) classified statistical methods for stability analysis into four groups:

- Group A: based on deviation from average genotype effect (DE) – represents sums of squares;
- Group B: based on GEI – represents sums of squares;
- Group C: based on either DE or GEI – represents regression coefficient against environment mean; and
- Group D: based on either DE or GEI – represents deviations from regression.

In Group A (Type 1 stability) which is equivalent to biological stability, a genotype is regarded as stable if its among-environment variance is small. In Groups B and C (Type 2 stability), which is equivalent to agronomic stability, a genotype is regarded as stable if its response to environments is parallel to the mean response of all genotypes in a test. In Group D (Type 3 stability), a genotype is regarded as stable if the residual mean square following regression of genotype performance or yield on environmental index is small. Lin and Binns (1988) proposed a Type 4 stability on the basis of predictable and unpredictable non-genetic variation. They suggested the use of a regression approach for the predictable portion. The mean square for years-within-locations for each genotype as a measure of the unpredictable variation was referred to as Type 4 stability.

The stability of cultivar performance across environments is influenced by the genotype of individual plants and the genetic structure of the plants. The terms homeostasis and individual buffering have been used to describe the stability in performance of individual plants or groups of plants over different environments (Allard and Bradshaw, 1964; Briggs and Knowles, 1967). It has been shown that heterozygous individuals, such as F_1 hybrids, are more stable than their homozygous parents. The stability of heterozygous individuals seems to be related to their ability to perform better under stress conditions than homozygous plants. The terms genetic homeostasis and population buffering were used to describe the stability of a group of plants that exceeds that of its individual members (Lerner, 1954; Allard and Bradshaw, 1964). Heterogeneous cultivars generally have higher stability than homogeneous cultivars.

A number of statistical procedures have been developed to enhance our understanding of GEI and to select genotypes that perform consistently well across many environments. The earliest approach was the linear regression analysis. Finlay and Wilkinson (1963), Eberhart and Russell (1966) and Tai (1971) popularized variations of the regression approach, assuming an expected linear response of yield to environments. Other statistical methods that have received significant attention are pattern analysis (DeLacy *et al.*, 1996), the AMMI model (Gauch and Zobel, 1996), the shifted multiplicative model (SHMM) (Cornelius *et al.*, 1996; Crossa *et al.*, 1996), linear–bilinear and mixed models (Crossa *et al.*, 2004) and non-parametric methods of Hühn (1996). The methods of Hühn (1996) and Kang (1988, 1993) investigate yield and stability into one statistic that can be used as a selection criterion. Flores *et al.* (1998) and Hussein *et al.* (2000) conducted comparative evaluation of 22 and 15 stability statistics/methods, respectively. Flores *et al.* (1998) classified 22 univariate and multivariate methods into three main groups. Group 1 statistics are mostly associated with yield level and show little or no correlation with stability parameters. In Group 2, both yield and stability of performance are considered simultaneously to reduce the effect of GEI. Group 3 statistics emphasize only stability. Recently, mixed model approaches have become increasingly important in GEI and stability analyses.

10.3.1 Linear–bilinear models for studying GEI

Statistical methods for detecting and quantifying COI and for forming subsets of environments and/or genotypes with negligible COI have been based on fixed effect linear–bilinear models. Several classes of these models have been developed, some of which are widely used. In this section, linear–bilinear model development will be discussed, mainly based on Crossa *et al.*'s (2005) review.

An early approach towards the analyses of GEI included the conventional fixed effect two-way (FE2W) ANOVA model with the sum to zero constraints running over indices as shown in Eqn 10.1. Yates and Cochran (1938) proposed to relate the GEI term in Eqn 10.1 linearly to the environmental main effect, that is, $(\tau\delta)_{ij} = \xi_i\delta_j + d_{ij}$, where ξ_i is the linear regression coefficient of the *i*th genotype on the environmental mean and d_{ij} is a deviation. This approach was later used by Finlay and Wilkinson (1963) and modified by Eberhart and Russell (1966). William (1952) linked the FE2W model with principal component analysis (PCA) by considering the model $\bar{y}_{ij} = \mu + \tau_i + \lambda\alpha_i\gamma_j + \bar{\varepsilon}_{ij}$, where λ is the largest singular value of $\boldsymbol{ZZ}'$ and $\boldsymbol{ZZ}$ (for $\boldsymbol{Z} = \bar{y}_{ij} - \bar{y}_{i.}$) and α_i and γ_j are the corresponding eigenvectors.

Gollob (1968) and Mandel (1969, 1971) extended William's (1952) work by considering the bilinear GEI term as $(\tau\delta)_{ij} = \sum_{k=1}^{t}\lambda_k\alpha_{ik}\gamma_{jk}$. Thus, the general formulation for the linear–bilinear model is

$$\bar{y}_{ij} = \mu + \tau_i + \delta_j + \sum_{k=1}^{t}\lambda_k\alpha_{ik}\gamma_{jk} + \bar{\varepsilon}_{ij} \qquad (10.4)$$

where the constant λ_k is the singular value of the *k*th multiplicative component (*k*th PCA axis), that is ordered $\lambda_1 \geq \lambda_2 \geq \ldots \geq \lambda_t$; α_{ik} corresponds to the left singular vector of the *k*th component and represents genotypic sensitivities to hypothetical environmental factors represented by the right singular vector of the *k*th component, γ_{jk}. The α_{ik} and γ_{jk} satisfy the ortho-normalization constraints $\Sigma_i\alpha_{ik}\alpha_{ik'} = \Sigma_j\gamma_{jk}\gamma_{jk'} = 0$ for $k \neq k'$ and $\Sigma_i\alpha^2_{ik} = \Sigma_j\gamma^2_{jk} = 1$ for $k = k'$. When Eqn 10.4 is saturated the number of bilinear terms is $t = \min(I-1, J-1)$ and for any smaller value, the model is said to be truncated. The interaction parameters λ_k, α_{ik} and γ_{jk} of the GEI subspace are estimated from the data themselves. The linear–bilinear model of Eqn 10.4 is a generalization of the regression on the mean model with more flexibility for describing GEI because more than one genotypic and environmental dimension is considered.

Several classes of linear–bilinear models, described by Cornelius *et al.* (1996), which are generally derived from Eqn 10.4, are Genotypes Regression Model (GREG) $\bar{y}_{ij} = \mu_i + \sum_{k=1}^{t}\lambda_k\alpha_{ik}\gamma_{jk} + \bar{\varepsilon}_{ij}$, the Sites (environments) Regression Model (SREG) $\bar{y}_{ij} = \mu_j + \sum_{k=1}^{t}\lambda_k\alpha_{ik}\gamma_{jk} + \bar{\varepsilon}_{ij}$, the Completely Multiplicative Model (COMM) $\bar{y}_{ij} = \sum_{k=1}^{t}\lambda_k\alpha_{ik}\gamma_{jk} + \bar{\varepsilon}_{ij}$ and the Shifted Multiplicative Model (SHMM) $\bar{y}_{ij} = \beta + \sum_{k=1}^{t}\lambda_k\alpha_{ik}\gamma_{jk} + \bar{\varepsilon}_{ij}$.

Two linear and bilinear models, SHMM and SREG, have been used for studying GEI and for clustering genotypes or sites into groups with statistically negligible COI (Cornelius *et al.*, 1992, 1993; Crossa and Cornelius, 1997, 2002; Crossa *et al.*, 1993, 1995). Only the SREG model permits the detection of COIs (Bernardo, 2002).

The SREG model has been used for grouping environments without genotypic rank change (Crossa and Cornelius, 1997). The interaction parameters α_{ik} and γ_{jk} of these linear–bilinear models define the behaviour of the genotypes and the environments and when α_{i1}, α_{i2} and γ_{j1}, γ_{j2} are plotted together in the bi-plot (Gabriel, 1978) useful interpretations of the relationships between genotypes, environments and GEI are obtained. In the bi-plot, the interaction between the *i*th genotypes and the *j*th environment is obtained from the projection of either vector on to the other. Crossa *et al.* (2002) used SREG_1 analysis (a reduced SREG model) to examine GEI among the 20 environments ranged from −0.41 to 0.43 and, consequently, the ranking of the nine genotypes differed among environments (Fig. 10.5). The primary effect for a given environment depends on the other environments included in the

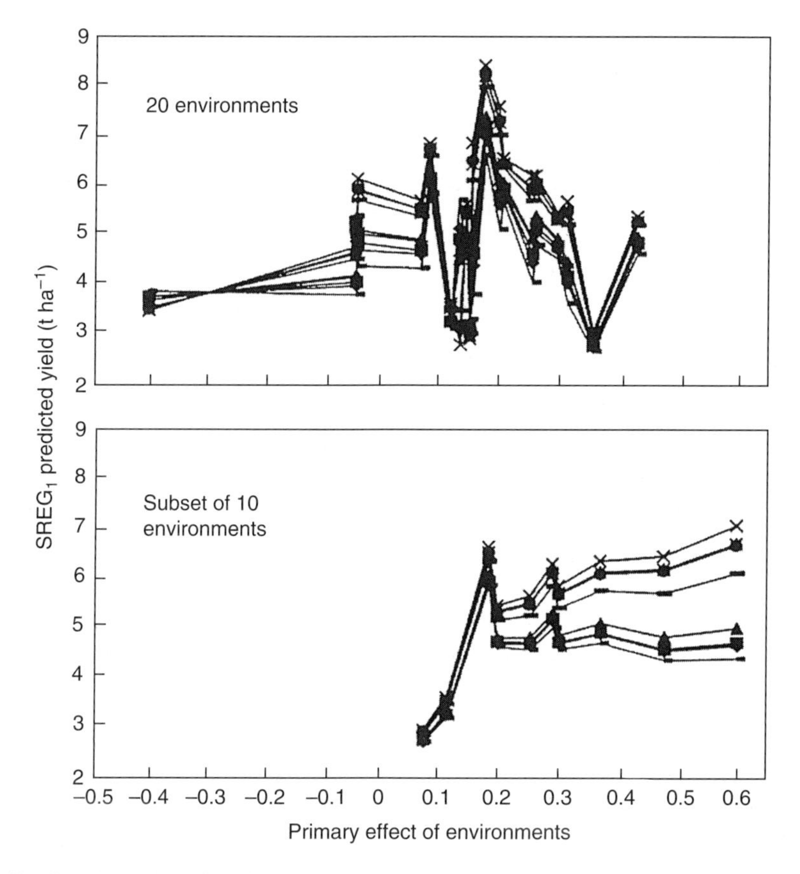

Fig. 10.5. Predicted yield from SREG$_1$: analysis of nine maize genotypes. Data from Crossa *et al.* (2002) courtesy of R. Bernardo (2002).

analysis. A subset of ten environments, in which the resulting primary effects are all positive and COIs were absent, was found (Fig. 10.5).

Gabriel (1978) described the least square fit of Eqn 10.4 and explained how the residual matrix of the GEI term, $\mathbf{Z} = \bar{y}_{ij} - \bar{y}_{i.} - \bar{y}_{.j} + \bar{y}_{..}$, is subjected to a singular value decomposition (SVD) after adjusting the additive (linear) terms. Zobel *et al.* (1988) and Gauch (1988) called Eqn 10.4 additive main effects and multiplicative interaction (AMMI) and proposed a cross-validation procedure for determining the number of important bilinear components. The AMMI model separates the multiplicative portion of GEI into specific patterns of response of genotypes and environments. In this analysis the information about GEI after taking out the main effects of genotypes and environments from METs is used for PCA to extract patterns of GEI or residual variation to understand the underlying causes of such interactions. It is thus a combination of ANOVA and the PCA.

In AMMI analysis, the least square estimates of the parameters along with mean values of genotypes and environments are interpreted to classify genotypes and environments for their stability. A bi-plot is developed by placing both genotype and environment means on the *x*-axis and placing the first PCA scores of the genotypes and environments on the *y*-axis. This 'bi-plot' can be used to facilitate identification of any pattern of GEI, i.e. specific interactions of individual genotypes and environments based on the sign and magnitude of PCA values especially, assuming that the first PCA accounts for the most important pattern of the GEI. The bi-plot helps to visualize the relationship between eigen values for PCA 1 and genotypic and environment means. Any genotype with a PCA 1 value close to zero shows general adaptation to the tested environments (Fox *et al.*, 1997). However, it should be noted that a genotype that is poor everywhere would have a zero PCA 1 score too. So the PCA 1 value should be used along with the average performance across the tested environments. A large genotypic PCA 1, with a high average performance, reflects more specific adaptation to environments with a PCA 1 score of the same size. The genotypes and environments with a PCA values of the same sign show positive interactions and suggest specific adaptations. The reverse sign of the PCA value of genotypes and environments depicts negative interaction, i.e. poor performance of genotypes in such environments.

10.3.2 GGE bi-plot analysis

As one of the major statistical methods for GED analysis, GGE bi-plot analysis has been developed by Yan *et al.* (2000), Yan and Kang (2003) and Yan and Tinker (2006). The method is based on the bi-plot originally developed by Gabriel (1971), which is a popular data visualization tool in many scientific research areas, including psychology, medicine, business, sociology, ecology and agricultural sciences. The bi-plot tool has become increasingly popular among plant breeders and agricultural researchers since its use in cultivar evaluation and mega-environment investigation. Yan *et al.* (2000) referred to bi-plots based on singular value decomposition (SVD) of environment-centred or within-environment standardized GED as 'GGE bi-plots,' because these bi-plots display both G and GE, the two sources of variation that are relevant to cultivar evaluation.

The GGE bi-plot is based on the SREG linear–bilinear (multiplicative) model (Cornelius *et al.*, 1996), which can be written as

$$\bar{y}_{ij} - \mu_j = \sum_{k=1}^{t} \lambda_k \alpha_{ik} \gamma_{jk} + \bar{\varepsilon}_{ij} \qquad (10.5)$$

The model is subjected to the constraint $\lambda_1 \geq \lambda_2 \geq \ldots \lambda_t \geq 0$ and to ortho-normality constraints on the α_{ik} scores, as indicated for Eqn 10.4.

Least squares solution for μ_j is the empirical mean ($\bar{y}_{.j}$) for the jth environment and the least squares solutions for parameters in the term $\lambda_k \alpha_{ik} \gamma_{jk}$ (for $i = 1, \ldots, I$; $j = 1, \ldots, J$) are obtained from the kth PC of the SVD of the matrix $\mathbf{Z} = [z_{ij}]$, where $z_{ij} = \bar{y}_{ij} - \bar{y}_{.j}$. The maximum number of PCs available for estimating the model parameters is $p = \text{Rank}(\mathbf{Z})$. In general, $p \leq \min(J, I - 1)$, with equality holding in most cases. For $k = 1, 2, 3, \ldots$, α_{ik} and γ_{jk} have also been characterized as primary, secondary, tertiary, etc., multiplicative effects of the ith cultivar/genotype and jth environment. Thus, Eqn 10.5 may be described as modelling the deviations of the cell means from the environment means as a sum of PCs, each of which is the product of a cultivar score (α_{ik}), an environment score (γ_{jk}) and a scale factor (the singular value, λ_k).

The GGE bi-plot is constructed from the first two PCs from the SVD of $\mathbf{Z}$ with 'markers,' one for each cultivar, plotted with $\hat{\lambda}_1^f \hat{\alpha}_{i1}$ as abscissa and $\hat{\lambda}_2^f \hat{\alpha}_{i2}$ as ordinate. Similarly, markers for environments are plotted with $\hat{\lambda}_1^{1-f} \hat{\gamma}_{j1}$ as abscissa and $\hat{\lambda}_2^{1-f} \hat{\gamma}_{j2}$ as ordinate. The exponent f, with $0 \leq f \leq 1$, is used to rescale the cultivar and environment scores to enhance visual interpretation of the bi-plot for a particular purpose. Specifically, singular values are allocated entirely to cultivar scores if $f = 1$ ('cultivar-focused' scaling), or entirely to environment scores if $f = 0$ ('environment-focused' scaling); and $f = 0.5$ will allocate the square roots of the $\hat{\lambda}_k$ values to cultivar scores and also to environment scores ('symmetric' scaling). Mathematically, a GGE bi-plot is a graphical representation of the rank 2 least squares approximation of the rank p matrix $\mathbf{Z}$. This representation is unique except for possible simultaneous sign changes on all $\hat{\alpha}_{i1}$ and $\hat{\gamma}_{j1}$ and/or all $\hat{\alpha}_{i2}$ and $\hat{\gamma}_{j2}$. An important property of the bi-plot is that the rank 2 approximation of any entry in the original matrix **Z** can be computed by taking the inner product of the corresponding genotype and environment vectors, i.e. $(\hat{\lambda}_1^f\hat{\alpha}_{i1}, \hat{\lambda}_2^f\hat{\mu}_{i2})(\hat{\lambda}_1^{1-f}\hat{\gamma}_{j1}, \hat{\lambda}_2^{1-f}\hat{\gamma}_{j2})' = \hat{\lambda}_1\hat{\alpha}_{i1}\hat{\gamma}_{j1} + \hat{\lambda}_2\hat{\alpha}_{i2}\hat{\gamma}_{j2}$. This is known as the inner-product property of the bi-plot.

The GGE bi-plot methodology consists of a set of bi-plot interpretation methods, whereby important questions regarding genotype evaluation and test-environment evaluation can be visually addressed. Within a single mega-environment, cultivars should be evaluated for their mean performance and stability across environments.

Figure 10.6 is the 'Average Environment Coordination' (AEC) view, which is based on genotype-focused singular value partitioning (SVP), that is, the singular values are entirely partitioned into the genotype scores (GGE bi-plot option 'SVP = 1'). This AEC view with SVP = 1 is also referred to as the 'Mean versus Stability' view because it facilitates genotype comparisons based on mean performance and stability across environments within a mega-environment. Since GGE represents G + GEI and since the AEC abscissa approximates the genotypes' contributions to G, the AEC ordinate must approximate the genotypes' contributions to GEI, which is a measure of their stability or instability. Thus, G4 in the figure was the most stable genotype, as it was located almost on the AEC abscissa and had a near-zero projection on to the AEC ordinate. This indicates that its rank was highly consistent across environments within this mega-environment. In contrast, G17 and G6 were two of the least stable genotypes with above average mean performance.

Several recent articles reviewed and compared the two statistical approaches discussed above, AMMI and GGE bi-plot analyses. For their pros and cons, readers are referred to Gauch (2006), Yan *et al.* (2007) and Gauch *et al.* (2008).

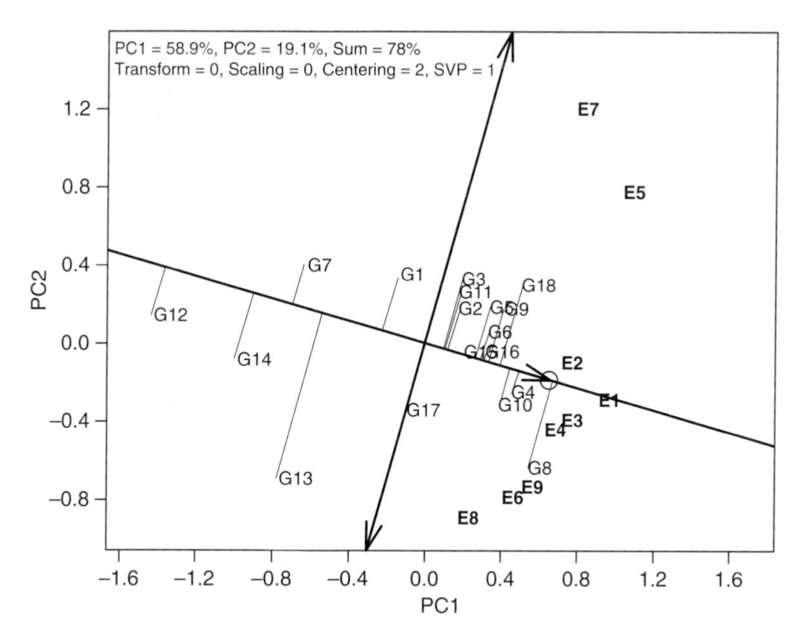

Fig. 10.6. The 'mean versus stability' view of the GGE bi-plot based on a subset of the G × E data in Table 10.1. The data were not transformed ('Transform = 0'), not scaled ('Scaling = 0'), and were environment-centred ('Centering = 2'). The bi-plot was based on genotype-focused singular value partitioning ('SVP = 1') and therefore is appropriate for visualizing the similarities among genotypes. It explained 79.5% of the total G + GE for the subset. From Yan *et al.* (2007) with permission.

10.3.3 Mixed model

A useful and statistically efficient approach would be to generate disjoint subsets of environments and genotypes with no significant COI within the linear mixed model framework and to detect COI in that context. Linear mixed models and the factor analytic (FA) VCOV structure offer a more realistic and effective approach for quantifying COI and forming subsets of environments and genotypes without COI.

Based on Crossa *et al.* (2004), the mixed model fitted to the data from s sites is (see Eqn 10.6 at bottom of page) where y_j is the vector of the response variable (i.e. grain yield) in the jth site ($j = 1, 2, ..., s$); **1** is a vector of ones; μ_j is the population mean of the jth site; Z_{Rj} and Z_{Gj} are the incidence matrices of the random effects of replicates and genotypes within the jth site, respectively; r, g, e are the vectors that contain the random effects of replicates within sites, genotypes within sites and residuals within sites, respectively, and are assumed to be random and normally distributed with zero mean vectors and VCOV matrices **R**, **G** and **E**, respectively, such that (see Eqn 10.7 at bottom of page). **R** and **E** are assumed to have the simple variance component structure:

$$\mathbf{R} = \mathrm{var}(rep) = \Sigma_{rep} \otimes \mathbf{I}_r = [diag(\sigma^2_{r_j}, j = 1, 2, ..., s)] \otimes \mathbf{I}_r \quad (10.8)$$

and

$$\mathbf{E} = \mathrm{var}(error) = \Sigma_{error} \otimes \mathbf{I}_{rg} = [diag(\sigma^2_{e_j}, j = 1, 2, ..., s)] \otimes \mathbf{I}_{rg} \quad (10.9)$$

where r is the number of replicates, g is the number of genotypes and $\mathbf{I}_r$ and $\mathbf{I}_{rg}$ are the identity matrices of orders r and $r \times g$, respectively; $\Sigma_{rep} = diag(\sigma^2_{r_j}, j = 1, 2, ..., s)$ and $\Sigma_{error} = diag(\sigma^2_{e_j}, j = 1, 2, ..., s)$ are the $s \times s$ replicate and error VCOV matrices among pairs of s sites, respectively; $\sigma^2_{r_j}$ and $\sigma^2_{e_j}$ are the replicate and residual variances within the jth site, respectively, and $\otimes$ is the Kronecker (or direct) product of the two matrixes.

The VCOV matrix **G** can be represented as (see Eqn 10.10 at bottom of page) where

$$\begin{bmatrix} y_1 \\ y_2 \\ \cdot \\ \cdot \\ \cdot \\ y_s \end{bmatrix} = \begin{bmatrix} 1\mu_1 \\ 1\mu_2 \\ \cdot \\ \cdot \\ \cdot \\ 1\mu_s \end{bmatrix} + \begin{bmatrix} Z_{R_1} & 0 & \cdot & \cdot & \cdot & 0 \\ 0 & Z_{R_2} & \cdot & \cdot & \cdot & 0 \\ \cdot & \cdot & \cdot & \cdot & \cdot & \cdot \\ \cdot & \cdot & \cdot & \cdot & \cdot & \cdot \\ \cdot & \cdot & \cdot & \cdot & \cdot & \cdot \\ 0 & \cdot & \cdot & \cdot & \cdot & Z_{R_s} \end{bmatrix} r + \begin{bmatrix} Z_{G_1} & 0 & \cdot & \cdot & \cdot & 0 \\ 0 & Z_{G_2} & \cdot & \cdot & \cdot & 0 \\ \cdot & \cdot & \cdot & \cdot & \cdot & \cdot \\ \cdot & \cdot & \cdot & \cdot & \cdot & \cdot \\ \cdot & \cdot & \cdot & \cdot & \cdot & \cdot \\ 0 & \cdot & \cdot & \cdot & \cdot & Z_{G_s} \end{bmatrix} g + e \quad (10.6)$$

$$\begin{bmatrix} r \\ g \\ e \end{bmatrix} \sim N\left(\begin{bmatrix} 0 \\ 0 \\ 0 \end{bmatrix}, \begin{bmatrix} \mathbf{R} = \mathrm{var}(rep) & 0 & 0 \\ 0 & \mathbf{G} = \mathrm{var}(genotype) & 0 \\ 0 & 0 & \mathbf{E} = \mathrm{var}(error) \end{bmatrix} \right) \quad (10.7)$$

$$\mathbf{G} = \mathrm{var}(genotype) = \sum\nolimits_g \otimes \mathbf{I}_g = \begin{bmatrix} \sigma^2_{g1} & \rho_{12}\sigma_{g_1}\sigma_{g2} & \cdot & \cdot & \cdot & \rho_{1s}\sigma_{g_1}\sigma_{g_s} \\ \rho_{12}\sigma_{g_1}\sigma_{g2} & \sigma^2_{g2} & \cdot & \cdot & \cdot & \cdot \\ \cdot & \cdot & \cdot & \cdot & \cdot & \cdot \\ \cdot & \cdot & \cdot & \cdot & \cdot & \cdot \\ \cdot & \cdot & \cdot & \cdot & \cdot & \cdot \\ \rho_{s1}\sigma_{g_s}\sigma_{g_1} & \cdot & \cdot & \cdot & \cdot & \sigma^2_{g_s} \end{bmatrix} \otimes \mathbf{I}_g \quad (10.10)$$

the jth diagonal element of the $s \times s$ matrix $\mathbf{\Sigma}_g$ is the genetic variance $\sigma^2_{g_j}$ within the jth site and the ijth element is the genetic covariance $\rho_{ij}\, \sigma_{gi}\, \sigma_{gj}$ of genotypic effects in sites i and j; thus ρ_{ij} is the correlation of genotypic effects in sites i and j. The hypothesis of interest is that the ρ_{ij} for pairs of sites within subsets of sites (or genotypes) previously identified as non-COI subsets by SHMM or SREG clustering are all unity (Crossa *et al.*, 2004).

The approach proposed by Crossa *et al.* (2004) is a step in the right direction for incorporating the linear mixed model methodology into the quantification of COI. However, detection of COI using SREG (and SHMM) has generally been done within the fixed effect linear–bilinear framework (Cornelius and Seyedsadr, 1997), that is, the differences between any two genotypic effects in any two environments are linear functions of least squares solutions for model parameters regarded as fixed. Recently, Yang (2007) recognized that in statistical analyses of METs, either genotypes or environments, or both, should be considered as random effects and, therefore, the detection of COI must consider that the difference between genotypic effects in a random environment is a predictable function that involves Best Linear Unbiased Estimators (BLUEs) as well as Best Linear Unbiased Predictions (BLUPs).

As a development of Crossa *et al.*'s (2004) approach, Burgueño *et al.* (2008) presented an integrated methodology for clustering environments and genotypes with negligible COI based on results obtained from fitting FA to MET data, which was used to detect COI using predictable functions based on the linear mixed model with FA and BLUP of genotypes.

For the genotype factor, the identity matrix $\mathbf{I}_g$ (of order g), described above, is used when it is assumed that the genotypes are not related and the breeding value of each genotype will be predicted only by the value of the empirical responses of the genotype itself. The genotypic component of $\mathbf{G}$ can be modelled by the identity matrix $\mathbf{I}_g$, which assumes no relationship among genotypes.

The genetic environmental component ($\mathbf{\Sigma}_g$) of the variance of the random effect vector, $\mathbf{g}$, can be modelled by the FA, which expresses the random effect of the ith genotype in the jth environment as a linear function of latent variables x_{ik} with coefficients δ_{jk} for $k = 1, 2, \ldots t$, plus a residual η_{ij}. Then

$$\mathbf{G} = (\mathbf{\Delta\Delta}' + \mathbf{\Psi}) \otimes \mathbf{I}_g = \mathrm{FA}(k) \otimes \mathbf{I}_g \quad (10.11)$$

where $\Sigma\Delta = \begin{bmatrix} \delta_{11} & \delta_{12} & . & . & \delta_{1t} \\ \delta_{21} & \delta_{22} & . & . & \delta_{2t} \\ . & . & . & . & . \\ . & . & . & . & . \\ \delta_{s1} & \delta_{s2} & . & . & \delta_{st} \end{bmatrix}$ is a matrix of order $s \times t$ with the kth column containing the environment loadings for the tth latent factor ($k = 1, \ldots, t$). The FA model can be interpreted as the linear regression of genotype and GEI on latent environmental covariates (environmental loadings, δ_{jk}), with each genotype having a separate slope (genotypic scores, x_{ik}) but a common intercept (if main effects of genotypes are not distinguished from GEI). The slopes of genotypes measure the sensitivity of the genotypes to hypothetical environmental factors represented by the loadings of each environment.

As an example, two CIMMYT maize international METs were used to illustrate the method for searching for subsets of environments and genotypes with negligible COI. Results from both data sets showed that the proposed method formed subsets of environments and/or genotypes with negligible COI. The main advantage of the integrated approach is that one unique linear mixed model, the FA model, can be used for: (i) modelling the association among environments; (ii) forming subsets of environments without COI; (iii) grouping genotypes into non-COI subsets; and (iv) detecting COI using the appropriate predictable function.

The multivariate approaches help to identify patterns of variation for genotypes on the basis of their multi-dimensional response to many environments. Such grouping, however, does not represent any specific types of stability index, but it can be used to draw meaningful conclusion in relation to the response of a cultivar that has been known for its ability. All the new genotypes grouped along with a well-known cultivar or falling near it are considered to have the same type of stability, since groups of genotypes that are known for their stability have been quite stable over years. Along with these known cultivars, a large number of new lines can be screened for their stability from 1-year experiments.

In general the multivariate methods do not provide a simple measure of stability for a specific genotype which could be used as a trait in breeding programmes. A detailed description of all such techniques is beyond the scope of this book and the reader is encouraged to consult Freeman (1973), Kang (1990) and Fox *et al.* (1997). Commercial statistical software packages such as SAS can be used for different models described in this section. For example, mixed models can be fitted using SAS PROC MIXED.

10.4 Molecular Dissection of GEI

Recent advances in molecular biology have provided some of the best tools for obtaining insights into the molecular mechanisms associated with GEI. Molecular markers can be employed to find genomic regions with stable responses. Marker-assisted QTL-by-environment interaction (QEI) analysis will ultimately provide a better genetic understanding and possible regulation of this phenomenon. Regions of plant genomes that provide stable responses across diverse environments can be identified. Experimental strategies have been proposed for resolving environmental factors into several components that affect specific quantitative traits so their effects can be either estimated or controlled (Xu, Y., 2002). Various statistical methods have been developed for mapping quantitative trait loci (QTL) involved in GEIs. In this section, genetic models involving GEI and molecular dissection of GEI will be discussed.

10.4.1 Partition of environmental factors

METs involve various environmental factors and some of them can be partitioned into several key components. Environment partition can be used to understand the effect of each environmental component, the response of a genotype to specific environmental factors and the genetic control of environment-dependent traits such as temperature or photoperiod-induced male sterility.

Genetic analysis in general involves extracting a genetic signal from many sources of 'noise', such as those from external environments and internal genetic backgrounds. For accurate genetic analysis, the 'noise' must be minimized or eliminated. 'Controlled' environments or genetic backgrounds are usually created for filtering the 'noise'. In Chapter 4, we described the development of a set of individuals such as near-isogenic lines (NILs) that have homogeneous genetic background. Similarly, Xu, Y. (2002) proposed the concept of near-iso environments (NIEs). Plant populations used for genetic analysis can be evaluated in either natural or controlled environments or both. Controlled environments can be compared with each other or with natural environments. If two environments mainly differ in one macro-environmental factor, they are considered contrasting or NIEs, if the standard plot-to-plot variation and other residual micro-environmental effects can be neglected. A relative trait value is then derived from two direct trait values measured in each environment to ascertain the sensitivity of plants to the stress (see for example Ni *et al.*, 1998).

Xu, Y. (2002) provided an example of how rice plants respond to photoperiod and temperature. Using Zhaiyeqing 8/Jingxi 17 doubled haploids (DHs), days-to-heading (DTH) and photo-thermo sensitivity (PTS) were measured in two environments (Beijing and Hangzhou) that mainly differ in day-length and temperature. At the photo-thermo sensitive stage, Beijing has long day-length (14.5–15 h) and low temperature (20–27°C), whereas Hangzhou has short day-length (13–13.5 h) and high temperature (25.5–30°C). Rice is considered a short-day plant and development from vegetative to reproductive stages is promoted under short day-length and high-temperature conditions. Differences in photoperiod and temperature in the two locations resulted in differences in DTH of 0–39 days for individual DH lines (Fig. 10.7). Using the relative difference, ((DTH in Beijing – DTH in Hangzhou)/DTH in Beijing × 100), genes associated with PTS were mapped with 155 restriction fragment length polymorphism (RFLP) and 92 simple sequence repeat (SSR) markers. Four chromosomal regions were identified significantly associated with DTH in either or both locations, whereas likelihood of odds (LOD) scores for the PTS in these regions were much lower than the threshold. A region on chromosome 7 (G397A–RM248) was significantly associated with PTS (LOD = 4.47), where LOD scores for DTH in both locations were much lower than the threshold (Fig. 10.7), indicating that this PTS QTL is independent of the QTL for heading date. As the rice breeding programme has been accelerated by growing rice in an off-season or an off-location where it is not a targeted environment, marker-assisted selection (MAS) for these types of traits would be important as they can only be identified under NIEs.

A second example in rice is from CO39/Moroberekan recombinant inbred lines (RILs), grown under greenhouse conditions and exposed to two different photoperiod regimes (Maheswaran *et al.*, 2000). Days-to-flowering (DTF) of individual lines was evaluated under 10-h and 14-h day-lengths and loci associated with photoperiod sensitivity were identified based on the delay in flowering less than the 14-h photoperiod

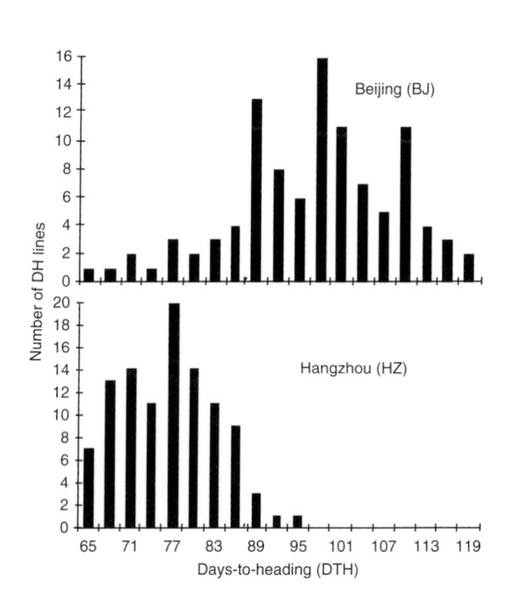

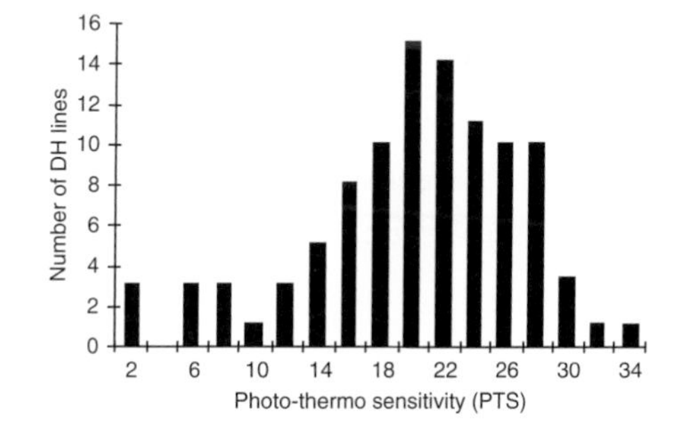

LODs for DTH and PTS

Chr	Marker interval	DTH(BJ)	DTH(HZ)	PTS
1	RG400–RM84	2.41*	1.68	0.87
7	G379A–RM248	1.21	0.56	4.47*
8	RG885–RM44	7.35*	6.56*	1.07
10	C16–RM228	2.67*	3.04*	0.41
12	RG463–RG323	2.30	2.55*	0.82

Fig. 10.7. QTL mapping for photo-thermo sensitivity (PTS) in rice under two environments (Beijing and Hangzhou). Left: Days-to-heading (DTH) distribution in Zhaiyeqing 8/Jingxi 17 DH population planted in Beijing and Hangzhou. Top right: PTS distribution in the population when PTS was measured by the difference of DTHs in the two environments divided by the DTH in Beijing. Bottom right: QTL identified for DTH in Beijing and Hangzhou and for PTS (*LOD > 2.4). Modified from Xu, Y. (2002). Chr, chromosome.

(DTF at 14 h – DTF at 10 h). In total, 15 QTL were associated with DTF. Only four of them were also identified as influencing response to photoperiod. None of these QTL is allelic to the PTS QTL on chromosome 7.

Genetic mapping performed on environmental sensitivity has provided much better quantitative evaluations of QEI and have been used successfully to investigate plasticity and GEI for agriculturally relevant traits in animals and plants.

10.4.2 QTL mapping across environments

Do genes function similarly in different environments? The answer is negative. Phenotypic expression of quantitative traits is affected by external environmental factors such as day-length, temperature, moisture and soil conditions, which can greatly modify the phenotype of quantitative traits. In many cases, external environments act as a regulator of expression of the traits. It was found that when the same mapping population was phenotypically evaluated in different environments, some QTL could be detected in all environments tested but others could be detected only in some of them (Paterson *et al.*, 1991; Stuber *et al.*, 1992; Lu *et al.*, 1996; Veldboom *et al.*, 1996), indicating that QTL detection depends on the specific environment. These QTL can be defined as environment-dependent (sensitive) QTL. For improving mapping power and efficient QTL cloning, therefore, the specific conditions highly suitable for expression of the quantitative trait of interest should be identified. The results from QTL mapping across multiple environments provide some evidence for QEIs, in addition to the information of how QTL detection depends on the environments and which traits are more environment-dependent.

QTL can be studied under adverse environments (abiotic stress), NIEs or a uniform environment by replicating DH or RIL populations and splitting tillers or ratooning a segregating population. Xu, Y. (2002) summarized the QTL mapping experiments in rice that have been done in two or more environments by using permanent populations. For the convenience of comparison, rice QTL mapped in two environments were selected for sharing analysis (Table 10.5). A total of 159 QTL was identified in ten QTL mapping reports for 11 categories of quantitative traits. For different traits, QTL-sharing frequencies between the two

Table 10.5. Comparison of QTL mapped in two environments using the same populations in rice (Xu, Y., 2002).

	Number of QTL		Mean VE[b] (%)		
Trait[a]	Total	Shared (%)	Total	Shared QTL	Unshared QTL
Yield	15	2 (13.3)	8.7	12.8	8.1
Panicle per plant	7	3 (42.9)	7.1	6.7	7.4
Grain per panicle	16	4 (25.0)	11.7	12.9	11.3
1000-grain weight	17	9 (52.9)	12.7	14.0	11.1
Root	30	9 (30.0)	11.8	15.0	10.4
Drought tolerance	21	2 (9.5)	9.8	10.2	9.8
Flood tolerance	12	3 (25)	24.4	48.8	16.3
Al tolerance	4	2 (50)	12.5	16.0	9.0
Disease resistance	17	7 (41.2)	10.4	11.0	10.1
Seedling vigour	13	3 (23.1)	16.0	19.5	14.9
Paste viscosity	7	2 (28.6)	19.5	37.7	11.9
Total	159	46 (30.0)	12.6	16.7	10.9

[a]Traits in each category: Yield, grains (t ha^{-1}); Root, root number, root length and thickness; Drought tolerance, leaf rolling and relative water content; Flood tolerance, initial plant height, plant height increment, internode increment and leaf-length increment; Seedling vigour, shoot length, root length, coleoptile length and mesocotyl length; Paste viscosity, peak viscosity, hot paste viscosity and cool paste viscosity.
[b]VE, Variance explained.

environments ranged from 9.5% for drought tolerance to 52.9% for 1000-grain weight and, for all traits, on average, 46 (30%) of them are shared or are common between the two environments. For all shared QTL, the mean variance explained is 16.7%, whereas for the unshared QTL, it is 10.9%. QTL with large effect (higher proportion of the variance explained) are shared more frequently. Major-gene-related QTL (for flooding tolerance and paste viscosity) had the highest QTL-sharing frequencies. When compared across three or more environments, QTL-sharing frequencies become lower. For example, a total of 22 QTL for six agronomic traits were identified in Zhaiyeqing 8/Jingxi 17 DHs, only seven of which were shared in all three tested environments (Lu *et al.*, 1996). In three trials using Tesanai 2/CB F_2 and its two equivalent F_3s, eight QTL were identified, two of which were detected in all three trials (Zhuang *et al.*, 1997). In another report, three of 11 QTL identified for leaf rolling were shared in the three trials with different drought-stress intensities (Courtois *et al.*, 2000).

With grain yield and test weight evaluated in four trials and for grain yield components evaluated in eight trials, Blanco *et al.* (2001) detected a total of 52 QTL in durum wheat that were significant in at least one environment at $P < 0.001$ or in at least two environments at $P < 0.01$. Paterson *et al.* (2003) described the impact of well-watered versus water-limited growth conditions on the genetic control of fibre quality, a complex suite of traits that collectively determine the utility of cotton. Fibre length, length uniformity, elongation, strength, fineness and colour (yellowness) were influenced by 6, 7, 9, 21, 25 and 11 QTL, respectively, that could be detected in one or more treatments. The genetic control of cotton fibre quality was markedly affected both by general differences between growing seasons (years) and by specific differences in water management regimes. Seventeen QTL were detected only in the water-limited treatment while only two were specific to the well-watered treatment.

Inconsistent QTL detection across environments may also be the result of QEI, which presumably represent the genetic factors underlying the GEI observed in line-based phenotypes (Beavis and Keim, 1996). QEI has been predicted by comparing the QTL detected separately in different environments in many crops. That a QTL can be detected in one environment but not in others, as discussed earlier, could result from experimental noise, sampling error or experimental error and thus does not necessarily indicate QEI. As indicated by Jansen *et al.* (1995), the chance for simultaneous detection of QTL in multiple environments is small. On the other hand, sharing QTL among environments does not necessarily mean lack of QEI. This is supported by the fact that QEI was identified for some sharing QTL by incorporating QE_{ij} into QTL analysis (e.g. Yan *et al.*, 1999) and by the fact that QTL effects estimated across environments could be very different.

10.4.3 QTL mapping with incorporated GEI

There are two approaches for the analysis of QEIs (Leflon *et al.*, 2005). The first approach deduced interactions by comparing QTL detected separately in different environments as described in the previous section: in many cases, an interaction was merely detected and no estimate made of the interaction itself. In other cases, QEIs were assessed by co-localization between QTL detected for the main effect and QTL detected for stability statistics (Emebiri and Moody, 2006). The second approach takes interaction effects into account in the analysis of multi-environment trials by introducing QTL main effects and QEI effects, like studies of GEI (see, for instance, Crossa *et al.*, 1999; Campbell *et al.*, 2003, 2004; Groos *et al.*, 2003). These methods are powerful but a large number of environmental measurements is necessary for their application.

With data collected from multiple location trials on a core set of genotypes, GEI can be detected by ANOVA and various statistical procedures that measure genotype stability (Lin *et al.*, 1986; Kang, 1993) as described in previous sections. To determine genetic factors responsible for GEI, QEI can be evaluated on the basis of agronomic data collected on a mapping population in multiple location trials and comparing QTL detection across environments by ANOVA to test marker locus × environment interactions. More recent efforts in QTL mapping involving GEI have proven far more effective, largely due to the incorporation of a QEI component by integrating this interaction component into actual mapping algorithms (Jiang and Zeng, 1995; Wang, D.L. *et al.*, 1999).

To analyse GED produced in METs, various statistical models have been proposed that differ in the extent to which additional genetic, physiological and environmental information is incorporated into the model formulation. The simplest model is the additive two-way ANOVA model, without GEI and with parameters whose interpretation depends strongly on the set of included genotypes and environments. The most complicated model is a synthesis of a multiple QTL model and an eco-physiological model to describe a collection of genotypic response curves. Among these models, factorial regression models allow direct incorporation of explicit genetic, physiological and environmental co-variables on the levels of the genotypic and environmental factor. They are also very suitable for the modelling of QTL main effects and QEI (van Eeuwijk *et al.*, 2005).

In the framework of factorial regression, modelling of QEI is a natural extension of modelling main effect QTL, i.e. QTL that are supposed to have constant expression across environments. A model with a QTL main effect and QEI at the same location in the genome can be written as

$$\mu_{ij} = \mu + x_i\rho + G_i^* + E_j + x_i\rho_j + (GE)_{ij}^* \quad (10.12)$$

The $(GE)_{ij}$ from the ANOVA model is partitioned in part due to a differential QTL expression, $x_i\rho_j$ and a residual, $(GE)_{ij}^*$, that is usually taken to be random and for that reason then disappears from the expression for the expectation. In the light of QEI, the parameter ρ_j adjusts the average QTL expression across environments, ρ_j, to a more appropriate level for the individual environment *j*. The QEI parameters, ρ_j, can themselves be regressed on an environmental co-variable, *z*, in an attempt to link differential QTL expression directly to key environmental factors. The QEI term $x_i\rho_j$ is replaced by a regression term $x_i(\lambda z_j)$ and a residual term $x_i\rho_j^*$ that again disappears from the expectation when ρ_j^* is assumed to be random. The parameter λ is a proportionality constant that determines the extent to which a unit change in the environmental co-variable *z*, influences the effect of a QTL allele substitution.

Mixed models

Using mixed models, several papers have been dedicated to the incorporation of the genetic basis of GEI: differential expression of QTL in relation to changing environmental conditions, or QEI. Early work on QEI was done by Jansen *et al.* (1995), Jiang and Zeng (1995) and Korol *et al.* (1998), who used a mixed model approach. Regression based approaches were presented by Sari-Gorla *et al.* (1997), Caliñski *et al.* (2000), Hackett *et al.* (2001) and van Eeuwijk *et al.* (2001, 2002). Piepho (2000) and Verbyla *et al.* (2003) presented other relevant work on QEI. These authors developed QTL mapping methods for the analysis of METs using mixed model theory, thereby giving special attention to the modelling of heterogeneity of variance across environments and correlations between environments, where the latter correlations may be due to undetected QTL.

Jansen *et al.* (1995) developed an analytic approach, multiple QTL mapping, which accommodates both the mapping of multiple QTL and GEI. This approach was compared to interval mapping in the mapping of QTL for flowering time in *Arabidopsis thaliana* under various photoperiod and vernalization conditions. Procedures developed by Jiang and Zeng (1995) for estimating the effect of QTL for multiple traits can be used to test the significance of QEI.

A least squares interval mapping approach developed by Sari-Gorla *et al.* (1997) allows inclusion in the model of the parameters describing the experimental and environmental situation so that the QEI can be tested. The analysis was performed on data concerning two components of maize pollen competitive ability, obtained from an experiment over 2 years. The method, in comparison with the traditional single marker approach, has been shown to be more powerful in detecting QTL and more precise in determining their map position. The analysis has identified QTL expressed across years, putative QTL with major effects and QTL accounting for GEI.

Piepho (2000) proposed a mixed model method to detect QTL with significant mean effect across environments and to characterize the stability of effects across multiple environments. He treated environment main effects as random, which meant that both environment main effects and QEI effects were random.

Verbyla *et al.* (2003) developed an approach for multi-environment QTL analysis. To accommodate a multi-environment analysis, the size of a QTL effect was assumed to be a random effect. The approach resulted in a multiplicative mixed model for QEI of the factor analytic type. The full genetic model may also include a factor analytic model for the residual GEI, whereas the environmental model for the non-genetic variation involves local, global and extraneous variation. The approach was used to determine QTL for yield in the Arapiles × Franklin DH population of the National Barley Molecular Marker Program.

Malosetti *et al.* (2004) presented a strategy for modelling QEI using mixed model methodology in combination with regression ideas. They proposed a simple interval mapping approach that consists of fitting along the genome a mixed model with both a fixed QTL main effect and a fixed QEI term and for the random part, the residual genetic variation, a factor analytic model with one multiplicative term and residual heterogeneity. For chromosome positions with identified QTL expression and QEI, a second modelling step regresses the QEI on one or more environmental co-variables. To illustrate the approach, they analysed grain yield data stemming from the North American Barley Genome Project (NABGP) (http://barleyworld.org/NABGP.html). QEI for yield, as identified at the 2H chromosome could be described as QTL expression in relation to the magnitude of the temperature range during heading.

Factorial regression model

If climatic data are available for precipitation, temperature and solar radiation, factorial regression models (van Eeuwijk *et al.*, 1996) and partial least squares models (Aastveit and Martens, 1986) can be used to determine the degree to which each of these factors influence GEI and QEI (Crossa *et al.*, 1999). Hence, just as molecular markers are commonly used to model the effects of chromosomal segments (QTL) on a particular quantitative trait, climatic data can also be used to model particular aspects of the environment that contribute to the differential performance of genotypes across a range of testing environments. Using factorial regression models, Crossa *et al.* (1999) were the first to explain QEI and found that temperature differences across environments accounted for a large portion of the QEI detected in a tropical maize (*Zea mays* L.) mapping population. They showed how regression methods such as the partial least squares regression and the factorial regression models, together with genetic markers and environmental co-variables (such as maximum and minimum temperature and sun hours), could be used to: (i) detect relevant sets of correlated markers and environmental variables that explain a significant proportion of the total GEI; and (ii) study the influence of environmental variables on the expression of QTL with the objective of assessing and interpreting the QEI that accounts for GEI. Vargas *et al.* (2006) used factorial regression and partial least squares methods for mapping QTL and QEI for the CIMMYT maize drought stress programme.

Van Eeuwijk (2001, 2002) extended the factorial regression models for GEI and QEI developed by Crossa *et al.* (1999) from the original marker-based regressions to interval mapping and composite interval mapping. The authors presented: (i) a randomization test for controlling the genome-wise error rate, following the logic introduced by

Churchill and Doerge (1994); and (ii) a partial least square (PLS) strategy to deal with the problem of multi-collinearity among multiple cofactors. The PLS strategy consisted of: (i) taking all the markers outside the chromosome being evaluated as cofactors; (ii) regressing the phenotypic responses on this set of markers using multivariate PLS; (iii) calculating the fitted values for the phenotypic responses; and (iv) using the corrected phenotypic observations, i.e. the residuals from the PLS regression, in a simple interval mapping (SIM) procedure for the chromosome being evaluated.

Structural equation model

Most agronomically important traits are the result of a number of genetic, molecular and physiological mechanisms that affect the trait of interest either directly or indirectly through other intermediate traits. The GEI of each trait in the network among variables will be influenced either directly or indirectly, by a number of QEI and GEI of other traits, which may, in turn, be influenced by other factors (Campbell *et al.*, 2003). A single dependent variable quantitative approach cannot describe the complicated relationship between traits, QTL and environments where some traits function simultaneously as both dependent variables to be predicted by other genetic and environmental factors and as independent predictor variables of other traits.

Dhungana *et al.* (2007) developed a systematic approach for understanding GEI of complex interrelated traits by combining chromosome institution lines that allowed studying the effects of genes on a single chromosome with a structural equation model (SEM) that approximated the complex process involving genes, environmental conditions and traits. Structural equation modelling is a generation of path analysis proposed by Wright (1921) and is used to quantitatively analyse the causal structure among a number of variables where each may function as a dependent variable in some equations and an independent variable in others (Bollen, 1989). Because of this, SEM is ideal for characterizing the complex relationships of GEI where many traits function as not only dependent variables to be predicted by environmental and genetic factors, but also as independent predictor variables of other traits further downstream (Dhungana *et al.*, 2007). To use SEM to analyse GEI, prior knowledge of the direction of the causal relationships is assumed and specified through a path diagram and the model is then algebraically specified by a system of regression-type equations where each variable is adjusted to contain only GEI effects. A final model is then developed by fitting successive models and retaining significant QEI variables which result in a better fitting model. The final model yields path coefficients and a path diagram that contains only significant paths thus giving insight into important relationships between traits, QTL and the environmental variables.

The approach was applied to recombinant inbred chromosome wheat lines grown in multiple environments. The final model explained 74% of the yield GEI variation and it was found that spikes per square metre GEI had the highest direct effect on yield GEI and that the genetic markers were mostly sensitive to temperature and precipitation during the vegetative and reproductive periods. In addition, a number of direct and indirect causal relationships were identified that described how genes interact with environmental factors to affect GEI of several important agronomic traits.

QEI mapping examples

There are numerous examples now available for QEI mapping using some of the approaches described above. Only a few examples will be discussed here to represent different approaches.

Romagosa *et al.* (1996) assessed AMMI's value in QTL mapping. This was done through the analysis of a large two-way table of GED of barley (*Hordeum vulgare* L.) grain yields. Grain yield data of 150 DHs derived from the Steptoe × Morex cross and the two parental lines, were taken by the NABGP at 16 environments throughout the barley production areas of the USA and Canada.

Four regions of the genome were identified to be responsible for most of differential genotypic expressions across environments. They accounted for approximately 50% of the genotypic main effect and 30% of the GEI sums of squares. The magnitude and sign of AMMI scores for genotypes and sites facilitate inferences about specific interactions. The parallel use of classification (cluster analysis of environments) and ordination (PCA of the GE matrix) techniques allowed most of the variation present in the GE matrix to be summarized in just a few dimensions, specifically four QTL showing differential adaptation to four clusters of environments.

An illustration of the uncertainties occurring when attempting to find specific genetic factors for yield is presented by Reyna and Sneller (2001). They attempted to introgress 'alleles for yield' into elite soybean material in the southern soybean area of the USA. As variation was scarce, the authors tried to exploit beneficial alleles for yield identified as QTL in cv. Archer from the northern soybean area of the country (Orf *et al.*, 1999). Reyna and Sneller (2001) built up four NILs for each QTL and tested them under different environmental conditions. They found that the QTL for yield identified in a particular cultivar and environmental condition did not contribute significantly to improved yields in a different genetic background and different environmental conditions. The authors concluded: 'It may be difficult to capture the value assigned to QTL alleles ... when the alleles are introgressed into populations with different genetic backgrounds, or when tested in different environments.'

Ungerer *et al.* (2003) examined inflorescence development patterns in *Arabidopsis* under different, ecologically relevant photoperiod environments for two RIL mapping populations (Ler × Col and Cvi × Ler) using a combination of quantitative genetics and QTL mapping. Plasticity and GEI were regularly observed for the majority of 13 inflorescence traits. These observations can be attributable (at least partly) to variable effects of specific QTL. Pooled across traits, 12/44 (27.3%) and 32/62 (51.6%) of QTL exhibited significant QEIs in the Ler × Col and Cvi × Ler lines, respectively. These interactions were attributable to changes in magnitude of effect of QTL more often than to changes in rank order (sign) of effect. Multiple QEIs (in Cvi × Ler) clustered in two genomic regions on chromosomes 1 and 5, indicating a disproportionate contribution of these regions to the phenotypic patterns observed.

By using factorial regression models, agronomic and molecular genotype data and three environmental covariates (daily mean temperature, precipitation and solar radiation) recorded in each test environment, Campbell *et al.* (2004) investigated to: (i) detect which of these three environmental covariates may account for GEI by testing individual genotype × environmental covariate interactions; and (ii) detect marker × environmental covariate interactions that provide explanations of variable QTL genotypic differences across environments. Agronomic performance and molecular marker data available for a population of chromosome 3A recombinant inbred chromosome lines (RICLs-3A) in seven environments were used along with environmental covariate data to construct individual factorial regressions to explain GEI and QEI. Precipitation and temperature before anthesis had the greatest influence on agronomic performance traits for the RICLs-3A and explained a sizeable portion of the total GEI for those traits. Individual molecular marker × environmental covariate interactions explained a large portion of the total marker × environment interactions for several agronomic traits.

Laperche *et al.* (2007) used three methodologies to reveal QTL × nitrogen interactions: (i) QTL detected separately under both types of N supply; (ii) QTL detected for global interaction variables assessed as N^+/N^- and N^-/N^+; and (iii) QTL considered for factorial regression slope and ordinate parameters, which represent a plant's sensitivity to N stress and plant performance under a limited N supply. In total, 233 QTL were detected for the traits measured in each combination of environment and N supply (N^+: high supply; N^-: low supply). Comparison of QTL detected under N^+ and N^- levels identified 13 non-specific QTL, eight N^+ specific loci and seven N^- specific loci. For QTL for global interaction variables, four adaptive loci were validated and eight constitutive loci were found to be involved in G × nitrogen interaction. Nine interactive loci were validated and three new loci detected using factorial regression variables.

10.4.4 Utilization of MET and genotypic data

While the international METs have been used effectively to exchange germplasm there have been only limited analyses of the large data sets that they generate. Further, many of these analyses have focused on a specific international MET conducted in 1 year. Analyses which integrate the information from international METs across years have been attempted in only a few cases. The strength of these studies is that they integrate large quantities of data on spatial and temporal GEIs and provide a basis for identifying repeatable interactions. However, their weakness is that in most cases there is only a limited information base for explanation of these interactions. There are great opportunities for synergy between the statistical, genetical and biophysical modelling methodologies (Cooper and Hammer, 1996). The complete data set which contains genotype, phenotype and environment information for numerous genetics and breeding materials opens the door for comprehensive use of them in both genetics and plant breeding including GEI through genome-wide association mapping. In Chapter 6, we provided an example of the use of cultivars in METs and their phenotypic and genotypic data in linkage disequilibrium mapping (Crossa *et al.*, 2007).

10.5 Breeding for GEI

How can a breeder deal with GEI? Eisemann *et al.* (1990) listed three ways of dealing with GEI in a breeding programme: (i) ignoring them, i.e. using genotypic means across environments even when GEI exists; (ii) avoiding them; or (iii) exploiting them. Kang (2002) discussed these three ways. Interactions should not be ignored when they are significant and of the crossover type. The second way of dealing with these interactions, i.e. avoiding them, involves minimizing the impact of significant interactions. One approach is to group similar environments (forming mega-environments) via a cluster analysis as discussed in previous sections. With environments being more or less homogeneous, genotypes evaluated in them would not be expected to show COIs. By clustering environments, potentially useful information may be lost. International research centres such as CIMMYT, aim to identify maize and wheat genotypes with broad adaptation (i.e. stable performance across diverse environments) at many international sites. If the subgrouping is used to eliminate the environments that share the same factors and are identical to each other (redundant test environments), optimization of the environment sites will also help determine the broad adaptation by using as few environment sites as possible. The third approach encompasses stability of performance across diverse environments by analysing and interpreting genotypic and environmental differences. This approach allows researchers to select genotypes with consistent performance, identify the causes of GEI and provide the opportunity to correct the problem. When the cause for the unstable performance of a genotype is known, either the genotype can be improved by genetic means or a proper environment (inputs and management practices) can be provided to enhance its productivity.

The best approach for breeders and geneticists would be to understand the nature and causes of GEI and to try to minimize its deleterious implications and exploit its beneficial potential through appropriate breeding, genetic and statistical methodologies (Singh *et al.*, 1999). Appropriate analyses of data can provide an opportunity for exploiting GEI through applied analytical methods, such as AMMI and GGE bi-plot, using climatic factors to explain GEI, evaluating risk of production and optimizing allocation of land resources to various genotypes for selection in heterogeneous environments.

10.5.1 Breeding for resource-limited environments

Breeding for resource-limited environments is one of the major objectives for many international breeding programmes. Commonly the performance of a genotype in an environment is a function of the influence of many interacting factors and environments differ in the type, intensity and timing of these challenges.

The performance level is important. Where productivity is low due to an overriding environmental limitation, it may be comparatively easy to accomplish improvement in performance through genetic and/or environmental change (Kang, 2002). A relatively simple genetic change may have quite a fundamental influence on performance and hence adaptation; for example, use of an early maturity, vernalization requirement or even a morphological character to avoid frost damage, genetic resistance to a specific disease, genetic tolerance of a nutritional disorder, etc. Equally, environmental modification to overcome the limitation may be possible. In these situations, the key to plant improvement is the recognition of the nature of the stress or challenge and of the adaptive response (Cooper and Byth, 1996).

In order to increase crop productivity through enhanced yield potential, heterosis, modified plant types, improved yield stability, gene pyramiding and exotic and transgenic germplasm, it is important to identify the factors that are responsible for GEI. Brancourt-Hulmel (1999) used crop diagnosis with the analysis of interaction by factorial regression in wheat. She provided an agronomic explanation of GEI and defined the responses or parameters for each genotype and each environment. Earliness at heading, susceptibility to powdery mildew and susceptibility to lodging were the major factors responsible for GEI. In the same study, factorial regression revealed that water deficits during the formation of grain number and N level were also associated with GEI.

To alleviate GEI concerns caused by stresses, breeders need to know as much about the various characteristics of genotypes as possible. They also need to characterize environments as fully as possible (Kang, 2002). Knowledge of soil characteristics and ranges of weather variables and stresses that plant materials will be exposed to is a prerequisite to exploiting the beneficial potentials of the genotypes and environments and to targeting appropriate cultivars to specific environments.

Although biotic stresses and interactions among them and/or with abiotic factors remain poorly understood, they have significant relevance to GEI in plants. Plants may respond to pathogen infection by inducing a long-lasting, broad-spectrum, systemic resistance to subsequent infections. Induced disease resistance has been referred to as physiological acquired immunity, induced resistance or systemic acquired resistance. Differences in insect and disease resistance among genotypes can be associated with stable or unstable performance. It is highly desirable to identify QTL for a complex trait that is expressed in a number of environments. Crossa *et al.* (1999) found that higher maximum temperature in low- and intermediate-altitude sites affected the expression of some QTL, whereas minimum temperature affected the expression of other QTL, in tropical maize.

10.5.2 Breeding for adaptation and stability

An understanding of the genetic basis of adaptation and stability and their physiological and environmental causes is of fundamental importance for understanding GEI, for assessing the association between phenotypic and genotypic values and for enhancing the selection of superior and stable genotypes. The presence of COIs has important implications for breeding strategies that aim to improve either broad or specific adaptation or some combination of both components of adaptation.

The broad adaptation concept – the need to minimize GEIs (and maximize G) – was successful for the rapid adoption of the seed-based technology of the Green Revolution. But is it appropriate for the green evolution? Cultivars must be diversified and matched with the diversity of pest systems to ensure effective and durable pest management. Genotypes will need to be matched with a less predictable water supply in the irrigated system in rice. Scientists may also need to match genotypes with radiation levels (again unpredictable) to address the challenge of increasing the yield by 50%.

When we look at plant adaptation, which exploits spatial GEI, are we interested in this primarily as a proxy for temporal GEIs to help ensure the stability of performance of our chosen genotypes over time? We need to have clarity in our objectives in pursuing this particular topic. Are we searching for adaptability per se in order to exploit technological spillover, or are we using spatial GEIs to ensure stability of performance over time for farmers using improved cultivars in particular locations? Tools like modelling and simulation can complement spatial genotype-by-environment experimentation and analysis aimed at addressing either adaptability or stability issues.

More commonly, however, there may be a range of responses to the environmental challenge and the same adaptive response can result from different challenges. In these circumstances, the factors influencing adaptation are multivariate, quantitative and complex and may vary in an undefined manner between different genotypes. Thus it is more difficult to recognize the nature of the challenge and explain the adaptive response. In advanced testing programmes where genotypes exhibit reasonably high levels of performance, relative differences in adaptation and the specific nature of the GEI become increasingly important in defining breeding objectives and strategies. However, where adaptation is a function of response to environments differing quantitatively in time and degree for a number of uncontrollable factors, analysis of genetic differences becomes complex (Cooper and Byth, 1996).

With a MET, a breeder can identify cultivars with specific adaptation as well as those with broad adaptation, which will not be possible from testing in a single environment. Broad adaptation provides stability against the variability inherent in an ecosystem, but specific adaptation may provide a significant yield advantage in particular environments as discussed in Section 10.1. MET makes it possible to identify cultivars that perform consistently from year to year (small temporal variability) and those that perform consistently from location to location (small spatial variability). There is a need for developing cultivars with broad adaptation to a number of diverse environments (adaptability) and a need for farmers to use new cultivars with reliable or consistent performance from year to year (reliability) (Evans, 1993). Genetic improvement for low-input conditions would require capitalizing on GEI and slower or limited gains in low-input or stress environments suggested that conventional high-input management of breeding nurseries and evaluation trials might not effectively select genotypes with improved performance at low-input levels (Smith, M.E. *et al.*, 1990). Because of the success in favourable environments, plant breeders have tried to solve the problems of poor farmers living in unfavourable environments by simply extending the same methodologies and philosophies applied to favourable, high-potential environments, without considering the possible limitations associated with the presence of a large GEI (Ceccarelli *et al.*, 2001). Responses to selection under stress and non-stress environments and to selection at high- and low-input levels need to be compared theoretically and practically.

As indicated by Kang (2002), stability of cultivars would be enhanced if multiple resistances/tolerances to stress factors were incorporated into the germplasm used for cultivar development. If every cultivar (different genotypes) possessed equal resistance/tolerance to every major stress encountered in diverse target environments, GEI would be reduced. Conversely, if genotypes possessed differential levels of resistance (a heterogeneous group) and, somehow, we could make all target environments as homogeneous as possible, GEI would again be reduced. Since we do not have any control over unpredictable environments from year to year, the only approach would be the former.

Plants have incorporated a variety of environmental signals into their development pathways that have provided for their wide range of adaptive capacities over time. In response to severe environmental changes, a genome can respond by selectively regulating (increasing, decreasing or even shutting down) the expression of specific genes. Jiang *et al.* (1999) used molecular markers to investigate adaptation differences between highland and lowland tropical maize. They concluded that breeding for broad thermal adaptation should be possible by pooling genes showing adaptation to specific thermal regimes, albeit at the expense of reduced progress for specific adaptation.

10.5.3 Measurement of GEI in breeding programmes

Measure interaction at intermediate growth stages

A crop is exposed to variable environmental factors throughout the developmental stages and the growing season. Generally, researchers investigate the causes of GEI for a quantitative trait such as yield that are phenotyped at the final harvest stages. To critically investigate GEI, one may need to record environmental variables and plant-growth measurements at specific time intervals throughout the growing season, as suggested by Xu (1997) for dynamic QTL mapping. This would help determine what effect, if any, the environmental variables from an earlier period had on GEI at intermediate stages and on the final yield. This may provide a better understanding of the dynamic development process of a quantitative trait.

Multi-environmental testing at early stages of breeding

Kang (2002) proposed early multi-environment testing. Usually, there is a shortage of seed at the earliest stages of breeding, which prevents extensive testing at multiple locations. However, in a clonally propagated crop, such as sugarcane or potato, one stalk of sugarcane or one tuber of potato can be divided into at least two pieces and planted in more than one environment. Similarly, in other crops, if only 20 kernels are available, one could plant ten seeds each in two diverse environments. In the absence of GEI, one would obtain a better evaluation of the genotypes, but, if GEI was present, one would obtain information about the consistency or inconsistency of performance of genotypes early in the programme. This strategy would prevent gene loss or genetic erosion, which could occur if testing was done in only one environment and would also result in an increased breeding effort without a corresponding increase in expenditure of resources.

Unbalanced data

Plant breeders often deal with unbalanced data. Searle (1987) classified unbalancedness as planned unbalanced data and missing observation. When a set of genotypes is grown in a specific set of environments, oftentimes a balanced data set (without any missing scores) is not possible, especially when a wide range of environments is used, or long-term trials are conducted. Hybrids/cultivars are continually replaced year after year. Also the number of replications may not be equal for all genotypes because experimental plots may be discarded for one reason or another. In such cases, plant breeders must deal with unbalanced data.

Researchers have used different approaches for studying GEI in unbalanced data (Kang, 2002). Usually environmental effects are considered as random and cultivar effects as fixed. Inference on random effects using least squares, in the case of unbalanced data, is not appropriate because information on variation among random effects is not incorporated (Searle, 1987). For this reason, mixed model equations are recommended (Henderson, 1975). The restricted maximum likelihood methodology is generally preferred to maximum likelihood estimates because it considers the degrees of freedom for fixed effects for calculating error. The calculation of restricted maximum likelihood stability variances for unbalanced data allow one to obtain a reliable estimate of stability parameters and overcomes the difficulties of manipulating unbalanced data (Kang and Magari, 1996).

10.5.4 MAS for QEI

Genotype means across test environments are used to select lines, populations, hybrids and cultivars for target environments and to select marker and QTL alleles for MAS across target environments. Test environments – samples of years, locations and other factors – are selected to maximize the speed of a selection cycle while minimizing the cost of testing and maximizing selection gains for target environments. GEIs can cause test environments to fail to maximize selection gains for target environments, with equivalent consequences for selection with and without markers. At the extreme, this happens when differences between genotypes are observed across test environments and there are no differences across target environments. Or when differences between genotypes are not observed across test environments and there are differences across target environments. The consequences are either to fix unfavourable alleles, or, for MAS only, to fix alleles at QTL which have no mean effect across test environments, but which had an effect across the sample of test environments used. The root of the problem with GEIs is differences between test and target environments. Nothing can be done about the outcome of selection if test and target environments are fixed (Knapp, 1994).

Additional methods are needed to determine the nature of QEIs. These methods are hardly necessary for practising MAS. Only the means of QTL genotypes across test environments need to be estimated to select QTL for MAS. This only fails if every QTL manifests COI and the test environments did not uncover these interactions, both of which are very unlikely for carefully selected test and target environments. Suppose, for example, a QTL manifests a COI which is not observed among the QTL genotype by test environment means, but which leads to no overall effect of the QTL within target environments, then putting selection pressure against this QTL is equivalent to selecting a neutral locus (Knapp, 1994). This diminishes the selection response by decreasing selection intensity, as do many errors (Edwards and Page, 1994).

Differentiating between non-crossover and crossover QEIs might be important for optimizing MAS. Crossover QEIs could affect the outcome of MAS, whereas non-crossover QEIs should be of no consequence to the outcome of MAS. Understanding and characterizing the nature of QEIs is useful for optimizing MAS or conventional selection.

10.6 Future Perspectives

QTL mapping so far has provided strong evidence that in addition to some consistent additive QTL effects across genetic background and environments, genetic architecture in elite breeding populations involves important components of epistasis, GEI and pleiotropy. However, this type of information has not been used to enhance breeding strategies. The realized progress from selection has been usually considerably lower than the predicted response, which in most cases might have been associated with GEI. To examine the potential of molecular-enhanced breeding strategies to enhance the predictive response, many attempts have been made and are currently underway to construct relevant gene-to-phenotype models for traits to assist the plant breeding process.

As such a example, Cooper *et al.* (2005, 2007) have tackled this as a genetic modelling problem by developing a flexible quantitative framework for studying the genetic architecture of traits in terms of gene networks. This framework, called the *E(NK)* model, is an extension of the *NK* gene network model that was introduced and used by Kauffman (1993) to study the behaviour of gene networks and their influences on organism development and evolutionary processes. When the *E(NK)* model is applied to the study of issues relevant to plant breeding processes, it allows for the property that the influence of a gene network on the expression of a trait can differ in varying environmental conditions. Thus, *E* identifies different environment-types within the context of a defined TPE, *N* identifies the different genes and *K* identifies the degree of connection between subsets of the total set of *N* genes, *i.e.* the gene network topology (Kauffman, 1993; Cooper *et al.*, 2005). Thus, in the terminology of quantitative genetics the *E(NK)* model is a finite locus polygenic model that can be defined to include effects of epistasis and GEIs. The parentheses around the *NK* term are used to indicate that the *N* genes can interact in *K* different ways to determine the trait phenotype in *E* different environment-types. $K = 0$ indicates the *N* genes act independently in the model and a larger *K* indicate increased levels of interactions among the *N* genes.

Kauffmann's landscape concept can be used in combination with the *E(NK)* model to examine how the shape of the phenotype landscape changes with the genetic architecture of a trait, as determined by changes in the levels of *E*, *N* and *K*. The simple additive finite locus model is defined by the case where $E = 1$ and $K = 0$, thus $E(NK) = 1(N{:}0)$ (Fig. 10.8). As *E* and *K* are increased for a given level of *N*, the effects of the alternative alleles for the *N* genes become increasingly context dependent on the genotypes of other genes and on the range of environment-types in the target population of environments. Thus, context dependent effects of genes due to epistasis and GEI can be simulated (Cooper and Podlich, 2002). Building on the landscape metaphor (Fig. 10.8), it is

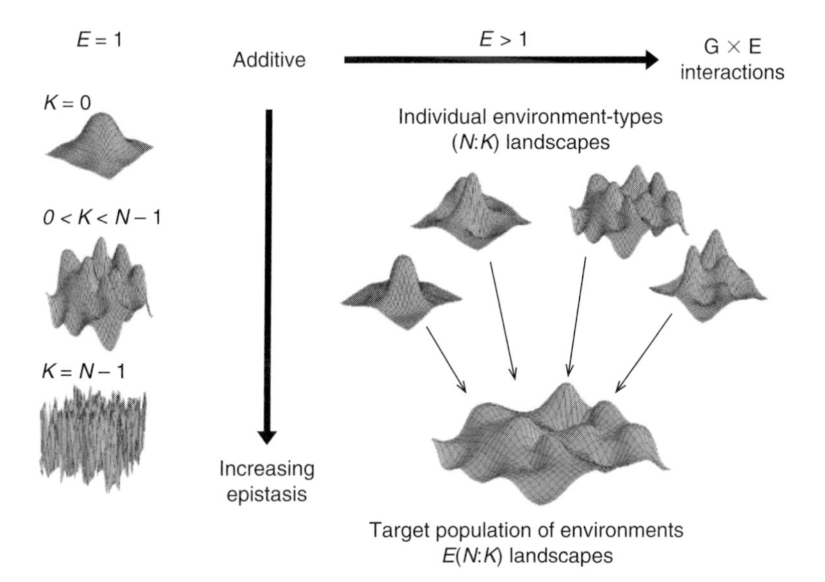

Fig. 10.8. Schematic three-dimensional representation of the phenotypic state-space performance landscapes for gene-to-phenotype (G → P) models simulated using the *E*(*NK*) model. The additive $E(NK) = 1(N : 0)$ G → P model is depicted as a single-peak landscape. Models with increasing levels of epistasis (i.e. from $K = 1$ to $K = N - 1$) are depicted by an increasingly more rugged landscape surface. Models with GEIs are depicted as a series of different landscape surfaces for different environment-types (*E*). The G → P response surface for the target population of environments (TPE) is depicted as a mixture of the response surfaces from the different environment-types. From Cooper *et al.* (2005) ©CSIRO 2009.

observed that as E and K are increased we move from a single peaked additive landscape for the $E(NK) = 1(N{:}0)$ case to a multiple peaked landscape and ultimately a random landscape when $K = N - 1$ and $E > 1$ (Cooper *et al.*, 2002b).

Recent advances in genome sequencing and high-throughput technologies, such as DNA and protein chips as described in Chapter 3, allow us to measure the spatiotemporal expression levels of thousands of genes or proteins. As gene expression in microarray experiments becomes increasingly popular in genetic studies of quantitative traits, it is possible to detect the relative signals of many genes simultaneously, allowing better understanding of genetic networks associated with specific developmental stages and/or environmental factors. Combining large-scale microarray experiments with genetic network simulation, it can be expected that GEI will be further revealed at the whole genome level and in the context of genetic networks.

11

Isolation and Functional Analysis of Genes

One of the challenges in molecular breeding is to understand how thousands of gene products interact with each other to control development and the ability of an organism to respond to its environment. Gene isolation and its functional analysis are not only for development of functional markers but also for manipulating plants through genetic transformation. For sequenced plant species identification of function for each gene has become a major focus in the era of functional genomics. For example, the *Arabidopsis* community has developed an initiative to empirically identify the function of all *Arabidopsis* genes by the year 2010.

To isolate and characterize all the genes in plants, it is important to first define what we mean by a gene. A gene was initially defined as the nucleic acid sequence that codes for a peptide. The definition now is extended to encompass many more features including the presence of gene families within a plant, alternative splicing, RNA that functions without translation into a protein and other confounding factors that together make a simple universal definition more difficult (Cullis, 2004). A gene, when defined as a transcribed (and translated) unit, is usually split into coding pieces (exons) that are separated by intervening sequences (introns) in the eukaryotic genomes.

All techniques for gene isolation exploit one or more of the four characteristics that define genes (Gibson and Somerville, 1993): they have a defined primary structure (sequence); they occupy a particular location within the genome; they encode an RNA with a particular expression pattern; and many genes encode protein or mRNA products with a defined function. Therefore, identification and functional analysis of genes can start at various points in the process of gathering information about genomes: they can be identified from their locations relative to closely linked markers on a genetic map, from their presence in populations of RNAs, from an analysis of genomic sequence data with gene finding programs, from comparisons with the genomic sequence data from related organisms, or from their disruption with the subsequent appearance of a phenotypic variant (Fig. 11.1; Cullis, 2004).

In a simple organism such as *E. coli*, one can readily isolate individual genes associated with a particular function. In a few Petri dishes one might select among millions of individuals to identify mutants in the function of interest, then artificially introduce wild-type (e.g. non-mutant) DNA into the mutants and identify those segments which restore normal function. Having a gene in hand, one might determine the sequence of genetic information comprising the gene,

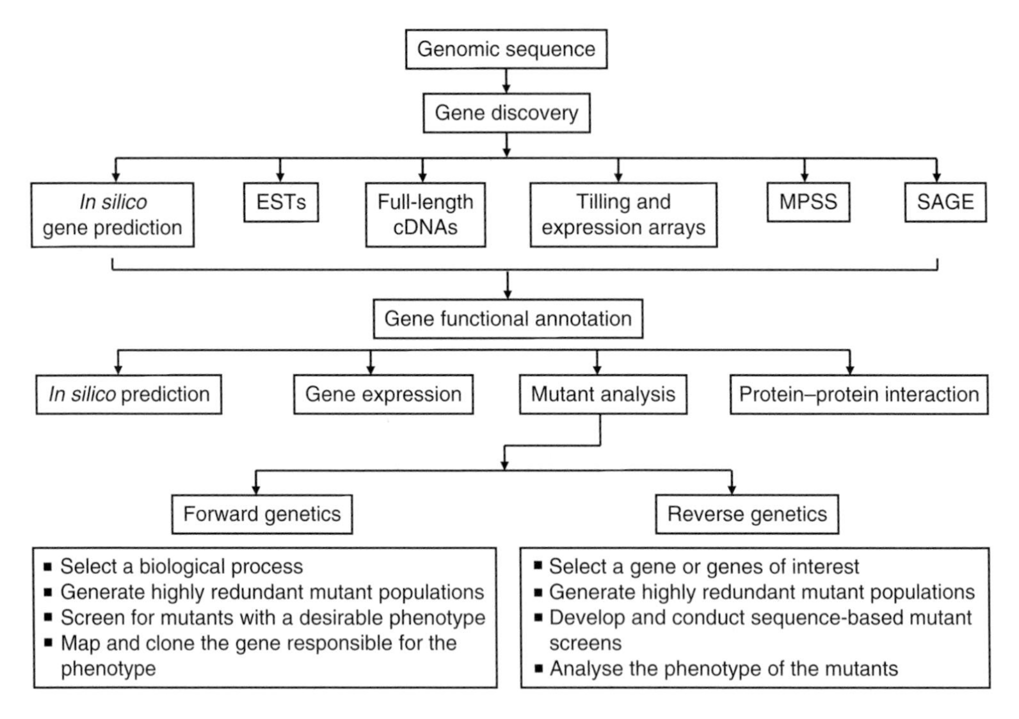

Fig. 11.1. From genomic sequence to gene function. Steps and experimental approaches that are used in the functional annotation of the genome. MPSS, massively parallel signature sequencing; SAGE, serial analysis of gene expression. Modified from Alonso and Ecker (2006).

the protein product encoded by this information, the function of the protein, the regulation of activity of the gene or protein by environmental factors and so on. Such elegant schemes for gene cloning are now being used for plants, but with significant challenges. Higher plants tend to have relatively large quantities of DNA in their genome and more non-coding DNA than coding DNA, making it difficult to identify particular genes of interest. Further, higher plants have relatively long generation times, months to years (versus a few minutes for *E. coli*). Based on phenotypes, one is seldom able to study millions of individuals or know enough about the function of a gene in order to isolate it from the many thousands of other genes in the organism. Many major genes have been cloned based on various approaches, but for traits affected by several genes or quantitative trait loci (QTL), the effects of any one are often partly masked by others and/or by environments. Thus, it is difficult to discern the effect of a single gene (QTL) by merely looking at the appearance (phenotype).

The main source of empirical information about gene function and structure has been the capture and characterization of mRNA transcripts (Fig. 11.1). A variety of high throughput methodologies have been successfully used for the model plant, *Arabidopsis thaliana* (Alonso and Ecker, 2006), including expressed sequence tags (ESTs) and full-length cDNA sequencing, whole genome tiling microarrays and gene-expression arrays, a massively parallel signature sequencing (MPSS) technique and serial analysis of gene expression (SAGE). Such methods provide information on gene splicing and transcribed units. The most widely used methods for isolating genes based on their functions involve protein purification, complementation of mutant phenotypes, positional cloning using genetic maps and mutagenesis-based gene identification. The major limitation to gene cloning based on its function is that for about half of the genes in most organisms the functional or physiological properties of their gene products are unknown or the corresponding proteins cannot be purified in sufficient quantities to permit amino acid sequence determination or preparation of antibodies. As described in this chapter, however, there are many elegant strategies for cloning genes in plants.

Objectives of this chapter are to review some basic methods that have been used for isolation and functional analysis of plant genes. These methods include those based on *in silico* prediction, comparative genomics, cDNA sequencing, microarray, map-based cloning and mutagenesis. For more comprehensive discussion, readers are recommended to refer to Gibson and Somerville (1993), Foster and Twell (1996), Jenks and Feldmann (1996), Paterson (1996b), Weigel *et al.* (2000), Davuluri and Zhang (2003), Ramakrishna and Bennetzen (2003), Cullis (2004), Jeon *et al.* (2004), Seki *et al.* (2005), Windsor and Mitchell-Olds (2006), Gibrat and Marin (2007), Nicolas and Chiapello (2007), Candela and Hake (2008) and Jung *et al.* (2008).

11.1 *In Silico* Prediction

Recent improvement in sequencing technologies has made large-scale DNA sequencing practical and widely accessible, which has enabled the development of sequence-based methods for identifying genes through exon discovery in genomic sequence data (Nunberg *et al.*, 1996). Ideally analytical methods should not depend only on a gene from another species that has already been isolated and sequenced. For organisms with compact genomes such as bacteria and yeast, exons tend to be large and the introns are either non-existent or short, so that the identification of genes by computational approaches is relatively straightforward. However, the challenge is much greater for larger genomes such as those of plants, because the exonic 'signal' is buried under non-genic 'noise'.

Currently, computational sequence analysis methods are usually exploited as a complement to and component of other functional genomics approaches. There are two main categories of computational methods that detect genes: extrinsic approaches relying essentially on comparison with other related sequences and intrinsic approaches based only on the local properties of the sequence under scrutiny (nucleotide composition and sequence motifs) (Windsor and Mitchell-Olds, 2006). Methods using information intrinsic to the sequence (Fig 11.2A) are based on the analysis of sequence, without referring to other sequences stored in the databases. These methodologies are commonly encountered as *ab initio* tools and are, by definition, not comparative (Davuluri and Zhang, 2003). *Ab initio* gene prediction algorithms display high sensitivity, but a low specificity in their output models. Extrinsic gene prediction methods are based on extrinsic data (including expression evidence) and/or sequence similarity (Fig. 11.2B, C, D), which supplement *ab initio* prediction by providing improved specificity and complementary sensitivity.

The first two features to be placed on newly acquired assembled genomic sequences are the possible open reading frames (ORFs) and splice sites. These two sets of data are combined to identify both already known and putative genes. Other available information that can be added to the sequence includes markers for genetic maps, submitted genes and ESTs and other EST data from different species. Comparisons of gene family members within and between species yielded the expected result that the genes were most highly conserved between the most closely related species. Moreover, sequence conservation was greatest in the protein-coding portions of the exons. All these make computational methods for gene identification practical.

11.1.1 Evidence-based gene prediction

Gene prediction frameworks using expression data, also called evidence-based gene

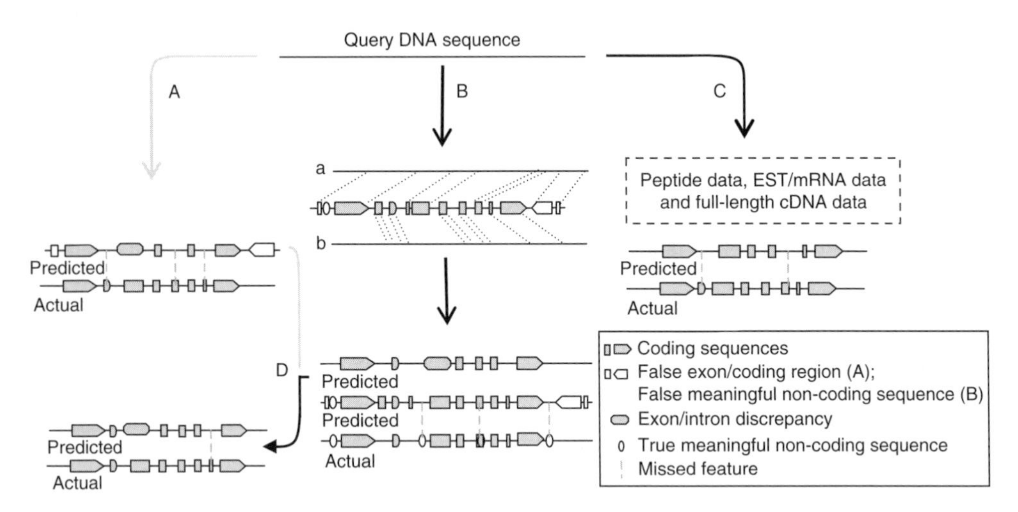

Fig. 11.2. Intrinsic (light grey arrows) and extrinsic (dark grey arrows) methods for gene prediction. Thick black lines represent query sequence data. Watson–Crick-strand coding sequences are indicated above or below sequence strands. Path (A), *ab initio* gene prediction algorithms model gene content with data from the query sequence itself. These methods can miss features such as small ORFs and small introns. Exons are also missed, but *ab initio* methods can erroneously identify exons or whole coding sequences. Generally, these methods are not applicable to the prediction of functional non-coding sequences. Path (B), similarity-based gene prediction. These methods are comparative and incorporate data from the alignment of one or more syntenic DNA sequences. Similarity-based methods display improved sensitivity and specificity for coding and non-coding sequences over *ab initio* methods. The ability to predict genes or conserved features is a function of the number of sequences compared, the evolutionary distance of these sequences, and the degeneracy and size of the features in the homologous sequences. Path (C), evidence-based gene prediction. These methods can be computational or experimental and display high specificity but low sensitivity. The efficacy of the prediction is contingent on the quality/extent of available expression data. Path (D), combinatorial approaches. In the example presented, similarity evidence is combined with an *ab initio* prediction to improve the overall prediction of gene content. Reprinted from Windsor and Mitchell-Olds (2006) with permission from Elsevier.

prediction, integrate empirical transcription and protein expression data with genome sequence to produce gene models (Fig. 11.2C) and facilitate annotation. Such data provide high specificity to gene model prediction, but sensitivity is contingent on the extent of the expression data sets. This property has negative impacts on the identification of sequences with tightly regulated or low-abundance transcripts or of RNA species that are not translated. The incorporation of expression data from multiple species can overcome some of these limitations and allows the use of species with no or partial genome sequence in comparative analyses (Windsor and Mitchell-Olds, 2006).

The analysis of the transcriptome and proteome data sets for the kingdom *Plantae* suggested that approximately 19,000 gene functions are encoded in the green plant lineage (Vandepoele and Van de Peer, 2005), nearly 6500 of which are encoded by orphan genes, novel genes that cannot be definitively assigned to a characterized homologue or gene family.

11.1.2 Homology-based gene prediction

Sequence homology or similarity is a very powerful type of evidence for detecting functional elements in genomic sequences. The homology-based methods to detect genes use either intraspecies or interspecies sequence comparison in different ways

(Davuluri and Zhang, 2003; Windsor and Mitchell-Olds, 2006; Nicolas and Chiapello, 2007). Most of the algorithms use one of three types of external information: protein sequences, mRNA sequences or DNA sequences.

Comparison with EST/cDNA databases

It has been demonstrated that homology-based cloning was a very effective way to identify tissue-, organ- and developmental stage-specific expressed genes by assembling EST sequences derived from the same library and performing homology searches in databases. In an early report, about 18% of 5000 human clones were assigned probable function (Adams *et al.*, 1991) by simply picking random cDNA clones, obtaining partial sequence from the 5' end and comparing the six possible translations of the partial cDNA sequences with the sequences of known proteins in the various databanks. The sequencing of mRNA through the sequencing of cDNA is an experimental technique to determine the sequence of genes that are transcribed. As sequencing takes places after splicing, cDNA sequencing also allows precise determination of the intron–exon structure of genes. By directly comparing a genomic DNA sequence (query) with ESTs or cDNA, regions of the query sequence that correspond to processed mRNA can be identified.

BLASTN is a common program that identifies similar nucleotide sequences that exist in databases (nr/EST) to the query sequence (see Basic Local Alignment Search Tool (BLAST) help at http://www.ncbi.nlm.nih.gov/BLAST for further details about BLASTN and other programs). The similarity between the sequences is estimated by aligning the sequences as closely as possible. The BLASTN algorithm finds similar sequences by generating an indexed table or dictionary of short subsequences called words for both the query and the database. However, determination of the DNA structure of genes from ESTs is not trivial, because these sequences are generally incomplete (often sequenced at the 3' end), of poor quality (sequenced only once), redundant (as a function of the abundance of mRNA) and of alternative splicing. The problem is more severe when the mRNA sequence has been obtained in a different organism (Nicolas and Chiapello, 2007). If the query sequence is very long, MEGABLAST is a better choice, which is specifically designed to efficiently find long alignments between very similar sequences. MEGABLAST is also optimized for aligning sequences that differ slightly as a result of sequencing errors. Davuluri and Zhang (2003) suggested the use of an expected value (e-value) of 0.1 and filtering for low complexity repeats. When larger word size (with the default value 28) is used, it increases the search speed and limits the number of database hits. For BLASTN, the word size can be reduced from the default value of 11 to a minimum of 7 to increase the sensitivity. Algorithms that can be used for DNA–cDNA and DNA–EST alignments include SIM4 (Florea *et al.*, 1998) and GENESEQUER (Usuka *et al.*, 2000).

Similarity-based methods (e.g. BLASTN, BLASTX) are perhaps the best to determine whether a given region of the genome is transcribed or not. A BLASTN match to a cDNA/EST or BLASTX match to a protein is good evidence that the region belongs to a gene. However, these methods have their own limitations (Davuluri and Zhang, 2003). Even the most comprehensive cDNA projects will miss low copy number transcripts and those transcripts whose expression is low, cell- or tissue-specific, or expressed only under unusual conditions. cDNA or ESTs can contain one or more introns, if the mRNA was partially spliced, which could lead to misclassification of intron regions as exons. Some cDNA sequences may result in incorrect protein prediction. Partial BLASTX alignment to a target protein should not be considered, as the protein may not be a true orthologue of the source gene and only shares some domains, although it would still give some information for gene prediction.

Comparison with protein sequence databases

DNA–protein similarity can be compared to predict the protein coding sequences that resemble proteins already present in the databases. BLASTX (Gish and States, 1993) and FASTX (Pearson *et al.*, 1997) are two programs for similarity analysis by local alignment between a DNA sequence translated in six reading frames and a protein sequence library. Subsequently, spliced alignment programs such as GENE-WISE, GENESEQUER, or PROCUSTES can be used to find gene structure by comparing the genomic sequence to the target protein sequences. These programs derive an optimal alignment based on sequence similarity score of the predicted gene product to the protein sequence and intrinsic splice site strength of the predicted introns. However, to predict the structure of a coding sequence composed of multiple exons, the DNA sequence and the protein sequence must be aligned taking into account the presence of introns that constitute long fragments of the DNA sequence that do not match with the protein sequence (Nicolas and Chiapello, 2007). PROCUSTES (Gelfand *et al.*, 1996) and PAIRWISE (Birney *et al.*, 1996) were the first two programs developed to resolve this spliced alignment problem.

Protein coding DNA from closely related plant species, such as sorghum and maize, show considerable sequence similarity. VISTA/AVID and PIGMAKER can be used to compare large genomic sequences to find orthologous genomic sequences from closely related species. For example, sequence analysis of orthologous genes from rice, maize and sorghum showed that the exons are more conserved than introns (Schmidt, 2002). The degree of sequence conservation, in terms of sequence identity, across species has been shown to be consistent with the divergence times of the respective species. For gene prediction programs, it would be best to compare two genomes that are very closely related, but distant enough that their intergenic repeat elements differ significantly. As a rule of thumb, two species are considered closely related if they diverged within the last 25 million years. For example, maize and sorghum are closely related species as they were diverged 15–20 million years ago. If homologous genomic sequences from two species are known, then a recently developed gene prediction tool called SGP-1 can be used to find protein-coding genes (Davuluri and Zhang, 2003).

Comparison of a translated genomic sequence with translated nucleotide database

TBLASTX can be used to identify similarities among protein coding regions by selecting 'Nucleotide query – Translated db [tblast]' option from the BLAST web page. TBLASTX takes a nucleotide query sequence, translates it in all six frames and compares the translations to nucleotide database sequences that are dynamically translated in all six frames.

Comparison of homologous genomic sequences

Often scientists want to isolate from a particular organism, a gene orthologous to one already isolated from another organism while others may be interested in isolating from the same organism other members of a gene family (paralogues) for which at least one cloned member is available. The homologous genes (orthologues or paralogues) may be used as probes to isolate the target genes from a library or the degenerate primer method. As the full genomes of plant species such as *Arabidopsis* and rice are more precisely annotated, the finding and isolation of potential genes in other, less well-defined systems may be possible with reference to the position of the sequence in a particular cluster of genes through synteny. However, these predictions are likely to be complicated by the presence of multiple copies of genes, the divergence between paralogues and orthologues in other species (see Chapter 1 in Cullis, 2004) and the micro- and macro-rearrangements of the chromosomes over evolutionary time. Therefore, any candidates will need to be extensively characterized to demonstrate that they are performing the same function in both time and space. Comparison of rice and *Arabidopsis* genome sequences showed that 90% of the *Arabidopsis* genes had a putative homologue in rice, where only 71% of rice genes had a putative homology in *Arabidopsis* (IRGSP, 2005). The creative application of similarity-based analyses has allowed the identification of novel coding sequences. Several thousands of conserved unannotated regions were recognized in *A. thaliana* relative to the partial genome sequences of *Brassica oleracea* (Ayele *et al.*, 2005; Katari *et al.*, 2005). In these approaches, conserved genomic regions, as identified by TwinScan, with physical proximity in the *A. thaliana* reference genome were chained together to produce novel gene models.

Homology-based cloning has been effective when the gene of interest is a known member of a multi-gene family. Often, amino acid sequence alignments of family members reveal particularly conserved regions. Consequently, degenerate oligonucleotides can be designed and used in library screening or directly for PCR cloning of the gene.

11.1.3 *Ab initio* gene prediction

Comparison of sequence differences between coding and non-coding regions has encouraged the development of prediction methods based on the probabilistic modelling of DNA sequences, which helps overcome the limitations of homology-based methods. *Ab initio* gene finding programs recognize signals of compositional features in an input genomic sequence by pattern matching or statistical methods. The performance of a gene-finding program is typically measured in terms of the sensitivity, defined as the proportion of true signals (e.g. donor signals, exons) that are correctly predicted and specificity, defined as the proportion of predicted signals that are correct. A program is considered accurate if its sensitivity and specificity are simultaneously high. A comprehensive review of these programs can be found at http://linkage.rockefeller.edu/wli/gene/. Stein (2001) reviewed various genome annotation methods.

Splice site prediction

As most plant genes have several exons, precise gene structure prediction in plants very much depends on correct splice site prediction. Although nearly all introns begin with GT and finish with AG, this information is not enough for the spliceosome to choose the splicing sites. Other sequence signals around these dinucleotides are used. Most splicing site recognition methods are based on the evaluation of the sequences of potential sites by using a probabilistic model that describes position by position the nucleotide composition of actual splicing sites. Many first generation gene prediction programs used simple position weight matrix methods to model the compositional biases present in the 5' and 3' splice sites. A limitation of such models is that they do not account for dependencies between positions that are not consecutive (O'Flanagan *et al.*, 2005). Most recent programs have investigated the correlations between different positions by using Markov models, maximal dependence decomposition models, decision tree models and artificial neural networks (as reviewed by Davuluri and Zhang, 2003). GeneSplicer, NetPlantGene, NetGene2 and SplicePredictor are some of the splice site prediction programs that use splice site models. Other models such as Bayesian networks that account for correlation between non-adjacent positions have also been proposed (Chen *et al.*, 2005).

Exon prediction

Exons are defined by what is retained in a spliced mRNA, which includes untranslated regions (UTRs) and the protein coding regions. The protein coding exons typically are of four types: (i) initial exons (ATG to first donor site); (ii) internal exons (acceptor site to donor site); (iii) terminal exons (acceptor site to stop codon); and (iv) single exons (ATG to stop codon without introns). Most of the gene prediction programs have been developed to predict protein coding exons. The accuracy of splice site prediction and hence exon prediction, by second generation programs (e.g. Genscan, GeneMark.hmm, mzef or

SPL) is significantly higher than simple splice site prediction programs described above, because these programs integrate splice site models with additional types of information, such as compositional features of exons and introns. MZEF, based on quadratic discriminant analysis, was specifically trained to predict internal exons (Davuluri and Zhang, 2003). It was shown to perform better than FGENESP, GRAIL, GENSCAN and GENEMARK.HMM in predicting internal exons for the *Arabidopsis* genome. For predicting initial and terminal exons, GENSCAN and GENEMARK.HMM are the best options, even though the accuracy of predicting these exons is significantly lower than that of internal exon prediction.

Gene modelling

The accuracy of individual exon prediction can be further improved by combining the compatibility of the reading frames of adjacent exons to make a full coding transcript. Probabilistic models, such as Hidden Markov Models, have been used to incorporate this information in GENSCAN and GENEMARK.HMM, which model different states (exon, intron, intergenic regions, etc.) of a gene. The GENEMARK program implements a sliding window strategy. The window slides along the sequence and at each position the program computes the probability of the sequence contained in the window under seven models: non-coding and coding on each of the two strands in each of three reading frames, to obtain a probability of the locally coding nature of the sequence in each reading frame. A simple alternative to sliding windows is implemented by GLIMMER (Nicolas and Chiapello, 2007). The program first extracts ORFs longer than a certain threshold and, secondly, attempts to classify them according to their coding or non-coding nature.

11.1.4 Gene prediction by integrated methods

Gene prediction by homology-based methods is perhaps the most efficient way of finding genes in genomic sequences, since the evidence of support (mRNA, EST, protein) was already derived experimentally (Davuluri and Zhang, 2003). *Ab initio* gene-prediction programs do not rely on such data, but miss some known genes (false negatives) and predict some that are not real (false positives). A combination of *ab initio* gene prediction programs and homology-based approaches has been automated in several programs such as GENOMESCAN and RICEGAAS to produce more reliable predictions of protein-coding regions. GENOMESCAN incorporates protein homology information (BLASTX hits) with the exon–intron predictions of GENSCAN. It first masks the interspersed repetitive elements in the genomic sequences with REPEATMASKER and then combines the GENSCAN predicted peptides with BLASTX hits. The program determines the most likely 'parse' (gene structure), conditional on the given similarity information under a probabilistic model of the gene structural and compositional properties of genomic DNA for the given organism. There are two major ways to integrate different approaches to improve the prediction (Nicolas and Chiapello, 2007). The first way is to use the programs separately, then carry out a post-treatment of the results. JIGSAW (Allen and Salzberg, 2005) is an example of program developed for this task. It uses dynamic programming algorithms to automatically combine predictions made with independent programs. The second way is to develop programs that have their predictions simultaneously based on intrinsic and extrinsic criteria.

Despite great progress, gene prediction by computational approaches alone is still far from perfect. Comparing the performances of different approaches is a careful and difficult task as indicated by Nicolas and Chiapello (2007). The difficulties arise partly from the diversity of information taken into account as well as the diversity of the predictions made (complete coding sequences, exons, splicing sites). Just as the quality of intrinsic methods depends clearly on their adjustment to a given species, that of extrinsic methods depends on the degree of similarity between the sequences compared.

Last but not least is the problem of a gold standard that could serve as a reference for the comparison of the approaches.

Before running any gene-finding program, Davuluri and Zhang (2003) suggested the use of programs such as RepeatMasker, which identifies known classes of interspersed repeats and long and short interspersed nuclear elements (LINEs and SINEs), which exist in non-coding regions of the genome. Almost all gene finding programs can predict only protein coding regions and have not been trained to predict untranslated exons and untranslated portions of first and last coding exons. In addition, identifying the exact boundaries of all the exons and assembly of the exons into different genes is not possible by computational approaches alone. As indicated by Davuluri and Zhang (2003), however, even the partial predictions are of immense value to design the experiments that can determine the complete gene structure faster than would be possible by experimental methods alone.

11.1.5 Detecting protein function from genomic sequences

There are three major classes of *in silico* methods used to obtain information on protein function: methods using information intrinsic to the sequence; homology search methods; and methods based on the context of genes (Gibrat and Marin, 2007). Homology search techniques provide precise information and thus occupy a central place, while others give only general information.

Intrinsic methods

The methods using information intrinsic to the sequence detect recognisable protein structure such as transmembrane segments, zones of low complexity, 'coiled coils' and cellular sorting signals (Gibrat and Marin, 2007). As transmembrane segments are mostly made up of hydrophobic residues, the detection methods are based on the search for segments that have an appropriate size and present a marked hydrophobic character. Most of the transmembrane segments are made up of α helices and around 25–30 residues are required for the polypeptide chain to cross the membrane in the form of an α helix. Some segments of protein sequences have an over-representation of a small number of amino acids or even show a more or less regular repetition of a particular peptide. They do not have a conventional three-dimensional globular structure. These zones rich in specific amino acids supply practically no information on the function of proteins and they must be masked before using homology search methods, because their abnormal amino acid composition disturbs the statistics associated with these techniques and often results in false inference of homology. Coiled coils zones are formed by a bundle of two or three α helices and they can be detected based on statistical techniques that take into account the probability of observing a particular amino acid at each of several characteristic positions (Lupas, 1996). Cellular sorting of proteins towards organelles that are their final destination depends on signals present in the primary structure. Techniques based on the overall composition of amino acids can be used to predict the location of proteins in various organelles.

Homology search methods

Homology search methods play a central role in *in silico* functional analysis to discover similar proteins in databases. There are two principal categories of homologous proteins: orthologous proteins (which result from a process of speciation) and paralogous proteins (which result from a process of duplication). Gibrat and Marin (2007) evaluated different means by which a relationship of homology between two proteins can be inferred, including sequence comparison, profile detection, motif detection and fold recognition.

As with DNA sequence comparison, protein sequence comparison is the most natural and the oldest method of indicating a relationship of homology between two proteins. BLAST and FASTA are sequence comparison methods based on the principle that two sequences with a common ancestor should maintain some traces of that relationship in the sequences. Profile detection, based on a multiple alignment of similar sequences, can be used to estimate the variability of each of the positions along the sequence of a protein. Multiple alignments can be made by using PSI-BLAST (Altschul *et al.*, 1997), which constructs the multiple alignments iteratively during the search by comparing with profiles of families of proteins from databases. Motif detection is to search for motifs that correspond to a functional signature, or even to residues necessary to maintain the correct geometry of the active site of a protein. As these residues are crucial for the function of the protein they are thoroughly conserved. Some programs, such as SCANREGEXP and PFSCAN, can be used to search for motifs characteristic of particular proteins in the Prosite library of motifs (Hofmann *et al.*, 1999). Fold recognition methods are based on the alignment of a sequence on a three-dimensional structure to indicate relationships of homology, which can be used to reveal the distant homologues that cannot be detected by sequence comparison methods.

Gene context methods

The methods based on the context of a gene depend on studying the co-location of genes in different genomes. Unlike homology search techniques that often provide information on the molecular function of proteins, this type of technique generally provides information on the interactions between proteins and thus their cellular role. With gene context methods protein function can be detected based on three different concepts: gene fusion, gene proximity and gene co-occurrence (Gibrat and Marin, 2007).

Gene fusion has been evident by observation that a metabolic pathway made up of independent enzymes in the prokaryotes is catalysed by a multi-enzyme system. Two proteins that exist in the independent form in one genome and that are fused in another genome have a high chance of being in close interaction. The gene fusion of functional models can be classified into those that occurred in an ancestor common to most of the present lines of organisms, those concerning the genetically mobile domain that are found in different proteins and those similar to cytochrome P450, providing some information on interaction between proteins (Gibrat and Marin, 2007). Gene proximity methods are based on the observation that functionally linked genes are co-regulated and have a tendency to be close together in the genome and that the position of a gene in the genome may provide information about its function. Protein function can be predicted, e.g. by measuring a local proximity that involves the conservation of nearby gene pairs in different genomes being compared. Gene co-occurrence is based on the concept that genes that are implicated in a particular cellular process, common to genomes of this set, can share an identical phylogenetic profile. Thus, an unknown protein that shares a phylogenetic profile with proteins that are known to be implicated in a particular cellular pathway has a high chance of playing a role in that pathway.

11.2 Comparative Approaches for Gene Isolation

Comparative approaches for gene isolation include *in silico* methods that have been described in the previous section, experimental procedures involving map-based cloning using comparative analysis between closely related species and major-gene assisted QTL mapping by comparing traits controlled by genes with different functions. These approaches have been used at different scales due to their various levels of practicability.

11.2.1 Genomic bases of comparative approaches

It is expected that the functional genomics of model plants will contribute to the understanding of basic plant biology as well as to the exploitation of genomic information for crop improvement. This is because a large number of gene functions are conserved across species, either directly or after identifying the functional homologues. Perhaps the most exciting application of comparative genomics will be the identification of different versions of genes for a target species from other related species. Orthologous genes in related species will be similar in sequence and function to those in the target species but could result in markedly different phenotypes (Xu *et al.*, 2005).

Conservation of gene content and gene order among closely related plant species greatly assists in gene identification and annotation. Even in closely related plant genomes, whose ancestors diverged from each other more than 10 million years ago (mya), only genes are conserved in orthologous regions. All of the plant species with large genomes studied have been evolved by the moment of retrotransposons within the last 6 million years and these sequences vary greatly between species (Ramakrishna and Bennetzen, 2003). Hence, plant species that diverged from each other more than 50 mya only have exonic regions conserved among genes. This feature has been used to improve gene annotation with great success (Tikhonov *et al.*, 1999; Dubcovsky *et al.*, 2001; Ramakrishna *et al.*, 2002). Gene structure can be predicted more accurately using comparative sequence analysis than by the combined use of ESTs, homology to entries in protein databases and gene prediction programs as described in the previous section. Conservation of genomic collinearity, gene content and order among plant genomes greatly assists in gene isolation from cross-species comparisons.

Differences in gene content are sometimes observed in otherwise collinear regions of plant genomes (Tikhonov *et al.*, 1999; Tarchini *et al.*, 2000; Ramakrishna *et al.*, 2002; Bennetzen and Ramakrishna, 2002). This phenomenon can complicate gene isolation, but does not completely invalidate the approach. Under almost all circumstances, a small genome species will provide numerous DNA markers on a single bacterial artificial chromosome (BAC), which permits more detailed mapping in the large genome species. Chromosome walking, as a basic step in map-based cloning to be described in detail later, is often difficult with large genomes such as barley, maize and wheat. In these cases, related plant species with small genomes such as rice, which show genomic collinearity with the large genome species, can be used to identify and isolate desired genes. This approach has potential pitfalls, especially with respect to some disease resistance genes (Kilian *et al.*, 1997; Leister *et al.*, 1998; Pan *et al.*, 2000). Resistance gene regions often undergo rapid rearrangement that results in a lack of micro-collinearity caused by deletion or translocation of the target loci. However, at the very least, the comparative genomic approach provides numerous probes from one species, which can be used for gene mapping and isolation in another species (Ramakrishna and Bennetzen, 2003).

Comparative genetics has been facilitated by the development of massive databases, efficient querying and comparison software and ever-improving computers. Many of the first genes sequenced in rice and other grasses were represented by abundant mRNAs (e.g. those encoding storage proteins and photosynthetic proteins). Members of the same gene families (e.g. paralogues), including those that were mapped to the same genomic position and thus were derived by vertical descent from a common ancestral gene (i.e. orthologues), were often cloned and analysed in multiple species (Bennetzen and Ma, 2003).

When sequences from two separate parts of the gene are moderately conserved, degenerate oligonucleotides based on the two sequences can be used to attempt to amplify the intervening sequence by PCR. An amplified RT-PCR product or a degenerate oligonucleotide may be used in nucleic acid hybridization to screen the colonies or plaques of a cDNA/genomic library for clones containing the gene of interest. Positive clones from a genomic library need to prove that the clone actually contains the gene and to be further examined to identify where on the clone the gene is located.

11.2.2 Experimental procedures involved in comparative analysis

Ramakrishna and Bennetzen (2003) described methods for plant gene isolation based on comparative genetic map and/or genomic sequence information. This technique involves identification of collinear regions, followed by clone selection and finally, sequence analyses to identify the gene of interest.

Basic procedures

IDENTIFICATION OF COLLINEAR REGIONS. The genetic map position of the targeted locus in the plant species with a large genome size must be determined accurately by segregation analysis of the locus with tightly linked markers. These markers should map to a collinear region in a related plant species with a small genome to enable isolation of the targeted locus. Comparative genetic linkage maps with common molecular markers serve as the best starting point. For example, the maize genome is about 2400 Mb in size, corresponding to a genetic map of about 2500 cM. This translates to an average of 1 Mb/cM for the maize genome. A large mapping population of 5000 gametes with no recombinants in the segregating progeny makes it likely that the targeted gene is present within a 500-kb region. The rice genome has a size of 380–450 Mb and a genetic map of about 1600 cM. This makes map-based gene isolation much easier in rice than in maize.

In cases where the gene of interest is absent in the small genome, we can still use markers from the orthologous region in the small genome to fine-map in the large genome, as markers are often the limiting factor for fine-mapping in some crop species. The complete genome sequence in crop plants provides abundant information for choosing suitable markers. For example, the maize BAC libraries can then be screened with suitable markers from rice to identify BACs that harbour the gene of interest. The next step is to look for the presence of flanking markers (tightly linked to the targeted gene) on continuous BACs in maize. The results of such studies will show whether overall collinearity is maintained in the region.

CLONE SELECTION AND MAPPING. Several thousand clones from the small genome BAC libraries are screened for individual clones that show homology to DNA markers mapped in the collinear regions in different plant species.

CONSTRUCTION OF SHOTGUN LIBRARIES AND SEQUENCING. These two steps can follow a standard procedure described in Chapter 3.

SEQUENCE ANALYSES AND ANNOTATION. The first step in the sequence analysis of collinear BACs (for instance, when a collinear sorghum BAC is sequenced to isolate a gene based on the genetic map location in maize) is the delimitation of regions that are conserved and not conserved relative to rice. Conserved regions are usually or always genes, while the unconserved regions are usually not genes. Complete sequences from orthologous BACs are then compared using the program DOTTER to identify the conserved regions. Genes can be predicted as described in the previous section.

CONFIRMATION OF CANDIDATE GENES. The possible functions of candidate genes can be investigated using several independent approaches. Sequence analyses and annotation, as described above, using comparative sequence analyses, gene-finding programs and BLAST searches, identify putative genes. Sequence variations and gene structure analysis of the gene identified in the region, for instance in susceptible and resistant lines in case of disease resistance genes, can help verify a candidate gene. As an example, preliminary mapping, cloning, sequencing, gene finding and BLAST searches identified two candidate genes for barley *Rpg1*. These were differentiated by segregation analysis in 8518 gametes and by sequence analysis in barley lines susceptible or resistant to stem rust (Brueggeman *et al.*, 2002).

Additional experimental analyses can be performed to evaluate candidate gene function. Several approaches can be used including mutation analysis and expression analysis.

Mutation analysis can follow these steps (Ramakrishna and Bennetzen, 2003):

- Analysis of knock-out mutations (i.e. transfer DNA (T-DNA) or transposon insertions).
- Wild-type lines that either have a non-functional or an overexpressed gene of interest can be generated by transforming wild-type plants with antisense or sense gene constructs.
- RNA interference can be employed, where homologous double-stranded RNA (dsRNA) is used to suppress a gene, generally resulting in a null phenotype (same as above, combine these two).
- Complementation studies, where a wild-type copy of the gene of interest is transformed into the mutant to see if the T1 progeny yields wild-type phenotype and whether complementation co-segregates with the transgene in subsequent generations.
- Searching for point mutations by targeting induced local lesions in genomes (TILLING) to provide an allelic series of mutations.

Tissue-specific expression of the genes can be studied using Northern blot analysis, microarrays, reporter constructs, or reverse transcription-PCR to see if the expression patterns agree with the predicted biology of the targeted gene.

Examples

Probably the most comprehensive application of collinearity in plants was the attempt to clone specific barley disease resistance genes by chromosome walking using rice. The collinearity provided numerous DNA markers from rice that facilitated the chromosome walk in barley, leading to the isolation of the desired stem-rust resistance gene *Rpg1*, although the synteny with rice failed to yield the gene because it does not seem to exist in rice (Brueggeman *et al.*, 2002). Another example is the *Lr21* leaf-rust-resistance gene of bread wheat that was successfully isolated using a strategy of shuttle-mapping between diploid wheat as a model and bread wheat (Huang *et al.*, 2003). Most of the time, however, there are breakages in microsynteny that prevent the straightforward identification of a candidate gene for a given trait. This was the case when attempts were made to isolate the leaf-rust-resistance gene *Rph7* (Brunner *et al.*, 2003) or the photoperiod response gene *Phd-H1* (Dunford *et al.*, 2002) from barley. A similar story was reported for the *Rfo* restorer genes isolated from radish: markers flanking these genes in radish are collinear with the *Arabidopsis* sequence, but the gene itself is not present in *Arabidopsis* although many homologues are present elsewhere in the *Arabidopsis* genome (Brown *et al.*, 2003; Desloire *et al.*, 2003).

Examples for the use of a shuttle-mapping strategy have to be evaluated on a case-by-case basis. The present information, from both successes and failures, strongly suggests that the development of efficient tools for isolating genes of agronomic importance within each important family should continue to be a priority and that, as indicated by Delseny (2004), restricting ourselves to use of several model species would be unwise, although collinearity has been useful in providing additional markers with which to saturate fine genetic and physical maps.

11.2.3 Cloning QTL facilitated by related major genes

Robertson (1985) presented evidence that qualitative and quantitative traits may be the result of different types of variation of genic DNA at the loci involved. At any given locus, variation of a minor nature may result in wild-type alleles responsible for gene products with different efficiencies (quantitative alleles) while major genic rearrangements or changes in the region of the gene essential for a normal functioning gene product may result in qualitative mutant alleles. Based on this hypothesis, Robertson proposed a possible approach to cloning QTL. It is apparent from previous work that the alleles for quantitative variation assume possible allelic interactions and have a smaller individual effect than alleles from qualitative variation. However, it is possible that alleles for qualitative mutants are simply loss-of-function alleles at the same loci underlying quantitative variation. Consider, for instance, a trait such as plant height. In maize, at least 17 known qualitative mutants that affect plant height have been identified (cf. Robertson, 1985). These are non-allelic mutants, all of which have been placed on chromosomes. In rice, over 50 loci responsible for semi-dwarfism or dwarfism have been found and mapped (Kinoshita, 1995). If all these loci had two or more 'wild-type' alleles responsible for quantitative variation, the combination of these would come close to being sufficient to explain the quantitative inheritance pattern observed by breeders. Theoretically, QTL mapping studies can provide a test of this hypothesis. If a gene contributing to quantitative variation is allelic to a gene controlling qualitative variation, then these genes should map to the same locus along the chromosome. For some organisms (e.g. maize and *Drosophila*), many of the major qualitative loci controlling morphological variation have been mapped with a high degree of precision on genetic maps and these locations should be predictive of the locations of QTL for the same character. As indicated by Robertson (1985), it seems unreasonable, to say nothing of wasteful, to assume that a living organism would have two sets of loci, one for qualitative traits and one for quantitative traits, when one set could account for both patterns.

Robertson (1989) gave two examples to support his hypothesis. One of them is that a difference in gibberellin deficiency, controlled by a major gene, resulted in a quantitative difference of plant height. He also listed a series of qualitative traits which are related to the quantitative traits of same kind. This hypothesis has been tested in maize. Beavis and colleagues attempted to test the relationship of qualitative mutants to quantitative variation by mapping QTL for plant height in four maize F_2 populations and comparing the map position of these QTL with previously known positions of qualitative variations for the same character (Beavis *et al.*, 1991). The results showed a general concordance in map positions of QTL and major genes affecting height, which is consistent with the hypothesis.

With the development of practical QTL mapping, the similar location of QTL and major genes is supported by many research results and only several early examples will be discussed here. In maize, many QTL affecting plant height were located at known major loci (Edwards *et al.*, 1992), indicating that some QTL may be allelic to the major genes. In genetic analysis of rice blast resistance, a major gene was located on chromosome 8 by randomly amplified polymorphic DNA (RAPD) analysis on resistant and susceptible plants of a double haploid (DH) population (Zhu *et al.*, 1994). A QTL controlling quantitative resistance was found in the same chromosomal region when using molecular markers to map the resistance gene with quantitative phenotype data (Wang *et al.*, 1994). In *A. thaliana*, five QTL affecting flowering time were identified in a cross between two ecotypes, H51 and Landsberg *erecta* and four of them were located in regions containing mutations or loci previously identified as conferring a late flowering phenotype (Clarke *et al.*, 1995). Generally, associations between qualitative mutants and QTL are more often than expected by chance. In maize, for example, 75% of chromosome intervals harbouring discrete height mutants also harboured height QTL and 43% of intervals harbouring QTL also harboured mutants (cf. Lin *et al.*, 1995), although the association is by no means absolute. A report on QTL mapping in five rice populations detected 23 plant height QTL. According to linkage relationships determined with restriction fragment length polymorphism (RFLP) markers, all of the 13 major dwarfing or semi-dwarfing genes were found to be in close proximity to these plant-height QTL (Huang *et al.*, 1996). In *Drosophila*, the map positions of bristle QTL in every case corresponded approximately to those of candidate neurogenic loci or loci with major bristle phenotypes

(Long *et al.*, 1995). However, the QTL were located on the map with a low degree of resolution in most cases mentioned above, raising the possibility that the QTL are linked, but not identical to the qualitative loci, as indicated by Tanksley (1993). Until QTL are mapped to higher degrees of precision and/or cloned, it will be difficult to prove that the particular QTL actually correspond to known loci defined by macromutant alleles. As more and more QTL are cloned, whether there are any corresponding macromutant alleles can be tested.

As suggested by Helentjaris *et al.* (1992), the theory proposed by Robertson (1985) provides a possible approach to identification and cloning of important quantitative genes. For QTL cloning based on Robertson's proposal, both major and minor genes should be examined based on their relative contribution to expression of traits and their interaction with each other. It can be expected that genetic relationships can be established between known major genes and QTL, or QTL can be paralleled with the extreme mutants and one can verify these relationships and facilitate QTL cloning through cloning the related major genes.

11.3 Cloning Based on cDNA Sequencing

One way to identify genes is to clone and sequence RNAs. Short stretches of cDNA sequences, derived from mRNA, are referred to as expressed sequence tags (ESTs). ESTs are usually 200–500 nucleotides long and generated by sequencing either one or both ends of an expressed gene. cDNA sequence-based approaches for gene cloning have been extensively applied in humans, *Caenorhabditis elegans* and plants. The precise nature of the sequence information obtained by EST analysis and the ever increasing number of gene sequences of known function make it possible and productive to identify specific genes by sequence similarity as discussed in the previous section. The idea is to sequence cDNA that represents genes expressed in certain cells, tissues, or organs from different organisms and use these 'tags' to fish a gene out of a portion of chromosomal DNA by matching base pairs. The challenge associated with identifying genes from genomic sequences varies among organisms and is dependent upon genome size as well as the presence or absence of introns, the intervening DNA sequences interrupting the protein coding sequence of a gene. A large number of new genes have been identified in many species by randomly sequencing cDNA clones to produce ESTs. In view of the efficiency of this approach as a mechanism for establishing relationships between plant phenotypes and the large amount of sequence information available for other organisms, it is desirable to obtain large numbers of partial sequences.

11.3.1 Generation of ESTs

To generate EST sequences, the mRNA is isolated and reverse transcribed into cDNA. The cDNA clones are sequenced from either the 5' or 3' ends of the cDNA or from both ends (Chapter 3). The sequences are then clustered to identify a series of tentative unique genes (TUGs) or tentative contigs (TCs) and an estimate of the number of different RNAs present in the initial sample. The TUGs/TCs can then be compared with the current databases to identify which of these have already been described in the species under consideration and which are still absent from the current databases. Where hits occur to ESTs from other organisms, a possible function may be ascribable to the sequence (Cullis, 2004). The sequencing of any given sample is continued until the rate of finding new sequences drops below an acceptable level. Although a huge redundancy of highly abundant RNAs will be produced, low-abundance RNAs and those genes that are only expressed in specialized cells are still likely to be missed. Therefore, techniques facilitating the isolation of specific tissues or cells, such as laser capture microscopy and RNA amplification, may help identify genes that are expressed at low levels or in very few cells.

Because a gene can be transcribed into mRNA many times, ESTs ultimately derived from this mRNA may be redundant. That is, there may be many identical, or similar, copies of the same EST. Such redundancy and overlap means that when someone searches dbEST for a particular EST, they may retrieve a long list of tags, many of which may represent the same gene. Searching through all of these identical ESTs can be very time consuming. To resolve the redundancy and overlap problem, National Center for Biotechnology Information (NCBI) investigators developed the UniGene database. UniGene automatically partitions GenBank sequences into a non-redundant set of gene-oriented clusters. The clustering and assembly of individual ESTs into TUGs/TCs will result in decreased sequence redundancy and a final consensus sequence that should be both more accurate and longer than any of the underlying individual ESTs in the database. The clustering algorithms will identify all the transcripts from a gene family and generate a consensus sequence from the EST data. An alternative way of reducing redundant sequencing is to enrich the RNA populations for low-abundance transcripts. Cullis (2004) described a number of normalization and subtraction methodologies for enrichment of these low-abundance RNAs before cloning. Abundant cDNA clones can be removed before sequencing by screening high-density cDNA filters with labelled RNA. The clones that have strong hybridizations are eliminated and the minimally-hybridizing clones are re-arrayed and sequenced. The ultimate goal of EST projects is to develop a UniGene set that eventually contains all the genes for the organism.

When a cDNA library is used in the process of gene cloning, the sequenced cDNAs are used as molecular markers for genetic mapping, generating a saturated cDNA marker-based genetic map to locate the gene and eventually determine which cDNA is related to the gene effects. Additionally, a cDNA can be verified as the gene by a transformation/complementation test. With the development of reversed genetic techniques, one can also study the genetic effect of a mutated cDNA and its effect on the quantitative trait by site-directed mutation and genetic engineering, which can help determine which cDNAs are related to a given trait (Xu, 1997).

Although it is widely recognized that the generation of ESTs constitutes an efficient strategy to identify genes, it is important to acknowledge that there are several limitations associated with the EST approach. One is that it is very difficult to isolate mRNA from some tissues and cell types. This results in a paucity of data on certain genes that may only be found in these tissues or cell types. Second is that important gene regulatory sequences may be found within an intron, as well as in untranscribed regions of the gene (promoter). Because ESTs are small segments of cDNA, generated from an mRNA in which the introns have been removed, much valuable information may be lost by focusing only on cDNA sequencing. Despite these limitations, ESTs continue to be invaluable in characterizing plant genomes.

11.3.2 Generation of full-length cDNAs

Construction of full-length cDNAs is a central focus in the post-sequence era of the various genome projects. As an essential resource for the functional analysis of plant genes, full-length cDNAs can be used in many fields such as genome annotation including splice sites, expression profiling, protein structure determination using X-ray crystallography and transgenic analysis (Cullis, 2004). By validating with a full-length cDNA, predicted transcription units from genomic sequence data and the occurrence of alternative splicing events can be confirmed. The full-length cDNA can be used in both homologous and heterologous expression systems to generate large amounts of protein for functional and structural studies to determine gene function. In addition, sequencing of the full-length transcripts will allow the identification of RNAs from different members of gene families.

Full-length cDNA library construction is more technically challenging compared with EST generation. A full-length first-strand cDNA is not efficiently produced by reverse transcription, especially if the mRNA has a stable secondary structure. Libraries made from cDNAs, therefore, can contain both full-length and partial cDNAs. One method for constructing cDNA libraries with a high content of full-length clones involves starting from the first transcribed nucleotide. A number of critical issues pertaining to synthesis and cloning of full-length cDNAs have been recently identified. Most important is the purity and integrity of the starting material. mRNA is often contaminated with heterogenous nuclear RNA (hnRNA) due to the difficulty to exclusively isolate cytoplasmic RNA from plant tissues. True full-length cDNAs will yield sequence information from both 5' and 3' non-coding regions as well. A full-length cDNA should encompass all sequences from the CAP site to the poly (A) addition site. However, it is generally agreed upon that a cDNA comprising the entire coding sequence of a protein should be considered worthy for full-length sequencing at high accuracy.

Cullis (2004) described a process of contraction of full-length cDNA. A biotin label for the CAP structure has been developed based on the principle that the CAP site and 3' end of mRNA are the only sites that carry the diol structure. The diol groups at each end of the mRNA are biotinylated and then the first-strand of cDNA is synthesized. This synthesis is primed with a degenerate primer (XTTTTTTTT (restriction site)). The reaction mixture is then digested with RNase I, which cleaves the single-stranded RNA molecules at any sites, to destroy RNA molecules or part of them unpaired with their cDNA. Therefore, the 5' ends of all the mRNAs not protected by their partial cDNAs and exposed as single-stranded are removed (along with the biotinylated CAP structure) as are all the biotinylated 3' ends. The full-length cDNAs are captured on streptavidin-coated magnetic beads and the cDNA is released from the beads and the mRNA is destroyed by treatment with RNase H and alkaline hydrolysis. The cDNA is then tailed with oligo(dG) that is used to prime the second-strand synthesis. Again, this primer also has an extension that includes a restriction enzyme site. After the second-strand synthesis the full-length cDNA is cloned with the restriction sites inserted with the first- and second-strand primers.

The full-length cDNA for a desired gene can be obtained using 5' and 3' RACE (rapid amplification of cDNA ends) technique. RACE results in the production of a DNA copy of the RNA sequence of interest, produced through reverse transcription, followed by PCR amplification of the DNA copy. The amplified DNA copy is then sequenced to obtain a partial sequence of the original RNA. RACE can provide the sequence of an RNA transcript from a small known sequence within the transcript to the 5' end (5' RACE-PCR) or 3' end (3' RACE-PCR) of the RNA.

11.3.3 Full-length cDNA sequencing

The wide availability and usefulness of cDNA clones has spurred an interest in using a high-throughput approach to obtaining the complete sequence of full-length clones. The approach to obtaining the sequence of a full-length cDNA clone is different from that used to generate EST data. Many of the full-length cDNAs are likely to be longer than the reads resulting from sequencing both the 3' and the 5' ends of the insert. Therefore, additional sequencing strategies are necessary to obtain the full-length cDNA sequence. There are three possible strategies for full-length sequencing (Cullis, 2004):

- Transposon mutagenesis (Kimmel *et al.*, 1997): a transposon is randomly inserted, *in vitro*, into the cDNA insert and primers designed from both sides of the transposon are used for sequencing. Typically, each cDNA clone is subjected to a transposon reaction to produce a population of subclones, each harbouring a transposon at a distinct location. The subclones are then sequenced using transposon-specific primers, most often from each end of the inserted transposon. Sequencing a number of independent transposon sites will be sufficient to assemble the complete cDNA sequence.

- Concatenated cDNA sequencing (CCS) (Yu, W. *et al.*, 1997): multiple cDNA inserts are isolated, pooled and enzymatically concatenated. The entire population of concatenated cDNAs is then subjected to shotgun sequencing (as if it was a single large-insert genomic clone), with the individual cDNA sequences then derived by computer analysis following sequence assembly. This approach is well suited for existing high-throughput sequencing environments. However, it is also associated with a number of technical challenges, including the initial construction of the cDNA concatemers and shotgun libraries, the computational de-convolution of the cDNA sequences and problems associated with the uneven molar representation of individual cDNAs.
- Primer walking: primers are designed from 5' and 3' end sequences and used for a second round of sequencing. Additional primers are then made and used until the whole contiguous cDNA sequence is obtained. This method has been applied to the large-scale, full-insert sequencing of cDNA clones (Wiemann *et al.*, 2001); however, it is also associated with several limitations. First, the extensive use of synthetic oligonucleotides adds considerably to the overall costs. Secondly, the sequencing of larger cDNA clones is associated with many iterative walking steps, in some cases requiring a protracted effort. Thirdly, there are logistical demands of ensuring correct primer–template associations, especially when applied on a large scale.

11.3.4 Directed EST screens to identify specific genes

When genes are to be isolated from a particular organism based on marginal similarities to known genes or predicted similarities, the similarities are frequently too distant for hybridization or PCR-based methods to be useful. In this case, a directed EST screen may prove useful. EST data is obtained and searched for characteristic sequences to identify a specific EST as a candidate gene. Maximal information about the characteristic sequence motifs and their approximate locations in the target protein sequence is critical to the success of this effort.

The success of direct EST screens for identification of specific genes depends on two factors (Nunberg *et al.*, 1996): (i) adequate sequence information from related species allowing reliable prediction on the conservation of specific motifs and the ability to recognize these and overall sequence similarities even with the relatively limited sequence data available; and (ii) the mRNAs of the target genes are moderately abundant and are enriched in the specific tissues. There are many successful examples of direct EST screens in plants and the direct EST approach can be the method of choice for the isolation of new genes based on sequence motif conservation or the isolation of poorly conserved known genes from new species. In these situations, directed EST screens should be intrinsically more rapid and reliable than hybridization- or PCR-based screens.

11.3.5 Full-length cDNAs for the discovery and annotation of genes

Sequencing the full-length cDNA solves the following problems (Seki *et al.*, 2005). First, it permits accurate identification of the 5' and 3' UTRs. Secondly, in comparison with complete genomic sequence, it enables one to identify the precise locations of all introns. Thirdly, it aids in the discovery of new genes.

In all gene predictions from genomic DNA the precise identity of the gene boundaries and exon–intron structure is hindered by the lack of supporting experimental evidence. Full-length cDNA sequences and bioinformatics software can produce insights on the structure of genes in chromosomal DNA. Therefore, full-length cDNA sequences are essential for confirmation of the predicted genes within a sequenced genome. Having a full-length cDNA enables the checking of both the extent of the coding region of the gene as well as the sequences immediately 5' upstream and 3' downstream of the coding sequence. In addition, having a full length cDNA makes it possible to train the gene finding programs so that the unknown regions of the genome can be more accurately annotated as far as the presence of genes is concerned (Cullis, 2004). The availability of many full-length cDNAs and trained gene-finding programs from a small number of model plants will also ease the identification of genes in partial genomic sequences of more exotic plant species.

11.4 Positional Cloning

High density linkage maps of molecular markers provide an alternative gene cloning approach, positional cloning or map-based cloning. Positional cloning usually consists of identifying the markers that flank and show tight genetic linkage to the target gene, walking to the gene by using various genomic libraries constructed in, for example, yeast artificial chromosome (YAC) vectors and confirming the gene effects by the comparison of the isolated gene with a wild-type allele or complementation of the recessive phenotype by transformation (Meyer *et al.*, 1996; Paterson, 1996b). Theoretically, positional cloning methods permit the isolation of any gene which can be precisely mapped.

11.4.1 Theoretical considerations of positional cloning

For a plant species with no complete sequence available, one can obtain the genic DNA corresponding to the target locus through walking from one marker to the other by identifying two DNA markers which flank a target locus and using these markers to identify a series of contiguous DNA clones ('contigs'), which are fundamental to a tremendous amount of research. On a local scale, chromosome walks in particular regions provide the means to isolate genes which have been assigned to the region by genetic mapping. On a global scale, assembly of contigs for entire chromosomes can effectively provide a resource which will reduce the future need for small-scale investigations of particular locations.

Chromosome walking is a technique to clone a gene (e.g. a disease gene) from its known closest markers. The closest linked marker (e.g. EST or a known gene) to the gene is used to probe a genomic library. A restriction fragment isolated from the end of the positive clones is used to re-probe the genomic library for overlapping clones. This process is repeated several times to walk across the chromosome and reach the gene of interest. In the diagram (Fig. 11.3A), the chromosome walk begins with a clone containing *mkrB*. The ends of the

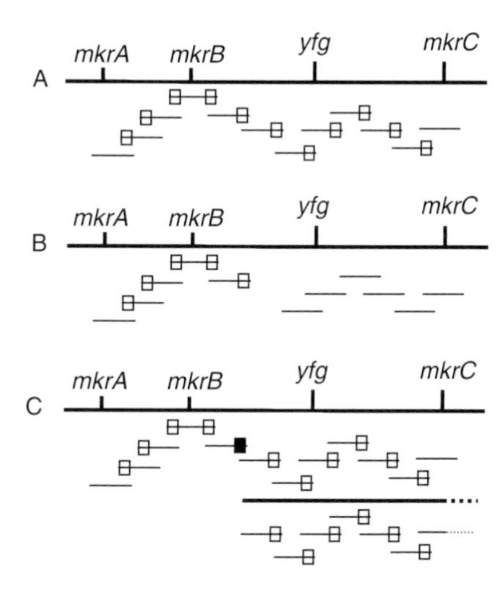

Fig. 11.3. Chromosome walking. (A) Chromosome walking begins with a clone containing *mkrB*. The ends of the clone (boxed) are used to probe a genomic library until a clone contains either *mkrA* or *mkrC* sequences. Clones between these two markers (*mkrB* and *mkrC*) are then evaluated for the presence of the target gene *yfg*. (B) Chromosome walking is interrupted because a segment of the region to be walked through is non-clonable. (C) Chromosome walking is detoured because one of the clone end probes (filled box) is a repeated sequence.

clone (boxed) are used to probe a genomic library. Clones from adjacent genome segments are thus identified and isolated. The distal ends of those clones are used to reprobe the library. These steps are continued until a clone contains either *mkrA* or *mkrC* sequences. Clones between two markers, *mkrB* and *mkrC*, must then be evaluated for the presence of the target gene *yfg*.

If the target species is a species whose genome has been completely molecularly mapped, an ordered set of YACs, P1-derived artificial chromosomes (PACs), BACs or cosmid clones will be available. Knowing which molecular markers are adjacent to the target gene automatically identifies the YACs and/or cosmids that need to be tested.

One difficulty of chromosome walking is recognizing where the gene is located between the two markers. Zoo (or garden) blots where DNAs of a variety of species have been restricted, electrophoresed and Southern blotted can be useful. Gene sequences are more likely to be conserved during evolution than intergenic sequences. The identification of GC islands or the use of exon trapping can also be useful. There are other problems with walking. Chromosome walks can be interrupted if a segment of the region to be walked through is non-clonable (for example if it is toxic to the host cell) (Fig. 11.3B). Chromosome walks can be detoured in many directions if one of the clone end probes (filled box in the Fig. 11.3C) is a repeated sequence.

By using a complete genetic map to estimate the total 'genetic length' of a genome, one can calculate the average quantity of DNA corresponding to a 'genetic distance' of 1% recombination (i.e. 1 cM). The physical quantity of DNA corresponding to 1 cM varies widely among higher plants, from about 280 kb in *Arabidopsis* to more than 7000 kb in barley. Curiously, despite gross differences in the average amount of DNA per cM in different taxa, the genetic (recombinational) distance between orthologous loci tends to be remarkably similar. This tends to suggest that the largely repetitive DNA elements which account for the differences in physical size of plant genomes may be relatively inactive in recombination. On the other hand, the correspondence between genetic and physical distance varies widely at different locations within a genome. In tomato, an organism with an average of about 750 kb cM^{-1}, individual regions have been estimated to show from as little as 50 kb cM^{-1} to as much as over 4000 kb cM^{-1} (Pillen *et al.*, 1996). It has been known that centromeric regions tend to be subject to recombination suppression and many genetic maps show clustering of DNA markers near the centromere. Factors other than repetitive DNA, such as introgressed chromatin or recombinational hotspots can also markedly influence the relationship between genetic and physical distance (Paterson, 1996a).

Positional cloning suffers from unpredictability in terms of the number of post-meiotic progeny that a research study can expect to genotype to narrow a candidate chromosomal region to a small number of candidate genes. For example, in rice, only 1600 gametes were genotyped to narrow the *Pi36(t)* allele to a resolution of 17 kb (Liu, X.Q. *et al.*, 2005), whereas 18,944 gametes were genotyped to map the *Bph15* allele to a lower resolution of 47 kb (Yang *et al.*, 2004). Dinka *et al.* (2007) described a detailed methodology to improve this prediction using rice as a model system. They derived and/or validated and then fine tuned equations that estimate the mapping population size by comparing these theoretical estimates to 41 successful positional cloning attempts. Then they used each validated equation to test whether neighbourhood meiotic recombination frequencies extracted from a reference RFLP map can help researchers predict the mapping population size.

A primary consideration in contig assembly is the size of each 'step' – i.e. the amount of DNA which can be held by the cloning vector used to construct the library. This consideration is a 'two-edged sword' – larger steps afford faster progress in assembling the contig, but yield lower resolution because target genes must be identified from a larger DNA segment. Chromosome walking has used different cloning vectors that can carry from 10–20 kb of exogenous

DNA by bacteriophage lambda (λ) up to 400–700 kb by YAC vectors.

While chromosome walking is straight forward in organisms with small genomes, it is more difficult to apply in most plant species with large and complex genomes. The strategy of chromosome walking is based on the assumption that it is difficult and time-consuming to find DNA markers that are physically close to a gene of interest. Technological developments have invalidated this assumption for many species. As a result, the mapping paradigm has changed such that it is often possible to isolate one or more DNA marker(s) at a physical distance from the targeted gene that is less than the average insert size of the genomic library being used for clone isolation. The DNA marker is then used to screen the library and isolate (or 'land' on) the clone containing the gene, without any need for chromosome walking and its associated problems (Tanksley *et al.*, 1995). Through this chromosome landing approach, Martin *et al.* (1993) isolated the tomato gene *Pto*, conferring resistance to the bacterial pathogen *Pseudomonas syringae* pv. This exemplifies the advantages of chromosome landing, in that initial emphasis on the isolation of many closely linked DNA markers eliminated the need for chromosome walking and the development of a high-resolution linkage map expedited the identification of candidate cDNAs. This approach has become the main strategy by which positional cloning is applied to isolate both major genes and QTL in plant species.

Contig assembly, or chromosome walking/landing, facilitate positional cloning – the isolation of genes based on genetic map information. Positional cloning has proven an effective means of isolation of genes in higher plants, but can be complicated by physically large genomes, prominent repetitive DNA fractions and polyploidy. Positional cloning has several basic requirements (Paterson, 1996b): (i) delineation of a target gene to a small chromosomal interval, preferably flanked by two DNA markers and spanned by a single megabase DNA clone, or by a contig of several megabase DNA clones; (ii) a means for identifying transcripts in the megabase DNA clone; and (iii) an efficient transformation system for introducing exogenous DNA into the plant species of interest, permitting identification of the target gene by mutant complementation. Currently, an essential requirement for positional cloning is the availability of comprehensive genomic libraries of relatively large DNA fragments, typically in YAC vectors. The chromosome landing paradigm can readily be applied to cloning QTL that can be mapped with a high degree of resolution. Progress has already been made in high-resolution mapping of QTL in plants and chromosome landing has been used to clone genes for quantitative traits, as exemplified in the following section. As anticipated more than a decade ago by Xu (1997), QTL that have been cloned so far are those with large effect and can be easily verified by transformation.

Genome sequence information has reshaped the procedures of positional cloning as the chromosome-aligned genome sequence information allows several of the steps in positional cloning to be skipped (Jander *et al.*, 2002). With a larger number of sequence-based molecular markers available, a certain level of genetic mapping may be able to quickly associate the target trait to a specific genomic region and further fine mapping effort may narrow the target genomic region to several candidate genes based on sequence information. This would be followed by cloning, complementation by transformation and high quality *de novo* determination of the sequence of the entire region of interest without a previously determined wild-type DNA sequence as a guide. Fig. 11.4 provides a comparison of map-based cloning in *Arabidopsis* between 1995 and 2002 (with complete genomic sequence available). The total effort required for map-based cloning reduced from 3–5 person-years to less than 1 person-year.

Methods have been proposed for cloning of multiple QTL and QTL with small effects. Peleman *et al.* (2005) proposed a method to fine map multiple QTL in a single population. As a first step, a rough mapping analysis is performed on a small part of the population. Once the QTL have been mapped to a

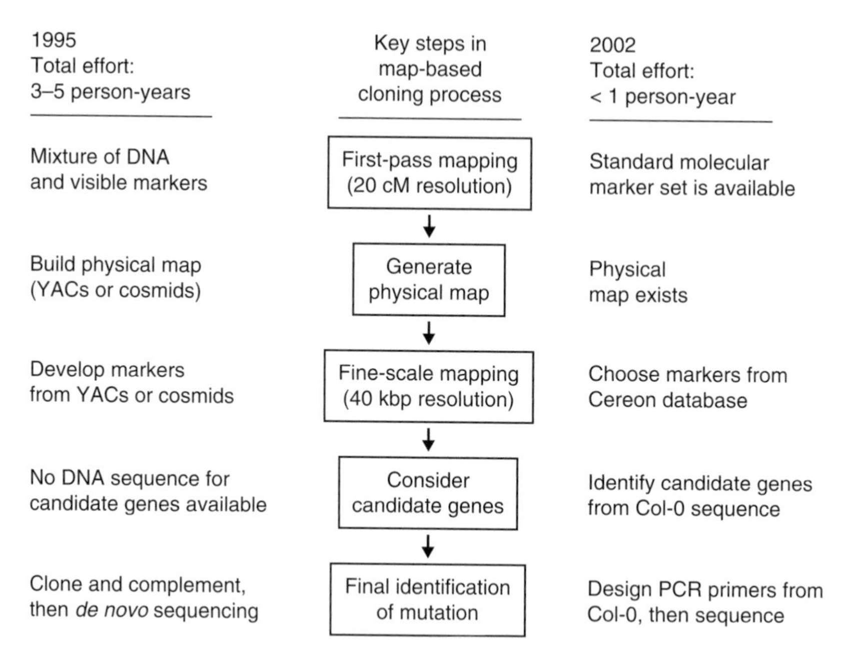

Fig. 11.4. Comparison of effort involved in map-based cloning in *Arabidopsis*. The key steps that have become easier between 1995 and 2002 are presented. From Jander *et al.* (2002) reproduced with permission of the American Society of Plant Biologists.

chromosomal interval by standard procedures, a large population of 1000 plants or more is analysed with markers flanking the defined QTL to select QTL isogenic recombinants (QIRs). QIRs bear a recombination event in the QTL interval of interest, while other QTL have the same homozygous genotype. Only these QIRs are subsequently phenotyped to fine map the QTL. By focusing at an early stage on the informative individuals in the population only, the efforts in population genotyping and phenotyping are significantly reduced as compared to prior methods. Linkage disequilibrium methods for fine mapping may also offer improved accuracy of QTL detection (Bink and Meuwissen, 2004; Grapes *et al.*, 2004).

For QTL with small effects, fine-scale mapping and positional cloning will be very difficult in the absence of a whole genome sequence. However, in these cases, reverse genetics may offer a solution, through functional genomics analysis of candidate genes that underlie QTL. For example, Liu *et al.* (2004) identified five candidate defence response (DR) genes that co-located with QTL for resistance to blast disease and were associated with the level of blast resistance.

11.4.2 Examples of positional cloning

Positional/candidate gene isolation has been very successful. Some early and significant examples include: (i) identification of genes underlying qualitative phenotypes using mutant analysis and a sequenced genome (Jander *et al.*, 2002) and no mutant in a sequenced or unsequenced genome (Buschhes *et al.*, 1997); (ii) identification of genes underlying quantitative phenotypes in an unsequenced genome (Frary *et al.*, 2000) and using positional analysis and structure/function interpretation in a sequenced genome (Yano *et al.*, 2000); and (iii) comparing the function of the gene with its orthologous counterparts in other species and exploring how the gene interacts with other genes in a pathway (Izawa *et al.*, 2003).

With advances made in rice genomics, several QTL associated with the same traits or trait components have been cloned. These include four QTL for heading date – *Hd1*, *Hd3a*, *Hd6* and *Ehd1* (Yano *et al.*, 2000; Takahashi *et al.*, 2001; Kojima *et al.*, 2002; Doi *et al.*, 2004) and QTL for grain number (*Gn1a*) and grain size (*GS3*) (Ashikari *et al.*, 2005; Fan *et al.*, 2006). More recently, the first QTL with significant pleiotropic effects has been isolated (Xue *et al.*, 2008). The QTL *Gdh7*, isolated from an elite hybrid rice and encoding a CONSTANS, CONSTANS-LIKE, TOC1 (CCT) domain protein, has major effects on an array of traits in rice, including number of grains per panicle, plant height and heading date. Enhanced expression of *Gdh7* under long-day conditions delays heading and increases plant height and panicle size. Sakamoto and Matsuoka (2008) summarized the genes identified in rice grain yield and its component trait including grain number, grain weight, grain filling, plant height and tillering, In this section, cloning of *fw2.2* (Frary *et al.*, 2000) will be discussed in detail as an example for identifying a gene(s) underlying a quantitative phenotype in an unsequenced genome.

Preliminary genetic mapping

1. Several QTL (~11) associated with tomato fruit weight were identified in primary interspecific mapping populations: *Lycopersicon pennellii* (small fruit) × *Lycopersicon esculentum* (large fruit).
2. All wild *Lycopersicon* spp. contain small-fruit alleles at the locus *fw2.2*; modern cultivars have large-fruit alleles, which suggested that this locus is a domestication locus and partially recessive mutations lead to large fruit. The alleles from modern cultivars at *fw2.2* increase fruit weight by 5–30% in segregating populations but 47% in near-isogenic line (NIL) populations.
3. NILs were developed with a total of 41.9cM of *L. pennellii* DNA (containing *fw2.2*) in an *L. esculentum* background (Fig. 11.5A).
4. Fine mapping narrowed the region to two YAC clones (150kb region) containing *fw2.2* (Alpert and Tanksley, 1996) (Fig. 11.6; Fig. 11.5A).

Fine mapping and candidate gene identification

1. YAC clones containing *fw2.2* were used as templates to screen the cDNA library that contains the dominant allele (*Fw2.2*), which allows a positionally targeted search for candidate genes.
2. 100 positive cDNA clones and four unique transcripts were identified.
3. 3472 F_2 individuals (derived from NIL × recurrent parent (RP) cross) were screened with four markers to establish the marker order of the cDNAs along the YACs (Fig. 11.5B). An alternative would be to sequence the YACs, which, however, is more expensive.
4. cDNAs were used to identify four cosmid clones (Fig. 11.5B) from an *L. pennellii* cosmid library consisting of 15–50kb genomic clones, which are large enough to contain more than one gene per clone including enhancer/promoter, 5' and 3' UTRs, introns and exons.

Complementation tests

1. Identified cosmid candidate clones were used in transformation experiments with two cultivated genetic backgrounds (Mogeor, TG496). In the hemizygous R_0 generation, the fruit weight of transformants was not significantly different from the controls due to partial dominance of the *L. pennellii*. Thus, R_0 plants were selfed and homozygous R_1 individuals with and without transgenes were compared for phenotype.
2. Significant differences in fruit weight were observed only for COS50 transformants and the differences were significant in both Mogeor and TG496 genetic backgrounds.
3. By sequencing the COS50 clone, two ORFs were identified: cDNA44 (used as probe) and ORFX (Fig. 11.5C).
4. Recombination with COS50 (XO33) delimited *fw2.2* to a region containing ORFX.

Exploration of ORFX identity

1. A significantly higher level of ORFX transcript in small-fruited NILs (*Fw2.2*) was found, compared to large-fruited (cultivated) RP (*fw2.2*). No ORFX transcript was

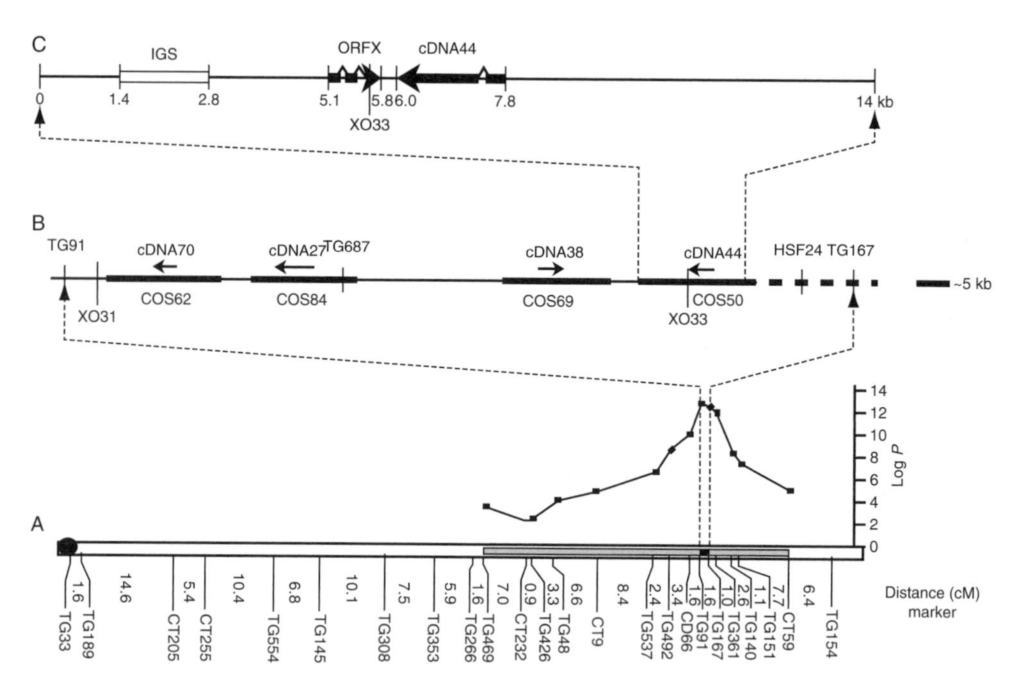

Fig. 11.5. High-resolution mapping of the *fw2.2* QTL. (A) The location of *fw2.2* on tomato chromosome 2 in a cross between *Lycopersicon esculentum* and a NIL containing a small introgression (grey area) from *Lycopersicon pennellii* (Alpert and Tanksley, 1996). (B) Contig of the *fw2.2* candidate region, delimited by recombination events at XO31 and XO33 (from Alpert and Tanksley, 1996). Arrows represent the four original candidate cDNAs (70, 27, 38 and 44), and heavy horizontal bars are the four cosmids (COS62, 84, 69 and 50) isolated with these cDNAs as probes. The vertical lines are positions of RFLP or cleaved amplified polymorphism (CAPs) markers. (C) Sequence analysis of COS50, including the positions of cDNA44, ORFX, the A-T-rich repeat region, and the 'rightmost' recombination event, XO33. From Frary *et al.* (2000). Reprinted with permission from AAAS.

detected in the *L. pennellii* cDNA library. The transcript was only detected at low levels with RT-PCR in mRNA extracted from pre-anthesis carpels.

2. Carpels were heavier in RP but cell size was the same in NILs and RP, which suggested that ORFX controls carpel cell number before anthesis.

3. ORFX was found to encode a 163-aa polypeptide of ~22 kDa.

4. BLASTP showed matches only with plant genes in GenBank (dicots, monocots, gymnosperm); and none of the putative homologues have known function.

5. The three-dimensional shape of the predicated protein was similar to heterotrimeric guanosine triphosphate-binding proteins in rat, which is associated with control of cell division.

Characterization of ORFX

1. ORFX represents a previously uncharacterized plant-specific multi-gene family and it has at least four paralogues in tomato and eight homologues in *Arabidopsis* (organized in two- or three-gene clusters).

2. Sequence comparison of ORFX alleles from *L. pennellii* and *L. esculentum* (830-nt fragment, which included 95 nt from 5' UTR, 55 nt from 3' UTR) showed 42-nt difference.

3. ORFs are highly conserved with only 35-nt differences in introns; four silent changes; three causing amino acid changes. It was concluded that functional differences in alleles may be due to a combination of sequence changes in coding and upstream regions of ORFX.

4. Reduction of cell division in the small-fruited NIL correlates with higher levels of

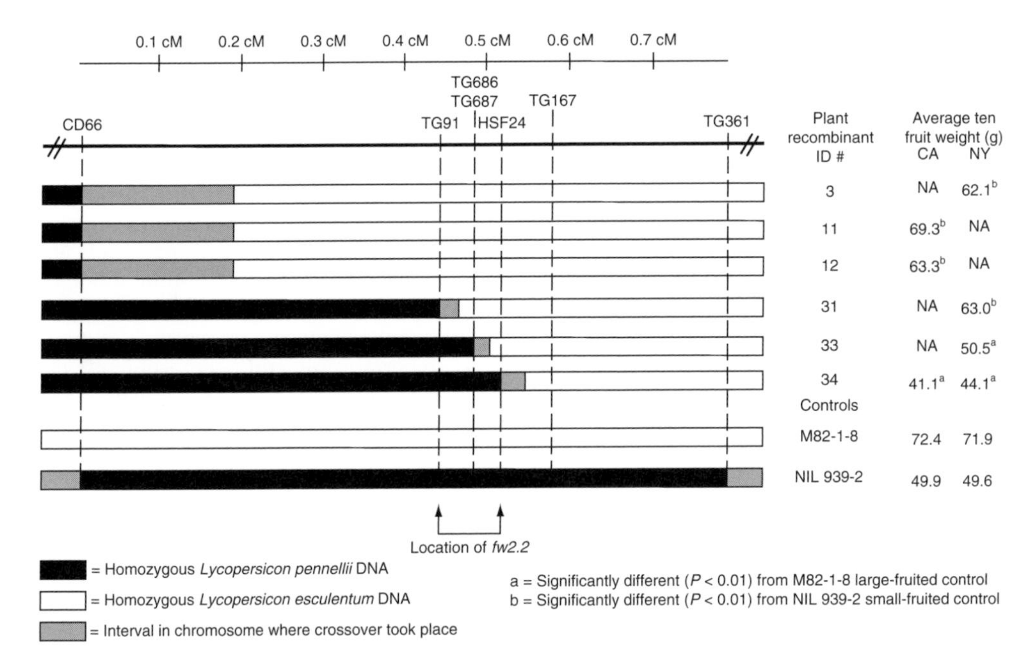

Fig. 11.6. Graphical genotypes of homozygous recombinants in the *fw2.2* region of chromosome 2. Five replications of each recombinant plant were grown in California (CA) and New York (NY). The average gram (g) weight of ten fruit from each recombinant was compared with the large-fruited, M82-1-8, and the small-fruited, NIL 939-2, controls. Recombinants #3, #11, #12, and #31 were significantly larger (b; $P < 0.01$) for average ten fruit weight in comparison to the small-fruited control, NIL 939-2, while recombinants #33 and #34 were significantly smaller (a; $P < 0.01$) for average ten fruit weight in comparison to the large-fruited control, M82-1-8. Recombinants #31 and #33 delineate the *fw2.2* region (bracketed by arrows), based on the smallest region demonstrating statistical significance. Plants for which few or no fruit were harvested due to pest infection were not available (NA) for fruit weight analysis. The black and white boxes indicate the homozygous condition for *Lycopersicon pennellii* (NIL 939-2) and *Lycopersicon esculentum* (M82-1-8) at the molecular markers, respectively. The grey boxes indicate the approximate position between two molecular markers where the genetic recombination event took place. The genetic distance between molecular markers (separated by dashed lines) is indicated by the scale shown in centiMorgans (cM). From Alpert and Tanksley (1996) © National Academy of Sciences, USA 1996.

ORFX transcript, suggesting a role as a negative regulator of cell division.

11.5 Identification of Genes by Mutagenesis

Phenotypic variation that has been used in genetic analysis and plant breeding comes from either natural variation or induced mutations. Natural phenotypic variation is observed in germplasm collections and exists as a random collection of diverse 'mutations' throughout the genome, although natural selection has led to maintenance of these mutations. Mutagenesis has been used as a tool in plant breeding for many years with numerous cultivars released. As described in Chapter 1, various chemical and physical mutagens have been used to create a wide variety of unique plant mutants to increase the amount of variation. Mutagenesis approaches have attracted the attention of plant molecular biologists as they provide a means for identifying desired genes (Xu *et al.*, 2005). Whole genome mutagenesis brings an opportunity of mutating every gene contained in a plant species.

In functional genomics, mutant populations or libraries that cover all possible genes become an increasingly important tool. Mutant libraries can be constructed using chemical and physical mutagenesis, T-DNA insertion and transposon tagging (e.g. Joen *et al.*, 2000; Leung *et al.*, 2001; Xue and Xu, 2002; An *et al.*, 2003; Hirochika, 2003; Hirochika *et al.*, 2004). These libraries can be used for functional analysis based on loss-of-function analyses. Gain-of-function approaches such as T-DNA activation tagging and gene overexpression are powerful complements to insertional mutagenesis for the successful identification of mutant phenotypes. Libraries of enhancer trap lines have also been developed which facilitate the detection and isolation of regulatory elements (Wu, C. *et al.*, 2003).

An alternative to nucleic acid sequence characterization is the direct demonstration that the sequence has a function. This can be done by the re-introduction of the sequence into the appropriate plant or by knock-out of the gene through mutagenesis (Cullis, 2004). The technology used to mutagenize genes and identify those mutants includes both insertional mutagenesis (Azpiroz-Leehan and Feldmann, 1997) and the TILLING methodology (McCallum *et al.*, 2000), which identifies single base changes in a gene of interest. A third method for disrupting gene function is by gene silencing through RNA interference (RNAi) (Cogoni and Macino, 2000).

Perhaps the most popular mutagens in use by plant molecular biologists for gene cloning are the insertion mutagens, T-DNAs and transposable elements (transposons). The advantage of these mutagens is that they both disrupt the gene and also serve as vehicles for recovery of the plant genic DNA. This process, known as gene tagging, provides a means to examine the biochemical and developmental consequences of mutations in the gene of interest, as well as a relatively facile means for isolating the affected genes.

The nature of the damage that is caused by mutagenesis determines the functional class of genetic alternations that are produced. Deletions, insertions and rearrangements are more likely to result in loss-of-function alleles, whereas point mutations can lead to a broader range of effects, including hypomorphic, hypermorphic and neomorphic effects (that is, alleles of reduced, enhanced or novel gene function, in corresponding order) (Alonso and Ecker, 2006).

11.5.1 Generation of mutant populations

The most reliable method to ascertain gene function is to disrupt the gene and to determine the phenotype change in the resulting mutant individual. A large collection of mutants for an organism will be extremely valuable to the scientific community and will accelerate the speed of gene function analysis. Table 11.1 lists mutagenic agents and related mutations. Two popular methods for mutagenesis are insertion and deletion.

Mutagenized populations that are derived from a single homozygous genotype, i.e. an inbred line, share a common genetic background and individuals differ only at one or a few 'mutagenized' loci in the genome ('isolines'). Transgenic mutagenized lines carry molecular tags that facilitate rapid identification of candidate loci. Chemically mutagenized lines carry only small insertion/deletion or point mutations (Table 11.1) which are difficult to find unless the candidate

Table 11.1. Mutagenic agents and related mutations.

Mutagenic agents	Type of mutation
Ethylmethane sulfonate (EMS)	CG >> AT transitions (point mutations)
Di-epoxy butane (DEB)	Point mutations, small deletions (6–8 bp)
Fast neutrons (FN)	Deletions (up to 1 kb)
X-rays	Chromosome breaks, rearrangements
T-DNA	Insertion
Transposable elements (TE)	Insertion/deletion
RNAi constructs	Insertion where transcribed product causes gene silencing

region has been defined. Mutation rates as high as 10^{-3} alleles per gene have been reported in maize, suggesting that alleles of any given gene might be found by screening as few as 3000 M_2 families, or 3000 M_1 plants in the case of non-complementation screen (Candela and Hake, 2008).

Knock-out mutagenesis including most that are chemically induced has the following limitations: (i) redundancy – a high level of gene duplication in plants provides 'genetic buffering' (backup/second copy) such that knocking out one member of a gene family may not affect phenotype; and (ii) lethality – some genes confer essential functions; disruption will lead to lethality so knock-outs at those loci will never be retrieved in generated plants or in offspring. Conditional lethals may be retrieved, if necessary conditions, such as temperature sensitivity, are met.

Resource populations in model systems provide genome-wide resources for all biologists so they do not have to develop them for each experiment. These populations allow results of independent experiments to build on each other, because data on the same set of mutants can be maintained in a common database, facilitating worldwide collaboration.

11.5.2 Insertional mutagenesis

Insertional mutagenesis occurs naturally in a number of plant species through the excision and reintegration of endogenous transposable elements. The insertion of a known DNA segment into a gene of interest has been an extremely valuable genomic tool for a number of systems in mammals and plants. Insertional events can be classified as T-DNA tagging, transposon tagging, retrotransposon tagging, or entrapment tagging, depending on the type of element used. Insertion events can also be labelled according to their sites and types of insertions (Jeon *et al.*, 2004). The 'knock-out' is a null mutation with an insertion in the coding or regulatory region of a gene. 'Knock-down' mutations cause reduced expression due to an insertion in the promoter or 3' UTRs. The 'knock-on' (or activation tagging) mutation has an insertion element carrying a constitutive promoter, such as the cauliflower mosaic virus (CaMV) 35S promoter, that is capable of driving the expression of genes adjacent to the insertions. The 'knock-about' mutation is an insertion that does not inhibit normal functioning of the gene. The 'knock-knock' mutation has more than one insertion, causing multiple knock-outs. Finally, the 'knock-worst' mutation includes insert events that lead to large-scale chromosomal rearrangement.

Gene knock-outs imply that the activity of a gene has been eliminated. In plants the two major methods for generating these are by inserting either a T-DNA or a transposon sequence (Azpiroz-Leehan and Feldmann, 1997). The unique advantage of using foreign DNA as a mutagen is that the inserted fragment not only disrupts the gene function but also tags the affected gene with known sequences, which greatly facilitates gene isolation. The DNA has a defined sequence and acts as a marker for the location of the mutation. Thus, by using oligonucleotide screening or specialized PCR, the mutagenized gene can be identified and sequenced easily. This method was first illustrated by the cloning of the 'white eye' locus of *Drosophila* (Bingham *et al.*, 1981).

As a principle of insertion tagging, an endogenous or an engineered DNA fragment (with known sequence) is allowed to insert at random into the genome. When it lands in a gene, it generally causes a recessive, loss of function mutation. For insertion mutagenesis to be useful for isolating all genes from a plant genome, it will be necessary to saturate the genome with insertions so that every single gene has been mutated. The probability that an insertion will be found within a given gene can be estimated based on the size of the gene, the size of the genome and number of inserts distributed among the population (Krysan *et al.*, 1999, 2002). Assuming random chromosomal insertion, tagging efficiency can be calculated according to the formula

$$P = 1 - [1 - (L/C)]^{nf}$$

where P is the probability of finding an insertion within a given gene, L is an average length for the gene, C is the haploid genome size, n is the number of independent insertional lines and f is the average number of loci inserted per line.

Consider an example that: (i) the rice haploid genome size is 3.8×10^8 bp; (ii) the average rice gene is 3.0 kb long; and (iii) the mean number of insertion loci per line is 1.4. A total of 417,000 tagging lines would be required for establishing a population in which a T-DNA insertion could be found within a given gene at 99% probability.The number of tagging lines required for saturation mutagenesis of a genome is highly dependent on the length of the target genes. A group of 1 kb genes in rice requires 1,250,000 lines to achieve 99% probability of being mutated, whereas 5 kb genes need 250,000 lines in the T-DNA tagging population. Jung *et al.* (2008) summarized the rice insertional mutants generated by different mutagens including T-DNA, Ac/Ds, Spm/dSpm, T-DNA with enhancer, full-length cDNA over-expresser (FOX) system and Tos17.

Insertion tagging has the following advantages: (i) insertion tagging generally inactivates a gene which simplifies phenotypic evaluation (disrupts an ORF, interrupts promoter, interferes with intron-splicing); (ii) it marks the gene for isolation via inverse-PCR, TAIL-PCR (thermal asymmetric interlaced), transposon display, AIMS (amplification of insertion-mutagenized sites), cDNA-AFLP, etc; and (iii) it can be used for both forward and reverse genetics. In forward genetics, it can be used to screen for an interesting phenotype and uses the 'tag' to isolate the gene. In reverse genetics, it can be used to identify insertion in a gene sequence of interest and to figure out the phenotypic consequences of the insertion; three-dimensional pools of insertion line DNA can be used for efficient screening.

Insertional mutagenesis as a tool for gene cloning has its limitations: (i) redundancy and lethality (the same as chemical mutagenesis); (ii) some species lack endogenous TEs or cannot mobilize them efficiently; (iii) engineered systems require the ability to make and propagate large numbers of transformants; (iv) the predominance of loss-of-functional alleles; (v) the biased distribution of insertion in the genome; (vi) the inability to characterize lethal mutations; and (vii) the difficulty of generating populations that are large enough to reach complete saturation of the genome.

T-DNA tagging

The transfer DNA (T-DNA) is a defined segment of the tumor-inducing (Ti) plasmid of *Agrobacterium tumefaciens* and delimited by short (25 bp) imperfect-repeat border sequences called left and right T-DNA borders. The insertion of a T-DNA element into a chromosome can lead to many different outcomes: insertion into the coding region can lead to partial or complete inactivation of the gene; while insertion into the promoter region can lead to complete inactivation of the gene, reduced expression of the gene, or increased expression of the gene.

Several methods have been developed for introducing T-DNA into *Arabidopsis*. These include various tissue culture and whole plant techniques. However, most tissue culture-based transformation protocols developed for *Arabidopsis* were not directed toward insertion mutagenesis. The vast majority of T-DNA tagged genes have been isolated from populations of transformants generated with whole plant transformation protocols (Jenks and Feldmann, 1996). A computer database has been established for *Arabidopsis* that contains the precise genomic locations of over 50,000 T-DNA insertions. Any gene of interest can quickly be found, if the collection contains a mutation in that gene, by performing a simple BLAST search. The database of these insertions can be found at http://signal.salk.edu/cgi-bin/tdnaexpress and the *Arabidopsis* Knock-out Facility at the University of Wisconsin. A number of other crop plants have similar resources, especially rice, which is already well served with T-DNA insertion lines (Parinov and Sundaresan, 2000; Ramachandran and Sundaresan, 2001). In rice, a mutant population containing 55,000 lines was generated and about 81% of the population carried one to two T-DNA copies per line. T-DNA was preferentially (~80%) integrated into genic regions (Hsing *et al.*, 2007).

T-DNA insertion is the most generally applicable method because it can be used for any plant that can be transformed and regenerated. Because each transformant is an independent event with the T-DNA being relatively randomly inserted into the genome a large number of independent transformation events are needed to inactivate every gene. The need for the generation of large numbers of independent transformants therefore limits this technology to plant species or particular lines that are capable of being transformed in a high-throughput manner (Cullis, 2004).

T-DNA tagging has the advantages of effective interruption of genes, low copy number (1.5) and preferential insertion into genic regions. Low copy number per line makes it easier to clone the tagged gene but requires a larger number of transformants to hit every gene with a T-DNA insertion. Disadvantages include the need for time-consuming and high efficiency transformation methods, somatic variation in tissue culture and a high percentage of untagged mutants. The T-DNA does not always segregate with mutation ('hit and run'); approximately 35–40% of mutants are actually tagged with the T-DNA. Dipping of whole plants in *Agrobacterium* avoids the tissue culture step and somaclonal variation that results.

Transposon tagging

Transposable elements (TEs) or transposons are sections of DNA (sequence elements) that move, or transpose, from one site in the genome to another and have been used as a tool for gene isolation. The insertion and excision of transposable elements result in changes to the DNA at the transposition site. The transposition can be identified when a known DNA sequence or selection markers are inserted within the elements.

Three major families of endogenous transposable elements have been used for gene tagging in maize (Candela and Hake, 2008). All transposable elements fall into one of the following two classes: DNA elements and retroelements. DNA elements, such as Ac/Ds and Spm in plants, P elements in animals and Tn in bacteria, transpose via DNA intermediates. A common feature of DNA elements is the flanking of the element by short inverted repeat sequences. The enzyme transposase recognizes these sequences, creates a stem/loop structure and then excises the loop from the region of the genome. The excised loop can then be inserted into another region of the genome. Retroelements transpose via RNA intermediates. The RNA is copied by reverse transcriptase into DNA and the DNA integrates into the genome. Two types of endogenous transposable elements have been identified in plants: autonomous and non-autonomous. TEs that are autonomous code for a transposase that cleaves the element from its insertion site in the chromosome. TEs that are non-autonomous lack transposase function but are competent to transpose if transposase is supplied.

TEs can be either low copy (e.g. Ac/Ds in maize, Tos17 in rice): typically one to three copies per genome or high copy (e.g. Mu in maize): up to 100s of copies per genome, which means very small populations are needed to ensure saturation tagging. It can be difficult to isolate a mutagenized gene by forward genetics because many genes are 'hit' simultaneously with a TE insertion and one or many may affect phenotype but TEs are very good for reverse genetics.

There are two major advantages in using the so-called Class II transposons such as Ac/Ds, En/Spm and Mu elements of maize, Tam elements of snapdragon and dTph1 from petunia (Jeon *et al.*, 2004). First, unlike T-DNA, transposons can be excised from the disrupted gene in the presence of transposase, resulting in functional revertants that can confirm the phenotypic consequences of the mutation. Secondly, excision often leads to insertion close to the original site, which can be exploited for local mutagenesis through targeted insertion into specific domains of a gene. To follow the excision and insertion of transposons elements, phenotypic assay systems have also been developed using marker genes containing transposon inserts, in which excision can be monitored by restoration of marker–gene activity. Transposon excision can be selected by employing visualized markers such as beta-glucuronidase (GUS), luciferase, streptomycin resistance or green fluorescent protein (GFP).

If the maize transposon Ac is used, the movement of the transposon is likely to be to relatively close sites on the same chromosome as the original insertion point. Therefore, a number of starter lines can be constructed with the Ac present at various known chromosomal locations across the genome (Cullis, 2004). Then the appropriate starter line can be chosen that will have a high probability of generating an insertion in a closely linked gene of interest.

As with T-DNA insertions, construction of the transposon to function as an enhancer trap is also possible. Such 'engineered' transposons can be introduced into the plant genome using T-DNA-mediated transformation. Once inserted the transposon can 'hop' from one chromosome location to another as long as an active transposase is present (Smith *et al.*, 1996). Although most transposons tend to hop to linked sites, a strategy has been devised to select for transposons that land at unlinked loci (Sundaresan *et al.*, 1995).

An Ac/Ds tagging system in plants is shown in Plate 3. The Ac/Ds tagging system has both advantages and disadvantages. It is an efficient and cost-effective method to generate a large mutant population, although secondary transposition complicates gene identification and this transposon system is not available in many species. The advantage of using transposons is that they can then be activated and moved into many regions of the genome. Therefore, after the generation of a small number of lines with the transposon present, the transposons can be launched to move around the genome and generate insertions in every gene. This technique is most easily applied to maize because this is the plant from which most of the transposable elements have been isolated (Cullis, 2004). The initial lines are, in essence, always available. The engineering of two-component transposon systems that include an inducible promoter will make this particular technique more widely applicable to a wide variety of other plant species.

Retrotransposon tagging

Class I retrotransposable elements are a group of mobile elements that transpose through reverse transcription of RNA intermediates by reverse transcriptase, RNaseH and integrase enzymes, leaving a copy of the retrotransposon at the original site. This replicative mode allows the retrotransposons to generate genetic diversity by producing insertional mutations. Retrotransposons are abundant in species with large genomes, however, only a small fraction of them are active. When the frequency of transposition in the original or heterologous hosts is considered, three Tyl-copia retrotransposons, Tnt1 and Tto1 in tobacco and Tos17 in rice, seem suitable for gene tagging. These retrotransposons prefer low-copy, gene-rich regions (Jeon *et al.*, 2004). Retrotransposon mutagenesis is shown in Fig. 11.7.

Of the rice genome 17% is estimated to consist of retrotransposons. Tos17 is an endogenous, copia-like element that can be activated by tissue culture in rice. The sequenced cultivar from *japonica* subspecies, Nipponbare, has only two copies of Tos17. A total of 47,196 Tos17-induced insertion mutant lines were generated and this population carried 500,000 insertions. Tos17 was three times more likely to be found in genic regions containing introns and exons than in intergenic regions. Frequency was low in centromeres and pericentromeric regions that were not in other types of retrotransposons. A total of 78% of insertions were in hotspots (clustered), with an average of 6.5 insertions/hotspot (Miyao *et al.*, 2003).

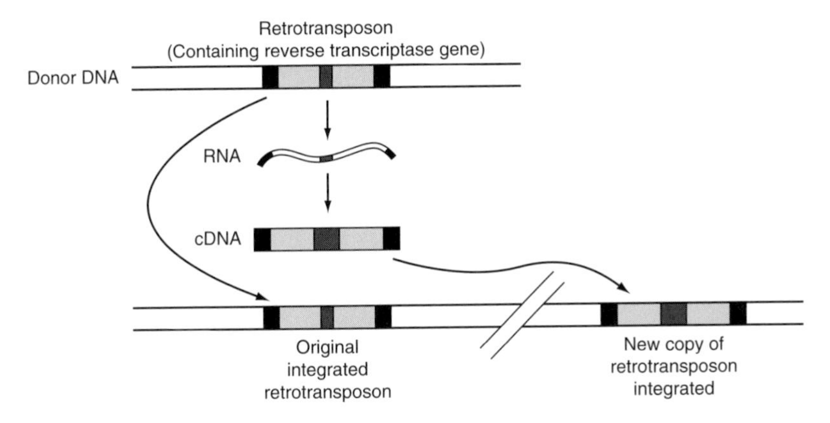

Fig. 11.7. Retrotransposon transposition. Modified from Buchanan *et al.* (2002).

Activation tagging

T-DNA mutagenesis can be adapted to generate gain-of-function alleles by activation tagging (Weigel *et al.*, 2000). To achieve this, several copies of a strong transcriptional enhancer are introduced into the T-DNA. On integration, the enhancers stimulate the transcription of a nearby gene and cause its ectopic expression. Conventional T-DNA mutagenesis shares with transposon tagging the disadvantage that it is not efficient for tagging genes that are functionally redundant because a phenotype will not be observed. In addition, neither T-DNA tagging nor transposon tagging will identify genes that are required during multiple stages of a life cycle and where loss of function results in early embryonic or in gametophytic lethality. To overcome these drawbacks, an activation-tagging system consisting of T-DNA vectors that contain strong transcriptional enhancers was developed in *Arabidopsis* and successfully used in gene cloning and the analysis of the function of redundant genes (Weigel *et al.* 2000). Multiple transcriptional enhancers from the CaMV 35S gene are positioned near the right T-DNA border. Genes immediately adjacent to the inserted CaMV 35S enhancers are over-expressed. A transposon-mediated activation tagging system has also been developed on the basis of a self-stabilizing Ac transposon derivative, using a Ds element that carries the tetramerized CaMV 35S enhancer (Suzuki *et al.*, 2001).

To overcome the tedious transformation process for many crop plants, a new strategy that combines activation tagging and Ac/Ds transposon tagging was proposed. In this system, the T-DNA carries the Ds element, which contains the Bar gene and a tetramer of the 35S enhancer. The T-DNA and the Ds element promote constitutive expression of genes in their vicinity after integration or transposition.

Inserted (T-DNA or TE) tags contain a selectable marker and multiple enhancer elements (CaMV 35S enhancers) and may contain a reporter gene (GUS or GFP) with a weak promoter as well (Weigel *et al.*, 2000). When the T-DNA construct inserts in or near a gene (within about 3.5 kb either upstream or downstream), transcriptional signals (enhancers) on the T-DNA construct interact with the native promoter and enhance gene expression producing dominant, gain of function mutations. The reporter gene may report the original expression pattern, or the new pattern. Selection can be applied to primary transformants and the resulting transformants analysed for desired phenotypes, or insertion events. The T-DNA tag is then cloned and nearby genes can be characterized.

As an early example, Weigel *et al.* (2000) characterized over 30 dominant, morphological mutants with various phenotypes from the T-DNA activation-tagging pools of *Arabidopsis*. T-DNA activation has also been used as a tool for isolation of the regulators of a complex metabolic pathway from a genetically non-tractable plant species (van der Fits *et al.*, 2001), where hundreds of thousands of *Catharanthus roseus* suspension cells (transformed with T-DNA that carried constitutive enhancer elements) were screened relatively easily for their resistance to a toxic substrate. In a recent example, Wan *et al.* (2008) generated about 50,000 individual transgenic rice plants by an *Agrobacterium*-mediated transformation approach with the pER38 activation tagging vector. The vector contains tandemly arranged double 35S enhancers next to the right border of T-DNA. Comparative field phenotyping of the activation tagging and enhancer trapping populations in two generations (6000 and 6400 lines, respectively, in the T_0 generation and 36,000 and 32,000 lines, respectively, in the T_1 generation) identified about 400 dominant mutants, indicating that the activation tagging pool is a valuable alternative tool for functional analysis of the rice genome.

Activation lines can be used to identify function of specific members of gene families where over-expression is diagnostic but loss-of-function provides no phenotype; or to identify the function of genes where knock-outs are lethal. Activation lines may be used to clarify new functions of previously characterized genes where phenotype depends on expression differences.

Entrapment/enhancer/promoter tagging

By creating fusions between tagged genes and a reporter gene such as GUS and GFP, an entrapment-tagging system allows one to monitor gene activity. Insertion of the promoterless reporter or the reporter with minimal promoter not only destroys normal gene function but also activates expression of the reporter gene. There are three commonly used entrapment systems: promoter trap, enhancer trap and gene trap (Springer, 2000). Promoter trap elements contain a promoterless reporter gene. The reporter gene is expressed when it is inserted into an exon and forms a translational fusion between the endogenous gene and the reporter. Enhancer trap elements carry a reporter gene with a minimal promoter. Reporter gene expression is activated by a chromosomal enhancer element located near the insertion point. Gene trap elements contain a reporter gene with an intron that carries multiple splicing donor and acceptor sites. The reporter gene can be expressed when the element is inserted into the transcribed region.

The use of gene traps consists of placing a reporter gene in a vector whereby the reporter gene is only activated when inserted within a functional gene. The reporter gene has a visual phenotype, so the tissue specificity of the promoter region (and therefore the endogenous gene itself) can be identified directly. The reporter activation demonstrates the spatial and temporal expression of the disrupted gene. Because expression levels can be monitored in heterozygous plants, the gene trap system is useful for studying the patterns of most plant genes, including essential genes that cause lethal mutations when homozygous. A finer dissection of various patterns within an organ has been demonstrated for enhancer trap GUS fusions in *Arabidopsis* roots (Cullis, 2004).

Devices for entrapment can be transferred into plant cells as part of T-DNA or transposable elements. This approach has been applied to genes that are difficult to identify by traditional methods. Because genes are identified based on their reporter gene expression, a mutant phenotype is not required. With this advantage, both functionally redundant genes and genes that have functions at multiple developmental stages can be identified (Jeon *et al.*, 2004).

11.5.3 Non-tagging mutagenesis

While insertional mutants offer obvious advantages in gene cloning and reverse genetics, there are a number of limitations associated with transformation-mediated mutagenesis. These include: (i) the high initial investment in producing the mutant lines or starter lines; (ii) the transgenic nature of the mutants prevents large-scale cultivation in the field and exchange of mutation collections in different countries; (iii) parts of the genome may not be accessible to insertional inactivation (e.g. hot spot insertion by Ac/Ds), thus preventing complete genome coverage; and (iv) in some organisms, insertional mutagenesis has never reached the efficiency needed for large-scale mutagenesis. Due to these issues, other types of mutations are also used in plant species, including deletion and chemical mutagenesis.

Chemical and radiation-induced mutations have been widely used for random mutagenesis in plants, resulting in a broader spectrum of mutation alleles that occurs randomly in the genome. Chemical mutagenesis produces a broad range of mutant alleles such as loss-of-function, gain-of-function, reduction-of-function and novel functions, in contrast to insertion and deletion mutagenesis that causes mainly loss-of-function mutations. When an efficient transformation tool is not available, it is not possible to adopt a gene tagging strategy, but these random, non-tagging systems can be utilized to create a mutagenized library.

Ionizing radiation has been widely used to induce mutations for plant breeding and classical genetic analysis, but the consequences of ionizing radiation have only recently been examined closely at the molecular level. Several genes have been identified in animals and plants using deletion mutants. Fast neutron, gamma ray, X-ray and UV radiations have been used in different systems. Usually, fast neutrons produce large deletions while the other three radiations yield small deletions or point mutations.

Besides ionizing radiation, a number of chemicals have been used to generate large mutant collections. Many chemicals can be used as mutagens but di-epoxy butane (DEB), N-ethyl-N-nitrosourea (ENU), ethylmethane sulfonate (EMS), di-epoxy octane (DEO), ultraviolet-activated trimethylpsoralen (UVTMP) and hexamethyllphosphoramide (HMPA) are common mutagens used in animals and plants. Generally, deletions caused by chemical mutagens are relatively small, ranging from point mutations to mutations of several kilobases.

Chemical agents such as EMS and nitrosomethylurea (NUM) are extremely efficient mutagens in *A. thaliana*. Under optimal conditions, EMS treatment of seeds can generate about 4000 mutations per genome, compared with an average of 1.5 insertions per transferred DNA (T-DNA) mutants (Alonso *et al.*, 2003; Till *et al.*, 2003). Chemical agents generate a broader range of DNA alternations; these are predominantly single base-pair substitution, but also induce small insertions and deletions. Importantly, the distribution of EMS-induced mutations is unbiased (Alonso and Ecker, 2006).

Point mutations

EMS, a base-alkylating agent that generates point mutations (of which the vast majority are G/C-A/T transitions, which often lead to the creation of stop codons/nonsense mutations), has been used most commonly because of its ease of use and the diversity of potential mutants. As EMS causes a high density of mutations, fewer plants need to be screened in order to target all genes, compared with other mutagenesis systems.

However, point mutations induced by EMS are subtle changes whose detection can be challenging. Once a phenotypic mutant is identified, it is necessary to determine the locus in the genome by a positional cloning strategy in order to clone the corresponding gene, as discussed in the previous section. If a mutant that exhibits an identical or similar phenotype has already been identified, complementation crosses are a first step to determine whether the new mutation is allelic. The existence of multiple alleles can give information on gene function and be useful for breeding.

Strategies have been developed recently so that subtle changes like point mutation can be detected easily. For efficient adaptation chemical induced mutagenesis for reverse genetics in *Arabidopsis* and other plants, McCallum *et al.* (2000) developed a large-scale screening system, targeting induced local lesions in genomes (TILLING), which allows a point mutation to be identified. In the basic TILLING method, seeds are mutagenized by treatment with EMS. The resulting M_1 plants are self-fertilized and DNA is prepared from the M_2 individuals. To screen many individuals a pooling strategy is used. DNA samples are pooled and pools are arrayed on microtitre plates and subjected to gene-specific PCR. High-throughput TILLING (Colbert *et al.*, 2001; Till *et al.*, 2003) uses the CEL I mismatch cleavage enzyme, which recognizes base-pair mismatches (Oleykowski *et al.*, 1998). The PCR is performed using a mixture of labelled and unlabelled primers. One primer is labelled with the IR Dye 700 and the other with IR Dye 800. Melting and re-annealing of PCR products is followed by CEL I treatment, which preferentially cleaves mismatches in heteroduplexes between wild-type and mutant DNA sequences. CEL I-treated PCR products are applied to slab gel electrophoresis, then detected in two separate channels by LI-COR scanners. Mutations are indicated by shorter, cleaved PCR products. If a mutation is detected in a pool, the individual DNA samples that went into the pool can be analysed separately to identify the individual that carries the mutation. Once this individual has been identified, its phenotype can be determined. This screening procedure can locate a mutation to within a few base pairs for PCR products of up to 1 kb in size. A potential problem with this method is that any one individual will carry multiple mutations. Genetic analysis is therefore necessary to confirm that any observed phenotypic alteration is associated with the mutation in the target gene and not with another mutation elsewhere in the genome. However, TILLING often results in a number of allelic mutations in different lines, which can help to confirm phenotype as well as provide information on protein function. An important advantage of the TILLING method is that it can be applied to any species for which a gene sequence is known.

TILLING has moved from proof-of concepts to production with the establishment of publicly available services for *Arabidopsis*, maize, lotus and barley. Pilot-scale projects have been completed on several other plant species, including wheat. The protocols developed for TILLING have been adapted for discovery of natural nucleotide variation linked to important phenotypic traits, a process termed EcoTILLING (Comai *et al.*, 2004). Till *et al.* (2007) reviewed the current TILLING and EcoTILLING technologies and discussed the process that has been made in applying these methods to many different plant species.

In addition, new methods for efficient genome-wide detection of point mutations are appearing on the horizon, including a mismatch-repair detection on tag arrays (Faham *et al.*, 2005). Mismatch-repair detection allows > 1000 amplicons to be screened for variations in a single laboratory reaction. This approach can be scaled up to allow sequence comparison in whole-genome coding regions among large sets of lines and controls at a reasonable cost.

Deletion mutagenesis

Ionizing radiation mutagenesis causes deletions and other types of chromosomal alternations. In plants, fast-neutrons are well-established, very effective deletion mutagens (Koornneef *et al.*, 1982; Li, X. *et al.*, 2001). Approximately ten genes are randomly deleted in each line when treated with fast neutrons at a dose of 60 Gy (Koornneef *et al.*, 1982). As fast neutron-deletion mutagenesis can be performed on numerous dry seeds and plant transformation is not necessary, it is easy to produce a great number of mutants with a high probability of finding a mutation in every gene.

Cloning a gene mutated by a deletion requires chromosome walking, as with chemical mutagenesis. However, deletion mutants can also be effective for reverse genetics. Deletion libraries have been established that contain knock-out mutants in *Arabidopsis* and rice (Li, X. *et al.*, 2001). Deletions can be identified by gene-specific PCR screening of pooled DNA, where PCR extension time is shortened so that amplification of the longer wild-type fragment is suppressed and only mutant lines yield products (Joen *et al.*, 2004). Experimental approaches for identifying DNA deletions in pools of mutants that are generated by high-energy ionizing radiation have been developed (Li and Zhang, 2002), which is called Deleteagene. Deleteagene can be applied to plants in which transformation is inefficient; it might also provide a means of simultaneously mutate (delete) tandem duplicated genes (Li and Zhang 2002; Zhang, S. *et al.*, 2003). In contrast to insertion mutants where the probability of finding a mutant is proportional to the size of the target gene, meaning that identification is difficult in a small gene (Krysan *et al.*, 1999), it is easier to find a knock-out of a small gene from a deletion mutant pool.

11.5.4 RNA interference

All gene disruption approaches have some inherent limitations. For example, it is difficult to identify the function of redundant genes or the functions of genes required in early embryogenesis or gametophyte development. One way to overcome the redundant gene problem is to simultaneously inhibit all the members of a gene family through gene silencing. RNA interference (RNAi) is a mechanism that inhibits gene expression by causing the degradation of specific RNA molecules or hindering the transcription of specific genes (Fire *et al.*, 1998). RNAi refers to the function of homologous double-stranded RNA (dsRNA) to specifically target a gene's product, resulting in null or hypomorphic phenotypes. As long as the interference is targeted to a region of the gene that is conserved within all the members of the gene family, all members of the family will be similarly inhibited (Tang *et al.*, 2003).

The most interesting aspects of RNAi are the following (Cullis, 2004):

- dsRNA, rather than single-stranded antisense RNA, is the interfering agent.
- It is highly specific.
- It is remarkably potent (only a few dsRNA molecules per cell are required for effective interference).
- The interfering activity (and presumably the dsRNA) can cause interference in cells and tissues far away from the site of introduction.

The RNAi pathway is initiated by the enzyme DICER, which cleaves long, dsRNA molecules into short fragments of 20–25 bp. One of the two strands of each fragment, known as the guide strand, is then incorporated into an RNA-induced silencing complex and pairs with complementary sequences. The most well-studied outcome of this recognition event is post-transcriptional gene silencing. In this way all the RNA transcripts from any of the members of a gene family can be simultaneously silenced if highly homologous regions are used. Any resulting phenotype can then be attributed to the functioning of that gene family, but it will still need to be determined whether the family members contribute redundant functions or whether only one of the members of the gene family actually conditions the particular phenotype observed.

Artificial microRNAs (amiRNAs), which are designed to target one or several genes of interest, provide a new and highly specific approach for effective post-transcriptional gene silencing in plants. Warthmann *et al.* (2008) devised an amiRNA-based strategy for both *japonica* and *indica* types of cultivated rice. Using an endogenous rice miRNA precursor and customized 21mers, amiRNA constructs were designed to target three different genes (*Phytoene desaturase -Pds, Spotted leaf -Spl11* and *elongated uppermost internode-Eui1/CYP714D1*). Upon constitutive expression of these amiRNAs in the cultivar Nipponbare (*japonica*) and IR64 (*indica*), the target genes were down-regulated by amiRNA-guided cleavage of the transcripts, resulting in the expected mutant phenotypes. The effects were highly specific to the target gene, the transgenes were stably inherited and they remained effective in the progeny. Ossowski *et al.* (2008) reviewed various strategies for small RNA-based gene silencing, described the design and application of artificial miRNAs for gene silencing in many plant species and compared the small RNA pathways mediating transgene-induced gene silencing, including post-transcriptional gene silencing, transcriptional gene silencing and virus-induced gene silencing.

11.5.5 Gene isolation via mutagenesis

There are two main approaches for disrupting gene function on the basis of its DNA sequence: using one of the targeted techniques such as RNAi or ectopic expression, or screening a collection of randomly generated mutants for a knock-out. The amplification and sequencing of genomic DNA next to an inserted transposon or T-DNA is an essential step in identifying a mutation within a gene. Several steps are required to isolate the disrupted flanking DNA, incorporating different possible techniques (Jenks and Feldmann, 1996), including:

1. Screen for the mutant: this can be done through the generation of genomic libraries from the mutants and screening with sequences homologous to the right or left border regions. When the screen is based on visible phenotypes, all the mutants are grown under regular growth conditions if screening for morphological variation; or they are grown under a special condition if screening for conditional mutants such as those to biotic or abiotic stresses.
2. Confirm co-segregation: because a large proportion of mutant lines are untagged in T-DNA or transposon mutagenized collections, co-segregation analysis of the T-DNA sequence or selection marker with the phenotype is the first step towards cloning the gene. It is estimated that 35–40% of the mutants in *Arabidopsis* are possibly due to deletion, rearrangement or somatic mutation during transformation. Once co-segregation is established for a given mutant, isolation of the mutated gene may be achieved by several methods such as plasmid rescue, IPCR or TAIL-PCR.
3. Plasmid rescue: this involves utilizing bacterial selectable markers and origin of replication from a linearized bacterial plasmid incorporated into the T-DNA to isolate T-DNA-plant junctions in *E. coli*. It includes the following procedures:

- Restriction enzymes are present in the T-DNA at the ends of the bacterial plasmid sequence.
- After extracting purified genomic DNA, the genomic DNA is digested with the appropriate restriction enzyme. After removal of the enzyme, samples are ligated.
- The ligated DNA is precipitated and transformed by electroporation into recombination-deficient *E. coli* cells to maximize the stability of the multimerized (CaMV 35S enhancers). Recovered plasmids can then be sequenced to identify captured flanking sequences.

4. Inverse PCR (IPCR): utilizing primers made from the left or right border sequences on circularized genomic fragments. IPCR has been implemented to isolate DNA segments of the genome that flank the inserted molecular in transgenic plants tagged by T-DNA, transposons or retrotransposons. The technique involves digestion by appropriate restriction enzymes containing the known sequence and its flanking region (Joen *et al.*, 2004). Many thousands of restriction fragments are circularized by self-ligation with T4 DNA ligase and the circularized DNA is then used as a template in PCR. The unknown flanking DNA segment is amplified by two primers located at the ends of the known sequence. The first primer is designed to locate near the junction point between the insert and plant sequences and the second primer is located near the enzyme site that is used for digestion of the mutagenized DNA. At least 50 nucleotides should be left between the primer sites and the junctions for nested PCR to isolate specific amplification products and DNA sequencing.
5. Thermal asymmetric interlaced (TAIL)-PCR using nested border specific primers and arbitrary degenerate primers. The TAIL-PCR strategy has been used to isolate insert-end sequences from P1 and YAC clones (Liu and Whittier, 1995), genomic sequences that flank T-DNA insertions from transgenic lines of *Arabidopsis* (Liu *et al.*, 1995) and genomic DNA flanking Tos17 in rice (Yamazaki *et al.*, 2001). TAIL-PCR depends on amplification between a set of three nested primers for the known sequences and shorter, arbitrary degenerate primers with low *Tm* values. Accordingly, the PCR programme is set to thermally control specific and non-specific products. In the primary reaction, five high-stringency cycles are used to specifically amplify linear products from the target flanking sequences by the known

insert-specific primer. This is followed by one low-stringency cycle to allow annealing by the arbitrary degenerate primers in the flanking sequence. The next cycles alternate between two high-stringency cycles and one reduced-stringency cycle, resulting in logarithmic amplification of the target sequences. Secondary and tertiary reactions by the nested primers decrease non-specific amplification. TAIL-PCR depends on the accidental position of an arbitrary primer sequence in the flanking sequence. Therefore, it is important to design optimal primers in order to successfully amplify a given insertion site.

6. Adaptor-ligated PCR: this was an early method for isolating the flanking region of a known DNA and now a modified method, PCR walking, has been developed for isolating genomic sequences that flank T-DNA borders (Balzergue *et al.*, 2001; Cottage *et al.*, 2001). The method comprises three major steps: restriction, adaptor ligation and PCR amplification. The genomic DNA carrying the T-DNA or transposon is digested with blunt-end restriction enzymes. An asymmetric adaptor cassette is ligated to the digested DNAs. By using a primer specific for the adaptor cassette and a primer specific to the T-DNA or transposon, unknown target DNAs that flank the insertions can be amplified.

7. Complementation test: to prove the isolated gene is the one responsible for the mutant phenotype, molecular complementation should be performed by introducing the wild-type allele into the mutant background and restoring the wild-type phenotype. Alternatively, good evidence can be obtained by molecular characterization of several alleles at the locus. Transposon mutants can sometimes be reverted by excision of the transposon, which can also be used as evidence that the insertion was the cause of the phenotype.

All approaches which produce lesions in unpredictable chromosomal locations can be used both in forward and in reverse genetic approaches for gene discovery. There are, however, three main requirements that must be met for a collection of random mutations to be efficiently used in a reverse genetic screen (Alonso and Ecker, 2006). First, the number of mutations in the collection should exceed (by five- to ten-fold) the number of genes in the genome. This redundancy is necessary to ensure that mutations in a particular gene will be found with a sufficiently high probability. Secondly, each individual plant in the mutagenized populations should be catalogued, propagated and pooled so that it can then be effectively screened. The number of individual mutants that need to be screened tends to be very high, making it necessary to pool individual mutants before testing for the presence of the mutation of interest. This requirement has promoted the development of more sophisticated pooling strategies that minimize the number of assays required but still allow the identification of an individual mutant line in one-step or two-step screens (Winkler and Feldman, 1998; Alonso *et al.*, 2003). The optimal strategy for screening the DNA pools depends on the type of mutagen used or, more specifically, the nature of the DNA lesion. Thirdly and perhaps most important, a DNA sequence-based screening approach needs to be developed that is sensitive enough for a single plant with a specific sequence alteration to be detected within a pool of wild-type individuals. Different types of DNA lesion (deletions, insertions and point mutations) require different screening methodologies as described by Alonso and Ecker (2006).

A major drawback of reverse genetic methods is that the screen has to be repeated for every gene. However, high-throughput identification of genome insertion sites in mutagenized populations can be achieved by taking advantage of the known sequence of the inserted DNA and the availability of a complete genome sequence in sequenced plant species. Various PCR-based strategies have been adapted for this purpose, including TAIL-PCR (Sessions *et al.*, 2002) and the adaptor ligation approach (Alonso *et al.*, 2003).

One of the most exciting uses of the near complete collection of gene-indexed mutations for a given species is the ability to carry out whole-genome forward genetic screens (Carpenter and Sabatini, 2004). This will enable researchers to test simultaneously the role of all genes in the genome for involvement in a particular biological process (Alonso and Ecker, 2006). The first step towards this goal involves generating a non-redundant collection of homozygous mutants. From more than 300,000 gene-indexed mutant lines in *A. thaliana*, ideally two independent lines per gene need to be selected, the mutations need to be confirmed and homozygous plants obtained. The hypothetical end product for this step will be a collection of 521 96-well plates that correspond to 50,000 mutant lines, two lines for each of the 25,000 genes. This seed library could then be systematically screened to study the role of each one of the 25,000 represented genes in any given biological process. The identification of mutants affected in the selected biological process allows the immediate identification of the underlying genes. By having two independent mutant lines per gene, false positives and the need for experimental replicates are substantially reduced.

Several important advances towards gene-function analysis in *Arabidopsis* are on the horizon (Alonso and Ecker, 2006), which should provide some guidelines for all plant species: the ability to do systematic forward genetics using reverse genetic tools (simultaneous phenotypic analysis of all gene-indexed mutants), the development of new phenomic platforms, improvements in targeted mutagenesis (specifically, homologous recombination) and the utilization of natural variation in gene function studies. Induced mutations (point mutations, deletions and transposon- and T-DNA-generated mutations) and other means of reducing gene expression (such as the use of small inhibitory RNAs (siRNAs), amiRNAs and artificial repressor proteins) will continue to be used for some time. However, the importance of natural allelic variation to study gene function in plants is likely to increase. The rapid advancement of ultra high-throughput sequencing (UHTS) technologies, which allow 1 gigabase of sequence in 48 h for a cost of US$3000 (Service, 2006), is likely to have a profound effect in two areas of gene-function analysis (Alonso and Ecker, 2006). First, induced alleles with phenotypes that have only been observed in a specific genetic background (Sanda and Amasino, 1996) point to the need for the creation and sequencing of insertions in large mutant populations using various accessions or ecotypes; a process that will be facilitated by UHTS. Secondly, UHTS technology will allow complete genome re-sequencing of many hundreds, or even thousands, of accessions. With the concomitant development of more phenotyping platforms and corresponding community phenotyping databases, whole-genome association studies that link genotype and phenotype will become the approach of choice to interrogate plant gene function and the role of natural allelic variation in plant adaptation to a range of local growth habitats (Weigel and Nordborg, 2005).

11.6 Other Approaches for Gene Isolation

As described in Chapter 3, DNA chip and microarray technology make it possible to do gene isolation in a high-throughput way. Gene isolation through microarrays is to identify target gene(s) from a genome. There are two different approaches: (i) parallel analysis of gene expression by comparison of expressions among different species or different individuals within a species, or expressions of the same individuals at different growth or developmental stages or under different environments. Microarray-based gene expression analysis can be used to detect the type and abundance of mRNA in the cell by hybridization, which needs few samples and is highly automatable. (ii) Genes can be isolated from cDNA or EST microarrays by using homologous probes.

11.6.1 Gene expression analysis

One of the major applications of DNA microarrays is gene expression profiling (Tessier *et al.*, 2005). Gene profiling via microarrays involves determining the expression of genes under specific conditions. Similarly, microarrays have enabled genome-wide class comparisons such as organs, genotypes or conditions. Several studies have identified genes that are consistently differentially expressed between two or more predefined classes, the degree to which a gene is active in a certain organ or tissue can be measured by the amount of mRNA found in the cells, although the correlation between mRNA and active protein is not always absolute due to post-transcriptional regulation. Such approaches hope to find the complete set of genes that differentiate between cellular states and shed light on the underlying differences between these classes at the molecular level.

Initial strategies for detecting differentially expressed genes between two classes are straightforward. Essentially they involve two-sample comparisons of the differences between mean log expressions of the classes. The significance of the differences is estimated using *t*-tests modified specifically for array data (simple *t*-tests are almost never used – usually more complex *t*-tests are needed due to the nature of the data) or its non-parametric analogues. When more than two classes are involved, *F*-statistics and non-parametric analogues can be applied.

A wide variety of statistical techniques is available for class prediction (Finak *et al.*, 2005), including linear discriminant analysis, weighted voting, nearest-neighbour classifiers, support vector machines, neural nets and Bayesian methods. At the centre of all of these methods is the issue of feature selection. The goal is to select a subset of features (genes) that best distinguish between known classes and can predict new, unseen samples.

Class discovery experiments attempt to determine biologically relevant subclasses of a particular cellular state. Several methods are used for this purpose and the most popular techniques are *k*-means clustering, hierarchical clustering, self-organizing maps and principal component analysis. The goal of clustering is to organize similar sample data together. Either the gene or the array dimensions can be clustered according to similarity indices, enabling one to see both genes with similar expression profiles across arrays and arrays that have underlying similarities across genes (Finak *et al.*, 2005).

Class prediction experiments aim to find subsets of genes that can best distinguish between two or more classes of sample. First do sequence analysis. Probes are designed using gene-specific DNA fragments and oligonucleotide or DNA arrays are made for all genes of an organism. All mRNA probes reverse transcribed under different classes from the organism are then hybridized with the DNA microarray. Based on the intensity of hybridization signal, differential gene expressions or co-expression under different classes can be detected. Class-dependent gene expression can be identified by comparison of expression profiles across all genes between different classes. The set of genes can be mapped on to the gene ontology or to metabolic pathways. In this way, physiological functions of a gene can be analysed and related functional genes can be determined. There are two early examples (Lockhart *et al.*, 1996; Wang, X. *et al.*, 1999). In addition, several tools for exploring mRNA expression data with known protein–DNA and protein–protein interaction databases have been reported. An example is CYTOSCAPE, which maps expression data on to the protein interaction network (Shannon *et al.*, 2003). This approach can yield important insights into the protein complexes perturbed in a given experiment and establish functional roles for the genes distinguished.

11.6.2 Using homologous probes

Availability of adequate quantities of sufficiently pure protein for production of specific antibodies or for partial peptide sequencing opens the door to cloning of the gene specifying that protein. Physiological and biochemical investigations may lead to the identification of a protein responsible for the phenotype or biological property of interest. If the purified polypeptide has a non-blocked N-terminus, the N-terminal sequence may be determined directly. Sequences from the interior of the protein can be obtained by producing and analysing proteolytic fragments of the polypeptide. If sufficient quantities of purified protein are available, the protein may be used to immunize animals (commonly rabbits or mice). The animals usually produce and secrete into the serum, antibodies specifically recognizing the protein. The antisera may be used to detect the immunizing protein specifically.

Peptides of ten to 30 residues in length can be chemically synthesized efficiently. Small peptides chemically coupled to larger carriers can be effective immunogens. Antisera can be used to recognize clones of an expression library that are synthesizing the cognate antigen. The antibodies bind to protein from the colony or plaque. Bound antibodies can be detected by any of a variety of methods such as radioimmune precipitation and enzyme-linked immunosorbent assay (ELISA).

Nucleotide sequences that could code for the determined sequence of amino acids can be deduced from the genetic code. Since the genetic code is redundant, multiple nucleotide sequences can encode the same peptide sequence. To be sure that the actual nucleotide sequence is present in a probe oligonucleotide, the oligonucleotide is synthesized incorporating, where needed, multiple nucleotides. The product is called a degenerate oligonucleotide.

When amino acid sequences from two separated parts of the polypeptide chain are available, degenerate oligonucleotides based on the two sequences can be used to attempt to amplify the intervening sequence by PCR. An amplified RT-PCR product may be used in nucleic acid hybridization to screen the colonies or plaques of a cDNA library for clones containing complementary sequences.

Positive clones obtained from the cDNA library can be used as nucleic acid hybridization probes to screen a library of genomic DNA to identify clones containing the gene that can produce the corresponding mRNA. Positively-reacting genomic library clones need to be further examined to identify where on the clone the gene is located and prove that the clone actually contains the gene.

Functional genomics has been broadly applied to include many endeavours aimed at determining functions of genes on a genome-wide scale, such as transcriptional profiling to determine gene expression patterns; and yeast two-hybrid and other interaction analyses to help identify pathways, networks and protein complexes (Chapter 3; Henikoff and Comai, 2003). Although a daunting task, several approaches have already been established, including the use of T-DNA knock-out lines and over-expression studies. In contrast to the previously prevalent gene-by-gene approaches, new high-throughput methods are being developed for expression analysis as well as for the recovery and identification of mutants. The experimental approach is consequently changing from hypothesis-driven to non-biased data collection and an archiving methodology that makes these data available for analysis by bioinformatics tools. Reverse genetics (sequenced gene to mutant and function) may play a more prominent role in functional genomics studies in the future (Xu *et al.*, 2005).

In true directed mutagenesis, researchers choose the gene to be perturbed. The most elegant and precise targeted mutagenesis approach relies on homologous recombination to target foreign DNA on a homologous sequence in the host genome. This is rarely possible in plants and therefore alternative approaches have been developed to alter the expression of selected genes. There are two main variants of direct mutagenesis: gene-silencing (RNAi) and zinc-finger nucleases. In these strategies, specific sequences that are unique for each gene to be disrupted must be engineered *in vitro* and then introduced into the plant. It has been shown that the expression of a sequence-specific zinc-finger nuclease in *A. thaliana* generates mutations in the target gene *in planta* (Lloyd *et al.*, 2005). The large battery of well-characterized zinc-fingers, each with different and specific DNA-recognition sequences, should allow the use of this methodology to target almost any gene in the *A. thaliana* genome (Alonso and Ecker, 2006).

Developments in high-throughput genomics will facilitate the process of dissecting the genetic basis of complex traits, including defining genetic intervals, identifying candidate genes and verifying an allele's contribution to the phenotype. Technologies such as microarrays, fluorescence polarization, mass spectrometry and molecular barcodes could achieve throughputs of 10,000 markers, which shows promise for high resolution association studies based on natural variation/natural populations and of high-throughput map-based cloning.

12

Gene Transfer and Genetically Modified Plants

As described in previous chapters, intra-specific transfer of genes is easily performed by cross-hybridization in all plants with a sexual cycle. Gene transfer by cross-hybridization becomes more difficult or impossible with increasing phylogenetic distance and as a result, the inter-generic gene transfer is very rare. By genetic transformation, DNA from any organism can be transferred into other species' genomes. The inserted gene sequence (known as the transgene) may come from another unrelated plant, or from a completely different species: transgenic *Bt* maize, for example, which produces its own insecticide, contains a gene from a bacterium. This powerful tool enables plant breeders to do what they have always done – generate more useful and productive crop cultivars containing new combinations of genes – but it expands the possibilities beyond the limitations imposed by traditional cross-pollination and selection techniques. Plants containing transgenes are often called genetically modified- or GM-crops, although in reality all crops have been genetically modified from their original wild state by domestication, selection and controlled breeding over long periods (Chapter 1). In this book, the term transgenic is used to describe a crop plant that has transgenes inserted.

Issues in gene transfer and GM-crops have been a hot topic for many years and some early reviews can be found in McElroy and Brettell (1994), Christou (1996) and McElroy (1996) among many other books and journal articles. Only some basic concepts will be introduced in this chapter. For a full coverage, readers are recommended to seek information in recent books, including Liang and Skinner (2004), Parekh (2004), Peña (2004), Skinner *et al.* (2004) and the *Transgenic Crops* series by Springer.

12.1 Plant Tissue Culture and Genetic Transformation

12.1.1 Plant tissue culture

Plant tissue culture exploits the *in vitro* plasticity of plant growth and development because whole plants can be regenerated from a wide range of plant cells (totipotency). For the majority of species gene transfer is carried out using explants competent of regeneration to obtain complete, fertile plants. Cell division and callus (dedifferentiated tissue) formation, embryogenesis and organogenesis can be induced using combinations of plant growth regulators. Auxins like 2,4-dichlorophenoxy-acetic acid (2,4-D), picloram and dicamba and cytokinins like benzylaminopurine (BAP), kinetin and zeatin are usually used in the tissue culture media. There are no universally applicable methods of plant tissue culture and thus, protocols need to be modified for each genus, species, cultivar and tissue. Within individual cereal species the 'elite' germplasm is usually least amenable to tissue culture.

12.1.2 Genetic transformation

Goals of plant transformation for crop improvement are to produce fertile transgenic plants with integrated transgenes at reasonable frequencies from 'elite' backgrounds. Once a gene has been isolated through one of the approaches as described in Chapter 11 and cloned (amplified in a bacterial vector), it must undergo several modifications before it can be effectively inserted into a plant. Components of any successful plant transformation system include delivery of DNA to the plant genome without compromising cell viability, selection of transformed cells, regeneration to produce intact fertile plants and the transmission of transgenes into subsequent generations. A simplified representation of a constructed transgene, containing the necessary components, which need to be developed in parallel, for successful integration and expression is as follows:

1. The promoter is the on/off switch that controls gene expression at different developmental stages and in response to certain environmental changes, or specific to certain tissues and organs. On the other hand, promoters like the most commonly used cauliflower mosaic virus (CaMV) 35S are constitutive. The genes under constitutive promoters are expected to be expressed throughout the life cycle of the plant in most tissues and organs.
2. The gene of interest is modified to achieve greater expression in a plant. For example, the *Bt* gene for insect resistance is of bacterial origin and has a higher percentage of A-T nucleotide pairs compared to plants, which prefer G-C nucleotide pairs. In a clever modification, researchers substituted A-T nucleotides with G-C nucleotides in the *Bt* gene without significantly changing the amino acid sequence. The result was the enhanced production of the gene product in plant cells.
3. The termination sequence signals to the cellular machinery that the end of the gene sequence has been reached.
4. A selectable marker gene in the gene construct is to identify plant cells with the integrated transgene. This is necessary because achieving incorporation and expression of transgenes in plant cells is a rare event, occurring in just a few of the targeted tissues or cells. Selectable marker genes encode proteins that provide resistance to agents that are normally toxic to plants, such as, metabolic inhibitors, antibiotics or herbicides. As explained below, only plant cells that have the integrated selectable marker gene will survive when grown on a medium containing the appropriate antibiotic or herbicide. Similar to the gene of interest, marker genes also require promoter and termination sequences for their proper function.

Conventional plant breeding represents the principal approach to crop improvement. It employs methods such as hybridization, introgression breeding, induced mutagenesis and somatic hybridization to randomly modify genomes and, as a result, create genetic variation (Fig. 12.1a). Genetic engineering is different from the traditional methods in that any modification can be designed and tailored to achieve the desired effect. This method often fuses promoters and genes to produce expression cassettes that are introduced into plants using bacterial transfer DNAs (T-DNAs) (Fig.12.1b). It excludes the transfer of known allergen- or toxin-encoding genes and analyses the sequence of insertion sites. The ability to identify rapidly and eliminate plants containing inadvertent fusions or disruptions of genes is not available to traditional plant breeding, where genes can be inactivated through unpredictable transposition of resident mobile elements. The second advantage of transgenic applications is that it generally takes less than a year to transform an existing cultivar with one or several traits.

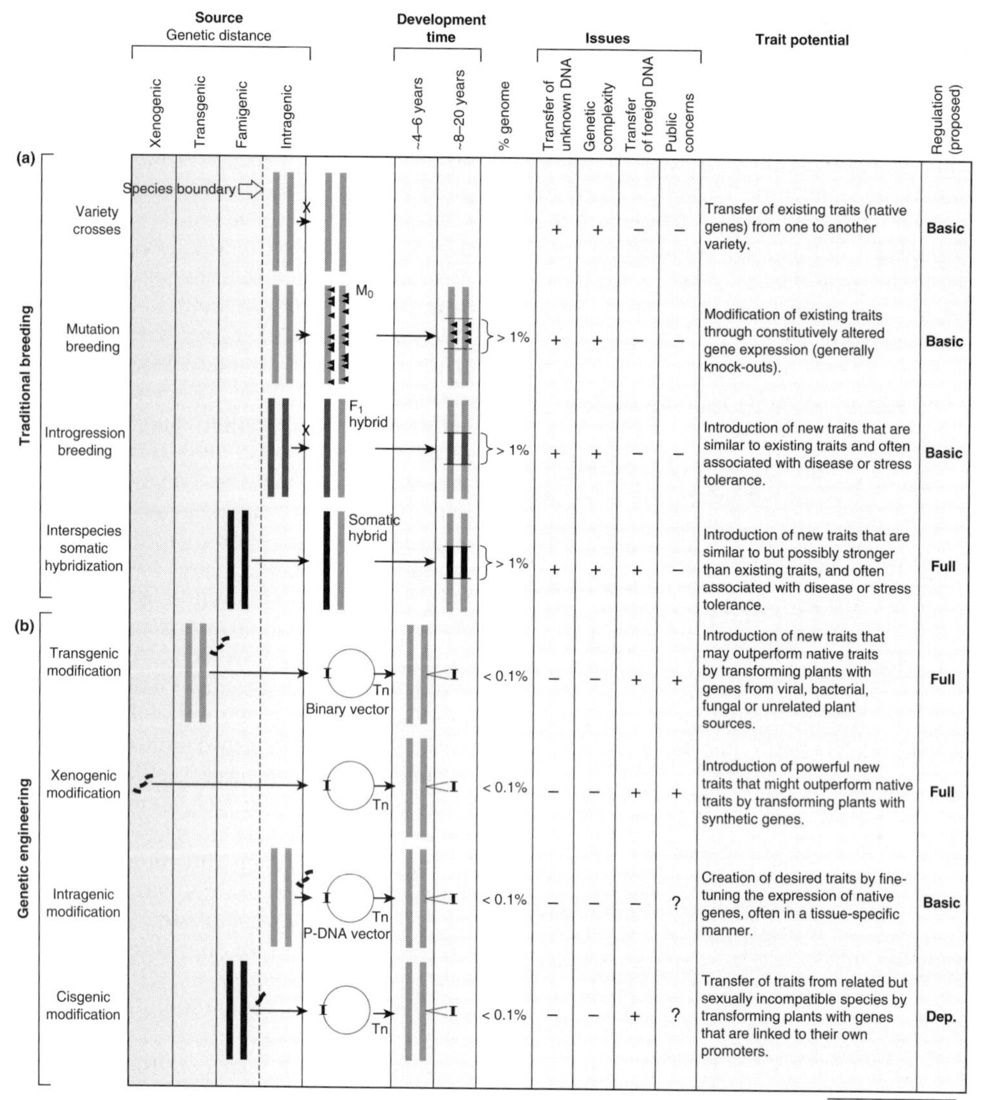

Fig. 12.1. Summary of various methods for crop improvement. The genetic distance between DNA source and target crop is indicated in the left four columns, including 'foreign' and 'sexually compatible'. The species barrier is shown as a dotted vertical line. Xenogenic, synthetic DNA; transgenic, DNA from unrelated species, such as viruses, bacteria, fungi and plants that belong to different families; famigenic, DNA from plants that belong to the same family; and intragenic, DNA from within the same sexual compatibility group. The '% genome' column shows the estimated size of the introduced DNA as a percentage of the entire genome. Proposed regulatory requirements are shown in bold letters with 'Basic' implying multi-year field tests on agronomic performance and an assessment of the nutritional profile, and 'Full' indicating more extensive studies, which include biosafety assessments of foreign proteins as well as environmental studies. Regulatory requirements for cisgenic applications are dependent on the trait ('Dep.'). In these cases, the transfer of traits that resemble native traits, such as those associated with disease resistance, should be considered for the basic regulatory assessment described above. However, traits that are new to the sexual compatibility group would require more extensive analyses. (a) Methods in traditional breeding. 'M_0' stands for an original plant derived from induced mutagenesis. Random mutations are shown as triangles, and can represent hundreds of point mutations/chromosome induced by ethylmethane sulfonate (EMS) or deletions of up to 100 kb pairs triggered by di-epoxy butane (DEB) or low linear energy transfer radiation (LET). (b) Methods in genetic engineering. Tn, plant transformation. Reprinted from Rommens *et al.* (2007) with permission from Elsevier.

There are several important fields in plant transformation that will not be discussed in detail in this chapter but are worthy of brief mention here: (i) high-throughput transformation, by which all candidate genes can be used for transformation prior to functional analysis; (ii) plastid transformation with a major advantage that in many plant species plastid DNA is not inherited, preventing gene flow from the GM-plant to other plants; and (iii) chromosome construction and transformation, by which high molecular weight DNA and multiple genes can be delivered into plant cells. Ogawa *et al.* (2008) established a large-scale, high-throughput protocol to construct *Arabidopsis thaliana* suspension-cultured cell lines, each of which carries a single transgene, using *Agrobacterium*-mediated transformation. They took advantage of RIKEN *Arabidopsis* full-length (RAFL) cDNA clones and the Gateway cloning system for high-throughput preparation of binary vectors carrying individual full-length cDNA sequences. Throughout all cloning steps, multiple-well plates were used to treat 96 samples simultaneously in a high-throughput manner. They evaluated the protocol by generating transgenic *Arabidopsis* T87 cell lines carrying individual 96 metabolism-related RAFL cDNA fragments and showed that the protocol was useful for high-throughput and large-scale production of gain-of-function lines for functional genomics. Plastid transformation is suitable only for certain crop species. For example, Ruf *et al.* (2007) studied genetically modified tobacco in which the transgene was integrated in chloroplasts. In a large screen, they detected low-level paternal inheritance of transgenic plastids in tobacco. Mini-chromosomes will be briefly discussed in Section 12.2.2.

12.1.3 Development of core plant transformation facilities

Transformation, particularly for cereals, is presently a technically demanding activity that is usually carried out as part of an industrial research process at large seed or agrochemical companies. Most university-based research groups do not have access to the physical or human resources necessary to establish a cereal transformation effort for their own target crop. These limitations have led to the establishment of core plant transformation facilities (PTFs) at a number of academic institutions. Examples of North American PTFs include Cornell University (tomato, algae, fungi), Iowa State University (maize and soybean), University of Nebraska at Lincoln (wheat and soybean), Texas A&M University (cotton, rice, sorghum, banana, conifers), University of Wisconsin at Madison (lucerne) and the National Research Council (NRC) of Canada (canola, wheat and pea). Core PTFs have advantages to exploit the economies of scale associated with the centralization of a labour-intensive activity, i.e. to assemble a critical mass of transformation specialists working on related problems with continuity of activity over long time frames, to provide an 'in-house' resource dedicated to fulfilling the exclusive transformation needs of the institution's own community, to eliminate the need to compete for limited collaborative opportunities elsewhere, to offer on-site teaching resources in plant tissue culture and transformation, to facilitate funding from local grower groups and to generate funds from private enterprises that contract out transformation activities to public sector organizations as a matter of economy. The core PTFs have been working well at big companies and international centres. A problem with core PTFs is that they may forget their reason for being and go off on tangents, thus being of limited use to the wider community.

12.2 Transformation Approaches

Transformation is the heritable change in a cell or organism brought about by the uptake and establishment of introduced DNA. Different methods have been developed to introduce foreign genes into plants. A common feature is that the foreign DNA first has to enter the plant cell by penetrating the plant cell wall and the plasma membrane and then must reach the nucleus and integrate into the resident chromosomes. There are two major techniques for introducing foreign genetic material into an organism (IBRD/World Bank, 2006). One is based on *Agrobacterium tumefaciens,* a bacterium that is able to insert its own or other genes into a plant genome. This method is accomplished through a plasmid (an autonomous piece of DNA) from the bacterium. The plasmid is used as the basis for the construction of a vector that incorporates the genes that are to be transferred to the plant cells. Another one is based on the direct, physical transfer of foreign genes into target plant cells. The most common example is particle bombardment (biolistics), in which metal particles are used as carriers of plasmids and are introduced into target plant tissue at high velocity. In this section, these two major approaches will be introduced along with a brief discussion of other direct gene transfer methods.

12.2.1 *Agrobacterium*-mediated transformation

Agrobacterium *strains*

Agrobacterium tumefaciens is a remarkable species of soil-dwelling bacteria that has the ability to infect plant cells with a piece of its DNA. When the bacterial DNA is integrated into a plant chromosome, it effectively hijacks the plant's cellular machinery and uses it to ensure the proliferation of the bacterial population. Many gardeners and orchard owners are unfortunately familiar with *A. tumefaciens*, because it causes crown gall diseases in many ornamental and fruit plants.

The DNA in an *A. tumefaciens* cell is contained in the bacterial chromosome as well as in another structure known as a Ti (tumour-inducing) plasmid. The Ti plasmid contains a stretch of DNA termed T-DNA (~20 kb long) that is transferred to the plant cell in the infection process and a series of *vir* genes that direct the infection process.

To harness *A. tumefaciens* as a transgene vector, the oncogenes (gall-forming sequences) have been moved from the T-DNA and in their place engineered expression cassettes with genes from virtually any source may be substituted, usually by convenient insertion into multiple cloning sequences that have been incorporated into these plasmids. The transgene and the selectable markers are inserted in the vector between two unique sequences, called the left border (LB) and the right border (RB). Only T-DNA is expected to be transferred to the plant cell and becomes integrated into the plant's chromosomes (Wong, 1997).

At the present time gene transfer by *Agrobacterium* is the established method of choice for the genetic transformation of most plant species. It has been successfully practised in both dicots (broadleaf plants like soybeans and tomatoes) and monocots (banana, grasses and their relatives). A general scheme for *Agrobacterium*-mediated transformation is outlined in Fig. 12.2. It is perceived to have several advantages over other forms of transformation (such as biolistics), including the ability to transfer large segments of DNA with minimal rearrangement and with fewer copies of inserted genes at higher efficiencies with lower cost (reviewed by Hiei *et al.*, 1997). In addition, *Agrobacterium* transformation may facilitate the removal of plant-selectable marker genes by segregation (Komari, 1996; Matthews *et al.*, 2001).

Application in cereals

As a vector-mediated transformation system, *Agrobacterium* was thought to have a restricted range of interacting hosts. For example, cereal cells were initially thought to be recalcitrant to *Agrobacterium*-mediated transformation because most monocot species are outside the natural host range of *A. tumefaciens*. Current *Agrobacterium*-mediated transformation protocols for cereals still involve a callus initiation stage. There are still issues like genotype-dependency and somaclonal variation. Development of reliable and efficient protocols is of great importance for the

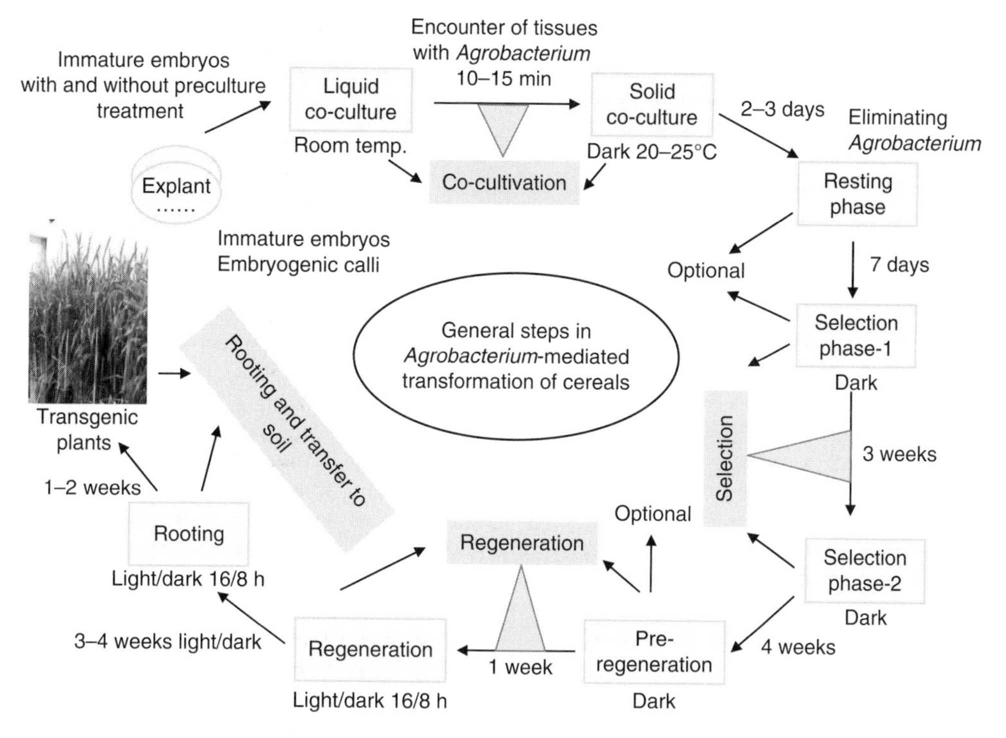

Fig. 12.2. General scheme for *Agrobacterium*-mediated transformation of cereal plants. From Shrawat and Lörz (2006) with permission from Wiley-Blackwell.

successful application of transformation technology in crop species.

Several factors influencing *Agrobacterium*-mediated transformation of monocots have been investigated and elucidated (as reviewed by Cheng *et al.*, 2004; Jones *et al.*, 2005), including the screening of the most responsive genotype and explant, *Agrobacterium* strain, binary vector, selectable marker gene and promoter, inoculation and co-culture conditions and tissue culture and regeneration medium. Of these factors, genotype and explant are considered the major limiting factors in *Agrobacterium*-mediated transformation of cereals, especially in extending the host range to commercial cultivars.

The explant type, explant quality and source of the explant have been found to be correlated with successful reports on the *Agrobacterium*-mediated genetic transformation of cereals (reviewed by Repellin *et al.*, 2001). For example, freshly isolated immature embryos with and without pretreatment have been part of the majority of successful reports on the genetic transformation of cereals and are considered the best explant type (Cheng *et al.*, 1997; Wu, H. *et al.*, 2003). Embryogenic callus derived from mature seeds has been reported to be the best explant for *Agrobacterium*-mediated transformation of rice as a result of active cell division (Hiei *et al.*, 1994). In conclusion, any explants at a vigorously dividing stage can be good for transformation.

Host plant genes involved in Agrobacterium*-mediated transformation*

The identification and molecular characterization of the plant genes involved in successful *Agrobacterium*-mediated transformation have opened up new avenues for a better understanding of the plant response to *Agrobacterium* infection (Veena *et al.*, 2003).

Such information may help to develop methods to enhance the transformation frequency of economically important plant species. In addition, the in-depth studies and evaluation of the genes responsible for stimulating plant cell division and the competency of plant cells to *Agrobacterium* may increase not only the extension of transformation protocols to elite genotypes but also the transformation efficiency in cereals (Shrawat and Lörz, 2006).

In recent years, efforts have also been made to understand the interactions of host plants with *Agrobacterium* at the molecular level (Ditt *et al.*, 2001; Veena *et al.*, 2003). Hwang and Gelvin (2004) have identified four *Arabidopsis* proteins that interact with the main T-pilus protein, VirB2 and have shown that the presence of these proteins is required for efficient transformation.

12.2.2 Particle bombardment

The microprojectile bombardment method, also known as 'gene gun' or biolistics, uses fine metal particles (typically tungsten or gold) coated with DNA that are usually accelerated with helium gas under pressure (Fig. 12.3). Particle bombardment involves the acceleration of DNA coated microparticles into cells and tissues. In biolistic bombardment, the primary delivering systems are the Biolistic PDS-1000/He helium-powered gun and similar designs, or the particle inflow gun. Parameters involved in the biolistic gun include pressure (ranging from 900 to 1300 psi), particle size (0.6–1.1 μm) and type of material (gold and tungsten), target distance (7.5–10 cm) and target material (cell suspension, callus, meristem, protoplast, immature embryo). Particle acceleration is rapid enough to penetrate the cell wall

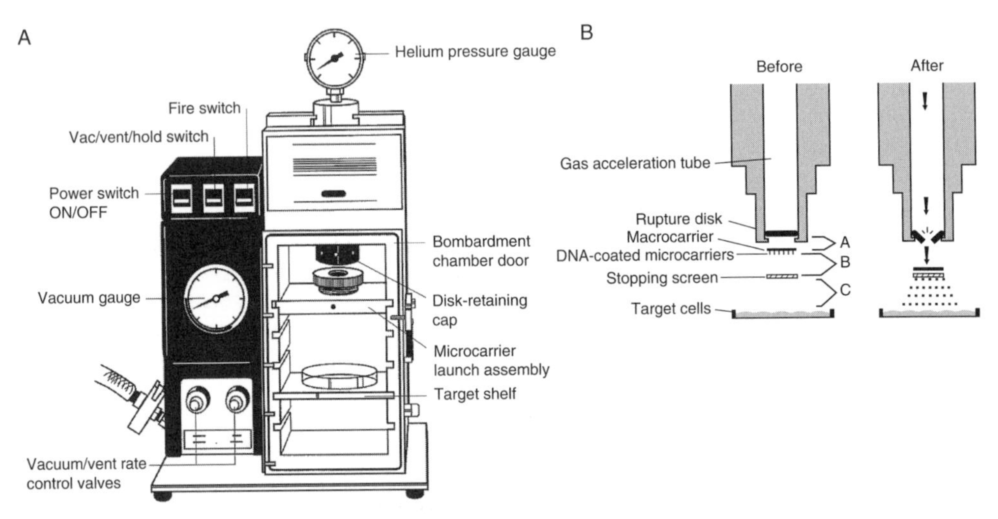

Fig. 12.3. Gene gun and system. (A) The biolistic system. The Biolistic PDS-1000/He instrument consists of the bombardment chamber (main unit), connective tubing for attachment to vacuum source, and all components necessary for attachment and delivery of high pressure helium to the main unit (helium regulator, solenoid valve, etc.). (B) Biolistic process. The Biolistic PDS-1000/He system uses high pressure helium, released by a rupture disk and partial vacuum to propel a macrocarrier sheet loaded with millions of microscopic tungsten or gold microcarriers towards target cells at high velocity. The microcarriers are coated with DNA or other biological materials for transformation. The macrocarrier is halted after a short distance by a stopping screen. The DNA-coated microcarriers continue travelling towards the target to penetrate and transform the cells. The launch velocity of microcarriers for each bombardment is dependent upon the helium pressure (rupture disk selection), the amount of vacuum in the bombardment chamber, the distance from the rupture disk to the macrocarrier, the macrocarrier travel distance to the stopping screen, and the distance between the stopping screen and target cells.

without causing excessive damage. For transformation, DNA is coated on to the surface of micron-sized tungsten or gold particles by precipitation with calcium chloride and spermidine and DNA must be delivered into cells from which whole plants can be generated. If the foreign DNA reaches the nucleus then transient expression is likely to result and the transgene may become incorporated in a stable manner into host chromosomes. Stanford developed the original bombardment concept (Stanford *et al.*, 1987; Stanford, 2000) and coined the term 'biolistics' (short for 'biological ballistics') for both the process and the device. Particle bombardment has been especially useful in transforming monocot species as it has no biological constraints or host limitations, can target diverse cell types and is the most convenient way for achieving organelle transformation. However, it is widely believed that particle bombardment produces large, multi-copy and highly complex transgenic loci that are prone to further recombination, instability and silencing.

Particle bombardment facilitates a wide range of transformation strategies

The Biolistic and Helios systems can be used to circumvent the need to maintain viruliferous populations of insect vectors, allowing direct introduction of infectious viral nucleic acids into a range of plant species. An attractive feature of such systems is the flexibility by which co-infections can be achieved with different viral species and genomic components, generating a powerful tool for investigating mechanisms of pathogenicity and host resistance. Particle bombardment was utilized both to produce transgenic cassava plants and to challenge them by simultaneous inoculation with two species of geminiviruses (Chellappan *et al.*, 2004). Particle bombardment also has an important role to play in extending virus-induced gene silencing (VIGS) into economically important crop plants (Fofana *et al.*, 2004).

Particle bombardment has no biological constraints or host limitations

Particle bombardment overcomes the boundaries defined by classical host ranges of *Agrobacterium*-mediated transformation by exploiting physical principles to introduce the DNA into the plant cell and then relying on factors that are common to all plants (i.e. DNA repair mechanisms) to enable stable transgene integration. For stable transformation and the recovery of transgenic plants, particle bombardment is restricted only by the requirement to deliver DNA into regenerable cells (Altpeter *et al.*, 2005a). By removing almost all the incidental biological constraints that limit other transformation methods, particle bombardment has facilitated the transformation of some of the most recalcitrant plant species.

The ability to transform diverse cell types by particle bombardment facilitates a broad range of applications that are difficult or impossible to achieve by other transformation methods. This is critical when the rapid analysis of large numbers of constructs in a specific tissue or cell type is required.

Diverse cell types can be targeted efficiently for foreign DNA delivery

Particle bombardment does not depend on any particular cell type as long as the DNA can be introduced into the cell without killing it. In rice, the range of suitable tissues includes immature embryos (7–8 days after anthesis), embryogenic callus derived from either immature embryos or mature seeds and suspension culture cells (Datta *et al.*, 1998, 2001; Tu *et al.*, 1998a, b; Baisakh *et al.*, 2001). Transient expression has even been achieved using the intact immature seed endosperm following bombardment with a vector carrying the *gus*A reporter gene (Grosset *et al.*, 1997; Clarke and Appels, 1998).

Vectors are not required for particle bombardment

The exogenous DNA used in transformation experiments typically comprises a plant expression cassette inserted in a vector based on a high-copy number bacterial cloning plasmid. Neither of these components is required for DNA transfer and only the expression cassette is required for transgene expression. The expression cassette typically comprises a promoter, open reading frame and polyadenylation site that are functional in plant cells, although other components may be present, such as a protein-targeting signal (Altpeter *et al.*, 2005a). Once this plasmid has been isolated from the bacterial culture it can be purified and used directly for transformation.

During *Agrobacterium*-mediated transformation, the T-DNA is naturally excised from the vector during the transformation process. This frequently, although not always, prevents the integration of vector backbone sequence into the plant genome (Fang *et al.*, 2002; Popelka and Altpeter, 2003), necessitating time-consuming sequence analysis of transgene insertion sites following *Agrobacterium*-mediated gene transfer. In contrast, particle bombardment involves no such processing. Cloning vectors are used in particle bombardment for convenience rather than necessity. Consequently, Fu *et al.* (2000) devised a clean DNA strategy in which all vector sequences were removed prior to particle loading. A standard plasmid vector was used to clone the plant expression cassette and transgene of interest in bacteria and then the cassette was excised from the plasmid and purified by agarose gel electrophoresis. This minimal, linear cassette was then used to coat the metal particles and carry out transformation.

High molecular weight DNA delivery into plant cells

Until recently, one serious limitation to plant transformation technology was the inability to introduce large intact DNA constructs into the plant genome. Such large constructs could incorporate multiple transgenes, or could comprise a segment of genomic DNA to facilitate the map-based cloning of plant genes. In *Agrobacterium*-mediated transformation, this limitation has been addressed by the development of binary bacterial artificial chromosome (BIBAC) and transformation-competent artificial chromosome (TAC) vectors (Shibata and Liu, 2000). The transfer of yeast artificial chromosome (YAC) DNA by particle bombardment was first demonstrated by Vaneck *et al.* (1995) using cell suspensions of two tomato cultivars. Only one of the cultivars yielded YAC transformants and initial studies suggested that the integrated YAC was 'fairly intact' in four of the five transformants recovered, based on the presence of two marker genes. The most promising way of introducing high molecular weight DNA into plant cells is to create engineered mini-chromosomes in maize and genes to those mini-chromosomes (Yu *et al.*, 2007). Mini-chromosomes are able to function in many of the same ways as chromosomes but allow for genes to be stacked on them. The technique developed in maize should be transferable to other plant species.

Particle bombardment is the most convenient way to achieve organelle transformation

Thus far, most genetically engineered plants have been subject to nuclear transformation. An alternative approach is to introduce transgenes into the chloroplast genome. This strategy offers advantages such as very high levels of transgene expression, uniparental plastid gene inheritance in most crop plants (preventing pollen transmission of transgenes), the absence of gene silencing and position effects, integration via a homologous recombination process that facilitates targeted transgene insertion, elimination of vector sequences, precise transgene control and sequestration of foreign proteins in the organelle, which prevents adverse interactions within the cytoplasmic environment (as reviewed by Altpeter *et al.*, 2005a).

Comparison with other methods

In addition to the properties discussed for particle bombardment, Altpeter *et al.* (2005a) provided a comprehensive review of this method by comparing it with other transformation methods. Transgene integration, mediated by either *A. tumefaciens* or particle bombardment, is a random process that appears to correlate with the position of naturally occurring chromosome breaks. Transcriptionally active regions of the genome are favoured, particularly the subterminal regions of the chromosomes, perhaps because the DNA is more accessible in these areas. It is possible, although still a matter of speculation, that further breaks may be caused by particle bombardment since the microprojectiles may shear the ends of DNA loops in the nucleus (Abranches *et al.*, 2000; Kohli *et al.*, 2003), which may partially explain the relative efficiency of bombardment in terms of stable transformation compared to other techniques.

Compared to biolistic techniques, *Agrobacterium*-mediated transformation offers several advantages (Tzfira and Citovsky, 2006), such as simpler integration patterns resulting in lower mutational consequences for the transgenic plant and limited transgene silencing via co-suppression. In addition, the option for fine tuning the *Agrobacterium*-based transformation protocols renders more and more cereal species amenable for efficient genetic engineering (Shrawat and Lörz, 2006; Conner *et al.*, 2007).

12.2.3 Electroporation and other direct gene transfer approaches

There are several less popular means of gene transfer that may be effective in specific cases: polyethylene glycol (PEG)-facilitated protoplast fusion, microinjection, sonication, *in planta* transformation and electroporation. The mechanism is to cause transient micro-wounds in the cell wall and the plasma membrane, allowing DNA in the medium to enter the cytoplasm before repair or fusion of the damaged cellular structures. The direct transfer of DNA to protoplasts using PEG, or electroporation resulting in the transient permeabilization of the cell membrane using high-voltage electric fields, has been shown to be possible in various plants. Leaf tissue or embryogenegic calli are often used to isolate protoplasts by enzymatic treatment. Using protoplasts as the starting material for transformation in cereals often employs callus induction, suspension culture initiation and maintenance, protoplast isolation, callus formation and plant regeneration. It is important to generate and maintain a cell suspension culture for its embryogenic capacity. Extensive time in tissue culture often results in low reproducibility and poor regeneration capacity.

A breakthrough in *Arabidopsis* research was the invention of the vacuum-infiltration procedure, a simple and reliable method of obtaining transformants at high efficiency while avoiding the use of tissue culture (Bent, 2000). *In planta* transformation involves floral dip, vacuum infiltration and spraying. They yield transformants at frequencies ranging up to several percent, with the most common frequency being 0.1–1%.

Electroporation utilizes short, high-intensity electric fields to permeabilize reversely the liquid bilayers of the cell membrane. It is widely believed that the electric pulse causes extensive compression and thinning of the plasmalemma. The resulting transient formation of pores permits free diffusion of various classes of macromolecules including dyes, antibodies, RNA and viral particles and DNA. Transient expression from electroporated plant cells has been used to define functional elements within a promoter, to examine the effects of antisense RNA on gene expression, to study the translocation of proteins into both plasmids and nuclei of intact protoplasts, to examine cell-cycle-specific gene expression and to study responses to plant hormones.

As a method of DNA transfer, electroporation is convenient and the results are consistently duplicated as a daily routine. In most cases it is more efficient than other methods designed for the same purpose, such as particle bombardment. In addition, it does not suffer from host-range limitations imposed by biology-based systems such as those employing *A. tumefaciens* or toxicity problems sometimes encountered using a PEG-based procedure. Finally, electroporation coupled with a transient expression assay is rapid, allowing for the reproducible detection of gene products within hours of the introduction of DNA. This is in contrast to a stable transformation strategy that involves months to regenerate transformants and suffers from uncontrollable larger variations in gene expression because of 'positional effects'. In the context of a transformation programme in which stable integration of genetic materials is required, transient expression may be used to rapidly demonstrate functionality of the new transgene sequences before they are used to generate transformants by some other methods of DNA introduction.

An electroporation-based transfection system consists of a number of potentially important variables, including methods of protoplast preparation, electric pulse strength and duration, ionic concentration and composition of the electroporation buffer and DNA purity, concentration and topology. Fisk and Dandekar (2004) have analysed the importance of these variables in addition to a few others with the goal of identifying and optimizing the parameters necessary to increase expression frequency among a population of tobacco protoplasts. They described optimized conditions for an electroporation-based transient expression assay that routinely results in nearly 90% expression frequency.

12.3 Expression Vectors

The progress in plant genetic engineering could not have been as productive as it is today without the development of small, easy-to-manipulate and simple-to-use *Agrobacterium* binary vectors (Komari *et al.*, 2006, 2007). The first generation of plant transformation binary vectors were rather simply designed, lacking cloning and expression versatility and offered very little flexibility for their manipulation for specific research or application purposes (Fig. 12.4; Tzfira *et al.*, 2007).

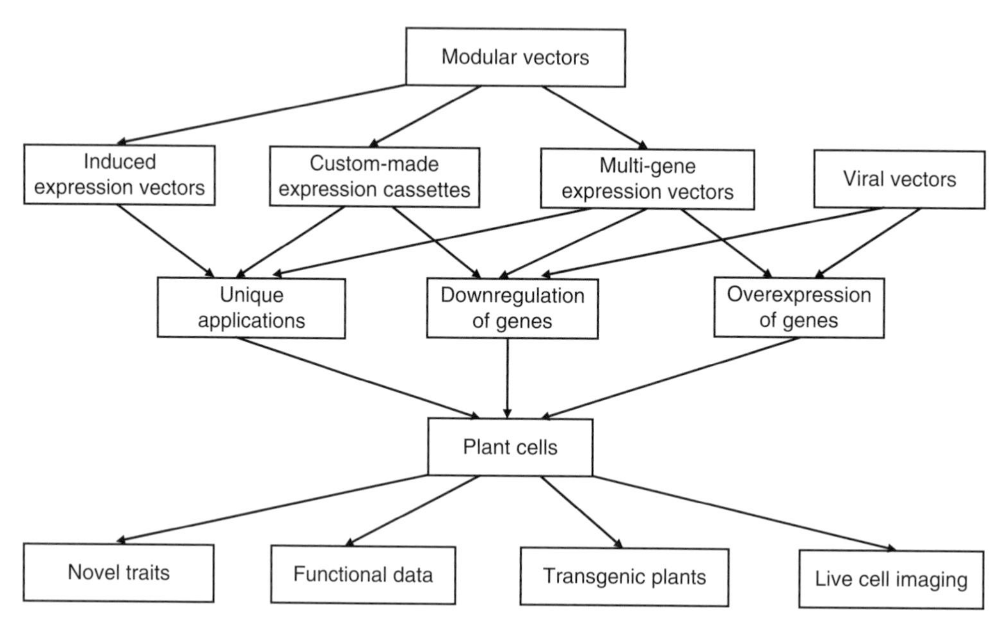

Fig. 12.4. From vectors to applications to cellular functions. Introduction of genetic information into target plant cells and acquisition of new data as a result of transgene expression may require a network of modular vectors, flexible gene cloning and expression systems, and specialized plasmids that result in different modes of transgene expression. Modular vectors may represent a starting point for assembly of custom-made expression vectors, multi-gene expression vectors, and other types of plant transformation vectors. These vectors in turn provide the users with the abilities to overexpress and downregulate genes, as well as with the capacity for specific, and often unique, applications, useful for obtaining novel traits and functional data, protein imaging in living plant cells, and generating transgenic plants for plant research and biotechnology.

A crucial improvement made to the first generation of binary plasmids was the introduction of an empty plant expression cassette, a feature that allowed the plant biologists a simple and more direct route for cloning their gene of interest under the control of a plant-expressing constitutive promoter. The constant improvements in binary vectors even included the most famous binary vector, one of which has been dominating the landscape of binary plasmids for several decades: pBin19 (Bevan, 1984; Komori *et al.*, 2007). This plasmid offers several features, including incorporation of the *lacZ* gene into the multiple cloning site (MCS) to facilitate identification of recombinant plasmids using a colorimetric assay, a bacterial kanamycin-resistance gene, an *E. coli* origin of replication, a complete plant selection marker expression cassette and an extended MCS.

New vectors were designed and constructed to provide users with a more specialized set of tools suitable for carrying out various tasks in plant cells, e.g. the transfer of extremely long DNA molecules (Hamilton, 1997), the expression of fluorescent protein fusions (Goodin *et al.*, 2002) and the detection of protein–protein interactions (Bracha-Drori *et al.*, 2004), while others were specifically designed for versatility and simplicity, allowing plant biologists not only a choice but also the ability to manipulate these vectors for their own needs. The latter group of vectors are typically constructed as families of plasmids and include, for example, the pCB mini-binary vector series that featured a collection of extremely small pBin19-derivative vectors (Xiang *et al.*, 1999) and the pGreen series of plasmids featuring versatile and flexible series of binary vectors (Hellens *et al.*, 2000b). These and many other families of vectors provide the plant research community with a vast number of versatile tools for various plant expression analyses. Some well-known binary and superbinary vectors are listed in Table 12.1.

Table 12.1. Well-known binary and superbinary vectors (from Komori *et al.* (2007) reproduced with permission of the American Society of Plant Biologists).

Vector	Plant selection marker[a]	Bacterial selection marker[b]	Replication origin for *A. tumefaciens*	Replication origin for *E. coli*	Mobilization	Reference	Frequency of use in recent literature[c]
pBin19	Kan	Kan	IncP	IncP	Yes	Bevan (1984)	40%
pBI121	Kan	Kan	IncP	IncP	Yes	Jefferson (1987)	40%
pCAMBIA series	Kan or Hyg	Cm or Kan	pVS1	ColE1	Yes	www.cambia.org	30%
pPZP series	Kan or Gen	Cm or Sp	pVS1	ColE1	Yes	Hajdukiewicz *et al.* (1994)	30%
pGreen series	Kan, Hyg, Sul, or Bar	Kan	IncW	pUC	Yes	Hellens *et al.* (2000)	3%
pGA482	Kan	Tc, Kan	IncP	ColE1[d]	Yes	An *et al.* (1985)	3%
pSB11[e]	None	Sp	None	ColE1	Yes	Komari *et al.* (1996)	3%
pSB1[e]	None	Tc	IncP	ColE1[d]	Yes	Komari *et al.* (1996)	3%
pPCV001	Kan	Ap	IncP	ColE1[d]	Yes	Koncz and Schell (1986)	1%
pCLD04541	Kan	Tc, Kan	IncP	IncP	Yes	Tao and Zhang (1998)	1%

Continued

Table 12.1. *Continued.*

Vector	Plant selection marker[a]	Bacterial selection marker[b]	Replication origin for *A. tumefaciens*	Replication origin for *E. coli*	Mobilization	Reference	Frequency of use in recent literature[c]
pBIBAC series	Kan or Hyg	Kan	pRi	F factor	Yes	Hamilton (1997)	0%
pYLTAC series	Kan or Bar	Kan	pRi	Phage P1	No	Liu *et al.* (1999)	0%

[a]Kan, Kanamycin; Hyg, hygromycin; Gen, gentamycin; Sul, sulfonylurea; Bar, phosphinothricin.
[b]Kan, Kanamycin; Cm, chloramphenicol; Sp, spectinomycin; Tc, tetracycline; Ap, ampicillin.
[c]From issues between 2005 and 2007 of 12 leading plant journals, 180 papers in which plant transformation mediated by *A. tumefaciens* is described were randomly chosen and surveyed.
[d]Although IncP is also active in *E. coli*, it is likely that the plasmid is replicated mainly by the ColE1 system.
[e]pSB11 and pSB1 are an intermediate vector and an acceptor vector of the superbinary vector system, respectively.

The 2007 Focus Issue of *Plant Physiology* presented a collection of original articles describing the development of new vector systems useful for plant research and biotechnology, as well as a compilation of short review articles that highlight some of the major developments in vector-assisted plant research technologies (Tzfira *et al.*, 2007). It includes papers describing an extensive collection of MultiSite Gateway-based plant expression vectors (Karimi *et al.*, 2007), a guide to vectors for chloroplast transformation (Lutz *et al.*, 2007) and a system of transformation vectors with the superpromoter (Lee, L.-Y., *et al.*, 2007). For a recent update on binary vectors, the reader is referred to an update by Komori *et al.* (2007).

12.3.1 Binary vectors

The binary vector was invented soon after it had been elucidated that crown gall tumorigenesis was caused by genetic transformation of plant cells with a piece of T-DNA from a Ti plasmid (tumour-inducing plasmid) harboured by *A. tumefaciens* (Fraley *et al.*, 1986). A key finding was that the virulence genes, which are involved in the transfer of T-DNA, could be placed on a replicon separate from the one with T-DNA (Hoekema *et al.*, 1983). Thus, combination of a 'disarmed' strain, which carries a Ti plasmid without the wild-type T-DNA and an artificial T-DNA within a plasmid that can be replicated both in *E. coli* and *A. tumefaciens* turned out to be fully functional in plant transformation. The term 'binary vector' literally refers to the entire combination, but the plasmid that carries the artificial T-DNA is usually called a binary vector.

A binary vector consists of T-DNA and the vector backbone (Fig. 12.5). T-DNA is the segment delimited by the border sequences, the right border (RB) and the left border (LB) and may contain MCS, a selectable marker gene for plants, a reporter gene and other genes of interest. The vector backbone carries plasmid replication functions for *E. coli* and *A. tumefaciens*, selectable marker genes for the bacteria, and optionally a function for plasmid mobilization between the bacteria and other accessory components (Komori *et al.*, 2007).

The RB and the LB are imperfect, direct repeats of 25 bases and said to be the only essential *cis*-elements for T-DNA transfer (Yadav *et al.*, 1982). The RB and the LB are integrated in binary vectors as DNA fragments cloned from well-known Ti plasmids, of either the octopine or nopaline type.

Insertion of genes of interest into appropriate locations of a binary vector is traditionally carried out by standard subcloning techniques. MCS, which are similar or identical to those in pUC, pBluescript and other standard vectors, are still very useful in this regard, but recently constructed vectors are

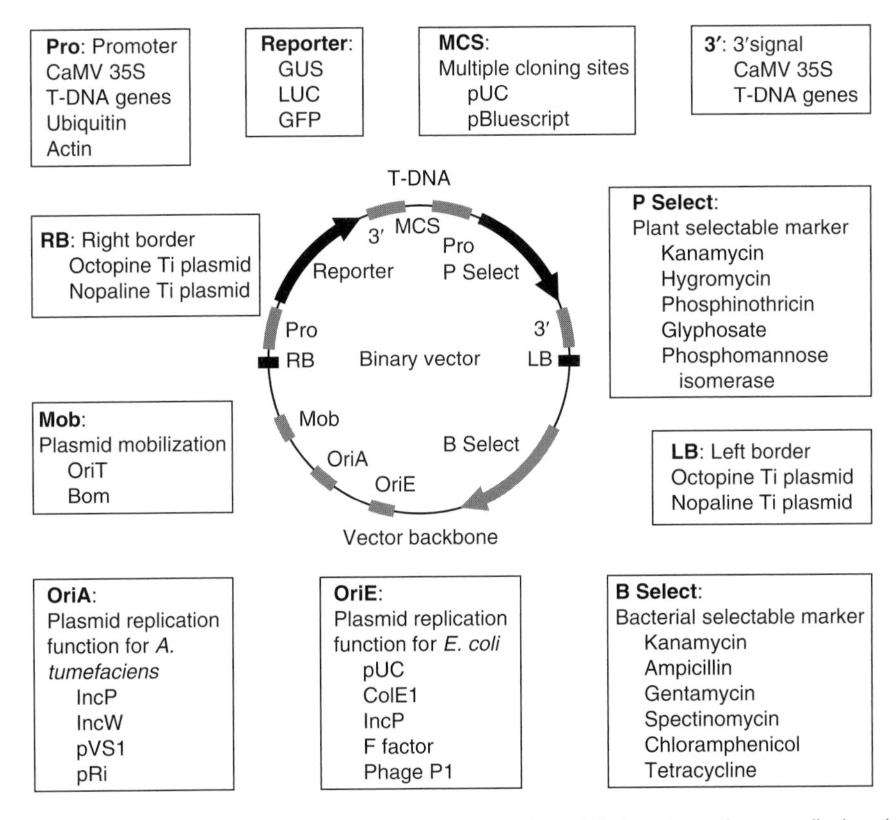

Fig. 12.5. Typical structure of a binary vector. Key components and their major options are displayed. From Komori *et al.* (2007) reproduced with permission of the American Society of Plant Biologists.

more user-friendly. Recognition sites for 'rare cutters', which are restriction enzymes with long recognition sequences, are very convenient in this respect because the DNA fragments that are to be inserted scarcely have such sites. In some of the recently created vectors termed modular vectors, a series of these rare sites are placed in the T-DNA (Chung *et al.*, 2005). An extensive set of auxiliary plasmids, which have full sets or subsets of these rare sites and other restriction sites, are provided and some of the plasmids also carry frequently used promoters, marker genes and/or 3' signals. Various types of expression units may be constructed in auxiliary plasmids and then the units may be inserted into the modular binary vectors. Thus, several expression cassettes could easily be assembled in a binary vector.

Until the early 1990s, *Agrobacterium*-mediated transformation had been used mainly in dicotyledons and it had been difficult to apply the method to cereals. The finding that some of the virulence genes exhibited gene dosage effects led to the development of a superbinary vector, which carried additional virulence genes. The superbinary vector has been highly efficient in the transformation of various plants and especially useful in the transformation of recalcitrant plants, such as important cereals.

A superbinary vector was developed and successfully used for the transformation of monocotyledons, such as rice and maize (Hiei *et al.*, 1994; Ishida *et al.*, 1996). The superbinary vector is an improved version of a binary vector and carries the 14.8-kb KpnI fragment that contains the *virB*, *virG* and *virC* genes derived from pTiBo542, which is responsible for the supervirulence phenotype of an *A. tumefaciens* strain, A281 (Jin *et al.*, 1987; Komari, 1990).

12.3.2 Gateway-based binary vectors

Binary vectors used for generation of transgenic cereal species are typically cumbersome due to their large size and the rather limited number of useful restriction sites. To bypass laborious preparation of constructs, Gateway technology (Invitrogen) is used especially for binary vectors generating knock-down lines. Gateway-derived cloning systems are based on the site-specific recombination system from bacteriophage 1 (Landy, 1989) and circumvent traditional cloning methods involving restriction and ligation of DNA sequences. A number of Gateway-based binary vector sets for plant functional genomics have been developed, thereby allowing overexpression or knock-down of effector genes, expression of fusion proteins (as reviewed by Earley *et al.*, 2006) and transformation of multiple genes (Chen *et al.*, 2006).

The Gateway system provides another user-friendly feature. A DNA fragment flanked by a pair of short, specific sequences may easily be replaced with another DNA fragment by the Gateway system. Thus, introduction of DNA fragments into a binary vector with the sites for the Gateway system is a straightforward step and is useful in many applications. Combination of the modularity based on rare-cutting restriction enzymes and the Gateway recombination sites provides an extensively versatile cloning system, which is especially useful in the production of T-DNA with multiple genes (Chen *et al.*, 2006).

Gateway-based binary vectors have been developed for dicotyledonous plants (e.g. Wesley *et al.*, 2001; Curtis and Grossniklaus, 2003; Tzfira *et al.*, 2005). However, these are typically not useful for monocotyledons, mainly because of the limited functionality of promoters that are used to drive either the gene of interest or the plant selection marker. However, other specific vector elements, such as the plant-selectable marker and the origin of replication, may impede the amenability of a binary vector.

Himmelbach *et al.* (2007) provided a set of generic binary vectors that is made available for phenotypic studies in stably transformed cereal species. Its modular configuration permits convenient insertion of promoter and effector sequences, as well as of plant selection marker cassettes of choice. The insertion of effector sequences into the binary overexpression and knock-down vector series is facilitated by the highly efficient Gateway recombination system. The spectrum of applications is further extended by the options to test constructs in transient expression assays (e.g. in barley) prior to starting the laborious stable transformation procedure and by the option to transform monocotyledonous and dicotyledonous plants using the same binary vector. Vector derivatives with strong, constitutive promoters, such as the maize ubiquitin promoter (ZmUbi1; Furtado and Henry, 2005), the double-enhanced CaMV 35S promoter (d35S; Furtado and Henry, 2005), or the rice actin promoter (OsAct1; McElroy *et al.*, 1990; Vickers *et al.*, 2006), are provided. In addition, the wheat glutathione S-transferase promoter (TaGstA1; Altpeter *et al.*, 2005b) permits the expression of transgenes confined to leaf epidermis in a constitutive manner. With the availability of a combination of the highly efficient Gateway cloning system, a selection of cereal promoters controlling the expression of genes of interest, different plant selection markers and the option of further convenient vector modifications, the functional characterization of DNA sequences in cereal species will be greatly facilitated.

12.3.3 Choice of transformation vectors

As a wide range of binary vectors and superbinary vectors is available now, helpful guidance for selection of the vectors is needed and has already been provided in the literature (Hellens *et al.*, 2000a; Komari *et al.*, 2006). Unfortunately there is no vector that is good for all purposes, but, fortunately, many of the vectors currently available are quite versatile. They may be used in various types of experiments and there is a good chance that vectors being routinely used can be employed in the experiment of interest. If this is not the case or better options are worthwhile searching for, a series of questions needs to be asked about the size and nature of the DNA fragments, the strains of *A. tumefaciens* to be employed, the species of plants to be transformed and the purposes of the experiments. If the DNA fragments are larger than 15 kb, IncP, BIBAC and TAC vectors are recommended. Otherwise, high-copy-number plasmids are very convenient and a wide range of vectors varying in restriction sites, selectable markers and Gateway sites is available. A series of vectors designed for specific purposes, e.g. vectors for suppression of plant genes by RNA interference (RNAi) technology (Miki and Shimamoto, 2004) may also be chosen.

Newer generations of plant transformation vectors provide us with improved strategies for cloning and delivering their genes of interest into plant cells, typically using *Agrobacterium* as a vehicle for the transformation process. Some of these vectors were developed as families of plasmids and others represented single constructs designed for specific purposes. One can find a plasmid for every task, including such relatively unique applications as activation tagging (e.g. the pSKI015 and pSKI074 binary vectors; Weigel *et al.*, 2000) or dexamethasone-inducible expression (e.g. the pOp/LhGR transcription activation system; Samalova *et al.*, 2005). In addition, vectors have been constructed that allow us to take advantage of radically new cloning methodologies and utilize new gene expression technologies. In addition, new vector systems are being produced to utilize transgenic technologies in an ever-expanding range of plant species, such as forest trees and transformation-recalcitrant crops (e.g. Meyer *et al.*, 2004; Coutu *et al.*, 2007). Furthermore, vectors for systemic gene expression without permanent genetic modification of the plant are being developed based on different plant viruses (e.g. Gleba *et al.*, 2005; Marillonnet *et al.*, 2005).

12.4 Selectable Marker Genes

The use of selectable marker gene systems facilitates the transformation process and allows the relatively straightforward recovery of transgenic crop plants (Ramessar *et al.*, 2007). Without them, the few plant cells that take up and stably integrate the foreign DNA would simply be lost in an ocean of wild-type cells, which would certainly overgrow these transformed cells in the absence of effective selection against them. However, under certain conditions, selectable marker genes may not be necessary and it may be feasible to get transgenic plants without selection of a marker gene.

12.4.1 Functions of selectable marker genes

Once a plant cell has incorporated the introduced DNA in a stable manner (i.e. covalently integrated within the host plant's genome), the next step is to regenerate plants from the transformed cells. Position, frequency and scope of regeneration events are critical to the isolation of transgenic plants. Most often, the major limiting step in the isolation of transgenic plants is a lack of regeneration occurring from within the transformed cell populations. There is a large amount of variability in the frequency and scope of regeneration among different angiosperm species as well as among different cultivars of any one species.

A critical step in the regeneration of transgenic plants is the ability to distinguish between transformed plant cells with an integrated transgene and the bulk of non-transformed cells. The traditional way to achieve this goal is to use marker genes within the transgene and to select for their expression. Genes conferring resistance to various antibiotics or herbicides are commonly used in laboratory transformation research. Selective marker genes act by expressing an enzyme that inactivates the selective agent (detoxification) and a resistant variant of a selective agent's target enzyme (tolerance). For example, the aminoglycoside antibiotics, such as kanamycin, neomycin and G418 kill cells by inhibiting protein translation. The *E. coli nptII* gene, encoding neomycin phosphotransferase, inactivates these antibiotics by phosphorylation, thus allowing preferential growth of plant cells transformed with this gene on media containing these selection agents. The herbicide phosphinothricin is an analogue of glutamine and acts by irreversibly inhibiting glutamine synthetase, a key enzyme for ammonium assimilation and the regulation of nitrogen assimilation in plants. The *bar* gene, cloned from the bacterium *Streptomyces hygroscopicus*, encodes phosphinothricin acetyltransferase, which converts phosphinothricin into the non-toxic acetylated form and allows growth of transformed plant cells in the presence of phosphinothricin, or commercial glufosinate ammonium-based herbicides.

All systems in general have low transformation efficiencies in the absence of selectable markers. However, in the presence of a selectable marker, in systems such as tobacco, rice and maize cells, transformation frequencies are extremely high. With high co-transformation frequencies selectable markers facilitate the identification of plants containing co-transformed transgenes. The utility of the individual selectable marker genes is a function of both the properties of the respective resistance protein they encode and the relative sensitivity of the target tissue to their corresponding selective agent. The timing of selective agent application is critical to its successful utilization and transformed cells need to recover and compete. The relative insensitivity of monocots to high levels of the antibiotic kanamycin (commonly used in dicot transformation) led to attempts to replace this antibiotic with other selective agents. The features of a particular transformation system (especially the nature of the material to be transformed and the route of transgenic plant regeneration) should be considered when choosing the resistance mechanism and the individual marker gene to be employed in any selection scheme. Patent and freedom to operate (FTO) issues often influence the choice of selectable marker gene.

Following the gene insertion process, plant tissues are transferred to a selective medium containing an antibiotic or herbicide, depending on which selectable marker was used. When grown on selective media, only plant tissues that have successfully integrated the transgene construct and express the selectable marker gene will survive. It is assumed that these plants will also possess the transgene of interest. Thus, subsequent steps in the process will only use these surviving plants.

12.4.2 Selectable marker genes for plants

There are two major classes of selectable marker genes, antibiotic and herbicide resistance genes. Antibiotic resistance genes are used in two important phases of transgenic plant production: (i) pre-plant transformation to select bacteria during routine molecular biology operations to manipulate transgenes and create expression vectors; and (ii) during the transformation process itself, to select cells and plants that have stably integrated introduced transgenes (selectable markers and gene(s) of interest) (Ramessar *et al.*, 2007). There are two issues frequently raised with respect to antibiotic resistance genes: (i) effects on the therapeutic efficacy of clinically used antibiotics, i.e. concerns that antibiotic resistance gene products in transgenic crops or products might render clinically important therapeutic antibiotics ineffective; and (ii) potential for horizontal gene transfer, i.e. concerns about the potential transfer of the antibiotic resistance marker gene to intestinal and soil microorganisms. For herbicide resistance genes the issues are: (i) gene flow – by which new genes can spread by normal outcrossing to wild or weedy relatives of the engineered crops; (ii) weediness – the potential for a crop or its sexually compatible wild relatives to become established and to persist and spread into new habitats as a result of newly introduced genes; and (iii) toxicity and allergenicity – an issue associated with human health and the safety of novel foods and potential negative effects on non-target organisms.

Choice of selectable marker genes is a key factor in plant transformation. Genes that give resistance to antibiotics or herbicides, such as kanamycin, hygromycin, phosphinothricin and glyphosate, are very popular. Kanamycin resistance has been most frequently employed in the transformation of many dicotyledonous plants. If the development of herbicide-resistant plants is aimed at, a trait gene could also be a selectable marker gene. Because of concerns over antibiotic resistance genes in commercial transformants, genes to add metabolic capabilities have been drawing considerable attention. For example, plant cells expressing a phosphomannose isomerase can grow on media with mannose as the sole carbon source. Such markers are referred to as positive selection markers (Joersbo *et al.*, 1998). Table 12.2 provides a list for selectable marker genes used in plant transformation. Although to date more than 20 selectable marker genes have been reported in the transformation of higher plants, many of them were tested only in a limited number of plant species on a limited scale. Therefore, further studies of marker genes may contribute to improvement of the transformation of certain plant species (Komori *et al.*, 2007).

Selectable marker genes are driven by constitutive promoters. The promoters of the CaMV 35S transcript (Odell *et al.*, 1985) and the nopaline synthase of *A. tumefaciens*

Table 12.2. Selectable marker genes for plant transformation.

Selectable marker gene	Gene product	Source	Selection
*npt*II	Neomycin phosphotransferase	Tn5	Kanamicin; G418 paromomycin; neomycin
ble	Bleomycin resistance	Tn5 and *Streptoalloteichus hindustanus*	Bleomycin; phleomycin
dhfr	Dihydrofolate reductase	Plasmid R67	Methotrexate
cat	Chloramphenicol acetyltransferase	Phage p1Cm	Chloramphenicol
*aph*IV	Hygromycin phosphotransferase	*E. coli*	Hygromycin B
ept	Streptomycin phosphotransferase	Tn5	Streptomycin
*aac*C3, *aac*C4	Gentamycin-3-N acetyltransferase	*Serratia marcescens*, *Klebsiella pneumoniae*	Gentamycin
bar	Phosphinothricin acetyltransferase	*Streptomyces hygroscopicus*	Phosphinothricin; bialophos
epsp	5-enolpyruvylshikimate-3-phosphate synthase	*Petunia hydrida*	Glyphosate
bxn	Bromoxynil specific nitrilase	*Klebsiella ozaenae*	Bromoxynil
*psb*A	Q_a protein	*Amaranthus hydridus*	Attrazine
*Ffd*A	2,4-D monooxygenase	*Alcaligenes eutrophus*	2,4 Dichlorophenoxyacetic acid
dhps	Dihydrodipicolinate synthase	*E. coli*	S-Aminoethyl; L-cystein
ak	Aspirate kinase	*E. coli* of lysine and threonine	High concentrations
sul	Dihydropteroate synthase	Plasmid R46	Sulfonamide
*Csr*1-1	Acetolactate synthase	*Arabidopsis thaliana*	Sulfonylurea herbicides
Tdc	Tryptophan decarboxylase	*Catharanthus roseus*	4-Methyl trytophan

(Depicker *et al.*, 1982) are very popular in dicotyledons and the promoters of the ubiquitin gene of maize (Christensen *et al.*, 1992) and the actin gene of rice are popular in monocotyledons (Zhang *et al.*, 1991). Selectable marker genes are followed by a DNA fragment, the so-called 3' signal. The 3' regions of the CaMV 35S transcript and the nopaline synthase gene in the wild-type T-DNA of *A. tumefaciens* are frequently used as a 3' signal.

Antibiotic resistance genes

Aminoglycoside antibiotics are bacterial inhibitors of prokaryotic, mitochondrial and chloroplast protein synthesis. Kanamycin, gentamycin/geneticin® (G418) and paromomycin bind the 30S ribosomal subunit to inhibit translation initiation. Hygromycin interacts with the elongation factor EF-2 to inhibit peptide chain elongation. Exposure of plants to these antibiotics leads to an inhibition of chlorophyll biosynthesis and leaf bleaching. The most widely used selectable markers in cereal transformation are the genes encoding neomycin phosphotransferase (*npt*II), hygromycin phosphotransferase (*hpt*) and phosphinothricin acetyltransferase (*bar*) (Cheng *et al.*, 2004). These genes confer resistance to kanamycin and some related aminoglycosides (such as G418 and paromomycin), hygromycin and PPT, respectively. Transformed cells in these systems are able to survive and non-transformed cells are killed by the selective agents. This type of selection is referred to as negative selection. Cereals have proven to be insensitive to relatively high concentrations of kanamycin. Paromomycin has been used for selection and regeneration of rice, maize, wheat, oats and barley transformed with the *npt*II gene. Resistance to hygromycin is encoded by the *aph*IV gene (commonly referred to as the *hpt* gene) of *E. coli*, which codes for hygromycin phosphotransferase (HPT). Rice showed relatively high sensitivity to hygromycin. There has been a move away from antibiotic marker genes in commercial cereal biotechnology because of associated regulatory concerns, especially in Europe and difficulty in using in breeding programmes for transgene identification.

Herbicide tolerance genes

A number of herbicides have been used as selective agents in cereal transformation. Markers have been developed by engineering tolerance to herbicides that inhibit amino acid biosynthesis. Both herbicides and antibiotics can be used to select materials by addition to the tissue culture media or by spraying the full-grown plants. They both can be readily used in breeding programmes to select for the inheritance of linked transgenes. However, herbicides have a more serious intellectual property problem than antibiotics. A problem with herbicide resistance genes is that we end up with plants that are herbicide resistant although this may not be a desired goal. Several strategies have been developed for engineering herbicide tolerance in transgenic cereals, by introducing a herbicide tolerant variant of an amino acid biosynthetic enzyme, e.g. a mutant *als* gene for sulfometuron methyl (Qust®) tolerance and by introducing an enzyme which inactivates the herbicide, e.g. the *bar* gene for phosphinothricin (PPT, Liberty®) tolerance. Resistance to PPT-based herbicides using the *bar* gene from *S. hygroscopicus* has been used for the selection of fertile transgenic cereals, e.g. rice, maize, wheat and barley, while Monsanto uses 5-enol-pyruvylshikimate-3-phosphate synthase (EPSPS) and DuPont uses imidazolinone, chlorsulfuron or acetolactate synthase (ALS).

Engineering detoxification of herbicides that inhibit glutamine synthase

The enzyme glutamine synthase (GS) catalyses the synthesis of glutamine from glutamate and free ammonium. PPT is a glutamate analogue that acts by inhibiting GS activity resulting in a cytotoxic accumulation of ammonium. Inactivation of PPT and PPT-containing herbicides (Liberty®) is conferred by the bar gene from *S. hygroscopicus,* which encodes a phosphinothricin acetyltransferase.

Engineering tolerance to and detoxification of herbicides that inhibit 5-enol-pyruvylshikimate-3-phosphate synthase

The chloroplast-localized enzyme EPSPS catalyses a common step in aromatic amino acid biosynthesis. Glyphosate, the active ingredient in the herbicide Roundup®, inhibits the plastid enzyme EPSPS and thus prevents the synthesis of chorismate-derived aromatic amino acids and secondary metabolites in plants. Dominant mutations that confer resistance to glyphosate-containing herbicides (Roundup®) have been shown to result from base pair substitutions within *epsps* genes. Transformation of plants with a mutant *epsps* gene renders them tolerant to glyphosate. Inactivation of glyphosate-containing herbicides is conferred by a bacterial *gox* gene that encodes a glyphosate oxidase. Howe *et al.* (2002) have described the development of an efficient selectable marker system for the production of transgenic maize plants using genes that confer resistance to the herbicide glyphosate.

12.4.3 Positive selection

In most cases, 'negative' selection markers are used to select transformed cells from a population of non-transformed cells, as described above. As a result of increasing concern worldwide regarding the use of antibiotic or herbicide markers in transgenic crop plants, although it is completely unfounded scientifically, several positive selection systems have been developed in recent years and successfully used for the production of transgenic plants. In the case of positive selection, a transformed cell acquires the ability to metabolize a substrate that it previously could not use (or not use efficiently) and thereby it grows out of the mass of non-transformed tissue. In contrast with negative selection, positive selection does not kill the non-transgenic cells, but gives clear advantages to the transformed cells. Positive selection systems include benzyladenine *N*-3-glucuronide (Joersbo and Okkels, 1996), xylose (Haldrup *et al.*, 1998a, b) and mannose (Miles and Guest, 1984). These selection systems allow transgenic plants to be produced without antibiotic or herbicide resistance genes in many plant species (e.g. Negrotto *et al.*, 2000; Lucca *et al.*, 2001; Reed *et al.*, 2001; He, Z. *et al.*, 2004; Gao *et al.*, 2005).

Positive selection can be of many types, from inactive forms of plant growth regulators that are then converted to active forms by the transformed enzyme, to alternative carbohydrate sources that are not utilized efficiently by the non-transformed cells that become available upon transformation with an enzyme that allows them to be metabolized. Non-transformed cells either grow slowly in comparison to transformed cells or not at all. Using positive selection, non-transformed cells may die, but, typically, production of phenolic compounds observed with negative selection markers does not occur.

The first example of positive selection was provided by Joersbo and Okkels (1996); these researchers demonstrated that transgenic tobacco plants could be obtained from a leaf disc transformed with β-glucuronidase (GUS) when a cytokinin glucuronide was provided as a substrate and cytokinin was absent from the media. Only cells expressing GUS could metabolize the cytokinin glucuronide. These cells could then proliferate and differentiate into shoots, whereas cells without GUS activity could not. This concept of positive selection was further expanded to include not only plant hormones, but also carbohydrate and nitrogen sources (Okkels and Whenham, 1994). Carbohydrates represent one of the most readily used aspects of positive selection because plant cells in culture require the presence of a carbohydrate source. Typically sucrose, glucose or maltose is incorporated into plant culture media. However, if another carbohydrate such as mannose is introduced instead into the media, in the majority of cases studied, the plant cells will be unable to proliferate and may die. In the case of mannose and many other carbohydrates, the compound is metabolized but the product of that step cannot be further metabolized.

Other examples of the use of alternative carbohydrate sources as a means of selection for transgenic cells have utilized deoxyglucose, xylose and ribitol. The best documented of these carbohydrate sources is the use of mannose combined with the phosphomannose isomerase (PMI) gene of *E. coli* (*man*A). PMI catalyses the conversion of mannose-6-phosphate to fructose-6-phosphate, which can be utilized as a carbohydrate source. The PMI system has been shown to be effective for sugarbeet, maize, rice, wheat, *Arabidopsis* and many other dicot and monocot species (reviewed by Wenck and Hansen, 2004).

12.4.4 Elimination of selectable marker genes from transgenic plants

Selectable marker genes are required for efficient generation of transgenic plants in nearly all transformation procedures, but serve no purpose once plants have been obtained that are homozygous for the transgene. On the contrary, their continued presence can pose technological problems because it precludes retransformation with the same marker systems.

Use of co-transformation

A simple approach is to co-transform plant cells with two separate pieces of T-DNA, one with a selective marker gene and the other with genes of interest and to select marker-free progeny segregated from the co-transformants (Hohn *et al.*, 2001). Unlinked integrations of the two T-DNAs lead to the segregation of the marker gene from the gene of interest in the T_1 generation. Co-transformation can be performed using either two strains, or a single strain, of *A. tumefaciens*. A mixture of two strains, each harbouring a binary vector (Komari *et al.*, 1996), or a co-integrate and a binary vector (De Buck *et al.*, 2000), have been used to study the factors that influence co-transformation frequencies. An alternative method for co-transformation using a single strain of *Agrobacterium* has been shown to yield higher co-transformation frequencies (Komari *et al.*, 1996). However, the convenience of the method will largely depend on the suitability of the analytical method required. Moreover, Hohn *et al.* (2001) suggested that the elimination of marker genes by co-transformation may be especially useful when using *Agrobacterium*-mediated transformation. In combination with 'twin T-DNA' vectors, marker-free transgenic plants can be produced by carefully designing the transformation vectors.

To carry out co-transformation in the superbinary vector system, a T-DNA with a selectable marker was located in an acceptor vector. For example, pSB4 and pSB6 were constructed by locating a T-DNA carrying the hygromycin resistance gene and the phosphinothricin resistance gene, respectively, and have been tested in a number of plant species (Komari *et al.*, 1996; Ishida *et al.*, 2004). The frequency of co-transformation, which is the ratio of transformants with the genes of interests among the number of plants with the selective marker gene, has been quite high, ranging typically between 50% and 80%. Marker-free transformants have then been obtained from more than 50% of the co-transformants. Co-transformation may be carried out using other types of vectors. For example, Huang *et al.* (2004) placed a marker gene in the vector backbone in a regular binary vector and observed that plants were co-transformed with one T-DNA processed from the right border and another 'T-DNA' processed from the left border.

Another method for producing marker-free transgenic plants was proposed by Vain *et al.* (2003) using a new dual binary vector system pGreen/pSoup (Hellens *et al.*, 2000a). pGreen is a small Ti binary vector unable to replicate in *Agrobacterium* without the presence of another binary plasmid, pSoup, in the same strain. Co-transformation with pGreen, carrying the gene of interest and pSoup, carrying the selectable marker, may lead to the production of marker-free transgenic plants in subsequent progeny (Vain *et al.*, 2003).

Removal of marker genes and other unnecessary segments by recombination

Recombinases from phages and yeasts, such as *cre*, *FLP* and *R*, which recombine specific sites *loxP*, *FRT* and *RS*, respectively, are powerful tools to remove selectable marker genes (Ow, 2001) and effective for a few model systems. A DNA segment placed between two of the specific recombination sites may be excised from the plant chromosome if the corresponding recombinase is somehow expressed in the plant cell. For example, transgenic lines that contained the *loxP* sites were crossed with lines that expressed the *cre* recombinase gene (Moore and Srivastava, 2006). Various sophisticated vector configurations and means to express the recombinases were reported to exploit this system (Wang, Y. *et al.*, 2005; Jia *et al.*, 2006). The recombinases may be able to cut out not only marker genes but also other unnecessary DNA segments. For example, tandem integration of two or more copies of T-DNA in a single locus has been observed quite frequently (Krizkova and Hrouda, 1998); it is a cumbersome phenomenon because clean, single-copy transformants are generally preferred. If a recombination site is possessed by the T-DNA, a segment between two of the sites in the tandem T-DNA could be deleted so that a clean, single T-DNA integration pattern could be generated.

The multi-auto transformation (MAT) vector system uses recombinase-based excision to enable the production of marker-free transgenic plants (Sugita *et al.*, 2000). An *Agrobacterium* isopentenyltransferase (*ipt*) gene provides a positive visual selectable marker for transformation by catalysing cytokinin synthesis and inducing a 'shooty' phenotype on hormone-free medium. After selection, subsequent excision via the *R/RS* system produces marker-free transgenic plants with a normal phenotype, allowing *ipt* and MAT to be used again for another round of transformation. Recent improvements to the method have increased its efficiency and have allowed it to be applied to species that do not regenerate through cytokinin-dependent organogenesis, but rather via somatic embryogenesis (Endo *et al.*, 2002b).

Verweire *et al.* (2007) presented a vector system to obtain homozygous marker-free transgenic plants without the need of extra handling and within the same period as transformation methods in which the marker is not removed. By introducing a germline-specific auto-excision vector containing a *cre* recombinase gene under the control of a germline-specific promoter, transgenic plants become genetically programmed to lose the marker when its presence is no longer required (i.e. after the initial selection of primary transformants). Using promoters with different germline functionality, two modules of this genetic programme were developed. In the first module, the promoter, placed upstream of the *cre* gene, confers CRE functionality in both the male and the female germline or in the common germline (e.g. floral meristem cells). In the second module, a promoter conferring single germline-specific CRE functionality was introduced upstream of the *cre* gene.

Recently, Mlynarova *et al.* (2006) and Luo *et al.* (2007) showed that it was possible to remove transgenes (selectable markers and others) efficiently by using an auto-excision vector in which a promoter that was specifically functional during microsporogenesis, in pollen or in seed, was placed upstream of a site-specific recombinase gene. More efficient transmission of the recombined allele to the progeny was observed compared to previously described auto-excision strategies that rely on chemical or physical induction of the recombinase. The results presented by Verweire *et al.* (2007), together with the results obtained by Mlynarova *et al.* (2006) and Luo *et al.* (2007), clearly indicate that germline-specific auto-excision is an efficient, flexible and versatile system to remove selectable markers from transgenic plants.

Use of transposons

The maize Ac/Ds transposable element system has been used to create novel T-DNA vectors for separating genes that are linked together on the same T-DNA after insertion into plants. The expression of the

Ac transposase from within the T-DNA can induce the transposition of the gene of interest from the T-DNA to another chromosomal location (Shrawat and Lörz, 2006). This results in the separation of the gene of interest from the T-DNA and selectable marker gene.

Use of homologous recombination

Homologous recombination between direct repeats provides a method for excising marker genes after transgenic cells and shoots have been isolated. The strategy uses native plant enzymes and is simple because it avoids the need for foreign site-specific DNA recombinases (Corneille *et al.*, 2001; Hajdukiewicz *et al.*, 2001). Efficient implementation of the method requires high rates of homologous recombination relative to illegitimate recombination pathways. The procedure works well in plasmids where homologous recombination predominates. Marker genes are flanked by engineered direct repeats. The number and length of direct repeats flanking a marker gene influence the excision rate. Excision is automatic and loss of the marker gene is controlled by selection alone. After transgenic cells have been isolated, selection is removed allowing loss of the marker genes. Excision is a unidirectional process resulting in the rapid accumulation of high levels of marker-free plastid genomes. Cytoplasmic sorting of marker-free plastids from marker-containing plastids leads to the isolation of marker-free plants. Marker-free plants can be isolated following vegetative propagation or among the progeny of sexual crosses.

Use of positive markers

Ebinuma *et al.* (2001) developed removal systems combined with a positive marker, which are called MAT vectors. The MAT vector system is designed to use the oncogenes (*ipt, iaaM/H, rol*) of *Agrobacterium*, which control the endogenous levels of plant hormones and the cell response to plant growth regulators, to differentiate transgenic cells and to select marker-free transgenic plants. The oncogenes are combined with the site-specific recombination system (*R/RS*). At transformation, the oncogenes regenerate transgenic plants and then are removed by the *R/RS* system to generate marker-free transgenic plants. The choice of a promoter for the oncogenes and the recombinase (*R*) gene, the state of plant materials and the tissue culture conditions greatly affect efficiency of both the regeneration of transgenic plants and the generation of marker-free plants (Ebinuma *et al.*, 2004). These conditions have been evaluated in several plant species to increase their generation efficiency and the MAT system has been applied to tobacco and rice (Endo *et al.*, 2002a, b).

As discussed above, marker-free transgenic cereal plants can be generated at varying efficiencies using different approaches and techniques, followed by segregation of the genes in the subsequent sexual generation. However, there are limitations associated with these techniques (Shrawat and Lörz, 2006). For example, co-transformation technology is not suitable for all plant species and its efficiency is clearly dependent on a number of variables, including the *Agrobacterium* strain and the plant tissue being transformed. In addition, this technique is labour intensive, requiring the production of a large number of transgenic plants to isolate the plant of interest. Although site-specific recombinases hold the greatest promise for the excision of selectable marker genes, concerns also exist about pleiotropic effects induced by the action of recombinase on cryptic excision sites in the plant genomes. A transposon to separate the selectable marker gene and gene of interest (Goldsbrough *et al.*, 1993) is of limited use. Homologous recombination approaches, although interesting from a scientific point of view, are only effective for a few model systems.

12.5 Transgene Integration, Expression and Localization

Once whole plants are generated and produce seeds, evaluation of the progeny begins. The transgenic plants should be evaluated for transgene integration, expression and localization.

12.5.1 Transgene integration

As a part of the regulatory process associated with commercial release of a transgenic plant product, transgene integration events must be fully characterized. For transgene technology to be useful, transgenes must have predictable and stable expression. Technologies have been sought that would enhance our ability to create transgenic plants with the desired expression characteristics. One of these technologies involves the use of matrix attachment regions (MARs). MARs are DNA sequences that bind specifically to a network of proteinaceous fibres, called the nuclear matrix, which permeates the nucleus. These MAR–matrix interactions are thought to organize chromatin into a series of independent loop domains. When MARs are positioned at the 5'- and 3'-ends of a transgene more predictable expression of the transgene results (Allen *et al.*, 2000).

Transgenic plants often contain complex integration structures at an undetermined genomic location, which may cause variations in gene expression. It has been demonstrated that the precise integration of a transgene in a pre-determined genomic location can reduce the variation in transgene expression (Day *et al.*, 2000). The integration of transgenes in a pre-determined genomic locus can be achieved by the use of site-specific recombinase systems, such as *cre/lox* and *FLP/frt* (Ow, 2002). Integration by homologous recombination would favour the establishment of a simple integration pattern and allow the insertion of a transgene into a known and stable region of the genome.

Individual transgenic lines with complex integration patterns are generally considered undesirable. There has been a drive to achieve cereal transformation using *Agrobacterium* and other target recombination/integration systems. *Agrobacterium*-mediated DNA integration is a defined process that generally results in low copy integration of defined T-DNAs, often into transcriptionally active sites. Gene targeting has the potential to place foreign gene sequences in predetermined regions of the genome thus potentially overcoming so-called position effects on transgene expression. Transposons can be used to deliver recombination targets for subsequent site-specific integration.

12.5.2 Transgene expression

Transformation technologies can be used for characterizing expression elements using reporter genes, utilizing transgene expression to modify endogenous metabolic activities, introducing transgenes conferring novel phenotypic characteristics, inactivating genes using anti-sense or co-suppression technologies and identifying genes by complementation. The characterization of constitutive and non-constitutive promoter elements has advanced the most in cereal transformation, but there are other non-promoter elements that regulate and control gene expression in transgenic plants, which include transcript termination, transcript stability, post-transcriptional modification, translation efficiency and protein targeting.

Transgenes currently used in cereal transformation have a relatively simple structure. They usually contain: (i) a promoter, usually of plant, bacterial or viral origin, which may be constitutive (Act1), inducible (Hsp70) or tissue-specific (Amy1) and which may have been modified for optimal activity; (ii) a coding sequence, which may have been modified for optimal expression in transgenic plants, e.g. translation initiation site modification, targeting information, glycosylation site modification and codon usage modification; and (iii) a transcript termination sequence.

12.5.3 Confirmation of transgene and analysis of gene expression in transgenic plants

Commonly used methods to confirm the putative transgenic plants, as discussed in

Chapter 3, are PCR, Southern blotting, Western blotting, Northern blotting, enzyme-linked immunosorbent assay (ELISA), functional assay (testing the presence of selectable marker and the target gene), *in situ* hybridization and progeny analysis (segregation of the target gene). In Southern blotting, whether an introduced gene is indeed present in the plant DNA and whether multiple transgenic plants carry the introduced genes on the same size of DNA fragment (suggesting a single transformation event) or on different sized fragments (suggesting independent transformation events) can be determined using the cloned gene as a probe. Northern blotting is used to determine whether the introduced gene has been transcribed into mRNA and accumulates in the transgenic plant. Western blotting, ELISA and specific techniques must be used for analysis of protein (enzyme) activity. The Western blotting detects the protein of the transgene in an extract of protein prepared from various parts of the transgenic plants and is, therefore, an assay for a functional transgene. When the selectable markers used are antibiotic or herbicide-resistant genes, a functional assay can be made by spraying antibiotics or smearing herbicide on the leaves of those putative transgenic seedlings or plants in later segregating populations. With stable transformed genes, progeny testing should show the presence and activity of the selectable marker and target genes, such as the gene *gfp* encoding green fluorescent protein (GFP), or *bar* and disease resistance.

When the PCR method is used, two primers specific for the selectable marker (*bar* or *cah* gene, for example) are used in a PCR reaction with genomics DNA extracted from the transgenic plants. The DNA fragment yielded should have the predicted size (the length equal to the number of base pairs between the two primers in the transgene). The presence of additional transgenes in the same plants carrying the selectable marker can be detected with a different set of primers using the same template DNA. It should be noted that organization of the transgenic locus with inverted repeats of multiple copies results in silencing while concatemeric head-to-tail multiple copies are good for high levels of stable expression. Numerous medium-to-high throughput analytical techniques are available to quantify the levels of mRNA transcripts of large numbers of genes. RNA-based techniques for the analysis of transgene expression begins with the extraction of RNA and its evaluation in terms of purity and integrity by spectrophotometry and gel electrophoresis, respectively, followed by the use of Northern blotting, reverse transcription- (RT-) PCR and quantitative or real-time RT-PCR (QRT-PCR) for quantifying RNA and *in situ* hybridization for studying tissue-level expression patterns. QRT-PCR is a highly sensitive technique for quantifying mRNA copy numbers of specific genes. The method permits a direct measurement of products during the log-linear phase of the PCR reaction via the incorporation of a fluorescent probe in the PCR reaction mix and the use of a thermocycler equipped with an optical sensor for fluorescence quantification. The details on the selection and suitability of and a comparison of various techniques for analysis of gene expression in general can be found in Jones (1995) and Bartlett (2002).

The stable inheritance and expression of foreign genes are of critical importance in the application of GM-plants to agriculture. The perfect transformation would contain a single copy of the transgene that would segregate in a Mendelian fashion, with uniform expression from one generation to the next. However, studies on transgene behaviour indicated that segregation in transformed lines of cereals does not always follow the typical Mendelian fashion but an aberrant segregation (Barro *et al.*, 1998; Vain *et al.*, 2003; Wu, H. *et al.*, 2006). This is a highly undesirable character when it occurs for transgenes encoding a useful trait. Many factors can contribute to variation in transgene expression, including tissue culture-induced variation or chimerism in the primary integration site (position effects), transgene copy number (dosage effects), transgene mutation and epigenetic gene silencing (as reviewed by Shrawat and Lörz, 2006). Gene silencing, the decline or loss of gene expression in subsequent generations of primary transformants, can occur at the transcriptional or post-transcriptional level and the phenomenon has often been associated with a high transgene copy number (Matzke and Matzke, 1995; Matzke *et al.*, 2000). Studies have indicated that the problem of transgene silencing raises serious concerns regarding the selection of transgenic lines for crop improvement with specific trait(s). Therefore, it now appears imperative that transgenic lines carrying gene(s) of economic importance need to be carefully tested for gene expression levels over many generations.

Particle bombardment has featured strongly in the burgeoning field of cereal functional genomics, specifically through the development of transposon-tagged plant lines for the systematic functional characterization of plant genes. For example, Kohli *et al.* (2001, 2004) produced a large population of transgenic rice plants tagged with the maize Ac transposon. They found that this population was suitable for saturation mutagenesis and the rapid PCR-based cloning of interrupted genes using unique barcode elements present in the DNA cassette used for transformation (Kohli *et al.*, 2001). Callus induced from specific transposon-tagged rice plants was maintained in a dedifferentiated state prior to regeneration into clonal transgenic lines, prolonging the developmental phase characterized by hypomethylation of genomic DNA (Kohli *et al.*, 2004). This resulted in a dramatically increased frequency of secondary transposition events compared to seed-derived plants, thus increasing the rate of genome saturation.

As detailed more fully in Chapter 6 of Cullis (2004), the use of tagged full-length cDNAs in transgenic plants can be a first step in isolating and identifying the protein complexes that exist *in vivo*. Genetic transformation can also be used to develop a protein atlas of where in the cell each of the genes is expressed. A full-length cDNA can be tagged with a dye and the tagged probe transformed back into the plant under the control of its native promoter. The site of the fluorescence will indicate the organ or tissue where the gene is expressed, as well as the cellular localization of the protein. The reintroduction of the full-length cDNA into a plant can also result in either overexpression or silencing of that gene. The subsequent phenotype that is observed provides clues as to the function of the gene. In addition, overexpression of such a gene, for which a full-length cDNA is available, can be accomplished in a heterologous system, such as yeast or *E. coli*, followed by *in vitro* studies of the protein function.

Transformation of allelic series into isogenic backgrounds can confirm the function of individual sequence motifs. However, current plant transformation protocols based on non-homologous end joining result in random genomic integration of transgenic DNA, position effects, multiple insertions of the transgene and transgene alterations (Xu, 1997; Hanin and Paszkowski, 2003), obscuring quantitative phenotypic differences between alleles. This can be circumvented using homologous recombination-based, locus-targeted integration of alleles. Recently, 1% of insertion events in rice were found to result from homologous recombination (Terada *et al.*, 2002). If this finding can be confirmed, rice genomics-genetics will be revolutionized. Further, if the method can be applied to other species, a similar advance in genomics of all plants would occur.

Virus-based vectors can be efficiently used for high levels of transient expression of foreign proteins in transfected plants and permit non-*Agrobacterium* bacterial species to be employed for the production of transgenic plants (reviewed by Chung *et al.*, 2006). Viral vectors hold great promise as efficient tools for transient recombinant protein expression in plant cells because of their ability to replicate in host cells autonomously (Marillonnet *et al.*, 2004, 2005). These viral vectors are built on the backbones of plus-sense RNA viruses, such as tobacco mosaic virus (TMV) or potato virus and have been used for the expression of foreign sequences in plants (Porta and Lomonossoff, 2002; Gleba *et al.*, 2004).

The recent development of reliable and efficient *Agrobacterium*-mediated transformation technologies for cereals (for review, see Shrawat and Lörz, 2006; Goedeke *et al.*, 2007) has stimulated a variety of strategies towards functional gene characterization, thereby paving the way for deeper understanding of crop plant biology in cereals (Himmelbach *et al.*, 2007). Comprehensive analyses of gene function include stable transformation with sequences for overexpression or knock-out of plant genes.

12.5.4 Reporter genes

Reporter genes, whose expression can be easily monitored, are useful in many ways in plant transformation. Strength and temporal, spatial and other types of regulation of promoters and other elements may be conveniently assayed by connecting these elements to the reporter genes. Genes for GUS (Jefferson, 1987), luciferase (Ow *et al.*, 1986) and GFP (Pang *et al.*, 1996) are popular examples. Gene fusions of the reporters and proteins of interest may be employed to examine the subcellular localization of the proteins.

Reporter genes that are connected to constitutive promoters may be used to monitor the process of transformation. The establishment of genetic transformation procedures has relied on, among other factors, the use of efficient reporter genes, which easily allows the detection of transgenic events after a transformation experiment, in either a transient or stable expression assay. Expression of the reporter genes soon after the inoculation of plant cells with *A. tumefaciens*, is referred to as 'transient expression'. Expression of the reporter genes later in a cluster of cells growing on selection media is a piece of evidence for integration of the T-DNA in plant chromosomes. A binary vector that carries a constitutive selectable marker and a constitutive reporter is very useful as a control vector both in transformation experiments and in assays of gene expression (Komori *et al.*, 2007).

It should also be mentioned that gene reporter systems have played a key role in many gene expression and regulation studies, in which expression of a reporter gene under, for instance, the direction of different promoters or the presence of different transcription factors may be investigated. Reporter genes are used in cereal transformation for analysing gene function, monitoring selection efficiency in both transformed tissue and transgenic plants and following the inheritance of foreign genes in subsequent plant generations.

Transient expression assays using promoter–reporter fusion genes may be used to analyse gene regulation and function. There can be incongruity between results obtained from transient assays and those observed in stably transformed plants. The utility of different reporter genes in cereal transformation is a function of the properties of the respective protein products they encode. The required properties a good reporter gene should have include: (i) expression in plant cells; (ii) low background activity in transgenic cereals; (iii) no detrimental effects on plant metabolism; (iv) only moderate stability *in vivo* so as to detect downregulation of gene expression as well as gene activation; and (v) coming with an assay system that is non-destructive, quantitative, sensitive, versatile, simple to carry out and inexpensive. The coral-derived red fluorescent protein DsRed is one of the reporter systems currently used in cereal transformation that have all these desired properties.

β-Glucuronidase

β-Glucuronidase (GUS) catalyses the hydrolysis and cleavage of a wide range of fluorometric and histochemical *β*-glucuronide substrates. Since GUS gene (*gus*, *gus*A, or *uid*A) was first isolated from *E. coli*, many efforts have been made to develop the *E. coli uid*A gene as a reporter system for plant transformation. Indeed, it has become the most widely used marker system, mainly because of the enzyme stability and high sensitivity and amenability of the assay to detection by fluorometric, spectrophotometric, or histochemical techniques. In addition, there is little or no detectable GUS activity in almost any higher plant tissues. The expression of *gus* gene fusions can be quantified by fluorometric assay.

Histochemical analysis can be used to localize gene activity in transgenic tissues.

There are a number of problems associated with the use of *gus* reporter genes. The expression assays of the *gus* gene are destructive. The GUS protein shows high *in vivo* stability, leading to problems when used to monitor gene inactivation. Histochemical localization of GUS enzyme activity can be 'leaky'. Dependence on the use of *gus* genes to monitor the efficiency of cereal transformation protocols has often been misleading.

Luciferase

The product of the firefly (*Photinus pyralis*) luciferase gene (*luc*) catalyses the oxidation of D(–)-luciferin in the presence of ATP to generate oxyluciferin and yellow-green light. The activity of luciferase gene fusions can be assayed in transformed cereal tissue non-destructively. There are a number of problems associated with the use of *luc* reporter genes. First, penetration of the luciferin substrate can be limiting in whole plant material. Secondly, detection equipment presently needed to monitor luciferase gene expression is relatively expensive. *luc* genes are widely used as an internal standard with *gus* fusions constructed to study gene expression in transient assays and in transgenic plants.

Anthocyanin biosynthetic pathway genes

C1, *B* and *R* genes code for trans-acting factors that regulate the anthocyanin biosynthetic pathway in maize seeds. Introduction of these regulatory genes, with constitutive promoters, into cereal cells induces cell autonomous pigmentation in non-seed tissues. This reporter system does not require the application of external substrates for its detection.

Green fluorescent protein

For a monitoring system to be effective, the genetic marker technology should be accurate with few false positives or negatives, detectable throughout the life cycle of the plant and able to inform on the status of genetically linked or fused transgenes of interest. Green fluorescent protein (GFP) has been proposed as a whole-plant marker for field-level applications. The GFP was isolated from a jellyfish (*Aequorea victoria*) in 1992 and has since been modified for specific applications and transformed into many different organisms. GFP monitoring has the potential to track transgenes under large spatial scales utilizing visual or instrumental detection of the characteristic green fluorescence of transgenic materials. There are other versions of GFP fluorescing at different wavelengths that allow detection of multiple proteins. GFP expression in mammalian cells yields a green fluorescence when excited by blue light, which does not require additional gene products or exogenous substrates for activity and detection is non-destructive.

GFP showed relatively weak activity in transformed plant cells and a number of modifications have been made to increase GFP expression in plants. The modifications include: (i) point mutations to increase signal intensity and shift excitation peak; (ii) mutations to alter codon usage for efficient translation and increased mRNA stability; (iii) mutation to remove cryptic intron splice junctions to increase mRNA processing and stability; (iv) subcellular localization, targeting to the oestrogen receptor, to reduce mild phytotoxicity; and (v) mutation to inhibit thermosensitive protein misfolding. The *mgfp5-er* variant gene has been shown to be a feasible transgene monitor in plants under field conditions (Haseloff *et al.*, 1997; Harper *et al.*, 1999). GFP has also been shown to be a feasible qualitative marker for the presence of a linked synthetic *Bt* cryl*Ac* endotoxic transgene (Harper *et al.*, 1999; Halfhill *et al.*, 2001). With these beneficial characteristics, the next step in the development of a GFP monitoring system is to better describe the system and resolve weaknesses that could limit the utility of the monitoring system (Halfhill *et al.*, 2004b).

12.5.5 Promoters

Promoters for constitutive transgene expression

The uses of constitutive promoters in transgenic cereals include: (i) the expression of reporter genes to monitor transformation protocols; (ii) expression of marker genes for transgenic cell selection; (iii) expression of herbicide tolerance genes; (iv) repression of endogenous and/or pathogenic gene expression through antisense and co-suppression technologies; (v) overproduction of biomolecules; and (vi) overexpression of disease and stress tolerance genes prior to the employment of targeted gene expression strategies. The constitutive promoter of the CaMV 35S RNA transcript (35S) was initially used for constitutive gene expression in cereal transformation. CaMV 35S promoter is used extensively in dicot transformation and is available at the onset of cereal transformation and used to express selectable marker genes in rice and maize. There are problems associated with the CaMV 35S promoter in transgenic cereals. It has relatively low activity in transient assays and is not completely constitutive in transgenic cereal plants. In general, the CaMV 35S promoter is not a strong promoter for cereals. In order to get it expressed at high levels, an intron or other enhancers are needed.

A number of strategies have been used to increase gene expression in monocots. These strategies are: (i) enhancement of the CaMV 35S promoter, e.g. e35S, 2 × 35S; (ii) incorporation of an intron into the foreign gene transcript unit to elevate mRNA abundance; (iii) modification of a monocot promoter for high-level constitutive activity in cereals, e.g. the modified maize *Adh1* sequence in the Emu promoter; and (iv) isolation of monocot promoter that show high level constitutive activity in cereals, e.g. the rice actin (*Act1*) and maize ubiquitin (*Ubi1*) promoters.

Most overexpression studies employ a strong, constitutive promoter, such as the CaMV 35S promoter, followed by phenotypic analysis of the transgenic plant. In many cases, ectopic expression experiments gave important insight into gene function (Jack *et al.*, 1994). However, as a possible consequence of ubiquitous overexpression and misdirection of gene products, undesirable pleiotropic effects on the plant may be caused. In addition, strong accumulation of unnecessary proteins leads to wasteful energy consumption, which could, in turn, generate phenotypes that are not directly correlated with the recombinant protein itself. To avoid such unwanted pleiotropic effects that occlude phenotypic analysis, as reviewed by Himmelbach *et al.* (2007), transgene expression can be controlled temporally and spatially by the use of cell- and tissue-specific or chemically inducible promoters.

Promoters for non-constitutive transgene expression

Progress in rice transformation has facilitated the study of non-constitutive promoters in transgenic cereals using *gus* reporter gene fusions. Non-constitutive promoters used in rice transformation include cereal promoters (maize *Adh1*, wheat *His3* and rice *rbcS*), dicot promoters (tomato *rbcS*, potato *pinII*), bacterial promoters (*Agrobacterium rhizogenes rolC*) and viral promoters (rice tungro bacilliform virus major transcript). A number of general conclusions can be drawn from these studies in rice: (i) inclusion of the promoter region alone is (usually) enough to give the expected pattern of reporter gene expression; (ii) signal transduction pathways are usually conserved between cereals, e.g. barley α-amylase promoter activity in transgenic rice; (iii) cereal promoters can show higher activity than their dicot homologues, e.g. rice versus tomato *rbcS-gus* gene expression; and (iv) intron-mediated enhancement of gene expression in cereals cells does not generally alter their pattern of activity. The isolation and utilization of cereal promoters will continue because of difference in the biochemistry, physiology and/or morphology between monocots and dicots (e.g. aleurone-specific expression) and the need to obviate potential problems with the use of non-cereal genetic elements in transgenic cereals.

12.5.6 Transgene inactivation

What causes transgene inactivation is not very conclusive. Although it might be associated with high copy/complex integration events and nucleic acid interactions between multiple copies of homologous DNA sequences, there are many examples where single copy events, simple interaction patterns and heterologous sequences also cause transgene inactivation. Two things matter in this context: one is organization of the transgenic locus, with inverted repeats of multiple copies causing silencing; the other is the integrity of the inserted DNA sequence, where partial or rearranged copies cause problems.

Inactivation can act at different steps in transgene expression: (i) transcription inactivation, by *de novo* methylation (often of promoter regions) and/or heterochromatin formation, e.g. homology-dependent trans-inactivation of the maize *AI* gene in transgenic petunia and natural paramutation at the maize *B* locus; and (ii) post-transcriptional inactivation, by increased RNA turnover, antisense and/or defective transcript effects (e.g. overproduction of untranslatable tobacco etch virus coat protein transcripts in transgenic tobacco) and co-suppression, such as *de novo* methylation of coding regions.

The potential 'signals' for transgene inactivation include: (i) DNA structure/integration site, the recognition of an introduced gene as 'foreign' and its subsequent *de novo* methylation, e.g. maize *AI* gene in petunia (but not the related gerbera *A1* gene); (ii) DNA–DNA association between genes leading to a transmission of chromatin-based transcription states and/or *de novo* methylation, e.g. repeat-induced transgene methylation in *Arabidopsis*; and (iii) DNA–RNA association – RNA (antisense/aberrant) accumulation causing a feedback 'signal' to reduce gene expression via DNA methylation or increased RNA turnover, e.g. potato spindle tuber viroid DNA methylation during viroid RNA–RNA replication.

There are several steps one can take to minimize transgene inactivation, which have been very useful for *Arabidopsis* or tobacco. These are: (i) transgene integration by development of recombination systems for transgene targeting to suitable genomic locations; (ii) transgene structure, the elimination of repeated elements from transgenes, including optimizing codon usage to produce less 'foreign'-looking transgenes and the isolation and utilization of 'insulating' sequences; (iii) transgene expression, the regulation of transgene transcription rates and/or transcript structure to reduce excess (aberrant read through/antisense) RNA-mediated transcript turnover; (iv) the use of site-directed gene targeting (e.g. *cre/lox*, *FLP/frp*) to target transgenes into chromosomal regions that provide an optimal sequence environment for stable expression; (v) use of double-haploid systems to rapidly evaluate transgene stability in homozygous plants; (vi) stress mediated induction of hypermethylation in tissue culture using stress mimics such as propionic or butyric acid; and (vii) evaluation of transgene expression under different field conditions and different genetic backgrounds. RNA silencing and associated RNAi, as a fundamental mechanism of gene regulation in plants, has great potential application in plant science including regulating transgene expression (Eamens *et al.*, 2008).

12.6 Transgene Stacking

Multiple gene transfer to plants is necessary for sophisticated genetic manipulation strategies, such as the stacking of transgenes specifying different agronomic traits, the expression of different polypeptide subunits making up a multimeric protein, the introduction of several enzymes acting sequentially in a metabolic pathway or the expression of a target protein and one or more enzymes required for specific types of post-translational modification. As discussed in Chapter 1, most agronomic traits are multigenic in nature. Plant genetic improvement will require manipulation of complex metabolic or regulatory pathways involving multiple genes. A plant breeder tries to assemble a combination of genes in a crop plant that will make it as useful and productive as possible. Combining the best genes in one plant is a long and difficult process, especially as traditional plant breeding has been limited to artificially crossing plants within the same species or with closely related species to bring different genes together. The growing interest in dissecting and analysing complex metabolic pathways and the need to exploit the full potential of multi-gene traits for plant biotechnology (for review, see Halpin and Boerjan, 2003; Tyo *et al.*, 2007) provide a mandate for the development of new methods and tools for the integration of multiple transgenes into the plant genome (multi-transgene pyramiding or stacking) and coordinated expression of these transgenes in transformed plants.

Several approaches can be considered when using single-gene vectors for the delivery of multiple genes into plant cells (Halpin *et al.*, 2001; Daniell and Dhingra, 2002; Halpin and Boerjan, 2003). Some of the approaches used for the production of transgenic plants carrying multiple new traits include: (i) re-transformation (Singla-Pareek *et al.*, 2003; Seitz *et al.*, 2007), the stacking of several transgenes by successive delivery of single genes into transgenic plants; (ii) co-transformation (Li, L. *et al.*, 2003; Altpeter et al., 2005a), the combined delivery of several transgenes in a single transformation experiment; and (iii) sexual crosses (Ma *et al.*, 1995; Zhao *et al.*, 2003; Lucker *et al.*, 2004) between transgenic plants carrying different transgenes.

In this section, several transgene-stacking/pyramiding methods will be discussed, which are mainly based on two reviews by Francois *et al.* (2002a) and Dafny-Yelin and Tzfira (2007) and revision of the different multi-transgene-pyramiding methods. Table 12.3 summarizes the advantages, disadvantages and examples of the different multi-transgene-stacking methods in plants.

12.6.1 Sexual crosses

In a crossing experiment, two plants are crossed to obtain progeny that consists of the traits of the two parents. In the case of transgenic plants, a first gene is introduced in one of the parents and a second gene in the other. Crossing both transgenic parental lines results in progeny of which 25% (in case both parents were hemizygous for the transgenes) or all (in case both parents were homozygous for the transgenes) contain the two transgenes.

The main advantage of the crossing-based method for transgene stacking is that the method is technically simple. It only involves transfer of pollen from one parent to the female reproductive organ of the other. One other advantage is that transgenic populations of each parent can be screened for optimal expression of each transgene, thus facilitating the combination of two optimally expressed transgenes. However, the procedure is relatively time-consuming, certainly if more than two transgenes need to be combined by sequential crossing. The two transgenes in the lines resulting from the cross will most probably reside on different chromosomal loci that complicate further breeding through conventional methods. Furthermore, for some agronomically important crops like potato and cassava, the high level of heterozygosity in the species makes crossing approaches difficult and time-consuming. Crossing is very difficult to apply to plants that are vegetatively propagated (e.g. perennial fruit crops and many ornamentals) since the (desired) heterozygous nature of the genetic background will be altered due to recombination during meiosis (Gleave *et al.*, 1999).

Sexual crosses among transgenic plants make it possible to exploit powerful ‘supertraits’ that are not attainable through traditional methods. One example of a crop carrying such new characteristics is Monsanto’s multi-stacked maize, which was produced via conventional crossing of three inbred transgenic maize lines: MON863, MON810 and NK603. The elements incorporated into this multi-stack include five loci, four of which carry a synthetic gene linked to combinations of strong regulatory elements from viruses, bacteria and unrelated plants. Expression of the first two synthetic genes produces an EPSPS that resembles the EPSPS from *E. coli* and is, unlike most plant versions, not inactivated by herbicides containing glyphosate. The third synthetic

Table 12.3. Summary of advantages, disadvantages and examples of multi-transgene-stacking methods in plants.

Techniques	Advantages	Disadvantages	Example and reference
Crossing	Technically simple; pre-selection of parents with optimal gene expression	Time-consuming; difficulties in further breeding; not applicable to vegetatively propagated plants	Mercury detoxification (Bizily *et al*, 2000); antibody engineering (Hiatt *et al.*, 1989); antimicrobial resistance (Zhu *et al.*, 1994)
Sequential transformation	Applicable to vegetatively propagated plants; allows maintenance of elite genotype	Time-consuming; necessity for different selection markers	Plant fertility restoration system (Hird *et al.*, 2000); removal of selectable marker gene (Gleave *et al.*, 1999)
Co-transformation			
Single plasmid	Linked integration[a]; single transformation event	Technically demanding; linked integration[a]	Reporter gene expression (Christou and Swain, 1990)
Multiple plasmid	Technically simple; single transformation event	Dependence on co-transformation frequency	Production of vitamin A-enriched rice ('Golden Rice') (Ye *et al.*, 2000); polyhydroxyalkanoate production (Slater *et al.*, 1999)
IRES[b]-based approach	Use of only a single set of promoter/ terminator sequences; single transformation event; natural system for co-expression of multiple genes; linked integration[a]	Tissue-specificity; developmental regulation; lower expression levels; linked integration[a]	Reporter gene expression (Urwin *et al.*, 2000)
Transplastomic technology	Use of only a single set of promoter/terminator sequences; single transformation event; natural system for co-expression of multiple genes; linked integration[a]; high expression levels; avoiding positional effects	Chloroplast containment of gene products; linked integration[a]; number of plant species to which transplastomic technology is applicable is limited; low success rate of gene insertion into chloroplast genome	Insect resistance (De Cosa *et al.*, 2001)
Polyprotein approach			
Potyviral system	Use of only a single set of promoter/ terminator sequences; coordinated expression	Energetically wasteful; aberrant cleavage of endogenous plant substrates by viral protease	Expression of mannityl opine biosynthetic pathway (Beck von Bodman *et al.*, 1995)
Cleavage dependent on plant protease	Single transformation event; possible targeting of gene products	Necessity for that plant protease	Nematode resistance (Francois *et al.*, 2002c); antifungal resistance (Francois *et al.*, 2002b)
2A system	Linked integration[a]	Addition of 2A originating sequences to the gene products	Reporter gene expression (Urwin *et al.*, 1998)

[a]Linked integration of the transgenes can be advantageous when the transgenic line is to be used in traditional breeding, whereas linked integration can be undesirable when one of the transgenes (e.g. the selectable marker gene) is to be removed via outcrossing.
[b]IRES, internal ribosome entrysite.

gene encodes the insecticidal cry3Bb1 protein with activity against specific Coleoptera, whereas the fourth gene product, cry1Ab, provides tolerance against certain Lepidopteran insects. The fifth gene is a bacterial kanamycin resistance gene encoding neomycin phosphotransferase (*npt*II). The pentuple stack maize currently occupies millions of hectares in the USA and supports a substantial reduction in pesticide usage.

12.6.2 Co-transformation via plasmids

Co-transformation is defined as the simultaneous introduction in a cell of multiple genes followed by the integration of the genes in the cell genome. The genes are either present on the same plasmid used in transformation ('single-plasmid co-transformation') or on separate plasmids ('multiple-plasmid co-transformation'). The main advantage of co-transformation for transfer of multiple genes into a plant is that a single transformation event can result in the integration of multiple transgenes as opposed to sequential transformation which requires multiple, time-consuming transformation events.

Theoretically speaking, however, co-transformation has some technical limitations. For single-plasmid co-transformation, the main technical limitation is the difficulty to assemble complex plasmids with multiple gene cassettes (François *et al.*, 2002a). Standard transformation vectors are not really up to such a task. A major problem is that their multiple cloning sites consist merely of hexa-nucleotide restriction sites, which are often present within one or more of the sequences that one wishes to insert in the vector. Insertion of more than one or two expression cassettes often requires inefficient partial digests or the use of linkers to convert one restriction site to another or the use of inefficient blunt-end cloning. When plant transformation vectors with multiple expression cassettes are eventually finalized, it is often not possible to move or replace the cassettes in single cloning steps, due to the presence of restriction sites at undesired locations.

Co-transformation with multiple plasmids has the obvious advantage that assembly of the different expression cassettes is technically easier as it is done independently on different plasmids (Komari *et al.*, 1996). The success of this technique depends on the frequency with which two (or more) independent transgenes are both transferred to the plant cell and integrated into the cell genome (= co-transformation frequency). Agrawal *et al.* (2005) transformed rice simultaneously with five minimal cassettes, each containing a promoter, coding region and polyadenylation site but no vector backbone. They found that multi-transgene co-transformation was achieved with high efficiency using multiple cassettes, with all transgenic plants generated containing at least two transgenes and 16% containing all five. They concluded that gene transfer using minimal cassettes is an efficient and rapid method for the production of transgenic plants containing and stably expressing several different transgenes. Their results facilitate effective manipulation of multi-gene pathways in plants in a single transformation step.

12.6.3 Co-transformation via particle bombardment

Particle bombardment is the most convenient method for multiple gene transfer to plants since DNA mixtures comprising any number of different transformation constructs can be used, with no need for complex cloning strategies, multiple *Agrobacterium* strains or sequential crossing (Altpeter *et al.*, 2005a). Many studies describe successful integration of two or three different transgenes, in addition to the selectable marker, into plants by particle bombardment.

Wu, L. *et al.* (2002) examined the co-transformation of rice with nine transgenes via particle bombardment and documented the levels of transgene expression. They found that non-selected transgenes were present along with the selectable marker in about 70% of the plants and that 56% carried seven or more genes. This was much higher than expected given the independent integration frequencies, agreeing with a model proposing that the integration of one gene into a specific locus in the rice genome could mediate the insertion of other genes into the same locus (Kohli *et al.*, 1998). This phenomenon is important when large numbers of genes are considered, since a much larger transgenic population would be required if each integration event were independent. Wu, L. *et al.* (2002) also found that all of the nine transgenes were expressed and that the expression of one gene was independent of each other. These findings are very useful in designing multiple plasmid transformation experiments such as those required for plant metabolic engineering.

One of the most interesting recent developments of particle bombardment is the combination of multiple gene transfer and clean DNA techniques, i.e. the simultaneous transfer of multiple gene cassettes into rice plants. Three coat protein genes from the same virus were introduced simultaneously to generate rice plants with pyramidal resistance against a single pathogen (Sivamani *et al.*, 1999). Similarly, Maqbool *et al.* (2001) have shown how the same transformation strategy can provide pyramidal insect resistance in rice. Datta *et al.* (2003) have succeeded in the development of 'Golden' *indica* rice lines containing four genes, i.e. those required to extend the existing carotenoid metabolic pathway (*psy*, *crtI* and *lcy*) in addition to the selectable marker gene, either phosphomannose isomerase (*pmi*) or hygromycin phosphotransferase (*hpt*). Romano and colleagues synthesized polyhydroxyalkanoates (PHAs) in transgenic potatoes by simultaneously introducing the *phaG* and *phaC* genes encoding acyl-CoA trans-acylase and PHA polymerase along with the neomycin-phosphotransferase selectable marker in three separate constructs (Romano *et al.*, 2005).

In addition to applications in metabolic engineering and multi-gene resistance strategies, the direct transfer of multiple genes has also become a practical strategy for generating crops that produce multimeric proteins. For example, Nicholson *et al.* (2005) produced full-sized multimeric antibodies in transgenic plants. These proteins comprise at least two components, the heavy and light chains, but more complex antibody forms such as secretory antibodies (sIgA) also require a joining chain and a secretory component. Nicholson *et al.* (2005) simultaneously delivered all four genes, together with a fifth gene encoding a selectable marker, into rice by particle bombardment.

For many applications of transgenesis, production of different heterologous proteins and hence introduction of multiple transgenes (multi-transgene-stacking), is highly desired. During the last decade, the number of approaches for multi-transgene-stacking in plants using transgenesis has significantly increased. For all the benefits and simplicity of combining co-transformation, retransformation and crosses while using single-gene vectors for the delivery of multiple genes into plant species, these methods suffer from several drawbacks. These include the undesirable incorporation of a complex T-DNA integration pattern, often observed during integration of T-DNA molecules from multiple sources (De Neve *et al.*, 1997; De Buck *et al.*, 1999) and the time needed for retransformation or crosses between transgenic plants. More importantly, transgenes derived from different sources typically integrate at different locations in the plant genome, which may lead to various expression patterns and possible segregation of the transgenes in the offspring.

Except for those discussed above, other approaches for transgene stacking include vector assembly, internal ribosome entry site (IRES), transplastomic technology and polyprotein approach. Therefore, after evaluation of the pros and cons of the different methods, one should be able now to select an appropriate approach for most purposes. Moreover, the potential of the different methods can be significantly increased by combining approaches. For example, for the delivery of different antimicrobial protein (AMP) genes, it has been able to double the capacity of modular plant transformation vector by combining it with a polyprotein strategy (Goderis *et al.*, 2002). For this purpose, single transgene units of the original vector were replaced by poly-AMP encoding expression cassettes and transformed to *A. thaliana*. Single biologically active AMPs could be demonstrated in the resulting transgenic plants.

12.7 Transgenic Crop Commercialization

Genetic transformation has the potential to address some of the most challenging biotic and abiotic constraints faced by farmers in non-industrialized agriculture, which are not easily addressed through conventional plant breeding alone. The major constraints include insect pests and viruses, as well as drought. A second advantage of genetic transformation is that it can add an economically valuable trait while maintaining other desirable characteristics of the host cultivar. For example, enhanced product quality or micronutrients can be added to a well-adapted cultivar that already yields well under local conditions. This feature is particularly attractive for semi-commercial, small-holder farmers in non-industrialized agriculture, who are more likely to consume as well as sell their farm products. The poor of the developing world should benefit from the deployment of desirable transgenic crops that follows scientifically-sound biosafety and food safety standards and appropriate intellectual property management and stewardship (Ortiz and Smale, 2007).

Intrinsic to the production of transgenic plants is an extensive evaluation process to verify whether the inserted gene has been stably incorporated without detrimental effects to other plant functions, product quality, or the intended agroecosystem. Initial evaluation includes attention to activity of the introduced gene, stable inheritance of the gene and unintended effects on plant growth, yield and quality.

If a plant passes these tests, it may not be used directly for crop production, but will be crossed with improved cultivars of the crop. This is because not all cultivars of a given crop can be efficiently transformed and these generally do not possess all the producer and consumer qualities required of modern cultivars. The initial cross to the improved cultivar must be followed by several cycles of repeated backcrosses to the improved parent. The goal is to recover as much of the improved parent's genome as possible, with the addition of the transgene from the transformed parent.

The next step in the process is multi-location and multi-year evaluation trials in greenhouse and field environments, as described in Chapter 10, to test the effects of the transgene and overall performance. This phase also includes evaluation of environmental effects and food safety.

12.7.1 Commercial targets

Commercialization of transgenic products is influenced by markets, i.e. consumer demand for improved processes and new products are dependent on technology – scientific discoveries in molecular genetics and biochemistry. Some examples of commercial targets include:

1. Hybrid seed systems for heterosis and intellectual property protection, such as nuclear male sterility systems for inbred line production.
2. Pest and disease tolerance genes: *Bt* genes, α-amylase inhibitors, viral coat proteins.
3. Stress tolerance genes: barley *Hva1,* maize *ZmPLC1.*
4. Herbicide resistant crops: mutation screens for resistance to sethoxydim (Poast®), an acetyl-CoA carboxylase (ACCase) inhibitor; transgenic plants for glyphosate (Roundup®) resistance.
5. Genes for commercially valuable oils, proteins and starches: fatty acid biosynthetic gene modification in high oil corn; modification of seed storage proteins; generation of transgenic corn with improved amino acid profiles; manipulation of carbon-partitioning genes for novel starch production.
6. Genes for improved plant performance: generation of dwarf cultivars of wheat and rice; *PhyA* expression for narrow-row crop production.

Most of these potential products generate revenue by lowering the costs (financial and/or environmental) of plant production, e.g. reducing the level of chemical inputs such as insecticides for both pests and viral vectors. Following are examples for the trangenes of agronomic importance that have been introduced into transgenic cereals:

Maize:

1. Insect resistance: synthetic truncated version of the CrylA(b) protein from *Bacillus thuringiensis* for tolerance to European corn borer.
2. Virus resistance: coat protein-mediated tolerance to maize dwarf mosaic virus.
3. Herbicide resistance: the *bar* gene for PPT (Liberty®) tolerance; mutant *epsps* synthase genes for glyphosate (Roundup®) tolerance; and mutant *als* gene for sulfonylurea (Glean®) tolerance.

Rice:

1. Resistance to bacterial pathogens: chitinase gene conferring enhanced tolerance to sheath blight; *Xa-21* bacterial blight resistance gene.
2. Virus resistance: coat protein-mediated rice stripe virus tolerance; coat protein-mediated rice dwarf phytoreovirus tolerance.
3. Insect resistance: *Bt* CrylA(b) gene expression for leaf folder and stem borer tolerance.
4. Improved grain quality: bean β-phaseolin seed storage gene expression in endosperm for improved lysine and isoleucine levels.

Barley:

1. Virus resistance: coat protein-mediated barley yellow dwarf virus tolerance.
2. Improved malting/brewing characteristics: hybrid bacterial β-glucanase gene expression for enzyme thermotolerance.

Wheat:

1. Improved bread-making characteristics: chimeric Dy10-Dx5 high molecular weight gluten gene expression in endosperm.
2. Transgenes conferring herbicide resistance: the *bar* gene for PPT (Liberty®) tolerance; mutant EPSPS synthase genes for glyphosate (Roundup®) tolerance.

12.7.2 Current status of transgenic crop commercialization

Commercial adoption by farmers of transgenic crops has been one of the most rapid cases of technology diffusion in the history of agriculture (Borlaug, 2000). Commercialization of transgenic crops started in 1996. Fig. 12.6 provides data on the global areas of biotech/GM-crops grown

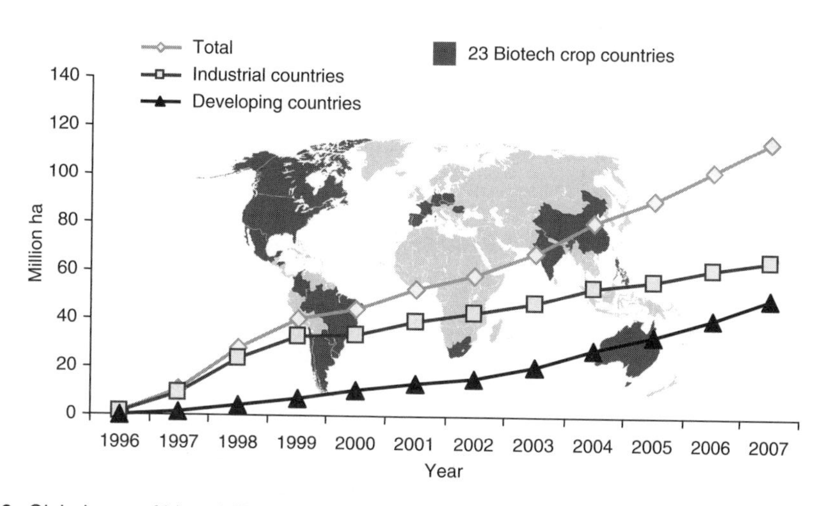

Fig. 12.6. Global area of biotech/GM crops (1996–2007). From James (2008) with permission.

over the last 12 years (1996–2007) (James, 2008). As a result of consistent and substantial benefits during the first dozen years of commercialization, farmers have continued to plant more biotech/GM-crops every single year. In 2007, the global area of biotech/GM-cOrops reached 114.3 million ha with an unprecedented 67-fold increase between 1996 and 2007, making it the fastest adopted crop technology in recent history. The proportion of the global area of biotech/GM crops grown by developing countries has increased consistently and by 2007, 43% of the global biotech crop area, equivalent to 49.4 million ha, was grown in developing countries (Table 12.4). The USA, followed by Argentina, Brazil, Canada, India and China are the principal adopters of biotech/GM crops globally, with the USA retaining its top world ranking with 57.7 million ha (50% of global biotech area) (Table 12.4). Notably, 63% of biotech/GM-maize, 78% of biotech/GM cotton and 37% of all biotech/GM crops in the USA in 2007 were stacked products containing two or three traits that delivered multiple benefits.

Soybean is the principal biotech/GM-crop, occupying 58.6 million ha (51% of global biotech/GM area), followed by fast-growing maize (35.2 million ha at 31%), cotton (15.0 million ha at 13%) and canola (5.5 million ha at 5% of global biotech/GM-crop area) (Fig. 12.7A; James, 2008). Since the genesis of commercialization in 1996, herbicide tolerance has consistently been the dominant trait (Fig. 12.7B). In 2007, herbicide tolerance, deployed in soybean, maize, canola, cotton and lucerne occupied 63% or 72.2 million ha of the global biotech/GM-crops.

The most recent survey of the global impact of biotech/GM-crops for the period 1996–2006, estimates that the global net economic benefits to biotech/GM-crop farmers in 2006 was US$7 billion, and US$34 billion (US$16.5 billion for

Table 12.4. Global area of biotech/GM-crops in 2007 by country (from James (2008) with permission).

Rank	Country	Area (million hectares)	Biotech/GM crops
1[a]	USA	57.7	Soybean, maize, cotton, canola, squash, papaya, lucerne
2[a]	Argentina	19.1	Soybean, maize, cotton
3[a]	Brazil	15	Soybean, cotton
4[a]	Canada	7	Canola, maize, soybean
5[a]	India	6.2	Cotton
6[a]	China	3.8	Cotton, tomato, poplar, petunia, papaya, sweet pepper
7[a]	Paraguay	2.6	Soybean
8[a]	South Africa	1.8	Maize, soybean, cotton
9[a]	Uruguay	0.5	Soybean, maize
10[a]	Philippines	0.3	Maize
11[a]	Australia	0.1	Cotton
12[a]	Spain	0.1	Maize
13[a]	Mexico	0.1	Cotton, soybean
14	Colombia	< 0.1	Cotton, carnation
15	Chile	< 0.1	Maize, soybean, canola
16	France	< 0.1	Maize
17	Honduras	< 0.1	Maize
18	Czech Republic	< 0.1	Maize
19	Portugal	< 0.1	Maize
20	Germany	< 0.1	Maize
21	Slovakia	< 0.1	Maize
22	Romania	< 0.1	Maize
23	Poland	< 0.1	Maize

[a]Thirteen biotech mega-countries growing 50,000 ha or more of biotech/GM crops.

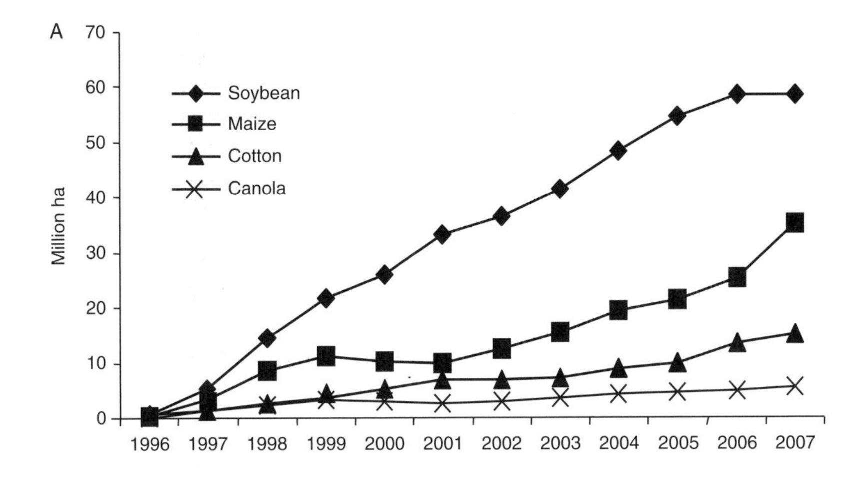

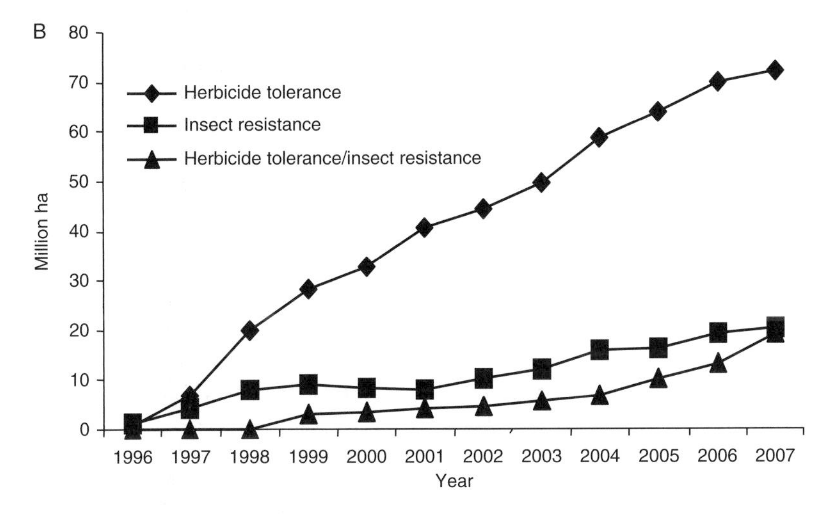

Fig. 12.7. Global area of biotech/GM crops (1996–2007). (A) By crops. (B) By traits. From James (2008) with permission.

developing countries and US$17.5 billion for industrial countries) for the accumulated benefits during the period 1996–2006; these estimates include the very important benefits associated with the double cropping of biotech/GM-soybean in Argentina (Brookes and Barfoot, 2008). The accumulative reduction in pesticides for the period 1996–2006 was estimated at 289,000t of active ingredient, which is equivalent to a 15.5% reduction in the associated environmental impact of pesticide use on these crops, as measured by the Environmental Impact Quotient (EIQ) – a composite measure based on the various factors contributing to the net environmental impact of an individual active ingredient.

While 23 countries planted commercialized biotech/GM-crops in 2007, an additional 29 countries, totalling 52, have granted regulatory approvals for biotech/GM-crops for import for food and feed use and for release into the environment since 1996. A total of 615 approvals have been granted for 124 events for 23 crops. Thus, biotech/GM-crops have been accepted for import for food and feed use and for release into the environment in 29 countries, including major food importing countries like Japan, which do not have plant biotech/GM-crops.

The most important potential contribution of biotech/GM-crops will be their contribution to the humanitarian Millennium Development Goals (MDG) of reducing poverty and hunger by 50% by 2015. With a dozen years of accumulated knowledge and significant economic, environmental and socio-economic benefits, biotech crops are poised for even greater growth in coming years, particularly in developing countries that have the greatest need for this technology. The number of biotech/GM-crop countries, crops and traits and hectarage are projected to double between 2006 and 2015, the second decade of commercialization (James, 2008).

Despite globally organized opposition, few innovations in agriculture have spread so rapidly as GM-crops. Still, much remains to be done – particularly the expansion of disease-resistant cultivars, increased yields, biofortification of food for poor consumers, substitution of plant-produced targeted endotoxins for broad-band pesticides and, perhaps most crucially, drought-tolerant and salt-tolerant cultivars (Herring, 2008). Disaggregating the concept of the genetically modified organism (GMO) is a necessary condition for confronting misconceptions that constrain the use of biotechnology in addressing imperatives of development and escalating challenges from nature. There are still several problems associated with commercialization. There is a high investment cost associated with the long lead time (8–10 years) for products to reach the marketplace; profitability of some 'technology push' products at the onset of product development is uncertain; intellectual property issues limit freedom to operate with key technologies; and uncertainty associated with regulatory and consumer acceptance issues inhibit trade and investment.

12.7.3 Regulating transgenic crops

There are many reasons why governments regulate and oversee processes and products for transgenic crops (Jaffe, 2004). One major concern about the use of GM-foods is that the molecular alterations designed to produce a beneficial trait could also result in unintentionally hazardous effects. Individual transgenic crops could potentially present risks to humans or the environment, although there is no evidence that it has happened over the 10 years of commercialization. In general, a strong, but not stifling, regulatory system needs to be established and properly implemented to ensure safe crops to humans and the environment. Other reasons include fraud avoidance and social, ethical and public concerns.

Risk assessment

All transgenic plants are required to undergo thorough and vigorous safety and risk assessments before commercialization. A risk assessment consists of hazard identification, hazard characterization, exposure assessment and risk characterization (Codex Alimentarious Commission, 2001; Craig *et al.*, 2008; Nickson, 2008). Regulatory justifications for these assessments differ between countries. In most countries there are two kinds of regulations that govern research and development of transgenic plants: (i) contained-use rules governing genetic modification in the laboratory, concentrating mainly on worker health and safety issues; and (ii) field-release regulations focusing on environmental risk assessment appropriate to the nature and final use of the transgenic plant. Each release is considered case by case to build up experience with particular crop and transgene combinations.

The United States National Research Council identified four categories of potential environmental hazards from the release of a transgenic crop:

> (i) hazards associated with the movement of the transgene itself with subsequent expression in a different organism or species, (ii) hazards associated directly or indirectly with the transgenic plant as a whole, (iii) non-target hazards associated with the transgene product outside the plant and (iv) resistance evolution in the targeted pest population.
>
> (NRC, 2002)

The potential human hazards from transgenic crops are also generally recognized by the scientific and regulatory communities (Jaffe, 2004). The potential risks generally related to:

> the possibility of introducing new allergens or toxins into food-plant varieties, the possibility of introducing new allergens into pollen, or the possibility that previously unknown protein combinations now being produced in food plants will have unforeseen secondary or pleiotropic effect.
> (NRC, 2001)

However, not all tests currently being applied to assessing allergenicity have a sound scientific basis (Goodman *et al.*, 2008). Therefore, factors to be borne in mind in the risk assessment should include, but are not limited to: (i) the function of the gene in the donor organism; (ii) the effect of the transgene on the phenotype of the transgenic plant; (iii) evidence of toxicity and/or allerginicity, e.g. Brazil nut seed storage protein; (iv) persistence in agricultural habitats (weediness); (v) invasiveness in natural habitats; (vi) impact on non-target organisms, e.g. *Bt* maize, regulation requires extensive analysis to identify any potential problem; and (vii) the likelihood/consequence of transgene movement to other plants by cross-pollination or to other (pathogenic) organisms by horizontal gene transfer, e.g. sexual compatibility between cultivated oats (*Avena sativa*) and wild oats (*Avena fatua*), viral host range extension by transgene-encoded coat protein transencapsidation or genetic recombination. More recent discussion on several specific issues can be found in Craig *et al.* (2008), Nickson (2008) and Romeis *et al.* (2008) and Tabashnik *et al.* (2008).

The risk assessment process for transgenic plants consists of two steps: (i) a comparative analysis (substantial equivalence) to identify potential differences with their non-engineered counterpart(s); followed by (ii) an assessment of the environmental and food/feed safety or nutritional impact of any identified differences (Ramessar *et al.*, 2007).

Regulatory systems

There are two kinds of regulation systems (the data assessment in both is similar): (i) horizontal – a process-based system that applies to all plants produced by transformation methods, e.g. Europe; and (ii) vertical – a product-based system that defines the characteristic of modified plants that require them to be regulated, e.g. the USA.

In the USA, transgenic plants are regulated by three federal agencies – the Department of Agriculture (USDA), the Environmental Protection Agency (EPA) and the Food and Drug Administration (FDA). The USDA controls permits for inter-state movement of transgenic materials, assesses the pest character of transgenic plants, determines when transgenic plants can be field-grown without notification or permits. The FDA determines whether a transgenic plant has been adequately evaluated in accordance with its biotechnology food and feed policy, e.g. for the safety of antibiotic selectable marker genes. The EPA regulates plants with pesticide properties, e.g. *Bt* plants and registers herbicides to be used on herbicide-resistant plants.

In the European Union (EU), EU Directive 2001/18/EC (European Parliament, 2001) sets forth regulations governing the deliberate release into the environment of GMOs. The Directive 'put in place a step-by-step approval process on a case-by-case assessment of the risks to human health and the environment before any GMO or products consisting of or containing GMOs can be released into the environment or placed on the market' (European Parliament, 2003). There are no European Community (EC)-wide regulations governing novel feed and foods, and some countries have established their own national regulations.

In Japan, transgenic plants are regulated by the Ministry of Agriculture, Forestry and Fisheries and the Ministry of Health and Welfare. In Canada, transgenic plants are regulated by Ag. & Agri-Foods Canada and Health Canada. Novel products are regulated in the same way whether they are generated by mutation or transformation.

There is no international harmonization of regulations to ensure that transgenic plant cultivars released in one country will be accepted in another. Antibiotic resistance genes in food products might inhibit international trade in transgenic products.

12.7.4 Product release and marketing strategies

In order to recover their substantial research investment the developer of a potential commercial product must either generate and market the transgenic seed directly or negotiate a royalty with a seed company/companies, e.g. universities, government agencies, technology development companies, large agrochemical companies. The marketing strategies for any one particular trait will be influenced by the nature of the product. Herbicides, disease or stress tolerance traits which enhance yield or reduce inputs will help increase market shares and/or increase seed sale premiums, e.g. *Bt* maize. Herbicide tolerant crops will help benefit from increased chemical sales, e.g. Roundup Ready® maize. Improved grain quality, for on-farm/downstream processing uses, will influence a seed sale premium, e.g. high lysine maize.

12.7.5 Monitoring transgenes

One of the principal concerns of GM-crops is the likelihood and possible consequence of the introduced transgenes being transferred through pollen dispersal to wild relatives or non-transgenic crops (Chandler and Dunwell, 2008). For pollen-mediated gene flow to occur among plant populations, dispersal of pollen to a different population must occur with successful fertilization of an ovule. Although the movement of pollen is a critical step in transgene escape, there are currently few systems for the direct monitoring of transgenic pollen movement under field conditions. Previous attempts to measure gene flow have evolved around the analyses of genetic markers (Slatkin, 1985) as discussed in Chapter 13. These systems have limitations because they are species-specific, requiring the use of expensive assays that hardly yield results in real time or in the field. Shi *et al.* (2008) reported on the gene flow between foxtail millet (*Setaria italica*), an autogamous crop and its weedy relative, *Setaria viridis*, growing within or beside fields containing three kinds of inherited herbicide resistance, dominant, recessive, or maternal. Over the 6-year study, in the absence of herbicide selection, the maternal chloroplast-inherited resistance was observed at a 2×10^6 frequency in the weed populations. Resistant weed plants were observed 60 times as often, at 1.2×10^4 in the case of the nuclear recessive resistance and 190 times as often, at 3.9×10^4 in the case of the dominant resistance. The results indicated that the hereditary mode of transmission of transgenes played a major role in interspecific gene flow. More recently visual markers such as GFP have been proposed for use, using whole plant expression to monitor gene flow under agricultural conditions. This method has been used successfully to assess outcrossing events in canola (*Brassica napus*) under field conditions (Halfhill *et al.*, 2004a). A direct method could be the use of GFP-tagged pollen to monitor pollen movement under field conditions. This system would allow the quantification of pollen flow directly from a group of individuals in the field and would determine the distance and directional patterns of pollen dispersal within a plant population. In Hudson *et al.*'s (2004) report, a pollen specific promoter was used to express the GFP gene in tobacco (*Nicotiana tabacum* L.). GFP was visualized in pollen and growing pollen tubes using fluorescence microscopy. Furthermore, the goal of the research was to compare the dynamics of pollen movement with that of gene flow by using another method of whole plant expression of GFP to estimate the outcrossing rate by progeny analysis. Pollen movement and gene flow were quantified under field conditions. Pollen was collected in traps and screened for the presence of GFP-tagged pollen using fluorescence microscopy. Progeny from wild-type plants were screened with a hand-held ultraviolet light for detection of the GFP phenotype. It should be noted that the GFP gene is from an animal or fly and thus it should be handled very carefully. The examples given here are only proposals by researchers and they are not used for commercial purposes.

A built-in strategy was developed to create selectively terminable transgenic rice, where the transgenic rice plants mixed in the conventional rice could be selectively eliminated by a spray of bentazon, a herbicide commonly used for rice weed control (Lin *et al.*, 2008). The gene(s) of interest is tagged with a RNAi cassette, which specifically suppresses the expression of the bentazon detoxification enzyme CYP81A6 and thus renders transgenic rice to be sensitive to bentazon. Transgenic rice plants were generated by this method using a new glyphosate resistant EPSPS gene from *Pesudomonas putida* as the gene of interest and it was demonstrated that these transgenic rice plants were highly sensitive to bentazon but tolerant to glyphosate, which is exactly the opposite of conventional rice. Field trials of these transgenic rice plants further confirmed that they could be selectively killed at 100% by one spray of bentazon at a regular dose used for conventional rice weed control. Furthermore, it was found that the terminable transgenic rice created showed no difference in growth, development and yield compared to its non-transgenic control. Therefore, this method of creating transgenic rice constitutes a novel strategy of transgene containment, which appears simple, reliable and inexpensive for implementation.

Metabolomics are being developed to assess the safety of GM-foods (Kuiper *et al.*, 2003). Because society demands that producers demonstrate 'substantial equivalence' between transgenic and non-transgenic crop plants, metabolite profiling is expected to provide a reliable means of detecting differences in metabolite levels between the two types of plants and identifying potential problems.

Monitoring of transgenic crops has been a wide concern among the centres of the Consultative Group on International Agricultural Research (CGIAR). However, decisions, policies and procedures about monitoring should be science-based and this requires education, an area where the CGIAR centres can play an important role. There will be a need to continue to evaluate the need for and type of monitoring, as new (and unique) products are developed and released in the emergent economies of the world (Hoisington and Ortiz, 2008).

Regarding future developments in genetic transformation, new techniques for producing transgenic plants will improve the efficiency of the process and will help resolve some of the environmental and health concerns. Among the expected changes are the following: (i) more efficient transformation, that is, a higher percentage of plant cells will successfully incorporate the transgene; (ii) better marker genes to replace the use of antibiotic resistance genes; (iii) better control of gene expression through more specific promoters, so that the inserted gene will be active only when and where needed; and (iv) transfer of multi-gene DNA fragments to modify more complex traits.

12.8 Perspectives

Improvement of crop plants through genetic transformation, also called transgenic breeding, is one of the two major approaches in molecular breeding. It can utilize the genes from any organisms, by which obstacles associated with sexual hybridization can be overcome. On the other hand, transgenic breeding provides a quick approach to pyramid genes of different sources into one genetic background. There are numerous examples of transgenic plants with single genes or traits incorporated. Typical examples include GM-crops containing genes for pest and disease resistance and for improved quality as discussed in Section 12.7.2. Examples of GM-crops with multiple traits that have been improved are Monsanto's multi-stacked maize as discussed in Section 12.6.1 and an ongoing project in China to develop green super hybrid rice by stacking genes for many traits including insect resistance, disease resistance, nutrient efficiency, drought tolerance, grain quality and yield (Zhang, 2007).

Transformation-based gene transfer should be integrated with genomics and other molecular approaches. For example, functional genomics, discussed in Chapters 3 and 11, will bring new frontiers and horizons to transgenic breeding by providing more genes with well-characterized function and tools optimized for transgene expression. Molecular markers can be used to facilitate the transformation process, to transfer the transgenes to a different genetic background and to identify and select transgenic plants as discussed in Chapter 13. Genetic transformation will also be increasingly combined with conventional breeding approaches, which will contribute to improved breeding efficiency.

As technology develops in transgenic breeding, transformation technologies that limit many steps in transgenic breeding will become less demanding compared to the discovery and characterization of genes and commercialization of transgenic products. It can be expected that transgenic breeding will become increasingly important by producing good-quality and high-yielding agricultural products. All regulatory and biosafety issues, both of which are man-made and currently slow or stop the adoption of transgenic crops by farmers in many countries, will be brought under control at a reasonable level.

13

Intellectual Property Rights and Plant Variety Protection

Intellectual property rights (IPR), especially patents, are become increasingly important as crop improvement increasingly becomes an industrial process with large private sector investments driven by an expectation of high economic returns. Biotechnological inventions, particularly in the field of agribiotechnology, are increasing with the worldwide expansion of the cultivation of transgenic plants and the increasing reliance of commercial breeding programmes on genomics tools such as marker-assisted selection (MAS). Proprietary control over many new breeding technologies and plant variety protection (PVP) of the products of crop improvement is increasingly influencing the nature and extent of public funding available for applied research in this area. Genomics tools also provide a means for DNA fingerprinting plants in order to obtain information on the relationship between new breeding lines and commercial cultivars which can be used in the protection of plant cultivars once they have been registered.

Intellectual property (IP) can take the form of genes, markers, technological processes, information and concepts and in some countries even whole plants. Major criteria for granting patents require that the IP be novel and non-obvious, useful and not previously disclosed. A particular piece of IP may be owned exclusively or jointly by an individual, a group of individuals, an organization/company, or a government. An exclusive owner of a particular piece of IP may choose to do nothing with it; use or practise it directly; or sell, license or gift it to others (Boyd, 1996). For a detailed coverage of IPR, please refer to Krattiger *et al.* (2006).

As has been chronicled extensively elsewhere, patents covering modified living organisms were first approved in the US Supreme Court in 1980. Since then, IPR covering organisms and/or their components have become commonplace in many countries and expanded internationally through treaties such as the International Convention for the Protection of New Varieties of Plants (known by its French acronym UPOV) and the World Trade Organization's (WTO) 1994 Agreement on Trade-Related Aspects of Intellectual Property Rights (TRIPS). This chapter is devoted to various aspects of PVP including its needs, impacts, strategies and related international agreements. It also covers IPR that affect development and application of molecular breeding techniques in crop improvement. Publications that serve as key references for this chapter include Heitz (1998), the proceedings from a seminar on the use of molecular techniques for plant variety protection organized by Canadian Food Inspection Agency and National Forum on Seed (CFIA/NFS,

2005), Chan (2006), Louwaars *et al.* (2006), a publication by The International Bank for Reconstruction and Development and The World Bank that is cited as IBRD/World Bank (2006), Tripp *et al.* (2006) and Henson-Apollonio (2007).

13.1 Intellectual Property and Plant Breeders' Rights

13.1.1 Basic aspects of intellectual property

As a specialized area of law, IPR covers 'all things which emanate from the exercise of the human brain' (Walden, 1998). Classically IPR have been divided into two groups: (i) industrial property (patents for inventions, trade secrets, trademarks, special rights for integrated circuits, etc.); and (ii) literary and artistic property (copyright, rights of performers, etc.; Walden, 1998; Ng'etich, 2005) although molecular breeding programmes may pursue rights under either group. National governments have established IPR laws in order to achieve several goals: (i) to create incentives to stimulate new technological advances by providing mechanisms to ensure capture of financial returns on investments; (ii) to reward inventors with exclusive rights for a certain period of time thereby ensuring that others do not capture financial returns on the inventor's investments; and (iii) to create avenues for public disclosure of new technologies, which provides stimulus for further advances. In nearly all national legal systems, individuals and companies are usually able to acquire IPR, protect the economic returns on their investment and sell, lease or license those IPR to third parties.

National IPR laws vary in terms of the specific details of registration, scope and duration of protection. General IPR laws are often considered inadequate for certain industrial sectors, leading to the development of so called '*sui generis*' systems to protect, for example, integrated computer circuits, databases and plant cultivars (e.g. plant breeders' rights, PBR). It is important to realize that IPR are only enforceable within the national territory within which they have been registered. Moreover, the level and nature of enforcement of IPR laws varies considerably across geographical regions. These factors significantly and differentially influence the product development and deployment strategies of commercial companies operating in different countries.

13.1.2 Intellectual property rights in plant breeding

Various attempts have been made to establish IPR systems for crop cultivars, but the concept of PVP and the possibility of plant patents emerged just under a century ago. Different IPR are now available to the plant breeding sector: PBRs, patents, trade secrets and trademarks are already highly important. In addition, copyright and database protection are likely to play an increasingly important role in molecular breeding. Very few countries (e.g. the USA and Japan) offer patent protection to cultivars so most plant breeders must still rely on PBRs for conventionally bred material. However, the increasing range of biotechnology interventions used in plant breeding can be patented, thus providing increasing scope for use of these patents in protecting new cultivars (Louwaars *et al.*, 2006).

An IPR regime in plant breeding should perform two basic roles (IBRD/World Bank, 2006). First, in the interest of the public, the IPR regime should ensure that knowledge and materials enter the public domain at the earliest possible point and it should stimulate improvements and innovations that increase the choices available to farmers and consumers. Secondly, in the interest of the rights-holder, the IPR regime should provide opportunities for breeders to recover their investments, which may include the rights to recover royalties from farmers who save seed for planting the next season, directly for themselves or for sharing with neighbours through informal sale of the seed. In addition, the IPR regime should keep competing commercial seed producers from multiplying and marketing the protected cultivar without a licence. Many breeding companies would like to keep competing plant breeders from using a protected cultivar or technology in the development of a new cultivar. In this case, they must use the patent system as PBRs have been explicitly structured to encourage rather than exclude this type of activity. The degree to which IPR and PVP systems in developing countries are able to limit practices depends on economic, administrative and political factors (Tripp *et al.*, 2006). A general prohibition on saving seed of protected cultivars is an unlikely strategy in most developing countries. Although coordinated systems of cleaning and dressing farmer-saved seed while collecting royalties have been successful in some Organisation for Economic Co-operation and Development (OECD) countries.

Crop cultivars present several important challenges for an IPR system (IBRD/World Bank, 2006). First, they are biological products that are easily reproduced and whose very use entails multiplication. Secondly, the users (and potential 'copiers') of the technology are millions of individual farmers whose compliance with any protection regime is difficult and expensive to monitor, particularly in developing countries. Thirdly, the agricultural sector involves cultural values and food security issues that in many countries affect the livelihoods and even potential survival of the rural poor, making the imposition of any controls a sensitive political issue. Fourthly, the inherent diversity of crop cultivars makes it difficult to apply the narrow technical criteria of novelty and reproducibility used in the conventional patent system, whereas the use of standard breeding methodologies may frustrate the application of the 'inventive step' criterion. Fifthly, the development of new crop cultivars has always relied to some extent on public research, partly in response to the traditional public goods nature of crop-related biodiversity. Thus, the application of IPR to the products of a publicly funded endeavour can be problematic. Sixthly, the increasing use of biotechnologies has brought additional challenges for the application of IPR in plant breeding.

Analysis of the historical seed-saving practices of soybean farmers in the USA indicates that large US farms have consistently saved seed – as much as 60% in some years. However, with the introduction of Roundup Ready® soybeans the nature of seed saving was drastically changed. The combination of an expanding array of IPR on technologies used in the development of new breeding materials, 'new' genetically modified (GM) technologies that in some countries have led to whole plant patents and the increasing application of industrial concepts to plant breeding has brought huge new private sector investments to plant breeding that are dramatically changing the nature of the business worldwide (Mascarenhas and Busch, 2006).

Not only do some countries allow the use of patents to protect plants, cultivars and genes, but the majority of the tools and processes of molecular biology and genetic transformation can be patented as well. Many of the biotechnology techniques, which are becoming increasingly important in conventional plant breeding, are also protected, thereby raising implications for the ownership of any cultivar resulting from their use. In addition, because biotechnology allows a much more precise understanding of the genetic make-up of any crop cultivar, it opens the door to sophisticated screening and reverse engineering techniques, which in turns offer new possibilities for utilizing protected cultivars, leading to pressure for more stringent protection.

Although a range of attempts have been made to provide a set of IPR for crop cultivars, only within the last few decades has a mechanism for PVP firmly taken hold in industrialized countries. International treaties such as UPOV and TRIPS, are attempting to establish common features of certain IPR although most developing country signatories are slow to ratify and implement these agreements through their national laws. One of the most controversial of these features is contained in Article 27.3(b) of the TRIPS Agreement, which requires all WTO member states to provide IP protection, through patents or an effective *sui generis* system or both, for crop cultivars.

13.2 Plant Variety Protection: Needs and Impacts

13.2.1 Needs for protection of crop cultivars

The agriculture and biotechnology sector as a whole has become a huge business with very large research and development (R&D) investments. For example, Syngenta has total sales in these areas of US$6340 million in 2004, while its total R&D investment was US$1738 billion, which means about 27.4% of the total sale was used for R&D. This compares with about 10% of profits reinvested into research for other crop science companies such as Monsanto (9.4%), BASF (8%), Pioneer Hi-Bred (10.9%), Bayer CropScience (11.4%) and Dow AgroSciences (9.9%). This R&D investment in 2004 ranged from US$350 million (Dow AgroSciences) to US$926M (Bayer CropScience), compared to US$428 million for the entire multidisciplinary R&D budget across all Consultative Group on International Agricultural Research (CGIAR) research centres in 2004, of which only 5–10% is spent on biotechnology (Spielman *et al.*, 2006).

Clearly, the CGIAR and many other public sector research organizations have little hope of competing with the private-sector agrobiotechnology investments. Instead, the public sector should strengthen their IP management capabilities in order to establish effective private sector partnerships that capitalize on these private sector investments and generate synergy with their own product development programmes. Seeking patents is the most frequently used approach in IP protection. For example, Pioneer alone submitted 463 applications to the USA and 176 to Australia in fiscal year 1999–2000. Considering all submitted patent applications, the percentage of patent applications that were granted by the end of 2004 varied greatly by country: the USA granted 81.4% of the 1040 applications while Australia approved about 30% of the 399 applications (Chan, 2006).

Plant breeding and crop biotechnology product development is a long-term process. It usually takes 7–12 years to develop a product from concept initiation to delivering a product to the market (Fig. 13.1), where this involves MAS or use of a transgene in cultivar development. Initial advances in biotechnology were heralded as offering substantial savings in time to bring new products to market. However, as the discipline has matured, the added value of the products possible through use of biotechnology tools has become increasingly important. Thus, the creation of a new crop cultivar continues to require a substantial investment in terms of skills, time, labour, resources and finances. The value of biotechnology investments for commercial agriculture, e.g. in OECD countries, now seems to be well established. However, the great remaining challenge is to apply these advances to the development of the new crop cultivars that are an essential tool for the improvements in sustainable agriculture and food security in developing countries. A new cultivar, once released to the market place, can in many cases be readily reproduced by others thereby depriving the original breeders of recovering the full profit on the investment. The granting of exclusive rights to a breeder of a new cultivar encourages future investments in plant breeding while the ability for other breeders to use that new cultivar in their breeding programmes contributes to the development of agriculture, horticulture and forestry. However, as companies make increasing investments in specific traits or processes there is an increasing pressure for them to protect these investments through patents.

Improved cultivars developed by breeders routinely replace old cultivars because they provide higher yield, better quality and stronger adaptability to ever-changing environments and market demands. The new cultivar is available on the market at a cost that is broadly similar to the cost of the cultivar(s) from the previous generation

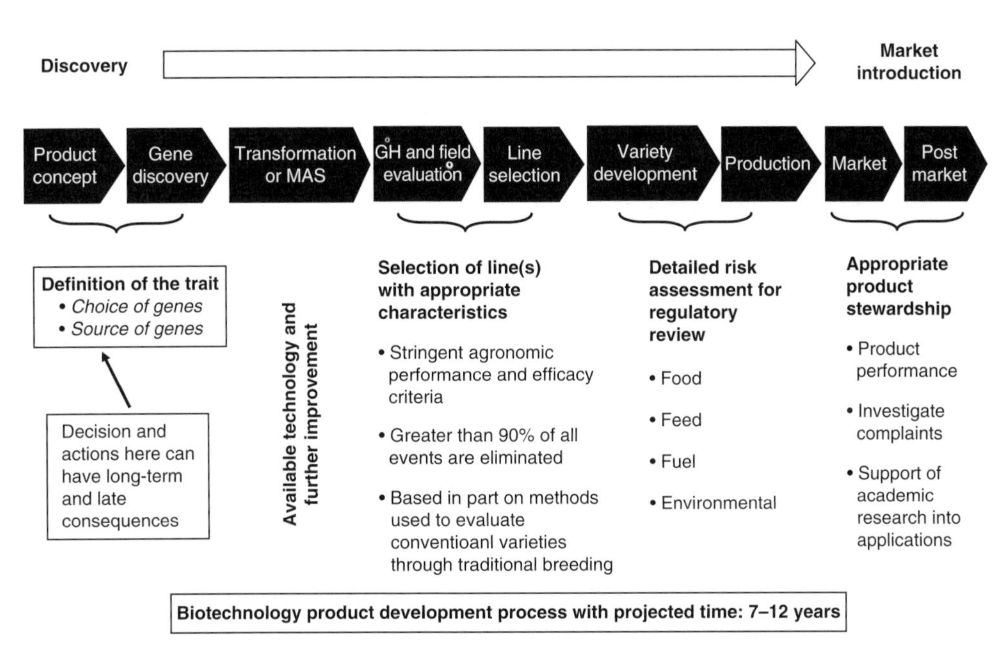

Fig. 13.1. Steps involved in crop biotechnology product development using transformation or marker-assisted selection (MAS). GH, greenhouse.

and thus the marginal cost to the farmer of shifting cultivar is low. The experience of Argentina shows that the advent of protected cultivars did not increase the price of the seed: breeders and seed producers, to remain competitive on the seed market with unprotected cultivars, rationalized the production of and trade with their seed and the royalty was taken from the savings. However, the situation is not always comparable in developing countries where many farmers may be currently obtaining seed through informal systems and are thus faced with significant initial investment requirements in order to shift to a new cultivar.

Many countries have strong investments in public research of relevance to plant breeders. But experience shows that this approach is not enough: public funds cannot adequately cover the needs of every crop, every agroclimatic zone, every market preference, etc. In addition, the interaction between strategic research and product development, between public plant breeding and private-sector product deployment, is frequently deficient (Heitz, 1998). The recognition of these difficulties is one of the main reasons for the interest shown by many developing countries in PVP: PVP is therefore an indispensable element of the new seed policies that give an important role to both public and private-sector plant breeding.

The exercise of the breeders' rights

The way in which plant breeders choose to exercise their right depends upon many factors and the scope of the protection conferred to them is only one of the options on hand. The breeder of seed crops will seek to organize the production of commercial (certified) seed in a fairly loose manner and seek to collect a royalty at each multiplication stage (to spread the risks); he will apply a very open licence policy. The breeder of an ornamental plant will seek to organize the production and sale of cut flowers and not just propagating material.

Heitz (1998) gave two reasons why economic theories and constructions based upon the notion of 'monopoly' are totally inappropriate in the case of crop cultivars: (i) breeders are bound to associate with others – in effect partners – to exploit their cultivars; the success of a particular cultivar – and of the commercial strategy of its breeder – is the result of many individual decisions; and (ii) the breeder of protected cultivars is almost always bound to compete with other breeders and their cultivars. Another relevant factor is the existence and scope of a 'farmer's privilege' (the need to make commercial seed competitive with farm-saved seed).

The derived benefits

There are three main reasons for national agricultural research institutes (NARIs) to embrace IPR: recognition, technology access and transfer and revenue (Louwaars *et al.*, 2006). In commercial breeding, the last reason prevails; IPR create additional value for the crop cultivar by providing a legal basis for licence contracts between the breeder and seed producers, which commonly includes a royalty payment that serves as an important tool to recoup the investment in research. In public research, however, cultivar development is funded from public sources and research managers tend to put some emphasis on other research objectives. IPR formally link the cultivar to the institute and individual breeders. Furthermore, IPR may facilitate seed production when only an exclusive market will entice an individual seed producer to take a new cultivar into its product range (to facilitate technology transfer) and technology may be more easily acquired if patents can be traded.

The direct effect of PVP is to promote plant breeding. All countries which have become a member of UPOV and whose agricultural sector is of a size that justifies investments in plant breeding have reported increases in the volume of plant breeding activities in developing countries with direct effects on their agriculture. However, the merges and acquisitions keep reducing the number of players in the seed sector in developed countries. The others have reported increases in the assortment of cultivars made available by foreign breeders. Given the declining public funding of agricultural research in many countries, revenue generation is an attractive option for many public institutions. Income from IPR can support the institution to cover operational costs or hire additional staff and provides managers with a financial tool to support particularly innovative researchers or research groups. Public cultivars can generate a ready income, especially if cultivars bred in the past can be protected.

13.2.2 Impacts of plant variety protection

Plant breeding

MAINTENANCE BREEDING. The PVP system is not only there to encourage creative plant breeding activities. The full benefit of a new and improved cultivar can only be drawn if, first of all, the cultivar is properly maintained and, secondly, if authentic propagating material of the cultivar is made available to users. The PVP system ensures that the breeder has a lasting interest in ensuring these activities (at least as long as the cultivar is commercially successful).

GENETIC DIVERSITY. The increased number of breeding programmes which enter into competition, if this happens, implies a diversification of the programmes that result an increasing probability of obtaining superior and genetically diverse cultivars. This scenario provides a strong counterbalance to the trend for uniformity that may be generated by the market demand in products (Heitz, 1998). However, the trend in the seed market for a few to be private providers of bred-seed worldwide contributes to a few crop cultivars that dominate the market, reducing genetic diversity. The pressure on the natural ecosystems can be lessened through: (i) providing uniformity for use of a cultivar in single fields (to allow rational and efficient production) but diversity for use across fields; and (ii) contributing to the widespread use of a relatively narrow, but improved, gene pool (to maximize agricultural production) and to the preservation of the large gene pool that serves as a genetic reservoir for further progress.

PUBLIC RESEARCH. PVP also benefits public breeding. Public institutions use the system to generate income and optimize the exploitation of their cultivars. The income can be used as an argument to resist cutbacks in their programmes. In addition, PVP helps to organize optimal distribution of the tasks between the various partners, for example, by sharing the tasks in an organized manner and handling the competition between public and private breeding.

UPOV (2005) provided a report that stated the introduction of the UPOV system of PVP and membership of the UPOV can open a door to economic development, particularly in the rural sector as shown by their individual research in Argentina, China, Kenya, Poland and the Republic of Korea. This demonstrated that PVP can produce benefits in a range of ways that differ from country to country reflecting the specific circumstances of each. Hence, a key conclusion is that the UPOV system of PVP provides an effective incentive for plant breeding in many different situations and in various sectors, resulting in the development of new, improved cultivars of benefit to farmers, growers and consumers.

Agricultural production and trade

Plant breeders draw their revenue, at least in the case of the major crops, from the trade in seeds and plant material by choosing serious partners and driving less serious people out of the market. In many countries, breeders have created licensing societies, so that seed production is also organized on a fairly uniform basis across cultivars and species (Heitz, 1998). Seed markets respond to the changes because of the involvement of plant breeders and their products, cultivars that are protected by IPR. As a primary user, the farmer does not just receive genetic potential in a high quality seed. In the highly competitive markets, breeders also offer fringe benefits, for example in the form of crop insurance. They undertake extension activities, in particular comparative trials and provide farmers with detailed technical information on crop management.

International trade becomes even more important with the extension of international trade and the reinforcement of IPR. For example, a producer of a tropical fruit may be refused access to the consumer market on the basis of the breeder's right granted in the country concerned. Conversely, the producer operating on the basis of a licence concluded under the breeder's right valid in the production country will secure, in the licence agreement, the right to enter the consumer market (Heitz, 1998). Subsidies, as measures taken by national governments to protect local farmers from market forces or to ensure abundant availability of agricultural products, will affect the trade of other nations by artificially interfering with the global markets. The issue of PVP generally mirrors the debate on trade barriers created from lack of IP rights. That is, lack of IPR for innovations in plant breeding results in depriving the property holder of rightful royalties to which they would be entitled to otherwise (Ragavan, 2006). The TRIPS Agreement was signed in 1994 to address the wider issue of facilitating international trade that was affected by the lack of IP protection in some countries. While introducing PBR provides one facet of the analysis concerning the reduction of distortions to trade, the discussion on subsidies provides a contextual understanding of all barriers, including those that are unrelated to IP, affecting agricultural trade.

Transfer of technology and know-how

PVP plays an essential role in the introduction of foreign cultivars and other novel genetic materials (and the associated technology) that enriches the assortment available to the farmer. It also contributes to the improvement of agricultural production, both in permitting such introduction and in speeding up the introduction.

Foreign plant breeders also take a more direct interest in breeding activities in specific regions, in particular by breeding plants for a specific environment so that the resulting cultivars grow better in that environment. This is done both through the organization of subsidiaries and through partnership or licensing agreements. In both cases there is a flow of technology and know-how, in both directions, as subsidiaries have to rely on the local seed trade.

Breeding strategies

Introducing the concept of revenue generation in public plant breeding is likely to have an impact on the distribution of funds within the NARI and on the breeding strategies applied. Louwaars *et al.* (2006) discussed the impact of IPR on plant breeding strategies which is summarized as follows.

First, IPR can be generated in plant breeding relatively easily compared with other agricultural research undertakings. As a result, the pursuit of revenue could lead to important disciplines such as soil science, socio-economics and plant pathology being marginalized or downgraded to supporting only breeding efforts.

A second possible impact is that funds will be distributed more to crops with a high value in seed production. These include, in general, crops that are produced for the market (where investment in seed is common), which are difficult to reproduce on-farm (e.g. cross-pollinated crops) and that have a low seed rate. In practical terms, this means that maize-breeding programmes will get priority over those for open-pollinated small grains, most pulses and root crops. The latter crops, however, may be important for the nutrition security of most of the population.

The third level of impact is within breeding programmes themselves, where researchers have to choose which ecological areas or client groups to target. Revenue generation will focus breeding on commercial farmers and hybrids rather than on resource-poor farmers who cannot afford to buy hybrid seed and they have to use open-pollinated cultivars instead. In the latter situation, the seed industry is unlikely to generate profits and pay royalties to the breeder.

The shift to commercial crops and farmers may be consistent with recent changes in national agricultural policy and trends of commercialization of public entities. In some countries, however, the public task of a NARI is to support both equity and national agricultural production. The trend towards crop diversification and breeding for low-input agriculture, which means better yield stability, may be reversed when NARIs focus on using IPR for revenue generation. Another strategy of a NARI may be to secure a choice of cultivars for farmers in a market that may otherwise be dominated by large commercial companies owing to IPR. However, this latter option also may shift research priorities away from smallholder farmers' needs (Louwaars *et al.*, 2006). Policy makers and research managers need to carefully consider the impact of the use of IPR in public breeding before including protection in their research strategies. If national public organizations are not supposed to protect their inventions, governments will have to provide the necessary funds for their research.

NARI organizations

Louwaars *et al.* (2006) discussed how PVP-related issues would greatly affect the NARI organizations. When a NARI intends to commercialize its cultivars using IPR, it must realize that the right holders are responsible for implementing their rights and that the NARI needs capacities to design commercialization strategies and licence contracts, as well as to follow up on these contracts. In addition, research managers have to be aware that there are many costs involved in IP protection such as those for additional personnel, IPR acquisition, implementation, application and maintenance fees. Commercial decisions have to be made on which rights to apply for and when to surrender them. A significant cost can arise when the rights have to be defended, especially against experienced negotiators of commercial companies with significant resources.

While crop cultivars are in almost all cases freely used as parents in further breeding, this is not the case in patented biotechnologies. Hence, NARIs will need to develop ways and means to observe rights on technologies and materials that they use in breeding. Most countries have a fairly liberal 'research exemption' in their patent laws. This situation commonly leads to a licence contract in which the patent holder can specify the uses, the ways of commercialization and benefit sharing (royalty payment) (Louwaars *et al.*, 2006). A NARI needs therefore to identify possible risks associated with the use of patented technologies. An IP plan needs to be developed for each project, in which it is decided when and how contact will be established with the technology provider.

The introduction of IPR brings new tasks and responsibilities to the NARI. It requires not just access to lawyers, IP specialists, negotiators and marketers, but more importantly, it calls for a shift in 'culture' among the researchers. All researchers will have to be aware of the potential impact of IPR on their work, when they commonly prefer to concentrate on their own science and not be bothered by 'administrative rules'. Senior management will have to lead the way in this gradual shift, assisted by well-designed capacity-building initiatives and support systems.

International agricultural research

The same considerations are important in international agricultural research (Louwaars *et al.*, 2006). Strategies for protecting inventions by the International Agricultural Research Centres (IARCs) concentrate on the technology transfer argument on the one hand and the original objective to develop international public goods on the other. Several IARCs are developing agribusiness parks or other mechanisms to link them directly with the private sector to provide additional routes for technology transfer. There are, however, also cases in which an IARC may obtain research funds for developing cultivars jointly with the private sector or obtain royalties on the commercialization of such cultivars.

Another challenge IARCs face is to get access to protected technologies without diminishing their primary task of poverty alleviation. Materials and tools may not be used in research if the products cannot be made available to the target groups (i.e. the 'resource poor') without restrictions. Humanitarian licences and cooperation agreements should at least contain such provisions.

A less debated result of the spread of IPR on IARCs is the impact of the commercialization of some NARIs on the capabilities of IARCs to reach the resource poor (Louwaars *et al.*, 2006). A NARI that will concentrate their strategy on revenue generation through IPR and thus move away from producing solutions for resource-poor farmers in favour of commercial production may not always be suitable partners of IARCs for reaching the poor. The latter may need to look for other ways, for example, through non-governmental organizations and in some cases, direct contacts with seed producers. All the IARCs have IPR policies, although most of these are still subject to adjustment and elaboration. The increased use of IPR has caused IARCs to re-evaluate their modes of interaction with both NARI organizations and seed companies. Various approaches have been taken to ensure that NARI germplasm reaches the farmers for whom it is intended.

13.3 International Agreements Affecting Plant Breeding

The international agreements related to regulatory systems that affect plant breeding include the TRIPS, the Convention on Biological Diversity (CBD), the International Treaty on Plant Genetic Resources for Food and Agriculture and the discussions in the Intergovernmental Committee on Intellectual Property and Genetic Resources, Traditional Knowledge and Folklore of the World Intellectual Property Organization (WIPO). In addition, the development of a Substantive Patent Law Treaty (SPLT) as discussed within WIPO is likely to reduce the current flexibility to protect plant cultivars (Wallø Tvet, 2005). These agreements that affect plant breeding are shown in Fig. 13.2.

13.3.1 The UPOV Convention and UPOV

After decades of attempts to obtain patent protection for their achievements, plant breeders, together with a segment of the IP specialists, requested that consideration be given to a specially designed protection system. The request was taken up by the French government, through the conferences and meetings it hosted between 1957 and 1961, leading to the signing on 2 December 1961 of the International Convention for the Protection of New Varieties of Plants (also known as the UPOV Convention).

The UPOV system – revised in 1972, 1978 and 1991 – has gradually strengthened the rights of plant breeders. The last revision, 30 years after the initial adoption, was substantial. The revisions were made: (i) to clarify certain provisions in the light of the experience of the UPOV member states in operating the Convention since 1961; (ii) to strengthen the protection offered to plant breeders in certain specific ways; and (iii) to reflect technological changes. The rights defined under UPOV are known as 'plant variety protection' (PVP). The UPOV system is considered as the most straightforward choice for countries wishing to comply with the TRIPS Agreement. The UPOV Convention is the only model for a PVP system. It is not only an IP treaty, but also an instrument in the field of agricultural policies.

The breeder is defined in the 1991 Act of the UPOV Convention as the 'person who bred, or discovered and developed, a variety' (cultivar in this book). Protection has thus to be afforded not only where a cultivar has originated from 'breeding' in the somewhat restricted sense of crossing parent plants and selecting from within the progeny, but also where a person identifies a mutation or a variation, of known or

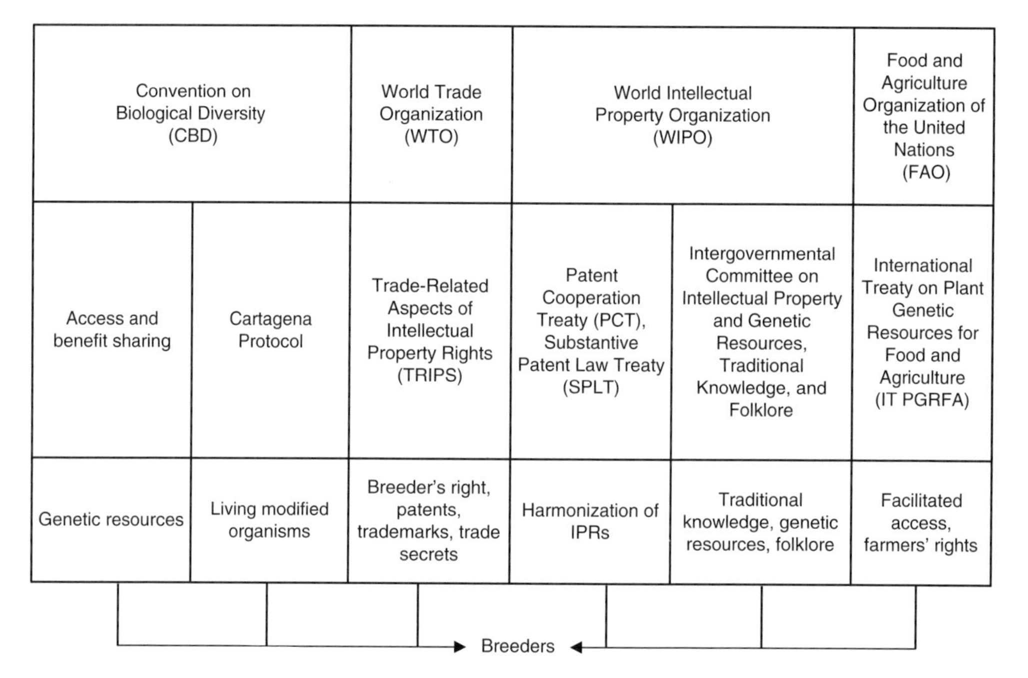

Fig. 13.2. International agreements that affect plant breeding. From IBRD/World Bank (2006) © The World Bank 2006.

unknown origin, in existing plant material and ensures that the mutation or variation is isolated and propagated as a new cultivar (Heitz, 1998).

The UPOV Convention has been very successful in impressing upon the crop cultivars and seed sector, in particular in the UPOV member states, a notion of variety (cultivar) that, from the technical point of view, is identical with 'protectable variety'. According to Article 1(vi) of the 1991 Act of the UPOV Convention, a 'variety' basically is a plant grouping that meets the conditions of distinctness, uniformity and stability, but not necessarily to the degree required for protection.

Distinctness, uniformity and stability (DUS)

There are five required conditions for protection.

1. Novelty – The cultivar to be protected must not have been the subject of commercial acts before certain dates determined on the basis of the date of application.
2. Distinctness – (Article 7): 'The variety shall be deemed to be distinct if it is clearly distinguishable from any other variety whose existence is a matter of common knowledge at the time of the filing of the application.' Distinctness is established on the basis of individual characteristics (descriptors in genetic resources parlance) that are botanical in nature and are not necessarily related to the agricultural or technological properties or value of the cultivar.
3. Uniformity (or homogeneity) – (Article 8): 'The variety shall be deemed to be uniform if, subject to the variation that may be expected from the particular features of its propagation, it is sufficiently uniform in its relevant characteristics.'
4. Stability – (Article 9): 'The variety shall be deemed to be stable if its relevant characteristics remain unchanged after repeated propagation or, in the case of a particular cycle of propagation, at the end of each such cycle.'
5. Denomination – The cultivar must be given a denomination under which it will be commercialized.

UPOV provides protocols for assessing and describing the unique characteristics of a new cultivar, ensuring that it is distinct, uniform and stable (DUS). These standards are adapted to the mode of reproduction of the protected species: cross-fertilizing crops admit a wider tolerance than the relatively strict requirements for uniformity in vegetatively propagated crops. Any cultivar that fulfils the DUS criteria and that is 'new' (in the market) is eligible for protection and there is no need to demonstrate an inventive step or industrial application, as required under a patent regime. A DUS examination involves growing the candidate cultivar together with the most similar cultivars as per common knowledge, usually for at least two seasons and recording a comprehensive set of morphological (and in some cases agronomic) descriptors (IBRD/World Bank, 2006).

Characteristics are used to assess DUS and include descriptors such as flower colour, or leaf shape. A characteristic must meet a number of basic requirements for it to be used for DUS testing or for producing a cultivar description. The characteristic must:

1. Result from a given genotype or combination of genotypes.
2. Be sufficiently consistent and repeatable in a particular environment.
3. Exhibit sufficient variation between cultivars to be able to establish distinctness.
4. Be capable of precise definition and recognition.
5. Allow uniformity and stability requirements to be fulfilled.

Characteristics may have direct commercial relevance or no commercial relevance. For example, using the criteria above may eliminate some commercially important traits, for example, yield. Chemical constituents may be acceptable characteristics, provided they meet the criteria. It is important that characteristics based on chemical constituents be well defined and supported by an appropriate method for examination.

UPOV test guidelines have been developed for individual species or cultivar groupings to provide guidance related to growing cycles, number of plants, material to be tested, or characteristics to be examined. The DUS test may be undertaken directly by the authority of the UPOV member, by a party designated by the authority (e.g. an institute, the breeder), or the authority may take into account the results from previous tests or trials conducted by, for example, other UPOV members. There can be therefore a high level of cooperation in DUS testing, including, for example, the purchase of DUS test reports, bilateral arrangements to avoid duplication of testing and centralized DUS testing at regional or global levels. Cooperation between authorities can minimize the time for DUS testing, minimize costs and optimize examination of characteristics in growing trials.

Essentially derived varieties/cultivars

Under the 1978 Act of the UPOV Convention, any protected cultivar may be freely used as a source of initial variation to develop further cultivars. Any such cultivar may itself be protected and, what is more important, exploited without any obligation on the part of its breeder and users towards the breeder of the cultivar that was used as a source of initial variation. These rules have with certain exceptions worked well in practice and have been reaffirmed in the 1991 Act.

However, the rules did not prevent a person finding a mutation within a crop cultivar (such mutations are for a few traits in some species), or selecting some other minor variant from within a cultivar, from exploiting the mutant or variant with no authorization from, or recognition of the contribution of, the original breeder to the final result. The lack of recognition of that contribution in such circumstances was generally considered to be improper (Heitz, 1998). Modern biotechnology has greatly increased the likelihood of such situations; it may take 12 years to develop a new cultivar but a mere 3 months to modify it by adding a transgene or genes introduced through genetic engineering in the laboratory.

This situation indeed can be a disincentive to the continued pursuit of 'classical' plant breeding (and also of genetic engineering since it suffices to add yet another gene to escape the protection of the cultivar taken as host for that gene). The concept of essentially derived variety (EDV) embodied in Article 14(5) of the 1991 Act of the UPOV Convention is designed to ensure that the Convention continues to provide an adequate incentive for plant breeding. Under that Article, a cultivar that is essentially derived from a protected cultivar may be the subject of protection (if it fulfils the normal protection criteria of DUS and novelty), but cannot be exploited without the authorization of the breeder of the protected cultivar. For practical purposes, cultivars will only be essentially derived when they are developed in such a way that they retain virtually the whole genetic structure of the earlier variety.

Farmer's privilege

The most prominent issues in the *sui generis* systems involve the so-called 'farmer's privilege' and 'breeder's exemption'. The traditional right of farmers to save seed from their harvests to plant the following season is an important aspect of *sui generis* systems and is one of the most contentious aspects of IPR in plant breeding. Although this practice is often described as a 'farmer's right', it is referred to here by the UPOV term of 'farmer's privilege' to distinguish it from the broader concept of 'farmers' rights'.

The 1978 UPOV Convention assumed that farmers were permitted to save and reuse seed of protected cultivars as part of 'private and non-commercial use'. However, Article 15(2) of the 1991 UPOV Convention rules that on-farm seed saving is not permitted without the consent of the breeder, although it allows member states to specify crops for which the use of farm-saved seed is permitted, 'taking into account the legitimate interests of the breeder'. In the European Union (EU), this provision is interpreted as the right of smallholder farmers to save seed for specific crops and the right of the breeder to collect royalties on farm-saved seed used on larger farms. The 1991 Convention also prohibits any transfer of seed of protected cultivars (through sale, barter or gift) between farmers. Utility patents on plant cultivars are even more rigid and a patented cultivar normally cannot be saved for subsequent use as seed on the farm or traded or exchanged with other farmers.

Various interpretations of farmer's privilege have favoured the adoption of laws based on the more liberal 1978 UPOV Convention in many developing countries. In most cases in these countries, restrictions on saving seed of food crops on the farm are neither administratively feasible nor politically acceptable. Making the transfer of seed from farmer to farmer illegal is widely considered incompatible with the traditions of small-scale farming.

The issue of seed saving is a good example of how IPR in plant breeding must be tailored to the conditions of national seed systems. Even within a single country, the requirements and conditions of different crop production systems are not uniform and countries could consider legal options that address this variability. For instance, earlier seed law in the Netherlands included severe restrictions on saving planting materials for ornamental crops, while field crops were regulated on the basis of the more liberal UPOV 1978 Convention. Many vegetatively propagated commercial flower species can be multiplied very rapidly by farmers, which would considerably reduce revenues for breeders and provide inadequate incentives for innovation in a sector that is very important for Dutch agriculture. Thus an amendment to the law made the farm-level propagation of such species illegal. This example emphasizes that countries need to design appropriate levels of protection for different types of commodities, in accord with the domestic agricultural economy and plant breeding capacities.

The breeders' rights

A completely new approach taken in the 1991 Act defines the scope of the breeders' rights. The basic right pertains to seven acts.

1. Production or reproduction (multiplication).
2. Conditioning for the purpose of propagation.
3. Offering for sale.
4. Selling or other marketing.
5. Exporting.
6. Importing.
7. Stocking for any of these purposes.

The purpose of this more detailed enumeration is not so much to give a more extensive right to the breeder, than to give him or her a more effective one. Conditioning, for instance, is essentially one (technical) step of seed or plant production.

The right applies to two classes of material to which such acts must relate and one class to which they may relate: (i) the propagating material; (ii) the harvested material (including whole plants and parts of plants), provided this has been obtained through the unauthorized use of propagating material and that the breeder has had no reasonable opportunity to exercise his or her right in relation to the propagating material; and (iii) optionally (at the discretion of the member state), products made directly from harvested material, provided this has been obtained through the unauthorized use of harvested material and that breeders have had no reasonable opportunity to exercise their rights in relation to the harvested material.

In the case under (ii), for instance, the breeder gets a more extensive right in relation to certain imported harvested material. Where harvested material has been produced with illegal seed, he or she now has another opportunity to exercise their right.

Furthermore, the 1991 Act specifies four subject matters to which the breeder's right extends:

1. The protected cultivar itself.
2. Cultivars that are not clearly distinguishable from the protected cultivar.
3. Cultivars that are essentially derived from the protected cultivar.
4. Cultivars whose production requires the repeated use of the protected cultivar.

The addition of cultivars that are not clearly distinguishable is designed to make the breeder's right more effective; to prevent an infringer from claiming that he was not exploiting the protected cultivar, but a very similar one falling outside the 'protection perimeter'.

The 1991 Act establishes three compulsory exceptions to the breeder's right and one optional exception. The three compulsory exceptions are: (i) acts done privately and for non-commercial purposes (in particular the reproduction of a protected cultivar by a subsistence farmer or by an amateur gardener); (ii) acts done for experimental purposes; and (iii) acts done for the purpose of breeding other cultivars and (provided protection has not been specifically extended to them, as for instance in the case of an EDV) for the purpose of exploiting such other cultivars.

The optional exception relates to farm-saved seed. States that are party to the 1991 Act of the UPOV Convention may exempt farm-saved seed from the breeder's right, within reasonable limits and subject to safeguarding the legitimate interests of the breeder. Each member state will exercise this option in the light of its own national conditions. Some states have chosen to give farmers an unconditional right to replant seed from their previous harvest while others have limited this right to certain crops or to small farmers.

Breeder's exemption

As plant breeding is generally considered as incremental, breeders have built on existing cultivars to develop improved ones. To make progress, contrary to the situation in mechanics or chemistry, the description of the invention is not enough, as it is not to rebuild a whole genome starting from nucleotides. That is why the UPOV Convention included an exception to breeders' rights: 'The utilisation [by others] of the [protected] new cultivar as an initial source of variation for the purpose of creating other new cultivars and the marketing of such cultivars' (Art. 5.3 of the 1961 Act). This exception, widely known as the breeder's exemption, has been one of the engines of the breeding industry since the late 1960s. It stems from the traditionally unrestricted use of seed by farmers and breeders. It provides that any person is allowed to use a protected cultivar for further breeding without requiring the consent of the rights holder.

It is seen as a way of promoting the development of the best cultivars for farmers, limiting the development of long-term commercial advantages, improving opportunities for smaller breeding companies and thus promoting competition in the sector. Unlike the farmer's privilege, the breeder's exemption has not dramatically changed in later UPOV Conventions, prompting some companies in the USA to look to the patent system for protecting their germplasm. The only modification in the 1991 Convention is the limitation on EDVs, which may fall under the rights of the original breeder.

13.3.2 The 1983 International Undertaking on Plant Genetic Resources

In 1983, the Food and Agriculture Organization of the United Nations (FAO) established a Commission on Plant Genetic Resources (later renamed the Commission on Genetic Resources), the first permanent intergovernmental forum devoted to germplasm conservation and development. The Commission's first major action was to adopt a non-binding resolution known as the International Undertaking on Plant Genetic Resources (hereafter 'the Undertaking'), which is based on the principle that 'plant genetic resources are a common heritage of mankind to be preserved and to be freely available for use, for the benefit of present and future generations.' The purpose of 'the Undertaking' is to ensure that genetic resources will be explored, preserved, evaluated and made available for breeding and science. It is based on the following underlying principles:

- Genetic resources are a heritage of humanity and should be available without restriction.
- Establishes 'farmers' rights': farmers should be compensated for development and conservation of genetic resources.
- Sovereign rights of nations to preserve, protect and be compensated for innovative utilization of their native genetic resources.

Article 5 of 'the Undertaking' (Availability of Plant Genetic Resources) provides as follows:

> It will be the policy of adhering Governments and institutions having plant genetic resources under their control to allow access to samples of such resources and to permit their export, where the resources have been requested for the purposes of scientific research, plant breeding or genetic resources conservation. The samples should be made available free of charge, on the basis of mutual exchange, or on mutually agreed terms.

13.3.3 The 1992 Convention on Biological Diversity

In 1992, in Rio de Janeiro, the United Nations hosted an Earth Summit to consider the state of the world's environment. In addition to producing a number of non-binding declarations of international environmental policy, the Earth Summit gave birth to the Convention on Biological Diversity (CBD). The specific concern of the CBD is biological diversity, which the convention defines as 'the variability among living organisms from all sources including diversity within species, between species and of ecosystems'. The CBD has the following objectives:

- to establish conserving biological diversity as an international priority;
- to promote fair and equitable sharing of benefits from genetic resources;
- to maintain appropriate access and transfer of relevant technology among countries;
- to reaffirm sovereign rights of states over natural resources, including genetic resources; and
- to promote international agreements, efforts in technology transfer, licensing, protection, sharing of R&D, cooperative training.

The CBD marked the end of the 'common heritage of mankind' conception of genetic resources. The CBD does not refer to a 'common heritage' and its preamble states only that conservation of biodiversity is a 'common concern of humankind'. The CBD is based on a new set of principles:

> *Affirming* that the conservation of biological diversity is a common concern of humankind,
>
> *Reaffirming* that States have sovereign rights over their own biological resources,
>
> *Reaffirming* also that States are responsible for conserving their biological diversity and for using their biological resources in a sustainable manner, [...]

The CBD must be viewed as a framework Convention that needs implementing measures. The analysis of the true scope of the obligations under the CBD is very difficult for its text is cluttered with limitations such as 'as far as possible and as appropriate' and 'subject to national legislation'.

It is nevertheless clear that there is no contradiction or conflict between the UPOV Convention and 'the Undertaking' under the FAO aegis, on the one hand and the CBD on the other. The measures that may be taken to implement the CBD, however, may create such a contradiction or conflict if they do not properly consider the prior legal instruments (and their objective background and rationale) and may even run counter to the CBD's stated objectives.

In particular, respect for IPR is expressly called for under Article 16(2), relating to access to and transfer of technology:

> Access to and transfer of technology... to developing countries shall be provided and/or facilitated under fair and most favourable terms, including on concessional and preferential terms where mutually agreed and, where necessary, in accordance with the financial mechanism established by Articles 20 and 21. In the case of technology subject to patents and other intellectual property rights, such access and transfer shall be provided on terms which recognize and are consistent with the adequate and effective protection of intellectual property rights.

With respect to the 'fair and equitable sharing of the benefits arising out of the utilization of genetic resources', it should also be obvious that it implies, first, the creation of benefits and, secondly, the identification of a person who would be called upon to share the benefits which he and his partners have created. All agreements that have been publicized so far – and follow the pattern created by the Merck-INBio agreement (http://www.american.edu/projects/mandala/TED/MERCK.HTM) – include as a major component the sharing of royalties derived from patents.

13.3.4 The 1994 TRIPS Agreement

The Uruguay Round of multilateral trade negotiations held under the framework of the General Agreement on Tariffs and Trade was concluded on 15 December 1993. The agreement embodying the results of those negotiations, the Agreement Establishing the World Trade Organization (WTO Agreement), was adopted on 15 April 1994, in Marrakech, Morocco.

The result of those negotiations, contained in an Annex to the WTO Agreement, was the Agreement on Trade-Related Aspects of Intellectual Property Rights (the TRIPS Agreement). The WTO Agreement, including the TRIPS Agreement (which is binding on all WTO members), came into force on 1 January 1995. The former agreement established a new organization, the World Trade Organization (WTO), which began its work on 1 January 1995.

The purpose and objective of the WTO Agreement is described in its preamble:

> *Recognizing* that their relations in the field of trade and economic endeavour should be conducted with a view to raising standards of living, ensuring full employment and a large and steadily growing volume of real income and effective demand and expanding the production of and trade in goods and services, while allowing for the optimal use of the world's resources in accordance with the objective of sustainable development, seeking both to protect and preserve the environment and to enhance the means for doing so in a manner consistent with their respective needs and concerns at different levels of economic development,
>
> *Recognizing* further that there is need for positive efforts designed to ensure that developing countries and especially the least developed among them, secure a share in the growth in international trade commensurate with the needs of their economic development,
>
> *Being desirous* of contributing to these objectives by entering into reciprocal and mutually advantageous arrangements directed to the substantial reduction of tariffs and other barriers to trade and to elimination of discriminatory treatment in international trade relations, [...]

Article 27.3 provides for an obligation to protect crop cultivars which became effective for developed countries on 1 January 1996 and became effective for developing countries on 1 January 2000 (1 January 2006 for least-developed countries):

> 3. Members may also exclude from patentability:
> [...]
> (b) plants and animals other than micro-organisms and essentially biological processes for the production of plants or animals other than non-biological and microbiological processes. However, Members shall provide for the protection of plant cultivars either by patents or by an effective *sui generis* system or by any combination thereof. The provisions of this subparagraph shall be reviewed four years after the date of entry into force of the WTO Agreement.

It is clear that WTO members enforced this obligation through the adoption of a *sui generis* protection system. At the Fourth Extraordinary Session of the FAO Commission on Genetic Resources for Food and Agriculture (Rome, 1–5 December 1997), the FAO Legal Adviser commented as follows:

> In fact, the concept of a *sui generis* system in the TRIPS Agreement is a very general concept that allows States to exercise ample discretion. The TRIPS Agreement does not give any direct indication on the elements or components that should be included in the *sui generis* system; nor does it require to follow the criteria of UPOV, which is already a *sui generis* system of plant cultivar protection although not the only possible one. Nevertheless, it is possible to infer, from the general context

of the TRIPS Agreement, some of the minimum requirements of the *sui generis* system, namely: (i) it should be, at least in the broad sense, a system to protect intellectual property rights; (ii) it should be applicable, in principle, to all traded plant cultivars; (iii) it should be effective, that is, enforceable; (iv) it should be non-discriminatory as regards the country of origin of the applicant (principle of national treatment); and (v) it should accord the most-favoured-nation treatment.

13.3.5 The 2001 International Treaty on Plant Genetic Resources for Food and Agriculture

On 3 November 2001 in Rome and after more than 15 sessions of the FAO Commission on Genetic Resources and its subsidiary bodies, representatives of 116 nations approved a new International Treaty on Plant Genetic Resources for Food and Agriculture (hereafter 'the Treaty'). The Treaty applies only to plant genetic resources useful for food and agriculture. It establishes the following objectives (Sullivan, 2004; Fowler and Lower, 2005):

- to encourage the conservation of plant genetic resources in order to preserve and enhance the genetic diversity of plant species and cultivars of value to food or agriculture;
- to provide a workable, juridical basis for rewarding farmers for their contribution in conserving, improving and making available plant genetic resources;
- to further develop the system of national sovereignty over plant genetic resources first established in the CBD, while ensuring that such exercise of sovereignty does not hinder international exchange of such resources; and
- to establish a multilateral system (MS) of access and benefit-sharing (ABS) that will coordinate exchanges of plant genetic resources and in some cases, require payments by persons or entities who commercially exploit such resources, to the nations from which such resources originated.

The MS of ABS applies to an initial Annex (1) of 35 food crops and 29 genera of forages, which include important staples such as wheat, rice, maize and potato. Collectively Annex 1 lists crops representing 80% of the world's calorie intake.

'The Treaty' is an ambiguous document. It seeks to vindicate the interests of parties that previously were underrepresented in international legal policy relating to plant genetic resources. At the same time, it seeks to assure industrial users of such resources that their economic interests will not be harmed. Although 'the Treaty' is more specific in some aspects than the CBD, its policies are stated broadly and often with significant practical detail (Sullivan, 2004). For these and other reasons, 'the Treaty' can be thought of as an international policy to be built for plant genetic resources. What that structure will look like when it is completed will depend upon a variety of political, economic and scientific influences that are already at work to shape the policies of the future.

'The Treaty' came into force on 29 June 2004, i.e. 90 days after 40 governments had ratified it. Governments that have ratified it will make up its Governing Body. At its first meeting, held in Madrid in June 2006, this Governing Body addressed important questions, such as the level, form and manner of monetary payments on commercialization, mechanisms to promote compliance with 'the Treaty', the funding strategy and an approved standard material transfer agreement (SMTA) for plant genetic resources for food and agriculture. Each country that ratifies will then develop the legislation and regulations it needs to implement 'the Treaty'.

Different sides in the MS bargained hard for an equal position before and after 'the Treaty'. On one hand, developed countries wanted 'the Treaty' to guarantee access to all crops. On the other hand, developing countries tend to believe that they were being exploited, as modern cultivars protected by IPR have been marketed to them at high prices when, in fact, these cultivars are based on genetic materials donated to developed countries' breeders. Developing countries also saw themselves as donors, not as recipients, of germplasm (Fowler and Lower, 2005).

A tremendous asset associated with 'the Treaty' is the genetic resources, mostly of the world's major food crops, that are held at the centres of the CGIAR. Historically, these have been considered as an international heritage and have been freely available to everyone, most recently under the terms of a formal agreement between FAO and the centres in which it is agreed that the centres are holding the materials 'in trust' for the benefit of the international community. The agreements signed by the centres with FAO on behalf of the Governing Body of 'the Treaty' on 16 October 2006 oblige the centres to deal differently with the plant genetic resources for food and agriculture (PGRFA) they hold and have brought under 'the Treaty', depending on whether or not the PGRFA is listed in Annex 1 of the Treaty. All transfers of PGRFA of crops listed in Annex 1 must be under the SMTA. It is assumed that the SMTA's prohibition against recipients acquiring IPR on the 'germplasm and related information' refers to the access and use of the material with few onerous restrictions. It therefore encourages use and development of the materials while keeping it available for use in the future by others (Fowler *et al.*, 2005).

13.4 Plant Variety Protection Strategies

A number of mechanisms are available to protect the interests of plant breeders and contribute to the development of a competitive and dynamic national seed sector. In addition to PVP (through the granting of plant breeders' rights) and patents, additional options include biological processes (such as the hybrid cultivar system), national seed laws, contract law, brand protection and other IPR (such as trademarks), as well as trade secrets. As with patents and PVP, the effectiveness of these alternatives depends on the local capacity for enforcement (IBRD/World Bank, 2006).

13.4.1 Plant variety protection or plant breeders' rights

UPOV is the most widely used system for PVP, currently with 63 member states (http://www.upov.int/). Most countries of the OECD and some developing countries are members of one of the UPOV conventions, although that is not the only *sui generis* option under the WTO's 1994 TRIPS Agreement. Countries wishing to join UPOV must present legislation compatible with the 1991 Convention. UPOV membership offers a number of advantages, including a source of technical backstopping for cultivar testing and the assurance of a PVP system recognized and respected by foreign investors. On the other hand, the 1991 Convention imposes potential restrictions on farmer seed management practices that may be politically unacceptable, a potential threat to food security and impossible to enforce in some circumstances. For these reasons some developing countries have declined to join UPOV. Only in specific cases where seed saving might threaten a market (e.g. export markets for flowers) or seed exchange would reduce incentives for plant breeding (e.g. informal seed sale by larger farmers or sales by grain merchants in competition with the commercial seed sector) would restrictions be justified in most developing countries (Tripp *et al.*, 2006).

The UPOV mission statement is 'to provide and promote an effective system of PVP, with the aim of encouraging the development of new cultivars of plants, for the benefit of society'. PVP provides the opportunity for breeders to gain a return on the investment made in breeding a new cultivar. For large-scale commercial farmers, market forces under UPOV schemes will generally lead to largely positive scenarios. However, the situation is very different and substantially more complex for farmers in developing countries. Cultivar protection systems may not be inappropriate in developing countries as long as resource-poor farmers continue to have choices through access to public cultivars or the right to save seed for their own purposes from commercial cultivars. PVP or PBR systems will benefit commercial farmers and agricultural productivity in all countries by stimulating private sector investment, increasing cultivar choice for farmers and facilitating technology transfer and agricultural development efforts, including the acquisition of crop biotechnology. However, for subsistence farmers, who are with different systems in developing countries, the benefits become complicated.

The PVP system has been largely beneficial for breeders and farmers in OECD countries over the past few decades. However, some believe that PVP rights are too burdensome to acquire in relation to the relatively limited protection they provide for returns on investment (Janis and Kesan, 2002). Naseem *et al.* (2005) investigated this issue for the case of cotton in the USA, first by examining trends in cotton cultivars planted and then by quantifying the effect of PVP cultivars on cotton yields. The analysis suggested that PVP had led to the development of more cultivars and that through these cultivars PVP had an overall impact on cotton yields.

Chiarrolla (2006) addressed the question of whether *sui generis* PVP legislation is becoming redundant due to the growing use of patents (to be discussed in the next section) for the protection of plant-related inventions. However, to be a fully functional standalone system, there would need to be modifications to the patent system in order to prevent agricultural exemptions, enjoyed by plant breeders and farmers under *sui generis* PVP systems, from being overridden by patent claims, particularly those related to entire plant cultivars. There is a danger that if *sui generis* PVP regimes continue to focus on broad societal objectives and promoting sustainable agriculture that a two tiered system will emerge where PVP-only cultivars fall behind in the private-sector-funded technology race. Alternatively the IPR system could be refined to support the needs of commercialization, research, further breeding and developing country agriculture. Important issues of global social relevance include the diversification of cropping systems, the promotion of development and commercialization of under-utilized crops and species, the development of new markets for local cultivars and maintenance of 'diversity-rich' products.

13.4.2 Patents

A patent is a legal right, granted by a government to the original and first discoverer or 'inventor' of a new IP, to exclude others from making, using or selling the subject IP 'invention' for a defined period of time. A patent is allowed by the grantor only if the claimed IP is deemed useful, novel and unobvious to others 'skilled in the art' (Boyd, 1996). A patent application is a written document which must fully disclose and describe the claimed invention in sufficient detail and completeness to allow 'one skilled in the art' to use or practise the invention. Patents are granted for inventions that are new, involve a creative step and can be applied in industry. Since the late 1980s, private companies, universities and federal governments all increased patenting in agricultural biotechnology particularly rapidly and they now hold a greater proportion of agricultural biotechnology patents than they do of patents in general. Private companies tend to dominate patenting in plant technologies and molecular level agricultural biotechnology. As Heisey *et al.* (2005) indicated, differences in patterns of patent production suggest not only differences in agricultural research investment but also differences in motivations for patenting.

PVP through the utility patent system is provided only in a few countries (for example, the USA, Australia and Japan). The US Utility Patent law designates four broad categories of patentable subject matter: composition, machines, articles of manufacture and processes. Plants and biological subject matter are not explicitly included. However, in 1980, the Supreme Court decision in *Diamond* v. *Chakrabarty*, construed section 101 to encompass genetically modified organisms (GMOs). This case undoubtedly helped to open the door for ensuing patents for genetically engineered biological materials and plant/plant cultivars. Only plant cultivars invented or discovered 'in a cultivated area' are eligible for patents, thus limiting the possibility of patents on wild relatives. Under the 1978 UPOV Convention, a cultivar could not be protected by both a patent and PVP, but the 1991 Convention allows this double protection. Table 13.1 provides a comparison of three major IP systems for plant cultivars. In Japan, the patent system is used only for plant cultivars that are considered innovative and not merely a product of normal plant breeding.

Utility patents for plant cultivars are not considered a reasonable option for developing country IPR systems (IBRD/World Bank, 2006). Nevertheless, aspects of patents for plant cultivars become increasingly important because of the pressure from some parts of the seed industry to move in this direction and because this option is included in some of the bilateral trade negotiations between the USA and several Latin American countries.

Patent protection first became available in 1985 and companies used both PVP and patent systems for some years; recently a

Table 13.1. Comparison of major intellectual property systems for plant cultivars (varieties) (from IBRD/World Bank (2006) © the World Bank).

Criterion	UPOV 1978	UPOV 1991	Utility patents (USA)
Protection	Varieties of species or genera as listed	Varieties of all genera and species	Sexually reproduced plants (and genes, tools, methods to produce varieties)
Exclusion	Unlisted species	None	First-generation hybrids, uncultivated varieties
Requirements	Novelty (in trade) Distinctness Uniformity Stability	Novelty (in trade) Distinctness Uniformity Stability	Novelty (in public knowledge) Utility Non-obviousness Industrial application
Disclosure	Description (DUS)	Description (DUS)	Enabling disclosure Best mode disclosure Deposit of novel materials
Rights	Prevent others from commercializing propagating materials	Prevent others from commercializing propagating materials and, under certain conditions, using harvested materials	Prevent others from making, using or selling the claimed invention or selling a component of the invention
Seed saving	Allowed for private and non-commercial use	For use on own holding only (for listed crops only)	Not allowed without consent of patent holder
Seed exchange	Allowed when non-commercial	Not allowed without consent of rights holder	Not allowed without consent of patent holder
Breeder's exemption	Use in breeding allowed	Use in breeding allowed (but sharing rights in case of EDV)	Not allowed without consent of patent holder
Duration	15–20 years (depending on crops)	20–25 years (depending on crops)	20 years from filing or 17 years from granting (prior to June 1995)
Double protection (PVP and patent)	Not allowed	Allowed	Allowed

reliance on patents has dominated. There are reasons for that choice, despite the higher cost of utility patents (Lesser, 2005). PVP allows farmers reuse of seed (although not an issue for F_1 hybrids) as well as open breeding access. Patents allow neither. Moreover, underfunding and resultant delays in issuing certificates reduced the value of PVP for breeders.

13.4.3 Biological protection

The oldest mechanism for protecting a plant cultivar is hybridization. The discovery of the phenomenon of hybrid vigour (heterosis) in the early 20th century opened new possibilities for producing high-yielding and uniform cultivars of cross-fertilizing crops and offered two distinct advantages for protecting the interests of commercial seed provision. First, seed of hybrid origin will lose some yield potential and other valuable characteristics (such as uniformity) in subsequent generations, which reduces farmers' incentives for saving seed. Secondly, competing seed companies cannot duplicate a particular hybrid cultivar if they do not have access to the inbred lines used to develop the hybrid cultivar. If the inbreds can be physically protected, they have the character of a trade secret. Hybrids from self-pollinated species including rice were first commercialized in China in the 1970s using genetic male sterility and now over 50% of rice land is planted with hybrid rice that has a huge seed market in China and South-east Asia. The use of hybrids thus provides a steady demand for seed, overcoming much of the uncertainty in the conventional seed market, where factors such as the weather determine how much seed is saved on the farm and hence the demand for fresh seed. In China a thriving and diverse commercial seed sector has existed for more than two decades because of the development of hybrid rice.

A more recent example of biological protection mechanisms is the introduction of genetic use restriction technologies, operating at the cultivar level (V-GURTS) (Louwaars *et al.*, 2002). In the absence of special treatments, plants containing these technologies produce sterile seed, thereby ensuring that farmers cannot save commercial seed of self-fertilizing crops (e.g. wheat and beans) for subsequent planting. The technologies also make it difficult for other breeders to use the protected germplasm. Companies are using the methods of genetic transformation to develop several such protection mechanisms including the so called 'terminator technology', a colloquial name given to proposed methods for restricting the use of GM-plants by genetically switching off a plant's ability to germinate a second time as next-generation seed. None is commercially viable yet, but the possibility of this technology has led to widespread debate and concern in the popular press (e.g. http://www.banterminator.org/; Guidetti, 1998) and has caused the technology to be specifically banned in India's Protection of Plant Varieties and Farmers' Rights Act (IBRD/World Bank, 2006).

13.4.4 Seed laws

Plant breeding and seed production are already subjected to a set of national regulations on cultivar release and seed quality control. These regulations are related to seed saving, seed exchange, the scope of protection, the breadth of coverage and the relation of PVP and patents to the concerns of farmers' rights. They have played an important part in determining the current evolution of seed systems. The following discussion on seed laws is based on IBRD/World Bank (2006).

Conventional seed laws can provide opportunities for controlling access to plant cultivars, even in the absence of IPR legislation. They determine what cultivars may be produced and establish regulations for seed certification and quality control. They can also limit the production and sale of seed by competitors and can perform some of the functions expected of PVP. Seed laws usually specify the extent to which seed must be certified and define the types of cultivar that may be offered for sale. Where seed certification is compulsory, the breeder may determine who is to produce seed by controlling access to breeders' (or pre-basic) seed. Any unauthorized multiplication will not be acceptable to the certification agency. A public or private breeder can establish an exclusive contract with a seed company for the production of specified cultivars. When a cultivar is not protected by PVP, the authorities can assign one or more maintainers to meet the continued demand for seed. Seed certification requirements can also be used to limit informal seed sales, especially when they occur on a large scale.

Where seed law specifies that a cultivar must be approved through a registration process or on the basis of performance tests before entering commercial seed production, this provision can also prohibit the sale of a released cultivar under a different name. In this way, the law limits the extent to which a competing company can market seed of a protected or an essentially derived version of a released cultivar, including the unauthorized use of a transgene.

Commercial seed systems usually begin with products that are difficult for farmers to save (hybrid cultivars or small seeded vegetables) and that generally require little IP protection. As the seed industry matures and farmers recognize the value of commercial seed, companies will offer a wider range of products, some of which may require attention to IPR. Seed industry development usually parallels the growth of agribusiness and markets for particular commodities may demand specific attention to IPR. Seed companies can sell seed to farmers who recognize the quality and convenience of commercial seed, on the basis of reputation and branding as is the case for small-size vegetable seeds in various countries.

13.4.5 Contract law

A contract is a legally binding exchange of promises or agreement between parties that the law will enforce. Breach of a contract is recognized by the law and remedies can be provided. Contract law can be classified, as is habitual in civil law systems, as part of a general law of obligations. Various types of contracts can be effective in providing legally enforceable agreements that restrict the use of a breeder's cultivar and offer complements or substitutes to IPR. Some contracts are aimed primarily at preventing seed saving and multiplication, whereas others are aimed at protecting the germplasm from being used in competitors' breeding programmes.

One type of contract that is increasingly prevalent in the US seed market is the grower contract, or 'bag tag'. This simple (unsigned) agreement restricts the farmer from using or disposing of any part of the harvest as seed. Farmers are considered to comply with the provisions of such contracts when they open the seed bag. If controlling the market for the harvested product becomes possible, another type of contract can be enforced. The breeder can oblige a grower to use crop cultivars in certain ways and can impose restrictions on the saving or multiplication of planting material. In the cut flower industry, for example, the vast majority of the output is sold in a limited number of wholesale markets. If a flower cultivar is protected in the country where a major wholesale market is located, growers in other countries may have to sign contracts limiting multiplication or unauthorized sale of that cultivar, or they risk being denied further access to the major market.

Access to germplasm may also be controlled through material transfer agreements (MTA), which may be seen as another form of contract regulating the use of plant germplasm. Such an example involving MTA includes the Agreements signed by the CGIAR centres with the FAO as discussed in Section 13.3.5. MTA and other contractual arrangements can be used by private companies to control access to genes or transgenic cultivars that are protected by IPR in one country, even if the recipient country does not recognize the particular IPR. For example, when a national agricultural research organization contracts with a major biotechnology company to use particular proprietary transgenes, the contract may specify how the national organization is to use the genes, the rights to any technologies that are produced and the company's obligations (for example, to provide training or other assistance). Access to various tools and processes of biotechnology, such as genetic transformation techniques or diagnostic methods, is also usually subject to contracts specifying limitations on their use and the rights of the provider in relation to commercial products.

13.4.6 Brands and trademarks

As a symbol such as a name, logo, slogan and design scheme, which embodies all the information connected to a company, product or service, brands and trademarks are part of IP law, but their utility in the seed industry is often overlooked in the policy debate about IPR (IBRD/World Bank, 2006). A minor point to remember is that terms such as AFLP® and 'Breeding by Design™', both trademarks of Keygene, Inc., should carry the ® or ™ designation. Seed companies frequently register their brands and trademarks as a way of distinguishing their products from those of their competitors and building up a loyal customer base. In the absence of other IP instruments, the development of a strong brand image and reputation can protect a company from some types of competition. While trademarks can be effective in communication with customers (farmers), they do not protect a breeder from competitors who 'steal' the cultivar and include it in their own (branded) product portfolio.

As there is usually a prohibition against using a cultivar name registered under PVP as a trademark, it is much less common for crop cultivars to be trademarked. However, in some cases a trademarked cultivar name may be very useful (IBRD/World Bank, 2006). For example, flower breeders often register a cultivar through PVP under one name but market it under a second, trademarked name, which can be used and protected long after the PVP expires. Some countries prohibit the use of separate trade names and prescribe that the name registered in the PVP or seed law lists is to be used in commerce.

13.4.7 Trade secrets

A trade secret can be considered a formula, practice, process, design, instrument, pattern, or compilation of information used by a business to obtain an advantage over competitors within the same industry or profession. In some instances, secrecy is an effective way to protect certain technologies and the choice between patenting and secrecy may depend on the type of technology and the size of the company. Trade secrets may not be included in a separate body of law but come under standard trade law. In plant breeding, the primary example of a trade secret is the protection of the inbred lines used to produce a hybrid. The ability to exploit this type of secrecy depends to an important extent on the degree of physical security that can be provided to plant breeding facilities and seed multiplication plots. Registration requirements (under PVP or seed law) may require the breeder to provide information on the pedigree (e.g. the specific inbred lines) or even deposit samples of the different parent lines. This requirement can nullify the trade secret unless the registration authority can keep the information and materials confidential. Advances in biotechnology make this type of secrecy more difficult to maintain, as reverse engineering of new cultivars becomes easier. Even though such actions might be covered by the enforcement of provisions on EDVs, they help to explain the pressure from some parts of the seed industry for further limitations on the breeder's exemption (IBRD/World Bank, 2006). Trade secrets are also useful for protecting certain aspects of plant biotechnology, particularly procedures or techniques that cannot be detected in the final product, such as markers and regeneration methods.

13.5 Intellectual Property Rights Affecting Molecular Breeding

As more and more plant patents are granted, IPR will increasingly affect each procedure of molecular breeding, which include methods for generation, identification and transfer and selection of genetic variation. Genetic materials (DNA, markers, genes and sequences) and methodologies (marker detection, MAS, genetic transformation and plant generation) that are keys to molecular breeding are heavily affected by biotechnology patents.

13.5.1 Genetic transformation technologies

Since the late 1980s, the contributions of biotechnology have transformed the science of plant breeding. The most visible (and controversial) aspect of plant biotechnology is the ability to transfer segments of DNA from one organism to another, resulting in GM plants. The range of commercial transgenic crop cultivars is still quite narrow (the majority feature herbicide tolerance or insect resistance) and about one-third of the global area planted to transgenic crops is in developing countries (the majority in Argentina, Brazil, India, China, India, Paraguay and South Africa) (James, 2006).

One of the recurring themes of the debates concerning the application of genetic transformation technology has been the role of IPR. This term covers both the content of patents and the confidential expertise usually related to methodology. The possession of appropriate genes and sequences is obviously not sufficient to produce a transgenic plant cultivar. As described in Chapter 12, there are two major techniques for introducing foreign genetic material into an organism. One is based on *Agrobacterium tumefaciens*, a species that is able to insert its own or other genes into a plant genome. The second is based on the direct, physical transfer of foreign genes into target plant cells, e.g. through biolistics. Both major techniques for introducing foreign genetic material into an organism are complex procedures characterized by a wide range of modifications and improvements. For instance, the *Agrobacterium* methodology was initially unsuccessful at transforming monocotyledons (such as cereals), but several recent advances have overcome this limitation. Similarly, the success of biolistics depends on a number of engineering considerations governing particle delivery. Hence both technologies are subject to a large number of broad and specific patent claims that make their utilization (and any claims on the resulting cultivars) far from straightforward. The particle gun technique was developed by US public researchers and licensed exclusively to a multinational company, thus providing an exclusive right to use and sublicense the technique. A development that may address restricted access to transformation methodology was the recent announcement of the discovery of transformation methods based on several genera of bacteria other than *Agrobacterium* and the establishment of an 'open source' licensing facility for these techniques (Broothaerts *et al.*, 2005). However, whether this methodology proves as efficient as the older methods remains to be seen.

Plant regeneration is a process of growing an entire plant from a single cell or group of cells through tissue culture in which fragments of tissue from a plant are transferred to an artificial environment in which they can continue to survive and function. In genetic transformation, the transformed plant cells (the products of *Agrobacterium* or biolistic methods) must be regenerated to produce whole GM plants (Chapter 12). Various techniques, which are mostly from tissue culture, are used to accomplish this goal and each regeneration protocol is appropriate to particular species or even cultivars. The majority of these transformation methods are described in the published literature and hence are available to all researchers. However, modifications that provide higher efficiency or that are appropriate to specific species may be kept secret by individual laboratories, because it is impossible to detect their utilization in the final product.

Dunwell (2005) reviewed the wide range of existing patents that cover all aspects of transgenic technology, from selectable markers and novel promoters to methods of gene introduction. Although few of the patents in this area have any real commercial value, there are a small number of key patents that restrict the 'freedom to operate' of new companies seeking to exploit the methods. Since the late 1980s, these restrictions have forced extensive cross-licensing between agricultural biotechnology companies and have been one of the driving forces behind the consolidation of these companies.

During the period since the production of the first transgenic plants a wide diversity of patents have been sought on all aspects of the process, ranging from the underlying tissue culture methods through to the means of introducing the heterologous DNA and to the composition of the DNA construct so introduced. The summary of granted US utility patents in the category 'genetic transformation' from 1976 to 2000 is available at http://www.ars.usda.gov/data/AgBiotechIP/. For detailed analysis of several of the key areas under discussion, the reader is referred to detailed summaries published elsewhere, for example in the series of comprehensive CAMBIA White Papers (http://www.cambia.org/daisy/bios/home.html). Frequently, the main point of interest in these discussions is the coverage of the patent(s) in question.

Transformation methods

As described in Chapter 12, there are several techniques for the introduction of recombinant vectors containing heterologous genes of interest into plant cells and the subsequent regeneration of plants from such cells. Some of the patents covering these techniques are summarized in Table 13.2.

Table 13.2. Selection of patents/applications covering plant transformation methods (from Dunwell (2005) with permission from Wiley-Blackwell).

Method	Company/institution	Patent/application number
Agrobacterium	University of Toledo	US 5177010, WO 02/102979
	Texas A&M University	US 5104310, WO 03/048369
	Leiden University	EP 120516, 159418, 176112
		US 5149645, 5469976, 5464763
		US 4940838, 4693976
	Max Planck	EP 116718, 290799, 320500
	Japan Tobacco	US 5591616
		EP 604662, 627752
	Ciba-Geigy	EP 267159, 292435
	Washington University	US 6051757
	Calgene	US 5463174, 4762785
	Agracetus	US 5004863, 5159135
	Monsanto	WO 03/007698
	BASF	WO 03/017752
	Purdue University	WO 01/020012
Particle bombardment	Cornell University	US 4945050
	DowElanco	US 5141131
	Dekalb	US 5538877, 5538880
	Agracetus	US 5015580, 5120657
Electroporation	Boyce Thompson Institute	WO 87/06614
	Dekalb	US 5472869, 5384253
	PGS	US 5679558, 5641664
		WO 92/09696, 93/21335
Whiskers	Zeneca	US 5302523, 5464765
Protoplasts	Ciba-Geigy	US 5231019

Most of these methods involve a tissue culture step and many of these enabling protocols are also the subject of patent claims.

The most extensive publication in this area is the 360-page CAMBIA White Paper (Roa-Rodriguez and Nottenburg, 2003a) on *Agrobacterium*-mediated transformation. This document focuses on the patents directed to methods and materials used for transformation, mainly of plants, but also of other organisms such as fungi.

Genes and DNA sequences

Much of the debate in this area concerns the ability to apply for patents on DNA sequences of unproven function. There have been several attempts to do so and the decisions on such applications have not been finalized. However, the fact remains that there is much useful sequence information available in patent databases and much of it is ignored by academic research scientists. Specifically, it is estimated that some 30–40% of all DNA sequences are only available in patent databases, since there is of course no obligation for commercial (or other) applicants to submit their sequences to public databases. Possibly, the best way to access this information is via the GENESEQ system, a commercial (Derwent) service.

As described in Chapter 12, transgenic crops are distinguished by the presence of several types of 'foreign' genetic material. These include: (i) functional genes (that is, genes that code for insect resistance, herbicide tolerance, or other desired characteristics); (ii) selectable marker genes (which have characteristics easily identifiable in the laboratory and, when linked to a functional gene, facilitate the detection of transformed cells); (iii) promoters (which regulate the timing and location of the expression of functional or marker genes); and (iv) end sequences (portions of DNA that terminate transcription). These different types of genes, sequences and techniques used in developing transgenic crops, as well as the diagnostic tools and processes of marker-assisted breeding used to produce conventional crop cultivars, are all candidates for patent protection (IBRD/World Bank, 2006). Almost all the significant components of the constructs used in plant transformation have been the subject of patent coverage. These include the 'effect gene' as well as its associated regulatory sequences, the selectable or screenable marker and additional sequences that might be required for the subsequent excision of the transgene.

It is important to recognize that patents on plant genes affect more than just the production of transgenic cultivars. It is possible to identify and protect genes that are used in more conventional breeding procedures. For instance, several herbicide-tolerant crop cultivars commercially available in North America incorporate patented genes that have been identified through techniques such as mutagenesis or whole cell selection and then incorporated in new crop cultivars through conventional breeding. Another example is imidazolinone-resistant maize, which is being tested in sub-Saharan Africa to control the weed *Striga*. The key to patent protection in these cases is the definition of novelty – that is, some countries prohibit patent protection on substances found in nature, which are considered to be discoveries rather than innovations. In most cases, a discovery must be further developed in order to be considered an innovation and eventually gain a patent that may effectively include the discovery. However, genes discovered and developed in the course of conventional breeding can be patented in several countries. IBRD/World Bank (2006) provided an example of the resistance to aphid (*Nasonovia ribisnigri*) in lettuce, patented by a Dutch breeding company in the USA and Europe. The European patent is, however, under appeal from various sides, including some important vegetable seed companies. So far, the US Patent and Trademark Office (USPTO) and the European Patent Office (EPO) have treated isolated and purified nucleotide sequences as if they were the same as man-made chemicals (Doll, 1998). Andrews (2002) argued that the useful properties of a gene sequence (such as its ability to bind to a complementary strand of DNA for diagnostic purposes) are not ones that scientists have invented, but instead, are natural, inherent properties of the genes themselves. Moreover, gene patents do not meet the criteria of non-obviousness, because, through *in silico* analysis, the function of genes can now be predicted on the basis of their homology to other genes.

Although the possibility of patenting genes is controversial, the concept itself seems straightforward. Even so, several issues contribute to making this area a particularly complex one for patent law. One problem is related to broad patent claims, which may cut a swathe as wide as 'all genetically engineered cotton plants'. Although such comprehensive claims may be more difficult to make now than in the early years of biotechnology, the issue of broad patents remains a concern for many areas of research, including the plant breeding industry (Barton, 2000). Another issue that affects gene patenting is the degree to which claims are allowed for genetic material whose functions are incompletely understood. For instance, the Human Genome Project witnessed a rush towards patents for a wide range of DNA sequences without any corresponding characterization and although such practices are more prevalent in the pharmaceutical industry than in plant breeding, they illustrate that there is not yet a widely accepted definition of how genetic material qualifies for a patent. This issue is related to a third issue, which concerns the type of genes or DNA sequences that might be patented. Claims have been made for protecting DNA that does not constitute a complete gene, including promoters, nucleic acid probes (used to identify DNA sequences) and polymorphisms. On the other hand, patents have been sought for collections of genes, from bacterial cloning vectors to entire genomes (IBRD/World Bank, 2006). Both the EPO and USPTO now have stronger guidelines concerning claims on genes: there must be a good knowledge and description of the gene's function.

A fourth issue that complicates the granting and defence of gene patents is the variable nature of the genes themselves. A good example of the difficulties in identifying what precisely is eligible for protection is provided by the *Bt* genes that are used for insect resistance in cotton, maize and other crops (IBRD/World Bank, 2006). The *Bt* bacterium produces certain insecticidal proteins and has been used as a source of 'natural' insecticide for many years. The techniques of biotechnology have allowed the identification and transfer of the genes that code for these crystalline ('Cry') proteins; the nomenclature describes a series of different *cry* genes (found in different strains of the bacterium), each coding for a distinct Cry protein that is effective against specific insects. Thus the *cry1Ac* gene codes for the Cry1Ac protein that is effective against the cotton bollworm and is the basis of most versions of *Bt*-cotton. Not only are there various claims on genes that code for specific Cry proteins; the *cry* genes that are used in transgenic plants are synthetic and significantly different from the original 'wild' genes found in the bacterium. In most cases, the *Bt* genes are 'codon modified' because part of the code that functions in a bacterium must be changed to be more effective in a plant. So although the insecticidal protein that is produced by the transgenic plant may be essentially identical to that produced by the bacterium, the governing gene may look somewhat different and patent claims can be made on the modified gene and the techniques used for its modification. A *cry* gene may be further altered by eliminating certain portions to produce a truncated form of the gene (which may prove more effective) and research has also created 'fusion' genes that code for novel proteins combining parts of two different Cry proteins. The various types of *cry* genes must be linked with specific promoters as well. The potential patent claims on various aspects of the process and disputes over definitions of novelty explain why *Bt* technology causes considerable uncertainty among scientists in developing countries and it is the subject of continuing legal disputes among the major biotechnology multinational corporations. Although the *Bt* example is particularly complex, it illustrates that genetic modification is rarely a case of simply identifying and moving a gene from one organism to another and it demonstrates how patent claims on genes may cover a range of issues.

Selection and identification of transformants

The production of transgenic organisms, including plants, involves the delivery of a gene of interest and the use of a selectable marker that enables the selection and recovery of transformed cells. This is necessary because only a minor fraction of the treated cells become transgenic while the majority remain untransformed. It has been estimated recently (Miki and McHugh, 2004) that approximately 50 marker genes used for transgenic and transplastomic plant research or crop development have been assessed for efficiency, biosafety, scientific applications and commercialization.

Selectable marker genes (see Table 13.3 for selected patents) can be divided into several categories depending on whether they confer positive or negative selection and whether selection is conditional or non-conditional on the presence of external substrates. The most common strategy currently used for selection is negative selection, the elimination of non-transformed cells in conditions where the transformed cells are allowed to thrive. Elimination is often affected by treatment of cells with chemicals, (e.g. antibiotics or herbicides) in conjunction with a transgene that confers resistance or tolerance to the chemical through detoxification or modification of the chemical. Much of the original work was conducted using antibiotic resistance marker (ARM) genes, which confer resistance to antibiotics such as neomycin, kanamycin and hygromycin. Roa-Rodriguez and Nottenburg (2003b) provided a summary of the most important scientific aspects of such resistance genes, together with an analysis of selected patents that relate to the most widely used ARM. Many of these marker genes are covered by patents or patent applications (Table 13.3) with a thorough IP analysis available on antibiotic markers and Basta resistance (Mayer *et al.*, 2004). As an alternative, or addition, to the use of selectable markers, transformants are often identified through the use of reporter or visualization molecules.

Promoters and other regulatory elements

Regulatory elements are crucial to gene expression in all organisms. The patent landscape of transcriptional regulators that are constitutively active, spatially active (e.g. tissue-specific) and temporally active (e.g. induced or active in response to a certain chemical or physical stimulus) has been well summarized (Roa-Rodriguez, 2003).

Table 13.3. Selection of patents covering selectable marker genes (Pardey *et al.*, 2003).

Selectable marker	Company	Region/country	Patent number
Phosphinothricin, Basta	Aventis/AgrEvo	Europe, USA *et al.*	EP 531716 *et al.* US 5767371 *et al.*
Kanamycin	Monsanto	Europe, USA *et al.*	EP 131623 US 6174724 *et al.*
Hygromycin	Novartis	Europe, USA *et al.*	EP 186425 *et al.* US 5668298 *et al.*
Sulfonamide	Rhone Poulenc	USA	US 5714096
Cyanamide	Syngenta Mogen	Europe, USA	EP 97201140 US 6660910
Aldehyde	Calgene	Europe, USA *et al.*	EP 0800583 US 5633153
Mannose/xylose	Novartis	Europe, USA *et al.*	US 5767378 *et al.*
Glucosamine	Danisco	USA *et al.*	US 6444878
2,4D	Unknown	Europe, USA	EP 0738326 US 5608147

Although the inventions protected by individual patents cannot be exactly the same, in certain cases, there are patents that due to the breadth of their scope may encompass other protected inventions or there may be patents which share common features. Where that is the case, Dunwell (2006) pointed out the juxtaposition of the different inventions and the possible room left to manoeuvre around the different entities in the field. It also needs to be taken into account that there are patents that while not totally directed to promoters may have an effect on gene expression control. This is the case for the restrictive reproductive technologies, for example, those termed as 'terminator' technologies, which may have a great impact on the use and development of methods to regulate the expression of genes related to plant reproduction and seed generation.

'Golden Rice' as an example for freedom-to-operate

One of the issues of over-riding importance to all companies is whether or not they are free to commercialize any particular product. Such 'freedom-to-operate' is determined by the status of any IPR that might cover the product in question and analysis of such IPR requires continuous (and therefore expensive) surveillance.

A well known example that can be used to demonstrate the complexity of this issue is 'Golden Rice', a transgenic line that is enhanced for β-carotene (pro-vitamin A) (Ye *et al.*, 2000). Vitamin A deficiency causes symptoms ranging from night blindness to those of xerophthalmia and keratomalacia, leading to total blindness. In developing countries, 500,000 children year^{-1} go blind and up to 600 day^{-1} die from vitamin A malnutrition (Potrykus, 2005). As oral delivery of vitamin A is problematic, mainly due to the lack of infrastructure, alternatives might be found in supplementation of the major staple food with pro-vitamin A. As a table food for many countries, rice is usually milled to remove the oil-rich aleurone layer that turns rancid upon storage. The remaining edible part of rice grains, the endosperm, lacks several essential nutrients including pro-vitamin A. Thus, predominant rice consumption promotes vitamin A deficiency. A combination of transgenes enabled biosynthesis of pro-vitamin A in the endosperm (Ye *et al.*, 2000). GM-rice that produces β-carotene (pro-vitamin A) in the endosperm shows the yellow colour of the grain that is visible after milling and polishing, from which the generic name 'Golden Rice' is derived. 'Golden Rice' could be used in food-based approaches and complement others, in reducing the persistent problem of vitamin A deficiency in rice-dependent populations. The 'Golden Rice' technology was developed by I. Potrykus and P. Beyer with their co-workers and was funded by the Rockefeller Foundation, the Swiss Federal Institute of Technology, the EU and the Swiss Federal Office for Education and Science.

'Golden Rice' and its use in grain production has involved a lot of controversies. It has been suggested that extensive patenting has hampered delivery of this rice to those in need since about 40 organizations hold 72 patents on the technology underlying its production (Kryder *et al.*, 2000). The range of patents covering various components of the pBin 19hpc plasmid used in the production of this rice include ones on the phytoene trait genes, the promoter sequences, the selectable marker and the transit peptide. Table 13.4 shows the product clearance profile detailing the possible required licences and/or agreements for 'Golden Rice'. Table 13.5 lists the tangible property received by ETH-Zurich, including the apparatuses used in the transformation. Some components were obtained under research-only licences or research-only MTA whereas others included use licences.

The challenges to freedom-to-operate for 'Golden Rice' at national and international levels include: (i) the technology is quite complex with many sophisticated components and processes; (ii) many potential IP owners or assignees; (iii) the range of potential producers and consumers of 'Golden Rice' is wide; (iv) a rapidly evolving global

Table 13.4. Product clearance profile: possible required licences and/or agreements for GoldenRice™ (from Kryder *et al.* (2000) with permission).

Company/institution[a]	Patent number
AMOCO	US 5545816, EP 0471056, US 5530189, WO 9113078, US 5530188, US 5656472
Bio-Rad Inc.	US 5186800
Biotechnica	WO 8603516
Calgene	WO 9907867, WO 9806862
Centra National de la RSK	WO 9636717
Cetus	WO 8504899, US 4965188, EP 0258017
Columbia University	US 4399216, US 4634665, WO 8303259
DuPont	WO 9955889, WO 995588, WO 9955887
Eli Lilly	US 5668298
Hoffman-La-Roche	US 4683202, EP 0509612, EP 0502588, US 4889818
ICI, Ltd	WO 9109128
Japan Tobacco	EP 0927765, US 5591616, EP 0604662, EP 0672752, US 5731179, EP 0687730, WO 9516031
Kirin Brewery	JP 3058786, US 5429939, US 5589581, EP 0393690, US 5350688
Max Planck Gesell.	EP 0265556, EP 0270822, EP 0257472
Monsanto	US 5352605, US 5858742, WO 8402913
National Foods RI	JP 63091085
NRC Canada	WO 9419930
Nederlandse OVT	EP 0765397, WO 9535389
Phytogen	US 4536475
Plant Genetic Systems	US 5717084, US 5778925, WO 8603776, WO 9209696
Promega	US 4766072
Rhone-Poulenc Agro	US RE36449, WO 9967357
Stanford University	US 4237224
Stratagene	US 5128256, US 5188957, US 5286636, EP 0286200, WO 880508
University of Maryland	WO 9963055
University of California	US 4407956, WO 9916890
Yissum RDC	US 5792903, EP 0820221, WO 9628014
Zeneca Corp.	US 5750865, EP 0699765A1

[a]Note that these are the names of the owners or assignees of the rights under the relevant patents. Because of possible subsequent licensing or assignment, these are not necessarily the current entities to approach for licences.

IP landscape; and (v) 'Golden Rice' may have significant commercial values (Kryder *et al.*, 2000). This issue has been overcome by a coordinated international programme designed to streamline the production and distribution of this material (http://www.goldenrice.org/). However, perceived problems with access to 'Golden Rice' and essential medicines have stimulated debate within the USA on the obligations of US universities to facilitate the provision of goods for the public benefit (Kowalski and Kryder, 2002; Phillips *et al.*, 2004).

The deal for 'Golden Rice' has the following clauses:

> The inventors have reached an agreement with Greenovation and Zeneca (now Syngenta)...to enable the delivery of this technology free-of-charge for humanitarian purposes in the developing world. Inventors (Beyer and Potrykus) assigned their rights exclusively...to [Syngenta] for all uses; [Syngenta] licensed inventors for humanitarian uses, with right to sublicense public research institutes and poor farmers in developing countries; the technology is to be made freely available, poor farmers can trade Golden Rice locally; [Syngenta] will support inventors in this task; and [Syngenta] retains commercial rights. In the Golden Rice Deal, Syngenta's role is to

Table 13.5. Material transfer agreements (MTAs), licences, documents and agreements relevant to Golden Rice™ (from Kryder *et al.* (2000) with permission).

Product	Company holding the licence/agreement
Rice germplasm transformed with Research Institite (IRRI) gene construct(s)	Taipei 309, obtained from International Rice
PGEM4	Promega
PbluescriptKS	Stratagene
PCIB900	Ciba-Geigy Limited (now Novartis Seeds AG)
CaMv35S promoter (component of pCIB900)	Monsanto
CaMv35S terminator (component of pCIB900)	Monsanto
AphIV gene: hygromycin phosphotransferase (component of pCIB900)	Ciba-Geigy Limited (now Novartis Seeds AG)
pKSP-1	Thomas Okita, Washington State University
GT1 promoter: glutelin storage protein	Thomas Okita, Washington State University (component of pKSP-1)
pUCET4	N. Misawa, Kirin Brewery Co., Ltd
Pea Rubisco transit peptide (component of pUCET4)	N. Misawa, Kirin Brewery Co., Ltd
CrtI gene: phytoene desaturase (component of pUCET4)	N. Misawa, Kirin Brewery Co., Ltd
PPZP100	Pal Maliga, Rutgers University
pYPIET4	Clontech, but now marketed by Life Technologies
Electroporation apparatus	Bio-Rad Corp., Gene Pulser II System
Miroprojectile bombardment apparatus	Bio-Rad Corp.

help the inventors in the management of Golden Rice deployment for humanitarian purposes and with other companies and universities obtained 'FTO' for humanitarian use; provide biosafety expertise; and share available regulatory data. Here 'Humanitarian Use' means (research leading to): developing country use (FAO list); resource poor farmer use (<US$10,000 pa from farming); in public germplasm (= seed); there must be no charge for technology (normal costs can be recovered; no premium); local sales are allowed by such farmers (... urban needs); and replanting is allowed. Other license terms include regulatory requirements – national sovereignty (or international standards); no export of grain allowed (or seed, expect for research, to other licenses) – liability, trade, biosafety approvals; and obliged to fulfil all regulatory requirements.

'Golden Rice 1' (SGR1) and 'Golden Rice 2' (SGR2), as the second generation of 'Golden Rice', were developed by Syngenta as a part of their commercial pipe-line and their pro-vitamin A (β-carotenoid) levels are up to 8.0 and 36.7 $\mu g\ g^{-1}$, respectively, compared to 1.2–1.8 $\mu g\ g^{-1}$ for 'Golden Rice' (Plate 4, colour photograph from Paine *et al.*, 2005). Consistent with Syngenta's support of the Humanitarian Project for Golden Rice, SGR2 transgenic events will be donated for further research and development. The use of the SGR2 events will be governed by the strategic directions of the Golden Rice Humanitarian Board and full regulatory compliance. It is expected that the third generation of 'Golden Rice' will be the rice with a high level of pro-vitamin A and the normal colour of polished rice grain, which is more acceptable to most rice consumers. Looking into the future, an interesting issue has been pointed out by Potrykus (2005), that is, experience with the 'Humanitarian Golden Rice' project has shown that 'extreme precautionary regulation' – not IPR – prevents use of the GMO potential to the benefit of the poor and that the public domain is incompetent and unwilling to deliver products. But a decade after its invention, 'Golden Rice' is still stuck in the laboratory. Well-organized opposition and a thicket of regulations on transgenic crops have prevented the plant from appearing on Asian farms within 2–3 years (Enserink, 2008). Although the first filed trial of 'Golden Rice' in Asia started in 2008, no farmer will plant the rice before 2011.

Although there are also some successful stories in other crops such as metabolic engineering of potato carotenoid content through tuber-specific overexpression of a bacterial mini-pathway (Diretto *et al.*, 2008), whether justified or not, the turmoil over 'Golden Rice' has shaped other efforts to improve the nutritional value of crops. For example, Harvest Plus has a US$13 million annual budget that aims to boost levels of three key nutrients, vitamin A, iron and zinc, and it relies almost entirely on conventional breeding with only some efforts involving marker-assisted breeding.

13.5.2 Marker-assisted plant breeding

'Marker-assisted selection' (MAS) has been described in detail in Chapters 8 and 9. The standard steps employed in MAS generally include: (i) selecting individuals to be tested; (ii) harvesting materials; (iii) extracting DNA from the materials, amplifying the DNA by PCR to enrich for gene sequences or DNA fragments associated with a particular trait or phenotype; and (iv) separating these fragments, visualizing and identifying the DNA fragments and interpreting and utilizing the information. MAS techniques are subject to varying degrees of protection. They rely on DNA sequences and probes, some of which may be available publicly whereas others are covered by patents. Jorasch (2004) provided an example for IPR in the field of molecular marker analysis including a figure showing some of the patents that claimed different steps of this typical marker experiment and to indicate which step of the experiment is claimed (Fig. 13.3). Henson-Apollonio (2007) reviewed the impacts of IPR on MAS research and application for agriculture in developing countries, providing some examples for patents, copyrighted software and trademarks that are relevant to MAS. It can be certain that many new patents have been claimed since then, particularly when the third generation of molecular markers such as single nucleotide polymorphism (SNP) are used, readers can update the figure with those related to the multiple procedures.

Selection of microsatellite primers and the PCR

There are some patents that claim primers for PCR analysis. Figure 13.3 shows two typical patents claiming primers for microsatellite (simple sequence repeat, SSR) markers. One is patent No. 1 (Röder *et al.*, 1997) which claims specific microsatellite markers for wheat, and the other is patent No. 2 (Nagaraju, 2003), which claims a certain class of SSR primers, the inter-simple sequence repeat-PCR primers (ISSR). Patents claiming specific primer sequences for marker analysis have become rare recently as primer sequences are increasingly treated as a business secret that is licensed to users.

After selecting primers, the PCR experiment will be affected by the basic patents in PCR that were registered in 1985 including patents Nos 3–5 (Mullis, 1992; Fig. 13.3). Meanwhile, there are many other patents that claim special polymerases or methods like reverse transcription PCR (RT-PCR) and quantitative PCR.

There are patents that claim both the primer sequences and the PCR methods such as patents No. 6 (Morgante and Vogel, 1997), 7 (Kuiper *et al.*, 1997) and 8 (van Eijk *et al.*, 2001; Fig.13.3). These patents' specifications claim processes for detecting polymorphisms between nucleic acid segments using defined primers sequences, sometimes starting with the previous restriction of the DNA sample by restriction endonucleases and the ligation of certain adaptor sequences similar to the amplified fragment length polymorphism (AFLP) approach.

Analysis of PCR products

Figure 13.3 shows three different methods for the analysis of the resulting PCR

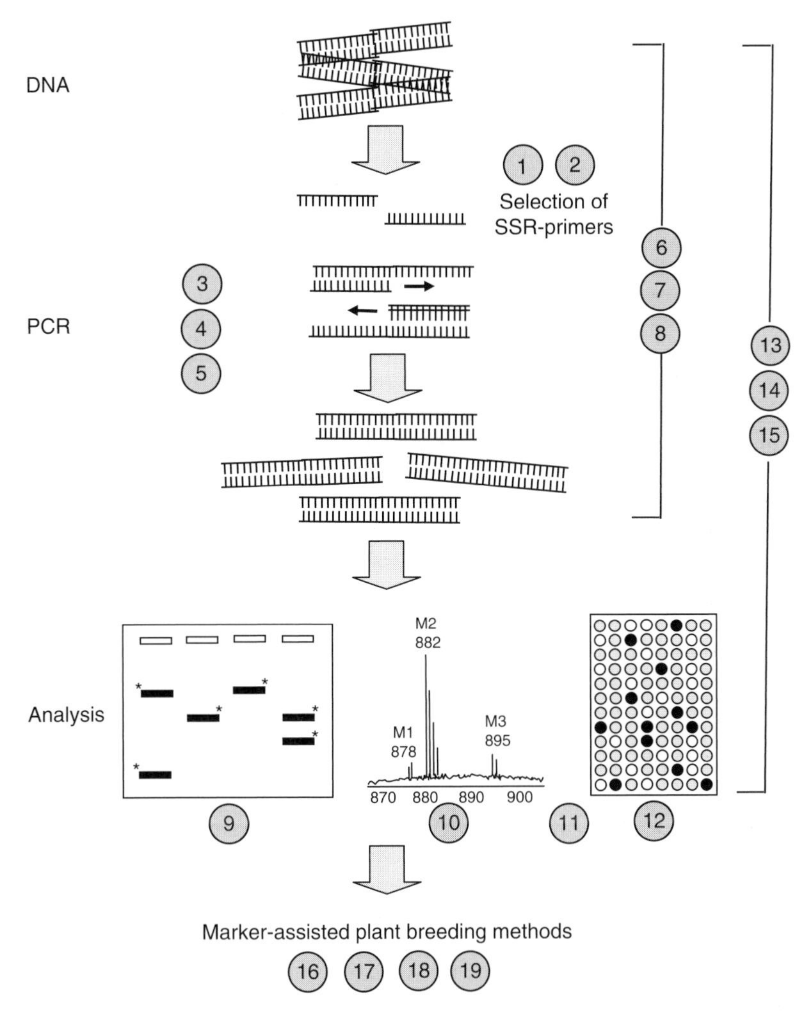

Fig. 13.3. An overview of a typical experiment involving SSR markers and some sample patents that are relevant for the different steps as described in Chapters 2 and 3. Patents are indicated by numbers. Starting with DNA isolation, the DNA is sometimes cut by restriction enzymes. After selection of specific SSR primers, a PCR reaction is carried out. There are different possible methods for the analysis of the PCR product including gel electrophoresis (the fluorescent label of the PCR product is indicated by *), mass spectrometry, and microarray analysis. The result from molecular marker analysis can be used in marker-assisted plant breeding, which involves several patents. From Jorasch (2004) with kind permission of Springer Science and Business Media.

products. The most common one, the analysis by gel electrophoresis, can also be claimed by patents, if for example special fluorescent labels for detection are used. Such a method is claimed by patent No. 9 (Shuber and Pierceall, 2002). The claimed method comprises the PCR with fluorescent primers, the detection of the labelled extension products and the comparison of the PCR product sizes. A second method of analysis, mass spectrometry, is claimed by patent No. 10 (Hillenkamp and Köster, 1999). This patent generally claims the analysis of nuclei acids by mass spectrometry in general. The microarray technique that can be used for high-throughput analysis of probe is protected by a patent of Affymetrix, patent No. 12 (Fodor *et al.*, 1998). This specification not only protects the detection of microsatellite markers by microarray analysis, but also the detection of nuclei acid sequences in general which comprises microsatellites. Another high-throughput technique described in patent No. 11 (Olek, 1996) combines the methods of mass spectrometry and microarray analysis of microsatellite markers. In addition, patent specifications No. 13 (Caskey and Edwards, 1992), No. 14 (Perlin, 1995) and No. 15 (Saint-Louis and Paquin, 2003) summarize the complete experimental process from DNA extraction to the analysis of PCR products, in which different PCR methods are combined, e.g. use of certain labelled nucleotide triphosphates and different analytical tools such as mass spectrometry or computer analytical tools.

Marker-assisted breeding methods

The most comprehensive patent specifications claim complete plant breeding methods in which molecular marker analysis is used. Examples are patent specifications No. 16 (Byrum and Reiter, 1998), No. 17 (Beavis, 1999), No. 18 (Openshaw and Bruce, 2001) and No. 19 (Jansen and Beavis, 2001) (Fig. 13.3). These comprise previously mentioned experimental steps in which they claim the association of genotype with phenotypic traits of interest for molecular marker analysis. The patents differ in the selection of plan populations that are the basis for the analysis, statistical methods applied in the analysis and the integration of molecular biological techniques such as expression profiling of genes.

This section just describes how biotechnology patents would affect marker-assisted plant breeding using microsatellites as an example. As SNP markers and gene-based markers become increasingly feasible, more claimed patents will be associated with their application in plant breeding. Although MAS techniques are used in conventional plant breeding and no foreign DNA sequences become part of any resulting cultivar, the use of patented diagnostic technology may have implications for a plant breeder's ability to claim ownership of the final product. The exact situation will depend on the conditions under which the technology was acquired and the wording of any contract with the supplier. So-called 'reach through claims' have not seemed to play a significant role in this area to date. Patent offices have also become aware of the negative effect of these claims and are very critical in granting wide claims.

National patent systems have been unable to keep pace with the rapid development of plant biotechnology, leaving many areas of uncertainty and dispute. In developing countries, only a small minority of patent offices have begun to consider applications related to plant biotechnology, while in several industrialized countries a number of claims to basic technologies are still the subject of complex court cases. It is therefore impossible to chart an unambiguous course for the development of effective IPR regimes for plant breeding-related biotechnology, but it is important to recognize the major parameters and to identify the issues that will affect IPR policy in the coming years. Areas of particular concern include the protection of genes and other sequences, the methods used for genetic transformation, information in bioinformatics databases and the diagnostic techniques that biotechnology offers conventional plant breeding (IBRD/World Bank, 2006).

13.5.3 Product development and commercialization

There are multiple steps involved in the procedure of developing biotechnology and breeding products, each of which might be associated with specific IPR issues. For example, the development of *Bt* maize involves multiple steps that are associated with specific patents and IPR issues: (i) gene ownership (*CryIF*, *PAT* marker gene); (ii) enabling technologies (microprojectile bombardment, herbicide selection, backcrossing, production of fertile transgenic plants); (iii) enhanced expression (chimeric genes using viral promoters, enhanced expression, enhanced transcription efficiency, selective gene expression); and (iv) developing elite maize inbreds and hybrids (patented inbreds, hybrids and patents for associated traits and genes).

There are many IPR issues involved in delivering a transformation product from research to the farmer's field. For example, in *Bt*-maize IPR issues would include: (i) research agreements among major players allowing forward movement in plant biotechnology; and (ii) cross-licences for Roundup Ready® (RR) YieldGard. Monsanto licenses Herculex 1 whereas Pioneer licenses RR for maize, soybean and canola, or Pioneer needs to deal on germplasm issues with Monsanto. Likewise, there was competition for developing basic technologies to most effective use of technologies to develop improved products. In addition, payment for technology or germplasm research is ultimately dependent on farmer purchases of seed.

It can be expected that a large number of new techniques will be developed and associated patents will be claimed in the field of molecular breeding in the near future. The new patents that may add to the current patent list and will affect molecular breeding include:

- High-throughput automated molecular marker profiling.
- High-throughput gene expression assays using DNA on silicon chips.
- High-throughput proteomics assays.
- High-throughput DNA sequencing facilities.
- The ability to DNA profile both the female and the male parents of hybrids without accessing either parent per se via use of maternally inherited tissue.
- The ability to conduct genome-wide gene–trait association studies involving hundreds or thousands of genotypes, including heterogeneous complexes such as landraces.
- The ability to conduct genome-wide scans comparing domesticated cultivars or landraces and to compare them with wild relatives to identify potentially useful loci and novel genetic diversity.

It would not be surprising at all if several years later many other fields and associated techniques and knowledge will have to add to and modify the best list we can generate so far.

13.6 Use of Molecular Techniques in Plant Variety Protection

Molecular techniques, particularly molecular markers, have been widely used in all procedures involved in plant breeding and some fields of PVP as well. For example, on 16 and 17 June 2005, the Plant Production Division of the Canadian Food Inspection Agency (CFIA) and the National Forum on Seed jointly held a seminar on UPOV plant variety protection and the use of molecular techniques. The objectives of the seminar were: (i) to provide information to Canadian plant breeders and other stakeholders on PVP and the use of molecular markers under the UPOV Convention; and (ii) to facilitate discussion on the potential application of molecular techniques to PBR, cultivar registration and seed certification in Canada. Information from this seminar is available at http://www.inspection.gc.ca/english/plaveg/pbrpov/molece.shtml (CFIA/NFS, 2005).

There are a number of advantages to the use of DNA marker techniques in plant breeding as described in Chapter 8, most of which are also applicable to PBR. Among all available molecular markers discussed in Chapter 2, SNP is the most prolific and is very efficient and inexpensive to use once developed. The technology holds enormous potential for multiplexing and high throughput. SNP technology is being used to characterize germplasm and in breeding programmes. Broad adoption of this technology would be useful to the plant protection regulatory systems, especially for cultivar identification and protection purposes. While SSRs are now generally accepted in the courts, there are some inherent limitations that would be overcome through the use of SNP.

Traditional methods based on morphological observations take time to complete and results are influenced by the environment. Molecular tools can play a complementary role to traditional methods. In the USA, except for complete descriptions of similar cultivars, colour chart references, statistical analyses, photographs, or plant specimens, the applicant may also provide, as supplementary information, results of isozyme analysis, restriction fragment length polymorphism (RFLP), SSR, SNP or other genetic fingerprinting testing results. Grant protection now can be based on molecular marker differences if those differences meet the definition of distinctness, that is, if they are 'clear'.

13.6.1 DUS testing

As tools for DUS testing, molecular techniques offer the potential for more rapid and cost-effective results that are less influenced by the environment, year, growth stage and other factors. The UPOV Working Group on Biochemical and Molecular Techniques and DNA-Profiling in Particular (BMT) reviewed three options for introducing molecular techniques into the UPOV system. For some applications, such as the use of gene specific markers to identify a phenotypic characteristic (option 1) or the use of molecular markers for the management of reference collections (option 2), molecular techniques would be acceptable within the terms of the UPOV Convention and would not undermine the effectiveness of the protection it provides. A third proposal (option 3), which would create a new system, raises a key concern that molecular techniques could potentially provide the opportunity to use a limitless number of markers to find differences between cultivars. There is also concern that differences would be found at the genetic levels which are not reflected in morphological characteristics.

GEVES (Groupe d'étude et de contrôle des varieties et des semences) in France and the Department for Environment, Food and Rural Affairs (Defra) in the UK have used molecular techniques for DUS testing. GEVES used fingerprinting (ISSR, AFLP, SSR, sequence tagged sites (STS)) and gene specific tests (GMO, species-specific reference genes or resistance genes). The National Institute of Agricultural Botany (NIAB, Cambridge, UK) carries out some of the technical work in this area for Defra and has been very active in research into the use of molecular markers for DUS testing. NIAB has undertaken research relating to the three UPOV BMT options. For example, Defra funded a project to develop a set of SSR primer pairs for wheat that could be used independently of the detection platform. The results indicated that in principle DUS could all be assessed using well-characterized SSR markers.

Early investigation of molecular techniques clearly indicated the potential of molecular markers for discrimination between sets of cultivars, confirmation of cultivar identity or measurement of diversity, etc., as discussed in Chapter 5. Overall, marker technology has the potential to assist the PBR process in determining distinctness and uniformity. Further studies are required to determine marker type, number of markers, quality of markers, distribution of markers on a genome and considerations pertaining to seed source and sampling (CFIA/NFS, 2005).

There was potential for molecular techniques for PBR DUS testing of some specific crops or species, rather than generalized broad-based use throughout the sector. Molecular techniques were considered to be particularly applicable to plants with novel traits and GMOs. However, for some crops, molecular techniques may be 'an unnecessary complication' (CFIA/NFS, 2005). It was noted that molecular techniques are especially effective for identifying 'newness' and therefore 'distinctness'. Generally, molecular techniques, particularly molecular markers, are considered to be more useful for 'distinctness' over 'uniformity' and 'stability'.

While the potential for molecular data to demonstrate stability was considered to be low, there is also potential regarding uniformity. As breeder seed samples are really populations of genotypes, genotypic non-uniformity is very typical. Some variability in a cultivar 'is not a bad thing'. What is important is that the amount of variability that is acceptable is defined (CFIA/NFS, 2005). When molecular markers are used for a uniformity test, a number of questions are raised:

- If genotypic data is used in DUS-like testing, how is an heterogeneous locus interpreted in a bulk seed sample?
- What is an appropriate sample size for single seed analysis?
- Genotypic stability will not be perfect in heterogeneous mixtures. Will the guidelines account for registration of deliberate mixtures of lines?

To have a cultivar that can be identified as unique only by markers is of no value if the physical appearance is not unique. If there is no physical inspection, someone could, theoretically, apply and receive a PBR for an obsolete cultivar. A library of DNA fingerprints could prevent this from occurring, but there would be significant costs to create it. It seems a system that uses both botanical and molecular measurements to protect cultivars provides the best protection.

It was noted that molecular techniques would be particularly advantageous for crops with few variable morphological traits or where cultivars are very similar. It was also suggested that the preferred approach would be for the use of molecular techniques to be on a voluntary, case-by-case and crop-based approach. Molecular markers would be considered by the Plant Breeders' Rights Office to be an 'additional descriptor' or supplement (CFIA/NFS, 2005). However, if there is a move towards creating a new platform in which molecular techniques would be a standalone tool, then all elements should be re-addressed, such as thresholds, definition of distinctness, purpose of protection, sampling methods, etc.

Finally, there are a number of issues to be considered and questions to be raised about the use of DNA markers in DUS testing:

- The variation measured may not have a genetic basis.
- If markers are used in conjunction with phenotypic traits, how will the two systems be weighted?
- Marker thresholds for distinctness will need to be established.
- On the issue of uniformity, it may be difficult to find cultivars that are fixed at all marker loci.
- There may be laboratory-to-laboratory variations in genotypic scores for some types of markers.

13.6.2 Essentially derived varieties

An EDV is distinct and predominantly derived from a protected initial cultivar, while retaining the essential characteristics of that initial cultivar. To avoid genetic erosion in breeding germplasm and to support creative, additive plant breeding, the principle of 'breeders' exemption' allows breeders to use protected cultivars for developing new cultivars. However, a newly derived cultivar may be developed from an already protected initial cultivar owned by a competitor, by applying breeding methods that are regarded as illegitimate. The potential occurrence of plagiarism has increased with the development of biotechnologies that enable the introduction of a single gene into a cultivar, such as genetic engineering or marker-assisted backcrossing, which facilitates deliberate selection of lines that retain a large amount of genome of the parental line.

The EDV provision is meant to limit the possibility of 'cosmetic breeding', which produces a cultivar that is only slightly different from the original. Introduction of the concept of EDV in the UPOV Convention improved the protection of the breeder of the initial cultivar. A cultivar is deemed to be essentially derived from another cultivar (the initial cultivar) when it: (i) is predominantly derived from the initial cultivar; (ii) is clearly distinguishable from the initial cultivar; and (iii) conforms to the initial cultivar in the expression of essential characteristics that result from the genotype of the initial cultivar (UPOV, 1991). Fowler *et al.* (2005) discussed a series of possible definitions of 'derivatives', each based on a different approach. These definitions are based on allelic differences, allelic frequencies, phenotype, breeding action, or composite definitions including two of the above. An alternative approach would be to employ a definition which, while allowing for IPR, aims to ensure the designated accessions or the designated accession and its components remain available for use by other recipients in defined ways. The goal of this approach would be to keep the materials in the public domain while encouraging research on designated accessions.

An EDV can be protected when it is DUS and new, but to commercialize the cultivar, the breeder of the EDV must have the consent of the person or entity that holds the rights to the initial cultivar. The EDV concept is susceptible to different interpretations, however, and is the subject of an ongoing debate among breeders (ISF, 2004). Under some laws, a newly bred cultivar may be considered an EDV and the IPR to that cultivar then depend on the rights to the initial cultivar (IBRD/World Bank 2006).

Although the EDV concept offers protection against piracy and violation of PBR, breeding companies have not yet agreed upon a catalogue of specific breeding procedures considered to yield EDVs. As a consequence, official guidelines for the estimation of genetic conformity between an initial cultivar and an EDV, as well as crop-specific thresholds to distinguish between independently derived cultivars and an EDV are important and should be established. Some larger companies would like to introduce the concept of 'genetic distance' in the definition of an EDV, but others fear that this step could lead to the monopolization of certain gene pools. After much debate, seed company representatives agreed upon arbitration rules for EDV disputes (ISF, 2005).

Historic samples can be used by a government agency related to PVP to establish thresholds regarding distinctness. Crops could be reviewed to determine what might be considered 'similar' and what might be considered 'dissimilar'. On the other hand, stakeholders may create working groups to develop marker thresholds. The underlying problem is the functionality of PVP as protecting the entire plant and not specific traits as is possible with patents (Lesser, 2005). As a result, we have to consider what portion of the germplasm was derived from the purported initial cultivar – something we can refer to as relatedness. In addition, the allegedly EDV must express the essential characteristics of the initial cultivar. The complexity arises because traits transferred into a cultivar may involve few genes for a simple trait or multiple genes for a complex trait. The complication arises when the degree of relatedness is set. If it is set high (e.g. 95% or higher), then there is an incentive when the trait is discrete to engage in unproductive 'cosmetic' breeding to circumvent the dependency standard. Conversely, if the relatedness is low (e.g. 90% or lower) and the trait discrete, then it is possible that an independently discovered trait will be identified as dependent, greatly expanding the control of the initial cultivar owner over cultivars he or she made no contribution to. This could happen when highly polymorphic molecular markers are used. Lesser and Mutschler (2004) concluded that a single relatedness requirement for a species cannot equitably be applied to both discrete and complex traits.

Among possible tools to assess genetic conformity between putative EDVs and their initial cultivars including morphological traits, agronomic descriptions and heterosis, molecular markers seem to be the most promising for establishing genetic conformity because of the reasons that have been described previously. Genetic distance estimates based on molecular markers represent a key to the assessment of essential derivation and effective determination of genetic conformity between a protected initial cultivar and a putative EDV. Highly polymorphic markers such as AFLP, SSR and SNP are useful in genotypic analyses because they are amenable to distinguish closely related cultivars.

A statistical test for the identification of EDVs with genetic distances from molecular markers was proposed by Heckenberger *et al.* (2005a). For a progeny line derived from a biparental cross, the genetic distance to each of the parents depends on the genetic distance between the parents and p, the parental genome contribution transmitted to the progeny. The treatise provides estimates of p and the variances of p (σ_i^2; Wang and Bernardo, 2000). Morphological distances based on 25 traits and midparent heterosis for 12 traits were observed for a total of 58 European maize inbred lines comprising 38 triplets. A triplet consisted of one homozygous line derived from an F_2, BC_1 or BC_2 population and both parental inbreds. All inbreds were genotyped with 100 uniformly distributed SSR markers and 20 AFLP primer combinations in companion studies for calculation of genetic distances. Correlations between the coancestry coefficient, genetic distances and morphological distances and midparent heterosis were significant and high for the majority of traits. However, thresholds for EDVs to discriminate between F_2- and BC_1-derived, or BC_1- and BC_2-derived progenies using only morphological distances or heterosis yielded a considerably higher probability of error than observed with genetic distances based on SSRs and AFLPs (Heckenberger *et al.*, 2005b). Consequently, morphological traits and heterosis are less suited for identification of EDVs in maize than molecular markers.

Heckenberger *et al.* (2006) observed considerable differences between AFLP- and SSR-based mean genetic distance estimates for unrelated inbred lines. With each marker system, the genetic distance between progeny lines and parents was little affected by the variation in genetic distance between the parents. Substantial differences in Type I and Type II errors were detected between flint and dent maize germplasm pools with different marker systems and when fixed EDV thresholds were considered. It was suggested that threshold levels should be crop-specific. With a crop, thresholds should be germplasm pool-specific. In addition, thresholds should also be molecular marker system-specific because marker systems vary in the way of generating polymorphism. Heckenberger *et al.* (2005a) reported that correlation between true and estimated genetic distances was considerably lower for randomly distributed markers, particularly with medium-to-low marker densities.

13.6.3 Cultivar identification

The identification of cultivars is an important aspect of plant production systems and is central to the protection of IPR through PBR. Preston *et al.* (1999) discussed the application of a range of molecular marker technologies to three points in the PBR registration process: (i) the analysis of the genetic distance between a candidate cultivar and the existing pool of cultivars in order to define a set of comparison cultivars; (ii) the contribution to the generation of a description of the cultivar for PBR registration; and (iii) the use of DNA markers to investigate and resolve the identity for cultivars in cases where infringement of PBR is claimed. Molecular techniques may be particularly useful in resolving the dispute relating to cultivar infringement (i.e. someone selling another's cultivar) dealt with by breeders in the courts. For example, molecular techniques are used in Canada by the Grain Research Laboratory (GRL) for cultivar identification testing of wheat and barley. The GRL has used two methods: acidic polyacrylamide gel electrophoresis (acid PAGE) and high performance liquid chromatography (HPLC). Protein fingerprinting works well, there are however limitations. With acid PAGE, estimates of sample composition are based on single kernels and large numbers of kernels can be necessary for statistically reliable estimates. HPLC can be used on ground samples, but it is not suitable for complex mixtures. Both methods are limited by finite protein diversity; there are a limited number of protein differences among cultivars and not all cultivars are distinguishable.

Quantitative DNA methods are now being developed using SNP and insertion/deletion polymorphisms (Indels). The goal is to be able to look at ground samples of grain to determine the cultivars present in a mixture and their proportions. Key challenges ahead include the development of accurate and sensitive quantification methods and, ultimately, the development of portable technologies that are capable of delivering rapid results.

Japanese barberry (*Berberis thunbergii*) is an ornamental shrub desired for its hardiness and attractiveness. However, because it is a host of black stem rust of wheat, it has been banned from importation. With permitted importation of 11 rust-resistant cultivars, molecular identification methods were used to assist CFIA inspectors with identification of the permitted cultivars, as their appearance is not always consistent with the morphological criteria, particularly for plants imported in the dormant state. AFLP test results identified 33 reference polymorphic bands. A sample is of the same cultivar if 31 or more polymorphic bands are shared, whereas if the number of shared bands is 28 or fewer, it is not considered to be of the same cultivar. Should results show 29 or 30 shared bands, the DNA is re-extracted and more primer sets are used to set the reference bands to 64 (CFIA/NFS, 2005).

While one gene or one trait may be sufficient for identifying a cultivar in one species, it may not be sufficient in all species. It may be appropriate to take a case-by-case and crop-by-crop approach.

13.6.4 Seed certification

The purpose of seed certification is to provide high quality seed to consumers by maintaining the cultivar identity and purity of seed and ensuring high standards of germination, seed health and mechanical purity. For example, in Canada for seed to be certified, it must be a recognized cultivar, multiplied according to strict rules that include process standards and cultivar purity standards established and monitored by the Canadian Seed Growers' Association (CSGA).

'Common seed' must meet germination, disease and mechanical purity standards, but there are no cultivar identity or purity guarantees associated with the purchase or use of the seed. For most crop kinds, common seed may not be sold by cultivar name. Common seed is less costly than certified seed and includes farm-saved seed.

'Certified (pedigreed) seed' is used by farmers who want additional assurance on seed quality, cultivar purity and performance. It is derived from a crop that has been issued a crop certificate from a seed association like CSGA indicating it has been granted Breeder, Select, Foundation, Registered or Certified status. Production of certified seed involves the planting of known seed stocks, previous land use restrictions, minimum isolation distances and field inspections.

CFIA seed laboratories use a number of test methods for the determination of cultivar purity and identity, depending on the crop kind and other factors. CFIA seed laboratories are International Organization for Standardization (ISO)-accredited and therefore carry responsibilities regarding the use of validated methods. The methods they use are classified as routine or non-routine and range from field, growth chamber and greenhouse grow outs to PCR.

Certified seed must be processed by an approved conditioner or by the grower of the seed and it must be sampled, tested and graded by accredited industry personnel. Molecular markers are not currently used in CFIA seed laboratories because seed certification has traditionally been based on phenotypic traits observable during crop inspection. However, molecular markers have the potential to be used as a control tool to ensure seed certification is working as it should. This could expand the level of public confidence regarding the purity level and security of certified seed or grain.

13.6.5 Seed purification

Breeder seed purification is a 3-year process. For example in wheat, the first year involves head selection from seed increases of material also being tested in first-year collaborative trials. Year 2 includes growing single-head-derived breeder lines in hill or short row plots with line discards based on visual and in some cases chemical phenotype. During the third year the remaining breeder lines are grown as individual breeder long rows with further line discarding based on visual phenotype with all rows of the remaining single-head-derived lines bulked as the first breeder seed. In addition during year 3, the cultivar is visually described in order to be registered and for purposes of future pedigree seed production (CFIA/NFS, 2005). Purification for molecular characterization is the same, with line discards during year 3 also based on molecular characterization and purification, but the molecular characterization process occurs in the laboratory rather than in the field.

An important application of molecular techniques in seed purification is seed purity testing, particularly for hybrid crops. Using cucumber as an example, Staub (1999) illustrated the usefulness of genetic markers in hybrid seed production including purity testing. Xu, Y. (2003) discussed the use of molecular markers in seed quality assurance including identification of off-types and false hybrids in rice seed production. When a two-line hybrid system is involved, the false hybrids in rice usually come from the selfing of the female parents due to the sterility instability of environment genic male sterility lines caused by temperature fluctuation beyond their critical temperature for fertility conversion. The false hybrids that co-exist with real hybrid seeds can also happen to other hybrid crops because of various reasons.

13.7 Plant Variety Protection Practice

13.7.1 Plant variety protection in the EU

The European Community's plant variety rights system (CPVO) was established in 1994. The IPR granted under this system are valid throughout the 25 member states of the EU. Most of the members of the EU are members of UPOV. The system is in line with UPOV 1991. It provides a one application, one procedure, one examination, one decision approach to the granting of rights.

The system follows UPOV's DUS requirements. The cultivar must also be 'novel', i.e. commercialized for less than 1 year within the EU and commercialized for less than 5 years outside the EU (6 years for trees). Protection is granted for 25 years (30 years for trees, vines and potatoes) and provides that authorization of the right-holder is required for the multiplication, sale or international trade of the cultivar.

There is currently no use of molecular techniques in CPVO DUS testing protocols, however CPVO funds research and development projects on the potential use of molecular techniques and supports ongoing discussions and consultations on implications and issues related to molecular techniques. CPVO received requests from breeders to add a 'genetic fingerprint' to the official cultivar description to facilitate the enforcement of European Community plant variety rights.

13.7.2 Plant variety protection in the USA

In the USA, IPR for plants are provided through plant patents, PVP and utility patents. Plant patents provide protection for asexually reproduced (by vegetation) cultivars excluding tuber crops. PVP provides protection for sexually (by seed) reproduced cultivars including tuber crops, F_1 hybrids and EDVs. Utility patents currently offer protection for any plant type or plant parts. A plant cultivar can also receive double protection under a utility patent and PVP.

The US Plant Variety Protection Office is responsible for administering the PVP Act, which provides plant cultivar owners with exclusive marketing rights within the USA. The requirements of protection are that the cultivar be new, uniform, stable and distinct from all other cultivars. The PVP Act states that a novel cultivar is distinct when it 'clearly differs by one or more identifiable morphological, physiological, or other characteristics...from all prior cultivars of public knowledge'. The meanings of 'characteristic' and 'identifiable' are purposefully vague in this definition to allow for future advances in knowledge and methodology.

PVP Office protection applies to cultivars that are sexually (seed) reproduced or tuber propagated and F_1 hybrids. Cultivars sold or used in the USA for longer than 1 year or more than 4 years in a foreign country are ineligible for protection. Fungi and bacteria are specifically excluded by the PVP Act. Asexually propagated crops fall under the purview of the US Patent Office.

A Certificate of Protection remains in effect for 20 years from the date of issue, or 25 years in the case of vines or trees. There are two exemptions to the rights granted. One exists to allow farmers to save seed for use on their own farm. Another exemption allows research to be conducted using the cultivar. This allows for the free exchange of germplasm within the research community.

Important events in the US history of IPR for plants and agriculture include: (i) hybrid cultivars could be protected through trade secrecy (1930s); (ii) the Plant Patent Act (1930), administered through the US Patent Office, provided protection for asexually propagated plants only (plants reproduced through buds or grafting) including horticultural crops and nursery stocks, with potato excluded; (iii) the Plant Variety Protection Act (1970), with the goal to promote commercial investments in plant breeding, provided 'patent-like' protection for plants reproduced by seed; and (iv) the Utility Patent of 'living organisms' (1980), as shown in *Diamond* v. *Chakrabarty* Supreme Court decision in 1980, established that 'anything under the sun made by man' is patentable, broadened patent law to encompass living organisms and established ownership of plant cultivars, traits, parts and processes.

13.7.3 Plant variety protection in Canada

The Canadian Plant Breeders' Rights Act came into force on 1 August 1990. The legislation makes it possible for plant breeders to legally protect new cultivars of plants. Crop cultivars may be protected under the legislation for a period of up to 18 years. All plant species are eligible for protection.

The owners of new cultivars who receives a 'Grant of Rights' will have exclusive rights over the use of the cultivar and will be able to protect their new cultivars from exploitation by others. To be protected, a cultivar must be new, distinct, uniform and stable.

The Plant Breeders' Rights Office, which is part of the CFIA, functions to secure the rights of plant breeders by granting protection for their new cultivars. It reviews and accepts applications, conducts site examinations, reviews data and comparative descriptions, publishes descriptions of cultivars and comparative photographs and grants rights.

13.7.4 Plant variety protection in developing countries

Systems for IPR have been recognized for more than a century, yet until recently IPR have not been an issue in the plant breeding and seed sector in most developing countries. Developing countries are being urged to strengthen IPR to foster innovation and expand trade. The field of agriculture is no exception and the TRIPS Agreement requires all WTO members to provide either patent or *sui generis* protection for plant cultivars. Developing countries will almost certainly look towards *sui generis* options for PVP to meet their TRIPS obligations (Tripp *et al.*, 2006). IPR are being introduced or strengthened in developing countries as a result of the TRIPS Agreement of the WTO, bilateral trade negotiations and pressure from export-oriented sectors in agriculture.

Most developing countries are in the early stages of implementing and/or enforcing IPR related to plant cultivars. The use of IPR in plant breeding in developing countries raises a number of important issues, including smallholders' access to technology, the role of public agricultural research, the growth of the domestic private seed sector, the status of farmer-developed cultivars and the growing north–south technology divide that restricts access to plant germplasm and research tools (IBRD/World Bank, 2006).

Relatively few developing countries have any significant experience with protecting cultivars. In systems where there is heavy emphasis on hybrid cultivars and considerable commercial competition, such as those in China and India, most interest centres on PVP for parent lines and hybrids, particularly in rice and maize. In countries where the production of ornamental plant materials is important, these materials dominate PVP applications.

The protection of transgenic crops has proven particularly difficult in developing countries. Most experience with transgenic crops resolves around Roundup Ready® soybean and *Bt* cotton. IBRD/World Bank (2006)'s report shows that the presence of IPR systems is not necessarily correlated with the effectiveness of controlling access to seed of transgenic cultivars.

Pressure to strengthen IPR in plant breeding in developing countries presents both immediate and long-term challenges to policy makers and development investors. The immediate challenges are related to framing and implementing appropriate legislation that is consistent with TRIPS and that supports national agricultural development goals. The long-term challenges are derived from the fact that an IPR regime, on its own, is not likely to provide the incentives that elicit the emergence of a robust plant breeding and seed sector; attention to other institutions and the provision of an enabling environment are also necessary (IBRD/World Bank, 2006). Collaboration and understanding between the 'south' and the 'north' should be strengthened for a better worldwide PVP.

13.7.5 Participatory plant breeding and plant variety protection

Participatory plant breeding (PPB) is the development of a plant breeding programme in collaboration between breeders and farmers, marketers, processors, consumers and policy makers (food security, health and nutrition, employment). In the context of plant breeding in the developing world, PPB is breeding that involves close farmer–researcher collaboration to bring about plant genetic improvement within a species.

PPB is seen as a way to overcome the limitations of conventional breeding by offering farmers the possibility to choose, in their own environment, which cultivars better suit their needs and conditions. PPB exploits the potential gains of breeding for specific adaptation through decentralized selection, defined as selection in the target environment and is the ultimate conceptual consequence of a positive interpretation of genotype-by-environment interactions (Ceccarelli and Grando, 2007). As one of the models, selection is conducted jointly by breeders, farmers and extension specialists in a number of target environments and the best selections are used in further cycles of recombination and selection.

In developing countries, plant breeding in the public sector is seldom a profit-making activity. Public sector plant breeders rarely make financial gains from their released products. This is unlikely to change if plant breeders' rights are introduced. Hence the issue of how to reward farmers is not complicated by a need to divide profits. Farmers participating in breeding programmes benefit from early access to new material, gain recognition from the community and learning new techniques. In Nepal, farmers involved in PPB have gained all of these benefits and have sold seed of the new cultivar at a higher price than the local landrace (Witcombe, 1996).

13.8 Future Perspectives

13.8.1 Extension and enforcement

A PVP system will not meet its goals unless it is supported by the full range of stakeholders. Breeders, seed producers, traders and farmers need to understand the objectives of the system in order to comply with it. The development of a PVP system should thus include an extensive information campaign involving all stakeholders, including the legal profession. One of the major challenges for a PVP system is providing effective enforcement. Establishing elaborate restrictions on seed use is counterproductive if there is no enforcement capacity. Private companies and public institutes that lobby for the establishment of PVP must be made aware that most enforcement responsibilities will fall on their shoulders. Likewise, identifying offenders is of little use if the court system is unable to understand or interpret PVP legislation. Developing judicial experience in PVP may take some time (Tripp *et al.*, 2006).

The following 'next steps' should be taken in moving forward as suggested for Canada by CFIA/NFS (2005), which should be applicable to other countries:

1. Develop standardized protocols.
2. Update marker systems.
3. Develop stakeholder agreement on threshold levels and techniques to be used.
4. Work towards harmonization of crop-specific protocols as they relate to PBR both nationally and internationally.
5. Develop a means for validation of tests and accrediting laboratories.
6. Review current national and international crop specific projects to determine all possible available markers.
7. Initiate research projects for selected species.
8. Canada should establish and lead a BMT subgroup on barley and possibly one on peas; and should participate in existing soybean, wheat and canola BMT subgroups.
9. Improve Canadian involvement in and feedback to Canadian experts and stakeholders from UPOV BMT meetings.
10. Explore further collaboration with the National Forum on Seed.

13.8.2 Administrative challenges for implementing PVP

In addition to establishing a framework for PVP legislation, there are administrative challenges for implementing PVP, including decisions on where to house the new authority, how to establish eligibility of a new cultivar for protection, which crops to protect first, how to recruit personnel with requisite technical and legal capacities and how the authority can pursue cost recovery while ensuring that small players can afford to apply for protection (Tripp *et al.*, 2006). Research managers and policy makers responsible for public research, who are commonly in favour of using IPR in public sector breeding, have to consider the potential impact on breeding strategies and on the costs and benefits before giving their unconditional support to IPR in plant breeding and their use in public agricultural research (Louwaars *et al.*, 2006).

Developing molecular techniques for IPR in plant breeding requires greater attention to strengthening capacities in national patent offices. As new methods of cultivar identification become available, the PVP Office should consult with the plant breeding community and research experts to best use these procedures. On the other hand, new tools also raise some concerns, including legal considerations relating to conformity with the UPOV Convention and the potential impact on the strength of protection. For example, countries that use transgenic cultivars will need to ensure adequate protection, although in many cases credible enforcement of the right combination of biosafety regulations, seed laws and PVP may offer adequate protection for transgenic cultivars, at least in the early stages of their availability in developing countries (IBRD/World Bank, 2006).

Not all crops need to be covered by PVP initially and choices should be made about which crop-breeding efforts would benefit most from IPR. With respect to public plant-breeding efforts, policy makers must distinguish between situations in which PVP will help stimulate the deployment of crop cultivars developed by public institutes and those in which PVP may turn national research institutes away from their public mandate. A further decision involves the protection afforded to extant (usually public) cultivars. Given that the rationale for IPR is to provide incentives for future breeding, rather than to reward past achievement, it seems reasonable to limit the protection periods for extant cultivars (Tripp *et al.*, 2006).

The general concept of a PVP-type system is appropriate and important to provide affordable IP for plant breeders while retaining the availability of germplasm as an initial source of variation in breeding. PVP remains especially important to provide IP for successful breeders who, either because of the incredible and still largely incomprehensible complex biology of their crop species or through lack of expensive technology cannot describe an individual gene and its agronomic impact, but who, none the less, develop improved cultivars that are needed in agriculture, horticulture, or forestry. Other forms of IP (trade secrets, contracts, patents) are also important.

The use of molecular techniques for cultivar registration needs to be harmonized with its use for PBR. To do this, international agreement on methods and procedures need to be established. There may be legal problems associated with the use of molecular markers that may require third party verification. Related government agencies should act as coordinators or verifiers of molecular markers and accredit or certify laboratories that wish to perform molecular techniques. The PBR Office may not be responsible for establishing standards or thresholds or for review of molecular markers. Instead, it would be handled in a similar manner as botanical descriptions.

13.8.3 The need to update UPOV

UPOV was updated once due to changes in technology. It is time to update the provisions once again to accommodate advances in technology that have occurred since 1991, in order to encourage continued infusions of new germplasm into breeding pools. As suggested by Donnenwirth *et al.* (2004), these UPOV updates should include:

- Providing compensation for and limits on saved seed in all countries.
- Making the EDV system more effectively defined to avoid technological loopholes.
- Revising the breeders' exemption to include a period of 'x' years from the date of a PVP application during which the breeders exemption would not be available for UPOV-protected material including commercialized cultivars.
- Requiring a seed deposit for all UPOV-related applications.
- Requiring the disclosure of all material deposited with PVP applications at the end of 'x' years and making all material deposited available for research under the breeder's exemption at the end of 'x' years unless the disclosure and availability would be in conflict with a utility patent on the same material.
- Placing all UPOV-related deposits (excepting parents and synthetics) into the public domain following expiration of UPOV protection.
- Creating a PCT (Patent Cooperation Treaty)-like system to facilitate filing of PVP applications on an international basis.
- Providing for and facilitating under UPOV global benefit sharing consistent with the International Treaty on Plant Genetic Resources for Food and Agriculture.

Janis and Smith (2007) made two novel and provocative claims in 'Obsolescence in intellectual property regimes'. They first argued that the legal regime for protecting new plant cultivars has become hopelessly outdated in light of recent changes in technology. They next asserted that the fate of the PVP system illustrates a broader and more disturbing phenomenon in IP law – the potential for *sui generis*, industry-specific IP regimes to become increasingly ineffective over time. Helfer (2006) believed that 'Obsolescence in intellectual property regimes' offered an insightful legal analysis of PVP, one of IP law's least understood *sui generis* regimes and that the article also made a persuasive case that the lynchpin of the PVP system is outdated and needs to be replaced with more flexible unfair competition principles. International and domestic policy makers interested in advancing innovation in the plant breeding industry and legal scholars concerned with the ever-evolving relationship between law and technological change would do well to consider the arguments from Janis and Smith (2007).

13.8.4 Collaboration in use of genetic resources

Historically, there has been excellent collaboration between the US Land Grant Institutions and publicly supported IARCs in crop improvement efforts. A hallmark of the collaboration has been the free exchange of plant germplasm and information. Now there are increasing restrictions to use and exchange of germplasm from the USA to the IARCs and from the private sector to the public sector, although the reverse cannot happen due to international public good nature of the CGIAR centres as well as their agreement with the International Treaty on Plant Genetic Resources for Food and Agriculture. This situation results in the following consequences:

- restricted access and use of germplasm;
- legal costs and enforcement;
- restrictions on progeny and publications;
- joint ownership of progenies and discoveries;
- complication caused by biotech patents on single genes and processes;
- public programmes increasingly being unable to access and use technology; and
- companies becoming increasingly restrictive and demanding.

On the other hand, international collaboration in the use of genetic resources becomes increasingly important and IPR issues are worth more attention. The recent revolution in the field of biotechnology has triggered off another round of controversy between the developed countries of the 'north' and the developing countries of the 'south' concerning access to genetic resources and equitable sharing of its benefits. Developed countries, as the genetic resource poor, assert ownership claims on associated technologies, while developing countries, as the genetic resource rich, claim ownership of genetic resources. The heart of the matter, however, lies in the application of conflicting conventions and protocols in respect of genetic resources and biotechnology: genetic resources are treated as public goods, while biotechnology is treated as a private good (Adi, 2006). Developing countries that claim ownership to a large reserve of the Earth's pool of genetic resource feel that this exposes them to the exploitative tendencies of multinational corporations (MNC) that are mainly owned by developed countries of the 'north', considering 74% of agbiotech patents held by six 'gene giants' (Monsanto, Dupont, Syngenta, Dow, Aventis and Grupo Pulsar) (http://www.etcgroup.org/upload/publication/247/01/com_globilization.pdf). MNC are sometimes regarded as exploiting the advantages as well as the weaknesses in the various conventions increasingly to monopolize the seed and germplasm industry, without due consideration for farmers and developing countries (Adi, 2006). It will take a long time to introduce a more even playing field that is mutually favourable to both parties and to establish a better regime of benefit sharing that recognizes farmers' or indigenous rights alongside patents and plant breeders' rights.

Public–private partnerships will need to be established to manage IP issues related to the transfer of information, material or technologies from private companies to developing countries (Naylor *et al.*, 2004). The African Agricultural Technology Foundation is one initiative that has been established to deal with such issues. Several private corporations with major investments in MAS in maize have agreed to provide access to germplasm and knowledge for African countries (Naylor *et al.*, 2004; Delmer, 2005).

13.8.5 Technology and intellectual property interaction

Technology can be a two-edged sword with respect to the effective level of IPR and the utilization of genetic resources. While technology can facilitate the use of genetic resources, it can also be used in a fashion that threatens to undermine existing levels of IPR. Donnenwirth *et al.* (2004) gave the following examples:

- Molecular marker technologies can be used to attack trade secrets by rapid identification of female parent inbred line contaminants in bags of hybrid seed. These inbred lines might then be used directly as parents of hybrids or as parents for further breeding.
- Molecular marker technology can be used to identify segregating molecular characteristics in an otherwise uniform cultivar and thus to select a distinct 'new' cultivar from the segregating source without any breeding effort being expended.
- An existing cultivar could be transformed by genetic engineering and thus achieve cultivar status by virtue of its distinctness but without any effort expended to change the genetic base of the cultivar.
- An existing cultivar could be changed just sufficiently and even only cosmetically using marker-assisted breeding so that it retains the important agronomic attributes of the initial cultivar but would evade the dependency resulting from its status as an EDV through selection for a molecular marker profile that is 'sufficiently different' from the initial cultivar.
- An existing cultivar could be changed dramatically in its overall DNA marker profile yet contain some or all of the key genetics impacting important agronomic traits due to targeted selection of its genetics using molecular marker or genomics data.
- An inbred containing the key genetics of the female parent of a hybrid can be rapidly recreated using one or a suite of technologies including di-haploidy, molecular markers, genomics, winter nurseries and high-throughput laboratory genetic profiling and screening. The inbred can then either be used as a parent of a hybrid or as a parent for further breeding.
- An inbred containing the key genetics of the male parent of a hybrid (hitherto essentially impossible to access via a hybrid) can similarly be recreated and used.

13.8.6 Seed saving and plant variety protection

Seed saving is a historical cultural phenomenon that dates back to the beginning of agriculture itself. It helps farmers control their enterprises and maintain their independence; it allows them to predict how well a crop will perform in the following season; it allows them to participate in maintaining the crop; it serves as insurance against inadequate supplies of seed; it helps to maintain food security; and it creates a viable market that ensures that seed prices remain affordable (Mascarenhas and Busch, 2006). Because a seed contains within itself the means for its own reproduction, seeds have offered a particularly large stumbling block to 'capital accumulation'. In the USA, IPR legislation and Supreme Court decisions have played a profound role in overcoming these unique characteristics. According to the ETC Group (2005) the top ten seed companies including Monsanto, Dupont and Syngenta now account for an estimated market value of US$21 billion for commercial seed sales worldwide and about 50% of the global seed market. Mascarenhas and Busch (2006) argued that the combination of expanding IPR, 'new' GM technology and the ideology of the technological treadmill have successfully overcome seeds' inherent obstacles to capitalist accumulation. As a result, US farmers are facing further loss of control of the farm production process.

For example, US large soybean farmers have consistently saved seed in the USA – as much as 60% in some years. However, with the introduction of Roundup Ready® soybeans the nature of seed saving was drastically changed. Savings rates have ranged from a peak of 63% in 1960 to 33% in 1991. The decline in saved soybean seed from 1955 to 1974 – before GM soybean – was approximately 1.4% year^{-1}. However, with the introduction in 1996 of Monsanto's Roundup Ready® soybean, a genetically modified herbicide tolerant cultivar, the rate of decline in soybean seed saving increased to 2.3% year^{-1} from 1996 to 2002.

More remarkable perhaps has been the intensive adoption of Monsanto's Roundup Ready® soybean since its introduction, e.g. Monsanto accounted for 91% of the worldwide GM-soybean area in 2004 (ETC Group, 2005). However, Ervin *et al.* (2000) suggest that when examined worldwide, all currently available transgenic crops account for a yield increase of no more than 2%. On the contrary, government data sources reveal that in some areas seed saving has all but ceased (USDA, 2002a). In order to explain the apparent contradiction, Mascarenhas and Busch (2006) invoked the theory of 'technological treadmill' (Cochrane, 1993) and they considered the rapid adoption of Roundup Ready® soybeans a classic example of the 'technological treadmill'. As the theory suggests, given the inability of farmers to affect the prices they receive for their commodity crops, farmers can only increase their profits by adopting new technologies that decrease their costs. However, only early adopters of new technology gain because the 'efficiencies' – usually in terms of increased profits – gained from widespread adoption itself pushes the prices received by all farmers downwards, thus abolishing any comparative advantage. When confronted with the rapidly expanding technologies of nature's production farmers are left with few options: loyalty to the 'technological treadmill' or exiting the industry all together, the latter being an option few are willing to consider.

For as rapidly as seed saving has been declining, the cost of seed has been rising. For example, in 1975 a bushel of soybean seed cost US$7.34. Twenty years later, in 1994, it was US$12.21. However, in 1997, 1 year after the introduction of Roundup Ready® soybeans, the price of soybean seed jumped to US$17.40 and 6 years later, in 2003, sold for US$24.20 per bushel. The decline in seed saving has shifted a significant portion of the value of bin-run seed from farmers to commercial seed retailers and their parent owners. The value of bin-run seed in 2000 was about US$170 million or approximately half its value before the introduction of Roundup Ready® soybeans. This decline in bin-run seed amounted to approximately US$374 million in additional profits in 2001 to commercial seed retailers (Mascarenhas and Busch, 2006).

The above information is not to say that farmers who adopted Roundup Ready® seed necessarily lost money. The major draw of Roundup Ready® soybeans was that they required less farmer labour and management time. This point is significant, particularly when one recalls that the persistence of family farms has been through their ability to self-exploit farm labour. Furthermore, GM soybeans are relatively simple to use and increased flexibility in herbicide application – provided that one used a glyphosate herbicide such as Roundup® – allows spraying to occur throughout most of the crop cycle. This flexibility also fits well with conservation tillage and other production inputs currently in practice (USDA, 2002b).

The decline in seed saving that has happened in the USA as shown by soybeans and would also be expected elsewhere in the world bring up two important issues. First, to prepare for the natural disaster or civil disturbance (such as particularly severe weather and global climate change), it is essential to domestic food security that farmers: (i) already have some saved seed on hand; and (ii) have the requisite skills needed to properly save that seed – skills that are only maintained if seed can be regularly saved. The second issue is associated with genetic diversity that might be narrowed down because of the decline in seed saving. If this narrowing continues it will result in great homogeneity in domestic crops that are planted across expansive contiguous areas. And, as demonstrated in the case of the 1972–1973 southern corn leaf blight, this lack of planted biodiversity can prove to be very costly.

13.8.7 Other plant products

As discussed in Chapters 1 and 5, plants provide various resources for humans, and crop cultivars developed by breeders through various breeding programmes are mainly for food, feed, fibre, fuel and other needs. In many cases, breeding efforts are combined with collection and selection of natural products from plants to meet human demands. As an example, traditional medicine has been an important part of human health care in many developing countries, and developed countries as well, for many years. It is estimated that between 25,000 and 75,000 plant species are used for traditional medicine, 1% of which are known by scientists and accepted for commercial purposes (Aguilar, 2001). The world market for herbal medicines has been estimated at US$60 billion with annual growth rates of between 5 and 15%. At the moment, the mechanisms for IPR are not able to protect traditional knowledge and indigenous peoples (Kartal, 2007). Local communities believe that they are subject of bio-piracy, which is unauthorized use of traditional knowledge or biological resources (Aguilar, 2001). Researchers or companies may claim IPR over biological resources and traditional knowledge, after slightly modifying them. However, the IP laws do not protect traditional knowledge adequately. A harmonized system of traditional knowledge protection should be effectively implemented on both national and international levels (Kartal, 2007). A golden triangle consisting of traditional knowledge, modern medicine and modern science with systems orientation will converge to form an innovative discovery engine for newer, safer and more affordable and effective therapies (Patwardhan, 2005). Countries and peoples providing the resources for natural products research and drug development now have well-defined benefits and rights. This will have a direct impact on the sharing of benefits accruing from collaborations (Gurib-Fakim, 2006). Increased legislation, rather than assisting traditional knowledge holders may in fact be to their detriment and also discourage companies from investing in bio-prospecting activities. Of greater concern to the world is the loss of traditional knowledge brought about by rapid sociocultural changes and inhibition of research by creating bureaucratic minefields that may prevent this knowledge from being documented and transmitted (Dutfield, 2003).

The book edited by Biber-Klemm and Cottier (2006) discussed the basic issues and perspectives on rights to plant genetic resources and traditional knowledge. It covers the means, instruments and institutions needed to create incentives to promote the conservation and sustainable use of traditional knowledge and plant genetic resources for food and agriculture, within the framework of the existing world trade order.

14

Breeding Informatics

In previous chapters, we have discussed genetic variation as it relates to plant breeding and the molecular tools used for the dissection, transfer and selection of novel traits and genes. Using these molecular techniques may result in the generation of a large amount of data. Extracting useful information from this ocean of data requires the integration of different sources of data and the ability to analyse and visualize the data in effective and efficient ways. The genomics projects of the last decade are most often carried out within universities, research institutes and companies, closely allied with laboratories producing large quantities of data. While bioinformatics has been involved with the primary data, it has yet to become focused significantly on applied areas such as plant breeding. However, this situation is beginning to change. Advances in plant breeding will depend heavily on how well we can manage and utilize all relevant information (Xu *et al.*, 2009b) In this chapter, breeding-related informatics will be discussed, including information collection, storage, integration and mining.

14.1 Information-driven Plant Breeding

As in other disciplines, plant breeding and in particular molecular plant breeding, has been driven by the availability and accessibility of various types of information. The first computer network, ARPAnet, was developed in the late 1950s as a product of the Cold War. By the 1980s, universities throughout North America and Western Europe were connected via countrywide networks such as the UK's Joint Academic Network (JANet). Molecular biologists were regularly logging in to central servers to run sequence analysis programs and transferring data from one machine to another. In the early 1990s, the World Wide Web (WWW) was invented and turned the Internet into the worldwide cultural phenomenon that it is today. The WWW has made the concept of a 'global village' developed by Marshal McLuhan decades ago into a reality. In 1991, Tim Berners-Lee and Robert Caillou, scientists working at CERN (the Organisation Européenne pour la Recherche Nucléaire: *European Organization for Nuclear Research*) in Geneva, developed the Hypertext Transfer Protocol (HTTP) as a way of linking and cross-referencing documents held on different computers. Many professionals, including plant breeders, now make regular use of the Internet as an integral part of their work.

Through these networks, information has been transferred at an increasing rate across the world and the quantity of information has been increasing exponentially.

Considering only DNA sequence data, the volume of biological information is doubling roughly every 6 months. This is faster than the exponential rate of increase in computing power, as suggested by Moore's law (an empirical observation made long ago that has held until today: the doubling of processor power every 12 months) (Sobral, 2002). With the development of high-throughput technologies, genotypic information including genetic polymorphisms and gene expression profiling, will increase exponentially. Since the first molecular genetic maps based on restriction fragment length polymorphism (RFLP) markers were developed in the 1980s, a significant amount of molecular marker and genetic map information has been generated and become available for many plant species. Genotype information is now generated primarily using PCR-based markers such as simple sequence repeats (SSRs) and single nucleotide polymorphisms (SNPs) and high-throughput systems. These molecular polymorphisms can be accurately sized and readily compared across laboratories and experiments.

Historically, plant breeding has been driven by phenotypic information and a large amount of phenotype and pedigree data has been accumulating for many decades in plant breeding programmes. A typical example is the use of multi-environment trials (METs), which have been in general practice for most plant breeding programmes for years. Yield trial data in many private breeding companies can often be traced back to the very beginning of each specific breeding programme. Most breeding institutions/companies have extensive facilities and expertise in collecting phenotype data for various agronomic traits. Integrating this type of information with other sources of information from genetics and genomics will lead to more efficient use of both types of information for plant breeding.

14.1.1 Basics of informatics

Bioinformatics can be considered a combination of several scientific disciplines including biology, biochemistry, statistics and computer and information science. It involves the use of computer technologies and statistical methods to manage and analyse a huge volume of biological data. Bioinformatics provides a common conceptual framework for molecular biologists, biochemists, molecular evolutionists, statisticians, computer scientists, information technologists and many others to work together.

Databases allow people to organize and manipulate large amounts of data and to quickly translate and deliver that information in useful summaries and formats. A database can be defined as a structured collection of records or data which is stored on a computer. The database can be queried and the records retrieved can be used to make decisions. The computer software used to manage and query a database is known as a database management system (DBMS).

The structural description of a database is known as a schema. The schema describes the objects that are represented in the database and the relationships among them. There are several different database models (or data models) and the most common in use today is the relational model, which represents information in the form of data records in different sets of tables and the relationships between them.

There are four main components to any database application: (i) a method for entering or editing data – usually data entry screens or import functions; (ii) a data storage mechanism – a way of storing the data on the computer; (iii) a query mechanism to allow users to filter and summarize data in structured ways; and (iv) a report generator to extract and interpret information from the stored data.

The first basic concept to understand about databases is the difference between data and information. What we call data is really a collection of facts in a specified domain; the facts may be measured values, observations, responses or even pictures. Data by itself is meaningless, but once it is organized in useful ways, it becomes meaningful information. Therefore, essentially a database is nothing more than a tool to organize and access large amounts of data so that people can turn it into useful information.

The content of a database determines its type. The main types of databases include those listed below and a combination of any of them:

- Bibliographic: examples are library catalogues and an article index. A library catalogue is a database that describes what the library owns. Each item in the catalogue describes a book or other item in the library. An article index is a database that describes the contents of a particular set of journals, magazines, newspapers and/or other documents.
- Full text: a full-text database provides the full text of a publication. For instance, the research library in GALILEO (Georgia Library Learning Online) provides not only the citation to a journal article, but often the entire text of the article as well.
- Numeric databases: examples are Census Bureau databases and databases for stock market information, each containing primarily numeric data (statistics, census data, economic indicators, etc.).
- Image databases: these collect only image information (EBSCO host image collection, www.ebscohost.com).
- Audio databases: those containing MP3 or wav files, etc.

Meta-databases contain information about databases. They allow users to search for content that is indexed by other databases. For example, the Genomes Online Database (GOLD, http://www.genomesonline.org/) is an Internet resource for access to information regarding complete and ongoing genome projects around the world and JAKE (jointly administered knowledge environment, http://jake.openly.com) is a meta-database of bibliographic databases. If you find a citation for an article in one of the bibliographic databases and want to determine if the article is available in full text in another database, you could do a search for the journal in JAKE to get a list of all the databases that index that specific publication and whether those databases include it in full text.

14.1.2 Gaps between bioinformatics and plant breeding

There has been some delay in the uptake of bioinformatics within the plant breeding community. Most bioinformatics databases are lacking information on phenotypes, traits and other organism data, largely because bioinformatics grew out of the fields of molecular biology and biochemistry. When applied to plant breeding, bioinformatics data must be combined with other types of information, including plant phenotype and information on the environment where the phenotype is measured. Therefore, breeding informatics focuses on the development of breeding-centric databases and algorithms and statistical tools to analyse, interpret and mine these datasets (Xu *et al*., 2009b).

Although the recent explosion of genetic and genomic data for a wide range of plant species has led to a proliferation of publicly available plant databases, this wealth of knowledge has not yet found its way into mainstream plant breeding. There may be several explanations for this. First, it is not obvious to many plant breeders how or if much of the primary information generated in plant genomics can be applied to real-life breeding situations. Secondly, breeding requires the integration of information from different sources, usually stored in different databases and managed by different groups of scientists (for example, pedigree, genotype and phenotype). Thirdly, many of the publicly available tools and interfaces available for bioinformatic data are oriented at the cellular/molecular level, while most breeders are working and thinking at the organism level. Fourthly, until recently, much of the genomic research and therefore the publically available data has concentrated on the comparison of genes between species rather than the gene diversity within species required for plant breeding. Therefore, there is a need to re-orient the tools and information so that crop researchers and biologists in general can query and use them properly. As in many informatics projects, an essential factor for success in plant bioinformatics will be the ability to integrate related information and to view and analyse it with tools that support decision-making functions. As the volume of information continues to increase, the need for such tools grows.

Bioinformatics data typically includes cDNA and genomic sequence data, genetic maps of mutants, DNA markers and maps, candidate genes and quantitative trait loci (QTL), physical maps based on chromosome breakpoints, gene expression data and libraries of large inserts of DNA such as bacterial artificial chromosomes and radiation hybrids. Information flow from molecular markers to genetic maps to sequences and to genes has been established. However, there is a gap between the sequence-based information and breeding-related information such as germplasm, pedigree and phenotype. We will depend on phenotyping as the basis for the functional analysis of about 40% of genes, even though a complete sequence is available. Therefore, the integration of breeding-related information with genomics databases is required for genomics-based breeding programmes.

Mayes *et al.* (2005) discussed how genetic information can be integrated into plant breeding programmes to produce cultivars from molecular variation using bioinformatics and what crop scientists might want from bioinformatics. They examined how bioinformatics tools might be used to track down the underlying genes controlling sustainability traits and how these may then be exploited in plant breeding programmes using marker-assisted selection (MAS).

14.1.3 A universal system for information management and data analysis

Modern plant breeding needs a standardized and widely accepted system for information acquisition, deposition, classification, integration, interpretation and utilization. Three major types of information – genotypic, phenotypic and environmental – should be brought under a single umbrella, with comprehensive tools for integrating, extracting and analysing useful information.

The success story from the data-hosting institutions and those for managing large-scale genome sequencing programs provide us with a clue that a web-based information management system is a better bet for modern plant breeding programmes than local, stand-alone systems. The web-based information management systems, once fully developed as stand-alone systems, are anticipated to offer several advantages, including the following:

- They provide highly efficient cutting edge technology solutions for breeding institutions/programmes looking to dramatically improve quality and reduce costs of information management.
- They provide a universal information management system that is suitable for all breeding programmes so that end-user application setup and maintenance is simplified in the Internet computing environment because there is nothing to install, configure, or maintain on the user's computer. Each institution has no need to maintain, manage, and integrate their data using their own facilities and personal.
- They accelerate breeding procedures by providing a much more affordable information management system to breeders. The application needs only to be installed, configured and modified on the web server, reducing the risk of inconsistent configurations and incompatible versions of software between client and the server machines.
- They create a knowledge base for typical customer support questions; a single integrated source for all customer support inquiries; and ability to analyse information more effectively.
- They stimulate collaborations by providing more accessible approaches to sharing data and sharing intellectual property based on mutual interests.
- They provide effective, flexible and competitive approaches to converting data into knowledge that is critical to companies and institutions continuously seeking ways to improve their products and services.

With such a system, customers can use the power of data management and analysis, along with computational biology and comparative genomics, to create an intellectual property portfolio associated with their particular cultivars and hybrids. The International Crop Information System (ICIS) to be discussed later in this chapter is under development towards providing such a universal system to worldwide plant breeding programmes, with a great potential but a long way to go.

14.1.4 Transforming information to new cultivars

A challenge to modern plant breeding is how to best utilize all relevant information efficiently and comprehensively, harnessing the power of informatics to support molecular breeding. Integrated exploration of genotypic, phenotypic and environmental information is critical for more efficient and predictable plant breeding.

A database would organize genetic information and give breeders the opportunity to pose specific questions through a software interface, helping them make selections and identify desired parents and progeny. Breeders will be able to look for particular traits they want to breed for by going back through breeding history and pedigrees to see where traceable characteristics could come.

As a result of years of research on genetic mapping, allele mining and molecular and functional diversity analysis of germplasm collections, we have amassed a large body of knowledge regarding the genomic location of factors/alleles that affect specific agronomic traits and the allelic variation available for utilization in plant breeding, but often it is not in an easy format for all researchers to use. For most of the traits, it is very difficult to detect the presence of particular alleles when the lines are only examined phenotypically in the field; however, by examining the lines at the DNA or sequence level, this becomes possible. Xu, Y. (2002) provided an example to illustrate the importance of this effort in information-driven plant breeding. A rice double haploid (DH) population derived from IR64/Azucena has been shared and used worldwide for the genetic mapping of many different traits with hundreds of genes/QTL identified, largely based on the first generation RFLP map. However the original phenotypic data has never been shared. The phenotypic data collected across laboratories should have been analysed through a meta-analysis and fine mapped using the updated genetic map consisting of about a thousand SSR markers, rather than individual efforts using the first version of the molecular map consisting of only 175 RFLP markers. The same story can be found in almost all well-studied crop plants. As both parental lines involved in the rice example have been used widely for breeding yield and adaptive traits, a collective effort bringing all related information on one page and mining it through an integrated data analysis would help transform them into new cultivars.

14.2 Information Collection

14.2.1 Data collection procedures

Planning the research and developing data collection strategies is the first step in research management. Before data collection begins, the following questions should be clearly answered: What hypothesis is to be tested? What data needs to be collected? How will this data be collected? What equipment or supplies are needed for the data collection?

Plant breeding information comes from many different sources and in many different forms, including a description of the plant itself, its genotype and phenotype and a depiction of the environment (Fig. 14.1). What data should be in the repository depends on many factors, however, if human and computing resources are not limiting it is advisable to preserve all the historical data, so that it can reanalysed for new hypotheses and guide new research. Whatever system will be built, it should be flexible, because there are clients

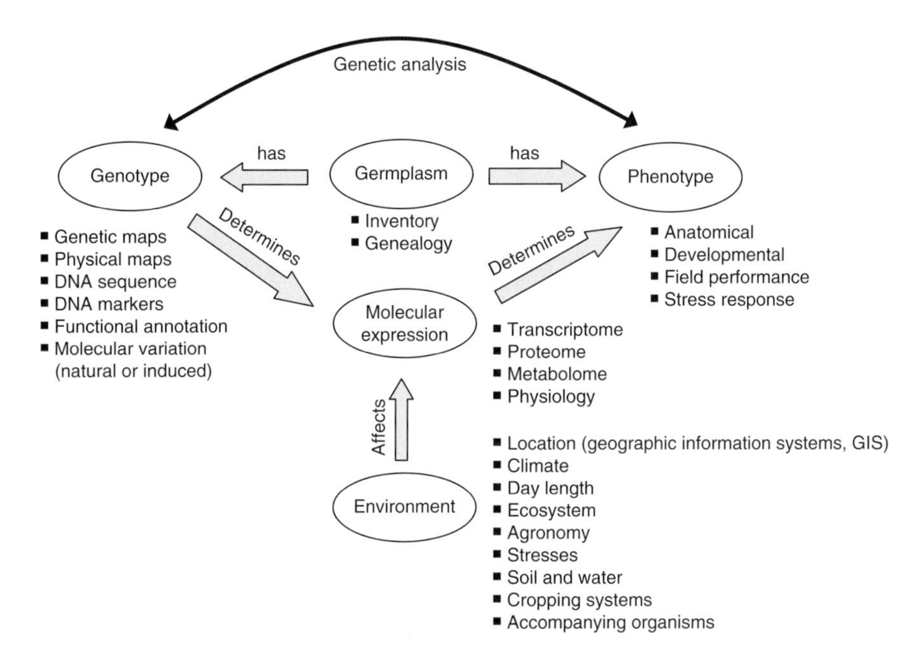

Fig. 14.1. Crop biological concepts, relationships and breeding related information. Modified from Richard Bruskiewich (ICIS Workshop, 2005, http://www.icis.cgiar.org).

who require minimum data but there are others who require all. In general, data should include germplasm information (passport data, pedigree and genealogy, genetic stocks), genotypic information (DNA markers, sequences, and expression information), phenotypic information and environmental information.

A reliable data-collection technique will ensure that information is systematically collected in a manner compatible with other existing information and it should take into account the following considerations: controls or checks to be used in data collection, sampling method, sample size, testing sites, replications and previously used data-collection techniques. Bias during data collection could come from defective instruments, biased observation, sampling errors, etc. Quality control for data collection can be done by checking relevant data and comparing and contrasting the collected information with expectations, controls and hypotheses. In addition, before the data is entered into a database, some preliminary organization and analysis might be needed. As phenotyping procedures are affected by multiple factors including environmental and measurement errors, multiple replications are usually required for most quantitatively inherited traits. To check the data quality and phenotyping reliability, data collected from multiple replications within a trial can be analysed for between-replication correlation. When a relatively large genetic variation exists within the tested population or cultivars, correlation coefficients should be high, e.g. 0.6 and 0.8 or higher, for traits with medium and high heritability, respectively.

Often there is relevant information that has already been collected by others, although it may not necessarily have been analysed or published. Taking the effort to locate and review this information is a good starting point and can help in planning a more efficient experiment.

14.2.2 Germplasm information

Information for a specific germplasm accession could include passport data, pedigree and genealogy information and all other measurements at a genotypic or phenotypic level. For a germplasm collection, the information can include genetic relationships and structure within the collection and other characteristics determined through population genetic analysis.

As a major effort in characterizing genetic resources for several important crop species, the Generation Challenge Program (GCP) has undertaken molecular characterization of thousands of accessions and is to cross-link the information to facilitate in the discovery of novel alleles and germplasm for crop improvement, with a focus on stress traits (drought in particular). This project contains a series of informatics development components, aimed at the development of an integrated web-enabled platform for crop research (Bruskiewich *et al.*, 2008). Obviously this kind of effort has been going on internally in most large breeding companies as well. Some selected Internet resources on germplasm resources are provided in Table 14.1.

Passport data

Passport data includes accession number and/or other numerical identifiers, attributes describing the origin (country of origin, collection site, collection expedition, donor institute), botanical classification (scientific name, taxonomic system, crop, regeneration method) and breeding information (institute, method and stage). Most plant germplasm databases contain passport data for their maintained accessions.

Pedigree and genealogy

A pedigree represents the ancestral history of a strain and shows how a particular accession/strain was derived from its parents, including crossing, selfing, backcrossing and selection. Genealogy is the ancestral relationships among sets of germplasm, which is a more general description of the breeding history of an accession/strain. For most crop plants, there is a registration system which requires pedigree and genealogy information. Names are assigned when a

Table 14.1. Selected Internet resources on germplasm resources.

Intergovernmental organizations
- Commission on Genetic Resources for Food and Agriculture (FAO): http://www.fao.org/ag/cgrfa/
- Consultative Group on International Agricultural Research (CGIAR): http://www.cgiar.org/
- Convention on Biological Diversity Secretariat: http://www.biodiv.org/
- FAO Plant Genetic Resources: http://www.fao.org/ag/cgrfa/PGR.htm
- Bioversity International: www.bioversityinternational.org
- CGIAR's System-wide Information Network for Genetic Resources (SINGER): http://singer.grinfo.net/
- System-wide Information Network for Genetic Resources (SINGER): http://singer.cgiar.org/

National/regional activities
- Asian Vegetable Research and Development Center: http://www.avrdc.org/
- Information System Genetic Resources: http://www.genres.de/genres-e.htm
- Centre for Genetic Resources, The Netherlands: http://www.cgn.wur.nl/UK/
- UK Plant Genetic Resource Group: http://ukpgrg.org/
- N.I. Vavilov Research Institute of Plant Industry, Russia: http://www.vir.nw.ru/
- Southern African Development Community (SADC) Plant Genetic Resources Project: http://www.ngb.se/sadc/sadc.html
- United States Department of Agricutlure (USDA) Genetic Resources Information Network: http://www.ars-grin.gov/
- Chinese Crop Germplasm Information System: http://icgr.caas.net.cn/cgris_english.html

Non-governmental organizations
- Conservation International: http://www.conservation.org/
- Global Biodiversity Forum: http://www.gbf.ch/
- World Resources Institute: http://www.wri.org/
- Genetic Resources Action International (GRAIN): http://www.grain.org/

cultivar is released or an accession/strain is collected; however, there is no standard for translating the name from one language to another. As a result, a given cultivar developed in a country like China can have several different names when it is translated into English, depending on which Chinese pronunciation system is used and which rules are used to separate multiple Chinese words in the cultivar name. Also, the way names are recorded by different breeders varies with dashes, spaces and codes.

Genealogy management systems have been established in several international research organizations. For example, the International Center for Agricultural Research in the Dry Areas (ICARDA) has such systems for barley and chickpea. Creation of a genealogy ontology is required for a standardized genealogy management system. Figure 14.2 provides an example of using the concept of ontology to name a germplasm accession.

Genetic stocks

A genetic stock is a plant sample that expresses a specific variation (or a specific small set of variations). Thus, genetic stocks are living examples of their underlying genetic variation. A database with genetic stock information helps scientists interested in a particular variation or locus to find and acquire living tissue that contains a desired variation.

The most frequently used types of plant genetic stocks include near-isogenic lines (NILs), single, series or genome-wide

- By an accession number
  - externally given
  - locally given
- Is a name for a commercial cultivar
  - heterozygous
    - propagated asexually
      - is a homogenous collection of a single heterozygous genotype
      - is a clonal traditional cultivar
    - propagated by crossing
      - of a small number of inbred lines
      - by farmer selection
      - by mass selection
      - of inbred lines
  - homozygous
    - derived from a single plant of a highly inbred population
    - is a traditional cultivar
    - is a deliberate mixture of fixed lines
- Is the name of a (collection of) homozygous individual(s)
  - from a breeding population
  - collected
    - from a bulk of many plants
      - is a weed
      - not a weed
    - from a single plant
- Is the name of a population (collection of heterogeneous /heterozygous individuals)
  - used for breeding
  - a recurrent selection cycle name
  - a recognized genetic stock
  - a deliberate mixture of populations
  - named after place of collection
  - collected off farm

Fig. 14.2. Genealogy ontology: how a germplasm accession is named. From the International Crop Information System (ICIA) Workshop (2005).

mutants, populations with segregating genotypes (recombinant inbred lines (RILs), DHs, introgression lines (ILs)), cytogenetic material (primary trisomics, translocation lines, etc), cell culture lines and gene and DNA clones. Genetic stock availability may vary greatly from one crop species to another. For example, the National Institute of Genetics (Japan) provides information on about 11,000 genetic stocks developed in Japan. These resources include marker gene testers, mutant lines, isogenic lines, autotetraploid lines, primary trisomics, reciprocal translocation homozygote lines, cytoplasm substitution lines and cell cultured lines.

14.2.3 Genotypic information

It is genotypic information that fundamentally drives the breeding informatics and products. Genotypic scores are determined by the genotype of an individual or a pooled DNA sample of multiple individuals. A genotype is determined by DNA sequences, the genes encoded by the sequences and the gene products translated from the sequences. Therefore, genotypic information is based on underlying DNA polymorphisms, which can be detected with many different techniques (Fig. 14.1). The action of these genes is not always additive, thus epistasis, the particular combination of alleles and genotype-by-environment interaction can be of great importance.

Molecular markers

Many molecular breeding projects involve a large number of molecular markers (hundreds to thousands) that cover the whole plant genome. These markers are used to genotype, or fingerprint, a large number of accessions (an entire or core collection, individuals derived from a selected cross, a set of landraces or a population). This will create a valuable database of information that can be used to determine which crosses or individuals may be more valuable than others.

Information for DNA markers (e.g. primers for PCR-based markers) includes characteristics describing the markers and how to best apply them to plant breeding. As an example, information about a PCR-based marker is given in Table 14.2, which is what a molecular database could manage and provide.

There are some databases that have been developed with features for displaying marker-related information. Despite the large numbers involved with SNPs and polymorphism data, presenting SNPs on a genome browser has become fairly straightforward. For example, Ensembl can show SNP locations as a track in ContigView displays, with colour coding to highlight those located in coding, intronic or upstream areas of genes. Clicking on an SNP produces a SNPView page with further details including, where appropriate, primers, validation status, heterozygosity, strain differences and links to entries in variation databases such as dbSNP and HGVBase (Hammond and Birney, 2004) and Panzea for plants.

Table 14.2. Information associated with PCR-based DNA markers.

Marker per se
Marker name and synonyms
Repeat motif/enzyme and repeat length
Primer sequence
PCR protocol (i.e. annealing temperature and number of cycles)
Expected allele size in a control cultivar or a group of cultivars
Number of alleles
Allele frequency (most/least frequent allele)
Signal strength
Allele size/range
Polymorphic information content
Chromosome location
Linkage to other markers
Images and gel pictures
References
Source (inventory)
Patent information
Historical data (e.g. associated with a trait)
Project data (date, title, germplasm, reports, etc.)
Marker-derived
Genetic maps (including haplotype block)
Physical maps
Consensus maps
Comparative maps

As discussed in the previous chapters, molecular markers are required for a broad spectrum of gene screening approaches, ranging from gene mapping within a traditional forward-genetics approach, to QTL identification studies, to genotyping and haplotyping studies. As we enter the post-genomics era, the need for genetic markers does not diminish, even in the species with fully sequenced genomes. Regardless of the ultimate reason for the application of molecular markers, there is a general need for the characterization of molecular markers and high density, uniform maps that represent whole genomes. The availability of complete genomic sequences in several plant species has led to the creation of several genome-based resources to expedite and facilitate traditional breeding and mapping approaches. Now, large expressed sequence tag (EST) collections are available for more and more plant species, with more than 1800 species having over 63 million ESTs in dbEST (25 September 2009; http://www.ncbi.nlm.nih.gov/dbEST). EST sequence information along with the underlying redundancy and parental associations has been used in predictive approaches for SNPs, SSRs and conserved orthologue set (COS) markers and the speed and success of marker characterization and validation typically exceeds traditional approaches. Given the technical ability to detect molecular markers *in silico*, there is truly a vast potential collection of molecular markers present with these sequences. A database resource, PlantMarkers, has been created for the prediction, analysis and display of plant molecular markers (Rudd *et al.*, 2005). Techniques have been developed to identify putative SNP, SSR and COS markers from the available sequence collections. A systematic approach to identify a broad range of putative markers has been undertaken by screening the available openSputnik unigene consensus sequences from over 50 plant species. Putative markers have been anchored to available protein-coding sequences where possible. Underlying sequence annotations that relate to clone library, strain or cultivar have been retained, allowing for the selection of putative polymorphic sequences that are segregating between different collections. A web presence at http:///markers.btk.fi provides functionality allowing a user to search for species-specific markers on the basis of many specific criteria, not limited to non-synonymous SNPs segregating between different cultivars or measured polymorphic SSRs.

Sequences

The typical data for sequences are strings of nucleic (or amino) acid residues. Each DNA or protein sequence database entry has the following information: an assigned accession number, source organism, name of locus, reference, key words that apply to the sequence, features in the sequence such as coding regions, intron splice sites and mutations and finally the sequence itself. Amino acid sequences are derived from the translation of cDNA sequences or predicted gene structures in genomic DNA sequences. Partial sequences are also derived by the translation of EST sequences or genomic DNA sequences in all six reading frames. There are many structural features that can be used to describe a specific protein, including active site, sequence and structural motifs, domains, fingerprints, primary structure, sequence and structural profiles, three-dimensional structure and family classification.

Plant genome sequence data have been accumulating from three major sources: (i) whole genome sequencing and assembly (e.g. *Arabidopsis thaliana*, rice, *Medicago truncatula*); (ii) genome survey sequencing (maize); and (iii) ESTs (for all target species). This data flow will likely continue, with a focus on the complete sequencing of 'reference species' (*Arabidopsis*, rice, sorghum, maize, *M. truncatula*, tomato), draft sequencing of other selected species and further EST and full-length cDNA sequencing.

A challenge facing the plant biotechnology and bioinformatics research community is the translation of complete genome sequence data into protein structures and predicted functions. Such a step will provide a vital link between the genetics of an organism and its expressed phenotype. Proteomes have a strong influence on the measured phenotype of the plant, either directly through protein content or function, or indirectly through the relationship of a protein with the metabolome.

Expression information

In contrast to DNA sequences which may show variation across a population but invariance within individuals, gene expression is highly dynamic and it is this variation in the expression patterns that researchers use to study the relationships between genes. In isolation, gene expression data contains little inherent information. The value or meaning of gene expression information comes from the context of the experiment (Sobral, 2002). What were the taxonomy, sex and developmental stages of the organism? What were its growth conditions? From which organ and tissue was the sample extracted? What protocols were used in sample preparation?

Microarrays, designed from sequence data, have been used to measure changes in gene expression in response to changes in ecologically relevant variables. Microarrays provide high-throughput identification of the transcriptional activity of the cell. This capacity places them at the centre of the revolution of plant functional genomics, just as high-throughput sequencing once was. There are many computational challenges relevant to gaining new insights from the analysis of the patterns of gene expression in response to environmental stress and responses. These include new methods for functional sequence annotation by combining sequence and expression data, better probabilistic techniques that make use of variation in sequence and expression data at the population and species level and the means to use genomic data in models that improve our explanatory and predictive capabilities.

The application of microarrays and sequence-based methods to expression profiling has added an extra dimension to current genomic data and has led to the development of several statistics-based disciplines within bioinformatics. Microarray technology continues to expand: cDNA arrays are being produced for gene expression analysis in many plant species and complete oligonucleotide-based UniGene arrays are being developed for the major plant species. Thus, the volume of plant breeding-related data being generated continues to expand. With continued developments in the field of microarray data production, significant improvements in data analysis and integration are necessary before these data can be structured for efficient interrogation.

14.2.4 Phenotypic information

The definition and scoring of phenotypes has a key role in genomics and plant breeding. Phenotypic data consist of all data collected in various genomics and breeding programmes, either for basic research or product delivery, which describe a distinguishable feature, characteristic, quality or physical feature of a developing or mature individual. Examples are concepts of gluten endosperm, disease resistance, plant height, photosensitivity, male sterility, etc. Phenotypes result from the expression of the genotype in specific environments. Here environment can be thought of in very general terms, including not only external growing conditions but also internal conditions such as 'other gees', regulators, organ and growth stage. In these terms, gene expression data are phenotypic.

Phenotypes are important for several reasons. They allow us to observe genetically inherited traits and events and aid in genetic manipulations. Genetic changes that alter a trait that can be scored physically have been exploited to great advantage. Phenotypes are also crucial because they are the expression of genotypes and reveal gene function. In this regard, phenotypes are an essential intermediate in the pathway from basic genetics to biological understanding.

Traditionally, plant breeders generate and collect large amounts of phenotypic data. These data are associated with different breeding procedures or stages. The most systematically collected information relating to plant breeding is probably yield trial data, which has been accumulating for many years and in many plants, since controlled plant breeding programmes started. As more and more breeding objectives are added and advanced instruments are developed, many additional phenotypic traits are now measured, such as nutritional characteristics, chemical responses, stress tolerance, etc.

Categories of phenotypes of interest to plant breeders include yield and yield components, product quality and biochemical characteristics, morphological characteristics (e.g. plant height), physiological characteristics (e.g. flowering time) and abiotic and biotic stress tolerance. A more comprehensive list can be derived from the breeding objectives as listed in Section 1.7.

In addition to phenotypic data generated by plant breeders, more phenotypes are being generated by physiologists, geneticists, pathologists and other biologists for both model and non-model organisms. New technologies such as RNA interference (RNAi) now make genome-wide knock-down studies feasible and have already been applied in a high-throughput manner for many novel characteristics. As microarray techniques become widely used in transcriptomics and metabolomics, molecular phenotypic data will keep widening the definition of phenotype.

We need efficient, precise and comprehensive large-scale phenotyping techniques. This presents a difficult challenge because phenotypes are numerous and diverse and they can be observed and annotated at the molecular, cellular and organism levels. Bochner (2003) described the efforts to develop new and efficient technologies for assessing cellular phenotypes for simple microbial-cell model organisms such as *E. coli* and *Saccharomyces cerevisiae*. Such a system could be exploited for the characterization of *in vitro* culture of any plant species. Phenotypic profiling through a whole plant procedure will facilitate phenotypic data collecting and processing.

14.2.5 Environmental information

Environmental informatics may be viewed as a merging of biodiversity and ecological informatics with geographic information systems (GIS) and other environmental data (Fig. 14.1). Environmental data include all the environmental factors that contribute to crop growth and development, including soil type as well as chemical, moisture and nutritional components in the soil, daily, monthly and annual temperature, humidity and precipitation profiles, day-length and even winds and other climatic factors as listed in Table 14.3, plus many environmental factors, such as drought and cold temperature, that cause stress to crop plants.

GIS has proven to be of great utility to predict the environments in which wild

Table 14.3. Environmental factors affecting plants and plant breeding.

Soil
Texture
Water content
Fertility
Nutrient content
Production index
Air
Pollutants
Emission of CO_2
Light
Light intensity
Day length
Temperature
Average, maximum, minimum daily temperature
Effective temperature
Length of available growing period
Water
Humidity
Precipitation
Ground water
Water quality
Potential evapotranspiration
Cropping systems
Intercropping
Previous crop
Accompanying organisms
Root microorganisms
Weeds
Pathogens
Insects

ancestors of crop plants can be expected to flourish, through a mathematical comparison of climatic data at known collection sites with that at all other sites. In the strictest sense, GIS is a computer system capable of integrating, storing, editing, analysing, sharing and displaying geographically referenced information. In a more generic sense, GIS is a tool that allows users to create interactive queries, analyse spatial information, edit data and maps and present the results of all these operations. GIS technology can be used for scientific investigations, resource management, asset management, environmental impact assessment, urban planning, cartography, criminology, history, sales, marketing and route planning. For example, Canadian Geographic Information Systems was used to store, analyse and manipulate data collected for the Canada Land Inventory (CLI) – an initiative to determine the land capability for rural Canada by mapping information about soils, agriculture, recreation, wildlife, waterfowl, forestry and land use.

While breeding programmes in high-income countries may employ real-time GIS information to more accurately weight information from METs (Podlich *et al.*, 1999), those opportunities rarely exist in low-income countries as there is a lack of both real-time GIS information and the resources for conducting a large number of METs. In the 1997 season, the International Maize and Wheat Improvement Center (CIMMYT) initiated a programme targeted at improving maize for the drought-prone mid-altitudes of southern Africa. The breeding programme was product-oriented and therefore simultaneously addressed several high-priority constraints including drought, low N and major leaf and ear diseases. To develop breeding strategies that increase productivity in highly variable drought-prone environments, cluster analysis applied to the most prominent genotype-by-environment interactions grouped trial sites into eight mega-environments (Bänziger *et al.*, 2004), mainly distinguished by GIS information available for seasonal rainfall, maximum temperature, subsoil pH and N application (Hodson *et al.*, 2002).

14.3 Information Integration

For genomics to deliver the promise of making agricultural systems more efficient, sustainable and environmentally friendly, various types of biological data must be integrated to reveal the functional relationships between DNA, RNA, proteins, environment and phenotypes (Sobral *et al.*, 2001). Such integrative approaches will rely heavily on the development of information systems and analytical methods that will require the same rigour and resources that have been applied to the development of laboratory experimental protocols. The new challenges facing the field of bioinformatics are to provide complex data integration between traditional genetics – through the genome, transcriptome, proteome and metabolome ('omics' disciplines) – and the observed phenotypes (Edwards and Batley, 2004). The major challenges for the plant science community will be how to extend genomics from models to crops and how to integrate various types of data.

It is increasingly recognized that many, if not most, biological functions result from interactions or networking among many components. This suggests that biological disciplines will need to incorporate the concept of systems, as is typical in engineering. As indicated by Sobral (2002), it will be necessary to provide integration for at least the following types of molecular data: (i) structural genomics: DNA sequences (to complete genomes) and maps (genetic, physical or cytological); (ii) gene expression: mRNA profiling and single gene profiles (Northern); and (iii) biochemistry: pathways (metabolic and signalling), metabolites, proteomics. When considering breeding-related information, the integration should be extended to include all kinds of phenotypic and environmental information.

Meeting the demands and challenges of an ideal plant improvement strategy remains a matter of combining traditional breeding concepts and genomic tools through rigorous phases of experimentation, including information integration. Logical connections among various types of information will enhance the intrinsic value of raw data of all types and facilitate new biological discoveries. For a given gene, for example, a database could horizontally link sequence, structure, map position and associated germplasm accessions and could include related elements pertaining to the expression profile of the gene, its protein structure, example phenotypes and environmental factors that affect gene expression. All this information should be correlated with the genetic resources available for a given crop.

At the level of databases, there are three main ways to integrate information, referred to as link integration, view integration and data warehousing. For link integration, researchers begin their query with one data source and then follow links to related information in other data sources. View integration leaves the information in its source database, but builds an environment around the databases that makes them appear to be part of one large system. A data warehouse brings all the data together under one 'roof' in a single database. Information integration will be promoted by standardized data collection, shared vocabularies/terms (ontology) and the development of database tools that help cross-database querying and parallel analysis of related data.

14.3.1 Data standardization

Genomics and plant breeding are generating a large and heterogeneous set of data. Efficient sharing, computational integration and accurate scientific interpretation of research outputs will require some agreement about the format and semantics of the basic data. A common set of biological domain models is essential to achieve this goal. A standardized nomenclature will facilitate database searches, comparisons and extrapolations throughout model biological systems from bacteria to *Arabidopsis* and rice.

To achieve some of the benefits possible from an integrative approach to biological questions and information, it is necessary to create standards for data interpretation and comparison and its transformation into information and knowledge (Sobral, 2002). Standardization of databases has been receiving more and more attention because it is required for integration across databases. Currently the emphasis is in functional genomics but is expanding to other fields including plant breeding. A number of initiatives have proposed standard reporting guidelines for functional genomics experiments. Associated with these are data models that may be used as the basis of the design of software tools that store and transmit experiment data in standard formats.

Data standards and the formal data descriptions that underlie them may yield a range of benefits including the following (Jenkins *et al.*, 2005): (i) consideration and development of best practice and standard operating procedures, which, in turn enable proper interpretation of experimental results, principled dataset comparison and experiment repetition; (ii) standardized reporting of experiments and deposition and archiving of data associated with publications or other standard pieces of work; and (iii) development of databases and verifiable transmission mechanisms for storage, collection and dissemination of results.

When phenotypes are characterized as a whole (phenome), which are characteristics of organisms that arise via the interaction of the genome with the environment, there is a need for phenotypic standardization that has been recognized by breeding and stock centres. Several projects associated with handling genetic mutants have begun to develop a standardized approach to developing annotation and databases. Jenkins *et al.* (2005) described the collection of datasets that conform to the recently proposed data model for plant metabolomics known as ArMet (architecture for metabolomics) and illustrated a number of approaches to robust data collection that have been developed in collaboration between software engineers and biologists.

Database curators are working to convert raw datasets contributed by researchers from their original format, including structure, syntax, assumptions, naming rules and conventions, into a format compatible and contextually consistent with respective genome databases, while maintaining accuracy of fact and interpretation. In addition, curators help users access and query the data and cooperate with other groups to improve the software and data distribution infrastructure.

14.3.2 Development of generic databases

The increasing number and types of databases and software applications make it more and more difficult for researchers to determine which databases to use for various types of information. A universal or updatable database system is required and so is an automated updating system. In addition, the different ways in which data are accessed and presented create an additional burden on researchers who seek to apply the available resources to their research. Using biodiversity as an example, the difficulties in finding, accessing and using biodiversity data include the long history of the 'bottom-up' evolution of scientific biodiversity information, the mismatch between the distribution of biodiversity itself and the distribution of information describing it and most importantly, the inherent complexity of biodiversity and ecological data. This stems from numerous data types, the non-existence of a common underlying language and the multiple perceptions of different researchers/data recorders across spatial or temporal distance or both (Lane *et al.*, 2000).

Emerging technologies to solve these problems have been proposed, such as the BioMOBY initiative (Wilkinson *et al.*, 2005). BioMOBY is an international research project involving biological data hosts, service providers and coders whose aim is to explore various methodologies for biological data representation, distribution and discovery. In addition, the National Human Genome Research Institute has funded a collaborative project called the Generic Model Organism Database (GMOD; http://www.gmod.org) to promote the development and sharing of software, schemas and standard operating procedures. The project's major aim is to build a generic organism database toolkit to allow researchers to set up a genome database 'off the shelf'. Recent developments in the Ensembl system include access to inter-species sequence levels and improvements to the display of polymorphism data while users can display their own data in the context of other annotation (Hammond and Birney, 2004).

14.3.3 Use of controlled vocabularies and ontologies

The diverse databases reflect the expertise and interests of the groups that maintain them. A current limitation of complex annotation and integration is the lack of agreed-upon formats across databases. There are many integration challenges. One of the most difficult is the one that might seem the most minor: how do you assign and maintain the correct names of biological objects across databases?

A more subtle problem is the clash of concepts as users move from one database to another. An extreme example, first noted by Michael Ashburner, considers the use of the term 'pseudogene' by different researchers and research communities. To some, a pseudogene is a gene-like structure that contains in-frame stop codons or evidence of reverse transcription. To others, the definition of a pseudogene is expanded to include gene structures that contain full open reading frames (ORFs) but are not transcribed. Some members of the *Neisseria gonorrhea* research community, meanwhile, use pseudogene to mean a transposable cassette that is rearranged in the course of antigenic variation (Stein, 2003).

There are also more subtle disagreements. The human genetics community uses the term allele to refer to any genomic variant, including silent nucleotide polymorphisms that lie outside of genes, whereas members of many model organism communities prefer to reserve the term allele to refer to variants that change genes. Even the concept of the gene itself can mean radically different things to different research communities because it has been refined as the field of genetics moves forward. Some researchers may treat the gene as the transcriptional unit itself, whereas others extend this definition to include up- and downstream regulatory elements and still others use the classical definition of cistron and genetic complementation. Plant breeders may consider a gene to be a manipulable unit during the breeding process, which can be as big as a gene complex that is transferred in conventional backcross breeding, or as small as a single nucleotide difference that can be detected in MAS.

It will be increasingly desirable for inter-database queries to be performed to exploit comparative genomic and phenomic strategies in order to elucidate functional aspects of plant biology and study synteny. However, terms used to describe comparative objects within and between databases are sometimes quite variable and limit the ability to accurately and successfully query information in and across different databases. To solve this problem, controlled vocabularies and ontologies become increasingly important. Unique identifiers that are associated with each concept in biological ontologies (bio-ontologies) can be used for linking and querying databases (Bard and Rhee, 2004). Natural language processing (NLP) techniques are increasingly being used to automate the capture of new biological discoveries described in text. A novel representational schema, PGschema, was developed that enables translation of phenotypic, genetic and other related information found in textual narratives to a well-defined data structure comprising phenotypic and genetic concepts taken from established ontologies along with modifiers and relationships (Friedman *et al.*, 2006).

Shared ontologies can help bioinformaticians agree on how to describe biological objects, but they do not necessarily help them agree on how to name them. The same biological object might have multiple names and the same name might denote multiple objects. One approach is to establish globally unique identifiers to standardize the description. There are two main lines of thought among groups that are interested in globally unique identifiers. One line holds that object identifiers should point to the objects themselves and use a Uniform Resource Locator (URL) syntax. The other decouples the notion of the location of a resource from its authoritative source.

Dictionaries, encyclopedias and database schemas are examples of ontologies, as are many web-based entities, such as the search engines Yahoo! and Google. One approach to this problem is to have biologists describe and conceptualize common biological elements and produce a dynamic, controlled vocabulary that can be applied to certain types of organisms. An ontology is simply an organized set of concepts about a specified domain. It generally consists of two components: (i) an indexed controlled vocabulary of terms (the 'concept'); and (ii) information about semantic relationships between these terms. As sophisticated types of controlled vocabularies that attempt to capture the main concepts in knowledge domains, ontologies are important facilitators but they do not, by themselves, lead to the integration of biological databases. The existence of a shared ontology allows two databases to be merged with some guarantee that a term used in one database corresponds to the same term in the other.

There are many ontology projects that range from descriptions of mutant phenotypes in plants to anatomical structures in vertebrates (details can be found at the Global Open Biological Ontologies web site). Imperfect ontologies in biology are the 'gene ontology' of terms for protein and gene sequences, the minimum information about a microarray experiment (MIAME) (Brazma *et al.*, 2001) and plant ontologies for broader plant-based information (Plant Ontology™ Consortium, 2002). Once an ontology is established, databases need to be annotated under the agreed terms. At present, only a few plant genome databases, such as *Arabidopsis* and rice, have primary gene ontology annotation.

In an effort to address the need for consistent descriptions of gene products in different databases, the Gene Ontology (GO) project began as a collaboration between three model organism databases (Gene Ontology

Consortium, 2000): FlyBase (*Drosophila*), the *Saccharomyces* Genome Database (SGD) and the Mouse Genome Database (MGD) in 1998. Since then, the GO Consortium has grown to include many databases, including several of the world's major repositories for plant, animal and microbial genomes (see the GO web page http://www.geneontology.org for a full list of member organizations). The GO collaborators are developing three structured, controlled vocabularies (ontologies) that describe gene products in terms of their associated biological processes, cellular components and molecular functions in a species-independent manner.

As an extended paradigm for the GO Consortium, the Plant Ontology™ Consortium (POC), funded by the National Science Foundation, aims to develop, curate and share controlled vocabularies (ontologies) that describe plant structures and growth/developmental stages providing a semantic framework for meaningful cross-species queries across databases (Plant Ontology™ Consortium, 2002; Avraham *et al.*, 2008). The first task of the POC project is to efficiently integrate diverse vocabularies currently in use to describe anatomy, morphology and growth and developmental stages in *Arabidopsis*, maize and rice. In coming years, POC will extend this controlled vocabulary to encompass the *Fabaceae, Solanaceae* and other plant families. The description of plant phenotypes has become one of the key issues in plant ontology that is based on complex knowledge and phenotyping protocols which are either to be developed or vary greatly from one species to another.

More recently, the Plant Structure Ontology (PSO), the first generic ontological representation of anatomy and morphology of a flowering plant, was created (Ilic *et al.*, 2007). The PSO is intended for the broad plant research community, including bench scientists, curators of genomic databases and bioinformaticians. The initial release of the PSO integrated existing ontologies for *Arabidopsis*, maize and rice; more recent versions of the ontology encompass terms relevant to the *Fabaceae, Solanaceae*, additional cereal crops and poplar.

At lower taxonomic levels the development of phenotypic trait ontologies promises to provide a formal framework for navigating between crop phenotype and genome. The Trait Ontology initiative is at an early stage of development and considerable work is required to match the plant development-focused hierarchical phenotype to relevant agronomic traits. Initial progress (http://www.gramene.org) in matching traits associated with rice to underlying biological entries provides a focus for discussion among the relevant research communities. Although this is starting to be achieved by associating rice trait ontologies with trait definitions established by the International Crop Information System (ICIS), definition of a broader range of vocabularies is required (King, 2004). As an example of Plant Ontology™, maize leaf morphology and specifically the ligule, can be explored as shown in Figure 2 of the Plant Ontology™ Consortium (2002) through the Ontology Search link at Gramene.

The problems associated with establishing satisfactory vocabularies, especially across species, are substantial. Even in taxonomy, where there has been an established formalism for centuries, there is disagreement on vocabulary. The situation is much less settled for such recent and dynamic domains as developmental biology, plant and animal pathologies, gene and protein naming conventions, metabolic relationships or laboratory protocols, all of which will play a major role in determining the utility of gene expression systems. There is a need for communities to collaborate and share restricted vocabularies. An example of successful collaboration in this respect is the development of the System-wide Information Network for Genetic Resources (SINGER). SINGER is a common gateway to the genebank databases managed in 12 centres of the Consultative Group on International Agricultural Research (CGIAR). In developing SINGER it was necessary to adopt a common structure based on agreed taxonomy and other descriptors, while retaining the identity and independence of the individual databases contributing to SINGER in terms of their software and hardware platforms and structure (Sobral, 2002).

14.3.4 Interoperable query system

Scientists have realized that there is a need to make existing data from different organisms simultaneously searchable, visible and, most importantly, comparable. To look for genes involved in a particular trait one has to search different databases and manually figure out the orthology relationships among the relevant genes. These species-specific databases are widely dispersed and tailored to different objectives and they store phenotypic data in different formats. Considerable handwork is therefore necessary to compare the phenotype of the same gene in different organisms. A simple meta-search engine or an interoperable query system for these databases alone does resolve this kind of problem.

Productive utilization of databases requires interoperability: that is, the precise yet flexible interrelating of information from one database to another. There are, at present, two major impediments to achieving wide-scale interoperability: the state of database protection legislation and computer security issues (Greenbaum *et al.*, 2005). While most non-commercial/academic databases may not be overly concerned with the protection of their intellectual property, they still put up barriers to entrance and consequently interoperability, due to concerns regarding the security of their computing infrastructure.

For rice and maize, cross-database querying and display of text objects can be implemented using a web-based object-oriented query system called the OPM (Object-Protocol Model) data management tools of Gene Logic Inc. These tools are unique in their capacity to impose a uniform object-oriented data model on an existing relational database framework where users can explore and assemble biological information from heterogeneous databases. This query system promotes direct analysis of collinearity at the nucleotide level in cereal species and may also be applied for exploring multiple crop databases. The further incorporation of datasets from different studies will close the loop and create the foundation for meta-integrated databases that facilitate queries across whole systems.

14.3.5 Redundant data condensing

Just as redundant accessions are found in germplasm collections, redundant data are unavoidable because duplicate datasets using the same genotypes are generated by different research groups or for different purposes. The availability of complete genome sequences, as well as the flood of other sequence data, is leading to alternative views on how these data can be organized and interrogated. The high level of redundancy in gene discovery programmes is being condensed through reference to consensus or complete genome sequences. If a complete genome sequence is unavailable for a specific crop, closely related syntenic genomes can be used. The ever-increasing size of DNA sequence databases continues to push bioinformatic capabilities and there is a growing need to condense redundant data (Edwards and Batley, 2004). Information integration and redundant data condensing often are two procedures that can interact and support each other.

14.3.6 Database integration

Sequence databases that evolve from rigorous and systematic sequencing efforts should not merely function as warehouses for millions of bases or amino acids. Of particular importance is the ability to attach substantial genomic information to the sequences. Studies will follow on identifying genes and predicting the proteins they encode, determining when and where the proteins are expressed and how they interact and how these expression and interaction profiles are modified in response to environmental signals. Emphasis on the underlying value of genotypic and genomic elements must be balanced with a phenocentric approach. One way to address this need is to link the resources containing the various types of information, such as genomic data, phenotypic or expression data and genetic resources.

The different source databases all use different gene loci description systems (i.e. gene indices) and the orthology relationships are not always obvious, so many important phenotypic relationships may be difficult to discover. Therefore, a common data model combining the data with a common gene index is required. Orthology data must be available and a case-oriented user interface should facilitate access to phenotypic data. There are some difficulties in integrating phenotypic terminology that varies significantly between each organism-specific research community.

Once integration from the molecular to the organism levels has occurred, the next frontier becomes integration of ecosystems information (environment and interactions among organisms and populations) as the concluding step in bridging from molecules to ecosystems. There are various efforts underway to tackle linking environmental information to the organism.

One such effort is represented by FloraMap (http://www.floramap-ciat.org), which is a computer tool for predicting the distribution of plants and other organisms in the world. Approaches like FloraMap could enable scientists to link disparate studies based on environmental characteristics (Sobral, 2002). Conceptually, it should be possible to perform a similar function with crop plants using GIS, to predict which environments are similar and can be expected to give a similar genotypic response, although in practice the process is quite distinct.

14.3.7 Tool-based information integration

In order to address biological questions more fully and to extract more information from various databases, researchers require tools that allow them to integrate different datasets in a dynamic, hypothesis-driven fashion. For example, researchers need efficient and intuitive tools to help identify common genomic regions and where possible specific genes, influencing the expression of target traits across diverse germplasm and growing conditions. To address these needs, Sawkins *et al.* (2004) developed a Comparative Map and Trait Viewer (CMTV) as a part of the ISYS™ platform that can help serve as an intuitive and extensible framework for the integration of various kinds of genomic data. Another example of this is PhenoMap® (http://www.deltaphenomics.com), which can create consensus maps, overlay all QTL on the consensus map, identify regions of synteny between species, connect common markers across maps, etc. GeneFlow® (http://www.geneflowinc.com/) also provides curated, annotated databases containing mapping and QTL information for the *Poaceae*, *Fabaceae* and *Solanaceae*.

A multi-species genotype/phenotype database, PhenomicDB (http://www.phenomicDB.de), was created by merging public genotype/phenotype data from a wide range of model organisms and *Homo sapiens* (Kahraman *et al.*, 2005). This wealth of data was compiled into a single integrated resource by coarse-grained semantic mapping of the phenotypic data fields, by including common gene indices (NCBI Gene) and by the use of associated orthology relationships. With its use-case-oriented interface, PhenomicDB allows scientists to compare and browse known phenotypes for a given gene or a set of genes from different organisms simultaneously.

Although it is tempting to treat the integration of biological databases as a technological problem, in fact the main impediment to achieving this goal is not technological but sociological. As Stein (2002) indicated, a meaningful scalable integration cannot be achieved without the cooperation of the data providers. This also includes the cooperation of the data generators. As long as the data providers continue to produce online databases without regard for the way in which the information will be aggregated, integration will be a monumental task (Stein, 2003).

14.4 Information Retrieval and Mining

14.4.1 Information retrieval

Information retrieval (IR) can encompass searching for documents themselves and information within documents and metadata which describe documents, or searching a database including stand-alone databases or hypertextually networked databases such as the WWW. In this section, we will discuss data retrieval, document retrieval and text retrieval, each of which has its own body of literature, theory, practice and technologies.

Automated IR systems are used to reduce information overload. Many universities and public libraries use IR systems to provide access to books, journals and other documents. IR systems are often based on objects and queries. An object is an entity that keeps or stores information in a database. User queries are matched to objects stored in the database. Web search engines such as Google, Live.com, or Yahoo! search are some of the current IR applications.

Bibliographic databases

Use of the literature is fundamental to the pursuit of all knowledge including plant breeding. Through searching and reading, we learn what our colleagues are doing, develop a broader perspective on our field of interest, get ideas and confirm our discoveries. The scientific literature has become an expanding knowledge base that represents the collective archive of the work carried out by the international scholarly community (Trawick and McEntyre, 2004) and bibliographic databases have become part of the daily life of scientists.

Traditionally, the term 'bibliographic databases' referred to the 'abstracting and indexing service' for scholarly literature. As technology has advanced, this has expanded to include full-text articles, original data, images and books. For the average biologist, mining the literature usually means a keyword search in Pubmed. However, methods for extracting biological facts from the scientific literature have improved considerably and the associated tools will probably soon be used to automatically annotate and analyse the growing number of system-wide experimental data sets. Thanks to the increasing body of text and the open-access policies of many journals, literature mining has also become useful for both hypothesis generation and biological discovery (Jensen *et al.*, 2006).

Abstracts

Searching an online collection of abstracts is often the first approach to investigating a new topic or seeking an update on a known research area. There are several abstracting and indexing services available online, many of which require a subscription. There is considerable content overlap among major bibliographic databases. There are three major databases for abstracts: Web of Science, Current Contents and BIOSYS Previews.

The Institute for Scientific Information (ISI) produces the 'Web of Science' – an interface to the ISI Citation Database that contains more than 5300 scientific journals, dating from 1980, which is updated weekly. One of the most valuable features of the Web of Science is the inclusion of the citations associated with each abstract. It is possible to: (i) view the abstracts of all articles cited in the original (parent) article; (ii) find all articles published since the original (parent) article and those that have cited it; and (iii) find all the articles that have cited a particular author.

The Web of Science also interfaces with the ISI Current Contents databases. Current Contents divides into broad subject categories, among which the Life Science category (coverage of about 1400 journals) is the most relevant to plant breeding. The Current Contents database can be searched, abstracts of articles found can be viewed and from these the table of contents of the journal issue can be displayed and browsed.

BIOSIS Previews is made up of two databases: Biological Abstracts, which contains about 12 million records from more than 5000 journals and Biological Abstracts/RRM, which covers reports, reviews and meetings – information not formally published in scientific research journals. This includes references to items from meetings, symposia, workshops, review articles, books, book chapters, software and US patents related to life sciences. It covers the biological sciences, from biochemistry to zoology, including almost all the areas related to plant breeding.

As a fully searchable abstracts database of internationally published research,

Plant Breeding Abstracts is available at http://www.cabi.org. The Plant Breeding Abstracts database contains the most up-to-date, relevant information about all aspects of plant breeding and genetics, including: (i) plant breeding for specific traits, genetic resources, cultivar trials and cultivar descriptions; (ii) plant genetics, both classical and molecular, cytogenetics, genetics of specific traits; (iii) plant biotechnology, genetic engineering, transgenic plants; (iv) taxonomy and evolution; (v) *in vitro* culture; (vi) pest, disease and pesticide resistance; (vii) stress tolerance; (viii) breeding for and genetics of crop production, botany, stability and quality traits; and (ix) reproductive behaviour. Each week Plant Breeding Abstracts Online delivers all the new highly targeted, searchable summaries covering key English and non-English language journal articles, reports, conferences and books about plant breeding, genetics and plant biotechnology. Over 16,000 records are added to the database each year.

Full text of research articles

Several thousand biology journals are now available in electronic form, most of which are online counterparts to the paper editions. Some journals are only available online. More and more journals have made articles from back issues freely available and more publishers offer free access to articles.

There are several common routes to online journal articles. Full articles can be accessed through abstract databases. Many databases have links between abstracts and the corresponding online full-text article allowing the user seamless access to the full text if the following is true: (i) the journal (more specifically, the journal issue) is published online; (ii) the publisher of the journal has agreed with the database to make the article available via this route; and (iii) the individual or individual's library subscribes to the journal, or the publisher makes the article freely available.

Full articles can also be accessed by logging on directly to the publisher web sites. Many publishers have developed their own online interfaces to their journal databases. Science Direct (http:///www.sciencedirect.com/) and Link (http://link.springer.de) are two of the major publishers that have a collection of full-text articles. Open-access journals (freely available online for reading, downloading, copying, distributing and using) have become increasingly popular. PLoS (Public Library of Science, http://www.plos.org/), as one of the open-access publishers, is now publishing seven online peer-reviewed journals. Another example is Hindawi Publishing Corporation (http://www.hindawi.com/), which publishes over 100 open-access journals in science, technology and medicine.

Finally, HighWire Press (http://highwire.stanford.edu/) works with scientific societies and publishers to create online counterparts to their print journals. It hosts the largest repository of high-impact, peer-reviewed content, with 1036 journals and 6,100,549 full-text articles from over 130 scholarly publishers. HighWire-hosted publishers have collectively made 1,940,665 articles free. With its partner publishers, HighWire produces 71 of the 200 most-frequently cited journals. It allows a basic search across all the journals with which they collaborate.

Books and text-rich web sites

While books have been slower than journals to make the transition from paper to electronic form, there are more and more books becoming available online. A growing trend is for books to be associated with web sites for further information and corrections. A project to put biomedical textbooks online, make them searchable and integrate them with PubMed and other data resources has recently begun at the National Center for Biotechnology Information (NCBI). Book publishers also provide information and brief descriptions for new and recently published books on their web sites.

Any search engine can be used to search for molecular biology information. Many publishers, biotech companies, research laboratories, teachers and others display information that can be browsed freely. As indicated by Trawick and McEntyre (2004), information found in this way should be carefully evaluated. Be aware that anyone can publish almost anything on the Internet, so a key factor in assessing the validity of online information is the reliability of its source. It is important to assess what qualifies the individual or organization to publish the information and what their motivation for doing so might be. As with any literature research, the information found should be cross-checked and critically evaluated.

A search engine provided by Google™ (http://scholar.google.com/) has recently become popular for literature searches. Searching by authors, key words or authors' affiliation will bring up all related publications (articles, books, etc.) with article title, author list, the number of citations, etc. A full article can be browsed from a provided link. All the articles that cite a specific article can be browsed. Therefore, a key-word search for a specific topic would provide a series of linked information.

14.4.2 Information mining

The first major goal for plant biologists in the post-genome era is to understand the function of every gene and how individual gene products interact and contribute to major plant processes. This new challenge for plant functional genomics is destined to become the most difficult hurdle in plant biology and requires the systematic application of global molecular approaches integrated through bioinformatics. Several tools are now required to decipher gene function including the traditional methods of random mutagenesis, gene knock-out and silencing, as well as high-throughout omics disciplines of transcriptomics, proteomics and metabolomics. Mining this genomics information and effectively applying it to plant breeding is a significant challenge indeed.

Data mining

Data mining, or knowledge discovery in databases (KDD), has been described as 'the nontrivial extraction of implicit, previously unknown and potentially useful information from data' (Frawely *et al.*, 1991). The majority of data mining exercises in bioinformatics at present are founded on the requirement to screen through large, usually sequence-based, datasets searching for 'homology'. Bioinformatics has traditionally helped to identify the molecular constituents of the cell and their functions, often described in relation to a biochemical activity. This has included gene finding, motif recognition, similarity searches, multiple sequence alignment, protein structure prediction, phylogenetic analysis and other related methods.

Compared to sequence databases, extracting information related to plant breeding is not an easy task. It is like finding a tiny bit of gold in the voluminous portion of ore taken from a gold mine. Breeding related data consist of many inter-related, complex data types and therefore require complex queries to search, retrieve and analyse them. Classical multivariate and discriminant statistics are relevant to many biological data mining exercises. Significant progress has yet to be made in carrying out systematic or integrated data mining for the disparate and complex information now available to plant scientists.

Plant breeders may want to mine for the following information: (i) germplasm information collected across worldwide institutions; (ii) marker–trait associations reported for traits of interest to specific breeding programmes; (iii) genes that are required for the improvement of traits of agronomic importance through transformation and introgression; and (iv) molecular markers and marker-related information for the development of MAS tools.

Comparative informatics

As comparative genetics is now viewed as a key component to expanding existing knowledge on plant genomes and genes, comparative bioinformatics remains an essential strategy of this pursuit. Comparative informatics facilitates linking the genomes of various crop species and will provide keys to understanding how genes and genomes are structured and how they evolve. Through the identification of synteny, it will be possible to isolate genes from crop plants with large genomes using information about homologous genes in related crops with smaller genomes. Linkages and interactions should also be promoted between databases of plants and non-plant species.

Horan *et al.* (2005) clustered all protein sequences from *Arabidopsis* and rice into similarity groups, calculated their corresponding alignments, localized their conserved domains and generated distance trees. The resulting datasets provide comprehensive information about the similarities and dissimilarities between a monocotyledon and dicotyledon representative with regard to the size, quantity and composition of their family and singlet proteins. The provided datasets represent a foundation for future studies of orthologous and paralogous sequences of the two species. The user-friendly Genome Cluster Database (GCD; http://bioinfo.ucr.edu/projects/GCD) was designed to provide an efficient cluster mining tool for *Arabidopsis* and rice, to perform various intraspecific and interspecific comparisons and also to retrieve related sequences from other organisms.

There are four basic comparative bioinformatics analyses, including:

- DNA–DNA conservation: the alignment of complete DNA sequence from two plant species to determine DNA–DNA conservation is computationally demanding and the algorithms that perform this are under active development.
- Syntenic blocks: the identification of segments of the genome in which the order of particular genes is conserved between two species (syntenic blocks) is of interest not only for studying the evolution of chromosome structure but also for helping to predict and identify pairs of genes between species that are (or are not) orthologues.
- Orthologues: another type of comparative analysis focuses on genes and proteins and attempts to identify the orthologous genes in different plants. In most cases, the orthologous genes can be expected to be functionally equivalent.
- Phenomic similarity: phenotypic and physiological similarities can be used to identify different genes for the same or similar phenotype.

Sequence similarity analysis

DNA sequence similarity analysis can be used to trace allele, gene or chromosomal fragments, identify similarities between sequences or genes and align multiple sequences. Protein sequence analysis includes searching for protein similarity and looking at primary, secondary and tertiary structure.

BLAST (Basic Local Alignment Search Tool) is a set of similarity search programs designed to explore all of the available sequence databases regardless of whether the query is a protein or DNA sequence. The BLAST services available include: Nucleotide Blast; Protein Blast; Translated Blast; Genomic Blast pages (human genome, eukaryotes, microbial genomes); and specialized Blast pages (VecScreen, a BLAST-based detection of vector contamination; IgBlast, for analysis of immunoglobulin sequences in GenBank®; Geo Blast, for gene expression data; and Snp Blast). These services are produced and made available on the Internet by the US NCBI.

14.5 Information Management Systems

The core components of information management tools for plant breeding should ideally support the acquisition, storage and analysis of information on genomes, proteins, biochemical pathways, cellular systems, organism models, ecological systems, geographic biodiversity, germplasm evaluation, field trial measurements, phenotypic variation and environmental interactions. In addition, the systems should adapt to emerging information infrastructure, such as: (i) scalable computing; (ii) distributed sensors, data, people and computers; (iii) web- and object-oriented software architecture; and (iv) decentralization of content and software authoring (Sobral, 2002).

Currently available information management and data analysis systems have their pitfalls:

- There is a big concept gap between breeding and molecular biology in terms of what information is available and how it can be used.
- Many of the systems have been developed independently for phenotypic and genotypic data and used by two different groups of scientists including breeders/agronomists and geneticists/ molecular biologists.
- Breeding information has been managed in most breeding programmes by using relatively simple tools such as MS Access and agrobase (http://www.agronomix.mb.ca/), for which less training is required. However, these tools are not suitable for data management and statistical analysis when molecular data and multiple-resource data are incorporated.
- Insufficient communication between breeders/agronomists and geneticists/ molecular biologists has contributed to the paucity of tools suitable for both groups. As a result, hardware, database and software support in most breeding institutions is very limited or very different from those established in the biotechnology and IT industries.
- Understanding and communication between IT scientists and breeders and between facility developers and breeders, is also lacking. This contributes to underdevelopment of information systems designed for plant breeding and the limited use of those currently available in genomics.
- Many breeding companies and institutions, especially in developing countries, are lacking the personnel and facilities for information management which are well developed in the biotechnology industry.

14.5.1 Laboratory information management systems

To handle the constant flow of data from the lab to the breeder and to integrate information from molecular markers, genetic mapping and phenotyping, many information management tools are needed. High-throughput laboratories, often required by molecular breeding programmes, make Laboratory Information Management Systems (LIMS) a necessity. LIMS manage data, samples, laboratory users, instruments, standards and support laboratory functions such as invoicing, plate management, sample tracking and work flow automation. Taking a typical genotyping project as an example, the LIMS may include tracking the samples from field to plate and to storage, managing the data flow from plate to genotyping facilities and to computers and organizing and optimizing experiments both internally and externally.

Today's trend is to move the whole process of information collection, management, analysis, decision making, review and release into the workplace. The goal of a LIMS is to create a seamless organization in which:

- Instruments are integrated in the lab network where they receive instructions and work lists from the LIMS and return finished results, including raw data, back to a central repository where the LIMS can update relevant information to external systems.
- Laboratory personnel perform calculations, review and document results using online information from connected instruments, reference databases and other resources using electronic lab notebooks connected to the LIMS.
- Management can supervise the lab process, react to bottlenecks in workflow and ensure regulatory requirements are met.
- Laboratory participants can place work requests and follow up on progress, review results and other documentation.

With several thousand data points flowing out of the laboratory every day, timely scoring and delivery of the results to breeders are basic requirements for an efficient breeding system. Well-trained assistants for genotyping and scoring, coupled with research scientists who can analyse data in meaningful ways, are the key components for a data management and delivery system. A laboratory with well-equipped facilities must also be well equipped with qualified personnel and the appropriate software for data integration, manipulation, analysis and mining. Timely delivery of data is equally important, because in many cases the window of time the breeder has to make selections is very limited. With high-throughput genotyping and data management systems currently available, it takes about a week to generate and analyse data for a breeding population consisting of several hundred individuals. This includes activities ranging from leaf tissue harvesting to DNA extraction, genotyping, data scoring, analysing, summarizing and reporting.

14.5.2 Breeding information management systems

Organizations involved in molecular breeding should establish information management systems capable of handling information from multiple sources, including institution-owned information and that in public databases. The data model serves as a foundation for data transfer between different entities in the field of plant breeding. Legacy data collected from different sources should be cleaned and organized according to this specification while new data is fed from users through a standard interface to ensure the format of the data. The collected data are stored in a centralized database. Different data warehouses can be built to meet the needs of data analysis and decision making. Raw phenotypic and genotypic information generated within the breeding institution can be stored in two separate databases; however, a knowledge base should be created which combines data from the two data sources.

The data management system serves as a bridge connecting genotypic and phenotypic information and provides tools for gathering data, integrating public information into the breeders' data warehouse and extracting useful information for breeding. The components of such a system include information support, IT infrastructure support, data scoring, acquisition and data formatting, data hosting, data integration and data mining.

Most bioinformatics data and tools are available through the Internet. Post-genomics experiments require access to dozens of data types for tens of thousands of data points simultaneously. This cannot be achieved with common web-based tools; such analyses require programmatic access to web interfaces such that large quantities of data can be pipelined from one interface to the next. A biological web service interoperability initiative, BioMOBY (http://www.biomoby.org/), was established to prove a simple, extensive platform through which the myriad of online biological databases and analytical tools can offer their information and analytical services in a fully automated and interoperable way (Wilkinson *et al.*, 2005).

The components of an information management system designed for molecular plant breeding are database modules that link gene location, function and allele value data with target environment characterization data and the germplasm units used in the breeding programme, together with the tools for querying the database. The importance and enormity of this task cannot be overstated. As discussed previously, web-based breeding information management systems provide several advantages over local and stand-alone systems because most public breeding institutions have limited IT and service support.

14.5.3 International Crop Information System

Informatics has become a prerequisite to molecular breeding because the volume of breeding-related information is increasing at such a high rate that collecting, storing, mining and manipulating this information for selection decisions is not possible without appropriate statistical, biometrical and informatics tools. An integrated breeding tool is therefore needed to rapidly collect, analyse and represent breeding-related data, uniquely identify the germplasm units, document their co-ancestry and associate gene information with these units in the short window of time available for most selection decisions. This is critical to practical implementation of any molecular-based breeding strategy. In addition, computational tools are required to translate and integrate research outputs into a usable form for plant breeding programmes (Dwivedi *et al.*, 2007). The International Crop Information System (ICIS) is identified as the key component that can link the gene, phenotype and environment data with uniquely identified germplasm units used and manipulated in breeding programmes.

ICIS is a database system already prototyped since 1996 by a CGIAR multicentre group of biologists and information scientists (www.icis.cgiar.org) to manage and integrate all research data on genetic resources, crop improvement and resource management and to link this information to global environmental and genomic data resources (McLaren *et al.*, 2005). ICIS is attempting to level the information playing field between developed and developing nations and it addresses the CGIAR mandate to share research information as well as germplasm and technology. Modest resources with strong commitments have so far produced an innovative prototype for tracking and recording generically all the processes in germplasm collection, characterization, evaluation and development. The system has been used or evaluated for rice, wheat, maize, barley, cowpea and common bean and is used by private and public breeding programmes.

ICIS has a modular structure with a core consisting of The Genealogy Management System (GMS) which manages data on nomenclature, origin, development and deployment of germplasm and the Data Management System (DMS) which manages and documents characterization and evaluation data. Specialized user interfaces deliver data views and decision support tools to crop scientists from different disciplines which access the same data resources leading to efficient use and re-use of research data. The development of distinct crop databases (separate ICIS implementations) are resulting in focused data and information management for each crop. Meanwhile, a common structure ensures that huge economies are gained by shared commitments to training in national agricultural systems and through collaboration in terms of intellectual development, programming, testing and maintenance.

Linkages between the GMS and DMS provide biological scientists with powerful querying functionality. The querying capabilities of ICIS will not place sensitive data at risk. To permit researchers to manage their own data in parallel with those from other sources, ICIS has a parallel structure of central and local versions. This structure provides local read/write capabilities, allowing data generated locally to be merged and harmonized with the central database at the local user's discretion.

ICIS must have seamless links to other information technologies used in agriculture. The System-wide Genetic Resources Program (SGRP) has endorsed ICIS as a critical initiative in germplasm information systems. A current project with SGRP ensures that ICIS and the System-wide Information Network for Genetic Resources (SINGER) exchange data smoothly; another project with the Collaborative Research Centre for Molecular Plant Breeding in Australia targets linkages between conventional evaluation data and molecular marker data within ICIS. In addition, ICIS is in many ways very complementary and becoming increasingly dependent upon the content and technologies of plant genome databases. As ICIS finds itself increasingly drawn towards the integration of breeding and field evaluations with associated molecular data, this requirement has inspired ongoing collaborations to integrate ICIS data with external genetic, genomic, transcriptomic and proteomic datasets, such as species-specific plant genomic databases, the United States Department of Agriculture (USDA) Gramene comparative genomics database, the European PlaNet group and others.

Although ICIS has created some fundamental components required for molecular breeding, there are several general needs for plant breeding, which still require a great deal of development: (i) databasing for all breeding-related information such as climate, soil and phenotype data for selection and target environments; (ii) data mining for specific breeding purposes such as environment classification, genotype-by-environment interaction and identification of novel alleles and genetic variation; (iii) modelling breeding processes and selection schemes using multiple sources of breeding information to eliminate some field and lab tests required for making selection decisions, which may be critical for complex traits; and (iv) extracting useful information by an integrated exploration of the information created in a specific breeding programme with all related information from public databases.

Phenotypic and genetic data should be stored in a generic database such as ICIS or where necessary in other databases compatible with a standard informatics platform. For example the CIMMYT MAIZE FIELDBOOK, currently used for capturing phenotypic and trial data in maize, is being integrated with ICIS. The functionality developed in the MAIZEFINDER software is being utilized as a data warehouse and an interface for queries on data in ICIS. This integration will allow breeders to continue to use the functions provided by both MAIZE FIELDBOOK and MAIZEFINDER, but also allows access to the functionalities of ICIS, such as pedigree management and storage for genomic data. ICIS also provides integration with the GCP informatics platform, called Pantheon (http://pantheon.generationcp.org). This software includes a web-based search engine, standalone Java graphical user interfaces and integration to other third party software such as ISYS™ (http://www.ncgr.org/cmtv), the Genomic Diverstiy and Phenotype Connection (GDPC) (http://www.maizegenetics.net/gdpc/index.html) and BioMOBY web service compliant tools (http://www.biomoby.org/). Through GDPC the platform has access to the TASSEL software used for association analysis and through ISYS™ access to the visualization tools such as the Comparative Map and Trait Viewer (http://www.ncgr.org/cmtv) that are useful for QTL mapping and MAS.

The data should be cross-linked with other publicly available data, including possibly several of the following data types: (i) QTL and (comparative) genetic mapping data from both specific projects and public sources of such information as in Gramene, GrainGenes and MaizeGDB; (ii) additional public sequence and annotation data, as it becomes available, possibly including molecular marker data of various types; (iii) crop mutant data; and (iv) information from other pertinent international plant databases, in particular, The Arabidopsis Information Resource (TAIR) and equivalent model organism databases.

To help users choose the most appropriate experimental design and data analysis methods and to provide them with a regularly updated selection of appropriate options, the system under development by the GCP project targets to provide automatic transition for data flow between all permutations and combinations of software to be used. This integrated decision support system for marker-assisted plant breeding (analogous to AGROBASE), called iMAS, will facilitate an *integrated*, *error-free* and *appropriate* data analysis from the beginning to the end of the molecular breeding pathway. As an integrated decision support system for marker-assisted plant breeding, iMAS was developed to seamlessly facilitate marker-assisted plant breeding by integrating freely available quality software involved in the journey from phenotyping and genotyping of individuals to identification and application of trait-linked markers and providing simple-to-understand-and-use online decision guidelines to correctly use these software programs and interpret and use their product. Potential useful software identified include those for generation of experimental design, biometric analysis of phenotypic data, building a linkage map, marker identification through QTL analysis, marker identification through association analysis and determination of sample sizes required for foreground and background selection. ICIS should finally integrate with iMAS to make all the software available under one single umbrella.

Other statistical tools and software should also be incorporated into the same platform through ICIS. One such tool is CropStat, a computer program for data management and basic statistical analysis of experimental data. CropStat is freely available from www.irri.cgiar.org and has been developed primarily for the analysis of data from agricultural field trials, but many of the features can be used for analysis of data from other sources. The main modules and facilities are:

- data management with a spreadsheet;
- text editor;
- summary statistics and scatter plot graphics;
- analysis of variance;
- regression and correlation;
- mixed model analysis;
- single site analysis of plant breeding cultivar trials;
- cross site and additive main effect and multiplicative interaction (AMMI) analysis;
- pattern analysis of genotype-by-environment interaction;
- generalized linear models;
- log linear models;
- QTL analysis;
- randomization and layout of experimental designs;
- display of linear forms for general factorial expected mean squares (EMS); and
- generation of coefficients for orthogonal polynomials.

Although CropStat is an easy-to-use software package, it is not suitable for analysing large-scale data sets.

14.5.4 Other informatics tools

There are several informatics tools available from either private or public sectors. Some of them have multiple functions relevant to plant breeding, while others only provide specific applications in plant breeding. Only some representative tools will be described here.

AGROBASE Generation II™ (http://www.agronomix.com/) is a comprehensive database management and analysis system for agronomists, plant breeders and plant researchers. Its Basic System offers data management, experiment management and statistical analysis. The Varietal Comparisons Module compares relative performance of cultivars or treatments within a trial or across all trials, locations and years and also analyses genotype-by-environment interactions. The Advanced Statistics Module supports the randomization and analysis of more advanced experimental designs, spatial analyses of yield trials, multivariate analyses and other advanced statistical analyses. The Pedigree Data Management Module supports the plant breeding needs of many types of crops. The Image Display Module supports the display of images of cultivars or treatments including the growth stages, flower colour or shape, plant components and characteristics, or molecular markers, for any cultivar or genotype.

GERMINATE (Lee *et al.*, 2005), developed by the Scottish Crop Research Institute and the John Innes Centre (http://germinate.scri.sari.ac.uk/), is a generic plant data management system designed to hold a diverse variety of data types, ranging from molecular to phenotypic and to allow querying between such data for any plant species. Data are stored in GERMINATE in a technology-independent manner, such that new technologies can be accommodated in the database as they emerge, without modification of the underlying schema.

The Plabsoft database (Heckenberger *et al.*, 2008) is a comprehensive database management system (DBMS) for integrating phenotypic and genomic data in academic and commercial plant breeding programmes. The database structure is capable of managing the following types of data observed in breeding programmes of all major crops: (i) germplasm data of any species including pedigree data; (ii) phenotypic data of any traits and trait complexity; (iii) trial management data for any field and trial design; (iv) molecular marker data for all common types of markers; and (v) project and study management data. By implementing the database structure into the DBMS, functions have been developed for data import, data retrieval and data transfer from and to commonly used statistical analysis software.

Compared to the above informatics tools, GeneFlow® (http://www.geneflowinc.com/geneflow.html) is a comprehensive tool more relevant to molecular breeding. The system integrates pedigree, genotype and phenotype data – information traditionally kept in separate databases but which gains considerable value and power by being linked together. This allows the researcher to study the inheritance of a trait, explore the relationship between genetic make-up and observed phenotype, look for genetic components associated with a trait, track genetic fingerprints for a set of individuals and identify ancestors that are the likely source of a gene or trait, all from a single platform. The primary components of the system include:

- a pedigree module to use as a pedigree-based display and to support the overlay and analysis of genetic and phenotypic data within the context of known family relationships;
- a genotype module to provide a detailed, chromosome-level view of the genetic content and organization of various individuals;
- a population module to analyse structured populations, ranking the progeny and producing a detailed report and display; and
- a report module to generate a large number of key reports and graphs, shedding light on the structure of genetic diversity and the relationship between genes and traits.

14.5.5 Future needs for informatics tools

A comprehensive statistical analysis system needs to be developed. This system should provide both basic and advanced statistical methods, analytical tools and web-based analysis and visualization software for managing, analysing and mining all kinds of data including those related to phenotype, genotype, sequence and expression. The following are examples of functional components that should be included in the system:

- multi-year trial data: environment classification and yield stability analysis;
- heterotic pattern analysis and heterotic pooling;
- genotype-by-environment interaction;
- germplasm characterization and classification;
- marker characterization and genetic mapping;
- genetic mating systems;
- identification and quantification of novel variation at both phenotypic and genotypic levels;
- molecular and functional analysis of genetic diversity and evolutionary process;
- association of genotypes with phenotypes through linkage genetics;
- statistical methods for fine mapping using multiple populations, multiple environments, comparative mapping and linkage disequilibrium mapping;
- simulation and modelling of gene networks and biological interactions including major gene interaction, major gene–QTL interaction and QTL–QTL interaction;
- heterosis and combining ability analysis and hybrid and inbred performance prediction; and
- statistical methods for molecular data of multiple sources.

Currently, bioinformatics is conducted by a specialized group of individuals. The majority of biologists use only basic bioinformatics tools and there is little involvement by plant breeders. This has been considered a major limitation of bioinformatics today (Rhee, 2005). In the decades to come, the majority of biologists will need skills such as programming, database development and management of large datasets and quantitative and statistical analysis of data. The richness and enormity of available information, such as understanding the function of every gene in an organism, will shift research into more theoretical biology using informatics approaches. Another current issue with bioinformatics includes the heterogeneity of data and how it is analysed, annotated and displayed and the lack of connectivity among the available data. Recent movements towards the creation of a scientific society for database curators (http://www.biocurator.org) and projects that bring together the efforts of different model organism databases (http://www.gmod.org) provide early hints that bioinformatics is developing into a more coherent discipline of biology. With the maturation of genomics has come the adoption of standard data formats and schemata for crop genome information, and it is likely that future databases will be designed with cross-connectivity capabilities as a priority. The availability of complete genome sequences enables further mining for novel promoter sequences and other regulatory features such as micro-RNA. This tertiary annotation provides links to both the phenotype and the complex regulatory mechanisms that govern development and response to the environment (Edwards and Batley, 2004).

One of the more significant changes to crop genome databases has been the move towards graphical user interfaces that provide a more user-friendly search environment. The Ensembl database schema, which has a strong emphasis on graphical user interaction, is used in the cereal comparative genomic database Gramene (Liang *et al.*, 2008). No single database can attempt to store all of the possible information about an organism. Therefore, a key role of genome browsers is to provide a rich variety of links to external databases. As genome sequences become available from more organisms, projects such as Ensembl are attempting to provide access to genome-wide inter-species comparisons of genomic and protein sequences. The same strategy is needed to develop inter-species crop informatics resources aimed at serving the plant breeding community.

The rapid evolution of the field of phenomics – the genome-wide study of gene dispensability by quantitative analysis of phenotype – has resulted in an increasing demand for new data analysis and visualization tools. Most of the valuable phenotypic data reside in the public literature, not captured in databases. Effective text mining is needed to gather these data as well. A prerequisite for text mining, however, is the availability of specific thesauri to catalogue and validate terms.

To meet this demand, a public resource for mining, filtering and visualizing phenotypic data – the PROPHECY database – was designed to allow easy and flexible access to physiologically relevant quantitative data for the growth behaviour of mutant strains in the yeast deletion collection during conditions of environmental challenges (Fernandez-Ricaud *et al.*, 2005). We would expect a similar effort in crop plants.

Some informatics tools are discussed in Chapter 15.

14.6 Plant Databases

A list of currently available molecular biology databases can be found at http://www.oxfordjournals.org/nar/database/a/. The list is updated in the January issue of *Nucleic Acid Research* each year. The 2008 update contains 1078 databases, which can be classified into 14 categories (Table 14.4; Galperin, 2008). In addition, a comprehensive list of databases is available at ExPASy Life Science Directory (http://expasy.ch/alinks.html).

Many attempts are being made to understand biological subjects at a systems level. A major resource for these approaches are biological databases, storing large volumes of information about DNA, RNA and protein sequences, including their functional and structural motifs, molecular markers, mRNA expression levels, metabolite concentrations, protein–protein interactions, phenotypic traits or taxonomic relationships. As an example of a comprehensive resource, the NCBI provides analysis and retrieval tools for the data in GenBank® and other biological data made available through NCBI's web site, in addition to maintaining the GenBank® nucleic acid sequence database (Wheeler *et al.*, 2007). NCBI resources include Entrez, the Entrez Programming Utilities, My NCBI, PubMed, PubMed Central, Entrez Gene, the NCBI Taxonomy Browser, BLAST, BLAST LINK (BLINK), Electronic PCR, OrfFinder, Spidey, Splign, RefSeq, UniGene, HomoloGene, ProtEST, dbMHC, dbSNP, Cancer Chromosomes,

Table 14.4. Molecular biology databases: categories and numbers. Summarized from http://www.oxfordjournals.org/nar/database/a/; the number in parentheses is the number of databases in the category.

Category	Subcategory
Nucleotide sequence databases	International Nucleotide Sequence Database Collaboration (3)
	Coding and non-coding DNA (41)
	Gene structure, introns and exons, splice sites (25)
	Transcriptional regulator sites and transcription factors (60)
RNA sequence databases (63)	
Protein sequence databases	General sequence databases (15)
	Protein properties (16)
	Protein localization and targeting (22)
	Protein sequence motifs and active sites (22)
	Protein domain databases; protein classification (38)
	Databases of individual protein families (65)
Structure databases	Small molecules (15)
	Carbohydrates (9)
	Nucleic acid structure (16)
	Protein structure (75)
Genomics databases (non-vertebrate)	Genome annotation terms, ontologies and nomenclature (12)
	Taxonomy and identification (10)
	General genomics databases (44)
	Viral genome databases (25)
	Prokaryotic genome databases (61)
	Unicellular eukaryotes genome databases (15)
	Fungal genome databases (32)
	Invertebrate genome databases (51)
Metabolic and signalling pathways	Enzymes and enzyme nomenclature (12)
	Metabolic pathways (19)
	Protein–protein interactions (70)
	Signalling pathways (5)
Human and other vertebrate genomes	Model organisms, comparative genomics (63)
	Human genome databases, maps and viewers (19)
	Human ORFs (28)
Human genes and diseases	General human genetics databases (13)
	General polymorphism databases (28)
	Cancer gene databases (22)
	Gene-, system- or disease-specific databases (50)
Microarray data and other gene expression databases (65)	
Proteomics resources (18)	
Other molecular biology databases (41)	Drugs and drug design (23)
	Molecular probes and primers (9)
Organelle databases	Mitochondrial genes and proteins (18)
Plant databases	General plant databases (38)
	Arabidopsis thaliana (26)
	Rice (17)
	Other plants (18)
Immunological databases (27)	

Entrez Genome, Genome Project and related tools, the Trace and Assembly Archives, the Map Viewer, Model Maker, Evidence Viewer, Clusters of Orthologous Groups (COGs), Viral Genotyping Tools, Influenza Viral Resources, HIV-1/Human Protein Interaction Database, Gene Expression Omnibus (GEO), Entrez Probe, GENSAT, Online Mendelian Inheritance in Man (OMIM), Online Mendelian Inheritance in Animals (OMIA), the Molecular Modelling Database (MMDB), the Conserved Domain Database (CDD), the Conserved Domain Architecture Retrieval Tool (CDART) and the PubChem suite of small molecule databases.

Information relevant to plant breeding may be housed in nucleotide, RNA and protein sequence databases, structure databases, genomics databases, metabolic and signalling pathways, microarray data and other gene expression databases, proteomics resources, organelle databases and plant databases. Some of the databases to be discussed in this section are not for plant species, however, they may be useful in comparative genomics, genetics and phenomics.

14.6.1 Sequence databases

Nucleotide sequence databases

The most important DNA sequence databases are listed in Table 14.5 with their URLs

Table 14.5. DNA and protein sequence databases.

Database name	Uniform Resource Locator (URL)	Database description
DDBJ	http://www.ddbj.nig.ac.jp	DNA Data Bank of Japan (DDBJ), one of the three major databases for the International Nucleotide Sequence Database Collaboration
EMBL Nucleotide Sequence Database	http://www.ebi.ac.uk/embl	The EMBL Nucleotide Sequence Database is maintained at the European Bioinformatics Institute (EBI) in an international collaboration with DDBJ and GenBank® at the NCBI (USA)
GenBank®	http://www.ncbi.nlm.nih.gov	A comprehensive sequence database that contains publicly available DNA sequences for more than 170,000 different organisms, obtained primarily through the submission of sequence data from individual laboratories and batch submissions from large-scale sequencing projects
EXProt	http://www.cmbi.kun.nl/EXProt	A non-redundant protein database containing a selection of entries from genome annotation projects and public databases, aimed at including only proteins with an experimentally verified function
MIPS	http://mips.gsf.de	Databases at Munich Information Center for Protein Sequences
NCBI Protein	http://www.ncbi.nlm.nih.gov/entrez/query.fcgi?db=Protein	The NCBI Entrez Protein database comprises sequences taken from a variety of sources, including Swiss-PROT, the Protein Information Resource, the Protein Research Foundation, the Protein Data Bank, and translations from annotated coding regions in the GenBank® and RefSeq databases
Patome	http://www.patome.org	Biological sequence data disclosed in patents and published applications, as well as their analysis information
PIR-PSD	http://pir.georgetown.edu	The Protein Information Resource (PIR) is an integrated public bioinformatics resource that supports genomic and proteomic research and scientific studies. PIR has provided many protein databases and analysis tools to the scientific community, including the PIR-International Protein Sequence Database (PSD) of functionally annotated protein sequences
PRF	http://www.prf.or.jp/en/index.shtml	Protein Research Foundation database of peptides: sequences, literature and unnatural amino acids
RefSeq	http://www.ncbi.nlm.nih.gov/RefSeq	The NCBI Reference Sequence (RefSeq) database provides curated non-redundant sequence standards for genomic regions, transcripts (including splice variants), and proteins
Swiss-PROT	http://www.expasy.org/sprot	The UniProt/Swiss-PROT Protein Knowledgebase is a curated protein sequence, providing a high level of annotation (such as the description of protein function, domains structure, post-translational modifications, variants, etc.), a minimal level of redundancy and high level of integration with other databases. It is part of the Universal Protein Knowledgebase (UniProtKB)

Continued

Table 14.5. *Continued.*

Database name	Uniform Resource Locator (URL)	Database description
TCDB	http://www.tcdb.org	The Transporter Classification Database (TCDB) is a curated, relational database containing sequence, classification, structural, functional and evolutionary information about transport systems from a variety of living organisms
UniProt	http://www.uniprot.org	UniProt (Universal Protein Resource) is the world's most comprehensive catalogue of information on proteins. It is a central repository of protein sequences and functions created by joining the information contained in Swiss-PROT, TrEMBL and PIR. UniProt has three components, each optimized for different uses. The UniProt Knowledgebase (UniProtKB) is the central access point for extensive curated protein information, including function, classification and cross-reference. The UniProt Reference Clusters (UniRef) databases combine closely related sequences into a single record to speed searches. The UniProt Archive (UniParc) is a comprehensive repository, reflecting the history of all protein sequences

and a brief description. The International Nucleotide Sequence Database Collaboration is a joint effort of the European Bioinformatics Institute (EBI), the DNA Data Bank of Japan (DDBJ) and the US NCBI. The nucleotide sequence databases are data repositories, accepting nucleic acid sequence data from the community and making it freely available.

Each entry in a database has a unique identifier, which is a string of letters and numbers corresponding to that record. This unique identifier, known as the Accession Number, can be quoted in the scientific literature. As the Accession Number is permanent, another code is used to indicate the number of changes that a particular sequence has undergone. This code is known as the Sequence Version and is composed of the Accession Number followed by a period and a number indicating the specific version.

Since their inception in the 1980s, the nucleic acid sequence databases have experienced exponential growth, with archives doubling in size about every 18 months, reflecting advances in sequencing technologies.

Protein sequence databases

The protein sequence databases are the most comprehensive source of information on proteins, some of which are listed in Table 14.5. They can be classified into universal databases, covering proteins from all species and specialized data collections storing information about specific families or groups of proteins, or about the proteins of a specific organism. Two categories of universal protein sequence databases can be discerned: (i) simple archives of sequence data; and (ii) annotated databases where additional information has been added to the sequence record.

The Protein Information Resource (PIR) is the oldest protein sequence database. It was established in 1984 by the National Biomedical Research Foundation (NBRF) and has been maintained since 1988 by PIR.

Swiss-PROT, established in 1986 as an annotated universal protein sequence database and maintained collaboratively by the Swiss Institute of Bioinformatics (SIB) and the EBI, provides a high level of annotation, a minimal level of redundancy, a high level of integration with other biomolecular databases and extensive external documentation. Each entry in Swiss-PROT is thoroughly analysed and annotated to ensure a high standard of annotation and maintain the quality of the database. In Swiss-PROT two classes of data can be distinguished: the core data and the annotation. The core data consists of the sequence data, the citation information (bibliographical references) and the taxonomic data (description of the biological source of the protein). The annotation describes the function(s) of the protein, post-transcriptional modification (carbohydrates, phosphorylation, acetylation, glycosylphosphatidylinositol-anchor, etc.), domains and sites (calcium binding regions, ATP-binding sites, zinc fingers, homeobox, kringle, etc.), secondary structure, quaternary structure (homodimer, heterotrimer, etc.), similarities to other proteins, disease(s) associated with deficiencies in the protein and sequence conflicts, variants, etc.

TrEMBL (Translation of EMBL nucleotide sequence database), the supplement of Swiss-PROT, was created in 1996 to make new sequences available as quickly as possible, since maintaining the high quality of Swiss-PROT is a time-consuming process that involves extensive sequence analysis and detailed curation by expert annotators. TrEMBL consists of computer-annotated entries derived from the translation of all coding sequences in the EMBL nucleotide sequence database, except for those already included in Swiss-PROT.

Searches in protein databases have become a standard research tool in the life sciences. To produce valuable results, the source databases should be comprehensive, non-redundant, well-annotated and up-to-date. However, the lack of a single protein sequence database satisfying all four criteria has forced users to search multiple databases. By unifying the PIR, Swiss-PROT and TrEMBL database activities, PIR International and its partners, EBI and SIB, have produced a single worldwide database of protein sequence and function, UniProt, which is the central repository of protein sequence and function data, created by joining the information contained in Swiss-Prot, TrEMBL and PIR.

14.6.2 General genomics and proteomics databases

Table 14.6 provides some general genomics and proteomics databases. Databases of two-dimensional gel electrophoresis data, like Swiss-2DPAGE which is maintained collaboratively by the Central Clinical Chemistry Laboratory of the Geneva University Hospital and SIB, has been considered one of the classical proteomics databases.

A number of databases address some aspect of genome or proteome comparisons. The Kyoto Encyclopedia of Genes and Genomes (KEGG) is a knowledge base for systematic analysis of gene functions, linking genomic information with higher order functional information. KEGG mainly addresses regulation and metabolic pathways, although the KEGG scheme is being extended to include a number of non-metabolism-related functions. Clusters of Orthologous Groups of proteins (COGs) is a phylogenetic classification of proteins encoded in completely sequenced genomes. COGs group together related proteins with similar but sometimes non-identical functions.

The Proteome Analysis Initiative at EBI has the more general aim of integrating information from a variety of sources that will together facilitate the classification of the proteins in complete proteome sets. These proteome sets are built from the Swiss-PROT and TrEMBL protein sequence databases that provide reliable, well-annotated data as the basis for the analysis. The Proteome Analysis Initiative provides a broad view of the proteome data classified according to signatures describing particular sequence motifs or sequence similarities. At the same time it affords the option of examining various specific details like structure or searchable functional classification. The InterPro (http://www.ebi.ac.uk/interpro) and CluSTr (http://www.ebi.ac.uk/clustr) resources have been used to classify the data by sequence similarity. Structural information includes amino acid composition for each of the proteomes, the Homology derived Secondary Structure of Proteins (HSSP) classification and links to the Protein Data Bank (PDB). A searchable functional classification using the Gene Ontology (GO) is also available. The Proteome Analysis Database contains statistical and analytical data for the proteins from completely sequenced genomes.

14.6.3 General plant databases

Table 14.7 lists the databases that generally cover multiple plant species or have multiple functions. Information contained in these databases includes genetic and physical mapping, sequencing, clustering, microarray analysis, functional annotation, signal transduction analysis, etc.

Some databases contain relatively simple information such as *cis*-element motifs, plant genome sizes (C-values), promoter sequences, small nucleolar RNAs (snoRNAs) or non-coding RNAs (ncRNAs), mitochondrial protein, *cis*-acting regulatory elements/enhancers/repressors and clusters of predicted plant proteins.

In addition to data integration and visualization, some databases provide specific tools required for managing and mining the data, such as plant EST clustering and functional annotation, signal transduction analysis, functional analysis of agricultural plant and animal gene products, classification of repetitive sequences, phylogeny-based tools for comparative genomics and retrieval of plant protease inhibitors (PIs) and their genes.

Some databases cover a group of plant species, such as GrainGenes for wheat, barley, rye, triticale and oats, TropGENE DB for tropical crops and PLANTS Database for the vascular plants, mosses, liverworts, hornworts and lichens of the USA and its territories.

Some databases provide multiple categories of information and also tools

Table 14.6. General genomics and proteomics databases.

Database name	Uniform Resource Locator (URL)	Database description
TIGR Gene Indices	http://compbio.dfci.harvard.edu/tgi	Databases to identify and classify transcribed sequences in eukaryotic species using available EST and gene sequence data
GO	http://www.geneontology.org	Gene Ontology Consortium database
KEGG	http://www.genome.ad.jp/kegg	KEGG (Kyoto Encyclopedia of Genes and Genomes) is the primary database resource of the Japanese GenomeNet service for understanding higher order functional meanings and utilities of the cell or the organism from its genome information
Swiss-2DPAGE	http://www.expasy.org/ch2d	Maintained collaboratively by the Central Clinical Chemistry Laboratory of the Geneva University Hospital and the Swiss Institute of Bioinformatics (SIB), the database contains data on proteins identified on various two-dimensional PAGE and SDS-PAGE reference maps from human, mouse, *Arabidopsis thaliana*, *Dictyostelium discoideum*, *Escherichia coli*, *Saccharomyces cerevisiae* and *Staphylococcus aureus*
COGs	http://www.ncbi.nlm.nih.gov/COG	Clusters of Orthologous Groups of proteins (COGs) were delineated by comparing protein sequences encoded in complete genomes, representing major phylogenetic lineages. Each COG consists of individual proteins or groups of paralogues from at least three lineages and thus corresponds to an ancient conserved domain
ERGO	http://www.ergo-light.com	ERGO, formerly WIT database, provides links to information about the functional role of enzymes (via links to data in KEGG); links to NCBI Medline entries for each enzyme; and links to enzymes and metabolic pathways records for each enzyme. The database also provides access to thoroughly annotated genomes within a framework of metabolic reconstructions, connected to the sequence data; protein alignments and phylogenetic trees, and data on gene clusters, potential operons and functional domains
wwPDB	http://www.wwpdb.org	The Worldwide Protein Data Bank (wwPDB) consists of organizations that act as deposition, data processing and distribution centres for PDB data. The mission of the wwPDB is to maintain a single Protein Data Bank Archive of macromolecular structural data that is freely and publicly available to the global community
Genome Project Database	http://www.ncbi.nlm.nih.gov/entrez/query.fcgi?CMD=search&DB=genomeprj	The NCBI Entrez Genome Project Database is intended to be a searchable collection of complete and incomplete (in-progress) large-scale sequencing, assembly, annotation, and mapping projects for cellular organisms

Continued

Table 14.6. *Continued.*

Database name	Uniform Resource Locator (URL)	Database description
Entrez Gene	http://www.ncbi.nlm.nih.gov/entrez/query.fcgi?db=gene	Entrez Gene is NCBI's database for gene-specific information with focus on the genomes that have been completely sequenced, that have an active research community to contribute gene-specific information, or that are scheduled for intense sequence analysis
Entrez Genomes	http://www.ncbi.nlm.nih.gov/sites/entrez?db=genome	NCBI's collection of databases for the analysis of complete and unfinished viral, pro- and eukaryotic genomes
ACeDB	http://www.acedb.org	*Caenorhabditis elegans*, *Schizosaccharomyces pombe*, and human sequences and genomic information
FlyBase	http://flybase.org	An integrated resource for genetic, molecular and descriptive data concerning the Drosophilidae, including interactive genomic maps, gene product descriptions, mutant allele phenotypes, genetic interactions, expression patterns, transgenic constructs and their insertions, anatomy and images, and genetic stock collections

Table 14.7. General plant databases.

Database name	Uniform Resource Locator (URL)	Database description
AgBase	http://www.agbase.msstate.edu	A curated, open-source, web-accessible resource for functional analysis of agricultural plant and animal gene products
BarleyBase	http://www.barleybase.org	An online database for plant microarrays with integrated tools for data visualization and statistical analysis
Cereal Small RNA Database	http://sundarlab.ucdavis.edu/smrnas	An integrated resource for small RNAs expressed in rice and maize that includes a genome browser and a smRNA-target relational database as well as relevant bioinformatic tools
CR-EST – Crop ESTs	http://pgrc.ipk-gatersleben.de/cr-est	A publicly available online resource providing access to sequence, classification, clustering, and annotation data of crop EST projects at IPK Gatersleben, Germany
CropNet	http://ukcrop.net	The UK Crop Plant Bioinformatics Network (UK CropNet) established to harness the extensive work in genome mapping in crop plants in the UK. The resource facilitates the identification and manipulation of agronomically important genes by laying a foundation for comparative analysis among crop plants and model species. A number of software tools have been developed to facilitate data visualization and analysis
FLAGdb++	http://urgv.evry.inra.fr/projects/FLAGdb++/HTML/index.shtml	Dedicated to the integration and visualization of data for high-throughput functional analysis of a fully sequenced genome, as illustrated for *Arabidopsis*
GénoPlante-Info	http://www.genoplante.com	Integrated and made publicly available the data have been generated for genomics sequence, transcriptome, proteome, allelic variability, mapping and synteny, mutation data) and tools (databases, interfaces, analysis software) through a collaboration between public French institutes and private companies that aims at developing genome analysis programs for crop species (maize, wheat, rapeseed, sunflower and pea) and model plants (*Arabidopsis thaliana* and rice)
GeneFarm	http://urgi.versailles.inra.fr/Genefarm	Expert annotation of *Arabidopsis* gene and protein families
GrainGenes	http://wheat.pw.usda.gov	Molecular and phenotypic information on wheat, barley, rye, triticale and oats
Gramene	http://www.gramene.org	A comparative genome mapping database for grasses with both automatic and manual curation performed to combine and interrelate information on genomic and EST sequences, genetic, physical and sequence-based maps, proteins, molecular markers, mutant phenotypes and QTL, and publications
MIPSPlantsDB	http://mips.gsf.de/proj/plant/jsf	The MIPS (Plant Genome Bioinformatics at the Institute for Bioinformatics) plant Genomics group focuses on the bioinformatics of plant genomes. It developed from the Arabidopsis Genome Annotation Group and currently provides the following databases: the MIPS *Arabidopsis thaliana* genome database, the maize genome, the rice genome (MOsDB), the Medicago Genome database, the Lotus Genome database, the Tomato Genome database, *Cis*-Regulatory Element Detection Online (CREDO), mips Repeat Element database (mips-REdat), mips Repeat Element catalogue (mips-REcat), and MotifDB

(*Continued*)

Table 14.7. *Continued.*

Database name	Uniform Resource Locator (URL)	Database description
MPIM	http://www.plantenergy.uwa.edu.au/applications/mpimp/index.html	A database containing information on the mitochondrial protein import apparatus from a wide range of organisms, including yeast, human, rat, mouse, *Drosophila*, *Danio rerio*, *Cenorhabtidis elegans*, *Arabidopsis*, rice and *Plasmodium falciparum*
PathoPlant®	http://www.pathoplant.de	A database on plant–pathogen interactions and components of signal transduction pathways related to plant pathogenesis
ICIS	http://www.icis.cgiar.org	The International Crop Information System (ICIS) is a database system for the management and integration of global information on genetic resources and crop improvement for any crop
Phytome	http://www.phytome.org	A comparative genomics database designed to facilitate functional plant genomics, molecular breeding, and evolutionary studies. It contains predicted protein sequences, protein family assignments, multiple sequence alignments, phylogenies, and functional annotations for proteins from a large, phylogenetically diverse set of plant taxa
PHYTOPROT	http://urgi.versailles.inra.fr/phytoprot	Clusters of predicted plant proteins
dbEST	http://www.ncbi.nlm.nih.gov/dbEST	A division of GenBank® that contains sequence data and other information on 'single-pass' cDNA sequences, or ESTs, from a number of organisms
PLACE	http://www.dna.affrc.go.jp/htdocs/PLACE	A database containing *cis*-element motifs found in plant genes
Plant DNA C-values database	http://www.kew.org/genomesize/homepage.html	A one-stop, user-friendly database for plant genome sizes. The most recent release (release 4.0, October 2005) contains genome size data for 5150 species comprising 4427 angiosperms, 207 gymnosperms, 87 monilophytes and lycopods, 176 bryophytes and 253 algal species
Plant Genome Central	http://www.ncbi.nlm.nih.gov/genomes/PLANTS/PlantList.html	Providing access to data from large-scale sequencing projects, genetic maps, and large-scale EST sequencing projects
Plant MPSS	http://mpss.udel.edu	MPSS (Massively Parallel Signature Sequencing) is a sequencing-based technology that uses a unique method to quantify gene expression level, generating millions of short sequence tags per library. The Plant MPSS databases are the largest publicly available set of tag-based gene expression data
Plant Ontology™ database	http://www.plantontology.org	The Plant Ontology™ (PO) is a collaborative effort among several plant databases and experts in plant systematics, botany and genomics to develop simple yet robust and extensible controlled vocabularies that accurately reflect the biology of plant structures (morphology and anatomy) and developmental stages

Plant snoRNA DB	http://bioinf.scri.sari.ac.uk/cgi-bin/plant_snorna/home	Small nucleolar RNA (snoRNA) genes in plant species
PLANT-PIs	http://bighost.area.ba.cnr.it/PLANT-PIs	A database for facilitating retrieval of information on plant protease inhibitors (PIs) and their genes
PlantGDB	http://www.plantgdb.org	A database for plant genomic sequences, in particular ESTs that correspond to fragments of genes that are actively transcribed under particular conditions (currently with data for 48 plant species)
PlantProm	http://mendel.cs.rhul.ac.uk/mendel.php?topic=plantprom	A database for plant promoter sequences
PlantsP/PlantsT	http://plantsp.sdsc.edu	PlantsP and PlantsT are plant-specific curated databases that combine sequence derived information with experimental functional genomics data. PlantsP focuses on proteins involved in the phosphorylation process (i.e. kinases and phosphatases), whereas PlantsT focuses on membrane transport proteins
POGs/PlantRBP	http://plantrbp.uoregon.edu	A relational database that integrates data from rice, *Arabidopsis*, and maize by placing the complete *Arabidopsis* and rice proteomes and available maize sequences into 'putative orthologous groups' (POGs)
TAED	http://www.bioinfo.no/tools/TAED	TAED (The Adaptive Evolution Database) is a phylogeny-based tool for comparative genomics
TIGR plant repeat database	http://www.tigr.org/tdb/e2k1/plant.repeats	Classification of repetitive sequences in plant genomes
TIGR Plant Transcript Assembly Database	http://plantta.tigr.org	The database uses expressed sequences collected from the NCBI GenBank® Nucleotide database for the construction of transcript assemblies. The sequences collected include ESTs and full-length and partial cDNAs, but exclude computationally predicted gene sequences
TropGENE DB	http://tropgenedb.cirad.fr	A database that manages genetic and genomic information about tropical crops
PLEXdb	http://www.plexdb.org	PLEXdb (Plant Expression Database) is a unified public resource for gene expression for plants and plant pathogens, serving as a bridge to integrate new and rapidly expanding gene expression profile data sets with traditional structural genomics and phenotypic data
PLANTS Database	http://plants.usda.gov	The PLANTS Database provides standardized information about the vascular plants, mosses, liverworts, hornworts, and lichens of the USA and its territories
UK CropNet Databases	http://ukcrop.net/db.html	Contains six databases (Arabidopsis Genome Resource, BarleyDB, BrassicaDG, CropSeqDB, FoggDB, MilletGenes) and mirrors many other plant-related databases

for multiple functions. A typical example is ICIS, which has been described in the previous section. As another example, Phytome is an online comparative genomics resource that is built upon publicly available sequence and map information from a diverse set of plant species, with a focus on the angiosperms, or flowering plants. It provides an interface to the results from a variety of phylogenomic analyses. Phytome is designed to facilitate functional genomics, molecular breeding and evolutionary studies in model and non-model plant species. Currently, Phytome contains phylogenetic and functional information for predicted protein sequences ('Unipeptides'). Future development will incorporate data and tools for analysis of sequence-based comparative maps.

There are some databases supporting functions for comparative biology. Gramene is one such database for comparative genome mapping of grasses, with both automatic and manual curation performed to combine and interrelate information on genomic and EST sequences, genetic, physical and sequence-based maps, proteins, molecular markers, mutant phenotypes and QTL and publications. As an information resource, Gramene's purpose is to provide added value to data sets available within the public sector, which will facilitate researchers' ability to understand the rice genome and leverage the rice genomic sequence for identifying and understanding corresponding genes, pathways and phenotypes in other crop grasses. This is achieved by building automated and curated relationships between rice and other cereals. The automated and curated relationships are queried and displayed using controlled vocabularies and web-based displays. The controlled vocabularies (ontologies) currently being utilized include Gene ontology, Plant Ontology™, Trait ontology, Environment ontology and Gramene Taxonomy ontology. The web-based displays for phenotypes include the Genes and Quantitative Trait Loci (QTL) modules. Sequence based relationships are displayed in the Genomes module using the genome browser adapted from Ensembl, in the Maps module using the comparative map viewer (CMAP) from GMOD and in the Proteins module displays. BLAST is used to search for similar sequences.

Some sites host their own databases and also mirror other related databases. The UK Crop Plant Bioinformatics Network (UK CropNet), established to harness the extensive work in genome mapping in crop plants in the UK, is one such example. The UK CropNet contains six databases (Arabidopsis Genome Resource, BarleyDB, BrassicaDG, CropSeqDB, FoggDB, MilletGenes). It also mirrors many other plant-related databases (Table 14.7).

14.6.4 Individual plant databases

Tables 14.8, 14.9 and 14.10 list databases for specific plants. As model plants, *Arabidopsis* and rice databases are listed in separate tables. Although most plant databases differ considerably in content, the general subject matter for a species-specific database may include:

- genetic and cytogenetic maps;
- genomic probes, nucleotide sequences;
- genes, alleles and gene products;
- phenotypes, quantitative traits and QTL;
- genotypes and pedigrees of cultivars, genetic stocks and other germplasm;
- pathologies and the corresponding pathogens, insects and abiotic stresses;
- taxonomy of the crops and related species;
- addresses and research interests of colleagues; and
- relevant bibliographic citations.

As a model crop, rice has highly diversified databases, which results in each database containing specific information, such as mutants and T-DNA insertions, or serving as a specific function, such as annotation and proteomic analysis (Table 14.9). There are two annotation related databases, one oriented around contigs for high-quality manual annotation (RAD) and the other providing a system to integrate programs for prediction and analysis of protein-coding gene structure (RiceGAAS). The evidence

Table 14.8. *Arabidopsis thaliana* databases.

Database name	Uniform Resource Locator (URL)	Database description
AGNS	http://wwwmgs.bionet.nsc.ru/agns	AGNS (*Arabidopsis* GeneNet supplementary) database provides access to description of the functions of the known *Arabidopsis* genes at various levels-the levels of mRNA, protein, cell, tissue, and ultimately at the levels of organs and the organism in both wild-type and mutant backgrounds
AGRIS	http://arabidopsis.med.ohio-state.edu	*Arabidopsis* Gene Regulatory Information Serve (AGRIS) is an information resource of *Arabidopsis* promoter sequences, transcription factors and their target genes, currently containing two databases, AtTFDB (*Arabidopsis thaliana* transcription factor database) and At*cis*DB (*Arabidopsis thaliana cis*-regulatory database)
Arabidopsis Mitochondrial Protein Database	http://www.plantenergy.uwa.edu.au applications/ampdb/index.html	Experimentally identified mitochondrial proteins in *Arabidopsis*
Arabidopsis MPSS	http://mpss.udel.edu/at	*Arabidopsis* gene expression detected by massively parallel signature sequencing
Arabidopsis Nucleolar Protein Database	http://bioinf.scri.sari.ac.uk/cgi-bin/atnopdb/proteome_comparison	Comparative analysis of nucleolar proteomes of human and *Arabidopsis*
ARAMEMNON	http://aramemnon.botanik.uni-koeln.de	A curated database for *A. thaliana* transmembrane (TM) proteins and transporters
ARTADE	http://omicspace.riken.jp/ARTADE	A database containing transcriptional structures elucidated by ARTADE which estimates exon/intron structures of structurally unknown genes based on both tiling array data and genomic sequence data
ASRP	http://asrp.cgrb.oregonstate.edu	A database for *A. thaliana* small RNA project
AtGDB	http://www.plantgdb.org/AtGDB	A part of 'PlantGDB – Plant Genome Database and Analysis Tools' to provide a convenient sequence-centred genome view for *A. thaliana*, with a narrow focus on gene structure annotation
AthaMap	http://www.athamap.de	Genome-wide map of putative transcription factor binding sites in *A. thaliana*
ATTED-II	http://www.atted.bio.titech.ac.jp	A database providing co-regulated gene relationships based on co-expressed genes deduced from microarray data and the predicted *cis* elements
DATF	http://datf.cbi.pku.edu.cn	The database of Arabidopsis Transcription Factors (DATF) collects all *Arabidopsis* transcription factors (total of 1922 loci and 2290 gene models) and classifies them into 64 families
GABI-Kat	http://www.gabi-kat.de	A Flanking Sequence Tag (FST)-based database for T-DNA insertion mutants generated by the GABI-Kat project

(*Continued*)

Table 14.8. *Continued.*

Database name	Uniform Resource Locator (URL)	Database description
MAtDB	http://mips.gsf.de/proj/thal/db	MIPS (Plant Genome Bioinformatics at the Institute for Bioinformatics) *A. thaliana* database
NASCarrays	http://affymetrix.arabidopsis.info	Nottingham Arabidopsis Stock Centre microarray database
PLprot	http://www.pb.ipw.biol.ethz.ch/proteomics	*A. thaliana* chloroplast protein database
RARGE	http://rarge.gsc.riken.jp	RIKEN Arabidopsis Genome Encyclopaedia (RARGE) contains *Arabidopsis* cDNAs, mutants and microarray data
SeedGenes	http://www.seedgenes.org	Genes essential for *Arabidopsis* development
SUBA	http://www.suba.bcs.uwa.edu.au	The Arabidopsis Subcellular Database (SUBA) contains publicly available protein subcellular localization data from a variety of sources from the model plant *Arabidopsis*
TAIR	http://www.arabidopsis.org	The Arabidopsis Information Resource (TAIR) contains data for *A. thaliana* genome

Table 14.9. Rice databases.

Database name	Uniform Resource Locator (URL)	Database description
Oryza Tag Line	http://urgi.versailles.inra.fr/OryzaTagLine	A database to organize data resulting from the phenotypic characterization of a library of T-DNA insertion lines of rice (*Oryza sativa* L. cv. Nipponbare)
BGI-RISe	http://rise.genomics.org.cn	Beijing Genomics Institute Rice Information System (BGI-RISe), containing comprehensive data from *O. sativa* L. ssp. *indica*, genome information from *O. sativa* L. ssp. *japonica* and EST sequences available from other cereal crops. Sequence contigs of *indica* (93-11) have been further assembled into Mbp-sized scaffolds and anchored on to the rice chromosomes referenced to physical/genetic markers, cDNAs and BAC-end sequences. The rice genomes have been annotated for gene content, repetitive elements, gene duplications (tandem and segmental) and SNPs between rice subspecies
WhoGA	http://rgp.dna.affrc.go.jp/whoga	WhoGA is a rice genome annotation viewer using the GBrowse web-server application. In addition to predicted genes, WhoGA also includes gene models for pseudogenes with or without EST/full-length cDNA support, regions wherein genes could not be modelled although showing significant homology to known genes, and ORFs predicted by a single gene prediction program
IRIS	http://www.iris.irri.org	The International Rice Information System (IRIS) is the rice implementation of the International Crop Information System (ICIS, www.cgiar.org/icis), a database system for the management and integration of global information on genetic resources and crop improvement for any crop
MOsDB	http://mips.gsf.de/proj/plant/jsf/rice/index.jsp	A resource for publicly available sequences of the rice (*O. sativa* L.) genome to provide all available data about rice genes and genomics, including mutant information and expression profiles
OryGenesDB	http://orygenesdb.cirad.fr	A database for rice genes, T-DNA and transposable elements flanking sequence tags
Oryzabase	http://www.shigen.nig.ac.jp/rice/oryzabase	A comprehensive rice science database with the original aim to gather as much knowledge as possible ranging from classical rice genetics to recent genomics and from fundamental information to hot topics
RAP-DB	http://rapdb.lab.nig.ac.jp	Rice Annotation Project Database (RAP-DB) provides access to the annotation data. By connecting the annotations to other rice genomics data, such as full-length cDNAs and *Tos17* mutant lines, the RAP-DB serves as a hub for rice genomics

(*Continued*)

Table 14.9. *Continued.*

Database name	Uniform Resource Locator (URL)	Database description
RetrOryza	http://www.retroryza.org	RetrOryza is a database that aims at providing the research community with the most complete resource on long terminal repeat-retrotransposon for rice
RAD	http://golgi.gs.dna.affrc.go.jp/SY-1102 rad/index.html	The Rice Annotation Database (RAD) is a contig-oriented database for high-quality manual annotation of the Rice Genome Project, which can present non-redundant contig analyses by merging the accumulated PAC/BAC clones
RMD	http://rmd.ncpgr.cn	The Rice Mutant Database (RMD) contains the information of approximately 129,000 rice T-DNA insertion (enhancer trap) lines generated by an enhancer trap system
Rice Pipeline	http://cdna01.dna.affrc.go.jp/PIPE	A unification tool which dynamically collects and compiles data from scientific databases in National Institute of Agrobiological Sciences (NIAS) to provide a unique scientific resource of rice that pools publicly available data
Rice Proteome Database	http://gene64.dna.affrc.go.jp/RPD/main_en.html	Rice proteome database
RiceGAAS	http://RiceGAAS.dna.affrc.go.jp	Rice Genome Automated Annotation System (RiceGAAS)
RMD	http://www.ricefgchina.org/mutant	Rice mutant database

for database diversification within a single species is also apparent from the list of *Arabidopsis* databases (Table 14.8).

For other plant databases (Table 14.10), two will be briefly described here. MaizeGDB – the Maize Genetics and Genomics Database – provides a central repository for public maize information and presents it in a way that creates intuitive biological connections for the researcher with minimal effort. It also provides a series of computational tools that directly address the questions of the biologist in an easy-to-use form. Its data centre contains the following information: data centres; bacterial artificial chromosomes (BACs); ESTs; gene products; locus/loci; maps; metabolic pathways; microarrays; overgos; people/organizations; phenotypes; probes; QTL; references; sequences; SSRs; stocks; variations. At CIMMYT, two crop-specific databases for wheat and maize (http://iwis.cimmyt.org/ICIS5/) have been developed.

Dendrome is a collection of forest tree genome databases and information resources for the international forest genetics community. Dendrome is part of a larger collaborative effort to construct genome databases for major crop and forest species. The primary genome database of Dendrome is called TreeGenes. TreeGenes provides curated information about genetic maps, DNA sequences, germplasm, markers, QTL and ESTs. The goal of this effort is to provide an improved interface for comparison between maps and to integrate expression and EST data.

14.7 Future Prospects for Breeding Informatics

Plant breeding in the future will be largely driven by molecular biology and informatics. Breeding efficiency will depend on how much information breeders can access and how wisely and effectively they can use it in their breeding programmes.

Breeding related databases and information systems have to be improved so they are more user-friendly to breeders. The lack of mutual understanding between breeders and scientists in other disciplines will continue to be a major limiting factor, so information management systems and tools should be enhanced so that they can be accessed and used by breeders more easily.

The great proliferation of relevant databases and informatics tools makes them less accessible to most breeders. The use of these resources is often hampered by the fact that they are designed for specific application areas and thus lack universality. As users, breeders have to visit many different databases and use different tool packages for specific purposes, depending on which crop species the breeder works with, the types of information the breeder wants to retrieve and the different functions the breeder wants to perform. As a result, knowing how to access and use these databases demands a significant investment of time and effort.

Data stored in central databases such as KEGG, BRENDA or SABIO-RK is often limited to read-only access. If researchers want to store their own data, they must either develop their own information system for managing that data, which can be time-consuming and costly, or they must store their data in existing systems, which is often restricted. Hence, an out-of-the-box information system for managing breeding-related data is needed. As an example of such effort, Weise *et al.* (2006) designed META-ALL, an information system that allows the management of metabolic pathways, including reaction kinetics, detailed locations, environmental factors and taxonomic information. Data can be stored together with quality tags and in different parallel versions.

As many information systems and databases are developed through specially funded projects, which generally only run for a specific period of time, they become outdated and ill supported. They may also be abandoned completely. Maintaining the databases and tools that have been developed requires continuous funding and technical support, which is almost impossible if the number of databases and informatics tools keep growing at the present rate. One

Table 14.10. Databases for other plants excluding *Arabidopsis* and rice.

Database name	Uniform Resource Locator (URL)	Database description
Brassica BASC	http://bioinformatics.pbcbasc.latrobe.edu.au	The BASC system provides tools for the integrated mining and browsing of genetic, genomic and phenotypic data, hosting information on *Brassica* species supporting the Multinational Brassica Genome Sequencing Project
Diatom EST Database	http://www.biologie.ens.fr/diatomics/EST	ESTs from two diatom algae, *Thalassiosira pseudonana* and *Phaeodactylum tricornutum*
ForestTreeDB	http://foresttree.org/ftdb	A resource that centralizes large-scale EST sequencing results from several tree species
Legume Information	http://www.comparative-legumes.org	The Legume Information System (LIS), formerly the Medicago Genome System Initiative (MGI), is an EST sequence database and analysis system that supports EST sequencing at the Noble Foundation Center for Medicago Genome Research
MaizeGDB	http://www.maizegdb.org	The Maize Genetics and Genomics Database (MaizeGDB) is a central repository for maize sequence, stock, phenotype, genotypic and karyotypic variation, and chromosomal mapping data. In addition, MaizeGDB provides contact information for over 2400 maize cooperative researchers, facilitating interactions among members of the rapidly expanding maize community
MtDB	http://www.medicago.org/MtDB	*Medicago truncatula* genome database
NRESTdb	http://genome.ukm.my/nrestdb	Natural Rubber EST Database (NRESTdb) serving as a molecular resource for functional genomics of the rubber tree
Panzea	http://www.panzea.org	The Panzea Database contains the genotype, phenotype, and polymorphism data produced by the Molecular and Functional Diversity in the Maize Genome project
PoMaMo	https://gabi.rzpd.de/PoMaMo.html	PoMaMo (Potato Maps and More), established within the German Plant Genome Project 'GABI', harbours information on molecular maps of all 12 potato chromosomes with about 1000 mapped elements, sequence data, putative gene functions, results from BLAST analysis, SNP and Indel information from different diploid and tetraploid potato genotypes, publication references, and links to other public databases like NCBI or SGN (see below) for example
SGMD	http://psi081.ba.ars.usda.gov/SGMD/default.htm	Soybean genomics and microarray database
SoyGD	http://soybeangenome.siu.edu	The Soybean Genome Database (SoyGD) genome browser integrates the publicly available physical map, BAC sequence database and genetic map-associated genomic data

TED	http://ted.bti.cornell.edu	Tomato expression database
TIGR Maize database	http://maize.tigr.org	A repository of publicly available maize genomic sequences
TomatEST DB	http://biosrv.cab.unina.it/tomatestdb/	A secondary database integrating EST/cDNA sequence informationindex2.php from different libraries of multiple tomato species collected from dbEST
Soybean Genome	http://www.soybeangenome.org	Dedicated to the sharing and dissemination of public information on all aspects of soybean genomics and the application of genome information to soybean
BarleyBase	http://www.plexdb.org/plex.php?	BarleyBase is a MIAME-compliant and Plant Ontology enhanced database=Barley expression database for plant microarray data
Dendrome	http://dendrome.ucdavis.edu	Dendrome is a collection of forest tree genome databases and other forest genetic information resources for the international forest genetics community
TropGENE	http://tropgenedb.cirad.fr	A database that manages genetic and genomic information about tropical crops studied by the Agricultural Research Centre for International Development (known by its French acronym, CIRAD), including banana, cocoa, coconut, coffee, cotton, oil palm, rice, rubber tree and sugarcane
Cotton	http://www.cottondb.org	A database that contains genomic, genetic and taxonomic information for cotton (*Gossypium* spp.). It serves both as an archival database and as a dynamic database which incorporates new data and user resources
CyanoBase	http://bacteria.kazusa.or.jp/cyano	CyanoBase provides an easy way of accessing the sequences and all-inclusive annotation data on the structures of the cyanobacterial genomes
BeanGenes	http://beangenes.cws.ndsu.nodak.edu	A plant genome database currently containing information relevant to *Phaseolus* and *Vigna* species
SGN	http://sgn.cornell.edu	The SOL Genomics Network (SGN) is a Clade Oriented Database (COD) containing genomic, genetic and taxonomic information for species in the Euasterid clade, including the families *Solanaceae* (e.g. tomato, potato, eggplant, pepper, petunia) and *Rubiaceae* (coffee)
RAPESEED	http://rapeseed.plantsignal.cn	Shanghai RAPESEED database contains information collected on ESTs, full-length cDNA, unique serial analysis of gene expression (SAGE) tags, and EMS mutants for *Brassica napus*
ICIS	http://www.icis.cgiar.org	A database system that provides integrated management of global information on crop improvement and management both for individual crops and for farming systems

approach is to develop databases and tools that need minimum maintenance or that can be upgraded or updated automatically. Another way is to develop a universal database and informatics tool package for information-driven plant breeding, which needs a worldwide collaboration through a global scientific programme in a way similar to the human genome sequencing project.

Developing a universal database or a database of all databases would require a universal language that can be shared across all plant species. Gene Ontology and Plant Ontology projects represent a good beginning of such effort. Another universal language is also needed that can be used for communications among breeders, database curators, bioinformaticians, molecular biologists and tool developers. Breeders should be a major player rather than an observer in the development of such a universal database or language.

15

Decision Support Tools

Molecular breeding involves identification of beneficial genetic variation and selection of desirable recombinants, and it manages and utilizes genetic variation more effectively and efficiently through molecular technology including two major procedures, marker-assisted-selection (MAS) and genetic transformation. In general, MAS relies on the reliable identification and application of simply inherited markers that are inside of, or in close proximity to, genetic factors affecting simple, oligogenic and multi-genic traits of importance to crop improvement. The journey from phenotyping and genotyping individuals from genetic populations to identifying marker–trait associations and finally applying markers in molecular breeding programmes depends on a sequential use of a number of decision support tools that facilitate communication and collaboration between molecular biologists, geneticists, bioinformaticians, trait specialists and breeders towards effective interdisciplinary decision making. Ultimately, molecular breeding programmes will combine MAS with a diverse range of technology-assisted interventions including whole genome scans, advanced biometrical analyses and quantitative genetics modelling that will require increasingly complex facilitating software.

Effective molecular breeding requires a careful balance of many diverse elements in order to provide the best compromise between time, cost and genetic gain:

- Identify new sources of beneficial genetic variation and develop robust marker–trait associations.
- Manage and manipulate large amounts of genotype, pedigree and phenotype data.
- Select desirable recombinants through an optimum combination (in time and space) of phenotypic and genotypic information.
- Develop breeding systems that minimize population sizes, number of generations and overall costs while maximizing genetic gain for traditional and novel target traits.

Figure 15.1 summarizes the forward and reverse genetics approaches associated with molecular breeding product delivery. Decision support tools are required to manage and optimize many components of molecular plant breeding. Many decision support tools are in the form of software. From the UK's gateway to high-quality Internet resources in biology and biomedical research (http://bioresearch.ac.uk/browse/mesh/D012984.html), there is a list of all software available, some of which are related to molecular breeding. The Laboratory of Statistical Genetics at Rockefeller University also provides web resources of genetic

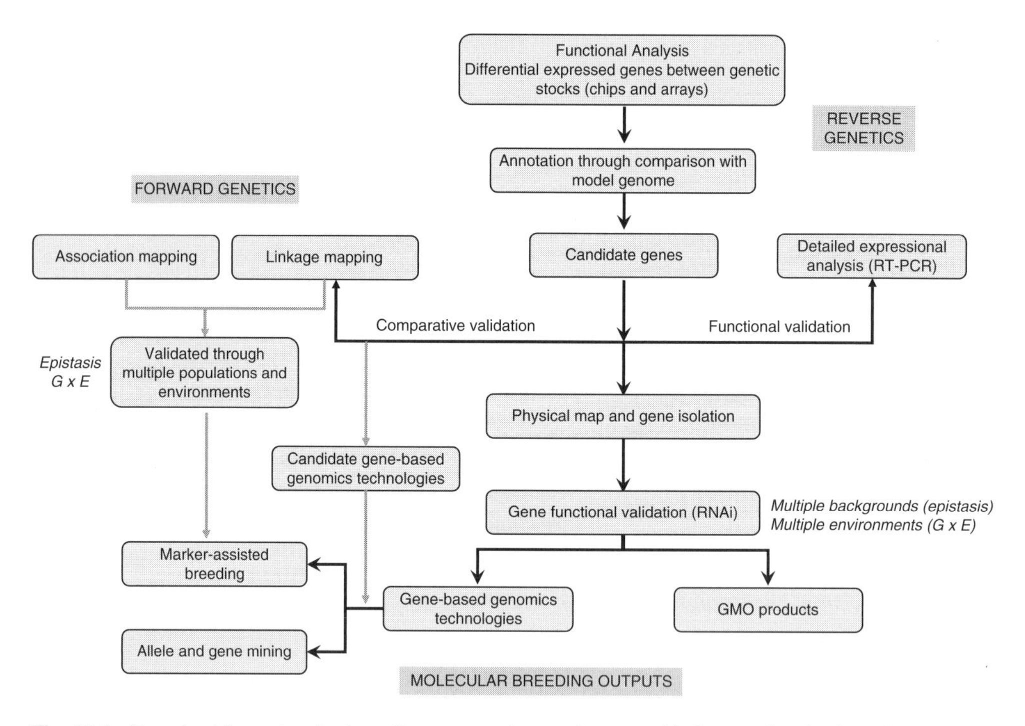

Fig. 15.1. Flowchart for molecular breeding approaches and outputs. Various molecular breeding approaches discussed in this book are summarized, including forward and reverse genetics approaches and their associated breeding outputs. Decision support tools may be needed in each step. G × E, genotype-by-environment interaction.

linkage analysis with various software listed in alphabetic order (http://linkage.rockefeller.edu/soft/list.html).

This chapter provides an overview of key decision support tools that need to support molecular breeding programmes, including germplasm evaluation, breeding population management, genotype-by-environment interaction (GEI), genetic map construction, marker–trait linkage and association analysis, MAS and breeding system design and simulation. Plant variety protection and breeding information management are discussed in Chapters 13 and 14, respectively.

15.1 Germplasm and Breeding Population Management and Evaluation

Germplasm collections and breeding populations are basic materials required in crop improvement. Decision support tools are needed to manage and evaluate crop genetic resources and breeding materials including genetic diversity and variation analysis, population structure evaluation and for hybrid crops, use of the genetic diversity to define heterotic groups and predict hybrid performance.

15.1.1 Germplasm management and evaluation

Genetic resources provide the foundation of any plant breeding programme. Efficient germplasm utilization requires well-founded sampling strategies. Genetic diversity analysis and its relationship with functional variation of the target trait is the fundamental basis of germplasm evaluation (Chapter 5). Novel alleles and new genes from both cultivated and wild relatives provide the engine of germplasm enhancement. However, marker technologies offer to dramatically increase the pace, precision and efficiency with which that genetic variation can drive new product development.

As discussed in Chapter 5, genebank curators use a variety of methodologies for guiding their germplasm collection and management strategies. Geographical information systems (GIS) data associated with the site from which germplasm was collected is an important component of the standard description (passport data) of an accession. Plant breeders can use GIS data to access valuable information about the ecological environment where new genetic resources were growing. Thus, genetic resources collected from drought prone areas or hot spots for important diseases can be rich sources of new beneficial genetic variation for crop improvement programmes. GIS data alone can be misleading in this respect where the site of collection is not related to the location. However, DNA marker analysis provides a valuable complement to GIS data, as it can assist in establishing generic relationships and estimating genetic distances between genetic resources. Thus, the combined use of GIS and DNA marker data can help to prioritize emphasis of germplasm screening efforts.

As the size of germplasm collections has increased, genebank curators have attempted to stratify genetic resources in order to provide breeders and researchers with a small number of accessions that represent a large proportion of the overall genetic variation. This has led to the establishment of so-called core collections (Chapter 5). Breeders and researchers will then typically first evaluate the core collection to identify accessions with high levels of their desired target trait and then move on to screen closely related germplasm from the main collection. Unfortunately, a core collection can only be as good as the data on which it is based, which often constitutes a relatively small number of taxonomic descriptors and agronomic traits. More robust core collections could be generated through combined analysis of phenotype, genotype and GIS data. However, a tool is needed that can be used to combine these different criteria. As more and more genetic information from functionally characterized genes and genomic regions becomes available, the construction of core collections can also utilize this functional diversity rather than the neutral information that is revealed by molecular markers used in most current studies. Development of core collections based on all possible types of data will require powerful new computational tools. Ultimately, germplasm users need to be provided with on-line dynamic selector tools that allow them to tailor the selection of a subset of germplasm based upon analysis of all available data using their own unique criteria.

Marker-assisted germplasm evaluation (MAGE) will play an important role in the procedures related to the acquisition/distribution, maintenance and use of germplasm (Bretting and Widrlechner, 1995; Xu, Y. 2003; Chapter 5). Efficient marker-assisted germplasm management relies on the availability of several key resources (Xu, Y., 2003), including: (i) suitable genetic markers characterized for the number of alleles, polymorphic information content (PIC value), allele sizes and ranges, signal strength, working conditions and the necessary information for multiplexing; (ii) high-density molecular maps allowing the selection of markers evenly spread over the whole genome or densely spread over the specific region of interest; (iii) established marker–trait associations for traits of agronomic importance; (iv) high-throughput genotyping systems; and (v) an efficient data management and analysis system. In addition, germplasm collection should provide a large number of relevant accessions for the target research purpose. Although computational programs are available for all relevant analyses, as discussed in Chapter 5, including computer simulation and re-sampling (Xu *et al.*, 2004), a fully integrated, user-friendly graphical program is needed to bring all these functions together to facilitate decisions through all aspects of germplasm evaluation.

The recent focus on molecular characterization of germplasm collections and the subsequent use of those collections in crop improvement has led to a number of bioinformatics projects developing new tools to improve the power and scope of such analyses. Ambiguous germplasm identification, difficulty in tracing pedigree information and lack of integration of databases across genetic resources, characterization, evaluation and utilization have been identified as the major constraints to developing knowledge-led germplasm enhancement programmes.

Visualization tools enable us to simultaneously view large quantities of these data and to identify underlying patterns in our data sets. We also need analytical tools to help search for association between the target trait and individual markers or marker haplotypes and to look for patterns of genetic diversity with our germplasm collections. GENE-MINE (http://www.gene-mine.org; Davenport *et al.*, 2004) was developed to bring together experts in database development, data querying and visualization, quantitative methods and computational methods, to develop novel tools for the analysis of germplasm collections characterized by molecular markers. Within the GENE-MINE project, a generic information system was developed for studying the relationships between large-scale collections databases from genebank material such as molecular marker data, trait phenotype, passport and environmental data. The system uses a generic model for associating properties, such as traits, genetic data and molecular data, with accession data. Users are able to make queries using terms from germplasm query language (GQL), an extension of structured query language. GQL allows the definition of specialist query terms that are not held with the database. For germplasm analysis, these may include pedigree terms such as grandparent or ancestor, geographical terms such as neighbouring country and marker terms such as haplotype.

Graphical tools for germplasm analysis that are considered to be essential to the GENE-MINE and other similar tools include: (i) a geographical tool that could show the origin of accessions and the distribution of genetic diversity; (ii) a haplotype tool showing genotype information; (iii) a graph drawing tool that could show both pedigrees and phylogenetic trees or networks (graphs containing closed loops, which can be used to represent genetic exchange between organisms); (iv) a map tool showing the distribution of genetic markers on the relevant linkage maps; and (v) a plotting tool that could show scatter plots of, for example, diversity distances between pairs of accessions and principal components (Davenport *et al.*, 2004).

A high-throughput platform for identifying single feature polymorphisms (SFPs) in complex genomes has been developed by Borevitz *et al.* (2003). This is based on hybridizing *Arabidopsis* genomic DNA against an RNA expression GeneChip. Their informatics analysis involved development of analytical tools to identify 4000 SFPs by comparing the reference ecotype Columbia against Landsberg *erecta*. A linear clustering algorithm enabled identification of SFPs representing potential deletions in 111 transposons, disease resistance genes and genes involved in secondary metabolism, at 5% error rate. In crop plants, a genome-wide rice DNA polymorphism database has been constructed based on the genomic sequences from two subspecies, *indica* (93-11) and *japonica* (Nipponbare). This database contains 1,703,176 single nucleotide polymorphisms (SNPs) and 479,406 insertion/deletions (Indels), approximately one SNP every 268 bp and one Indel every 953 bp in rice genome (Shen *et al.*, 2004).

Several commercialized or freely available software packages, such as STATISTICA, JMP, SAS, NTSYS, GENEFLOW, STRUCTURE and POWERMARKER, can be used for germplasm evaluation including principal component or coordinate analysis to identify distinct groups or populations, cluster or structure analysis to find population structure. STRUCTURE, developed by Pritchard *et al.* (2000a), uses multi-locus genotype data to investigate population structure, which can be used to infer the presence of distinct populations, assign individuals to populations, study hybrid zones, identify migrants and admixed individuals and estimate population allele frequencies in situations where many individuals are migrants or admixed. It can be applied to most of the commonly used genetic markers.

PowerMarker (http://www.powermarker.net), as a software package to perform statistical analysis of marker data collected from a set of germplasm accessions, delivers a data-driven, integrated analysis environment (IAE) for marker data. The IAE integrates the data management, analysis and visualization in a user-friendly graphic interface. It accelerates the analysis process and enables users to maintain data integrity throughout the process. PowerMarker handles a variety of data from most of the commonly used genetic markers including simple sequence repeat (SSR), SNP and restriction fragment length polymorphism (RFLP). The results can be exported as frequency, distance and tree. Various data analyses can be performed by PowerMarker. Its summary statistics include basic statistics, allele and genotype frequencies, haplotype frequencies, Hardy–Weinberg disequilibrium, two-locus linkage disequilibrium and multi-locus linkage disequilibrium. Structure analysis includes population differentiation test, classic *F*-statistics, population-specific *F*-statistics and co-ancestry matrix. In phylogenetic analysis, tree construction can be made after computing frequencies and frequency-based distances with bootstrapping implemented. Association analysis can be done through a single-locus case control test, single-locus *F*-test and haplotype trend regression.

There are several software packages for treatment and analysis of data collected for germplasm accessions. Graphical Genotypes (ggt) software, developed by van Berloo (1999), allows the user to transform molecular marker data into simple colourful chromosome drawings. Besides graphical representation, ggt can also be used for selection or filtering of marker data. popdist calculates a number of different genetic identities, phylogeny reconstructing measures and distance reconstructing measures (http://genetics.agrsci.dk/~bg/popgen/). adegenet is a package dedicated to the handling of molecular marker data for multivariate analysis (http://pbil.univ-lyon1.fr/software/adegenet/). This package is related to ade4, an R package for multivariate analysis, graphics, phylogeny and spatial analysis.

The exploration of DNA sequence variation for making inferences on evolutionary processes in populations has become increasingly important recently and requires the coordinated implementation of a Suite of Nucleotide Analysis Programs (SNAP; http://www.cals.ncsu.edu/plantpath/people/faculty/carbone/snap.html), each bound by specific assumptions and limitations. A workbench tool was developed to make existing population genetic software more accessible and to facilitate the integration of new tools for analysing patterns of DNA sequence variation, within a phylogenetic context. Collectively, SNAP tools can serve as a bridge between theoretical and applied population genetic analysis (Aylor *et al.*, 2006).

15.1.2 Breeding population management

Decision support tools for the management of breeding populations are needed to assist in the choice of parental lines, types of crosses and the nature of breeding system. Computational tools may also assist in the establishment and maintenance of heterotic groups, selection of lines for creation of a synthetic cultivar, prediction of progeny and hybrid performance; and monitoring of genomic profiles during population improvement.

Establishing heterotic patterns

Generating highly heterotic hybrids is highly dependent on having sufficient genetic diversity in the germplasm pool of potential parents. However, it is still not possible in many crops to predict the level of hybrid vigour from analysis of parental lines. For example, commercial maize hybrids are typically generated from crosses between inbreds from complementary heterotic groups. Therefore, construction or development of heterotic groups has been one of the key strategies in hybrid breeding for many crops. However, moving to a more definitive system to predict which genotypes in each heterotic group should be crossed to maximize heterosis, is still not possible in many crops.

Genotyping parental lines on a genome-wide scale, especially when gene-based markers are available, may provide an opportunity for establishing parent–hybrid performance relationships at the molecular level. Genome-wide heterozygosity and specific combinations of alleles (linkats) may be useful determinants in some crops for maximizing heterosis and hybrid vigour. Melchinger and Gumber (1998) suggested a multi-stage procedure to identify heterotic groups (Chapter 9). Determining heterotic patterns is a continual process, each cycle of which consists of three steps: (i) cluster analysis to identify broad heterotic groups; (ii) combining ability and heterosis analysis to define the heterotic pattern; and (iii) update and maintain heterotic groups. Tools for heterotic group identification are usually the same as those that have been used in germplasm classification and grouping.

Predicting hybrid performance

A successful hybrid development process depends on a full understanding of the parental genotypes and the consequences of their genetic combinations and interactions in the hybrid. Hybrid breeding includes two major procedures: breeding parental lines and selection of the best combinations of those parental lines for hybrid production. These procedures involve a large amount of work for field evaluation, testcrossing and progeny tests. Breeders continually have to decide which experimental single crosses to test, which advanced hybrids to recommend for further testing or commercialization and which inbred parents to cross to form new base populations for inbred/population development (Bernardo, 1999). As a result, large-scale testcrossing is required for all hybrid-related inbred development. Testcrossing might be carried out at many stages in the breeding process often beginning from the very first generations. This 'trying all to find the best' process consumes a large proportion of the breeding effort but there is currently no alternative as hybrid performance is highly unpredictable in most crops. Therefore, predicting hybrid performance has always been a primary objective in all hybrid-breeding programmes.

Methods for predicting the performance of single crosses would greatly enhance the efficiency of hybrid breeding programmes. Development of a reliable method for predicting hybrid performance or heterosis without generating and testing hundreds or thousands of single cross combinations has been the goal of numerous studies using marker data and combinations of marker and phenotypic data, particularly in maize and rice. Considering that hybrid performance must be governed by many genes, genotyping parental lines on a genome-wide scale, especially when gene-based markers are used, provides an opportunity of establishing parent–hybrid performance relationships at the molecular level. Genome-wide heterozygosity and allele combination analysis may provide some clue for breeding more heterotic and vigorous hybrids. Therefore, using parental genotyping may reduce the required level of testcross-based phenotyping analysis.

The best linear unbiased prediction (BLUP) procedure has been used for decades for evaluating the genetic merit of animals, especially dairy cattle. Intrapopulation, additive genetic models have traditionally been used for BLUP in animal breeding (Henderson, 1975). Bernardo (1994, 1996) used BLUP in maize breeding with interpopulation genetic models that involve both general combining ability and specific combining ability and found that BLUP is useful for routine prediction of single-cross performance. Results have indicated that BLUP is useful for routine prediction of single-cross performance. The predicted performance of single crosses may subsequently be used to predict the performance of F_2 × tester combinations, three-way crosses, or double crosses. Along with the pedigree relationship, the BLUP method can use trait data, or both trait and marker data, for prediction.

In some specific cases within a breeding programme, tools are needed for selective genotyping and pooled DNA analysis as described in Chapter 7 and by Xu *et al.* (2008). GenePool (http://genepool.tgen.org/) is such a software package that provides analytical tools for the detection of shifts in relative allele frequency between pooled genomic DNA from cases and controls using SNP-based genotyping microarrays. GenePool supports genotyping platforms from Affymetrix and Illumina (Pearson *et al.*, 2007). Another package is PDA, Pooled DNA Analyser, a tool for analysis of pooled DNA data (http://www.ibms.sinica.edu.tw/~csjfann/first%20flow/programlist.htm; Yang, H.-C. *et al.*, 2006b).

In addition to the tools for germplasm and breeding population management and evaluation described above, decision support tools are needed for intellectual property rights and plant variety protection. Chapter 13 provides a section on how molecular markers can be used for this purpose.

15.2 Genetic Mapping and Marker–Trait Association Analysis

Construction of genetic maps using molecular markers (Chapter 2) and use of these maps in marker–trait association analysis (Chapters 6 and 7) are two prerequisite steps required for MAS. There is a large number of methodologies and tools currently available for various types of populations and markers. In this section, only some of these tools will be discussed and at the same time, we expect more tools will be developed for new mapping strategies and markers and new types of populations.

15.2.1 Genetic map construction

Genetic maps can be constructed using segregating populations of different types for species with different levels of ploidy as described in Chapter 2. The first and most frequently used software for map construction is MAPMAKER/EXP, which was developed by the Whitehead Institute (Lander *et al.*, 1987). Almost all molecular maps based on the first generation of molecular markers, RFLPs, were constructed using this software. As an alternative, MAP MANAGER CLASSIC is a graphic, interactive program to map Mendelian loci using intercrosses with co-dominant markers, backcrosses or recombinant inbred lines (RILs) in experimental plants or animals (Manly, 1993; http://www.mapmanager.org/mapmgr.html).

Some special statistical modifications may be needed to construct a map using markers with severe distortion of segregation. MAPDISTO (web/ftp: http://mapdisto.free.fr/) is such a program for mapping genetic markers in case of segregation distortion using experimental segregating populations such as backcross, double haploid (DH) and RIL populations. It can: (i) compute and draw genetic maps through a graphical interface; and (ii) facilitate the analysis of marker data showing segregation distortion due to differential viability of gametes or zygotes.

Maps or data from multiple populations derived from different crosses can be combined into single or consensus maps through joint mapping. JOINMAP is a software package for construction of genetic linkage maps for several types of mapping populations: BC_1, F_2, RIL, F_1- and F_2-derived DH and out-breeder full-sib family (http://www.kyazma.nl/index.php/mc.JoinMap/). It can combine ('join') data derived from several sources into an integrated map, with several other functions including linkage group determination, automatic phase determination for out-breeder full-sib family, several diagnostics and map charts (van Ooijen and Voorrips, 2001).

A software package with comparative function is CMap, which was developed as a web-based tool to allow users to view comparisons of genetic and physical maps. The package also includes tools for curating map data (http://www.gmod.org/cmap; Ware *et al.*, 2002).

15.2.2 Linkage-based QTL mapping

Demonstrated linkages/associations between target traits/genes and molecular markers are based on genetic linkage and linkage-disequilibrium (LD) mapping experiments (Chapters 6 and 7). Decision support tools required for genotype–phenotype association include: (i) statistical methods and tools to establish, validate and compare genotype–phenotype associations through linkage mapping, LD or association mapping and *in silico* mapping, using single populations, multiple populations or all genetic resources with information available from multiple trials across years, seasons and locations; (ii) statistical methods and tools for identification of genetic background effects, quantitative trait loci (QTL) alleles at multiple loci and multiple alleles at a locus; (iii) tools facilitating the process from linked markers to functional markers and candidate genes; and (iv) tools facilitating management of genetic populations, maps and related marker and phenotypic data.

There are many commercial or freely available software packages for establishing association between marker genotypes and trait phenotypes. The most commonly used are QTL CARTOGRAPHER, MAPQTL, PLABQTL and QGENE. All of these only handle bi-allelic populations, while MCQTL (Jourjon *et al.*, 2005) also performs QTL mapping in multi-allelic situations, including bi-parental populations made from segregating parents, or sets of bi-parental, bi-allelic populations. The most frequently used software during the 1980s and 1990s was MAPMAKER/QTL, which is a sister software package to MAPMAKER/EXP, developed by Lander *et al.* (1987) (http://www-genome.wi.mit.edu/genome_software). This software is based on maximum likelihood estimation of linkage between marker and phenotype using interval mapping, which deals with simple QTL and several standard populations. Another early software package, MAPL (MAPping and QTL analysis; http://lbm.ab.a.u-tokyo.ac.jp/software.html; Ukai *et al.*, 1995) allows a user to get results on segregation ratio, linkage test, recombination value, group markers, order markers by metric multi-dimensional scaling, draw a map and graphical genotype and map QTL through interval mapping and analysis of variance (ANOVA).

A currently widely used QTL mapping software is QTL CARTOGRAPHER (http://statgen.ncsu.edu/qtlcart/cartographer.html), which implements several statistical methods using multiple markers simultaneously including composite interval mapping and multiple composite interval mapping. Interaction between identified QTL can also be estimated. PLABQTL uses composite interval mapping with many functions similar to QTL CARTOGRAPHER. QTL can be localized and characterized in populations derived from a biparental cross by selfing or production of DHs. Simple and composite interval mapping are performed using a fast multiple regression procedure. As an additional function to many other software packages, it can be used for QTL × environment interaction analysis (Utz and Melchinger, 1996).

QGENE (http://www.qgene.org/) is intended for doing comparative analyses of QTL mapping data sets in computationally efficient ways that are of maximum use to analysts. It is also written with a plug-in architecture for ready extensibility. QGENE was begun in about 1991 as a map and population simulation program, to which QTL analyses were added on. Recently QGENE has been rewritten in the Java language, allowing it to run on any computer operating system. It offers most conventional QTL-mapping methods and allows their side-by-side comparison. Its interface can be rendered in any human language desired; the conversion requires only that the interested user writes a translation file. QGENE can be used for analysis of trait, QTL and permutation and simulation of populations and traits as well.

Several software packages can be used for constructing linkage maps in outcrossing plant species. ONEMAP provides such an environment using full-sib families derived from two outbreed (non-inbreeding) parent plants (http://www.ciagri.usp.br/~aafgarci/OneMap/; Garcia *et al.*, 2006). Another is MAPQTL (software for the calculation of QTL positions on genetic maps, http://www.mapqtl.nl), which can be used for several types of mapping populations including BC_1, F_2, RILs, (doubled) haploids and full-sib family of out-breeders. It can be used for QTL mapping through interval mapping, composite interval mapping and non-parametric mapping, with functions for automatic cofactor selection and permutation test.

A few mapping software programs consider epistasis in QTL mapping. EPISTACY is an SAS program designed to test all possible two-locus combinations for epistatic (interaction) effects on a quantitative trait. The program is really an SAS program template that users must modify to suit their own data sets. In the simplest cases, users will need only to change the names of the files containing their data. However, the program uses least squares methods and does not employ interval mapping methods (Holland, 1998).

Bayesian QTL mapping has received a lot of attention in recently years with several software packages developed. For example, BQTL, Bayesian Quantitative Trait Locus mapping, was developed for the mapping of genetic traits from line crosses and RILs (http://hacuna.ucsd.edu/bqtl; Borevitz *et al.*, 2002). It performs: (i) maximum likelihood estimation of multi-gene models; (ii) Bayesian estimation of multi-gene models via Laplace Approximations; and (iii) interval mapping and composite interval mapping of genetic loci. BLADE, Bayesian LinkAge DisEquilibrium mapping, was developed for Bayesian analysis of haplotypes for LD mapping (http://www.people.fas.harvard.edu/~junliu/TechRept/03folder/; Liu *et al.*, 2001; Lu, X. *et al.*, 2003). MULTIMAPPER is a Bayesian QTL mapping software for analysing backcross, DH and F_2 data from designed crossing experiments of inbred lines (Martinez *et al.*, 2005). MULTIMAPPER/OUTBRED extended this to the populations derived from out-bred lines (http://www.rni.helsinki.fi/~mjs/).

Several mapping software packages were developed for QTL mapping required in some specific situations. MCQTL was developed for simultaneous QTL mapping in multiple crosses and populations (http://www.genoplante.com; Jourjon *et al.*, 2005). It allows the analysis of the usual populations derived from inbred lines and can link the families by assuming that the QTL locations are the same in all of them. Moreover, a diallel modelling of the QTL effects is allowed when using multiple related families. MAPPOP was developed for selective mapping and bin mapping by choosing good samples from mapping populations and for locating new markers on pre-existing maps (Vision *et al.*, 2000). In addition, QTLNETWORK was developed for mapping and visualizing the genetic architecture underlying complex traits for experimental populations derived from a cross between two inbred lines (http://ibi.zju.edu.cn/software/qtlnetwork; Yang *et al.*, 2008).

As web-based tools become increasingly important, web-based QTL analytical tools become available. As such an example, WEBQTL was developed as an interactive web site useful for exploring the genetic modulation of thousands of phenotypes gathered over a 30-year period by hundreds of investigators using reference panels of recombinant inbred strains of mice (http://www.webqtl.org/search.html). WEBQTL includes dense error-checked genetic maps, as well as extensive gene expression data sets (Affymetrix) acquired across more than 35 strains of mice. As a web-based user-friendly package to map QTL in out-bred populations, QTL EXPRESS (http://qtl.cap.ed.ac.uk; Seaton *et al.*, 2002) was developed for line crosses, half-sib families, nuclear families and sib-pairs. It provides two options for QTL significance tests: permutation tests to determine empirical significance levels and bootstrapping to estimate empirical confidence intervals of QTL locations. Fixed effects/covariates can be fitted and models may include single or multiple QTL.

15.2.3 eQTL mapping

With the availability of whole genome sequences in many plant species, linkage analysis, positional cloning and microarray are gradually becoming powerful tools for revealing the links between phenotype and genotype or genes. To display the myriad of relationships between eTraits, markers and genes, we need a convenient bioinformatics tool to visualize eQTL mapping results at a variety of scales ranging from a single locus to the entire genome. Additionally, researchers need quick and straightforward ways to integrate these results with the extra information from previous studies on the organism. To address these needs, eQTL Explorer was developed (Mueller *et al.*, 2006) to store expression profiles, linkage data and information from external sources in a relational database, enabling simultaneous visualization and intuitive interpretation of the combined data via a Java graphical interface. Zou *et al.* (2007) developed eQTL Viewer, a web-based tool that plots eQTL mapping results. The resulting plot displays eQTL for thousands of eTraits in a single view, which makes patterns such as *cis*- and *trans*-regulations readily identifiable. They also empowered such a plot with the ability to present annotations, highlight features and organize eTraits in biological groups, such as biochemical pathways. All these characteristics make eQTL Viewer an intuitive and information rich environment to discover and understand genome-wide transcriptional regulation patterns.

A web site developed by Bhave *et al.* (2007), PhenoGen, can be used to search for candidate genes that control a complex trait based on the co-occurrence of differentially expressed genes in microarray experiments and phenotypic QTL or co-occurrence of phenotypic QTL and expression QTL. PhenoGen needs to know how many candidate genes exist within the QTL region according to known literature reports and detailed information of those candidates and related reports indicating their candidacy. Xiong *et al.* (2008) developed a software tool, PGMAPPER, for automatically matching phenotype to genes from a defined genome region or a group of given genes by combining the mapping information from the Ensembl databases and gene function information from the OMIM (http://www.ncbi.nlm.nih.gov/sites//entrez?db=omim) and PubMed databases (http://www.ncbi.nlm.nih.gov/sites//entrez). PGMAPPER is currently available for candidate gene search of human, mouse, rat, zebrafish and 12 other species.

15.2.4 Linkage-disequilibrium based QTL mapping

Association or LD mapping has become increasingly popular (Chapter 6). It uses unstructured populations that consist of unrelated individuals, germplasm accessions, or randomly selected cultivars. Before LD mapping, genotyped units are subjected to statistical analysis to remove the most important factor, population structure, which can cause false positive associations due to circumstantial correlations rather than real linkage. For example, the STRUCTURE software (Pritchard *et al.*, 2000a) can be used for this purpose. Some software packages have been developed for LD mapping with the population structure analysis functionality included. STRAT, as a companion program to STRUCTURE, uses a structured association method for LD mapping, enabling valid case-control studies even in the presence of population structure (http://pritch.bsd.uchicago.edu/software/STRAT.html; Pritchard *et al.*, 2000b).

LD-based QTL mapping

TASSEL is a comprehensive software package for trait analysis by association, evolution and linkage, which performs a variety of genetic analyses including LD mapping, diversity estimation and calculating LD (http://sourceforge.net/projects/tassel/; Zhang, Z. *et al.*, 2006). The LD analysis between genotypes and phenotypes can be performed by either a general linear model or a mixed linear model. The general linear model allows users to analyse complex field designs, environmental interactions and epistasis. The mixed model is specially designed to handle polygenic effects at multiple levels of relatedness including pedigree information. These analyses should permit LD analysis in a wide range of plant and animal species.

Other software packages include Multiallelic Interallelic Disequilibrium Analysis Software (MIDAS), which was designed for analysis and visualization of interallelic disequilibrium between multiallelic markers (http://www.genes.org.uk/

software/midas; Gaunt *et al.*, 2006) and PedGenie (http://bioinformatics.med.utah.edu/PedGenie/index.html; Allen-Brady *et al.*, 2006), which was developed as a general-purpose tool to analyse association and transmission disequilibrium (TDT) between genetic markers and traits in families of arbitrary size and structure. With PedGenie, any size pedigree may be incorporated into this tool, from independent individuals to large genealogies. Independent individuals and families may be analysed together.

GeneRecon (http://www.daimi.au.dk/~mailund/GeneRecon/) is another software package for LD mapping using coalescent theory. It is based on a Bayesian Markov-chain Monte Carlo method for fine-scale LD mapping using high-density marker maps in animals. GeneRecon explicitly models the genealogy of a sample of the case chromosomes in the vicinity of a disease locus. Given case and control data in the form of genotype or haplotype information, it estimates a number of parameters, most importantly, the disease position (Mailund *et al.*, 2006).

Genome-wide association mapping

Genome-wide association (GWA) studies are now being widely undertaken to find the link between genetic variations and common diseases in humans and agronomic traits in plants. Ideally, a well-powered GWA study will involve the measurement of hundreds of thousands of SNPs in thousands of individuals. The sheer volume of data generated by these experiments creates very high analytical demands. There are a number of important steps during the analysis of such data, many of which may present several bottlenecks. The data need to be imported and reviewed to perform initial quality control before proceeding to LD testing. Evaluation of results may involve further statistical analysis, such as permutation testing, or further quality control of associated markers, for example, reviewing raw genotyping intensities. Finally, significant associations need to be prioritized using functional and biological interpretation methods, browsing available biological annotation, pathway information and patterns of LD (Pettersson *et al.*, 2008). GoldSurfer2 (gs2), a comprehensive tool for the analysis and visualization of GWA studies, was developed by Pettersson *et al.* (2008). gs2 is an interactive and user-friendly graphical application that can be used in all steps in GWA projects from initial data quality control and analysis to biological evaluation and validation of results. The program is implemented in Java and can be used on all platforms. With gs2, very large data sets (e.g. 500K markers and 5000 samples) can be quality assessed, rapidly analysed and integrated with genomic sequence information. Candidate SNPs can be selected and functionally evaluated.

Other tools that are developed for GWA studies include Genomizer (a platform-independent Java program for the analysis of GWA experiments; http://www.ikmb.uni-kiel.de/genomizer), plink (a whole-genome LD analysis toolset; http://pngu.mgh.harvard.edu/purcell/plink; Purcell *et al.*, 2007), MapBuilder (for chromosome-wide LD mapping; http://bios.ugr.es/BMapBuilder; Abad-Grau *et al.*, 2006) and power Calculator for Association with Two Stage design (cats), which calculates the power and other useful quantities for two-stage GWA studies (http://www.sph.umich.edu/csg/abecasis/CaTS) (Skol *et al.*, 2006).

The results of large GWA studies are being deposited in public databases with increasing frequency. But the currently available software to analyse and interpret GWA data sets can be difficult to use (Buckingham, 2008). User-friendly software is urgently needed to provide new ways of making GWA data sets easy to explore and share among researchers and to design analysis packages that deal with the increasing computational demands posed by these data sets.

Integrated haplotype and LD analysis

The analysis of large amounts of SNP data creates difficulties for the analysis of haplotypes and their association to traits of interest. Commonly fairly simple methods, such as two- or three-SNP sliding windows are used to create haplotypes across large regions, but these may be of limited value when adjacent SNPs are in strong LD and provide redundant information. Genetic analysis of SNP data and haplotypes have received more and more attention recently and various software packages have been developed for haplotype analysis and these are sometimes integrated with LD analysis.

HAPLOBUILD (http://snp.bumc.bu.edu/modules.php?name=HaploBuild), was created for constructing and testing haplotypes for SNPs in close physical proximity to one another but which are not necessarily contiguous (Laramie *et al.*, 2007). The number of SNPs contained in the haplotype is not restricted, thereby permitting the evaluation of complex haplotype structures.

HAPLOVIEW (http://www.broad.mit.edu/personal/jcbarret/haploview) was designed to simplify and expedite the process of haplotype analysis by providing a common interface to several tasks relating to such analyses. HAPLOVIEW currently allows users to examine block structures, generate haplotypes in these blocks, run association tests and save the data in a number of formats. All functionalities are highly customizable (Barrett *et al.*, 2005).

HAPSTAT (http://www.bios.unc.edu/~lin/hapstat/) is a user-friendly software interface for the statistical analysis of haplotype-disease association. HAPSTAT allows the user to estimate or test haplotype effects and haplotype–environment interactions by maximizing the (observed-data) likelihood that properly accounts for phase uncertainty and study design. The current version considers cross-sectional, case-control and cohort studies.

Other related software packages include:

- DPPH (Direct method for Perfect Phylogeney Haplotyping; http://wwwcsif.cs.ucdavis.edu/~gusfield/dpph.html; Bafna *et al.*, 2003);
- EHAP (detecting association between haplotypes and phenotypes; http://wpicr.wpic.pitt.edu/WPICCompGen/ehap_v1.htm);
- HAPLOBLOCK, a software package which provides an integrated approach to haplotype block identification, haplotype resolution and LD mapping, suitable for high-density phased or unphased SNP data (http://bioinfo.cs.technion.ac.il/haploblock);
- HAPLOT, a simple program for graphical presentation of haplotype block structures, tagSNP selection and SNP variation (Gu *et al.*, 2005);
- HAPLOREC, population-based haplotyping software (Eronen *et al.*, 2004); and
- HAP, a haplotype analysis system which is aimed at helping to perform disease association studies and a phasing method which is based on the assumption of imperfect phylogeny (http://research.calit2.net/hap).

15.2.5 Genotype-by-environment interaction analysis

To better separate the genetic effects from the environmental effects and their interaction, statistical methods are of paramount importance in a traditional as well as molecular breeding programme (Chapter 10). These methods become even more essential when developing MAS systems for abiotic stress tolerance where germplasm must be tested under drought or low nitrogen conditions, for example. Under such stress conditions the soil where the plants are grown becomes extremely variable and patchy so that the separation of genetic effects from environmental effects is much more difficult than under normal conditions.

Various processes contribute to the characterization of a genotype–environment system (Cooper *et al.*, 1999). There is a great need for integrated decision support tools for genotype-by-environment interaction (GEI) analysis: (i) developing field experimental designs; defining the target population of environments (TPE) and genotypes; (ii) assessing GEI for various field conditions and determining subsets of genotypes and sites with negligible crossover interaction effects from which subgroups of sites and genotypes with similar response can be identified in order to maximize responses to selection; (iii) mapping QTL and QTL-by-environment interaction (QEI) of component traits important for the target traits; (iv) developing a selection index for phenotypic as well as molecular marker data in order to select the best genotypes to be used in the next cycle of selection; (v) incorporating environmental and/or genotypic variables into statistical models to explain the causes of GEI (physical and chemical soil conditions may be of importance under drought and may be the main cause of GEI); (vi) studying genetic diversity of crop genotypes associated with the target traits; (vii) performing LD mapping of those traits; and (viii) studying gene expression of genes under target conditions from microarray experiments.

Decision support tools are required for classifying the most important testing environments into mega-environments that will then define the appropriate TPE. Based on these environmental classifications breeding strategies can be developed and established for a more efficient and rapid realization of genetic gains targeting those specific environments. Furthermore, the incorporation of climatic variables (attributes of environments) and molecular markers (attributes of genotypes such as QTL) into statistical models facilitate the identification of the causes of GEI and therefore help explain QEI. This allows interpreting, understanding and exploiting GEI and QEI and it allows identification of the regions of the chromosomes affecting a trait that are highly affected by external climatic conditions. This also facilitates grouping of environments with negligible genetic crossover effects as well as clustering genotypes with no genotypic crossover GEI.

Podlich and Cooper (1998) developed the QuGene software for carrying out quantitative genetic analyses of GEI in crop breeding and this has become an increasingly widely utilized decision support tool in breeding programmes. More recently, a statistical model developed by Crossa *et al.* (2006) incorporates pedigree information (through the coefficient of parentage) for test genotypes when modelling GEI. This model can be used to perform more efficient LD mapping studies as well as *in silico* QTL detection. Among various models, mixed linear models are fundamental in the process of *in silico* QTL linkage and LD mapping. These decision support tools are being further refined through the integration of whole-plant physiology models.

15.2.6 Comparative mapping and consensus maps

In the past few decades, a wealth of genomic data has been produced in a wide variety of species using a diverse array of functional and molecular marker approaches. In order to unlock the full potential of the information contained in these independent experiments, researchers need efficient and intuitive means to identify common genomic regions and genes involved in the expression of target phenotypic traits across diverse conditions. Experimenters who seek to apply many diverse studies on QTL face complex problems in summarizing, interrelating and integrating them. Tools for QTL consensus map building offer extensive analysis or meta-analysis of data prior to assigning a consensus QTL location for a trait (Sawkins *et al.*, 2004; Arcade *et al.*, 2004).

CMTV (Comparative Map and Trait Viewer; Sawkins *et al.*, 2004) was developed as a software component to help serve as an intuitive and extensible framework for the integration of various kinds of genomic data. The software components use the ISYS (Integrated SYStem) integration platform developed by the National Center for Genetic Resources (Siepel *et al.*, 2001) to access and visualize map data and related information such as germplasm pedigree relationships. CMTV is based on algorithmically determining correspondences between sets of objects on multiple genomic maps, and can display syntenic regions across taxa, combine maps from separate experiments into a consensus map, or project data from different maps into a common coordinate framework. As such an example, Schaeffer *et al.* (2006) used a strategy for consensus QTL maps that leverages the highly curated data in MaizeGDB, in particular, the numerous QTL studies and maps that are integrated with other genome data on a common coordinate system. In addition, they exploited a systematic QTL nomenclature and a hierarchical categorization of over 400 maize traits developed in the mid-1990s; the main nodes of the hierarchy are aligned with the trait ontology at Gramene, a comparative mapping database for cereals (http://www.gramene.org). Consensus maps are presented for one trait category, insect response (80 QTL); two traits, grain yield (71 QTL) and kernel weight (113 QTL), representing over 20 separate QTL map sets of ten chromosomes each. The strategy is germplasm-independent and reflects any trait relationships that may be chosen.

A systematic approach for associating genes and phenotypic characteristics that combines literature mining with comparative genome analysis has also been practised. The underlying principle is that species sharing a phenotype may share orthologous genes associated with the same biological process and thus correlations between the presence and absence of both genes and traits across species should indicate relevant genotype–phenotype associations (Korbel *et al.*, 2005). In a global analysis involving 92 prokaryotic genomes, 323 clusters containing a total of 2700 significant gene–phenotype associations were retrieved from the MEDLINE literature database that reflect phenotypic similarities of species. Some clusters contain mostly known relationships, such as genes involved in motility or plant degradation, often with additional hypothetical proteins associated with those phenotypes. Other clusters comprise unexpected associations: for example, a group related to food and spoilage is linked to genes predicted to be involved in bacterial food poisoning. Among the clusters, an enrichment of pathogenicity-related associations was observed, suggesting that this approach revealed many novel genes likely to play a role in infectious diseases (Korbel *et al.*, 2005).

The explosion of interest in marker–trait association studies has led to numerous reports in plants, each based on its own experimental population(s). Each experiment is limited in size and usually restricted to a single population or a cross planted in a specific environment. As suggested by Xu, Y. (2002), it is important for researchers to follow general rules for naming and reporting genes and traits. This will then facilitate the combination of information from several studies, for example, through meta-analysis of results of QTL studies (Goffinet and Gerber, 2000) or joint analysis of the raw data (Haley, 1999). Extension of current databases to include raw data from gene mapping projects will stimulate this effort. On the other hand, many permanent populations have been shared internationally for genomic studies and the raw phenotype and genotype data should also be shared at the same time. A rice RFLP map based on a DH population from the cross between IR64 and Azucena has been saturated with about a thousand SSR markers (Chen *et al.*, 1997; Temnykh *et al.*, 2000; McCouch *et al.*, 2002). However, researchers involved in QTL mapping continued to use a molecular map consisting of only 175 RFLP markers for many years after the SSR markers were developed. Clearly, sharing marker and phenotype information through a well-established database such as Gramine or GrainGenes has made all sources of data more valuable.

A standard reporting system is also critical for comparative genomics, QTL allelism tests, data sharing and mining and the association between major genes and QTL. As discussed by Xu, Y. (2002), a standard system for marker–trait association should include allele characterization data such as allele sizes, gene effects, variation explained by each gene or all genes in the model, gene interaction if more than one gene is identified and GEI if more than one environment is involved. Genetic information should be shared and combined with data generated by plant breeding programmes, for example, germplasm diversity, mapping populations, pedigrees, graphical genotypes, mutants and other genetic stocks.

Finally, as a comprehensive tool, The Rosetta Syllego System (http://www.rosettabio.com/products/syllego/; Broman *et al.*, 2003b) was developed as a genetic data management and analysis system to advance whole genome linkage, LD and eQTL studies. Designed for biologists, statistical geneticists and investigators responsible for generating genotyping data, the Syllego system provides us with an easy-to-use project workspace so that we can organize, analyse and share genotype and phenotype data along with analysis results. With the Syllego system, generating high quality analysis data and meaningful results becomes simple. It automates all tedious data management and data formatting tasks so that genetic analysis workflows can be streamlined using analysis methods of choice. Managing all genetic data and reference information is straightforward. The Syllego system converts public and private genotype data sets and reference annotations, such as dbSNP (http://www.ncbi.nlm.nih.gov/projects/SNP/) and HapMap (http://www.hapmap.org/), as well as individual (sample) information into a single, consistent repository for fast, convenient access.

15.3 Marker-assisted Selection

MAS is one of the major activities in molecular breeding (Chapters 8 and 9). It needs various decision support tools including those for foreground and background selection and identification of the recombinants with favourable alleles and allele combinations. However, only a few tools are available so far for some procedures of MAS. Development of decision support tools for fully functional MAS still faces a lot of challenges.

A huge amount of data will be generated with large-scale MAS and this set of data needs to be analysed and also integrated with other types of data to make selection decisions in a short time window, e.g. 4 weeks during vegetative to flowering stages, or harvest to planting the next season. Thus, decision support tools are essential to accelerate this process while maintaining accuracy and precision. Although many tools have been developed for assisting germplasm evaluation, genetic mapping and MAS, they either work independently, depending on different operating systems, or require different data formats which makes it impossible to complete a comprehensive data analysis to make the results available to breeders for decision making in such a short time window.

15.3.1 MAS methodologies and implementation

There are many factors that affect the efficiency of MAS. In theory, MAS is expected to be more efficient than phenotypic selection when the heritability of a trait is low, where there is tight linkage between the QTL and the DNA markers (Dudley, 1993; Knapp, 1998), with larger population sizes (Moreau *et al.*, 1998) and in earlier generations of selection before recombinational erosion of marker–trait associations (Lee, 1995). Edwards and Page (1994) proposed that the distance between the markers and the QTL was the single largest constraining factor for gains from MAS. An example in Lande and Thompson (1990) demonstrated that on a single trait the potential selection efficiency, using a combination of molecular and phenotypic information, depends on the heritability of the trait, the proportion of additive genetic variance associated with marker loci and the selection scheme. The relative efficiency of MAS is greatest for traits with low heritability if a large fraction of the additive genetic variance is associated with marker loci.

Decision support tools are required for the following procedures related to MAS: (i) determining minimum sample size for foreground/background selection; (ii) estimation of genetic gains (response to selection); (iii) construction of selection indices for multiple traits and whole genome selection; (iv) estimation and graphical display of recipient genome content of selected individuals at each generation of introgression; (v) identification of desirable plants based on both phenotype and genotype; (vi) cost–benefit analysis; and (vii) software for

MAS and simulations (using all available information).

There has been much interest in the development of software that simulates MAS using genetic models. Early efforts had somewhat limited value, for example, GREGOR simulates MAS based only on predefined genetic linkage maps and is thus restricted in its value for simulation of MAS in breeding programmes (Tinker and Mather, 1993). The program GREGOR implements the basic principles, but the interactive use and the fact that it simulates only some predefined genetic linkage maps restricts its value for simulation of breeding programmes.

Frisch *et al.* (2000) present PLABSIM, a tool for simulation of MAS programmes. The software can be used to investigate the effect of varying population size, marker density and positions and selection strategies on the genetic composition of the breeding product and on the required number of marker data points. It has the following features: (i) simulations can be made for any diploid genome with an arbitrary number of loci at arbitrary positions on an arbitrary number of chromosomes; (ii) the implemented reproduction schemes include all common breeding methods; (iii) an arbitrary number of selection steps can be combined with a selection strategy; (iv) selection can be carried out for genotypes at defined loci, or for selection indices calculated from allele frequencies at several loci; and (v) the simulated data can be analysed for a broad range of genetic parameters including population size, marker density and positions and selection strategies for the genetic composition of the breeding product and on the required number of marker data points.

To integrate various tools into a common platform to assist their effective deployment in crop improvement, iMAS (www.generationcp.org) is a preliminary attempt to create a publicly available computational platform to assist the development and application of marker-assisted breeding. iMAS currently integrates freely available software for the journey from phenotyping-and-genotyping of individuals to identification and application of trait-linked markers. iMAS also provides simple-to-understand-and-use on-line decision-support guidelines to help the user correctly operate the software and correctly interpret the outputs.

Other MAS tools include: (i) POPMIN, a program for the numerical optimization of population sizes in marker-assisted backcross programmes (Hospital and Decoux, 2002); (ii) BCSIM, backcross simulation software for evaluation of marker-assisted backcross programmes (http://www.plantbreeding.wur.nl/UK/software_bcsim.html); and (iii) the GGT, GRAPHICAL GENOTYPES software, allowing the user to transform molecular marker data into simple colourful chromosome drawings (van Berloo, 1999).

15.3.2 Marker-assisted inbred and synthetic creation

For open-pollinated crops, a synthetic cultivar is developed by inter-crossing selected clones or inbred lines, with seed production of the cultivar through open pollination. For self-pollinated crops, a synthetic cultivar is a mix of different inbred lines. The breeding procedures used to develop a synthetic cultivar depend on the feasibility of developing superior inbred lines and clones. For species such as maize, inbred lines for synthetic cultivars are developed by the same procedures used for the development of hybrid cultivars. For many forage crops, inbreeding depression is too severe to permit the formation of inbred lines, but the parent can be maintained and reproduced readily by cloning. The factors to consider in the development of a synthetic cultivar include: (i) formation of a population; (ii) evaluation of individual inbreds/clones per se; (iii) evaluation of the combining ability of the inbreds/clones; (iv) evaluation of experimental synthetics; and (v) preparation of seed for commercial use (Fehr, 1987).

Synthetic cultivars can be developed by mixing inbred lines that have been bred by MAS or by mixing individual plants derived from any stage of MAS (Dwivedi *et al.*, 2007). With genotypic information available across the whole genome for all the selected individuals or inbred lines, support tools are needed to facilitate developing synthetic cultivars to contain complementary genotypes, fixed heterozygosity and the best combinations of genetic structure.

15.4 Simulation and Modelling

Along with the fast development in molecular biology and biotechnology, a large amount of biological data becomes increasingly available for important breeding traits, which in turn allows selection based on information of multiple sources. As discussed in the previous sections, however, available information has not been effectively used in crop improvement due to the lack of appropriate tools. In this section, plant breeding through simulation and modelling will be discussed including utilizing the vast and diverse information by incorporating simulation and modelling into breeding programmes to develop and upgrade various decision support tools.

15.4.1 Importance of simulation and modelling

The accumulation in genomics information for breeding traits has made simulation and modelling more and more practical and important, as computer simulation can help to investigate many 'what if' crossing and selection scenarios, allowing many scenarios to be tested *in silico* in a short period of time, which in turn helps breeders make important decisions to minimize and optimize highly resource demanding field experiments. As the number of published genes and QTL for various traits continues to increase, for example, plant breeders face a challenge to determine how to best utilize this multitude of information for crop improvement. Although quantitative genetics provides much of the framework for the design and analysis of selection methods used within breeding programmes (Falconer and Mackay, 1996; Lynch and Walsh, 1998; Goldman, 2000), various assumptions are made in quantitative genetics to render theories mathematically or statistically tractable. Some of these assumptions can be easily tested or satisfied by certain experimental designs; others, such as the assumptions of no linkage, no multiple alleles and no GEI, can seldom, if ever, be met. Other assumptions, like the presence or absence of epistasis and pleiotropy, are statistically difficult to define and test. Computer simulation provides a tool to investigate the implications of relaxing some of the assumptions and the effect it has on the implementation of a breeding programme (Kempthorne, 1988). Computer simulation provides an opportunity to lessen the impact of these assumptions by accommodating these factors, thereby improving the validity of genetic models for use in plant breeding. This approach would be very helpful when the breeders want to compare breeding efficiencies from different selection strategies, to predict the cross performance with known gene information and to utilize efficiently identified major genes and QTL in breeding.

As agronomically important traits are significantly affected by the environment, whole-plant physiology modelling is becoming increasingly important for partitioning complex traits into their components and understanding how those components interact with each other and contribute to the overall trait expression in different environmental conditions. With a commitment to genomic analysis of component traits, whole-plant physiology modelling provides a critical link between molecular genetics and crop improvement. Crop models with generic approaches to underlying physiological processes (Wang *et al.*, 2002) provide a means to link phenotype and genotype, through simulation analysis, of an *in silico* or virtual plant (Tardieu, 2003). In this way it is possible to dissect the physiological basis of adaptive traits and determine their control at whole-plant level through modelling.

A plant requires information about its environment and its interaction with that environment and uses that information to dictate its adaptive responses that result in the plant phenotype. Significant endeavours in the field of whole-plant modelling are now being directed at understanding genetic regulation and aiding crop improvement (Cooper *et al.*, 2002a; Chapman *et al.*, 2002, 2003; Hammer *et al.*, 2002; Yin *et al.*, 2003; Wang *et al.*, 2004; Yin, X. *et al.*, 2004; Wang, J. *et al.*, 2005). There are three areas in which crop modelling could assist in assessing *in silico* the multitude of options to improve the efficiency of plant breeding (Cooper *et al.* 2002a): (i) characterizing environments to define the target population of environments; (ii) assessing the value of specific putative traits in improved plant types; and (iii) enhancing integration of molecular genetic methodologies. Hence, plant breeders can pose questions that range from how to better utilize field performance data to how knowledge of gene action or function can be utilized for selection in a complex TPE.

15.4.2 Genetic models used in simulation

Multiple mathematical formalisms have been used to model genetic and, more generally, metabolic networks. Examples include: (i) Boolean (ON/OFF) networks; (ii) Petri (concurrent information flow) nets; (iii) S-systems (continuous time models motivated by chemical kinetics); (iv) differential equation models; (v) neutral network models; and (vi) Bayesian networks (Welch *et al.*, 2004). Despite this extensive effort, little attention has focused on predicting phenotypes of interest to plant breeders or on integrating the effect of multiple environmental factors.

Simulation, using relatively simple genetic models, has been used for many special studies in plant breeding (Casali and Tigchelaar, 1975; Reddy and Comstock, 1976; van Oeveren and Stam, 1992; van Berloo and Stam, 1998; Frisch and Melchinger, 2001). When it is used for genetic models with complex traits involved, however, the result is uncertain. Coors (1999) summarized many of the published recurrent selection studies for grain yield in maize. The synopsis we can take from Coors's synthesis of published studies strongly suggested that the realized progress from selection for this trait is considerably lower than the predicted response. For most involved in applied breeding this result is not surprising. However, this quantified observation forces us to consider the possible reasons for the discrepancies between the predictions made from classical quantitative genetic theory and the realized responses from applied breeding.

A crop can be analysable for processes at various scales: community, population, plant, organ, tissues, cell and downwards to molecular levels. White and Hoogenboom (2003) identified six levels of genetic details for simulation to elucidate differences in plant growth and development among cultivars:

1. Genetic model with no reference to species.
2. Species-specific model with no reference to genotypes.
3. Genetic differences represented by cultivar-specific parameters.
4. Genetic differences represented by specific alleles, with gene action and gene effects presented through linear effects on model parameters.
5. Genetic differences represented by genotypes with gene action explicitly simulated based on knowledge of regulation of gene expression and effects of gene products.
6. Genetic differences represented by genotypes, with gene action simulated at the level of interactions of regulators, gene products and other metabolites.

The first two levels are found in early models of crops and are still used for models where only genetic representations of species are required. Most current crop models are at level 3. Level 4 corresponds to the approach used in GeneGro Version 1 (White and Hoogenboom, 1996) and linear models of gene effects and level 5 is partially represented in the phenology routines of GeneGro Version 2 (Hoogenboom and White, 2003) and based on knowledge of gene action. The feasibility of level

6 is implicitly considered for unicellular organisms in models such as E-CELL (Tomita *et al.*, 1999), which can advance our understanding of cell biochemistry and gene regulation, but current application are far from providing the capacity of simulating growth of a plant, even if simplified to a few key cell types and maintained in a constant environment (White and Hoogenboom, 2003). The last three levels represent a continuum of approaches involving greater levels of genetic and biochemical details.

To study the behaviour of gene networks and their influences on organism development and evolutionary processes, Cooper *et al.* (2005) developed the *E*(*NK*) model, as discussed in Chapter 10, which is an extension of the *NK* gene network model introduced and used by Kauffman (1993). van Eeuwijk *et al.* (2004) presented various statistical models for the analysis of multi-environment trial data that differ in the extent to which additional genetic, physiological and environmental information is incorporated into the model formulation. Their models range from a simplest one with only the additive two-way ANOVA model to a complex one involving a synthesis of a multiple QTL model and an eco-physiological model to describe a collection of genotypic response curves. Between these extremes, they discussed linear–bilinear models, whose parameters can only indirectly be related to genetic and physiological information and factorial regression models that allow direct incorporation of explicit genetic, physiological and environmental co-variables on the levels of the genotypic and environmental factors.

Hammer *et al.* (2005) explored whether physiological dissection and integrative modelling of complex traits could link the complexity of the phenotype to underlying genetic systems in a way that could enhance the power of molecular breeding strategies in sorghum. This approach was applied to four key adaptive traits (phenology, osmotic adjustment, transpiration efficiency and stay-green). It was assumed that the three to five genes associated with each trait, had two alleles per locus acting in an additive manner. The results indicated that use of a crop growth and development modelling framework can link phenotype complexity to underlying genetic systems in a way that enhances the power of molecular breeding strategies. The environmental characterization and physiological knowledge helped to dissect and explain gene and environment context dependencies in the data and based on estimated gene effects to simulate a range of MAS breeding strategies.

QTL mapping allows the dissection of a phenotype into underlying genetic factors but it has limited ability to predict how QTL detected in one set of environmental factors or management practices will behave in a new set of conditions (Stratton, 1998). Eco-physiological modelling provides an insight into the factors influencing GEI (Tardieu, 2003), but it does help define the genetic basis for differences in response to environmental changes. Combining eco-physiological modelling with genetic mapping provides the opportunity for creating a QTL-based crop physiology model that could be powerful tool for resolving the genetic basis of complex environment-dependent yield-related traits. For example, using this approach researchers predicted specific leaf area in barley (Yin *et al.*, 1999), stay-green response to nitrogen in sorghum (Borrell *et al.*, 2001), leaf growth response to temperature and water deficit in maize (Reymond *et al.*, 2003) and pre-flowering duration in barley (Yin *et al.*, 2005). By removing gene and environment context dependencies, it was possible to devise breeding strategies that generated an enhanced rate of yield improvement over several cycles of selection. Messina *et al.* (2006) combined an eco-physiological model (CROPGRO-Soybean) with a linear model that predicted cultivar-specific parameters as a function of E-loci. This approach predicted 75% of the variance in time to maturity and 54% of the variance in yield, demonstrating that agricultural genomics data can be effectively used for predicting cultivar performance and refining crop breeding systems.

The genotype-to-phenotype (GP) model, as a key component of breeding design, describes how different genotypes interact with environments to produce different phenotypes (Cooper *et al.*, 2005). Using information from genes, core germplasm collections and cornerstone parents, when combined with the biological characteristics and breeding objectives for the target environments, breeding procedure and selection methods can be simulated and optimized and desirable genotypes and the probability of breeding new cultivars can be predicted.

Comparisons among genomes of different crop species reveal high levels of similarity and it appears likely that models of gene action in one crop can be extrapolated to other crops in the same botanical family (e.g. among legumes or among cereals). However, Helentjaris and Briggs (1998) noted that efforts to identify maize homologues for genes described in other species have proven more difficult than originally anticipated. One problem is that a single species may have multiple genes with similar sequences but different functions.

In the future it will be possible to build more realistic genetic models if advances in genomics improve our understanding of the GP relationship and GEIs (Bernardo, 2002; Cooper *et al.*, 2005). Conclusions on the relative merits of breeding strategies based on simple GP models may have to be re-evaluated in the context of an exponentially growing knowledge base. This information will aid in determining gene number and gene effects on phenotype. In addition, conventional plant breeding provides a wealth of information about trait heritabilities and correlations. This information, once determined, will help define errors, linkage and pleiotropic effects. In addition, crop physiological models may also help fine-tune the genetic models for breeding modelling (Reymond *et al.*, 2003; Yin, X. *et al.*, 2004; Hammer *et al.*, 2005). White and Hoogenboom (2003) discussed several practical issues in gene-based modelling, including how to access genetic and molecular data, which species, traits and what scale and level of detail to model, the relevance of results from animal systems to plant biology and how to ensure effective collaboration among crop modellers, geneticists and molecular biologists.

15.4.3 A simulation module for genetics and breeding: QuLine

Typically, breeding is done by crossing and selecting from progeny. With the opportunity to make predictions of crop performance and to explicitly model *in silico* the desired genotype × environment × breeding scheme combinations, breeding shifts in its character. Breeders become model testers themselves while model systems (or other information rich systems) become useful tools for model building. Once phenotype × genotype × environment models are verified through explicit breeding experiments, the task is to move the models themselves around through breeding programmes of different crop species. One of the interesting efforts pursuing this type of paradigm shift is the QUantitative GENEtics (QuGene; www.pig.ag.uq.edu.au/qu-gene) system.

QuGene is a simulation platform for quantitative analysis of genetic models, which provides the opportunity to develop a general simulation program for actual breeding programmes through its two-stage architecture (Podlich and Cooper, 1998). The first stage is the engine, which has two roles: (i) to define the genotype-by-environment (GE) system (i.e. all the genetic and environmental information of the simulation experiment); and (ii) to generate the starting population of individuals (base germplasm). The second stage includes the application modules, whose role is to investigate, analyse, or manipulate the starting population of individuals within the GE system defined by the engine. The application module usually represents the operation of a breeding programme. The core model within the engine can incorporate many of the features for the architecture of traits that are revealed by the characterization of GE system. It includes multiple traits and QTL with different effects, genome positional information such as that provided by molecular maps, epistasis within gene networks, differential gene expression, GEIs and structure within the TPE. Cooper *et al.* (1999) provided an example of this approach for comparisons between conventional phenotypic and MAS strategies.

Using QuGene software, a breeding module was developed for sorghum by incorporating physiological constraints and was implemented by linking QuGene to the Agricultural Production System Simulator (APSIM) cropping systems model (Keating *et al.*, 2003; http://www.apsru.gov.au). This module can be used to simulate breeding line performance in a given environment and extrapolate the effects of long-term selection over many breeding cycles and seasons. Another project supported by the Generation Challenge Programme links QuGene/APSIM with QTL data on maize leaf growth under drought. These projects aim to deliver modelling tools into the hands of molecular breeders and other researchers to extend the scope and impact of their use, particularly with respect to molecular breeding of complex traits such as drought tolerance (Dwivedi *et al.*, 2007).

As a QuGene application module, QuLine was developed at the International Maize and Wheat Improvement Center (CIMMYT) specifically for wheat-breeding programme simulation. It is a computer tool capable of defining a range of genetic models from simple to complex and simulating breeding processes for developing final advanced lines. Simulation indicated that it can be used to optimize breeding methodology and improve breeding efficiency. QuLine can be used to integrate various genes with multiple alleles functioning within epistatic networks and differentially interacting with the environment and predict the outcomes from a specific cross following the application of a real selection scheme (Wang *et al.* 2003, 2004). The breeding methods that can be simulated by QuLine are mass selection, pedigree system (including single seed descent), bulk population system, backcross breeding, top cross (or three-way cross) breeding, DH breeding, MAS and many combinations and modifications of these methods.

QuLine has the potential to provide a bridge between the vast amount of biological data and breeders' queries on optimizing selection gain and efficiency. It has been used to compare two selection strategies (Wang *et al.*, 2003), study the effects on selection of dominance and epistasis (Wang *et al.*, 2004), predict cross performance using known gene information (Wang, J. *et al.*, 2005) and optimize MAS to efficiently pyramid multiple genes (Kuchel *et al.*, 2005; Wang, J. *et al.*, 2007).

By defining breeding strategy, QuLine translates the complicated breeding process into a way that the computer can understand and simulate. QuLine allows for several breeding strategies to be defined simultaneously. The programme then starts with the same virtual crosses for all the defined strategies at the first breeding cycle, including the same initial population, crosses and genotype and environment systems, allowing appropriate comparisons. A breeding strategy in QuLine is defined to include all activities involved in an entire breeding cycle such as crossing, seed propagation and selection (Wang and Pfeiffer, 2007). A breeding cycle begins with crossing and ends at the generation when the selected advanced lines are returned to the crossing block as new parents. The genotypic value of a genotype is calculated based on the definition of gene actions. The phenotypic value and family mean is derived from the genotypic value and its associated error (environmental deviation). With all defined phenotypic and genotypic values, QuLine then makes within-family selection from phenotypic values and among-family selection from family means.

To simulate in QuLine, the seed propagation type must be defined to describe how the selected plants in a retained family from the previous selection round or generation are propagated to generate the seed for the current selection round or generation. Wang and Pfeiffer (2007) defined nine options for seed propagation, which can be presented in the order of increasing genetic diversity (the F_1 excluded) as: (i) *clone* (asexual reproduction); (ii) *DH* (doubled haploid); (iii) *self* (self-pollination);

(iv) *singlecross* (single crosses between two parents); (v) *backcross* (backcrossed to one of the two parents); (vi) *topcross* (crossed to a third parent, also known as a three-way cross); (vii) *doublecross* (crossed between two F_1s); (viii) *random* (random mating among the selected plants in a family); and (ix) *noself* (random mating but self-pollination is eliminated). The seed for the F_1 is derived from crossing among the parents in the initial population (or crossing block). QuLine randomly determines the female and the male parents for each cross from a defined initial population, or alternatively, one may select some preferred parents from the crossing block. The selection criteria used to identify such preferred parents can be defined in terms of among-family and within-family selection descriptors within the crossing block (referred to as the F_0 generation). By using the parameter of seed propagation type, most if not all methods of seed propagation in self-pollinated crops can be simulated in QuLine.

15.4.4 The future of simulation and modelling

There are several practical implementation issues in simulation and modelling to be solved, including: (i) communications and training required to combine modelling and simulation with real breeding programmes through involvement of other scientists including breeders, agronomists and geneticists; (ii) standardization and documentation of data collection for phenotypic, environmental and genomic information needs to be enforced through the project; (iii) unexpected and great variation within selection and target environments requires much more comprehensive data collection, compared to other breeding environments with much less stressful factors; and (iv) when more and more factors are involved in modelling and simulation, data generation and collection should be done with more data dimensions including more locations, samples and replications.

Practical applications often oblige crop modellers to emphasize simulation of economic yield. A set of traits that are involved in stress response is also worthy considering. While allowing precise control of plant response and gene expression, specific stress responses may largely be survival mechanisms. Thus, whereas their study could improve the simulation of plant survival, the results might prove harder to relate to the simulation of basic processes of growth and partitioning (White and Hoogenboom, 2003). Innovative simulation models will bridge the gap between molecular and conventional plant breeding and will inform both strategic research and tactical breeding decisions (www.generationcp.org/sccv10/sccv10_upload/modelling_links.pdf). Simulation models integrate molecular information about interaction between genes and simpler traits to allow realistic predictions for more complex traits such as drought tolerance and yield.

Developing and implementing a design-led breeding system for complex traits requires enhanced attention to precision phenotyping, eco-physiological modelling and marker validation to ensure robustness and selective power. These approaches require the iterative and systemic integration of a range of scientific disciplines including modellers, physiologists, geneticists, breeders and molecular biologists. Nevertheless, the first preliminary studies reviewed in this section suggest that a new paradigm in knowledge-led, design-driven plant breeding is a feasible option and that for the first time, genomics may finally realize its potential impact on breeding complex traits (Dwivedi *et al.*, 2007).

Although many public databases on genes, alleles, gene and genomic sequences and related information are maintained by geneticists and molecular biologists, physiologists and modellers may find these databases less useful than expected. The user interfaces assume familiarity with bioinformatics. Databases of gene sequences and protein structure lack information on actual gene function in most cases. The number of lines or cultivars characterized for a given gene is usually limited to the parents used in describing the gene and few data on field performance are found (White and Hoogenboom, 2003). For example, the Arabidopsis Information Resource (2000) purports to provide more phenotypic data for *Arabidopsis* as a model plant, but still falls short of meeting the requirements of whole plant model. The same is true for rice and its related databases.

15.5 Breeding by Design

The advances in applied genomics and the possibility of generating large-scale marker data sets provide us with the tools to determine the genetic basis for all traits of agronomic importance. In addition, methods for assessing the allelic variation at these agronomically important loci are now available. This combined knowledge will eventually allow the breeder to combine the most favourable alleles at all these loci in a controlled manner to design superior cultivars *in silico*. This concept is called 'breeding by design' (Peleman and van der Voort, 2003) and has been generalized to breeding design using genome-wide QTL–marker associations identified through all types of effort due to the fast development in molecular marker technology (Bernardo, 2002; Peleman and van der Voort, 2003). The goal can be reached following a three-step approach: (i) mapping loci involved in all agronomically relevant traits; (ii) assessing the allelic variation at those loci; and (iii) following the 'breeding by design' approach. Because the positions of all loci of importance are mapped precisely, recombinant events can be accurately selected using flanking markers to collate the different favourable alleles next to each other. Software tools should enable us to determine the optimal route for generation of those mosaic genotypes by crossing lines and using markers to select for the specific combinations that will eventually combine all those alleles. The prerequisites for this approach include extremely saturated marker maps available to enable the generation of high-resolution chromosome haplotypes, extensive phenotyping of all agronomic traits for both the mapping populations and the inbred lines that are used for chromosome haplotyping and allele assessment. 'Breeding by design' involves the integrative, complementary application of technological tools and the materials currently available to develop superior cultivars. During this process, an enormous resource of knowledge is generated and accumulated that should enable breeders to deploy more rational and refined breeding strategies in the future. The developments in high-throughput genotyping and genetic mapping with associated statistical methodology have now brought this strategy within reach. The optimal exploitation of the naturally available genetic resources should create unsurpassed possibilities to generate new traits and crop performance.

15.5.1 Parental selection

Selecting parents to make crosses is the first and essential step in plant breeding (Fehr, 1987). Due to incomplete gene information (i.e. some resistance genes and their effects on phenotype are known, while other genes and most genes for other agronomic traits are unknown), many seemingly good crosses are discarded during the segregating phase of a breeding programme. Almost all agronomic traits including disease resistance, stress tolerance and yield involve complex genetics. It makes sense to understand as much as we can about the plant parents, including genotype, before we make decisions about crossing one parent with another. In most plant breeding programmes, less than 1% of all the crosses made end up in a cultivar. To a layperson, that may seem incredibly inefficient, but that's the nature of the beast. What is most important in plant breeding is to pick the right parents so that breeders would have fewer crosses to deal with and would be able to spend more time and attention on the crosses that will result in superior material.

Generally speaking, the cross with the highest progeny mean and largest genetic variance has the most potential to produce the best lines (Bernardo, 2002). Under an additive genetic model, the mid-parent value is a good predictor of the progeny mean, but the variance cannot be deduced from the performance of the parents alone. The best way to estimate the progeny variance is to generate and test the progeny. Breeders normally use one of two types of parental selection: one based on parental information, such as parental performance or the genetic diversity among parents; the other based on parental and progeny information. In the first case, previous studies found that both high × high and high × low crosses have the potential to produce the best lines. In the second case, the progeny needs to be grown and tested, which precludes parental selection. Due to complicated intra-genic, inter-genic and GEIs, no method has given a precise prediction of cross performance (Wang *et al.*, 2005).

Breeders are already aware of what parents are available, but often breeders' phenotypic and field data comes in spreadsheets with numerous columns and reams of data without much association with other types of data generated in genetics and genomics. Once software becomes available to show a full genome genotype of all possible parents, one can ask, for example, which parents will provide high yield and resistance to a specific disease. The informatics tool will indicate to the breeder what genes will be traceable in the progeny and which are the best sets of molecular markers for tracking these genes.

15.5.2 Breeding product prediction

Designing effective breeding systems requires information about target genes, donor germplasm and proposed elite recurrent parents. This can then be combined with evaluation data on the target biological characteristics, breeding objectives for the TPE, in order to optimize the breeding procedure and selection methods through modelling and simulation analysis. This type of analysis will also predict the desirable target genotype and the probability of successfully generating new cultivars through the proposed breeding system.

Cross performance can be accurately predicted when information about the genes controlling the traits of interest is known. If progeny arrays after selection in a breeding programme could be predicted, then the efficiency of plant breeding would be greatly increased. Take wheat as an example. For the majority of economically important traits in wheat breeding the genes controlling their expression remain unknown. However, for wheat quality this information is known, though incompletely, for certain aspects of wheat quality (Eagles *et al.*, 2002, 2004). Wang and Pfeiffer (2007) demonstrated how cross performance, following selection, can be predicted in wheat quality breeding by using QuLine, under the condition that all the gene information of key selection traits is known.

Plant breeders have been always confronted with the problem of predicting the expected phenotypic performance of new individuals with untested gene combinations (new genotypes) with limited information on the GP architecture for traits. The success of molecular breeding relies on an effective prediction of phenotypic variation based on allelic variation. There are opportunities to apply molecular technologies to further refine the pedigree-based breeding strategies used today. Ultimately it will not be sufficient to demonstrate that we can predict phenotypic variation and the phenotypic changes that result from selection using genetic information, but this knowledge allows us to improve on the outcomes that are currently being achieved by conventional selection on phenotype alone.

15.5.3 Selection method evaluation

To develop new genotypes that are genetically superior to those currently available for a specific target environment, plant breeders employ a range of selection methods. Many field experiments have been conducted to compare the efficiencies of different breeding methods. However, because of the time and effort spent in conducting field experiments, the concept of modelling and prediction has always been of interest to plant breeders.

Taking the bread wheat breeding at CIMMYT as an example, breeders spend great efforts in choosing parents to make the targeted crosses and approximately 50–80% of crosses are discarded in generations F_1 to F_8, following selection for agronomic traits (e.g. plant height, lodging tolerance, tillering, appropriate heading date and balanced yield components), disease resistance (e.g. stem rust, leaf rust and stripe rust) and end-use quality (e.g. dough strength and extensibility, protein quantity and quality). Then, after two cycles of yield trials (i.e. preliminary yield trial in F_8 and replicated yield trial in F_9), only 10% of the initial crosses remain, among which 1–3% of the crosses originally made are released as cultivars from CIMMYT's international nurseries (Wang *et al.*, 2003, 2005). This fact is true across plant breeding programmes of different species, which calls for a more efficient breeding system.

Two selection methods are commonly used in CIMMYT's wheat breeding programmes. Pedigree selection was used primarily from 1944 until 1985. From 1985 until the second half of the 1990s the main selection method was a modified pedigree/bulk method (MODPED), which resulted in many widely adapted wheat cultivars and was replaced in the late 1990s by the selected bulk method (SELBLK) (van Ginkel *et al.*, 2002). The MODPED method begins with pedigree selection of individual plants in the F_2 followed by three times of bulk selection from F_3 to F_5 and pedigree selection in the F_6; hence the name modified pedigree/bulk. In the SELBLK method, spikes of selected F_2 plants within a cross are harvested in bulk, resulting in one F_3 seed lot per cross. This process continues from F_3 to F_5, while pedigree selection is used only in the F_6. A major advantage of SELBLK compared with MODPED is that fewer seed lots need to be harvested, threshed and visually selected for seed appearance, leading to significant savings in time, labour and costs associated with nursery preparation, planting and plot labelling (van Ginkel *et al.*, 2002).

Before simulation, the breeders already knew that SELBLK can save costs compared with MODPED. Some small-scale field experiments have been conducted comparing the efficiencies of MODPED and SELBLK (Singh *et al.*, 1998), but the relative efficiency of the two methods remains untested on a larger scale. Wang and Pfeiffer (2007) illustrated the simulation principles by using the QuLine module with CIMMYT's wheat breeding programme as an example. They developed the genetic models accounting for epistasis, pleiotropy and GEI. For each selection method, the simulation experiment comprised the same 1000 crosses derived from 200 parents with an assumption that a total of 258 advanced lines remained following ten generations of selection. The tests for the two methods were each repeated 500 times on 12 GE systems. The simulation not only provided a clear answer that the adoption of SELBLK would not cause a yield-gain penalty, but also indicated a fact that CIMMYT's breeders did not realize, i.e. SELBLK can retain more crosses in the final selected population than MODPED.

15.6 Future Perspectives

The use of appropriate experimental design and data analysis is a critical component for successful development and application of molecular breeding approaches, in particular, marker-assisted breeding systems. Figure 15.2 shows an information flowchart from data to outputs through use of various analytical tools. Making these choices correctly is a highly specialized function. There is a lack of proper and simple-to-use guidelines for non-specialists, which makes it difficult for them to confidently choose the appropriate design and analysis methods offered by various types of software. Having a centralized and evolving resource offering biometric inputs required for molecular breeding would be a tremendously valuable asset to the research and breeding community.

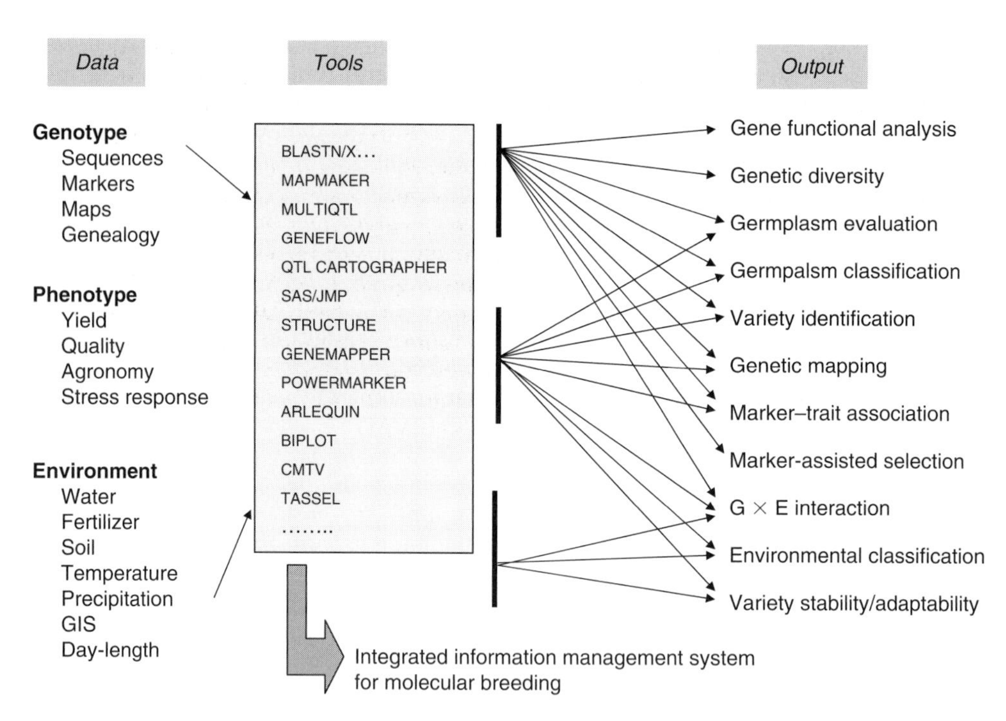

Fig. 15.2. Analytical tools and outputs associated with procedures in plant breeding. Three types of data from genotype (G), phenotype (P) and environment (E) are analysed using various tools, and outputs will be delivered to breeders for decision making.

There is an urgent need for integrated molecular tools including those for facilitating molecular breeding design, integrated mapping and MAS and communications between genomics scientists, geneticists, bioinformaticians and breeders.

There is also a need to develop molecular breeding decision support tools that can use modelling and simulation analysis of all pre-existing and project generated data. These tools will help breeders design and implement the most efficient breeding schemes (including cost- and time-related factors) based on the optimum combination of MAS (for both foreground and background) and phenotypic selection. Other decision support tools that are needed in molecular breeding include: (i) those for sample colleting, depositing, retrieving and tracking; (ii) those for data acquiring, collecting, processing and mining; and (iii) databases.

Independent of the platform and the analysis methods used, the result of microarray experiments is, in most cases, a list of differentially expressed genes. An automatic ontological analysis approach using Gene Ontology has been proposed to help with the biological interpretation of such results (Khatri *et al.*, 2002). Currently this approach is the *de facto* standard for the secondary analysis of high-throughput experiments and a large number of tools have been developed for this purpose. Khatri and Drăghici (2005) provided a detailed comparison of 14 such tools using the following criteria: scope of the analysis, visualization capabilities, statistical model(s) used, correlation for multiple comparisons, reference microarray available, installation issues and sources of annotation data. This detailed analysis of the capabilities of these tools will help researchers choose the most appropriate tool for a given type of analysis. More importantly, in spite of the fact that this type of analysis has been generally adopted, this approach has several intrinsic drawbacks. These drawbacks are associated with all tools discussed and represent conceptual limitations of the current state-of-the-art in ontological analysis. These are challenges for the next generation of secondary data analysis tools. It would be more beneficial if future tools expand the current approach by trying to address some of these limitations rather than providing endless variations of the same idea.

There is a need for systematic construction of biological relationship graphs from the integration of gene, protein, metabolite and phenotype data (Blanchard, 2004). The challenge is now to use the large-scale data sets in a meaningful way to weed out the high false positive rates associated with high-throughput techniques and to encapsulate knowledge to validate and extend existing models (Blanchard, 2004). Graphical models represent a union between probability theory and graph theory and thus represent a natural extension of the work on relationship graphs.

References

Aastveit, H. and Martens, H. (1986) ANOVA interactions interpreted by partial least squares regression. *Biometrics* 42, 829–844.

Abad-Grau, M.M., Montes, R. and Sebastiani, P. (2006) Building chromosome-wide LD maps. *Bioinformatics* 22, 1933–1934.

Able, J.A., Langridge, P. and Milligan, A.S. (2008) Capturing diversity in the cereals: many options but little promiscuity. *Trends in Plant Sciences* 12, 71–79.

Abranches, R., Santos, A.P., Williams, S., Wegel, E., Castilho, A., Christou, P., Shaw, P. and Stoger, E. (2000) Widely-separated multiple transgene integration sites in wheat chromosomes are brought together at interphase. *The Plant Journal* 24, 713–723.

Acosta-Gallegos, J.A., Kelly, J.D. and Gepts, P. (2007) Prebreeding in common bean and use of genetic diversity from wild germplasm. *Crop Science* 47(S3), S44–S59.

Adams, M.D., Kelley, J.M., Gocayne, J.D., Dubnick, M., Polymeropoulos, M.H., Xiao, H., Merril, C.R., Wu, A., Olde, B., Moreno, R.E., Kerlavage, A.R., Combie, W.R. and Venter, J.C. (1991) Complementary DNA sequencing: expressed sequence tags and human genome project. *Science* 252, 1651–1653.

Adams, R.P. (1997) Conservation of DNA: DNA banking. In: Callow, J.A., Ford-Lloyd, B.V. and Newbury, H.J. (eds) *Biotechnology and Plant Genetics Resources Conservation and Use.* CAB International, Wallingford, UK, pp.163–174.

Adi, B. (2006) Intellectual property rights in biotechnology and the fate of poor farmers' agriculture. *The Journal of World Intellectual Property* 9, 91–112.

Aebersold, R. and Goodlett, D.R. (2001) Mass spectrometry in proteomics. *Chemical Reviews* 101, 269–295.

Aebersold, R. and Mann, M. (2003) Mass spectrometry-based proteomics. *Nature* 422, 198–207.

Agrawal, P.K., Kohli, A., Twyman, R.M. and Christou, P. (2005) Transformation of plants with multiple cassettes generates simple transgene integration patterns and high expression levels. *Molecular Breeding* 16, 247–260.

Aguilar, G. (2001) Access to genetic resources and protection of traditional knowledge in the territories of indigenous peoples. *Environmental Science and Policy* 4, 241–256.

Ahmadi, N., Albar, L., Pressoir, G., Pinel, A., Fargette, D. and Ghesquiere, A. (2001) Genetic basis and mapping of the resistance to rice yellow mottle virus. III. Analysis of QTL efficiency in introgressed progenies confirmed the hypothesis of complementary epistasis between two resistance QTL. *Theoretical and Applied Genetics* 103, 1084–1092.

Ahmadian, A., Gharizadeh, B., Gustafsson, A.C., Sterky, F., Nyren, P., Uhlen, M. and Lundeberg, J. (2000) Single nucleotide polymorphism analysis by pyrosequencing. *Analytical Biochemistry* 280, 103–110.

Ahn, S.N. and Tanksley, S.D. (1993) Comparative linkage maps of the rice and maize genomes. *Proceedings of the National Academy of Sciences of the United States of America* 90, 7980–7984.

Ahn, S.N., Anderson, J.A., Sorrells, M.E. and Tanksley, S.D. (1993) Homoeologous relationships of rice, wheat and maize chromosomes. *Molecular and General Genetics* 241, 483–490.

Ajmone Marson, P., Castiglioni, P., Fusari, F., Kuiper, M. and Motto, M. (1998) Genetic diversity and its relationship to hybrid performance in maize as revealed by RFLP and AFLP markers. *Theoretical and Applied Genetics* 96, 219–227.

Akaike, H. (1969) Fitting autoregressive models for prediction. *Annals of the Institute of Statistical Mathematics* 21, 243–247.

Alan, A.R., Mutchler, M.A., Brants, A., Cobb, E. and Earle, E.D. (2003) Production of gynogenic plants from hybrids of *Allium apa* L. and *A. roylei* Stearn. *Plant Science* 165, 1201–1211.

Allard, R.W. (1956) Formulas and tables to facilitate the calculation of recombination values in heredity. *Hilgardia* 24, 235–278.

Allard, R.W. (1988) Genetic changes associated with the evolution of adaptedness in cultivated plants and their progenitors. *Journal of Heredity* 79, 225–238.

Allard, R.W. (1999) *Principles of Plant Breeding*, 2nd edn. John Wiley & Son, Inc., New York, 254 pp.

Allard, R.W. and Bradshaw, A.D. (1964) Implications of genotype–environmental interactions in applied plant breeding. *Crop Science* 4, 503–507.

Allen, G.C., Spiker, S. and Thompson, W.F. (2000) Use of matrix attachment regions (MARs) to minimize transgene silencing. *Plant Molecular Biology* 43, 361–376.

Allen-Brady, K., Wong, J. and Camp, N.J. (2006) PedGenie: an analysis approach for genetic association testing in extended pedigrees and genealogies of arbitrary size. *BMC Bioinformatics* 7, 209.

Allison, D.B., Cui, X., Page, G.P. and Sabripour, M. (2006) Microarray data analysis: from disarray to consolidation and consensus. *Nature Reviews Genetics* 7, 55–65.

Alonso, J.M. and Ecker, J.R. (2006) Moving forward in reverse: genetic technologies to enable genome-wide phenomic screens in *Arabidopsis*. *Nature Reviews Genetics* 7, 524–536.

Alonso, J.M., Stepanova, A.N., Leisse, T.J., Kim, C.J., Chen, H., Shinn, P., Stevenson, D.K., Zimmerman, J., Barajas, P., Cheuk, R., Gadrinab, C., Heller, C., Jeske, A., Koesema, E., Meyers, C.C., Parker, H., Prednis, L., Ansari, Y., Choy, N., Deen, H., Geralt, M., Hazari, N., Hom, E., Karnes, M., Mulholland, C., Ndubaku, R., Schmidt, I., Guzman, P., Aguilar-Henonin, L., Schmid, M., Weigel, D., Carter, D.E., Marchand, T., Risseeuw, E., Brogden, D., Zeko, A., Crosby, W.L., Berry, C.C. and Ecker, J.R. (2003) Genome-wide insertional mutagenesis of *Arabidopsis thaliana*. *Science* 301, 653–657.

Alpert, K.B. and Tanksley, S.D. (1996) High-resolution mapping and isolation of a yeast artificial chromosome contig containing *fw2.2*: a major fruit weight quantitative trait locus in tomato. *Proceedings of the National Academy of Sciences of the United States of America* 93, 15503–15507.

Altpeter, F., Baisakh, N., Beachy, R., Bock, R., Capell, T., Christou, P., Daniell, H., Datta, K., Datta, S., Dix, P.J., Fauquet, C., Huang, N., Kohli, A., Mooribroek, H., Nicholson, L., Nguyen, T.H., Nugent, G., Raemakers, K., Romano, A., Somers, D.A., Stoger, E., Taylor, N. and Visser, R. (2005a) Particle bombardment and the genetic enhancement of crops: myths and realities. *Molecular Breeding* 15, 305–327.

Altpeter, F., Varshney, A., Abderhalden, O., Douchkov, D., Sautter, C., Kumlehn, J., Dudler, R. and Schweizer, P. (2005b) Stable expression of a defense-related gene in wheat epidermis under transcriptional control of a novel promoter confers pathogen resistance. *Plant Molecular Biology* 57, 271–283.

Altschul, S., Madden, T., Schaffer, A., Zhang, J., Zhang, Z., Miller, W. and Lipman, D. (1997) Gapped BLAST and PSI-BLAST: a new generation of protein database search programs. *Nucleic Acids Research* 25, 3389–3402.

Álvarez-Castro, J.M. and Carlborg, Ö. (2007) A unified model for functional and statistical epistasis and its application in quantitative trait loci analysis. *Genetics* 176, 1151–1167.

Amratunga, D. and Cabrera, J. (2004) *Exploration and Analysis of DNA Microarray and Protein Array Data*. John Wiley & Sons, Inc., New York.

An, G., Watson, B.D., Stachel, S. and Gordon, M.P. (1985) New cloning vehicles for transformation of higher plants. *EMBO Journal* 4, 277–284.

An, G., Jeong, D.-H., An, S., Kang, H.-G., Moon, S., Han, J., Park, S., Lee, H. S. and An, K. (2003) Activation tagged mutants to discover novel rice genes. In: Mew, T.W., Brar, D.S., Peng, S., Dawe, D. and Hardy, B. (eds) *Rice Science: Innovations and Impact for Livelihood. Proceedings of the International Rice Research Conference*, 16–19 September 2002, Beijing, China, International Rice Research Institute, Chinese Academy of Engineering and Chinese Academy of Agricultural Sciences, pp. 195–204.

Andersen, J.R. and Lübberstedt, T. (2003) Functional markers in plants. *Trends in Plant Science* 8, 554–560.

Anderson, J.A., Churchill, G.A., Autrique, J.E., Tanksley, S.D. and Sorrells, M.E. (1993) Optimizing parental selection for genetic linkage maps. *Genome* 36, 181–186.

Andrews, L.B. (2002) Genes and patent policy: rethinking intellectual property rights. *Nature Reviews Genetics* 3, 803–808.

Anido, F.L., Cravero, V., Asprelli, P., Firpo, T., García, S.M. and Cointry, E. (2004) Heterotic patterns in hybrids involving cultivar-groups of summer squash, *Cucurbita pepo* L. *Euphytica* 135, 355–360.

Annicchiarico, P., Bellah, F. and Chiari, T. (2005) Defining subregions and estimating benefits for a specific-adaptation strategy by breeding programs: a case study. *Crop Science* 45, 1741–1749.

Annicchiarico, P., Bellah, F. and Chiari, T. (2006) Repeatable genotype × location interaction and its exploitation by conventional and GIS-based cultivar recommendation for durum wheat in Algeria. *European Journal of Agronomy* 24, 70–81.

Antonio, B.A., Inoue, T., Kajiya, H., Nagamura, Y., Kurata, N., Minobe, Y., Yano, M., Nakagahra, M. and Sasaki, T. (1996) Comparison of genetic distance and order of DNA markers in five populations of rice. *Genome* 39, 946–956.

Arabidopsis Information Resource (2000) The Arabidopsis Information Resource (TAIR). TAIR, Stanford, California. Available at: http://www.arabidopsis.org (accessed 17 November 2009).

Aranzana, M.J., Kim, S., Zhao, K., Bakker, E., Horton, M., Jakob, K., Lister, C., Molitor, J., Shindo, C., Tang, C., Toomajian, C., Traw, B., Honggang Zheng, H., Bergelson, J., Dean, C., Marjoram, P. and Nordborg, M. (2005) Genome-wide association mapping in *Arabidopsis* identifies previously known flowering time and pathogen resistance genes. *PLoS Genetics* 1, e60.

Arcade, A., Labourdette, A., Falque, M., Mangin, B., Chardon, F., Charcosset, A. and Joets, J. (2004) BioMercator: integrating genetic maps and QTL towards discovery of candidate genes. *Bioinformatics* 20, 2324–2326.

Arcelllana-Panlilio, M. (2005) Principles of application of DNA microarrays. In: Sensen, C.W. (ed.) *Handbook of Genome Research, Genomics, Proteomics, Metabolomics, Bioinformatics, Ethical and Legal Issues.* Wiley-VCH Verlag GmbH & Co. KGaA, Weinheim, Germany, pp. 239–260.

Arumuganathan, K. and Earle, E.D. (1991) Nuclear DNA content of some important plant species. *Plant Molecular Biology Reporter* 9, 208–219.

Ashikari, M., Sakakibara, H., Lin, S., Yamamoto, T., Takashi, T., Nishimura, A., Angeles, R.E., Qian, Q., Kitano, H. and Matsuoka, M. (2005) Cytokinin oxidase regulates rice grain production. *Science* 309, 741–745.

Ashman, K., Moran, M.F., Sicheri, F., Pawson, T. and Tyers, M. (2001) Cell signalling – the proteomics of it all. *Science's STKE.* Available at: http://stke.sciencemag.org/cgi/content/full/sigtrans;2001/103/pe33 (accessed 17 November 2009).

Ashmore, S. (1997) *Status Report on the Development and Application of* in vitro *Techniques for the Conservation and Use of Plant Genetic Resources.* Engelmann, F. (vol. ed.) International Plant Genetic Resources Institute, Rome.

Auger, D.L., Gray, A.D., Ream, T.S., Kato, A., Coe, E.H., Jr and Birchler, J.A. (2005) Nonadditive gene expression in diploid and triploid hybrids of maize. *Genetics* 169, 389–397.

Auzanneau, J., Huyghe, C., Julier, R. and Barre, P. (2007) Linkage disequilibrium in synthetic varieties of perennial ryegrass. *Theoretical and Applied Genetics* 115, 837–847.

Avise, J.C. (1986) Mitochondrial DNA and the evolutionary genetics of higher animals. *Philosophical Transactions of the Royal Society of London B* 312, 325–342.

Avise, J.C. (2004) *Molecular Markers, Natural History and Evolution*, 2nd edn. Sinauer Associates, Inc., Sunderland, Massachusetts.

Avraham, S., Tung, C.-W., Ilic, K., Jaiswal, P., Kellogg, E.A., Susan McCouch, S., Pujar, A., Reiser, L., Rhee, S.Y., Sachs, M.M., Schaeffer, M., Stein, L., Stevens, P., Vincent, L., Zapata, F. and Ware, D. (2008) The Plant Ontology Database: a community resource for plant structure and developmental stages controlled vocabulary and annotations. *Nucleic Acids Research* 36, D449–D454.

Ayele, M., Haas, B.J., Kumar, N., Wu, H., Xiao, Y., Van Aken, S., Utterback, T.R., Wortman, J.R., White, O.R. and Town, C.D. (2005) Whole genome shotgun sequencing of *Brassica oleracea* and its application to gene discovery and annotation in *Arabidopsis. Genome Research* 15, 487–495.

Aylor, D.L., Price, E.W. and Carbone, I. (2006) SNAP: combine and map modules for multilocus population genetic analysis. *Bioinformatics* 22, 1399–1401.

Ayoub, M., Armstrong, E., Bridger, G., Fortin, M.G. and Mather, D.E. (2003) Marker-based selection in barley for a QTL region affecting α-amylase activity of malt. *Crop Science* 43, 556–561.

Ayres, N.M., Mclung, A.M., Larkin, P.D., Bligh, H.F.J., Jones, C.A. and Park, W.D. (1997) Microsatellites and a single-nucleotide polymorphism differentiate apparent amylose classes in an extended pedigree of US rice germ plasm. *Theoretical and Applied Genetics* 94, 773–781.

Azpiroz-Leehan, R. and Feldmann, K.A. (1997) T-DNA insertion mutagenesis in *Arabidopsis*: going back and forth. *Trends in Genetics* 13, 152–156.

Babar, M.A., Reynolds, M.P., van Ginkel, M., Klatt, A.R., Raun, W.R. and Stone, M.L. (2006) Spectral reflectance to estimate genetic variation for in-season biomass, leaf chlorophyll and canopy temperature in wheat. *Crop Science* 46, 1046–1057.

Babar, M.A., van Ginkel, M., Klatt, A.R., Prasad, B. and Reynold, M.P. (2007) The potential of using spectral reflectance indices to estimate yield in wheat grown under reduced irrigation. *Euphytica* 150, 155–172.

Babu, R., Nair, S.K., Prasanna, B.M. and Gupta, H.S. (2004) Integrating marker assisted selection in crop breeding – prospects and challenges. *Current Science* 87, 607–619.

Babu, R., Nair, S.K., Kumar, A., Venkatesh, S., Sekhar, J.C., Singh, N.N., Srinivasan, G. and Gupta, H.S. (2005) Two-generation marker-aided backcrossing for rapid conversion of normal maize lines to Quality Protein Maize (QPM). *Theoretical and Applied Genetics* 111, 888–897.

Bachem, C.W.B., van der Hoeven, R.S., de Bruijn, S.M., Vreugdenhil, D., Zabeau, M. and Visser, G.R.F. (1996) Visualization of differential gene expression using a novel method of RNA fingerprinting based on AFLP: analysis of gene expression during potato tuber development. *The Plant Journal* 9, 745–753.

Bafna, V., Gusfield, D., Lancia, G. and Yooseph, S. (2003) Haplotyping as perfect phylogeny: a direct approach. *Journal of Computational Biology* 10, 323–340.

Bagge, M. and Lübberstedt, T. (2008) Functional markers in wheat: technical and economic aspects. *Molecular Breeding* 22, 319–328.

Bagge, M., Xia, X. and Lübberstedt, T. (2007) Functional markers in wheat. *Current Opinion in Plant Biology* 10, 211–216.

Baginsky, S. and Gruissem, W. (2004) Choroplast proteomics: potentials and challenges. *Journal of Experimental Botany* 55, 1213–1220.

Baginsky, S. and Gruissem, W. (2006) *Arabidopsis thaliana* proteomics: from proteome to genome. *Journal of Experimental Botany* 57, 1485–1491.

Baieri, A., Bogdan, M., Frommlet, F. and Futschik, A. (2006) On locating multiple interacting quantitative trait loci in intercross designs. *Genetics* 173, 1693–1703.

Baisakh, N., Datta, K., Oliva, N., Ona, I., Rao, G.J.N., Mew, T.W. and Datta, S.K. (2001) Rapid development of homozygous transgenic rice using anther culture harboring rice chitinase gene for enhanced sheath blight resistance. *Plant Biotechnology* 18, 101–108.

Baker, R.J. (1986) *Selection Indices in Plant Breeding.* CRC Press, New York.

Bal, U. and Abak, K. (2007) Haploidy in tomato (*Lycopersicon esculenttum* Mill.): a critical review. *Euphytica* 158, 1–9.

Balint-Kurti, P.J., Zwonitzer, J.C., Wisser, R.J., Carson, M.L., Oropeza-Rosas, M.A., Holland, J.B. and Szalma, S.J. (2007) Precise mapping of quantitative trait loci for resistance to southern leaf blight, caused by *Cochliobolus heterostrophus* race O and flowering time using advanced intercross maize lines. *Genetics* 176, 645–657.

Balzergue, S., Dubreucq, B., Chauvin, S., Le-Clainche, I., Le Boulaire, F., de Rose, R., Samson, F., Biaudet, V., Lecharny, A., Cruaud, C., Weissenbach, J., Caboche, M. and Lepiniec, L. (2001) Improved PCR-walking for large-scale isolation of plant T-DNA borders. *Biotechniques* 30, 496–503.

Bänziger, M., Setimela, P.S., Hodson, D. and Vivek, B. (2004) Breeding for improved drought tolerance in maize adapted to southern Africa. In: *New Directions for a Diverse Planet*, Proceedings of the 4th International Crop Science Congress, 26 September–1 October 2004, Brisbane, Australia. Published on CD-ROM. Available at: http://www.cropscience.org.au/icsc2004 (accessed 17 November 2009).

Bänziger, M., Setimela, P.S., Hodson, D. and Vivek, B. (2006) Breeding for improved abiotic stress tolerance in maize adapted to southern Africa. *Agricultural Water Management* 80, 212–224.

Bao, J.B., Lee, S., Chen, C., Zhang, X.-Q., Zhang, Y., Liu, S.-Q., Clark, T., Wang, J., Cao, M.-L., Yang, H.-M., Wang, S.M. and Yu, J. (2005) Serial analysis of gene expression study of a hybrid rice strain (*LYP9*) and its parental cultivars. *Plant Physiology* 138, 1216–1231.

Barclay, I.R. (1975) High frequencies of haploid production in wheat (*Triticum aestivum*) by chromosome elimination. *Nature* 256, 410–411.

Bard, J.B.L. and Rhee, S.Y. (2004) Ontologies in biology: design, applications and future challenges. *Nature Reviews Genetics* 5, 213–222.

Bar-Hen, A., Charcosset, A., Bourgoin, M. and Guiard, J. (1995) Relationship between genetic markers and morphological traits in a maize inbred lines collection. *Euphytica* 84, 145–154.

Barrett, J.C., Fry, B., Maller, J. and Daly, M.J. (2005) Haploview: analysis and visualization of LD and haplotype maps. *Bioinformatics* 21, 263–265.

Barrett, S.C.H and Kohn, J.R. (1991) Genetic and evolutionary consequences of small population size in plants: implications for conservation. In: Falk, D.A. and Holsinger, K.E. (eds) *Genetics and Conservation of Rare Plants*. Oxford University Press, Oxford, UK, pp. 3–30.

Barro, F., Cannell, M.E., Lazzeri, P.A. and Barcelo, P. (1998) The influence of auxins on transformation of wheat and tritordeum and analysis of transgene integration patterns in transformants. *Theoretical and Applied Genetics* 97, 684–695.

Bartlett, J.M.S. (2002) Approaches to the analysis of gene expression using mRNA – a technical overview. *Molecular Biotechnology* 21, 149–160.

Barton, J. (2000) Reforming the patent system. *Science* 287, 1933–1934.

Barton, N.H. and Keightley, P.D. (2002) Understanding quantitative genetic variation. *Nature Reviews Genetics* 3, 11–21.

Barua, U.M., Chalmers, K.J., Hackett, C.A., Thomas, W.T., Powell, W. and Waugh, R. (1993) Identification of RAPD markers linked to a *Rhynchosporium secalis* resistance locus in barley using near-isogenic lines and bulked segregant analysis. *Heredity* 71, 177–184.

Beaujean, A., Sangwan, R.S., Hodges, M. and Sangwan-Norreel, B.S. (1998) Effect of ploidy and homozygosity on transgene expression in primary tobacco transformants and their androgenetic progenies. *Molecular and General Genetics* 260, 362–371.

Beavis, W.D. (1994) The power and deceit of QTL experiments: lessons from comparative QTL studies. In: *49th Annual Corn and Sorghum Industry Research Conference*. American Seed Trade Association, Washington, DC, pp. 250–266.

Beavis, W.D. (1998) QTL analyses: power, precision and accuracy. In: Paterson, A.H. (ed.) *Molecular Dissection of Complex Traits*. CRC Press, Boca Raton, Florida, pp. 145–162.

Beavis, W.D. (1999) QTL mapping in plant breeding populations. Patent EP 1042507.

Beavis, W.D. and Keim, P. (1996) Identification of QTL that are affected by environment. In: Kang, M.S. and Gaugh, H.G. (eds) *Genotype-by-Environment Interaction*. CRC Press, Boca Raton, Florida, pp. 123–149.

Beavis, W.D., Grant, D., Albertson, M. and Fincher, R. (1991) Quantitative trait loci for plant height in four maize populations and their associations with qualitative genetic loci. *Theoretical and Applied Genetics* 83, 141–145.

Beck von Bodman, S., Domier, L.L. and Farrand, S.K. (1995) Expression of multiple eukaryotic genes from a single promotor in *Nicotiana*. *BioTechnology* 13, 587–591.

Beckert, M. (1994) Advantages and disadvantages of the use of *in vitro/in situ* produced DH maize plants. In: Bajaj, Y.P.S. (ed.) *Biotechnology in Agriculture and Forestry*, Vol. 25. Springer-Verlag, Berlin, pp. 201–213.

Beckmann, J.S. and Soller, M. (1986a) Restriction fragment length polymorphisms in plant genetic improvement. *Oxford Surveys of Plant Molecular and Cell Biology* 3, 196–250.

Beckmann, J.S. and Soller, M. (1986b) Restriction fragment length polymorphisms and genetic improvement of agricultural species. *Euphytica* 35, 111–124.

Bedell, J.A., Budiman, M.A., Nunberg, A., Citek, R.W., Robbins, D., Jones, J., Flick, E., Rohlfing, T., Fries, J., Bradford, K., McMenamy, J., Smith, M., Holeman, H., Roe, B.A., Wiley, G., Korf, I.F., Rabinowicz, P.D., Lakey, N., McCombie, W.R., Jeddeloh, J.A. and Martienssen, R.A. (2005) Sorghum genome sequencing by methylation filtration. *PLoS Biology* 3, 0103–0115.

Beer, S.C., Siripoonwiwat, W., O'Donoughue, L.S., Sousza, E., Matthews, D. and Sorrells, M.E. (1997) Associations between molecular markers and quantitative traits in a germplasm pool: can we infer linkages? *Journal of Agricultural Genomics* 3. Available at: http://www.ncgr.org/research/jag/papers97/paper197/indexp197.html (last accessed 31 December 2007).

Bekaert, S., Storozhenko, S., Mehrshahi, P., Bennett, M.J., Lambert, W., Gregory, J.F. III, Schubert, K., Hugenholtz, J., van der Straeten, D. and Hanson, A.D. (2008) Folate biofortification in food plants. *Trends in Plant Science* 13, 28–35.

Benchimol, L.L., de Souza, C.L., Jr, Garcia, A.F.F., Kono, P.M.S., Mangolin, C.A., Barbosa, A.M.M., Coelho, A.S.G. and de Souza, A.P. (2000) Genetic diversity in tropical maize inbred lines: heterotic group assignment and hybrid performance determined by RFLP markers. *Plant Breeding* 119, 491–496.

Benjamini, Y. and Hochberg, Y. (1995) Controlling the false discovery rate: lessons from comparative QTL approach to multiple testing. *Journal of the Royal Statistical Society, Series B* 57, 289–300.

Bennet, S.T., Barnes, C., Cox, A., Davies, L. and Brown C. (2005) Toward the 1,000 dollar human genome. *Pharmacogenomics* 6, 373–382.

Bennett, M.D., Finch, R.A. and Barclay, I.R. (1976) The time rate and mechanism of chromosome elimination in *Hordeum* hybrids. *Chromosoma* 54, 175–200.

Bennetzen, J.L. (1996) The use of comparative genome mapping in the identification, cloning and manipulation of important plant genes. In: Sobral, B.W.S. (ed.) *The Impact of Plant Molecular Genetics.* Birkhäuer, Boston, Massachusetts, pp. 71–85.

Bennetzen, J.L. and Ma, J. (2003) The genetic colinearity of rice and other cereals on the basis of genomic sequence analysis. *Current Opinion in Plant Biology* 6, 128–133.

Bennetzen, J.L. and Ramakrishna, W. (2002) Numerous small rearrangements of gene content, order and orientation differentiate grass genomes. *Plant Molecular Biology* 48, 821–827.

Benson, E.E. (1990) *Free Radical Damage in Stored Plant Germplasm.* International Board for Plant Genetic Resources (IBPGR), Rome.

Bent, A.F. (2000) *Arabidopsis in planta* transformation. Uses, mechanisms and prospects for transformation of other species. *Plant Physiology* 124, 1540–1547.

Bernacchi, D., Beck-Bunn, T., Emmatty, D., Eshed, Y., Inai, S., Lopez, J., Petiard, V., Sayama, H., Uhlig, J., Zamir, D. and Tanksley, S.D. (1998a) Advanced backcross QTL analysis of tomato: II. Evaluation of near-isogenic lines carrying single-donor introgressions for desirable wild QTL-alleles derived from *Lycopersicon hirsutum* and *L. pimpinellifolium. Theoretical and Applied Genetics* 97, 170–180.

Bernacchi, D., Beck-Bunn, T., Eshed, Y., Lopez, J., Petiard, V., Uhlig, J., Zamir, D. and Tanksley, S.D. (1998b) Advanced backcross QTL analysis in tomato. I. Identification of QTLs for traits of agronomic importance from *Lycopersicon hirsutum. Theoretical and Applied Genetics* 97, 381–397.

Bernardo, R. (1991) Retrospective index weights used in multiple trait selection in a maize breeding program. *Crop Science* 31, 1174–1179.

Bernardo, R. (1992) Relationship between single-cross performance and molecular marker heterozygosity. *Theoretical and Applied Genetics* 83, 628–634.

Bernardo, R. (1993) Estimation of coefficient of coancestry using molecular markers in maize. *Theoretical and Applied Genetics* 85, 1055–1062.

Bernardo, R. (1994) Prediction of maize single-cross performance using RFLPs and information from related hybrids. *Crop Science* 34, 20–25.

Bernardo, R. (1996) Best linear unbiased prediction of maize single-cross performance. *Crop Science* 36, 50–56.

Bernardo, R. (1999) Best linear unbiased predictor analysis. In: Coors, J.G. and Pandey, S. (eds) *Genetics and Exploitation of Heterosis in Crops.* ASA-CSSA-SSSA, Madison, Wisconsin, pp. 269–276.

Bernardo, R. (2001) What if we knew all the genes for a quantitative trait in hybrid crops? *Crop Science* 41, 1–4.

Bernardo, R. (2002) *Breeding for Quantitative Traits in Plants.* Stemma Press, Woodbury, Minnesota, 369 pp.

Bernardo, R. (2004) What proportion of declared QTL in plants are false? *Theoretical and Applied Genetics* 109, 419–424.

Bernardo, R. (2008) Molecular markers and selection for complex traits in plants: learning from the last 20 years. *Crop Science* 48, 1649–1664.

Bernardo, R. and Yu, J. (2007) Prospects for genomewide selection for quantitative traits in maize. *Crop Science* 47, 1082–1090.

Bernot, A. (2004) *Genome, Transcriptome and Protein Analysis.* John Wiley & Sons, Ltd, Chichester, UK.

Betrán, F.J., Ribaut, J.M., Beck, D. and Gonzalez de León, D. (2003) Genetic diversity, specific combining ability and heterosis in tropical maize under stress and nonstress environments. *Crop Science* 43, 797–806.

Bevan, M. (1984) Binary *Agrobacterium* vectors for plant transformation. *Nucleic Acids Research* 12, 8711–8721.

Bhave, S.V., Hombaker, C., Phang, T.L., Saba, L., Lapadat, R., Kechris, K., Gaydos, J., McGoldrick, D., Dolbey, A., Leach, S., Soriano, B., Ellington, A., Ellington, E., Jones, K., Mangion, J., Belknap, J.K., Williams, R.W., Hunter, L.E., Hoffman, P.L. and Tabakoff, B. (2007) The PhenoGen informatics website: tools for analyses of complex traits. *BMC Genetics* 8, 59.

Bhojwani, S.S. (ed.) (1990) *Plant Tissue Culture: Applications and Limitations.* Elsevier Science Publishers, The Netherlands.

Biber-Klemm, S. and Cottier, T. (2006) (eds) *Rights to Plant Genetic Resources and Traditional Knowledge: Basic Issues and Perspectives.* CAB International, Wallingford, UK, 448 pp.

Bidinger, F.R., Serraj, R., Rizvi, S.M.H., Howarth, C., Yadav, R.S. and Hash, C.T. (2005) Field evaluation of drought tolerance QTL effects on phenotype and adaptation in pearl millet (*Pennisetum glaucum* (L.) R. Br.) top cross hybrids. *Field Crops Research* 94, 14–32.

Bijlsma, R., Allard, R.W. and Kahler, A.L. (1986) Nonrandom mating in an open-pollinated maize population. *Genetics* 112, 669–680.

Bingham, P.M., Levis, R. and Rubin, G.M. (1981) Cloning of DNA sequences from the white locus of *Drosophila melanogaster* by a general and novel method. *Cell* 25, 693–704.

Bink, M.C.A.M. and Meuwissen, T. (2004) Fine mapping of quantitative trait loci using linkage disequilibrium in inbred plant populations. *Euphytica* 137, 95–99.

Birchler, J.A., Auger, D.L. and Riddle, N.C. (2003) In search of the molecular basis of heterosis. *The Plant Cell* 15, 2236–2239.

Birney, E., Thompson, J.D. and Gibson, T.J. (1996) PairWise and SearchWise: finding the optimal alignment in a simultaneous comparison of a protein profile against all DNA translation frames. *Nucleic Acids Research* 24, 2730–2739.

Biswas, S., Storey, J.D. and Akey, J.M. (2008) Mapping gene expression quantitative trait loci by singular value decomposition and independent component analysis. *BMC Bioinformatics* 9, 244.

Bizily, S.P., Rugh, C.L. and Meagher, R.B. (2000) Phytodetoxification of hazardous organomercurials by genetically engineered plants. *Nature Biotechnology* 18, 213–217.

Blakeslee, A.F. and Avery, A.H. (1937) Methods of inducing chromosome doubling in plants. *Journal of Heredity* 28, 393–411.

Blanc, G., Charcosset, A., Mangin, B., Gallais, A. and Moreau, L. (2006) Connected populations for detecting quantitative trait loci and testing for epistasis: an application in maize. *Theoretical and Applied Genetics* 113, 206–224.

Blanchard, J.L. (2004) Bioinformatics and systems biology, rapidly evolving tools for interpreting plant response to global change. *Field Crops Research* 90, 117–131.

Blanco, A., Lotti, C., Simeone, R., Signorile, A., De-Santis, V., Pasqualone, A., Troccoli, A. and Di-Fonzo, N. (2001) Detection of quantitative trait loci for grain yield and yield components across environments in durum wheat. *Cereal Research Communications* 29, 237–244.

Bligh, H.F.J., Till, R.I. and Jones, C.A. (1995) A microsatellite sequence closely linked to the waxy gene of *Oryza sativa*. *Euphytica* 86, 83–85.

Blow, N. (2008) Mass spectrometry and proteomics: hitting the mark. *Nature Methods* 5, 741–747.

Bochner, B.R. (1989) Sleuthing out bacterial identifies. *Nature* 339, 157–158.

Bochner, B.R. (2003) New technologies to assess genotype–phenotype relationships. *Nature Reviews Genetics* 4, 309–314.

Boer, M.P., ter Braak, C.J.F and Jansen, R.C. (2002) A penalized likelihood method for mapping epistatic quantitative trait loci with one-dimensional genome searches. *Genetics* 162, 951–960.

Bogyo, T.P., Lance, R.C.M., Chevalier, P. and Nilan, P.A. (1988) Genetic models for quantitatively inherited endosperm characters. *Heredity* 60, 61–67.

Bohanec, B., Jakse, M. and Havey, M.J. (2003) Genetic analysis of gynogenetic haploid production in onion. *Journal of American Horticulture Science* 128, 571–574.

Bollen, K.A. (1989) *Structural Equations with Latent Variables*. John Wiley & Sons, New York.

Bonnet, D.G., Rebetzke, G.J. and Spielmeyer, W. (2005) Strategies for efficient implementation of molecular markers in wheat breeding. *Molecular Breeding* 15, 75–85.

Boppenmaier, J., Melchinger, A.E., Seitz, G., Geiger, H.H. and Herrmann, R.G. (1993) Genetic diversity for RFLPs in European maize inbreds. III. Performance of crosses within versus between heterotic groups for grain traits. *Plant Breeding* 111, 217–226.

Borevitz, J.O. and Ecker, J.R. (2004) Plan genomics: the third wave. *Annual Review of Genomics and Human Genetics* 5, 443–477.

Borevitz, J.O., Maloof, J.N., Lutes, J., Dabi, T., Redfern, J.L., Trainer, G.T., Werner, J.D., Asami, T., Berry, C.C., Weigel, D. and Chory, J. (2002) Quantitative trait loci controlling light and hormone response in two accessions of *Arabidopsis thaliana*. *Genetics* 160, 683–696.

Borevitz, J.O., Liang, D., Plouffe, D., Chang, H.S., Zhu, T., Weigel, D., Berry, C.C., Winzeler, E. and Chory, J. (2003) Large-scale identification of single-feature polymorphisms in complex genomes. *Genome Research* 13, 513–523.

Borevitz, J.O., Hazen, S.P., Michael, T.P., Morris, G.P., Baxter, I.R., Hu, T.T., Chen, H., Werner, J.D., Nordborg, M., Salt, D.E., Kay, S.A., Chory, J., Weigel, D., Jones, J.D.G. and Ecker, J.R. (2007)

Genome-wide patterns of single-feature polymorphism in *Arabidopsis thaliana. Proceedings of the National Academy of Sciences of the United States of America* 104, 12057–12062.
Borlaug, N.E. (1972) *The Green Revolution, Peace and Humanity.* CIMMYT Reprint and Translation Series No. 3, International Maize and Wheat Improvement Center, Mexico DF.
Borlaug, N.E. (2000) Ending world hunger. The promise of biotechnology and the threat of antiscience zealotry. *Plant Physiology* 124, 487–490.
Borlaug, N.E. (2001) Feeding the world in the 21st century: the role of agricultural science and technology. Speech given at Tuskegee University, April 2001. Available at: http://www.agbioworld.org/biotech-info/topics/borlaug/borlaugspeech.html (accessed 17 November 2009).
Borrell, A.K., Hammer, G.L. and van Oosterom, E. (2001) Staygreen: a consequence of the balance between supply and demand for nitrogen during grain filling? *Annals of Applied Biology* 138, 91–95.
Botstein, D.R., White, R.L., Skolnick, M. and Davis, R.W. (1980) Construction of a genetic linkage map in man using restriction fragment length polymorphisms. *American Journal of Human Genetics* 32, 314–331.
Boumedine, K.S. and Rodolakis, A. (1998) AFLP allows the identification of genomic markers of ruminant *Chlamydia psittaci* strains useful for typing and epidemiological studies. *Research in Microbiology* 149, 735–744.
Bourgault, R., Zulak, K.G. and Facchini, P.J. (2005) Applications of genomics in plant biology. In: Sensen, C.W. (ed.) *Handbook of Genome Research, Genomics, Proteomics, Metabolomics, Bioinformatics, Ethical and Legal Issues.* Wiley-VCH Verlag GmbH & Co. KGaA, Weinheim, Germany, pp. 59–80.
Bowers, J.E., Abbey, C., Anderson, S., Chang, C., Draye, X., Hoppe, A.H., Jessup, R., Lemke, C., Lennington, J., Li, Z., Lin, Y.-R., Liub, S.-C., Luo, L., Marler, B.S., Ming, R., Mitchell, S.E., Qiang, D., Reischmann, K., Schulze, S.R., Skinner, D.N., Wang, Y.-W., Kresovich, S., Schertz, K.F. and Paterson, A.H. (2003a) A high-density genetic recombination map of sequence-tagged site for *Sorghum*, as a framework for comparative structural and evolutionary genomics of tropical grains and grasses. *Genetics* 165, 367–386.
Bowers, J.E., Chapman, B.A., Rong, J. and Paterson, A.H. (2003b) Unravelling angiosperm genome evolution by phylogenetic analysis of chromosomal duplication events. *Nature* 422, 433–438.
Bowman, J.G.P., Blake, T.K., Surber, L.M.M., Habernicht, D.K. and Bockelman, H. (2001) Feed-quality variation in the barley core collection of the USDA National Small Grains Collection. *Theoretical and Applied Genetics* 41, 863–870.
Boyd, M.R. (1996) The position of intellectual property rights in drug discovery and development from natural products. *Journal of Ethnopharmacology* 51, 17–27.
Boyer, J.S. (1982) Plant productivity and environment. *Science* 218, 443–448.
Bracha-Drori, K., Shichrur, K., Katz, A., Oliva, M., Angelovici, R., Yalovsky, S. and Ohad, N. (2004) Detection of protein–protein interactions in plants using bimolecular fluorescence complementation. *The Plant Journal* 40, 419–427.
Bradshaw, A.D. (1965) Evolutionary significance of phenotypic plasticity in plants. *Advances in Genetics* 13, 115–155.
Bradshaw, H.D., Jr and Settler, R.F. (1995) Molecular genetics of growth and development in populus. IV. Mapping QTLs with large effects on growth, form and phenology traits in a forest tree. *Genetics* 139, 963–973.
Brancourt-Hulmel, M. (1999) Crop diagnosis and probe genotypes for interpreting genotype environment interaction in winter wheat trials. *Theoretical and Applied Genetics* 99, 1018–1030.
Branton, D., Deamer, D.W., Marziali, A., Bayley, H., Benner, S.A., Butler, T., Ventra, M.D., Garaj, S., Hibbs, A., Huang, X., Jovanovich, S.B., Krstic, P.S., Lindsay, S., Ling, X.S., Mastrangelo, C.H., Meller, A., Oliver, J.S., Pershin, Y.V., Ramsey, J.M., Riehn, R., Soni, G.V., Tabard-Cossa, V., Wanunu, M., Wiggin, M. and Schloss, J.A. (2008) The potential and challenges of nanopore sequencing. *Nature Biotechnology* 26, 1146–1153.
Brazma, A., Hingamp, P., Quackenbush, J., Sherlock, G., Spellman, P., Stoeckert, C., Aach, J., Ansorge, W., Ball, C.A., Causton, H.C., Gaasterland, T., Glenisson, P., Holstege, F.C., Kim, I.F., Markowitz, V., Matese, J.C., Parkinson, H., Robinson, A., Sarkans, U., Schulze-Kremer, S., Stewart, J., Taylor, R., Vilo, J. and Vingron, M. (2001) Minimum information about a microarray experiment (MIAME) – toward standards for microarray data. *Nature Genetics* 29, 365–371.
Breitling, R., Pitt, A.R. and Barrett, M.P. (2006) Precision mapping of the metabolome. *Trends in Biotechnology* 24, 543–548.
Brem, R.B. and Kruglyak, L. (2005) The landscape of genetic complexity across 5,700 gene expression traits in yeast. *Proceedings of the National Academy of Sciences of the United States of America* 102, 1572–1577.

Brem, R.B., Yvert, G., Clinton, R. and Kruglyak, L. (2002) Genetic dissection of transcriptional regulation in budding yeast. *Science* 296, 752–755.

Brenner, S., Johnson, M., Bridgham, J., Golda, G., Lloyd, D.H., Johnson, D., Luo, S., McCurdy, S., Foy, M., Ewan, M., Roth, R., George, D., Eletr, S., Albrecht, G., Vermaas, E., Williams, S.R., Moon, K., Burcham, T., Pallas, M., DuBridge, R.B., Kirchner, J., Fearon, K., Mao J.-I. and Corcoran, K. (2000) Gene expression analysis by massively parallel signature sequencing (MPSS) on microbead arrays. *Nature Biotechnology* 18, 630–634.

Breseghello, F. and Sorrells, M.E. (2006a) Association mapping of kernel size and milling quality in wheat (*Triticum aestivum* L.) cultivars. *Genetics* 172, 1165–1177.

Breseghello, F. and Sorrells, M.E. (2006b) Association analysis as a strategy for improvement of quantitative traits in plants. *Crop Science* 46, 1323–1330.

Bretting, P. and Duvick, D. (1997) Dynamic conservation of plant genetic resources. *Advances in Agronomy* 61, 1–51.

Bretting, P.K. and Goodman, M.M. (1989) Genetic variation in crop plants and management of germplasm collections. In: Stalker, H.T. and Chapman, C. (eds) *Scientific Management of Germplasm: Charaterization, Evaluation and Enhancement.* International Board for Plant Genetic Resources (IBPGR) Training Courses: Lecture Series 2. Department of Crop Science, North Carolina State University, Raleigh, North Carolina and IBPGR, Rome, pp. 41–54.

Bretting, P.K. and Widrlechner, M.P. (1995) Genetic markers and plant genetic resource management. *Plant Breeding Reviews* 13, 11–86.

Brick, M.A., Byrne, P.F., Schwartz, H.F., Ogg, J.B., Otto, K., Fall, A.L. and Gilbert, J. (2006) Reaction to three races of *Fusarium* wilt in the *Phaseolus vulgaris* core collection. *Crop Science* 46, 1245–1252.

Briggs, R.N. and Knowles, P.F. (1967) *Introduction to Plant Breeding.* Reinhold Books, New York.

Broman, K.W. (1997) Identifying quantitative trait loci in experimental crosses. PhD thesis, Department of Statistics, University of California, Berkeley.

Broman, K.W. (2005) The genomes of recombinant inbred lines. *Genetics* 169, 1133–1146.

Broman, K.W., Churchill, G.A., Yandell, B.S. and Zeng, Z.B. (2003a) Statistical methods for mapping quantitative trait loci in experimental crosses. Available at: http://www.stat.wisc.edu/~yandell/statgen (accessed 17 November 2009).

Broman, K.W., Wu, H., Sen, S. and Churchill, G.A. (2003b) R/qtl: QTL mapping in experimental crosses. *Bioinformatics* 19, 889–890.

Brondani, C., Rangel, N., Brondani, V. and Ferreira, E. (2002) QTL mapping and introgression of yield-related traits from *Oryza glumaepatula* to cultivated rice (*Oryza sativa*) using microsatellite markers. *Theoretical and Applied Genetics* 104, 1192–1203.

Brookes, G. and Barfoot, P. (2008) *GM Crops: Global Socio-economic and Environmental Impacts 1996–2006.* PG Economics, Dorchester, UK.

Broothaerts, W., Mitchell, H.J., Weir, B., Kaines, S., Smith, L.M.A., Yang, W., Mayer, J.E., Roa-Rodriguez, C. and Jefferson, R.A. (2005) Gene transfer to plants by diverse species of bacteria. *Nature* 433, 629–633.

Brown, A.D.H. (1989a) The case for core collections. In: Brown, A.D.H., Frankel, O.H., Marshall, R.D. and Williams, J.T. (eds) *The Use of Plant Genetic Resources.* Cambridge University Press, Cambridge, UK, pp. 136–156.

Brown, A.D.H. (1989b) Core collection: a practical approach to genetic resources management. *Genome* 31, 818–824.

Brown, A.H.D. and Brubaker, C.L. (2002) Indicators for sustainable management of plant genetic resources: how well are we doing? In: Engels, J.M.M., Ramanatha Rao, V., Brown, A.H.D. and Jackson, M.T. (eds) *Managing Plant Genetic Diversity.* International Plant Genetics Resources Institute (IPGRI), Rome, pp. 249–262.

Brown, A.H.D. and Weir, B.S. (1983) Measuring genetic variability in plant populations. In: Tanksley, S.D. and Orton, T.J. (eds) *Isozymes in Plant Genetics and Breeding*, Vol. 1A. *Developments in Plant Genetics and Breeding* 1. Elsevier, Amsterdam, pp. 219–240.

Brown, G.G., Formanova, N., Jin, H., Wargachuk, R., Dondy, C., Patil, P., Laforest, M., Zhang, J., Cheung, W.Y. and Landry, B.S. (2003) The radish *Rfo* restorer gene of Ogura cytoplasmic mole sterility encodes a protein with multiple pentatricopeptide repeat. *The Plant Journal* 35, 262–272.

Brown, P.J., Rooney, W.L., Franks, C. and Kresovich, S. (2008) Efficient mapping of plant height quantitative trait loci in a sorghum association population with introgressed dwarfing genes. *Genetics* 180, 629–637.

Brown, S.D. and Peters, J. (1996) Combining mutagenesis and genomics in the mouse – closing the phenotype gap. *Trends in Genetics* 12, 433–435.

Brown, S.M. and Kresovich, S. (1996) Molecular characterization for plant genetic resources conservation. In: Paterson, A.H. (ed.) *Genome Mapping in Plants.* R.G. Landes Co., Austin, Texas, pp. 85–93.

Brown, T.A. (2002) *Genomics*, 2nd edn. Wiley-Liss, Wilmington, Delaware, pp. 125–159.

Brownstein, M.J., Carpten, J.D. and Smith, J.R. (1996) Modulation of non-templated nucleotide addition by *Taq* DNA polymerase: primer modifications that facilitate genotyping. *Biotechniques* 20, 1004–1006.

Brueggeman, R., Rostoks, N., Kudrna, D., Kilian, A., Han, F., Chen, J., Druka, A., Steffenson, B. and Kleinhofs, A. (2002) The barley stem rust-resistance gene *Rpg1* is a novel disease-resistance gene with homology to receptor kinases. *Proceedings of the National Academy of Sciences of the United States of America* 99, 9328–9333.

Brummer, E.C. (1999) Capturing heterosis in forage crop cultivar development. *Crop Science* 39, 943–954.

Brummer, E.C. (2006) Breeding for cropping systems. In: Lamkey, K.R. and Lee, M. (eds) *Plant Breeding: the Arnel R. Hallauer International Symposium.* Blackwell Publishing, Oxford, UK, pp. 97–106.

Brunner, S., Keller, B. and Feuillet, C. (2003) A large rearrangement involving genes and low copy DNA interrupts the micro-collinearity between rice and barley at the *Rph7* locus. *Genetics* 164, 673–683.

Bruskiewich, R., Senger, M., Davenport, G., Ruiz, M., Rouard, M., Hazekamp, T., Takeya, M., Doi, K., Satoh, K., Costa, M., Simon, R., Balaji, J., Akintunde, A., Mauleon, R., Wanchana, S., Shah, T., Anacleto, M., Portugal, A., Ulat, V.J., Thongjuea, S., Braak, K., Ritter, S., Dereeper, A., Skofic, M., Rojas, E., Martins, N., Pappas, G., Alamban, R., Almodiel, R., Barboza, L.H., Detras, J., Manansala, K., Mendoza, M.J., Morales, J., Peralta, B., Valerio, R., Zhang, Y., Gregorio, S., Hermocilla, J., Echavez, M., Yap, J.M., Farmer, A., Schiltz, G., Lee, J., Casstevens, T., Jaiswal, P., Meintjes, A., Wilkinson, M., Good, B., Wagner, J., Morris, J., Marshall, D., Collins, A., Kikuchi, S., Metz, T., McLaren, G. and van Hintum, T. (2008) The Generation Challenge Programme platform: semantic standards and workbench for crop science. *International Journal of Plant Genomics*, Article ID 369601, 6 pages. Available at: http://www.hindawi.com/journals/ijpg/2008/369601.html (accessed 17 November 2009).

Buchanan, B., Gruissem, W. and Jones, R.L. (eds) (2002) *Biochemistry and Molecular Biology of Plants.* John Wiley & Sons Inc., Chichester, UK.

Buckingham, S.D. (2008) Scientific software: seeing the SNPs between us. *Nature Methods* 5, 903–908.

Buckler, E.S. IV and Thornsberry, J.M. (2002) Plant molecular diversity and applications to genomics. *Current Opinion in Plant Biology* 5, 107–111.

Buckler, E.S., Holland, J.B., Bradbury, P.J., Acharya, C.B., Brown, P.J., Browne, C., Ersoz, E., Flint-Garcia, S., Garcia, A., Glaubitz, J.C., Goodman, M.M., Harjes, C., Guill, K., Kroon, D.E., Larsson, S., Lepak, N.K., Huihui Li, H., Mitchell, S.E., Pressoir, G., Pfeiffer, J.A., Oropeza Rosas, M., Rocheford, T.R., Cinta Romay, M., Romero, S., Salvo, S., Sanchez-Villeda, H., Sofia da Silva, H., Qi Sun, Q., Tian, F., Upadyayula, N., Ware, D.,Yates, H., Yu, J., Zhang, Z., Kresovich, S. and McMullen, M.D. (2009) The genetic architecture of maize flowering time. *Science* 325, 714–718.

Burgueño, J., Crossa, J., Cornelius, P.L. and Yang, R.-C. (2008) Using factor analytic models for joining environment and genotypes without crossover genotype × environment interaction. *Crop Science* 48, 1291–1305.

Burns, J., Fraser, P.D. and Bramley, P.M. (2003) Identification and quantification of carotenoids, tocopherols and chlotophylls in commonly consumed fruits and vegetables. *Phytochemistry* 62, 939–947.

Burr, B. and Burr, F.A. (1991) Recombinant inbreds for molecular mapping in maize: theoretical and practical considerations. *Trends in Genetics* 7, 55–60.

Burr, B., Burr, F.A., Thompson, K.H., Albertson, M.C. and Stuber, C.W. (1988) Gene mapping with recombinant inbreds in maize. *Genetics* 118, 519–526.

Burton, G.W. (1981) Meeting human needs through plant breeding: past progress and prospects for the future. In: Frey, K.J. (ed.) *Plant Breeding II.* Iowa State University Press, Ames, Iowa, pp. 433–466.

Busch, W. and Lohmann, J.U. (2007) Profiling a plant: expression analysis in *Arabidopsis. Current Opinion in Plant Biology* 10, 136–141.

Büschhes, R., Hollricher, K., Ranstruga, R., Simons, G., Wolter, M., Frijters, A., van Daelen, R., van der Lee, T., Diergaarde, P., Groenendijk, J., Töpsch, S., Vos, P., Salamini, F. and Schulze-Lefert, P. (1997) The barley *Mio* gene: a novel control element of plant pathogen resistance. *Cell* 88, 695–705.

Busso, C.S., Liu, C.J., Hash, C.T., Witcombe, J.R., Devos, K.M., deWet, J.M.J. and Gale, M.D. (1995) Analysis of recombination rate in female and male gametogenesis in pearl millet (*Pennisetum glaucum*) using RFLP markers. *Theoretical and Applied Genetics* 90, 242–246.

Bustamam, M., Tabien, R.E., Suwarmo, A., Abalos, M.C., Kadir, T.S., Ona, I., Bernardo, M., VeraCruz, C.M. and Leung, H. (2002) Asian rice biotechnology network: improving popular cultivars through marker-assisted backcrossing by the NARES. Abstract of International Rice Congress, 16–22 September 2002, Beijing China. Available at: http://www.irri.org/irc2002/index.htm (last accessed 31 December 2007).

Butlin, R.K. and Tregenta, T. (1998) Levels of genetic polymorphism: marker loci versus quantitative traits. *Philosophical Transactions of the Royal Society of London B* 353, 1–12.

Byrum, J. and Reiter, R. (1998) A method for identifying genetic marker loci associated with trait loci. Patent EP 0972076.

Caetano-Anollés, G., Bassam, B.J. and Gresshoff, P.M. (1991) DNA amplification fingerprinting using very short arbitrary oligonucleotide primers. *Bio/Technology* 9, 553–557.

Caicedo, A.L. and Purugganan, M.D. (2005) Comparative plant genomics. Frontiers and prospects. *Plant Physiology* 138, 545–547.

Caliñski, T., Kaczmarek, Z., Krajewski, P., Frova, C. and Sari-Gorla, M. (2000) A multivariate approach to the problem of QTL localization. *Heredity* 84, 303–310.

Campbell, B.T., Baezinger, P.S., Gill, K.S., Eskridge, K.M., Budak, H., Erayman, M., Dweikat, I. and Yen, Y. (2003) Identification of QTLs and environmental interactions associated with agronomic traits on chromosome 3A of wheat. *Crop Science* 43, 1493–1505.

Campbell, B.T., Baenziger, P.S., Eskridge, K.M., Budak, H., Streck, N.A., Weiss, A., Gill, K.S. and Erayman, M. (2004) Using environmental covariates to explain genotype × environment and QTL × environment interactions for agronomic traits on chromosome 3A of wheat. *Crop Science* 44, 620–627.

Campbell, M.A., Zhu, W., Jiang, N., Lin, H., Ouyang, S., Childs, K.L., Haas, B.J., Hamilton, J.P. and Buell, C.R. (2007) Identification and characterization of lineage-specific genes within the Poaceae. *Plant Physiology* 145, 1311–1322.

Candela, H. and Hake, S. (2008) The art and design of genetic screens: maize. *Nature Reviews Genetics* 9, 192–203.

Cardon, L.R. and Bell, J.I. (2001) Association study designs for complex diseases. *Nature Reviews Genetics* 2, 91–98.

Carlborg, Ö. and Andersson, L. (2002) Use of randomization testing to detect multiple epistatic QTLs. *Genetical Research* 79, 175–184.

Carlborg, Ö. and Haley, C.S. (2004) Epistasis: too often neglected in complex trait studies? *Nature Reviews Genetics* 5, 618–625.

Carlborg, Ö., Andersson, L. and Kinghorn, B. (2000) The use of a genetic algorithm for simultaneous mapping of multiple interacting quantitative trait loci. *Genetics* 155, 2003–2010.

Carlborg, Ö., Brockmann, G.A. and Haley, C.S. (2005) Simultaneous mapping of epistatic QTL in DU6i × DBA/2 mice. *Mammalian Genome* 16, 481–494.

Carninci, P. and Hayashizaki, Y. (1999) High-efficiency full-length cDNA cloning. *Methods in Enzymology* 303, 19–44.

Carpenter, A.E. and Sabatini, D.M. (2004) Systematic genome-wide screens of gene function. *Nature Reviews Genetics* 5, 11–22.

Cartwright, D.A., Troggio, M., Velasco, R. and Gutin, A. (2007) Genetic mapping in the presence of genotyping errors. *Genetics* 176, 2521–2537.

Casali, V.W.D. and Tigchelaar, E.C. (1975) Computer simulation studies comparing pedigree, bulk and single seed descent selection in self-pollinated populations. *Journal of American Society of Horticulture Science* 100, 364–367.

Casasoli, M., Derory, J., Morera-Dutrey, C., Brendel, O., Porth, I., Guehl, J.-M., Villani, F. and Kremer, A. (2006) Comparison of quantitative trait loci for adaptive traits between oak and chestnut based on an expressed sequence tag consensus map. *Genetics* 172, 533–546.

Caskey, T. and Edwards, A. (1992) DNA typing with short tandem repeat polymorphisms and identification of polymorphic short tandem repeats. Patent EP 0639228.

Castle, W.E. (1921) On a method of estimating the number of genetic factors concerned in cases of blending inheritance. *Science* 54, 93–96.

Causier, B., Graham, J. and Davis, B. (2005) Large-scale yeast two-hybrid analysis. In: Leister, D. (ed.) *Plant Functional Genomics*. Food Products Press, New York, pp. 119–135.

Causse, M.A., Fulton, T.M., Cho, Y.G., Ahn, S.N., Chunwongse, J., Wu, K., Xiao, J., Yu, Z., Ronald, P.C., Harrington, S.E., Second, G., McCouch, S.R. and Tanksley, S.D. (1994) Saturated molecular map of the rice genome based on an interspecific backcross population. *Genetics* 138, 1251–1274.

Cavalli-Sforza, L.L. and Edwards, A.W.F. (1967) Phylogenetic analysis: models and estimation procedures. *American Journal of Human Genetics* 19, 233–257.

Ceccarelli, S. and Grando, S. (2007) Decentralized participatory plant breeding: an example of demand driven research. *Euphytica* 155, 349–360.

Ceccarelli, S., Grando, S., Amri, A., Asaad, F.A., Benbelkacem, A., Harrabi, M., Maatougui, M., Mekni, M.S., Mimoun, H., El-Einen, R.A., El-Felah, M., El-Sayed, A.F., Shreidi, A.S. and Yahyaoui, A. (2001) Decentralized and participatory plant breeding for marginal environments. In: Cooper, H.D., Spillane, C. and Hodgkins, T. (eds) *Broadening the Genetic Bases of Crop Production.* CAB International. Wallingford, UK, pp. 115–135.

Cerna, F.J., Cianzio, S.R., Rafalski, A., Tingey, S. and Dyer, D. (1997) Relationship between seed yield heterosis and molecular marker heterozygosity in soybean. *Theoretical and Applied Genetics* 95, 460–467.

CFIA/NFS (Canadian Food Inspection Agency/National Forum on Seed) (2005) Seminar on the use of molecular techniques for plant variety protection. Available at: http://www.inspection.gc.ca/english/plaveg/pbrpov/molece.shtml (last accessed 30 June 2008).

Chagné, D., Batley, J., Edwards, D. and Forster, J.W. (2007) Single nucleotide polymorphisms genotyping in plants. In: Oraguzie, N.C., Rikkerink, E.H.A., Gardiner, S.E. and De Silva, H.N. (eds) *Association Mapping in Plants.* Springer, Berlin, pp.77–94.

Chahal, G.S. and Gosal, S.S. (2002) *Principles and Procedures of Plant Breeding, Biotechnological and Conventional Approaches.* Alpha Science International Ltd, Pangbourne, UK.

Chaïb, J., Lecomte, L., Buret, M. and Causse, M. (2006) Stability over genetic backgrounds, generations and years of quantitative trait loci (QTLs) for organoleptic quality in tomato. *Theoretical and Applied Genetics* 112, 934–944.

Chan, E.K.F., Rowe, H.C. and Kliebenstein, D.J. (2009) Understanding the evolution of defense metabolites in *Arabidopsis thaliana* using genome-wide association mapping. *Genetics* (in press).

Chan, H.P. (2006) International patent behaviour of nine major agricultural biotechnology firms. *AgBioForum* 9, 59–68.

Chandler, P.M., Marrion-Poll, A., Ellis, M. and Gubler, F. (2002) Mutants at the *Slender1* locus of barley cv Himalaya. Molecular and physical characterization. *Plant Physiology* 129, 181–190.

Chandler, S. and Dunwell, J.M. (2008) Gene flow, risk assessment and the environmental release of transgenic plants. *Critical Reviews in Plant Sciences* 27, 25–49.

Chapman, S.C., Hammer, G.L., Podlich, D.W. and Cooper, M. (2002) Linking biophysical and genetic models to integrate physiology, molecular biology and plant breeding. In: Kang, M.S. (ed.) *Quantitative Genetics, Genomics and Plant Breeding.* CAB Internationl, Wallingford, UK, pp. 167–187.

Chapman, S., Cooper, M., Podlich, D.W. and Hammer, G.L. (2003) Evaluating plant breeding strategies by simulating gene action and dryland environment effects. *Agronomy Journal* 95, 99–113.

Charcosset, A. and Essioux, L. (1994) The effect of population structure on the relationship between heterosis and heterozygosity at marker loci. *Theoretical and Applied Genetics* 89, 336–343.

Charcosset, A. and Gallais, A. (1996) Estimation of the contribution of quantitative trait loci (QTL) to the variance of a quantitative trait by means of genetic markers. *Theoretical and Applied Genetics* 93, 1193–1201.

Charcosset, A., Lefort-Buson, M. and Gallais, A. (1991) Relationship between heterosis and heterozygosity at marker loci: a theoretical computation. *Theoretical and Applied Genetics* 81, 571–575.

Charcosset, A., Causse, M., Moreau, L. and Gallais, A. (1994) Investigation into the effect of genetic background on QTL expression using three recombinant inbred lines (RIL) populations. In: van Ooijen, J.W. and Jansen, J. (eds) *Biometrics in Plant Breeding: Applications of Molecular Markers.* Centre for Plant Breeding and Reproduction Research, Wageningen, The Netherlands, pp. 75–84.

Charcosset, A., Mangin, B., Moreau, L., Combes, L., Jourjon, M.F. and Gallais, A. (2000) Heterosis in maize investigated using connected RIL populations. In: *Quantitative Genetics and Breeding Methods: the Way Ahead.* Institut National de la Recherche Agronomique (INRA), Paris, pp. 89–98.

Chardon, F., Virlon, B., Moreau, L., Falque, M., Joets, J., Decousset, L., Murigneux, A. and Charcosset, A. (2004) Genetic architecture of flowering time in maize as inferred from quantitative trait loci meta-analysis and synteny conservation with the rice genome. *Genetics* 168, 2169–2185.

Charmet, G., Robert, N., Perretant, M.R., Gay, G., Sourdille, P., Groos, C., Bernard, S. and Bernard, M. (1999) Marker-assisted recurrent selection for cumulating additive and interactive QTLs in recombinant inbred lines. *Theoretical and Applied Genetics* 99, 1143–1148.

Chase, S.S. (1969) Monoploids and monoploid derivatives of maize (*Zea mays* L.). *The Botanical Review* 35, 117–167.

Chavarriaga-Aguirre, P., Maya, M.M., Tohme, J., Duque, M.C., Iglesias, C., Bonierbale, M.W., Kresovich, C. and Kochert, G. (1999) Using microsatellites, isozymes and AFLPs to evaluate genetic diversity and

redundancy in the cassava core collection and to assess the usefulness of DNA-based markers to maintain germplasm collections. *Molecular Breeding* 5, 263–273.
Chellappan, P., Masona, M.V., Vanitharani, R., Taylor, N.J. and Fauquet, C.M. (2004) Broad spectrum resistance to ssDNA viruses associated with transgene-induced gene silencing in cassava. *Plant Molecular Biology* 56, 601–611.
Chen, H., Wang, S., Xing, Y., Xu, C., Hayes, P.M. and Zhang, Q. (2003) Comparative analyses of genomic locations and race specificities of loci for quantitative resistance to *Pyricularia grisea* in rice and barley. *Proceedings of the National Academy of Sciences of the United States of America* 100, 2544–2549.
Chen, J., Griffey, C.A., Chappell, M., Shaw, J. and Pridgen, T. (1999) Haploid production in twelve wheat F_1 by wheat × maize hybridization method. In: *Proceedings of National Fusarium Head Blight Forum*, December 1999, Sioux Falls, South Dakota, pp.147–149.
Chen, J.Q., Zhou, H.M., Chen, J., and Wang, X.C. (2006) A GATEWAY-based platform for multiple plant transformation. *Plant Molecular Biology* 62, 927–936.
Chen, L. and Storey, J.D. (2006) Relaxed significance criteria for linkage analysis. *Genetics* 173, 2371–2381.
Chen, M., Presting, G., Barbazuk, W.B., Goicoechea, J.L., Blackmon, B., Fang, G., Kim, H., Frisch, D., Yu, Y., Sun, S., Higingbottom, S., Phimphilai, J., Phimphilai, D., Thurmond, S., Gaudette, B., Li, P., Liu, J., Hatfield, J., Main, D., Farrar, K., Henderson, C., Barnett, L., Costa, R., Williams, B., Walser, S., Atkins, M., Hall, C., Budiman, M.A., Tomkins, J.P., Luo, M., Bancroft, I., Salse, J., Regad, F., Mohapatra, T., Singh, N.K., Tyagi, A.K., Soderlund, C., Dean, R.A. and Wing, R.A. (2002) An integrated physical and genetic map of the rice genome. *The Plant Cell* 14, 537–545.
Chen, S., Lin, X.H., Xu, C.G. and Zhang, Q. (2000) Improvement of bacterial blight resistance of 'Minghui 63', an elite restorer line of hybrid rice, by molecular marker-assisted selection. *Crop Science* 40, 239–244.
Chen, T.M., Lu, C.C. and Li, W.H. (2005) Prediction of splice sites with dependency graphs and their expanded Bayesian networks. *Bioinformatics* 21, 471–482.
Chen, X., Temnykh, S., Xu, Y., Cho, Y.G. and McCouch, S.R. (1997) Development of microsatellite framework map providing genome-wide coverage in rice (*Oryza sativa* L.). *Theoretical and Applied Genetics* 95, 553–567.
Chen, Y., Lu, C., He, P., Shen, L., Xu, J., Xu, Y. and Zhu, L. (1997) Gametic selection in a doubled haploid population derived from anther culture of *indica/japonica* cross of rice. *Acta Genetica Sinica* 24, 322–329.
Cheng, M., Fry, J.E., Pang, S., Zhou, H., Hironaka, C., Duncan, D.R., Conner, T.W. and Wan, Y. (1997) Genetic transformation of wheat mediated by *Agrobacterium tumefaciens*. *Plant Physiology* 115, 971–980.
Cheng, M., Lowe, B.A., Spencer, T.M., Ye, X. and Armstrong, C.L. (2004) Factors influencing *Agrobacterium*-mediated transformation of monocotyledonous species. *In Vitro Cellular and Development Biology – Plant* 40, 31–45.
Chiarrolla, C. (2006) Commodifying agricultural biodiversity and development-related issues. *The Journal of World Intellectual Property* 9 (1), 25–60.
Chin, H.E. and Roberts, E.H. (eds) (1980) *Recalcitrant Crop Seeds*. Tropical Press Sdn. Bhd., Kuala Lumpur, Malaysia.
Cho, Y.G., Ishii, T., Temnykh, S., Chen, X., Lipovich, L., McCouch, S.R., Park, W.D., Ayres, N. and Cartinhour, S. (2000) Diversity of microsatellites derived from genomic libraries and GeneBank sequences in rice. *Theoretical and Applied Genetics* 100, 713–722.
Choisne, N., Samain, S., Demange, N., Orjeda, G., Michelet, L., Pelletier, E., Salanoubat, M., Weissenbach, J. and Quetier, F. (2007) The sequencing of plant nuclear genomes. In: Morot-Gaudry, J.F., Lea, P. and Briat, J.F. (eds) *Functional Plant Genomics*. Science Publishers, Enfield, New Hampshire, pp. 23–51.
Choo, T.M., Reinbergs, E. and Park, S.J. (1982) Comparison of frequency distribution of doubled haploid and single seed descent lines in barley. *Theoretical and Applied Genetics* 61, 215–218.
Choo, T.M., Reinbergs, E. and Kasha, K.J. (1985) Use of haploids in breeding barley. *Plant Breeding Reviews* 3, 219–252.
Christensen, A.H., Sharrock, R.A. and Quail, P.H. (1992) Maize polyubiquitin genes: structure, thermal perturbation of expression and transcript splicing and promoter activity following transfer to protoplasts by electroporation. *Plant Molecular Biology* 18, 675–689.
Christiansen, M.J., Anderson, S.B. and Ortiz, R. (2002) Diversity changes in an intensively bred wheat germplasm during the 20th century. *Molecular Breeding* 9, 1–11.

Christou, P. (1996) Transformation technology. *Trends in Plant Science* 1, 423–431.

Christou, P. and Swain, W.F. (1990) Cotransformation frequencies of foreign genes in soybean cell cultures. *Theoretical and Applied Genetics* 79, 337–341.

Chung, S.M., Frankman, E.L. and Tzfira, T. (2005) A versatile vector system for multiple gene expression in plants. *Trends in Plant Science* 10, 357–361.

Chung, S.-M., Vaidya, M. and Tzfira, T. (2006) *Agrobacterium* is not alone: gene transfer to plants by viruses and other bacteria. *Trends in Plant Science* 11, 1–4.

Churchill, G.A. and Doerge, R.W. (1994) Empirical threshold values for quantitative trait mapping. *Genetics* 138, 963–971.

Clark, R.L., Shands, H.L., Bretting, P.K. and Eberhart, S.A. (1997) Germplasm regeneration: developments in population genetics and their implications. *Crop Science* 37, 1–6.

Clark, R.M., Schweikert, G., Toomajian, C., Ossowski, S., Zeller, G., Shinn, P., Warthmann, N., Hu, T.T., Fu, G., Hinds, D.A., Chen, H., Frazer, K.A., Huson, D.H., Schölkopf, B., Nordborg, M., Rätsch, G., Ecker, J.R. and Weigel, D. (2007) Common sequence polymorphisms shaping genetic diversity in *Arabidopsis thaliana*. *Science* 317, 338–342.

Clarke, B.C. and Appels, R. (1998) A transient assay for evaluating promoters in wheat endosperm tissue. *Genome* 41, 865–871.

Clarke, J.H., Mithen, R., Brown, J.K.M. and Dean, C. (1995) QTL analysis of flowering time in *Arabidopsis thaliana*. *Molecular and General Genetics* 248, 278–286.

Coburn, J., Temnykh, S., Paul, E. and McCouch, S.R. (2002) Design and application of microsatellite marker panels for semi-automated genotyping of rice (*Oryza sativa* L.). *Crop Science* 42, 2092–2099.

Cochrane, W. (1993) *The Development of American Agriculture*. University of Minnesota Press, Minneapolis, Minnesota.

Codex Alimentarious Commission (2001) *Codex Guidelines (ALINORM 01/22)*. FAO/WHO, Rome. Available at: http:/www.codexalimentarious.net (accessed 17 November 2009).

Coe, E.H. (1959) A line of maize with high haploid frequency. *American Naturalist* 93, 381–382.

Cogoni, C. and Macino, G. (2000) Post-transcriptional gene silencing across kingdoms. *Genes and Development* 10, 638–643.

Cokcerham, C.C. and Zeng, Z.-B. (1996) Design III with marker loci. *Genetics* 143, 1437–1456.

Colbert, T., Till, B.J., Tompa, R., Reynolds, S., Steine, M.N., Yeung, A.T., McCallum, C.M., Comai, L. and Henikoff, S. (2001) High-throughput screening for induced point mutations. *Plant Physiology* 126, 480–484.

Collard, B.C.Y. and Mackill, D.J. (2008) Marker-assisted selection: an approach for precision plant breeding in the twenty-first century. *Philosophical Transactions of The Royal Society* B 363, 557–572.

Collard, B.C.Y., Jahufer, M.Z.Z., Brouwer, J.B. and Pang, E.C.R. (2005) An introduction to markers, quantitative trait loci (QTL) mapping and marker-assisted selection for crop improvement: the basic concepts. *Euphytica* 142, 169–196.

Collard, B.C.Y., Vera Cruz, C.M., McNally, K.L., Virk, P.S. and Mackill, D.J. (2008) Rice molecular beeding laboratories in the genomics era: current status and future considerations. *International Journal of Plant Genomics* Article ID 524847, 25 pp. Available at: http://www.hindawi.com/journals/ijpg/2008/524847.html (accessed 17 November 2009).

Collins, W.W. and Qualset, C.O. (1999) *Biodiversity in Agroecosystems*. CRC Press, Boca Raton, Florida.

Comai, L., Young, K., Till, B.J., Reynolds, S.H., Greene, E.A., Codomo, C.A., Enns, L.C., Johnson, J.E., Burtner, C., Odden, A.R. and Henikoff, S. (2004) Efficient discovery of DNA polymorphisms in natural populations by Ecotilling. *The Plant Journal* 37, 778–786.

Complex Trait Consortium (2004) The Collaborative Cross, a community resource for the genetic analysis of complex traits. *Nature Genetics* 36, 1133–1137.

Comstock, R.E. and Robinson, H.F. (1952) Estimation of average dominance of genes. In: Gowen, J.W. (ed.) *Heterosis*. Iowa State College Press, Ames, Iowa, pp. 494–516.

Comstock, R.E., Robinson, H.F. and Harvey, P.H. (1949) A breeding procedure designed to make maximum use of both general and specific combining ability. *Agronomy Journal* 41, 360–367.

Concibido, V.C., Denny, R.L., Lange, D.A., Orf, J.H. and Young, N.D. (1996) RFLP mapping and molecular marker-assisted selection of soybean cyst nematode resistance in PI 209332. *Crop Science* 36, 1643–1650.

Concibido, V.C., La Vallee, B., Mclaird, P., Pineda, N., Meyer, J., Hummel, L., Yang, J., Wu, K. and Delannay, X. (2003) Introgression of a quantitative trait locus for yield from *Glycine soja* into commercial soybean cultivars. *Theoretical and Applied Genetics* 106, 575–582.

Cone, K.C., McMullen, M.D., Bi, I.V., Davis, G.L., Yim, Y.-S., Gardiner, J.M., Polacco, M.L., Sanchez-Villeda, H., Fang, Z., Schroeder, S.G., Havermann, S.A., Bowers, J.E., Paterson, A.H., Soderlund, C.A., Engler, F.W., Wing, R.A. and Coe, E.H. (2002) Genetic, physical and informatics resources for maize. On the road to an integrated map. *Plant Physiology* 130, 1598–1605.

Conner, A.J., Barrell, P.J., Baldwin, S.J., Lokerse, A.S., Cooper, P.A., Erasmuson, A.K., Nap, J.P. and Jacobs, J.M.E. (2007) Intragenic vectors for gene transfer without foreign DNA. *Euphytica* 154, 341–353.

Cooper, M. and Byth, D.E. (1996) Understanding plant adaptation to achieve systematic applied crop improvement – a fundamental challenge. In: Cooper, M. and Hammer, G.L. (eds) *Plant Adaptation and Crop Improvement.* CAB International, Wallingford, UK, pp. 5–23.

Cooper, M. and Hammer, G.L. (1996) Synthesis of strategies for crop improvement. In: Cooper, M. and Hammer, G.L. (eds) *Plant Adaptation and Crop Improvement.* CAB International, Wallingford, UK, pp. 591–623.

Cooper, M. and Podlich, D.W. (2002) The *E*(*NK*) model: extending the NK model to incorporate gene-by-environment interactions and epistasis for diploid genomes. *Complexity* 7, 31–47.

Cooper, M., Podlich, D.W. and Chapman, S.C. (1999) Computer simulation linked to gene information databases as a strategic research tool to evaluate molecular approaches for genetic improvement of crops. Workshop on Molecular Approaches for the Genetic Improvement of Cereals for Stable Production in Water-Limited Environments, Cento Internacional de Mejoramiento de Maiz y Trigo (CIMMYT), Mexico, 21–25 June 1999. Available at: http://www.cimmyt.org/ABC/map/research_tools_results/wsmolecular/workshopmolecular/WorkshopMolecularcontents.htm (accessed 30 June 2008).

Cooper, M., Chapman, S.C., Podlich, D.W. and Hammer, G.L. (2002a) The GP problem: quantifying gene-to-phenotype relationships. *In Silico Biology* 2, 151–164.

Cooper, M., Podlich, D.W., Micallef, K.P., Smith, O.S., Jensen, N.M., Chapman, S.C. and Kruger, N.L. (2002b) Complexity, quantitative traits and plant breeding: a role for simulation modelling in the genetic improvement of crops. In: Kang, M.S. (ed.) *Quantitative Genetics, Genomics and Plant Breeding.* CAB International, Wallingford, UK, pp. 143–166.

Cooper, M., Smith, O.S., Graham, G., Arthur, L., Feng, L. and Bodlich, D.W. (2004) Genomics, genetics and plant breeding: a private sector perspective. *Crop Science* 44, 1907–1914.

Cooper, M., Podlich, D.W. and Smith, O.S. (2005) Gene-to-phenotype and complex trait genetics. *Australian Journal of Agricultural Research* 56, 895–918.

Cooper, M., Podlich, D.W. and Luo, L. (2007) Modelling QTL effects and MAS in plant breeding. In: Varshney, R.K. and Tuberosa, R. (eds) *Genomics-Assisted Crop Improvement.* Volume 1. *Genomics Approaches and Platforms.* Springer, Dordrecht, Netherlands, pp. 57–95.

Coors, J.G. (1999) Selection methodologies and heterosis. In: Coors, J.G. and Pandey, S. (eds) *The Genetics and Exploitation of Heterosis in Crops.* ASA-CSSA-SSSA, Madison, Wisconsin, pp. 225–245.

Coque, M. and Gallais, A. (2006) Genomic regions involved in response to grain yield selection at high and low nitrogen fertilization in maize. *Theoretical and Applied Genetics* 112, 1205–1220.

Corneille, S., Lutz, K., Svab, Z. and Maliga, P. (2001) Efficient elimination of selectable marker genes from the plastid genome by the CRE-lox site-specific recombination system. *The Plant Journal* 27, 171–178.

Cornelius, P.L. and Seyedsadr, M.S. (1997) Estimation of general linear–bilinear models for two-way tables. *Journal of Statistical Computation and Simulation* 58, 287–322.

Cornelius, P.L., Seyedsadr, M. and Crossa, J. (1992) Using the shifted multiplicative model in search for 'separability' in corn cultivar trials. *Theoretical and Applied Genetics* 84, 161–172.

Cornelius, P.L., van Sanford, D.A. and Seyedsadr, M.S. (1993) Clustering cultivars into groups without rank-change interactions. *Crop Science* 33, 1193–1200.

Cornelius, P.L., Crossa, J. and Seyedsadr, M.S. (1996) Statistical tests and estimates of multiplicative models for GE interaction. In: Kang, M.S. and Hauch, H.G., Jr (eds) *Genotype-by-Environment Interaction.* CRC Press, Boca Raton, Florida, pp. 199–234.

Correns, C. (1901) Bastarde zwischen Maisrassen, mit besonderer Berucksichtigung der Xenien. *Bibliotheca Botanica* 53, 1–161.

Cottage, A., Yang, A.P., Maunders, H., de Lacy, R.C. and Ramsay, N.A. (2001) Identification of DNA sequences flanking T-DNA insertion by PCR walking. *Plant Molecular Biology Reporter* 19, 321–327.

Courtois, B. (1993) Comparison of single seed descent and anther culture-derived lines of three single crosses of rice. *Theoretical and Applied Genetics* 85, 625–631.

Courtois, B., McLaren, G., Sinha, P.K., Prasad, K., Yadav, R. and Shen, L. (2000) Mapping QTL associated with drought avoidance in upland rice. *Molecular Breeding* 6, 55–66.

Coutu, C., Brandle, J., Brown, D., Brown, K., Miki, B., Simmonds, J. and Hegedus, D.D. (2007) pORE: a modular binary vector series suited for both monocot and dicot plant transformation. *Transgenic Research* 16, 771–781.

Craig, W., Tepfer, M., Degrassi, G. and Ripandelli, D. (2008) An overview of general feature of rick assessments and genetically modified crops. *Euphytica* 164, 853–880.

Cravatt, B.F., Simon, G.M. and Yates, J.R. (2007) The biological impact of mass-spectrometry-based proteomics. *Nature* 450, 991–1000.

Cregan, P.B., Shoemaker, R.C. and Specht, J.E. 1999) An integrated genetic linkage map of the soybean genome. *Crop Science* 39, 1464–1490.

Cresham, D., Dunham, M.J. and Botstein, D. (2008) Comparing whole genomes using DNA microarrays. *Nature Reviews Genetics* 9, 291–302.

Crosbie, T.M., Eathington, S.R., Johnson, G.R., Edwards, M., Reiter, R., Stark, S., Mohanty, R.G., Oyervides, M., Buehler, R.E., Walker, A.K., Dobert, R., Delannay, X., Pershing, J.C., Hall, M.A. and Lamkey, K.R. (2006) Plant breeding: past, present and future. In: Lamkey, K.R. and Lee, M. (eds) *Plant Breeding: the Arnel R. Hallauer International Symposium*. Blackwell Publishing, Oxford, UK, pp. 3–50.

Croser, J.S., Lulsdorf, M.M., Davies, P.A., Clarke, H.J., Dayliss, K.L., Mallikarjuna, N. and Siddique, K.H.M. (2006) Toward doubled haploid production in the Fabaceae: progress, constraints and opportunities. *Critical Reviews in Plant Sciences* 25, 139–157.

Crossa, J. and Cornelius, P.L. (1997) Sites regression and shifted multiplicative model clustering of cultivar trial sizes under heterogeneity of error variances. *Crop Science* 37, 406–415.

Crossa, J. and Cornelius, P. (2002) Linear–bilinear models for the analysis of genotype-environment interaction. In: Kang, M.S. (ed.) *Quantitative Genetics, Genomics and Plant Breeding*. CAB International, Wallingford, UK, pp. 305–322.

Crossa, J. and Franco, J. (2004) Statistical methods for classifying genotypes. *Euphytica* 137, 19–37.

Crossa, J., Cornelius, P.L., Seyedsadr, M. and Byrne, P. (1993) A shifted multiplicative model cluster analysis for grouping environments without genotypic rank change. *Theoretical and Applied Genetics* 85, 577–586.

Crossa, J., Cornelius, P.L., Sayre, K. and Ortiz-Monasterio, R.J.I. (1995) A shifted multiplicative model fusion method for grouping environments without cultivar rank change. *Crop Science* 35, 54–62.

Crossa, J., Cornelius, P.L. and Seyedsadr, M.S. (1996) Using the shifted multiplicative model cluster methods for crossover GE interaction. In: Kang, M.S. and Hauch, H.G., Jr (eds) *Genotype-by-Environment Interaction*. CRC Press, Boca Raton, Florida, pp. 175–198.

Crossa, J., Vargas, M., van Eeuwijk, F.A., Jiang, C., Edmeades, G.O. and Hoisington, D. (1999) Interpreting genotype × environment interaction in tropical maize using linked molecular markers and environmental covariables. *Theoretical and Applied Genetics* 99, 611–625.

Crossa, J., Cornelius, P.L. and Yan, W. (2002) Biplots of linear–bilinear models for studying crossover genotype × environment interaction. *Crop Science* 42, 619–633.

Crossa, J., Yang, R.-C. and Cornelius, P.L. (2004) Studying crossover genotype × environment interaction using linear–bilinear models and mixed models. *Journal of Agricultural Biological and Environmental Statistics* 9, 362–380.

Crossa, J., Burgueño, J., Autran, D., Vielle-Calzada, J.-P., Cornelius, P.L., Garcia, N., Salamanca, F. and Arenas, D. (2005) Using linear–bilinear models for studying gene-expression × treatment interaction in microarray experiments. *Journal of Agricultural, Biological and Environmental Statistics* 10, 337–353.

Crossa, J., Burgueño, J., Cornelius, P.L., McLaren, G., Trethowan, R. and Krischnamachari, A. (2006) Modeling genotype × environment interaction using additive genetic covariance of relatives for predicting breeding values of wheat genotypes. *Crop Science* 46, 1722–1733.

Crossa, J., Burdueno, J., Dreisigacker, S., Vargas, M., Herrera-Foessel, S.A., Lillemo, M., Singh, R.P., Trethowan, R., Warburton, M., Franco, J., Reynolds, M., Crouch, J.H. and Ortiz, R. (2007) Association analysis of historical bread wheat germplasm using additive genetic covariance of relatives and population structure. *Genetics* 177, 1889–1013.

Crow, J.F. (1999) Dominance and overdominance. In: Coors, J.G. and Pandey, S. (eds) *Genetics and Exploitation of Heterosis in Crops*. ASA-CSSA-SSSA, Madison, Wisconsin, pp. 49–58.

Crow, J.F. (2000) The rise and fall of overdominance. *Plant Breeding Reviews* 17, 225–257.

Cui, Y. and Wu, R. (2005) Statistical model for characterizing epistatic control of triploid endosperm triggered by maternal and offspring QTLs. *Genetical Research* 86, 65–75.

Cullis, C.A. (2004) *Plant Genomics and Proteomics*. John Wiley & Sons, Inc., Chichester, UK.

Curtis, J.J., Brunson, A.M., Hubbard, J.E. and Earle, F.R. (1956) Effect of the parent on oil content of the corn kernel. *Agronomy Journal* 48, 551–555.
Curtis, M.D. and Grossniklaus, U. (2003) A Gateway cloning vector set for high-throughput functional analysis of genes *in planta*. *Plant Physiology* 133, 462–469.
Dafny-Yelin, M. and Tzfira, Z. (2007) Delivery of multiple transgenes to plant cells. *Plant Physiology* 145, 1118–1128.
D'Amato, F. (1975) The problem of genetic stability in plant tissues and cell cultures. In: Frankel, O. and Hawkes, J.G. (eds) *Crop Genetic Resources for Today and Tomorrow.* Cambridge University Press, Cambridge, UK, pp. 333–348.
Damude, H.G. and Kinney, A.J. (2008) Enhancing plant seed oils for human nutrition. *Plant Physiology* 147, 962–968.
Daniell, H. and Dhingra, A. (2002) Multigene engineering: dawn of an exciting new era in biotechnology. *Current Opinion in Biotechnology* 13, 136–141.
Dargie, J.D. (2007) Marker-assisted selection: policy considerations and options for developing countries. In: Guimarães, E.P., Ruane, J., Scherf, B.D., Sonnino, A. and Dargie, J.D. (eds) *Marker-Assisted Selection, Current Status and Future Perspectives in Crops, Livestock, Forestry and Fish.* Food and Agriculture Organization of the Unites Nations, Rome, pp. 441–471.
Darrah, L.L. and Zuber, M.S. (1986) 1985 United States maize germplasm base and commercial breeding strategy. *Crop Science* 26, 1109–1113.
Darvasi, A. and Soller, M. (1992) Selective genotyping for determination of linkage between a molecular marker and a quantitative trait. *Theoretical and Applied Genetics* 85, 353–359.
Darvasi, A. and Soller, M. (1994) Selective DNA pooling for determination of linkage between a molecular marker and a quantitative trait. *Genetics* 138, 1365–1373.
Darvasi, A. and Soller, M. (1995) Advanced intercross lines, an experimental population for fine genetic mapping. *Genetics* 141, 1199–1207.
Darvasi, A. and Soller, M. (1997) A simple method to calculate resolving power and confidence interval of QTL map location. *Behavior Genetics* 27, 125–132.
Darvasi, A., Weinreb, A., Minke, V., Weller, J.I. and Soller, M. (1993) Detecting marker-QTL linkage and estimating QTL gene effect and map location using a saturated genetic map. *Genetics* 134, 943–951.
Datta, K., Vasquez, A., Tu, J., Torrizo, L., Alam, M.F., Oliva, N., Abrigo, E., Khush, G.S. and Datta, S.K. (1998) Constitutive and tissue-specific differential expression of cryIA(b) gene in transgenic rice plants conferring resistance to rice insect pest. *Theoretical and Applied Genetics* 97, 20–30.
Datta, K., Tu, J., Oliva, N., Ona, I., Velazhahan, R., Mew, T.W., Muthukrishnan, S. and Datta, S.K. (2001) Enhanced resistance to sheath blight by constitutive expression of infection-related rice chitinase in transgenic elite *indica* rice cultivars. *Plant Science* 160, 405–414.
Datta, K., Baisakh, N., Thet, K.M., Tu, J. and Datta, S.K. (2002) Pyramiding transgenes for multiple resistance in rice against bacterial blight, yellow stem borer and sheath blight. *Theoretical and Applied Genetics* 106, 1–8.
Datta, K., Baisakh, N., Oliva, N., Torrizo, L., Abrigo, E., Tan, J., Rai, M., Rehana, S., Al-Babili, S., Beyer, P., Potrykus, I. and Datta, S.K. (2003) Bioengineered 'golden' *indica* rice cultivars with beta-carotene metabolism in the endosperm with hygromycin and mannose selection systems. *Plant Biotechnology Journal* 1, 81–90.
Davenport, C.B. (1908) Degeneration, albinism and inbreeding. *Science* 28, 454–455.
Davenport, G., Ellis, N., Ambrose, M. and Dicks, J. (2004) Using bioinformatics to analyse germplasm collections. *Euphytica* 137, 39–54.
Davuluri, R.V. and Zhang, M.Q. (2003) Computer software to find genes in plant genomic DNA. In: Grotewold, E. (ed.) *Methods in Molecular Biology*, Vol. 236: *Plant Functional Genomics: Methods and Protocols.* Humana Press, Inc., Totowa, New Jersey, pp. 87–107.
Day, C.D., Lee, E., Kobayashi, J., Holappa, L.D., Albert, H. and Ow, D.W. (2000) Transgene integration into the same chromosome location can produce alleles that express at a predictable level, or alleles that are differentially silenced. *Genes and Development* 14, 2869–2880.
Day Rubenstein, K., Heisey, P., Shoemaker, R., Sullivan, J. and Frisvold, G. (2005) *Economic Information Bulletin* No. (EIE2), p. 47. Available at: http://www.ers.usda.gov/publications/eib2/ (accessed 17 November 2009).
De Buck, S., Jacobs, A., Van Montagu, M. and Depicker, A. (1999) The DNA sequences of T-DNA junctions suggest that complex T-DNA loci are formed by a recombination process resembling T-DNA integration. *The Plant Journal* 20, 295–304.

De Buck, S., De Wilde, C., Van Montagu, M. and Depicker, A. (2000) T-DNA vector backbone sequences are frequently integrated into the genome of transgenic plants obtained by *Agrobacterium* mediated transformation. *Molecular Breeding* 6, 459–468.

De Cosa, B., Moar, W., Lee, S.B., Miller, M. and Daniell, H. (2001) Overexpression of the *Bt cry2Aa2* operon in chloroplasts leads to formation of insecticidal crystals. *Nature Biotechnology* 19, 71–74.

De Groote, H., Wangare, L., Kanampiu, F., Odendo, M., Diallo, A., Karaya, H. and Friesen, D. (2008) The potential of a herbicide resistant maize technology for Striga control in Africa. *Agricultural Systems* 97, 83–94.

De Hoog, C.L. and Mann, M. (2004) Proteomics. *Annual Review of Genomics and Human Genetics* 5, 267–293.

de Koning, D.J. and Haley, C.S. (2005) Genetical genomics in humans and model organisms. *Trends in Genetics* 21, 377–381.

De Neve, M., De Buck, S., Jacobs, A., Van Montagu, M. and Depicker, A. (1997) T-DNA integration patterns in co-transformed plant cells suggest that T-DNA repeats originate from co-integration of separate T-DNAs. *The Plant Journal* 11, 15–29.

De Silva, H.N. and Ball, R.D. (2007) Linkage disequilibrium mapping concepts. In: Oraguzie, N.C., Rikkerink, E.H.A., Gardiner, S.E. and De Silva, H.N. (eds) *Association Mapping in Plants*. Springer, Berlin, pp. 103–132.

De Vicente, M.C. and Tanksley, S.D. (1991) Genome-wide reduction in recombination of backcross progeny derived from male versus female gametes in an interspecific cross of tomato. *Theoretical and Applied Genetics* 83, 173–178.

De Vicente, M.C. and Tanksley, S.D. (1993) QTL analysis of transgressive segregation in an interspecific tomato cross. *Genetics* 134, 585–596.

Dean, R.E., Dahlberg, J.A., Hopkins, M.S. and Kresovich, S. (1999) Genetic redundancy and diversity among 'Orange' accessions in the U.S. national sorghum collection as assessed with simple sequence repeat (SSR) markers. *Crop Science* 39, 1215–1221.

Deimling, S., Röber, F.K. and Geiger, H.H. (1997) Methodik und Genetik der *in-vivo*-Haploideninduktion bei Mais. *Vortr. Pflanzenzüchtg.* 38, 203–224.

DeLacy, I.H. and Cooper, M. (1990) Pattern analysis for the analysis of regional variety trials. In: Kang, M.S. (ed.) *Genotype-by-Environment Interaction and Plant Breeding*. Louisiana State University Agricultural Center, Baton Rouge, Louisiana, pp. 301–334.

DeLacy, I.H., Cooper, M. and Basford, K.E. (1996) Relationships among analytical methods used to study genotype-by-environment interactions and evaluation of their impact on response to selection. In: Kang, M.S. and Hauch, H.G., Jr (eds) *Genotype-by-Environment Interaction*. CRC Press, Boca Raton, Florida, pp. 51–84.

DellaPenna, D. and Last, R.L. (2008) Genome-enabled approaches shed new light on plant metabolism. *Science* 320, 479–481.

Delmer, D.P. (2005) Agriculture in the developing world: connecting innovations in plant research to downstream applications. *Proceedings of the National Academy of Sciences of the United States of America* 102, 15739–15746.

Delseny, M. (2004) Re-evaluating the relevance of ancestral shared synteny as a tool for crop improvement. *Current Opinion in Plant Biology* 7, 126–131.

Delvin, B. and Risch, N. (1995) A comparison of linkage disequilibrium measures for fine-scale mapping. *Genomics* 29, 311–322.

Dempster, A.P., Laid, N.M. and Rubin, D.B. (1977) Maximum likelihood from incomplete data via the EM algorithm. *Journal of the Royal Statistical Society Series B* 39, 1–38.

Depicker, A., Stachel, S., Dhaese, P., Zambryski, P. and Goodman, H.M. (1982) Nopaline synthase: transcript mapping and DNA sequence. *Journal of Molecular and Applied Genetics* 1, 561–573.

Dereuddre, J., Blandin, S. and Hassen, N. (1991) Resistance of alginate-coated somatic embryos of carrot (*Daucus carota* L.) to desiccation and freezing in liquid nitrogen: 1. Effects of preculture. *Cryo-Letters* 12, 125–134.

Desloire, S., Gherbi, H., Laloui, W., Marhadour, S., Clouet, V., Cattolico, L., Falentin, C., Giancola, S., Renard, M., Budar, F., Small, I., Caboche, M., Delourme, R. and Bendahmane, A. (2003) Identification of the fertility restoration locus, *Rfo*, in radish, as a member of the pentatricopeptide-repeat protein family. *EMBO Reports* 4, 588–594.

Devaux, P. and Zivy, M. (1994) Protein markers for anther culturability in barley. *Theoretical and Applied Genetics* 88, 701–706.

Devaux, P., Kilian, A. and Kleinhofs, A. (1995) Comparative mapping of the barley genome with male and female recombination-derived, doubled haploid populations. *Molecular and General Genetics* 249, 600–608.

DeVerna, J.W., Chetelat, R.T., Rick, C.M. and Stevens, M.A. (1987) Introgression of *Solanum lycopersicoides* germplasm. In: Nevins, D.J. and Jones, R.A. (eds) *Tomato Biotechnology*. Proc. Seminar, University of California, Davis, California, 20–22 August 1986. *Plant Biology* Vol.4, Alan R. Liss, New York, pp. 27–36.

DeVerna, J.W., Rick, C.M., Chetelat, R.T., Lanini, B.J. and Alpert, K.B. (1990) Sexual hybridization of *Lycopersicon esculentum* and *Solanum rickii* by means of a sesquidiploid bridging hybrid. *Proceedings of the National Academy of Sciences of the Unites States of America* 87, 9486–9490.

Dhillon, B.S., Boppenmaier, J., Pollmer, W.G., Hermann, R.G. and Mechinger, A.E. (1993) Relationship of restriction fragment length polymorphisms among European maize inbreds with ear dry matter yield of their hybrids. *Maydica* 38, 245–248.

D'hoop, B.B., Paulo, M.J., Mank, R.A., van Eck, H.J. and van Eeuwijk, F.A. (2008) Association mapping of quality traits in potato (*Solanum tuberosum* L.). *Theoretical and Applied Genetics* 161, 47–60.

Dhungana, P., Eskridge, K.M., Baenziger, P.S., Champbell, B.T., Gill, K.S. and Dweikat, I. (2007) Analysis of genotype-by-environment interaction in wheat using a structural equation model and chromosome substitution lines. *Crop Science* 47, 477–484.

Dias, A.P., Brown, J., Bonello, P. and Brotewold, E. (2003) Metabolite profiling as a functional genomics tool. In: Grotewold, E. (ed.) *Methods in Molecular Biology 236. Plant Functional Genomics: Methods and Protocols*. Humana Press, Totowa, New Jersey, pp. 415–425.

Diatchenko, L., Lau, Y.-F.C., Campbell, A.P., Chenchik, A., Moqadam, F., Huang, B., Lukyanov, S., Lukyanov, K., Gurskaya, N., Sverdlov, E.D. and Siebert, P.D. (1996) Suppression subtractive hybridization: a method for generating differentially regulated or tissue-specific cDNA probes and libraries. *Proceedings of the National Academy of Sciences of the United States of America* 93, 6025–6030.

Dijkhuizen, A., Dudley, J.W., Rocheford, T.R., Haken, A.E. and Eckhoff, S.R. (1998) Comparative analysis for kernel composition using near infrared reflectance and 100g Wetmill Analysis. *Cereal Chemistry* 75, 266–270.

Dilday, R.H. (1990) Contribution of ancestral lines in the development of new cultivars of rice. *Crop Science* 30, 905–911.

Dinka, S.J., Campbell, M.A., Demers, T. and Raizada, M.N. (2007) Predicting the size of the progeny mapping population required to positionally clone a gene. *Genetics* 176, 2035–2054.

Diretto, G., Al-Babili, S., Tavazza, R., Papacchioli, V., Beyer, P. and Giiliano, G. (2008) Metabolic engineering of potato carotenoid content through tuber-specific over-expression of a bacterial mini-pathway. *PLoS ONE* 2(4), e350. doi:10.1371/journal.pone.0000350. Available at: http://www.plosone.org (accessed 17 November 2009).

Ditt, R.F., Nester, E.W. and Comai, L. (2001) Plant gene expression to *Agrobacterium tumefaciens*. *Proceedings of the National Academy of Sciences of the United States of America* 98, 10954–10959.

Dixon, A.L., Liang, L., Moffatt, M.F., Chen, W., Heath, S., Wong, K.C., Taylor, J., Burnett, E., Gut, I., Farrall, M., Lathrop, G.M., Abecasis, G.R. and Cookson, W.O.C. (2007) A genome-wide association study of global gene expression. *Nature Genetics* 39, 1202–1207.

Dodds, J.H. (1991) Introduction: conservation of plant genetic resources – the need for tissue culture. In: Dodds, J.H. (ed.) In Vitro *Methods for Conservation of Plant Genetic Resources*. Chapman & Hall, London, pp. 1–9.

Doebley, J. (1992) Molecular systematics and crop evolution. In: Soltis, D.E., Soltis, P.S. and Doyle, J.J. (eds) *Molecular Systematics of Plants*. Chapman & Hall, New York, pp. 202–222.

Doebley, J., Stec, A. and Gustus, C. (1995) Teosinte branched1 and the origin of maize: evidence for epistasis and the evolution of dominance. *Genetics* 141, 333–346.

Doerge, R.W. and Churchill, G.A. (1996) Permutation tests for multiple loci affecting a quantitative character. *Genetics* 142, 285–294.

Doi, K., Izawa, T., Fuse, T., Yamanouchi, U., Kubo, T., Shimatani, Z., Yano, M. and Yoshimura, A. (2004) *Ehd1*, a B-type response regulator in rice, confers short-day promotion of flowering and controls *FT*-like gene expression independently of *Hd1*. *Genes and Development* 18, 926–936.

Doll, J. (1998) The patent of DNA. *Science* 280, 689–690.

Dong, Y.S., Cao, Y.S., Zhang, X.Y., Liu, S.C., Wang, L.F., You, G.X., Pang, B.S., Li, L.H. and Jia, J.Z. (2003) Establishment of candidate core collections in Chinese common wheat germplasm. *Journal of Plant Genetic Resources* 4, 1–8.

Donnenwirth, J., Grace, J. and Smith, S. (2004) Intellectual property rights, patents, plant variety protection and contracts: a perspective from the private sector. *IP Strategy Today*, No. 9.

Doumas, P., Al-Ghazi, Y., Rothan, C. and Robin, S. (2007) DNA microarrays in plants. In: Morot-Gaudry, J.F., Lea, P. and Briat, J.F. (eds) *Functional Plant Genomics.* Science Publishers, Enfield, New Hampshire, pp. 165–190.

Dreher, K., Khairallah, M., Ribau, J.M. and Morris, M. (2003) Money matters (I): cost of field and laboratory procedures associated with conventional and marker-assisted maize breeding at CIMMYT. *Molecular Breeding* 11, 221–234.

Dubcovsky, J. (2004) Marker-assisted selection in public breeding programs. The wheat experience. *Crop Science* 44, 1895–1898.

Dubcovsky, J., Ramakrishna, W., SanMiguel, P.J., Busso, C.S., Yan, L., Shiloff, B.A. and Bennetzen, J.L. (2001) Comparative sequence analysis of colinear barley and rice BACs. *Plant Physiology* 125, 1342–1353.

Dudley, D.N., Saghai Maroof, M.A. and Rufener, G.K. (1991) Molecular markers and grouping of parents in a maize breeding program. *Crop Science* 31, 718–723.

Dudley, J.W. (1977) Seventy six generations of selection for oil and protein percentage in maize. In: Pollak, E., Kempthorne, O. and Bailey, T.B. (eds) *Proceedings of International Conference on Quantitative Genetics.* Iowa State University Press, Ames, Iowa, pp. 459–473.

Dudley, J.W. (1993) Molecular markers in plant improvement: manipulation of genes affecting quantitative traits. *Crop Science* 33, 660–668.

Dudley, J.W. (1997) Quantitative genetics and plant breeding. *Advances in Agronomy* 59, 1–23.

Dudley, J.W. (2007) From means to QTL: the Illinois Long-Term Selection Experiment as a case study in quantitative genetics. *Crop Science* 47(S3), S20–S31.

Dudley, J.W. (2008) Epistatic interactions in crosses of Illinois High Oil × Illinois Low Oil and of Illinois High Protein × Illinois Low Protein corn strains. *Crop Science* 48, 59–68.

Dudley, J.W. and Lambert, R.J. (1992) Ninety generations of selection for oil and protein in maize. *Maydica* 37, 81–87.

Dudley, J.W. and Lambert, R.J. (2004) 100 generations of selection for oil and protein in corn. *Plant Breeding Reviews* 24 (Part 1), 79–110.

Dudley, J.W., Lambert, R.J. and Alexander, D.E. (1974) Seventy generations of selection for oil and protein concentration in the maize kernel. In: Dudley, J.W. (ed.) *Seventy Generations of Selection for Oil and Protein in Maize.* Crop Science Society of America, Madison, Wisconsin, pp. 181–212.

Dudley, J.W., Lambert, R.J. and de la Roche, I.A. (1977) Genetic analysis of crosses among corn strains divergently selected for percent oil and protein. *Crop Science* 17, 111–117.

Dunford, R.P., Yano, M., Kurata, N., Sasaki, T., Huestis, G., Rocheford, T. and Laurie, D.A. (2002) Comparative mapping of the barley *Phd-H1* photoperiod response gene region, which lies close to a junction between two rice linkage segments. *Genetics* 161, 825–834.

Dunn, G. and Everitt, B.S. (1982) *An Introduction to Mathematical Taxonomy.* Cambridge Studies in Mathematical Biology. Vol. 5. Cambridge University Press, Cambridge, UK.

Dunning, A.M., Durocher, F., Healey, C.S., Teare, M.D., McBride, S.E., Carlomagno, F., Xu, C.-F., Dawson, E., Rhodes, S., Ueda, S., Lai, E., Luben, R.N., Van Rensburg, E.J., Mannermaa, A., Kataja, V., Rennart, G., Dunham, I., Purvis, I., Easton, D. and Ponder, B.A.J. (2000) The extent of linkage disequilibrium in four populations with distinct demographic histories. *American Journal of Human Genetics* 67, 1544–1554.

Dunninnton, E.A., Haberefeld, A., Stallard, L.G., Siegel, P.B. and Hillel, J. (1992) Deoxyribonucleic-acid fingerprint bands linked to loci coding for quantitative traits in chicken. *Poultry Science* 71, 1251–1258.

Dunwell, J.M. (2005) Intellectual property aspects of plant transformation. *Plant Biotechnology Journal* 3, 371–384.

Dunwell, J.M. (2006) Patents and transgenic plants. In: Fári, M.G., Holb, I. and Bisztray, G.D. (eds) *Proceedings of Vth International Symposium on In Vitro Culture and Horticultural Breeding.* International Society for Horticultural Science. *Acta Horticulturae* 725, 719–732.

Dutfield, G. (2003) Protecting Traditional Knowledge and Folklore, Issue Paper 1. International Centre on Trade and Sustainable Development and United Nations Conference on Trade and Development Project on Intellectual Property Rights and Sustainable Development, Geneva.

Duvick, D.N. (1977) Major USA crops in 1976. *Annals of the New York Academy of Sciences* 287, 86–96.

Duvick, D.N. (1984) Genetic contribution to yield grains of U.S. hybrid maize, 1930–1980. In: Fehr, W.R. (ed.) *Genetic Contributions to Yield Grains of Five Major Crop Plants.* Crop Science Society of America

(CSSA) Spec. Publ. 7. CSSA and American Society of Agronomy (ASA), Madison, Wisconsin, pp. 15–47.
Duvick, D.N. (1990) Genetic enhancement and plant breeding. In: Janick, J. and Simon, J.E. (eds) *Advances in New Crops.* Proc. First National Symposium on New Crops: Research, Development, Economics. Timber Press, Portland, Oregon, pp. 90–96.
Duvick, D.N. (1999) Heterosis: feeding people and protecting natural resources. In: Coors, J.G. and Pandey, S. (eds) *Genetics and Exploitation of Heterosis in Crops.* ASA-CSSA-SSSA, Madison, Wisconsin, pp. 19–29.
Duvick, D.N., Smith, J.S.C. and Cooper, M. (2004) Long-term selection in commercial hybrid maize breeding programs. *Plant Breeding Reviews* 24 (Part 2), 109–151.
Dwivedi, S.L., Blair, M., Upadhyaya, H.D., Serraj, R., Balaji, J., Buhariwalla, H.K., Ortiz, R. and Crouch, J.H. (2005) Using genomics to exploit grain legume biodiversity in crop improvement. *Plant Breeding Reviews* 26, 176–357.
Dwivedi, S.L., Crouch, J.H., Mackill, D.J., Xu, Y., Blair, M.W., Ragot, M., Upadhyaya, H.D. and Ortiz, R. (2007) The molecularization of public sector crop breeding: progress, problems and prospects. *Advances in Agronomy* 95, 163–318.
Eagles, H.A., Bariana, H.S., Ogbonnaya, F.C., Rebetzke, G.J., Hollamby, G.J., Henry, R.J., Henschke, P.H. and Carter, M. (2001) Implementation of markers in Australian wheat breeding. *Australian Journal of Agricultural Research* 52, 1349–1356.
Eagles, H.A., Hollamby, G.J., Gororo, N.N. and Eastwood, R.F. (2002) Estimation and utilization of glutein gene effects from the analysis of unbalanced data from wheat breeding programs. *Australian Journal of Agricultural Research* 53, 367–377.
Eagles, H.A., Eastwood, R.F., Hollamby, G.J., Martin, E.M. and Cornish, G.B. (2004) Revision of the estimates of glutenin gene effects at the *Glu-B1* locus form southern Australian wheat breeding programs. *Australian Journal of Agricultural Research* 55, 1093–1096.
Eamens, A., Wang, M.-B., Smith, N.A. and Waterhouse, P.M. (2008) RNA silencing in plants: yesterday, today and tomorrow. *Plant Physiology* 147, 456–468.
Earley, K.W., Haag, J.R., Pontes, O., Opper, K., Juehne, T., Song, K. and Pikaard, C.S. (2006) GATEWAY-compatible vectors for plant functional genomics and proteomics. *Plant Journal* 45, 616–629.
East, E.M. (1908) Inbreeding in corn. *Rep. Connecticut Expt. Stat. Years 1907–1908*, pp. 419–428.
Eathington, S.R. (2005) Practical applications of molecular technology in the development of commercial maize hybrids. In: *Proceedings of the 60th Annual Corn and Sorghum Seed Research Conferences*. American Seed Trade Association, Washington, DC.
Eathington, S.R., Crosbie, T.M., Edwards, M.D., Reiter, R.S. and Bull, J.K. (2007) Molecular markers in a commercial breeding program. *Crop Science* 47(S3), S154–S163.
Eberhart, S.A. and Russell, W.A. (1966) Stability parameters for comparing varieties. *Crop Science* 6, 36–40.
Ebinuma, H.K., Sugita, K., Matsunaga, E., Endo, S., Yamada, K. and Komamine, A. (2001) Systems for removal of a selection marker and their combination with a positive marker. *Plant Cell Reports* 20, 383–392.
Ebinuma, H.K., Sugita, E., Endo, S., Matsunaga, E. and Yamada, K. (2004) Elimination of markers genes from transgenic plants using MAT vector system. In: Peña, L. (ed.) *Methods in Molecular Biology*, vol. 286: *Transgenic Plants: Methods and Protocols.* Humana Press Inc., Totowa, New Jersey, pp. 237–253.
Eder, J. and Chalyk, S. (2002) *In vivo* haploid induction in maize. *Theoretical and Applied Genetics* 104, 703–708.
Edmeades, G.O., Bänziger, M. and Ribaut, J.M. (2000) Maize improvement for drought-limited environments. In: Otegui, M.E. and Slafer, G.A. (eds) *Physiological Bases for Maize Improvement.* Food Products Press, New York, pp. 75–111.
Edwards, D. and Batley, J. (2004) Plant bioinformatics: from genome to phenome. *Trends in Biotechnology* 22, 232–237.
Edwards, D., Forster, J.W., Chagné, D. and Batley, J. (2007a) What is SNPs? In: Oraguzie, N.C., Rikkerink, E.H.A., Gardiner, S.E. and De Silva, H.N. (eds) *Association Mapping in Plants.* Springer, Berlin, pp. 41–52.
Edwards, D., Forster, J.W., Cogan, N.O.I., Batley, J. and Chagné, D. (2007b) Single nucleotide polymorphism discovery. In: Oraguzie, N.C., Rikkerink, E.H.A., Gardiner, S.E. and De Silva, H.N. (eds) *Association Mapping in Plants.* Springer, Berlin, pp. 53–76.

Edwards, J.D., Janda, J., Sweeney, M.T., Gaikwad, A.B., Liu, B., Leung, H. and Galbraith, D.W. (2008) Development and evaluation of a high-throughput, low-cost genotyping platform based on oligonucleotide microarrays in rice. *Plant Methods* 4, 13.
Edwards, M. and Johnson, L. (1994) RFLPs for rapid recurrent selection. In: *Proceedings of Symposium on Analysis of Molecular Marker Data*. American Society of Horticultural Science and Crop Science Society of America, Corvallis, Oregon, pp. 33–40.
Edwards, M.D. and Page, N.J. (1994) Evaluation of marker-assisted selection through computer simulation. *Theoretical and Applied Genetics* 88, 376–382.
Edwards, M.D., Stuber, C.W. and Wendel, J.F. (1987) Molecular-marker-facilitated investigations of quantitative trait loci in maize. I. Numbers, genomic distribution and types of gene action. *Genetics* 116, 113–125.
Edwards, M.D., Helentjaris, T., Wright, S. and Stuber, C.W. (1992) Molecular-marker-facilitated investigations of quantitative trait loci in maize. 4. Analysis based on genome saturation with isozyme and restriction fragment length polymorphism markers. *Theoretical and Applied Genetics* 83, 765–774.
Eisemann, R.L., Cooper, M. and Woodruff, D.R. (1990) Beyond the analytical methodology, better interpretation and exploiting of GE interaction in plant breeding. In: Kang, M.S. (ed.) *Genotype-by-Environment Interaction and Plant Breeding*. Louisiana University Agricultural Center, Baton Rouge, Louisiana, pp. 108–117.
Eitan, Y. and Soller, M. (2004) Selection induced genetic variation. In: Wasser, S. (ed.) *Evolutionary Theory and Processes: Modern Horizon*. Papers in honour of Eviatar Nevo. Kluwer Academic Publishers, Dordrecht, Netherlands, pp. 154–176.
Elston, R.C. (1984) The genetic analysis of quantitative trait differences between two homozygous lines. *Genetics* 108, 733–744.
Emebiri, L.C. and Moody, D.B. (2006) Heritable basis for some genotype-environment stability statistics: inference from QTL analysis of heading date in two-rowed barley. *Field Crops Research* 96, 243–251.
Empig, L.T., Gardner, C.O. and Compton, W.A. (1972) Theoretical grains for different population improvement procedures. Nebraska Agricultural Experiment Station Miscellaneous Publications 26 (revised).
Emrich, S., Li, L., Wen, T.J., Ashlock, D., Aluru, S. and Schnable, P. (2007b) Nearly identical paralogs: implications for maize (*Zea mays* L.) genome evolution. *Genetics* 175, 429–439.
Endo, S., Kasahara, Y., Sugita, K. and Ebinuma, H. (2002a) A new GST-MAT vector containing both the *ipt* gene and *iaaM/H* genes can produce marker-free transgenic plants with high frequency. *Plant Cell Reports* 20, 923–928.
Endo, S., Sugita, K., Sakai, M., Tanaka, H. and Ebinuma, H. (2002b) Single-step transformation for generating marker-free transgenic rice using the ipt-type MAT vector system. *The Plant Journal* 30, 115–122.
Engelmann, F. and Engels, J.M.M. (2002) Technologies and strategies for *ex situ* conservation. In: Engels, J.M.M., Ramanatha Rao, V., Brown, A.H.D. and Jackson, M.T. (eds) *Managing Plant Genetic Diversity*. International Plant Genetic Resources Institute, Rome, pp. 89–103.
Engels, J.M.M. and Visser, L. (2003) A guide to effective management of germplasm collections. *IPGRI Handbook for Genebanks* No. 6. International Plant Genetic Resources Institute, Rome.
Enserink, M. (2008) Tough lessons from golden rice. *Science* 320, 468–471.
Eronen, L., Geerts, F. and Toivonen, H. (2004) A Markov chain approach to reconstruction of long haplotypes. *Pacific Symposium on Biocomputing* 9, 104–115.
Ervin, D., Batie, S., Welsh, R., Carpentier, C.L., Fern, J.I., Richman, N.J. and Schulz, M.A. (2000) *Transgenic Crops: an Environmental Assessment*. Henry A. Wallace Center for Agricultural and Environmental Policy at Winrock International, Arlington, Virginia.
Erwin, T. (1991) An evolutionary basis for conservation strategies. *Science* 253, 750–752.
Eshed, Y. and Zamir, D. (1994) A genomic library of *Lycopersicon pennellii* in *L. esculentum*: a tool for fine mapping of genes. *Euphytica* 79, 175–179.
Eshed, Y. and Zamir, D. (1995) An introgression line population of *Lycopersicon pennellii* in the cultivated tomato enables the identification and fine mapping of yield associated QTL. *Genetics* 141, 1147–1162.
Eshed, Y. and Zamir, D. (1996) Less-than-additive epistatic interactions of quantitative trait loci in tomato. *Genetics* 143, 1807–1817.
Esquinas-Alcázar, J.T. (1993) Plant genetic resources. In: Hayward, M.D., Bosemark, N.O. and Romagosa, I. (eds) *Plant Breeding: Principles and Prospects*. Chapman & Hall, London, pp. 33–51.
Esquinas-Alcázar, J. (2005) Protecting crop genetic diversity for food security: political, ethical and technical challenges. *Nature Reviews Genetics* 6, 946–953.

ETC Group (Action Group on Erosion, Technology and Concentration) (2005) Global seed industry concentration – 2005. *Communique* September/October 2005, pp. 1–12.
Etzel, C. and Guerra, R. (2003) Meta-analysis of genetic-linkage of quantitative trait loci. *American Journal of Human Genetics* 71, 56–65.
Eujayl, I., Sorrels, M.E., Baum, M., Wolters, P. and Powell, W. (2002) Isolation of EST-derived microsatellite markers for genotyping the A and B genomes of wheat. *Theoretical and Applied Genetics* 104, 399–407.
European Parliament (2001) Directive 2001/18/EC of the European Parliament and of the Council of 12 March 2001 on the deliberate release into the environment of genetically modified organisms and repealing Council Directive 90/220/EEC – Commission Declaration. *Official Journal of European Community L* 106, 1–39.
Evans, L.T. (1993) *Crop Evolution, Adaptation and Yield.* Cambridge University Press, New York.
Faham, M., Zheng, J., Moorhead, M., Fakhrai-Rad, H., Namsaraev, E., Wong, K., Wang, Z., Chow, S.G., Lee, L., Suyenaga, K., Reichert, J., Boudreau, A., Eberle, J., Bruckner, C., Jain, M., Karlin-Neumann, G., Jones, H.B., Willis, T.D., Buxbaum, J.D. and Davis, R.W. (2005) Multiplexed variation scanning for 1,000 amplicons in hundreds of patients using mismatch repair detection (MRD) on tag arrays. *Proceedings of the National Academy of Sciences of the United States of America* 102, 14717–14722.
Falconer, D.S. (1960) *Introduction to Quantitative Genetics.* Oliver & Boyd, Edinburgh, UK.
Falconer, D.S. (1981) *Introduction to Quantitative Genetics*, 2nd edn. Longman, London.
Falconer, D.S. (1989) *Introduction to Quantitative Genetics*, 3rd edn. Wiley, New York.
Falconer, D.S. and Mackay, T.F.C. (1996) *Introduction to Quantitative Genetics*, 4th edn. Longman Scientific & Technical Ltd, Harlow, UK.
Faleiro, F.G., Ragagnin, V.A., Moreira, M.A. and de Barros, E.G. (2004) Use of molecular markers to accelerate the breeding of common bean lines resistant to rust and anthracnose. *Euphytica* 138, 213–218.
Falque, M. and Santoni, S. (2007) Molecular markers and high-throughput genotyping analysis. In: Morot-Gaudry, J.F., Lea, P. and Briat, J.F. (eds) *Functional Plant Genomics.* Science Publishers, Enfield, New Hampshire, pp. 503–527.
Falque, M., Decousset, L., Dervins, D., Jacob, A.-M., Joets, J., Martinant, J.-P., Raffoux, X., Ribière, N., Ridel, C., Samson, D., Charcosset, A. and Murigneux, A. (2005) Linkage mapping of 1454 new maize candidate gene loci. *Genetics* 170, 1957–1966.
Falush, D., Stephens, M. and Pritchard, J.K. (2003) Inference of population structure using multilocus genotype data: linked loci and correlated allele frequencies. *Genetics* 164, 1567–1587.
Fan, C., Xing, Y., Mao, H., Lu, T., Han, B., Xu, C., Li, X. and Zhang, Q. (2006) GS3, a major QTL for grain length and weight and minor QTL for grain width and thickness in rice, encodes a putative transmembrane protein. *Theoretical and Applied Genetics* 112, 1164–1171.
Fang, Y.-D., Akula, C. and Altpeter, F. (2002) *Agrobacterium*-mediated barley (*Hordeum vulgare* L.) transformation using green fluorescent protein as a visual marker and sequence analysis of the T-DNA:genomic DNA junctions. *Journal of Plant Physiology* 159, 1131–1138.
FAO (Food and Agriculture Organization of the United Nations) (1998) *The State of the World's Plant Genetic Resources for Food and Agriculture.* FAO, Rome.
Faris, J.D., Laddomada, B. and Gill, B.S. (1998) Molecular mapping of segregation distortion loci in *Aegilops tauschii. Genetics* 149, 319–327.
Faris, J.D., Fellers, J.P., Brooks, S.A. and Gill, B.S. (2003) A bacterial artificial chromosome contig spanning the major domestication locus *Q* in wheat and identification of a candidate gene. *Genetics* 164, 311–321.
Fashena, S.J., Serebriiskii, I. and Golemis, E.A. (2000) The continued evolution of two-hybrid screening approaches in yeast: how to outwit different preys with different baits. *Gene* 250, 1–14.
Fatokun, C.A., Menancio-Hautea, D.I., Danesh, D. and Young, N.D. (1992) Evidence for orthologous seed weight genes in cowpea and mung bean based on RFLP mapping. *Genetics* 132, 841–846.
Fauquet, C.M. and Tohme, J. (2004) The global cassava partnership for genetic improvement. *Plant Molecular Biology* 86, v–x (editorial).
Fehr, W.R. (1987) *Principles of Cultivar Development.* Vol. 1. *Theory and Techniques.* Macmillan Publishing Company, London.
Feltus, F.A., Singh, H.P., Lohithaswa, H.C., Schulze, S.R., Silva, T.D. and Paterson, A.H. (2006) A comparative genomic strategy for targeted discovery of single-nucleotide polymorphisms and conserved-noncoding sequences in orphan crops. *Plant Physiology* 140, 1183–1191.

Fenn, J.B., Mann, M., Meng, C.K., Wong, S.F. and Whitehouse, C.M. (1989) Electrospray ionization for the mass spectrometry of large biomolecules. *Science* 246, 64–71.

Fernandez-Ricaud, L., Warringer, J., Ericson, E., Pylvanainen, I., Kemp, G.J.L., Nerman, O. and Blomberg, A. (2005) PROPHECY – a database for high-resolution phenomics. *Nucleic Acids Research* 33, D369–D373.

Fernando, R.L. (2002) Methods to map QTL. Available at: http://meishan.ansci. iastate.edu/rohan/notes-dir/QTL.pdf (accessed 31 December 2007).

Fernando, R.L., Nettleton, D., Southey, B.R., Dekkers, J.C.M., Rothschild, M.F. and Soller, M. (2004) Controlling the proportion of false positives in multiple dependent tests. *Genetics* 166, 611–619.

Ferrie, A.M.R. (2007) Doubled haploid production in nutraceutical species: a review. *Euphytica* 158, 347–357.

Ferro, M., Salvi, D., Rivière-Polland, H., Vernat, T., Seigneurin-Berny, D., Grunwald, D., Garin, J., Joyard, J. and Rolland, N. (2002) Integral membrane proteins of the chloroplast envelope: identification and subcellular localization of new transporters. *Proceedings of the National Academy of Sciences of the United States of America* 99, 11487–11492.

Fiehn, O. (2002) Metabolomics – the link between genotypes and phenotypes. *Plant Molecular Biology* 48, 155–171.

Fiehn, O., Wohlgemuth, G., Scholz, M., Kind, T., Lee, D.Y., Lu, Y., Moon, S. and Nikolau. B. (2008) Quality control for plant metabolomics: reporting MSI-compliant studies. *The Plant Journal* 53, 691–704.

Fields, S. and Song, O. (1989) A novel genetic system to detect protein–protein interactions. *Nature* 340, 245–246.

Filipski, A. and Kumar, S. (2005) Comparative genomics in eukaryotes. In: Gregory, T.R. (ed.) *The Evolution of the Genome.* Elsevier Inc., Amsterdam, pp. 521–583.

Finak, G., Hallett, M., Park, M. and Pepin, F. (2005) Bioinformatics tools for gene-expression studies. In: Sensen, C.W. (ed.) *Handbook of Genome Research. Genomics, Proteomics, Metabolomics, Bioinformatics, Ethical and Legal Issues.* WILEY-VCH, Weinheim, Germany, pp. 415–434.

Finlay, K.W. and Wilkinson, G.N. (1963) The analysis of adaptation in a plant-breeding programme. *Australian Journal of Agricultural Research* 14, 742–754.

Fire, A., Xu, S., Montgomery, M., Kostas, S., Driver, S. and Mello, C. (1998) Potent and specific genetic interference by double-stranded RNA in *Caenorhabditis elegans. Nature* 391, 806–811.

Fisher, R.A. (1918) The correlation between relatives on the supposition of Mendelian inheritance. *Transactions of the Royal Society of Edinburgh, Earth Sciences* 52, 399–433.

Fisher, R.A. (1935) The detection of linkage with dominant abnormalities. *Annals of Eugenics* 6, 187–201.

Fisher, R.A. (1936) The use of multiple measurements in taxonomic problems. *Annals of Eugenics* 7, 179–188.

Fisk, H.J. and Dandekar, A.M. (2004) Electroporation. In: Peña, L. (ed.) *Methods in Molecular Biology*, Vol. 286. *Transgenic Plants: Methods and Protocols.* Humana Press Inc., Totowa, New Jersey, pp. 79–90.

Flint, J. and Mott, R. (2001) Finding the molecular basis of quantitative traits: successes and pitfalls. *Nature Reviews Genetics* 2, 437–445.

Flint-Garcia, S.A., Thornsberry, J.M. and Buckler, E.S. (2003) Structure of linkage disequilibrium in plants. *Annual Review of Plant Biology* 54, 357–374.

Florea, L., Hartzell, G., Zhang, Z., Rubin, G.G. and Miller, W. (1998) A computer program for aligning a cDNA sequence with a genomic DNA sequence. *Genome Research* 8, 967–974.

Flores, F., Moreno, M.T. and Cubero, J.I. (1998) A comparison of univariate and multivariate methods to analyze G × E interaction. *Field Crops Research* 56, 271–286.

Fodor, S., Dower, W. and Solas, D. (1998) Detection of nucleic acid sequences. Patent EP 0834576.

Fofana, I.B.F., Sangaré, A., Collier, R., Taylor, C. and Fauquet, C.M. (2004) A geminivirus-induced gene silencing system for gene function validation in cassava. *Plant Molecular Biology* 56, 613–624.

Foolad, M.R. and Jones, R.A. (1992) Models to estimate maternally controlled genetic variation in quantitative seed characters. *Theoretical and Applied Genetics* 83, 360–366.

Foolad, M.R. and Jones, R.A. (1993) Mapping salt-tolerance genes in tomato (*Lycopersicon esculentum*) using trait-based marker analysis. *Theoretical and Applied Genetics* 87, 184–192.

Forster, B.P. and Thomas, W.T.B. (2004) Doubled haploids in genetics and plant breeding. *Plant Breeding Reviews* 25, 57–88.

Forster, B.P., Ellis, R.P., Thomas, W.T.B., Newton, A.C., Tuberosa, R., This, D., El-Enein, R.A., Bahri, M.H. and Ben Salem, M. (2000) The development and application of molecular markers for abiotic stress. *Journal of Experimental Botany* 51, 19–27.

Forster, B.P., Herberle-Bors, E., Kasha, K.J. and Touraev, A. (2007) The resurgence of haploids in higher plants. *Trends in Plant Science* 12, 368–375.

Foster, G.D. and Twell, D. (eds) (1996) *Plant Gene Isolation: Principles and Practice.* John Wiley & Sons, Chichester, UK, 426 pp.

Fowler, C. and Hodgkin, T. (2004) Plant genetic resources for food and agriculture: assessing global availability. *Annual Review of Environment and Resources* 29, 143–179.

Fowler, C. and Lower, R.L. (2005) Politics of plant breeding. *Plant Breeding Reviews* 25, 21–55.

Fowler, C., Hawtin, G., Ortiz, R., Iwanaga, M. and Engels, J. (2005) The questions and derivatives: promoting use and ensuring availability of non-proprietary plant genetic resources. *The Journal of World Intellectual Property* 7, 641–663.

Fox, P.N., Crossa, J. and Romagosa, I. (1997) Multi-environment testing and genotype × environment interaction. In: Kempton, R.A. and Fox, P.N. (eds) *Statistical Methods for Plant Variety Evaluation.* Chapman & Hall, London, pp. 117–138.

Fraley, R. (2006) Presentation at Monsanto European Investor Day, 10 November 2006. Available at: http://www.monsanto.com (accessed 17 November 2009).

Fraley, R.T., Rogers, S.G. and Horsch, R.B. (1986) Genetic transformation in higher plants. *Critical Reviews in Plant Sciences* 4, 1–46.

Francia, E., Tacconi, G., Crosatti, C., Barabaschi, D., Bulgarelli, D., Dall'Aglio, E. and Vale, G. (2005) Marker assisted selection in crop plants. *Plant Cell, Tissue and Organ Culture* 82, 317–342.

Franco, J., Crossa, J., Taba, S. and Shands, H. (2005) A sampling strategy for conserving genetic diversity when forming core subsets. *Crop Science* 45, 1035–1044.

Franco, J., Crossa, J., Warburton, M.L. and Taba, S. (2006) Sampling strategies for conserving maize diversity when forming core subsets using genetic markers. *Crop Science* 46, 854–864.

François, I., Broekaert, W. and Cammue, B. (2002a) Different approaches for multi-transgene-stacking in plants. *Plant Science* 163, 281–295.

François, I.E.J.A., De Bolle, M.F.C., Dwyer, G., Goderis, I.J.W.M, Wouters, P.F.J., Verhaert, P., Proost, P., Schaaper, W.M.M., Cammue, B.P.A and Broekaert, W.F. (2002b) Transgenic expression in *Arabidopsis thaliana* of a polyprotein construct leading to production of two different antimicrobial proteins. *Plant Physiology* 128, 1346–1358.

François, I.E.J.A., Dwyer, G.I., De Bolle, M.F.C., Goderis, I.J.W.M, van Hemelrijck, W., Proost, P., Wouters, P.F.J., Broekaert, W.F. and Cammue, B.P.A. (2002c) Processing in transgenic *Arabidopsis thaliana* plants of polyproteins with linker peptide variants derived from the *Impatiens balsamina* antimicrobial polyprotein precursor. *Plant Physiology and Biochemistry* 40, 871–879.

Frankel, O. (1984) Genetic perspectives of germplasm conservation. In: Arber, W., Limensee, K., Peacock, W.J. and Starlinger, P. (eds) *Genetic Manipulation: Impact on Man and Society.* Cambridge University Press, Cambridge, UK, pp. 161–170.

Frankel, O.H. (1986) Genetic resources – museum or utility. In: Williams, T.A. and Wratt, G.S. (eds) *Plant Breeding Symposium, DSIR 1986.* Agronomy Society of New Zealand, Christchurch, pp. 3–7.

Frankel, O.H. and Brown, A.H.D. (1984) Current plant genetic resources: a critical appraisal. In: *Genetics: New Frontiers* (Vol. IV). Oxford & IBH, New Delhi.

Frankel, W.N. (1995) Taking stock of complex trait genetics in mice. *Trends in Genetics* 11, 471–477.

Frary, A., Nesbitt, T.C., Frary, A., Grandillo, S., van de Knaap, E., Cong, B., Liu, J., Meller, J., Elber, R., Alpert, K.B. and Tanksley, S.D. (2000) *fw2.2*: a quantitative trait locus key to the evolution of tomato fruit size. *Science* 289, 85–88.

Frascaroli, E., Canè, M.A., Landi, P., Pea, G., Gianfranceschi, L., Villa, M., Morgante, M. and Pè, M.E. (2007) Classical genetic and quantitative trait loci analyses of heterosis in a maize hybrid between two elite inbred lines. *Genetics* 176, 625–644.

Frawely, W.J., Piatetsky-Shapiro, G. and Matheus, C.J. (1991) Knowledge discovery in databases: an overview. In: Piatetsky-Shapiro, G. and Frawely, W.J. (eds) *Knowledge Discovery in Databases.* AAAI Press, Menlo Park, California and MIT Press, Cambridge, Massachusetts, pp. 1–27.

Freeman, G.H. (1973) Statistical methods for the analysis of genotype–environment interactions. *Heredity* 31, 339–354.

Freudenreich, C.H., Stavenhagen, J.B. and Zakian, V.A. (1997) Stability of CTG:CAG trinucleotide repeat in yeast is dependent on its orientation in the genome. *Molecular and Cell Biology* 4, 2090–2098.

Fridman, E., Pleban, T. and Zamir, D. (2000) A recombination hotspot delimits a wild-species quantitative trait locus for tomato sugar content to 484 bp within an invertase gene. *Proceedings of the National Academy of Sciences of the United States of America* 97, 4718–4723.

Fridman, E., Carrari, F., Liu, Y.S., Fernie, A.R. and Zamir, D. (2004) Zooming in on a quantitative trait for tomato yield using interspecific introgressions. *Science* 305, 1786–1789.

Friedman, C., Borlawsky, T., Shagina, L., Xing, H.R. and Lussier, Y.A. (2006) Bio-ontology and text: bridging the modelling gap. *Bioinformatics* 22, 2421–2429.

Frisch, M. (2004) Breeding strategies: optimum design of marker-assisted backcross programs. In: Lörz, H. and Wenzl, G. (eds) *Biotechnology in Agriculture and Forestry*, Vol. 55. *Molecular Marker Systems in Plant Breeding and Crop Improvement.* Springer-Verlag, Berlin, pp. 319–334.

Frisch, M. and Melchinger, A.E. (2001) Marker-assisted backcrossing for simultaneous introgression of two genes. *Crop Science* 41, 1716–1725.

Frisch, M. and Melchinger, A.E. (2005) Selection theory for marker-assisted backcrossing. *Genetics* 170, 909–917.

Frisch, M. and Melchinger, A.E. (2008) Precision of recombination frequency estimates after random intermating with finite population sizes. *Genetics* 178, 597–600.

Frisch, M., Bohn, M. and Melchinger, A.E. (1999a) Comparison of selection strategies for marker-assisted backcrossing of a gene. *Crop Science* 39, 1295–1301.

Frisch, M., Bohn, M. and Melchinger, A.E. (1999b) Minimum sample size and optimal positioning of flanking markers in marker-assisted backcrossing for transfer of a target gene. *Crop Science* 39, 967–975.

Frisch, M., Bohn, M. and Melchinger, A.E. (2000) PLABSIM: software for simulation of marker-assisted backcrossing. *Journal of Heredity* 91, 86–87.

Fu, H. and Dooner, H.K. (2002) Intraspecific violation of genetic colinearity and its implications in maize. *Proceedings of the National Academy of Sciences of the United States of America* 99, 9573–9578.

Fu, X.D., Duc, L.T., Fontana, S., Bong, B.B., Tinjuangjun, P., Sudhakar, D., Twyman, R.M., Christou, P. and Kohli, A. (2000) Linear transgene constructs lacking vector backbone sequences generate low-copy number transgenic plants with simple integration patterns. *Transgenic Research* 9, 11–19.

Fu, Y., Wen, T.J., Ronin, Y.I., Chen, H.D., Guo, L., Mester, D.I., Yang, Y., Lee, M., Korol, A.B., Ashlock, D.A. and Schnable, P.S. (2006) Genetic dissection of intermated recombinant inbred lines using a new genetic map of maize. *Genetics* 174, 1671–1683.

Fu, Y.B., Peterson, G.W., Williams, D., Richards, K.W. and Fetch, J.M. (2005) Patterns of AFLP variation in a core subset of cultivated hexaploid oat germplasm. *Theoetical and Applied Genetics* 111, 530–539.

Fulton, T.M., Beck-Bunn, T., Emmatty, D., Eshed, Y., Lopez, J., Petiard, V., Uhlig, J., Zamir, D. and Tanksley, S.D. (1997) QTL analysis of an advanced backcross of *Lycopersicon peruvianum* to the cultivated tomato and comparisons with QTLs found in other wild species. *Theoretical and Applied Genetics* 95, 881–894.

Fulton, T.M., van der Hoeven, R., Eannetta, N.T. and Tanksley, S.D. (2002) Identification, analysis and utilization of conserved ortholog set markers for comparative genomics in higher plants. *The Plant Cell* 14, 1457–1467.

Furtado, A. and Henry, R.J. (2005) The wheat Em promoter drives reporter gene expression in embryo and aleurone tissue of transgenic barley and rice. *Plant Biotechnology Journal* 3, 421–434.

Gabriel, K.R. (1971) The biplot graphic display of matrices with application to principal component analysis. *Biometrika* 58, 453–467.

Gabriel, K.R. (1978) Least squares approximation of matrices by additive and multiplicative models. *Journal of the Royal Statistical Society, Series B* 40, 186–196.

Gale, M.D. (1975) High α-amylase breeding and genetical aspects of the problem. *Cereal Research Communications* 4, 231–243.

Gale, M.D. and Devos, K.M. (1998) Comparative genetics in the grasses. *Proceedings of the National Academy of Sciences of the United States of America* 95, 1971–1974.

Galinat, W.C. (1977) The origin of corn. In: Sprague, G.F. (ed.) *Corn and Corn Improvement*, 2nd edn. American Society of Agronomy, Madison, Wisconsin, pp. 1–48.

Gallais, A. and Bordes, J. (2007) The use of doubled haploids in recurrent selection and hybrid development in maize. *Crop Science* 47(S3), S190–S201.

Gallais, A., Moreau, L. and Charcosset, A. (2007) Detection of marker–QTL associations by studying change in marker frequencies with selection. *Theoretical and Applied Genetics* 114, 669–681.

Galperin, M.Y. (2008) The molecular biology database collection: 2008 update. *Nucleic Acids Research* 36, D2–D4.

Galperin, M.Y. and Koller, E. (2006) New metrics for comparative genomics. *Current Opinion in Biotechnology* 17, 440–447.

Gao, S., Martinez, C., Skinner, D.J., Krivanek, A.F., Crouch, J.H. and Xu, Y. (2008) Development of a seed DNA-based genotyping system for marker-assisted selection in maize. *Molecular Breeding* 22, 477–494.

Gao, Z., Xie, X., Ling, Y., Muthukrishnan, S. and Liang, G.H. (2005) *Agrobacterium tumefaciens*-mediated sorghum transformation using a mannose selection system. *Plant Biotechnology Journal* 3, 591–599.

Garcia, A.A., Kido, E.A., Meza, A.N., Souza, H.M., Pinto, L.R., Pastina, M.M., Leite, C.S., Silva, J.A., Ulian, E.C., Figueira, A. and Souza, A.P. (2006) Development of an integrated genetic map of a sugarcane (*Saccharum* spp.) commercial cross, based on a maximum-likelihood approach for estimation of linkage and linkage phases. *Theoretical and Applied Genetics* 112, 298–314.

Gauch, H.G., Jr (1988) Model selection and validation for yield trials with interaction. *Biometrics* 44, 705–715.

Gauch, H.G. (2006) Statistical analysis of yield trials by AMMI and GGE. *Crop Science* 46, 1488–1500.

Gauch, H.G. and Zobel, R.W. (1988) Predictive and postdictive success of statistical analysis of yield trials. *Theoretical and Applied Genetics* 76, 1–10.

Gauch, H.G. and Zobel, R.W. (1996) AMMI analysis of yield trials. In: Kang, M.S. and Hauch, H.G., Jr (eds) *Genotype-by-Environment Interaction.* CRC Press, Boca Raton, Florida, pp. 85–122.

Gauch, H.G. and Zobel, R.W. (1997) Identifying mega-environments and targeting genotypes. *Crop Science* 37, 311–326.

Gauch, H.G., Piepho, H.-P. and Annicchiarico, P. (2008) Statistical analysis of yield trials by AMMI and GGE: further considerations. *Crop Science* 48, 866–889.

Gaunt, T.R., Rodriguez, S., Zapata, C. and Day, I.N.M. (2006) MIDAS: software for analysis and visualisation of interallelic disequilibrium between multiallelic markers. *BMC Bioinformatics* 7, 227.

Gaut, B.S. and Ross-Ibarra, J. (2008) Selection on major components of angiosperm genomes. *Science* 320, 484–486.

Gayen, P., Madan, J.K., Kumar, R. and Sarkar, K.R. (1994) Chromosome doubling in haploids through colchicine. *Maize Genetics Cooperation Newsletter* 68, 65.

Gebhardt, C., Ballvora, A., Walkemeier, B., Oberhagemann, P. and Schüler, K. (2004) Assessing genetic potential in germplasm collections of crop plants by marker–trait association: a case study for potatoes with quantitative variation of resistance to late blight and maturity type. *Molecular Breeding* 13, 93–102.

Gedil, M.A., Wye, C., Berry, S., Segers, B., Peleman, J., Jones, R., Leon, A., Slabaugh, M.B. and Knapp, S.J. (2001) An integrated restriction fragment length polymorphism-amplified fragment length polymorphism linkage map for cultivated sunflower. *Genome* 44, 213–221.

Geldermann, H. (1975) Investigations on inheritance of quantitative characters in animals by gene markers. I. Methods. *Theoretical and Applied Genetics* 46, 319–330.

Geleta, L.F., Labuschagne, M.T. and Viljoen, C.D. (2004) Relationship between heterosis and genetic distance based on morphological traits and AFLP markers in pepper. *Plant Breeding* 123, 467–473.

Gelfand, M.S., Mironow, A.A. and Pevzner, P.A. (1996) Gene recognition via spliced sequence alignment. *Proceedings of the National Academy of Sciences of the United States of America* 93, 9061–9066.

Gene Ontology Consortium (2000) Gene ontology: tool for the unification of biology. *Nature Genetics* 25, 25–29.

George, E.I. and McMulloch, R.E. (1993) Variable selection via Gibbs sampling. *Journal of The American Statistical Association* 91, 883–904.

Georgiady, M.S., Whitkus, R.W. and Lord, E.M. (2002) Genetic analysis of traits distinguishing outcrossing and self-pollinating forms of currant tomato, *Lycopersicon pimpinellifolium* (Jusl.) Mill. *Genetics* 161, 333–344.

Gepts, P. (2006) Plant genetic resources conservation and utilization: the accomplishments and future of a societal insurance policy. *Crop Science* 46, 2278–2292.

Gerdes, J.T. and Tracy, W.F. (1993) Pedigree diversity within the Lancaster Surecrop heterotic group of maize. *Crop Science* 33, 334–337.

Gerdes, J.T., Behr, C.F., Coors, J.G. and Tracy, W.F. (1993) *Compilation of North America Maize Breeding Programs.* Crop Science Society of America, Madison, Wisconsin.

Gernand, D., Rutten, T., Varshney, A., Rubtsova, M., Prodanovic, S., Brüß, C., Kumlehn, J., Matzk, F. and Houben, A. (2005) Uniparental chromosome elimination at mitosis and interphase in wheat and pearl millet crosses involves micronucleus formation, progressive heterochromatinization and DNA fragmentation. *The Plant Cell* 17, 2431–2438.

Gerry, N.P., Witowski, N.E., Day, J., Hammer, R.P., Barany, G. and Barany, F. (1999) Universal DNA microarray method for multiplex detection of low abundance point mutations. *Journal of Molecular Biology* 292, 251–262.

Gethi, J.G., Labate, J.A., Lamkey, K.R., Smith, M.E. and Kresovich, S. (2002) SSR variation in important U.S. maize inbred lines. *Crop Science* 42, 951–957.

Gibbon, B.C. and Larkins, B.A. (2005) Molecular genetic approaches to developing quality protein maize. *Trends in Genetics* 21, 227–233.

Gibrat, J.F. and Marin, A. (2007) Detecting protein function from genome sequences. In: Morot-Gaudry, J.F., Lea, P. and Briat, J.F. (eds) *Functional Plant Genomics*. Science Publishers, Enfield, New Hampshire, pp. 87–106.

Gibson, G. and Weir, B. (2005) The quantitative genetics of transcription. *Trends in Genetics* 21, 616–623.

Gibson, S. and Somerville, C. (1993) Isolating plant genes. *Trends in Biotechnology* 11, 306–313.

Gill, B.S., Appels, R., Botha-Oberholster, A.-M., Buell, C.R., Bennetzen, J.L., Chalhoub, B., Chumley, F., Dvořák, J., Iwanaga, M., Keller, B., Li, W., McCombie, W.R., Ogihara, Y., Quetier, F. and Sasaki, T. (2004) A workshop report on wheat genome sequencing: International Genome Research on Wheat Consortium. *Genetics* 168, 1087–1096.

Gimelfarb, A. and Lande, R. (1994a) Simulation of marker-assisted selection in hybrid populations. *Genetical Research* 63, 39–47.

Gimelfarb, A. and Lande, R. (1994b) Simulation of marker-assisted selection for non-additive traits. *Genetical Research* 64, 127–136.

Gimelfarb, A. and Lande, R. (1995) Marker-assisted selection and marker-QTL associations in hybrid populations. *Theoretical and Applied Genetics* 91, 522–528.

Giovannoni, J.J., Wing, R.A., Ganal, M.W. and Tanksley, S.D. (1991) Isolation of molecular markers from specific chromosome intervals using DNA pools from existing populations. *Nucleic Acids Research* 19, 6553–6558.

Gish, W. and States, D.J. (1993) Identification of protein coding regions by database similarity search. *Nature Genetics* 3, 266–272.

Gizlice, Z., Carter, T.E., Jr and Burton, J.W. (1993) Genetic diversity in North American soybean: II. Prediction of heterosis in F_2 populations of southern founding stock using genetic similarity measures. *Crop Science* 33, 620–626.

Glass, G.V. (1976) Primary, secondary and meta-analysis of research. *Educational Researcher* 5, 3–8.

Glazier, A.M., Nadeau, J.H. and Aitman, T.J. (2002) Finding genes that underlie complex traits. *Science* 298, 2345–2349.

Gleave, A.P., Mitra, D.S., Mudge, S.R. and Morris, B.A.M. (1999) Selectable marker-free transgenic plants without sexual crossing: transient expression of cre recombinase and use of a conditional lethal dominant gene. *Plant Molecular Biology* 40, 223–235.

Gleba, Y., Marillonnet, S. and Klimyuk, V. (2004) Engineering viral expression vectors for plants: the ‘full virus’ and the ‘deconstructed virus’ strategies. *Current Opinion in Plant Biology* 7, 182–188.

Gleba, Y., Klimyuk, V. and Marillonnet, S. (2005) Magnifection – a new platform for expressing recombinant vaccines in plants. *Vaccine* 23, 2042–2048.

Goderis, I.J.W.M., De Bolle, M.F.C., François, I.E.J.A., Wouters, P.F.J., Broekaert, W.F. and Cammue, B.P.A. (2002) A set of modular plant transformation vectors allowing flexible insertion of up to six expression units. *Plant Molecular Biology* 50, 17–27.

Godshalk, E.B., Lee, M. and Lamkey, K.R. (1990) Relationship of restriction fragment length polymorphisms to single-cross hybrid performance of maize. *Theoretical and Applied Genetics* 80, 273–280.

Goedeke, S., Hensel, G., Kapusi, E., Gahrtz, M. and Kumlehn, J. (2007) Transgenic barley in fundamental research and biotechnology. *Transgenic Plant Journal* 1, 104–117.

Goff, S.A., Ricke, D., Lan, T.H., Presting, G., Wang, R., Dunn, M., Glazebrook, J., Sessions, A., Oeller, P., Varma, H., Hadley, D., Hutchison, D., Martin, C., Katagiri, F., Lange, B.M., Moughamer, T., Xia, Y., Budworth, P., Zhong, J., Miguel, T., Paszkowski, U., Zhang, S., Colbert, M., Sun, W.L., Chen, L., Cooper, B., Park, S., Wood, T.C., Mao, L., Quail, P., Wing, R., Dean, R., Yu, Y., Zharkikh, A., Shen, R., Sahasrabudhe, S., Thomas, A., Cannings, R., Gutin, A., Pruss, D., Reid, J., Tavtigian, S., Mitchell, J., Eldredge, G., Scholl, T., Miller, R.M., Bhatnagar, S., Adey, N., Rubano, T., Tusneem, N., Robinson, R., Feldhaus, J., Macalma, T., Oliphant, A. and Briggs, S. (2002) A draft sequence of the rice genome (*Oryza sativa* L. ssp. *japonica*). *Science* 296, 92–100.

Goffinet, B. and Gerber, S. (2000) Quantitative trait loci: a meta-analysis. *Genetics* 155, 463–473.

Goldman, I.L. (1999) Inbreeding and outbreeding in the development of a modern heterosis concept. In: Coors, J.G. and Pandey, S. (eds) *Genetics and Exploitation of Heterosis in Crops*. ASA-CSSA-SSSA, Madison, Wisconsin, pp. 7–18.

Goldman, I.L. (2000) Prediction in plant breeding. *Plant Breeding Reviews* 19, 15–40.

Goldman, I.L., Rocheford, T.R. and Dudley, J.W. (1993) Quantitative trait loci influencing protein and starch concentration in the Illinois long term selection maize strains. *Theoretical and Applied Genetics* 87, 217–224.

Goldman, I.L., Rocheford, T.R. and Dudley, J.W. (1994) Molecular markers associated with maize kernel oil concentration in the Illinois High Protein × Illinois Low Protein Cross. *Crop Science* 34, 908–915.

Goldsbrough, A.P., Lastrella, C.N. and Yoder, J.I. (1993) Transposition mediated re-positioning and subsequent elimination of marker genes from transgenic tomato. *Bio/Technology* 11, 1286–1292.

Gollob, H.F. (1968) A statistical model which combines features of factor analytic and analysis of variance. *Psychometrika* 33, 73–115.

Goodin, M.M., Dietzgen, R.G., Schichnes, D., Ruzin, S. and Jackson, A.O. (2002) pGD vectors: versatile tools for the expression of green and red fluorescent protein fusions in agroinfiltrated plant leaves. *The Plant Journal* 31, 375–383.

Goodman, R.E., Vieths, S., Sampson, H.A., Hill, D., Ebisawa, M., Tyaler, S.L. and van Ree, R. (2008) Allergenicity assessment of genetically modified crops – what make sense? *Nature Biotechnology* 26, 73–81.

Goodnight, C.J. (2004) Gene interaction and selection. *Plant Breeding Reviews* 24 (Part 2), 269–291.

Gorg, A., Obermaier, C., Boguth, G. and Weiss, W. (1999) Recent developments in two-dimensional gel electrophoresis with immobilized pH gradients: wide pH gradients up to pH 12, longer separation distances and simplified procedures. *Electrophoresis* 20, 712–717.

Grandillo, S. and Tanksley, S.D. (1996) QTL analysis of horticultural traits differentiating the cultivated tomato from the closely related species *Lycopersicon pimpinellifolium*. *Theoretical and Applied Genetics* 92, 935–951.

Graner, A., Jahoor, A., Schondelmaier, J., Siedler, H., Pollen, K., Fischbeck, G., Wenzel, G. and Herrmann, R.G. (1991) Construction of an RFLP map of barley. *Theoretical and Applied Genetics* 83, 250–256.

Grapes, L., Dekkers, J.C.M., Rothschild, M.F. and Fernando, R.L. (2004) Comparing linkage disequilibrium-based methods for fine mapping quantitative trait loci. *Genetics* 166, 1561–1570.

Green, P.J. (1995) Reversible jump Markov chain Monte Carlo computation and Bayesian model determination. *Biometrika* 82, 711–732.

Greenbaun, D., Smith, A. and Gerstein, M. (2005) Impediments to database interoperation: legal issues and security concerns. *Nucleic Acids Research* 33, D3–D4.

Greene, S.L. and Guarino, L. (eds) (1999) *Linking Genetic Resources and Geography: Emerging Strategies for Conserving and Using Crop Biodiversity*. American Society of Agronomy (ASA) and Crop Science Society of America (CSSA), Madison, Wisconsin.

Gregory, B.D., Yazaki, J. and Ecker, J.R. (2008) Utilizing tiling microarrays for whole-genome analysis in plants. *The Plant Journal* 53, 636–644.

Groos, C., Robert, N., Bervas, E. and Charmet, G. (2003) Genetic analysis of grain protein-content, grain yield and thousand-kernel weight in bread wheat. *Theoretical and Applied Genetics* 106, 1032–1040.

Grosset, J., Alary, R., Gautier, M.F., Menossi, M., Martinez-Izquierdo, J.A. and Joudrier, P. (1997) Characterization of a barley gene coding for an alpha-amylase inhibitor subunit (CMd protein) and analysis of its promoter in transgenic tobacco plants and in maize kernels by microprojectile bombardment. *Plant Molecular Biology* 34, 331–338.

Grupe, A., Germer, S., Usuka, J., Aud, D., Belknap, J.K., Klein, R.F., Ahluwalia, M.K., Higuchi, R. and Peltz, G. (2001) *In silico* mapping of complex disease-related traits in mice. *Science* 292, 1915–1918.

Gu, S., Pakstis, A.J. and Kidd, K.K. (2005) HAPLOT: a graphical comparison of haplotype blocks, tagSNP sets and SNP variation for multiple populations. *Bioinformatics* 21, 3938–3939.

Guidetti, G. (1998) Seed terminator and mega-merger threaten food and freedom. Available at: http://www.sustainable-city.org/articles/terminat.htm (accessed 17 November 2009).

Guo, B., Sleper, D.A., Sun, J., Nguyen, H.T., Arelli, P.R. and Shannon, J.G. (2006) Pooled analysis of data from multiple quantitative trait locus mapping populations. *Theoretical and Applied Genetics* 113, 39–48.

Guo, M., Rupe, M.A., Zinselmeier, C., Habben, J., Bowen, B.A. and Smith, O.S. (2004) Allelic variation of gene expression in maize hybrids. *The Plant Cell* 16, 1707–1716.

Guo, M., Rupe, M.A., Yang, X., Crasta, O., Zinselmeier, C., Smith, O.S. and Bowen, B. (2006) Genome-wide transcript analysis of maize hybrids: allelic additive gene expression and yield heterosis. *Theoretical and Applied Genetics* 113, 831–845.

Gupta, P.K. and Rustgi, S. (2004) Molecular markers from the transcribed/expressed region of the genome in higher plants. *Functional and Integrated Genomics* 4, 139–162.

Gur, A. and Zamir, D. (2004) Unused natural variation can lift yield barriers in plant breeding. *PLoS Biology* 2(10), e245.

Gurib-Fakim, A. (2006) Medicinal plants: traditions of yesterday and drugs of tomorrow. *Molecular Aspects of Medicine* 27, 1–93.

Haanstra, J.P.W., Wye, C., Verbakel, H., Meijer-Dekens, F., Van den Berg, P., Odinot, P., van Heusden, A.W., Tanksely, S., Lindhout, P. and Peleman, J. (1999) An integrated high-density RFLP-AFLP map of tomato based on two *Lycopersicon esculentum* × *L. pennellii* F_2 populations. *Theoretical and Applied Genetics* 99, 254–271.

Haberer, G., Young, S., Bharati, A.K., Gundlach, H., Raymond, C., Fuks, G., Butler, E., Wing, R.A., Rounsley, S., Birren, B., Nusbaum, C., Mayer, K.F.X. and Messing, J. (2005) Structure and architecture of the maize genome. *Plant Physiology* 139, 1612–1624.

Hackett, C.A., Meyer, R.C. and Thomas, W.T.B. (2001) Multi-trait QTL mapping in barley using multivariate regression. *Genetical Research* 77, 95–106.

Hagberg, A. and Hagberg, G. (1980) High frequency of spontaneous haploids in the progeny of an induced mutation barley. *Hereditas* 93, 341–343.

Hahn, W.J. and Grifo, F.T. (1996) Molecular markers in plant conservation genetics. In: Sobral, B.W.S. (ed.) *The Impact of Plant Molecular Genetics.* Birkhäuer, Boston, Massachusetts, pp. 113–136.

Hajdukiewicz, P., Svab, Z. and Maliga, P. (1994) The small, versatile pPZP family of *Agrobacterium* binary vectors for plant transformation. *Plant Molecular Biology* 25, 989–994.

Hajdukiewicz, P.T.J., Gilbertson, L. and Staub, J.M. (2001) Multiple pathways for Cre/lox-mediated recombination in plastids. *The Plant Journal* 27, 161–170.

Haldane, J.B.S. (1919) The combination of linkage values and the calculation of distance between the loci of linkage factors. *Journal of Genetics* 8, 299–309.

Haldane, J.B.S. and Smith, C.A.B. (1947) A new estimate of the linkage between the genes for colour-blindness and haemophilia in man. *Annals of Eugenics* 14, 10–31.

Haldane, J.B.S. and Waddington, C.H. (1931) Inbreeding and linkage. *Genetics* 16, 357–374.

Haldrup, A., Petersen, S.G. and Okkels, F.T. (1998a) Positive selection: a plant selection principle based on xylose isomerase, an enzyme used in the food industry. *Plant Cell Reports* 18, 76–81.

Haldrup, A., Petersen, S.G. and Okkels, F.T. (1998b) The xylose isomerase gene from *Thermoanaerobacterium thermosulfurogenes* allows effective selection of transgenic plant cells using D-xylose as the selection agent. *Plant Molecular Biology* 37, 287–296.

Haley, C. (1999) Advances in quantitative trait locus mapping. In: Dekkers, J.C.M., Lamont, S.J. and Rothschild, M.F. (eds) *From Jay Lush to Genomics: Visions for Animal Breeding and Genetics.* Animal Breeding and Genetics Group, Department of Animal Science, Iowa State University, Ames, Iowa, pp. 47–59.

Haley, C.S. and Knott, S.A. (1992) A simple regression method for mapping quantitative trait loci in line crosses using flanking markers. *Heredity* 69, 315–324.

Haley, C.S., Knott, S.A. and Elsen, J.-M. (1994) Mapping quantitative trait loci in crosses between outbred lines using least squares. *Genetics* 136, 1195–1207.

Halfhill, M.D., Richards, H.A., Mabon, S.A. and Stewart, C.N., Jr (2001) Expression of GFP and Bt transgenes in *Brassica napus* and hybridization and introgression with *Brassica rapa. Theoretical and Applied Genetics* 103, 362–368.

Halfhill, M.D., Zhu, B., Warwick, S.I., Raymer, P.L., Millwood, R.J., Weissinger, A.K. and Stewart, C.N., Jr (2004a) Hybridization and backcrossing between transgenic oilseed rape and two related weed species under field conditions. *Environmental Biosafety Research* 3, 73–81.

Halfhill, M.D., Millwood, R.J. and Stewart, C.N., Jr (2004b) Green fluorescent protein quantification in whole plants. In: Peña, L. (ed.) *Methods in Molecular Biology,* Vol. 286. *Transgenic Plants: Methods and Protocols.* Humana Press Inc., Totowa, New Jersey, pp. 215–225.

Hall, J.G., Eis, P.S., Law, S.M., Reynaldo, L.P., Prudent, J.R., Marshall, D.J., Allawi, H.T., Mast, A.L., Dahlberg, J.E., Kwiatkowski, R.W., de Arruda, M., Neri, B.P. and Lyamichev, V.I. (2000) Sensitive detection of DNA polymorphisms by the serial invasive signal amplification reaction. *Proceedings of the National Academy of Sciences of the United States of America* 97, 8272–8277.

Hallauer, A.R. (1990) Methods used in developing maize inbreds. *Maydica* 35, 1–16.
Hallauer, A.R. (2007) History, contribution and future of quantitative genetics in plant breeding: lessons from maize. *Crop Science* 47(S3), S4–S19.
Hallauer, A.R. and Miranda, J.B. (1988) *Quantitative Genetics in Maize Breeding*, 2nd edn. Iowa State University Press, Ames, Iowa.
Hallauer, A.R., Russell, W.A. and Lamkey, K.R. (1988) Corn breeding. In: Sprague, G.F. and Dudley, J.W. (eds) *Corn and Corn Improvement*, 3rd edn. ASA-CSSA-SSSA, Madison, Wisconsin, pp. 463–564.
Hallauer, A.R., Ross, A.J. and Lee, M. (2004) Long-term divergent selection for ear length in maize. *Plant Breeding Reviews* 24 (Part 2), 153–168.
Halpin, C. and Boerjan, W. (2003) Stacking transgenes in forest trees. *Trends in Plant Science* 8, 363–365.
Halpin, C., Barakate, A., Askari, B.M., Abbott, J.C. and Ryan, M.D. (2001) Enabling technologies for manipulating multiple genes on complex pathways. *Plant Molecular Biology* 47, 295–310.
Hamilton, C.M. (1997) A binary-BAC system for plant transformation with high-molecular-weight DNA. *Gene* 200, 107–116.
Hamilton, C.M., Frary, A., Lewis, C. and Tanksley, S.D. (1996) Stable transfer of intact high molecular weight DNA into plant chromosomes. *Proceedings of the National Academy of Sciences of the United States of America* 93, 9975–9979.
Hammer, G.L., Kropff, M.J., Sinclair, T.R. and Porter, J.R. (2002) Future contribution of crop modeling: from heuristics and supporting decision making to understanding genetic regulation and aiding crop improvement. *European Journal of Agronomy* 18, 15–31.
Hammer, G.L., Chapman, S., van Oosterom, E. and Podlich, D.W. (2005) Trait physiology and crop modeling as a framework to link phenotypic complexity to underlying genetic systems. *Australian Journal of Agricultural Research* 56, 947–960.
Hammond, M.P. and Birney, E. (2004) Genome information resources – developments at Ensembl. *Trends in Genetics* 20, 268–272.
Han, B. and Xue, Y. (2003) Genome-wide intraspecific DNA-sequence variations in rice. *Current Opinion in Plant Biology* 6, 134–138.
Han, O.K., Kaga, A., Isemura, T., Wang, X.W., Tomooka, N. and Vaughan, D.A. (2005) A genetic linkage map for azuki bean [*Vigna angularis* (Willd.) Ohwi & Ohashi]. *Theoretical and Applied Genetics* 111, 1278–1287.
Han, X., Aslanian, A. and Yates, J.R. III (2008) Mass spectrometry for proteomics. *Current Opinion in Chemical Biology* 12, 483–490.
Hanash, S. (2003) Disease proteomics. *Nature* 422, 226–232.
Hanin, M. and Paszkowski, J. (2003) Plant genome modification by homologous recombination. *Current Opinion in Plant Biology* 6, 157–162.
Hanocq, E., Laperche, A., Jaminon, O., Lainé, A.-L. and Le Guis, J. (2007) Most significant genome regions involved in the control of earliness traits in bread wheat, as revealed by QTL meta-analysis. *Theoretical and Applied Genetics* 114, 569–584.
Hansen, B.G., Halkier, B.A. and Kliebenstein, D.J. (2008) Identifying the molecular basis of QTLs: eQTLs add a new dimension. *Trends in Plant Science* 13, 72–77.
Hansen, M., Kraft, T., Ganestam, S., Säll, T. and Nilsson, N.-O. (2001) Linkage disequilibrium mapping of the bolting gene in sea beet using AFLP markers. *Genetical Research* 77, 61–66.
Hanson, W.D. (1959) Early generation analysis of lengths of heterozygous chromosome segments around a locus held heterozygous with backcrossing or selfing. *Genetics* 44, 833–837.
Harding, K. (2004) Genetic integrity of cryopreserved plant cells: a review. *Cryo Letters* 25, 3–22.
Harlan, H.V. and Pope, M.N. (1922) The use and value of back-crosses in small grain breeding. *Journal of Heredity* 13, 319–322.
Harlan, H.V., Martini, M.L. and Stevens, H. (1940) A study of methods in barley breeding. *USDA Technical Bulletin* 720.
Harlan, J. (1965) The possible role of weed races in the evolution of cultivated plants. *Euphytica* 14, 173–176.
Harlan, J.R. (1971) Agricultural origins: centers and noncenters. *Science* 174, 468–474.
Harlan, J. (1992) *Crops and Man*, 2nd edn. Crop Science Society of America, Madison, Wisconsin.
Harlan, J.R. (1987) Gene centers and gene utilization in American agriculture. In: Yeatman, C.W., Kafton, D. and Wilkes, G. (eds) *Plant Genetic Resources: a Conservation Imperative*. Westview Press, Boulder, Colorado, pp. 111–129.

Harlan, J.R. and de Wet, J.M.J. (1971) Towards a rational classification of cultivated plants. *Taxon* 20, 509–517.

Harper, B.K., Mabon, S.A., Leffel, S.M., Halfhill, M.D., Richards, H.A., Moyer, K.A. and Stewart, C.N., Jr (1999) Green fluorescent protein as a marker for expression of a second gene in transgenic plants. *Nature Biotechnology* 17, 1125–1129.

Harris, S.A. (1999) Molecular approaches to assessing plant diversity. In: Benson, E.E. (ed.) *Plant Conservation Biotechnology*. Taylor & Francis Ltd, London, pp. 11–24.

Hart, G.E., Gale, M.D. and McIntosh, R.A. (1993) Linkage maps of *Triticum aestivum* (Hexaploid wheat, $2n = 42$, genome A, B and D) and *T. tauschii* ($2n = 14$, genome D). In: O'Brien, S.J. (ed.) *Genetic Maps: Locus Maps of Complex Genomes*. Cold Spring Harbor Laboratory Press, Cold Spring Harbor, New York, pp. 6.204–6.219.

Harushima, Y., Kurata, N., Yano, M., Nagamura, Y., Sasaki, T., Minobe, Y. and Nakagahra, M. (1996) Detection of segregation distortions in an *indica–japonica* rice cross using a high-resolution molecular map. *Theoretical and Applied Genetics* 92,145–150.

Harushima, Y., Yano, M., Shomura, A., Sato, M., Shimano, T., Kuboki, Y., Yamamoto, T., Lin, S.Y., Antonio, B.A., Parco, A., Kajiya, H., Huang, N., Yamamoto, K., Nagamura, Y., Kurata, N., Khush, G.S. and Sasaki, T. (1998) A high-density rice genetic linkage map with 2275 markers using a single F_2 population. *Genetics* 148, 479–494.

Haseloff, J., Siemering, K.P., Prasher, D. and Hodge, S. (1997) Removal of a cryptic intron and subcellular localization of green fluorescent protein are required to mark transgenic *Arabidopsis* plants brightly. *Proceedings of the National Academy of Sciences of the United States of America* 94, 2122–2127.

Havey, M.J. (1998) Molecular analyses and heterosis in the vegetables: can we breed them like maize? Lamkey, K.R. and Staub, J.E. (eds) *Concepts and Breeding of Heterosis in Crop Plants*. Crop Science Society of America (CSSA), Madison, Wisconsin, pp. 109–116.

Hawtin, G. (1998) Conservation of agrobiodiversity for tropical agriculture. In: Chopra, V.L., Singh, R.B and Varma, A. (eds) *Crop Productivity and Sustainability – Shaping the Future, Proceedings of the 2nd International Crop Science Congress*. Oxford & IBH Publishing Co., New Delhi, pp. 917–925.

Hayes, B. and Goddard, M.E. (2001) The distribution of the effects of genes affecting quantitative traits in livestock. *Genetics Selection Evolution* 33, 209–229.

Hazekamp, Th. (2002) The potential role of passport data in the conservation and use of plant genetic resources. In: Engels, J.M.M., Ramanatha Rao, V., Brown, A.H.D. and Jackson, M.T. (eds) *Managing Plant Genetic Diversity*. International Plant Genetic Resources Institute, Rome, pp. 185–194.

Hazekamp, Th., Serwinski, J. and Alercia, A. (1997) Mulit-crop passport descriptors. In: Lipmann, E., Jongen, M.W.M., Hintum, Th.J.L. van, Gass, T. and Maggioni, L. (compilers) *Central Crop Databases: Tools for Plant Genetic Resources Management*. Report of a Workshop, 13–16 October 1996, Budapest, Hungary. International Plant Genetic Resources Institute, Rome, Italy/CGN, Wageningen, Netherlands, pp. 35–39.

Hazen, S.P., Pathan, M.S., Sanchez, A., Baxter, I., Dunn, M., Estes, B., Chang, H.-S., Zhu, T., Kreps, J.A. and Nguyen, H.T. (2005) Expression profiling of rice segregating for drought tolerance QTL using a rice genome array. *Functional and Integrative Genomics* 5, 104–116.

He, P., Li, J.Z., Zheng, X.W., Shen L.S., Lu, C.F., Chen, Y. and Zhu, L.H. (2001) Comparison of molecular linkage maps and agronomic trait loci between DH and RIL populations derived from the same rice cross. *Crop Science* 41, 1240–1246.

He, X.H. and Zhang, Y.M. (2008) Mapping epistatic quantitative trait loci underlying endosperm traits using all markers on the entire genome in a random hybridization design. *Heredity* 101, 39–47.

He, Y., Chen, C., Tu, J., Zhou, P., Jiang, G., Tan, Y., Xu, C. and Zhang, Q. (2002) Improvement of an elite rice hybrid, Shanyou 63, by transformation and maker-assisted selection. In: *Abstracts of the Fourth International Symposium on Hybrid Rice*, 14–17 May 2002, Hanoi, Vietnam, p. 43.

He, Y., Li, X., Zhang, J., Jiang, G., Liu, S., Chen, S., Tu, J., Xu, C. and Zhang, Q. (2004) Gene pyramiding to improve hybrid rice by molecular marker technique. *4th International Crop Science Congress*. Available at: http://www.cropscience.org.au/icsc2004/ (accessed 17 November 2009).

He, Z., Fu, Y., Si, H., Hu, G., Zhang, S., Yu, Y. and Sun, Z. (2004) Phosphomannose-isomerase (*pmi*) gene as a selectable marker for rice transformation via *Agrobacterium*. *Plant Science* 166, 17–22.

Heckenberger, M., Bohn, M., Maurer, H.P., Frisch, M. and Melchinger, A.E. (2005a) Identification of essentially derived varieties with molecular markers: an approach based on statistical test theory and computer simulations. *Theoretical and Applied Genetics* 111, 598–608.

Heckenberger, M., Bohn, M., Klein, D. and Melchinger, A.E. (2005b) Identification of essentially derived varieties obtained from biparental crosses of homozygous lines: II. Morphological distances and heterosis in comparison with simple sequence repeat and amplified fragment length polymorphism data in maize. *Crop Science* 45, 1132–1140.

Heckenberger, M., Muminovic, J., van der Voort, J.R., Peleman, J., Bohn, M. and Melchinger, A.E. (2006) Identification of essentially derived varieties from biparental crosses of homogenous lines. III. AFLP data from maize inbreds and comparison with SSR data. *Molecular Breeding* 17, 111–125.

Heckenberger, M., Maurer, H.P., Melchinger, A.E. and Frisch, M. (2008) The Plabsoft database: a comprehensive database management system for integrating phenotypic and genomic data in academic and commercial plant breeding programs. *Euphytica* 161, 173–179.

Hedden, P. (2003) The genes of the green revolution. *Trends in Genetics* 19, 5–19.

Hedgecock, D., Lin, J.Z., DeCola, S., Haudenschild, C., Meyer, E., Manahan, D.T. and Bowen, B. (2002) Analysis of gene expression in hybrid Pacific oysters by massively parallel signature sequencing. *Plant & Animal Genome X Conference Abstract.* Available at: http://www.intl-pag.org/pag/10/abstracts/PAGX_W15.html (accessed 30 June 2007).

Hedges, L.V. and Olkin, I. (1985) *Statistical Methods for Meta-analysis.* Academic Press, Orlando, Florida.

Heisey, P.W., King, J.L. and Rubenstein, K.D. (2005) Patterns of public sector and private-sector patenting in agricultural biotechnology. *AgBioForum* 8, 73–82.

Heitz, A. (1998) Intellectual property rights and plant variety protection in relation to demands of the world trade organization and farmers in sub-Saharan Africa. In: *Proceedings of the Regional Technical Meeting on Seed Policy and Programmes for Sub-Saharan Africa*, Abidjan, Côte d'Ivoire, 23–27 November 1998. Available at: http://www.fao.org/ag/agp/AGPS/abidjan/tabcont.htm (accessed 17 November 2009).

Helentjaris, T. and Briggs, K. (1998) Are there too many genes in maize? *Maize Genetics Cooperation Newsletter* 72, 39–40.

Helentjaris, T., Cushman, M.A.T. and Winkler, R. (1992) Developing a genetic understanding of agronomy traits with complex inheritance. In: Dettee, Y., Dumas, C. and Gallais, A. (eds) *Reproductive Biology and Plant Breeding.* Springer-Verlag, Berlin, pp. 397–406.

Helfer, L.R. (2006) The demise and rebirth of plant variety protection: a comment on obsolescence in intellectual property. *Regimes. Public Law and Legal Theory* (Vanderbilt University Law School), Working Paper Number 06–28. Vanderbilt University, Nashville, Tennessee.

Hellens, R., Mullineaux, P. and Klee, H. (2000) Technical focus: a guide to *Agrobacterium* binary Ti vectors. *Trends in Plant Science* 5, 446–451.

Hellens, R.P., Edwards, E.A., Leyland, N.R., Bean, S. and Mullineaux, P.M. (2000) pGreen, a versatile and flexible binary Ti vector for *Agrobacterium*-mediated plant transformation. *Plant Molecular Biology* 42, 819–832.

Henderson, C.R. (1975) Best linear unbiased estimation and prediction under a selection model. *Biometrics* 31, 423–447.

Henikoff, S. and Comai, L. (2003) Single-nucleotide mutations for plant functional genomics. *Annual Review of Plant Biology* 54, 375–401.

Henry, Y., De Buyser, J., Agache, S., Parker, B.B. and Snape, J.W. (1988) Comparison of methods of haploid production and performance of wheat lines produced by doubled haploidy and single seed descent. In: Miller, T.E. and Koebner, R.M.D. (eds) *Proceedings of 7th International Wheat Genetics Symposium,* Cambridge, 13–19 July 1988. Institute of Plant Science Research, Cambridge, UK, pp. 1087–1092.

Henson-Apollonio, V. (2007) Impacts of intellectual property rights on marker-assisted selection research and application for agriculture in developing countries. In: Guimarães, E.P., Ruane, J., Scherf, B.D., Sonnino, A. and Dargie, J.D. (eds) *Marker-Assisted Selection, Current Status and Future Perspectives in Crops, Livestock, Forestry and Fish.* Food and Agriculture Organization of the United Nations, Rome, pp. 405–425.

Herring, R.J. (2008) Opposition to transgenic technologies: ideology, interests and collective action frames. *Nature Reviews Genetics* 9, 458–463.

Heun, M., Kennedy, A.E., Anderson, J.A., Lapitan, N.L.V., Sorrells, M.E. and Tanksley, S.D. (1991) Construction of a restriction fragment length polymorphism map for barley (*Hordeum vulgare*). *Genome* 34, 437–447.

Hiatt, A.C., Cafferkey, R. and Bowdish, K. (1989) Production of antibodies in transgenic plants. *Nature* 342, 76–78.

Hiei, Y., Ohta, S., Komari, T. and Kumashiro, T. (1994) Efficient transformation of rice (*Oryza sativa* L.) mediated by *Agrobacterium* and sequence analysis of the boundaries of the T-DNA. *The Plant Journal* 6, 271–282.

Hiei, Y., Komari, T. and Kubo, T. (1997) Transformation of rice mediated by *Agrobacterium tumefaciens. Plant Molecular Biology* 35, 205–218.

Hijmans, R.J., Guarino, L., Cruz, M. and Rojas, E. (2001) Computer tools for spatial analysis of plant genetic resources data. 1. DIVA-GIS. *Plant Genetic Resources Newsletter* 127, 15–19.

Hillel, D. and Rosenzweig, C. (2005) The role of biodiversity in agronomy. *Advances in Agronomy* 88, 1–34.

Hillel, J., Avner, R., Baxter-Jones, C., Dunnington, E.A., Cahaner, A. and Siegel, P.B. (1990) DNA fingerprints from blood mixes in chickens and turkeys. *Animal Biotechnology* 2, 201–204.

Hillenkamp, F. and Köster, H. (1999) Infrared matrix-assisted laser desorption/ionization mass spectrometric analysis of macro-molecules. Patent EP 1075545.

Himmelbach, A., Zierold, U., Hensel, G., Riechen, J., Douchkov, D., Schweizer, P. and Kumlehn, J. (2007) A set of modular binary vectors for transformation of cereals. *Plant Physiology* 145, 1192–1200.

Hintum, Th.J.L. van (1999) The Core Selector, a system to generate representative selections of germplasm accessions. *Plant Genetic Resources Newsletter* 118, 64–67.

Hird, D.L., Paul, W., Hollyoak, J.S. and Scott, R.J. (2000) The restoration of fertility in male sterile tobacco demonstrates that transgene silencing can be mediated by T-DNA that has no DNA homology to the silenced transgene. *Transgenic Research* 9, 91–102.

Hirochika, H. (2003) Insertional mutagenesis in rice using the endogenous retrotransposon. In: Mew, T.W., Brar, D.S., Peng, S., Dawe, D. and Hardy, B. (eds) *Rice Science: Innovations and Impact for Livelihood, Proceedings of the International Rice Research Conference*, 16–19 September 2002, Beijing, China. International Rice Research Institute, Chinese Academy of Engineering and Chinese Academy of Agricultural Sciences, pp. 205–212.

Hirochika, H., Guiderdoni, E., An, G., Hsing, Y.I., Eun, M.Y., Han, C.D., Upadhyaya, N., Ramachandran, S., Zhang, Q., Pereira, A., Sundaresan, V. and Leung, H. (2004) Rice mutant resources for gene discovery. *Plant Molecular Biology* 54, 325–334.

Hittalmani, S., Parco, A., Mew, T.V., Zeigler, R.S. and Huang, N. (2000) Fine mapping and DNA marker-assisted pyramiding of the three major genes for blast resistance in rice. *Theoretical and Applied Genetics* 100, 1121–1128.

Hodgkin, T. and Ramanatha Rao, V. (2002) People, plant and DNA: technical aspects of conserving and using plant genetic resources. In: Engels, J.M.M., Ramanatha Rao, V., Brown, A.H.D. and Jackson, M.T. (eds) *Managing Plant Genetic Diversity.* International Plant Genetic Resources Institute, Rome, pp. 469–480.

Hodson, D.P. and White, J.W. (2007) Use of spatial analyses for global characterization of wheat-based production systems. *Journal of Agricultural Science* 145, 115–125.

Hodson, D.P., Martinez-Romero, E., White, J.W., Corbett, J.D. and Bänziger, M. (2002) *Africa Maize Research Atlas* (v. 3.0), CD-ROM Publication. Centro Internacional de Mejoramiento de Maiz y Trigo (CIMMYT), Mexico, DF.

Hoekema, A., Hirsch, P.R., Hooykaas, P.J.J. and Schilperoort, R.A. (1983) A binary plant vector strategy based on separation of vir- and T-region of the *Agrobacterium tumefaciens* Ti-plasmid. *Nature* 303, 179–180.

Hoeschele, I. and VanRaden, P.M. (1993a) Bayesian analysis of linkage between genetic markers and quantitative trait loci. I. Prior knowledge. *Theoretical and Applied Genetics* 85, 953–960.

Hoeschele, I. and VanRaden, P.M. (1993b) Bayesian analysis of linkage between genetic markers and quantitative trait loci. II. Combining prior knowledge with experimental evidence. *Theoretical and Applied Genetics* 85, 946–952.

Hofmann, K., Bucher, P., Falquet, L. and Bairoch, A. (1999) The Prosite database, its status in 1999. *Nucleic Acids Research* 27, 215–219.

Hoheisel, J.D. (2006) Microarray technology: beyond transcript profiling and genotype analysis. *Nature Reviews Genetics* 7, 200–210.

Hohn, B., Levy, A.A. and Puchta, H. (2001) Elimination of selection markers from transgenic plants. *Current Opinion in Biotechnology* 12, 139–143.

Hoisington, D. and Ortiz, R. (2008) Research and field monitoring on transgenic crops by the Centro Internacional de Mejoramiento de Maiz y Trigo (CIMMYT). *Euphytica* 164, 893–902.

Holland, J.B. (1998) EPISTACY: a SAS program for detecting two-locus epistasis interactions using genetic marker information. *Journal of Heredity* 89, 374–375.

Holland, J.B. (2001) Epistasis and plant breeding. *Plant Breeding Reviews* 21, 29–32.
Holland, J.B. (2004) Implementation of molecular markers for quantitative traits in breeding programs – challenges and opportunities. In: *New Direction for a Diverse Planet*, Proceedings of the 4th International Crop Science Congress, 26 September–1 October 2004, Brisbane, Australia. Published on CD-ROM. Available at: http://www.cropscience.org.au/icsc 2004/ (accessed 17 November 2009).
Hopkins, C.G. (1899) Improvement in the chemical composition of the corn kernel. *Illinois Agricultural Experiment Station Bulletin* 55, 205–240.
Horan, K., Lauricha, J., Bailey-Serres, J., Raikhel, N. and Girke, T. (2005) Genome cluster database. A sequence family analysis platform for *Arabidopsis* and rice. *Plant Physiology* 138, 47–54.
Hori, K., Kobayashi, T., Shimizu, A., Sato, K., Takeda, K. and Kawasaki, S. (2003) Efficient construction of high-density linkage map and its application to QTL analysis in barley. *Theoretical and Applied Genetics* 107, 806–813.
Hori, K., Sato, K. and Takeda, K. (2007) Detection of seed dormancy QTL in multiple mapping populations derived from crosses involving novel barley germplasm. *Theoretical and Applied Genetics* 115, 869–876.
Hormaza, J.I., Dollo, L. and Polito, V.S. (1994) Identification of a RAPD marker linked to sex determination in *Pistacia vera* using bulked segregant analysis. *Theoretical and Applied Genetics* 89, 9–13.
Hospital, F. (2001) Size of donor chromosome segments around introgressed loci and reduction of linkage drag in marker-assisted backcross programs. *Genetics* 158, 1363–1379.
Hospital, F. (2002) Marker-assisted backcross breeding: a case study in genotype building theory. In: Kang, M.S. (ed.) *Quantitative Genetics, Genomics and Plant Breeding*. CAB International, Wallingford, UK, pp. 135–141.
Hospital, F. and Charcosset, A. (1997) Marker-assisted introgression of quantitative trait loci. *Genetics* 147, 1469–1485.
Hospital, F. and Decoux, G. (2002) Popmin: a program for the numerical optimization of population sizes in marker-assisted backcross breeding programs. *Journal of Heredity* 93, 383–384.
Hospital, F., Chevalet, C. and Mulsant, P. (1992) Using markers in gene introgression breeding programs. *Genetics* 231, 1199–1210.
Hospital, F., Moreau, L., Lacoudre, F., Charcosset, A. and Gallais, A. (1997) More on the efficiency of marker-assisted selection. *Theoretical and Applied Genetics* 95, 1181–1189.
Hospital, F., Goldringer, I. and Openshaw, S. (2000) Efficient marker-based recurrent selection for multiple quantitative trait loci. *Genetical Research* 75, 1181–1189.
Hoti, F. and Sillanpää, M.J. (2006) Bayesian mapping of genotype × expression interaction in quantitative and qualitative traits. *Heredity* 97, 4–18.
Howe, A.R., Gasser, C.S., Brown, S.M., Padgette, S.R., Hart, J., Parker, G.B., Fromn, M.E. and Armstrong, C.L. (2002) Glyphosate as a selective agent for the production of fertile transgenic maize (*Zea mays* L.) plants. *Molecular Breeding* 10, 153–164.
Howell, W.M., Jobs, M., Gyllensten, U. and Brooks, V. (1999) Dynamic allele-specific hybridization. A new method for scoring single nucleotide polymorphisms. *Nature Biotechnology* 17, 87–88.
Hsing, Y.-I., Chern, C.-G., Fan, M.-J., Lu, P.-C., Chen, K.-T., Lo, S.-F., Sun, P.-K., Ho, S.-L., Lee, K.-W., Wang, Y.-C., Huang, W.-L., Ko, S.-S., Chen, S., Chen, J.-L., Chung, C.-I., Lin, Y.-C., Hour, A.-L., Wang, Y.-W., Chang, Y.-C., Tsai, M.-W., Lin, Y.-S., Chen, Y.-C., Yen, H.-M., Li, C.-P., Wey, C.-K., Tseng, C.-S., Lai, M.-H., Huang, S.-C., Chen, L.-J. and Yu, S.-M. (2007) A rice gene activation/knockout mutant resource for high throughput functional genomics. *Plant Molecular Biology* 63, 351–364.
Hu, J. and Vick, B.A. (2003) Target region amplification polymorphism: a novel marker technique for plant genotyping. *Plant Molecular Biology Reporter* 21, 289–294.
Hua, J., Xing, Y., Wu, W., Xu, C., Sun, X., Yu, S. and Zhang, Q. (2003) Single-locus heterotic effects and dominance by dominance interactions can adequately explain the genetic basis of heterosis in an elite rice hybrid. *Proceedings of National Academy of Sciences of United States of America* 100, 2574–2579.
Hua, J.P., Xing, Y.Z., Xu, C.G., Sun, X.L., Yu, S.B. and Zhang, Q. (2002) Genetic dissection of an elite rice hybrid revealed that heterozygotes are not always advantageous for performance. *Genetics* 162, 1885–1895.
Huamán, Z., Ortiz, R., Zhang, D. and Rodríguez, F. (2000) Isozyme analysis of entire and core collection of *Solanum tuberosum* subsp. *andigena* potato cultivars. *Crop Science* 40, 273–276.
Huang, L., Brooks, S.H., Li, W., Fellers, J.P., Trick, H.N. and Gill, B.S. (2003) Map based cloning of leaf rust resistance gene *Lr21* from the large and polyploid genome in bread wheat. *Genetics* 164, 655–664.

Huang, N., Courtois, B., Khush, G.S., Lin, H., Wang, G., Wu, P. and Zheng, K. (1996) Association of quantitative trait loci for plant height with major dwarfing genes in rice. *Heredity* 77, 130–137.

Huang, N., Angeles, E.R., Domingo, J., Magpantay, G., Singh, S., Zhang, G., Kumaravadivel, N., Bennet, J. and Khush, G.S. (1997) Pyramiding of bacterial blight resistance genes in rice: marker-assisted selection using RFLP and PCR. *Theoretical and Applied Genetics* 95, 313–320.

Huang, S., Gilbertson, L.A., Adams, T.H., Malloy, K.P., Reisenbigler, E.K., Birr, D.H., Snyder, M.W., Zhang, Q. and Luethy, M.H. (2004) Generation of marker-free transgenic maize by regular two-border *Agrobacterium* transformation vectors. *Transgenic Research* 13, 451–461.

Huang, X., Feng, Q., Qian, Q., Zhao, Q., Wang, L., Wang, A., Guan, J., Fan, D., Wang, Q., Huang, T., Dong, G., Sang, T. and Han, B. (2009) High-throughput genotyping by whole-genome resequencing. *Genome Research* 19, 1068–1076.

Hudson, L.C., Halfhill, M.D. and Stewart, C.N., Jr (2004) Transgene dispersal through pollen. In: Peña, L. (ed.) *Methods in Molecular Biology*, Vol. 286. *Transgenic Plants: Methods and Protocols*. Humana Press Inc., Totowa, New Jersey, pp. 365–374.

Huelsenbeck, J.P., Ronquist, F., Nielsen, R. and Bollback, J.P. (2001) Bayesian inference of phylogeny and its impact on evolutionary biology. *Science* 294, 2310–2314.

Hühn, M. (1996) Nonparametric analysis of genotype × environment interactions by ranks. In: Kang, M.S. and Hauch, H.G., Jr (eds) *Genotype-by-Environment Interaction*. CRC Press, Boca Raton, Florida, pp. 235–271.

Hulden, M. (1997) Standardization of central crop databases. In: Lipmann, E., Jongen, M.W.M., Hintum, Th.J.L. van, Gass, T. and Maggioni, L. (compilers) *Central Crop Databases: Tools for Plant Genetic Resources Management.* Report of a Workshop, 13–16 October 1996, Budapest, Hungary. International Plant Genetic Resources Institute, Rome, Italy/CGN, Wageningen, Netherlands, pp. 26–34.

Hunt, M. (1997) *How Science Takes Stock: the Story of Meta Analysis*. Russell Sage Foundation, New York.

Hussein, M.A., Bjornstad, A. and Aastveit, A.H. (2000) SASG × ESTAB: a SAS program for computing genotype × environment stability statistics. *Agronomy Journal* 92, 454–459.

Hyne, V. and Kearsey, M.J. (1995) QTL analysis – further uses of marker regression. *Theoretical and Applied Genetics* 91, 471–476.

Hyten, D.L., Song, Q., Choi, I.-Y., Yoon, M.-P., Specht, J.E., Matukumalli, L.K., Nelson, R.L., Shoemaker, R.C., Young, N.D. and Cregan, P.B. (2008) High-throughput genotyping with the GoldenGate assay in the complex genome of soybean. *Theoretical and Applied Genetics* 116, 945–952.

IBPGR (International Board for Plant Genetic Resources) (1986) *Design, Planning and Operation of* In Vitro *Genebanks: Reports of a Subcommittee of the IBPGR Advisory Committee on* In Vitro *Storage*. IBPGR, Rome.

IBRD/World Bank (The International Bank for Reconstruction and Development/The World Bank) (2006) *Intellectual Property Rights: Designing Regimes to Support Plant Breeding in Developing Countries*. The World Bank, Washington, DC.

Ideta, O., Yoshimura, A. and Iwata, N. (1996) An integrated linkage map of rice. *Rice Genetics III. Proceedings of the Third International Rice Genetics Symposium*, 16–20 October 1995, Manila. International Rice Research Institute (IRRI), Manila, Phillipines.

Igartua, E., Casas, A.M., Ciudad, F., Montoya, L. and Romagosa, I. (1999) RFLP markers associated with major genes controlling heading date evaluated in a barley germ plasm pool. *Heredity* 83, 551–559.

Igartua, E., Edney, M., Rossnagel, B.G., Spaner, D., Legge, W.G., Scoles, G.L., Ecksteins, P.E., Penner, G.A.,Tinker, N.A., Briggs, K.G., Falk, D.E. and Mather, D.E. (2000) Marker-assisted selection of QTL affecting grain and malt quality in two-row barley. *Crop Science* 40, 1426–1433.

Ikeda, A., Ueguchi-Tanaka, M., Sonoda, Y., Kitano, H., Koshioka, M., Futsuhara, Y., Matsuoka, M. and Yamaguchi, J. (2001) *slender* rice, a constitutive gibberellin response mutant, is caused by a null mutation of the *SLR1* gene, an ortholog of the height-regulating gene *GAI/RGA/RHT/D8. The Plant Cell* 13, 999–1010.

Ilic, K., Kellogg, E.A., Jaiswal, P., Zapata, F., Stevens, P.F., Vincent, L.P., Avraham, S., Reiser, L., Pujar, A., Sachs, M.M., Whitman, N.T., McCouch, S.R., Schaeffer, M.L., Ware, D.H., Stein, L.D. and Rhee, S.Y. (2007) The Plant Structure Ontology, a unified vocabulary of anatomy and morphology of a flowering plant. *Plant Physiology* 143, 587–599.

International Human Genome Sequencing Consortium (2001) Initial sequencing and analysis of human genome. *Nature* 409, 860–921.

Ioannidis, J.P., Ntzani, E.E., Trikalinos, T.A. and Contopoulos-Ioannidis, D.G. (2001) Replication validity of genetic association studies. *Nature Genetics* 29, 306–309.

IRGSP (International Rice Genome Sequencing Project) (2005) The map-based sequence of the rice genome. *Nature* 436, 793–800.

ISF (International Seed Federation) (2004) *Protection of Intellectual Property and Access to Plant Genetic Resources*. Proceedings of an International Seminar, 27–28 May, 2004, Berlin, CD-ROM.

ISF (International Seed Federation) (2005) Essential derivation from a not-yet protected variety and dependency. ISF Position Paper, June 2005. Available at: http://www.worldseed.org/Position_papers/ED&Dependency.htm (accessed 30 June 2007).

Ishida, Y., Saito, H., Ohta, S., Hiei, Y., Komari, T. and Kumashiro, T. (1996) High efficiency transformation of maize (*Zea mays* L.) mediated by *Agrobacterium tumefaciens*. *Nature Biotechnology* 14, 745–750.

Ishida, Y., Murai, N., Kuraya, Y., Ohta, S., Saito, H., Hiei, Y. and Komari, T. (2004) Improved co-transformation of maize with vectors carrying two separate T-DNAs mediated by *Agrobacterium tumefaciens*. *Plant Biotechnology* 21, 57–63.

Ishimaru, K. (2003) Identification of a locus increasing rice yield and physiological analysis of its function. *Plant Physiology* 122, 1083–1090.

Ivandic, V., Hackett, C.A., Nevo, E., Keith, R., Thomas, W.T.B. and Forster, B.P. (2002) Analysis of simple sequence repeats (SSRs) in wild barley from the Fertile Crescent: associations with ecology, geography and flowering time. *Plant Molecular Biology* 48, 511–527.

Ivandic, V., Thomas, W.T.B., Nevo, E., Zhang, Z. and Forster, B.P. (2003) Association of SSRs with quantitative trait variation including biotic and abiotic stress tolerance in *Hordeum spontaneum*. *Plant Breeding* 122, 300–304.

Iwata, H., Uga, Y., Yoshioka, Y., Ebana, K. and Hayashi, T. (2007) Bayesian association mapping of multiple quantitative trait loci and its application to the analysis of genetic variation among *Oryza sativa* L. germplasms. *Theoretical and Applied Genetics* 114, 1437–1449.

Izawa, T., Takahashi, Y. and Yano, M. (2003) Comparative biology comes into bloom: genomic and genetic comparison of flowering pathways in rice and *Arabidopsis*. *Current Opinion in Plant Biology* 6, 113–120.

Jaccoud, D., Peng, K., Feinstein, D. and Kilian, A. (2001) Diversity arrays: a solid state technology for sequence information independent genotyping. *Nucleic Acids Research* 29, e25.

Jack, T., Fox, G.L. and Meyerowitz, E.M. (1994) *Arabidopsis* homeotic gene APETALA3 ectopic expression: transcriptional and posttranscriptional regulation determine floral organ identity. *Cell* 76, 703–716.

Jaffe, G. (2004) Regulation transgenic crops: a comparative analysis of different regulatory processes. *Transgenic Research* 13, 5–19.

Jain, S.M., Sopory, S.K. and Veilleux, R.E. (1996–1997) In Vitro *Haploid Production in Higher Plants*. Kluwer Academic Publishers, Dordrecht, Netherlands.

James, C. (2006) *Global Status of Commercialized Biotech/GM Crops: 2006*. *ISAAA Briefs* No. 35. International Service for the Acquisition of Agri-biotech Applications (ISAAA), Ithaca, New York.

James, C. (2008) *2007 ISAAA Report on Global Status of Biotech/GM Crops*. International Service for the Acquisition of Agri-biotech Applications (ISAAA). Available at: http://www.isaaa.org (accessed 17 November 2009).

Jander, G., Norris, S.R., Rounsley, S.D., Bush, D.F., Levin, I.M. and Last, R.L. (2002) *Arabidopsis* map-based cloning in the post-genome era. *Plant Physiology* 129, 440–450.

Janick, J. (1988) Horticulture, science and society. *HortScience* 23, 11–13.

Janick, J. (1998) Hybrids in horticulture crops. In: Lamkey, K.R. and Staub, J.E. (eds) *Concepts and Breeding of Heterosis in Crop Plants*. Crop Science Society of America (CSSA), Madison, Wisconsin, pp. 45–56.

Janis, M.D. and Kesan, J.P. (2002) U.S. plant variety protection: sound or furry ...? *Houston Law Review* 39, 727–778.

Janis, M.D. and Smith, S. (2007) Obsolescence in intellectual property regimes. University of Iowa Legal Studies Research Paper No. 05-48. Abstract available at: http://papers.ssrn.com/sol3/papers.cfm?abstract_id=897728 (accessed 17 November 2009).

Jannink, J.L. (2005) Selective phenotyping to accurately mapping quantitative trait loci. *Crop Science* 45, 901–908.

Jannink, J.L. and Jansen, R.C. (2000) The diallel mating design for mapping interacting QTLs. In: *Quantitative Genetics and Breeding Methods: the Way Ahead*. Institut National de la Recherche Agronomique (INRA), Paris, pp. 81–88.

Jannink, J.L. and Jansen, R. (2001) Mapping epistatic quantitative trait loci with one-dimensional genome searches. *Genetics* 157, 445–454.

Jannink, J.L. and Walsh, B. (2002) Association mapping in plant populations. In: Kang, M.S. (ed.) *Quantitative Genetics, Genomics and Plant Breeding.* CAB International, Wallingford, UK, pp. 59–68.

Jannink, J.L., Bink, M. and Jansen, R.C. (2001) Using complex plant pedigrees to map valuable genes. *Trends in Plant Science* 6, 337–342.

Jansen, C., Thomas, D.Y. and Pollock, S. (2005) Yeast two-hybrid technologies. In: Sensen, C.W. (ed.) *Handbook of Genome Research, Genomics, Metabolomics, Bioinformatics, Ethical and Legal Issues.* WILEY-VCH Verlag GmbH & Co., KGaA, Weinheim, Germany, pp. 261–272.

Jansen, J.P.A. (1996) Aphid resistance in composites. International application published under the patent cooperation treaty (PCT) No. WO 97/46080.

Jansen, R.C. (1996) A general Monte Carlo method for mapping multiple quantitative trait loci. *Genetics* 142, 305–311.

Jansen, R.C. and Beavis, W.D. (2001) MQM mapping using haplotyped putative QTL-alleles: a simple approach for mapping QTLs in plant breeding populations. Patent EP 1265476.

Jansen, R.C. and Nap, J.P. (2001) Genetical genomics: the added value from segregation. *Trends in Genetics* 17, 388–391.

Jansen, R.C. and Stam, P. (1994) High resolution of quantitative traits into multiple loci via interval mapping. *Genetics* 136, 1447–1455.

Jansen, R.C., Van-Ooijen, J.W., Stam, P., Lister, C. and Dean, C. (1995) Genotype-by-environment interaction in genetic mapping of multiple quantitative trait loci. *Theoretical and Applied Genetics* 91, 33–37.

Jansen, R.C., Jannink, J.-L. and Beavis, W.D. (2003) Mapping quantitative trait loci in plant breeding populations: use of parental haplotype sharing. *Crop Science* 43, 829–834.

Jarvis, A., Yeaman, S., Guarino, L. and Tohme, J. (2005) The role of geographic analysis in locating, understanding and using plant genetic diversity. In: Zimmer, E. (ed.) *Molecular Evolution: Producing the Biochemical Data*, Part B. Elsevier, New York, pp. 279–298.

Jarvis, D.I. and Hodgkin, T. (1999) Wild relatives and crop cultivars: detecting natural introgression and farmer selection of new genetic combinations in agroecosystems. *Molecular Ecology* 8, S159–S173.

Jayasekara, N.E.M. and Jinks, J.L. (1976) Effect of gene dispersion on estimates of components of generation means and variances. *Heredity* 36, 31–40.

Jefferson, R.A. (1987) Assaying chimeric genes in plants: the GUS gene fusion system. *Plant Molecular Biology Reporter* 5, 387–405.

Jenkins, H., Johnson, H., Kular, B., Wang, T. and Hardy, N. (2005) Toward supportive data collection tools for plant metabolomics. *Plant Physiology* 138, 67–77.

Jenkins, S. and Gibson, N. (2002) High-throughput SNP genotyping. *Comparative and Functional Genomics* 3, 57–66.

Jenks, M.A. and Feldmann, K. (1996) Cloning genes by insertion mutagenesis. In: Paterson, A.H. (ed.) *Genome Mapping in Plants.* R.G. Landes Company, Austin, Texas, pp. 155–168.

Jensen, C.J. (1974) Chromosome doubling techniques in haploids. In: Kasha, K.J. (ed.) *Haploids in Higher Plants: Advances and Potentials.* Guelph University Press, Guelph, Canada, pp. 153–190.

Jensen, L.J., Saric, J. and Bork, P. (2006) Literature mining for the biologist: from information retrieval to biological discovery. *Nature Reviews Genetics* 7, 119–129.

Jeon, J.-S., Kang, H.-G. and An, G. (2004) Tools for gene tagging and mutagenesis. In: Christou, P. and Klee, H. (eds) *Handbook of Plant Biotechnology.* John Wiley & Sons Ltd, Chichester, UK, pp. 103–125.

Jia, H., Pang, Y., Chen, X. and Fang, R. (2006) Removal of the selectable marker gene from transgenic tobacco plants by expression of Cre recombinase from a tobacco mosaic virus vector through agroinfection. *Transgenic Research* 15, 375–384.

Jiang, C. and Zeng, Z.B. (1995) Multiple trait analysis of genetic mapping for quantitative trait loci. *Genetics* 140, 1111–1127.

Jiang, C., Pan, X. and Gu, M. (1994) The use of mixture models to detect effects of major genes on quantitative characters in plant breeding experiment. *Genetics* 136, 383–394.

Jiang, C., Edmeades, G.O., Armstead, I., Lafitte, H.R., Hayward, M.D. and Hoisington, D. (1999) Genetic analysis of adaptation differences between highland and lowland tropical maize using molecular markers. *Theoretical and Applied Genetics* 99, 1106–1119.

Jiang, N., Bao, Z., Zhang, X., Hirochika, H., Eddy, S.R., McCouch, S.R. and Wessler, S.R. (2003) An active DNA transposon family in rice. *Nature* 421, 163–167.

Jin, C., Lan, H., Attie, A.D., Churchill, G.A., Bulutuglo, D. and Yandell, B.Y. (2004) Selective phenotyping for increased efficiency in genetic mapping study. *Genetics* 168, 2285–2293.

Jin, S., Komari, T., Gordon, M.P. and Nester, E.W. (1987) Genes responsible for the supervirulence phenotype of *Agrobacterium tumefaciens* A281. *Journal of Bacteriology* 169, 4417–4425.

Jinks, J.L. and Perkins, J.M. (1969) The detection of linked epistatic genes for a metrical trait. *Heredity* 24, 465–475.

Jinks, J.L. and Perkins, J.M. (1972) Predicting the range of inbred lines. *Heredity* 28, 399–403.

Joen, J.-S., Lee, S., Jung, K.-H., Jun, S.-H., Joeng, D.-H., Lee, J., Kim, C., Jang, S., Yang, K., Nam, J., An, K., Han, M.J., Sung, R.-J., Choi, H.-S., Yu, J.-H., Choi, J.-H., Cho, S.-S., Cha, S.-S., Kim, S.-I. and An, G. (2000) T-DNA insertional mutagenesis for functional genomics in rice. *The Plant Journal* 22, 561–571.

Joersbo, M. and Okkels, F.T. (1996) A novel principle for selection of transgenic plant cells: positive selection. *Plant Cell Reports* 16, 219–221.

Joersbo, M., Donaldson, I., Kreiberg, J., Petersen, S.G., Brunstedt, J. and Okkels, F.T. (1998) Analysis of mannose selection used for transformation of sugar beet. *Molecular Breeding* 4, 111–117.

Johannes, F. (2007) Mapping temporally varying quantitative trait loci in time-to-failure experiments. *Genetics* 175, 855–865.

Johnson, B., Gardner, C.O. and Wrede, K.C. (1988) Application of an optimization model to multi-trait selection programs. *Crop Science* 28, 723–728.

Johnson, G.R. (2004) Marker assisted selection. *Plant Breeding Reviews* 24, 293–310.

Johnson, H.E., Broadburst, D., Goodacre, R. and Smith, A.R. (2003) Metabolic fingerprinting of salt-stressed tomatoes. *Phytochemistry* 62, 919–928.

Johnson, H.W., Robinson, H.F. and Comstock, R.E. (1955) Estimates of genetic and environmental variability in soybeans. *Agronomy Journal* 47, 314–318.

Johnson, R. (2001) Marker-assisted sweet corn breeding: a model for special crops. In: *Proceedings of 56th Annual Corn and Sorghum Industry Research Conference* Chicago, Illinois, 5–7 December 2001. American Seed Trade Association, Washington, DC, pp. 25–30.

Jones, H. (ed.) (1995) *Plant Gene Transfer and Expression Protocols.* Humana Press, Totowa, New Jersey.

Jones, H.D., Doherty, A. and Wu, H. (2005) Review of methodologies and a protocol for the *Agrobacterium*-mediated transformation of wheat. *Plant Methods* 2005, 1–5.

Jorasch, P. (2004) Intellectual property rights in the field of molecular marker analysis. In: Lörz, H. and Wenzel, G. (eds) *Biotechnology in Agriculture and Forestry*, Vol. 55. *Molecular Marker Systems.* Springer-Verlag Berlin, pp. 433–471.

Jordaan, J.P., Engelbrecht, S.A., Malan, J.H. and Knobel, H.A. (1999) Wheat and heterosis. In: Coors, J.G. and Pandey, S. (eds) *Genetics and Exploitation of Heterosis in Crops.* ASA-CSSA-SSSA, Madison, Wisconsin, pp. 411–421.

Jordan, D., Tao, Y., Godwin, I., Henzell, R., Cooper, M. and McIntyre, C. (2004) Prediction of hybrid performance in grain sorghum using RFLP markers. *Theoretical and Applied Genetics* 106, 559–567.

Jorde, L.B. (2000) Linkage disequilibrium and the search for complex disease genes. *Genome Research* 10, 1435–1444.

Joseph, M., Gopalakrishnan, S., Sharma, R.K., Singh, V.P., Singh, A.K., Singh, N.K. and Mohapatra, T. (2003) Combining bacterial blight resistance and Basmati quality characteristics by phenotypic and molecular marker-assisted selection in rice. *Molecular Breeding* 13, 1–11.

Jourjon, M.F., Jasson, S., Marcel, J., Ngom, B. and Mangin, B. (2005) MCQTL: multi-allelic QTL mapping in multi-cross design. *Bioinformatics* 21, 128–130.

Jung, K.-H., An, G. and Ronald, P.C. (2008) Towards a better bowl of rice: assigning function to tens of thousands of rice genes. *Nature Reviews Genetics* 9, 91–101.

Kahler, A.L., Gardner, C.O. and Allard, R.W. (1984) Nonrandom mating in experimental populations of maize. *Crop Science* 24, 350–354.

Kahraman, A., Avramov, A., Nashev, L.G., Popov, D., Ternes, R., Pohlenz, H.-D. and Weiss, B. (2005) PhenomicDB: a multi-species genotype/phenotype database for comparative phenomics. *Bioinformatics* 21, 418–420.

Kahvejian, A., Quackenbush, J. and Thompson, J.F. (2008) What would you do if you could sequence everything? *Nature Biotechnology* 26, 1125–1133.

Kamujima, O., Tanisaka, T. and Kinoshita, T. (1996) Gene symbols for dwarfness. *Rice Genetics Newsletter* 13,19–25.

Kang, M.S. (1988) A rank-sum method for selecting high-yielding, stable corn genotypes. *Cereal Research Communications* 16, 113–115.

Kang, M.S. (1990) Understanding and utilization of genotype–environment interaction in plant breeding. In: Kang, M.S. (ed.) *Genotype-By-Environment Interactions and Plant Breeding*. Louisiana State University Agriculture Center, Baton Rouge, Louisiana, pp. 52–68.

Kang, M.S. (1993) Simultaneous selection for yield and stability in crop performance trials: consequences for growers. *Agronomy Journal* 85, 754–757.

Kang, M.S. (2002) Genotype–environment interaction: progress and prospects. In: Kang, M.S. (ed.) *Quantitative Genetics, Genomics and Plant Breeding*. CAB International, Wallingford, UK, pp. 221–243.

Kang, M.S. and Magari, R. (1996) New developments in selecting for phenotypic stability in crop breeding. In: Kang, M.S. and Gauch, H.G., Jr (eds) *Genotype-by-Environment Interaction*. CRC Press, Boca Raton, Florida, pp. 1–14.

Kantety, R.V., Rota, M.L., Mathews, D.E. and Sorrels, M.E. (2002) Data mining for simple-sequence repeats in expressed sequence tags from barley, maize, rice, sorghum and wheat. *Plant Molecular Biology* 48, 501–510.

Kao, C.H. (2004) Multiple-interval mapping for quantitative trait loci controlling endosperm traits. *Genetics* 167, 1987–2002.

Kao, C.H. (2006) Mapping quantitative trait loci using the experimental designs of recombinant inbred populations. *Genetics* 174, 1373–1386.

Kao, C.H. and Zeng, Z.B. (1997) General formulas for obtaining the MLEs and the asymptotic variance–covariance matrix in mapping quantitative trait loci when using the EM algorithm. *Biometrics* 53, 653–665.

Kao, C.H., Zeng, Z.B. and Teasdale, R.D. (1999) Multiple interval mapping for quantitative trait loci. *Genetics* 152, 1203–1216.

Karas, M. and Hillenkamp, F. (1988) Laser desorption ionization of proteins with molecular mass exceeding 10000 daltons. *Analytical Chemistry* 60, 2299–2301.

Karimi, M., Bleys, A., Vanderhaeghen, R. and Hilson, P. (2007) Building blocks for plant gene assembly. *Plant Physiology* 145, 1183–1191.

Karp, A. and Edwards, J. (1997) DNA markers: a global overview. In: Caetano-Anolles, G. and Gresshoff, P.M. (eds) *DNA Markers – Protocols, Applications and Overviews*. Wiley-Liss, Inc., New York, pp. 1–13.

Kartal, M. (2007) Intellectual property protection in the natural product drug discovery, traditional herbal medicine and herbal medicinal products. *Phytotherapy Research* 21, 113–119.

Kasha, K.J. (2005) Chromosome doubling and recovery of doubled haploid plants. In: Palmer, C.E., Keller, W.A. and Kasha, K.J. (eds) *Biotechnology in Agriculture and Forestry*, Vol. 56. *Haploids in Crop Improvement II*. Springer-Verlag, Berlin, pp. 123–152.

Katari, M.S., Balija, V., Wilson, R.K., Martienssen, R.A. and McCombie, W.R. (2005) Comparing low coverage random shotgun sequence data from *Brassica oleracea* and *Oryza sativa* genome sequence for their ability to add to the annotation of *Arabidopsis thaliana*. *Genome Research* 15, 496–504.

Kato, A. (2002) Chromosome doubling of haploid maize seedlings using nitrous oxide gas at the flower primordial stage. *Plant Breeding* 121, 370–377.

Kauffman, S.A. (1993) *The Origins of Order: Self-Organization and Selection in Evolution*. Oxford University Press, Oxford, UK.

Kaushik, N., Sirohi, M. and Khanna, V.K. (2004) Influence of age of the embryo and method of hormone application on haploid embryo formation in wheat x maize crosses. In: *New Directions for a Diverse Planet*, Proceedings of the 4th International Crop Science Congress, 26 September–1 October, 2004, Brisbane, Australia. Published on CD-ROM. Available at: http://www.cropscience.org.au/icsc2004/ (accessed 17 November 2009).

Kearsey, M.J. and Farquhar, A.G.L. (1998) QTL analysis in plants: where are we now? *Heredity* 80, 137–142.

Kearsey, M.J. and Hyne, V. (1994) QTL analysis: a simple 'marker regression' approach. *Theoretical and Applied Genetics* 89, 698–702.

Kearsey, M.J. and Jinks, J.L. (1968) A general method of detecting additive, dominance and epistasis variation for metrical traits. I. Theory. *Heredity* 23, 403–409.

Kearsey, M.J., Pooni, H.S. and Syed, N.H. (2003) Genetics of quantitative traits in *Arabidopsis thaliana*. *Heredity* 91, 456–464.

Keating, B.A., Carberry, P.S., Hammer, G.L., Probert, M.E., Robertson, M.J., Holzworth, D., Huth, N.I., Hargreaves, J.N.G., Meinke, H., Hockman, Z., McLean, G., Verburg, K., Snow, V., Dimes, J.P., Silburn,

M., Wang, E., Brown, S., Bristow, K.L., Asseng, S., Chapman, S., McCown, R.L., Freebairn, D.M. and Smith, C.J. (2003) An overview of APSIM, a model designed for farming system simulation. *European Journal of Agronomy* 18, 267–288.

Keightley, P.D. (2004) Mutational variation and long-term selection response. *Plant Breeding Reviews* 24(1), 227–247.

Keightley, P.D. and Bulfield, G. (1993) Detection of quantitative trait loci from frequency changes of marker alleles under selection. *Genetical Research* 62, 195–203.

Keller, E.R.J. and Korzun, L. (1996) Haploidy in onion (*Allium cepa* L.) and other *Allium* species. In: Jain, S.M., Sopory, S.M. and Veilleux, R.E. (eds) In Vitro *Haploid Production in Higher Plants*. Vol. 3: *Important Selected Plants*. Kluwer Academic Publisher, Dordrecht, Netherlands, pp. 51–75.

Kempthorne, O. (1957) *An Introduction to Genetics Statistics*. Wiley, New York.

Kempthorne, O. (1988) An overview of the field of quantitative genetics. In: Weir, B.S., Eisen, E.J., Goodman, M.M. and Namkoong, G. (eds) *Proceedings of the 2nd International Conference on Quantitative Genetics*. Sinauer Associates, Inc., Sunderland, Massachusetts, pp. 47–56.

Kennedy, B.G., Waters, D.L.E. and Henry, R.J. (2006) Screening for the rice blast resistance gene *Pi-ta* using LNA displacement probes and real-time PCR. *Molecular Breeding* 18, 185–193.

Kermicle, J.L. (1969) Androgenesis conditioned by a mutation in maize. *Science* 166, 1422–1424.

Kerns, M.R., Dudley, J.W. and Rufener, G.K. (1999) QTL for resistance to common rust and smut in maize. *Maydica* 44, 37–45.

Kersten, B., Berkle, L., Kuhn, E.J., Giavalisco, P., Konthur, Z., Lueking, A., Walter, G., Eickhoff, H. and Schneider, U. (2002) Large-scale plant proteomics. *Plant Molecular Biology* 48, 133–141.

Keurentjes, J.J., Bentsink, L., Alonso-Blanco, C., Hanhart, C.J., Blankestijn-De Vries, H., Effgen, S., Vreugdenhil, D. and Koornneef, M. (2007a) Development of a near-isogenic line population of *Arabidopsis thaliana* and comparison of mapping power with a recombinant inbred line population. *Genetics* 175, 891–905.

Keurentjes, J.J.B., Jingyuan Fu, L., Terpstra, I.R., Garcia, J.M., Ackerveken, G., Snoek, L.B., Peeters, A.J.M., Vreugdenhil, D., Koornneef, M. and Jansen, R.C. (2007b) Regulatory network construction in *Arabidopsis* by using genome-wide gene expression quantitative trait loci. *Proceedings of the National Academy of Sciences of the United States of America* 104, 1708–1713.

Keurentjes, J.J.B., Koornnef, M. and Vreugdenhil, D. (2008) Quantitative genetics in the age of omics. *Current Opinion in Plant Biology* 11, 123–128.

Khatkar, M.S., Thomson, P.C., Tammen, I. and Raadsma, H.W. (2004) Quantitative trait loci mapping in dairy cattle: review and meta-analysis. *Genetics Selection Evolution* 36, 163–190.

Khatri, P. and Drăghici, S. (2005) Ontological analysis of gene expression data: current tools, limitations and open problems. *Bioinformatics* 21, 3587–3595.

Khatri, P., Drăghici, S., Ostermeier, G.C. and Krawetz, S.A. (2002) Profiling gene expression using Onto-Express. *Genomics* 79, 266–270.

Khush, G.S. (1987) List of gene markers maintained in the Rice Genetic Stock Center, IRRI. *Rice Genetics Newsletter* 4, 56–62.

Khush, G.S. (1999) Green revolution: preparing for the 21st century. *Genome* 42, 646–655.

Kiesselbach, T.A. (1926) The immediate effect of gametic relationship and of parental type upon the kernel weight of corn. *Nebraska Agricultural Experiment Station Bulletin* 33, 1–69.

Kikuchi, K., Terauchi, K., Wada, M. and Hirano, Y. (2003) The plant MITE mPing is mobilized in anther culture. *Nature* 421, 167–170.

Kilian, A., Chen, J., Han, F., Steffenson, B. and Kleinhofs, A. (1997) Towards map-based cloning of the barley stem rust resistance gene *Rpg1* and *rpg4* using rice as an intergenomic cloning vehicle. *Plant Molecular Biology* 35, 187–195.

Kilian, A., Kudrna, D. and Kleinhofs, A. (1999) Genetic and molecular characterization of barley chromosome telomeres. *Genome* 42, 412–419.

Kilpikari, R. and Sillanpää, M.J. (2003) Bayesian analysis of multilocus association in quantitative and qualitative traits. *Genetic Epidemiology* 25, 122–135.

Kim, K.-W., Chung, H.-K., Cho, G.-T., Ma, K.-H., Chandrabalan, D., Gwag, J.-G., Kim, T.-S., Cho, E.-G. and Park, Y.-J. (2007) PowerCore: a program applying the advanced M strategy with a heuristic search for establishing core mining sets. *Bioinformatics* 23, 2155–2162.

Kimmel, A. and Oliver, B. (eds) (2006a) *DNA Microarrays Part A: Array Platforms and Wet-Bench Protocols*. Elsevier Inc., Amsterdam.

Kimmel, A. and Oliver, B. (eds) (2006b) *DNA Microarrays Part B: Databases and Statistics*. Elsevier Inc., Amsterdam.

Kimmel, B.E., Palazzolo, M.J., Martin, C.H., Boeke, J.D. and Devine, S.E. (1997) Transposon-mediated DNA sequencing. In: Birren, B., Green, E.D., Klapholz, S., Myers, R.M. and Roskams, J. (eds) *Genome Analysis: a Laboratory Manual*, Vol. 1. Cold Spring Harbor Laboratory Press, Cold Spring Harbor, New York, pp. 455–532.

Kimura, M. (1969) The number of heterozygous nucleotide sites maintained in a finite population due to steady flux of mutations. *Genetics* 61, 893–903.

King, G.L. (2004) Bioinformatics: harvesting information for plant and crop science. *Seminars in Cell and Developmental Biology* 15, 721–731.

King, J., Armstead, I.P., Donnison, I.S., Thomas, H.M., Jones, R.N., Kearseyc, M.J., Roberts, L.A., Thomas, A., Morgan, W.G. and King, I.P. (2002) Physical and genetic mapping in the grasses *Lolium perenne* and *Festuca pratensis*. *Genetics* 161, 315–324.

Kinoshita, T. (1995) Report of Committee on Gene Symbolization, Nomenclature and Linkage Groups. *Rice Genetics Newsletter* 12, 9–153.

Kinoshita, T. and Takahashi, M. (1991) The one hundredth report of genetical studies on rice plant: linkage studies and future prospects. *Journal of the Faculty of Agriculture, Hokkaido University* 65, 1–61.

Kirst, M., Myburg, A.A., De León, J.P.G., Kirst, M.E., Scott, J. and Sederoff, R. (2004) Coordinated genetic regulation of growth and lignin revealed by quantitative trait locus analysis of cDNA microarray data in an interspecific backcross of eucalyptus. *Plant Physiology* 135, 2368–2378.

Kisana, N.S., Nkongolo, K.K., Quick, J.S. and Johnson, D.L. (1993) Production of doubled haploids by anther culture and wheat × maize method in a wheat breeding programme. *Plant Breeding* 110, 96–102.

Kiviharju, E., Moisander, S. and Laurila, J. (2005) Improved green plant regeneration rates from oat anther culture and the agronomic performance of some DH lines. *Plant Cell, Tissue and Organ Culture* 81, 1–9.

Kjemtrup, S., Boyes, D.C., Christensen, C., McCaskill, A.J., Hylton, M. and Davis, K. (2003) Growth stage-based phenotypic profiling of plants. In: Grotewold, E. (ed.) *Methods in Molecular Biology*, Vol. 236. *Plant Functional Genomics: Methods and Protocols*. Humana Press, Totowa, New Jersey, pp. 427–441.

Klein, P.E., Klein, R.R., Cartinhour, S.W., Ulanch, P.E., Dong, J., Obert, J.A., Morishige, D.T., Schlueter, S.D., Childs, K.L., Ale, M. and Mullet, J.E. (2000) A high-throughput AFLP-based method for constructing integrated genetic and physical maps: progress toward a sorghum genome map. *Genome Research* 10, 789–807.

Klein, P.E., Klein, R.R., Vrebalov, J. and Mullet, J.E. (2003) Sequence-based alignment of sorghum chromosome 3 and rice chromosome 1 reveals extensive conservation of gene order and one major chromosomal rearrangement. *The Plant Journal* 34, 605–621.

Klose, J., Nock, C., Herrmann, M., Stühler, K., Marcus, K., Blüggel, M., Krause, E., Schalkwyk, L.C., Rastan, S., Brown, S.D.M., Büssow, K., Himmelbauer, H. and Lehrach, H. (2002) Genetic analysis of mouse brain proteome. *Nature Genetics* 30, 385–393.

Knapp, S.J. (1991) Using molecular markers to map multiple quantitative trait loci: models for backcross, recombinant inbred and doubled haploid progeny. *Theoretical and Applied Genetics* 81, 333–338.

Knapp, S.J. (1994) Mapping quantitative trait loci. In: Philip, R.I. and Vasil, I.K. (eds) *DNA-Based Markers in Plants*. Kluwer Academic Publishers, Dordrecht, Netherlands, pp. 58–96.

Knapp, S.J. (1998) Marker-assisted selection as a strategy for increasing the probability of selecting superior genotypes. *Crop Science* 38, 1164–1174.

Knapp, S.J., Holloway, J.L., Bridges, W.C. and Liu, B.-H. (1995) Mapping dominant markers using F_2 matings. *Theoretical and Applied Genetics* 91, 74–81.

Knoll, J. and Ejeta, G. (2008) Marker-assisted selection for early-season cold tolerance in sorghum: QTL validation across populations and environments. *Theoretical and Applied Genetics* 116, 541–553.

Knox, M.R. and Ellis, T.H. (2001) Stability and inheritance of methylation states at PstI sites in *Pisum*. *Molecular Genetics and Genomics* 265, 497–507.

Kobiljski, B., Quarrie, S., Denčić, S., Kirby, J. and Ivegeš, M. (2002) Genetic diversity of the Novi Sad wheat core collection revealed by microsatellites. *Cellular and Molecular Biology* 7, 685–694.

Koebner, R.M. and Summers, R.W. (2003) 21st century wheat breeding: plot selection or plate detection? *Trends in Biotechnology* 21, 59–63.

Koester, R.P., Sisco, P.H. and Stuber, C.W. (1993) Identification of quantitative trait loci controlling days to flowering and plant height in two near isogenic lines of maize. *Crop Science* 33, 1209–1216.

Kohli, A., Leech, M., Vain, P., Laurie, D.A. and Christou, P. (1998) Transgene organization in rice engineered through direct DNA transfer supports a two-phase integration mechanism mediated by the establishment of integration hot spots. *Proceedings of the National Academy of Sciences of the United States of America* 95, 7203–7208.

Kohli, A., Xiong, J., Greco, R., Christou, P. and Pereira, A. (2001) Transcriptome Display (TTD) in *indica* rice using Ac transposition. *Molecular Genetics and Genomics* 266, 1–11.

Kohli, A., Twyman, R.M., Abranches, A., Wegel, E., Christou, P. and Stoger, E. (2003) Transgene integration, organization and interaction in plants. *Plant Molecular Biology* 52, 247–258.

Kohli, A., Prynne, M.Q., Berta, M., Pereira, A., Cappell, T., Twyman, R.M. and Christou, P. (2004) Dedifferentiation-mediated changes in transposition behavior make the Activator transposon an ideal tool for functional genomics in rice. *Molecular Breeding* 13, 177–191.

Koizuka, N., Imai, R., Fujimoto, H., Hayakawa, T., Kimura, Y., Kohno-Murase, J., Sakai, T., Kawasaki, S. and Imamura, J. (2003) Genetic characterization of a pentatricopeptide repeat protein gene, *orf687*, that restores fertility in the cytoplasmic male-sterile Kosena radish. *The Plant Journal* 34, 407–415.

Kojima, S., Takahashi, Y., Kobayashi, Y., Monna, L., Sasaki, T., Araki, T. and Yano, M. (2002) *Hd3a*, a rice ortholog of the *Arabidopsis FT* gene, promotes transition to flowering downstream of *Hd1* under short-day conditions. *Plant Cell and Physiology* 43, 1096–1105.

Koller, A., Washburn, M.P., Lange, B.M., Andon, N.L., Deciu, C., Haynes, P.A., Hays, L., Schieltz, D., Ulaszek, R., Wei, J., Wolters, D. and Yates, J.R. III (2002) Proteomic survey of metabolic pathways in rice. *Proceedings of the National Academy of Sciences of the United States of America* 99, 11969–11974.

Komari, T. (1990) Transformation of cultured cells of *Chenopodium quinoa* by binary vectors that carry a fragment of DNA from the virulence region of pTiBo542. *Plant Cell Reports* 9, 303–306.

Komari, T., Hiei, Y., Saito, Y., Murai, N. and Kumashiro, T. (1996) Vectors carrying two separate T-DNAs for co-transformation of higher plants mediated by *Agrobacterium tumefaciens* and segregation of transformants free from selection markers. *The Plant Journal* 10, 165–174.

Komari, T., Takakura, Y., Ueki, J., Kato, N., Ishida, Y. and Hiei, Y. (2006) Binary vectors and super-binary vectors. In: Wang, K. (ed.) *Methods in Molecular Biology* 343: *Agrobacterium Protocols*, Vol. 1, 2nd edn. Humana Press, Totowa, New Jersey, pp. 15–41.

Komori, T., Ohta, S., Murai, N., Takakura, Y., Kuraya, Y., Suzuki, S., Hiei, Y., Imaseki, H. and Nitta, N. (2004) Map-based cloning of a fertility restorer gene, *Rf-1*, in rice (*Oryza sativa* L.). *The Plant Journal* 37, 315–325.

Komori, T., Imayama, T., Kato, N., Ishida,Y., Ueki, J. and Komari, T. (2007) Current status of binary vectors and superbinary vectors. *Plant Physiology* 145, 1155–1160.

Koncz, C. and Schell, J. (1986) The promoter of TL-DNA gene 5 controls the tissue-specific expression of chimaeric genes carried by a novel type of *Agrobacterium* binary vector. *Molecular and General Genetics* 204, 383–396.

Konieczny, A. and Ausubel, F. (1993) A procedure for mapping *Arabidopsis* mutations using co-dominant ecotype-specific PCR based markers. *The Plant Journal* 4, 403–410.

Konishi, T., Abe, K., Matsuura, S. and Yano, Y. (1990) Distorted segregation of the esterase isozyme genotypes in barley *Hordeum vulgare* L. *Japanese Journal of Genetics* 65, 411–416.

Konishi, T., Yano, Y. and Abe, K. (1992) Geographic distribution of alleles at the *ga2* locus for segregation distortion in barley. *Theoretical and Applied Genetics* 85, 419–422.

Koonin, E.V. (2005) Orthologies, paralogs and evolutionay genomics. *Annual Review of Genetics* 39, 309–338.

Koornneef, M., Dellaert, L.W.M. and van der Veen, J.H. (1982) EMS- and radiation-induced mutation frequencies at individual loci in *Arabidopsis thaliana* (L.) Heynh. *Mutation Research* 93, 109–123.

Korbel, J.O., Doerks, T., Jensen, L.J., Perez-Iratxeta, C., Kaczanowski, S., Hooper, S.D., Andrade, M.A. and Bork, P. (2005) Systematic association of genes to phenotypes by genome and literature mining. *PLos Biology* 3, e134.

Korol, A.B., Ronin, Y.I., Nevo, E. and Hayes, P. (1998) Multi-interval mapping of correlated trait complexes: simulation analysis and evidence from barley. *Heredity* 80, 273–284.

Korol, A.B., Ronin, Y.I., Itskovichi, A.M., Peng, J. and Nevo, E. (2001) Enhanced efficiency of quantitative trait loci mapping analysis based on multivariate complexes of quantitative traits. *Genetics* 157, 1789–1803.

Kosambi, D.D. (1944) The estimation of map distances from recombination values. *Annals of Eugenics* 12, 172–175.

Kota, R., Rudd, S., Facius, A., Kolesov, G., Theil, T., Zhang, H., Stein, N., Mayer, K. and Graner, A. (2003) Snipping polymorphisms from large EST collections in barley (*Hordeum vulgare* L.). *Molecular Genectics and Genomics* 270, 24–33.

Kowalski, S.P. and Kryder, R.D. (2002) Golden rice: a case study in intellectual property management and international capacity building. *RISK: Health, Safety and Environment* 13, 47–67.

Kraakman, A.T.W., Niks, R.E., van den Berg, P.M.M.M., Stam, P. and van Eeuwijk, F.A. (2004) Linkage disequilibrium mapping of yield and yield stability in modern spring barley cultivars. *Genetics* 168, 435–446.

Kraft, T., Hansen, M. and Nilsson, N.-O. (2000) Linkage disequilibrium and fingerprinting in sugar beet. *Theoretical and Applied Genetics* 101, 323–326.

Krapp, A., Morot-Gaudry, J.F., Boutet, S., Bergot, G., Lelarge, C., Prioul, J.L. and Noctor, G. (2007) Metabolomics. In: Morot-Gaudry, J.F., Lea, P. and Briat, J.F. (eds) *Functional Plant Genomics.* Science Publishers, Enfield, New Hampshire, pp. 311–333.

Krattiger, A., Mahoney, R.T., Nelsen, L., Bennett, A.B., Graff, G.D., Fernandez, C. and Kowalski, S.P. (eds) (2006) *Intellectual Property Management in Health and Agricultural Innovation, a Handbook of Best Practices.* Centre for the Management of Intellectual Property in Health R&D, Oxford, UK and Public Intellectual Property Resource for Agriculture, Davis, California.

Kresovich, S. and McFerson, J.R. (1992) Assessment and management of plant genetic diversity: consideration of intra- and interspecific variation. *Field Crops Research* 29, 185–204.

Kresovich, S., Luongo, A.J. and Schloss, S.J. (2002) 'Mining the gold': finding allelic variants for improved crop conservation and use. In: Engels, J.M.M., Ramanatha Rao, V., Brown, A.H.D. and Jackson, M.T. (eds) *Managing Plant Genetic Diversity.* International Plant Genetic Resources Institute, Rome, pp. 379–386.

Kriegner, A., Cervantes, J.C., Burg, K., Mwanga, R.O.M. and Zhang, D.P. (2003) A genetic linkage map of sweetpotato (*Ipomoea batatas* (L) Lam) based on AFLP markers. *Molecular Breeding* 11, 169–185.

Krishnan, P., Kruger, N.J. and Ratcliffe, R.G. (2005) Metabolite fingerprinting and profiling in plants using NMR. *Journal of Experimental Botany* 56, 255–265.

Krizkova, L. and Hrouda, M. (1998) Direct repeats of T-DNA integrated in tobacco chromosome: characterization of junction regions. *The Plant Journal* 16, 673–680.

Kruglyak, L. (2008) The road to genome-wide association studies. *Nature Reviews Genetics* 9, 314–318.

Kryder, R.D., Kowalski, S.P and Krattiger, A.F. (2000) The intellectual and technical property components of pro-vitamin A rice (GoldenRice™): a preliminary freedom-to-operate review. *ISAAA Briefs* No. 20. International Service for the Acquisition of Agri-biotech Applications (ISAAA), Ithaca, New York, 56 pp.

Krysan, P.J., Young, J.C. and Sussman, M.R. (1999) T-DNA as an insertional mutagen in *Arabidopsis. The Plant Cell* 11, 2283–2290.

Krysan, P.J., Young, J.C., Jester, P.J., Monson, S., Copenhaver, G., Preuss, D. and Sussman, M.R. (2002) Characterization of T-DNA insertion sites in *Arabidopsis thaliana* and the implications for saturation mutagenesis. *OMICS* 6, 163–174.

Kuchel, H., Ye, G., Fox, R. and Jefferies, S. (2005) Genetic and economic analysis of a targeted marker-assisted wheat breeding strategy. *Molecular Breeding* 16, 67–78.

Kuchel, H., Fox, R., Reinheimer, J., Mosionek, L., Willey, N., Bariana, H. and Jefferies, S. (2008) The successful application of a marker-assisted wheat breeding strategy. *Molecular Breeding* 20, 295–308.

Kuiper, H.A., Kok, E.J. and Engel, K.H. (2003) Exploitation of molecular profiling techniques for GM food safety assessment. *Current Opinion in Biotechnology* 14, 238–243.

Kuiper, M., Zabeau, M. and Vos, P. (1997) Amplification of simple sequence repeats. Patent EP 0805875.

Kumar, I. and Khush, G.S. (1986) Genetics of amylose content in rice (*Oryza sativa* L.). *Journal of Genetics* 65, 1–11.

Kumar, P.V.S. (1993) Biotechnology and biodiversity – a dialectical relationship. *Journal of Scientific and Industrial Research* 52, 523–532.

Kumpatla, S.P. and Mukhopadhyay, S. (2005) Mining and survey of simple sequence repeats in expressed sequence tags of dicotyledonous species. *Genome* 48, 985–998.

Kurata, N., Moore, G., Nagamura, Y., Foote, T., Yano, M., Minobe, Y. and Gale, M. (1994) Conservation of genome structure between rice and wheat. *Nature Biotechnology* 12, 276–278.

Kusterer, B., Piepho, H.P., Utz, H.F., Schön, C.C., Muminovic, J., Meyer, R.C., Altmann, T. and Melchinger, A.E. (2007) Heterosis for biomass-related traits in *Arabidopsis* investigated by a novel QTL analysis of the triple testcross design with recombinant inbred lines. *Genetics* 177, 1839–1850.

Lagercrantz, U. and Lydiate, D. (1995) RFLP mapping in *Brassica nigra* indicates different recombination rates in male and female meiosis. *Genome* 38, 255–264.

Laird, N.M. and Lange, C. (2006) Family-based designs in the age of large-scale gene-association studies. *Nature Reviews Genetics* 7, 385–394.

Lalonde, S., Ehrhardt, D.W., Loqué, D., Chen, J., Rhee, S.Y. and Frommer, W.B. (2008) Molecular and cellular approaches for the detection of protein–protein interactions: latest techniques and current limitations. *The Plant Journal* 53, 610–635.

Lamkey, K.R. and Edwards, J.W. (1999) Quantitative genetics of heterosis. In: Coors, J.G. and Pandey, S. (eds) *The Genetics and Exploitation of Heterosis in Crops*. American Society of Agronomy (ASA) and Crop Science Society of America (CSSA), Madison, Wisconsin, pp. 31–48.

Lamkey, K.R., Schnicker, B.J. and Melchinger, A.E. (1995) Epistasis in an elite maize hybrid and choice of generation for inbred development. *Crop Science* 35, 1272–1281.

Lan, H., Chen, M., Flowers, J.B., Yandell, B.S., Stapleton, D.S., Mata, C.M., Mui, E.T.-K., Flowers, M.T., Schueler, K.L., Manly, K.F., Williams, R.W., Kendziorski, C. and Attie, A.D. (2006) Combined expression trait correlations and expression quantitative trait locus mapping. *PLoS Genetics* 2(1), e6.

Lande, R. and Thompson, R. (1990) Efficiency of marker-assisted selection in the improvement of quantitative traits. *Genetics* 124, 743–756.

Landegren, U., Kaiser, R., Sanders, J. and Hood, L. (1988) A ligase-mediated gene detection technique. *Science* 241, 1077–1080.

Lander, E. and Kruglyak, L. (1995) Genetic dissection of complex traits: guidelines for interpreting and reporting linkage results. *Nature Genetics* 11, 241–247.

Lander, E.S. and Botstein, D. (1989) Mapping Mendelian factors underlying quantitative traits using RFLP linkage maps. *Genetics* 121,185–199.

Lander, E.S. and Green, P. (1987) Construction of multilocus genetic linkage maps in humans. *Proceedings of the National Academy of Sciences of the United States of America* 84, 2363–2367.

Lander, E.S., Green, P., Abrahamson, J., Barlow, A., Daly, M.J., Lincoln, S.E. and Newburg, L. (1987) MAPMAKER: an interactive computer package for constructing primary genetic linkage maps of experimental and natural populations. *Genomics* 1, 174–181.

Landy, A. (1989) Dynamic, structural and regulatory aspects of lambda-site-specific recombination. *Annual Review of Biochemistry* 58, 913–949.

Lane, M.A., Edwards, J.L. and Nielsen, E.S. (2000) Biodiversity informatics: the challenges of rapid development, large databases and complex data. In: *Proceedings of the 26th International Conference on Very Large Databases*, 10–14 September 2000, Cairo, Egypt. Very Large Data Base Endowment, Inc., USA.

Lang, N.T., Subudhi, P.K., Virmani, S.S., Brar, D.S., Khush, G.S., Li, Z. and Huang, N. (1999) Development of PCR-based markers for thermosensitive genetic male sterility gene *tms3(t)* in rice (*Oryza sativa* L.). *Hereditas* 131, 121–127.

Laperche, A., Brancourt-Hulmel, M., Heumez, E., Gardet, O., Hanocq, E., Devienne-Barret, F. and Le Gouis, J. (2007) Using genotype × nitrogen interaction variables to evaluate the QTL involved in wheat tolerance to nitrogen constraints. *Theoretical and Applied Genetics* 115, 399–415.

Laramie, J.M., Wilk, J.B., DeStefano, A.L. and Myers, R.H. (2007) HaploBuild: an algorithm to construct non-contiguous associated haplotypes in family based genetic studies. *Bioinformatics* 23, 2190–2192.

Larkin, P.J. and Scowcroft, W.R. (1981) Somaclonal variation – a novel source of variability from cell cultures for plant improvement. *Theoretical and Applied Genetics* 60, 197–214.

Lashermes, P. and Beckert, M. (1988) Genetic control of maternal haploidy in maize (*Zea mays* L.) and selection of haploid inducing lines. *Theoretical and Applied Genetics* 76, 405–410.

Lassner, M.W. and Orton, T.J. (1983) Detection of somatic variation. In: Tanksley, S.D. and Orton, T.J. (eds) *Isozymes in Plant Genetics and Breeding*. Vol. 1A. *Developments in Plant Genetics and Breeding*, 1. Elsevier, Amsterdam, Netherlands, pp. 209–217.

Laurie, C.C., Chasalow, S.D., LeDeaux, J.R., McCarroll, R., Bush, D., Hauge, B., Lai, C., Clark, D., Rocheford, T.R. and Dudley, J.W. (2004) The genetic architecture of response to long-term artificial selection for oil concentration in the maize kernel. *Genetics* 168, 2141–2155.

Laurie, D.A. and Bennett, M.D. (1986) Wheat and maize hybridization. *Canadian Journal of Genetics and Cytology* 28, 313–316.

Laurie, D.A. and Reymondie, S. (1991) High frequencies of fertilization and haploid seedling production in crosses between commercial hexaploid wheat varieties and maize. *Plant Breeding* 106, 182–189.

Laurie, D.A., Pratchett, N., Bezant, J.H. and Snape, J.W. (1994) Genetic analysis of a photoperiod response gene on the short arm of chromosome 2(2H) on *Hordeum vulgare* (barley). *Heredity* 72, 619–627.

Lebowitz, R.L., Soller, M. and Beckmann, J.S. (1987) Trait-based analysis for the detection of linkage between marker loci and quantitative trait loci in cross between inbred lines. *Theoretical and Applied Genetics* 73, 556–562.
Lee, E.A, Ash, M.J. and Good, B. (2007) Re-examining the relationship between degree of relatedness, genetic effects and heterosis in maize. *Crop Science* 47, 629–635.
Lee, J.M., Davenport, G.F., Marshall, D., Noel Ellis, T.H., Ambrose, M.J., Dicks, J., van Hintum, T.J.L. and Flavell, A.J. (2005) GERMINATE: a generic database for integrating genotypic and phenotypic information for plant genetic resource collections. *Plant Physiology* 139, 619–631.
Lee, L.-Y., Kononov, M.E., Bassuner, B., Frame, B.R., Wang, K. and Gelvin, S.B. (2007) Novel plant transformation vectors containing the superpromoter. *Plant Physiology* 145, 1294–1300.
Lee, M. (1995) DNA markers and plant breeding programs. *Advances in Agronomy* 55, 265–344.
Lee, M., Godshalk, E.B., Lamkey, K.R. and Woodman, W.L. (1989) Association of restriction length polymorphism among maize inbreds with agronomic performance of their crosses. *Crop Science* 29, 1067–1071.
Lee, M., Sharopova, N., Beavis, W.D., Grant, D., Katt, M., Blair, D. and Hallauer, A. (2002) Expanding the genetic map of maize with the intermated B73 × Mo17 (IBM) population. *Plant Molecular Biology* 48, 453–461.
Leflon, M., Lecomte, C., Barbottin, A., Jeuffroy, M.-H., Robert, N. and Brancourt-Hulmel, M. (2005) Characterization of environments and genotypes for analyzing genotype × environment interaction. Some recent advances in winter wheat and prospects for QTL detection. *Journal of Crop Improvement* 14, 249–298.
Leister, D.M., Kurth, J., Laurie, D.A., Yano, M., Sasaki, T., Devos, K., Graner, A. and Schulze-Lefert, P. (1998) Rapid re-organization of resistance gene homologues in cereal genomes. *Proceedings of the National Academy of Sciences of the United States of America* 95, 370–375.
Leng, E.R. (1962) Results of long-term selection for chemical composition in maize and their significance in evaluating breeding systems. *Zeitschrift für Pflanzenzüchtung* 47, 67–91.
Lerner, I.M. (1950) *Population Genetics and Animal Improvement.* Cambridge University Press, Cambridge.
Lerner, I.M. (1954) *Genetic Homeostasis.* Oliver and Boyd, London.
Lesser, W. (2005) Intellectual property rights in a changing political environment: perspectives on the types and administration of protection. *AgBioForum* 8, 64–72.
Lesser, W. and Mutschler, M.A. (2004) Balancing investment incentives and social benefits when protecting plant varieties: implementing initial systems. *Crop Science* 44, 1113–1120.
Leung, H., Wu, C., Baraoidan, M., Bordeos, A., Ramos, M., Madamba, S., Cabauatan, P., Vera Cruz, C., Portugal, A., Reyes, G., Bruskiewich, R., McLaren, G., Lafitte, R., Gregorio, G., Bennett, J., Brar, D., Khush, G., Schnable, P., Wang, G. and Leach, J. (2001) Deletion mutants for functional genomics: progress in phenotyping, sequence assignment and database development. In: Khush, G.S., Brar, D.S. and Hardy, B. (eds) *Rice Genetics IV. Proceedings of the Fourth International Rice Genetics Symposium*, 22–27 October 2000, Los Baños, Philippines. Science Publishers, Inc., New Delhi and International Rice Research Institute, Los Baños, Philippines, pp. 239–251.
Levinson, G. and Gutman, G.A. (1987) Slipped-strand mispairing: a major mechanism for DNA sequence evolution. *Molecular Biology and Evolution* 4, 203–221.
Lewin, B. (2007) *Genes IX.* Jones & Bartlett, Sudbury, Massachusetts, 892 pp.
Lewington, A. (2003) *Plants for People.* Eden Project Books, London.
Lewontin, R.C. (1964) The interaction of selection and linkage. I. General considerations; heterotic models. *Genetics* 49, 49–67.
Lewontin, R.C. and Berlan, J.P. (1990) The political economy of agricultural research: the case of hybrid corn. In: Carroll, C.R., Vandermeer, J.H. and Rosset, P. (eds) *Agroecology.* McGraw Hill, New York, pp. 613–628.
Li, C.C. (1955) *Population Genetics.* University of Chicago Press, Chicago, Illinois.
Li, H., Ye, G. and Wang, J. (2007) A modified algorithm for the improvement of composite interval mapping. *Genetics* 175, 361–374.
Li, H., Ribaut, J.M., Li, Z. and Wang, J. (2008) Inclusive composite interval mapping (ICIM) for digenic epistasis of quantitative traits in biparental populations. *Theoretical and Applied Genetics* 116, 243–260.
Li, L., Zhou, Y., Cheng, X., Sun, J., Marita, J.M., Ralph, J. and Chiang, V.L. (2003) Combinatorial modification of multiple lignin traits in trees through multigene cotransformation. *Proceedings of the National Academy of Sciences of the United States of America* 100, 4939–4944.

Li, R., Lyons, M.A., Wittenburg, H., Paigen, B. and Churchill, G.A. (2005) Combining data from multiple inbred line crosses improves the power and resolution of quantitative trait loci mapping. *Genetics* 169, 1699–1709.

Li, R., Tsaih, S.W., Shockley, K., Stylianou, I.M., Wergedal, J., Paigen, B. and Churchill, G.A. (2006) Structural model analysis of multiple quantitative traits. *PLoS Genetics* 2(7), e114.

Li, X. and Zhang, Y. (2002) Reverse genetics by fast neutron mutagenesis in higher plants. *Functional and Integrative Genomics* 2, 254–258.

Li, X., Song, Y., Century, K., Straight, S., Ronald, P.C., Dong, X., Lasser, M. and Zhang, Y. (2001) Deleagene™: a fast neutron mutagenesis-based reverse genetics system for plants. *The Plant Journal* 27, 235–242.

Li, Y., Shi, Y., Cao, Y. and Wang, T. (2004) Establishment of a core collection for maize germplasm preserved in Chinese national gene bank using geographic distribution and characterization data. *Genetic Resources and Crop Evolution* 51, 845–852.

Li, Z.K., Pinson, S.R., Stansel, J.W. and Park, W.D. (1995) Identification of quantitative trait loci (QTL) for heading date and plant height in cultivated rice (*Oryza sativa* L.). *Theoretical and Applied Genetics* 91, 374–381.

Li, Z.K., Luo, L.J., Mei, H.W., Wang, D.L., Shu, Q.Y., Tabien, R., Zhong, D.B., Ying, C.S., Stansel, J.W., Khush, G.S. and Paterson, A.H. (2001) Overdominance epistatic loci are the primary genetic basis of inbreeding depression and heterosis in rice: I. Biomass and grain yield. *Genetics* 158, 1737–1753.

Li, Z.K., Fu, B.-Y., Gao, Y.-M., Xu, J.-L., Ali, J., Lafitte, H.R., Jiang, Y.-Z., Rey, J.D., Vijayakumar, C.H.M., Maghirang, R., Zheng, T.-Q. and Zhu, L.-H. (2005) Genome-wide introgression lines and their use in genetic and molecular dissection of complex phenotypes in rice (*Oryza sativa* L.). *Plant Molecular Biology* 59, 33–52.

Liang, C., Jaiswal, P., Hebbard, C., Avraham, S., Buckler, E.S., Casstevens, T., Hurwitz, B., McCouch, S., Ni, J., Pujar, A., Ravenscroft, D., Ren, L., Spooner, W., Tecle, I., Thomason, J., Tung, C.-W., Wei, X., Yap, I., Youens-Clark, K., Ware, D. and Stein, L. (2008) Gramene: a growing plant comparative genomics resource. *Nucleic Acids Research* 36, D947–D953.

Liang, F., Deng, Q., Wang, Y., Xiong, Y., Jin, D., Li, J. and Wang, B. (2004) Molecular marker-assisted selection for yield-enhancing genes in the progeny of '9311 × *O. rufipogon*' using SSR. *Euphytica* 139, 159–165.

Liang, G.H. and Skinner, D.Z. (eds) (2004) *Genetically Modified Crops: Their Development, Uses and Risks.* Food Products Press, Binghamton, New York.

Lillemo, M., van Ginkel, M., Trethowan, R.M., Hernández, E. and Rajaram, S. (2004) Associations among international CIMMYT bread wheat yield testing locations in high rainfall areas and their implications for wheat breeding. *Crop Science* 44, 1163–1169.

Lilley, J.M., Ludlow, M.M., McCouch, S.R. and O'Toole, J.C. (1996) Locating QTL for osmotic adjustment and dehydration tolerance in rice. *Journal of Experimental Botany* 47, 1427–1436.

Lin, C., Fang, J., Xu, X., Zhao, T., Cheng, J., Tu, J., Ye, G. and Shen, Z. (2008) A built-in strategy for containment of transgenic plants: creation of selectively terminable transgenic rice. *PLoS ONE* 3, e1818. Available at: http://www.plosone.org (accessed 17 November 2009).

Lin, C.S. and Binns, M.R. (1988) A method of analyzing cultivar × location × year experiments: a new stability parameter. *Theoretical and Applied Genetics* 76, 425–430.

Lin, C.S., Binns, M.R. and Lefkovitch, L.P. (1986) Stability analysis: where do we stand? *Crop Science* 26, 894–900.

Lin, H.X., Yamamoto, T., Sasaki, T. and Yano, M. (2000) Characterization and detection of epistatic interactions of 3 QTLs, *Hd1*, *Hd2* and *Hd3*, controlling heading date in rice using nearly isogenic lines. *Theoretical and Applied Genetics* 101, 1021–1028.

Lin, Y.R., Schertz, K.F. and Paterson, A.H. (1995) Comparative analysis of QTLs affecting plant height and maturity across the Poaceae, in reference to an interspecific sorghum population. *Genetics* 140, 391–411.

Lippman, Z.B. and Zamir, D. (2007) Heterosis: revisiting the magic. *Trends in Genetics* 23, 60–66.

Liu, B., Zhang, S., Zhu, X., Yang, Q., Wu, S., Mei, M., Mauleon, R., Leach, J., Mew, T. and Leung, H. (2004) Candidate defense genes as predictors of quantitative blast resistance in rice. *Molecular Plant–Microbe Interaction* 17, 1146–1152.

Liu, B.H. (1998) *Statistical Genomics: Linkage, Mapping and QTL Analysis.* CRC Press, Boca Baton, Florida, 611 pp.

Liu, G., Zhang, Z., Zhu, H., Zhao, F., Ding, X., Zeng, R., Li, W. and Zhang, G. (2008) Detection of QTLs with additive effects and additive-by-environment interaction effects on panicle number in rice (*Oryza sativa* L.) with single-segment substitution lines. *Theoretical and Applied Genetics* 116, 923–931.

Liu, J.H., Xu, X.Y. and Deng, X.X. (2005) Intergeneric somatic hybridization and its application to crop genetic improvement. *Plant Cell, Tissue and Organ Culture* 82, 19–44.

Liu, J.S., Sabatti, C., Teng, J., Keats, B.J.B. and Risch, K. (2001) Bayesian analysis of haplotypes for linkage disequilibrium mapping. *Genome Research* 11, 1716–1724.

Liu, K., Goodman, M., Muse, S., Smith, J.S., Buckler, E.D. and Doebley, J. (2003) Genetic structure and diversity among maize inbred lines as inferred from DNA microsatellites. *Genetics* 165, 2117–2128.

Liu, S., Zhou, R., Dong, Y., Li, P. and Jia, J. (2006) Development, utilization of introgression lines using a synthetic wheat as donor. *Theoretical and Applied Genetics* 112, 1360–1373.

Liu, S.C., Kowalski, S.P., Lan, T.H., Feldmann, K.A. and Paterson, A.H. (1996) Genome-wide high-resolution mapping by recurrent intermating using *Arabidopsis thaliana* as a model. *Genetics* 142, 247–258.

Liu, X.C. and Wu, J.L. (1998) SSR heterotic patterns of parents for making and predicting heterosis in rice breeding. *Molecular Breeding* 4, 263–268.

Liu, X.Q., Wang, L., Chen, S., Lin, F. and Pan, Q.H. (2005) Genetics and physical mapping of *Pi36(t)*, a novel rice blast resistance gene located on rice chromosome 8. *Molecular Genetics and Genomics* 274, 394–401.

Liu, X.Z., Peng, Z.B., Fu, J.H., Li, L.C. and Huang, C.L. (1997) Application of RAPD in group classification studies. *Scientia Agricultura Sinica* 30, 44–51.

Liu, Y. and Zeng, Z.B. (2000) A general mixture model approach for mapping quantitative trait loci from diverse cross designs involving multiple inbred lines. *Genetical Research* 75, 345–355.

Liu, Y.G. and Whittier, R. (1995) Thermal asymmetric interlaced PCR: automatable amplification and sequencing of insert and fragments from P1 and YAC clones for chromosome walking. *Genomics* 25, 674–681.

Liu, Y.G., Mitsukawa, N., Oosumi, T. and Whittier, R. (1995) Efficient isolation and mapping of *Arabidopsis thaliana* T-DNA insert junctions by thermal asymmetric interlaced PCR. *The Plant Journal* 8, 457–463.

Liu, Y.-G., Shirano, Y., Fukaki, H., Yanai, Y., Tasaka, M., Tabata, S. and Shibata, D. (1999) Complementation of plant mutants with large genomic DNA fragments by a transformation-competent artificial chromosome vector accelerates positional cloning. *Proceedings of the National Academy of Sciences of the United States of America* 96, 6535–6540.

Lloyd, A., Plaisier, C.L., Carroll, D. and Drews, G.N. (2005) Targeted mutagenesis using zinc-finger nucleases in *Arabidopsis*. *Proceedings of the National Academy of Sciences of the United States of America* 102, 2232–2237.

Lockhart, D.J., Dong, H., Byrne, M.C., Follettie, M.T., Gallo, M.V., Chee, M.S., Mittmann, M., Wang, C., Kobayashi, M., Norton, H. and Brown, E.L. (1996) Expression monitoring by hybridization to high-density oligonucleotide arrays. *Nature Biotechnology* 14, 1675–1680.

Löffler, C.M., Wei, J., Fast, T., Gogerty, J., Langton, S., Bergman, M., Merrill, B. and Cooper, M. (2005) Classification of maize environments using crop simulation and geographic information systems. *Crop Science* 45, 1708–1716.

Lolle, S.J., Victoria, J.L., Young, J.M. and Pruitt, R.E. (2005) Genome-wide non-Mendelian inheritance of extra-genomic information in *Arabidopsis*. *Nature* 434, 505–509.

Long, A.D., Mullaney, S.L., Reid, L.A., Fry, J.D., Langley, C.H. and Mackay, T.F. (1995) High resolution mapping of genetic factors affecting abdominal bristle number in *Drosophila melanogaster*. *Genetics* 139, 1273–1291.

Longin, C.F.H., Utz, H.F., Reif, J.C., Schipprack, W. and Melchinger, A.E. (2006) Hybrid maize breeding with doubled haploids: I. One stage versus two-stage selection for testcross performance. *Theoretical and Applied Genetics* 112, 903–912.

Lonnstedt, I. and Speed, T.P. (2002) Replicated microarray data. *Statistica Sinica* 12, 31–46.

Lonosky, P.M., Zhang, X., Honavar, V.G., Dobbs, D.L., Fu, A. and Rodermel, S.R. (2004) A proteomic analysis of maize chloroplast biogenesis. *Plant Physiology* 134, 560–574.

Lörz, H. and Wenzel, G. (eds) (2005) *Molecular Marker Systems in Plant Breeding and Crop Improvement. Biotechnology in Agriculture and Forestry*, Vol. 55. Springer-Verlag, Berlin.

Louwaars, N.P., Visser, B., Eaton, D., Beekwilder, J. and van der Meer, I. (2002) Policy response to technological developments: the case of GURTs. In: Louwaars, N.P. (ed.) *Seed Policy, Legislation and Law: Widening a Narrow Focus.* Food Products Press, Binghamton, New York, pp. 89–102.

Louwaars, N.P., Tripp, R. and Eaton, D. (2006) Public research in plant breeding and intellectual property rights: a call for new institutional policies. *Agricultural and Rural Development Notes* Issue 13, p. 4. World Bank, Washington, DC.

Lu, C., Shen, L., Tan, Z., Xu, Y., He, P., Chen, Y. and Zhu, L. (1996) Comparative mapping of QTL for agronomic traits of rice across environments using a double haploid population. *Theoretical and Applied Genetics* 93, 1211–1217.

Lu, H., Romero-Severson, J. and Bernardo, R. (2003) Genetic basis of heterosis explored by simple sequence repeat markers in a random-mated maize population. *Theoretical and Applied Genetics* 107, 494–502.

Lu, H., Redus, M.A., Coburn, J.R., Rutger, J.N., McCouch, S.R. and Tai, T.H. (2005) Population structure and breeding patterns of 145 U.S. rice cultivars based on SSR marker analysis. *Crop Science* 45, 66–76.

Lu, L., Romero-Severson, J. and Bernardo, R. (2002) Chromosomal regions associated with segregation distortion in maize. *Theoretical and Applied Genetics* 105, 622–628.

Lu, X., Niu, T. and Liu, J.S. (2003) Haplotype information and linkage disequilibrium mapping for single nucleotide polymorphisms. *Genome Research* 13, 2112–2117.

Lu, X.G., Gu, M.H. and Li, C.Q. (eds) (2001) *Theory and Technology of Two-line Hybrid Rice.* China Science Press, Beijing.

Lu, X.G., Mou, T.M., Hoan, N.T. and Virmani, S.S. (2004) Two-line hybrid rice breeding in and outside of China. In: Virmani, S.S., Mao, C.X. and Hardy, B. (eds) *Hybrid Rice for Food Security, Poverty Alleviation and Environmental Protection.* International Rice Research Institute, Manila, Phillipines.

Lu, Y., Yan, J., Guimarães, C.T., Taba, S., Hao, Z., Gao, S., Chen, S., Li, J., Zhang, S., Vivek, B.S., Magorokosho, C., Mugo, S., Makumbi, D., Parentoni, S.N., Shah, T., Rong, T., Crouch, J.H. and Xu, Y. (2009) Molecular characterization of global maize breeding germplasm based on genome-wide single nucleotide polymorphisms. *Theoretical and Applied Genetics* 120, 93–115.

Lübberstedt, T., Klien, D. and Melchinger, A.E. (1998a) Comparative QTL mapping of resistance to *Ustilago maydis* across four populations of European flint-maize. *Theoretical and Applied Genetics* 97, 1321–1330.

Lübberstedt, T., Melchenger, A.E., Fähr, S., Klein, D., Dally, A. and Westhoff, P. (1998b) QTL mapping in test crosses of flint lines of maize: III. Comparison across populations for forage traits. *Crop Science* 38, 1278–1289.

Lucca, P., Ye, X.D. and Potrykus, I. (2001) Effective selection and regeneration of transgenic rice plants with mannose as selective agent. *Molecular Breeding* 7, 43–49.

Lucken, K.A. (1986) The breeding and production of hybrid wheat. In: Smith, E.L. (ed.) *Genetic Improvement in Yield of Wheat.* American Society of Agronomy (ASA) and Crop Science Society of America (CSSA), Madison, Wisconsin, pp. 87–107.

Lucken, K.A. and Johnson, K.D. (1988) Hybrid wheat status and outlook. In: International Rice Research Institute (IRRI) (ed.) *Hybrid Rice.* IRRI, Manila, Philippines, pp. 243–255.

Lucker, J., Schwab, W., van Hautum, B., Blaas, J., van der Plas, L.H., Bouwmeester, H.J. and Verhoeven, H.A. (2004) Increased and altered fragrance of tobacco plants after metabolic engineering using three monoterpene synthases from lemon. *Plant Physiology* 134, 510–519.

Luo, K., Duan, H., Zhao, D., Zheng, X., Deng, W., Chen, Y., Stewart, C.N., Jr, McAvoy, R., Jiang, X., Wu, Y., He, A., Pei, Y. and Li, Y. (2007) 'GM-Gene-deletor': fused loxP-FRT recognition sequences dramatically improve the efficiency of FLP or CRE recombinase on transgene excision from pollen and seed of tobacco plants. *Plant Biotechnology Journal* 5, 263–274.

Luo, L.J., Li, Z.K., Mei, H.W., Shu, Q.Y., Tabien, R., Zhong, D.B., Ying, C.S., Stansel, J.W., Khush, G.S. and Paterson, A.H. (2001) Overdominant epistatic loci are the primary genetic basis of inbreeding depression and heterosis in rice. II. Grain yield components. *Genetics* 158, 1755–1771.

Lupas, A. (1996) Prediction and analysis of coiled coil structures. *Methods in Enzymology* 266, 513–523.

Lush, J.L. (1937) *Animal Breeding Plans.* Iowa State College Press, Ames, Iowa.

Lush, J.L. (1945) *Animal Breeding Plans*, 3rd edn. Iowa State College Press, Ames, Iowa.

Lussier, Y.A. and Li, J. (2004) Terminological mapping for high throughput comparative biology of phenotypes. *Proceedings of the Pacific Symposium on Biocomputing*, 6–10 January 2004, Hawaii. PSB, Stanford, California, pp. 202–213.

Lutz, K.A., Azhagiri, A.K., Tungsuchat-Huang, T. and Maliga, P. (2007) A guide to choosing vectors for transformation of the plastid genome of higher plants. *Plant Physiology* 145, 1201–1210.

Lyamichev, V., Mast, A.L., Hall, J.G., Prudent, J.R., Kaiser, M.W., Takova, T., Kwiatkowski, R.W., Sander, T.J., de Arruda, M., Arco, D.A., Neri, B.P. and Brow, M.A.D. (1999) Polymorphism identification and quantitative detection of genomic DNA by invasive cleavage of oligonucleotide probes. *Nature Biotechnology* 17, 292–296.

Lyman, J.M. (1984) Progress and planning for germplasm conservation of major food crops. *Plant Genetic Resources Newsletter* 60, 3–21.

Lynch, M. and Walsh, B. (1998) *Genetics and Analysis of Quantitative Traits*. Sinauer Associates, Sunderland, Massachusetts, 980 pp.

Ma, C.X., Casella, G. and Wu, R.L. (2002) Functional mapping of quantitative trait loci underlying the character process: a theoretical framework. *Genetics* 61, 1751–1762.

Ma, J.K., Hiatt, A., Hein, M., Vine, N.D., Wang, F., Stabila, P., van Dolleweerd, C., Mostov, K. and Lehner, T. (1995) Generation and assembly of secretory antibodies in plants. *Science* 268, 716–719.

Ma, J.K.-C., Chikwamba, R., Sparrow, P., Fischer, R., Mahoney, R. and Twyman, R.M. (2005) Plant-derived pharmaceuticals – the road forward. *Trends in Plant Science* 10, 580–585.

MacBeath, G. and Schreiber, S.L. (2000) Printing proteins as microarrays for high-throughput function determination. *Science* 289, 1760–1763.

MacCoss, M.J., McDonald, W.H., Saraf, A., Sadygov, R., Clark, J.M, Tasto, J.J., Gould, K.L., Wolters, D., Washburn, M., Weiss, A., Clark, J.I. and Yates III, J.R. (2002) Shotgun identification of protein modifications from protein complexes and lens tissue. *Proceedings of the National Academy of Sciences of the United States of America* 99, 7900–7905.

MacDonald, J.A., Mackey, A.J., Pearson, W.R. and Haystead, T.A. (2002) A strategy for the rapid identification of phosphorylation sites in the phosphoproteome. *Molecular and Cellular Proteomics* 1, 314–322.

Mackay, I. and Powell, W. (2007) Methods for linkage disequilibrium mapping in crops. *Trends in Plant Science* 12, 57–63.

Mackay, T.F.C. (1995) The genetic basis of quantitative variation: number of sensory bristles of *Drosophila melanogaster* as a model system. *Trends in Genetics* 11, 464–470.

Mackay, T.F.C., Stone, E.A. and Ayroles, J.F. (2009) The genetics of quantitative traits: challenges and prospects. *Nature Reviews Genetics* 10, 565–577.

Mackill, D.J. and McNally, K.L. (2004) A model crop species: molecular markers in rice. In: Lörz, H. and Wenzel, G. (eds) *Molecular Marker Systems in Plant Breeding and Crop Improvement*. Springer Verlag, Heidelberg, pp. 39–54.

Mackill, D.J., Salam, M.A., Wang, Z.Y. and Tanksley, S.D. (1993) A major photoperiod-sensitivity gene tagged with RFLP and isozyme markers in rice. *Theoretical and Applied Genetics* 85, 536–540.

Mackill, D.J., Zhang, Z., Redoña, E.D. and Colowit, P.M. (1996) Level of polymorphism and genetic mapping of AFLP markers in rice. *Genome* 39, 969–977.

Macomber, R.S. (1998) *A Complete Introduction to Modern NMR Spectroscopy*. John Wiley & Sons, Chichester, UK.

Magnuson, V.L., Ally, D.S., Nylund, S.J., Karanjawala, Z.E., Rayman, J.B., Knapp, J.I., Lowe, A.L., Ghosh, S. and Collins, F.S. (1996) Substrate nucleotide-determined non-templates addition to adenine by *Taq* DNA polymerase: implication for PCR-based genotyping and cloning. *Biotechniques* 21, 700–709.

Maheswaran, M., Huang, N., Sreerangasamy, S.R. and McCouch, S.R. (2000) Mapping quantitative trait loci associated with days to flowering and photoperiod sensitivity in rice (*Oryza sativa* L.). *Molecular Breeding* 6, 145–155.

Mailund, T., Schierup, M.H., Pedersen, C.N.S., Madsen, J.N., Hein, J. and Schauser, L. (2006) GeneRecon – a coalescent based tool for fine-scale association mapping. *Bioinformatics* 22, 2317–2318.

Malakoff, D. (1999) Bayes offers a 'new' way to make sense of numbers. *Science* 286, 1460–1464.

Malmberg, R.L. and Mauricio, R. (2005) QTL-based evidence for the role of epistasis in evolution. *Genetical Research* 86, 89–95.

Malmberg, R.L., Held, S., Waits, A. and Mauricio, R. (2005) Epistasis for fitness-related quantitative traits in *Arabidopsis thaliana* grown in the field and in the greenhouse. *Genetics* 171, 2013–2027.

Malosetti, M., Voltas, J., Romagosa, I., Ullrich, S.E. and van Eeuwijk, F.A. (2004) Mixed models including environmental variables for studying QTL by environment interaction. *Euphytica* 137, 139–145.

Malosetti, M., van der Linden, C.G., Vosman, B. and van Eeuwijk, A. (2007) A mixed-model approach to association mapping using pedigree information with an illustration of resistance to *Phytophthora infestans* in potato. *Genetics* 175, 879–889.

Maluszynski, M., Kasha, K.J., Forster, B.P. and Szarejko, I. (eds) (2003) *Doubled Haploid Production in Crop Plants – a Manual.* Kluwer Academic Publishers, Dordrecht, Netherlands.

Mandel, J. (1969) The partitioning of interaction in analysis of variance. *Journal of Research of the National Bureau of Standards, Series B* 73, 309–328.

Mandel, J. (1971) A new analysis of variance model for nonadditive data. *Technometrics* 13, 1–18.

Manenti, G., Galvan, A., Pettinicchio, A., Trincucci, G., Spada, E., Zolin, A., Milani, S., Gonzalez-Neira, A. and Dragani, T.A. (2009) Mouse genome-wide association mapping needs linkage analysis to avoid false-positive loci. *PLoS Genetics* 5(1), e1000331.

Mangelsdorf, P.C. (1974) *Corn: Its Origin, Evolution and Improvement.* Harvard University Press, Cambridge, Massachusetts.

Manly, K.F. (1993) A Macintosh program for storage and analysis of experimental genetic mapping data. *Mammalian Genome* 4, 303–313.

Mannschreck, S. (2004) Optimierung der Methode zur Chromosomalen Aufdopplung von *in-vivo* induzierten Haploiden bei Mais (*Zea mays* L.). MSc thesis, Universität Hohenheim, Germany.

Maqbool, S.B., Riazuddin, S., Loc, N.T., Gatehouse, A.M.R., Gatehouse, J.A. and Christou, P. (2001) Expression of multiple insecticidal genes confers broad resistance against a range of different rice pests. *Molecular Breeding* 7, 85–93.

Marchini, J., Donnelly, P. and Cardon, L.R. (2005) Genome-wide strategies for detecting multiple loci that influence complex diseases. *Nature Genetics* 4, 413–417.

Margulies, M., Egholm, M., Altman, W.E., Attiya, S., Bader, J.S., Bemben, L.A., Berka, J., Braverman, M.S., Chen, Y.-J., Chen, Z., Dewell, S.B., Du, L., Fierro, J.M., Gomes, X.V., Godwin, B.C., He, W., Helgesen, S., Ho, C.H., Irzyk, G.P., Jando, S.C., Alenquer, M.L.I., Jarvie, T.P., Jirage, K.B., Kim, J.-B., Knight, J.R., Lanza, J.R., Leamon, J.H., Lefkowitz, S.M., Lei, M., Li, J., Lohman, K.L., Lu, H., Makhijani, V.B., McDade, K.E., McKenna, M.P., Myers, E.W., Nickerson, E., Nobile, J.R., Plant, R., Puc, B.P., Ronan, M.T., Roth, G.T., Sarkis, G.J., Simons, J.F., Simpson, J.W., Srinivasan, M., Tartaro, K.R., Tomasz, A., Vogt, K.A., Volkmer, G.A., Wang, S.H., Wang, Y., Weiner, M.P., Yu, P., Begley, R.F. and Rothberg, J.M. (2005) Genome sequencing in open microfabricated high density picoliter reactors. *Nature* 437, 376–380.

Marillonnet, S., Giritch, A., Gils, M., Kandzia, R., Klimyuk, V. and Gleba, Y. (2004) *In planta* engineering of viral RNA replicons: efficient assembly by recombination of DNA modules delivered by *Agrobacterium. Proceedings of the National Academy of Sciences of the United States of America* 101, 6852–6857.

Marillonnet, S., Thoeringer, C., Kandzia, R., Klimyuk, V. and Gleba, Y. (2005) Systematic *Agrobacterium tumefaciens*-mediated transfection of viral replicons for efficient transient expression in plants. *Nature Biotechnology* 23, 718–723.

Martienssen, R.A., Rabinowicz, P.D., O'Shaughnessy, A. and McCombie, W.R. (2004) Sequencing the maize genome. *Current Opinion in Plant Biology* 7, 102–107.

Martin, G.B., Williams, J.G.K. and Tanksley, S.D. (1991) Rapid identification of markers linked to a *Pseudomonas* resistance gene in tomato by using random primers and near-isogenic lines. *Proceedings of the National Academy of Sciences of the United States of America* 88, 2336–2340.

Martin, G.B., Brommonschenkel, S.H., Chunwongse, J., Frary, A., Ganal, M.W., Spivey, R., Wu, T., Earle, E.D. and Tanksley, S.D. (1993) Map-based cloning of a protein kinase gene conferring disease resistance in tomato. *Science* 262, 1432–1436.

Martin, J.M., Talbert, L.E., Lanning, S.P. and Blake, N.K. (1995) Hybrid performance in wheat as related to parental diversity. *Crop Science* 35, 104–108.

Martin, O.C. and Hospital, F. (2006) Two- and three-locus tests for linkage analysis using recombinant inbred lines. *Genetics* 173, 451–459.

Martinez, L. (2003) *In vitro* gynogenesis induction and doubled haploid production in onion (*Allium cepa* L.). In: *Doubled Haploid Production in Crop Plants.* Kluwer Academic Publisher, Dordrecht, Netherlands, pp. 275–281.

Martinez, O. and Curnow, R.N. (1992) Estimating the locations and the sizes of the effects of quantitative trait loci using flanking markers. *Theoretical and Applied Genetics* 85, 480–488.

Martinez, V., Thorgaard, G., Robison, B. and Sillanpää, M.J. (2005) An application of Bayesian QTL mapping to early development in double haploid lines of rainbow trout including environmental effects. *Genetical Research* 86, 209–221.

Mascarenhas, M. and Busch, L. (2006) Seeds of change: intellectual property rights, genetically modified soybeans and seed saving in the United States. *Sociologia Ruralis* 46, 122–138.

Mather, K. (1949) *Biometrical Genetics.* Chapman & Hall, London.

Mather, K. and Jinks, J.L. (1982) *Biometrical Genetics.* Chapman & Hall, London.

Mathesius, U., Imin, N., Natera, S.H.A. and Rolfe, B.G. (2003) Proteomics as a functional genomics tool. In: Grotewold, E. (ed.) *Methods in Molecular Biology*, Vol. 236. *Plant Functional Genomics: Methods and Protocols.* Humana Press, Totowa, New Jersey, pp. 395–413.

Matsumura, H., Ito, A., Saitoh, H., Winter, P., Kahl, G., Reuter, M., Kruger, D.H. and Terauchi, R. (2005) SuperSAGE. *Cell Microbiology* 7, 11–18.

Matthews, P.R., Wang, M.B., Waterhouse, P.M., Thornton, S., Fieg, S.J., Gubler, F. and Jacobsen, J.V. (2001) Marker gene elimination from transgenic barley, using co-transformation with adjacent 'twin T-DNA' on a standard *Agrobacterium* transformation vector. *Molecular Breeding* 7, 195–202.

Matus, I., Corey, A., Filichkin, T., Hayes, P.M., Vales, M.I., Kling, J., Riera-Lizarazu, O., Sato, K., Powell, W. and Waugh, R. (2003) Development and characterization of recombinant chromosome substitution lines (RCSLs) using *Hordeum vulgare* subsp. *spontaneum* as a source of donor alleles in a *Hordeum vulgare* subsp. *vulgare* background. *Genome* 46, 1010–1023.

Matzke, M.A. and Matzke, A.J.M. (1995) How and why do plants inactivate homologous (Trans) genes? *Plant Physiology* 107, 679– 685.

Matzke, M.A., Mette, M.F. and Matzke, A.J.M. (2000) Transgene silencing by the host genome defense: implications for the evolution of epigenetic control mechanisms in plants and vertebrates. *Plant Molecular Biology* 43, 401–415.

Maxted, N., Ford-Lloyd, B.V. and Hawkes, J.G. (1997) Complementary conservation strategies. In: Maxted, N., Ford-Lloyd, B.V. and Hawkes, J.G. (eds) *Plant Genetic Resources Conservation.* Chapman & Hall, London, pp. 15–39.

Mayer, J., Sharples, J. and Nottenburg, C. (2004) *Resistance to Phosphinothricin.* CAMBIA Intellectual Property, Canberra.

Mayer, J.E., Pfeiffer, W.H. and Beyer, P. (2008) Biofortified crops to alleviate micronutrient malnutrition. *Current Opinion in Plant Biology* 11, 166–177.

Mayes, S., Parsley, K., Sylvester-Bradley, R., May, S. and Foulkes, J. (2005) Integrating genetic information into plant breeding programmes: how will we produce varieties from molecular variation, using bioinformatics? *Annals of Applied Biology* 146, 223–237.

McCallum, C.M., Comai, L., Greene, E.A. and Henikoff, S. (2000) Targeting Induced Local Lesions IN Genomes (TILLING) for plant functional genomics. *Plant Physiology* 123, 439–442.

McCarthy, M.I., Abecasis, G.R., Cardon, L.R., Goldstein, D.B., Little, J., Ioannidis, J.P.A. and Hirschhorn, J.N. (2008) Genome-wide association studies for complex traits: consensus, uncertainty and challenges. *Nature Reviews Genetics* 9, 356–269.

McCouch, S.R., Teytelman, L., Xu, Y., Lobos, K.B., Clare, K., Walton, M., Fu, B., Maghirang, R., Li, Z., Xing, Y., Zhang, Q., Kono, I., Yano, M., Fjellstrom, R., DeClerck, G., Schneider, D., Cartinhour, S., Ware, D. and Stein, L. (2002) Development and mapping of 2240 new SSR markers for rice (*Oryza sativa* L.). *DNA Research* 9, 199–207.

McCouch, S.R., Sweeney, M., Li, J., Jiang, H., Thomson, M., Septiningsih, E., Edwards, J., Moncada, P., Xiao, J., Garris, A., Tai, T., Martinez, C., Tohme, J., Sugiono, M., McClung, A., Yuan, L.P. and Ahn, S.N. (2007) Through the genetic bottleneck: *O. rufipogon* as a source of trait-enhancing alleles for *O. sativa. Euphytica* 154, 317–339.

McElroy, D. (1996) The industrialization of plant transformation. *Nature Biotechnology* 14, 715–716.

McElroy, D. and Brettell, R.I.S. (1994) Foreign gene expression in transgenic cereals. *Trends in Biotechnology* 12, 62–68.

McElroy, D., Zhang, W.G., Cao, J. and Wu, R. (1990) Isolation of an efficient actin promoter for use in rice transformation. *The Plant Cell* 2, 163–171.

McLaren, C.G., Bruskiewich, R.M., Portugal, A.M. and Cosico, A.B. (2005) The International Rice Information System. A platform for meta-analysis of rice crop data. *Plant Physiology* 139, 637–642.

McMullen, M.M., Kresovich, S., Villeda, H.S., Bradbury, P., Li, H., Sun, Q., Flint-Garcia, S., Thornsberry, J., Acharya, C., Bottoms, C., Brown, P., Browne, C., Eller, M., Guill, K., Harjes, C., Kroon, D., Lepak, N., Mitchell, S.E., Peterson, B., Pressoir, G., Romero, S., Rosas, M.O., Salvo, S., Yates, H., Hanson, M., Jones, E., Smith, S., Glaubitz, J.C., Goodman, M., Ware, D., Holland, J.B. and Buckler, E.S. (2009) Genetic properties of the maize nested association mapping population. *Science* 325, 737–740.

McNally, K.L., Bruskiewich, R., Mackill, D., Buell, C.R., Leach, J.E. and Leung, H. (2006) Sequencing multiple and diverse rice varieties. Connecting whole-genome variation with phenotypes. *Plant Physiology* 141, 26–31.

Meaburn, E., Butcher, L.M., Schalkwyk, L.C. and Plomin, R. (2006) Genotyping pooled DNA using 100K SNP microarrays: a step towards genomewide association scans. *Nucleic Acids Research* 34, e28.

Meghi, M.R., Dudley, J.W., Lamkey, R.J. and Sprauge, G.F. (1984) Inbreeding depression, inbred and hybrid grain yields and other traits of maize genotypes representing three eras. *Crop Science* 24, 545–549.

Melchinger, A.E. (1993) Use of RFLP markers for analyses of genetic relationships among breeding materials and prediction of hybrid performance. In: Buxton, D.R., Shibles, R., Forsberg, R.A., Blad, B.L., Asay, K.H., Paulson, G.M. and Wilson, R.F. (eds) *International Crop Science I*. Crop Science Society of America (CSSA), Madison, Wisconsin, pp. 621–628.

Melchinger, A.E. (1999) Genetic diversity and heterosis. In: Coors, J.G. and Pandey, S. (eds) *The Genetics and Exploitation of Heterosis in Crops*. Crop Science Society of America (CSSA), Madison, Wisconsin, p. 54 (abstract).

Melchinger, A.E. and Gumber, R.K. (1998) Overview of heterosis and heterotic groups in agronomic crops. In: Lamkey, K.R. and Staub, J.E. (eds) *Concepts and Breeding of Heterosis in Crop Plants*. Crop Science Society of America (CSSA), Madison, Wisconsin, pp. 29–44.

Melchinger, A.E., Geiger, H.H. and Schnell, F.W. (1986) Epistasis in maize (*Zea mays* L.) I. Comparison of single and three-way cross hybrids among early flint and dent inbred lines. *Maydica* 31, 179–192.

Melchinger, A.E., Lee, M., Lamkey, K.R. and Woodman, W.L. (1990) Genetic diversity for restriction fragment length polymorphisms: relation to estimated genetic effects in maize inbreds. *Crop Science* 30, 1033–1040.

Melchinger, A.E., Messmer, M.M., Lee, M., Woodman, W.L. and Lamkey, K.R. (1991) Diversity and relationships among U.S. maize inbreds revealed by restriction fragment length polymorphism. *Crop Science* 31, 669–678.

Melchinger, A.E., Boppenmaier, J., Dhillon, B.S., Pollmer, W.G. and Herrmann, R.G. (1992) Genetic diversity for RFLPs in European maize inbreds. II. Relation to performance of hybrids within versus between heterotic groups for forage traits. *Theoretical and Applied Genetics* 84, 627–681.

Melchinger, A.E., Graner, A., Singh, M. and Messmer, M.M. (1994) Relationships among European germplasm: I. Genetic diversity among winter and spring cultivars revealed by RFLPs. *Crop Science* 34, 1191–1199.

Melchinger, A.E., Utz, H.F. and Schön, C.C. (1998) Quantitative trait locus (QTL) mapping using different testers and independent populations samples in maize reveals low power of QTL detection and large bias in estimates of QTL effects. *Genetics* 149, 383–403.

Melchinger, A.E., Utz, H.F. and Schön, C.C. (2000) From Mendel to Fisher. The power and limits of QTL mapping for quantitative traits. *Vortr Pflanzenzüchtg* 48, 132–142.

Melchinger, A.E., Utz, H.F. and Schön, C.C. (2004) QTL analyses of complex traits with cross validation, bootstrapping and other biometric methods. *Euphytica* 137, 1–11.

Melchinger, A.E., Longin, C.F., Utz, H.F. and Reif, J.C. (2005) Hybrid maize breeding with doubled haploid lines: quantitative genetic and selection theory for optimum allocation of resources. In: *Proceedings of the Forty First Annual Illinois Corn Breeders' School* 7–8 March 2005, Urbana-Champaign, Illinois. University of Illinois at Urbana-Champaign, pp. 8–21.

Melchinger, A.E., Utz, H.F., Piepho, H.P., Zeng, Z.-B. and Schön, C.C. (2007) The role of epistasis in the manifestation of heterosis – a system-oriented approach. *Genetics* 177, 1815–1825.

Melchinger, A.E., Utz, H.F. and Schön, C.C. (2008) Genetic expectations of quantitative trait loci main and interaction effects obtained with the triple testcross design and their relevance for the analysis of heterosis. *Genetics* 178, 2265–2274.

Menkir, A., Melake-Berhan, A., The, C., Ingelbrecht, I. and Adepoju, A. (2004) Grouping of tropical mid-altitude maize inbred lines on the basis of yield data and molecular markers. *Theoretical and Applied Genetics* 108, 1582–1590.

Menz, M.A., Klein, R.R., Mullet, J.E., Obert, J.A., Unruh, N.C. and Klein, P.E. (2002) A high-density genetic map of *Sorghum bicolor* (L.) Moench based on 2926 AFLP, RFLP and SSR markers. *Plant Molecular Biology* 48, 483–499.

Mertz, E.T., Bates, L.S. and Nelson, O.E. (1964) Mutant gene that changes protein composition and increases lysine content of maize endosperm. *Science* 145, 279–280.

Messina, C.D., Jones, J.W., Boote, K.J. and Vallejos, C.E. (2006) A gene-based model to simulate soybean development and yield response to environment. *Crop Science* 46, 456–466.

Messmer, M.M., Melchinger, A.E., Herrmann, R.G. and Boppenmaier, J. (1993) Relationships among early European maize inbreds. II. Comparisons of pedigree and RFLP data. *Crop Science* 33, 944–950.

Meudt, H.M. and Clarke, A.C. (2007) Almost forgotten or latest practice? AFLP applications, analyses and advances. *Trends in Plant Science* 12, 106–117.

Meuwissen, T.H.E., Hayes, B.J. and Goddard, M.E. (2001) Prediction of total genetic value using genome-wide dense marker maps. *Genetics* 157, 1819–1829.

Meyer, K., Benning, G. and Grill, E. (1996) Cloning of plant genes based on genetic map location. In: Paterson, A.H. (ed.) *Genome Mapping in Plants*. R.G. Landes Company, Austin, Texas, pp. 137–154.

Meyer, S., Nowak, K., Sharma, V.K., Schulze, J., Mendel, R.R. and Hansch, R. (2004) Vectors for RNAi technology in poplar. *Plant Biology* 6, 100–103.

Meyers, B.C., Scalabrin, S. and Morgante, M. (2004) Mapping and sequencing complex genomes: let's get physical! *Nature Reviews Genetics* 5, 578–588.

Michelmore, R.W. and Shaw, D.V. (1988) Character dissection. *Nature* 335, 672–673.

Michelmore, R.W., Paran, I. and Kesselli, R.V. (1991) Identification of markers linked to disease resistance genes by bulked segregant analysis: a rapid method to detect markers in specific genome regions using segregating populations. *Proceedings of the National Academy of Sciences of the United States of America* 88, 9828–9832.

Miernyk, J.A. and Thelen, J.J. (2008) Biochemical approaches for discovering protein–protein interactions. *The Plant Journal* 53, 597–609.

Miki, B. and McHugh, S. (2004) Selectable marker genes in transgenic plants: applications, alternatives and biosafety. *Journal of Biotechnology* 107, 193–232.

Miki, D. and Shimamoto, K. (2004) Simple RNAi vectors for stable and transient suppression of gene function in rice. *Plant and Cell Physiology* 45, 490–495.

Mikkilineni, V. and Rocheford, T.R. (2004) RFLP variant frequency differences among Illinois long-term selection protein strains. *Plant Breeding Reviews* 24(1), 111–131.

Miklas, P.N., Kelly, J.D. and Singh, S.P. (2003) Registration of anthracnose-resistant pinto bean germplasm line USPT-ANT-1. *Crop Science* 43, 1889–1890.

Miles, J.S. and Guest, J.R. (1984) Nucleotide sequence and transcriptional start point of the phosphomannose isomerase gene (mana) of *Escherichia coli*. *Gene* 32, 41–48.

Miller, W., Makova, K.D., Nekrutenko, A. and Hardison, R.C. (2004) Comparative genomics. *Annual Review of Genomics and Human Genetics* 5, 15–56.

Mitchell, A.A. and Chakravarti, A. (2003) Undetected genotyping errors cause apparent overtransmission of common alleles in the transmission/disequilibrium test. *American Journal of Human Genetics* 72, 598–610.

Miyahara, K. (1999) Analysis of LGC1, low glutelin mutant of rice. *Gamma Field Symposia* 38, 43–52.

Miyao, A., Tanaka, K., Murata, K., Sawaki, H., Takeda, S., Abe, K., Shinozuka, Y., Onosato, K. and Hirochika, H. (2003) Target site specificity of the Tos17 retrotransposon shows a preference for insertion within genes and against insertion in retrotransposon-rich regions of the genome. *The Plant Cell* 15, 1771–1780.

Miyata, M., Yamamoto, T., Komori, T. and Nitta, N. (2007) Marker-assisted selection and evaluation of the QTL for stigma exsertion under *japonica* rice genetic background. *Theoretical and Applied Genetics* 114, 539–548.

Mlynarova, L., Conner, A.J. and Nap, J.P. (2006) Directed microspore-specific recombination of transgenic alleles to prevent pollen-mediated transmission. *Plant Biotechnology Journal* 4, 445–452.

Mo, H. (1988) Genetic expression for endosperm traits. In: Weir, B.S., Eisen, E.J., Goodman, M.M. and Namkoog, S.N. (eds) *Proceedings of the 2nd International Conference of Quantitative Genetics*. Sinauer Associates, Sunderland, Massachusetts, pp. 478–487.

Mo, H. (1993a) Genetic analysis for qualitative–quantitative traits. I. The genetic constitution of generations and identification of major gene genotypes. *Acta Agronomica Sinica* 19, 1–6 (in Chinese with English abstract).

Mo, H. (1993b) Genetic analysis for qualitative–quantitative traits. II. Generation means and genetic variances. *Acta Agronomica Sinica* 19, 193–200 (in Chinese with English abstract).

Mockler, T.C. and Ecker, J.R. (2004) Application of DNA tiling arrays for whole-genome analysis. *Genomics* 85, 1–15.

Mohler, V. and Singrün, C. (2004) General considerations: marker-assisted selection. In: Lörz, H. and Wenzl, G. (eds) *Biotechnology in Agricultural and Forestry*, Vol. 55. *Molecular Marker Systems in Plant Breeding and Crop Improvement*. Springer-Verlag, Berlin, pp. 305–317.

Mohler, V. and Schwartz, G. (2005) Genotyping tools in plant breeding: from restriction fragment length polymorphisms to single nucleotide polymorphisms. In: Lörz, H. and Wenzel, G. (eds) *Molecular Marker Systems in Plant Breeding and Crop Improvement. Biotechnology in Agriculture and Forestry*, Vol. 55. Springer-Verlag, Berlin, pp. 23–38.

Moing, A., Deborde, C. and Rolin, D. (2007) Metabolic fingerprinting and profiling by proton NMR. In: Morot-Gaudry, J.F., Lea, P. and Briat, J.F. (eds) *Functional Plant Genomics.* Science Publishers, Enfield, New Hampshire, pp. 335–344.

Molloy, M.P. and Witzmann, F.A. (2002) Proteomics: technologies and applications. *Briefings in Functional Genomics and Proteomics* 1, 23–29.

Moncada, P., Martinez, C.P., Borrero, J., Chatel, M., Gauch, H., Guimaraes, E., Tohme, J. and McCouch, S.R. (2001) Quantitative trait loci for yield and yield components in an *Oryza sativa* × *Oryza rufipogon* BC_2F_2 population evaluated in an upland environment. *Theoretical Applied Geneics* 102, 41–52.

Monna, L., Lin, H.X., Kojima, S., Sasaki, T. and Yano, M. (2002) Genetic dissection of a genomic region for a quantitative trait locus, *Hd3*, into two loci, *Hd3a* and *Hd3b*, controlling heading date in rice. *Theoretical and Applied Genetics* 104, 772–778.

Mooers, C.A. (1921) The agronomic placement of varieties. *Journal of American Society of Agronomy* 13, 337–352.

Moore, S.K. and Srivastava, V. (2006) Efficient deletion of transgenic DNA from complex integration locus of rice mediated by *Cre/lox* recombination system. *Crop Science* 46, 700–705.

Moreau, L., Charcosset, A., Hospital, F. and Gallais, A. (1998) Marker-assisted selection efficiency in populations of finite size. *Genetics* 148, 1353–1365.

Moreau, L., Lemarie, S., Charcosset, A. and Gallais, A. (2000) Economic efficiency on one cycle of marker-assisted selection. *Crop Science* 40, 329–337.

Moreau, L., Charcosset, A. and Gallais, A. (2004) Experimental evaluation of several cycles of marker-assisted selection in maize. *Euphytica* 137, 111–118.

Moreno-Gonzalez, J., Dudley, J.W. and Lambert, R.J. (1975) A design II study of linkage disequilibrium for percent oil in maize. *Crop Science* 15, 840–843.

Morgante, M. and Vogel, J. (1997) Compound microsatellite primers for the detection of genetic polymorphisms. Patent EP 0804618.

Morris, M., Dreher, K., Ribau, J.M. and Khairallah, M. (2003) Money matters (II): cost of maize inbred line conversion schemes at CIMMYT using conventional and marker-assisted selection. *Molecular Breeding* 11, 235–247.

Morton, N.E. (1955) Sequential test for the detection of linkage. *American Journal of Human Genetics* 7, 277–318.

Moser, H. and Lee, M. (1994) RFLP variation and genealogical distance, multivariate distance, heterosis and genetic variance in oats. *Theoretical and Applied Genetics* 87, 947–956.

Mu, J., Zhou, H., Zhao, S., Xu, C., Yu, S. and Zhang, Q. (2004) Development of contiguous introgression lines covering entire genome of the sequenced *japonica* rice. In: *New Directions for a Diverse Planet*: Proceedings of the 4th International Crop Science Congress, 26 September–1 October 2004, Brisbane, Australia. Published on CD-ROM. Available at: http://www.cropscience.org.au/icsc2004/ (accessed 17 November 2009).

Muehlbauer, G.J., Specht, J.E., Thomas-Compton, M.A., Staswick, P.E. and Bernard, R.L. (1988) Near-isogenic lines – a potential resource in the integration of conventional and molecular marker linkage maps. *Crop Science* 28, 729–735.

Mueller, M., Goel, A., Thimma, M., Dickens, N.J., Aitman, T.J. and Mangion, J. (2006) eQTL Explorer: integrated mining of combined genetic linkage and expression experiments. *Bioinformatics* 22, 509–511.

Mukhambetzhanov, S.K. (1997) Culture of nonfertilized female gametophytes *in vitro. Plant Cell, Tissue and Organ Culture* 48, 111–119.

Mullis, K. (1992) Process for amplifying nucleic acid sequences. Patent EP 0201184B1.

Mumm, R.H. and Dudley, J.W. (1994) Classification of 148 U.S. maize inbreds. I. Cluster analysis based on RFLPs. *Crop Science* 34, 842–851.

Munafò, M.R. and Flint, J. (2004) Meta-analysis of genetic association studies. *Trends in Genetics* 20, 439–444.

Muranty, H. (1996) Power of tests for quantitative trait loci detection using full-sib families in different schemes. *Heredity* 76, 156–165.

Murigneux, A., Baud, S. and Beckert, M. (1993) Molecular and morphological evaluation of doubled-haploid lines in maize: 2. Comparison with single-seed-descent lines. *Theoretical and Applied Genetics* 87, 278–287.

Mýles, S., Peiffer, J., Brown, P.J., Ersoz, E.S., Zhang, Z., Costich, D.E. and Buckler, E.S. (2009) Association mapping: critical considerations shift from genotyping to experimental design. *The Plant Cell* 21, 2194–2202.

Nagaraju, J. (2003) Novel FISSR-PCR primes and method of identifying genotyping diverse genomes of plant and animal systems including rice varieties, a kit thereof. Patent WO 03085133.

Nakagahra, M. (1972) Genetic mechanism on the distorted segregation of marker gene belonging to the eleventh linkage group in cultivated rice. *Japanese Journal of Breeding* 22, 232–238.

Nakazaki, T., Okumoto, Y., Horibata, A., Yamahira, S., Teraishi, M., Nishida, H., Inoue, H. and Tanisaka, T. (2003) Mobilization of a transposon in the rice genome. *Nature* 421, 170–172.

Naqvi, S., Zhu, C., Farrea, G., Ramessara, K., Bassiea, L., Breitenbach, J., Conesa, D.P., Ros, G., Sandmann, G., Capell, T. and Christou, P. (2009) Transgenic multivitamin corn through biofortification of endosperm with three vitamins representing three distinct metabolic pathways. *Proceedings of the National Academy of Sciences of the United States of America* 106, 7762–7767.

Narayanan, N.N., Baisakh, N., Oliva, N.P., Vera Cruz, C.M., Gnanamanickam, S.S., Datta, K. and Datta, S.K. (2004) Molecular breeding: marker-assisted selection combined with biolistic transformation for blast and bacterial blight resistance in *indica* rice (cv. CO39). *Molecular Breeding* 14, 61–71.

Naseem, A., Oehmmke, J.F. and Schimmelpfennig, D.E. (2005) Does plant variety intellectual property protection improve farm productivity? Evidence from cotton varieties. *AgBioForum* 8, 100–107.

Navarro, R.L., Warrier, G.S. and Maslog, C.C. (2006) *Genes Are Gems: Reporting Agri-Biotechnology. A Sourcebook for Journalists.* International Crops and Research Institute for the Semi-Arid Tropics, Andhra Pradesh, India, 136 pp.

Naylor, R.L., Falcon, W.P., Goodman, R.M., Jahn, M.M., Sengooba, T., Tefera, H. and Nelson, R.J. (2004) Biotechnology in the developing world: a case for increased investments in orphan crops. *Food Policy* 29, 15–44.

Negrotto, D., Jolley, M., Beer, S., Wenck, A.R. and Hansen, G. (2000) The use of hosphomannose-isomerase as a selectable marker to recover transgenic maize plants (*Zea mays* L.) via *Agrobacterium* transformation. *Plant Cell Reports.* 19, 798–803.

Nei, M. (1972) Genetic distance between populations. *The American Naturalist* 106, 283–292.

Nei, M. (1973) Analysis of gene diversity in subdivided populations. *Proceedings of the National Academy of Sciences of the United States of America* 70, 3321–3323.

Nei, M., Tajima, F. and Tateno, Y. (1983) Accuracy of estimated phylogenetic trees from molecular data. II. Gene frequency data. *Journal of Molecular Evolution* 19, 153–170.

Nelson, O.E. (2001) Maize: the long trail to QTM. In: Reeve, E.C.R. and Black, I. (eds) *Encyclopedia of Genetics.* Fitzroy Dearborn, London, pp. 657–660.

Neuffer, M.G., Coe, E.H. and Wessler, S. (1997) *Mutants of Maize.* Cold Spring Harbor Laboratory Press, Cold Spring Harbor, New York.

Ng'etich, K.A. (2005) *Indigenous Knowledge, Alternative Medicine and Intellectual Property Rights Concerns in Kenya.* 11th General Assembly, 6–10 December 2005, Maputo, Mozambique. Egerton University, Njoro, Kenya.

Nguyen, B.D., Brar, D.S., Bui, B.C., Nguyen, T.V, Pham, L.N. and Nguyen, H.T. (2003) Identification and mapping of the QTL for aluminum tolerance introgressed from the new source *Oryza rufipogon* Griff. into *indica* rice (*Oryza sativa* L.). *Theoretical and Applied Genetics* 106, 583–593.

Nguyen, H.T., Chandra Babu, R. and Blum, A. (1997) Breeding for drought tolerance in rice: physiology and molecular genetics considerations. *Crop Science* 37, 1426–1434.

Nguyen, T.T.T., Klueva, N., Chamareck, V., Aarti, A., Magpantay, G., Millena, A.C.M., Pathan, M.S. and Nguyen, H.T. (2004) Saturation mapping of QTL regions and identification of putative candidate genes for drought tolerance in rice. *Molecular Genetics and Genomics* 272, 35–46.

Ni, J.J., Wu, P., Senadhira, D. and Huang, N. (1998) Mapping QTLs for phosphorus deficiency tolerance in rice (*Oryza sativa* L.). *Theoretical and Applied Genetics* 97, 1361–1369.

Ni, Z.F., Sun, Q.X., Liu, Z.Y. and Huang, T.C. (1997) Studies on heterotic grouping in wheat: II. Genetic diversity among common wheat, Tibet semi-wild wheat and spelt wheat. *Journal of Agricultural Biotechnology* (China) 5, 103–111.

Nicholas, F.W. (2006) Discovery, validation and delivery of DNA markers. *Australian Journal of Experimental Agriculture* 46, 155–158.

Nicholson, L., Gonzalez-Melendi, P., van Dolleweerd, C., Tuck, H., Perrin, Y., Ma, J.K.-C., Fischer, R., Christou, P. and Stoger, E. (2005) A recombinant multimeric immunoglobulin expressed in rice shows assembly dependent subcellular localization in endosperm cells. *Plant Biotechnology* 3, 115–127.

Nickson, T.E. (2008) Planning environmental risk assessment for genetically modified crops: problem formulation for stress-tolerant crops. *Plant Physiology* 147, 494–502.

Nicolas, P. and Chiapello, H. (2007) Gene prediction. In: Morot-Gaudry, J.F., Lea, P. and Briat, J.F. (eds) *Functional Plant Genomics.* Science Publishers, Enfield, New Hampshire, pp. 71–85.

Niebur, W.S., Rafalski, J.A., Smith, O.S. and Cooper, M. (2004) Applications of genomics technologies to enhance rate of genetic progress for yield of maize within a commercial breeding program. In: Fischer, T. (ed.) *New Directions for a Diverse Planet.* Proceedings of the 4th International Crop Science Congress, Brisbane, Australia. Available at: http://www.cropscience.org.au/icsc2004/ (accessed 17 November 2009).

Nilsson, M., Malmgren, H., Samiotaki, M., Kwiatkowski, M., Chowdhary, B.P. and Landegren, U. (1994) Padlock probes: circularization oligonucleotides for localized DNA detection. *Science* 265, 2085–2088.

Nobécourt, P. (1939) Sur la pérennite et l'augmentation de volume des cultures de tissus végétaux. *Comptes Rendus des Séances-Societe Biologie* 130, 1270–1271.

Noirot, M., Anthony, F., Dussert, S. and Hamon, S. (2003) A method for building core collections. In: Hamon, P., Seguin, M., Perrier, X. and Glaszmann, J.C. (eds) *Genetic Diversity of Cultivated Tropical Plants.* Science Publishers, Enfield, New Hampshire and CIRAD, Paris, pp. 65–75.

Nordborg, M. (2000) Linkage disequilibrium, gene trees and selfing: an ancestral recombination graph with partial self-fertilization. *Genetics* 154, 923–929.

Nordborg, M., Borevitz, J.O., Bergelson, J., Berry, C.C., Chory, J., Hagenblad, J., Kreitman, M., Maloof, J.N., Noyes, T., Oefner, P.J., Stahl, E.A. and Weigel, D. (2002) The extent of linkage disequilibrium in *Arabidopsis thaliana. Nature Genetics* 30, 190–193.

NRC (National Research Council) (2001) *Genetically Modified Pest-Protected Plants: Science and Regulation.* National Academy Press, Washington, DC.

NRC (National Research Council) (2002) *Environmental Effects of Transgenic Plants: the Scope and Adequacy of Regulation.* National Academy Press, Washington, DC.

Nunberg, A.N., Li, Z. and Thomas, T.L. (1996) Analysis of gene expression and gene isolation by high-throughput sequencing of plant cDNAs. In: Paterson, A.H. (ed.) *Genome Mapping in Plants.* R.G. Landes Company, Austin, Texas, pp. 169–177.

Nyquist, W.E. (1991) Estimation of heritability and prediction of selection response in plant populations. *Critical Review of Plant Science* 10, 235–322.

O'Brien, S.J. and Mayr, E. (1991) Bureaucratic mischief: recognizing endangered species and subspecies. *Science* 251, 1187–1188.

O'Flanagan, R.A., Paillard, G., Lavery, R. and Sengupta, A.M. (2005) Non-additivity in protein–DNA binding. *Bioinformatics* 21, 2254–2263.

Odell, J.T., Nagy, F. and Chua, N.H. (1985) Identification of DNA sequences required for activity of the cauliflower mosaic virus 35S promoter. *Nature* 313, 810–812.

Ogawa, Y., Dansako, T., Yano, K., Sakurai, N., Suzuki, H., Aoki, K., Noji, M., Saito, K. and Shibata, D. (2008) Efficient and high-throughput vector construction and *Agrobacterium*-mediated transformation of *Arabidopsis thaliana* suspension-cultured cells for functional genomics. *Plant and Cell Physiology* 49, 242–250.

Oka, H.I. (1988) *Origin of Cultivated Rice.* Japan Scientific Societies Press, Tokyo.

Okkels, T.F. and Whenham, R.J. (1994) Method for the selection of genetically transformed cells and compound for the used in the method. Patent EP 0601092B1.

Olek, A. (1996) Amplification of simple sequence repeats. Patent EP 0870062.

Oleykowski, C.A., Bronson Mullins, C.R., Godwin, A.K. and Yeung, A.T. (1998) Mutation detection using a novel plant endonuclease. *Nucleic Acids Research* 26, 4597–4602.

Oliver, S.G., Winson, M.K., Kell, D.B. and Baganz, F. (1998) Systematic functional analysis of the yeast genome. *Trends in Biotechnology* 16, 373–378.

Olufowote, J.O., Xu, Y., Chen, X., Park, W.D., Beachell, H.M., Dilday, R.H., Goto, M. and McCouch, S.R. (1997) Comparative evaluation of within-cultivar variation of rice (*Oryza sativa* L.) using microsatellite and RFLP markers. *Genome* 40, 370–378.

Openshaw, S. and Bruce, W.B. (2001) Marker-assisted identification of a gene associated with a phenotypic trait. Patent EP 1230385.

Openshaw, S.J. and Frascaroli, E. (1997) QTL detection and marker-assisted selection for complex traits in maize. *Proceedings of Corn and Sorghum Industrial Research Conference* 52, 44–53.

Oraguzie, N.C., Wilcox, P.L., Rikkerink, E.H.A. and De Silva, H.N. (2007) Linkage disequilibrium. In: Oraguzie, N.C., Rikkerink, E.H.A., Gardiner, S.E. and De Silva, H.N. (eds) *Association Mapping in Plants.* Springer, Berlin, pp. 11–39.

Orf, J.H., Chase, K., Jarvik, T., Mansur, L.M., Cregan, P.B., Adler, F.R. and Lark, K.G. (1999) Genetics of agronomic traits: I. Comparison of three related recombinant inbred populations. *Crop Science* 39, 1642–1651.

Ortiz, R. and Smale, M. (2007) Transgenic technology: pro-poor or pro-rich? *Chronica Horticulturae* 47, 9–12.

Ossowski, S., Schwab, R. and Weigel, D. (2008) Gene silencing in plants using artificial microRNAs and other small RNAs. *The Plant Journal* 53, 674–690.

Ouyang, Z., Mowers, R.P., Jensen, A., Wang, S. and Zeng, S. (1995) Cluster analysis for genotype × environment interaction with unbalanced data. *Crop Science* 33, 1300–1305.

Ow, D.W. (2001) The right chemistry for marker gene removal? *Nature Biotechnology* 19, 115–116.

Ow, D.W. (2002) Recombinase-directed plant transformation for the post-genomic era. *Plant Molecular Biology* 48, 183–200.

Ow, D.W., Wood, K.V., DeLuca, M., de Wet, J.R., Helinski, D.R. and Howell, S.H. (1986) Transient and stable expression of the firefly luciferase gene in plant cells and transgenic plants. *Science* 234, 856–859.

Owen, H.R. (1996) Plant germplasm. In: Hunter-Cevera, J.C. and Belt, A. (eds) *Maintaining Cultures for Biotechnology and Industry.* Academic Press, Inc., London, pp. 197–228.

Paine, J.A., Shipton, C.A., Chaggar, S., Howells, R.M., Kennedy, M.J., Vernon, G., Wright, S.Y., Hinchliffe, E., Adams, J.L., Silverstone, A.L. and Drake, R. (2005) Improving the nutritional value of Golden Rice through increased pro-vitamin A content. *Nature Biotechnology* 23, 482–487.

Palmer, C.E. and Keller, W.A. (2005) Overview of haploidy. In: Palmer, C.E., Keller, W.A. and Kasha, K.J. (eds) *Biotechnology in Agriculture and Forestry*, Vol. 56. *Haploids in Crop Improvement II.* Springer-Verlag, Berlin, pp. 3–9.

Palmer, C.E., Keller, W.A. and Kasha, K.J. (eds) (2005) *Biotechnology in Agriculture and Forestry*, Vol. 56. *Haploids in Crop Improvement II.* Springer-Verlag, Berlin.

Palmer, L.E., Rabinowicz, P.D., O'Shaughnessy, A.L., Balija, V.S., Nascimento, L.U., Dike, S., de la Bastide, M., Martienssen, R.A. and McCombie, W.R. (2003) Maize genome sequencing by methylation filtration. *Science* 302, 2115–2117.

Palmer, R.G. and Shoemaker, R.C. (1998) Soybean genetics. In: Hrustic, M., Vidic, M. and Jackovic, D. (eds) *Soybean Institute of Field and Vegetative Crops.* Novi Sad, Yugoslavia, pp. 45–82.

Palmiter, R.D., Norstedt, G., Gelinas, R.E., Hammer, R.E. and Brinster, R.L. (1983) Metallothionein – human GH fusion genes stimulate growth of mice. *Science* 222, 809–814.

Pan, Q.L., Liu, Y.S., Budai-Hadrian, O., Sela, M., Carmel-Goren, L., Zamir, D. and Fluhr, R. (2000) Comparative genetics of nucleotide binding size leucine-rich repeat resistance gene homologues in the genomes of two dicotyledons: tomato and *Arabidopsis*. *Genetics* 155, 309–322.

Panaud, O., Chen, X. and McCouch, S.R. (1996) Development of microsatellite markers and characterization of simple sequence length polymorphism (SSLP) in rice (*Oryza sativa* L.). *Molecular and General Genetics* 252, 597–607.

Pang, S.-Z., DeBoer, D.L., Wan, Y., Ye, G., Layton, J.G., Neher, M.K., Armstrong, C.L., Fry, J.E., Hinchee, M.A.W. and Fromm, M.E. (1996) An improved green fluorescent protein gene as a vital marker in plants. *Plant Physiology* 112, 893–900.

Para, R., Acosta, J., Delgado-Salinas, A. and Gepts, P. (2005) A genome-wide analysis of differentiation between wild and domesticated *Phaseolus vulgaris* from Mesoamerica. *Theoretical and Applied Genetics* 111, 1147–1158.

Paran, I. and Michelmore, R.W. (1993) Development of reliable PCR-based markers linked to downy mildew resistance genes in lettuce. *Theoretical and Applied Genetics* 85, 985–993.

Paran, I., Kesseli, R.V. and Michemore, R.W. (1991) Identification of RFLP and RAPD markers linked to downy mildew resistance genes in lettuce using near-isogenic lines. *Genome* 34,1021–1027.

Pardey, P.G., Wright, B.D., Nottenburg, C., Binenbaum, E. and Zambrano, P. (2003) Intellectual property and developing countries: freedom to operate in agricultural biotechnology. *Biotechnology and Genetic Resource Policies* Brief 3. International Food Policy Research Institute (IFPRI), Washington, DC.

Parekh, S.R. (ed.) (2004) *The GMO Handbook: Genetically Modified Animals, Microbes and Plants in Biotechnology.* Humana Press, Totowa, New Jersey.

Parinov, S. and Sundaresan, V. (2000) Functional genomics in *Arabidopsis*: large scale insertional mutagenesis complements the genome sequencing project. *Current Opinion in Biotechnology* 11, 157–161.

Parisseaux, B. and Bernardo, R. (2004) *In silico* mapping of quantitative trait loci in maize. *Theoretical and Applied Genetics* 109, 508–514.

Park, S.J., Walsh, E.J., Reinbergs, E., Song, L.S.P. and Kasha, K. (1976) Field performance of doubled haploid barley lines in comparison with lines developed by the pedigree and single seed descent methods. *Canadian Journal of Plant Science* 56, 467–474.

Parkin, I.A.P., Gulden, S.M., Sharp, A.G., Lukens, L., Trick, M., Osborn, T.C. and Lydiate, D.J. (2005) Segmental structure of the *Brassica napus* genome based on comparative analysis with *Arabidopsis thaliana*. *Genetics* 171, 765–781.

Paterson, A.H. (1996a) Mapping genes responsible for differences in phenotype. In: Paterson, A.H. (ed.) *Genome Mapping in Plants.* R.G. Landes Company, Austin, Texas, pp. 41–54.

Paterson, A.H. (1996b) Physical mapping and map-based cloning: bridging the gap between DNA markers and genes. In: Paterson, A.H. (ed.) *Genome Mapping in Plants.* R.G. Landes Company, Austin, Texas, pp. 55–62.

Paterson, A.H. (ed.) (1998) *Molecular Dissection of Complex Traits.* CRC Press, Boca Raton, Florida, 305 pp.

Paterson, A.H., Lander, E.S., Hewitt, J.D., Peterson, S., Lincoln, S.E. and Tanksley, S.D. (1988) Resolution of quantitative traits into Mendelian factors, using a complete linkage map of restriction fragment length polymorphisms. *Nature* 335, 721–726.

Paterson, A.H., Deverna, J.W., Lanini, B. and Tanksley, S.D. (1990) Fine mapping of quantitative trait loci using selected overlapping recombinant chromosomes, in an interspecific cross of tomato. *Genetics* 124, 735–742.

Paterson, A.H., Damon, S., Hewitt, J.D., Zamir, D., Rabinowitch, H.D., Lincoln, S.E., Lander, E.C. and Tanksley, S.D. (1991) Mendelian factors underlying quantitative traits in tomato: comparison across species, generation and environments. *Genetics* 127, 181–197.

Paterson, A.H., Lin, Y.R., Li, Z., Schertz, K.F., Doebley, J.F., Pinson, S.R.M., Liu, S.-C., Stansel, J.W. and Irvine, J.E. (1995) Convergent domestications of cereal crops by independent mutations at corresponding genetic loci. *Science* 269, 1714–1718.

Paterson, A.H., Saranga, Y., Menz, M., Jiang, C.X. and Wright, R.J. (2003) QTL analysis of genotype × environment interactions affecting cotton fiber quality. *Theoretical and Applied Genetics* 106, 384–396.

Patwardhan, B. (2005) Ethnopharmacology and drug discovery. *Journal of Ethnopharmacology* 100, 50–52.

Peacock, J. and Chaudhury, A. (2002) The impact of gene technologies on the use of genetic resources. In: Engels, J.M.M., Ramanatha Rao, V., Brown, A.H.D. and Jackson, M.T. (eds) *Managing Plant Genetic Diversity.* International Plant Genetic Resources Institute, Rome, pp. 33–42.

Peakall, R., Gilmore, S., Keys, W., Morgante, M. and Rafalski, A. (1998) Cross-species amplification of soybean (*Glycine max*) simple sequence repeats (SSRs) within the genus and other legume genera: implications for the transferability of SSRs in plants. *Molecular Biology and Evolution* 15, 1275–1287.

Pearson, J.V., Huentelman, M.J., Halperin, R.F., Tembe, W.D., Melquist, S., Homer, N., Brun, M., Szelinger, S., Coon, K.D., Zismann, V.L., Webster, J.A., Beach, T., Sando, S.B., Aasly, J.O., Heun, R., Jessen, F., Kölsch, H., Tsolaki, M., Daniilidou, M., Reiman, E.M., Papassotiropoulos, A.P., Hutton, M.L., Stephan, D.A. and Craig, D.W. (2007) Identification of the genetic basis for complex disorders by use of pooling-based genomewide single-nucleotide-polymorphism association studies. *American Journal of Human Genetics* 80, 126–139.

Pearson, W.R., Wood, T., Zhang, Z. and Miller, W. (1997) Comparison of DNA sequences with protein sequences. *Genomics* 15, 24–36.

Peleg, Z., Saranga, Y., Suprunova, T., Ronin, Y., Röder, M.S., Kilian, A., Korol, A.B. and Fahima, T. (2008) High-density genetic map of durum wheat × wild emmer wheat based on SSR and DArT markers. *Theoretical and Applied Genetics* 117, 103–115.

Peleman, J.D. and van der Voort, J.R. (2003) Breeding by design. *Trends in Plant Science* 8, 330–334.

Peleman, J.D., Wye, C., Zethof, J., Sorensen, A.P., Verbakel, H., van Oeveren, J., Gerats, T. and van der Voort, J.R. (2005) Quantitative trait locus (QTL) isogenic recombinant analysis: a method for high-resolution mapping of QTL within a single population. *Genetics* 171, 1341–1352.

Peña, L. (ed.) (2004) *Methods in Molecular Biology*, Vol. 286: *Transgenic Plants: Methods and Protocols.* Humana Press Inc., Totowa, New Jersey.

Peng, J., Richards, D.E., Hartley, N.M., Murphy, G.P., Devos, K.M., Flintham, J.E., Beales J., Fish, L.J., Wordland, A.J., Pelica, F., Sudhakar D., Christou, P., Snape, J.W., Gale, M.D. and Harberd, N.P. (1999) 'Green revolution' genes encode mutant gibberellin response modulators. *Nature* 400, 256–261.

Peng, Z.B., Liu, X.Z., Fu, J.H., Li, L.C. and Huang, C.L. (1998) Preliminary studies on the superior inbred groups and construction of heterosis mode. *Acta Agronomica Sinica* 24, 711–717.

Pereira, M.G., Lee, M.M. and Rayapati, P.J. (1994) Comparative RFLP and QTL mapping in sorghum and maize. In: *Second Internal Conference on the Plant Genome*. Scherago Int., New York, Poster 169.

Pérez, T., Albornoz, J. and Dominguez, A. (1998) An evaluation of RAPD fragment reproducibility and nature. *Molecular Evolution* 7, 1347–1358.

Pérez-Enciso, M. (2004) *In silico* study of transcriptome genetic variation in outbred populations. *Genetics* 166, 547–554.

Perkins, J.M. and Jinks, J.L. (1973) The assessment and specificity of environmental and genotype–environmental components of variability. *Heredity* 30, 111–126.

Perlin, M. (1995) Method and system for genotyping. Patent EP 0714537.

Perumal, R., Krishnaramanujam, R., Menz, M.A., Katilé, S., Dahlberg, J., Magill, C.W. and Rooney, W.L. (2007) Genetic diversity among sorghum races and working groups based on AFLPs and SSRs. C*rop Science* 47, 1375–1383.

Pesek, J. and Baker, R.J. (1969) Desired improvement in relation to selection indices. *Canadian Journal of Plant Science* 49, 803–804.

Peters, J.L., Cnudde, F. and Gerats, T. (2003) Forward genetics and map-based cloning approaches. *Trends in Plant Science* 8, 484–491.

Peterson, D.G., Schulze, S.R., Sciara, E.B., Lee, S.A., Bowers, J.E., Nagel, A., Jiang, N., Tibbitts, D.C., Wessler, S.R. and Paterson, A.H. (2002) Integration of Cot analysis, DNA cloning and high throughput sequencing facilitate genome characterization and gene discovery. *Genome Research* 12, 795–807.

Peterson, P.A. (1992) Quantitative inheritance in the era of molecular biology. *Maydica* 37, 7–18.

Pettersson, F., Morris, A.P., Barnes, M.R. and Cardon, L.R. (2008) Goldsurfer2 (Gs2): a comprehensive tool for the analysis and visualization of genome wide association studies. *BMC Bioinformatics* 9, 138.

Phillips, R.L. (2006) Genetic tools from nature and the nature of genetic tools. *Crop Science* 46, 2245–2252.

Phillips, R.L. (2008) Can genome sequencing of model plants be helpful for crop improvement? *Proceedings of 5th International Crop Science Congress*, 13–18 April 2008, Jeju, Korea. International Crop Science Society, Madison, Wisconsin.

Phillips, R.L., Chen, J., Okediji, R. and Burk, D. (2004) Intellectual property rights and the public good. *The Scientist* 18, 8.

Phizicky, E., Bastiaens, P.I.H., Zhu, H., Snyder, M. and Fields, S. (2003) Protein analysis on a proteomic scale. *Nature* 422, 208–215.

Pickering, R.A. and Devaux, P. (1992) Haploid production: approaches and use in plant breeding. In: Shewry, P.R. (ed.) *Barley: Genetics, Molecular Biology and Biotechnology.* CAB International, Wallingford, UK, pp. 511–539.

Picoult-Newberg, L., Ideker, T.E., Pohl, M.G., Taylor, S.L., Donaldson, M.A., Nickerson, D.A. and Boyce-Jacino, M. (1999) Mining SNPs from EST databases. *Genome Research* 9, 167–174.

Piepho, H.P. (2000) A mixed model approach to mapping quantitative trait loci in barley on the basis of multiple environment data. *Genetics* 156, 2043–2050.

Pillen, K., Pineda, O., Lewis, C.B. and Tanksley, S.D. (1996) Status of genome mapping tools in the taxon Solonaceae. In: Paterson, A.H. (ed.) *Genome Mapping in Plants.* R.G. Landes Company, Austin, TX, pp. 281–308.

Pineda, O., Bonierbale, M.W., Plaisted, R.L., Brodie, B.B. and Tanksley, S.D. (1993) Identification of RFLP markers linked to the *H1* gene conferring resistance to the potato cyst nematode *Globodera rostochiensis. Genome* 36, 152–156.

Plant Ontology™ Consortium (2002) The Plant Ontology™ Consortium and plant ontologies. *Comparative and Functional Genomics* 3, 137–142.

Plomion, C., Durel C.-E. and O'Malley, D.M. (1996) Genetic dissection of height in maritime pine seedlings raised under accelerated growth conditions. *Theoretical and Applied Genetics* 93, 849–858.

Plotsky, Y., Cahaner, A., Haberfeld, A., Lavi, U., Lamont, S.J. and Hillel, J. (1993) DNA fingerprint bands applied to linkage analysis with quantitative trait loci in chickens. *Animal Genetics* 24, 105–110.

Podlich, D.W. and Cooper, M. (1998) QU-GENE: a platform for quantitative analysis of genetic models. *Bioinformatics* 14, 632–653.

Podlich, D.W., Cooper, M.E. and Basford, K.E. (1999) Computer simulation of a selection strategy to accommodate genotype–environment interactions in a wheat recurrent selection programme. *Plant Breeding* 118, 17–28.

Podlich, D.W., Winkler, C.R. and Cooper, M. (2004) Mapping as you go: an effective approach for marker-assisted selection of complex traits. *Crop Science* 44, 1560–1571.

Poehlman, J.M. and Quick, J.S. (1983) Crop breeding in a hunger world. In: Wood, D.R., Rawal, K.M. and Wood, M.N. (eds) *Crop Breeding*. American Society of Agronomy and Crop Science Society of America, Madison, Wisconsin, pp. 1–19.
Pollak, L.M., Gardner, C.O., Kahler, A.L. and Thomas-Compton, M. (1984) Further analysis of the mating system in two mass selected populations of maize. *Crop Science* 24, 793–796.
Pooni, H.S., Kumar, I. and Khush, G.S. (1992) A comprehensive model for disomically inherited metrical traits expressed in triploid tissues. *Heredity* 69, 166–174.
Pooni, H.S., Kumar, I. and Khush, G.S. (1993) Genetical control of amylose content in selected crosses of *indica* rice. *Heredity* 70, 269–280.
Popelka, J.C. and Altpeter, F. (2003) *Agrobacterium tumefaciens*-mediated genetic ransformation of rye (*Secale cereale* L.). *Molecular Breeding* 11, 203–211.
Popelka, J.C., Xu, J. and Altpeter, F. (2003) Generation of rye plants with low copy number after biolistic gene transfer and production of instantly marker-free transgenic rye. *Transgenic Research* 12, 587–596.
Porceddu, A., Albertini, E., Barcaccia, G., Marconi, G., Bertoli, F. and Veronesi, F. (2002) Development of S-SAP markers based on an LTR-like sequence from *Medicago sativa* L. *Molecular Genetics and Genomics* 267, 107–114.
Porta, C. and Lomonossoff, G.P. (2002) Viruses as vectors for the expression of foreign sequences in plants. *Biotechnology and Genetic Engineering Reviews* 19, 245–291.
Portyanko, V.A., Hoffman, D.L., Lee, M. and Holland, J.B. (2001) A linkage map of hexaploid oat based on grass anchor DNA clones and its relationship to other oat maps. *Genome* 44, 249–265.
Potrykus, I. (2005) Golden Rice, vitamin A and blindness – public responsibility and failure. Available at: http://www.goldenrice.org/PDFs/Potrykus_Zurich_2005.pdf (accessed 17 November 2009).
Prasanna, B.M., Vasal, S.K., Kassahun, B. and Singh, N.N. (2001) Quality protein maize. *Current Science* 81, 1308–1319.
Preston, L.R., Harker, N., Holton, T. and Morell, M.K. (1999) Plant cultivar identification using DNA analysis. *Plant Varieties and Seeds* 12, 191–205.
Price, A.H. and Tomos, A.D. (1997) Genetic dissection of root growth in rice (*Oryza sativa* L.): II. Mapping quantitative trait loci using molecular markers. *Theoretical and Applied Genetics* 95, 143–152.
Primmer, C.R., Ellengren, H., Saino, N. and Moller, A.P. (1996) Directional evolution in germline microsatellite mutations. *Nature Genetics* 13, 391–393.
Primrose, S.B. (1995) *Principles of Genome Analysis*. Blackwell Science, Oxford, UK, pp. 14–37.
Pritchard, J.K. and Rosenberg, N.A. (1999) Use of unlinked genetic markers to detect population stratification in association studies. *American Journal of Human Genetics* 65, 220–228.
Pritchard, J.K., Stephens, M. and Donnelly, P. (2000a) Inference of population structure using multilocus genotype data. *Genetics* 155, 945–959.
Pritchard, J.K., Stephens, M., Rosenberg, N.A. and Donnelly, P. (2000b) Association mapping in structured populations. *American Journal of Human Genetics* 67, 170–181.
Purcell, S., Neale, B., Todd-Brown, K., Thomas, L., Ferreira, M.A.R., Bender, D., Maller, J., de Bakker, P.I.W., Daly, M.J. and Sham, P.C. (2007) PLINK: a toolset for whole-genome association and population-based linkage analysis. *American Journal of Human Genetics* 81, 559–575.
Qi, X., Stam, P. and Lindhout, P. (1998) Use of locus-specific AFLP markers to construct a high-density molecular map in barley. *Theoretical and Applied Genetics* 96, 376–384.
Qi, X., Pittaway, T.S., Lindup, S., Liu, H., Waterman, E., Padi, F.K., Hash, C.T., Zhu, J., Gale, M.D. and Devos, K.M. (2004) An integrated genetic map and a new set of simple sequence repeat markers for pearlmillet, *Pennisetum glaucum*. *Theoretical and Applied Genetics* 109, 1485–1493.
Qian, W., Sass, O., Meng, J., Li, M., Frauen, M. and Jung, C. (2007) Heterotic patterns in rapeseed (*Brassica napus* L.): I. Crosses between spring and Chinese semi-winter lines. *Theoretical and Applied Genetics* 115, 27–34.
Quarrie, S.A., Lazić-Jančić, V., Kovačević, D., Steed, A. and Pekić, S. (1999) Bulk segregant analysis with molecular markers and its use for improving drought resistance in maize. *Journal of Experimental Botany* 50, 1299–1306.
Rabinowicz, P.D., Schulz, K., Dedhia, N., Yordan, C., Parnemm, L.D., Parnell., L.D., Stein, L., McCombie, R. and Martienssen, R.A. (1999) Differential methylation of genes and retrotransposons facilitates shot gun sequencing of maize genome. *Nature Genetics* 23, 305–308.
Raboin, L.-M., Pauquet, J., Butterfield, M., D'Hont, A. and Glasmann, J.-C. (2008) Analysis of genome-wide linkage disequilibrium in the high polyploidy sugarcane. *Theoretical and Applied Genetics* 116, 701–714.

Rae, S.J., Macaulay, M., Ramsay, L., Leigh, F., Mathews, D., O'Sullivan, D.M., Donini, P., Morris, P.C., Powell, W., Marshall, D.F., Waugh, R. and Thomas, W.T.B. (2007) Molecular barley breeding. *Euphytica* 158, 295–303.

Rafalski, A. (2002) Applications of single nucleotide polymorphisms in crop genetics. *Current Opinion in Plant Biology* 5, 94–100.

Ragavan, S. (2006) Of plant variety protection, agricultural subsidies and the WTO. Available at: http://www.law.ou.edu/faculty/facfiles/OfPlantVarietyProtection.pdf (accessed 17 November 2009).

Ragot, M. and Lee, M. (2007) Marker-assisted selection in maize: current status, potential, limitations and perspectives from the private and public sectors. In: Guimarães, E.P., Ruane, J., Scherf, B.D., Sonnino, A. and Dargie, J.D. (eds) *Marker-Assisted Selection, Current Status and Future Perspectives in Crops, Livestock, Forestry and Fish.* Food and Agriculture Organization of the United Nations, Rome, pp. 117–150.

Ragot, M., Biasiolli, M., Delbut, M.F., Dell'Orco, A., Malgarini, L., Thevenin, P., Vernoy, J., Vivant, J., Zimmermann, R. and Gay, G. (1995) Marker-assisted backcrossing: a practical example. In: Bervillé, A. and Tersac, M. (eds) *Les Colloques,* No. 72. *Techniques et Utilisations des Marqueurs Moléculaires.* Institute National de la Recherche Agronomique (INRA), Paris, pp. 45–56.

Ragot, M., Gay, G., Muller, J.P. and Durovray, J. (2000) Efficient selection for the adaptation to the environment through QTL mapping and manipulation in maize. In: Ribaut, J.-M. and Poland, D. (eds) *Molecular Approaches for the Genetic Improvement of Cereals for Stable Production in Water-limited Environments*, Centro Internacional de Mejoramiento de Maiz y Trigo (CIMMYT), México, DF, pp. 128–130.

Rajaram, S., van Ginkel, M. and Fischer, R.A. (1994) CIMMYT's wheat breeding mega-environments (ME). In: *Proceedings of the 8th International Wheat Genetics Symposium*, 20–25 July 1993 Beijing, China. Agricultural Scientech Press, Beijing, pp. 1101–1106.

Ramachandran, S. and Sundaresan, V. (2001) Transposons as tools for functional genomics. *Plant Physiology and Biochemistry* 39, 243–252.

Ramage, R.T. (1983) Heterosis and hybrid seed production in barley. In: Frankel, R. (ed.) *Monographs on Theoretical and Applied Genetics*, Vol. 6. *Heterosis.* Springer-Verlag, Berlin, pp. 71–93.

Ramakrishna, W. and Bennetzen, J.L. (2003) Genomic colinearity as a tool for plant gene isolation. In: Grotewold, E. (ed.) *Methods in Molecular Biology*, Vol. 236. *Plant Functional Genomics: Methods and Protocols.* Humana Press, Inc., Totowa, New Jersey, pp. 109–121.

Ramakrishna, W., Dubcovsky, J., Park, Y.-J., Busso, C., Emberton, J., SanMiguel, P. and Bennetzen, J.L. (2002) Different types and rates of genome evolution detected by comparative sequence analysis of orthologous segments from four cereal genomes. *Genetics* 162, 1389–1400.

Ramessar, K., Peremarti, A., Gómez-Galera, S., Naqvi, S., Moralejo, M., Muñoz, P., Capell, T. and Christou, P. (2007) Biosafety and risk assessment framework for selectable marker genes in transgenic crop plants: a case of the science not supporting the politics. *Transgenic Research* 16, 261–280.

Ramlingam, J., Basharat, H.S. and Zhang, G. (2002) STS and microsatellite marker-assisted selection for bacterial blight resistance and waxy gene in rice, *Oryza sativa* L. *Euphytica* 127, 255–260.

Rao, K.E.P. and Rao, V.R. (1995) The use of characterization data in developing a core collection of sorghum. In: Hodgkin, T., Brown, A.H.D., van Hintum, Th.J.L. and Morales, E.A.V. (eds) *Core Collections of Plant Genetic Resources.* Wiley–Sayce, Chichester, UK, pp. 109–115.

Rappsilber, J., Siniossoglou, S., Hurt, E.C. and Mann, M. (2000) A generic strategy to analyze the spatial organization of multi-protein complexes by cross-linking and mass spectrometry. *Analytical Chemistry* 72, 267–275.

Rebai, A. and Goffinet, B. (1993) Power of test for QTL detection using replicated progenies derived from a diallel crosses. *Theoretical and Applied Genetics* 86, 1014–1022.

Rebai, A., Goffinet, B., Mangin, B. and Perret, D. (1994) QTL detection with diallel schemes. In: van Ooijen, J.W. and Jansen, J. (eds) *Biometrics in Plant Breeding: Applications of Molecular Markers.* Centre for Plant Breeding and Reproduction Research, Wageningen, Netherlands, pp. 170–177.

Reddy, B.V.S. and Comstock, R.E. (1976) Simulation of the backcross breeding method. I. Effect of heritability and gene number on fixation of desired alleles. *Crop Science* 16, 825–830.

Reed, J., Privalle, L., Powell, M.L., Meghji, M., Dawson, J., Dunder, E., Suttie, J., Wenck, A., Launis, K., Kramer, C., Chang, Y.-F., Hansen, G. and Wright, M. (2001) Phosphomannose isomerase: an efficient selectable marker for plant transformation. In Vitro *Cellular and Developmental Biology – Plant* 37, 127–132.

Reeves, T., Pinstrup-Anderson, P. and Randya-Lorch, R. (1999) Food security and role of agricultural research. In: Coors, J.G. and Pandey, S. (eds) *The Genetics and Exploitation of Heterosis in Crops.* ASA-CSSA-SSSA, Madison, Wisconsin, pp. 1–5.

Reif, J.C., Melchinger, A.E., Xia, X.C., Warburton, M.L., Hoisington, D.A., Vasal, S.K., Srinivasan, G., Bohn, M. and Frisch, M. (2003) Genetic distance based on simple sequence repeats and heterosis in tropical maize populations. *Crop Science* 43, 1275–1282.

Reif, J.C., Xia, X.C., Melchinger, A.E., Warburton, M.L., Hoisington, D.A., Beck, D., Bohn, M. and Frisch, M. (2004) Genetic diversity determined within and among CIMMYT maize populations of tropical, subtropical and temperate germplasm by SSR markers. *Crop Science* 44, 326–334.

Reif, J.C., Melchinger, A.E. and Frisch, M. (2005) Genetical and mathematical properties of similarity and dissimilarity coefficients applied in plant breeding and seed bank management. *Crop Science* 45, 1–7.

Reiter, R. (2001) PCR-based marker systems. In: Phillip, R.L. and Vasil, I.K. (eds) *DNA-Based Markers in Plants.* Kluwer Academic Publishers, Dordrecht, Netherlands, pp. 9–29.

Remington, D.L., Thornsberry, J.M., Matsuoka, Y., Wilson, L.M., Whitt, S.R., Doebley, J., Kresovich, S., Goodman, M.M. and Buckler IV, E.S. (2001) Structure of linkage disequilibrium and phenotypic associations in the maize genome. *Proceedings of the National Academy of Sciences of the United States of America* 98, 11479–11484.

Repellin, A., Båga, M., Jauhar, P.P. and Chibbar, R.N. (2001) Genetic enrichment of cereal crops via alien gene transfer: new challenges. *Plant Cell, Tissue and Organ Culture* 64, 159–183.

Reymond, M., Muller, B., Leonardi, A., Charcosset, A. and Tardieu, F. (2003) Combining quantitative trait loci analysis and an ecophysiological model to analyze the genetic variability of the responses of maize leaf growth to temperature and water deficit. *Plant Physiology* 131, 664–675.

Reyna, N. and Sneller, C.H. (2001) Evolution of marker-assisted introgression of yield QTL alleles into adapted soybean. *Crop Science* 41, 1317–1321.

Reynolds, J., Weir, B.S. and Cockerham, C.C. (1983) Estimation of the coancestry coefficient: basis for a short-term genetic distance. *Genetics* 105, 767–769.

Rhee, S.Y. (2005) Bioinformatics: current limitations and insights for the future. *Plant Physiology* 138, 569–570.

Ribaut, J.-M. and Betrán, J. (1999) Single large-scale marker-assisted selection (SLS-MAS). *Molecular Breeding* 5, 531–541.

Ribaut, J.-M. and Ragot, M. (2007) Marker-assisted selection to improve drought adaptations in maize: the backcross approach, perspectives, limitations and alternatives. *Journal of Experimental Botany* 58, 351–360.

Ribaut, J.-M., Hoisington, D.A., Deutsch, J.A., Jiang, C. and González-de-León, D. (1996) Identification of quantitative trait loci under drought conditions in tropical maize. I. Flowering parameters and the anthesis-silking interval. *Theoretical and Applied Genetics* 92, 905–914.

Ribaut, J.-M., Huu, X., Hoisington, D. and Gonzales de Leon, D. (1997) Use of STSs and SSRs as rapid and reliable preselection tools in marker-assisted selection backcross scheme. *Plant Molecular Biology Reporter* 15, 156–164.

Ribaut, J.-M., Edmeades, G., Perotti, E. and Hoisington, D. (2000) QTL analyses, MAS results and perspectives for drought-tolerance improvement in tropical maize. In: Ribaut, J.-M. and Poland, D. (eds) *Molecular Approaches for the Genetic Improvement of Cereals for Stable Production in Water-limited Environments.* Centro Internacional de Mejoramiento de Maiz y Trigo (CIMMYT), México, DF, pp. 131–136.

Ribaut, J.-M., Jiang, C. and Hoisington, D. (2002a) Simulation experiments on efficiencies of gene introgression by backcrossing. *Crop Science* 42, 557–565.

Ribaut, J.-M., Bänziger, M., Betran, J., Jiang, C., Edmeades, G.O., Dreher, K. and Hoisington, D. (2002b) Use of molecular markers in plant breeding: drought tolerance improvement in tropical maize. In: Kang, M.S. (ed.) *Quantitative Genetics, Genomics and Plant Breeding.* CAB International, Wallingford, UK, pp. 85–99.

Richardson, K.L., Vales, M.I., Kling, J.G., Mundt, C.C. and Hayes, P.M. (2006) Pyramiding and dissecting disease resistance QTL to barley stripe rust. *Theoretical and Applied Genetics* 113, 485–495.

Rick, C.M. (1974) High soluble-solids content in large-fruited tomato lines derived from a wild green-fruited species. *Hilgardia* 42, 493–510.

Rick, C.M. (1988) Tomato-like nightshades: affinities, autoecology and breeders' opportunities. *Economic Botany* 42, 145–154.

Rickert, A.M., Premstaller, A., Gebhardt, C. and Oefner, P.J. (2002) Genoptying of SNPs in a polyploid genome by pyrosequencing™. *Biotechniques* 32, 592–603.

Roa-Rodriguez, C. (2003) *Promoters Used to Regulate Gene Expression.* CAMBIA Intellectual Property, Canberra.

Roa-Rodriguez, C. and Nottenburg, C. (2003a) *Agrobacterium-mediated Transformation of Plants.* CAMBIA Intellectual Property, Canberra.

Roa-Rodriguez, C. and Nottenburg, C. (2003b) *Antibiotic Resistance Genes and Their Uses in Genetic Transformation, Especially in Plants.* CAMBIA Intellectual Property, Canberra.

Röber, F.K. (1999) Fortpflanzungsbiologische und genetische Untersuchungen mit RFLP-Markern zur *in-vivo*-Haploideninduktion bei Mais. Dissertation, University of Hohenheim. Grauer Verlag, Stuttgart.

Röber, F.K., Gordillo, G.A. and Geiger, H.H. (2005) *In vivo* haploid induction in maize – performance of new inducers and significance of doubled haploid lines in hybrid breeding. *Maydica* 50, 275–284.

Robert, V.J.M., West, M.A.L., Inai, S., Caines, A., Arntzen, L., Smith, J.K. and St-Clair, D.A. (2001) Marker-assisted introgression of blackmold resistance QTL alleles from wild *Lycopersicon chesmanii* to cultivated tomato (*L. esculentum*) and evaluation of QTL phenotypic effects. *Molecular Breeding* 8, 217–233.

Roberts, E.H. (1973) Predicting the viability of seeds. *Seed Science and Technology* 1, 499–514.

Roberts, J.K. (2002) Proteomics and a future generation of plant molecular biologists. *Plant Molecular Biology* 48, 143–154.

Robertson, D.S. (1985) A possible technique for isolating genomic DNA for quantitative traits in plants. *Journal of Theoretical Biology* 117, 1–10.

Robertson, D.S. (1989) Understanding the relationship between qualitative and quantitative genetics. In: Helentjaris, T. and Burr, B. (eds) *Development and Application of Molecular Markers to Problems in Plant Genetics.* Cold Spring Harbor Laboratory Press, Cold Spring Harbor, New York, pp. 81–87.

Rockman, M.V. and Kruglyak, L. (2006) Genetics of global gene expression. *Nature Reviews Genetics* 7, 862–872.

Rockman, M.V. and Kruglyak, L. (2008) Breeding designs for recombinant inbred advanced intercross lines. *Genetics* 179, 1069–1078.

Rockman, M.V. and Wray, G.A. (2002) Abundant raw material for *cis*-regulatory evolution in humans. *Molecular Biology and Evolution* 19, 1991–2004.

Röder, M., Plaschke, J. and Ganal, M. (1997) Microsatellite markers for plants of the species *Triticum aestivum* and tribe Triticeae and the use of said markers. Patent EP 0835324B1.

Rogers, J.S. (1972) Measures of genetic similarity and genetic distance. In: *Studies in Genetics* VII, Publ. 7213. University of Texas, Austin, Texas, pp. 145–153.

Romagosa, I. and Fox, P.N. (2003) Genotype × environment interaction and adaptation. In: Hayward, M.D., Bosemark, N.O. and Romagosa, I. (eds) *Plant Breeding, Principles and Prospects.* Chapman & Hall, London, pp. 373–390.

Romagosa, I., Ullrich, S.E., Han, F. and Hayes, P.M. (1996) Use of the additive main effects and multiplicative interaction model in QTL mapping for adaptation in barley. *Theoretical and Applied Genetics* 93, 30–37.

Romano, A., van der Plas, L.H.W., Witholt, B., Eggink, G. and Mooibroek, H. (2005) Expression of poly-3-(R)-hydroxyalkanoate (PHA) polymerase and acyl-CoA-transacylase in plastids of transgenic potato leads to the synthesis of a hydrophobic polymer, presumably medium-chain-length PHAs. *Planta* 220, 455–464.

Romeis. J., Bartsch, D., Bigler, F., Candolfi, M.P., Gielkens, M.M.C., Hartley, S.E., Hellmich, R.L., Huesing, J.E., Jepson, P.C., Layton, R., Quemada, H., Raybould, A., Rose, R.I., Schiemann, J., Sears, M.K., Shelton, A.M., Sweet, J., Vaituzis, Z. and Wolt, J.D. (2008) Assessment of risk of insect-resistant transgenic crops to nontarget anthropods. *Nature Biotechnology* 26, 203–208.

Rommens, C.M., Haring, M.A., Swords, K., Davies, H.V. and Belknap, W.R. (2007) The intragenic approach as a new extension to traditional plant breeding. *Trends in Plant Science* 12, 397–403.

Ron Parra, J. and Hallauer, A.R. (1997) Utilization of exotic maize germplasm. *Plant Breeding Reviews* 14, 165–187.

Rong, J., Feltus, F.A., Waghmare, V.N., Pierce, G.J., Chee, P.W., Draye, X., Saranga, Y., Wright, R.J., Wilkins, T.A., May, O.L., Smith, C.W., Gannaway, J.R., Wendel, J.F. and Paterson, A.H. (2007) Meta-analysis of polyploid cotton QTL shows unequal contributions of subgenomes to a complex network of genes and gene clusters implicated in lint fiber development. *Genetics* 176, 2577–2588.

Roos, E.E. (1984) Genetic shifts in mixed bean populations. I. Storage effects. *Crop Science* 24, 240–244.

Roos, E.E. (1988) Genetic changes in a collection over time. *HortScience* 23, 86–90.

Rostoks, N., Mudie, S., Cardle, L., Russell, J., Ramsay, L., Booth, A., Svensson, J.T., Wanamaker, S.I., Walia, H., Rodriguez, E.M., Hedley, P.E., Liu, H., Morris, J., Close, T.J., Marshall, D.F. and Waugh, R. (2005) Genome-wide SNP discovery and linkage analysis in barley based on genes responsive to abiotic stress. *Molecular Genetics and Genomics* 274, 515–527.

Rudd, S., Schoof, H. and Mayer, K. (2005) PlantMarkers – a database of predicted molecular markers from plants. *Nucleic Acids Research* 33, D628–632.

Ruf, S., Karcher, D. and Rock, R. (2007) Determining the transgene containment level provided by chloroplast transformation. *Proceedings of the National Academy of Sciences of the United States of America* 114, 6998–7002.

Sackville Hamilton, N.R. and Chorlton, K.H. (1997) Regenaration of accessions in seed collections: a decision guide. *Handbook for Genebanks* No. 5. International Plant Genetic Resources Institute, Rome.

Saghai Maroof, M.A., Yang, G.P., Zhang, Q. and Gravois, K.A. (1997) Correlation between molecular marker distance and hybrid performance in US southern long grain rice. *Crop Science* 37, 145–150.

Saha, S., Sparks, A.B., Rago, C., Akmaev, V., Wang, C.J., Vogelstein, B., Kinzler, K.W. and Velculescu, V.E. (2002) Using the transcriptome to annotate the genome. *Nature Biotechnology* 20, 508–512.

Saint-Louis, D. and Paquin, B. (2003) Method for genotyping microsatellite DNA markers by mass spectrometry. Patent WO 03035906.

Sakamoto, T. and Matsuoka, M. (2008) Identifying and exploiting grain yield genes in rice. *Current Opinion in Plant Biology* 11, 209–214.

Salathia, N., Lee, H.N., Sangster, T.A., Morneau, K., Landry, C.R., Schellenberg, K., Behere, A.S., Gunderson, K.L., Cavalieri, D., Jander, G. and Queitsch, C. (2007) Indel arrays: an affordable alternative for genotyping. *The Plant Journal* 51, 727–737.

Salse, J., Piegu, B., Cooke, R. and Delseny, M. (2004) New *in silico* insight into the synteny between rice (*Oryza sativa* L.) and maize (*Zea mays* SL.) highlights reshuffling and identifies new duplications in the rice genome. *The Plant Journal* 38, 396–409.

Salvi, S. and Tuberosa, R. (2005) To clone or not to clone plant QTLs: present and future challenges. *Trends in Plant Science* 10, 297–304.

Samalova, M., Brzobohaty, B. and Moore, I. (2005) pOp6/LhGR: a stringently regulated and highly responsive dexamethasone-inducible gene expression system for tobacco. *The Plant Journal* 41, 919–935.

San Noeum, L.H. (1976) Haploids of *Hordeum vulgare* L. from *in vitro* culture of unfertilized ovaries. *Annales de l'Amelioration des Plantes* 26, 751–754.

Sánchez-Monge, E. (1993) Introduction. In: Hayward, M.D., Bosemark, N.O. and Romagosa, I. (eds) *Plant Breeding, Principles and Prospects.* Chapman & Hall, London, pp. 3–5.

Sanda, S.L. and Amasino, R.M. (1996) Ecotype-specific expression of a flowering mutant phenotype in *Arabidopsis thaliana. Plant Physiology* 111, 641–644.

Sano, Y. (1990) The genic nature of gamete eliminator in rice. *Genetics* 125, 183–191.

Sant, V.J., Patankar, A.G., Sarode, N.D., Mhase, L.B., Sainani, M.N., Deshmukh, R.B., Ranjekar, P.K. and Gupta, V.S. (1999) Potential of DNA markers in detecting divergence and in analyzing heterosis in Indian elite chickpea cultivars. *Theoretical and Applied Genetics* 98, 1217–1225.

Saravanan, R.S., Bashir, S. and Rose, J.K.C. (2004) Plant proteomics. In: Christou, P. and Klee, H. (eds) *Handbook of Plant Biotechnology.* John Wiley & Sons Ltd, Chichester, UK, pp. 183–199.

Sari-Gorla, M., Calinski, T., Kaczmarek, Z. and Krajewski, P. (1997) Detection of QTL × environment interaction in maize by a least squares interval mapping method. *Heredity* 78, 146–157.

Sarkar, K.R., Pandey, A., Gayen, P., Mandan, J.K., Kumar, R. and Sachan, J.K.S. (1994) Stabilization of high haploid inducer lines. *Maize Genetics Cooperation Newsletter* 68, 64–65.

Satagopan, J.M., Yandell, B.S., Newton, M.A. and Osborn, T.G. (1996) A Bayesian approach to detect quantitative trait loci using Markov chain Monte Carlo. *Genetics* 144, 805–816.

Sauer, S., Gelfand, D.H., Boussicault, F., Bauer, K., Reichert, F. and Gut, I.G. (2002) Facile method for automated genotyping of single nucleotide polymorphisms by mass spectrometry. *Nucleic Acid Research* 30, e22.

Sawkins, M.C., Farmer, A.D., Hoisington, D., Sullivan, J., Tolopko, A., Jiang, Z. and Ribaut, J.M. (2004) Comparative Map and Trait Viewer (CMTV): an integrated bioinformatic tool to construct consensus maps and compare QTL and functional genomics data across genomes and experiments. *Plant Molecular Biology* 56, 465–480.

Sax, K. (1923) The association of size differences with seed coat pattern and pigmentation in *Phaseolus vulgaris. Genetics* 8, 552–560.

Scarascia-Mugnozza, G.T. and Perrino, P. (2002) The history of *ex situ* conservation and use of plant genetic resources. In: Engels, J.M.M., Ramanatha Rao, V., Brown, A.H.D. and Jackson, M.T. (eds) *Managing Plant Genetic Diversity.* International Plant Genetics Resources Institute (IPGRI), Rome, pp. 1–22.

Schadt, E.E., Monks, S.A., Drake, T.A., Lusis, A.J., Che, N., Colinayo, V., Ruff, T.G., Milligan, S.B., Lamb, J.R., Cavet, G., Linsley, P.S., Mao, M., Stoughton, R.B. and Friend, S.H. (2003) Genetics of gene expression surveyed in maize, mouse and man. *Nature* 422, 297–302.

Schaeffer, M., Byrne, P. and Coe, E.H., Jr (2006) Consensus quantitative trait maps in maize: a database strategy. *Maydica* 51, 357–367.

Schauer, N. and Fernie, A.R. (2006) Plant metabolomics: towards biological function and mechanism. *Trends in Plant Science* 11, 508–516.

Scheuring, C., Barthelson, R., Gailbraith, D., Betran, J., Cothren, J.T., Zeng, Z.-B. and Zhang, H.-B. (2006) Preliminary analysis of differential gene expression between a maize superior hybrid and its parents using the 57K maize gene-specific long-oligonucleotide microarray. In: *48th Annual Maize Genetic Conference*, 9–12 March 2006, Pacific Grove, California, 132 pp.

Schmid, K.J., Rosleff Sörensen, T., Stracke, R., Törjék, O., Altmann, T., Mithell-Olds, T. and Weisshaar, B. (2003) Large-scale identification and analysis of genome wide single nucleotide polymorphisms for mapping in *Arabidopsis thaliana. Genome Research* 13, 1250–1257.

Schmidt, R. (2002) Plant genome evolution: lessons from comparative genomics at the DNA level. *Plant Molecular Biology* 48, 21–37.

Schmierer, D.A., Kandemir, N., Kudrna, D.A., Jones, B.L., Ullrich, S.E. and Kleinhofs, A. (2004) Molecular marker-assisted selection for enhanced yield in malting barley. *Molecular Breeding* 14, 463–473.

Schön, C.C., Utz, H.F., Groh, S., Truberg, B., Openshaw, S. and Melchinger, A.E. (2004) Quantitative trait locus mapping based on resampling in a vast maize testcrosses experiment and its relevance to quantitative genetics for complex traits. *Genetics* 167, 485–498.

Schranz, M.E., Song, B.-H., Windsor, A.J. and Mitchell-Olds, T. (2007) Comparative genomics in the Brassicaceae: a family-wide perspective. *Current Opinion in Plant Biology* 10, 168–175.

Schüller, C., Backes, G., Fischbeck, G. and Jahoor, A. (1992) RFLP markers to identify the alleles on the *Mla* locus conferring powdery mildew resistance in barley. *Theoretical and Applied Genetics* 84, 330–338.

Schuster, S.C. (2008) Next-generation sequencing transforms today's biology. *Nature Methods* 5, 16–18.

Schwarz, G., Herz, M., Huang, X.Q., Michalek, W., Jahoor, A., Wenzel, G. and Mohler, V. (2000) Application of fluorescence-based semi-automated AFLP analysis in barley and wheat. *Theoretical and Applied Genetics* 100, 545–551.

Scott, K.D. (2001) Microsatellites derived from ESTs and their comparison with those derived by other methods. In: Henry, R.J. (ed.) *Plant Genotyping: the DNA Fingerprinting of Plants.* CAB International, Wallingford, UK, pp. 225–237.

Searle, S.R. (1987) *Linear Model for Unbalanced Data.* John Wiley & Sons, New York.

Seaton, G., Haley, C.S., Knott, S.A., Kearsey, M. and Visscher, P.M. (2002) QTL Express: mapping quantitative trait loci in simple and complex pedigrees. *Bioinformatics* 18, 339–340.

Seitz, C., Vitten, M., Steinbach, P., Hartl, S., Hirsche, J., Rathje, W., Treutter, D. and Forkmann, G. (2007) Redirection of anthocyanin synthesis in *Osteospermum hybrida* by a two-enzyme manipulation strategy. *Phytochemistry* 68, 824–833.

Seitz, G. (2005) The use of doubled haploids in corn breeding. In: *Proceedings of the Forty First Annual Illinois Corn Breeders' School,* 7–8 March 2005, Urbana-Champaign, Illinois. University of Illinois at Urbana-Champaign, pp. 1–8.

Seki, M., Narusaka, M., Satou, M., Fujita, M., Sakurai, T., Oono, Y., Akiyama, T., Yamaguchi-Shinozaki, K., Iida, K., Carninci, P., Ishisa, J., Kawai, J., Nakajima, M., Hayashizaki, Y., Enju, A. and Shinozaki, K. (2005) Full-length cDNAs for the discovery and annotation of genes in *Arabidopsis thaliana.* In: Leister, D. (ed.) *Plant Functional Genomics.* Food Products Press, Binghamton, New York, pp. 3–22.

Semagn, K., Bjørnstad, Å., Skinnes, H., Marøy, A.G., Tarkegne, Y. and William, M. (2006) Distribution of DArT, AFLP and SSR markers in a genetic linkage map of a doubled-haploid hexaploid wheat population. *Genome* 49, 545–555.

Sen, S. and Churchill, G.A. (2001) A statistical framework for quantitative trait mapping. *Genetics* 159, 371–387.

Septiningsih, E.M., Prasetiyono, J., Lubis, E., Tai, T.H., Tjubaryat, T., Moeljopawiro, S. and McCouch, S.R. (2003) Identification of quantitative trait loci for yield and yield components in an advanced backcross population derived from the *Oryza sativa* variety IR64 and the wild relative *O. rufipogon. Theoretical and Applied Genetics* 107, 1419–1432.

Service, R.F. (2006) Gene sequencing. The race for the $1000 genome. *Science* 311, 1544–1546.
Servin, B., Martin, O.C., Mézard, M. and Hospital, F. (2004) Toward a theory of marker-assisted gene pyramiding. *Genetics* 168, 513–523.
Sessions, A. Burke, E., Presting, G., Aux, G., McElver, J., Patton, D., Dietrich, B., Ho, P., Bacwaden, J., Ko, C., Clarke, J.D., Cotton, D., Bullis, D., Snell, J., Miguel, T., Hutchison, D., Kimmerly, B., Mitzel, T., Katagiri, F., Glazebrook, J., Law, M. and Goff, S.A. (2002) A high-throughput *Arabidopsis* reverse genetics system. *The Plant Cell* 14, 2985–2994.
Setimela, P., Chitalu, Z., Jonazi, J., Mambo, A., Hodson, D. and Bänziger, M. (2005) Environmental classification of maize-testing sites in the SADC region and its implication for collaborative maize breeding strategies in the subcontinent. *Euphytica* 145, 123–132.
Sham, P., Bader, J.S., Craig, I., O'Donovan, M. and Owen, M. (2002) DNA pooling: a tool for large-scale association studies. *Nature Reviews Genetics* 3, 862–871.
Shannon, P., Markiel, A., Ozier, O., Baliga, N.S., Wang, J.T., Ramage, D., Amin, N., Schwikowski, B. and Ideker, T. (2003) Cytoscape: a software environment for integrated models of biomolecular interaction networks. *Genome Research* 13, 2498–2504.
Sharopova, N., McMullen, M.D., Schultz, L., Schroeder, S., Sanchez-Villeda, H., Gardiner, J., Bergstrom, D., Houchins, K., Melia-Hancock, S., Musket, T., Duru, N., Polacco, M., Edwards, K., Ruff, T., Register, J.C., Brouwer, C., Thompson, R., Velasco, R., Chin, E., Lee, M., Woodman-Clikeman, W., Long, M.J., Liscum, E., Cone, K., Davis, G. and Coe, E.H., Jr (2002) Development and mapping of SSR markers for maize. *Plant Moelcular Biology* 48, 463–481.
Shatskaya, O.A., Zabirova, E.R., Shcherbak, V.S. and Chumak, M.V. (1994) Mass induction of maternal haploids in corn. *Maize Genetics Cooperation Newsletter* 68, 51.
Shen, J.H., Li, M.F., Chen, Y.Q. and Zhang, Z.H. (1982) Breeding by anther culture in rice improvement. *Scientia Agricultura Sinica* 2,15–19.
Shen, L., Courtois, B., McNally, K.L., Robin, S. and Li, Z. (2001) Evaluation of near-isogenic lines of rice introgressed with QTLs for root depth through marker-aided selection. *Theoretical and Applied Genetics* 103, 75–83.
Shen, Y.-J., Jiang, H., Jin, J.-P., Zhang, Z.-B., Xi, B., He, Y.-Y., Wang, G., Wang, C., Qian, L., Li, X., Yu, Q.-B., Liu, H.-J., Chen, D.-H., Gao, J.-H., Huang, H., Shi, T.-L. and Yang, Z.-N. (2004) Development of genome-wide DNA polymorphism database for map-based cloning of rice genes. *Plant Physiology* 135, 1198–1205.
Shendure, J. and Ji, H. (2008) Next-generation DNA sequencing. *Nature Biotechnology* 26, 1135–1145.
Shi, Y., Wang, T., Li, Y. and Darmency, H. (2008) Impact of transgene inheritance on the mitigation of gene flow between crops and their wild relatives: the example of foxtail millet. *Genetics* 180, 969–975.
Shibata, D. and Liu, Y.G. (2000) *Agrobacterium*-mediated plant transformation with large DNA fragments. *Trends in Plant Science* 5, 354–357.
Shimamoto, K. and Kyozuk, J. (2002) Rice as a model for comparative genomics of plants. *Annual Review of Plant Biology* 53, 399–419.
Shin, B.K., Wang, H., Yim, A.M., Naour, F.L., Brichory, F., Jang, J.H., Zhao, R., Puravs, E., Tra, J., Michael, C.W., Misek, D.E. and Hanash, S.M. (2003) Global profiling of the cell surface proteome of cancer cells uncovers an abundance of proteins with chaperone function. *Journal of Biological Chemistry* 278, 7607–7616.
Shizuya, H., Birren, B., Kim, U., Mancino, V., Slepak, T., Tachiiri, Y. and Simon, M. (1992) Cloning and stable maintenance of 300-kilobase-pair fragments of human DNA in *Escherichia coli* using an F-factor-based vector. *Proceedings of the National Academy of Sciences of the United States of America* 89, 8794–8797.
Shoemaker, J.S., Painter, I.S. and Weir, B.S. (1999) Bayesian statistics in genetics. A guide for the uninitiated. *Trends in Genetics* 15, 354–358.
Shrawat, A.K. and Lörz, H. (2006) *Agrobacterium*-mediated transformation of cereals: a promising approach crossing barriers. *Plant Biotechnology Journal* 4, 575–603.
Shuber, A. and Pierceall, W. (2002) Methods for detecting nucleotide insertion or deletion using primer extension. Patent EP 1203100.
Shull, G.H. (1908) The composition of a field of maize. *American Breeders' Association Report* 4, 296–301.
Siepel, A., Farmerm A., Tolopko, A., Zhuang, M, Mendes, P., Beavis, W. and Sobral, B. (2001) ISYS: a decentralized, component-based approach to the integration of heterogeneous bioinformatic resources. *Bioinformatics* 17, 83–94.
Sillanpää, M.J. and Arjas, E. (1998) Bayesian mapping of multiple quantitative trait loci from incomplete inbred line cross data. *Genetics* 148, 1373–1388.

Sillanpää, M.J. and Arjas, E. (1999) Bayesian mapping of multiple quantitative trait loci from incomplete outbred offspring data. *Genetics* 151, 1605–1619.

Sillanpää, M.J. and Bhattacharjee, M. (2005) Bayesian association-based fine mapping in small chromosomal segments. *Genetics* 169, 427–439.

Sillanpää, M.J. and Corander, J. (2002) Model choice in gene mapping: what and why. *Trends in Genetics* 18, 301–307.

Silver, J. (1985) Confidence limits for estimates of gene linkage based on analysis of recombinant inbred strains. *Journal of Heredity* 76, 436–440.

Simko, I., Costanzo, S., Haynes, K.G., Christ, B.J. and Jones, R.W. (2004a) Linkage disequilibrium mapping of a *Verticillium dahliae* resistance quantitative trait locus in tetraploid potato (*Solanum tuberosum*) through a candidate gene approach. *Theoretical and Applied Genetics* 108, 217–224.

Simko, I., Haynes, K.G., Ewing, E.E., Costanzo, S., Christ, B.J. and Jones, R.W. (2004b) Mapping genes for resistance to *Verticillium albo-atrum* in tetraploid and diploid potato populations using haplotype association tests and genetic linkage analysis. *Molecular Genetics and Genomics* 271, 522–531.

Simmonds, N.W. (1979) *Principles of Crop Improvement.* Longman, London.

Simmonds, N.W. (1982) The context of the workshop. In: Withers, L.A. and Williams, J.T. (eds) *Crop Genetic Resources – the Conservation of Difficult Material.* IUBS Series B42, International Union of Biological Sciences/International Board for Plant Genetic Resources/International Genetic Federation, Paris, pp. 1–3.

Singh, M., Ceccarelli, S. and Grando, S. (1999) Genotype × environment interaction of crossover type: detecting its presence and estimating the crossover point. *Theoretical and Applied Genetics* 99, 988–995.

Singh, R.P., Rajaram, S., Miranda, A., Huerta-Espino, J. and Autrique, E. (1998) Comparison of two crossing and four selection schemes for yield, yield traits and slow rusting resistance to leaf rust in wheat. *Euphytica* 100, 35–43.

Singla-Pareek, S.L., Reddy, M.K. and Sopory, S.K. (2003) Genetic engineering of the glyoxalase pathway in tobacco leads to enhanced salinity tolerance. *Proceedings of the National Academy of Sciences of the United States of America* 100, 14672–14677.

Sinha, S.K. and Swaminathan, M.S. (1984) New parameters and selection criteria in plant breeding. In: Vose, P.B. and Blixt, S.G. (eds) *Crop Breeding, a Contemporary Basis.* Pergamon Press, Oxford, UK.

Siripoonwiwat, W. (1995) Application of restriction fragment length polymorphism (RFLP) markers in the analysis of chromosomal regions associated with some quantitative traits for hexaploid oat improvement. MS thesis, Cornell University, Ithaca, New York.

Sivamani, E., Huet, H., Shen, P., Ong, C.A., DeKochko, A., Fauquet, C.M. and Beachy, R.N. (1999) Rice plants (*Oryza sativa* L.) containing three rice tungro spherical virus (RTSV) coat protein transgenes are resistant to virus infection. *Molecular Breeding* 5, 177–185.

Skinner, D.Z., Muthukrishnan, S. and Liang, G.H. (2004) Transformation: a powerful tool for crop improvement. In: Liang, G.H. and Skinner, D.Z. (eds) *Genetically Modified Crops: Their Development, Uses and Risks.* Food Products Press, Binghamton, New York, pp. 1–16.

Skol, A.D., Scott, L.J., Abecasis, G.R. and Boehnke, M. (2006) Joint analysis is more efficient than replication-based analysis for two-stage genome-wide association studies. *Nature Genetics* 38, 209–213.

Slater, S., Mitsky, T.A., Houmiel, K.L., Hao, M., Reiser, S.E., Taylor, N.B., Tran, M., Valentin, H.E., Rodriguez, D.J., Stone, D.A., Padgette, S.R., Kishore, G. and Gruys, K.J. (1999) Metabolic engineering of *Arabidopsis* and *Brassica* for poly(3-hydroxybutyrate-co-3-hydroxyvalerate) copolymer production. *Nature Biotechnology* 17, 1011–1016.

Slatkin, M. (1985) Gene flow in natural populations. *Annual Review of Ecology and Systematics* 16, 393–430.

Smith, D., Yanai, Y., Lui, Y.-G., Ishiguro, S., Okada, K., Shibata, D., Whitter, R.F. and Fedoroff, N.V. (1996) Characterization and mapping of Ds-GUS-T-DNA lines for targeted insertional mutagenesis. *The Plant Journal* 10, 721–732.

Smith, G.D. and Egger, M. (1998) Meta-analysis bias in location and selection of studies. *BMJ* 317, 625–629.

Smith, H.F. (1936) A discriminant function for plant selection. *Annals of Eugenics* 7, 240–250.

Smith, J.S.C. (1986) Genetic diversity within the corn belt dent racial complex of maize (*Zea mays* L.). *Maydica* 21, 349–367.

Smith, J.S.C. and Smith, O.S. (1992) Fingerprinting crop varieties. *Advances in Agronomy* 47, 85–140.

Smith, M.E., Coffman, W.R. and Barker, T.C. (1990) Environmental effects on selection under high and low input conditions. In: Kang, M.S. (ed.) *Genotype-By-Environment Interactions and Plant Breeding.* Louisiana State University Agriculture Center, Baton Rouge, Louisiana, pp. 261–272.

Smith, O.S., Smith, J.S.C., Bowen, S.L., Tenborg, R.A. and Wall, S.J. (1990) Similarities among a group of elite maize inbreds as measured by pedigree, F_1 grain yield, grain yield heterosis and RFLPs. *Theoretical and Applied Genetics* 80, 833–840.

Smith, O.S., Smith, J.S.C., Bowen, S.L. and Tenborg, R.A. (1991) Numbers of RFLP probes necessary to show associations between lines. *Maize Genetics Newsletter* 65, 66.

Smith, O.S., Hoard, K., Shaw, F. and Shaw, R. (1999) Prediction of single-cross performance. In: Coors, J.G. and Pandey, S. (eds) *The Genetics and Exploitation of Heterosis in Crops.* American Society of Agronomy (ASA) and Crop Science Society of America (CSSA), Madison, Wisconsin, pp. 277–285.

Smith, S. and Beavis, W. (1996) Molecular marker assisted breeding in a company environment. In: Sobral, B.W.S. (ed.) *The Impact of Plant Molecular Genetics.* Birkhäuer, Boston, Massachusetts, pp. 259–272.

Smith, S. and Helentjaris, T. (1996) DNA fingerprinting and plant variety protection. In: Paterson, A.H. (ed.) *Genome Mapping in Plants.* R.G. Landes Company, Austin, Texas, pp. 95–110.

Sneath, P. and Sokal, R.R. (1973) *Numerical Taxonomy*, 2nd edn. W.H. Freeman, San Francisco, California.

Sobral, B.W.S. (2002) The role of bioinformatics in germplasm conservation and use. In: Engels, J.M.M., Ramanatha Rao, V., Brown, A.H.D. and Jackson, M.T. (eds) *Managing Plant Genetic Diversity.* International Plant Genetics Resources Institute (IPGRI), Rome, pp. 171–178.

Sobral, B.W.S., Waugh, M. and Beavis W. (2001) Information systems approaches to support discovery in agricultural genomics. In: Phillips, R.L. and Vasil, I.K. (eds) *DNA-based Markers in Plants.* Kluwer Academic Publishers, Dordrecht, Netherlands.

Sobrino, B., Briona, M. and Carracedoa, A. (2005) SNPs in forensic genetics: a review on SNP typing methodologies. *Forensic Science International* 154, 181–194.

Sobrizal, K., Ikeda, K., Sanchez, P.L., Doi, K., Angeles, E.R., Khush, G.S. and Yoshimura, A. (1999) Development of *Oryza glumaepatulla* introgression lines in rice, *O. sativa* L. *Rice Genetics Newsletter* 16, 107.

Sokal, R.R. (1986) Phenetic taxonomy: theory and methods. *Annual Review of Ecological Systems* 17, 423–442.

Soller, M. and Beckmann, J.S. (1990) Marker-based mapping of quantitative trait loci using replicated progenies. *Theoretical and Applied Genetics* 80, 205–208.

Somers, D.J., Isaac, P. and Edwards, K. (2004) High-density microsatellite consensus map for bread wheat (*Triticum aestivum* L.). *Theoretical and Applied Genetics* 109, 1105–1114.

Song, J., Bradeen, J.M., Naess, S.K., Raasch, J.A., Wielgus, S.M., Haberlach, G.T., Liu, J., Austin-Phillips, S., Buell, C.R., Helgeson, J.P. and Jiang, J. (2003) Gene *RB* cloned from *Solanum bulbocastanum* confers broad spectrum resistance to potato late blight. *Proceedings of the National Academy of Sciences of the United States of America* 100, 9128–9133.

Song, R. and Messing, J. (2003) Gene expression of a gene family in maize based on noncollinear haplotypes. *Proceedings of the National Academy of the Sciences of the United States of America* 100, 9055–9060.

Song, R., Llaca, V. and Messing, J. (2002) Mosaic organization of orthologous sequences in grass genome. *Genome Research* 12, 1549–1555.

Sopory, S. and Munshi, M. (1996) Anther culture. In: Jain, S.M., Sopory, S.K. and Vielleux, R.E. (eds) In Vitro *Haploid Production in Higher Plants*, Vol. 1.Kluwer Academic Publisher, Dordrecht, Netherlands, pp. 145–176.

Sorensen, D. and Gianola, D. (2002) *Likelihood, Bayesian and MCMC Methods in Quantitative Genetics.* Springer-Verlag Inc., New York.

Sorrells, M.E. and Wilson, W.A. (1997) Direct classification and selection of superior alleles for crop improvement. *Crop Science* 37, 691–697.

Sorrells, M.E., La Rota, M., Bermudez-Kandianis, C.E., Greene, R.A., Kentety, R., Munkvold, J.D., Miftahudin, Mahmoud, A., Ma, X.F., Gustafson, P.J., Qi, L.L., Echalier, B., Gill, B.S., Matthews, D.E., Lazo, G.R., Chao, S., Anderson, O.D., Edwards, H., Linkiewicz, A.M., Dubcovsky, J., Akhunov, E.D., Dvorak, J., Zhang, D., Nguyen, H.T., Peng, J., Lapitan, N.L.V., Gonzalez-Hernandez, J.L., Anderson, J.A., Hossain, K., Kalavacharla, V., Kianian, S.F., Choi, D.-W., Close, T.J., Dilbirligi, M., Gill, K.S., Steber, C., Walker-Simmons, M.K., McGuire, P.E. and Qualset, C.Q. (2003) Comparative DNA sequence analysis of wheat and rice genomes. *Genome Research* 13, 1818–1827.

Sourdille, P., Singh, S., Cadalen, T., Brown-Guedira, G.L., Gay, G., Qi, L., Gill, B.S., Dufour, P., Murigneux, A. and Bernard, M. (2004) Microsatellite-based deletion bin system for the establishment of genetic-physical map relationships in wheat (*Triticum aestivum* L.). *Functional and Integrative Genomics* 4, 12–25.

Southern, E.M. (1975) Detection of specific sequences among DNA fragments separated by gel electrophoresis. *Journal of Molecular Biology* 98, 503–517.

Spielman, D., Cohen, J. and Zambrano, P. (2006) Will agbiotech applications reach marginalized farmers? Evidence from developing countries. *AgBioForum* 9, 23–30.

Spielman, R.S., McGinnis, R.E. and Ewens, W.J. (1993) Transmission test for linkage disequilibrium: the insulin gene region and insulin-dependent diabetes mellitus (IDDM). *American Journal of Human Genetics* 52, 506–516.

Spooner, D., van Treuren, R. and de Vicente, M.C. (2005) Molecular markers for genebank management. *IPGRI Technical Bulletin* No. 10. Available at: http://www.ipgri.cgiar.org/publications/pdf/1082.pdf (accessed 30 June 2007).

Sprague, G.F. and Tatum, L.A. (1942) General vs. specific combining ability in single crosses of corn. *Journal of American Society of Agronomy* 34, 923–932.

Sprague, G.F., Russell, W.A., Penny, L.H. and Horner, T.W. (1962) Effects of epistasis on grain yield of maize. *Crop Science* 2, 205–208.

Springer, P.S. (2000) Gene traps: tools for plant development and genomics. *The Plant Cell* 12, 1007–1020.

Stadler, L.J. (1928) Mutations in barley induced by X-rays and radium. *Science* 68, 186–187.

Stam, P. (1991) Some aspects of QTL analysis. *Proceedings of the Eighth Meeting of the Eucarpia Section Biometrics on Plant Breeding*, 1–6 July 1991, Brno, Czechoslovakia, pp. 24–32.

Stam, P. (1993) Construction of integrated genetic linkage maps by means of a new computer package: JoinMap. *The Plant Journal* 3, 739–744.

Stam, P. (1995) Marker-assisted breeding. In: Van Ooijen, J.W. and Jansen, J. (eds) *Biometrics in Plant Breeding: Applications of Molecular Markers. Proceedings of the 9th Meeting of EUCARPIA Section on Biometrics in Plant Breeding* (1994). Centre for Plant Breeding and Reproduction Research, Wageningen, Netherlands, pp. 32–44.

Stam, P. (2003) Marker-assisted introgression: speed at any cost? In: van Hintum, Th.J.L., Lebeda, A., Pink, D. and Schut, J.W. (eds) *Proceedings of the Eucarpia Meeting on Leafy Vegetables Genetics and Breeding*, 19–21 March 2003, Noordwijkerhout, Netherlands. Centre for Genetic Resources (CGN), Wageningen, Netherlands, pp. 117–124.

Stam, P. and Zeven, A.C. (1981) The theoretical proportion of the donor genome in near-isogenic lines of self-fertilizers bred by backcrossing. *Euphytica* 30, 227–238.

Stamatoyannopoulos, J.A. (2004) The genomics of gene expression. *Genomics* 84, 449–457.

Stanford, J.C. (2000) The development of the biolistic process. In Vitro *Cellular and Developmental Biology – Plant* 36, 303–308.

Stanford, J.C., Klein, T.M., Wolf, E.D. and Allen, N. (1987) Delivery of substances into cells and tissues using a particle bombardment process. *Particulate Science and Technology* 5, 27–37.

Staub, J.E. (1999) Intellectual property rights, genetic markers and the hybrid seed production. *Journal of New Seeds* 1, 39–64.

Stebbins, G.L. (1957) Self fertilization and population variability in the higher plants. *American Nature* 91, 337–354.

Stebbins, G.L. (1970) Adaptive radiation of reproductive characteristics in angiosperms: I. Pollination mechanisms. *Annual Review of Ecology and Systematics* 1, 307–326.

Steele, K.A., Price, A.H., Shashidhar, H.E. and Witcombe, J.R. (2006) Marker-assisted selection to introgress rice QTL controlling root traits into an Indian upland rice variety. *Theoretical and Applied Genetics* 112, 208–221.

Stein, L. (2001) Genome annotation: from sequence to biology. *Nature Reviews Genetics* 2, 493–503.

Stein, L.D. (2002) Creating a bioinformatics nation. *Nature* 417, 119–120.

Stein, L.D. (2003) Integrating biological databases. *Nature Reviews Genetics* 4, 337–345.

Stein, N., Perovic, D., Kumlehn, J., Pellio, B., Stracke, S., Streng, S., Ordon, E. and Graner, A. (2005) The eukaryotic translation initiation factor 4E confers multiallelic recessive Bymovirus resistance in *Hordeum vulgare* (L.). *The Plant Journal* 42, 912–922.

Stelly, D.M., Lee, J.A. and Rooney, W.L. (1988) Proposed schemes for mass-extraction of doubled haploids of cotton. *Crop Science* 28, 885–890.

Sterling, T.D. (1959) Publication decision and their possible effects on inferences drawn from tests of significance – or vice versa. *Journal of the American Statistical Association* 54, 30–34.

Stich, B. and Melchinger, A.E. (2009) Comparison of mixed-model approaches for association mapping in rapeseed, potato, sugar beet, maize, and *Arabidopsis*. *BMC Genomics* 10, 94.

Stich, B., Melchinger, A.E., Piepho, H.-P., Heckenberger, M., Maurer, H.P. and Reif, J.C. (2006) A new test for family-based association mapping using inbred lines from plant breeding programs. *Theoretical and Applied Genetics* 113, 1121–1130.

Stich, B., Yu, J., Melchinger, A.E., Piepho, H.P., Utz, H.F., Maurer, H.P. and Buckler, E.S. (2007) Power to detect higher-order epistatic interactions in a metabolic pathway using a new mapping strategy. *Genetics* 176, 563–570.

Stich, B., Möhring, J., Piepho, H.-P., Heckenberger, M., Buckler, E.S. and Melchinger, A.E. (2008) Comparison of mixed-model approaches for association mapping. *Genetics* 178, 1745–1754.

Stitt, M. and Fernie, A.R. (2003) From measurements of metabolites to metabolomics: an 'on the fly' perspective illustrated by recent studies of carbon–nitrogen interactions. *Current Opinion in Biotechnology* 14, 136–144.

Stoyanova, S.D. (1991) Genetic shifts and variations of gliadins induced by seed aging. *Seed Science and Technology* 19, 363–371.

Stratton, D.A. (1998) Reaction norm functions and QTL-environments for flowering time in *Arabidopsis thaliana*. *Heredity* 81, 144–155.

Stuber, C.W. (1992) Biochemical and molecular markers in plant breeding. *Plant Breeding Reviews* 9, 37–61.

Stuber, C.W. (1994a) Breeding multigenic traits. In: Phillips, R.L. and Vasil, I.K. (eds) *DNA Based Markers in Plants*. Kluwer Academic Publishers, Dordrecht, Netherlands, pp. 97–115.

Stuber, C.W. (1994b) Heterosis in plant breeding. *Plant Breeding Reviews* 12, 227–251.

Stuber, C.W. (1995) Mapping and manipulating quantitative traits in maize. *Trends in Genetics* 11, 477–481.

Stuber, C.W. (1999) Biochemistry, molecular biology and physiology of heterosis. In: Coors, J.G. and Pandey, S. (eds) *The Genetics and Exploitation of Heterosis in Crops*. American Society of Agronomy (ASA) and Crop Science Society of America (CSSA), Madison, Wisconsin, pp. 173–184.

Stuber, C.W. and Moll, R.H. (1972) Frequency changes of isozyme alleles in a selection experiment for grain yield in maize (*Zea mays* L.). *Crop Science* 12, 337–340.

Stuber, C.W. and Sisco, P.H. (1991) Marker-facilitated transfer of QTL alleles between elite inbred lines and responses of hybrids. *Proceedings of 46th Annual Corn and Sorghum Industry Research Conference* 46, 104–113.

Stuber, C.W., Moll, R.H., Goodman, M.M., Schaffer, H.E. and Weir, B.S. (1980) Allozyme frequency changes associated with selection for increased grain yield in maize (*Zea mays*). *Genetics* 95, 225–336.

Stuber, C.W., Goodman, M.M. and Moll, R.H. (1982) Improvement of yield and ear number resulting from selection at allozyme loci in a maize population. *Crop Science* 22, 737–740.

Stuber, C.W., Lincoln, S.E., Wolff, D.W., Helentjaris, T. and Lander, E.S. (1992) Identification of genetic factors contributing to heterosis in a hybrid from two elite maize inbred lines using molecular markers. *Genetics* 132, 823–839.

Stuber, C.W., Polacco, M. and Senior, M.L. (1999) Synergy of empirical breeding, marker-assisted selection and genomics to increase crop yield potential. *Crop Science* 39, 1571–1583.

Stuper, R.M. and Springer, N.M. (2006) *Cis*-transcriptional variation in maize inbred lines B73 and Mo17 lead to additive expression patterns in the F1 hybrid. *Genetics* 173, 2199–2210.

Subrahmanyam, N.C. and Kasha, K.J. (1975) Chromosome doubling of barley haploids by nitrous oxide and colchicine treatment. *Canadian Journal of Genetics and Cytology* 17, 573–583.

Subramanian, A., Tamayo, P., Mootha, V.K., Mukherjee, S., Ebert, B.L., Gillette, M.A., Paulovich, A., Pomeroy, S.L., Golub, T.R., Lander, E.S. and Mesirov, J.P. (2005) Gene set enrichment analysis: a knowledge-based approach for interpreting genome-wide expression profiles. *Proceedings of the National Academy of Sciences of the United States of America* 102, 15545–15550.

Sughrou, J.R. and Rockeford, T.R. (1994) Restriction fragment length polymorphism differences among the Illinois long-term selection oil strains. *Theoretical and Applied Genetics* 87, 916–924.

Sugita, K., Kasahara, T., Matsunaga, E. and Ebinuma, H. (2000) A transformation vector for the production of marker-free transgenic plants containing a single copy transgene at high frequency. *The Plant Journal* 22, 461–469.

Sullivan, S.N. (2004) Plant genetic resources and the law: past, present and future. *Crop Science* 135, 10–15.

Sumner, L.W., Mendes, P. and Dixon, R.A. (2003) Plant metabolomics: large-scale phytochemistry in the functional genomics era. *Phytochemistry* 62, 817–836.

Sun, D.J., He, Z.H., Xia, X.C., Zhang, L.P., Morris, C., Appels, R., Ma, W. and Wang, H. (2005) A novel STS marker for polyphenol oxidase activities in bread wheat. *Molecular Breeding* 16, 209–218.

Sun, Q.X., Huang, T.C., Ni, Z.F. and Procunier, D.J. (1996) Studies on heterotic grouping in wheat: I. Genetic diversity between varieties revealed by RAPD. *Journal of Agricultural Biotechnolgy* (China) 4, 103–110.

Sun, Q.X., Wu, L.M., Ni, Z.F., Meng, F.R., Wang, Z.K. and Lin, Z. (2004) Differential gene expression patterns in leaves between hybrids and their parental inbreds are correlated with heterosis in a diallelic cross. *Plant Science* 166, 651–657.

Sun, Y., Wang, J., Crouch, J.H. and Xu, Y. (2009) Efficiency of selective genotyping for complex traits and its innovative use in genetics and plant breeding. *Molecular Breeding* (in press)

Sundaresan, V., Springer, P., Volpe, T., Haward, S., Jones, J.D., Dean, C., Ma, H. and Martienssen, R. (1995) Patterns of gene action in plant development revealed by enhancer trap and gene trap transposable elements. *Genes and Development* 9, 1797–1810.

Suter, B., Kittanakom, S. and Stagljar, I. (2008) Two-hybrid technologies in proteomics research. *Current Opinion in Biotechnology* 19, 316–323.

Suzuki, Y., Uemura, S., Saito, Y., Murofushi, N., Schmitz, G., Theres, K. and Yamaguchi, I. (2001) A novel transposon tagging element for obtaining gain-of-function mutants based on a self-stablizing *Ac* derivative. *Plant Molecular Biology* 45, 123–131.

Swaminathan, M.S. (2006) An evergreen revolution. *Crop Science* 46, 2293–2303.

Swaminathan, M.S. (2007) Can science and technology feed the world in 2025? *Field Crops Research* 104, 3–9.

Swaminathan, M.S. and Singh, M.P. (1958) X-ray induced somatic haploidy in watermelon. *Current Science* 27, 63–64.

Swanson-Wagner, R.A., Jia, Y., DeCook, R., Borsuk, L.A., Nettleton, D. and Schnable, P.S. (2006) All possible modes of gene action are observed in a global comparison of gene expression in a maize F_1 hybrid and its inbred parents. *Proceedings of the National Academy of Sciences of the United States of America* 103, 6805–6810.

Syvänen, A.-C. (1999) From gels to chips: 'minisequencing' primer extension for analysis of point mutations and single nucleotide polymorphisms. *Human Mutation* 13, 1–10.

Syvänen, A.-C. (2001) Accessing genetic variation: genotyping single nucleotide polymorphisms. *Nature Reviews Genetics* 2, 930–942.

Syvänen, A.-C. (2005) Toward genome-wide SNP genotyping. *Nature Genetics* 37, S5–S10.

Syvänen, A.-C., Aalto-Setala, K., Harju, L., Kontula, K. and Soderlund, H. (1990) A primer-guided nucleotide incorporation assay in the genotyping of apolipoprotein E. *Genomics* 8, 684–692.

Szalma, S.J., Hostert, B.M., LeDeaux, J.R., Stuber, C.W. and Holland, J.B. (2007) QTL mapping with near-isogenic lines in maize. *Theoretical and Applied Genetics* 114, 1211–1228.

Szarejko, I. and Forster, B.P. (2007) Doubled haploidy and induced mutation. *Euphytica* 158, 359–370.

Tabashnik, B.E., Gassmann, A.J., Crowder, D.W. and Carrière, Y. (2008) Insect resistance to *Bt* crops: evidence verus theory. *Nature Biotechnology* 26, 199–202.

Taberner, A., Dopazo, J. and Castaäera, P. (1997) Genetic characterization of populations of a *de novo* arisen sugar beet pest, *Aubeonymus mariaefranciscae* (Coleopteram Curculionidae), by RAPD analysis. *Journal of Molecular Evolution* 45, 24–31.

Tai, G.C.C. (1971) Genotypic stability analysis and its application to potato regional trials. *Crop Science* 11, 184–190.

Taji, A., Kumar, P.P. and Lakshmann, P. (2002) In Vitro *Plant Breeding*. Food Products Press, Binghamton, New York, 167 pp.

Takahashi, Y., Shomura, A., Sasaki, T. and Yano, M. (2001) *Hd6*, a rice quantitative trait locus involved in photoperiod sensitivity, encodes the alpha subunit of protein kinase CK2. *Proceedings of the National Academy of Sciences of the United States of America* 98, 7922–7927.

Talbot, C.J., Nicod, A., Cherny, S.S., Fulker, D.W., Collins, A.C. and Flint, J. (1999) High-resolution mapping of quantitative trait loci in outbred mice. *Nature Genetics* 21, 305–308.

Tan, Y.F., Li, J.X., Yu, S.B., Xing, Y.Z., Xu, C.G. and Zhang, Q. (1999) The three important traits for cooking and eating quality of rice grain are controlled by a single locus in an elite rice hybrid, Shanyou 63. *Theoretical and Applied Genetics* 99, 642–648.

Tang, G.L., Reinhart, B.J., Bartel, D.P. and Zamore, P.D. (2003) A biochemical framework for RNA silencing in plants. *Genes and Development* 17, 49–63.
Tang, H., Bowers, J.E., Wang, X., Ming, R., Alam, M. and Paterson, A.H. (2008) Synteny and collinearity in plant genomes. *Science* 320, 486–488.
Tanksley, S.D. (1983) Molecular markers in plant breeding. *Plant Molecular Biology Reporter* 1, 1–3.
Tanksley, S.D. (1993) Mapping polygenes. *Annual Review of Genetics* 27, 205–233.
Tanksley, S.D. and McCouch, S.R. (1997) Seed banks and molecular maps: unlocking genetic potential from the wild. *Science* 277, 1063–1066.
Tanksley, S.D. and Nelson, J.C. (1996) Advanced backcross QTL analysis: a method for the simultaneous discovery and transfer of valuable QTLs from unadapted germplasm into elite breeding. *Theoretical and Applied Genetics* 92, 191–203.
Tanksley, S.D. and Rick, C.M. (1980) Isozyme gene linkage map of the tomato: applications in genetics and breeding. *Theoretical and Applied Genetics* 57, 161–170.
Tanksley, S.D., Miller, J., Paterson, A. and Bernatzky, R. (1988) Molecular mapping of plant chromosomes. In: Gustafson, J.P. and Appels, R. (eds) *Chromosome Structure and Function – Impact of New Concepts.* Proceedings of the 18th Stadller Genetics Symposium. Plenum Press, New York, pp. 157–173.
Tanksley, S.D., Young, N.D., Paterson, A.H. and Bonierbale, M.W. (1989) RFLP mapping in plant breeding: new tools for an old science. *Bio/Technology* 7, 257–263.
Tanksley, S.D., Ganal, M.W. and Martin, G.B. (1995) Chromosome landing: a paradigm for map based gene cloning in plants with large genomes. *Trends in Genetics* 11, 63–68.
Tanksley, S.D., Grandillo, S., Fulton, T.M., Zamir, D., Eshed, Y., Petiard, V., Lopez, J. and Beck-Bunn, T. (1996) Advanced backcross QTL analysis in a cross between an elite processing line of tomato and its wild relative *L. pimpinellifolium. Theoretical and Applied Genetics* 92, 213–224.
Tao, Q. and Zhang, H.-B. (1998) Cloning and stable maintenance of DNA fragments over 300 kb in *Escherichia coli* with conventional plasmid-based vectors. *Nucleic Acids Research* 26, 4901–4909.
Tarchini, R., Biddle, P., Wineland, R., Tingey, S. and Rafalski, A. (2000) The complete sequence of 340kb of DNA around the rice *Adh1-Adh2* region reveals interrupted colinearity with maize chromosome 4. *The Plant Cell* 12, 381–391.
Tardieu, F. (2003) Virtual plants: modeling as a tool for the genomics of tolerance to water deficit. *Trends in Plant Science* 8, 9–14.
Tauz, D. and Renz, M. (1984) Simple sequences are ubiquitous repetitive components of eukaryotic genomes. *Nucleic Acids Research* 12, 4127–4138.
Taylor, B.A. (1978) Recombinant inbred strains: use in gene mapping. In: Morse, H.C. (ed.) *Origin of Inbred Mice.* Academic Press, New York, pp. 423–438.
Tekeoglu, M., Rajesh, P.N. and Muehlbauer, F.J. (2002) Integration of sequence tagged microsatellites to the chickpea genetic map. *Theoretical and Applied Genetics* 105, 847–854.
Temnykh, S., Park, W.D., Ayres, N., Cartinhour, S., Hauck, N., Lipovich, L., Cho, Y.G., Ishii, T. and McCouch, S.R. (2000) Mapping and genome organization of microsatellite sequences in rice (*Oryza sativa* L.). *Theoretical and Applied Genetics* 100, 697–712.
Tenaillon, M.I., Sawkins, M.C., Long, A.D., Gaut, R.L., Doebley, J.F. and Gaut, B.S. (2001) Patterns of DNA sequence polymorphism along chromosome 1 of maize (*Zea mays* ssp. *mays* L.). *Proceedings of the National Academy of Sciences of the United States of America* 98, 9161–9166.
Tenhola-Roininen, T., Immonen, S. and Tanhuanpää, P. (2006) Rye doubled haploids as a research and breeding tool – a practical point of view. *Plant Breeding* 125, 584–590.
Terada, R., Urawa, H., Inagaki, Y., Tsugane, K. and Iida, S. (2002) Efficient gene targeting by homologous recombination in rice. *Nature Biotechnology* 20, 1030–1034.
Tessier, D.C., Arbour, M., Benoit, F., Hogues, H. and Rigby, T. (2005) A DNA microarray fabrication strategy for research laboratories. In: Sensen, C.W. (ed.) *Handbook of Genome Research. Genomics, Proteomics, Metabolomics, Bioinformatics, Ethical and Legal Issues.* WILEY-VCH, Weinheim, Germany, pp. 223–238.
The Arabidopsis Genome Initiative (2000) Analysis of the genome sequence of the flowering plant *Arabidopsis thaliana. Nature* 408, 796–815.
Therneau, T.M. and Grambsch, P.M. (2000) *Modeling Survival Data: Extending the Cox Model.* Springer, New York.
Thiel, T., Michalek, W., Varshney, R.K. and Graner, A. (2003) Exploiting EST data bases for the development and characterization of gene-derived SSR-markers in barley (*Hordeum vulgare* L.). *Theoretical and Applied Genetics* 106, 411–422.

Thoday, J.M. (1961) Location of polygenes. *Nature* 191, 368–370.
Thomas, C.D., Cameron, A., Green, R.E., Bakkenes, M., Beaumont, L.J., Collingham, Y.C., Erasmus, B.F.N., Ferreira de Siqueira, M., Grainger, A., Hannah, L., Hughes, L., Huntley, B., van Jaarsveld, A.S., Midgley, G.F., Miles, L., Ortega-Huerta, M.A., Peterson, A.T., Phillips, O.L. and Williams, S.E. (2004) Extinction risk from climate change. *Nature* 427, 145–148.
Thompson, J.A., Halewood, M., Engels, J. and Hoogendoorn, C. (2004) Plant genetic resources collections: a survey of issues concerning their value, accessibility and status as public goods. In: *New Directions for a Diverse Planet: Proceedings of the 4th International Crop Science Congress*, 26 September–1 October 2004, Brisbane, Australia. Published on CD-ROM. Available at: http://www.cropscience.org.au/icsc2004/ (accessed 17 November 2009).
Thomson, M.J., Tai, T.H., McClung, A.M., Hinga, M.E., Lobos, K.B., Xu, Y., Martinez, C. and McCouch, S.R. (2003) Mapping quantitative trait loci for yield, yield components and morphological traits in an advanced backcross population between *Oryza rufipogon* and the *Oryza sativa* cultivar Jefferson. *Theoretical and Applied Genetics* 107, 479–493.
Thomson, M.J., Edwards, J.D., Septiningsih, E.M., Harrington, S.E. and McCouch, S.R. (2006) Substitution mapping of *dth1.1*, a flowering-time quantitative trait locus (QTL) associated with transgressive variation in rice, reveals multiple sub-QTL. *Genetics* 172, 2501–2514.
Thorisson, G.A., Muilu, J. and Brookes, A.J. (2009) Genotype–phenotype databases: challenges and solutions for the post-genomic era. *Nature Reviews Genetics* 10, 9–18.
Thornsberry, J.M., Goodman, M.M., Doebley, J., Kresovich, S., Nielsen, D. and Buckler IV, E.S. (2001) *Dwarf8* polymorphisms associate with variation in flowering time. *Nature Genetics* 28, 286–289.
Tikhonov, A.P., SanMiguel, P.J., Nakajima, Y., Gorenstein, N.M., Bennetzen, J.L. and Avramova, Z. (1999) Colinearity and its exceptions in orthologous *adh* regions of maize and sorghum. *Proceedings of the National Academy of Sciences of the United States of America* 96, 7409–7414.
Till, B.J., Reynolds, S.H., Greene, E.A., Codomo, C.A., Enns, L.C., Johnso, J.E., Burtner, C., Odden, A.R., Young, K., Taylor, N.E., Henikoff, J.G., Comai, L. and Henikoff, S. (2003) Large-scale discovery of induced point mutations with high-throughput TILLING. *Genome Research* 13, 524–530.
Till, B.J., Comai, L. and Henikoff, S. (2007) TILLING and EcoTILLING for crop improvement. In: Varshney, R.K. and Tuberosa, R. (eds) *Genomics-Assisted Crop Improvement.* Vol.1: *Genomic Approaches and Platforms.* Springer, Dordrecht, Netherlands, pp. 333–350.
Tinker, N.A. and Mather, D.E. (1993) GREGOR: software for genetic simulation. *Journal of Heredity* 84, 237.
Tirosh, I., Bilu, Y. and Barkai, N. (2007) Comparative biology: beyond sequence analysis. *Current Opinion in Biotechnology* 18, 371–377.
Tomita, M., Hashimoto, K., Takahashi, K, Shimizu, T.S., Matsuzaki, Y., Miyoshi, F., Saito, K., Tanida, S., Yugi, K., Venter, J.C. and Hutchison, C.A. III (1999) E-CELL: software environment for whole-cell simulation. *Bioinformatics* 15, 72–84.
Trawick, B.W. and McEntyre, J.R. (2004) Bibliographic databases. In: Sansom, C.E. and Horton, R.M. (eds) *The Internet for Molecular Biologists.* Oxford University Press, Oxford, UK, pp. 1–16.
Trethewey, R.N. (2005) Metabolite profiling in plants. In: Leister, D. (ed.) *Plant Functional Genomics.* Food Products Press, Binghamton, New York, pp. 85–117.
Tripp, R., Louwaars, N.P. and Eaton, D. (2006) Intellectual property rights for plant breeding and rural development: challenges for agricultural policymakers. *Agricultural and Rural Development Notes* Issue 12.
Truco, M.J., Antonise, R., Lavelle, D., Ochoa, O., Kozik, A., Witsenboer, H., Fort, S.B., Jeuken, M.J.W., Kesseli, R.V., Lindhout, P., Michelmore, R.W. and Peleman, J. (2007) A high-density, integrated genetic linkage map of lettuce (*Lactuca* spp.). *Theoretical and Applied Genetics* 115, 735–746.
Tsien, R.Y. (1998) The green fluorescent protein. *Annual Review of Biochemistry* 67, 509–544.
Tu, J., Datta, K., Alam, M.F., Khush, G.S. and Datta, S.K. (1998a) Expression and function of a hybrid *Bt* toxin gene in transgenic rice conferring resistance to insect pests. *Plant Biotechnology* 15, 183–191.
Tu, J., Ona, I., Zhang, Q., Mew, T.W., Khush, G.S. and Datta, S.K. (1998b) Transgenic rice variety IR72 with Xa21 is resistant to bacterial blight. *Theoretical and Applied Genetics* 97, 31–36.
Turcotte, E.L. and Feaster, C.V. (1963) Haploids: high-frequency production from single-embryo seeds in a line of Pima cotton. *Science* 140, 1407–1408.
Turcotte, E.L. and Feaster, C.V. (1967) Semigamy in Pima cotton. *Journal of Heredity* 58, 54–57.
Tuvesson, S., Dayteg, C., Hagberg, P., Manninen, O., Tanhuanpää, P., Tenhola-Roininen, T., Kiviharju, E., Weyen, J., Förster, J., Schondelmaier, J., Lafferty, J., Marn, M. and Fleck, A. (2007) Molecular markers and doubled haploids in European plant breeding programmers. *Euphytica* 158, 305–312.

Tyo, K.E., Alper, H.S. and Stephanopoulos, G.N. (2007) Expanding the metabolic engineering toolbox: more options to engineer cells. *Trends in Biotechnology* 25, 132–137.

Tzfira, T. and Citovsky, V. (2006) *Agrobacterium*-mediated genetic transformation of plants: biology and biotechnology. *Current Opinion in Biotechnology* 17, 147–154.

Tzfira, T., Tian, G.W., Lacroix, B., Vyas, S., Li, J., Leitner-Dagan, Y., Krichevsky, A., Taylor, T., Vainstein, A. and Citovsky, V. (2005) pSAT vectors: amodular series of plasmids for autofluorescent protein tagging and expression of multiple genes in plants. *Plant Molecular Biology* 57, 503–516.

Tzfira, T., Kozlovsky, S.V. and Vitaly Citovsky, V. (2007) Advanced expression vector systems: new weapons for plant research and biotechnology. *Plant Physiology* 145, 1087–1089.

Ufaz, S. and Galili, G. (2008) Improving the content of essential amino acids in crop plants: goals and opportunities. *Plant Physiology* 147, 954–961.

Uga, Y., Fukuta, Y., Cai, H.W., Iwata, H., Ohsawa, R., Morishima, H. and Fujimura, T. (2003) Mapping QTLs influencing rice floral morphology using recombinant inbred lines derived from a cross between *Oryza sativa* L. and *Oryza rufipogon* Griff. *Theoretical and Applied Genetics* 107, 218–226.

Uga, Y., Nonoue, Y., Liang, Z.W., Lin, H.X., Yamamoto, S., Yamanouchi, U. and Yano, M. (2007) Accumulation of additive effects generates a strong photoperiod sensitivity in the extremely late-heading rice cultivar 'Nona Bokra'. *Theoretical and Applied Genetics* 114, 1457–1466.

Ukai, Y., Osawa, R., Saito, A. and Hayashi, T. (1995) MAPL: a package of computer programs for construction of DNA polymorphism linkage maps and analysis of QTL (in Japanese). *Breeding Science* 45, 139–142.

Ulloa, M., Saha, S., Jenkins, J.N., Meredith, W.R., Jr, McCarty, J.C., Jr and Stelly, D.M. (2005) Chromosomal assignment of RFLP linkage groups harboring important QTLs on an intraspecific cotton (*Gossypium hirsutum* L.) joinmap. *Journal of Heredity* 96, 132–144.

Ungerer, M.C., Halldorsdottir, S.S., Purugganan, M.D. and Mackay, T.F.C. (2003) Genotype–environment interactions at quantitative trait loci affecting inflorescence development in *Arabidopsis thaliana. Genetics* 165, 353–365.

Ünlü, M., Morgan, M.E. and Minden, J.S. (1997) Difference gel electrophoresis: a single gel method for detecting changes in protein extracts. *Electrophoresis* 18, 2071–2077.

Upadhyaya, H.D. and Ortiz, R. (2001) A mini core subset for capturing diversity and promoting utilization of chickpea genetic resources in crop improvement. *Theoretical and Applied Genetics* 102, 1292–1298.

Upadhyaya, H.D., Bramel, P.J., Ortiz, R. and Singh, S. (2002) Developing a mini core of peanut for utilization of genetic resources. *Crop Science* 42, 2150–2156.

Upadhyaya, H.D., Gowda, C.L.L., Pundir, R.P.S., Reddy, V.G. and Singh, S. (2006a) Development of core subset of fingermillet germplasm using geographical origin and data on 14 quantitative traits. *Genetic Resources and Crop Evolution* 53, 679–685.

Upadhyaya, H.D., Reddy, L.J., Gowda, C.L.L., Reddy, K.N. and Singh, S. (2006b) Development of a mini core subset for enhanced and diversified utilization of pigeonpea germplasm resources. *Crop Science* 46, 2127–2132.

UPOV (The International Union for the Protection of New Varieties of Plants) (1991) The 1991 Act of the UPOV Convention. Available at: http://www.upov.int/en/publications/conventions/1991/content.htm (accessed 17 November 2009).

UPOV (The International Union for the Protection of New Varieties of Plants) (2005) *UPOV Report on the Impact of Plant Variety Protection.* UPOV Publication No. 353 (E), UPOV, Geneva, December 2005, 98 pp.

Urwin, P.E., McPheron, M.J. and Atkinson, H.J. (1998) Enhanced transgenic plant resistance to nematodes by dual proteinase inhibitor constructs. *Planta* 204, 472–479.

Urwin, P., Yi, L., Martin, H., Atkinson, H. and Gilmartin, P.M. (2000) Functional characterization of the EMCV IRES in plants. *Plant Journal* 24, 583–589.

USDA (United States Department of Agriculture) (2002a) Statistical indicators. *Agricultural Outlook*, January–February 2002, Economic Research Service, USDA, Washington, DC, pp. 30–59.

USDA (United States Department of Agriculture) (2002b) Genetically engineered crops: US adoption and impacts. *Agricultural Outlook*, September 2002, Economic Research Service, USDA, Washington, DC, pp. 24–27.

Usuka, J., Zhu, W. and Brendel, V. (2000) Optimal sliced alignment of homologous cDNA to a genomic DNA template. *Bioinformatics* 16, 203–211.

Utz, H.F. and Melchinger, A.E. (1994) Comparison of different approaches to interval mapping of quantitative trait loci. In: van Ooijen, J.W. and Jansen, J. (eds) *Biometrics in Plant Breeding: Applications of*

Molecular Markers. Proceedings of the Ninth Meeting of the EUCARPIA Section Biometrics in Plant Breeding, 6–8 July 1994, Wageningen, Netherlands, pp. 195–204.

Utz, H.F. and Melchinger A.E. (1996) PLABQTL: a program for composite interval mapping of QTL. *Journal of Agricultural Genomics*. Available at: http://www.cabi-publishing.org/jag/papers96/paper196/indexp196.html (accessed 30 June 2007).

Utz, H.F., Melchinger, A.E. and Schön, C.C. (2000) Bias and sampling error of the estimated proportion of genotypic variance explained by quantitative loci determined from experimental data in maize using cross validation and validation with independent samples. *Genetics* 154, 1839–1849.

Vain, P., Afolabi, A.S., Worland, B. and Snape, J.W. (2003) Transgene behaviour in populations of rice plants transformed using a new dual binary vector system: pGreen/pSoup. *Theoretical and Applied Genetics* 107, 210–217.

Vallegos, C.E. and Chase, C.D. (1991) Linkage between isozyme markers and a locus affecting seed size in *Phaseolus vulgaris* L. *Theoretical and Applied Genetics* 81, 413–419.

van Berloo, R. (1999) GGT: software for the display of graphical genotypes. *Journal of Heredity* 90, 328–329.

van Berloo, R. and Stam, P. (1998) Marker-assisted selection in autogamous RIL populations: a simulation study. *Theoretical and Applied Genetics* 96, 147–154.

van Berloo, R. and Stam, P. (1999) Comparison between marker-assisted selection and phenotypical selection in a set of *Arabidopsis thaliana* recombinant inbred lines. *Theoretical and Applied Genetics* 98, 113–118.

van Berloo, R. and Stam, P. (2001) Simultaneous marker-assisted selection for multiple traits in autogamous crops. *Theoretical and Applied Genetics* 102, 1107–1112.

van Berloo, R., Aalbers, H., Werkman, A. and Niks, R.E. (2001) Resistance QTL confirmed through development of QTL-NILs for barley leaf rust resistance. *Molecular Breeding* 8, 187–195.

van der Fits, L., Hilliou, F. and Memelink, J. (2001) T-DNA activation tagging as a tool to isolate regulators of a metabolic pathway from a generally non-tractable plant species. *Transgenic Research* 10, 513–521.

van der Wurff, A.W., Chan, Y.L., Van Straalen, N.M. and Schouten, J. (2000) TE-AFLP: combining rapidity and robustness in DNA fingerprinting. *Nucleic Acids Research* 28, e105.

van Deynze, A.E., Nelson, J.C., O'Donoughue, L.S., Ahn, S.N., Siripoonwiwat, W., Harrington, S.E., Yglesias, E.S., Braga, D.P., McCouch, S.R. and Sorrells, M.E. (1995a) Comparative mapping in grasses. Oat relationships. *Molecular and General Genetics* 249, 349–356.

van Deynze, A.E., Nelson, J.C., O'Donoughue, L.S., Ahn, S.N., Siripoonwiwat, W., Harrington, S.E., Yglesias, E.S., Braga, D.P., McCouch, S.R. and Sorrells, M.E. (1995b) Comparative mapping in grasses. Wheat relationships. *Molecular and General Genetics* 248, 744–754.

van Eeuwijk, F.A., Denis, J.-B. and Kang, M.S. (1996) Incorporating additional information on genotype and environments in models for two-way genotype by environment tables. In: Kang, M.S. and Gaugh, H.G. (eds) *Genotype-by-Environment Interaction*. CRC Press, Boca Raton, Florida, pp. 15–50.

van Eeuwijk, F.A., Crossa, J., Vargas, M. and Ribaut, J.M. (2001) Variants of factorial regression for analysing QTL by environment interaction. In: Gallais, A., Dillmann, C. and Goldringer, I. (eds) *Eucarpia, Quantitative Genetics and Breeding Methods: the Way Ahead*. Institut National de la Rescherche Agronomique (INRA) Editions, Versailles. Les colloques 96, 107–116.

van Eeuwijk, F.A., Crossa, J., Vargas, M. and Ribaut, J.-M. (2002) Analysing QTL by environment interaction by factorial regression, with an application to the CIMMYT drought and low nitrogen stress programme in maize. In: Kang, M.S. (ed.) *Quantitative Genetics, Genomics and Plant Breeding*. CAB International, Wallingford, UK, pp. 245–256.

van Eeuwijk, F.A., Malosetti, M., Yin, X., Struik, P.C. and Stam, P. (2004) Modeling differential phenotypic expression. In: *New Directions for a Diverse Planet: Proceedings 4th International Crop Science Congress* (ICSC), 26 September–1 October 2004, Brisbane, Australia. ICSC, Brisbane, Australia. Available at: http://www.cropscience.org.au/icsc2004/ (accessed 17 November 2009).

van Eeuwijk, F.A., Malosetti, M., Yin, X., Struik, P.C. and Stam, P. (2005) Statistical models for genotype by environment data: from conventional ANOVA models to eco-physiological QTL models. *Australian Journal of Agricultural Research* 56, 883–894.

van Eijk, M., Peleman, J. and de Ruiter-Bleeker, M. (2001) Microsatellite-AFLP. Patent EP 1282729.

van Ginkel, M., Trethowan, R., Ammar, K., Wang, J. and Lillemo, M. (2002) Guide to bread wheat breeding at CIMMYT (rev). *Wheat special report* No. 5. Centro Internacional de Mejoramiento de Maiz y Trigo (CIMMYT), Mexico, DF.

van Oeveren, A.J. and Stam, P. (1992) Comparative simulation studies on the effects of selection for quantitative traits in autogamous crops: early selection versus single seed decent. *Heredity* 69, 342–351.
van Ooijen, A.J. and Voorrips, R.E. (2001) *JoinMap (tm) 3.0: Software for the Calculation of Genetic Linkage Maps.* Plant Research International, Wageningen, Netherlands.
van Ooijen, J.W. (1992) Accuracy of mapping quantitative trait loci in autogamous species. *Theoretical and Applied Genetics* 84, 803–811.
van Os, H., Andrzejewski, S., Bakker, E., Barrena, I., Bryan, G.J., Caromel, B., Ghareeb, B., Ishidore, E., de Jong, W., van Koert, P., Lefebvre, V., Milbourne, D., Ritter, E., van der Voort, J.N.A.M., Rousselle-Bourgeois, E., van Vliet, J., Waugh, R., Visser, R.G.F., Bakker, J. and van Eck, H.J. (2006) Construction of a 10,000-marker ultradense genetic recombination map of potato: providing a framework for accelerated gene isolation and a genomewide physical map. *Genetics* 173, 1075–1087.
van Treuren, R. (2001) Efficiency of reduced primer selectivity and bulked DNA analysis for the rapid detection of AFLP polymorphisms in a range of crop species. *Euphytica* 117, 27–37.
van Wijk, K.J. (2001) Challenges and prospects of plant proteomics. *Plant Physiology* 126, 301–308.
Vandepoele, K. and Van de Peer, Y. (2005) Exploring the plant transcriptome through phylogenetic profiling. *Plant Physiology* 137, 31–42.
Vaneck, J.M., Blowers, A.D. and Earle, E.D. (1995) Stable transformation of tomato cell-cultures after bombardment with plasmid and YAC DNA. *Plant Cell Reports* 14, 299–304.
Vane-Wright, R.I., Humphries, D.J. and Williams, P.H. (1991) What to protect? Systematics and the agony of choice. *Biological Conservation* 55, 235–254.
Varela, M., Crossa, J., Rane, J., Joshi, A. and Trethowan, R. (2006) Analysis of a three-way interaction including multi-attributes. *Australian Journal of Agricultural Research* 57, 1185–1193.
Vargas, M., van Eeuwijk, F.A., Crossa, J. and Ribaut, J.-M. (2006) Mapping QTL and QTL × environment interaction for CIMMYT maize drought stress program using factorial regression and partial least squares methods. *Theoretical and Applied Genetics* 122, 1009–1023.
Varshney, R.K., Graner, A. and Sorrells, M.E. (2005a) Genic microsatellite markers in plants: features and applications. *Trends in Biotechnology* 23, 48–55.
Varshney, R.K., Graner, A. and Sorrells, M.E. (2005b) Genomics-assisted breeding for crop improvement. *Trends in Plant Science* 10, 621–630.
Varshney, R.K., Nayak, S.N., May, G.D. and Jackson, S.A. (2009) Next-generation sequencing technologies and their implications for crop genetics and breeding. *Trends in Biotechnology* 27, 522–530.
Vavilov, N.I. (1926) Studies on the origin of cultivated plants. *Bulletin of Applied Botany, Genetics and Plant Breeding* 16, 1–248.
Veena, J.H., Doerge, R.W. and Gelvin, S. (2003) Transfer of T-DNA and Vir proteins to plant cells by *Agrobacterium tumefaciens* induces expression of host genes involved in mediating transformation and suppresses host defense gene expression. *The Plant Journal* 35, 219–236.
Velculescu, V.E., Zhang, L., Vogelstein, B. and Kinzler, K.W. (1995) Serial analysis of gene expression. *Science* 270, 484–487.
Veldboom, L.R., Lee, M. and Woodman, W.L. (1994) Molecular-marker-facilitated studies in an elite maize population: I. Linkage analysis and determination of QTL for morphological traits. *Theoretical and Applied Genetics* 88, 7–16.
Veldboom, L.R., Lee, M. and Woodman, W.L. (1996) Molecular-marker-facilitated studies in an elite maize population: I. Linkage analysis and determination of QTL for morphological traits. *Theoretical and Applied Genetics* 88, 7–16.
Venter, J.C., Adams, M.D., Myers, E.W., Li, P.W., Mural, R.J. *et al.* (2001) The sequence of the human genome. *Science* 291, 1304–1351.
Verbyla, A.P., Eckermann, P.J., Thompson, R. and Cullis, B.R. (2003) The analysis of quantitative trait loci in multi-environment trials using a multiplicative mixed model. *Australian Journal of Agricultural Research* 54, 1395–1408.
Verdonk, J.C., De Vos, C.H.R., Verhoeven, H.A., Harina, M.A., van Tunen, A.J. and Schuurink, R.C. (2003) Regulation of floral scent production in petunia revealed by targeted metabolomics. *Phytochemistry* 62, 997–1008.
Verhaegen, D., Plomion, C., Gion, J.-M., Poitel, M., Costa, P. and Kremer, A. (1997) Quantitative trait dissection analysis in *Eucalyptus* using RAPD markers: 1. Detection of QTL in interspecific hybrid progeny, stability of QTL expression across different ages. *Theoretical and Applied Genetics* 95, 597–608.

Verweire, D., Verleyen, K., Buck, S.D., Claeys, M. and Angenon, G. (2007) Marker-free transgenic plants through genetically programmed auto-excision. *Plant Physiology* 145, 1220–1231.

Veyrieras, J.-B., Goffinet, B. and Alain Charcosset, A. (2007) MetaQTL: a package of new computational methods for the meta-analysis of QTL mapping experiments. *BMC Bioinformatics* 8, 49.

Vickers, C., Xue, G. and Gresshoff, P.M. (2006) A novel *cis*-acting element, ESP, contributes to high-level endosperm-specific expression in an oat globulin promoter. *Plant Molecular Biology* 62, 195–214.

Vigouroux, Y., Mitchell, S., Matsuoka, Y., Hamblin, M., Kresovich, S., Smith, J.S.C., Jaqueth, J., Smith, O.S. and Doebley, J. (2005) An analysis of genetic diversity across the maize genome using microsatellites. *Genetics* 169, 1617–1630.

Villar, M., Lefevre, F., Bradshaw, H.D., Jr and du-Cros, E.T. (1996) Molecular genetics of rust resistance in poplars (*Melampsora larici-populina* Kleb/*Populus* sp.) by bulked segregant analysis in a 2 × 2 factorial mating design. *Genetics* 143, 531–536.

Virk, P.S., Ford-Lloyd, B.V., Jackson, M.T., Pooni, H.S., Clemeno, T.P. and Newbury, H.J. (1996) Predicting quantitative variation within rice germplasm using molecular markers. *Heredity* 76, 296–304.

Vision, T.J., Brown, D.G., Shmoys, D.B., Durrett, R.T. and Tanksley, S.D. (2000) Selective mapping: a strategy for optimizing the construction of high-density linkage maps. *Genetics* 155, 407–420.

Visscher, P.M. and Goddard, M.E. (2004) Prediction of the confidence interval of quantitative trait loci location. *Behavior Genetics* 34, 477–482.

Visscher, P.M., Thompson, R. and Haley, C.S. (1996) Confidence intervals in QTL mapping by bootstrapping. *Genetics* 143, 1013–1020.

Visscher, P.M., Hill, W.G. and Wray, N.R. (2008) Heritability in the genomics era – concepts and misconceptions. *Nature Reviews Genetics* 9, 255–266.

Vogl, C. and Xu, S. (2000) Multipoint mapping of viability and segregation distorting loci using molecular markers. *Genetics* 155, 1439–1447.

Vos, P., Hogers, R., Bleeker, M., Reijans, M., van de Lee, T., Hornes, M., Frijters, A., Pot, J., Peleman, H., Kuiper, M. and Zabeau, M. (1995) AFLP: a new technique for DNA fingerprinting. *Nucleic Acids Research* 23, 4407–4414.

Vuylsteke, M., Kuiper, M. and Stam, P. (2000) Chromosomal regions involved in hybrid performance and heterosis: their AFLP®-based identification and practical uses in prediction models. *Heredity* 85, 208–218.

Walden, I. (1998) Preserving diversity: the role of property rights. In: Swanson, T.M. (ed.) *Intellectual Property Rights and Biodiversity Conservation.* Cambridge University Press, Cambridge, UK, pp. 176 –197.

Walker, D., Boerma, H.R., All, J. and Parrott, W. (2002) Combining *cry1Ac* with QTL alleles from PI 229358 to improve soybean resistance to lepidopteran pests. *Molecular Breeding* 9, 43–51.

Walker, D.R., Narvel, J.M., Boerma, H.R., All, J.N. and Parrott, W.A. (2004) A QTL that enhances and broadens *Bt* insect resistance in soybean. *Theoretical and Applied Genetics* 109, 1051–1957.

Wallace, D.H. (1985) Physiological genetics of plant maturity, adaptation and yield. *Plant Breeding Reviews* 3, 21–158.

Wallace, R.B., Shaffer, J., Murphy, R.F., Bonner, J., Hirose, T. and Itakura, K. (1979) Hybridization of synthetic oligodeoxyribonucleotide to phi 174 DNA: the effect of single base pair mismatch. *Nucleic Acids Research* 6, 3543–3557.

Walling, G.A., Visscher, P.M., Andersson, L., Rothschild, M.F., Wang, L., Moser, G., Groenen, A.M., Bidanel, J.P., Cepica, S., Archibald, A.L., Geldermann, H., Koning, D.J., Milan, D. and Haley, C.S. (2000) Combined analysis of data from quantitative trait loci mapping studies: chromosome 4 effects on porcine growth and fatness. *Genetics* 155, 1369–1378.

Wallø Tvet, M.W. (2005) How will a Substantive Patent Law Treaty affect the public domain for genetic resources and biological material? *Journal of World Intellectual Property* 8, 311–344.

Walsh, B. (2001) Quantitative genetics in the age of genomics. *Theoretical Population Biology* 59, 175–184.

Walsh, B. (2004) Population- and quantitative-genetic models of selection limits. *Plant Breeding Reviews* 24 (Part 1), 177–225.

Wan, S., Wu, J., Zhang, Z., Sun, X., Lv, Y., Gao, C., Ning, Y., Ma, J., Guo, Y., Zhang, Q., Zheng, X., Zhang, C., Ma, Z. and Lu, T. (2008) Activation tagging, an efficient tool for functional analysis of the rice genome. *Plant Molecular Biology* 69, 69–80.

Wang, D.L., Zhu, J., Li, Z.K. and Paterson, A.H. (1999) Mapping QTLs with epistatic effects and QTL × environment interactions by mixed linear model approaches. *Theoretical and Applied Genetics* 99, 1255–1264.

Wang, E., Robertson, M.J., Hammer, G.L., Carberry, P.S., Holzworth, D., Meinke, H., Chapman, S.C., Hargreaves, J.N.G., Huth, N.I. and McLean, G. (2002) Development of a generic crop model template in the cropping system model APSIM. *European Journal of Agronomy* 18, 121–140.

Wang, G.-L., Mackill, D.J., Bonman, J.M., McCouch, S.R., Champoux, M.C. and Nelson, R.J. (1994) RFLP mapping of genes conferring complete and partial resistance to blast in a durably resistant rice cultivar. *Genetics* 136, 1421–1434.

Wang, G.W., He, Y.Q., Xu, C.G. and Zhang, Q. (2005) Identification and confirmation of three neutral alleles conferring wide compatibility in inter-subspecific hybrids of rice (*Oryza sativa* L.) using near-isogenic lines. *Theoretical and Applied Genetics* 111, 702–710.

Wang, G.W., He, Y.Q., Xu, C.G. and Zhang, Q. (2006) Fine mapping of *f5-Du*, a gene conferring wide-compatibility for pollen fertility in inter-subspecific hybrids of rice (*Oryza sativa* L.). *Theoretical and Applied Genetics* 112, 382–387.

Wang, H., Zhang, Y.M., Li, X., Masinde, G.L., Mohan, S., Baylink, D.J. and Xu, S. (2005) Bayesian shrinkage estimation of quantitative trait loci parameters. *Genetics* 170, 465–480.

Wang, J., van Ginkel, M., Podlich, D., Ye, G., Trethowan, R., Pfeiffer, W., DeLacy, I.H., Cooper, M. and Rajaram, S. (2003) Comparison of two breeding strategies by computer simulation. *Crop Science* 43, 1764–1773.

Wang, J., van Ginkel, M., Trethowan, R., Ye, G., DeLacy, I., Podlich, D. and Cooper, M. (2004) Simulating the effects of dominance and epistasis on selection response in the CIMMYT Wheat Breeding Program using QuCim. *Crop Science* 44, 2006–2018.

Wang, J., Eagles, H.A., Trethowan, R. and van Ginkel, M. (2005) Using computer simulation of the selection process and known gene information to assist in parental selection in wheat quality breeding. *Australian Journal of Agricultural Research* 56, 465–473.

Wang, J., Chapman, S.C., Bonnett, D.G., Rebetzke, G.J. and Crouch, J. (2007) Application of population genetic theory and simulation models to efficiently pyramid multiple genes via marker-assisted selection. *Crop Science* 47, 582–588.

Wang, J.K. and Bernardo, R. (2000) Variance and marker estimates of parental contribution to F_2 and BC_1-derived inbreds. *Crop Science* 40, 659–665.

Wang, J.K. and Pfeiffer, W.H. (2007) Simulation modeling in plant breeding: principles and applications. *Agricultural Sciences in China* 6, 908–921.

Wang, X., Rea, T., Bian, J., Gray, S. and Sun, Y. (1999) Identification of the gene responsive to etoposide-induced apoptosis: application of DNA chip technology. *FEBS Letters* 445, 269–273.

Wang, X., Hu, Z., Wang, W., Li, Y., Zhang, Y.M. and Xu, C. (2007) A mixture model approach to the mapping of QTL controlling endosperm traits with bulked samples. *Genetica* 132, 59–70.

Wang, X.Y., Chen, P.D. and Zhang, S.Z. (2001) Pyramiding and marker-assisted selection for powdery mildew resistance genes in common wheat. *Acta Genetica Sinica* 28, 640–646 (in Chinese; summary in English).

Wang, Y., Chen, B., Hu, Y., Li, J. and Lin, Z. (2005) Inducible excision of selectable marker gene from transgenic plants by the *Cre/lox* site-specific recombination system. *Transgenic Research* 14, 605–614.

Wang, Y.H., Liu, S.J., Ji, S.L., Zhang, W.W., Wang, C.M., Jiang, L. and Wan, J.M. (2005) Fine mapping and marker-assisted selection (MAS) of a low glutelin content gene in rice. *Cell Research* 15, 622–630.

Wang, Z., Zou, Y., Li, X., Zhang, Q., Chen, L., Wu, H., Su, D., Chen, Y., Guo, J., Luo, D., Long, Y., Zhong, Y. and Liu, Y.G. (2006) Cytoplasmic male sterility of rice with Boro II cytoplasm is caused by a cytotoxic peptide and is restored by two related PPR motif genes via distinct modes of mRNA silencing. *The Plant Cell* 18, 676–687.

Ware, D. and Stein, L. (2003) Comparison of genes among cereals. *Current Opinion in Plant Biology* 6, 121–127.

Ware, D.H., Jaiswal, P., Ni, J., Yap, I.V., Pan, X., Clark, K.Y., Teytelman, L., Schmidt, S.C., Zhao, W., Chang, K., Cartinhour, S., Stein, L.D. and McCouch, S.R. (2002) Gramene, a tool for grass genomics. *Plant Physiology* 130, 1606–1613.

Warthmann, N., Chen, H., Ossowski, S., Weigel, D. and Hervé, P. (2008) Highly specific gene silencing by artificial miRNAs in rice. *PLoS ONE* 3(3), e1829.

Wassom, J.J., Wong, J.C., Martinez, E., King, J.J., DeBaene, J., Hotchkiss, J.R., Mikkilineni, V., Bohn, M.O. and Rocheford, T.R. (2008) QTL associated with maize kernel oil, protein and starch concentrations; kernel mass; and grain yield in Illinois High Oil × B73 backcross-derived lines. *Crop Science* 48, 243–252.

Waugh, R., Mclean, K., Flavell, A.J., Pearce, S.R., Kumar, A. and Thomas, B.B.T. (1997) Genetic distribution of Bare-1 like retrotransposable elements in the barley genome revealed by sequence-specific amplification polymorphism (S-SAP). *Molecular and General Genetics* 253, 687–694.

Wayne, M.L. and McIntyre, L.M. (2002) Combining mapping and arraying: an approach to candidate gene identification. *Proceedings of the National Academy of Sciences of the United States of America* 99, 14903–14906.

Weber, A.L., Briggs, W.H., Rucker, J., Baltazar, B.M., Sánchez-Gonzalez, J.D.J., Feng, P., Buckler, E.S. and Doebley, J. (2008) The genetic architecture of complex traits in teosinte (*Zea mays* ssp. *parviglumis*): new evidence from association mapping. *Genetics* 180, 1221–1232.

Weckwerth, W. (2003) Metabolomics in systems biology. *Annual Review of Plant Biology* 54, 669–689.

Weckwerth, W., Wenzel, K. and Fiehn, O. (2004) Process for the integrated extraction, identification and quantification of metabolites, proteins and RNA to reveal their co-regulation in biochemical networks. *Proteomics* 4, 78–83.

Wehrhahn, C. and Allard, R.W. (1965) The detection and measurement of the effects of individual genes involved in the inheritance of a quantitative character in wheat. *Genetics* 51, 109–119.

Weigel, D. and Nordborg, M. (2005) Natural variation in *Arabidopsis*. How do we find the causal genes? *Plant Physiology* 138, 567–568.

Weigel, D., Ahn, J.H., Blazquez, M.A., Borevitz, J.O., Christensen, S.K., Fankhauser, C., Ferrandiz, C., Kardailsky, I., Malancharuvil, E.J., Neff, M.M., Nguyen, J.T., Sato, S., Wang, Z., Xia, Y., Dixon, R.A., Harrison, M.J., Lamb, C.J., Yanofsky, M.F. and Chory, J. (2000) Activation tagging in *Arabidopsis*. *Plant Physiology* 122, 1003–1013.

Weir, B.S. (1990) *Genetic Data Analysis, Methods for Discrete Population Genetic Data*. Sinauer Associates, Inc., Sunderland, Massachusetts, pp. 222–260.

Weir, B.S. (1996) *Genetic Data Analysis II*. Sinauer Associates, Inc., Sunderland, Massachusetts, 376 pp.

Weise, S., Grosse, I., Klukas, C., Koschützki, D., Scholz, U., Schreiber, F. and Junker, B.H. (2006) Meta-All: a system for managing metabolic pathway information. *BMC Bioinformatics* 7, 465.

Welch, R.M. and Graham, R.D. (2004) Breeding for micronutrients in staple food crops from a human nutrition perspective. *Journal of Experimental Botany* 55, 353–364.

Welch, S.M., Dong, Z. and Roe, J.L. (2004) Modeling gene networks controlling transition to flowering in *Arabidopsis*. In: *New Directions for a Diverse Planet: Proceedings 4th International Crop Science Congress* (ICSC), 26 September–1 October 2004, Brisbane, Australia. ICSC, Brisbane, Australia. Available at: http://www.cropscience.org.au/icsc2004/ (accessed 17 November 2009).

Welsh, J. and McClelland, M. (1990) Fingerprinting genomes using PCR with arbitrary primers. *Nucleic Acids Research* 18, 7231–7238.

Wenck, A. and Hansen, G. (2004) Positive selection. In: Peña, L. (ed.) *Methods in Molecular Biology*, Vol. 286. *Transgenic Plants: Methods and Protocols*. Humana Press Inc., Totowa, New Jersey, pp. 227–235.

Wenzel, W.G. and Pretorius, A.J. (2000) Heterosis and xenia in sorghum malt quality. *South-African Journal of Plant Soil* 17, 66–69.

Wenzl, P., Carling, J., Kudrna, D., Jaccoud, D., Huttner, E., Kleinhofs, A. and Kilian, A. (2004) Diversity array technology (DArT) for whole-genome profiling of barley. *Proceedings of the National Academy of Sciences of the United States of America* 101, 9915–9920.

Wenzl, P., Li, H., Carling, J., Zhou, M., Raman, H., Paul, E., Hearnden, P., Maier, C., Xia, L., Caig, V., Ovesná, J., Cakir, M., Poulsen, D., Wang, J., Raman, R., Smith, K.P., Muehlbauer, G.J., Chalmers, K.J., Kleinhofs, A., Huttner, E. and Kilian, A. (2006) A high-density consensus map of barley linking DArT markers to SSR, RFLP and STS loci and agricultural traits. *BMC Genomics* 7, 206.

Werner, K., Friedt, W. and Ordon, F. (2005) Strategies for pyramiding resistance genes against the barley yellow mosaic virus complex (BaMMV, BaYMV, BaYMV-2). *Molecular Breeding* 16, 45–55.

Wesley, S.V., Helliwell, C.A., Smith, N.A., Wang, M.B., Rouse, D.T., Liu, Q., Gooding, P.S., Singh, S.P., Abbott, D., Stoutjesdijk, P.A., Robinson, S.P., Gleave, A.P., Green, A.G. and Waterhouse, P.M. (2001) Construct design for efficient, effective and high-throughput gene silencing in plants. *The Plant Journal* 27, 581–590.

Wheeler, D.L., Barrett, T., Benson, D.A., Bryant, S.H., Canese, K., Chetvernin, V., Church, D.M., DiCuccio, M., Edgar, R., Federhen, S., Feolo, M., Geer, L.Y., Helmberg, W., Kapustin, Y., Khovayko, O., Landsman, D., Lipman, D.J., Madden, T.L., Maglott, V., Miller, D.R., Ostell, J., Pruitt, K.D., Schuler, G.D., Shumway, M., Sequeira, E., Sherry, S.T., Sirotkin, K., Souvorov, A., Starchenko, G., Tatusov, R.L., Tatusova, T.A., Wagner, L. and Yaschenko, E. (2007) Database resources of the National Center for Biotechnology Information. *Nucleic Acids Research* 36, D13–D21.

White, J.W. and Hoogenboom, G. (1996) Integrating effects of genes for physiological traits into crop growth models. *Agronomy Journal* 88, 416–422.

White, J.W. and Hoogenboom, G. (2003) Gene-based approaches to crop simulation: past experiences and future opportunities. *Agronomy Journal* 95, 52–64.

White, P.J. and Broadley, M.R. (2005) Biofortifying crops with essential mineral elements. *Trends in Plant Science* 10, 586–593.

White, P.R. (1934) Potentially unlimited growth of excised tomato root tips in a liquid medium. *Plant Physiology* 9, 585–600.

Whitelaw, C.A., Barbazuk, W.B., Pertea, G., Chan, A.P., Cheung, F., Lee, Y., Zheng, L., van Heeringen, S., Karamycheva, S., Bennetzen, J.L., SanMiguel, P., Lakey, N., Bedell, J., Yuan, Y., Budiman, M.A., Resnick, A., van Aken, S., Utterback, T., Riedmuller, S., Williams, M., Feldblyum, T., Schubert, K., Beachy, R., Fraser, C.M. and Quackenbush, J. (2003) Enrichment of gene-coding sequences in maize by genome filtration. *Science* 302, 2118–2120.

Whitelegge, J.P. (2002) Plant proteomics: BLASTing out of a MudPIT. *Proceedings of the National Academy of Sciences of the United States of America* 99, 11564–11566.

Whitesides, G.M. (2006) The origins and the future of microfluidics. *Nature* 442, 368–373.

Whittaker, J.C., Haley, C.S. and Thompson, R. (1997) Optimal weighting of information in marker-assisted selection. *Genetical Research* 69, 137–144.

Wiemann, S., Weil, B., Wellenreuther, R., Gassenhuber, J., Glassl, S., Ansorge, W., Bocher, M., Blocker, H., Bauersachs, S., Blum, H., Lauber, J., Düsterhöft, A., Beyer, A., Köhrer, K., Strack, N., Mewes, H.-W., Ottenwälder, B., Obermaier, B., Tampe, J., Heubner, D., Wambutt, R., Korn, B., Klein, M. and Poustka, A. (2001) Toward a catalog of human genes and proteins: sequencing and analysis of 500 novel complete protein coding human cDNAs. *Genome Research* 11, 422–435.

Wilkes, G. (1993) Germplasm collections: their use, potential, social responsibility and genetic vulnerability. In: Buxton, D.R., Shibles, R., Forsberg, R.A., Blad, B.L., Asay, K.H., Paulsen, G.M. and Wilson, R.F. (eds) *International Crop Science I*. Crop Science Society of America, Madison, Wisconsin, pp. 445–450.

Wilkinson, M., Schoof, H., Ernst, R. and Haase, D. (2005) BioMOBY successfully integrates distributed heterogeneous bioinformatics web services. The PlaNet Exemplar case. *Plant Physiology* 138, 5–7.

Wilkins-Stevens, P., Hall, J.G., Lyamichev, V., Neri, B.P., Lu, M., Wang, L., Smith, L.M. and Kelso, D.M. (2001) Analysis of single nucleotide polymorphisms with solid phase invasive cleavage reactions. *Nucleic Acids Research* 29, e77.

William, H.M., Morris, M., Warburton, M. and Hiosington, D.A. (2007a) Technical, economic and policy considerations on marker-assisted selection in crops: lessons from the experience at an international agricultural research center. In: Guimarães, E.P., Ruane, J., Scherf, B.D., Sonnino, A. and Dargie, J.D. (eds) *Marker-Assisted Selection, Current Status and Future Perspectives in Crops, Livestock, Forestry and Fish.* Food and Agriculture Organization of the Unites Nations, Rome, pp. 381–404.

William, H.M., Trethowan, R. and Crosby-Galvan, E.M. (2007b) Wheat breeding assisted by markers: CIMMYT's experience. *Euphytica* 157, 307–319.

Williams, C.E. and St Clair, D.A. (1993) Phenetic relationships and levels of variability detected by restriction fragment length polymorphism and random amplified polymorphic DNA analysis of cultivated and wild accessions of *Lycopersicon esculentum. Genom*e 36, 619–630.

Williams, E.J. (1952) The interpretation of interactions in factorial experiments. *Biometrika* 39, 65–81.

Williams, J.G.K., Kubelik, A.R., Livak, K.J., Rafalski, J.A. and Tingey, S.V. (1990) DNA polymorphisms amplified by arbitrary primers are useful as genetic markers. *Nucleic Acids Research* 18, 6531–6535.

Williams, J.S. (1962) The evaluation of a selection index. *Biometrics* 18, 375–393.

Wilson, J.A. (1968) Problems in hybrid wheat breeding. *Euphytica* 17 (Suppl.1), 13–33.

Wilson, L.M., Whitt, S.R., Ibanez, A.M., Rocheford, T.R., Goodman, M.M. and Buckler IV, E.S. (2004) Dissection of maize kernel composition and starch production by candidate gene association. *The Plant Cell* 16, 2719–2733.

Wilson, P. and Driscoll, C.J. (1983) Hybrid wheat. In: Frankel, R. (ed.) *Monographs on Theoretical and Applied Genetics*, Vol. 6. *Heterosis*. Springer-Verlag, Berlin, pp. 94–123.

Wilson, W.A., Harrington, S.E., Woodman, W.L., Lee, M., Sorrells, M.E. and McCouch, S.R. (1999) Inferences on the genome structure of progenitor maize through comparative analysis of rice, maize and the domesticated panicoids. *Genetics* 153, 453–473.

Windsor, A.J. and Mitchell-Olds, T. (2006) Comparative genomics as a tool for gene discovery. *Current Opinion in Biotechnology* 17, 1–7.

Wingbermuehle, W.J., Gustus, C. and Smith, K.P. (2004) Exploiting selective genotyping to study genetic diversity of resistance to *Fusarium* head blight in barley. *Theoretical and Applied Genetics* 109, 1160–1168.

Wink, M. (1988) Plant breeding: importance of plant secondary metabolites for protection against pathogens and herbivores. *Theoretical and Applied Genetics* 75, 225–233.

Winkler, R.G. and Feldman, K.A. (1998) PCR-based identification of T-DNA insertion mutants. *Methods in Molecular Biology* 82, 129–136.

Winzeler, E.A., Richards, D.R., Conway, A.R., Goldstein, A.L., Kalman, S., McCullough, M.J., McCusker, J.H., Stevens, D.A., Wodicka, L., Lockhart, D.J. and Davis, R.W. (1998) Direct allelic variation scanning of the yeast genome. *Science* 281, 1194–1197.

Wishart, D.S., Tzur, D., Knox, C., Eisner, R., Guo, A.C., Young, N., Cheng, D., Jewell, K., Arndt, D., Sawhney, S., Fung, C., Nikolai, L., Lewis, M., Coutouly, M.-A., Forsythe, I., Tang, P., Shrivastava, S., Jeroncic, K., Stothard, P., Amegbey, G., Block, D., Hau, D.D., Wagner, J., Miniaci, J., Clements, M., Gebremedhin, M., Guo, N., Zhang, Y., Duggan, G.E., Macinnis, G.D., Weljie, A.M., Dowlatabadi, R., Bamforth, F., Clive, D., Greiner, R., Li, L., Marrie, T., Sykes, B.D., Vogel, H.J. and Querengesser, L. (2007) HMDB: The Human Metabolome Database. *Nucleic Acids Research* 35(Database issue), D521–526.

Witcombe, J.R. (1996) Participatory approaches to plant breeding and selection. *Biotechnology and Development Monitor* 29, 2–6.

Witcombe, J.R. and Hash, C.T. (2000) Resistance gene deployment strategies in cereal hybrids using marker-assisted selection: gene pyramiding, three-way hybrids and synthetic parent populations. *Euphytica* 112, 175–186.

Withers, L.A. (1993) New technologies for the conservation of plant genetic resources. In: Buxton, D.R., Shibles, R., Forsberg, R.A., Blad, B.L., Asay, K.H., Paulsen, G.M. and Wilson, R.F. (eds) *International Crop Science I*. Crop Science Society of America, Madison, Wisconsin, pp. 429–435.

Withers, L.A. (1995) Collecting *in vitro* for genetic resources conservation. In: Guarino, L., Ramanatha Rao, V. and Reid, R. (eds) *Collecting Plant Genetic Diversity*. CAB International, Wallingford, UK, pp. 511–515.

Wold, B. and Myers, R.M. (2008) Sequence census methods for functional genomics. *Nature Methods* 5, 19–21.

Wolf, Y.I., Rogozin, I.B., Grishin, N.V. and Koonin, E.V. (2003) Genome-scale phylogenetic trees. In: *Frontiers in Computational Genomics*. Caister Academic Press, Wymondham, UK, pp. 241–260.

Wollenweber, B., Porter, J.R. and Lübberstedt, T. (2005) Need for multidisciplinary research towards a second green revolution. Commentary. *Current Opinion in Plant Biology* 8, 337–341.

Wong, D.W.S. (1997) *The ABCs of Gene Cloning*. Chapman & Hall, New York.

Worland, A.J. and Law, C.N. (1986) Genetic analysis of chromosome 2D of wheat. I. The location of genes affecting height, daylength insensitivity, hybrid dwarfism and yellow-rust resistance. *Zeitschrift für Pflanzenzüchtung* 96, 331–345.

Wouters, F.S., Verveer, P.J. and Bastiaens, P.I.H. (2001) Imaging biochemistry inside cells. *Trends in Cell Biology* 11, 203–221.

Wright, A.J. and Mowers, R.P. (1994) Multiple regression for molecular-marker, quantitative trait data from large F2 populations. *Theoretical and Applied Genetics* 89, 305–312.

Wright, S. (1921a) Correlation and causation. *Journal of Agricultural Research* 20, 557–585.

Wright, S. (1921b) Systems of mating I. The biometric relations between parent and offspring. *Genetics* 6, 111–123.

Wright, S. (1978) *Evolution and Genetics of Populations*, Vol. IV. The University of Chicago Press, Chicago, Illinois.

Wright, S.I., Bi, I.V., Schroeder, S.G., Yamasaki, M., Doebley, J.F., McMullen, M.D. and Gaut, B.S. (2005) The effects of artificial selection on the maize genome. *Science* 308, 1310–1314.

Wu, C., Li, X.J., Yuan, W.Y., Chen, G.X., Kilian, A., Li, J., Xu, C., Li, X.H., Zhou, D.-X., Wang, S. and Zhang, Q. (2003) Development of enhancer trap lines for functional analysis of the rice genome. *The Plant Journal* 35, 418–427.

Wu, H., Sparks C., Amoah, B. and Jones, H.D. (2003) Factors influencing successful *Agrobacterium*-mediated genetic transformation of wheat. *Plant Cell Reports* 21, 659–668.

Wu, H., Sparks, C. and Jones, H.D. (2006) Characterization of T-DNA loci and vector backbone sequences in transgenic wheat produced by *Agrobacterium*-mediated transformation. *Molecular Breeding* 18, 195–208.

Wu, L., Nandi, S., Chen, L., Rodriguez, R.L. and Huang, N. (2002) Expression and inheritance of nine transgenes in rice. *Transgenic Research* 11, 533–541.

Wu, M.S., Wang, S.C. and Dai, J.R. (2000) Application of AFLP markers to heterotic grouping of elite maize inbred lines. *Acta Agronomica Sinica* 26, 9–13.

Wu, R., Lou, X.Y., Ma, C.X., Wang, X., Larkins, B.A. and Casella, G. (2002a) An improved genetic model generates high-resolution mapping of QTL for protein quality in maize endosperm. *Proceedings of the National Academy of Sciences of the United States of America* 99, 11281–11286.

Wu, R., Ma, C.-S. and Casella, G. (2002b) Joint linkage and linkage disequilibrium mapping of qualitative trait loci in natural mapping populations. *Genetics* 160, 779–792.

Wu, R., Ma, C.X., Gallo-Meagher, M., Littell, R.C. and Casella, G. (2002c) Statistical methods for dissecting triploid endosperm traits using molecular markers: an autogamous model. *Genetics* 162, 875–892.

Wu, R., Ma, C.-X., Lin, M., Wang, Z. and Casella, G. (2004) Functional mapping of quantitative trait loci underlying growth trajectories using the transform-both-sides of the logistic model. *Biometrics* 60, 729–738.

Wu, R., Ma, C. and Casella, G. (2007) *Statistical Genetics of Quantitative Traits: Linkage, Maps and QTL (Statistics for Biology and Health).* Springer, Berlin.

Wu, R.L. and Lin, M. (2006) Functional mapping: how to map and study the genetic architecture of dynamic complex traits. *Nature Reviews Genetics* 7, 229–237.

Wu, R.L. and Zeng, Z.B. (2001) Joint linkage and linkage disequilibrium mapping in natural populations. *Genetics* 157, 899–909.

Wu, W., Zhou, Y., Li, W., Mao, D. and Chen, Q. (2002) Mapping of quantitative trait loci based on growth models. *Theoretical and Applied Genetics* 105, 1043–1049.

Wu, W.R. and Li, W.M. (1994) A new approach for mapping quantitative trait loci using complete genetic marker linkage maps. *Theoretical and Applied Genetics* 89, 535–539.

Wu, W.R. and Li, W.M. (1996) Model fitting and model testing in the method of joint mapping of quantitative trait loci. *Theoretical and Applied Genetics* 92, 477–482.

Wu, W.-R., Li, W.-M., Tang, D.-Z., Lu, H.-R. and Worland, A.J. (1999) Time-related mapping of quantitative trait loci underlying tiller number in rice. *Genetics* 151, 297–303.

Xi, Z.Y., He, F.H., Zeng, R.Z., Zhang, Z.M., Ding, X.H., Li, W.T. and Zhang, G.Q. (2006) Development of a wide population of chromosome single segment substitution lines in the genetic background of an elite cultivar of rice (*Oryza sativa* L.). *Genome* 49, 476–484.

Xia, L., Peng, K., Yang, S., Wenzl, P., de Vincente, M.C., Fregene, M. and Kilian, A. (2005) DArT for high-throughput genotyping of cassava (*Manihot esculenta*) and its wild relatives. *Theoretical and Applied Genetics* 110, 1092–1098.

Xia, X.C., Reif, J.C., Melchinger, A.E., Frisch, M., Hoisington, D.A., Beck, D., Pixley, K. and Warburton, M.L. (2005) Genetic diversity among CIMMYT maize inbred lines investigated with SSR markers: II. Subtropical, tropical mid-altitude and highland maize inbred lines and their relationships with elite U.S. and European maize. *Crop Science* 45, 2573–2582.

Xiang, C., Han, P., Lutziger, I., Wang, K. and Oliver, D.J. (1999) A mini binary vector series for plant transformation. *Plant Molecular Biology* 40, 711–717.

Xiao, J., Li, J., Yuan, L. and Tanksley, S.D. (1995) Dominance is the major genetic basis of heterosis in rice as revealed by QTL analysis using molecular markers. *Genetics* 140, 745–754.

Xiao, J., Grandillo, S., Ahn, S.N., McCouch, S.R., Tanksley, S.D., Li, J. and Yuan, L. (1996a) Genes from wild rice improve yield. *Nature* 384, 223–224.

Xiao, J., Li, J., Yuan, L., McCouch, S.R. and Tanksley, S.D. (1996b) Genetic diversity and its relationship to hybrid performance and heterosis in rice as revealed by PCR-based markers. *Theoretical and Applied Genetics* 92, 637–643.

Xiao, J., Li, J., Yuan, L. and Tanksley, S.D. (1996c) Identification of QTLs affecting traits of agronomic importance in a recombinant inbred population derived from subspecific rice cross. *Theoretical and Applied Genetics* 92, 230–244.

Xiao, J., Li, L., Grandillo, S., Yuan, L., Tanksley, S.D. and McCouch, S.R. (1998) Identification of trait-improving quantitative trait loci alleles from a wild rice relative, *Oryza rufipogon. Genetics* 150, 899–909.

Xiong, Q., Qiu, Y. and Gu, W. (2008) PGMapper: a web-based tool linking phenotype to genes. *Bioinformatics* 24, 1011–1013.

Xu, C., He, X. and Xu, S. (2003) Mapping quantitative trait loci underlying triploid endosperm traits. *Heredity* 90, 228–235.

Xu, S. (1996) Mapping quantitative trait loci using four-way crosses. *Genetical Research* 68, 175–181.

Xu, S. (1998) Mapping quantitative trait loci using multiple families of line crosses. *Genetics* 148, 517–524.

Xu, S. (2002) QTL analysis in plants. In: Camp, N.J. and Cox, A. (eds) *Methods in Molecular Biology,* Vol. 195. *Quantitative Trait Loci: Methods and Protocols.* Humana Press, Totowa, New Jersey, pp. 283–310.

Xu, S. (2003) Estimating polygenic effects using markers of the entire genome. *Genetics* 163, 789–801.

Xu, S. (2007) An empirical Bayes method for estimating epistatic effects of quantitative trait-loci. *Biometrics* 63, 513–521.

Xu, S. and Jia, Z. (2007) Genome-wide analysis of epistatic effects for quantitative traits in barley. *Genetics* 175, 1955–1963.

Xu, Y. (1994) Application of molecular markers in genetic improvement of quantitative traits in plants. In: *Proceedings of the Third Young Scientists Symposium on Crop Genetics and Breeding.* Publishing House of Agricultural Science and Technology of China, Beijing, pp. 38–49.

Xu, Y. (1997) Quantitative trait loci: separating, pyramiding and cloning. *Plant Breeding Reviews* 15, 85–139.

Xu, Y. (2002) Global view of QTL: rice as a model. In: Kang, M.S. (ed.) *Quantitative Genetics, Genomics and Plant Breeding.* CAB International, Wallingford, UK, pp. 109–134.

Xu, Y. (2003) Developing marker-assisted selection strategies for breeding hybrid rice. *Plant Breeding Reviews* 23, 73–174.

Xu, Y. and Crouch, J.H. (2008) Marker-assisted selection in plant breeding: from publications to practice. *Crop Science* 48, 391–407.

Xu, Y. and Luo, L. (2002) Biotechnology and germplasm resource management in rice. In: Luo, L., Ying, C. and Tang, S. (eds) *Rice Germplasm Resources.* Hubei Science and Technology Publisher, Wuhan, China, pp. 229–250.

Xu, Y.B. and Shen, Z.T. (1991) Diallel analysis of tiller number at different growth stages in rice (*Oryza sativa* L.). *Theoretical and Applied Genetics* 83, 243–249.

Xu, Y. and Shen, Z. (1992a) Detection and genetic analyses of the gene dispersed crosses: some theoretical considerations. *Acta Agricultura Zhejiangensis* 18, 109–117 (in English with Chinese abstract).

Xu, Y. and Shen, Z. (1992b) Detection and genetic analyses of the gene dispersed cross for tiller angle in rice (*Oryza sativa* L.). *Acta Agricultura Zhejiangensis* 4, 54–60.

Xu, Y. and Shen, Z. (1992c) Accumulation of the alleles with similar effects at four loci controlling tiller angle from gene dispersed crosses in rice (*Oryza sativa* L.). *Journal of Biomathematics* (Beijing) 7, 1–10.

Xu, Y.B. and Shen, Z.T. (1992d) Distorted segregation of waxy gene and its characterization in *indica–japonica* hybrids. *Chinese Journal of Rice Science* 6, 89–92 (in Chinese).

Xu, Y. and Zhu, L. (1994) *Molecular Quantitative Genetics* (in Chinese). China Agriculture Press, Beijing, China, 291 pp.

Xu, Y., Shen, Z., Chen, Y. and Zhu, L. (1995) A statistical technique and generalized computer software for interval mapping of quantitative trait loci and its application. *Acta Agronomica Sinica* 21, 1–8 (in Chinese with English abstract).

Xu, Y., Zhu, L., Xiao, J., Huang, N. and McCouch, S.R. (1997) Chromosomal regions associated with segregation distortion of molecular markers in F_2, backcross, doubled haploid and recombinant inbred populations of rice (*Oryza sativa* L.). *Molecular and General Genetics* 253, 535–545.

Xu, Y., McCouch, S.R. and Shen, Z. (1998) Transgressive segregation of tiller angle in rice caused by complementary action of genes. *Crop Science* 38, 12–19.

Xu, Y., Lobos, K.B. and Clare, K.M. (2002) Development of SSR markers for rice molecular breeding. In: *Proceedings of Twenty-Ninth Rice Technical Working Group Meeting,* 24–27 February 2002, Little Rock, Arkansas. Rice Technical Working Group, Little Rock, Arkansas, p. 49.

Xu, Y., Ishii, T. and McCouch, S.R. (2003) Marker-assisted evaluation of germplasm resources for plant breeding. In: Mew, T.W., Brar, D.S., Peng, S. and Hardy, B. (eds) *Rice Science: Innovations and Impact for Livelihood. Proceedings of the 24th International Rice Research Conference,* 16–19 September 2002, Beijing. International Rice Research Institute, Chinese Academy of Engineering and Chinese Academy of Agricultural Sciences, Beijing, pp. 213–229.

Xu, Y., Beachell, H. and McCouch, S.R. (2004) A marker-based approach to broadening the genetic base of rice (*Oryza sativa* L.) in the US. *Crop Science* 44, 1947–1959.

Xu, Y., McCouch, S.R. and Zhang, Q. (2005) How can we use genomics to improve cereals with rice as a reference genome? *Plant Molecular Biology* 59, 7–26.

Xu, Y., Wang, J. and Crouch, J.C. (2008) Selective genotyping and pooled DNA analysis: an innovative use of an old concept. In: *Proceedings of the 5th International Crop Science Congress,* 13–18 April 2008, Jeju, Korea. Published on CD-ROM. Available at: http://www.cropscience2008.com (accessed 30 June 2008).

Xu, Y., Babu, R., Skinner D.J., Vivek, B.S. and Crouch, J.H. (2009a) Maize mutant Opaque2 and the improvement of protein quality through conventional and molecular approaches. In: Shu, Q.Y. (ed.) *Induced Plant Mutations in the Genomics Era*. Food and Agriculture Organization of the United Nations, Rome, pp. 191–196.

Xu, Y., Lu, Y., Yan, J., Babu, R., Hao, Z., Gao, S., Zhang, S., Li, J., Vivek, B.S., Magorokosho, C., Mugo, S., Makumbi, D., Taba, S., Palacios, N., Guimarães, C.T., Araus, J.-L., Wang, J., Davenport, G.F., Crossa, J. and Crouch, J.H. (2009b) SNP-chip based genomewide scan for germplasm evaluation and marker–trait association analysis and development of a molecular breeding platform. *Proceedings of 14th Australasian Plant Breeding & 11th Society for the Advancement in Breeding Research in Asia & Oceania Conference*, 10–14 August 2009, Cairns, Tropical North Queensland, Australia. Distributed by CD-ROM.

Xu, Y., Skinner, D.J., Wu, H., Palacios-Rojas, N., Araus, J.L., Yan, J., Gao, S., Warburton, M.L. and Crouch, J.H. (2009c) Advances in maize genomics and their value for enhancing genetic gains from breeding. *International Journal of Plant Genomics* Volume 2009, Article ID 957602, 30 pages. Available at: http://www.hindawi.com/journals/ijpg/2009/957602.html (accessed 21 December 2009).

Xu, Y., This, D., Pausch, R.C., Vonhof, W.M., Coburn, J.R., Comstock, J.P., McCouch, S.R. (2009d) Water use efficiency determined by carbon isotope discrimination in rice: genetic variation associated with population structure and QTL mapping. *Theoretical and Applied Genetics* 118, 1065–1081.

Xue, W., Xing, Y., Weng, X., Zhao, Y., Tang, W., Wang, L., Zhou, H., Yu, S., Xu, C., Li, X. and Zhang, Q. (2008) Natural variation in *Gdh7* is an important regulator of heading date and yield potential in rice. *Nature Genetics* 40, 761–767.

Xue, Y. and Xu, Z. (2002) An introduction to the China Rice Functional Genomics Program. *Comparative and Functional Genomics* 3, 161–163.

Yadav, N.S., Vanderleyden, J., Bennett, D.R., Barnes, W.M. and Chilton, M.D. (1982) Short direct repeats flank the T-DNA on a nopaline Ti plasmid. *Proceedings of the National Academy of Sciences of the United States of America* 79, 6322–6326.

Yadav, R.S., Hash, C.T., Bidinger, F.R., Cavan, G.P. and Howarth, C.J. (2002) Quantitative trait loci associated with traits determining grain and stover yield in pearlmillet under terminal drought stress conditions. *Theoretical and Applied Genetics* 104, 67–83.

Yamada, K., Lim, J., Dale, J.M., Chen, H., Shinn, P., Palm, C.J., Southwick, A.M., Wu, H.C., Kim, C., Nguyen, M., Pham, P., Cheuk, R., Karlin-Newmann, G., Liu, S.X., Lam, B., Sakano, H., Wu, T., Yu, G., Miranda, M., Quach, H.L., Tripp, M., Chang, C.H., Lee, J.M., Toriumi, M., Chan, M.M.H., Tang, C.C., Onodera, C.S., Deng, J.M., Akiyama, K., Ansari, Y., Arakawa, T., Banh, J., Banno, F., Bowser, L., Brooks, S., Carninci, P., Chao, Q., Choy, N., Enju, A., Goldsmith, A.D., Gurjal, M., Hansen, N.F., Hayashizaki, Y., Johnson-Hopson, C., Hsuan, V.W., Iida, K., Karnes, M., Khan, S., Koesema, E., Ishida, J., Jiang, P.X., Jones, T., Kawai, J., Kamiya, A., Meyers, C., Nakajima, M., Narusaka, M., Seki, M., Sakurai, T., Satou, M., Tamse, R., Vaysberg, M., Wallender, E.K., Wong, C., Yamamura, Y., Yuan, S., Shinozaki, K., Davis, R.W., Athanasios Theologis, A. and Ecker, J.R. (2003) Empirical analysis of transcriptional activity in the *Arabidopsis* genome. *Science* 302, 842–846.

Yamagishi, M., Yano, M., Fukuta, Y., Fukui, K., Otani, M. and Shimada, T. (1996) Distorted segregation of RFLP markers in regenerated plants derived from anther culture of an F_1 hybrid of rice. *Genes & Genetic Systems* 71, 37–41.

Yamamoto, T., Takemori, N., Sue, N. and Nitta, N. (2003) QTL analysis of stigma exsertion in rice. *Rice Genetics Newsletter* 20, 33–34.

Yamazaki, M., Tsugawa, H., Miyao, A., Yano, M., Wu, J., Yamamoto, S., Matsumoto, T., Sasaki, T. and Hirochika, H. (2001) The rice retrotransposon Tos17 prefers low-copy-number sequences as integration targets. *Molecular and General Genetics* 265, 336–344.

Yan, J., Zhu, J., He, C., Benmoussa, M. and Wu, P. (1998a) Molecular dissection of developmental behavior of plant height in rice (*Oryza sativa* L.). *Genetics* 150, 1257–1265.

Yan, J., Zhu, J., He, C., Benmoussa, M. and Wu, P. (1998b) Quantitative trait loci analysis for the developmental behavior of tiller number in rice (*Oryza sativa* L.). *Theoretical and Applied Genetics* 97, 267–274.

Yan, J., Zhu, J., He, C., Benmoussa, M. and Wu, P. (1999) Molecular marker-assisted dissection of genotype × environment interaction for plant type traits in rice (*Oryza sativa* L.). *Crop Science* 39, 538–544.

Yan, J., Yang, X., Shah, T., Sánchez-Villeda, H., Li, J., Warburton, M., Zhou, Y., Crouch, J.H. and Xu, Y. (2009) High-throughput SNP genotyping with the GoldenGate assay in maize. *Molecular Breeding* (in press).

Yan, W. and Kang, M.S. (2003) *GGE Biplot Analysis: a Graphical Tool for Breeders, Geneticists and Agronomists*. CRC Press, Boca Raton, Florida.

Yan, W. and Rajcan, I. (2002) Biplot evaluation of test sides and trait relations of soybean in Ontario. *Crop Science* 42, 11–20.

Yan, W. and Tinker, N.A. (2006) Biplot analysis of multi-environment trial data: principles and applications. *Canadian Journal of Plant Science* 86, 623–645.

Yan, W., Hunt, L.A., Sheng, Q. and Szlavnies, Z. (2000) Cultivar evolution and mega-environment investigation based on GGE biplot. *Crop Science* 40, 596–605.

Yan, W., Rutger, J.N., Bockelman, H.E. and Tai, T. (2004) Development of a core collection from the USDA rice germplasm collection. In: Norman, R.J., Meullenet, J.-F. and Moldenhauer, K.A.K. (eds) *B. R. Wells Rice Research Studies 2003*. Arkansas Agricultural Expteriment Research Station Series No. 517, pp. 88–96. Available at: http://www.uark.edu/depts/agripub/publications/research (accessed 31 December 2007).

Yan, W., Kang, M.S., Ma, B., Woods, S. and Cornelius, P.L. (2007) GGE biplot vs. AMMI analysis of genotype-by-environment data. *Crop Science* 47, 643–655.

Yang, H., You, A., Yang, Z., Zhang, F., He, R., Zhu, L. and He, G. (2004) High-resolution genetic mapping at the *Bph5* locus for brown planthopper resistance in rice (*Oryza sativa* L.). *Theoretical and Applied Genetics* 110, 182–191.

Yang, H.-C., Liang, Y.-J., Huang, M.-C., Li, L.-H., Lin, C.H., Wu, J.-Y., Chen, Y.-T. and Fann, C.S.J. (2006a) A genome-wide study of preferential amplification/hybridization in microarray-based pooled DNA experiments. *Nucleic Acids Research* 34, e106.

Yang, H.-C., Pan, C.-C., Lin, C.-Y. and Fann, C.S.J. (2006b) PDA: pooled DNA analyzer. *BMC Bioinformatics* 7, 233.

Yang, J., Hu, C., Hu, H., Yu, R., Xia, Z., Ye, X. and Zhu, J. (2008) QTLNetwork: mapping and visualizing genetic architecture of complex traits in experimental populations. *Bioinformatics* 24, 721–723.

Yang, R. and Xu, S. (2007) Bayesian shrinkage analysis of quantitative trait loci for dynamic traits. *Genetics* 176, 1169–1185.

Yang, R.-C. (2004) Epistasis of quantitative trait loci under different gene action models. *Genetics* 167, 1493–1505.

Yang, R.-C. (2007) Mixed model analysis of crossover genotype-environment interactions. *Crop Science* 47, 1051–1062.

Yang, R.Q., Tan, Q. and Xu, S.Z. (2006) Mapping quantitative trait loci for longitudinal traits in line crosses. *Genetics* 173, 2339–2356.

Yang, X., Rupe, M., Bickel, D., Arthur, L., Smith, O. and Guo, M. (2006) Effects of *cis–trans*-regulation on allele-specific transcript expression in the meristems of maize hybrids. In: *48th Annual Maize Genetic Conference*, 9–12 March 2006, Pacific Grove, California, 132 pp.

Yang, X.R., Wang, J.R., Li, H.L. and Li, Y.F. (1983) Studies on the general medium for anther culture of cereals and increasing of the frequency of green pollen-plantlets-induction of *Oryza sativa* subsp. *hseni*. In: Shen, J.H., Zhang, Z.H. and Shi, S.D. (eds) *Studies on Anther-Cultured Breeding in Rice*. Agriculture Press, Beijing, pp. 61–69.

Yano, M., Harushima, Y., Nagamura, Y., Kurata, N., Minobe, Y. and Sasaki, T. (1997) Identification of quantitative trait loci controlling heading date in rice using a high-density linkage map. *Theoretical and Applied Genetics* 95, 1025–1032.

Yano, M., Katayose, Y., Ashikari, M., Yamanouchi, U., Monna, L., Fuse, T., Baba, T., Yamamoto, K., Umehara, Y., Nagamura, Y. and Sasaki, T. (2000) *Hd1*, a major photoperiod sensitivity quantitative trait locus in rice, is closely related to the *Arabidopsis* flowering time gene *CONSTANS*. *The Plant Cell* 12, 2473–2484.

Yano, M., Kojima, S., Takahashi, Y., Lin, H.X. and Sasaki, T. (2001) Genetic control of flowering time in rice, as short-day plant. *Plant Physiology* 127, 1425–1429.

Yao, Y., Ni, Z., Zhang, Y., Chen, Y., Ding, Y., Han, Z., Liu, Z. and Sun, Q. (2005) Identification of differentially expressed genes in leaf and root between wheat hybrid and its parental inbreds using PCR-based cDNA subtraction. *Plant Molecular Biology* 58, 367–384.

Yates, F. and Cochran, W.G. (1938) The analysis of groups of experiments. *Journal of Agricultural Science* 28, 556–580.

Ye, X., Al-Babili, S., Klöti, A., Zhang, J., Lucca, P., Beyer, P. and Potrykus, I. (2000) Engineering the provitamin A (β-carotene) biosynthetic pathway into (carotenoid-free) rice endosperm. *Science* 287, 303–305.

Yi, N. (2004) A unified Markov chain Monte Carlo framework for mapping multiple quantitative trait loci. *Genetics* 167, 967–975.

Yi, N. and Shriner, D. (2008) Advances in Bayesian multiple quantitative trait loci mapping in experimental crosses. *Heredity* 100, 240–252.

Yi, N. and Xu, S. (2002) Mapping quantitative trait loci with epistatic effects. *Genetical Research* 79, 185–198.

Yi, N., George, V. and Allison, D.B. (2003) Stochastic search variable selection for identifying multiple quantitative trait loci. *Genetics* 164, 1129–1138.

Yi, N., Yandell, B.S., Churchill, G.A., Allison, D.B., Eisen, E.J. and Pomp, D. (2005) Bayesian model selection for genome-wide epistatic quantitative trait loci analysis. *Genetics* 70, 1333–1344.

Yi, N., Zinniel, D.K., Kim, K., Eisen, E.J., Bartolucci, A., Allison, D.B. and Pomp, D. (2006) Bayesian analyses of multiple epistasis QTL models for body weight and body composition in mice. *Genetical Research* 87, 45–60.

Yi, N., Banerjee, S., Pomp, D. and Yandell, B.S. (2007) Bayesian mapping of genomewide interacting quantitative trait loci for ordinal traits. *Genetics* 176, 1855–1864.

Yin, T.M., DiFazio, S.P., Gunter, L.E., Riemenschneider, D. and Tuskan, G.A. (2004) Large-scale heterospecific segregation distortion in *Populus* revealed by a dense genetic map. *Theoretical and Applied Genetics* 109, 451–463.

Yin, X., Kropff, M.J. and Stam, P. (1999) The role of ecophysiological models in QTL analysis: the example of specific leaf area in barley. *Heredity* 82, 415–421.

Yin, X., Stam, P., Kropff, M.J. and Schapendonk, A.H.C.M. (2003) Crop modeling, QTL mapping and their complementary role in plant breeding. *Agronomy Journal* 95, 90–98.

Yin, X., Struik, P.C. and Kropff, M.J. (2004) Role of crop physiology in predicting gene-to-phenotype relationships. *Trends in Plant Science* 9, 426–432.

Yin, X., Struik, P.C., Tang, J., Qi, C. and Liu, T. (2005) Model analysis of flowering phenology in recombinant inbred lines of barley. *Journal of Experimental Botany* 56, 959–965.

Yoo, B.H. (1980) Long-term selection for a quantitative character in large replicate populations of *Drosophila melanogaster*. I. Response to selection. *Genetical Research* 35, 1–17.

Yoon, D.-B., Kang, K.-H., Kim, H.-J., Ju, H.-G., Kwon, S.-J., Suh, J.-P., Jeong, O.-Y. and Ahu, S.-N. (2006) Mapping quantitative trait loci for yield components and morphological traits in an advanced backcross population between *Oryza grandiglumis* and the *O. japonica* cultivar Hwaseongbyeo. *Theoretical and Applied Genetics* 112, 1052–1062.

Young, N.D. (1999) A cautiously optimistic vision for marker assisted breeding. *Molecular Breeding* 5, 505– 510.

Young, N.D. and Tanksley, S.D. (1989a) Restriction fragment length polymorphism maps and the concept of graphical genotypes. *Theoretical and Applied Genetics* 77, 95–101.

Young, N.D. and Tanksley, S.D. (1989b) RFLP analysis of the size of chromosomal segments retained around the *Tm-2* locus of tomato during backcross breeding. *Theoretical and Applied Genetics* 77, 353–359.

Young, N.D., Zamir, D., Ganal, M. and Tanksley, S.D. (1988) Use of isogenic lines and simultaneous probing to identify DNA markers tightly linked to the *Tm-2a* gene in tomato. *Genetics* 120, 579–585.

Yousef, G.G. and Juvik, J.A. (2001a) Comparison of phenotypic and marker-assisted selection for quantitative traits in sweet corn. *Crop Science* 41, 645–655.

Yousef, G.G. and Juvik, J.A. (2001b) Evaluation of breeding utility of a chromosomal segment from *Lycopersicon chmielewskii* that enhances cultivated tomato soluble solids. *Theoretical and Applied Genetics* 103, 1022–1027.

Yu, G.-X. and Wise, R.P. (2000) An anchored AFLP- and retrotransposon-based map of diploid *Avena*. *Genome* 43, 736–749.

Yu, J., Hu, S., Wang, J., Wong, G.K.S., Li, S., Liu, B., Deng, Y., Dai, L., Zhou, Y., Zhang, X., Cao, M., Liu, J., Sun, J., Tang, J., Chen, Y., Huang, X., Lin, W., Ye, C., Tong, W., Cong, L., Geng, J., Han, Y., Li, L., Li, W., Hu, G., Huang, X., Li, W., Li, J., Liu, Z., Li, L., Liu, J., Qi, Q., Liu, J., Li, L., Li, T., Wang, X., Lu, H., Wu, T., Zhu, M., Ni, P., Han, H., Dong, W., Ren, X., Feng, X., Cui, P., Li, X., Wang, H., Xu, X., Zhai, W., Xu, Z., Zhang, J., He, S., Zhang, J., Xu, J., Zhang, K., Zheng, X., Dong, J., Zeng, W., Tao, L., Ye, J., Tan, J., Ren, X., Chen, X., He, J., Liu, D., Tian, W., Tian, C., Xia, H., Bao, Q., Li, G., Gao, H., Cao, T., Wang, J., Zhao, W., Li, P., Chen, W., Wang, X., Zhang, Y., Hu, J., Wang, J., Liu, S., Yang, J., Zhang, G., Xiong, Y., Li, Z., Mao, L., Zhou, C., Zhu, Z., Chen, R., Hao, B., Zheng, W., Chen, S., Guo, W., Li, G., Liu, S., Tao, M., Wang, J., Zhu, L., Yuan, L. and Yang, H. (2002) A draft sequence of the rice genome (*Oryza sativa* L. ssp. *indica*). *Science* 296, 79–92.

Yu, J., Arbelbide, M. and Bernardo, R. (2005a) Power of *in silico* QTL mapping from phenotypic, pedigree and marker data in a hybrid breeding program. *Theoretical and Applied Genetics* 110, 1061–1067.

Yu, J., Wang, J., Lin, W., Li, S., Li, H., Zhou, J., Ni, P., Dong, W., Hu, S., Zeng, C., Zhang, J., Zhang, Y., Li, R., Xu, Z., Li, S., Li, X., Zheng, H., Cong, L., Lin, L., Yin, J., Geng, J., Li, G., Shi, J., Liu, J., Lv, H., Li, J., Wang, J., Deng, Y., Ran, L., Shi, X., Wang, X., Wu, Q., Li, C., Ren, X., Wang, J., Wang, X., Li, D., Liu, D., Zhang, X., Ji, Z., Zhao, W., Sun , Y., Zhang, Z., Bao, J., Han, Y., Dong, L., Ji, J., Chen, P., Wu, S., Liu, J., Xiao, Y., Bu, D., Tan, J., Yang, L., Ye, C., Zhang, J., Xu, J., Zhou, Y., Yu, Y., Zhang, B., Zhuang, S., Wei, H., Liu, B., Lei, M., Yu, H., Li, Y., Xu, H., Wei, S., He, X., Fang, L., Zhang, Z., Zhang, Y., Huang, X., Su, Z., Tong, W., Li, J., Tong, Z., Li, S., Ye, J., Wang, L., Fang, L., Lei, T., Chen, C., Chen, H., Xu, Z., Li, H., Huang, H., Zhang, F., Xu, H., Li, N., Zhao, C., Li, S., Dong, L., Huang, Y., Li, L., Xi, Y., Qi, Q., Li, W., Zhang, B., Hu, W., Zhang, Y., Tian, X., Jiao, Y., Liang, X., Jin, J., Gao, L., Zheng, W., Hao, B., Liu, S., Wang, W., Yuan, L., Cao, M., McDermott, J., Samudrala, R., Wang, J., Wong, G.K.-S. and Yang, H. (2005b) The genome of *Oryza sativa*: a history of duplications. *PLoS Biology* 3, E38.

Yu, J., Pressoir, G., Briggs, W., Bi, I.V., Yamasaki, M., Doebley, J.F., McMullen, M.D., Gaut, B.S., Nielsen, D.M., Holland, J.B., Kresovich, S. and Buckler, E.S. (2006) A unified mixed-model method for association mapping that accounts for multiple levels of relatedness. *Nature Genetics* 38, 203–207.

Yu, J., Hollan, J.B., McMullen, M.D. and Buckler, E.S. (2008) Genetic design and statistical power of nested association mapping in maize. *Genetics* 178, 539–551.

Yu, J., Zhang, Z., Zhu, C., Tabanao, D.A., Pressoir, G., Tuinstra, M.R., Kresovich, S., Todhunter, R.J. and Buckler, E.S. (2009) Simulation appraisal of the adequacy of number of background markers for relationship estimation in association mapping. *The Plant Genome* 2, 63–77.

Yu, J.K., La Rota, M., Kantety, R.V. and Sorrells, M.E. (2004) EST-derived SSR markers for comparative mapping in wheat and rice. *Molecular Genetics and Genomics* 271, 742–751.

Yu, K., Park, S.J. and Poysa, V. (2000) Marker-assisted selection of common beans for resistance to common bacterial blight: efficacy and economics. *Plant Breeding* 119, 411–415.

Yu, S.B., Li, J.X., Xu, C.G., Tan, Y.F., Gao, Y.J., Li, X.H., Zhang, Q.F. and Saghai Maroof, M.A. (1997) Importance of epistasis as the genetic basis of heterosis in an elite rice hybrid. *Proceedings of the National Academy of Sciences of the United States of America* 94, 9226–9231.

Yu, W., Andersson, B., Worley, K.C., Muzny, D.M., Ding, Y., Liu, W., Ricafrente, J.Y., Wentland, M.A., Lennon, G. and Gibbs, R.A. (1997) Large-scale concatenation cDNA sequencing. *Genome Research* 7, 353–358.

Yu, W., Han, F., Gao, Z., Vega, J.M. and Birchler, J. (2007) Construction and behavior of engineered minichromosomes in maize. *Proceedings of the National Academy of Sciences of the United States of America* 104, 8924–8929.

Yuan, L.P. (1992) Development and prospects of hybrid rice breeding. In: You, C.B. and Chen, Z.L. (eds) *Agricultural Biotechnology. Proceedings of the Asian Pacific Conference on Agricultural Biotechnology.* China Agricultural Press, Beijing, pp. 97–105.

Yuan, L.P. (2002) Future outlook on hybrid rice research and development. In: *Abstracts of the Fourth International Symposium on Hybrid Rice*, 14–17 May 2002, Hanoi, Vietnam. International Rice Research Institute (IRRI), Manila, Philippines, p.3.

Yuan, L.P. and Chen, H.X. (eds) (1988) *Breeding and Cultivation of Hybrid Rice.* Hunan Science and Technology Press, Changsha, China.

Yuan, Y., SanMiguel, P.J. and Bennetzen, J.L. (2003) High Cot sequence analysis of the maize genome. *The Plant Journal* 34, 249–255 (erratum: *The Plant Journal* 36, 430).

Zabeau, M. and Voss, P. (1993) Selective restriction fragment amplification: a general method for DNA fingerprinting. European Patent Application. 92402629.7 (Publ. Number 0 534 858 A1).

Zale, J.M., Clancy, J.A., Ullrich, S.E., Jones, B.L., Hays, P.M. and the North American Barley Genome Mapping Project (2000) Summary of barley malting QTL mapped in various mapping populations. *Barley Genetics Newsletter* 30, 1–4.

Zamir, D. (2001) Improving plant breeding with exotic genetic libraries. *Nature Reviews Genetics* 2, 983–989.

Zeng, R., Zhang, Z. and Zhang, G. (2000) Identification of multiple alleles at the *Wx* locus in rice using microsatellite class and G–T polymorphism. In: Liu, X. (ed.) *Theory and Application of Crop Research.* China Science and Technology Press, Beijing, pp. 202–205.

Zeng, Z.-B. (1993) Theoretical basis of separation of multiple linked gene effects on mapping quantitative trait loci. *Proceedings of the National Academy of Sciences of the United States of America* 90, 10972–10976.

Zeng, Z.-B. (1994) Precision mapping of quantitative trait loci. *Genetics* 136, 1457–1468.

Zeng, Z.-B. (1998) Mapping quantitative trait loci: interval mapping, composite interval mapping and multiple interval mapping. *Summer Institute for Statistical Genetics*, Module 7. Department of Statistics, North Carolina State University, Raleigh, North Carolina.

Zenkteler, M. and Nitzsche, W. (1984) Wide hybridization experiments in cereals. *Theoretical and Applied Genetics* 68, 311–315.

Zhang, H.B. and Wing, R.A. (1997) Physical mapping of the rice genome with BACs. *Plant Molecular Biology* 35, 115–127.

Zhang, J., Chandra Babu, R., Pantuwan, G., Kamoshita, A., Blum, A., Wade, L., Sarkarung, S., O'Toole, J.C. and Nguyen, N.T. (1999) Molecular dissection of drought tolerance in rice: from physio-morphological traits to field performance. In: Ito, O., O'Toole, J. and Hardy, B. (eds) *Genetic Improvement of Rice for Water-limited Environments*. International Rice Research Institute (IRRI), Manila, Philippines, pp. 331–343.

Zhang, J., Zheng, H.G., Aarti, A., Pantuwan, G., Nguyen, T.T., Tripathi, J.N., Sarial, A.K., Robin, S., Babu, R.C., Nguyen, B.D., Sarkarung, S., Blum, A. and Nguyen, H.T. (2001) Locating genomic regions associated with components of drought resistance in rice: comparative mapping within and across species. *Theoretical and Applied Genetics* 103, 19–29.

Zhang, J., Xu, Y., Wu, X. and Zhu, L. (2002) A bentazon and sulfonylurea sensitive mutant: breeding, genetics and potential application in seed production of hybrid rice. *Theoretical and Applied Genetics* 105, 16–22.

Zhang, J., Li, X., Jiang, G., Xu, Y. and He, Y. (2006) Pyramiding of *Xa7* and *Xa21* for the improvement of disease resistance to bacterial blight in hybrid rice. *Plant Breeding* 125, 600–605.

Zhang, J.F. and Stewart, J.McD. (2004) Semigamy gene is associated with chlorophyll reduction in cotton. *Crop Science* 44, 2054–2062.

Zhang, L.P., Lin, G.Y., Niño-Liu, D. and Foolad, M.R. (2003) Mapping QTLs conferring early blight (*Alternaria solani*) resistance in a *Lycopersicon esculentum* × *L. hirsutum* cross by selective genotyping. *Molecular Breeding* 12, 3–19.

Zhang, N., Xu, Y., Akash, M., McCouch, S. and Oard, J.H. (2005) Identification of candidate markers associated with agronomic traits in rice using discriminant analysis. *Theoretical and Applied Genetics* 110, 727–729.

Zhang, Q. (2007) Strategies for developing green super rice. *Proceedings of the National Academy of Sciences of the United States of America* 104, 16402–16409.

Zhang, Q. and Huang, N. (1998) Mapping and molecular marker-based genetic analysis for efficient hybrid rice breeding. In: Virmani, S.S., Siddiq, E.A. and Muralidharan, K. (eds) *Advances in Hybrid Rice Technology. Proceedings of the Third International Symposium on Hybrid Rice*, 14–16 November 1996, Hyderabad, India. International Rice Research Institute (IRRI), Manila, Philippines, pp. 243–256.

Zhang, Q., Gao, Y.J., Yang, S.H., Ragab, R.A., Saghai Maroof, M.A. and Li, Z.B. (1994) A diallel analysis of heterosis in elite hybrid rice based on RFLPs and microsatellites. *Theoretical and Applied Genetics* 89, 185–192.

Zhang, Q., Gao, Y.J., Saghai Maroof, M.A., Yang, S.H. and Li, J.X. (1995) Molecular divergence and hybrid performance in rice. *Molecular Breeding* 1, 133–142.

Zhang, S., Raina, S., Li, H., Li, J., Dec, E., Ma, H., Huang, H. and Fedoroff, N.V. (2003) Resources for targeted insertional and deletional mutagenesis in *Arabidopsis*. *Plant Molecular Biology* 53, 133–150.

Zhang, W., McElroy, D. and Wu, R. (1991) Analysis of rice *Act1* 5' region activity in transgenic rice plants. *The Plant Cell* 3, 1155–1165.

Zhang, X., Yazaki, J., Sundaresan, A., Cokus, S., Chan, S.W.-L., Chen, H., Henderson, I.R., Shinn, P., Pellegrini, M., Jacobsen, S.E. and Ecker, J.R. (2006) Genome-wide high-resolution mapping and functional analysis of DNA methylation in *Arabidopsis*. *Cell* 126, 1189–1201.

Zhang, Y.M. and Xu, S. (2004) Mapping quantitative trait loci in F2 incorporating phenotypes of F3 progeny. *Genetics* 166, 1981–1993.

Zhang, Y.M. and Xu, S. (2005) A penalized maximum likelihood method for estimating epistatic effects of QTL. *Heredity* 95, 96–104.

Zhang, Z., Bradbury, P.J., Kroon, D.E., Casstevens, T.M. and Buckler, E.S. (2006) TASSEL 2.0: a software package for association and diversity analyses in plants and animals. Poster presented at *Plant and Animal Genomes XIV Conference*, 14–18 January 2006, San Diego, California.

Zhao, J.Z., Cao, J., Li, Y., Collins, H.L., Roush, R.T., Earle, E.D. and Shelton, A.M. (2003) Transgenic plants expressing two *Bacillus thuringiensis* toxins delay insect resistance evolution. *Nature Biotechnology* 21, 1493–1497.

Zhao, M.F., Li, X.H., Yang, J.B., Xu, C.G., Hu, R.Y., Liu, D.J. and Zhang, Q. (1999) Relationships between molecular marker heterozygosity and hybrid performance in intra- and inter-subspecific crosses of rice. *Plant Breeding* 118, 139–144.

Zhao, S. and Bruce, W.B. (2003) Expression profiling using cDNA microarray. In: Grotewold, E. (ed.) *Methods in Molecular Biology*, Vol. 236. *Plant Functional Genomics: Methods and Protocols*. Humana Press, Totowa, New Jersey, pp. 365–380.

Zhao, W., Li, H., Hou, W. and Wu. R. (2007) Wavelet-based parametric functional mapping of developmental trajectories with high-dimensional data. *Genetics* 176, 1879–1892.

Zhao, Z., Wang, C., Jiang, L., Zhu, S., Ikehashi, H. and Wan, J. (2006) Identification of a new hybrid sterility gene in rice (*Oryza sativa* L.). *Euphytica* 151, 331–337.

Zheng, K., Qian, H., Shen, B., Zhuang, J., Liu, H. and Lu, J. (1994) RFLP-based phylogenetic analysis of wide compatibility varieties in *Oryza sativa* L. *Theoretical and Applied Genetics* 88, 65–69.

Zheng, X., Wu, E.J.G., Lou, X.Y., Xu, H.M. and Shi, C.H. (2008) The QTL analysis on maternal and endosperm genome and their environmental interactions for characters of cooking quality in rice (*Oryza sativa* L.). *Theoretical and Applied Genetics* 116, 335–342.

Zhou, P.H., Tan, Y.F., He, Y.A., Xu, C.G. and Zhang, A. (2003) Simultaneous improvement of four quality traits of Zhenshan 97, an elite parent of hybrid rice, by molecular marker-assisted selection. *Theoretical and Applied Genetics* 106, 326–331.

Zhu, C., Gore, M., Buckler, E.S. and Yu, J. (2008) Status and prospects of association mapping in plants. *The Plant Genome* 1, 5–20.

Zhu, H., Bilgin, M. and Snyder, M. (2003) Proteomics. *Annual Review of Biochemistry* 72, 783–812.

Zhu, J. and Weir, B.S. (1994) Analysis of cytoplasmic and maternal effects. II. Genetic models for triploid endosperm. *Theoretical and Applied Genetics* 89, 160–166.

Zhu, L., Xu, J., Chen, Y., Ling, Z., Lu, C. and Xu, Y. (1994) Location of unknown resistance gene to rice blast using molecular markers (in Chinese). *Science in China* (Ser. B) 24, 1048–1052.

Zhu, Q., Maher, A., Masoud, S., Dixon, R.A. and Lamb, C.J. (1994) Enhanced protection against fungal attack by constitutive co-expression of chitinase and glucanase genes in transgenic tobacco. *Bio/Technology* 12, 807–812.

Zhu, Y. Nomura, T., Xu, Y., Zhang, Y., Peng, Y., Mao, B., Hanada, A., Zhou, H., Wang, R., Li, P., Zhu, X., Mander, L.N., Kamiya, Y., Yamaguchi, S. and He, Z. (2006) *ELONGATED UPPERMOST INTERNODE* encodes a cytochrome P450 monooxygenase that epoxidizes gibberellins in a novel deactivation reaction in rice. *The Plant Cell* 18, 442–456.

Zhu, Z.F., Sun, C.Q., Jiang, T.B., Fu, Q. and Wang, X.K. (2001) The comparison of genetic divergences and its relationships to heterosis revealed by SSR and RFLP markers in rice (*Oryza sativa* L.). *Acta Genetica Sinica* 28, 738–745.

Zhuang, J.Y., Lin, H.X., Lu, J., Qian, H.R., Hittalmani, S., Huang, N. and Zheng, K.L. (1997) Analysis of QTL × environment interaction for yield components and plant height in rice. *Theoretical and Applied Genetics* 95, 799–808.

Zimmerli, L. and Somerville, S. (2005) Transcriptomics in plants: from expression to gene function. In: Leister, D. (ed.) *Plant Functional Genomics*. Food Products Press, Binghamton, New York, pp. 55–84.

Zivy, M., Joyard, J. and Rossignol, M. (2007) Proteomics. In: Morot-Gaudry, J.F., Lea, P. and Briat, J.F. (eds) *Functional Plant Genomics*. Science Publishers, Enfield, New Hampshire, pp. 217–244.

Zobel, R.W., Wright, M.J. and Gaugh, H.G., Jr (1988) Statistical analysis of a yield trial. *Agronomy Journal* 80, 388–393.

Zou, F., Yandell, B.S. and Fine, J.P. (2001) Statistical issues in the analysis of quantitative traits in combined crosses. *Genetics* 158, 1339–1346.

Zou, W. and Zeng, Z.-B. (2008) Statistical methods for mapping multiple QTL. *International Journal of Plant Genomics* 2008, Article ID 286561. Available at: http://www.hindawi.com/journals/ijpg/2008/286561.html (accessed 17 November 2009).

Zou, W., Aylor, D.L. and Zeng, Z.-B. (2007) eQTL Viewer: visualizing how sequence variation affects genome-wide transcription. *BMC Bioinformatics* 8, 7.

Index